Environmental Science and Engineering

The ultimate goal of this series is to contribute to the protection of our environment, which calls for both profound research and the ongoing development of solutions and measurements by experts in the field. Accordingly, the series promotes not only a deeper understanding of environmental processes and the evaluation of management strategies, but also design and technology aimed at improving environmental quality. Books focusing on the former are published in the subseries Environmental Science, those focusing on the latter in the subseries Environmental Engineering.

More information about this series at http://www.springer.com/series/7487

Junfu Lyu · Shuiqing Li
Editors

Clean Coal and Sustainable Energy

Proceedings of the 9th International Symposium on Coal Combustion

清華大學出版社
TSINGHUA UNIVERSITY PRESS

Springer

Editors
Junfu Lyu
Department of Energy and Power Engineering
Tsinghua University
Beijing, China

Shuiqing Li
Department of Energy and Power Engineering
Tsinghua University
Beijing, China

ISSN 1863-5520 ISSN 1863-5539 (electronic)
Environmental Science and Engineering
ISBN 978-981-16-1656-3 ISBN 978-981-16-1657-0 (eBook)
https://doi.org/10.1007/978-981-16-1657-0

Jointly published with Tsinghua University Press
The print edition is not for sale in the Mainland of China. Customers from the Mainland of China please order the print book from Tsinghua University Press.

This Springer imprint is published by the registered company Springer Nature Singapore Pte Ltd.
The registered company address is: 152 Beach Road, #21-01/04 Gateway East, Singapore 189721, Singapore

Preface

The International Symposium on Coal Combustion (ISCC), initiated in 1987 by Tsinghua University, has now evolved into a unique and influential series of forums every four years, which concentrates on the clean, efficient, flexible, low-carbon, and intelligent technologies for solid fuel-based energy.

The most recent 9th Symposium was successfully held at Qingdao, China in July 21–24, 2019, as co-hosted by the Institute of Thermal Engineering of Tsinghua University, and Shandong University of Science and Technology (local organizer). The renewed International Advisory Committee now consists of 59 distinguished experts from 15 countries (27 Chinese and 32 International).

The 9th ISCC receives a record-breaking number of full-paper submissions covering the latest research and development on the solid fuel-based energy technology. The colloquium chairs and reviewers dedicate themselves to a high-quality, two-round review procedure. In this symposium, 148 oral presentations and 122 posters were selected. Invited talks include 11 plenary and 6 keynote lectures. Among them about fifty prominent papers were further submitted to special issues of *Fuel* and *Frontiers in Energy*.

More than 260 delegates attended the symposium from 16 countries including Australia, China, Germany, India, Japan, Portugal, Russia, South Korea, UK, USA, Vietnam, etc. Five best paper awards were delivered for the first time to promote youth scholars for the ISCC. A lasting aim of the symposium is to establish and strengthen international bridges between universities, research institutes and the industry community. The 9th ISCC has seen more collaborative contributions, and launched novel sessions of panel discussions on international projects, encompassing Sino-US and Sino-Europe collaboration.

The proceedings of the 9th ISCC collect 5 invited speeches and 84 technical papers that are presented in oral or in poster. Papers are divided into 6 following sections including the number of papers.

- Basic Coal Quality & Combustion/14
- Pulverized Coal Combustion/15
- Fluidized Bed Combustion/11
- Low Carbon Energy/6

– Emissions Control/24
– Design and Operating Experiences/14

The proceedings can serve as a reference, and hopefully inspire new ideas for scientists, engineers, graduate students, equipment manufactures and operators, as well as technical managers who are working on the broad field of solid fuel-based energy technologies.

We would like to acknowledge our members of the International Advisory Committee who devoted their time and energy to review the manuscripts, and to express our appreciation to Dr. Qiang Li of Tsinghua University Press for his efforts in publishing the proceedings.

Beijing, China

Prof. Junfu Lyu
Prof. Shuiqing Li

Contents

About the Editors

Dr. Junfu Lyu Professor at Institute of Thermal Engineering, Tsinghua University, where he received his Bachelor Degree in 1991, Master Degree in 1996, and Ph.D. in 2005. His interests lie in coal combustion, coal gasification, circulating fluidized bed combustion technology, Hydrodynamic of boiler, and coal fired pollutant control. He has once organized and hold a series of the key technology research and development plant of China government, the foundation of China National Nature Science, as well as the projects from enterprises. His achievement work includes the 600 MW supercritical CFB boiler, the superheat steam generator by the salted waste water from heavy oil recovery, and the ultra-low emission circulating fluidized bed boiler technology by limestone injection into furnace and low NOx combustion. These developed technologies have been commercialized and used widely. He was elected as the New Century Excellent Talents in University, the Outstanding Academy. He won the BEST paper award twice in 15th and 23th International Conference on Fluidized Bed Combustion. He got the first-class National Scientific & Technological Progress Award in 2017 and the second-class National Scientific & Technological Progress Award in 2005. Besides, he also won the first-class prizes in scientific & technological progress or invention from the Ministries and Commissions. He published more than 300 Journal papers and 5 books. He has more than 70 authorized invention patents.

Shuiqing Li (born 10 December 1975), a Professor at the Department of Energy and Power Engineering, Tsinghua University, Beijing, China. He obtained both his B.S. and Ph.D. degree from Zhejiang University and has been a visiting scientist at the University of Leeds (2004–2005), the University of Iowa (2006), Princeton University (2010–2011), Yale University (2014), and RWTH Aachen University (2018). Current research interests of Prof. Li include adhesive particulate flow, combustion physics, gas-phase material synthesis, clean coal technology. Prof. Li has more than 140 peer-reviewed papers published on Prog. Eng. Combust. Sci., Phys. Rev. Lett., Soft Matter, J. Fluid Mech., etc., with about 2800 citations and H. Index of 30. He co-authored a monograph of *Adhesive Particle Flow* at Cambridge University Press, and holds 20 patents.

Prof. Li is a recipient of the National Young and Middle-aged Leading Scientists, Engineers, and Innovators (2018), the National Science Fund for Distinguished Young Scholars of China (2017), the National Award for New Century Excellent Talents (2009), and the Tsinghua University Award for Young Talents on Fundamental Studies (2011). Due to his outstanding achievements, he is elected a fellow of the Combustion Institute (2020).

Part I
Invited Talks

Chapter 1
Advances in Two-Fluid LES of Two-Phase Combustion

L. X. Zhou

Abstract Recently, large-eddy simulation (LES) was used to study spray-air and coal-air two-phase combustion. Most LES of two-phase combustion takes Eulerian–Lagrangian approach, which needs much more computational time than the Eulerian-Eulerian (E-E) or two-fluid approach. A few studies on E-E LESof two-phase combustion were reported in the literature. In this paper at first previous studies are reviewed. Then, the mathematical models for two-fluid LES of two-phase combustion, proposed by the present author are presented, including filtered controlling equations, specific for full two-fluid and two-fluid-Lagrangian approaches, a two-phase sub-grid scale (SGS) stress model and a SGS combustion model. For the SGS stress model, the present author proposes a two-phase SGS energy equation model, accounting for the interaction between two phases. For the gas-phase combustion model, a second-order moment SGS (SOM-SGS) turbulence-chemistry model, proposed by the present author, is suggested to simulate gas-phase (liquid-fuel vapor or CO and coal-volatile) combustion. Also, for coal combustion, the coal pyrolization and char combustion models are used. These sub-models are separately assessed by comparison with experiments.

Keywords Two-fluid modeling · Large-eddy simulation · Two-phase combustion

1.1 Introduction

In recent years large-eddy simulation (LES) of two-phase combustion, including spray combustion and pulverized-coal or other solid-fuel combustion, attracts more and more attention, since it can give instantaneous flow and flame structures and more accurate statistical results than those given by the Reynolds-averaged Navier–Stokes (RANS) simulation. Most LES of two-phase combustion takes Eulerian–Lagrangian (E-L) approach (Li and Kong 2008; Moin and Apte 2006; Patel and Menon 2008; Yan et al. 2008; Wen et al. 2017, 2018; Knappstein et al. 2018; Wan et al. 2017; Rabaçal

L. X. Zhou (✉)
Department of Engineering Mechanics, Tsinghua University, Beijing 100084, China
e-mail: zhoulx@mail.tsinghua.edu.cn

J. Lyu and S. Li (eds.), *Clean Coal and Sustainable Energy*, Environmental Science and Engineering,
https://doi.org/10.1007/978-981-16-1657-0_1

et al. 2018; Zhou et al. 2013). It is known from the experience in RANS modeling that the shortcomings of E-L approach are: (1) it is difficult to give the detailed continuous particle/droplet velocity and concentration distributions at each location of the 3-D space in a complex geometric configuration; (2) it frequently underpredicts the particle/droplet dispersion; (3) particles/droplets are difficult to enter or leave the recirculation zone; (4) to increase the accuracy it is needed to calculate a large number of trajectories, leading to a large computational time. On the other hand, the Eulerian-Eulerian or two-fluid modeling is successfully developed in RANS modeling of coal combustion (Zhou 2016). It can give the detailed particle/droplet information with less computation requirement than the E-L approach. Therefore it is needed to develop models for two-fluid LES of two-phase combustion.

The reasonability and accuracy of LES, no matter in E-L or E-E approach, are determined by three factors: (1) sufficiently fine grids; (2) more accurate difference scheme than that used in RANS modeling; (3) sub-grid scale (SGS) stress models and combustion models. For SGS stress models, in the E-L approach, the single-phase SGS stress models are adopted for the gas phase and there are no particle SGS stress models. The most widely used are the Smagorinsky eddy-viscosity model (Smagorinsky 1963), Germano dynamic eddy-viscosity model (Germano et al. 1991) and Kim SGS energy equation model (Kim and Menon 1995). In these SGS models the effect of particles on the gas-phase SGS stress is neglected. Subsequently, Yuu et al. (2001) and Zhou et al. (2004) separately proposed gas SGS stress models accounting for the effect of particles. Boileau et al. (2008) and Moreau et al. (2010) reported the two-fluid LES of two-phase combustion, using a Smagorinsky-type model for the particle phase, imitated from the gas-phase Smagorinsky model, without accounting for the interaction between two-phase SGS stresses. Hence, more reasonable two-phase SGS stress models for two-fluid LES of two-phase combustion remain to be developed. In this paper the present author proposes a two-phase SGS energy equation model for two-fluid LES of two-phase combustion, in which the interaction between two-phase SGS stresses is taken into account.

For SGS gas combustion models in LES, many investigators used different models. Frequently, the G-equation model and linear-eddy model were used for LES of premixed combustion; the laminar flamelet model was used for LES of non-premixed combustion. However, these models work well only for certain flame types and flame structures. The filtered PDF equation (FDF) model was used for both premixed and non-premixed combustion, but it needs a large computation time in LES. In recent years a second-order moment (SOM) combustion model was developed by the present author (Zhou et al. 2002), and it was successfully used in LES of single-phase non-premixed and premixed combustion and E-L LES of spray-air two-phase combustion (Zhou et al. 2016) with not much computational requirement. Now, it is suggested to be used for gas-phase homogeneous reaction in two-fluid LES of two-phase combustion. In this paper, at first a few studies on E-E LES of two-phase combustion reported in the literature are reviewed and then the mathematical models for two-fluid LES of two-phase combustion, suggested by the present author are presented, including filtered controlling equations, specific for full two-fluid and two-fluid-Lagrangian approaches, a two-phase sub-grid scale (SGS) stress model, a

SGS gas combustion model, as well as coal-particle pyrolization and char combustion models for the case of coal combustion. Separate assessment of these models by comparison with experiments is reported.

1.2 Previous Studies on Two-Fluid LES of Two-Phase Combustion

Boileau et al. (2008) adopted a two-fluid-LES (Eulerian-Eulerian LES) for an annular gas-turbine combustor. A Wale eddy-viscosity SGS stress model, similar to the Smagorinsky model, was used for both gas and dispersed phases. A dynamic thickened-flame combustion model was used in LES.

The SGS stresses are taken as

$$\tau_{gs,ij} = -\rho_g\left(\overline{u_{gi}\,u_{gj}} - \overline{u}_{gi}\,\overline{u}_{gj}\right)$$
$$\tau_{ps,ij} = -\rho_p\left(\overline{u_{pi}\,u_{pj}} - \overline{u}_{pi}\,\overline{u}_{pj}\right)$$

The particle SGS stress model given in Boileau et al. (2008) is taken as

$$\tau_{ps,ij} = C_s^2\Delta^2\left|\tilde{S}_p\right|\tilde{S}_{p,ij} + \frac{2}{3}C_Y\Delta^2\left|\tilde{S}_p\right|^2\delta_{ij}$$

It is similar to the gas Smagorinsky eddy-viscosity SGS stress model. No interaction between gas and particle SGS stresses is taken into account. The statistical results for isothermal swirling gas-particle flows are validated by the measurement results, given in references. The instantaneous fuel concentration, droplet size and temperature maps, and the time series for ignition and flame propagation were obtained. However, no experimental validation of the statistical results for the combustion case was reported.

Pedel et al. (2012) and Shen et al. (2018) proposed a so-called <Direct Quadrature Method of Moments> (DQMOM) method for <Eulerian LES> of coal combustion. It is a population balance method for treating the number densities of particles of different sizes, velocities, temperatures and masses. The filtered gas-phase equations are similar to those used in E-L LES approach. For the particle phase, the particle number density function (NDF) is used to describe the number of particles per volume as a function of the spatial location and so-called <internal coordinate> (independent variables of the particles, e.g. particle sizes, velocities, temperatures, etc.).The internal coordinates are contained in the coordinate vector and are dependent upon space and time. The particle SGS stress was neglected in this approach. In fact, the transport equation of PDF for the particle number density in the phase space, given by Williams and Zhou (Zhou 2018) is the same as that in the DQMOM method. The DQMOM method approximates the NDF by a summation of multidimensional Dirac-delta functions. For three particle sizes with different velocities and temperatures, six partial differential equations should be solved. However, for particle velocities,

Fig. 1.1 Gas temperature

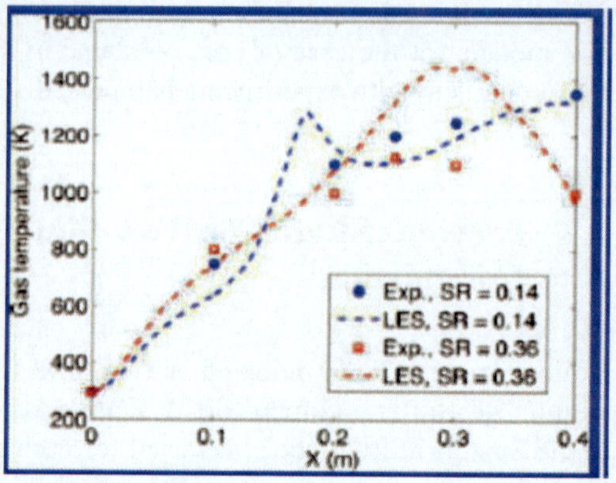

Fig. 1.2 Coal burnout

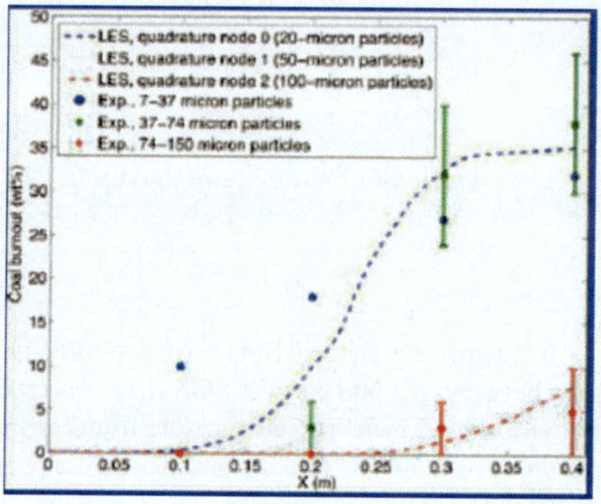

temperature and mass change in the physical space, the Lagrangian equations are still used. So, this method is actually a combined Eulerian–Lagrangian approach for the particle phase. The gas temperature change along the axial direction and coal burnout obtained by E-E LES (Figs. 1.1 and 1.2) are not in good agreement with experimental results.

1.3 Suggested Models for Two-Fluid LES of Two-Phase Combustion

The mathematical models for two-fluid LES of two-phase combustion proposed by the present author are given as follows. Taking the white-noise filtration (i.e., volume averaging over the computational cell of LES), the filtered gas and particle continuity, momentum, energy and gas species equations for two-fluid LES of two-phase combustion can be obtained as:

$$\frac{\partial}{\partial t}(\alpha_g \rho_g) + \frac{\partial}{\partial x_j}(\alpha_g \rho_g \overline{u_{gi}}) = S_g \tag{1.1}$$

$$\begin{aligned}&\frac{\partial}{\partial t}(\alpha_g \rho_g \overline{u_{gi}}) + \frac{\partial}{\partial x_j}(\alpha_g \rho_g \overline{u_{gi} u_{gj}})\\ &= -\frac{\partial \overline{p_g}}{\partial x_j} + \frac{\partial \tau_{g,ij}}{\partial x_j} + \frac{\partial \tau_{gs,ij}}{\partial x_j} + \frac{\alpha_g \rho_g}{\tau_r}(\overline{u_{gi}} - \overline{u_{pi}}) + \bar{u}_{gi} S_g\end{aligned} \tag{1.2}$$

$$\begin{aligned}&\frac{\partial}{\partial t}(\alpha_p \rho_p \overline{u_{pi}}) + \frac{\partial}{\partial x_j}(\alpha_p \rho_p \overline{u_{pi} u_{pj}})\\ &= \frac{\partial \tau_{p,ij}}{\partial x_j} + \frac{\partial \tau_{ps,ij}}{\partial x_j} + \frac{\alpha_p \rho_p}{\tau_r}(\overline{u_{gi}} - \overline{u_{pi}}) + \bar{u}_{gi} S_p\end{aligned} \tag{1.3}$$

$$\begin{aligned}&\frac{\partial(\alpha_g \rho_g \overline{h_g})}{\partial t} + \frac{\partial}{\partial x_j}(\alpha_g \rho_g \bar{h}_g \bar{u}_{gj})\\ &= \frac{\partial}{\partial x_j}\left(\frac{\mu_g}{Pr}\frac{\partial \overline{h_g}}{\partial x_j}\right) - \frac{\partial q_{gj,sgs}}{\partial x_j} - \bar{q}_r + \bar{S}_h\end{aligned} \tag{1.4}$$

$$\begin{aligned}&\frac{\partial(\alpha_p \rho_p \overline{h_p})}{\partial t} + \frac{\partial}{\partial x_j}(\alpha_p \rho_p \bar{h}_p \bar{u}_{pj})\\ &= \frac{\partial}{\partial x_j}\left(\frac{\mu_p}{Pr}\frac{\partial \overline{h_p}}{\partial x_j}\right) - \frac{\partial q_{pj,sgs}}{\partial x_j} - \bar{q}_{pr} + \bar{S}_{ph}\end{aligned} \tag{1.5}$$

$$\begin{aligned}&\frac{\partial \alpha_g \rho_g \bar{Y}_s}{\partial t} + \frac{\partial}{\partial x_j}(\alpha_g \rho_g \bar{u}_{gj} \bar{Y}_s)\\ &= \frac{\partial}{\partial x_j}\left(\frac{\mu}{Sc}\frac{\partial \bar{Y}_s}{\partial x_j}\right) - \bar{w}_s - w_{sgs} - \frac{\partial g_{gj,sgs}}{\partial x_j} + \bar{\alpha}_s \bar{S}_g\end{aligned} \tag{1.6}$$

In case of full two-fluid modeling for coal combustion, more continuity equations for the particle number density, particle total mass, particle dry-and-ash-free (daf) coal mass and particle moisture are needed as

$$\frac{\partial \bar{n}_p}{\partial t} + \frac{\partial}{\partial x_j}(\bar{n}_p \bar{u}_{pj}) = -\frac{\partial}{\partial x_j}(\overline{n_p u_{pj}}) - \frac{\partial g_{p,sgs}}{\partial x_j} \tag{1.7}$$

$$\begin{aligned}&\frac{\partial(\bar{n}_p m_p)}{\partial t} + \frac{\partial}{\partial x_j}(\bar{n}_p m_p \bar{u}_{pj})\\ &= -m_p \frac{\partial}{\partial x_j}(\overline{n_p u_{pj}}) - m_c \frac{\partial g_{p,sgs}}{\partial x_j} + n_p \dot{m}_p\end{aligned} \tag{1.8}$$

$$\frac{\partial(\bar{n}_p m_c)}{\partial t}+\frac{\partial}{\partial x_j}(\bar{n}_p m_c \bar{u}_{pj})$$
$$= -m_c \frac{\partial}{\partial x_j}\left(\overline{n_p u_{pj}}\right) - m_c \frac{\partial g_{p,sgs}}{\partial x_j} + n_p \dot{m}_c \tag{1.9}$$

$$\frac{\partial(\bar{n}_p m_w)}{\partial t}+\frac{\partial}{\partial x_j}(\bar{n}_p m_w \bar{u}_{pj})$$
$$= -m_w \frac{\partial}{\partial x_j}\left(\overline{n_p u_{pj}}\right) - m_c \frac{\partial g_{p,sgs}}{\partial x_j} + n_p \dot{m}_w \tag{1.10}$$

For coal particle combustion, the particle mass changing rate $\dot{m}_p$ due to moisture evaporation, pyrolyzation and char combustion is determined by the following equations

$$\dot{m}_p = \dot{m}_w + \dot{m}_v + \dot{m}_h \tag{1.11}$$

$$\dot{m}_w = \pi d_p Nu D \rho \ln\left(1+\frac{Y_{ws}-Y_{wg}}{1-Y_{ws}}\right),$$
$$Y_{ws} = B_w \exp\left(-\frac{E_w}{RT_p}\right) \tag{1.12}$$

$$\dot{m}_v = m_c \alpha B_v \exp\left(-\frac{E_v}{RT_p}\right) \tag{1.13}$$

$$\dot{m}_c = -m_c B_v \exp\left(-\frac{E_v}{RT_p}\right) \tag{1.14}$$

$$\dot{m}_p = \pi d_p Nu D \rho \ln\left(\frac{\dot{m}_s/\dot{m}_p - Y_s}{\dot{m}_s/\dot{m}_p - Y_{ss}}\right) \tag{1.15}$$

$$\dot{m}_s = \pi d_p^2 \rho Y_{ss} B_s \exp\left(-\frac{E}{RT_p}\right) \tag{1.16}$$

$$\dot{m}_h = \sum \dot{m}_s \tag{1.17}$$

To save the computational storage and time, a combined two-fluid-Lagrangian (TFL) approach can be introduced into large-eddy simulation. In the TFL approach, only Eqs. (1.3) and (1.7) are used to simulate particle velocity and particle number density by Eulerian-Eulerian or two-fluid approach, and particle mass and temperature change are simulated using a Lagrangian approach. The Eulerian particle energy equation (1.5) is not used. Instead, a Lagrangian particle energy equation is adopted as

$$
\begin{aligned}
m_p C \frac{dT_p}{dt} &= \pi d_p^2 \varepsilon \sigma \left(T_g^4 - T_p^4 \right) \\
&\quad + \dot{m}_p C_p (T_g - T_p) \left[\exp\left(\dot{m}_p C_p / (\pi d_p Nu \lambda) \right) - 1 \right]^{-1} \\
&\quad - \dot{m}_w L_w - \dot{m}_v q_v + \dot{m}_h Q_c
\end{aligned}
\tag{1.18}
$$

1.4 The Two-Phase SGS Energy Equation Model

In the proposed two-phase SGS energy equation model, the filtered gas and particle viscous forces are:

$$\tau_{g,ij} = \mu_{gl} \left(\frac{\partial u_{gi}}{\partial x_j} + \frac{\partial u_{gj}}{\partial x_i} \right) - \frac{2}{3} \mu_{gl} \frac{\partial u_{gj}}{\partial x_j} \delta_{ij};$$

$$\tau_{p,ij} = \mu_p \left(\frac{\partial u_{pi}}{\partial x_j} + \frac{\partial u_{pj}}{\partial x_i} \right) - \frac{2}{3} \mu_p \frac{\partial u_{pj}}{\partial x_j} \delta_{ij}$$

The SGS stresses of gas and particle phases are given by

$$\tau_{g,ij} = \mu_{gl} \left(\frac{\partial u_{gi}}{\partial x_j} + \frac{\partial u_{gj}}{\partial x_i} \right) - \frac{2}{3} \mu_{gl} \frac{\partial u_{gj}}{\partial x_j} \delta_{ij}$$

$$\tau_{p,ij} = \mu_p \left(\frac{\partial u_{pi}}{\partial x_j} + \frac{\partial u_{pj}}{\partial x_i} \right) - \frac{2}{3} \mu_p \frac{\partial u_{pj}}{\partial x_j} \delta_{ij}$$

The SGS kinetic energies for gas and particle phases k_g^{sgs} and k_p^{sgs} are defined as:

$$k_g^{sgs} = \frac{1}{2} \sum_{j=1}^{3} R_{gs,ii}$$

$$k_p^{sgs} = \frac{1}{2} \sum_{j=1}^{3} R_{ps,ii}$$

The dissipation rates of SGS kinetic energies for gas and particle phases ε and ε_p are given as:

$$\varepsilon = (k_g^{sgs})^{3/2} / \Delta$$

$$\varepsilon_p = (k_p^{sgs})^{3/2} / \Delta$$

where Δ is the sub-grid scale size. The two-phase SGS stresses can be assumed to behave isotropically, and they can be defined as

$$\tau_{gs,ij} = -2\alpha_g \rho_g \nu_g (\bar{S}_{g,ij} - \frac{1}{3}\bar{S}_{g,kk}\,\delta_{ij})$$

$$\tau_{ps,ij} = -2\alpha_p \rho_p \nu_p (\bar{S}_{p,ij} - \frac{1}{3}\bar{S}_{p,kk}\,\delta_{ij}) \quad (1.19)$$

$$\nu_g = c_\mu (k_{gc})^{1/2}\Delta$$
$$\nu_p = c_{\mu p} (k_{pc})^{1/2}\Delta \quad (1.20)$$

The two-phase SGS energies are determined by

$$\frac{\partial}{\partial t}(\rho_g k_{gs,ij}) + \frac{\partial}{\partial x_j}(\rho_g \overline{u_{gj}} k_{gs,ij}) = \frac{\partial}{\partial x_j}\left(\frac{\mu_g}{\sigma_k}\frac{\partial k_{gs,ij}}{\partial x_j}\right) + G_{gk}^{sgs} + G_{pg,ij}^{sgs} - \rho_g \varepsilon_g \quad (1.21)$$

$$\frac{\partial}{\partial t}(\rho_p k_{ps,ij}) + \frac{\partial}{\partial x_j}(\rho_p \overline{u_{pj}} k_{ps,ij}) = \frac{\partial}{\partial x_j}\left(\frac{\mu_p}{\sigma_p}\frac{\partial k_{ps,ij}}{\partial x_j}\right) + G_{pk}^{sgs} + \rho_p \varepsilon_p \quad (1.22)$$

where $G_{pg,ij}^{sgs} = \frac{m_p}{\tau_r}\left[n_p\left(\overline{u_{gi}u_{pi}} - 2k_g^{sgs}\right)\right]$

$$\varepsilon_p = -\frac{1}{\tau_r}\left[(\overline{u_{gi}u_{pi}} - 2k_p^{sgs}) - \frac{\nu_p}{\sigma_p}\frac{(\overline{u_{gi}} - \overline{u_{pi}})}{n_p}\frac{\partial n_p}{\partial x_i}\right]$$

The two-phase velocity correlation equation is

$$\frac{\partial}{\partial t}(\overline{u_{gi}u_{pi}}) + (\overline{u_k} + \overline{u_{pk}})\frac{\partial}{\partial x_k}(\overline{u_{gi}u_{pi}}) = \frac{\partial}{\partial x_k}\left[(\nu_e + \nu_p)\frac{\partial}{\partial x_k}(\overline{u_{gi}u_{pi}})\right] + \frac{1}{\rho_g \tau_{rp}}[\rho_g \overline{u_{pi}u_{pi}} + \rho_g \overline{u_{gi}u_{gi}} - (\rho_g + \rho_p)\overline{u_{gi}u_{pi}}] - \left(\overline{u_{pk}u_{gi}}\frac{\partial \overline{u_{pi}}}{\partial x_k} + \overline{u_{gk}u_{pi}}\frac{\partial \overline{u_{gk}}}{\partial x_k}\right) - \frac{\varepsilon_g}{k_g^{sgs}}\overline{u_{pi}u_{gi}}\delta_{ij} \quad (1.23)$$

1.5 The SGS Mass Flux and Heat Flux Models

To close the SGS mass-flux and heat-flux terms, the following gradient modeling is used as

$$q_{gjsgs} = \rho\left(\overline{u_{gj}h_g} - \bar{u}_{gj}\bar{h}_g\right) = \frac{\mu_g}{\sigma_T}\frac{\partial \bar{h}_g}{\partial x_j} \tag{1.24}$$

$$g_{gjsgs} = \rho\left(\overline{u_{gj}Y_s} - \bar{u}_{gj}\bar{Y}_s\right) = \frac{\mu_g}{\sigma_Y}\frac{\partial \bar{Y}_s}{\partial x_j} \tag{1.25}$$

$$q_{pjsgs} = \rho\left(\overline{u_{pj}h_p} - \bar{u}_{pj}\bar{h}_p\right) = \frac{\mu_p}{\sigma_T}\frac{\partial \bar{h}_p}{\partial x_j} \tag{1.26}$$

$$g_{pjsgs} = \rho\left(\overline{u_{gj}n_p} - \bar{u}_{gj}\bar{n}_p\right) = \frac{\mu_g}{\sigma_p}\frac{\partial \bar{n}_p}{\partial x_j} \tag{1.27}$$

1.6 The SOM-SGS Gas Combustion Model

When taking a global one-step chemical kinetics, the SOM-SGS gas (CO and coal-volatile) combustion model is expressed by

$$w_{sgs} = \rho^2 \begin{bmatrix} \bar{K}(\overline{Y_{ox}Y_{fu}} - \bar{Y}_{ox}\bar{Y}_{fu}) + \\ \bar{Y}_{ox}(\overline{KY_{fu}} - \bar{Y}_{fu}\bar{K}) + \\ \bar{Y}_{fu}(\overline{KY_{ox}} - \bar{Y}_{ox}\bar{K}) \end{bmatrix} \tag{1.28}$$

$$\bar{K} = B\int \exp(-E/R\bar{T})p(\bar{T})d\bar{T}$$

$$\overline{\Phi\Psi} - \bar{\Phi}\bar{\Psi} = c\mu_t\left(\frac{\partial\bar{\Phi}}{\partial x_j}\right)\left(\frac{\partial\bar{\Psi}}{\partial x_j}\right)\Big/\left[\rho\left(\frac{a}{\tau_T} + \frac{(1-a)}{\tau_C}\right)\right]$$

where Φ and Ψ denote Y_1 or Y_2 or K, and τ_C is the chemical reaction time, τ_T is the turbulent diffusion time, *a* and c are model constants. The reaction time and fluctuation time are given by

$$\tau_C = \left[B\rho(\bar{Y}_{O_2} + \beta\bar{Y}_{CH_4})\exp\left(-\frac{E}{R\bar{T}}\right)\right]^{-1}$$

$$\tau_T = 1/|\bar{S}|$$

where β is the stoichiometric coefficient.

1.7 Assessment of the Two-Phase SGS Energy Equation Model

To assess the two-phase SGS energy equation model, the co-axial gas-particle sudden-expansion flows are studied by a 2-D two-fluid LES (Zhou and Liu 2018). The simulated chamber is given in Fig. 1.3.

The flow parameters are given in Table 1.1.

For this 2-D LES, the grid sizes were taken as 1 mm × 1 mm, the grid number in the half 2-D plane was taken as 100,000. The time step was taken as 10^{-6} s. Figure 1.4 shows the instantaneous two-phase streamlines. It is seen that both gas and particle flows have coherent structures, which cannot be given by RANS modeling.

Figure 1.5 gives the particle statistically averaged and RMS fluctuation velocities. It is seen that in average the LES results are better than the RANS modeling results,

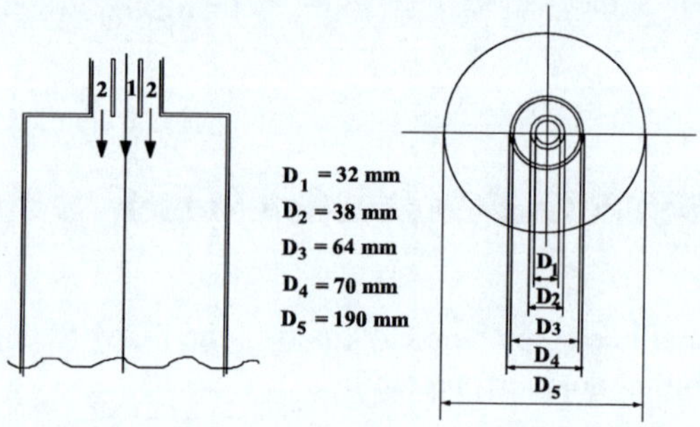

D_1=32 mm
D_2=38 mm
D_3=64 mm
D_4=70 mm
D_5=194mm

1 primary gas-particle flow

2 annular gas flow

Fig. 1.3 Simulated chamber for co-axial gas-particle flows

Table 1.1 Flow parameters

Annular gas flow rate	38.3 g/s
Inlet Reynolds number	53,256
Particle size	60 μm
Particle material density	2500 kg/m^3
Particle mass flow rate	0.34 g/s
Mass loading	0.034
Primary gas flow rate	9.9 g/s

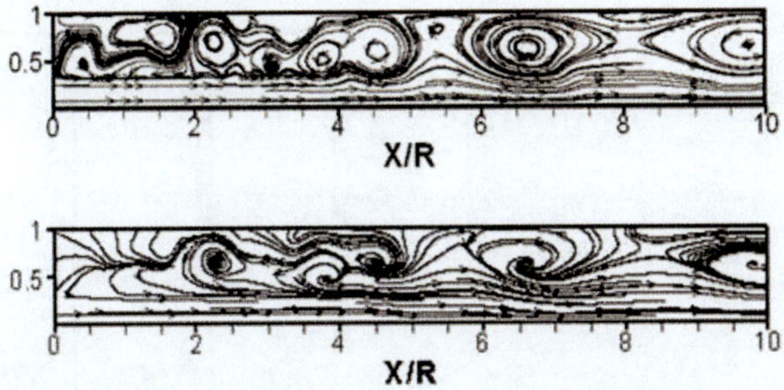

Fig. 1.4 Instantaneous two-phase streamlines (top-gas; bottom-particle)

although in some regions the LES still under-predicts the velocities due to the errors of 2-D LES.

1.8 Assessment of the SOM Combustion Model

To assess the SOM combustion model, the methane-air turbulent diffusion combustion in a swirl combustor, shown in Fig. 1.6 was studied by LES (Zhou et al. 2016).

For 3-D LES the grid sizes in the x, y, and z directions are 1–2 mm. The total number of cells is 839,595. The time step is taken as 0.005 s. For the numerical procedure, the pressure-implicit split-operator (PISO) algorithm is used for p–v corrections, the second-order implicit difference scheme is used for the time dependent term, and the central difference scheme is used for the convection and diffusion terms. Figure 1.7 shows the tangential RMS fluctuation velocity. Obviously, the LES can well simulate the peak value at the axis, whereas the RANS modeling using the Reynolds stress model cannot do. Figure 1.8 shows the time-averaged temperature. It is seen that the SOM SGS model gives the best results than those obtained by the widely used EBU model.

1.9 Assessment of the Two-Fluid RANS Modeling of Coal Combustion

To assess the coal combustion modeling, the coal combustion in a swirl combustor was simulated using two-fluid RANS modeling with a k-ε-kp two-phase turbulence model and the SOM turbulence-chemistry model (Zhou et al. 2003). The swirl coal combustor is shown in Fig. 1.9.

The flow parameters and coal proximate analysis are given in Table 1.2 and Table 1.3 respectively.

The predicted particle axial velocity and gas temperature are shown in Figs. 1.10

Fig. 1.5 Particle time-averaged (top) and RMS fluctuation (bottom) velocities

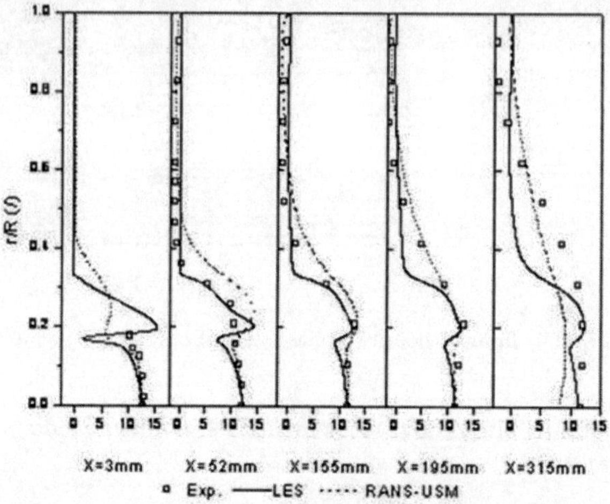

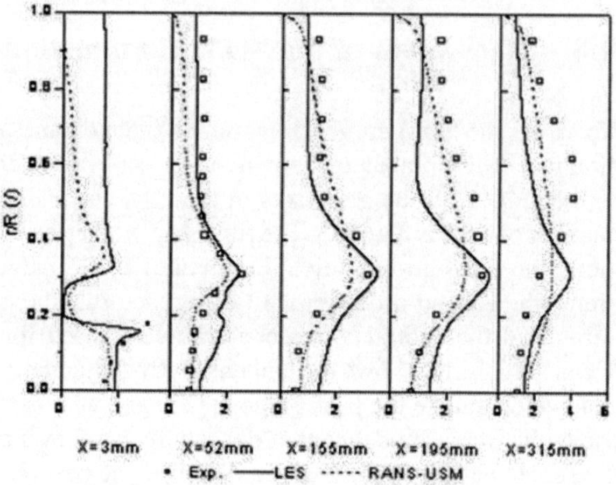

Fig. 1.6 A swirl gas combustor

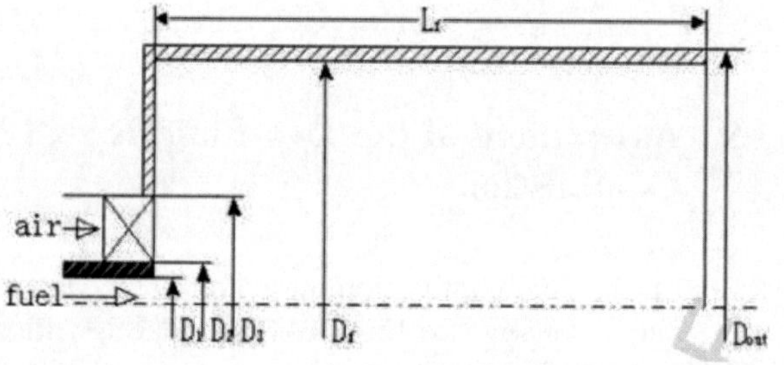

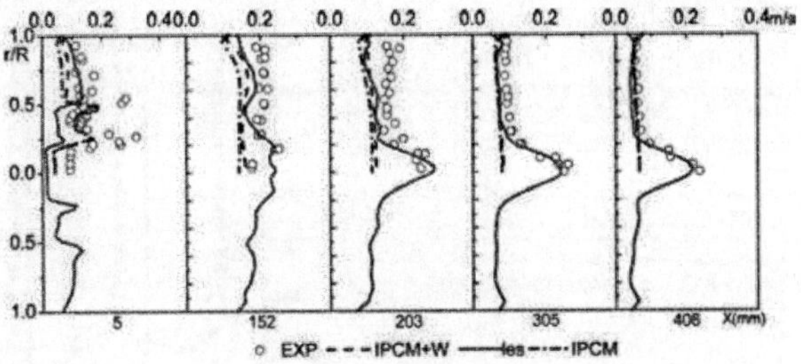

Fig. 1.7 Tangential RMS fluctuation velocity

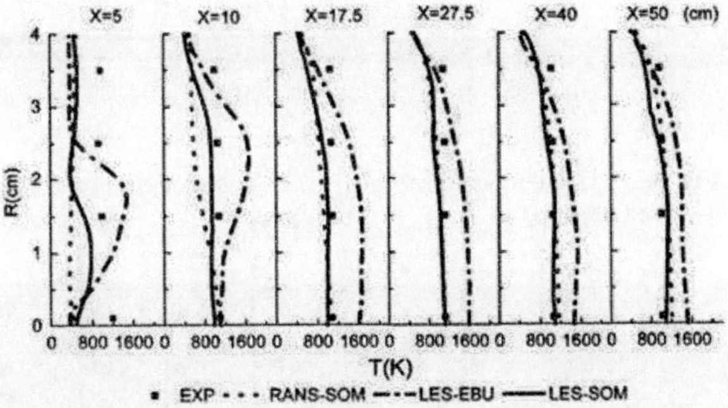

Fig. 1.8 Time-averaged temperature

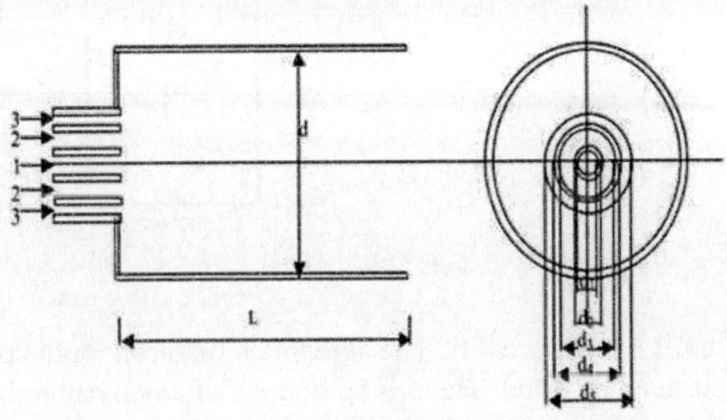

Fig. 1.9 A swirl coal combustor

Table 1.2 Flows parameters

	Air flow rate (kg/h)	Coal flow rate (kg/h)	Swirl number
Primary air	31.8	14	0
Secondary air	120.4	0	0.5, 0.8, 1.0, 1.4

Table 1.3 Proximate analysis of coal

Volatile (%)	Fixed carbon (%)	Moisture (%)	Ash (%)
35.8	53.7	6.3	4.2

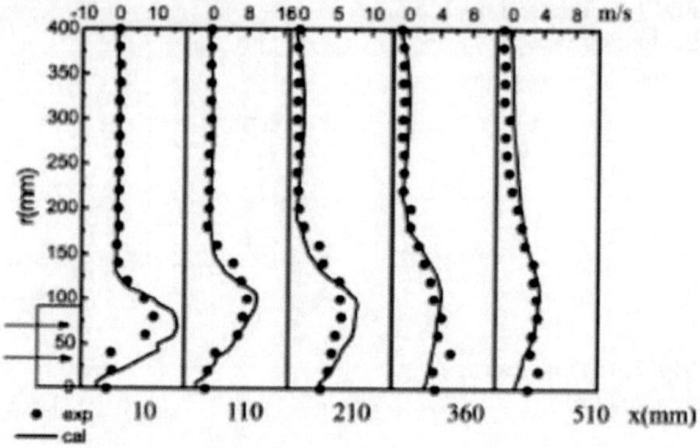

Fig. 1.10 Particle axial velocity (— Pred., • Exp.)

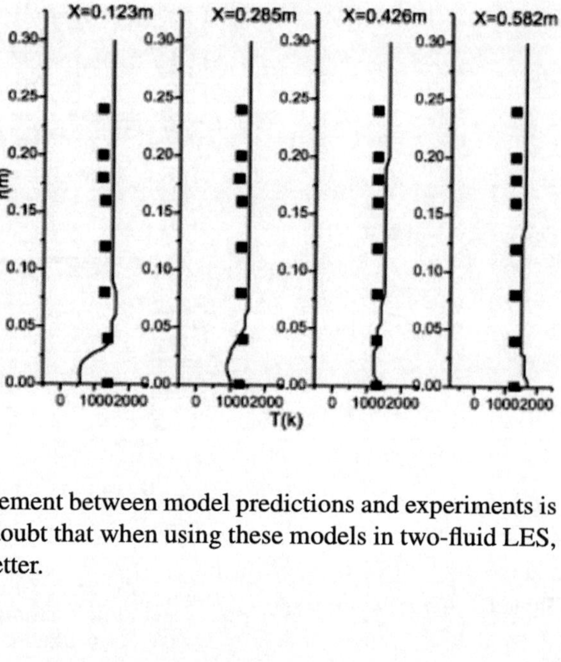

Fig. 1.11 Gas temperature (— Pred. ■ Exp)

and 1.11 respectively. The agreement between model predictions and experiments is sufficiently good. There is no doubt that when using these models in two-fluid LES, the agreement would be still better.

1.10 Conclusions

(1) The mathematical models for two-fluid LES of two-phase combustion are proposed, including the filtered governing equations specific for full two-fluid and two-fluid-Lagrangian modeling, the two-phase SGS stress model, the SOM-SGS gas combustion model, as well as the particle pyrolization model and char combustion model for coal combustion.

(2) The proposed two-fluid-Lagrangian LES model needs about the same computational cost for simulating three particle size groups as those of the DQMOM model, reported in references.

(3) The proposed two-phase SGS stress model was assessed in two-fluid LES of co-axial gas-particle flows.
(4) The proposed SOM-SGS combustion model was assessed in LES of gas swirling diffusion combustion.
(5) The two-fluid simulation of coal combustion was assessed in RANS modeling.
(6) The next step is applying these comprehensive mathematical models for two-fluid LES of coal combustion to a practical case for its assessment.

Acknowledgements This study was sponsored by the National Natural Science Foundation of China under the Grant 51390493.

References

Boileau M, Pascaud S et al (2008) Investigation of two-fluid methods for large eddy simulation of spray combustion in gas turbines. Flow Turbul Combust 80:291–321

Germano M, Piomelli U et al (1991) A dynamic sub-grid-scale eddy viscosity model. Phys Fluids A 3:1760–1765

Kim WW, Menon SS (1995) A new dynamic one-equation sub-grid-scale model for large eddy simulation, AIAA 95-0356. In: 33th Aerospace sciences meeting and exhibition, USA

Knappstein R, Kuenne G et al (2018) Large eddy simulation of a novel gas-assisted coal combustion chamber. Flow Turbul Combust 101:895–926

Li YH, Kong SC (2008) Diesel combustion modeling using LES turbulence model with detailed chemistry. Combust Theory Model 12:205–219

Moin P, Apte SV (2006) Large-eddy simulation of realistic gas turbine combustors. AIAA J 44:698–708

Moreau M, Simonin O et al (2010) Development of gas–particle Euler-Euler LES approach: a priori analysis of particle sub-grid models in homogeneous isotropic turbulence. Flow Turbul Combust 84:295–324

Patel N, Menon S (2008) Simulation of spray–turbulence–flame interactions in a lean direct injection combustor. Combust Flame 153:228–257

Pedel J, Thornock JN et al (2012) Large eddy simulation of pulverized coal jet flame ignition using the direct quadrature method of moments. Energy Fuels 26:6686–6694

Rabaçal M, Costa M et al (2018) A large eddy simulation study on the effect of devolatilization modeling and char combustion modeling on the structure of a large-scale, biomass and coal co-fired flame. J Combust Paper 7036425

Shen HS, Wu YX et al (2018) Eulerian LES simulation of coal jet flame with a simplified DQMOM model. Fuel 216:475–483

Smagorinsky J (1963) General circulation experiments with the primitive equation (I): the basic experiment. Monthly Weather Rev 91:99–164

Wan KD, Xia J et al (2017) Large-eddy simulation of pilot-assisted pulverized-coal combustion in a weakly turbulent jet. Flow Turbul Combust 99:531–550

Wen X, Luo YJ et al (2017) LES of pulverized coal combustion with a multi-regime flame-let model. Fuel 188:661–671

Wen X, Luo K et al (2018) Large-eddy simulation of multiphase combustion jet in cross-flow using flame-let model. Int J Multiph Flow 108:211–225

Yan YW, Zhao JX et al (2008) Large-eddy simulation of two-phase spray combustion for gas turbine combustors. Appl Therm Eng 28:1365–1374

Yuu S, Ueno T et al (2001) Numerical simulation of the high Reynolds number slit nozzle gas-particle jet using sub-grid-scale coupling large eddy simulation. Chem Eng Sci 56:4293–4307
Zhou LX (2018) Theory and modeling of dispersed multiphase turbulent reacting flows. Elsevier and Tsinghua University Press
Zhou LX (2016) A review for developing two-fluid modeling and LES of turbulent combusting gas-particle flows. Powder Technol 297:438–447
Zhou LX, Liu Y (2018) Two-fluid LES and RANS modeling of sudden-expansion gas-particle flows. Powder Technol 323:45–50
Zhou LX, Qiao L et al (2002) A USM turbulence-chemistry model for simulating NO formation in turbulent combustion. Fuel 81:1703–1709
Zhou LX, Zhang Y et al (2003) Simulation of swirling coal combustion using a full two-fluid model and an AUSM turbulence-chemistry model. Fuel 82:1001–1007
Zhou HS, Flamant G et al (2004) DEM-LES of coal combustion in a bubbling fluidized bed, Part I: Gas-particle turbulent flow structure. Chem Eng Sci 59:4193–4203
Zhou LX, Hu LY et al (2013) Measurement and simulation of swirling coal combustion. Particuology 11:189–197
Zhou LX, Wang F et al (2016) A review on studies of a SOM combustion model for single-and-two-phase combustion. Int J Heat Mass Transf 96:54–163

Chapter 2
Flexible Operation of High Efficiency Coal Power Plants to Ensure Grid Stability When Intermittent Renewables Are Included

Andrew Minchener

Abstract The introduction of intermittent renewables such as wind and solar has created a major integration problem that has to be addressed to ensure operational stability for power generation grid systems. While there is some expectation that large scale batteries may provide a solution, this still appears to be far from commercial viability. Consequently, in most cases coal fired power generation has needed to adapt to a new operating regime so that there is adequate energy system stability. This includes being able to achieve, at low minimum load, rapid ramp rates and cycling together with fast start-up. Flexible operation can have significant impact on a coal power plant with most components being affected. This is because of the increase in thermal and mechanical fatigue stresses in the different parts of the plant, which together with corrosion, differential expansion and other effects, often occurring in synergy, reduce the life time of many plant components. There has been considerable development work undertaken to counter these adverse effects while balancing the grid, including new technologies, processes and skills. There are several means to achieve low minimum load combustion, with the critical need to maintain stable combustion. These include the need to understand and mitigate the technical limitations to low burning rates, such as fire stability, flame monitoring, and minimising unburned coal and CO emissions. Fire stability itself depends on many factors, such as changes in firing rate or fuel quality, inaccurate fuel/air ratio or uneven coal flow. Another impact to consider is the effect of low load operation on downstream NOx control systems and connected equipment. Measures for achieving minimum load include ensuring coal quality, air/fuel flow optimisation and coal fineness, operation with a reduced number of mills or smaller mills, indirect firing, thermal energy storage for feedwater heating, tilting burners, reliable flame scanners and economiser modifications. Start-up procedures are complex and expensive as they usually require auxiliary fuel such as gas or oil, during burners the ignition period. Start-up times in power plants can be shortened through application of reliable ignition, turbine integration, reduced thickness of walls in boiler and turbine design, external heating of boiler thick wall components, measures in the turbine (sliding

A. Minchener (✉)
IEA Clean Coal Centre, Apsley House, 176 Upper Richmond Road, London SW15 2SH, UK
e-mail: andrew.minchener@iea-coal.org

J. Lyu and S. Li (eds.), *Clean Coal and Sustainable Energy*,
Environmental Science and Engineering,
https://doi.org/10.1007/978-981-16-1657-0_2

pressure, advanced sealings, steam cooling of the outer casting), proactive cleaning of boiler deposits plus more effective instrumentation and control. Measures to improve high ramp up rates include exploring mill storage capacity, use of a dynamic classifier instead of a static one; measures in the turbine such as opening of throttled main steam valves, condensate throttling, thermal storage for feedwater heater bypass and HP stage bypass, lower thicknesses of pressure parts, and increased number of headers. Preservation during standby periods is important including targeted plant chemistry management of the boiler and turbine. These and other issues are considered and examples presented.

Keywords Flexibility · Power generation stability · Impacts

2.1 Introduction

As intermittent renewables such as solar and wind increase their share of grid-based power generation, fossil fuel units are being called upon to operate more frequently in cycling modes, compared to the baseload modes for which many were designed. Coal power plant flexible operation, together with grid and demand side management, has become increasingly important for the integration of the intermittent energy sources on the grid to ensure adequate system stability (Henderson 2014; Sloss 2016). Without commercial-scale alternatives such as battery storage, this use of coal (or other fossil fuel) power sources is essential. Coal-fired units are required to now achieve fast start-up, very low minimum load, rapid ramp rates and on/off cycling. Such operation at off-design conditions increases the wear and tear of plant components, which bring new challenges. Consequently, with the expectation of further introduction of intermittent renewable energy sources, new strategies and effective management continue to be required to mitigate and/or avoid higher probability of equipment failure and consequent reduction in unit life, critical risk of process safety and increased costs (Hilleman 2018) (Fig. 2.1).

Issues include thermal and mechanical fatigue stresses together with corrosion, and differential expansion, often occurring in synergy, which can reduce the life time

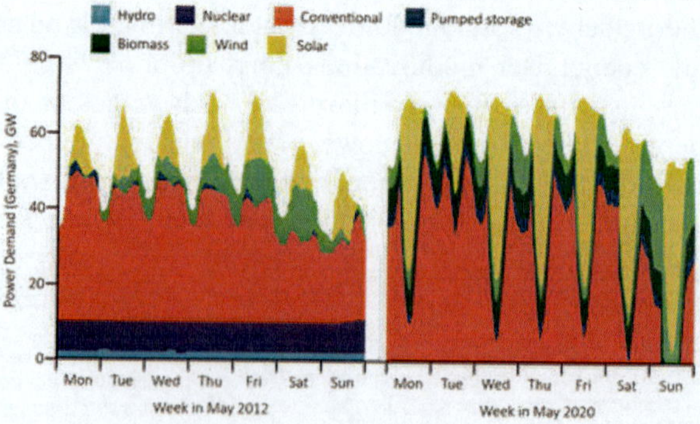

Fig. 2.1 Increasing grid instability with increasing introduction of intermittent renewable energy (Morris and Pehnt 2014)

of certain components (Daury 2018; Henderson 2014). There are also adverse impacts on coal power performance. For example, a reduction in load leads to a corresponding heat rate reduction, which together with higher auxiliary power consumption leads to specific CO_2 emissions increasing. When high levels of variable renewables occur on the grid, the operating costs for the fossil fuel plants can increase by an average 2–5% (Hilleman 2018).

For existing plants, there are several ways to improve flexibility (Henderson 2016). These include retrofitting new technology variants, modifying existing or adopting new operating procedures, introducing awareness training for both operating and managerial level personnel. Thus, different modes of cyclic operation of fossil plants and strategies for managing the negative impacts are identified. Options include new operating practices, use of advanced materials and installation of improved control systems. Such measures can improve heat rates and reduce the number of forced outages in existing fossil plants.

2.2 Instrumentation and Control

Modern control systems are vital for ensuring flexible power plant operation (Lockwood 2015). They allow navigation between different loads and ensure stable operation by adjusting all related process variables. The upgrade of Instrumentation and Control (I&C) systems improves accuracy, reliability and speed of change. For example, it allows operation of the plant closer to the material limitations of components such as the superheater headers without significantly reducing the lifetime of the component. Optimisation of I&C is the most cost effective way to improve plant flexibility and should be a precondition for other measures. In many plants, an upgrade of I&C is combined with plant engineering upgrades such as retrofits of the boiler, burners or turbine or other components (Agora Energiewende 2017) (Fig. 2.2).

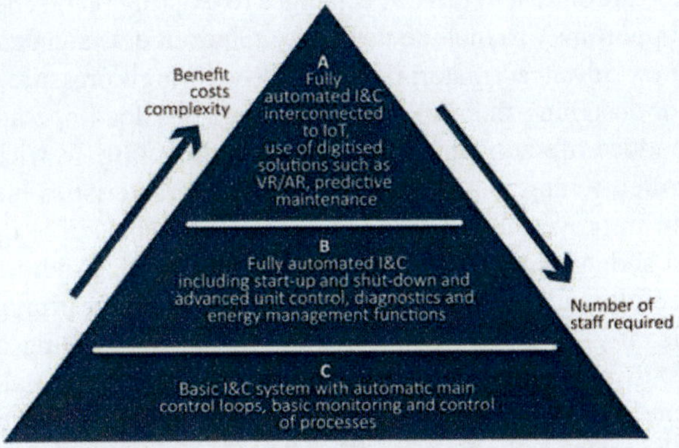

Fig. 2.2 Different levels of I&C systems (VGB 2018)

2.3 Flexibility Options

Operation with low minimum load limits the number of shut-downs required, which means less adverse impact on plant component life. There are several measures to achieve low minimum load, with stable combustion the key based on careful control of the boiler, fuel supply and combustion systems (Hamel and Nachtigall 2013). Such measures include ensuring consistent coal quality and particle fineness, operation with low excess air, use of flame monitoring, fuel/air flow control systems, tilting burners, auxiliary firing, operation with a lesser number of mills and only top-level burners, deploying smaller mills, thermal energy storage for feedwater heating, vertical internally rifled evaporators, a sliding pressure operation and economiser modifications (IEA 2018).

Today's optimised coal-fired power plants can operate at less than 20% of full-load capacity (Schiffer 2015), with indications that levels as low as 10% can be achieved if various measures are implemented together.

Start-up procedures are complex as well as expensive since they usually require auxiliary fuel during the burners' ignition stage. Consequently, they should be avoided if possible, and if that is not achievable then the start-up time should be shortened in combination with a fast ramp-up. Such changes can be achieved by several measures. These include: reliable ignition, integration of a gas turbine, reducing thickness of thick wall boiler components such as headers, including more headers, external heating of thick boiler components, and cleaning of boiler deposits (Martino 2013). Measures in the steam turbine include use of advanced seals, turbine bypass (HP or LP), internal cooling of the turbine casing.

Many of the improvements for start-up aid high ramp up rates, which allow dynamic adjustment to net power requirements. Additional measures include changing mill storage capacity, condensate throttling, and the use of an additional turbine valve. The change (i.e. ramp) between partial load and full load for power plants involves load changes of approximately three percentage points per minute, and the change in mode of operation can therefore be achieved at all plants in less than half an hour.

In contrast to possible changes to existing plants, designers of new plants have an opportunity to include flexibility requirements at an early stage. For example, use of new advanced materials for thick-wall high pressure components such as headers, or designing them based on a shorter baseload operational life have been shown to reduce life consumption during rapid cycling. Vertical evaporators with internally rifled tubing has shown good flow characteristics and flow stability, which helps to improve the rate of load change during flexible operation. Designing plant for a sliding pressure operation is also effective. Additionally, plants which include a condensate throttling system can increase their primary frequency response significantly. Other design futures include steam cooling of the inner turbine casing as well as HP and feedwater heaters bypass and thermal energy storage for feedwater preheating. The designers of new power plants, however, may face a conflict between flexibility and efficiency (Fig. 2.3).

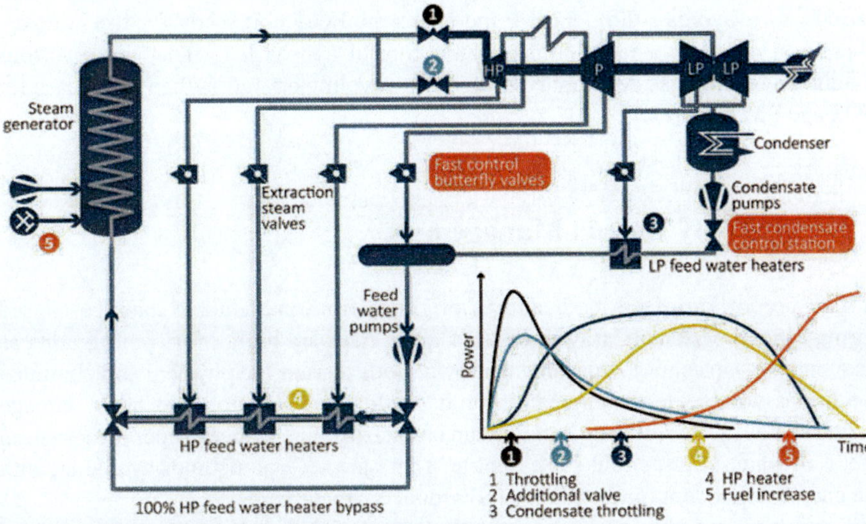

Fig. 2.3 Measures for fast load ramping (Chittora 2018)

2.4 Pollution Control Systems

The performance of some emission control systems can be affected by off-design conditions arising from the flexible operation of power plants since the temperature of the flue gas can change with the cycling regime. The removal of particulate matter either with ESPs or fabric filters can accommodate rapid load changes provided that the temperature does not fall below the dew point (~90 °C). At the latter condition, any moisture can lead to a build-up of dust, which can be difficult to remove (EPRI 2013). This can be countered with installation of a warming system. This could be in the form of a gas burner system to pre-heat the precipitators while the unit is being brought back on load. Generally, ESPs perform better at low loads because of the reduced proportion of unburnt carbon in ash and the increased residence time of the gases in the precipitator which allows more of the dust to be collected.

In contrast, maintaining the temperature at the required level is essential to ensure effective NOx control (Żmuda 2019; Boyle et al. 2015). For systems based on selective catalytic reduction, the inclusion of an additional flue gas heater prior to the SCR inlet has been used. For selective non-catalytic reduction systems, the use of a multiple zone injection approach, and the ability to bring injectors in and out of service as needed, allows for chemical release within the desired chemical and thermal environment (Davis et al. 2013).

For SO_2 removal with flue gas desulphurization, it is essential to minimize the number of shut-downs and start-ups to avoid slurry solidification and accumulation of start-up fuel oil residues on linings, as well as averting long warm-up periods. It is normal practice to keep the FGD unit in stand-by mode in case of short outage periods. This avoids solid deposits and keeps the FGD unit ready for fast start-up. There is a need for a more sophisticated control approach to meet these various conditions for flue gas desulphurisation during cycling operation than is the case for NOx and PM controls.

2.5 Flexibility Impact Management

A high proportion of on-load failures originate from preventable damage caused during offload periods (Caravaagio 2014). The risks are higher for cycling units as frequent start-ups/shutdowns and standby periods disrupt the physical and chemical conditions within the water/steam circuit, leading to corrosion and other damage during standby. The resulting damage can be catastrophic. Hence proper preservation of the all water-steam circuits is essential. There are several methods available, with the choice dependent on the plant's individual characteristics.

Flexible plant operation means more start-ups and shut-downs and more periods of standby ranging from a few hours to several days or more. Plant shut-downs and start-ups disrupt physical and chemical conditions within the water/steam circuit, leading to corrosion and other damage mechanisms during standby, which then affect plant operation unless proper lay-up procedures are applied (McCann 2018). Such damage can compromise start-up reliability as well as result in serious failures during service including potentially catastrophic LP turbine blade damage. Frequent, short-term outages from unit cycling increase by nearly an order of magnitude the percentage of operating life and annual hours for which components are stressed or imperfectly protected (Mathews 2013) (Fig. 2.4).

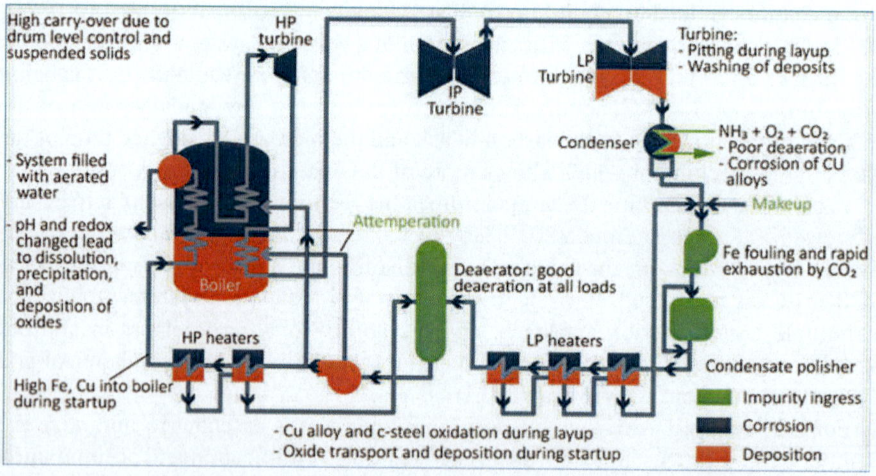

Fig. 2.4 Areas of the steam/water cycle affected by lay-up and start-up practices (Mathews 2013)

Choosing the most applicable practices depends on site-specific factors, and the entire unit must be considered. The practices applied may differ from outage to outage but should always focus on the most practical and beneficial techniques for minimising equipment damage during standby (EPRI 2014).

Cycling operation increases the number of start-ups/shut-downs and standby periods. These change the water/steam equilibrium and lead to corrosion and other damage mechanisms during standby periods, unless proper lay-up procedures are applied. Damage initiated during a standby period affects plant operation and reliability. All water/steam circuits need to be preserved. There is no one-size-fits-all solution for plant lay-up and the most appropriate method depends on site-specific conditions (EPRI 2014).

Wet storage of the water systems and often the boiler is considered the most practical approach for cycling units, for which pH adjustment and elimination of oxygen are essential. This procedure includes complete deaeration of the condensate and feedwater and prevention of air entering the boiler and superheater. The latter can be achieved by nitrogen blanketing and/or maintaining boiler pressure. The pH adjustments for all the liquid, including condensed steam in the superheater, must be equal to or higher than normal pH conditions.

The wet lay-up practices in all parts of the water/steam cycle can be enhanced using filming amines as a corrosion inhibitor. These are dosed to the entire circuit before the unit shutdown and their dosage needs to be controlled precisely (Moore 2018).

The best method for preserving the reheater and steam turbine is dry storage. Residual heat of the turbine can typically maintain 'dry' conditions for 24–36 h, but once a relative humidity either greater than 40% or equal to the 'dew point' temperature is reached, condensation and oxygen will initiate corrosion. Reheaters that are force-cooled require immediate purging of steam vapour as exclusion of oxygen laden air is difficult to achieve. Dry reheaters, like the turbine, are subject to condensation and aeration when cooling.

Preservation of condensers and the shell (steam) side of feedwater heaters is difficult. Both systems are often the major areas of corrosion during unit shutdown and the source of deposit forming corrosion products during start-up. Filming amines can provide corrosion protection for the reheater, turbine condenser and feedwater heater. Applied during operation in advance of a shut-down, this method enables quick return to service, and hence is applicable for frequently cycling units.

Monitoring of lay-up conditions is required to ensure the protective conditions are maintained (Figs. 2.5 and 2.6).

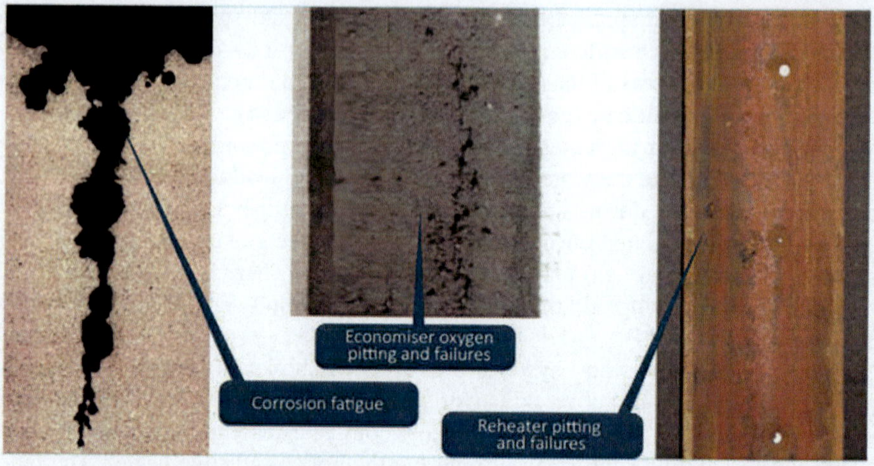

Fig. 2.5 Boiler tube failures influenced by off load corrosion (McCann 2018)

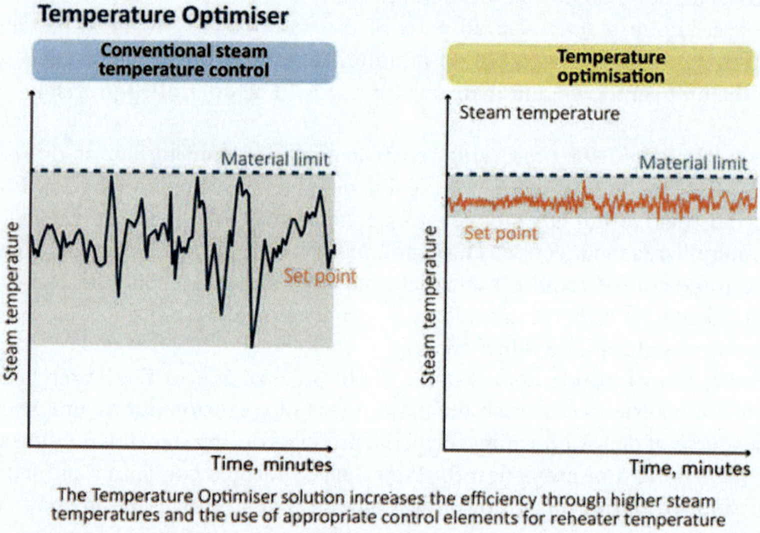

Fig. 2.6 Example of temperature optimisation using Siemens SPPA-P3000 system (Chittora 2018)

2.6 Conclusions

The inclusion of intermittent renewable energy sources within a grid-based power generation system provides power with zero carbon emissions but this must be backed up with other sources to ensure adequate grid stability. In most instances, this is provided by coal fired power plants, originally designed for baseload operations but now operating in cycling modes with faster ramp rates, very low load and on/off operation (Reischke 2012). While these coal-based units are meeting the stability challenge, they are operating mostly at off-design conditions, with increased wear and tear of plant components that brings new challenges. As more intermittent renewable sources are added to the grid worldwide, the need for flexible operation will increase (Kumar and Hillemann 2018). Further new strategies and effective management will be required to ensure that this operational approach can continue to be achieved (VGB 2018).

Acknowledgements I acknowledge with grateful thanks the information provided by my colleagues Dr. Colin Henderson and Dr. Maggie Wiatros-Motyka of the IEA Clean Coal Centre.

References

Agora Energiewende (2017) Flexibility in thermal power plants. https://www.agora-energiewende.de/fileadmin2/Projekte/2017/Flexibility_in_thermal_plants/115_flexibility-report-WEB.pdf. Accessed 05 Nov 2018

Boyle J, Stamatakis P, de Havilland P (2015) Advanced SNCR NOx reduction experience on multiple large utility boilers

Caravaagio M (2014) Layup practices for cycling units. https://www.power-eng.com/articles/print/volume-118/issue-8/features/layup-practices-for-cycling-units.html

Chittora S (2018) Flexible operation of thermal power plants in India—experience so far. Presented at Improving power plant flexibility—paving the way for greening the Grid, NTPC Power Management Institute, Noida, India 27 Sep 2018

Daury D (2018) Boiler design futures for flexible operation. Presented at Improving power plant flexibility—paving the way for greening the Grid, NTPC Power Management Institute, Noida, India 27 Sept 2018

Davis, Rummenhohl, Benisvy and Schultz (2013) Optimizing catalyst performance aids in lower operational & management costs. https://www.powermag.com/optimizing-catalyst-performance-lowers-om-costs/

EPRI (2013) Impact of cycling on the operation and maintenance cost of conventional and combined-cycle power plants, 208 pp. https://www.epri.com/#/pages/product/3002000817/?lang=en-US

EPRI (2014) Layup for cycling units: requirements, issues, and concerns—an EPRI white paper

Hamel S, Nachtigall C (2013) Measures for flexibility of steam generators in low load operation. In: VGB conference: power plants in competition, Neuss, Germany, 24–25 April 2013. VGB, Germany

Henderson C (2014) Increasing the flexibility of coal-fired power plants. CCC/242, London, UK, IEA Clean Coal Centre, 57 pp

Henderson (2016) Coal-fired power plants—flexibility options and challenge. Presented at UNECE/WCA workshop on the sustainability credentials of coal and its role in the UN, Geneva, 26 October 2016

Hilleman D (2018) Coal plants: American experience, modifying coal plants for generation flexibility. Presented at Greening the grid workshop, New Delhi, India

IEA (2018) Status of power system transformation advanced power plant flexibility 2018, 115 pp

Kumar N, Hilleman D (2018) Grid operations with high renewable generation. Presented at Greening the grid workshop, New Delhi, India

Lockwood T (2015) Advanced sensors and smart controls for coal-fired power plant. CCC/251, London, UK, IEA Clean Coal Centre, 104 pp

Martino J (2013) Advances in boiler cleaning technology. Power Eng 117 (6):68–74

Mathews J (2013) Layup practices for fossil plants. https://www.powermag.com/layup-practices-for-fossil-plants/?pagenum=1
McCann P (2018) Plant layup management during operational standby period presentation given during ExPPerts conference in Wroclaw, Poland, 26–27 Sep 2018
Moore W (2018) Failure mechanisms in steam turbine condensers. Presented at Power plant operation and flexibility conference, London, UK, 4–6 Jul 2018
Morris C, Pehnt M (2014) Energy transition the German Energiewende (An initiative of the Heinrich Böll Foundation Released on 28 November 2012 Revised January 2014). https://pl.boell.org/sites/default/files/german-energy-transition.pdf
Reischke A (2012) Increasing the flexibility of conventional coal-fired power plants in the context of the energy transition. In: Sixth workshop on power plant components 2012: intelligent power plant components for a flexible base load supply, Gelsenkirchen, Germany, 25 September 2012
Schiffer H-M (2015) The flexibility of German coal-fired power plants amid increased renewables. Cornerstone 2(4)
Sloss L (2016) Levelling the intermittency of renewables with coal. CCC/268, London, UK, IEA Clean Coal Centre, 71 pp
VGB PowerTech (2018) Flexibility toolbox. Compilation of measures for the flexible operation of coal-fired power plants, 60 pp
Żmuda R (2019) SBB Energy S.A. 45-324 opole, ul. Łowicka 1, Poland, personal communication

Chapter 3
Advances in Modeling Coal Pyrolysis, Char Combustion, and Soot Formation from Coal and Biomass Tar

Thomas H. Fletcher, David O. Lignell, Alex Josephson, Andrew Richards, and Troy Holland

Abstract Simulations of pulverized coal boilers and gasifiers have become easier and more complex as computational resources become more available. The improvements in simulations have generally focused on the fluid dynamics and grid resolution, with marginal improvements in treatments of the fundamental coal reactions. In this work, suggestions are made in several areas to improve boiler and gasifier simulations with only relatively small impacts on computational time. New correlations are presented for the elemental composition of coal char and tar as a function of parent coal characteristics, temperature, and heating rate. The importance and correlation of char oxidation effects are discussed. A new generalized model for soot formation, oxidation, and gasification is also discussed.

Keywords Coal combustion modeling · Tar · Soot · Char oxidation

This paper will discuss three topics that can improve simulations of coal combustors and gasifiers: (1) correlations of elemental compositions of coal tar and char; (2) modeling of soot from coal and biomass tar; and (3) generalized modeling of coal char annealing, with the eventual goal of a generalized coal-general model of coal char combustion and gasification.

3.1 Elemental Composition of Tar and Char

Modeling coal tar and subsequent reactions of coal tar have largely been ignored or treated lightly in large simulations. For example, some prominent groups use a coal gas mixture fraction approach, where the turbulence chemistry is treated with

T. H. Fletcher (✉) · D. O. Lignell · A. Richards
Chemical Engineering Department, Brigham Young University, Provo, UT 84602, USA
e-mail: tom_fletcher@byu.edu

A. Josephson · T. Holland
Los Alamos National Laboratory, Los Alamos, NM 87545, USA

J. Lyu and S. Li (eds.), *Clean Coal and Sustainable Energy*,
Environmental Science and Engineering,
https://doi.org/10.1007/978-981-16-1657-0_3

a probability density function combining equilibrium chemistry with fluctuations in the mixture fraction (Brewster et al. 1988; Smith et al. 1981; Zhou et al. 2019). This coal gas mixture fraction approach assumes that the elemental composition of all volatiles is the same as the coal and the char. Simulations with two coal gas mixture fractions, one mixture fraction for the volatiles and one for the char, have also been made, assuming that the char is 100% carbon (Flores and Fletcher 2000). Other approaches are to correlate char composition versus burnout (Smoot et al. 1984; Smoot and Smith 1985), or to use a light gas submodel and estimate the coal (and as a result, the tar and char) as excess hydro-carbons (Ahmed and Gupta 2013; Backreedy et al. 1999; Beck and Hayhurst 1990; Smoot et al. 1977), or to assume that the char only contains carbon (Al-Abbas et al. 2013; Chen and Ghoniem 2012; Chen et al. 2012). Some approaches have attempted to use a simple tar species with a 1:1 molar ratio of carbon and hydrogen, such as acetylene or benzene (Musarra et al. 1986), or assume that the tar composition was the same as that of the parent coal (Bradley et al. 2006). While these assumptions can be beneficial to reduce computational time in large simulations, these assumptions are not consistent with measured elemental compositions of tar and char (Perry et al. 2000; Pugmire et al. 1991).

To aid in the two coal gas mixture fraction approach, a correlation is presented here for the elemental composition of different components after pyrolysis, namely tar and char. These correlations can be used to improve simulations based on mixture fraction approaches.

A set of possible correlating variables was first determined, including parent coal composition (ultimate and proximate analysis), chemical structural parameters, heating rate, peak pyrolysis temperature, and residence time. The chemical structural parameters include those used by the CPD model (Fletcher et al. 1992) (molecular weight per cluster, molecular weight per side chain, number of attachments per cluster, and the fraction of attachments that are bridges). However, it seemed logical to include carbon aromaticity as a correlating parameter. Therefore, correlations of carbon aromaticity were reviewed and a new correlation for carbon aromaticity was developed based on the elemental coal composition and ASTM volatile matter. Results of this correlation are shown in Fig. 3.1 (Richards 2019).

The best form of the aromaticity correlation is shown in Eq. 3.1:

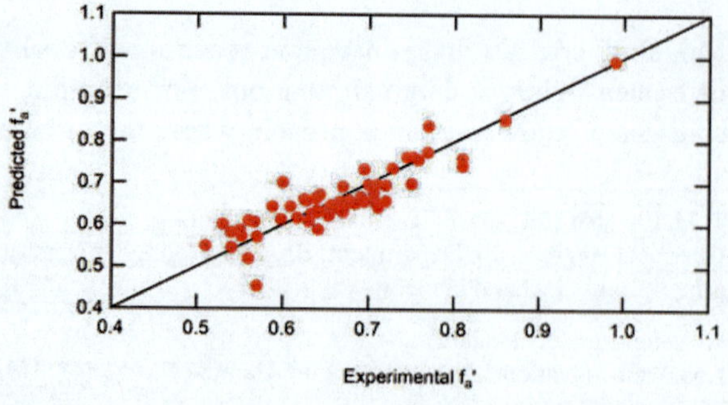

Fig. 3.1 Predicted versus measured carbon aromaticity

$$f'_a = c_1 + c_2 x_C + c_3 x_C^2 + c_4 x_H + c_5 x_H^2 + c_6 x_O + c_7 x_O^2 + c_8 x_{VM} + c_9 x_{VM}^2 \quad (3.1)$$

The elemental composition analysis uses experimental data (Perry et al. 2000; Fletcher and Hardesty 1992; Freihaut et al. 1989; Hambly et al. 1998; Parkash 1985; Tyler 1980; Watt et al. 1996) from literature based on several conditions: primary tar formation (temperatures below 1100 K, so as to limit secondary tar reactions), low residence times (no hold times) and heating rates of at least 1000 K/s. Data from fluidized bed systems were not used unless residence times were available. Compositions and yields are on a dry, ash-free basis (DAF). A detailed cross-validation procedure was used with several statistical measures of error to determine the best fits of elemental compositions of coal tar and char (Richards 2019). Examples of the curve fits for tar and char are shown in Figs. 3.2 and 3.3. Note that if the tar and coal compositions were equal, the data would all line up at a ratio of 1.0 in Fig. 3.2. The char H composition ratio starts at 1.0 and decreases to as low as 0.4, but not to 0.0.

The correlations at first included up to 17 coefficients. A study was made to determine the best correlation with the fewest parameters. The value of R^2/N_{coeff} (N_{coeff} being the number of coefficients in the correlation) was used as a screening test.

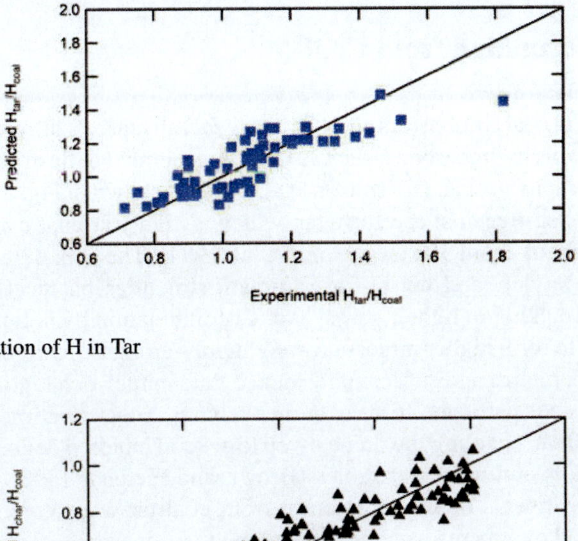

Fig. 3.2 Correlation of H in Tar

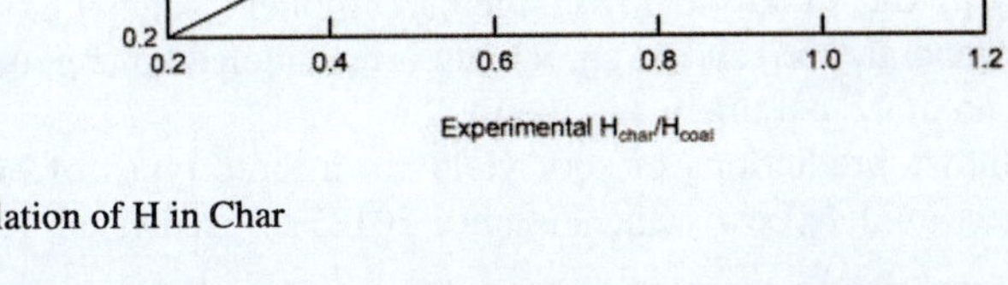

Fig. 3.3 Correlation of H in Char

The coefficients were removed one at a time to see if a better or almost equal fit could be determined with fewer coefficients. The final tar H correlation had 10 coefficients with 4 independent variables: maximum gas temperature ($T_{g,max}$), residence time, actual volatiles yield (V_{norm}), and the molecular weight of a side chain in the parent coal (m_δ):

$$\begin{aligned} H_{tar}/H_{coal} = \ & c_1 + c_2 T_{g,max} + c_3 t_{res} + c_4 t_{res}^2 + c_5 t_{res}^3 \\ & + c_6 V_{norm} + c_7 V_{norm}^2 + c_8 V_{norm}^3 + c_9 M_d + c_{10} m_d^2 \end{aligned} \quad (3.2)$$

The final char H correlation had 10 coefficients with 5 independent variables: $T_{g,max}$, residence time, volatiles yield (V_{norm}), coal H content ($x_{H,coal}$), and ASTM volatiles content (V_{ASTM}).

$$H_{char}/H_{coal} = c_1 + c_2 T_{g,max}^{c3} + c_4 t_{res}^{c5} + c_6 V_{norm}^{c7} + c_8 x_{H,coal}^{c9} + c_{10} V_{ASTM}^{c11} \quad (3.3)$$

Correlations of the composition of coal tar and char have also been developed for the other elements (C, O, N, S) using this data set (Richards 2019; Richards et al. 2019).

3.2 Soot Formation from Coal

Soot formation in coal combustors and gasifiers generally takes a different path than in flames from simple hydrocarbons, such as CH_4. A generalized illustration of this path proposed by Josephson et al. (Josephson et al. 2017) is shown in Fig. 3.4. As much as 30% of the dry ash-free coal can form tar, which is highly aromatic and has a mean molecular weight of about 350 amu (Smith et al. 1994). The bulk of these tars do not break down to acetylene and then re-form aromatic structures, but react by two general pathways: (1) cracking to lighter gases, and (2) combination by polymerization and other reactions to form higher molecular weight poly-aromatic hydrocarbons (PAH). With high temperatures and sufficient residence time in fuel-rich regions, these PAH form incipient soot particles. In near burner regions, soot radiation transfers heat away from the flame, cooling the flame by up to several hundred degrees and heating surrounding walls and nearby particles (Brown and Fletcher 1998; Fletcher et al. 1997). Early treatments of soot formation from coal tar were very empirical, but more generalized treatments have been developed (Josephson et al. 2017).

Figure 3.5 shows coal soot predictions for a flat-flame burner in comparison to experimentally measured data. Although not as good as a curve fit, these predictions are based on the predicted tar yield from the CPD model (Fletcher et al. 1992), and hence are more general. The relative agreement is considered quite good considering that there were no tuned parameters in the model.

Figure 3.6 shows predictions of soot yield from three types of biomass using the CPD-biomass model (Lewis and Fletcher 2013) to predict tar yield from the

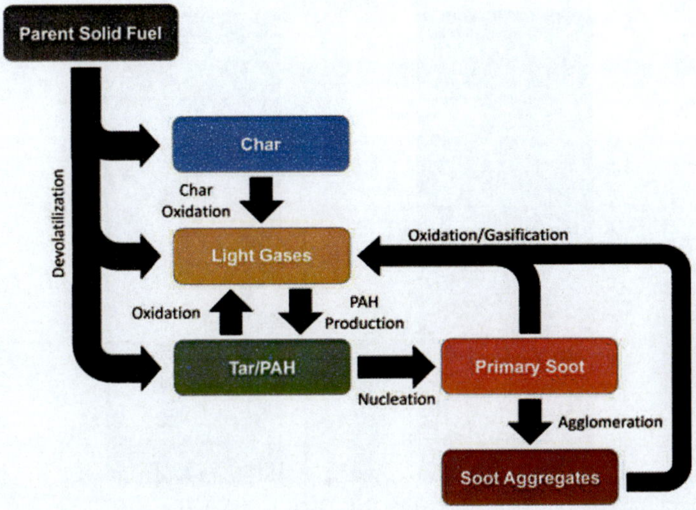

Fig. 3.4 General mechanism for soot formation in coal systems (Josephson et al. 2017). Reprinted with permission, copyright (2017) American Chemical Society

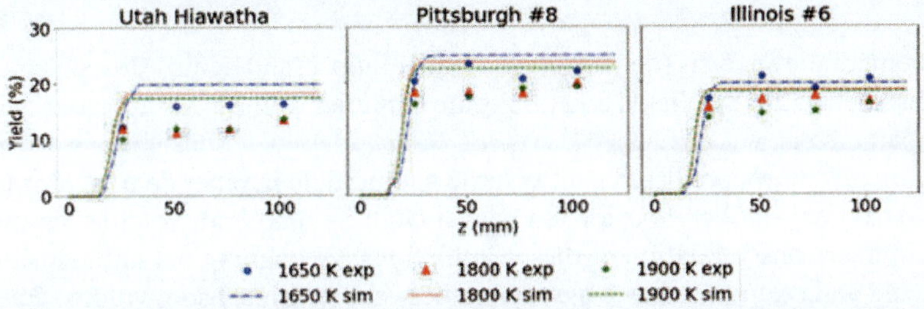

Fig. 3.5 Soot predictions versus coal data from Ma et al. (1996) made by Josephson (2018), Josephson et al. (2018) Reprinted with permission from Elsevier. Copyright 2018

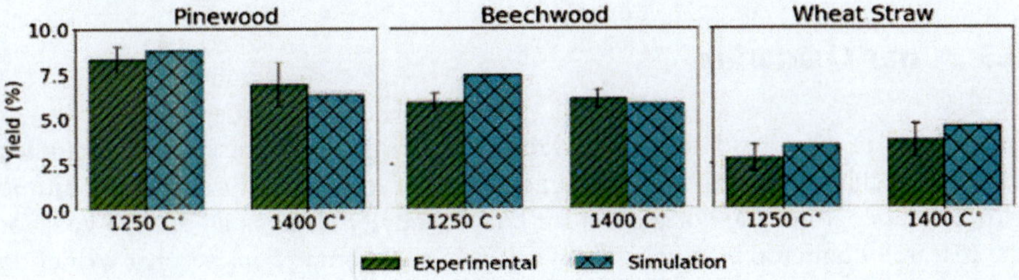

Fig. 3.6 Soot predictions versus biomass data made by Josephson (2018), Josephson et al. (2018). Reprinted with permission from Elsevier, Copyright 2018

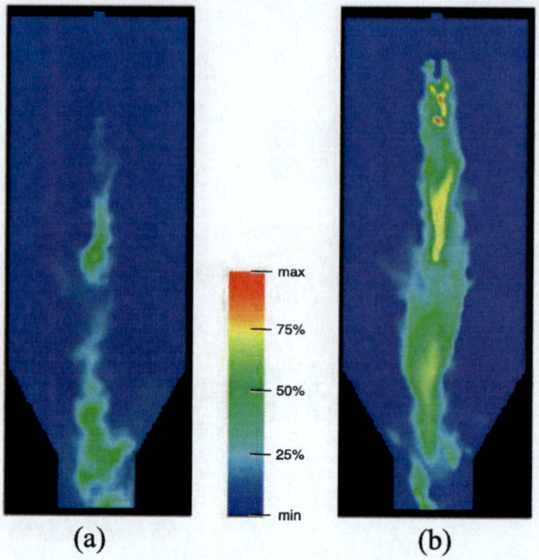

Fig. 3.7 **a** Soot volume fraction and **b** particle number density predictions for soot in a biomass-fed pilot-scale oxy-fuel combustor

basic component analysis (i.e., cellulose, hemicellulose, and lignin) (Josephson 2018; Josephson et al. 2018). Yields and trends are captured well by this approach with no tuned parameters.

Figure 3.7 shows predicted soot volume and particle number density profiles for a pilot-scale oxy-fuel combustor fed with a Utah Skyline high-volatile bituminous coal. Comparisons/validation of these simulations are ongoing but initial results are promising and seem to indicate agreement between predicted soot volume fractions and optically measured values. Radiative effects of these soot concentrations are currently being quantified.

3.3 Char Oxidation

Treatments of char oxidation range from simple global expressions to complex surface mechanisms with detailed treatments of pore structure and evolution. However, a coal-general correlation for char reactivity remains an elusive goal for the research community. A sensitivity analysis of a comprehensive char oxidation model revealed that the most important processes (other than the chemical kinetic coefficients) were annealing, the oxidation reaction order, particle swelling, and the mode of burning (i.e., diameter and density changes) (Holland and Fletcher 2016). Effects of char formation conditions such as heating rate, temperature, residence time, and pressure have been explored but are not yet clearly understood. Changes in intrinsic char reactivity with heating conditions and residence time are sometimes

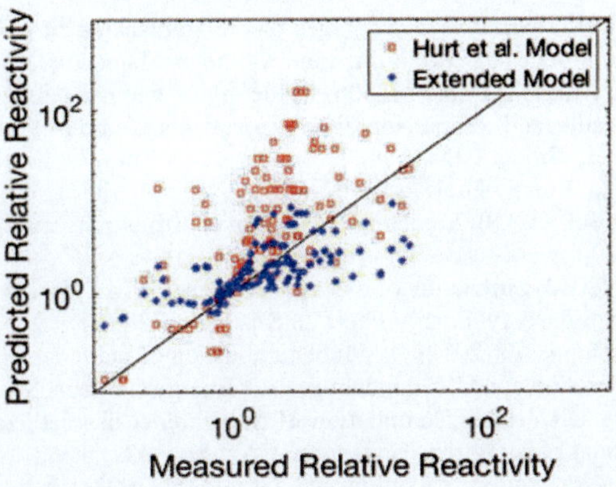

Fig. 3.8 Comparison of improved coal char annealing model with Hurt et al. model (from Holland et al. (2017)). Reprinted with permission, copyright (2017) American Chemical Society

treated with char annealing models. An attempt at a coal general annealing model was developed by Holland et al. (2017), Holland and Fletcher (2016). This model includes effects of coal type through a chemical structure parameter, heating rate, and peak particle temperature. The model attempts to describe the changes in char reactivity to O_2 during both the initial heatup and pyrolysis region as well as during char oxidation. Figure 3.8 shows the comparison of the improved char annealing model versus the previous annealing model developed by Hurt et al. (1998). In this figure, the intrinsic reactivity of a specific char is normalized to a reference char for that same set of experiments. The reference char may have been more reactive or less reactive than the char of interest. The error factor for the new extended model is a factor of four lower for the new model (Holland et al. 2017).

Even with the annealing model, it is currently not possible to make an accurate prediction of char oxidation from first principles without some tuning of coefficients based on experiments performed at the conditions of interest.

Acknowledgements This material is partially based upon work supported by the Department of Energy, National Nuclear Security Administration, under Award Number DE-NA0002375.

References

Ahmed II, Gupta AK (2013) Experiments and stochastic simulations of lignite coal during pyrolysis and gasification. Appl Energy 102:355–363

Al-Abbas AH, Naser J, Hussein EK (2013) Numerical simulation of brown coal combustion in a 550 MW tangentially-fired furnace under different operating conditions. Fuel 107:688–698

Backreedy RI, Habib R, Jones JM, Pourkashanian M, Williams A (1999) An extended coal combustion model. Fuel 78(14):1745–1754

Beck NC, Hayhurst AN (1990) The early stages of the combustion of pulverized coal at high-temperatures. 1. The kinetics of devola-tilization. Combust Flame 79(1):47–74
Bradley D, Lawes M, Park HY, Usta N (2006) Modeling of laminar pulverized coal flames with speciated devolatilization and comparisons with experiments. Combust Flame 144(1–2):190–204
Brewster BS, Baxter LL, Smoot LD (1988) Treatment of coal devolatilization in comprehensive combustion modeling. Energy Fuels 2(4):362–370
Brown AL, Fletcher TH (1998) Modeling soot derived from pulverized coal. Energy Fuels 12(4):745–757
Chen L, Ghoniem AF (2012) Simulation of oxy-coal combustion in a 100 kW(th) test facility using rans and les: a validation study. Energy Fuels 26(8):4783–4798
Chen L, Yong SZ, Ghoniem AF (2012) Oxy-fuel combustion of pulverized coal: characterization, fundamentals, stabilization and CFD modeling. Prog Energy Combust Sci 38(2):156–214
Fletcher TH, Hardesty DR (1992) Compilation of Sandia coal devolatilization data: Milestone report. Sandia National Laboratories, Livermore, CA, May 1992, p 362
Fletcher TH, Kerstein AR, Pugmire RJ, Solum MS, Grant DM (1992) Chemical percolation model for devolatilization. 3. Direct use of C-13 NMR data to predict effects of coal type. Energy Fuels 6(4):414–431
Fletcher TH, Ma J, Rigby JR, Brown AL, Webb BW (1997) Soot in coal combustion systems. Prog Energy Combust Sci 23(3):283–301
Flores DV, Fletcher TH (2000) The use of two mixture fractions to treat coal combustion products in turbulent pulverized-coal flames. Combust Sci Technol 150(1–6):1–26
Freihaut JD, Proscia WM, Seery DJ (1989) Chemical characteristics of tars produced in a novel low-severity, entrained-flow reactor. Energy Fuels 3(6):692–703
Hambly EM, Fletcher TH, Solum M, Pugmire RJ (1998) Solid-state 13-C NMR analysis of coal tar and char. Abstr Pap Am Chem Soc 215:U607–U607
Holland T, Fletcher TH (2016) Global sensitivity analysis for a comprehensive char conversion model in oxy-fuel conditions. Energy Fuels 30(11):9339–9350
Holland T, Bhat S, Marcy P, Gattiker J, Kress JD, Fletcher TH (2017) Modeling effects of annealing on coal char reactivity to O_2 and CO_2, based on preparation conditions. Energy Fuels 31(10):10727–10744
Hurt R, Sun J-K, Lunden M (1998) A kinetic model of carbon burnout in pulverized coal combustion. Combust Flame 113(1):181–197
Josephson A (2018) Modeling soot formation derived from solid fuels. PhD Dissertation, Chemical Engineering Department, Brigham Young University, Provo, UT
Josephson AJ, Gaffin ND, Smith ST, Fletcher TH, Lignell DO (2017) Modeling soot oxidation and gasification with Bayesian statistics. Energy Fuels 31(10):11291–11303
Josephson AJ, Linn RR, Lignell DO (2018) Modeling soot formation from solid complex fuels. Combust Flame 196:265–283
Lewis AD, Fletcher TH (2013) Prediction of sawdust pyrolysis yields from a flat-flame burner using the CPD model. Energy Fuels 27(2):942–953
Ma J, Fletcher TH, Webb BW (1996) Conversion of coal tar to soot during coal pyrolysis in a post-flame environment. Symp (Int) Combust 26(2):3161–3167
Musarra SP, Fletcher TH, Niksa S, Dwyer HA (1986) Heat and mass transfer in the vicinity of a devolatilizing coal particle. Combust Sci Technol 45(5–6):289–307
Parkash S (1985) True density and elemental composition of subbituminous coals. Fuel 64(5):631–634
Perry ST, Hambly EM, Fletcher TH, Solum MS, Pugmire RJ (2000) Solid-state C-13 NMR characterization of matched tars and chars from rapid coal devolatilization. Proc Combust Inst 28:2313–2319
Pugmire RJ, Solum MS, Grant DM, Critchfield S, Fletcher TH (1991) Structural evolution of matched tar-char pairs in rapid pyrolysis experiments. Fuel 70(3):414–423

Richards A (2019) Coal pyrolysis models for use in massively parallel oxyfuel-fired boiler simulations. PhD Dissertation, Chemical Engineering Department, Brigham Young University, Provo, UT, in preparation

Richards AP, Johnson C, Fletcher TH (2019b) Correlations of the elemental compositions of primary coal tar and char. Energy Fuels 33(10):9520–9537

Smith PJ, Thomas HF, Smoot LD (1981) Model for pulverized coal-fired reactors. Symp (Int) Combust 18(1):1285–1293

Smith KL, Smoot LD, Fletcher TH, Pugmire RJ (1994) The structure and reaction processes of coal. Plenum Press, New York, p 471

Smoot LD, Smith PJ (1985) Coal combustion and gasification. Plenum Press, New York

Smoot LD, Horton MD, Williams GA (1977) Propagation of laminar pulverized coal-air flames. Symp (Int) Combust 16(1):375–387

Smoot LD, Hedman PO, Smith PJ (1984) Pulverized-coal combustion research at Brigham Young University. Prog Energy Combust Sci 10(4):359–441

Tyler RJ (1980) Flash pyrolysis of coals. Devolatilization of bituminous coals in a small fluidized-bed reactor. Fuel 59(4):218–226

Watt M, Fletcher TH, Bai S, Solum MS, Pugmire RJ (1996) Chemical structure of coal tar during devolatilization. Symp (Int) Combust 26(2):3153–3160

Zhou M-M, Parra-Álvarez JC, Smith PJ, Isaac BJ, Thornock JN, Wang Y, Smith ST (2019) Large-eddy simulation of ash deposition in a large-scale laboratory furnace. Proc Combust Inst 37(4):4409–4418

Chapter 4
The Change in Bed Materials Size Distribution and Its Effect on CFB Boiler Operation

G. A. Ryabov and O. M. Folomeev

Abstract When burning solid fuels in circulating fluidized bed boilers (CFB), it is important to ensure a large multiplicity of particle circulation along the loop of the furnace–cyclone–return system. The high flow rate of circulating particles determines the uniformity of the temperature field along the height of the furnace, ensures temperature equalization along the bed section, which is especially important for large CFB boilers with a furnace section of 200 and more m^2. The size of circulating particles is determined by the efficiency of capture in the cyclone, the higher it is, the smaller the average particle size. It is believed that a well-operated CFB boiler has an average size of circulating particles less than 0.2 and even 0.15 mm. Another factor determining the bed material size distribution is the effective removal of bottom ash, which is especially important for high-ash fuels. The first in Russia large CFB boiler of unit #9 of Novocherkassk TPP has some problems with overly large part of course particles in the bed. The report presents the characteristics of the fuel, bed material and circulating ash compositions. The estimate of circulating flow rate was made. The main characteristics of the boiler, including the temperature field and estimation of circulating flow rate are presented. Special attention is paid to the regimes with the addition of limestone and fly ash. The operating date of changes in the chemical and particle size distribution (PSD) of ash streams is presented. According to results of the study of changes in ash composition, it can be concluded that the composition of fly and circulating ash varies quite rapidly and reaches design values after 10 h (according to CaO). In bottom ash the PSD varies slowly (the order of magnitude is 50 h). It is possible that the fractional composition of bed material also varies slowly. The simplified model for calculation of changes in bed material size distribution after start-up of the boiler is presented. It takes into account the initial composition of sand and coal supplied. The calculation is carried out to determine the specific removal of particles from the bed. The capture efficiency is calculated by dependences of VTI and corrected taking into account published data and results of experiments. According to experimental data, the coefficients of bed material size distribution and the removed bottom ash are introduced. According to the results of calculations, it is shown that for the available fuel fractional composition, the number of large fractions

G. A. Ryabov (✉) · O. M. Folomeev
All-Russia Thermal Engineering Institute (VTI), 14 Avtozavodskaya st., Moscow 116280, Russia

J. Lyu and S. Li (eds.), *Clean Coal and Sustainable Energy*,
Environmental Science and Engineering,
https://doi.org/10.1007/978-981-16-1657-0_4

increases over time. The addition of limestone leads to an increase in fractions with sizes of 0.1–0.2 mm. the Best results are obtained when feeding coal with sizes less than 5 mm.

Keywords Circulating fluidized bed boilers · Bed material size distribution · Circulation rate · Capture efficiency · Limestone and fly ash addition

4.1 Introduction

A key parameter for the proper operation of a circulating fluidized bed boiler (CFB) is the particle size distribution (PSD) of the bed inventory. The solid concentration distribution along the height of furnace takes an important role in coal combustion and heat transfer in CFB boiler, and it is mainly determined by the PSD and bed inventory (Blaszczuk et al. 2012; Redemann et al. 2009; Werner 2001). There are a lot of main parameters such as: particle size distribution (Lin et al. 2011), sphericity for bed inventory particles, gas velocity (Gao et al. 2009), density of fluidizing medium, and solid suspension density (Qi et al. 2008), which has an impact on properly CFB combustion behaviour. The high flow rate of circulating particles determines the uniformity of the temperature field along the height of the furnace, ensures temperature equalization along the bed, which is especially important for large CFB boilers with a furnace section of 200 and more m^2.

The PSD of the bed material is determined by a number of factors, among which the most important are the PSD of coal, its ability to crack and attrition, temperature and mixing in the bed (gas velocity, share of primary air), the addition of limestone, sand, ash recycling with its fractional compositions. The efficiency of particle capture in cyclones determines the size distribution of the circulating material, and the removal of bottom ash should reduce the proportion of coarse fractions. Reducing in bed the particle size, ceteris paribus, leads to an increase in the concentration at the cyclone entrance and improve the capture of small particles.

The entrainment flow rate from the bed depends on the bed inventory. Thus, to increase circulation it is necessary to have the maximum possible number of circulating particles in the bed. For CFB boilers such particles are typically in the range of sizes 0.1–0.3 mm. As shown in Redemann et al. (2009), the maximum residence time refers to particles with sizes 0.2–0.3 mm and is 4 and more hours. A typical PSD graph for large CFB boilers is given in Basu (2006) (Fig. 4.1).

In Guangxi (2015), the position on the effective composition of bed material with a maximum number of particles 0.1–0.3 mm, which allows you to go completely into the fast bed mode from turbulent bed mode. This is achieved by the appropriate composition of the feed material, increasing the capture efficiency of fine particles. In Fig. 4.2 the parameters of size distribution and capture efficiency are given (Guangxi 2015).

To calculate the concentration profile along the height of the furnace, heat exchange to the wall screen, particles entrainment from the bed there are many

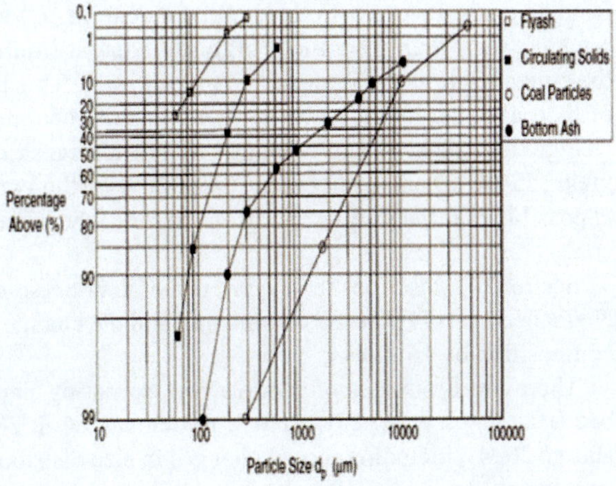

Fig. 4.1 Size distributions of coal particles and different solid streams in a large CFB boiler (Basu 2006)

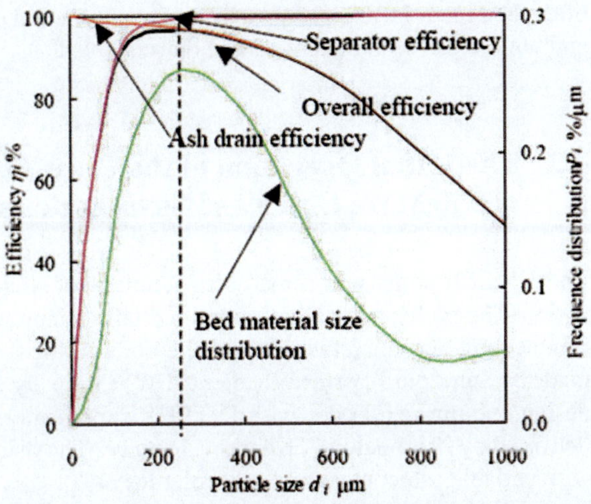

Fig. 4.2 The collection efficiency and the particles size distribution according to Guangxi (2015)

well-tested recommendations. This also applies to the determination of the cyclones capture efficiency, for example, Muschelknautz and Trefz (1993).

It is more difficult to take into account the factors of changes in the size of fuel particles during combustion, taking into account their tendency to fragmentation and attrition, as well as possible agglomeration processes. These issues cannot be resolved without pilot combustion of specific fuels. Primary and secondary particles fragmentation is distinguished and primary fragmentation for most fuels is of the greatest importance. Fragmentation is more prone to fuel with a large yield of volatile (brown coals, biomass).

Particle size changes during combustion and attrition depend on the internal ash structure, the gas velocity and temperature in the bed. Our studies of a number of

Russian coals have shown that anthracite calm (AC) with an extremely low yield of volatiles (less than 4%) and a high ash content (more than 30%) characterized by the almost complete absence of changes in size. This is due to the solid whole frame of the particles. In addition the fuel contains incombustible inclusions that do not change their sizes during combustion. For Kuznetsk coals with a moderate volatile yield (25–30%) and ash content up to 20% of the particles are strongly crushed. It is possible that the graphs in Fig. 4.1 refer to the same type of fuel as the Kuznetsk coals.

Section 4.2 describes the experience of the once-thought CFB boiler of unit #9 of Novocherkassk TPP at AC combustion and the change in the fractional and chemical composition of ash flows.

There are a number of models for estimating particle size distributions in the bed (Blaszczuk et al. 2012, 2013; Redemann et al. 2009; Yang 2003; Werther and Hartge 2004), including model changes in size distribution of bed material at star-up operation (Ma et al. 2015). In this paper, approaches (Ma et al. 2015) are used to determine the effect of size distribution of fuel, sand, limestone on the composition of the bed material. The comparison of experimental data on the PSD of circulating and bottom ash with the calculated ones is given.

4.2 The Initial Operation of the Once-Thought CFB Boiler of 330 MWe Unit #9 of Novocherkassk TPP

The first CFB boiler was constructed at unit #9 of Novocherkasskaya TPP of Rostov region. The boiler was delivered and partially manufactured by JSC "EMAlliance". Engineering and manufacturing of the boiler critical parts and its equipment were made by Sumitomo-Foster Wheeler (SFW) Company. It is a typical SFW "compact" design one through boiler with INTREX superheaters. The 330 MWe turbine was delivered by "Turboatom", Kharkov, Ukraine. The main fuel is Anrtacite Culm (AC) with variable ash content and low volatiles (4%).

Based on several AC probes analysis, actual range of the AC heating value was determined at level 4500–5100 kcal/kg within ash variety of 24–35% and moisture of 6–12%. Key components of the boiler are furnace, solid separators, INTREX chambers, cross-over ducts and convective cage. These elements are consisting of the gas-tight membrane walls. In the upper part of the furnace wingwalls are located; ECO and reheater I/II surfaces can be found within the convective cage area. Integrated bed ash heat exchangers (INTREX) are acting as the final stage of the superheater.

During first startup experience high deviation of the furnace bed temperature (up to 150 °C) was determined: maximal values were registered in the middle area of furnace cross section and the points of minimal temperature values located near the side walls (Fig. 4.3).

Flue gas duct temperature value analysis revealed good correspondence between calculated and actual values. Typical furnace temperature profile is given in Fig. 4.4.

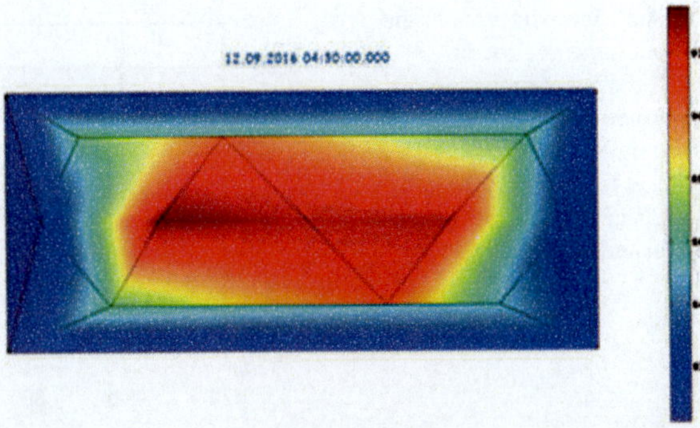

Fig. 4.3 Furnace bed temperature gradient

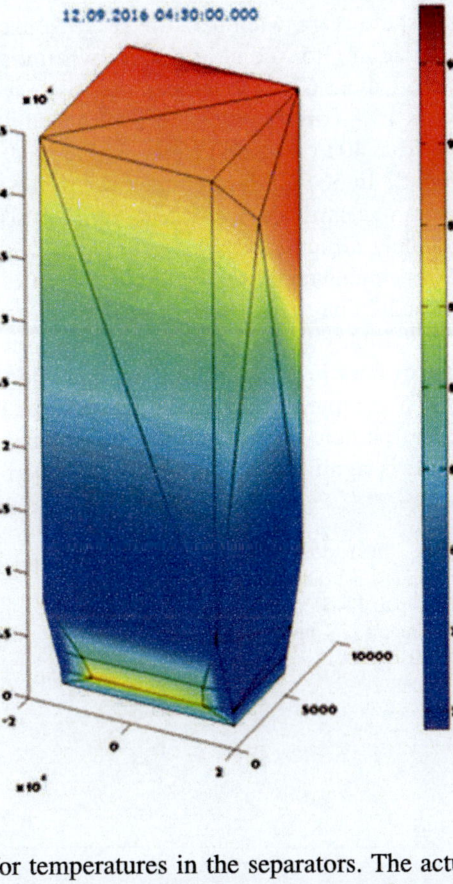

Fig. 4.4 Typical furnace temperature profile

A significant difference is observed for temperatures in the separators. The actual value of the temperature in the separators is much higher than the one at the outlet of the furnace, indicating the burning of fine particles in the separators. Also this phenomenon was registered at Chinese CFB boilers with anthracite combustion (Yue et al. 2006). Figure 4.5 shows the dependence of average temperature in cyclones

Fig. 4.5 Temperatures in the cyclones and at furnace outlet during combustion of low grade fuels. 1—AC firing boilers in China, 2—data for 70 and 90% BMCR cases, Novocherkassk TPP

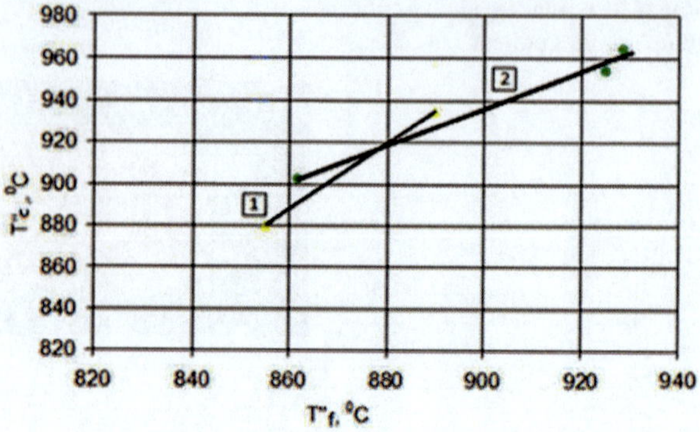

against the average temperature at the furnace outlet.

According to Yue et al. (2006) when a share of particles smaller than 0.4 mm reached about 63% the temperature rise exceeded 50 °C, and at the same share as low as 15% corresponding value dropped down to below 30 °C. Increasing of air flow from 40 to 44 m^3/h ceteris paribus leads to a slight increase in temperature rise (from 28 to 36 °C). It is pointed out in the paper that one of the main influencing factors was particles size distribution. It's also stated that on the existing boilers it's extremely difficult to completely avoid coal burn-out in the cyclones. Fang et al. (2006) implemented boiler mode, which allowed increasing boiler efficiency by almost 2% due to reduction of heat losses caused by unburned carbon.

A possible reason for the temperature growth in the cyclones could be a substantial share of fines in the fuel. Figure 4.6 provides requirements for coal PSD stipulated by SFW company. Both actual coal samples PSD curves are close to each other. The average particle size is about 1 mm, which is close to SFW design data. But a share of fines is significantly higher: a share of particles less than 0.2 mm is approximately

Fig. 4.6 The residue on the R sieves against the particle size of coal: 1—SFW requirements; 2—samples from 2016 to 2018

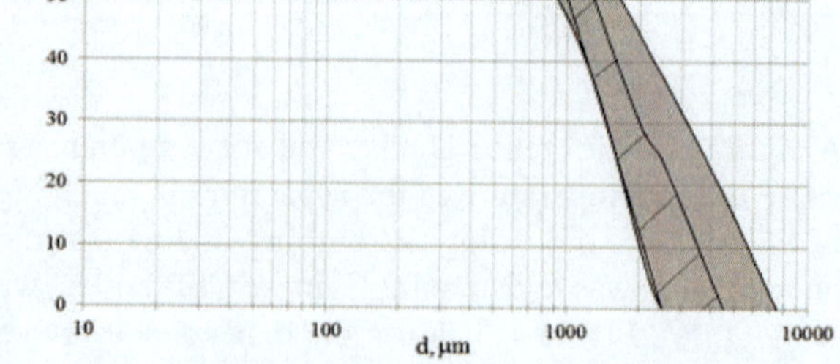

Table 4.1 Heat balance tests of the boiler

Name of quantity	Dimension	Data			
		310 MWe	310 MWe, SFW	193 MWe	193 MWe, SFW
Total heat of evaporating tubes	Gkal/h	231.1	232.9	167.5	183.4
Temperature difference	°C	621	539	506	527
Heat transfer coefficient	Kkal/m^{2}h°C	105.6	122.6	93.9	98.7

15%, whereas it should not exceed 5% according to SFW requirements. Also high (up to 10%) is a share of coarse particles (larger than 6 mm), which should not be present at all.

The heat balance tests of the boiler on the loads of 310, 260 and 194 MWe showed that the experimental values of heat fluxes in heating surfaces practically coincide with the design values (Ryabov et al. 2018). At the same time, the thermal capacity of the boiler was 96, 82 and 69% of the nominal (BMCR) in these modes. The boiler's efficiency, determined from the reverse balance, at these capacities was 92.30%, 91.58% and 90.24% respectively, which is higher than the design data.

When experimental and design values of heat flux in furnace wall are close to each other, the gas temperature in the furnace is substantially higher than the design temperature. A possible cause is an increased share of the radiant component of heat transfer to walls and a lower share of convective heat transfer than projected ones (Table 4.1).

It is known that the conductive component of the heat transfer coefficient in CFB boilers can reach 20%, depending on the particle size and their concentration near the furnace walls. Therefore, to increase the concentration of particles in the furnace over the bed and improve heat transfer to walls, it is advisable to transfer the boiler to the design mode of operation with the supply of limestone and recirculation ash, which will increase the amount of particles of the "right" size. Another factor that leads to a reduction the amount of circulating particles in comparison with the design one may be inefficient performance of solid particle separators.

The flow rate of the circulating material can be estimated from the thermal balance of the ash heat exchangers INTREX. The most thorough experiments were made on INTREX-4 with a capacity of about 300 MWe. This flow rate was at 130 kg/s, including the internal particles circulation from the furnace to the heat exchanger. The total consumption from four cyclones is estimated as 250–300 kg/s.

The PSD of the bed material can be changed by feeding limestone and ash recycling from the first stage of the precipitator. Unfortunately, the limestone supply system was not able to work for a long time due to malfunctions of its pneumatic supply to the boiler silo. Below are the results of testing operation for three days with the supply of limestone and one day with the supply of limestone and ash recycling.

The actual values of limestone particle sizes are close to the design ones. Almost 80% of these particles can escape from the furnace and be captured in separators, increasing the flow rate of circulating material.

In the one day tests flow rate of limestone was about 9.7 t/h. We could not to estimate of recycling ash flow. The average bed temperature during the experiments was within the usual limits (there was a slight decrease in the average temperature). However, there was a noticeable non-uniformity of bed temperatures. Thus according to the thermocouples at the front of the boiler the maximum difference in time-averaged experiments, the temperature dropped from 60 to 43 °C, rear—43–35 °C. At the same time, there was an increase of thermocouples with relatively low temperatures and a decrease of thermocouples with relatively high temperatures. This fact indirectly indicates an improvement in the mixing of the bed material. The temperature difference in the INTREX 4 was 120 °C before the experiment and 110 °C when feeding limestone and ash recycling. This resulted in an increase in circulation flow rate through a heat exchanger with 138–146 kg/s.

The PSD of circulating ash also changed in the direction of reducing the average size (Fig. 4.7).

Experiments with long-term supply of limestone were carried out in the typical operation mode with a daily electrical load of about 280 MWe and night—about 196 MWe. Coal flow rate was about 130 t/h during the day and 90 t/h at night. On the first day, the limestone flow rate was about 7 t/h, in the second it was increased to 9.4 t/h and further reduced to 5.4 t/h.

The average bed temperature slightly decreased (up to 15 °C at a load of about 280 MW). This value is close to the calculated by the reactions of limestone with sulfur in fuel—the heat of these reactions (taking into account the physical heat of about 5.3 Gkal/h) should lead to a bed temperature decreasing by 15 °C. At a load of about 280 MW, bed temperature is approximately equal or lower than the temperature in separators, and then when the load drops to 200 MW, the bed

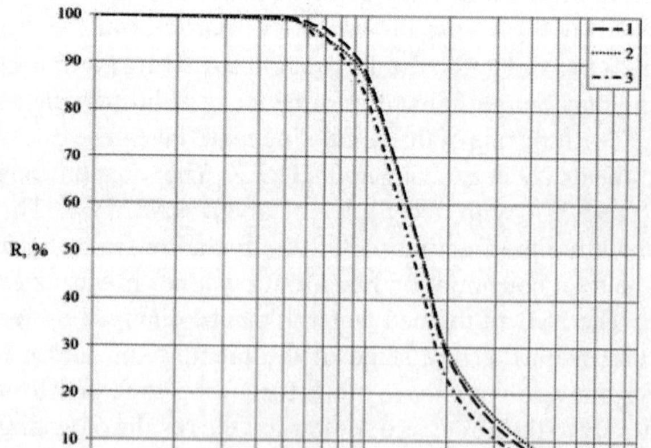

Fig. 4.7 PSD of circulation ash in one day experiments with limestone and ash return consumption. 1—Without of limestone); 2—9.7 t/h limestone; 3—9.7 t/h limestone + recirculated ash

temperature exceeds the temperature in separators. The thermocouple indications and averaged temperature values in the separators were significantly closer now (deviation from average temperature in each separator decreased from 26 to 13 °C).

Analysis of temperature regime in the ash return system and ash heat exchangers showed a steady tendency to increase the gas temperature after all ash exchangers in modes with limestone feed. The temperature in separators and standpipes (wall seals) even decreased somewhat. Based on the heat balance of INTREX 3 and 4 heat exchangers, the circulating material rate can be estimated. If it is assumed that this flow is approximately the same for each heat exchanger, then the range of total circulation rate in regimes without supply of limestone at a power of 280 MWe is 250–300 kg/s. Large values relate to the calculation of maximum temperatures for heat exchanger, the smaller values refer to the average. When feeding limestone, the circulating material flow rate increased by approximately 20%. The average circulation particle diameter by Sauter prior to limestone feeding was 0.241 mm, and in regimes with limestone feeding it decreased to 0.203–0.218 mm. The carbon content in samples does not exceed 1%, which indicates a good burnout of the fuel.

Chemical analysis of bottom, circulating and fly ash samples was performed during experiments with long limestone supply. The results were important for the assessment of changes in the chemical composition over time with the achievement of stable values. In the absence of limestone, the chemical composition of all samples is close to composition of fuel ash. All ash flows during the boiler operation differs from anthracite ash with an increasing concentration of CaO and SO_3 with lime supply of limestone while the components of other oxides are reduced. It can be noted that the design data of JSC VTI and the typical compositions of ash flows are satisfactory according to the data of boiler supplier. It is of interest to change the content of CaO, $CaSO_4$ and SO_3 in time from the beginning of the limestone supply. At thc samc timc, the supply of recirculating ash can somewhat accelerates the growth in concentrations of these components, since the fly ash contains more of it than other ash streams. Figures 4.8 and 4.9 shows curves for changes in concentrations of CaO and SO_3 in

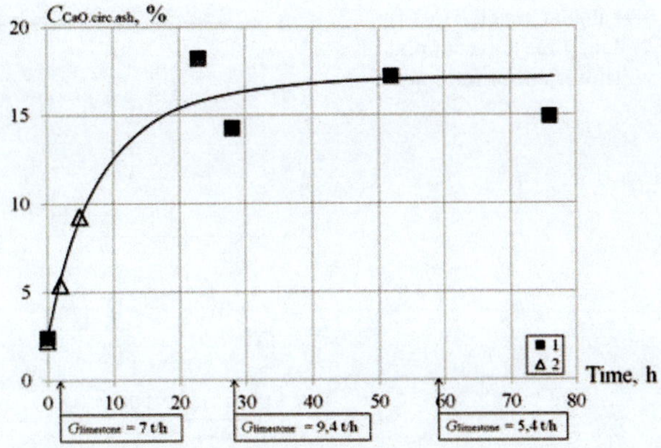

Fig. 4.8 Content of CaO in circulating ash when limestone is feeding in furnace: 1—long limestone supply; 2—short limestone supply

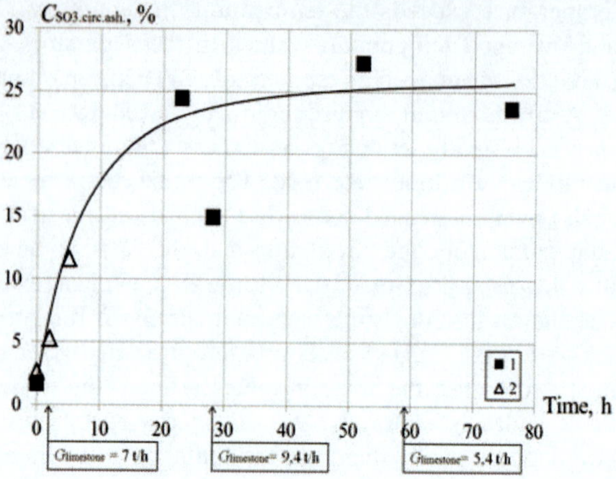

Fig. 4.9 SO_3 content in circulating ash when limestone is feeding in furnace: 1—long limestone supply; 2—short limestone supply

circulating ash. There are also points from the experiments with a short-term feeding of limestone.

From the consideration of these figures it follows that approximately 24 h after the beginning of limestone supply, the concentrations of these components stabilize. For bottom ash, the stabilization time was not reached (Fig. 4.10). Apparently, the bed size distribution is also slowly changing when limestone is fed. The order of the stabilization time is 50 h. Perhaps this is due to the fact that the reactions of the limestone compound with sulfur in fuel happen on the surface of the bed and in the upper part of furnace.

Thus, it can be concluded that the key issue for achieving the design performance of a CFB boiler is to coming the required composition of bed material with an

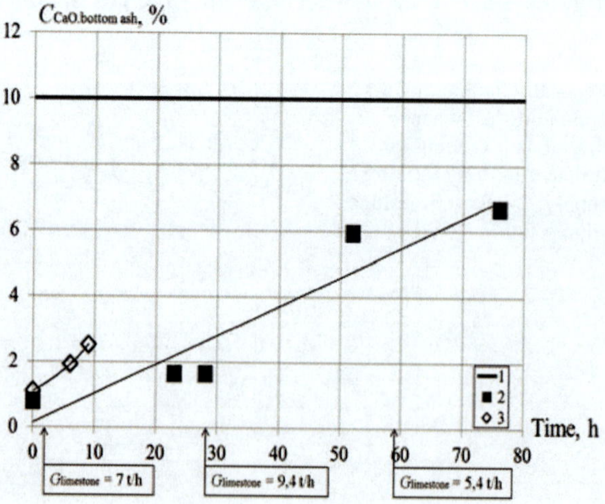

Fig. 4.10 CaO content in bottom ash when limestone is fed into the furnace: 1—calculated value (VTI); 2—long limestone supply; 3—short limestone supply

increase in the external ash circulation through the combustion circuit (furnace–cyclones–return systems). It can be provided by certain fractional fuel composition, limestone supply and recirculation ash. It will be possible to redistribute the primary and secondary air streams.

To understand the results of the initial operation of CFB boiler unit #9 of Novocherkassk TPP, the calculated estimates of the influence of various factors (fuel particle size, limestone and ash recirculation) on the change in the bed size distribution were made.

4.3 Model for Estimation of Changes in the PSD of the Bed Material

The change in the size distribution of the bed material can be estimated by material balance. The CFB boiler on the solid material balance is a system with one inlet (fuel, limestone, recirculation ash) and two outlets (bottom and fly ash not caught in cyclones). In Ma et al. (2015) considers the issues of changes in the PSD of bed material in time after boiler start-up when filling for the first bed of sand. In this work, it is believed that the reference time of the change in the composition of the bed material begins directly from the output to a certain constant load. That is, the processes of initial heating at start-up with gas or oil burners are neglected. To simplify, it is believed that the furnace already receives fuel ash with a fixed particle size distribution. It is commonly believed that the PSD of this ash is smaller than the fuel composition due to initial fragmentation and burnout.

The bed material is divided into N fractions according to the particle diameter, and the following material balance equation is written for each fraction:

$$M\frac{dX_i}{dt} = m_{fs}(i) - m_{fa}(i) - m_{ba}(i) + m_{att}(i) \tag{4.1}$$

where M is the bed inventory, kg; Xi—mass fraction of the bed material of class i; $m_{fs}(i)$—mass flow rate of coal ash class i, kg/s; $m_{fa}(i)$—mass flow rate of fly ash particles (cyclone entrainment) of class i, kg/s; $m_{ba}(i)$—mass flow rate of bottom ash particles of class i, kg/s; $m_{att}(i)$—change in the mass flow rate of the class i bed due to iteration, kg/s.

Flow rate of fly ash is determined using the saturated entrained flow rate (constants of outflow, E_i) and fractional separation efficiency (η_i) of:

$$m_{fa}(i) = (1 - \eta_i) \cdot E_i \cdot X_i \cdot A \tag{4.2}$$

In our calculations, we used a well-known empirical dependence for the outflow constant (Kunii and Levenspiel 1991; Leva 1959):

$$E_i = 0.011 \cdot \left(1 - U/U_o\right)^2 \cdot \rho_s \tag{4.3}$$

where U_o is the superficial gas velocity in the furnace, m/s; ρ_s–particle density, kg/m^3; U—transport velocity, m/s, calculated for each fraction by the formula:

$$U(i) = (d_i)^{1.14} \cdot (0.153 \cdot g^{0.71} \cdot \rho_s^{0.71}) \Big/ (\rho_g^{0.29} \cdot \mu_g^{0.43}) \tag{4.4}$$

where ρ_g is the gas density, kg/m^3, μ_g—dynamic viscosity of the gas, Ns/m^2.

Calculations according to the formulas (4.3), (4.4) at the velocity and temperature of gases in the furnace are characteristic of the modes of operation of the CFB boiler unit #9 of Novocherkassk TPP at load of 310–290 MWe showed that particles with dimensions of more than 0.6 mm are practically not entrained. For further calculations, the following range of fractions was adopted: 0–0.05 mm; 0.05–0.1 mm; 0.1–0.2 mm; 0.2–0.3 mm; 0.3–0.4 mm; 0.4–0.5 mm; 0.5–0.6 mm; 0.6–1 mm; 1–3 mm; 3–8 mm. Total 10 fractions characteristic of the bed material and fuel particles supplied.

Fractional efficiency of the cyclone was determined by the method of VTI (Ryabov et al. 2002). This method is based on the three-zone model proposed in Dietz (1981) and got its development in Abrahamson and Allen (1987) for high-temperature cyclones. In this model, three zones are considered: the entrance zone, the downflow and the lifting flow. Turbulent mixing is determined by the profile of radial concentrations in each zone, taking into account the exchange of particles. The proposed in Dietz (1981) system of equations for each zone under the appropriate assumptions and transformations performed by the authors (Ryabov et al. 2002) leads to the following set of dependencies for the calculation of fractional efficiency:

$$\eta_i = 1 - (k_0 - \sqrt{k_1^2 + k_2}) \cdot \exp(-f(d)) \tag{4.5}$$

where k_0, k_1 and k_2—depend on the ratio of the diameters of the cyclones and the exhaust pipe, as well as on the function of the particle diameter.

The function of the particle diameter $f(d)$ was determined on the basis of generalization of data on high-temperature cyclones and own experimental data. As a result of processing this data, the following dependence was obtained:

$$f(d) = 0.565 \cdot \left(\frac{h_e - \frac{a}{2}}{l}\right)^{0.44} \cdot \left(\frac{t_g + 273}{293}\right)^{0.3} \cdot \left(\frac{d_i}{d_p}\right)^{1.04} \tag{4.6}$$

where h_e is the height of the exhaust pipe, m; t_g—gas temperature in the cyclone, °C; d_i and d_p—current and calculated particle size, m.

The calculated particle size is determined by the formula:

$$d_p = \sqrt{\frac{9 \cdot_g \cdot a \cdot b}{\pi \cdot H_c \cdot_s \cdot U_e}} \tag{4.7}$$

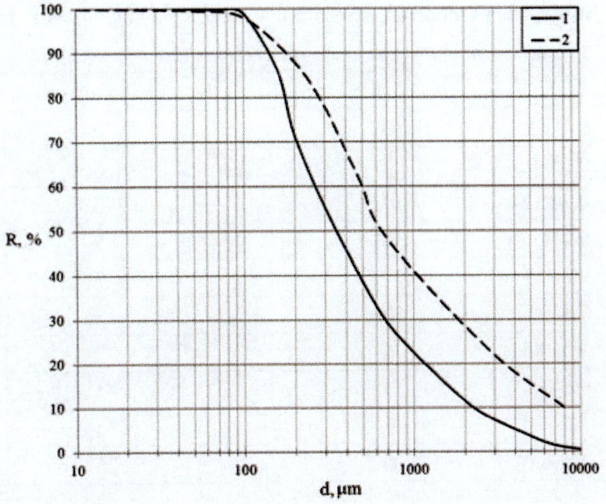

Fig. 4.11 Typical PSD curves of bed material (1) and bottom ash (2)

where H_c—height of the cylindrical part of the cyclone, m; U_e—gas velocity at the cyclone inlet, m/s; ***a***—the height of inlet duct, m; ***b***—the width of inlet duct, m.

Thus, for each fraction the flow rate of captured and entrained particles was calculated. This consumption was summed, and flow rate of bottom ash was determined by subtraction from the flow of incoming solid particles and the flow of entrainment.

In Ma et al. (2015) it was proposed to calculate the bottom flow rate of each fraction using the efficiency of removal of this fraction by dependencies (Yang 2003). These dependencies cannot be precisely moved to a specific object—CFB boiler of unit #9 of Novocherkassk TPP. The ratio between the fractions of different classes of the bed material and bottom ash was determined from the experimental data of Fig. 4.11.

Initial calculations were carried out without taking into account attrition. The purpose of these calculations was to estimate the change in fractions and to determine the effect of the addition of limestone and ash recycling on the PSD of the bed. Data on the fractional composition of solid particle fluxes are given in Table 4.2.

The bed inventory was 200,000 kg, the flow rate of fuel ash 8.16 kg/s (29.3 t/h, ash content 24%, coal flow rate 122 t/h), limestone flow rate 2.78 kg/s, recirculation ash flow rate 2.78 kg/c (10 t/h). The calculation was carried out with a step of 1000 s.

4.4 Results and Discussion

Figure 4.12 shows the change in the fraction of 0.1–0.2 mm and 3–8 mm in time when feeding typical for Novocherkassk TPP coal. The accumulation of large fractions in the bed is clearly noticeable, with the proportion of small fractions of the initial sand. The numerical values of large fractions are close to the experimental ones, but the calculated share of small fractions is significantly less experienced. With the supply

Table 4.2 Fractional composition of the flows of solid particles

Fractions (mm)	The share of fractions (%)			
	Sand	Fuel ash	Limestone	Recirculated ash
0–0.05	0	2	2	2
0.05–0.1	2	8	8	30
0.1–0.2	10	5	68	30
0.2–0.3	36	5	8	35
0.3–0.4	33	3	4	2
0.4–0.5	10	3	3	1
0.5–0.6	6	3	3	0
0.6–1.0	3	11	4	0
1.0–3.0	0	35	0	0
3.0–8.0	0	25	0	0

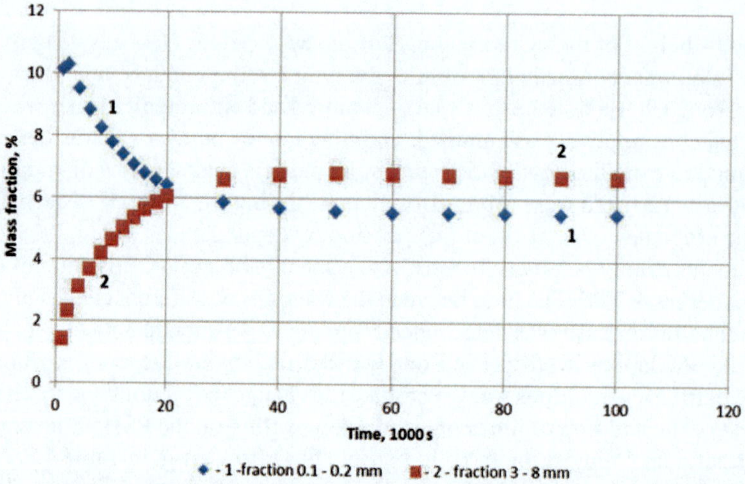

Fig. 4.12 The dynamic changes of fine and course fraction. Typical AC without limestone: 1—fraction 0.1–0.2 mm; 2—fraction 3—8 mm

of limestone with a flow rate of 10 t/h (Fig. 4.13), the proportion of fractions of 0.1–0.2 mm increases rapidly and almost 3 times.

The additional feed of recirculation ash increases the share of fine fractions to 18%. The feed efficiency of ash recirculation is noticeably lower than that of limestone.

Calculations were also performed for the design PSD fuel supply. In that case and the addition of limestone after 10^5 s, the share of large fractions is about 1% and small ones exceed 20% (Fig. 4.14).

In the variants of calculations without limestone and ash recirculation, the PSD differed from the experimental one, and the calculated fraction of 0.1–0.5 mm was

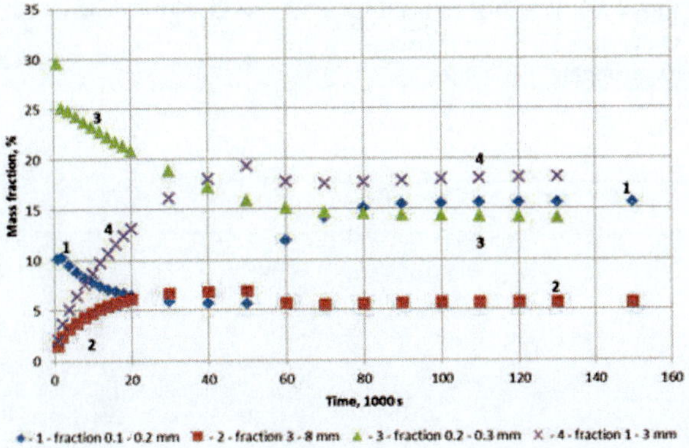

Fig. 4.13 The dynamic changes of four sizes fraction. Typical AC without limestone before 50,000 s and with limestone after 50,000 s

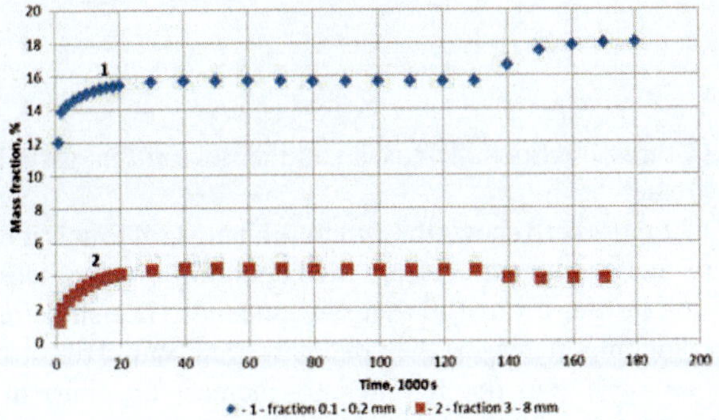

Fig. 4.14 The dynamic changes of fine and course fraction. Design AC with limestone after 10^5 s. 1—fraction 0.1–0.2 mm; 2—fraction 3—8 mm

lower than the experimental one. A possible reason for this is the fragmentation and attrition of the particles. To account for it, the method of calculation proposed in Yang (2003), Ma et al. (2015) was used.

The attrition constant depends on many factors and is determined experimentally. Reliable data on attrition of Rostov AC and limestone in the fluidized bed is not available, and they depend on the place of production. Therefore, it is not possible to perform an accurate calculation. However, a rough estimate can be done of variant calculations for different value of the constants, the same for all factions. If the calculated and experimental of PSD in the bed is a satisfactory match after the completion of the dynamic process, such a constant can be further used for further estimates. In Fig. 4.15 the comparison of the PSD curve obtained by calculation when the value of the attrition constant of the order of 0.001 and the experimental data from the combustion of typical coal without limestone.

It turned out that in the initial period of time the share of fine fractions (0.1–0.2 mm) in the calculation taking into account of attrition increases, reaches a maximum and then decreases and stabilizes. This is due to the fact that the initial share of these particles in the sand was small, then, due to the attrition of the ash fuel, this fraction increases, but then the effect of the entrainment with the increase in the proportion

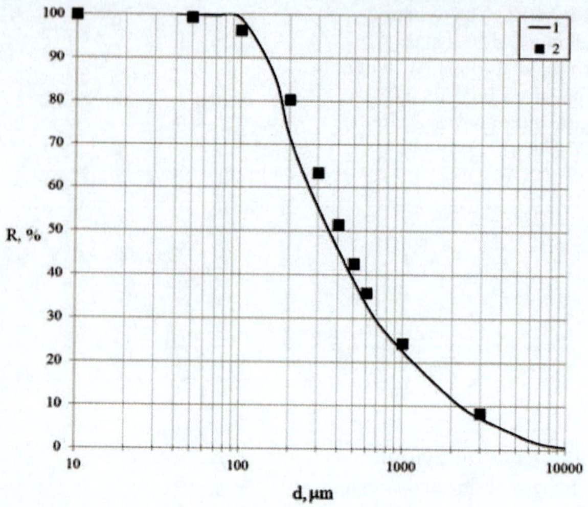

Fig. 4.15 The PSD of bed material. Calculated (1) and experimental (2) values

of these fractions increases and the flow of this particles into this fraction becomes greater.

Figure 4.16 shows the dynamic changes of four fractions with simultaneous supply of fuel and limestone from the beginning of the process, as well as the recirculation ash, starting from 100 thousand seconds. The share of fractions 0.1–0.2 mm in the beginning increases, then decreases and stabilizes at 21%. The addition of recirculation ash does not significantly increase the share of this fraction (by about 1%). The results of the calculations without attrition and fragmentation (Fig. 4.5) showed a slightly larger increase in these fractions (by 3%) when feeding recirculation ash. Recirculation ash feeding for other fractions is also not much changes of their share. This is due to the increased entrainment of small particles of recirculation ash.

The share of fractions 0.2–0.3 mm, which was a lot in the sand (35%) is reduced and stabilized at a value of about 16% (excluding attrition, this value was about 14%). The share of large fractions of 1–3 mm and 3–8 mm was 13 and 7%, respectively.

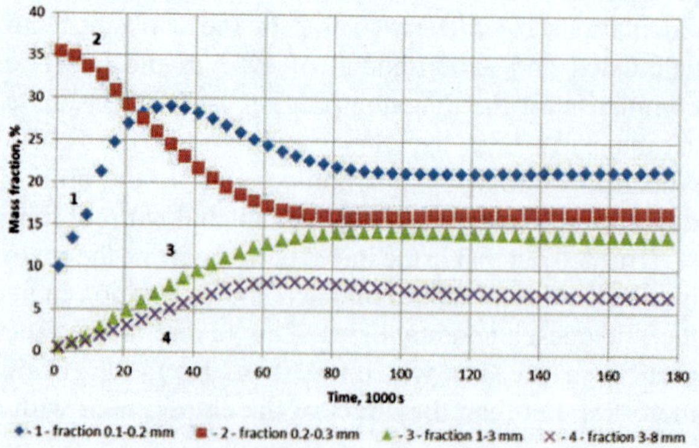

Fig. 4.16 The dynamic changes of four sizes fraction. Typical AC with limestone before 10^5 s and with recirculation ash after 10^5 s

It should be noted that the account of attrition gives a noticeable increase in the share of fly ash to 0.7–0.8, whereas without attrition this share was 0.4–0.5.

Of interest is the comparison of the time stabilization of fine fraction and stabilization of the calcium oxide concentration. Figure 4.17 shows the change in the concentration of CaO in circulating ash and the change in the ratio of fractions of 0.1–0.2 mm after beginning of limestone feeding.

It turned out that the stabilization of the share of fine fractions in the bed and the concentration of calcium oxide occurs approximately simultaneously. However, the stabilization times of the calcium oxide concentration in the bottom ash (Fig. 4.10) almost 5 times more. Apparently, such a long time is required for the complete stabilization of the ash composition of the bed.

Unfortunately, there is no reliable data on the fragmentation of Rostov AC particles. Perhaps influenced by the methodical simplification of the already fragmented particles and not changes in the composition of each fraction due to fragmentation in time. At the same time, the proposed method allows at least qualitatively predict the influence of a number of factors on the bed size distribution. This is especially important for decisions to optimize the fractional composition of fuel and limestone, as well as the mode of supply of limestone and ash recycling.

The calculations clearly showed that to increase the proportion of circulating particles it is necessary to work with limestone, which will lead to an increase in the number of particles of 0.1–0.3 mm to almost 40%. Experiments with long-term supply of limestone gave a positive effect in terms of reducing the temperature non-uniformity in the bed and cyclones, increasing the circulating particles flow rates.

Calculations have shown that an even greater effect will be obtained by reducing the proportion of large particles in the fuel. It is planned to carry out regime and reconstructive measures for the fuel preparation system, in particular: optimization of the operation of existing crushers, reducing the gap between the impact part of hammers and the armor plate to the minimum possible when operating crushers with

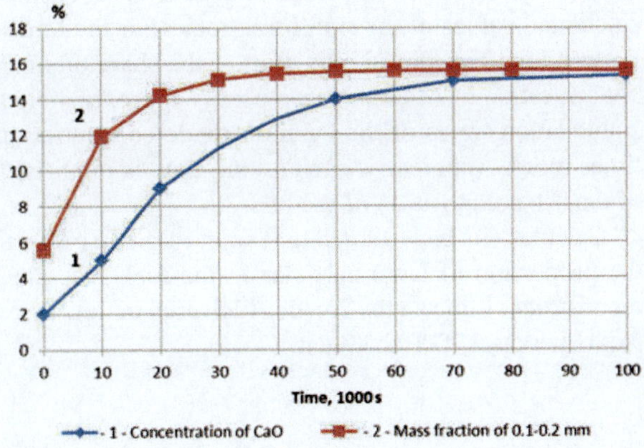

Fig. 4.17 The dynamic changes of experimental CaO concentration in circulation ash (1) and calculated mass fraction of 0.1–0.2 mm (2) in the bed after star-up of limestone feeding

a full load of up to 250 t/h, the development of the project of the installation of the screen after crushers with the return of large fraction.

4.5 Conclusions

Particle size distribution is more important factor of successful operation of CFB boilers. The proper share of particles 0.1–0.4 mm in the bed determines of high solid concentration in upper part of the furnace and high circulation flow rate. It is more difficult to take into account the factors of changes in the size of fuel particles during combustion, taking into account their tendency to fragmentation and attrition, as well as possible agglomeration processes. These issues cannot be resolved without pilot combustion of specific fuels.

The results of the experience of the initial operation of CFB boiler of 330 MWe unit #9 of Novocherkassk TPP were done. It is shown that the excessive number of small and large particles in the fuel leads to an increase in the temperature in the cyclone, uneven temperatures in the furnace and other problems. The results of changes in the PSD of ash and the concentration of calcium oxides in the supply of limestone and ash recycling were done.

To understand the results of the initial operation of CFB boiler, the calculated estimates of the influence of various factors (fuel particle size, limestone and ash recirculation) on the change in the bed size distribution were made. The simplified model for calculation of changes in bed material size distribution after start-up of the boiler is presented.

The calculations showed that to increase the proportion of circulating particles it is necessary to work with limestone, which will lead to an increase in the number of particles of 0.1–0.3 mm to almost 40%. Experiments with long-term supply of limestone gave a positive effect in terms of reducing the temperature non-uniformity in the bed and cyclones, increasing the circulating particles flow rates.

The experimental data on the change in the concentration of calcium oxide and the calculated values of the small fractions after the limestone feeding were compared. It turned out that the stabilization of the share of small fractions in the bed and the concentration of calcium oxide occurs approximately simultaneously. However, the stabilization times of the calcium oxide concentration in the bottom ash almost 5 times more. Apparently, such a long time is required for the complete stabilization of the ash composition of the bed.

Calculations have shown that an even greater effect will be obtained by reducing the proportion of large particles in the fuel. It is planned to carry out regime and reconstructive measures for the fuel preparation system on CFB boiler of unit #9 Novocherkassk TPP that year.

References

Abrahamson J, Allen RWK (1987) The efficiency of conventional return-flow cyclone at high temperatures. Institution of chemical engineers symposium series, № 99. Pergamon Press, pp 31–43

Basu P (2006) Combustion and gasification in fluidized bed. Taylor & Francis Group, LLC, p 460

BlaszczukA, Komorowski M, Nowak W (2012) Distribution of solids concentration and temperature within combustion chamber of SC-OTU CFB boiler. J Powder Technol 91:27–33

Blaszczuk A, Leszczynski J, Nowak W (2013) Simulation model of the mass balance in a supercritical circulating fluidized bed combustor. Powder Technol 246:317–326

Dietz PW (1981) Collection efficiency of cyclone separators. AJChE J 27(6):888–892

Fang M, Yang L, Zhang F, Ma Z, Lua Z, Cen K (2006) An experimental study of burnout characteristics of anthracite in CFB boiler. In: Proceedings of FBC 19, 21–24 May 2006, Vienna, Austria, Part 2

Gao J, Lan X, Fan Y, Chang J, Wang G, Lu Ch, Xu Ch (2009) Hydrodynamics of gas-solid fluidized bed of disparately sized binary particles. Chem Eng Sci 64:4302–4316

Guangxi Y (2015) The formation of the CFB design theory and its practice in China. In: Proceedings of the 22nd international conference on FBC, 14–17 June 2015, Finland, Turku, pp 12–22

Kunii D, Levenspiel O (1991) Fluidization engineering, 2nd edn. Butterworth-Heinemann, Stoneham

Leva M (1959) Fluidization. McGraw-Hill, New York

Lin ChL, Peng TH, Wang WJ (2011) Effect of particle size distribution on agglomeration/defluidization during fluidized bed combustion. Powder Technol 207:290–295

Ma S-X, Guo J, Chang W-M, Yue G-X, Zhang H (2015) The dynamic balance of bed material size distribution in start-up process for circulating fluidized bed boiler. In: Proceedings of international conference on FBC22, Turku, Finland, pp 788–796

Muschelknautz E, Trefz M (1993) Extended cyclone theory for gas flows with high solids concentrations. Chem Eng Technol 16:153–160

Qi XB, Zhu J, Huang WX (2008) A new correlation for predicting solids concentrations in the fully developed zone of circulating fluidized bed risers. Powder Technol 188:64–72

Redemann K, Hartge E-U, Werther J (2009) A particle population balancing model for a circulating fluidized bed combustion system. Powder Technol 191:78–90

Ryabov GA, Folomeev OM, Shaposhnik GA (2002) A study of systems for collection and returning ash of installation with circulated fluidized bed. Therm Eng 49(8):631–637

RyabovG, Kuchmistrov D, Antonenko E, Krutitskiy I, Folomeev O, Melnikov D, Beliaev A (2018) The first year experience of once through CFB boiler operation of 330 MWe unit. In: Proceedings of 23rd international conference on FBC, Seoul, Korea, 13–17 May 2018, pp 108–117

Werner A (2001) Solids distribution as a basis for modeling of heat transfer in circulating fluidized bed boilers. Exp Therm Fluid Sci 25:269–276

Werther J, Hartge E-U (2004) A population balance model of the particle inventory in a fluidized-bed reactor/regenerator system. Powder Technol 148:113–122

Yang H (2003) Study on the mass balance of circulating fluidized boiler. Doctoral dissertation, Peking, Tsinghua University

Yue GX, Yang HR, Zhang H, Lu JF (2006) Post combustion in circulating fluidized bed boilers. In: Proceedings of BC 19, 21–24 May 2006, Vienna, Austria, Part 2, pp 3–8

Chapter 5
Insights of MILD Combustion from High-Fidelity Simulations

Hang Zhou, Josh McConnell, Terry A. Ring, and James C. Sutherland

Abstract Simulation of coal combustion remains challenging due to the many physical processes involved which span a large range of length and time scales. Although detailed models exist for devolatilization, char oxidation and gas-phase kinetics, most simulation efforts simplify these models considerably to reduce the high cost of simulation. Three mixture fraction-based chemistry models are evaluated, namely, the steady laminar flamelet, equilibrium, and Burke-Schumann models in the scope of coal volatiles combustion. Coal volatiles are assumed to be composed of "light gasses" (CH_4, CO, etc.) as well as tar, which refers to the various large aromatic compounds released during the devolatilization process. Here, tar is treated as a single empirical species. Each mixture fraction-based model is evaluated by comparing predicted gas phase properties to computations using finite-rate chemistry with a detailed reaction model. The results indicate that the reconstructions for gas phase temperature and composition from steady flamelet model is the most accurate. The Burke-Schumann chemistry model performed very poorly for predicting the gas phase temperature and composition under stoichiometric conditions. We apply the steady laminar flamelet model to Moderate or Intense Low Oxygen Dilution (MILD) combustion. A key requirement for MILD combustion is that mixing rates are sufficiently fast that gas-phase chemistry occurs nearly volumetrically, eliminating visible flame structures. A Well-stirred reactor assumption is applied to MILD combustion due to its characteristic of volumetric reactions. The necessary conditions to achieve MILD combustion, including recirculation rate of flue gas and heat loss, are determined under various mixture fractions and mass fractions of light gas in the fuel stream. We conclude that the increasing the recirculation rate and heat loss are helpful for achieving MILD regime. Additionally, we observe that the recirculation rate and heat loss values required to achieve MILD combustion increase as the fuel stream is enriched in light gases. Steady flamelet computations reveal that

H. Zhou · J. McConnell · T. A. Ring · J. C. Sutherland (✉)
Department of Chemical Engineering, The University of Utah, Salt Lake City, UT, USA
e-mail: James.Sutherland@utah.edu

J. Lyu and S. Li (eds.), *Clean Coal and Sustainable Energy*,
Environmental Science and Engineering,
https://doi.org/10.1007/978-981-16-1657-0_5

MILD combustion can be achieved when reactants are not well-mixed as long as the scalar dissipation rate is sufficiently large. Our considerations indicate that the steady laminar flamelet model provides a reliable method to model MILD combustion in the absence of well-mixed reactants.

Keywords High-fidelity simulation · Coal combustion · Mixture fraction-based models · MILD

5.1 Introduction

Coal combustion involves a variety of highly-coupled, complex phenomena that span a large range of spatial and temporal scales such as thermochemistry and turbulence in the fluid phase, as well as vaporization, devolatilization, heterogeneous reactions in the particle phase. Direct numerical simulation (DNS) and large-eddy simulation (LES) are useful computational tools for studying the physical processes that occur during the in coal combustion process, and a number of computational studies have been undertaken using these two approaches (Rieth et al. 2018; Bai et al. 2016; Hara et al. 2015; Watanabe and Yamamoto 2015; Zhou 2019). However, simplified methods are often used for gas phase reaction kinetics, devolatilization, and char oxidization, such as the two-step devolatilization model (Rieth et al. 2018; Luo et al. 2012) and kinetic/diffusion model for char oxidization (Watanabe and Yamamoto 2015), in order to reduce the computational burden of performing simulations. Therefore, finding a high-fidelity model that is capable of balancing the accuracy of resolving complex coal combustion process with simulation cost becomes a critical step.

The One-Dimensional Turbulence (ODT) model proposed by Kerstein (1999) provides the ability to resolve the full range of length and time scales as in DNS at a substantially lower computational cost. The ODT model represents a line of sight through a three-dimensional turbulent flow field, in which the size and frequency of mixing events are determined by the local fluid dynamics. Past work has demonstrated that one-dimensional approaches to combustion simulation are capable of accurately predicting ignition delay (Goshayeshi and Sutherland 2014), flame standoff (Goshayeshi and Sutherland 2015a, b) as well as char burnout (McConnell et al. 2016, 2017), and have been used as a cost-effective method for combustion models spanning a wide range of physical fidelity. In this work, the ODT model is used as means to generate data using advanced combustion models.

To ensure that calculations are tractable, simplified chemistry models are often employed, typically parameterized by one or more mixture fractions in conjunction with other parameters such as normalized heat loss. Pedel et al. (2013) utilize a mixture fraction approach with equilibrium chemistry in a study of the ignition behavior in a semi-industrial coal furnace. In a study by Olenik and coworkers (2015), gas phase properties are determined by calculating the state among 6 species. A more recent development is the application of laminar flamelet models to coal combustion. Studies undertaken by Watanabe and Yamamoto (2015) and Rieth et al. (2016)

model gas phase chemistry using a flamelet model parameterized with two mixture fractions corresponding to the products of devolatilization char oxidation. Another work outlined in Watanabe et al. (2017) extends the flamelet approach described in Watanabe and Yamamoto (2015) to include consideration of a mixture fraction for moisture to account for the effects of water evaporation that occurs as coal is heated. More recently, Wen et al. (2019) perform an a priori analysis of the flamelet model for a laminar coal flame near an isothermal wall and find that the flamelet model is able to reproduce the thermochemical state accurately. Furthermore, a study by McConnell and Sutherland (2020) develop a practical approach to joining an empirical tar and soot model to a steady flamelet model and and examine the impact of the required modeling assumptions through an a priori comparison to data generated using a finite rate chemistry model. To evaluate the accuracy of various simplified chemistry models, three mixture fraction-based models are applied in this work.

MILD combustion has attracted increased attention in recent years because of its high efficiency and low emissions. To attain these characteristics, the main criteria of MILD combustion needed to be satisfied are that the inlet temperature of the reactant mixture is higher than mixture self-ignition temperature, and the maximum allowable temperature increase with respect to inlet temperature during combustion is lower than mixture self-ignition temperature (in Kelvin) (Cavaliere and De Joannon 2004). Strong turbulent intensity is always applied to increase the recirculation of high temperature flue gases to dilute and preheat the reactants prior to combustion. However, simulation of coal combustion with strong turbulent intensity using detailed kinetics for gas phase and coal devolatilization models remains difficult due to the high computational cost. In the present investigation, studies of MILD combustion using solid fuels, such as pulverized coal, are still limited (Weber et al. 2005; Li 2014; Smart and Riley 2012). Saha et al. (2013, 2014, 2015, 2016, 2017) have carried out several studies investigating the effect of particle size, turbulence, and coal type on MILD combustion using pulverized coal in a self-recuperative furnace. However, the existence of "ghost flames" or sparks caused by poor mixing between volatile matter and the surrounding hot combustion products was reported in all of these papers. This conflicts with the MILD combustion characteristic of invisible flames. This indicates that it is difficult to attain well-mixedness in a combustion reactor, especially around the pulverized coal particles. On the other hand, OH Planar Laser Induced Fluorescence (PLIF) images from experiments indicate the presence of thin reaction zones where MILD combustion is observed (Wunning and Wunning 1997; Özdemir and Peters 2001; Weber et al. 2005), conflicting with the characteristic of distributed reaction zone for MILD combustion. This observation demonstrates that the conditions required for MILD combustion, such as well-mixedness, are not well-characterized for solid fuels such as coal.

In this work, we first perform an a priori assessment of the predictive capability of three mixture fraction-based modeling approaches, specifically, the steady laminar flamelet, equilibrium, and Burke-Schumann chemistry models in the scope of coal volatiles combustion with consideration of a tar and soot formation model. We subsequently investigate the required conditions to obtain MILD combustion, including recirculation rate and heat loss. Requirements to attain MILD combustion are determined over a large range of oxidizer/fuel mixtures as well as a fuel composition with a various light gas to tar ratios. These requirements are investigated under the assumption of a perfectly-mixed system using a a well-stirred reactor model, whereas poorly-mixed conditions are investigated using the steady laminar flamelet model. Results from the steady laminar flamelet model are compared with the those from the well-stirred reactor model to evaluate the effects of the main parameters, including recirculation rate and heat loss on the achievement of MILD regime and its efficacy to model MILD combustion. The insight provided in this study should help guide the setup of experimental conditions to achieve MILD combustion.

5.2 Theoretical Formulation

5.2.1 *Governing Equations for Coal Combustion Simulation*

In each simulation, the gas and particle phases are evolved in Eulerian and Lagrangian frames of reference, respectively.

5.2.1.1 Gas Phase

The gas phase conservation equations are written as (Goshayeshi and Sutherland 2014, 2015a)

$$\frac{\partial \rho\varphi}{\partial t} = -\frac{\partial \rho\varphi v}{\partial y} - \frac{\partial \Theta_\varphi}{\partial y} + \omega_\phi + \sum_{j=1}^{n_p} \frac{S_{p_j\varphi}}{V_{cell}}, \tag{5.1}$$

where φ is an intensive quantity, Θ_φ is the diffusive flux of φ, ω_φ is the net rate of production of φ in the gas phase, V_{cell} is the quantity representing the volume of the control volume, and $S_{p_j\varphi}$ is gas-phase source term for φ from the particle phase. In this formulation, $\varphi = \{1, u, v, e_0, Y_i\}$ where ρ is the mass density, u and v are the x and y components of velocity, respectively, e_0 is the specific total internal energy, and Y_i are species mass fractions. For the continuity equation, $\varphi = 1$ and $\Theta_\rho = 0$. The set of equations defined by (5.1) are closed using the ideal gas equation, $P = \rho RT/M$, and the following relationships are used for the diffusive fluxes for momentum and energy:

$$\Theta_v = \tau_{yy} + P = -\frac{4}{3}\mu\frac{\partial v}{\partial y} + P, \tag{5.2}$$

$$\Theta_u = \tau_{yx} = -\mu\frac{\partial u}{\partial y}, \tag{5.3}$$

$$\Theta_{e_0} = -\kappa\frac{\partial T}{\partial y} + \sum_{i=1}^{n_s} h_i\Theta_{Y_i}, \tag{5.4}$$

where μ is the viscosity, κ is the thermal conductivity, and n_s is the number of species, and h_i is the enthalpy of species i.

To investigate the impact of differential diffusion on model accuracy, two different forms of the species diffusive flux are implemented. The first assumes uniform diffusivity, D, across all species:

$$\Theta_{Y_i} = -\rho D \frac{\partial Y_i}{\partial y}, \tag{5.5}$$

where D is set based on the assumption that all species have a unity Lewis number, i.e.

$$D_i = \frac{\kappa}{\mathrm{Le}_i \rho c_p} = \frac{\kappa}{\rho c_p}, \tag{5.6}$$

where c_p is the heat capacity. The second form of the species diffusive flux implemented is given as

$$\Theta_{Y_i} = -\frac{\rho Y_i}{X_i} D_i \frac{\partial X_i}{\partial y}, \tag{5.7}$$

where X_i and D_i are the mole fraction and mixture-averaged diffusivity of species i.

Turbulent mixing is modeled using the One-Dimensional Turbulence (ODT) model. ODT models mimic turbulent mixing through instantaneous rearrangement of adjacent fluid parcels (eddy events). By construction, the ODT model conserves momentum, energy, and mass over the interval of the eddy event (Kerstein 1999; Sutherland et al. 2019). The size, duration, and location of each eddy event are determined by the local properties of the flow field, which allows ODT to capture the $-5/3$ energy cascade inherent to isotropic turbulence (Kerstein 1999; Sutherland et al. 2019). For details of the ODT model, the reader is referred to Kerstein (1999), Sutherland et al. (2019).

A model based on the GRI 3.0 mechanism (Lu and Law 2008) consisting of 184 reactions among 30 species is used as a basis for calculating species source terms, ω_{Y_i}.

5.2.1.2 Particle Phase

In this work, only mass loss by devolatilization is considered, so conservation equations for particle phase quantities do not include terms for evaporation or char consumption. Therefore, conservation equations for the mass, velocity, and temperature of each particle are given as

$$\frac{\mathrm{d}m_\mathrm{p}}{\mathrm{d}t} = \frac{\mathrm{d}m_v}{\mathrm{d}t}, \tag{5.8}$$

$$\frac{\mathrm{d}u_\mathrm{p}}{\mathrm{d}t} = \frac{g_x(\rho_\mathrm{p} - \rho_\mathrm{g})}{\rho_\mathrm{p}} + S_{\mathrm{p,u}}, \tag{5.9}$$

$$\frac{\mathrm{d}v_\mathrm{p}}{\mathrm{d}t} = \frac{g_y(\rho_\mathrm{p} - \rho_\mathrm{g})}{\rho_\mathrm{p}} + S_{\mathrm{p,v}}, \tag{5.10}$$

$$\frac{\mathrm{d}T_\mathrm{p}}{\mathrm{d}t} = \frac{A_\mathrm{p}}{m_\mathrm{p} C_\mathrm{p}} \left[h_c\left(T_\mathrm{p} - T_\mathrm{g}\right) + \epsilon_\mathrm{p}\sigma\left(T_\mathrm{p}^4 - T_\mathrm{w}^4\right)\right], \tag{5.11}$$

where m_p and m_v are the total and volatile mass of the particle, respectively, u_p and v_p are the x and y components of the particle velocity, respectively, u_g and v_g are the x and y components of the gas velocity, respectively. T_p, A_p, ρ_p, C_p, and ϵ_p are the particle temperature, surface area, density, heat capacity, and emissivity, respectively, ρ_g is the gas density, T_wall is a furnace "wall" temperature, σ is the Stefan-Boltzmann constant, and $h_c = Nu\kappa/d_\mathrm{p}$ is a convective heat transfer coefficient where $Nu = 2 + 0.6Re_\mathrm{p}^{1/2}Pr^{1/3}$ and d_p is the particle diameter.

In this work, two devolatilization models are considered. The first is the Chemical Percolation and Devolatilization (CPD) model described in Jupudi et al. (2009), Goshayeshi and Sutherland (2014, 2015a). For devolatilization, we consider a modified form of the Chemical Percolation and Devolatilization (CPD) model described in Jupudi et al. (2009), Goshayeshi and Sutherland (2014, 2015a) in which speciation of coal volatiles is disabled, and referred to as CPD-NS henceforth. The CPD-NS model uses the following definition for species devolatilization rates:

$$\left(\frac{\mathrm{d}m_{v,i}}{\mathrm{d}t}\right)^{\text{CPD-NS}} = Y_{v,i}^0 \left(\frac{\mathrm{d}m_v}{\mathrm{d}t}\right)^{\text{CPD}}, \tag{5.12}$$

where $m_{v,i}$ is the mass of species i in the coal volatiles and $Y_{v,i}^0$ is the initial mass fraction of species i in the coal volatiles.

As implemented, the CPD-NS model assumes that the coal volatiles are composed of CH_4, CO, CO_2, H_2, and H_2O, and tar. NH_3 and HCN, which are additional products of devolatilization considered in the formulation of the CPD model, are not considered here as neither of these species are included in the gas phase kinetic mechanism implemented in this work.

5.2.1.3 Tar and Soot Treatment

In this work, we consider a treatment of coal-derived tar and soot based on the model developed by Brown and Fletcher (1997), in which tar is assumed to be a high molecular weight hydrocarbon and soot is assumed to be composed of carbon. We assume tar and soot do not affect the thermodynamic properties of the system directly and only interact with the system through reactive source terms. Herein, we assume that tar is dihydronaphthalene ($C_{10}H_{10}$) and further assume that soot has the same empirical formula as tar, which results in the following reaction scheme

$$\begin{aligned} \text{tar} &\rightarrow \text{soot}, \\ \text{tar} + O_2 &\rightarrow CO + H_2O, \\ \text{soot} + O_2 &\rightarrow CO + H_2O. \end{aligned} \tag{5.13}$$

The primary motivation for assuming soot has the same stoichiometry as tar is that the product composition resulting from (5.13) can be parametrized by a single stream, whereas three streams are required to adequately parameterize tar and soot reactions if soot is assumed to be carbon. This modification of the Brown-Fletcher tar and soot approach ensures that implementation of the mixture fraction-based models described in Sect. 5.2.2 remains tractable. It has been demonstrated that the steady laminar flamelet model is capable of providing an accurate a priori prediction of gas composition and temperature by assuming the same stoichiometry for tar and soot (McConnell and Sutherland 2020).

5.2.2 *Mixture Fraction Based Models*

One of the goals of this study is to assess the performance of three mixture fraction based chemistry models. What follows is a concise description of each of the considered modeling approaches.

All models are parameterized by Z_{lg}, Z_{tp}, and γ, where Z_{lg} and Z_{tp} are mixture fractions for volatile light gases and reaction products of tar and soot oxidation, respectively, and γ is the normalized heat loss. The flamelet model described in Sect. 5.2.2.3 is parameterized by an additional variable, χ_{max}, which is the scalar dissipation rate. The total mixture fraction, Z, is given by

$$Z = Z_{\text{lg}} + Z_{\text{tp}}, \tag{5.14}$$

where

$$Z_{\text{lg}} = \frac{M_{\text{lg}}}{M_{\text{lg}} + M_{\text{tar}} + M_{\text{ox}}}, \tag{5.15}$$

$$Z_{\text{tp}} = \frac{M_{\text{tp}}}{M_{\text{lg}} + M_{\text{tar}} + M_{\text{ox}}}. \tag{5.16}$$

M_{lg} and M_{tar} are masses originating from light gases in the coal volatiles and tar, respectively, M_{tp} is the mass of CO, H_2O, and N_2 supplied to the system through reaction of tar and soot and M_{ox} is the mass originating from the oxidizer, which for this study, is air. The mixture fractions Z_{lg} and Z_{tp} are obtained by performing a linear best-fit analysis on elemental mass balance equations:

$$E_i = Z_{\text{lg}} E_{i,\text{lg}} + Z_{\text{tp}} E_{i,\text{tp}} + \left(1 - Z_{\text{lg}} - Z_{\text{tp}}\right) E_{i,\text{ox}} \tag{5.17}$$

with $i = \{C, H, O\}$. E_i, $E_{i,\text{lg}}$, $E_{i,\text{tp}}$, and $E_{i,\text{ox}}$ are the mass fractions of element i obtained from simulation data, in the volatile light gas stream, in the tar product stream, and in the oxidizer stream, respectively. The parameter ζ is the fraction of fuel mass originating from volatile light gases given by

$$\zeta = \frac{Z_{\text{lg}}}{Z_{\text{lg}} + Z_{\text{tp}}}. \tag{5.18}$$

The variable ζ is used to parameterize fuel-side boundary for each model; $\zeta = 1$ indicates all fuel originates from volatile light gases and $\zeta = 0$ indicates all fuel originates from tar and soot combustion products. The normalized heat loss, γ, is determined through the following expression

$$\gamma = \frac{h_{\text{ad}} - h_{\text{sim}}}{h_{\text{ad}} - h(T_{\text{g,ref}})}, \tag{5.19}$$

where $h_{\text{ad}} = Z_{\text{lg}}h_{\text{lg}}(T_{\text{lg}}) + Z_{\text{tp}}h_{\text{tp}}(T_{\text{tp}}) + (1 - Z_{\text{lg}} - Z_{\text{tp}})h_{\text{ox}}(T_{\text{ox}})$ is the adiabatic mixture enthalpy where h_{lg}, h_{tp} and h_{ox} are the enthalpy of volatile light gas, tar products and oxidizer streams, respectively, and T_j, $j = \{\text{lg, tp, ox}\}$ are the stream temperatures. The values h_{sim} and $T_{\text{g,sim}}$ are the gas phase enthalpy and temperature, respectively, obtained from simulation data, $T_{\text{g,ref}} = 298$ K is the reference temperature, and $h(T) = Z_{\text{lg}}h_{\text{lg}}(T) + Z_{\text{tp}}h_{\text{tp}}(T)(1 - Z_{\text{lg}} - Z_{\text{tp}})h_{\text{ox}}(T)$ for $h(T_{\text{g,ref}})$.

5.2.2.1 Burke-Schumann Chemistry

The Burke-Schumann model (1928) assumes an infinitely-fast reaction between fuel and oxidizer, where CO_2, H_2O, and N_2 are the products of combustion. The gas composition according to the Burke-Schumann model is given by

$$Y_i = \begin{cases} Y_{i,\text{O}} + \frac{Z}{Z_{\text{st}}}\left(Y_{i,\text{st}} - Y_{i,\text{O}}\right) & Z \le Z_{st} \\ Y_{i,\text{F}} + \frac{1-Z}{1-Z_{\text{st}}}\left(Y_{i,\text{st}} - Y_{i,\text{F}}\right) & Z > Z_{\text{st}} \end{cases}, \tag{5.20}$$

where $Y_{i,\text{O}}$, $Y_{i,\text{F}}$, and $Y_{i,\text{st}}$ are the mass fractions of species i in the oxidizer stream, fuel stream, and at stoichiometric conditions, and Z_{st} is the stoichiometric mixture fraction.

The mixture enthalpy, h, is given in terms of the heat loss parameter as

$$h = (1 - \gamma)h_{\text{ad}} + \gamma h_{\text{ref}}. \tag{5.21}$$

Temperature is obtained using the Cantera software package (Goodwin et al. 2017) by specifying the enthalpy, pressure, and composition.

5.2.2.2 Equilibrium Chemistry

Like the Burke-Schumann model, the equilibrium model assumes infinitely-fast chemical kinetics. Rather than assuming complete combustion of fuel, the product composition for the equilibrium model is set based on a suitable reaction mechanism. In this work, equilibrium composition and temperature are obtained with the same chemical mechanism (Lu and Law 2008) used to generate simulation data using the Cantera software package (Goodwin et al. 2017).

5.2.2.3 Steady Laminar Flamelet

Unlike the Burke-Schumann and equilibrium models, the steady laminar flamelet model accounts for the effect of diffusion on the chemical state. The temperature and composition are determined by solving the following system of ODEs

$$\frac{\chi}{2\mathrm{Le}_i}\frac{\mathrm{d}^2 Y_i}{\mathrm{d}Z^2} = -\frac{1}{\rho}\omega_{Y_i}, \tag{5.22}$$

$$\rho c_p \frac{\chi}{2}\frac{\mathrm{d}^2 T}{\mathrm{d}Z^2} = \sum_{j=1}^{n_s} h_j \omega_{Y_j} - \omega_h, \tag{5.23}$$

where

$$\chi = \chi_{\max}\exp\left(-2\left[\mathrm{erf}^{-1}\left(2Z-1\right)\right]^2\right), \tag{5.24}$$

as suggested in Peters (1984), and Le_i is the Lewis number for species i. The quantity ω_h is an energy source term chosen included for the purpose of generating non-adiabatic flamelet solutions and is set to

$$\omega_h = H(T - T_{\mathrm{conv}}), \tag{5.25}$$

where T_{conv} is chosen to be $T_{\mathrm{ad}} + \Delta T$, T_{ad} is the adiabatic flamelet temperature, and ΔT is set to $(\pm)2000$ K for solutions corresponding to positive and negative values of γ, respectively. Boundary conditions for (5.22) and (5.23) are set using the following formulae

$$Y_i(Z=0) = Y_{\mathrm{ox},i}, \tag{5.26}$$

$$Y_i(Z=1) = \zeta Y_{\mathrm{lg},i} + (1-\zeta)\, Y_{\mathrm{tp},i}, \tag{5.27}$$

$$T(Z=0) = T_{\mathrm{ox}}, \tag{5.28}$$

$$T(Z=1) = T \text{ such that } h(T) = \zeta h_{\mathrm{lg}}(T_{\mathrm{lg}}) + (1-\zeta)\, h_{\mathrm{tp}}(T_{\mathrm{tp}}), \tag{5.29}$$

where the light gas (lg) and tar products (tp) temperatures are determined using

$$T_{\mathrm{lg}} = -\frac{\sum_{j}^{n_{\mathrm{p}}} \int_{0}^{\infty} \frac{\mathrm{d}m_{v,j}}{\mathrm{d}t} T_{\mathrm{p}} dt}{\sum_{j}^{n_{\mathrm{p}}} m_{v,j}}, \tag{5.30}$$

$$T_{\mathrm{tp}} = \frac{\int_{0}^{\infty} \left(\omega_{\mathrm{tar},ox} + \omega_{\mathrm{soot},ox}\right) T_{g} dt}{\int_{0}^{\infty} \left(\omega_{\mathrm{tar},ox} + \omega_{\mathrm{soot},ox}\right) dt}, \tag{5.31}$$

where $\omega_{\mathrm{tar},ox}$ and $\omega_{\mathrm{soot},ox}$ are the tar and soot oxidation rates, respectively.

5.2.3 Well-Stirred Reactor Models

In MILD combustion, high turbulent intensity is required for rapid dilution and high temperature of the reactants. This leads to high mixing rate and lower temperature in the reactor, which subsequently leads to lower chemical reaction rates. These two main characteristics suggest the use of well-stirred reactors to model MILD gas-phase combustion (Cavaliere and De Joannon 2004; Plessing et al. 1998), which provides a perfectly mixed, volumetric reaction environment.

The well-stirred reactor model in this work assumes an open, rigid perfectly-stirred spherical vessel with radius r at constant pressure p. In the present study, the volatiles, including the volatile light gases and reaction products of tar and soot oxidation, produced from coal particles are used as the fuel, and no particle is considered. The equations solved for this reactor with residence time $\tau_{\mathrm{mix}} = V/\dot{V}_{\mathrm{in}}$ are

$$\frac{dY_i}{dt} = \frac{\rho_{\mathrm{in}}}{\rho \tau_{\mathrm{mix}}} (Y_{i,\mathrm{in}} - Y_i) + \frac{\dot{\omega}_i}{\rho}, \tag{5.32}$$

$$\frac{dh}{dt} = \frac{\rho_{\mathrm{in}}}{\rho \tau_{\mathrm{mix}}} (h_{\mathrm{in}} - h) - \frac{3k}{\rho r} (T - T_{\mathrm{inf}}), \tag{5.33}$$

where ρ is the density, h is the enthalpy of the gas phase, Y_i is the species mass fraction, $\dot{\omega}_i$ is the volumetric species chemical reaction rate and k is the convective heat transfer coefficient, which considers the convective heat transfer between gas and surroundings with temperature T_{inf}. The "in" subscript indicates an inlet (feed) condition. The Dual-Time stepping method described in Hansen and Sutherland (2017) is used to provide an efficient, robust solution of these equations.

In industrial scale reactors, it is difficult to obtain the required turbulent intensity for perfectly mixed reactants throughout the whole reactor due to the limitations of the equipment. Different amounts of flue gas are recirculated and are mixed with the local reactants in various locations of the reactor to get well-mixed reactants locally. To mimic the reactions in various locations, well-stirred reactors with various mixture fractions, heat losses and recirculation rates are applied in this work. To analyze the degree of dilution of the reactants due to the recirculation of flue gas in MILD combustion, Wunning and Wunning (1997) define the recirculation rate (K_v) as

$$K_v = \frac{\dot{m}_e}{\dot{m}_o + \dot{m}_f} \tag{5.34}$$

where, $\dot{m}_e$, $\dot{m}_o$ and $\dot{m}_f$ are the mass flow rate of entrained flue gas, initial oxidizer and initial fuel respectively. To mimic the heat loss of the gas phase, the definition for normalized heat loss in (5.19) is applied. Considering the transient changing process of light gases and tar products, different values of ζ are used to get the oxidizer temperature and fuel temperature and composition based on Eq. (5.18).

5.3 Computational Configuration

In this work, the data used as a basis for a priori assessment of chemistry models was generated through simulation of a turbulent coal combustion simulation system consistent with high temperature air combustion (HiTAC) conditions (Suda et al. 2002; Tamura et al. 2015). For the ODT calculation, a domain with length 0.2 m is initialized to a temperature and composition consistent with air at 1200 K. Air at 1200 K as well as coal particles with a diameter of 50 μm and temperature of 350 K are injected though a nozzle with a diameter of 15.8 mm at the center of the domain at a velocity of 14.92 m/s. Velocity over the remainder of the domain is set to zero. The governing equations are solved using a fully-coupled scheme with a first-order explicit time integration scheme, a second-order spatial discretization, and characteristic boundary conditions Sutherland and Kennedy (2003). The spatial and temporal resolutions of 200 μm and 50 ns, respectively, are used to obtain a solution to the numerical system.

The Illinois #6 coal is considered in this work, with proximate and ultimate analyses given in Tables 5.1 and 5.2. The composition of coal volatiles is given in Table 5.3. The composition of the coal volatiles was determined using a 2D interpolation technique based on molar O/C and H/C ratios from the Illinois #6 coal and data from library coals (Zhao et al. 1994).

Table 5.1 Proximate analyses of the Illinois #6 coal

Proximate %			
Moisture	Ash	Volatiles	Fixed C
0.0	8.86	40.70	50.44

Table 5.2 Ultimate analyses of the Illinois #6 coal

Ultimate (dry) %				
C	H	O	N	S
78.11	5.49	9.81	1.36	4.83

Table 5.3 Composition of coal volatiles (%)

Tar	CH_4	CO	CO_2	H_2	H_2O
42.94	13.5	27.2	5.81	3.47	7.08

For the investigation of MILD combustion, the steady laminar flamelet and perfectly-stirred reactor models are applied. Air at 1200 K is used as undiluted oxidizer. The mixture of volatile light gases (with composition of CH_4, CO, CO_2, H_2 and H_2O as shown in Table 5.3) and reaction products of tar and soot oxidation (with composition of CO, H_2O and N_2) is used as fuel stream. Its composition and temperature are calculated using Eqs. (5.27) and (5.29). The adiabatic equilibrium product of stoichiometric air and fuel stream is used as the recirculated flue gas, and the recirculation rate in Eq. (5.34) is calculated based on the mass flow rate of stoichiometric air and fuel stream.

5.4 Results and Discussion

5.4.1 Comparison of Mixture Fraction-Based Models

In this section, the a priori performance of the Burke-Schumann (BS), equilibrium (EQ), and steady laminar flamelet (SLF) models is evaluated using data from a turbulent coal combustion simulation in which a detailed kinetics (DK) model for chemistry is implemented. Because soot is assumed to have the same empirical formula as tar and both differential diffusion and speciation are neglected, differences between simulation data and model reconstructions should be due only to variations in the chemical and diffusive time scales. Herein, use of "expected" or "observed" data refers to that originating from detailed kinetics calculations.

Figure 5.1 shows model reconstructions of T_g using the Burke-Schumann, equilibrium, and flamelet models in addition to T_g from simulation data plotted against equivalence ratio and colored by heat loss. As Fig. 5.1 indicates, the accuracy of model reconstructions of T_g greatly depends on model fidelity. The Burke-Schumann model yields temperature reconstructions that overestimate T_g by almost 600 K near stoichiometric conditions. Reconstruction of gas temperature by the equilibrium model also overestimates the expected temperature in the vicinity of $\phi = 1$, though is substantially more accurate the Burke-Schumann reconstruction of T_g. Additionally, flamelet reconstructions of T_g very closely resemble values obtained from simulation data. Figure 5.2 shows model reconstructions of Y_{O_2} as well as values obtained from simulation data plotted against equivalence ratio and colored by heat loss. As Fig. 5.2 indicates, reconstructions of Y_{O_2} mirror the expected behavior when $\phi < 0.5$ for all models considered. However, the behavior of Y_{O_2} is model-dependent for larger ϕ. For $\phi > 0.5$, the Burke-Schumann model underpredicts Y_{O_2} values from

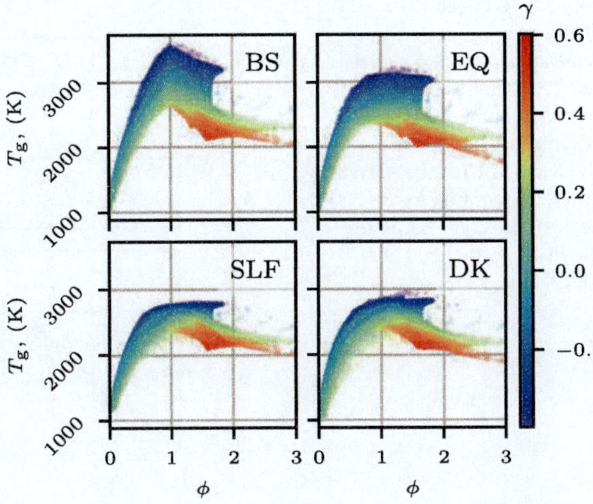

Fig. 5.1 Gas temperature, T_g, plotted against equivalence ratio, ϕ, for the detailed kinetics (DK), Burke-Schumann (BS), equilibrium (EQ), and flamelet (SLF) models. Data colored according to the scale of the color bar is set based on normalized heat loss, γ

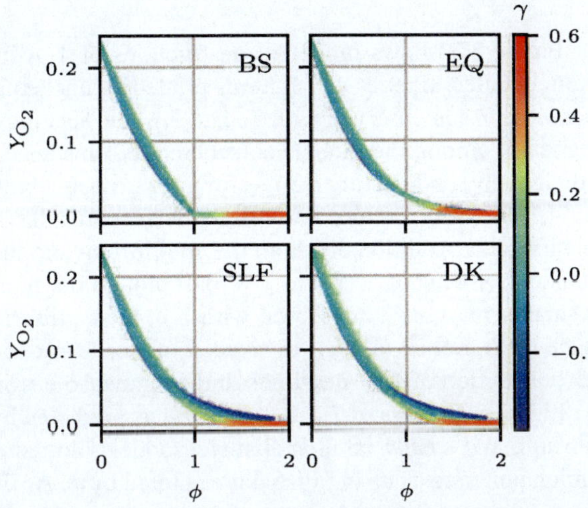

Fig. 5.2 Oxygen mass fraction, Y_{O_2}, plotted against equivalence ratio, ϕ, for the detailed kinetics (DK), Burke-Schumann (BS), equilibrium (EQ), and flamelet (SLF) models. Data colored according to the scale of the color bar is set based on normalized heat loss, γ

the simulation data. Underestimation of Y_{O_2} by the Burke-Schumann model is especially egregious under stoichiometric conditions where it predicts $Y_{O_2} = 0$, which is a consequence of the "mixed is burnt" assumption. Unlike the Burke-Schumann model, equilibrium chemistry produces a Y_{O_2} profile that is nearly identical to values obtained from simulation data for sufficiently lean conditions, ($\phi < 0.8$) and matches the general trend of Y_{O_2} for larger ϕ, but fails to capture the observed variance. Similar to the equilibrium model the flamelet reconstruction of Y_{O_2} mirrors values from simulation data under lean conditions. Additionally, the flamelet model is able to reproduce the expected behavior of Y_{O_2} at and near stoichiometric conditions which neither the Burke-Schumann nor equilibrium models are able to do.

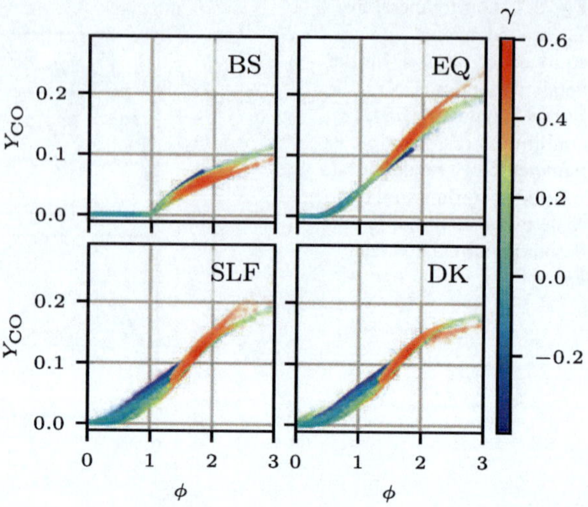

Fig. 5.3 Oxygen mass fraction, Y_{CO}, plotted against equivalence ratio, ϕ, for the detailed kinetics (DK), Burke-Schumann (BS), equilibrium (EQ), and flamelet (SLF) models. Data colored according to the scale of the color bar is set based on normalized heat loss, γ

Figure 5.3 shows model reconstructions of Y_{CO} in addition to data originating from detailed kinetics calculations plotted against equivalence ratio and colored by heat loss. As is observed with T_g and Y_{O_2}, the accuracy of predicted Y_{CO} varies substantially among the models considered. For instance, the Burke-Schumann model significantly underestimates Y_{CO} for nonzero ϕ, but both the equilibrium and flamelet models yield predictions that match the general behavior of Y_{CO} obtained from simulation data. Additionally, both the equilibrium and flamelet models tend to overestimate Y_{CO} when $\phi > 2$ and $\gamma > 0.4$, though the flamelet model overestimates by a smaller margin. As observed with Y_{O_2}, the equilibrium model is unable to reproduce the variance in Y_{CO} for fixed ϕ, under lean conditions, whereas the flamelet reconstruction of Y_{CO} are almost indistinguishable from expected values.

Figure 5.4 shows of T_g, Y_{O_2}, and Y_{CO} as predicted by the Burke-Schumann, equilibrium, and steady laminar flamelet models plotted against values obtained from simulation data. Data in Fig. 5.4 are colored by ϕ. As Fig. 5.4 illustrates, the flamelet model generally reproduces expected values of T_g, Y_{O_2}, and Y_{CO} more accurately than both the equilibrium and Burke-Schumann models. The only exception to this generalization is T_g at $\phi \approx 0$, where T_g as predicted by the flamelet model approaches 1200 K as ϕ approaches 0. As mentioned in Sect. 5.2.2.3, non-adiabatic flamelets are generated by including a source term on the right-hand side of the energy conservation equation (Sect. 5.2.2.3) while maintaining adiabatic boundary conditions. Consequently, this method of generating non-adiabatic flamelets cannot resolve heat loss at mixture fraction boundaries, which is the cause of the errant behavior of the flamelet temperature at $\phi \approx 0$. It is also evident from Fig. 5.4 that, for the dataset considered, the accuracy of equilibrium reconstructions is typically close to that of flamelet reconstructions. Furthermore, Fig. 5.4 illustrates the poor accuracy of Burke-Schumann the model for predicting T_g, Y_{O_2}, and Y_{CO}. Accurate Burke-Schumann

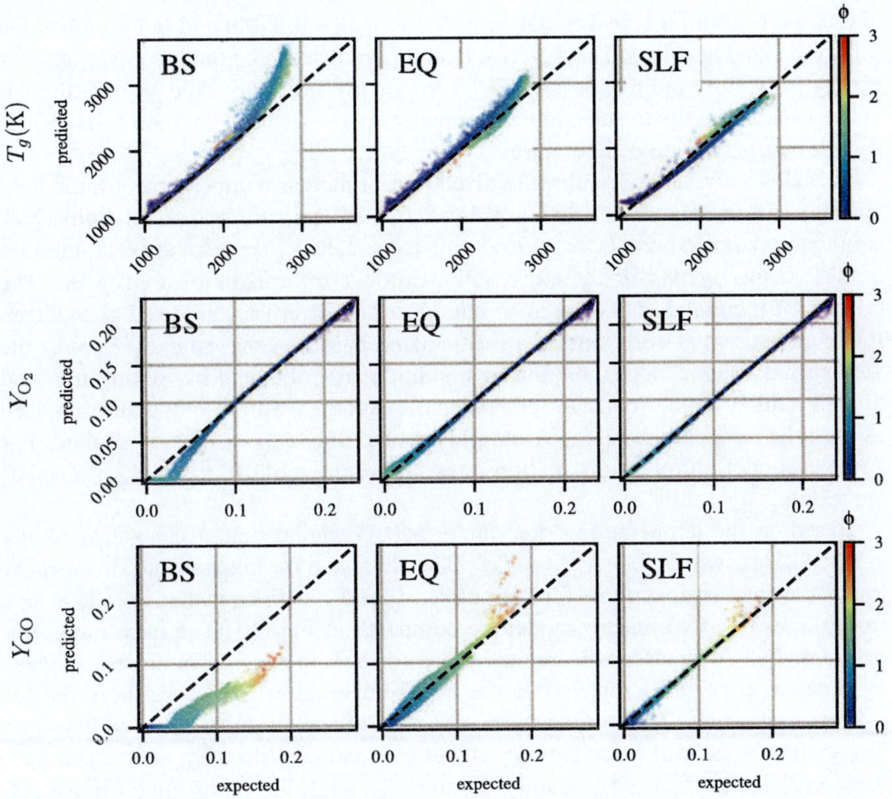

Fig. 5.4 Parity plots of a priori predictions of gas temperature (top row), O_2 mass fraction (middle row) and CO mass fraction (bottom row) given by the Burke-Schumann (left column), equilibrium (middle column), and steady laminar flamelet (right column) models

reconstructions for T_g and Y_{O_2} are limited to $\phi < 0.5$, while the Y_{CO} reconstruction is inaccurate for $\phi > 0$.

5.4.2 *Analysis of MILD Combustion Using Well-Stirred Reactor and Steady Flamelet Models*

From the analysis in Sect. 5.4.1, the steady flamelet model is significantly more accurate than both the equilibrium and Burke-Schumann models, especially considering predictions of T_g. Therefore, it is applied to investigate the MILD combustion behavior and is compared with the well-stirred reactor model.

To achieve MILD combustion and test the effect of dilution rate, we vary the recirculation rate (K_v). Five values of K_v, namely 0.0, 0.5, 1.0, 1.5 and 2.0, are used in this work. From the analysis in Sect. 5.4.1, $\gamma = -0.2$, 0.0 and 0.4 are obtained for normalized heat loss. The negative heat loss indicates that the gas phase absorbs energy from surroundings or particles. The positive heat loss indicates that the gas phase releases energy to surroundings or particles. $\zeta = 0.10, 0.25, 0.40$ and 0.80 are used for mass fraction of light gases.

In both steady flamelet and well-stirred reactor models, temperatures for the light gas (lg) and tar and soot reaction product (tp) streams are computed from ODT simulation data, and are $T_{\mathrm{lg}} = 1176K$ and $T_{\mathrm{tp}} = 2380K$, respectively. The mixture of the light gas (lg) and tar and soot reaction product (tp) streams are used as fuel. The mixture of recirculated flue gases and undiluted oxidizer (air) are used as oxidizer. Its temperature and composition are calculated based on the value of K_v. For the well-stirred reactor model, the inflow conditions are obtained by mixing fuel and diluted oxidizer under constant pressure and enthalpy, whose temperature is used as the inlet temperature, T_{inlet}. A long residence time, $\tau_{\mathrm{mix}} = 10^3$s, is applied. For flamelet model, different dissipation rates in the range of $10^{-2} \leq \chi_{\mathrm{max}} \leq 2 \times 10^4$ are tested.

Based on the temperature criterion for MILD combustion, $T_{\mathrm{inlet}} > T_{\mathrm{ignition}}$ and $\Delta T < T_{\mathrm{ignition}}$, with $\Delta T = T_{\mathrm{max}} - T_{\mathrm{inlet}}$ (Cavaliere and De Joannon 2004), the maximum and inlet temperatures for various K_v, γ and ζ with equivalence ratio $\phi = 1$ from the well-stirred reactor model are compared in Fig. 5.5. The maximum temperature, T_{max}, decreases with the increment of heat loss, γ, as indicated by three red lines in each sub-figure, while the inlet temperatures, T_{inlet}, are the same for various heat losses. This results in the decrease of the disparity between T_{max} and T_{inlet}, which is helpful for achieving MILD combustion. Also, T_{inlet} decreases as ζ increases because T_{lg} is significantly less than T_{tp}, while T_{max} does not change much for different ζ. This indicates that having more production of tar and soot oxidation in the fuel stream makes it easier to achieving MILD regime. For different recirculation rate K_v, T_{inlet} increases significantly for bigger K_v because of the high temperature of recirculated flue gas. For $\gamma = 0.0$, T_{max} keeps constant for various K_v because the recirculated flue gas has the same composition and temperature as the gas phase in the PSR without recirculation. The recirculated flue gas only increases T_{inlet} and does not have any effects in the reactor due to the well-mixed reactants within the reactor. When γ is not zero, the products, from the *adiabatic* stoichiometric reaction, are different from the gas phase in the PSR, leading to the variance of T_{max}. For large K_v, the adiabatic and reference enthalpy, h_{ad} and $h(T_{\mathrm{g,ref}})$, in Eq. (5.19) increase due to bigger amount of flue gas, which results in bigger heat loss ($h_{\mathrm{ad}} - h_{\mathrm{sim}}$) with the same γ. Therefore, T_{max} decreases for $\gamma = 0.4$ and increases for $\gamma = 0.2$ with the increment of K_v.

Figure 5.6 delineates the combustion regimes as the right bottom area based on the definition of MILD combustion. Three subplots respond to different heat loss values, while different symbols respond to different mass fractions of light gases in fuel stream. In this work, the undiluted oxidizer(air) is preheated to 1200 K. The combustion drops into the *High Temperature Combustion* area when $K_v = 0$. With the increment of K_v, the combustion moves from *High Temperature Combustion* area to *MILD Combustion* area. The heat loss value does have effect on the achievement

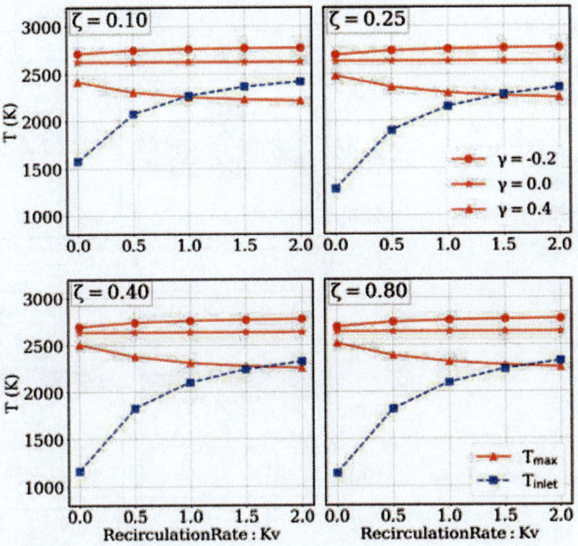

Fig. 5.5 Variances of inlet and maximum temperatures for different K_v, γ and ζ with equivalence ratio $\phi = 1$ from well-stirred reactor model

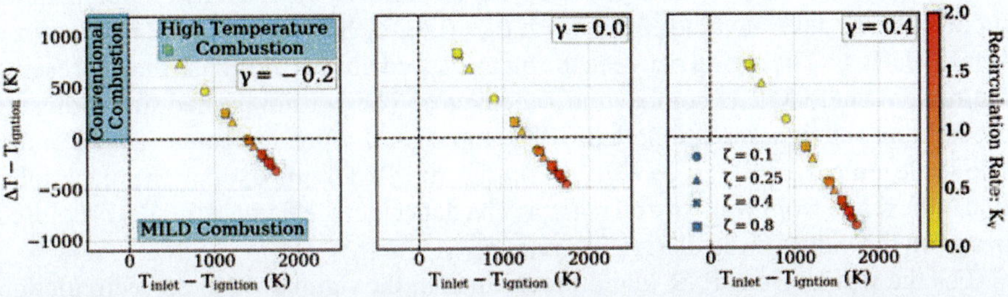

Fig. 5.6 The effect of recirculation rate on determining combustion regimes for different γ and ζ from well-stirred reactor model. Data colored according to the scale of the color bar is set based on recirculation rate, K_v

of MILD combustion. For adiabatic conditions ($\gamma = 0$), MILD is achieved for all ζ when $K_v \geq 1.0$, while it only requires $K_v \geq 0.5$ for positive heat loss, $\gamma = 0.4$. For negative heat loss, $\gamma = -0.2$, it requires $K_v \geq 1.0$ for $\zeta < 0.8$. The case with $\zeta = 0.8$ and $\gamma = -0.2$ is on the edge between *High Temperature Combustion* and *MILD Combustion* area for $K_v = 1.0$. When ζ is decreased for the same value of K_v, the state of the system moves closer to the *MILD Combustion* area due to the higher temperature of the production of tar and soot oxidation as indicated in Fig. 5.5. From the analysis of Figs. 5.5 and 5.6, it is concluded that higher recirculation K_v, elevated heat loss γ and lower mass fraction of light gas in fuel stream ζ are helpful to reach MILD regime.

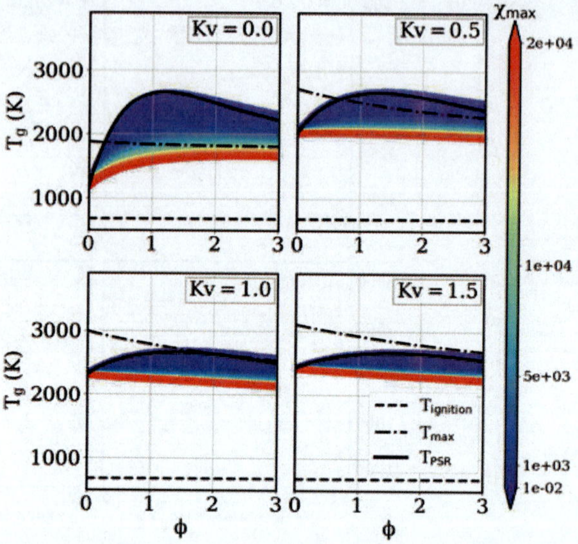

Fig. 5.7 Gas temperature, T_g, plotted against equivalence ratio, ϕ, for various recirculation rates, K_v, with no heat loss, $\gamma = 0$, and $\zeta = 0.4$. Data are colored according to the scale of the color bar based on dissipation rate, χ_{max}

Figures 5.7, 5.8 and 5.9 compare the results from well-stirred reactor and flamelet models for for several values of K_v with $\zeta = 0.4$. Each figure gives the results for a single heat loss. The region between the black dashed line and dot-dash line represents the region where MILD combustion is achieved. The bottom dashed line indicates the mixture self-ignition temperature. The top dash-dot line indicates the maximum allowable temperature, $T_{max} = T_{inlet} + T_{ignition}$, of MILD combustion. The black solid line is the result from well-stirred reactor. The data from the flamelet model is colored based on the value of the dissipation rate, χ_{max}.

For the results from PSR (solid black lines in the figures) with no recirculation ($K_v = 0.0$), the temperature far exceeds the allowable maximum temperature for the three heat loss values considered, especially near stoichiometric, indicating that MILD regime is not achieved. When ϕ is smaller than ~0.5, the temperature drops into the MILD regime between two black dashed lines. That is, only when the the stoichiometry is very lean can MILD be achieved in PSR with no recirculation. As K_v increases, the allowable maximum temperature, T_{max}, increases due to increasing inlet temperature, T_{inlet}. The MILD regime between two black dashed lines becomes much larger. The range of equivalence ratios for achieving MILD regime becomes bigger for larger K_v until the entire black line drops into the MILD regime with $K_v = 1.0$ for $\gamma = 0.0$ and with $K_v = 0.5$ for $\gamma = 0.4$. For $\gamma = -0.2$, cases with $\phi > 1.5$ cannot reach the MILD regime even for $K_v = 1.5$.

For the results from steady flamelet model, the temperatures are almost the same as the results from PSR for small dissipation rate, $\chi_{max} = 10^{-2}$. As dissipation rate increases, temperature decreases and drops into the MILD regime for all equivalence ratios of 0.0–3.0 when χ_{max} is sufficiently large. The required dissipation rate to achieve MILD combustion for all equivalence ratios of 0.0–3.0 is referred to as

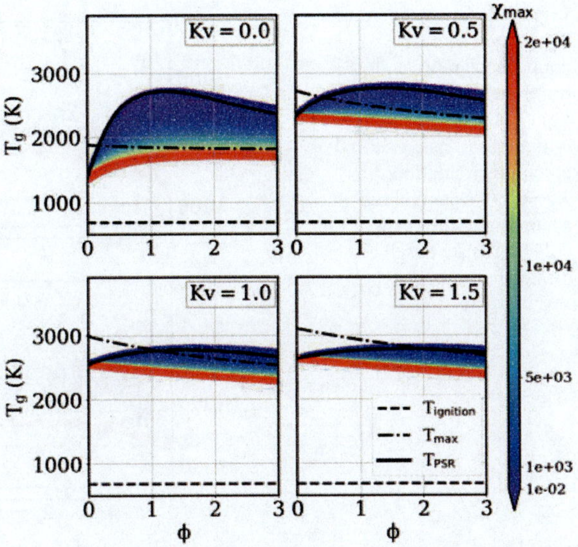

Fig. 5.8 Gas temperature, T_g, plotted against equivalence ratio, ϕ, for different recirculation rates, K_v, with negative heat loss, $\gamma = -0.2$, and $\zeta = 0.4$. Data are colored according to the scale of the color bar based on dissipation rate, $\chi_{\max}$

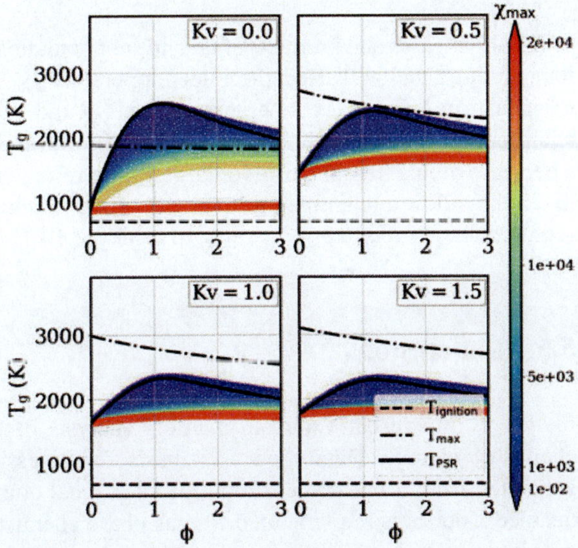

Fig. 5.9 Gas temperature, T_g, plotted against equivalence ratio, ϕ, for different recirculation rates, K_v, with positive heat loss, $\gamma = 0.4$, and $\zeta = 0.4$. Data colored according to the scale of the color bar is set based on dissipation rate, $\chi_{\max}$

$\chi_{\max,\text{MILD}}$. For $K_v = 0.0$, $\chi_{\max,\text{MILD}}$ is around 1.5×10^4 for the adiabatic case ($\gamma = 0$) and for $\gamma = -0.2$, while it is approximately 1.0×10^4 for $\gamma = 0.4$. However, extinction is observed for the case with $K_v = 0.0$ and $\gamma = 0.4$ when the dissipation rate reaches $\sim 1.5 \times 10^4$. That is, a maximum dissipation rate should also been considered to avoid extinction. $\chi_{\max,\text{MILD}}$ decreases with the increment of recirculation rate. When $K_v = 1.5$ for $\gamma = 0$ and $K_v = 0.5$ for $\gamma = 0.4$, MILD regime is achieved for all dissipation rates in $10^{-2} - 2 \times 10^4$.

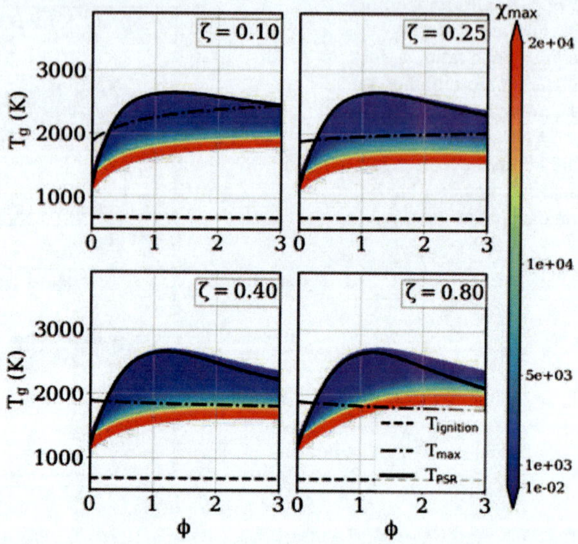

Fig. 5.10 Gas temperature, T_g, plotted against equivalence ratio, ϕ, for different mass fractions of light gas in fuel stream, ζ, with no heat loss, $\gamma = 0.0$ and no recirculation $K_v = 0.0$. Data colored according to the scale of the color bar is set based on dissipation rate, χ_{max}

The light gases and products of tar and soot oxidation are generated under different temperatures, and will affect the inlet temperature T_{inlet} as shown in Fig. 5.5. In order to evaluate its effect, gas temperatures from PSR and steady flamelet model are plotted against equivalence ratio in Fig. 5.10 with $K_v = 0.0$ and $\gamma = 0.0$. Figure 5.5 indicates a decrease of inlet temperature for large values of ζ, which causes the decrease of the maximum allowable temperature for the MILD combustion regime. As a result, the required dissipation rate, $\chi_{\text{max,MILD}}$, to achieve MILD regime increases for bigger ζ.

5.5 Conclusions

In this study, we carry out an a priori analysis of three mixture fraction-based chemistry models. Data used as a basis for assessing model performance was generated from a one-dimensional turbulent coal combustion simulation in which detailed kinetics is implemented for gas phase chemistry. This work only considers devolatilization; neither char combustion nor evaporation are accounted for. Furthermore, tar and soot are considered through an approach based on Brown-Fletcher tar and soot model. Soot is assumed to have the same empirical formula as tar in order to facilitate the two mixture fraction parameterization of the system composition.

Of the three mixture fraction-based approaches considered, the flamelet model performed the best. Flamelet reconstructions for T_g, Y_{O_2}, and Y_{CO} were typically very accurate. This is especially the case for T_g near stoichiometric conditions where the Burke-Schumann and equilibrium models struggle to be accurate. However, the accuracy of T_g predictions by the flamelet model declines for $\phi \approx 0$ because the approach used is to generate non-adiabatic flamelets is not capable of resolving heat loss at mixture fraction boundaries. The accuracy of equilibrium reconstructions was often close to those of the flamelet model, although the equilibrium predictions of temperature were typically off by 150 K at stoichiometric conditions. The Burke-Schumann model was the least accurate of the three models considered, performing very poorly for T_g, Y_{O_2}, and Y_{CO} under stoichiometric conditions.

To investigate the MILD combustion behavior, the steady flamelet model and well-stirred reactor model are applied in this work. The definition of MILD combustion based on temperature relationship, $T_{\text{inlet}} > T_{\text{ignition}}$ and $\Delta T < T_{\text{ignition}}$, with $\Delta T = T_{\max} - T_{\text{inlet}}$, is applied as the criteria for achieving MILD combustion. The effects of three main parameters, including recirculation rate, K_v, normalized heat loss, γ, and mass fraction of light gases in fuel stream, ζ, are evaluated for MILD combustion. Increasing recirculation rate, positive heat loss and decreasing mass fraction of light gases in fuel steam have positive influence on attaining of MILD combustion. Negative heat loss and increasing light gases mass fraction in fuel stream increases the critical value of recirculation rate needed to achieve MILD regime. The dissipation rate plays important role in achieving MILD combustion. For instance, the MILD regime can be achieved with higher dissipation rate under poorly-mixed combustion , and its critical value decreases as the recirculation rate or heat loss increase and as mass fraction of light gas in fuel stream decreases. In other words, flamelet model provides a reliable method to model MILD regime with high dissipation rate in diffusion combustion.

Acknowledgements This research was funded by the National Science Foundation under grant NSF1704141.

References

Bai Y, Luo K, Qiu K, Fan J (2016) Numerical investigation of two-phase flame structures in a simplified coal jet flame. Fuel 182:944–957. ISSN: 0016-2361

Brown AL, Fletcher TH (1997) Modeling soot derived from pulverized coal. Energy Fuels 4:745–757

Burke SP, Schumann TEW (1928) Diffusion flames. Ind Eng Chem 20:998–1004

Cavaliere A, De Joannon M (2004) Mild combustion. Prog Energy Combust Sci 30:329–366

Goodwin DG, Moffat HK, Speth RL (2017) Cantera: an object-oriented software toolkit for chemical kinetics, thermodynamics, and transport processes version 2.3.0. https://doi.org/10.5281/zenodo.170284. http://www.cantera.org

Goshayeshi B, Sutherland JC (2015a) Prediction of oxy-coal flame stand-off using high-fidelity thermochemical models and the one-dimensional turbulence model. Proc Combust Inst 35:2829–2837. ISSN: 15407489

Goshayeshi B, Sutherland JC (2015b) A comparative study of thermochemistry models for oxy-coal combustion simulation. Combust Flame 162:4016–4024. ISSN: 0010-2180

Goshayeshi B, Sutherland JC (2019) A comparison of various models in predicting ignition delay in single-particle coal combustion. Combust Flame 161:1900–1910. ISSN: 15562921 (2014). In: 9th International symposium on coal combustion (ISCC-9) Qingdao, China, 21–24 July 2019

Hansen MA, Sutherland JC (2017) Dual timestepping methods for detailed combustion chemistry. Combusti Theory Model Month 21:329–345

Hara T, Muto M, Kitano T, Kurose R, Komori S (2015) Direct numerical simulation of a pulverized coal jet flame employing a global volatile matter reaction scheme based on detailed reaction mechanism. Combust Flame 162:4391–4407. ISSN: 0010-2180

Jupudi R, Zamansky V, Fletcher T (2009) Prediction of light gas composition in coal devolatilization. Energy Fuels 23:3063–3067

Kerstein AR (1999) One-dimensional turbulence: model formulation and application to homogeneous turbulence, shear flows, and buoyant stratified flows. J Fluid Mech 392:277–334

Li P et al (2014) Moderate or intense low-oxygen dilution oxy-combustion characteristics of light oil and pulverized coal in a pilot-scale furnace. Energy Fuels 28:1524–1535. ISSN: 08870624

Lu T, Law CK (2008) A criterion based on computational singular perturbation for the identification of quasi steady state species: a reduced mechanism for methane oxidation with NO chemistry. Combust Flame 154:761–774. ISSN: 00102180

Luo K, Wang H, Fan J, Yi F (2012) Direct numerical simulation of pulverized coal combustion in a hot vitiated co-flow. Energy Fuels 26:6128–6136

McConnell J, Sutherland JC (2020) Assessment of various tar and soot treatment methods and a priori analysis of the steady laminar flamelet model for use in coal combustion simulation. Fuel 265:116775. ISSN: 00162361

McConnell J, Goshayeshi B, Sutherland JC (2016) The effect of model fidelity on prediction of char burnout for single-particle coal combustion. Proc Combust Inst

McConnell J, Goshayeshi B, Sutherland JC (2017) An evaluation of the efficacy of various coal combustion models for predicting char burnout. Fuel 201:53–64

Olenik G, Stein O, Kronenburg A (2015) LES of swirl-stabilised pulverised coal combustion in IFRF furnace No. 1. Proc Combust Inst 35:2819–2828. ISSN: 1540-7489

Özdemir IB, Peters N (2001) Characteristics of the reaction zone in a combustor operating at mild combustion. Exp Fluids 30:683–695. ISSN: 07234864

Pedel J, Thornock JN, Smith PJ (2013) Ignition of co-axial turbulent diffusion oxy-coal jet flames: experiments and simulations collaboration. Combust Flame 160:1112–1128. ISSN: 0010-2180

Peters N (1984) Laminar diffusion flamelet models in non-premixed turbulent combustion. Prog Energy Combust Sci 10:319–339. ISSN: 0360-1285

Plessing T, Peters N, Wünning JG (1998) Laseroptical investigation of highly preheated combustion with strong exhaust gas recirculation. Symp (Int) Combust 27:3197–3204. ISSN: 00820784

Rieth M et al (2016) Flamelet LES of a semi-industrial pulverized coal furnace. Combust Flame 173:39–56. ISSN: 0010-2180

Rieth M, Kempf A, Kronenburg A, Stein O (2018) Carrier-phase DNS of pulverized coal particle ignition and volatile burning in a turbulent mixing layer. Fuel 212:364–374. ISSN: 0016-2361

Saha M, Dally BB, Medwell PR, Cleary E (2013) An experimental study of MILD combustion of pulverized coal in a recuperative furnace 2 experimental setup. In: 9th Asia-Pacific conference on combustion, Gyeongju Hilton, Gyeongju, Korea, 19–22 May 2013, pp 3–6

Saha M, Dally BB, Medwell PR, Cleary EM (2014) Moderate or intense low oxygen dilution (MILD) combustion characteristics of pulverized coal in a self-recuperative furnace. Energy Fuels 28:6046–6057. ISSN: 15205029

Saha M, Chinnici A, Dally BB, Medwell PR (2015) Numerical study of pulverized coal MILD combustion in a self-recuperative furnace. Energy Fuels 29:7650–7669. ISSN: 15205029

Saha M, Dally BB, Medwell PR, Chinnici A (2016) Burning characteristics of Victorian brown coal under MILD combustion conditions. Combust Flame 172:252–270. ISSN: 15562921

Saha M, Dally BB, Medwell PR, Chinnici A (2017) Effect of particle size on the MILD combustion characteristics of pulverised brown coal. Fuel Process Technol 155:74–87. ISSN: 03783820

Smart JP, Riley GS (2012) Combustion of coal in a flameless oxidation environment under oxyfuel firing conditions: the reality. J Energy Inst 85:131–134

Suda T, Takafuji M, Hirata T, Yoshino M, Sato J (2002) A study of combustion behavior of pulverized coal in high-temperature air. Proc Combust Inst 29:503–509. ISSN: 1540–7489

Sutherland JC, Kennedy CA (2003) Improved boundary conditions for viscous, reacting, compressible flows. J Comput Phys 191:502–524. ISSN: 00219991

Sutherland J, Punati N, Kerstein A (2019) A unified approach to the various formulations of the one-dimensional-turbulence model tech. rep. (Inst. Clean Secure Energy, 2010). In: 9th International symposium on coal combustion (ISCC-9), Qingdao, China, 21–24 July 2019

Tamura M, Watanabe S, Komaba K, Okazaki K (2015) Combustion behaviour of pulverised coal in high temperature air condition for utility boilers. Appl Thermal Eng 75:445–450

Watanabe J, Yamamoto K (2015) Flamelet model for pulverized coal combustion. Proc Combust Inst 35:2315–2322. ISSN: 1540-7489

Watanabe J, Okazaki T, Yamamoto K, Kuramashi K, Baba A (2017) Large-eddy simulation of pulverized coal combustion using flamelet model. Proc Combust Inst 36:2155–2163. ISSN: 1540-7489

Weber R, Smart JP, Kamp WV (2005) On the (MILD) combustion of gaseous, liquid, and solid fuels in high temperature preheated air. Proc Combust Inst 30(II):2623–2629. ISSN: 15407489

Wen X, Wang H, Luo K, Fan J (2019) Analysis and flamelet modelling for laminar pulverised coal combustion considering the wall effect. Combust Theory Model 23:353–375

Wunning JA, Wunning JG (1997) Flameless oxidation to reduce thermal NO-formation. Prog Energy Combust Sci 23(23):81–94

Zhao Y, Serio MA, Bassilakis R, Solomon PR (1994) A method of predicting coal devolatilization behavior based on the elemental composition. Symp (Int) Combust 25:553–560. ISSN: 0082-0784

Zhou MM et al (2019) Large-eddy simulation of ash deposition in a large-scale laboratory furnace. Proc Combust Inst 37:4409–4418. ISSN: 15407489

Part II
Basic Coal Quality and Combustion

Chapter 6
Study on Alkali Metal Migration Characteristics in the Pyrolysis of Naomaohu Coal in CO Atmosphere

Peipei Gao, Junjie Fan, and Zhiyuan Ren

Abstract The four-step chemical extraction method was used for stepwise extraction to compare the occurrence of alkali metals in Naomaohu coal and the other four types of high alkali coal. The coal sample was placed in a vertical furnace with CO gas atmosphere to study the migration and release law of alkali metal with temperature changing at 400–800 °C. The results show that sodium in the Naomaohu coal is mainly in the form of water-soluble sodium, while potassium is mainly composed of insoluble potassium and organic potassium. The increase of pyrolysis temperature promotes the release of sodium and potassium, and the water-soluble sodium changes to insoluble sodium, and a substitution reaction occurs between sodium and potassium. At 700 °C, the amount of sodium and potassium released from the coal char is relatively reduced, resulting in enrichment.

Keywords Naomaohu coal · Alkali metal · Occurrence · Pyrolysis · Migration

The coal content of Naomaohu in Hami area of Xinjiang is extremely rich. The amount of resources in the central mining area has been approved as high as 7.641 billion tons, which is one of the main mining areas of "Western Coal East Transportation". However, due to the high content of sodium and potassium alkali metals in the Naomaohu coal, in the direct combustion, there are serious problems of slagging in the furnace and contamination of the heating surface of the tail (Zeng et al. 2019). Based on the characteristics of high oil content of Naomaohu coal, the use of coal-to-liquid technology can not only make up for the shortage of oil shortage in China, but also extract clean fuel such as hydrogen and methane from natural gas as a natural gas substitute to reduce the burning of coal to the environment. The resulting threat has broad prospects for development (Zhao et al. 2018). Therefore, in the process of pyrolysis of Naomaohu coal, it is of great significance to study the

P. Gao (✉) · J. Fan · Z. Ren
School of Environment and Architecture, University of Shanghai for Science and Technology, Shanghai 200093, China
e-mail: 18239194470@163.com

J. Lyu and S. Li (eds.), *Clean Coal and Sustainable Energy*, Environmental Science and Engineering,
https://doi.org/10.1007/978-981-16-1657-0_6

migration characteristics of alkali metals, improve the quality of gas and tar, and promote the clean utilization of high alkali coal.

Liu et al. conducted pyrolysis experiments of wuwenwan coal at different temperatures to study the occurrence forms and pyrolysis migration rules of alkali metals in coal, and analyzed the existence forms and evolution rules of Na in pyrolysis semi-coke under N_2 atmosphere (Liu et al. 2014); Song et al. and others studied the occurrence and precipitation characteristics of Na in the quasi-east high Na coal at 600–900 °C. The results show that the alkali metal Na is mainly precipitated in the form of NaCl (Song et al. 2015); Li et al. believe that during the pyrolysis process, bond cleavage and bond formation are repeated between alkali metal and alkaline earth metal and carbon matrix. The volatilization of alkali metal is determined by the bond strength between it and the carbon matrix. The temperature is The main factor determining the volatilization of alkali metals (Li et al. 2000a); Wang studied the release and transformation of Na during pyrolysis of Zhundong coal. It was found that water-soluble Na is the main chemical form in Zhundong coal. Most of the water-soluble Na is released into the gas phase as a volatile substance during pyrolysis, and the remaining part is converted into an insoluble form. The release of Na and the particle size of the coal show a non-monotonic change (Wang et al. 2015); Luo et al. studied the migration of Na in two N_2, CO_2 and H_2O atmospheres and Na at different temperatures in two high Na coals. The results show that sodium hydroxide vapor formed under water vapor can promote the volatilization of Na, and the CO_2 atmosphere promotes carbonation. Formation of sodium, inhibiting the volatilization of Na (Luo et al. 2018); Quyn et al. used thermogravimetric analyzer and fluidized bed/fixed bed reactor to study the volatilization and catalysis of alkali metal in the pyrolysis and gasification process of Victoria brown coal. It was concluded that NaCl in brown coal is mainly Na and The release of Cl separately concluded that sodium in the form of NaCl is more volatile than the form of the carboxylate in coal (Quyn et al. 2002; Wu et al. 2002).

At present, domestic and foreign scholars have done a lot of research on the migration of alkali metals in the process of coal pyrolysis, and there are few reports on the migration and release of alkali metals in the process of pyrolysis of Xinjiang high alkali coal under reducing atmosphere conditions. In this paper, the fixed bed pyrolysis experiment platform was built, and the migration law of Na and K alkali metals in the pyrolysis process of Naomaohu coal under CO atmosphere was studied, which provided a theoretical basis for the large-scale utilization of Xinjiang high alkali coal.

6.1 Experimental Part

6.1.1 Experimental Sample

The Naomaohu coal was selected as the experimental raw material. After the coal sample was dried, it was crushed and sieved, and the sample with the particle size of 0.5–1.0 mm was selected, dried and sealed for storage. Comparing the occurrence forms of alkali metals in four typical Xinjiang high alkali coals such as Shaerhu coal, Xiheishan coal, Shenhuo coal and surface high-chlorine coal. Proximate analysis, ultimate analysis and ash composition analysis of coal are shown in Tables 6.1 and 6.2.

6.1.2 Experimental Setup and Steps

Figure 6.1 shows the experimental device of vertical furnace pyrolysis. The whole experimental system consists of gas flow control system, electric heating reaction device, tar condensation collection device and exhaust gas treatment device. The

Table 6.1 Proximate and ultimate analysis of coal samples

Coal type	Proximate analysis (%)				Ultimate analysis (%)				
	FCad	Mad	Aad	Vad	Car	Har	Oar	Nar	Sar
Naomaohu coal	43.6	3.25	9.74	46.22	67.16	5.09	1.84	0.67	0.25
Shaerhu coal	43.6	10.32	8.05	36.34	59.34	3.936	17.434	0.57	0.35
Xiheishan coal	46.26	5.76	18.67	29.31	56.07	3.646	14.204	0.7	0.95
Shenhuo coal	51.16	6.55	5.63	36.66	71.16	3.771	12.219	0.53	0.14
Surface high chloride coal	48.8	9.14	6.12	35.86	62.62	3.862	17.538	0.49	0.23

Table 6.2 Composition of coal ash

Coal type	Ash composition analysis (%)								
	SiO_2	Al_2O_3	Fe_2O_3	CaO	MgO	K_2O	Na_2O	TiO_2	SO_3
Naomaohu coal	29.28	15.07	8.22	30.92	1.16	0.38	2.94	0.97	8.26
Shaerhu coal	16.68	9.4	5.51	44.16	16.67	0.51	2.02	1.56	5.51
Xiheishan coal	35.95	14.6	16.32	11.72	5.69	0.78	4.46	0.96	9.1
Shenhuo coal	30.13	8.44	7.67	23.34	9.37	0.68	5.45	0.53	13.72
Surface high chloride coal	24.88	13.58	7.87	35.74	6.71	0.21	7.16	0.77	2.35

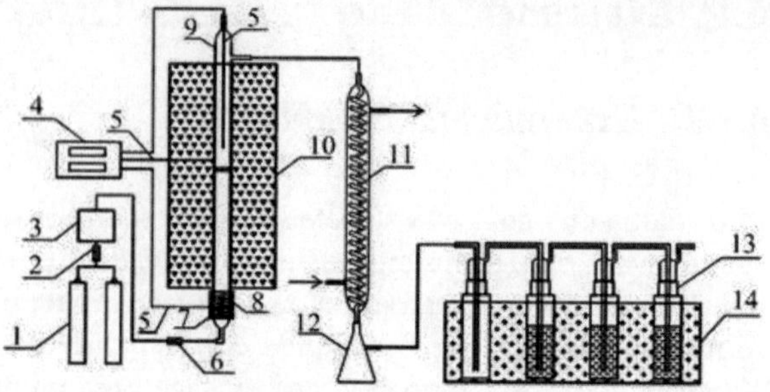

Fig. 6.1 Vertical furnace pyrolysis experimental device diagram

1-Cylinder gas (N2, CO); 2-flow meter; 3-premixer; 4-temperature controller; 5-thermocouple; 6-valve; 7-insulation cotton; 8-heating belt; ;10-resistance furnace; 11-condensation tube; 12-tar collection bottle; 13-wash bottle (diluted hydrochloric acid); 14-ice bath

quartz tube reactor has a total length of 750 mm and an inner diameter of 21 mm, and a quartz sand core is arranged in the middle. The maximum heating temperature of the resistance furnace is 1200 °C. In the experimental process, the resistance furnace reaches the set pyrolysis temperature, and 100 mL/min N_2 is applied for 10 min. After the air in the tube is removed, the pyrolysis atmosphere is adjusted to 20% CO + 80% N_2. The gas flow is stable, and 20 g is poured from the upper part of the reaction tube. The Xinjiang Naomaohu coal with a particle size of 0.5–1.0 mm is collected by a condensing collection device and an exhaust gas treatment device, respectively, and the absorption liquid is 1 mol/L HCl solution. After the end of the pyrolysis, the quartz tube was taken out and continuously passed through N_2, and after cooling, the coal char sample was collected.

6.1.3 Determination of the Form and Content of Alkali Metals

In this paper, the four-step chemical extraction method is used to extract the water-soluble, ammonium acetate-soluble, acid-soluble and insoluble alkali metals in coal. The distilled alkali is used as the extract to extract the inorganic alkali metal forms in the form of chloride crystals and hydrated forms. The ammonium acetate solution can extract the organic state in the form of a carboxylate in addition to the extractable water-soluble state; The organic state of the nitrogen-containing or oxygen-containing functional group present in the coal structure can only be extracted by dilute hydrochloric acid. The residual alkali metal can be considered to be insoluble. The experiment selected distilled water (H_2O), ammonium acetate (CH_3COONH_4, 1 mol/L), hydrochloric acid (HCl, 1 mol/L) three kinds of solution to extract the sample coal/coal coke, test and analysis of water-soluble state, ammonium acetate dissolved state and acid dissolution the content of alkali metal. Weigh 1 g sample into a conical flask, add 30 mL of extract, shake at room temperature for 24 h at 60 °C with a constant temperature oscillator, then centrifuge to separate the supernatant, and dilute to 50 ml. The Na/K content was measured by a coupled plasma emission spectrometer (ICP-OES). The insoluble state and the alkali metal content in the tar are determined by weighing a certain amount of coal, coal char, and tar, and the obtained digestion solution is measured by ICP-OES by constant volume and filtration. The method for determining the alkali metal content released into the gas phase during coal pyrolysis is to dilute and filter the dilute hydrochloric acid absorption liquid, and determine its concentration by ICP-OES to calculate the Na/K content.Its concentration by ICP-OES to calculate the Na/K content.

6.2 Results and Discussion

6.2.1 *Analysis of the Occurrence of Alkali Metals in Raw Coal*

Figure 6.2a shows the experimental results of the occurrence of alkali metal sodium in raw coal. The analysis found that Na in the coal mainly exists in the form of water soluble Na. The content of Na in the surface of high-chlorine coal is as high as 80.7%, which may be related to the fact that the raw coal contains more chlorine. The content of insoluble Na in Xiheishan coal and Shenhuo coal is higher, and the content of insoluble Na in Shaerhu coal and surface high-chlorine coal is lower, which may be related to the higher content of Si and Al. However, the insoluble Na content of Min Mao Lake coal is not high because it contains more Ca, and forms insoluble calcium salts with Si and Al, so the insoluble Na content is lower. There are certain differences in the forms of Na in different coal types. This difference has a certain relationship with the formation and evolution of coal. Coal-forming plants, coal-forming environments and geological environments all have certain effects on the content and morphology of Na (Li and Xiao).

Figure 6.2b shows that potassium in coal is mainly in the form of insoluble potassium and organic potassium, both of which exceed 50%. The insoluble potassium content in the coal of Minmao Lake and Xiheishan is 77% and 89%, respectively. It mainly exists in the form of aluminosilicate. Potassium hydride in organic potassium is significantly higher than potassium acetate in ammonium acetate, indicating that the organic potassium in the coal containing nitrogen or oxygen-containing functional groups in the coordination form is more than the organic potassium in the form of carboxylate.

Fig. 6.2 **a** Raw coal alkali metal sodium combined forms. **b** Raw coal alkali potassium deposit form

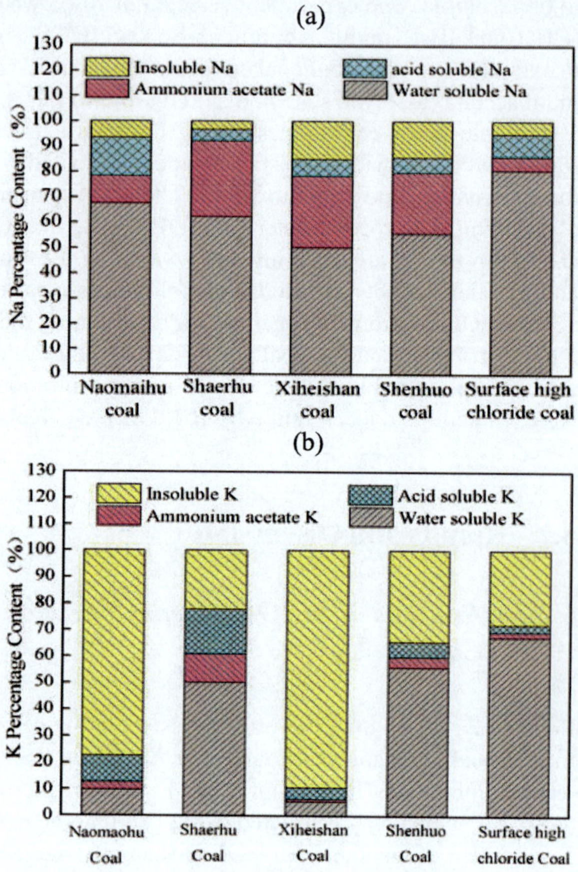

6.2.2 *Migration Characteristics of Alkali Metals at Different Temperatures*

6.2.2.1 Release of Alkali Metal

The effect of the pyrolysis temperature in the CO atmosphere on the form and residual amount of Na in the coal char is shown in Fig. 6.3a. It can be seen from the figure that as the pyrolysis temperature increases, the release rate of Na gradually increases. When the pyrolysis temperature is 400 °C, the content of water-soluble Na in the pyrolysis semi-coke decreases, and the content of insoluble Na and acid-soluble Na increases. As the pyrolysis temperature increases from 600 °C, the release of sodium becomes more and more obvious, and the water is dissolved. The Na content began to increase, the insoluble Na content began to decrease, and the ammonium acetate dissolved Na content remained basically unchanged.

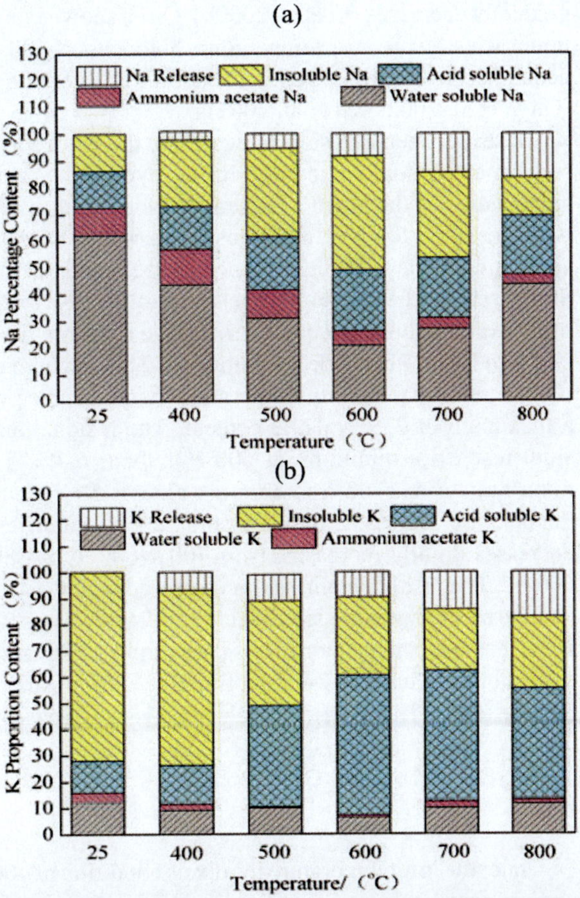

Fig. 6.3 **a** Effect of temperature on the form and residual ratio of Na in coal char. **b** Effect of temperature on the form and residual ratio of K in coal char

Many scholars have also obtained similar results. Wei et al. found the following mechanism for the conversion of water-soluble Na to insoluble Na (Wei et al. 2008):

$$2NaCl + 2SiO_2 + H_2O \rightarrow Na_2Si_2O_5 + 2HCl \quad (6.2.1)$$

$$Na_2SO_4 + 3SiO_2 \rightarrow Na_2O \cdot 3SiO_2 + SO_2 + 1/2O_2 \quad (6.2.2)$$

$$Na_2O \cdot 3SiO_2 + Al_2O_3 \rightarrow Na_2O \cdot Al_2O_3 \cdot 2SiO_2 + SiO_2 \quad (6.2.3)$$

Li et al. found that 600–800 °C is the main release form of water-soluble sodium. With the increase of temperature, the parts of water-soluble sodium, sodium acetate and sodium chloride are continuously released and released, and the residual amount gradually decreases (Li et al. 2000b). Quyn showed that the release process of ammonium acetate was at a temperature higher than 600 °C, the carboxylate was first decomposed to release CO_2, and sodium was combined with the coal char structure (Quyn et al. 2002; Wu et al. 2002).

The experimental results indicate that there is a conversion between water-soluble sodium and insoluble sodium during pyrolysis. As the temperature increases, the stabilization of the organic sodium and the reaction of the water-soluble sodium salt with the silica form an aluminosilicate which causes an increase in the insoluble sodium content. In addition, since the released Na atoms are combined with Cl^-, $SO4^{2-}$, etc. in the coal to form a water-soluble Na salt, to some extent, the reduction of the water-soluble Na content in the coal char is suppressed.

From Fig. 6.3b can be seen under CO atmosphere conditions, volatilization of K with increasing pyrolysis temperature increases, the pyrolysis temperature of 800 °C a maximum of 23% volatile content. The residual amount of water-soluble potassium reached a minimum at 600 °C, about 6.4%. The potassium acetate ammonium acetate reached a minimum value at 500 °C, about 0.3%. As the pyrolysis temperature increases, the acid-soluble potassium residue first increases and then decreases slightly, increasing from 400 to 600 °C by 38% and reaching a maximum at 600 °C. The insoluble potassium increased slightly with the increase of temperature, and the residue was the lowest at 700 °C, which was 23%. The release of potassium is due to the fact that a portion of the sodium volatilized by the pyrolysis displaces an equivalent amount of potassium from the aluminosilicate by a displacement reaction (Raask 1985). The reaction mechanism is as follows:

$$mK_2O \cdot xSiO_2 \cdot yAl_2O_3 + 2NaCl \rightarrow K_2O \cdot Na_2O \cdot xSiO_2 \cdot yAl_2O_3 + 2KCl \quad (6.2.4)$$

Under the low temperature pyrolysis condition of 400 °C, potassium in coal begins to be released in large quantities, insoluble potassium begins to decompose, and the content drops sharply, while acid-soluble K content shows an upward trend, while the content of ammonium acetate and water-soluble potassium has no significant change, which is consistent with the research results of Wang et al. (2014).

6.2.2.2 Alkali Metal Migration Occurs in Different Forms

Figure 6.4a shows the contents of sodium and potassium in coal char at different temperatures. When the pyrolysis temperature is between 400 and 600 °C, the water-soluble sodium content is significantly lower than that of the raw coal, the acid-soluble sodium content is gradually decreased, the ammonium acetate sodium is slightly reduced, and the insoluble sodium content is rapidly increased, which may be the conversion of water-soluble sodium. In addition, the alkali metal will volatilize during the pyrolysis process. During the heating and dehydration of coal, the water-soluble sodium in the coal is also brought to the surface of the coal particles by the water to precipitate in the form of NaCl (Raask 1985). Therefore, as the pyrolysis temperature

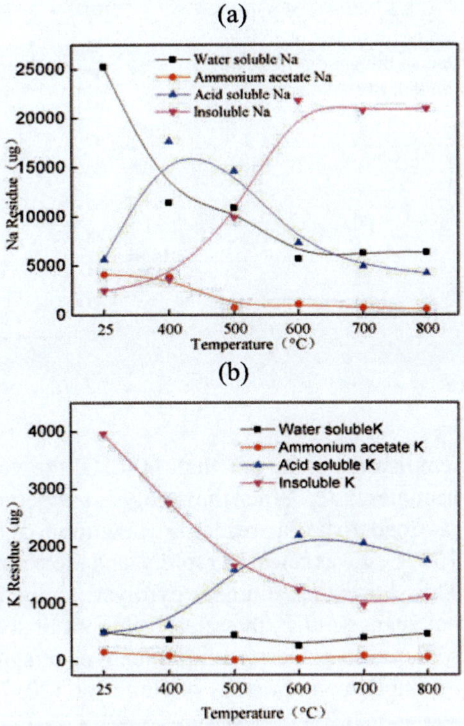

Fig. 6.4 **a** Effect of temperature on the form and residual amount of Na in coal char. **b** Effect of temperature on the form and residual amount of K in coal char

increases, the content of water-soluble sodium decreases overall, and the sodium in the mink lake coal has begun to release at a pyrolysis temperature of 400 °C.

It can be seen from Fig. 6.4b that the release of potassium before 600 °C is large. With the increase of pyrolysis temperature, the fluctuation of potassium content in each occurrence state is small, which can be considered as coal char after 600 °C. The potassium content is basically stable. Among them, the content of water-soluble potassium is relatively stable; while the potassium acetate of ammonium acetate decreases with temperature; the acid-soluble potassium shows a decreasing trend with temperature, but it does not change significantly within the range of 600–800 °C. The potassium in coal is mainly in the form of aluminosilicate, which is relatively stable and not volatile at high temperatures, so when the temperature is high, the amount of potassium released is relatively small.

6.2.2.3 Effect of Temperature on Coal Char Yield and Sodium and Potassium Content

It can be seen from Fig. 6.5 that, with the increase of pyrolysis temperature, the semi-coke yield decreases rapidly and then slowly. As the pyrolysis temperature increases, alkali metal sodium and potassium are continuously released. When the

Fig. 6.5 Effects of pyrolysis temperature on coke yield and sodium and potassium content in coal coke

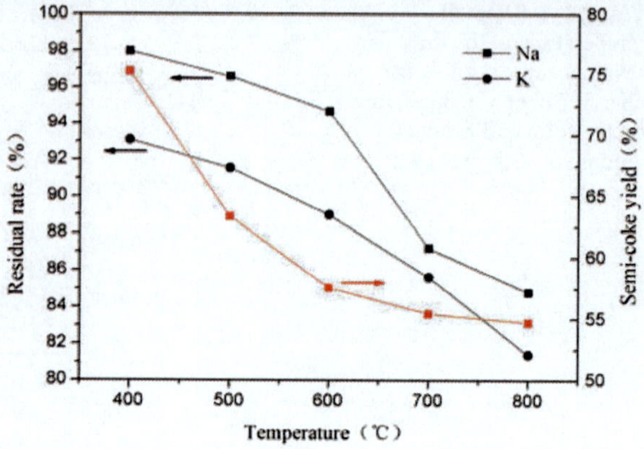

temperature is higher than 600 °C, the volatilization rate of sodium alkali metal becomes faster. When the pyrolysis temperature is further increased, the volatilization rate tends to be flat, reaching a maximum of 15.2% at 800 °C. When K was higher than 700 °C, it was released rapidly, and the K residual rate in the coke decreased sharply. This shows that at higher pyrolysis temperature, when the yield of semi-coke does not change much, the release amount of sodium and potassium is relatively low, so sodium and potassium enrichment occurs in the semi-coke, which is consistent with the experimental results of Guo et al. (2017), Jiang et al. (2012).

6.3 Conclusion

Through the extraction experiments of different coal types, it is found that sodium in Xinjiang high alkali coal mainly exists in the form of water-soluble sodium. The main forms of potassium are insoluble potassium and organic potassium. The content of ammonium acetate and insoluble potassium in ammonium sulphate coal is higher than other coal types, and the cause of the difference may be related to coal formation factors.

During the pyrolysis process of the Naomaohu coal under the CO atmosphere, the alkali metal sodium and potassium were partially released. As the pyrolysis temperature increases, the release rates of sodium and potassium increase gradually. The water-soluble sodium changed to insoluble sodium, and the release of sodium was most significant at 600 °C. Potassium begins to release in large amounts at 400 °C. The release of potassium is related to the partial exchange of potassium from the aluminosilicate by a displacement reaction.

As the temperature increases, the semi-coke yield decreases rapidly and then slowly decreases, and sodium and potassium are continuously released and released. At higher pyrolysis temperatures, the release of sodium and potassium is relatively reduced, resulting in enrichment.

References

Guo S, Jiang Y, Xiong Q et al (2017) Study on the regularity of different occurrence forms in the pyrolysis process of Zhundong coal. J Fuel Chem 45(3):257–264

Jiang L, Hu S, Xiang J et al (2012) Release characteristics of alkali and alkaline earth metallic species during biomass pyrolysis and steam gasification process. Bioresour Technol 116:278–284

Li CZ, Sathe C, Kershaw JR et al (2000a) Fates and roles of alkali and alkaline earth metals during the pyrolysis of a Victorian brown coal. Fuel 79(3):427–438

Li CZ, Sathe C, Kershaw JR et al (2000b) Fates and roles of alkali and alkaline earth metals during the pyrolysis of a Victorian brown coal. Fuel 79(3):427–438

Li Y, Xiao J (2005) Occurrence and migration of alkali metals in coal combustion process and related research progress. Clean Coal Technol (01):39–44. https://doi.org/10.13226/j.issn.1006-6772.2005.01.010

Liu D, Zhang S, Tu S et al (2014) Morphological changes during pyrolysis of Wucaiwan coal. J Fuel Chem 42(10):1190–1196

Luo A, Zhu P, Zhang J et al (2018) Effects of carbon dioxide and water atmosphere on migration during high coal conversion. J Fuel Chem (5)

Quyn DM, Wu H, Li C-Z (2002) Volatilisation and catalytic effects of alkali and alkaline earth metallic species during the pyrolysis and gasification of Victorian brown coal. Part I. Volatilisation of Na and Cl from a set of NaCI-loaded samples. Fuel 81(2):143–149

Raask E (1985) Mineral impurities in coal combustion, behavior, problems, and remedial measures. Hemisphere Publishing

Song W, Song G, Zhang H et al (2015) Experimental study on migration characteristics of high coal pyrolysis in Zhundong. J Fuel Chem 43(1):16–21

Wang Z, Li Q, Liu J et al (2014) Occurrence forms of alkali metals in Zhundong coal and their migration during pyrolysis. Chin J Electr Eng 34(1):130–135

Wang C, Jin X, Wang Y et al (2015) Release and transformation of sodium during pyrolysis of Zhundong coals. Energy Fuels 29:78–85

Wei X, Huang J, Liu T et al (2008) Transformation of alkali metals during pyrolysis and gasification of a lignite. Energy Fuels 22(3):1840–1844

Wu H, Quyn DM, Li CZ (2002) Volatilisation and catalytic effects of alkali and alkaline earth metallic species during the pyrolysis and gasification of Victorian brown coal. Part III. The importance of the interactions between volatiles and char at high temperature. Fuel 81(8):1033–1039

Zeng X, Yu D, Yu G et al (2019) Transformation of inorganic elements in different forms into ash particles during Zhundong coal combustion. J China Coal Soc 44(2):588–595

Zhao P, Zhang X, Li J et al (2018) Hydroliquefaction characteristics of Naomaohu coal and distribution of hydrogen in liquefaction products. Coal Convers 41(164)(04):45–50

Chapter 7
Precipitation Characteristics of Alkali/Alkaline Earth Metal in High Alkali Coal

Hang Shi, Yuxin Wu, Junfu Lyu, Man Zhang, and Hai Zhang

Abstract Serious problems of fouling and slagging occurred in combustion of high alkali coal. As the physical and chemical characteristics of coal ash may change at different ashing temperature (AT). The investigation on the precipitation characteristics of alkali/alkaline earth metal (AAEM) in high alkali coal is important for the safety and stable operation of the boilers. To understand the effect of AT on AAEM precipitation characteristics, four different coals were selected as samples. For each coal, three coal ashes were acquired at 200 °C/500 °C/815 °C respectively. X-ray fluorescence spectroscopy (XRF) and Inductively Coupled Plasma Optical Emission (ICP-OES) were applied for the elementary analysis of ash. The test results indicated that ash yields and alkali contents in coal ash decrease while alkaline earth metal content changes little as AT increases. The precipitation temperature of chlorine is lower than 500 °C, and chlorine is substantially completely precipitated before 815 °C. The precipitation characteristic of Na is closely related to the content of chlorine in the coal. Low-temperature ashing will cause carbon burnout problems for high ash coals while the relative error of alkali metal measurement is limited. The test results of ICP-OES on element content in same sample is lower than XRF. However, the elemental content changes with AT are similar between two measurement results. It is recommended to use the ICP-OES test method when measuring AAEM content due to its higher precision.

Keywords Precipitation characteristics · Alkali/alkaline earth metal (AAEM) · High alkali coal · Ashing temperature

---

H. Shi · Y. Wu (✉) · J. Lyu · M. Zhang · H. Zhang
Key Laboratory for Thermal Science and Power Engineering of Ministry of Education, Department of Energy and Power Engineering, Tsinghua University, Beijing 100084, China
e-mail: wuyx09@mail.tsinghua.edu.cn

J. Lyu and S. Li (eds.), *Clean Coal and Sustainable Energy*, Environmental Science and Engineering,
https://doi.org/10.1007/978-981-16-1657-0_7

7.1 Introduction

Coal plays an important role in the consumption of primary energy source in China (BP 2016). Nowadays, some high-alkali coal (the sodium content in ash is higher than 2%) was applied in power plant for power generation or heating supply. Zhundong coal field is the largest coal field in China and Zhundong coal was characterized by low ignition temperature, high burnout rate, high combustion economy and low pollutant emissions (Xu et al. 2017). However, the actual burning experiences indicate that serious problems of fouling, slagging occurred during the combustion of Zhundong coal (Li et al. 2017; Wang et al. 2015a). The high alkali/alkaline earth metal AAEM contents (mainly CaO, MgO, Na_2O) in coal ash were thought to be the cause (Wang et al. 2015b).

The previous studies indicated that AAEM tend to release to the flue gas and then condensation of AAEM vapor in heating surfaces thus to promote deposition. And the high AAEM contents in ash will lower the ash fusion temperatures thus to worsen slagging (Hurley and Benson 1995; Vorres 1977). Considering that high ashing temperature (AT) will decrease AAEM contents in coal ash so as to influence the physical and chemical characteristics of coal ash. It is necessary to study the precipitation characteristics of AAEM under different ashing temperature.

Since the organic matter in coal affects the accuracy of the mineral test results, the coal sample is usually first ashed at 815 °C (reference "Test method for analysis of coal ash", GB/T 1574–2007, China) to remove the organic matter and then conduct the component analysis. Li et al. (2000), Li (2007, 2013) and Quyn et al. (2002a) found that about 10% of the AAEM released from the coal when the pyrolysis temperature is 300–400 °C. When the pyrolysis temperature is higher than 600 °C, the release rate of Na increased rapidly. Quyn et al. (2002b) reported that the release rate of Na was positively correlated with the NaCl content in coal. It was also found that Na in ash gradually changed from water-soluble state to insoluble Na as the AT increased (Li et al. 2016; Xu et al. 2018). Li et al. (2000) reported that the temperature won't affect the release of Ca in the temperature range of 300–600 °C. The release rate of Ca increased with increasing temperature in the temperature range of 600–900 °C. Yang et al. (2015) and Fu et al. (2014) found that the increase of AT will promote the volatilization of alkali metals while the precipitation of alkaline earth metals were basically not affected by the AT.

Some scholar (Wang et al. 2008; Xiao et al. 2011) studied the influence of AT on the fusion characteristic of biomass. Wang et al. (2008) claimed that high-melting matters generated on seaweed biomass at high AT so as to affect the ash fusion temperature. and Xiao et al. (2011) reported that the optimal AT for analysis of ash properties is 600 °C.

At present, there are few studies on the precipitation characteristics of AAEM in coal when the AT is lower than 500 °C. X-ray fluorescence spectroscopy (XRF) and ICP-OES are two methods applied in previous studies. The measurement of AAEM contents by these two methods have been reported, but the measurement data is biased. Compared with the high pre-processing requirements and cumbersome operation of ICP, the sample preparation is much easier for XRF. However, XRF can only measure the element content on the surface of sample. And particle size and the unevenness of the sample will affect the accuracy of the results. In contrast, measurement accuracy of ICP-OES is higher, but high pre-processing requirements and cumbersome operation (Gao 2017; 24 elements in geochemical samples by low dilution ratio multi-lithium borate melting flake-XRF 2013).

In this paper, four coals with high AAEM content (all experiments using air-dry coal samples) were selected for experiments. The effects of AT on the precipitation of AAEM in coal and the influence of chlorine on alkali metal were investigated. The measurement results of two measurement methods were compared.

7.2 Experimental Approaches

7.2.1 Preparation of Coal Samples and Sample Properties

Four coals including Shaowu coal (SW coal) and three Zhundong coals: Wucaiwan coal (WCW coal), Wudong coal (WD coal) and Jiangjunmiao coal (JJM coal).

All experiments using air-dry coal samples. The dried coal was ground and sieved with a 200 mesh sieve. The particle size was less than 75 μm. The proximate and ultimate analysis results were shown in Table 7.1. It's clear that Zhunodng coals have lower ash, higher volatile and fixed carbon contents. For further component analysis, coals were first digested and analysed by ICP-OES. The test results were shown in Table 7.2. It can be indicated from Table 7.2 that Zhundong coals have a lower Si/Al contents and higher Ca content. Na/Ca/S were found to be the main components in the ash deposits (Song et al. 2017). But the primary Na content is even lower than SW coal and the Ca content in Zhundong coal is not that high. The ash deposition occurred due to the lower ash in Zhundong coal, which result in a much higher AAEM contents in coal ash.

Table 7.1 Proximate and ultimate analysis of coal

Coal samples		JJM	SW	WCW	WD
Proximate analysis	M_{ad}/%	2.47	4.74	11.22	3.98
	V_{ad}/%	29.67	14.96	29.27	29.34
	A_{ad}/%	13.92	64.85	3.54	14.56
	FC_a/%	53.94	15.45	55.97	52.13
Ultimate analysis	C_{ar}/%	60.63	23.76	65.31	61.01
	H_{ar}/%	4.06	2.54	4.41	4.28
	O_{ar}/%	17.12	2.33	14.66	11.45
	N_{ar}/%	1.15	0.85	0.37	0.37
	S_{ar}/%	0.65	0.94	0.49	0.43

Table 7.2 ICP-OES test results of metal elements in coal

Analyte	Unit	JJM	SW	WCW	WD
Si	%	3.11	21.77	0.23	9.72
Al	%	1.25	1.83	0.20	4.50
Ca	%	0.95	0.36	1.12	1.46
Mg	%	0.22	0.04	0.23	0.28
Na	%	0.8	0.61	0.19	0.19
K	%	0.17	1.12	0.01	0.54
Ti	%	0.06	0.37	0.01	0.20

7.2.2 *Experimental Equipment and Facilities*

In order to investigate the influence of AT on the ash properties, low temperature (100–200), 500 and 815 °C were selected as the final temperature of ashing. The coal ash was acquired in a low-temperature ashing machine (100–200 °C) and a muffle furnace (500 °C/815 °C). The specific ashing method are shown as follows:

(1) Preparation of low-temperature ash

YAMATO PR300 plasma ashing instrument was applied to prepare low-temperature ash. The size of the heating tube is Φ64 mm × 160 mm, and the maximum output power is 300 W. The organic matter in the coal is oxidized by oxygen plasma oxidation in a low temperature and low-pressure environment (100–200 °C, <0.2 kPa). coal samples were ashed until the mass didn't change. Lower AT (100–200 °C) can reduce the decomposition and volatilization of the original minerals during the ashing process of coal. It's beneficial to the analysis of coal ash properties.

(2) Preparation of medium-temperature ash

1 g coal samples were put into the muffle furnace when the furnace temperature is lower than 100 °C. A gap of about 15 mm was reserved for the furnace door. The furnace temperature was first raised to 300 °C in not less than 30 min and maintained for 30 min. Then the furnace temperature was raised to 500 °C in not less than 20 min and maintained for 1 h.

(3) Preparation of high-temperature ash

High-temperature coal ash were prepared at 815 °C (reference "Test method for analysis of coal ash", GB/T 1574–2007, China). 1 g coal samples were put into the muffle furnace when the furnace temperature is lower than 100 °C. A gap of about 15 mm was reserved for the furnace door. The furnace temperature was first raised to 500 °C in not less than 30 min and maintained for 30 min. Then the furnace temperature was raised to (815 ± 10) °C and maintained for 1 h. After the preparation of the coal ash, an inspection calcination was performed for 20 min each time until the mass change between two calcination did not exceed 0.001 g.

7.2.3 *Experimental Approaches*

The measurement of AAEM contents were conducted by XRF and ICP-OES. Coal ashes were ground to no more than 30 μm for XRF. For ICP-OES measurement, the coal ashes were digested by microwave digestion. The digestion reagent were 9 ml HNO3, 1 mlHF and 1 ml HCl (guarantee reagent).

7.3 Experimental Results and Discussion

7.3.1 *Influence of AT on Ash Content*

Figure 7.1 shows that the ash yield decrease from 73.1 to 20.3% (72.5% at 500 °C) for JJM coal, from 4.7 to 3.9% (4.6 at 500 °C) for WCW coal and from 37.1 to 31.6% (36.4 at 500 °C) for WD coal as the AT increase from 200 to 815 °C. Except for JJM coal, other coal's ash yield decrease more when the AT increase from 500 to 815 °C, which maybe caused by oxidation of the unburned carbon at low AT and the change of precipitation ratio of alkali and volatile matter at different AT.

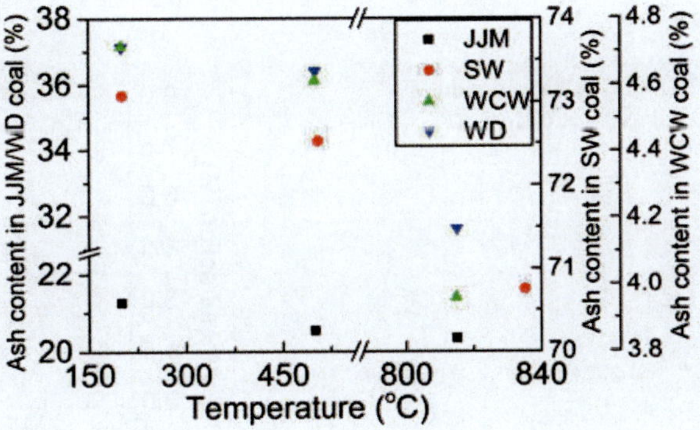

Fig. 7.1 Coal ash yield at different AT

7.3.2 Alkali and Chlorine Contents in Coal Ash at Different AT

The alkali content in Zhundong coals are similar to SW coal, however, it can be indicated from Fig. 7.2 that the Na content in JJM and WCW coal ash are much higher than SW and WD coal ash while the K content in coal ash shows the opposite result. WD coal is often applied for co-firing with other high-alkali Zhundong coal due to its relatively lower Na content and higher ash yield.

Similar to the influence of AT on ash yield, Fig. 7.2 shows that Na content in JJM/WCW coal ash decrease when the AT increase from 500 to 815 °C while it doesn't change much when the AT increase from 200 to 500 °C. Figure 7.1 shows that for JJM/WCW coal, Na/K content in coal ash decrease from 7.30/1.09% to 5.79/0.10% (7.53/1.09% at 500 °C) for JJM coal, from 5.36/0.26% to 3.77/0.15% (5.47/0.27% at 500 °C) for WCW coal.

For SW/WD coal, Na/K content in coal ash change a little. It changes from 0.83/3.06% to 0.78/3.10% (5.47/3.27% at 500 °C) for SW coal, from 0.70/1.75% to 0.79/1.91% (0.70/1.79% at 500 °C) for WD coal as the AT increase from 200 to 815 °C.

Compare the Na/K contents and ash yield of four coals, the alkali content is more sensitive to the AT as the ash yield increase.

Figure 7.3 shows the Cl content in coal ash. JJM/WCW coal have a relatively higher Cl content in coal ash. For JJM/WCW coal which have a higher Cl content, the Na/K content in coal ash decrease no more than 0.01% while the chlorine content in coal ash decrease by 0.16/0.03%. The Cl begin to precipitate from coal before 500 °C and nearly all release to air at 815 °C. For SW/WD coal, Cl content in coal ash is extremely low and it does not change with the increase of AT.

Fig. 7.2 Alkali content in coal ash at different AT

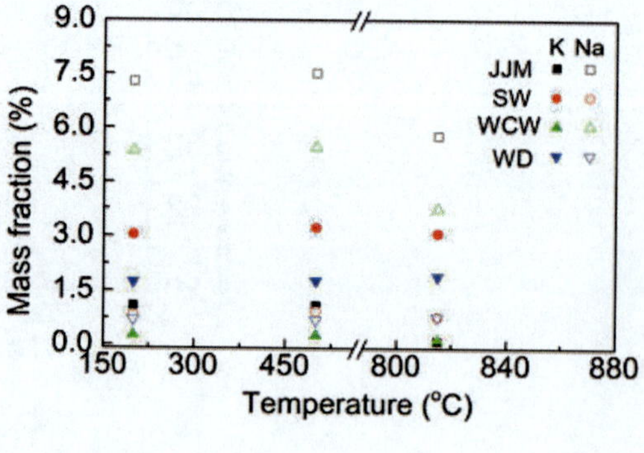

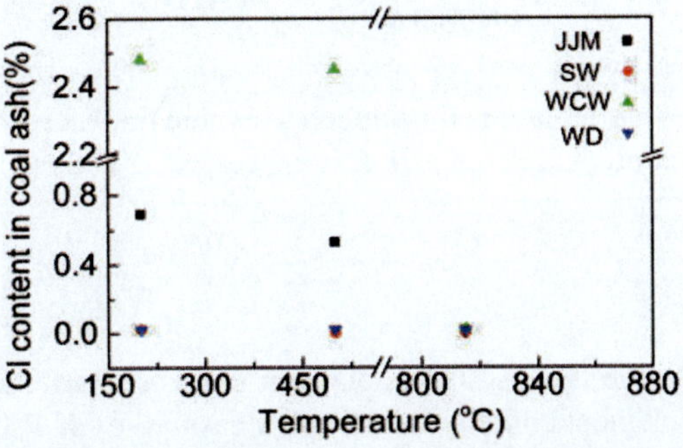

Fig. 7.3 Chlorine content in coal ash at different AT

7.3.3 *Alkaline Earth Metal Content in Coal Ash at Different AT*

The change of alkaline earth metal content is different from alkali/Cl content in coal ash. Ca/Mg content change a little in coal ash as the AF increase from 200 to 815 °C (Fig. 7.4).

Combine all the measurement results from Figs. 7.1, 7.2 and 7.3. The Na/K/Ca/Mg contents in coal ash decrease no more than 0.1% while the ash yield increase by 0.73/0.55/0.10/0.74% for JJM/SW/WCW/WD coal as the AT increase from 200 to 500 °C. It can be indicated that there is a certain amount of unburned carbon when the AT is lower than 500 °C. When the AT increase from 500–815 °C, the Na/K/Cl content in coal ash shows a sharp decrease for JJM/WCW coal. The decrease of ash yield is caused by the combustion of unburned carbon and precipitation of Na/K/Cl for JJM/WCW coal.

Therefore, unburned carbon is the main factor that influence the measurement of AAEM contents when the AT is lower than 500 °C. When the AT is higher than 500 °C, not only unburned carbon should be considered, but also the precipitation of AAEM should be evalued. The unburned carbon content is lower than 5% of air-dry coal, the influence on the measurement of alkali metal content is limited.

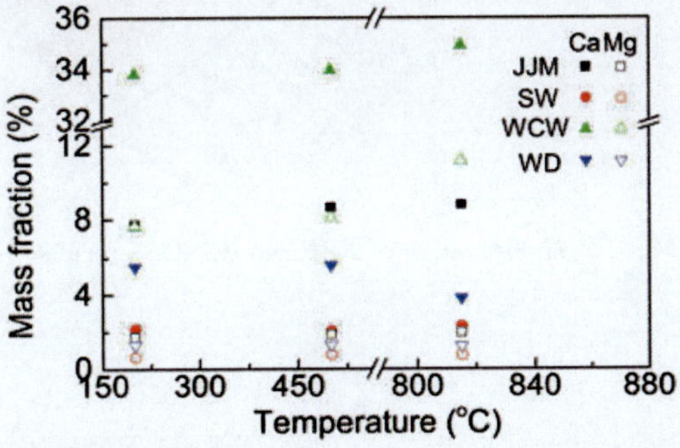

Fig. 7.4 Alkaline earth metal content in coal ash at different AT

7.3.4 Comparation of XRF and ICP-OES Results

A variable R is introduced to evalue the XRF and ICP-OES test results of the AAEM contents in coal ash. R is expressed as follows:

$$R = \frac{\left(W_{t,XRF} - W_{t,ICP\text{-}OES}\right)}{W_{t,ICP\text{-}OES}} * 100 \tag{7.1}$$

where $W_{t,XRF}$ is the element content in coal ash measured by XRF; $W_{t,XRF}$ is the element content in coal ash measured by ICP-OES.

As is shown in Fig. 7.5, the AAEM content in coal ash measured by XRF are higher than ICP-OES. But the change trend of AAEM at different ashing temperatures is similar. XRF test result is about 30–70% higher than ICP-OES.

It is recommended to study the precipitation characteristics of AAEM by ICP-OES due to its high accuracy. Both ICP-OES and XRF can be applied when carry out the qualitative analysis the influence of the ashing temperature on the precipitation law of AAEM.

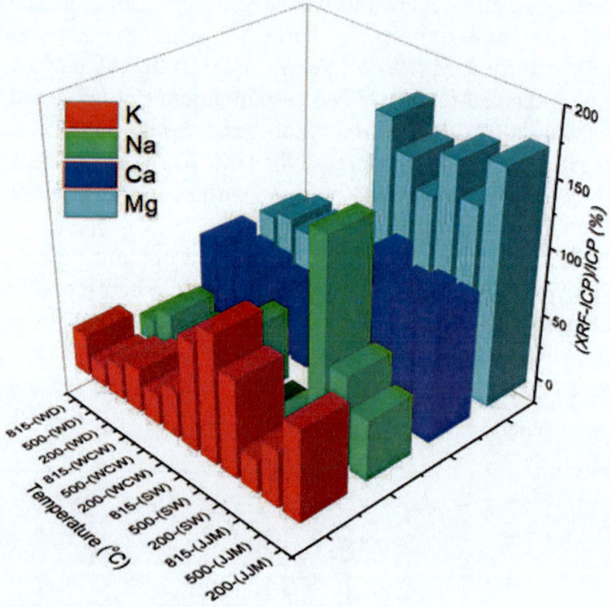

Fig. 7.5 Comparation of XRF and ICP-OES results

7.4 Conclusion

Based on the obtained results the following conclusions can be drawn:

Na content in JJM/WCW coal ash decrease when the AT increase from 500 to 815 °C while it doesn't change much when the AT increase from 200 to 500 °C. Ca/Mg content change a little in coal ash as the AF increase from 200 to 815 °C.

Cl precipitate from coal before 500 °C and nearly all release to air at 815 °C while most of Na precipitate from coal from 500 °C.

Unburned carbon is the main factor that influence the measurement of AAEM contents when the AT is lower than 500 °C. When the AT is higher than 500 °C, unburned carbon and the precipitation of AAEM should be considered. It is recommended to use the coal ash prepared at no more than 500 °C.

ICP-OES is recommended to measure the AAEM content. Both ICP-OES and XRF can be applied when carry out the qualitative analysis on the precipitation law of AAEM.

Acknowledgements This work was supported by the National Science & Technology Pillar Program (No. 2015BAA04B02).

References

(2013) 24 elements in geochemical samples by low dilution ratio multi-lithium borate melting flake-XRF. Jilin University (in Chinese)

BP (2016) BP Statistical review of world energy 2016, pp 30–34. http://www.bp.com/content/dam/bp/pdf/energy-economics/statistical-review-2016/bp-statistical-review-of-world-energy-2016-full-report.pdf

Fu Z, Wang C, Che D et al (2014) Experimental study on the effect of ashing temperature on physicochemical properties of Zhundong coal ashes. J Eng Thermophys 35(3):609–613 (in Chinese)

Gao X (2017) Practical X ray spectrum analysis. Chemical Industry Press, Beijing (in Chinese)

Hurley JP, Benson SA (1995) Ash deposition at low temperatures in boilers burning high-calcium coals 1. Problem definition. Energy Fuels 9:775–781

Li CZ (2007) Some recent advances in the understanding of the pyrolysis and gasification behaviour of Victorian brown coal. Fuel 86(12–13):1664–1683

Li CZ (2013) Importance of volatile–char interactions during the pyrolysis and gasification of lowrank fuels—a review. Fuel 112(3):609–623

Li CZ, Sathe C, Kershaw JR et al (2000) Fates and roles of alkali and alkaline earth metals during the pyrolysis of a Victorian brown coal. Fuel 79(3):427–438

Li GY, Wang CA, Yan Y et al (2016) Release and transformation of sodium during combustion of Zhundong coals. J Energy Inst 89(1):48–56

Li GY, Wang GCA, Wang PQ et al (2017) Ash deposition and alkali metal migration during Zhundong high-alkali coal gasification. Energy Procedia 105:1350–1355

Quyn DM, Wu H, Bhattacharya SP et al (2002a) Volatilisation and catalytic effects of alkali and alkaline earth metallic species during the pyrolysis and gasification of Victorian brown coal. Part II. Effects of chemical form and valence. Fuel 81(2):151–158

Quyn DM, Wu H, Li CZ (2002b) Volatilisation and catalytic effects of alkali and alkaline earth metallic species during the pyrolysis and gasification of Victorian brown coal. Part I. Volatilisation of Na and Cl from a set of NaCl-loaded samples. Fuel 81(2):143–149

Song G, Yang S, Song W, Qi X (2017) Release and transformation behaviors of sodium during combustion of high alkali residual carbon. Appl Therm Eng 122:285–296

Vorres KS (1977) Melting behavior of coal ash materials from coal ash composition. Quat Int 371(2):197–208

Wang S, Jiang X, Wang N et al (2008) Fusing characteristic analysis on seaweed biomass ash. Proc Chin Soc Electr Eng 28(5):96

Wang XB, Xu ZX, Wei B et al (2015a) The ash deposition mechanism in boilers burning Zhundong coal with high contents of sodium and calcium: a study from ash evaporating to condensing. Appl Therm Eng 80:150–159

Wang C, Jin X, Wang Y et al (2015b) Release and transformation of sodium during pyrolysis of Zhundong coals. Energy Fuels 29(1):78–85

Xiao R, Chen X, Wang F et al (2011) The physicochemical properties of different biomass ashes at different ashing temperature. Renew Energy 36(1):244–249

Xu L, Liu H, Fang H et al (2017) Effects of various inorganic sodium salts present in Zhundong coal on the char characteristics. Fuel 203:120–127

Xu L, Liu H, Zhao D et al (2018) Transformation mechanism of sodium during pyrolysis of Zhundong coal. Fuel 233:29–36

Yang Y et al (2015) Release of alkali/alkaline earth metal species in Zhundong coal at different ashing temperatures. J Combust Sci Technol 21(04):297–300 (in Chinese)

Chapter 8
Experimental Study on Removal of Low Concentration Coal Tar in Syngas by Mg-Ca Composite Catalyst

Liang Wenzheng, Yan Hao, Li Yanhui, Cui Weiwei, Cong Xiaochun, and Wang Cuiping

Abstract The catalytic cracking of low concentration coal tar in syngas was studied using Mg-Ca composite catalyst. The reaction was carried out in a two-stage fixed bed reactor. The effects of bed temperature (400–800 °C), catalyst particle size (400–600 μm, 200–400 μm, 100–200 μm, and <75 μm) and catalyst layer height in bed reactor on tar conversion, pyrolysis gas products (CO, H_2, CO_2, CH_4), and the CO_2 absorption by CaO in catalyst were investigated. The results showed that the cracking rate of coal tar vapor increased with the increasing of bed temperature, till reaching a maximum of 94% at 700 °C. The production of H_2 also increased with temperature, and the amount of CO_2 in the gaseous products decreased between 500 and 700°C responding to the CO_2 absorption capacity of CaO reaching its peak in this temperature range. The absorbed CO_2 subsequently released at higher temperatures. The catalytic cracking rate was highest when the catalyst particle size was 100–200 μm, and the catalytic cracking efficiency increased with the increasing of catalyst layer height and reaction time. These results demonstrate that when the catalyst particle size is too small (<75 μm) or the reaction temperature is too high (>700 °C), the catalyst activity for coal tar cracking is reduced.

Keywords Coal tar vapor · Mg-Ca composite catalyst · Cracking rate · Catalyst layer height · Carbon deposition

L. Wenzheng · Y. Hao · L. Yanhui · W. Cuiping (✉)
Electrical and Mechanical Engineering College, Qingdao University, Qingdao 266071, Shandong Province, China
e-mail: wangcuiping@tsinghua.org.cn

C. Weiwei · C. Xiaochun · W. Cuiping
College of Civil Engineering and Architecture, Shandong University of Science and Technology, Qingdao 266590, Shandong Province, China

J. Lyu and S. Li (eds.), *Clean Coal and Sustainable Energy*, Environmental Science and Engineering,
https://doi.org/10.1007/978-981-16-1657-0_8

8.1 Introduction

Gasification technology is one of the most efficient methods of clean coal conversion, and is widely used in industries such as chemical synthesis, synthetic ammonia, industrial gas, metallurgical reduction *gas production, and coal-based polygeneration (Franco et al. 2003; Antal and Allen 2000; Bridgwater 1995). However, the gasification process inevitably accompanies with the formation of by-product tar, which might be gradually condensed out of the gas mixture as it cools and bonds with water, coke powder or ash residue when the gasification products are transported in pipelines. The condensate coal tar is a major problem causing blockage and corrosion in pipeline and the downstream equipments, and even harmful to the stable operation of the gasification system. The tar concentration in the syngas has been significantly reduced by operation conditions in the gasifier are adjusted, but it is still higher than 2 g/m^3 (Nunes et al. 2008).

At present, the main methods of removing tar from gasification products include biological, physical and chemical methods (Han et al. 2008; Masurel et al. 2015; Xie et al. 2014). The biological method is of low efficiency, whereas the physical method is simple and effective with 40–90% removal rate of the tar; however, it is not effectively recycled and might cause secondary pollution. The third option (i.e. chemical methods) mainly includes the processes of thermal cracking and catalytic reforming, of which the latter is considered to be more promising method due to its high efficiency and cleanliness. Hence, the factors affecting the tar catalytic cracking process and the development of efficient and inexpensive catalysts have been studied extensively. Boot-Handford et al. (2018) simulated the downdraft gasification process and studied the tar catalysis by calcined limestone and dolomite catalysts. The results show that the tar pyrolysis efficiency after calcined dolomite layer is higher, so calcined dolomite is an effective tar pyrolysis catalyst for downdraft gasifier. Wang (2017) and Sun et al. (2011) used a small fluidized bed two-stage reactor to carry out the experiments of tar removal, by thermal cracking and semi-coke catalytic reforming respectively. The results showed that the coal tar removal efficiency increased with the increasing of reaction temperature, residence time, and with the specific surface area and the developed pore structure of catalyst. The semi-coke was better for the removal of light tar composites, whereas between 600 and 700 °C, the removal rate of heavy tar composites was enhanced. But the semi-coke catalyst was limited to be applied for its poor mechanical strength, hence, it is often compounded with metal Ni to enhance its low-temperature performance and mechanical strength. Luo et al. (2004) studied the catalytic cracking of tar in a fixed-bed reactor using limestone and Al_2O_3 as catalysts. The catalytic cracking efficiency is above 80% when the cracking temperature is of 900 °C; when the gas residence time is increased from 0.5 to 1 s, the tar removal efficiency increases by 7%. So the limestone is also a suitable catalyst. Hu et al. (2016) studied the catalytic performance of expanded perlite and olivine in granular bed using toluene as tar model compound. The conversion of toluene can reach 90% at 800 °C.

There are also studies on other types of catalysts, such as Ni-based catalysts (Lv et al. 2017), alkali metal catalysts (Moud et al. 2015), could achieve tar removal efficiency of more than 90% under appropriate operating conditions. And the recognized dolomite (Sun et al. 2012) is the cheaper catalyst with higher catalytic removal efficiency, but it is suitable for the light tar components pyrolysis and of poor performance for heavy tar components. The coal tar byproduct from high-temperature gasification is a multi-component mixture of benzene, phenol, aromatic hydrocarbons, etc., and it is necessary to apply a feasible catalyst for both light and heavy components pyrolysis to meet the operation index. Therefore, it is significant to find or composite the novel catalyst for coal tar pyrolysis.

The current study using a Ca-Mg composite catalyst prepared by mechanical mixing method, investigates the effects of different reaction temperatures, catalyst particle sizes, and catalyst heights in bed on the gas phase reaction of coal tar in a two-stage fixed-bed reactor, in order to improve the catalyst performance.

8.2 Experimental

8.2.1 Experimental Apparatus

The experimental system is shown in Fig. 8.1 and the reactor is divided into two parts: (1) the tar gasification part, and (2) the catalytic cracking part. In a typical experiment process, the coal tar was heated and vaporized to form a certain concentration of tar vapor in the lower-stage fixed bed at low temperature, which was carried into the

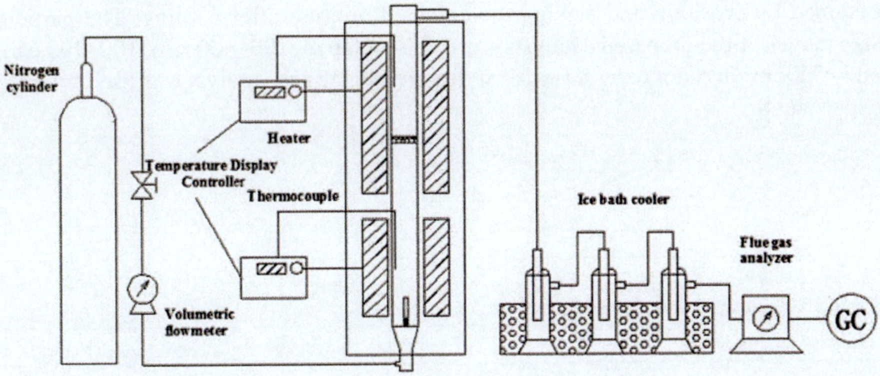

Fig. 8.1 Two-stage fixed bed unit for coal tar catalytic cracking

upper-stage high temperature catalytic cracker using N_2 as the carrier gas. Following the reaction, the products were passed through a multi-stage ice bath cooler, where the tar was separated out. The remaining gas was collected, and its composition was analyzed via a flue gas analyzer. The temperatures of the upper and lower reactors were each controlled by separate temperature controllers, with the temperature range of the high-temperature catalytic cracking stage varied between 400 and 800 °C. The gas phase reaction time was controlled to be within 0.25–1.25 s by changing the thickness of the catalyst material layer, and the gas flow rate was set to 300 ml/min.

The two-stage fixed bed reactor was 700 mm high with an inner diameter of 32 mm. The main reactor tube was constructed from a steel alloy material, with the outer heating unit wrapped with a refractory material (asbestos) to achieve good thermal insulation. A fine wire mesh was installed at the bottom of the bed to evenly distribute the gas throughout the bed and support the material inside.

8.2.2 Experimental Materials

The coal tar used in this study was obtained from the Qingdao Iron and Steel Company. It is a black and viscous liquid at room temperature, with an irritating odor and poor fluidity. In order to add tar to the fixed bed reactor, the coal tar was heated at 60 °C for 10 min in an air blast dryer to enhance its fluidity. The industrial and elemental analyses of the coal tar are shown in Table 8.1. As to the data in industrial analysis, M is the moisture, V is the volatile, A means ash and FC is the fixed carbon.

The composite catalyst was prepared using a mechanical mixing method, as follows: 200 mesh CaO and MgO powders were evenly mixed in a 7:3 ratio. Sesbania gum was then added as a binder and the mixture was stirred to a wet gel by adding water, then extruded, granulated, and dried at 100 °C in a blast drying box for 5 h. Finally, the dried mixture was calcined at 800 °C in a muffle furnace for 2 h to improve its mechanical strength. Catalyst products with different particle sizes were obtained by crushing and sieving the calcined mixture after cooling. The particle size ranges of the prepared catalysts were 400–600 μm, 200–400 μm, 100–200 μm, and <75 μm. In order to avoid catalyst deactivation, fresh catalyst was used for each experiment.

Table 8.1 Industrial and elemental analyses of coal tar

Industry analysis/%				Elemental analysis/%					Q_{net}/(MJ/kg)
M	V	A	FC	C	H	O	N	S	37.44
1.55	81.3	0.58	16.57	88.66	5.44	4.35	0.93	0.62	

8.3 Results and Discussion

During the catalytic cracking of gasified coal tar, the effect of changing various factors including the reaction temperature, catalyst particle size, and gas phase reaction time were investigated. The catalytic cracking rate of coal tar, η, is defined as follows:

$$\eta = \frac{M_{in} - M_{out}}{M_{in}} \tag{8.1}$$

where M_{in} is the quantity of gasified tar fed into the reactor, and M_{out} is the quantity of tar collected by the tar collector.

8.3.1 Effect of Temperature on Tar Conversion Rate and Gas Products

The tar conversion efficiency and the concentration distribution of gaseous products are shown in Figs. 8.2 and 8.3, respectively, varying with the temperature.

From Fig. 8.2, it can be seen that the conversion rate of tar increased from 78 to 94% with the increasing of reaction temperature between 400 and 700 °C. However, there was very little change between 700 and 800 °C. It is evident that temperature has a big influence on the cracking rate of tar, as the reaction is endothermic. The slowdown in conversion between 700 and 800 °C may be attributed to a small amount of carbon deposit to weaken the activity of the catalyst, thereby slowing down the reaction rate.

Figure 8.3 shows that the total gas production follows a very similar trend to that of the tar conversion rate shown in Fig. 8.2. It is evident that CO is the most abundant

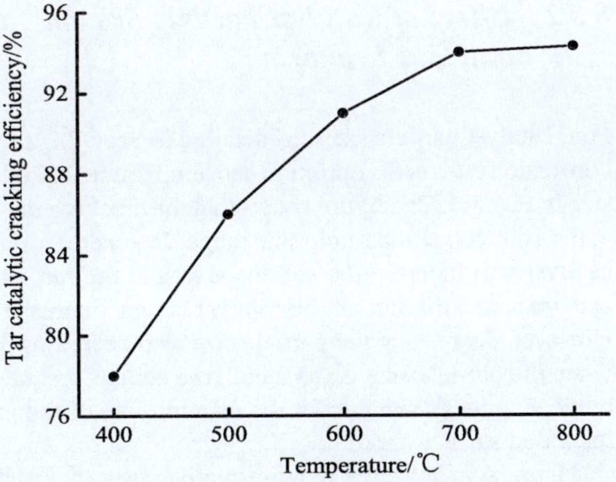

Fig. 8.2 Conversion rate of tar at different temperatures

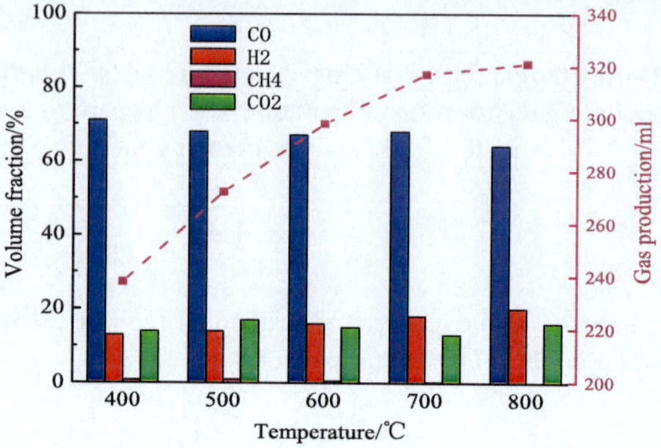

Fig. 8.3 Gas component distributions at different temperatures

product at each temperature, although its volume fraction decreases slightly with increasing temperature. The amount of H_2 increases monotonically with temperature increasing, and the proportion of CH_4 in the mixture is very small (<1%). Above 600 °C, the amount of CH_4 becomes negligible. The CO_2 volume fraction shows a less straightforward temperature dependence compared to the other gases. Initially, the CO_2 production increases with temperature increasing, however, there is a decrease between 500 and 700 °C, which is likely due to some of the CO_2 being consumed by reaction with CaO to form $CaCO_3$. However, between 700 and 800°C, the decomposition of the formed $CaCO_3$ occurs, thereby increasing the CO_2 volume fraction yet again. On the whole, the results presented in Figs. 8.2 and 8.3 show that when the temperature exceeded 700 °C, the catalytic cracking performance of tar vapor was not significantly improved. Therefore, the optimum reaction temperature is 700 °C.

8.3.2 *Effect of Catalyst Particle Size on Tar Conversion Rate and Gas Products*

Four catalyst particle sizes, as detailed in Sect. 8.2.2 were used in this study. The tar conversion rates at the optimum bed temperature of 700 °C are shown for each particle size in Fig. 8.4. The figure shows that the cracking degree of tar showed a maximum at the 100–200 μm particle size range. In general, decreasing the particle size of the catalyst will increase the surface area and the rate of out-diffusion, accelerate the catalytic cracking rate of the catalyst tar, and increase the conversion rate of the tar. However, decreasing the particle size also decreases the porosity of the bed; hence, when the particle size is too small, the contact frequency of the reactants is greatly reduced, thus slowing down the reaction rate of the tar faster than the effect of the increased surface area.

Figure 8.5 shows the relationship between the catalyst particle size and the total

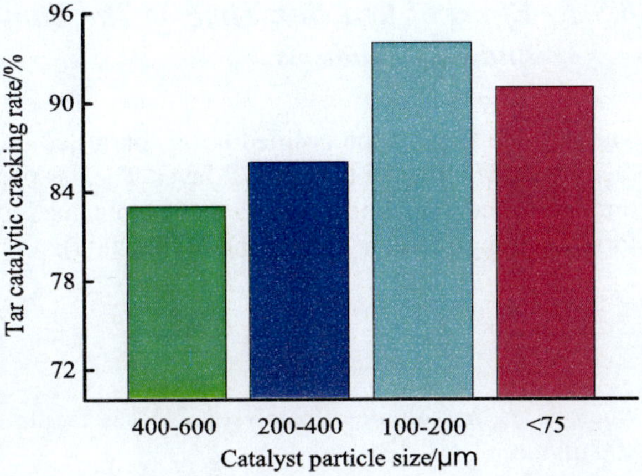

Fig. 8.4 Tar conversion rate at 700 °C with different catalysts particle sizes

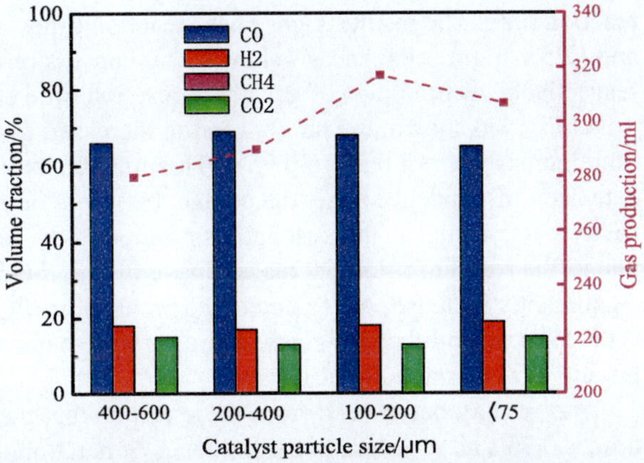

Fig. 8.5 The effects of particle size on gas composition and gas production at 700 °C

gas yield and volume fraction of each product gas. The total gas production shows a similar trend to the catalytic cracking rate from Fig. 8.4. This is due to the lower particle size increasing the surface area of the catalyst, resulting in greater contact frequency between the surface active sites of the catalyst and the tar molecules. However, the catalyst particle size has no significant effect on the product distribution ratio of gas components. Based on these results, it can be inferred that the optimum catalyst particle size lies in the 100–200 μm range.

8.3.3 *Effect of Reaction Time on Tar Conversion Rate and Gas Products*

The reaction time of the gasified tar is controlled by changing the thickness (and thereby the volume) of the catalyst layer. The gas phase reaction time is estimated through the contact time, *t*, which may be obtained using the following theoretical formula (Yong-Bin et al. 2004; Zhou et al. 2003):

$$\mathrm{t}=\frac{\mathrm{V_s}}{\mathrm{V_g}} \tag{8.2}$$

where V_s is the volume of catalyst and V_g is the flow rate of N_2 into the catalyst, 300 ml/min.

As the catalyst particles have a certain void density, and a void fraction of 0.2, the contact time should be multiplied by the void fraction to obtain the gas phase reaction time. The resulting gas phase reaction times were 0.25, 0.37, 0.625, 0.75, 1 and 1.25 s. Figures 8.6 and 8.7 show the tar conversion rate and product distributions, respectively, as functions of gas phase reaction time at 700 °C and 100–200 μm.

As Fig. 8.6 shown, the tar conversion increased from 87 to 96% as the residence time increased from 0.25 to 0.75 s. This may be explained by noting that with the extension of residence time, the contact between the gaseous tar molecules and the catalyst active sites is increased, thus promoting the reaction. Figure 8.7 also shows that as the residence time increases above 0.75 s, the conversion rate of tar does not significantly increase. As expected, the proportion of CO in the gaseous components is the highest, and shows a gentle decrease with the increase of residence time, while H_2 and CO_2 increase slightly.

Thus, it may be inferred from these results that the optimum gas phase reaction time is 0.75 s here, although the influence of reaction time on the overall conversion

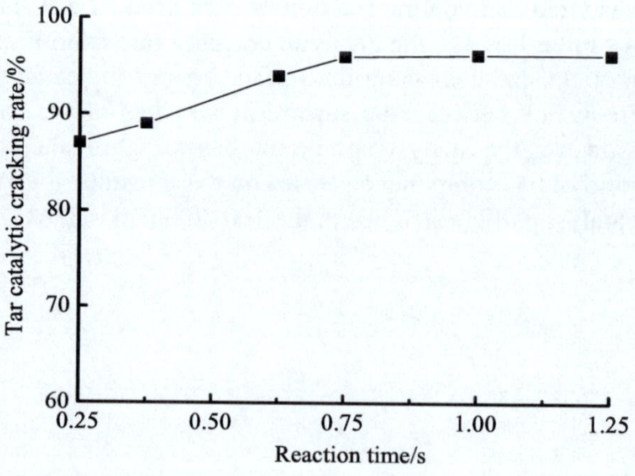

Fig. 8.6 Effect of reaction time on tar conversion rate at 700 °C and 100–200 μm

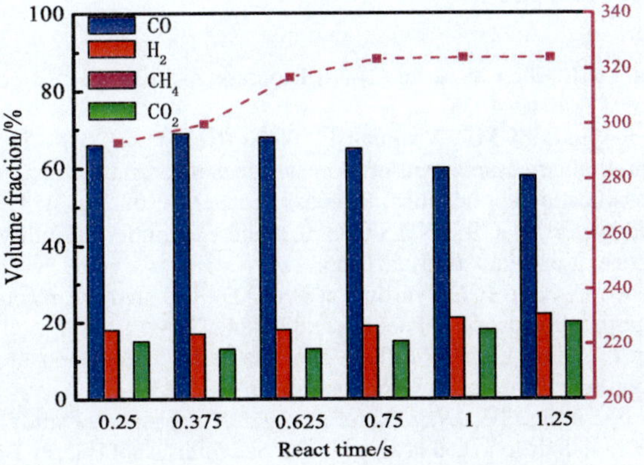

Fig. 8.7 Effect of reaction time on gas composition and gas yield at 700 °C and 100–200 μm

of tar oil is not as pronounced as the other factors (i.e. temperature and catalyst particle size) that were investigated herein.

8.4 Conclusions

Experimental results for tar catalytic cracking in a two-stage fixed-bed reactor with Ca-Mg composite catalyst showed that reaction temperature, catalyst particle size, and gas-phase reaction time all influence the overall tar conversion and the concentration distribution of gaseous products. The temperature plays the most important role, with the conversion rate reaching as high as 94% at 700 °C. The largest component in the gaseous products is CO, the volume fraction of which is not significantly affected by temperature. The next most abundant products are H_2 and CO_2, with the volume fraction of former increasing with temperature, while the latter shows a more complex trend due to absorption and reaction with CaO in the catalyst between 500 and 700 °C and the subsequent decomposition of $CaCO_3$ between 700 and 800 °C. The conversion of tar vapor, on the other hand, shows a maximum value for the catalyst particle size range of 100–200 μm. Finally, increasing the residence time in the reactor promotes the conversion of tar vapor. Based on the above analysis, it may be concluded that the optimum reaction conditions for the highest tar conversion of 96%, lowest CO_2 content, and highest H_2/CO ratio consist of a temperature of 700 °C, a catalyst particle size of 100–200 μm and a gas phase reaction time of 0.75 s.

Acknowledgements This work was supported by the National Natural Science Foundation of China (51676102). The authors greatly acknowledge the support from the Foundation of State Key Laboratory of Coal Clean Utilization and Ecological Chemical Engineering (Grant No. 2016-07) and Taishan Scholar Program of Shandong Province (201511029).

References

Antal MJ, Allen SG et al (2000) Biomass gasification in supercritical water. Ind Eng Chem Res 39(11):4040–4053

Boot-Handford ME, Virmond E, Florin NH et al (2018) Simple pyrolysis experiments for the preliminary assessment of biomass feedstocks and low-cost tar cracking catalysts for downdraft gasification applications. Biomass Bioenergy 108:398–414

Bridgwater AV (1995) The technical and economic feasibility of biomass gasification for power generation. Fuel 74(5):631–653

Franco C, Pinto F, Gulyurtlu I et al (2003) The study of reactions influencing the biomass steam gasification process. Fuel 82(7):835–842

Han P, Li DZ, Liu XW (2008) The discussion on tar removal method in biomass gasification for power system. Renew Energy Resour 21(1):25–36

Hu FX, Yang GH, Ding GZ et al (2016) Experimental study on catalytic cracking of model tar compounds in a dual layer granular bed filter. Appl Energy Barking Then Oxford 170(15):47–57

Luo ZY, Zhang XD, Zhou JS et al (2004) Experimental study on catalytic and thermal cracking of tar from biomass pyrolysis. J Chem Eng Chin Univ (china) 18(2):162–167

Lv X, Xiao J, Sun T et al (2017) Steam reforming of α-methylnaphthalene as a model compound of biomass tar over Ni-based catalyst for hydrogen-rich gas. Korean J Chem Eng 11:1–15

Masurel E, Authier O, Castel C et al (2015) Screening method for solvent selection used in tar removal by absorption process. Environ Technol 36(20):25–56

Moud PH, Andersson KJ, Lanza R et al (2015) Effect of gas phase alkali species on tar reforming catalyst performance: initial characterization and method development. Fuel 154:95–106

Nunes SM, Paterson N, Herod AA et al (2008) Tar formation and destruction in a fixed bed reactor simulating downdraft gasification: optimization of conditions. Energy Fuels 22(3):1955–1964

Sun Q, Sang Y, Wang F et al (2011) Decomposition and gasification of pyrolysis volatiles from pine wood through a bed of hot char. Fuel 90(3):1041–1048

Sun Y, Jiang J, Kantarelis E et al (2012) Development of a bimetallic dolomite based tar cracking catalyst. Catal Commun 20(3):36–40

Wang F (2017) Characteristics of catalytic tar removal from semi-coke in two-stage fluidized bed. J Chem Eng 10:77–81

Xie QL, Borges FC, Cheng YL et al (2014) Fast microwave-assisted catalytic gasification of biomass for syngas production and tar removal. Bioresour Technol 156(2):291–296

Yong-Bin J, Jie-Jie H, Yang W (2004) Effect of residence time on tar catalytic cracking behavior with calcium oxide. J Combust Sci Technol 10(6):549–553

Zhou JS, Wang TZ, Luo ZY et al (2003) Catalytic cracking of biomass tar. J Fuel Chem Technol 10(1):112–134

Chapter 9
Experimental Investigation on Sodium Migration and Mineral Transformation in Ash Deposit During Gasification of Zhundong Coal Using a Drop Tube Furnace

Chengchang Liu, Chang'an Wang, Chaowei Wang, Yongbo Du, Guantao Tang, Guangyu Li, and Defu Che

Abstract Zhundong coal has attracted an ever-increasing concern due to its super-huge reserve but high content of alkali metals. Volatilized into the gaseous phase during the gasification process, alkali metals are extremely unfavorable for coal utilization. Gasification technology can promote the large-scale utilization of high-alkali coal. However, few efforts, if any, have been conducted on gasification of Zhundong coal. The present study aimed to elucidate the sodium migration and mineral transformation characteristics in coal ash deposition process under different gasification conditions, while the behavior of sodium migration and mineral transformation were further analyzed using Inductively Coupled Plasma-Atomic Emission Spectrometer (ICP-OES), X-ray diffraction (XRD) and X-ray fluorescence (XRF) techniques. In this paper, the sodium migration characteristics in gasification process under different conditions were investigated by adding three sodium salts (NaCl, Na_2CO_3 and Na_2SO_4), kaolin and diatomite into coal, respectively. The effects of ash deposition time on sodium migration and mineral transformation characteristics in ash were also explored. The experimental results showed that the addition of sodium salt could increase the sodium content in the ash to a large extent. Kaolin and diatomite could solidify alkali metals in the gaseous phase during gasification process. As the ash accumulation time increased, the minerals such as chlorine, sulphur, and iron in the gaseous phase were easily bonded to the surface of the ash and reacted to form other crystal phases. The present study can provide guidance for the utilization of Zhundong coal and benefit the development of clean coal technology.

Keywords Gasification · Ash deposition · Sodium migration · Zhundong coal

---

C. Liu · C. Wang (✉) · C. Wang · Y. Du · G. Tang · G. Li · D. Che
State Key Laboratory of Multiphase Flow in Power Engineering, School of Energy and Power Engineering, Xi'an Jiaotong University, Xi'an 710049, China
e-mail: changanwang@mail.xjtu.edu.cn

J. Lyu and S. Li (eds.), *Clean Coal and Sustainable Energy*, Environmental Science and Engineering,
https://doi.org/10.1007/978-981-16-1657-0_9

9.1 Introduction

Coal occupies a dominant position in China's energy. With a large reserve, high volatile content and high calorific value, Zhundong coal has attracted an ever-increasing attention from many scholars and research institutes (Li and Li 2018). However, some ash-related problems, such as ash deposition and slagging which results from the characteristics of low ash melting point and high content of alkali metals, have greatly hindered the safe and efficient utilization of Zhundong coal (Wang and Xu 2015; Song and Song 2016; Liu and Li 2019). Coal gasification technology is one of the clean coal technologies, and it is also foundamental for the development of coal-based chemicals, coal-based liquid fuels, hydrogen production, fuel cells, and other process industries (Ji and Tabassum 2016). Compared with combustion, coal gasification technology can significantly reduce pollution emissions (Liang and Xie 2019). However, there are still many problems in its application. The deposition and adhesion characteristics of fly ash caused by high content of alkali metals can corrode the equipment and decrease the heat transfer efficiency during the gasification process (Huang and Li 2019; Jiang and Wei 2019), which is likely to cause great environmental pollution, decrease the utilization efficiency of energy, and affect the operation safety in the production process (Xu and Liu 2018). Therefore, it is necessary to further elucidate the mechanisms of ash deposition and slagging caused by alkali metal migration and mineral transformation, which is of great practical significance for protecting the environment, ensuring safe operation of boilers, and improving economic efficiency.

In order to explore the characteristics of deposited ash under different gasification conditions, an experimental study on the multi-functional test bench was conducted. Three kinds of sodium salts (NaCl, Na_2CO_3 and Na_2SO_4), kaolin and diatomite were added to Zhundong coal to study the effects on the migration of sodium, respectively. The characteristics of sodium migration and mineral transformation in deposited ash under different duration time were also studied.

9.2 Experimental

The gas composition, the gas volume of components, the feeding powder rate, and other impacted parameters in the process of gasification can be controlled by using the multi-function test bench as shown in Fig. 9.1. The experimental device mainly includes five subsystems: the gas supply system, steam generation system, the feeding powder system, gasification reaction system, data acquisition and analysis system. The gasification reaction system is the main part of the experimental system and can simulate the pulverized coal gasification reaction process in the actual gasification device. It consists of heating elements, temperature-controlled thermocouples, furnace body, refractory layers, insulation layers, and control cabinets. In order to make the temperature of the system uniform in the cross section, the cavity of the

Fig. 9.1 Schematic diagram of the experimental system

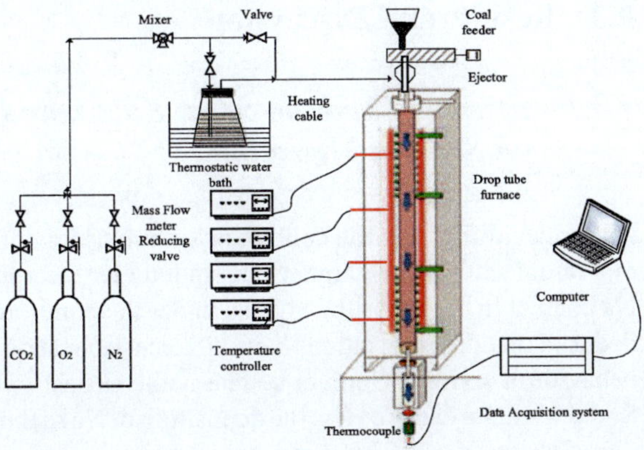

electric heating system has a small cross-sectional area and is well insulated and insulated in the experiment. The designed heating rate is up to 10 °C min^{-1}. High temperature corundum tube in the reaction system has a total length of 1500 mm, an inner diameter of 40 mm which can withstand the temperature of 1600 °C. The present gasification reaction system can simulate the pulverized coal gasification reaction process in the actual gasification furnace.

In this experiment, the fuel selected was Wucaiwan coal with a particle size of 75–150 μm, the gasification temperature in the gasification reactor was controlled to 1500 °C, the coal-supply was kept as 0.4 g min^{-1}, and O_2/coal and steam/coal were 0.8 and 0.1, respectively. Three kinds of sodium salts (NaCl, Na_2CO_3, Na_2SO_4), kaolin and diatomite accounting for 5% of the total fuel mass, were mixed with Wucaiwan coal. The deposited ash under different conditions was collected at the temperature of 500 and 700 °C in the gasifier tail. The ash sample was analyzed by an Inductively Coupled Plasma-Atomic Emission Spectrometer (ICP-OES). The mass content of major metal elements with various sodium salt additives and additive-free ash samples was compared and analyzed under the experimental conditions. Similarly, the ash after gasification of pure Wucaiwan coal and the addition of kaolin and diatomite was also analyzed.

Experimental study focused on the migration of sodium and the transformation characteristics of principal minerals in deposited ash with ash deposition time. In the present study, the deposition time of the ash on the stainless-steel surface were 30 min, 120 min, and 180 min, respectively. The X-ray diffraction (XRD) and X-ray fluorescence (XRF) analyses were performed on the ash samples. The effects of the ash deposition time in the gasification process were further demonstrated by analyzing the elements and crystal phase composition.

9.3 Results and Discussion

9.3.1 *Effects of Sodium Salts, Kaolin and Diatomite on Sodium Migration*

As shown in Fig. 9.2, the content of sodium (Na) in the deposited ash with addition of sodium salt is much higher than that under the additive-free condition, while the Na content in the ash is the highest under the condition with addition of $NaSO_4$. It is deduced that different anions have different effects on the deposition and coagulation behavior of sodium. Sodium sulfate is the easiest to bond to the surface of the ash. Sodium sulfate can promote the deposition of Na in the deposited ash, which increases the adhesion of ash. The mass content of Na in the ash at 500 °C under the condition of adding NaCl is much higher than that at 700 °C. The differences between the remaining two sodium salt additions are unobvious. The deposition of NaCl in the flue gas occurs principally at 500–700 °C and the condensation of Na_2SO_4 and Na_2CO_3 mainly occurs above 700 °C.

From Fig. 9.3, it can be seen that the mass contents of Na and potassium (K) in the deposited ash with addition of kaolin and diatomite are higher than those in the rawcoal (Xing and Liu 2019; Zhang and Liu 2019). Hence, kaolin and diatomite can absorb alkali metals in the gas phase, reduce the corrosion and ash adhesion caused by alkali metals. Adsorption of gaseous phase alkali metals by kaolin and diatomite is more easily performed at high temperature.

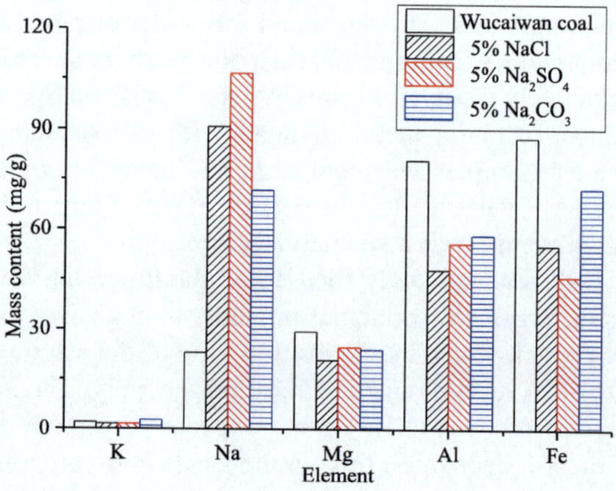

Fig. 9.2 The element contents of ash deposit at 500 °C with various additive containing sodium

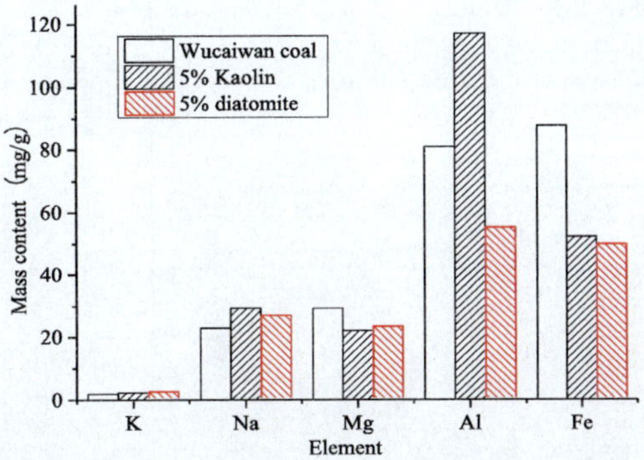

Fig. 9.3 The element contents of ash deposit at 500 °C with various additive of Si–Al series

9.3.2 *Effect of Ash Deposition Time on Sodium Migration and Mineral Transformation*

As shown in Fig. 9.4, only the crystal phases of CaO, MgO, and SiO_2 are detected in the deposited ash sample when the ash deposition time is 30 min. As the ash deposition time is prelonged to 120 min, more kinds of crystalline phases, such as Fe_2O_3 and $CaSO_4$, are present, and there are more and higher intensity diffraction peaks of Fe_2O_3. As the duration time increases from 30 to 120 min, the content of iron in the deposited ash increases significantly. With the ash deposition occurring near the heating surface, iron could easily migrate to the surface of the ash and react with the ash to form Fe_2O_3.

It can be found out from Fig. 9.5 that, the content of sulphur in the ash increases obviously with an increase in the ash deposition time. The sulphur content is raised with the deposition duration time. The enrichment degree of sulphur in the deposited ash on the heating surface is greater, and the sulphur compound in the gaseous phase is more likely to deposit into the ash and form metal sulfate. The X-ray diffraction (XRD) analysis results of the two ash samples in the comparison of the ash deposition

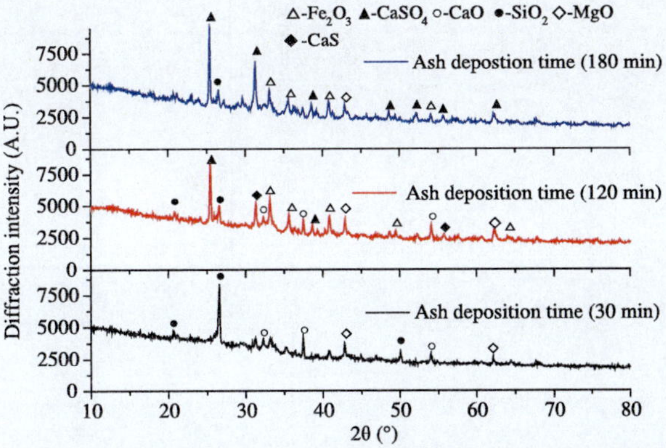

Fig. 9.4 The XRD test results of gasification ash in different times

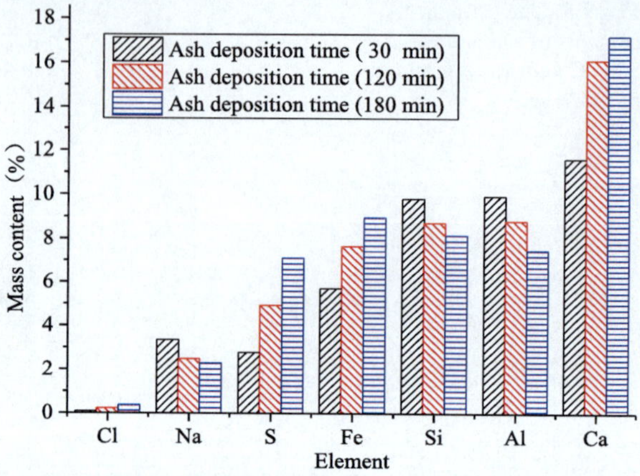

Fig. 9.5 Element distribution of gasification ash deposit in different times

time of 120 and 180 min showed that the crystal phases did not change greatly, but the diffraction intensity of $CaSO_4$ further increased. The contents of sulphur and calcium in the ash are obviously increased, indicating that more $CaSO_4$ formed in the ash after the further increase in the ash deposition time.

From the X-ray fluorescence (XRF) analysis results in Fig. 9.5, it can be observed that the content of Na in the ash decreases with the ash deposition time. The same conclusion can be drawn from Fig. 9.6. As the ash deposition time increases, the mass content of Na in the ash is reduced. It is indicated that the compound of Na is more likely to adhere to the heating surface at the beginning of the ash deposition process, and it is not deposited in the form of NaCl, but bonded to the heating surface in the form of its sulfate. It exists in the form of viscous underlayer.

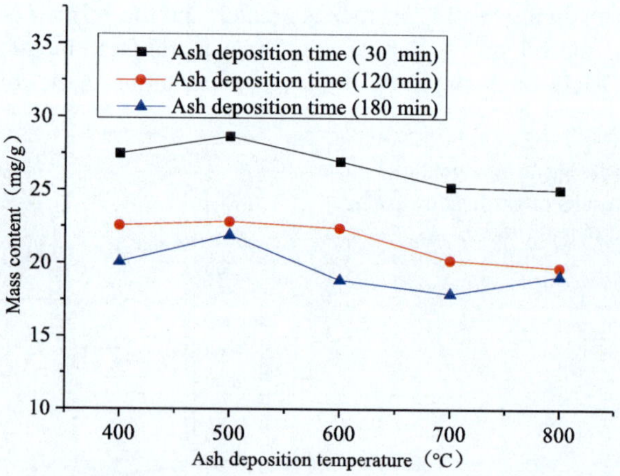

Fig. 9.6 Variation of Na content in ash deposit with ash deposition time and temperature

9.4 Conclusions

In this paper, an experimental study on sodium migration and mineral transformation during coal gasification under different conditions was carried out on a coal gasification multi-functional experimental device system. The migration of sodium in Wucaiwan coal under the conditions with addition of sodium salt, kaolin and diatomite was studied. The effect of the deposition duration time on the ash accumulation properties was elucidated. The main conclusions are as follows:

Among the three sodium salts of NaCl, Na_2CO_3 and Na_2SO_4, Na_2SO_4 is most likely to deposit into the ash. As the temperature of the flue gas decreases, the condensation of NaCl is the strongest, and the addition of NaCl can promote the release of potassium in coal. Kaolin and diatomite have adsorption effects on Na and K in the gaseous phase, and the addition of kaolin and diatomite during coal gasification can absorb alkali metals and reduce their corrosion and ash adhesion to the heating surface. The adsorption performance of kaolin is better than that of diatomite.

In the initial stage of ash deposition, only some alkali metal sulfates and composite salts can adhere to the heating surface to form a viscous underlayer, and minerals in other gaseous phases are less likely to migrate onto the surface. As the ash deposition time increases, the minerals such as chlorine, sulphur, and iron in the gaseous phase are easily bonded to the ash surface and other crystal phases are formed. Hence, the adhesion of ash onto the heating surface becomes more and more serious.

Acknowledgements The authors acknowledge financial support from the National Key R&D Program of China (2017YFB0602003) and the National Natural Science Foundation of China (51506163).

References

Huang Z, Li N (2019) A comparative study of the pyrolysis and combustion characteristics of sodium-rich Zhundong coal in slow and rapid processes. Energy Sci Eng 7(1):98–107

Ji Q, Tabassum S (2016) A review on the coal gasification wastewater treatment technologies: past, present and future outlook. J Clean Prod 126:38–55

Jiang G, Wei L (2019) Depolymerization model for flash pyrolysis of Zhundong coal: competition and coordination reaction mechanisms between the bridge scission and condensation. Thermochim Acta 675:44–54

Li X, Li J (2018) Clean and efficient utilization of sodium-rich Zhundong coals in China: behaviors of sodium species during thermal conversion processes. Fuel 218:162–173

Liang D, Xie Q (2019) Evolution of structural and surface chemistry during pyrolysis of Zhundong coal in an entrained-flow bed reactor. J Anal Appl Pyrol 140:331–338

Liu Z, Li J (2019) An experimental investigation into mineral transformation, particle agglomeration and ash deposition during combustion of Zhundong lignite in a laboratory-scale circulating fluidized bed. Fuel 243:458–468

Song G, Song W (2016) Transformation characteristics of sodium of Zhundong coal combustion/gasification in circulating fluidized bed. Energy Fuels 30(4):3473–3478

Wang X, Xu Z (2015) The ash deposition mechanism in boilers burning Zhundong coal with high contents of sodium and calcium: a study from ash evaporating to condensing. Appl Therm Eng 80:150–159

Xing H, Liu H (2019) Enhanced sodium adsorption capacity of kaolinite using a combined method of thermal pre-activation and intercalation-exfoliation: alleviating the problems of slagging and fouling during the combustion of Zhundong coal. Fuel 239:312–319

Xu L, Liu H (2018) Transformation mechanism of sodium during pyrolysis of Zhundong coal. Fuel 233:29–36

Zhang X, Liu H (2019) Correlations between the sodium adsorption capacity and the thermal behavior of modified kaolinite during the combustion of Zhundong coal. Fuel 237:170–177

Chapter 10
Study on Characteristics and Influencing Factors of Coal-Water Slurry Pyrolysis

Fan Feng, Boyang Li, Juan Yu, Yao Zhang, Chen Lin, and Zhongxiao Zhang

Abstract The distribution, composition and yield of coal-water slurry pyrolysis products have an important impact on the efficient combustion/gasification of coal-water slurry. In this paper, the coal-water slurry made of Shenmu coal (bituminous coal) is rapidly pyrolyzed by a high-frequency heating furnace, and the yield, composition and composition of pyrolysis gas were measured and analyzed. The effects of pyrolysis temperature, heating rate and residence time on the pyrolysis characteristics of coal-water slurry were studied. The results have shown that as the temperature increases, the yields of volatile matters and pyrolysis gas continue to increase. The compositions of pyrolysis gas are mainly H_2, CO, CH_4 and CO_2. With the increase of temperature, the yields of H_2, CO and CH_4 increase first and then decrease, and peaks appear at around 1100 °C. The CO yield continues to increase with increasing temperature. The rate of temperature increase affects the yield of volatiles. The research results provide a reference for understanding and mastering the formation characteristics of primary pyrolysis products of coal-water slurry.

Keywords Coal-water slurry · Gasification · High-frequency heating furnace · Pyrolysis · Volatiles

10.1 Introduction

Coal-water slurry is a new, efficient and clean coal-based fuel. It is a mixture of 65–70% coal with different particle size distribution, 29–34% water and about 1% chemical additive. Many countries have researched and developed it as coal-based fuel technologies based on long-term energy strategy considerations.

Niu (2004) used a quartz bed reactor to conduct fuel nitrogen release experiments for ordinary coal-water slurry, black liquor coal-water slurry and its raw coal. Meng et al. (2006) used a tube furnace to perform pyrolysis experiments on coal-water slurry at 800, 1000, 1200 °C respectively and analyzed the evolution of HCN and

F. Feng · B. Li · J. Yu (✉) · Y. Zhang · C. Lin · Z. Zhang
School of Mechanical Engineering, Shanghai Jiao Tong University, Shanghai 200240, China
e-mail: yujuan@sjtu.edu.cn

J. Lyu and S. Li (eds.), *Clean Coal and Sustainable Energy*,
Environmental Science and Engineering,
https://doi.org/10.1007/978-981-16-1657-0_10

NH_3 with temperature by chemical analysis. It was found that the release of HCN in raw coal and coal-water slurry increased with increasing temperature, but the change was slow and tended to be stable. As the temperature increased, the yield of NH_3 from the coal-water slurry increased. However, the chemical analysis and experiments of Derun Meng et al. did not eliminate the interference of the secondary reaction, so the experiment has space for optimization.

Wang et al. (2006) explored the pyrolysis characteristics of coal-water slurry at different heating rates, and concluded that the experimental coal-water slurry would lose weight at room temperature. At about 110–180 °C, free water in coal-water slurry and the internal water had been basically evaporated. When the temperature reached about 380–430 °C, the volatile matter in the coal-water slurry began to release. Wang et al. (2013) used the Curie point pyrolysis instrument to crack the coal-water slurry made from Shanxi Datong bituminous coal at 770 °C, and found that the pyrolysis rate of the coal-water slurry after reaching the activation energy of pyrolysis was faster than the raw coal and the final output was slightly higher.

Wu (2013) used a high-frequency heating furnace to rapidly pyrolyze coal-water slurry to analyze the characteristics of coke. It was found that as the pyrolysis temperature increased, the yield of coke in bituminous coal decreased, that is, the yield of pyrolysis gas increased. Wang et al. (2005) used a horizontal tube furnace to carry out high-temperature rapid pyrolysis of different coal types. It was found that for Xiezhuang coal (bituminous coal), the content of pyrolysis gas increased with the increase of temperature, and after 900 °C, it was basically stable. In the temperature range of 500–900 °C, the volume concentration of H_2 increased with the increase of temperature in the pyrolysis gas, and the CO volume concentration also increased but it was basically stable after 700 °C, and the CH_4 volume concentration decreased.

Shi (2010) studied the pyrolysis characteristics of typical coal types and found that the pyrolysis of coal experienced three stages of cracking, depolymerization and decomposition, and polycondensation. The results showed that the yield of coke decreased gradually with the increase of temperature at 850–1000 °C, and the yield of pyrolysis gas increased gradually. The concentration of CH_4 in pyrolysis gas increased with the increase of temperature. In the research, Li (2016) found that the yields of H_2, CO and CH_4 in the circulating fluidized bed increased with the increase of temperature in the temperature range of 500–900 °C.

Yuan et al. (2011) used high frequency to rapidly pyrolyze coal, and explored the law of pyrolysis gas generation in the range of 600–1200 °C. It was concluded that the yield of pyrolysis gas increased with the pyrolysis temperature. The rise was reduced as the coal rank increased. The pyrolysis gas components were mainly H_2 and CO, and the contents of CH_4 and CO_2 were low. As the pyrolysis temperature increased, the contents of H_2 and CO increased, and the contents of CH_4 and CO_2 decreased. The increase of pyrolysis temperature could effectively increase the contents of H_2+CO and H_2/CO in pyrolysis gas.

According to previous studies, the main components of pyrolysis gas for coal pyrolysis are H_2, CH_4, CO and CO_2, but the research on pyrolysis gas components for rapid pyrolysis of coal-water slurry is still insufficient. In this paper, a rapid pyrolysis of a coal-water slurry with a concentration of about 50% made of Shenmu coal

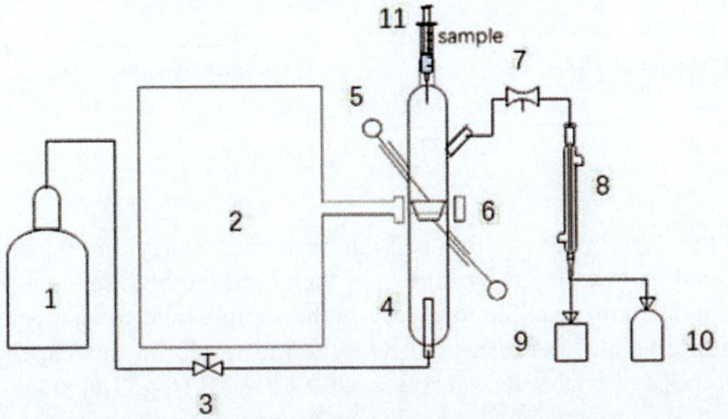

Fig. 10.1 Experimental system diagram. 1: Gas cylinder (N_2) 2: High frequency heating furnace 3: Manual gate valve 4: Quartz tube 5: Thermocouple 6: Molybdenum crucible 7: Flow meter 8: Condenser 9: Bag (liquid) 10: Bag (gas) 11: Syringe

(bituminous coal) in a high frequency furnace is carried out to investigate the volume concentration of H_2, CH_4, CO and CO_2 in pyrolysis gas at different temperatures and its influencing factors, and explore the factors affecting the yield of four gases and their causes (Fig. 10.1).

10.2 Experiment

The heating device used in this paper is a high-frequency heating furnace with an output oscillating power of 25 kW, an output oscillating frequency of 30–100 kHz, and a heating current of 200–1000 A. The high-frequency furnace pyrolysis device has the advantages of high heating rate, fast cooling, accurate quantification, etc., and can carry out research on rapid pyrolysis of coal-water slurry. The heating mechanism is using an induction coil to induce eddy current in the metal crucible, so that the crucible is heated rapidly to heat the coal-water slurry. A quartz tube and a molybdenum crucible are used as the reactor. The inner diameter of the quartz tube is 40 mm, 4 mm thick, and can withstand a high temperature of 1350 °C. The reaction carrier is the molybdenum crucible, with an upper diameter of 30 mm, a bottom diameter of 18 mm, a height of 13 mm, and a volume of about 6.5 ml, which is placed in the middle of the induction coil. Only the molybdenum crucible is in a high temperature state throughout the experimental apparatus, which is safer. The experimental carrier gas is high purity nitrogen with a flow rate of 0.5 L/min.

Use Shenmu coal (bituminous coal) to prepare a coal-water slurry with a concentration of about 50%. Because the slurryability of Shenmu coal is not good, and the high concentration will cause the coal powder to assemble into spherical during the mixing process, it cannot be made into high concentration slurry.

Table 10.1 Analysis of Shenmu coal

Coal	Proximate, w_d/%				Ultimate, w_{daf}/%			
	M	Ash	VM	FC	C	H	N	S
Shenmu	5.24	5.05	31.2	63.8	82.25	4.53	1.03	0.19

During the experiment, we dropped the coal-water slurry into the crucible from the upper feed port through a syringe first, then used the high frequency furnace to heat the molybdenum crucible to pyrolysis the sample, and the temperature of the molybdenum crucible was measured by a thermocouple. In this experiment, each feed was 3 drops of coal-water slurry, about 0.15 ml, 0.16 g. The inner diameter of the bottom of the molybdenum crucible was 18 mm. After the water coal slurry was dripped into the molybdenum crucible, a film with a thickness of about 0.6 mm was formed at the bottom, which ensured that the coal-water slurry was heated evenly and substantially consistent with the temperature at the bottom of the crucible. The generated pyrolysis gas was blown out of the quartz tube by nitrogen gas, and the purging speed was 6.63 mm/s, and all the gas in the quartz tube could be blown out of the quartz tube in 30 s. Considering that the high temperature zone in the tube existed only in the heating section of the crucible, about 20 mm in length, the carrier gas could blow the gas away from the high temperature reaction zone in about 3 s, effectively reducing the influence of the secondary reaction. The pyrolysis gas was separated from the quartz tube through the condenser to separate the gas phase and the liquid phase. The liquid phase was mainly water, tar, and the like. The gas phase was collected in a gas bag and analyzed by gas chromatography.

The proximate and ultimate analyses of Shenmu coal are shown in the following Table 10.1.

The main purpose of the experiment was to measure the concentrations of H_2, CH_4, CO, and CO_2 in the pyrolysis gas, and analyzed the effects of temperature, heating rate, and reaction time on the composition of the pyrolysis gas.

10.3 Experimental Results and Analysis

The obtained by the gas chromatograph is the volume concentration of the components measured in the mixed gas, and that in the pyrolysis gas can be converted according to the ratio of the respective components to the N_2 volume concentration. This calculation requires neglecting the effects of N_2 in the pyrolysis gas. In fact, N_2 reacts generally above 1200 °C, so the error is negligible (Niu 2004).

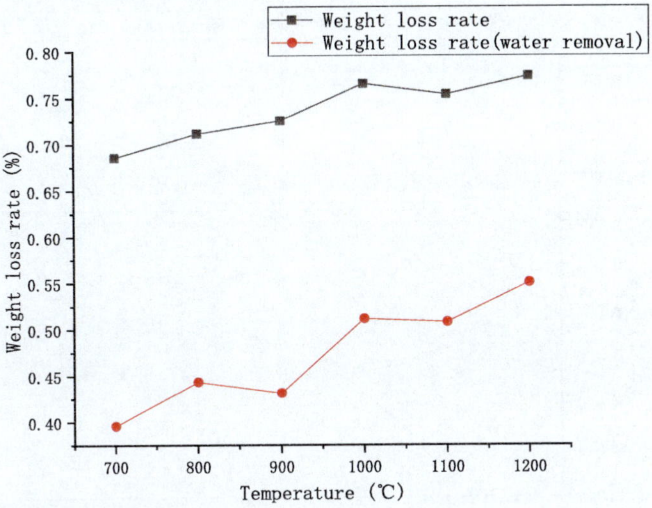

Fig. 10.2 Weight loss rate with temperature

10.3.1 Effect of Temperature on Weight Loss Rate and Pyrolysis Gas Yield

The Fig. 10.2 above shows the relationship between temperature and weight loss rate and weight loss rate (water removal). The heating rate was 400 °C/min, the nitrogen purge rate was 0.5L/min, and the gas velocity in the reactor was about 6.63 mm/s. The weight loss rate is the ratio of the mass lost after pyrolysis to the total mass of coal-water slurry. The weight loss rate (water removal) is the ratio of the mass lost by coal in coal-water slurry to the mass of coal. It can be seen from the figure that the trends of the two are consistent, that is, as the temperature increases, the weight loss rate and the weight loss rate (water removal) also increase. This indicates that the degree of pyrolysis of coal-water slurry increases with the increase of temperature. This is consistent with previous research conclusions (Lei 2013; Wang et al. 2005). And with the increase of temperature, the yield of pyrolysis gas of coal-water slurry also shows an upward trend, as shown in Fig. 10.3.

10.3.2 Effect of Temperature on Combined Gas Volume Concentration

As the temperature increases, as seen from Fig. 10.4, the volumetric concentration of H_2, CH_4, CO_2 and CO combined gas volume in the total pyrolysis gas decreases in the range of 700–900 °C. It remains stable in the range of 900–1100 °C. Above 1100 °C it continues to decline. This indicates that the volume concentration of other gas

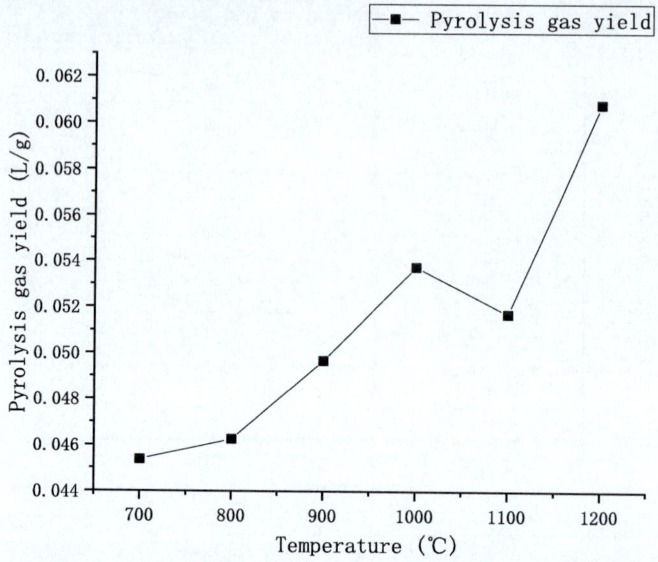

Fig. 10.3 Prolysis gas yield with temperature

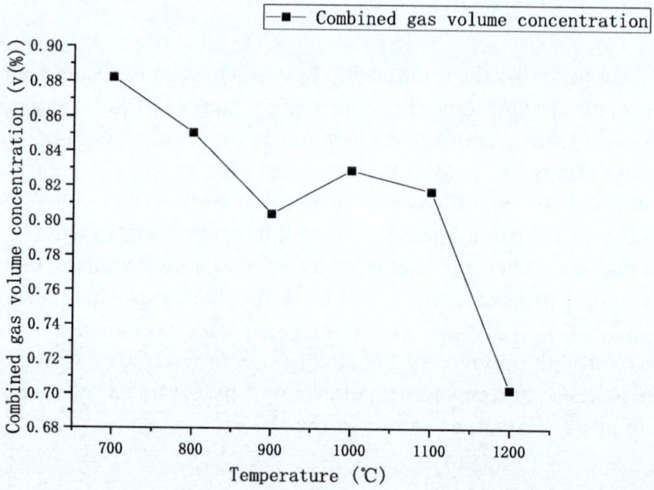

Fig. 10.4 The volumetric concentration of H_2, CH_4, CO_2 and CO combined gas with temperature

components will increase in the range of 700–900 °C and above 1100 °C. According to references, the other gas are components containing N(NO, NO_2, HCN, etc.) and compounds containing sulfur (SO_2, CS_2). High temperature promotes the formation of HCN, and increases the concentration of NO, and leads to the decomposition of sulfate to form SO_2 (Wang et al. 2005).

10.3.3 Effect of Temperature on Gas Yield and Volume Concentration

From Fig. 10.5, we can see the generation of H_2 is mainly divided into different stages.

In the temperature range of 700–900 °C a large amount of hydrogen has produced. Because as the pyrolysis temperature increases, the aromatic ring both inside the coal and in the volatile products produced by pyrolysis will undergo a second polycondensation reaction to generate a large yield of H_2, which is different from the H_2 generation in the range 400–600 °C where H_2 is mainly from the dehydrogenation of the fat chain and the production of hydrogen radicals.

Other investigations (Wang et al. 2005; Li 2016; Lu et al. 2016) also confirmed a significant increase in the yield of H_2 from 700 to 900°C. In the pyrolysis reaction of coal, as the thermal decomposition reaction continues, the thermal decomposition temperature continuously increases. And in the later stage of the thermal decomposition reaction, a certain polycondensation reaction occurs, and finally an aromatic hydrocarbon compound is formed, accompanied by a large yield of hydrogen being released.

The increase in H_2 release during the pyrolysis of coal-water slurry continues until 1100 °C. The specific surface area and pore diameter of the coal char when the coal-water slurry is rapidly pyrolyzed, are much higher than the coal pyrolysis coke, which is more conducive to the release of volatile gases (Lei 2013; Dun et al.

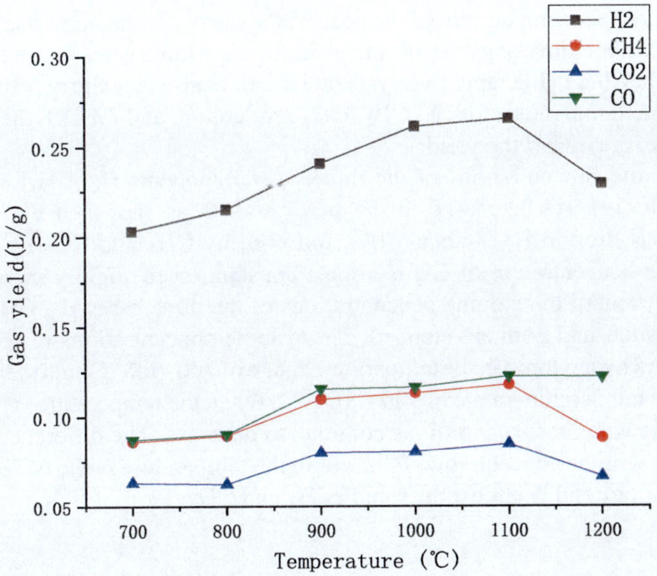

Fig. 10.5 Gas yields with temperature

2019). The yield of H_2 released began to decrease after reaching a maximum at about 1100 °C.

CH_4 is mostly derived from a large number of side chains and branches in the macromolecular structure of coal. Since the hydrocarbon is at a lower temperature, the –CH_3 of the aliphatic hydrocarbon side chain is broken and the hydrogen radical generates CH_4. As the temperature continues to increase, the degree of methyl group breakage increases, and long-chain aliphatic hydrocarbons are cleaved to form a methyl group. Due to the polycondensation of the aromatic structure, the carbon released from the fat structure also forms methane with hydrogen radicals (Chen et al. 2017), and a small amount of CH_4 is derived from the decomposition of partially volatile hydrocarbons and the free hydrogenation of coke. At the same time, more H_2 components also make CH_4 easier to form.

The CO and CO_2 in the pyrolysis gas mainly come from the oxygen-containing functional groups in the coal. At the lower pyrolysis temperature, the pyrolysis gases mainly come from the decomposition of the oxygen-containing functional groups. After the temperature rises in the late stage of pyrolysis, the slight increase in CO content may come from the chemical reaction of oxygen-containing heterocyclic rings such as ether bonds and helium–oxygen bonds in coal.

As the temperature increases, only the yield of CO increases with the increase of temperature. The trends of yield of the other three gases are basically the same as that of H_2, which continuously increase in the range of 700–1100 °C, reach maximums at 1100 °C and then begin to decrease. Among them, the change of H_2 is the most obvious. The coal-water slurry is rapidly heated to 1100 °C and maintained this temperature for pyrolysis. At this time, the yields of H_2, CH_4 and CO_2 reached maximums. As the temperature of the coal-water slurry continued to rise and stayed at a higher temperature for pyrolysis, the yields of these three gases began to decrease. This indicates that in the rapid pyrolysis process of coal-water slurry, 1100 °C is the most suitable temperature for H_2, CH_4, CO_2 production, and for CO, the higher the temperature, the higher the yield.

The volume concentrations of the three gas components H_2, CH_4, and CO_2 in the pyrolysis gas are decreased in the pyrolysis gas as shown in Fig. 10.6. The most obvious drop in H_2 is about 10%, followed by CH_4 and CO_2 by about 5%. The volume concentration of CO is almost constant, even slightly increased. The decreasing trend of the volume concentrations of the three gases H_2, CH_4 and CO_2 are very similar and both are stepped. The volume concentrations of H_2 and CH_4 showed a downward trend in the temperature range of 700–900 °C, and then remained stable in the temperature range of 900–1100 °C. When the temperature is higher than 1100 °C, the volume concentrations continue to decrease. The difference in CO_2 is that the first volume concentration falls within the temperature range of 700–800 °C, while the latter trend is almost the same as H_2 and CH_4.

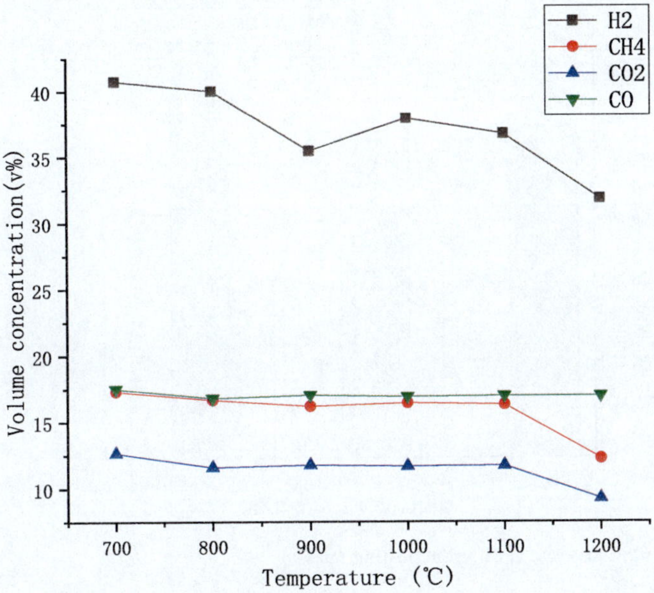

Fig. 10.6 Volume concentrations with temperature

10.3.4 Effect of Heating Rate and Residence Time on Weight Loss Rate

As seen from Fig. 10.7, at the reaction temperature of 1000 °C, as the heating rate increases, the weight loss rate also increases but the degree of change is not large. This is consistent with the findings of Wang et al. (2007). When the temperature is raised rapidly, the release rate of volatiles in the coal is fast, and it is easy to form a positive pressure inside the particles, which promotes the rich pore structure of the coal char, which is beneficial to the gasification activity of the coal char. Therefore, when the heating rate is fast, the more pyrolysis gas is generated, the greater the weight loss rate (Lei 2013).

In Fig. 10.8, at a reaction temperature of 800 °C, the rate of temperature rise was 400 °C/min, and after 2 min, the residence time had little effect on the weight loss rate. Because the sample quality of the experiment was considered to be small, between 0.15 and 0.2 g, the reactions would almost finish in 2 min at high temperature.

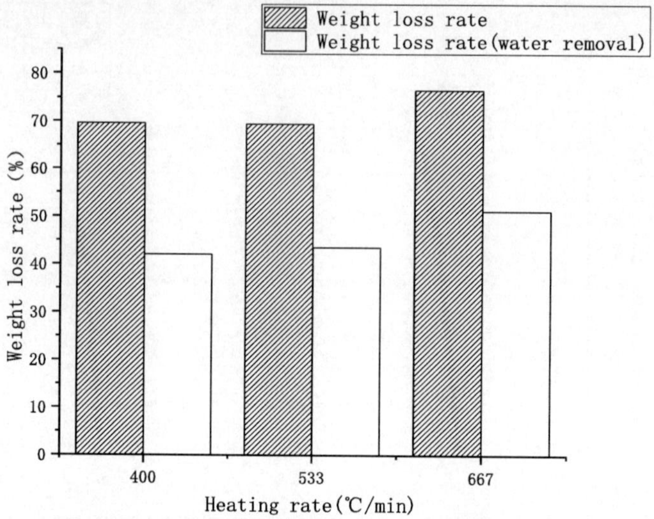

Fig. 10.7 Weight loss rate at different heating rates

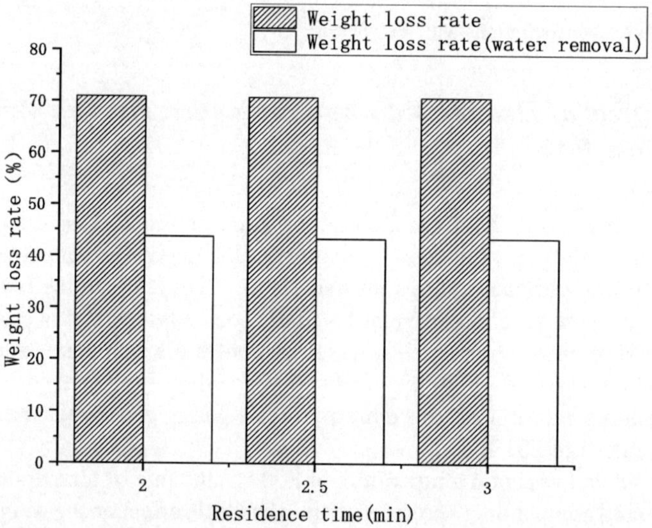

Fig. 10.8 Weight loss rate at different residence times

10.4 Conclusion

(1) With the increase of temperature, the degree of pyrolysis of coal-water slurry gradually increases and the yield of pyrolysis gas of coal-water slurry also shows an upward trend.
(2) As the temperature increases, the volume concentration of H_2, CH_4, CO_2 and CO combined gas decreases stepwise, and the proportion of other gas components becomes larger and larger.
(3) The yields of H_2, CH_4, CO_2 continuously increase in the range of 700–1100 °C, reach maximums at 1100 °C and then begin to decrease. The yield of CO continues to increase as the temperature increases.
(4) The volume concentrations of H_2, CH_4 and CO_2 decrease stepwise as the temperature increases. The most obvious decline in H_2 is about 10%, followed by CH_4 and CO_2, which each reduces by about 5%. Although CO decreased slightly, the decline is not significant. The yield of other gases is generated in large quantities as the temperature increases. The volume concentration of CO is almost constant, even slightly increased.
(5) As the heating rate increases, the degree of pyrolysis of the coal-water slurry also increases but does not change much.

Acknowledgements The authors gratefully acknowledge the financial support from National Key R&D Program of China, Grant No. 2018YFB0604202.

References

Chen W, Lu K, Zhao W (2017) Study on the characteristics of methane precipitation during pyrolysis of typical bituminous coal. Sci Technol Innov Herald 14(34):79–80 (in Chinese)
Li C (2016) Research on bituminous coal pyrolysis behavior in a fluidized bed reactor and release of volatile products. Zhejiang University (in Chinese)
Lu Z, Wang Y, Li J, Wang F, Yu G (2016) Evolution characteristics of gas from pyrolysis of typical coals. J East China Univ Sci Technol (Natural Science) 42(06):764–770 (in Chinese)
Meng D, Zhao X, Zhou J, Qi K (2006) Analysis of release characteristics of HCN and NH_3 during pyrolysis of coal water slurry. Thermal Power Eng 04:394–396 + 400 + 437–438 (in Chinese)
Niu Z (2004) Study on fuel-N and fuel-NOx produce characteristics of coal and coal water slurry. Zhejiang University (in Chinese)
Shi J (2010) Characteristics study on pyrolysis and gasification of typical coals. Huazhong University of Science and Technology (in Chinese)
Wang P, Wen F, Xuepeng Bu, Liu Y, Bian W, Deng Y (2005) Study on coal pyrolysis characteristics. Coal Convers 01:8–13 (in Chinese)
Wang H, Ma Y, Jiang X, Wan P (2006) Study on the mechanism of pyrolysis and volatilization of coal water slurry. Coal Convers 03:50–53 (in Chinese)
Wang H, Jiang X, Liu J, Lijun Yu, Zhang Q (2007) Analysis of pyrolysis characteristics of coal water slurry under various heating rates. Power Eng 02:263–266 (in Chinese)
Wang H, Jiang X, Liu J (2013) Experimental study on comparison of coal water slurry and parent pulverized coal with Curie point pyrolyser. Chem Eng 2007(03):19–22 (in Chinese)

Wu L (2013) Study on the rapid pyrolysis and the char gasification reactivity of coal water slurry. East China University of Science and Technology (in Chinese)

Wu D, Zhang W, Fu B, Hu G (2019) Chemical Structure and gas products of different rank coals during pyrolysis. J Therm Anal Calorim 136(5):2017–2031

Yuan S, Chen X, Li J, Dai Z, Zhou Z, Wang F (2011) Formations of solid and gas phase products during rapid pyrolysis of coal. J Chem Indus Eng (China) 62(05):1382–1388 (in Chinese)

Chapter 11
Soot Formation in High-Temperature Pyrolysis of Various Coals

Shengjie Bai, Yongbing Wang, Gaofeng Dai, Peng Li, and Xuebin Wang

Abstract Soot formed during high temperature pyrolysis or incomplete combustion, has strong effects on human health and the environment, as well as radiation heat transfer during combustion. Most studies on soot formation are focused on gas combustion, while the studies of soot formation during solid fuel combustion are rarely reported. In this study, the formation and properties of soot particles from coal pyrolysis are investigated in a drop tube furnace at 1000–1300 °C. The soot morphology, number size distribution, ignition characteristics, gas composition and soot-ash composition were characterized with transmission electron microscopy, scanning mobility particle sizer, thermogravimetric analyzer, gas chromatography, and energy disperse spectroscopy. The effect of coal rank on soot formation and yield is mainly discussed. The results show that with the increase of pyrolysis temperature, the soot yield increases. The soot yields show no dependent relationship with the coal ranks. Among the five kinds of coal tested (anthracite, lean coal, bituminous coal, lignite, and high sodium coal), bituminous coal has a much higher soot yield (>7.5%) than that of other coals (0.5–2%). The oxidation reactivity of soot particles from high sodium coal pyrolysis is much higher than that from other coals, because of the high contents of alkali and alkaline earth metals in soot particles.

Keywords Soot formation · Pyrolysis · Coal types

S. Bai · G. Dai · X. Wang (✉)
MOE Key Laboratory of Thermo, Fluid Science and Engineering, Xi'an Jiaotong University, Xi'an 710049, China
e-mail: wxb005@mail.xjtu.edu.cn

Y. Wang · P. Li
Xinjiang Uygur Autonomous Region Special Equipment Inspection and Research Institute, Uygur 830000, China

J. Lyu and S. Li (eds.), *Clean Coal and Sustainable Energy*, Environmental Science and Engineering,
https://doi.org/10.1007/978-981-16-1657-0_11

11.1 Introduction

Soot formation in solid fuel combustion is regarded as a result of secondary pyrolysis of volatile matter especially tar based on polycyclic aromatic hydrocarbons hypothesis, inception, surface growth and surface oxidation (Frenklach 2002). Soot is one of the fine particle pollutants but also important to combustion facilities because of its strong radiation heat transfer effect, that the near-burner coal flame temperature could be lowered by several hundred degree due to the radiation heat transfer to surrounding wall by soot (Fletcher et al. 1997; Cheung et al. 2004; Bi et al. 2014). Therefore, it is important to study the properties and mechanisms for soot formation in solid fule combustion systems. There have been extensive experimental studies on soot formation process in diesel engines and in diffusion flames at various combustion conditions. Detailed mechanisms of soot have been developed for simple hydrocarbon fuels. However, little attention has been given to the soot mechanisms from solid fuel combustion. The compositions of solid fuels are much more complex compared with gas or liquid fuels, which might result in a big difference in soot yields and properties during combustion or pyrolysis process. The effect of coal ranks on the formation of soot from coal pyrolysis has not been reported before.

This paper aims to compare the yields and properties of soot particles generated from the pyrolysis of varied coals at high temperatures (1000–1300 °C). Five coals of different ranks (Zhundong coal, bituminous, lignite, lean coal and anthracite) were pyrolyzed in a drop tube furnace at 1300 °C. The soot particles are sampled and weighted for yield measuring, and the soot particle size is measured on line. The gas composition is measured to demonstrate the soot formation mechanism. The oxidation reactivity and mineral compositions of soot are also measured and compared for different coals.

11.2 Experimental Section

11.2.1 Experimental Setups

The pyrolysis experiment was carried out in a drop tube furnace as is shown in Fig. 11.1 错误!未找到引用源。, which is adapted from our previous work (Li et al. 2018). The sampling system consists of three parts, namely gas composition, char and soot collection and $PM_{1.0}$ size distribution. For gas composition analysis, passing through a filter, the flue gas was collected in air bag, and then its composition was analyzed using gas chromatography (GC, GC-2014, SHIMADZU). For $PM_{1.0}$ size distribution, particles larger than 1 μm were separated first with cyclones, then $PM_{1.0}$ together with flue gas was diluted using diluter (Dekati Diluter DI-1000) and $PM_{1.0}$ concentration size distribution was obtained with scanning mobility particle sizer (SMPS, 10-1000 nm, model 3082, TSI). To avoid tar condensation during sampling process, all parts of sampling unit is heated up to 200 °C.

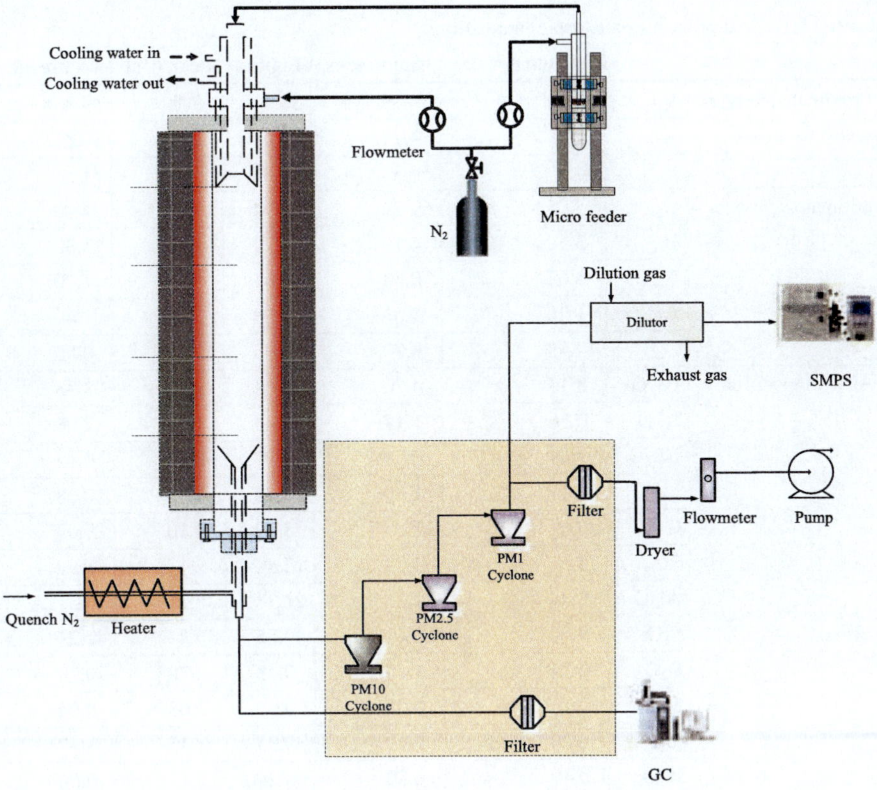

Fig. 11.1 Drop tube furnace and sampling system (Bi et al. 2014)

For all the experiments, the coals were pyrolyzed in N_2 atmosphere, in the temperature range of 1000–1300 °C. Soot and carbonaceous materials are seperated and collected using cyclones and filters according to their size. In total, solid phase products were collected, namely $PM_{1.0}$, $PM_{1.0-2.5}$, $PM_{2.5-10}$ and PM_{10+}.

11.2.2 Fuel Property

Five coals of different ranks are used, their proximate and ultimate analysis, together with ash composition is shown in Table 11.1. From the table, it can be seen that the contents of Na_2O and CaO in zhundong coal ash reaches 7.41% and 30.9%, respectively. In contrast, for the other coals, the contents of Si and Al in coal ash are high.

Table 11.1 Coal properties and ash compositions

		Zhundong Coal	Bituminous	Lignite	Lean coal	Anthracite
Proximate analysis	V_d	31.0	34.39	38.9	12.1	9.4
	A_d	4.7	7.29	13.3	22.6	19.3
	FC_d	64.3	58.3	47.8	65.3	71.3
Ultimate analysis	C_d	75.05	69.24	53.74	61.36	74.00
	H_d	3.58	4.19	4.21	1.48	3.30
	N_d	0.62	0.88	0.81	0.95	1.20
	S_d	1.28	1.61	1.29	1.00	1.40
	O_d*	14.78	16.80	26.65	12.60	0.80
Ash composition	Fe_2O_3	3.11	4.70	5.32	4.17	3.33
	K_2O	0.56	1.14	1.64	1.27	2.08
	Na_2O	7.41	0.51	3.47	0.78	0.45
	MgO	10.1	1.60	3.28	0.50	0.68
	CaO	30.9	11.8	8.18	4.10	5.93
	SO_3	25.5	4.48	12.42	2.17	1.97
	Al_2O_3	5.85	22.7	21.33	38.2	36.51
	SiO_2	7.20	50.4	42.65	46.8	47.21
	P_2O_5	0.14	1.45	0.48	0.28	0.28
	MnO	0.10	0.08	0.06	0.03	0.04
	TiO_2	0.24	0.89	0.93	1.45	1.47
	BaO	0.49	0.10	0.240	0.12	0.00
	Cl	0.63	0.01	0.00	0.01	0.01
	SrO	1.68	0.09	0.00	0.15	0.05

*Calculated by difference

11.3 Results and Discussion

11.3.1 Soot Yield and Number Size Distribution

The yield of soot and carbonaceous materials is shown in Fig. 11.2. In this study, $PM_{1.0}$ is considered as soot. The soot yield from zhundong coal, lignite, bituminous, lean coal and anthracite are 0.6%, 1.0%, 7.7%, 1.2% and 2.0%, respectively. The yield of $PM_{1.0\text{–}2.5}$ is very low mainly because the size range is narrow. The yield of $PM_{2.5\text{–}10}$ is the same order as $PM_{1.0}$. While for all kinds of coals, the yield of PM_{10+} is much higher, ranging from 30 to 65%, depending on coal types. This shows that PM_{10+} is the main product during fragmentation under high temperature. Note that the soot yield from the pyrolysis of bituminous coal is 7.5%, which is much higher than other coals.

The $PM_{1.0}$ concentration size distribution is shown in Fig. 11.3. $PM_{1.0}$ produced

Fig. 11.2 Soot yield of different coals

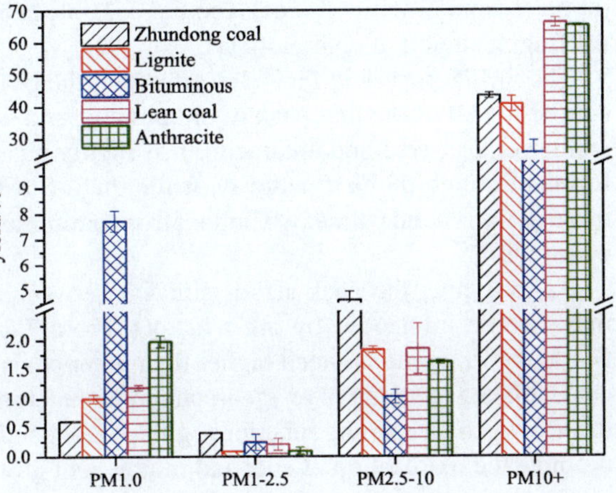

Fig. 11.3 $PM_{1.0}$ number size distribution

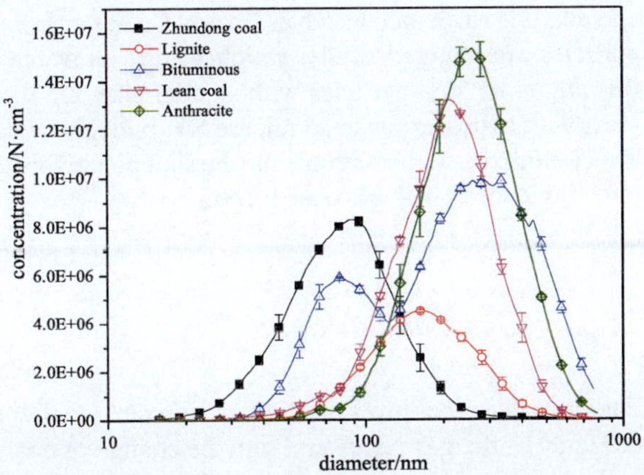

from zhundong coal, lignite, lean coal and anthracite shows a unimodal distribution, and the peak size are 80 nm, 150 nm, 200 nm and 300 nm, respectively. While $PM_{1.0}$ from bituminous has two peaks, the first peak is around 80 nm and the other showed up around 300 nm. As to the number concentration, the result is lignite < zhondong coal < lean coal < anthracite < bituminous, that is consistent with mass yield of $PM_{1.0}$.

There are two routes of soot formation during coal pyrolysis, the first is that during the secondary pyrolysis of volatile mater, tar crack under high temperature and form small moleculars, those small moleculars react and form aromatic rings first and through a series of reactions, forming soot with smaller size. The other pathway is that the PAHs in tar act as soot precursors and directly form soot, so the size could be larger (Frenklach 2002; Wang et al. 2018). As to the large peak of soot particles from bituminous coal, another possible explanation is that it might be from the fragmentation of char particles.

The SMPS curves in Fig. 11.3 show that the peak size of zhundong coal soot is similar to that of the first peak of bituminous, and the second peak of bituminous is similar to lean coal and anthracite. It is highly possible that zhundong coal soot is formed through the first pathway, while that of lean coal and anthracite is formed through the second pathway. The soot from bituminous pyrolysis is governed by both pathways.

As to lignite, the peak size is situated between the two peaks, indicates that it may also be influenced by other factors. From Table 11.1 it can be seen that the O content in lignite is much higher than other coals and it is apparently a low rank coal with large amount of oxygen-containing functional groups. During its pyrolysis, those oxygen-containing functional groups react and form oxidizing radicals, which promote the oxidization of soot and inhibit soot growing up.

For zhundong coal, the Na and Ca contents are high in ash, during soot formation, the nucleation rate is accelerated, thus, the soot formation is promoted. Besides, alkali and alkaline earth metal such as Na and Ca has smaller ionization energy, so the soot particles are charged easily, which in turn prevents the soot agglomeration, thus forming more soot particles with smaller size. On the other hand, elements like Ca react with hydrocarbons and release OH radicals, which have a strong oxidizability. Those oxidizing radicals consume the soot precursors and accelerate soot oxidization rate (Bockhorn 1994; Kaskan 1965).

11.3.2 Gas Compositions

For all coals, either low rank coal or high rank coal, the main gas phase products are CO and H_2, the gas yield vary with the change of coal types. Zhundong coal has the highest CO and H_2 yield, reaching 25 mol/kg and 32.5 mol/kg respectively. Lignite has second highest CO and H_2 yield (15 mol/kg and 17 mol/kg). The CO yield of bituminous, lean coal and anthracite are 2.5 mol/kg, almost the same, while H_2 yield are 5.8 mol/kg, 13.3 mol/kg and 14.4 mol/kg respectively.

There was hardly any C_2H_2 during the pyrolysis of zhundong and lignite, and CH_4 during the pyrolysis of lignite, bituminous and lean coal was not detected. All the coals had the same C_2H_4 yield, 0.05 mol/kg.

Since there is no C_2H_2 observed during zhundong coal and lignite pyrolysis, and the H_2 yield is obviously higher, it can be perceived that during the pyrolysis of these two kind of coals "HACA" route should dominate the soot formation (Frenklach and Hai 1991). This means the soot formed from zhundong coal and lignite pyrolysis was through the first pathway as mentioned above. Small moleculars like CO and CO_2 are considered to be the products of oxygen-containing functional groups like carbonyl, carboxyl during high temperature cracking. CO_2 yield of lignite and bituminous are relatively high, that is consistent with the high O content in these two coals (Fig. 11.4).

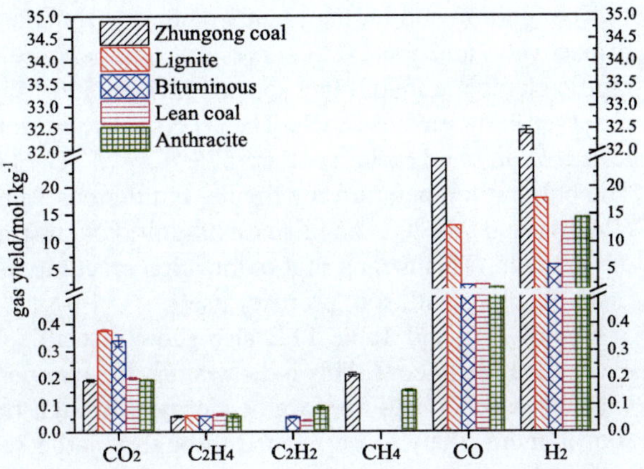

Fig. 11.4 Pyrolysis gas composition

11.3.3 Soot Oxidization

TGA was applied to study soot ignition and burnout characteristics, TG and DTG curves are shown in Fig. 11.5, the ignition and burnout characteristics are shown in Table 11.2.

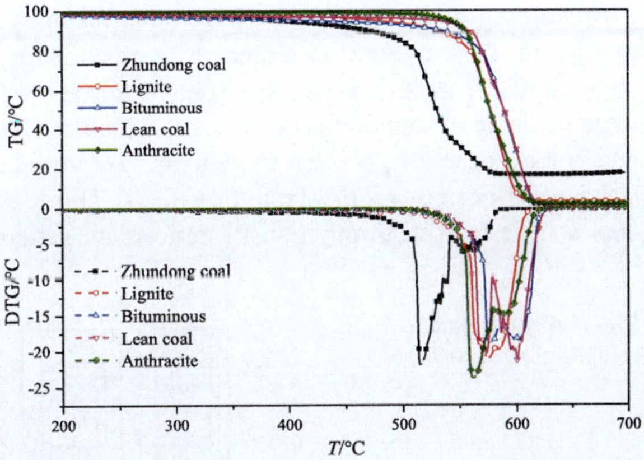

Fig. 11.5 TG-DTG curves during soot combustion

Table 11.2 Soot ignition and burnout characteristics

	T_i/°C	T_{max}/°C	T_b/°C	Residual mass/%
Zhundong coal	505	514	534.5	17.55
Lignite	555.1	575.5	598.7	2.98
Bituminous	561.8	575.8	624.3	0
Lean coal	560.5	600.1	615.2	0.04
Anthracite	551.4	561.5	599.8	1.48

The ignition and burnout temperature for zhundong coal is 505°C and 534.5 °C respectively, and the DTG curve peak temperature is around 514 °C. While the ignition temperature of other soot are all over 550 °C, and no obvious difference was observed between those coals. The DTG peak temperatures of for lignite, bituminous, lean coal and anthracite soot are 575 °C, 575 °C, 600 °C and 561 °C respectively. The burnout temperature for lignite, bituminous, lean coal and anthracite are 599, 624, 615 and 599 °C. The increase of ignition temperature is not consistent with that of coal rank. This means soot oxidization reactivity is not so closely related to coal rank, but more with soot property itself.

Figure 11.5 and Table 11.2 also show that all soot have very low ash content except zhundong coal. This indicates that the composition of soot is quite different with that of char, and they are not formed through fragmentation. Contrast to that, soot is more likely to be formed through volatile secondary reaction during high temperature pyrolysis. As to the soot of zhundong coal, its ash content is up to 17.55%, much higher that any other soot. This indicates that there might exist a different soot formation mechanism.

EDS was applied to analyze soot ash composition. The ratio of major elements is shown in Fig. 11.6. The soot ash of zhundong coal has high Ca, S, Si and Na contents, especially Ca and S content are much higher than other soot ash. While the soot ash of lignite, bituminous, lean coal and anthracite has high Al and Si content, just like their corresponding coal ash. Besides, the soot ash of lignite and lean coal has high Na content, while that of bituminous and anthracite has high Fe content.

According to ash contents and ash compositions, there might be three reasons to explain the low ignition temperature of zhundong coal soot. First, the size of soot particle is small with a large specific surface area, making it easier to contact oxygen. Second, zhundong coal soot has a high ash content compared to other coals, and amongst the ash, Na and Ca contents are very high, which may facilitate soot oxidization because of the catalytic effects. Third, as mentioned above, zhundong coal soot is formed through small moleculars generating aromatic rings, thus the

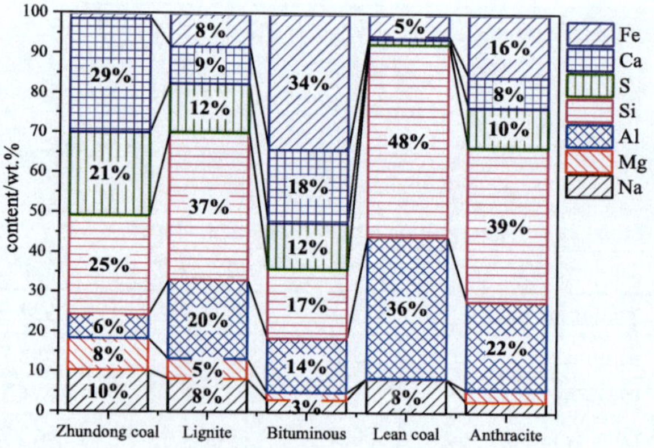

Fig. 11.6 Elemental compositions of soot ash

soot is younger, which means there remain some side-chain hydrocarbons with high reactivity.

11.4 Conclusions

Five coals with different ranks were pyrolyzed in a drop tube furnace, the soot yield, soot particle size distribution, gas composition, and ignition characteristics were studied. The main conclusions are as follows:

(1) Bituminous has the highest soot yield, up to 7.7%, while the other coals are all lower than 2%. The soot yield is not consistent with coal rank.
(2) The number size distribution of bituminous soot shows bimodal, while that of other coals only has one peak, which is ascribed into different soot formation mechanisms.
(3) Zhundong coal soot has the lowest ignition temperature, because of its small particle size and high Na/Ca contents.

Acknowledgements This work was supported by the National Key Development Plan of China (No. 2016YFC0801904) and the Natural Science Foundation of Xinjiang Uygur Autonomous Region (Nos. 2017D01A69 and 2017D01B41), Natural Science Basic Research Plan in Shaanxi Province of China (No. 2017JZ010).

References

Bi X, Liu H, Huo M, Shen C, Qiao X, Lee CF (2014) Experimental and numerical study on soot formation and oxidation by using diesel fuel in constant volume chamber with various ambient oxygen concentrations. Energy Convers Manage 84:152–163

Bockhorn H (1994) Soot formation in combustion

Cheung SC, Yuen RK, Yeoh G, Cheng GW (2004) Contribution of soot particles on global radiative heat transfer in a two-compartment fire. Fire Saf J 39:412–428

Fletcher TH, Ma J, Rigby JR, Brown AL, Webb BW (1997) Soot in coal combustion systems. Prog Energy Combust Sci 23:283–301

Frenklach M (2002) Reaction mechanism of soot formation in flames. Phys Chem Chem Phys 4(11):2028–2037

Frenklach M, Hai W (1991) Detailed modeling of soot particle nucleation and growth. Symp (Int) Combust 23 (1):1559–1566

Kaskan WE (1965) The reaction of alkali atoms in lean flames. Symp Combust 10(1):41–46

Li Y, Tan HZ, Wang XB, Bai SJ et al (2018) Characteristics and mechanism of soot formation during the fast pyrolysis of biomass in an entrained flow reactor. Energy Fuels 32(11):11477–11488

Wang XB, Li Y, Bai SJ et al (2018) Nano-scale soot particle formation during the high-temperature pyrolysis of waste plastics in an entrained flow reactor. Waste Biomass Valorizat 3:1–10

Chapter 12
Effect of Microstructures on Char Combustion Reactivity

Jingyuan Li, Lele Feng, Yiqun Huang, Yang Zhang, Lyu Junfu, and Man Zhang

Abstract Reaction of char dominates the combustion of coal in terms of the time scale and heat release, thus study on the reactivity of char is essential. Some researchers found that the coal char combustion reactivity is the function of its microstructure and the devolatilization temperature. However, in the related reported studies, the coal types used in the experiments were limited. Therefore, in this paper series of systematic experiments were conducted with 12 types of coal, including anthracite, bituminous, lean coal and lignite, and the volatile matter content ranges from 5 to 55%. The char samples were prepared in the muffle furnace in ambient atmosphere at 900 °C. The thermogravimetry analyzer (TGA) was employed to evaluate the char combustion reactivity. Moreover, the microstructure of the char was measured by X-Ray Diffraction (XRD). The experimental results indicate that the combustion reactivity of chars decreases with increasing graphitization degree. And an obvious liner relationship between f_a and T_{max}/E was found.

Keywords Char · Microstructure · Combustion reactivity · Kinetics · XRD

12.1 Introduction

Coal combustion generally includes three steps: (i) pyrolysis or devolatilization to release volatiles; (ii) combustion of volatile; (iii) combustion of char particles. Compared with the rapid devolatilization process, the combustion processes of char is much slower. Therefore, char combustion is the rate-controlling step of the combustion process of coal (Miura et al. 1989). After heat treatment, the combustion reactivity of char decreases. With the increase of heat treatment temperature and time, this phenomenon is becoming more obvious (Zhang et al. 2004). The microstructure of coal also changes in the process of char formation, including the loss of functional groups and the carbon microstructure changing to graphite. The change of

J. Li · L. Feng · Y. Huang · Y. Zhang · L. Junfu · M. Zhang (✉)
Key Laboratory for Thermal Science and Power Engineering of Ministry of Education, Department of Energy and Power Engineering, Tsinghua University, Beijing, China
e-mail: zhangman@mail.tsinghua.edu.cn

J. Lyu and S. Li (eds.), *Clean Coal and Sustainable Energy*, Environmental Science and Engineering,
https://doi.org/10.1007/978-981-16-1657-0_12

char microstructure after heat treatment is recognized as the major reason for the decrease of char combustion reactivity (Lu et al. 2002). Therefore, many studies have reported the relation between combustion reactivity and microstructure.

So far, many techniques have been employed to measure the microstructure of char, such as X-ray diffraction, transmission electron microscopy, and Raman spectroscopy (Oluwadayo 2010). X-ray diffraction does not cause destruction to the sample, so it was widely used to obtain the structural information, considering the crystalline carbon and amorphous carbon for carbon structure. Quantitative analysis of X-ray diffraction yields several parameters, including crystallite size (L_a, L_c, d_{002}, g) and aromaticity (f_a) (Lu et al. 2001).

Lu et al. (2002) used low volatile bituminous coal to prepare char samples. They found that even at very low temperature, char structure was more ordered during combustion. The amorphous concentration, which can reflect char structure, was decreased, while the aromaticity and average crystallite size was increased. As a result, char becomes less reactive. They also defined an index of disorder, *IOD*, to quantitatively describe char structure.

Zhu and Sheng (2010) used a non-isothermal TGA method to measure char combustion reactivity, and correlated the results with the first-order Raman parameters. A lignite was pyrolyzed from 773 to 1673 K. A good linear correlations were found between combustion reactivity indexes and the band area ratios of I_{D1}/I_G and I_G/I_{ALL}.

Similar work had also been carried out by Everson et al. (2013). They found that the reactivity was depended on aromaticity. The intrinsic activity decreased significantly and activation energy increased with the increase of aromaticity.

However, in the related reported studies, the coal types used in the experiments were limited. Therefore, this work was addressed to the evolution of the microstructure of char under heat treatment and its influence on combustion reactivity for more different char samples. 12 samples, including anthracite, bituminous, lean coal and lignite, were used for char preparation at 900 °C. The combustion reactivity of char was measured with the thermogravimetric analyzer based on a non-isothermal method. The microstructure was measured by X-ray diffraction. The relationship between the char combustion reactivity and microstructure were discussed.

12.2 Experimental

12.2.1 Coal Samples and Preparation of Char

The experiment using chars from different coals. There are 12 kinds of coal coming different regions, including anthracite, bituminous, lean coal and lignite. And the particle size was 75–100 μm. The properties of the samples are presented in Tables 12.1 and 12.2.

Table 12.1 Proximate analysis of coal

Sample ID		Ma	Va	Aa	Fca	Vd	Ad	Vdaf
Anthracite	XR	3.66	4.85	17.45	74.04	5.03	18.11	6.15
	YQ	2.56	8.34	10.02	79.08	8.56	10.28	9.54
	HN	0.81	5.65	16.05	77.49	5.70	16.18	6.80
Lean coal	QB	1.52	10.09	24.41	63.98	10.25	24.79	13.62
	HD	2.70	12.28	25.89	59.13	12.62	26.61	17.20
Bituminous	QJY	1.63	19.28	35.72	43.37	19.60	36.31	30.77
	ZY	4.43	25.65	33.23	36.69	26.84	34.77	41.15
	PS	5.00	26.24	32.91	35.85	27.62	34.64	42.26
Lignite	XL	27.45	28.75	15.71	28.09	39.63	21.65	50.58
	HLH	23.88	27.20	26.80	22.12	35.73	35.21	55.15
	CY	19.96	25.26	26.13	28.65	31.56	32.65	46.86
	SZ	16.62	41.34	8.84	33.20	49.58	10.60	55.46

Table 12.2 Ultimate analysis of coal

Sample ID		C%	H%	O%	N%	S%
Anthracite	XR	72.74	2.50	0.82	0.26	1.66
	YQ	79.83	3.09	2.04	0.27	0.78
	HN	78.30	0.16	1.68	0.11	2.09
Lean coal	QB	68.88	3.06	0.01	0.46	0.51
	HD	59.90	3.42	4.56	0.44	0.61
Bituminous	QJY	51.16	3.19	5.98	1.35	0.98
	ZY	43.86	3.33	13.43	1.00	0.71
	PS	47.72	3.54	7.91	1.15	1.76
Lignite	XL	33.42	2.65	22.31	0.48	0.86
	HLH	35.43	2.53	9.60	0.74	0.45
	CY	11.18	0.80	38.15	0.13	0.34
	SZ	49.08	4.06	19.35	1.01	1.06

The char samples were prepared in the muffle furnace at ambient atmosphere in 900 °C. Before preparation, the sample was placed in a dry box and dried at 105 °C for two hours to remove moisture.

12.2.2 Char Characterization

In order to test the properties of char, proximate analysis has been carried out. And the result is showed in Table 12.3. All samples should be stored in dry condition.

Table 12.3 Proximate analysis of char

Sample ID		Ma	Va	Aa	FCa
Anthracite	XR	0.81	3.98	16.24	78.97
	YQ	0.72	5.80	8.95	84.53
	HN	0.01	4.48	18.13	77.38
Lean coal	QB	0.61	6.03	23.60	69.76
	HD	0.43	5.46	29.07	65.03
Bituminous	QJY	0.35	3.27	43.55	52.83
	ZY	1.95	4.76	42.61	50.68
	PS	0.58	3.52	43.25	52.65
Lignite	XL	0.82	6.50	19.21	73.48
	HLH	0.78	5.07	50.90	43.26
	CY	0.77	5.68	45.09	48.46
	SZ	0.87	5.22	15.12	78.79

12.2.3 X-Ray Diffraction Analysis

D8-Discover X-ray diffractometer was used to record X-ray intensities scattered from char samples. Cu Kα radiation (40 kV, 40 mA) was used as the X-ray source. Samples were scanned over the angular 2θ range of 5° to 80°, and in a step-scan mode with a step size of 0.02°. Two peaks were expected over the examined 2θ range, the (002) peak and (10) peak. The position of the peak in the low angle corresponds to the (002) peak of graphite, which is attributed to the stacking of the graphitic basal plans of char crystallites. The peak in the high angle called (10) peak, which is attributed to hexagonal ring structure in char crystallites. Theoretically, the (002) peak symmetric, but the pattern shows obvious asymmetry. This is due to the existence of γ band on the left of (002) peak. And it is associated with packing of saturated structure such as aliphatic side chains (Watanable et al. 2002).

We can obtain four structural parameters based on XRD, viz. crystallite height (L_c), crystallite diameter (L_a), interlayer spacing (d_{002}) and aromaticity (f_a). They can be calculated by Eqs. (12.1)–(12.4).

$$d_{002} = \frac{\lambda}{2\sin\theta_{002}} \tag{12.1}$$

$$L_c = \frac{0.89\lambda}{\beta_{002}\cos\theta_{002}} \tag{12.2}$$

$$L_a = \frac{1.84\lambda}{\beta_{10}\cos\theta_{10}} \tag{12.3}$$

$$f_a = \frac{A_{002}}{A_{002} + A_\gamma} \times 100\% \tag{12.4}$$

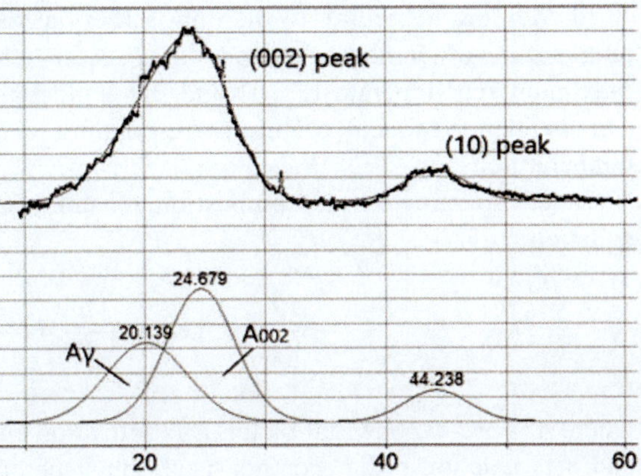

Fig. 12.1 X-ray diffraction pattern

where λ is the wavelength of the incident X-rays (Å). For copper Kα radiation, the λ value is 1.5406 Å. The θ is the peak position, β is the full width of the peak at half maximum intensity. And A is the area under the corresponding peak respectively.

As a function of heat treated conversion process, these parameters would be gradually evolved with the coal oxidation process. With the structure of char tending to be ordered, L_c, L_a and f_a will increase with decreasing d_{002} (Fig. 12.1).

12.2.4 Char Combustion Reactivity Measurement

Char combustion reactivity was measured by a thermogravimetric analyzer (Mettler-Toledo TGA/DSC1/1600HT) in non-isothermal method. The reaction was performed in the air atmosphere with 21% O_2. About 10 mg of char samples was placed in the TGA pan, and heat from room temperature to 1000 °C at a heating rate of 15 K/min. And then the chars were cooled to the programed temperature. The gas flow rate is 100 ml/min. And the temperature and weight of sample can be recorded (Nie et al. 2001).

The conversion of the char at any time can be calculated by Eq. (12.5).

$$X = \frac{m_0 - m}{m_0 - m_{ash}} \times 100\% \tag{12.5}$$

where m_0 is the initial weight of sample and m_{ash} is the weight at the end of the reaction.

Define R as the combustion rate:

$$R = \left(\frac{dX}{dt}\right)_{\max} \tag{12.6}$$

In addition, according to the non-isothermal method, we can obtain the initial temperature of combustion reaction (T_i), the temperature corresponding to the maximum reaction rate (T_{max}) and the temperature at the end of reaction. We can also define the temperature corresponding to 20% conversion as characteristic temperature.

Kinetic parameters of combustion reaction can be obtained from Arrhenius equation

$$\frac{dX}{dt} = A\exp(-\frac{E}{RT})(1-X)^n \tag{12.7}$$

where A is pre-exponential factor, E is activation energy, R is universal gas constant and n is reaction order. For non-isothermal method, a constant heating rate β is introduced into the equation. By transforming the equation, can get the following equation (Xie 2002)

$$\ln\left[\frac{1-\ln(1-X)^{1-n}}{T^2(1-n)}\right] = \ln\left(\frac{AR}{\beta E}\right) - \frac{E}{RT}\text{n} \neq 1$$
$$\ln\left[-\frac{\ln(1-X)}{T^2}\right] = \ln\left(\frac{AR}{\beta E}\right) - \frac{E}{RT}\text{n} = 1 \tag{12.8}$$

Therefore, the activation energy of the reaction can be used to characterize the combustion reactivity of char, and it was derived by fitting the Arrhenius plot with the reaction rate in the conversion range of 20–80% for each sample. The conversion of 20% was chosen as the low limit to avoid the measurement uncertainty at very low conversion.

12.3 Results and Discussion

12.3.1 X-Ray Diffraction Carbon Crystallite Analyses

The XRD diffraction pattern of char is shown in the Fig. 12.2. The curves show the representative samples selected from each kind of coal. The 002 peak is sharper and narrower with the increase of coal rank (Li 2013). This is because of a more compact, more ordered and more orientated structure.

The aromaticity (f_a), inter layer spacing (d_{002}), crystallite diameter (L_a) and crystallite height (L_c) are shown in Table 12.3. The small spikes on the diffractograms are peaks of traces of minerals still remaining in the samples (Table 12.4).

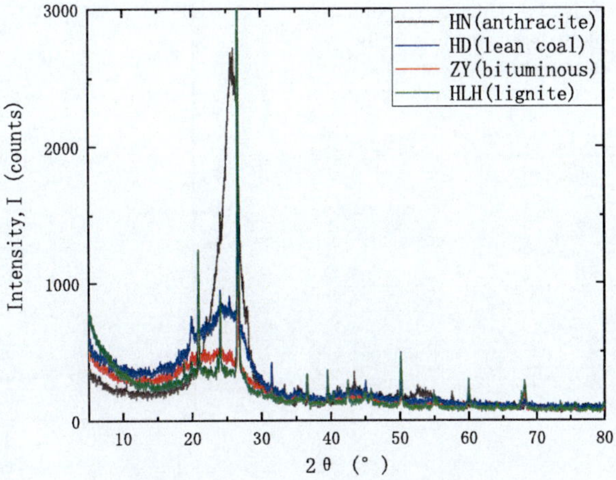

Fig. 12.2 Diffractograms of different chars

Table 12.4 Carbon crystallite properties of char used for combustion experiments

Sample ID		Lc/Å	La/Å	fa/%	d_{002}/Å
Anthracite	XR	15.26	40.87	71.31	1.83
	YQ	12.54	35.79	75.35	1.91
	HN	28.47	42.62	76.84	1.77
Lean coal	QB	15.95	31.61	66.49	1.83
	HD	14.27	36.45	67.78	1.82
Bituminous	QJY	15.58	31.34	71.70	1.80
	ZY	12.83	26.48	57.60	1.84
	PS	11.13	31.09	71.37	1.89
Lignite	XL	11.84	35.97	49.74	1.91
	HLH	11.32	38.79	59.91	1.90
	CY	13.90	37.14	55.12	1.92
	SZ	10.34	40.73	62.37	1.93

12.3.2 Char Combustion Reactivity

The combustion reactivity of char is affected by many factors, including the process of getting char, the nature of the original sample and so on. Several representative thermogravimetric curves of char are listed in Fig. 12.3. The parameters of char combustion reactivity are listed in Table 12.5.

The combustion reactivity of chars made from different kinds of coal under the same conditions is quite different. The initial temperature of combustion reaction (T_i) is at about 400 °C, and the temperature corresponding to the maximum reaction rate (T_{max}) is at about 600 °C. Actually, it was observed that all chars were completely

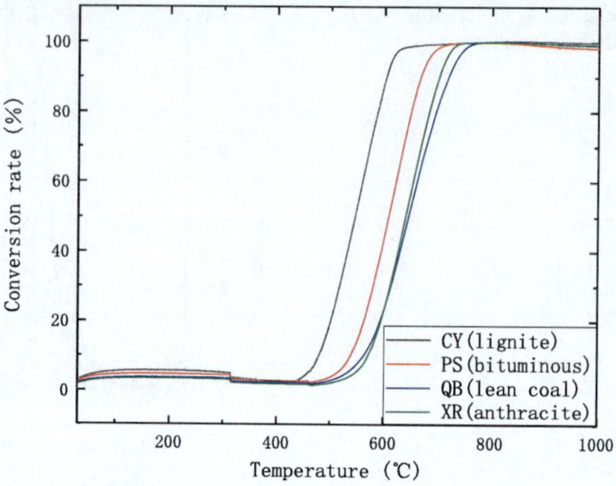

Fig. 12.3 Char conversion is plotted as a function of the temperature

Table 12.5 The parameters of char combustion reactivity

Sample ID		Tmax (°C)	Ti (°C)	Tf (°C)	T20 (°C)	R	E (J/mol)
Anthracite	XR	638.00	468.31	777.14	600.61	0.19	2.44E+05
	YQ	652.90	468.17	790.71	609.63	0.16	2.48E+05
	HN	720.86	467.83	830.54	610.36	0.14	2.48E+05
Lean coal	QB	638.89	468.16	804.85	598.87	0.17	2.45E+05
	HD	638.24	468.18	829.34	605.65	0.17	2.47E+05
Bituminous	QJY	634.11	468.28	791.68	600.56	0.18	2.44E+05
	ZY	557.06	418.42	723.64	512.05	0.17	2.21E+05
	PS	617.39	468.00	752.41	568.68	0.19	2.35E+05
Lignite	XL	537.29	334.94	710.87	480.89	0.15	2.13E+05
	HLH	536.60	357.63	757.75	487.09	0.21	2.14E+05
	CY	547.74	419.53	711.15	502.82	0.19	2.18E+05
	SZ	598.94	419.05	777.55	524.54	0.14	2.26E+05

burned before 850 °C. The burnout temperature of each char was lower than the corresponding treatment temperature.

12.3.3 *Correlations Between Char Combustion Reactivity and X-Ray Diffraction*

It is clear that the aromaticity, inter layer spacing (d_{002}), crystallite diameter (L_a) and crystallite height (L_c) are interrelated. So that any of these parameters can be chosen for correlating XRD properties with char combustion reactivity. And an obvious liner

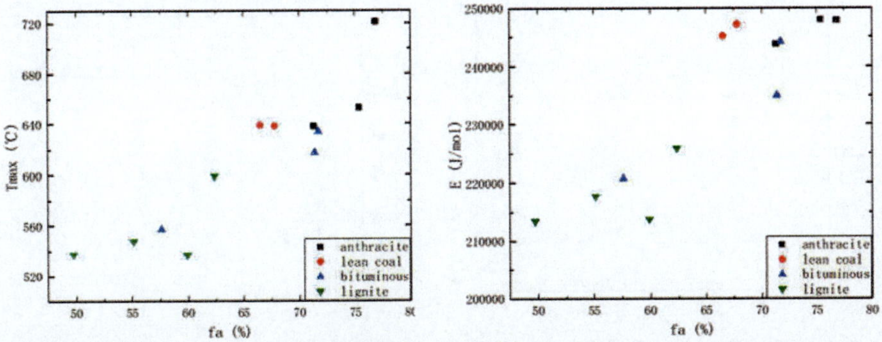

Fig. 12.4 Effect of aromaticity on char combustion reactivity

relationship between f_a and T_{max}/E exists and is presented in Fig. 12.4. The increase of T_{max} and E indicates that the reactivity of char decreases.

Figure 12.4 shows that T_{max} and E increase with the increasing aromaticity. Aromaticity is the fraction of aromatic carbon atoms within the char crystalline structure. The aromaticity in certain cases was as high as 0.75 has a very orderly structure. Therefore, the increase of aromaticity leads to more carbon microcrystals with ordered structure and orientation, which tend to graphite structure and with decreasing edge defects. This is consistent with previous studies.

Figure 12.5 shows that the effect of layer spacing (d_{002}) on char combustion reactivity. The decrease of d_{002} indicates the ordering of char structure. Figure 12.5 shows that with the decrease of d_{002}, T_{max} and E increase.

Figure 12.6 shows the effect of L_c and L_a on char combustion reactivity. Theoretically, with the increase of L_c and L_a, the combustion reactivity of char decreases, and correspondingly, T_{max} and E increase. But from the experimental results, the effect of L_c and L_a is not very obvious. At the same time, the change of L_c and L_a in the heat treatment process is staged and very complicated (Zubkova 2005; Li 2018).

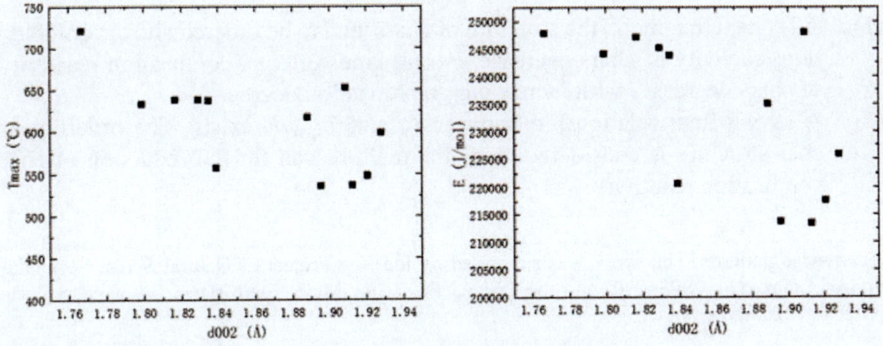

Fig. 12.5 Effect of layer spacing (d_{002}) on char combustion reactivity

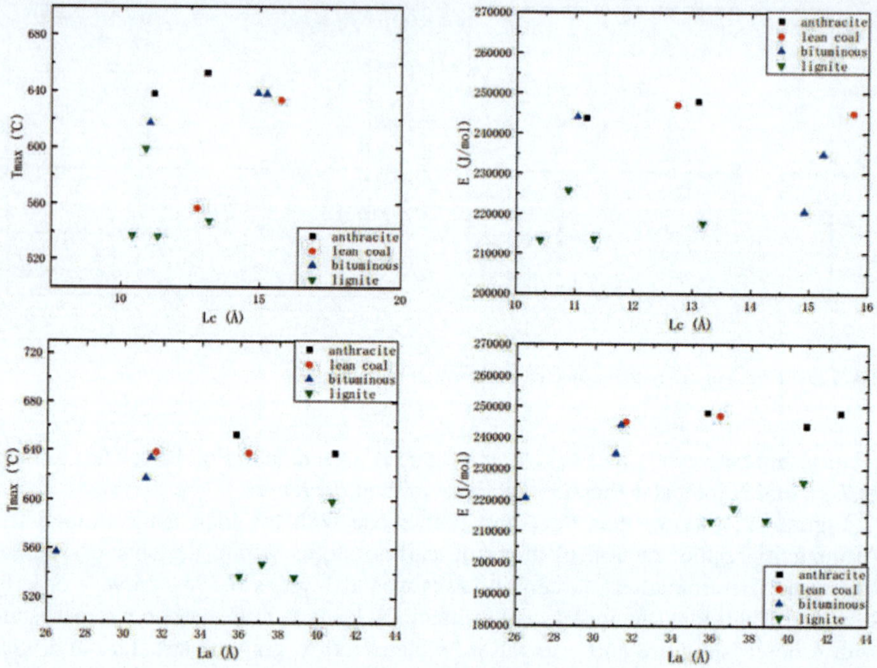

Fig. 12.6 Effect of crystallite diameter (L_a) and crystallite height (L_c) on char combustion reactivity

12.4 Conclusions

In present work, the microstructure of chars and their combustion reactivity had been studied using thermogravimeteic analyzer (TGA) and X-ray diffraction (XRD) analyzer, respectively. The correlations between structural parameters and combustion reactivity had been analyzed in detail. The key points of the present research are summarized below:

(1) After heat treatment, the structure of chars tend to be ordered and the combustion reactivity of chars decrease. For all kinds of coal, combustion reactivity of chars decreases with increasing graphitization degree.

(2) A strong liner relationship between f_a and T_{max}/E exists. The ordering of char structure is considered to be the main reason for the reduction of char combustion reactivity.

Acknowledgements This work was supported by the Key Project of Natural Science Foundation of China (No. 29936090) and the Special Funds for Major State Basic Research projects (G1999022102).

References

Everson RC, Okolo GN, Neomagus HW, Dos Santos JM (2013) X-ray diffraction parameters and reaction rate modeling for gasification and combustion of chars derived from inertinite-rich coals. Fuel 109:148–156

Li Q (2018) Investigation on the structure evolution of pre and post explosion of coal dust using X-ray diffraction. Int J Heat Mass Transf 120:1162–1172

Li M et al (2013) Aggregate structure evolution of low-rank coals during pyrolysis by in-situ X-ray diffraction. Int J Coal Geol 116–117:262–269

Lu L, Sahajwalla V, Kong C, Harris D (2001) Quantitative X-ray diffraction analysis and its application to various coals. Carbon 39(12):1821–1833

Lu L, Kong C, Sahajwalla V, Harris D (2002) Char structural ordering during pyrolysis and combustion and its influence on char reactivity. Fuel 81:1215–1225

Miura K, Hashimoto K, Silveston PL (1989) Factors affecting the reactivity chars during gasification, representing reactivity. Fuel 68:1461–1475

Nie QH, Sun SZ, Li ZQ, Zhang XJ, Wu SH, Qin YK (2001) Thermogravimetric analysis on the combustion characteristics of brown coal blends. J Combust Sci Technol 7:72–76

Oluwadayo O (2010) Sonibare: structural characterization of Nigerian coals by X-ray diffraction, Raman and FTIR Spectroscopy. Energy 35:5347–5353

Watanabe I, Sakanishi K, Mochida I (2002) Changes in coal aggregate structure by heat treatment and their coal rank dependency. Energy & Fuels 16:18–22

Xie KC (2002) Coal structure and its reactivity. Science Press, Beijing, pp 221–222

Zhang S, Lyu J et al (2004) Effect of heat treatment on the reactivity and microstructure of coal-char. J Fuel Chem Technol 32(6):673–678

Zhu X, Sheng C (2010) Evolution of the char structure of lignite under heat treatment and its influences on combustion reactivity. Energy Fuels 24:152–159

Zubkova VV (2005) Some aspects of structural transformation taking placc in organic mass of Ukrainian coals during heating. Part 1. Study of structural transformations when heating coals of different caking capacity. Fuel 84:741–754

Chapter 13
Effect of K_2CO_3 Addition on CO_2 Gasification Characteristics and Ash Sintering Behaviour of a Chinese Lignite at Different Temperatures and Pressures as Examined Using a High-Pressure Thermogravimetric Analyser

Jianbo Li, Zhezi Zhang, Jian Hao, Jiguang Zhang, Mingming Zhu, and Dongke Zhang

Abstract The gasification characteristics and the morphology of the residue ash from pressurised K_2CO_3-catalysed gasification of a Chinese lignite in CO_2 was investigated using a High-Pressure Thermogravimetric Analyser operating at pressure of 2.0 or 3.5 MPa and temperature between 750 and 900 °C for at least 2 h, after being heated from room temperature at 10 °C/min. The K_2CO_3 addition was varied from 0 to 10% w/w. Gasification characteristics of the K_2CO_3-doped lignite was determined by analysing the weight loss and conversion rate as a function of time whereas the ash morphology was analysed by using SEM–EDS. Results showed that at 3.5 MPa the in-situ weight loss of the lignite increased as K_2CO_3 addition ratio increased, suggesting that K_2CO_3 addition promoted lignite gasification. The conversion rate of the lignite correspondingly increased from 61 to 92% as the temperature elevated to 750 °C. An increase in the final temperature to 900 °C significantly promoted lignite gasification when K_2CO_3 was less than 5%, however this was not obvious for lignite with 10% K_2CO_3 addition. This is because the conversion rate of the

J. Li (✉)
Key Laboratory of Low-Grade Energy Utilization Technologies and Systems, Ministry of Education of PRC, Chongqing University, 174 Shazheng Street, Shapingba District, Chongqing 400044, China
e-mail: jianbo.li@cqu.edu.cn

Z. Zhang · M. Zhu · D. Zhang
Centre for Energy (M473), The University of Western Australia, 35 Stirling Highway, Crawley, WA 6009, Australia

J. Zhang
Huadian Electric Power Research Institute Co. Ltd, No.10 Xiyuan 1st Road, Xihu District, Hangzhou City 310030, Zhejiang, China

J. Hao
State Key Laboratory of High-Efficiency Utilization of Coal and Green Chemical Engineering, Ningxia University, Yinchuan 750021, China

J. Lyu and S. Li (eds.), *Clean Coal and Sustainable Energy*, Environmental Science and Engineering,
https://doi.org/10.1007/978-981-16-1657-0_13

lignite with 10% K_2CO_3 addition had exceeded 90% before the final temperature of 900 °C was reached. Furthermore, as pressure decreased from 3.5 to 2.0 MPa, the lignite gasification rate slowed down, with or without K_2CO_3 addition. Conversion rate of the lignite decreased from 61 to 42% while the temperature initially elevated to 750 °C. SEM–EDS analysis revealed that sintering of the lignite ash was not observed at 750 °C, but became apparent at 1% K_2CO_3 addition. The degree of ash sintering further aggravated at 5 and 10% K_2CO_3 addition. As the temperature increased from 750 to 900 °C, the ashes of the raw lignite and 1% K_2CO_3 doped-lignite remained largely similar, whereas the sizes of the 10% K_2CO_3-doped lignite ash was increased and the particle surfaces became smooth, suggesting an enhanced sintering of the ash. The formation of K-aluminosilicate and Ca-aluminosilicate of low-melting points in the ash was responsible for possible deactivation of the doped catalyst K_2CO_3 and the observed ash behaviour.

Keywords Ash morphology · Catalytic gasification · Gasification rate · HP-TGA · K_2CO_3 · Pressure

13.1 Introduction

Clean coal technologies including integrated gasification combined cycle (IGCC) and pressurised fluidised bed combustor (PFBC) are regarded as the most viable alternatives for clean coal utilisation and have gained increased technological and scientific interests (Wall and Liu 2002; Jing and Wang 2013a; Ye and Agnew 1998). To increase the throughput of coal and promote reaction intensity of gasification, higher operating pressures and the use of catalytic additives have been applied (Wall and Liu 2002; Ye and Agnew 1998; Sharma and Takanohashi 2008; Zhang and Zhang 2015a; Fan and Yuan 2016). For instance, operating PFBC at 10–15 atm and IGCC at 15–25 atm have been reported in the literature (Wall and Liu 2002; Gupta 2005). Moreover, the use of catalytic additives not only promote gasification reaction at relatively low temperatures, but also shift the product gas to the desired direction (Akyurtlu and Akyurtlu 1995). However, the gasification characteristics of coal are affected by the operating procedures including pressure, temperature and catalyst addition. The ash residue might also lead to ash sintering, bed agglomeration and ash deposition on gas circuits and heat exchangers, further influencing the system stability (Jing and Wang 2011a, 2013a, b; Jing and Zhu 2016). Investigating the gasification characteristics and ash sintering behaviour during pressurised catalytic gasification at varying operating conditions are therefore essential.

Gasification characteristics of solid fuels including coal and biomass have been intensively investigated (Liang and Xie 2018; Liu and Guan 2018a, b; Zhao and Zhang 2018; Takematsu and Maude 1991; Irfan 2011; Nzihou and Stanmore 2013; Lee and Kim 1995). It is known that the gasification reactivity of coal is dependent on coal rank and char structure, the inherent inorganic matter, and operation conditions. Typically, low-rank coals have widely spread reactivity, while high-rank coals have low reactivities. The gasification characteristics could be promoted by the inorganic mineral matter mainly alkali and alkali earth (Lee and Kim 1995). Thus, additives are used to promoted gasification and reform composition of the gas yield (Wu and Wang 2011; Zhang and Kudo 2013; Jaffri and Zhang 2008). Among the catalysts reported in the literature and used in industrial practices, potassium carbonate is believed to have superior catalytic properties, which not only promotes gasification reactivity, but also shifts the products to methane-rich or hydrogen-rich gases (Wang and Jiang 2009; Hattingh and Everson 2011). Moreover, the gasification characteristics are also dependent on temperature and pressure of the reactors (Wall and Liu 2002). As gasification temperature increases, the gasification kinetics are promoted, promoting gasification of coal or char. Likewise, pressure has been found to significantly influence the volatiles yield and coal swelling during devolatilisation, changing the structure and morphology of the char generated. At higher pressures, char combustion and gasification reaction rate would be enhanced as more char particles of high porosity are formed at higher pressures. However, comprehensive study during pressurized K_2CO_3-catalytic gasification of typical Chinese lignite is still required for better comprehension its gasification characteristics.

Sintering characteristics of the ash during gasification have also been widely studied (Bai and Li 2011; Chen and Dai 2012; Zhang and Zhang 2015b). It has been reported that sintering characteristics of the ash are dependent on the chemistry and mineral composition of the ash, atmosphere, temperature as well as pressure (Jing and Wang 2011a, b, 2013a, b; Jing and Zhu 2016; Zhang and Zhang 2015b; Li and Huang 2010). Jing and Wang (2011b) reported that the presence of acidic oxides such as SiO_2 and Al_2O_3 might increase ash sintering temperature yet the increase in Fe_2O_3, CaO, Na_2O, and K_2O decreased ash sintering temperature monotonously. In addition, the effect of MgO on the sintering temperature is not obvious when MgO is increased above a certain limit. Meanwhile, Fe in the ash during gasification would be in the form of Fe^{2+} and react with other oxides to form liquid phases (van Li and Huang 2010), promoting ash sintering between ash particles. Besides, FeO appears to react with CaO, SiO_2, and Al_2O_3 to produce the low-temperature eutectics (Dyk and Waanders 2007). As the catalyst K_2CO_3 are added into coal during gasification, a portion of potassium would be retained the ash, increasing the presence of K and the mineral interactions associated with K, thereby increasing the propensity of ash sintering (Erickson and Benson 1992; Kühn and Plogmann 1983; Bruno et al. 1986). Different gasification atmospheres such as CO_2 and steam atmosphere would also lead to varied mineral interaction and thus different ash sintering temperature (Lee and Kim 1995; Erickson and Benson 1992; Formella and Leonhardt 1986; Duchesne and Hall 2014). In addition, the sintering characteristics of the ash is also dependent on gasification temperature: an increase in gasification temperature might promote mineral interactions as well as the release of the alkali and alkali earth contents (Zhao et al. 2016). As for pressure, the increase in pressure would influence mineral transformation by affecting reactions between minerals and accelerates the formation of high-temperature minerals. However, supplemental study on the sintering characteristics of the ash during both pressurized and K_2CO_3-catalysed conditions are still essential.

This paper presents an investigation into the gasification characteristics and the ash sintering behaviour during K_2CO_3-catalysed gasification of lignite with varying pressures, temperatures, and K_2CO_3 addition ratios examined. The lignite with or without K_2CO_3 addition was gasified in a high pressure Thermogravimetric Analyser (HP-TGA) at varying pressure, and temperatures. The weight loss curve as a function of temperature was analysed and the morphology and chemistry of the obtained ash samples were analysed. It is expected that this paper would provide better comprehension on the effect of operating procedures and K_2CO_3 addition on gasification characteristics and ash sintering behaviour during pressurised K_2CO_3-catalysed gasification.

13.2 Experimental

13.2.1 Materials

A Chinese lignite, namely Wangjiata (WJT for short) lignite, was used during experimentation. The lignite was pulverised and sieved to particles less than 200 μm. Proximate and ultimate analysis of the lignite, chemistry of the ash prepared at 815 °C in air, and the ash fusion temperatures are shown in Table 13.1. Note that the volatile matter of WJT was 30.85% while its ash content was 26.5%. Moreover, the carbon content of the lignite was 75% and H was less than 4.10%. In terms of the ash chemistries, the WJT ash was rich in acidic oxides mainly SiO_2 (58.1%) and Al_2O_3 (12.1%), accounting for 60.2% of the total ash. Basic oxides such as CaO (11.5%) and Fe_2O_3 (6.3%) were also presented, and might contribute to ash sintering. As a consequence, the fusion temperatures of the ash were relatively low compared with other lignite or bituminous ash (Li and Zhu 2017).

K_2CO_3 was used as the catalyst for catalytic gasification. K_2CO_3 was analytical potassium carbonate with an impurity higher than 99.99%. During sample preparation, the lignite sample was firstly mixed with deionized water at a ratio of 10 ml (water)/1 mg (lignite) to form water-lignite slurry. K_2CO_3 with addition ratios of 1, 5 and 10 wt% (relative to lignite), respectively, was then dissolved into the water-lignite slurry. The mixture was then stirred by using a magnetic stirrer with a speed of 500r/min for 6 h. This was followed by drying in an oven at 60 °C to obtain the dried K_2CO_3-doped sample. These samples were denoted as WJT, WJT+1%K_2CO_3, WJT+5%K_2CO_3, and WJT+10%K_2CO_3 throughout this paper.

13.2.2 Experimental Setup

A high pressure Thermogravimetric Analyser (HP-TGA-150S, TA, US) as shown in Fig. 13.1 was used during experimentation. The HP-TGA was comprised of a

Table 13.1 Proximate and ultimate analysis of WJT lignite and its ash chemistry and fusion temperatures

	Proximate analysis (wt%)			Ultimate analysis (wt%)				
	A	V	FC	C	H	O	N	S
WJT	26.5	30.9	42.7	75.3	4.1	8.2	0.9	0.4

Ash composition (wt%)							Fusion temperatures (°C)			
	SiO_2	Al_2O_3	CaO	Fe_2O_3	K_2O	Na_2O	SO_3	DT	ST	HT	FT
WJT	58.1	12.1	11.5	6.3	2.1	2.7	2.3	1130	1212	1260	1311

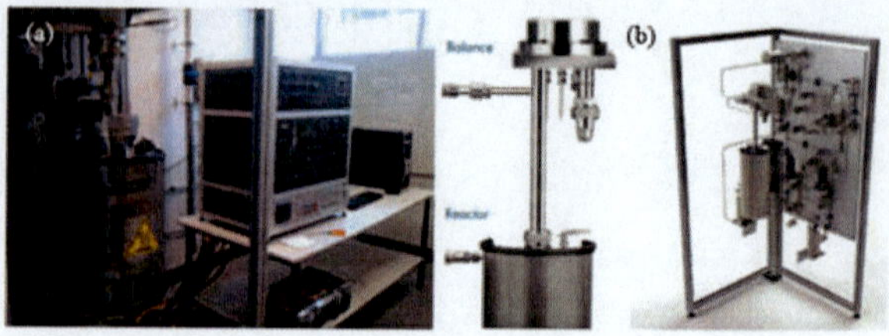

Fig. 13.1 Experimental setup of **a** the HP-TGA-150S system and **b** its main components

reactor, control system, and gas system. The reactor was made of stainless steel, which could be heated to 1100 °C and stand for 5.0 MPa. Moreover, five gas inlets were designed for input of various gases. The key components of the system are illustrated in Fig. 13.1b. A Rubotherm magnetic balance was used to measure the weight of the sample (and crucible), and the accuracy of the balance was ±30 μg.

During experimentation, the balance was firstly elevated above the reactor and ca. 0.2 g K_2CO_3-doped lignite was loaded into the crucible. The balance was then sent back to the reactor. For gasification at 2.0 MPa, pure CO_2 with a flow rate of 1L/min was used to elevate the pressure to 2.0 MPa. Whereas for gasification at 3.5 MPa, N_2 with a flow rate of 0.75 L/min was also provided at the same CO_2 flow rate. The reactor was then heated with a heating rate of 10 °C/min from ambient temperature to final temperatures of 750 to 900 °C, respectively. The experimentation maintained at the final temperatures for at least 2 h, and the weight of the sample (and crucible) was simultaneously recorded. Once the experimentation was finished, the reactor was cooled down to room temperature at N_2 atmosphere and the ash sample was afterwards collected.

13.2.3 *Data Processing and Analysis*

The in-situ weight loss curve as a function of time of each sample was corrected to eliminate the buoyancy effect. This correction included the correction of the magnet assembly, the sample holder, and the sample volume, i.e., Eq. 13.1,

$$\mathrm{Wt} = \mathrm{W_0} + \mathrm{Wt_1} + \mathrm{Wt_2} + \mathrm{Wt_3} \tag{13.1}$$

where Wt the final weight of the lignite (g), W_0 the in-situ recorded sample weight (g), Wt_1, Wt_2, and Wt_3 the corrected weight (g) of the magnet assembly, sample holder, and sample volume, respectively.

Wt_1 was obtained based on the temperature, pressure and the corrected factor as shown in Eq. 13.2. P is the pressure of the system (bar), MW the mole weight of CO_2 or CO_2/N_2 mixture (g/mol), Z_1 the correction factor of the magnet assembly, and T_0 the set temperature of the assembly, 50 °C in this case.

$$\mathrm{Wt_1} = (2.29 \times \mathrm{P} \times \mathrm{MW})/(\mathrm{Z_1} \times 83.144 \times (273.16 + \mathrm{T_0})) \tag{13.2}$$

The correction factor Z_1 was determined based on the properties of the mixtures, as shown in Eq. 13.3, where ψ the mole fraction of the gas in the mixture, P_{N2} the partial pressure of N_2 in the system (bar), ρ_{N2} and ρ_{CO2} are the densities of the pure N_2 and CO_2 (g/cc) determined based on the pressure and temperature of the gas.

$$\begin{aligned} Z_1 = Z_{N2} + Z_{CO2} = \psi_{N2} \times (P_{N2} \times 28)/(83.144 \times \rho_{N2} \times (273.16 + T_0)) \\ + \psi_{CO2} \times (P_{CO2} \times 44)/(83.144 \times \rho_{CO2} \times (273.16 + T_0)) \end{aligned} \quad (13.3)$$

Wt_2 was corrected based on Eq. 13.4 as below. Among Eq. 13.4, P is the pressure of the system (bar), MW the mole weight of the CO_2 or CO_2/N_2 mixture (g/mol), Z_2 the correction factor of the sample holder, and T the temperature as measured by the thermocouple (°C)

$$Wt_2 = (0.668 \times P \times MW)/(Z_2 \times 83.144 \times (273.16 + T)) \quad (13.4)$$

The correction factor of the sample holder Z_2 was determined based on the properties of the gas mixtures, as shown in Eq. 13.5. ψ is the mole fraction of the gas in the mixture, P_{N2} the partial pressure of N_2 in the system (bar), ρ_{N2} and ρ_{CO2} the densities of the pure N_2 and CO_2 (g/cc) determined based on the pressure and temperature of the gas, and T the temperature as measured by the thermocouple (°C)

$$\begin{aligned} Z_2 = Z_{N_2} + Z_{CO_2} = \psi_{N2} \times \left(P_{N_2} \times 28\right)/\left(83.144 \times \rho_{N_2} \times (273.16 + T)\right) \\ + \psi_{CO_2} \times \left(P_{CO_2} \times 44\right)/\left(83.144 \times \rho_{CO_2} \times (273.16 + T)\right) \end{aligned} \quad (13.5)$$

The sample weight after correction of magnet assembly and sample holder (Wt′) was thus obtained as shown in Eq. 13.6, and the volume of the sample V_1 was obtained as Eq. 13.7

$$Wt' = W_0 + Wt_1 + Wt_2 \quad (13.6)$$

$$V_1 = Wt'/2.2 \quad (13.7)$$

Wt_3 was correction of sample volume based on Eq. 13.7, where V_1 the volume of the sample, P the pressure of the system (bar), MW the mole weight of CO_2 or CO_2/N_2 mixture (g/mol), Z_3 the correction factor of sample volume, the same as that of sample holder Z_2.

$$Wt_3 = (V_1 \times P \times MW)/(Z_3 \times 83.144 \times (273.16 + T)) \quad (13.8)$$

Once the weight of the lignite was corrected, the conversion rate as a function of time was also obtained. The conversion rate (α) of the lignite was determined as the amount of weight loss relative to the overall weight loss after gasification (Eq. 13.9), where m_0 represents the initial sample weight, m_1 the final sample weight, and

m_i the sample weight during gasification at time i. Moreover, t_{50} during conversion, representing the time when 50% conversation rate was achieved, was also determined,

$$\alpha = (m_0 - m_i)/(m_0 - m_1) \tag{13.9}$$

13.2.4 SEM–EDS analysis

A TESCAN Vega 3 scanning electronic microscopy (SEM) coupled with X-ray energy dispersive spectroscopy (EDS) was used to analyse the morphological features and the chemistries of the obtained ashes. The chemistries of the ash particles were semi-quantitatively determined by the detector coupled in EDS.

13.3 Results and Discussion

13.3.1 Gasification Characteristics

13.3.1.1 Effect of K_2CO_3 Addition

Figure 13.2 illustrates the effect of K_2CO_3 addition on the weight loss and conversion rate of the K_2CO_3-doped lignite as a function of time during CO_2 gasification at 3.5 MPa in HP-TGA. The temperature profile during gasification was also recorded and are shown in Fig. 13.2. Based on the temperature profile as recorded, the weight loss as a function of time were divided into two periods. The first period was non-isothermal gasification during which temperature was increased from ambient temperature (<100 °C) to 750 °C. Note that the recorded temperature might be higher than 750 °C due to system response. The second period was the isothermal gasification period at 750 °C.

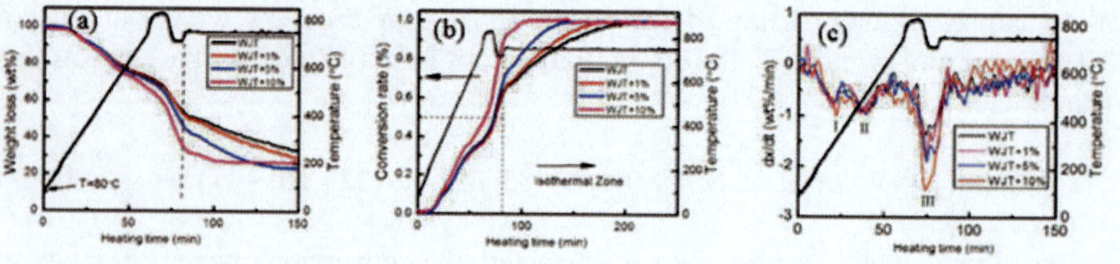

Fig. 13.2 Effect of K_2CO_3 addition on the **a** weight loss and **b** conversion rate of WJT during CO_2 gasification at 3.5 MPa and 750 °C

During non-isothermal gasification as shown in Fig. 13.2a, the weight of WJT decreased to 55 wt% as temperature increased to 750 °C. The sample weight further decreased slowly and reached to ca. 40 wt% after gasification at 750 °C for 150 min. This indicates that gasification of WJT without catalysts might be quite slow. Moreover, Gasification at this moment was believed to be incomplete as the ash content (Table 13.1) as determined ca. 26.5% was much less than the current of 40 wt%. As 1% K_2CO_3 was added into the lignite, the weight loss of WJT+1% K_2CO_3 was similar to that of WJT before the temperature reached 750 °C, which was however decreased faster at isothermal temperature of 750 °C and to ca. 30 wt% after gasification for 150 min. This indicated that the addition of 1% K_2CO_3 might not significantly promoted gasification at temperatures less than 750 °C, but would be apparent at 750 °C. As K_2CO_3 addition ratio further increased to 5% and 10%, the weight loss of the sample at the same time (and temperature) further increased, revealing that addition of 5 and 10% K_2CO_3 further promoted gasification. It is noted that the weight of WJT+10% K_2CO_3 decreased to less than 40 wt% while the temperature initially elevated to 750 °C. This enlightens that gasification was greatly promoted by the incorporating K_2CO_3. Moreover, the weight of WJT+10% K_2CO_3 at isothermal temperature of 750 °C was relatively stable after 20 min's gasification, while those with less K_2CO_3 required longer time. This again confirmed that addition of 10% K_2CO_3 promoted gasification of WJT.

Looking into the conversion rate of the lignite samples as shown in Fig. 13.2b, it is found that t_{50} of WJT, WJT+1% and WJT+5% were all at ca. 75 min. This illustrated that the addition of K_2CO_3, might not promoted gasification at lower temperatures. However, when the temperature was higher than 750 °C, the conversion rate was increased significantly. This implies that for gasification at temperatures lower than 750 °C, the effect of K_2CO_3 on gasification was not obvious when the addition ratio was in the range of 1–5%. However, at isothermal temperature 750 °C, the conversion rate became faster as K_2CO_3 addition increased from 0 to 5%, suggesting that the effect of K_2CO_3 on gasification at this range became significant. In comparison, t_{50} of WJT+10% was ca. 60 min, which was much earlier than those of lignites with less K_2CO_3 addition. This again indicates that addition of 10% K_2CO_3 promoted gasification even at lower temperatures.

13.3.1.2 Effect of Temperature

Figure 13.3 shows the effect of final temperature on the weight loss of WJT, WJT+1% K_2CO_3, and WJT+10% K_2CO_3 during CO_2 gasification at 3.5 MPa. It can be seen that for gasification of WJT (Fig. 13.3a), the sample weight decreased to 55 wt% as temperature elevated to 750 °C. The weight loss curves during heating stage was largely the same, confirming the repeatability of the experiments. During isothermal gasification, the sample weight at 850 and 900 °C decreased faster than that at 750 °C, indicating that the increase in temperature from 750 to 900 °C promoted gasification. Similar findings were also observed for lignite with 1% K_2CO_3 addition. However, when K_2CO_3 addition was increased to 10% as shown in Fig. 13.3c, the sample

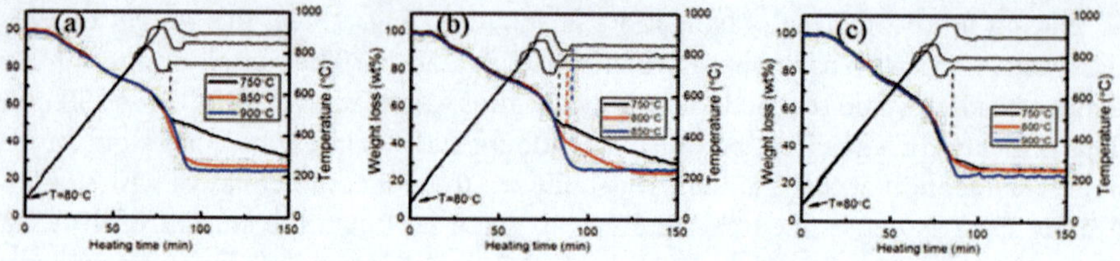

Fig. 13.3 Effect of Temperature on the weight loss of **a** WJT, **b** WJT+1% K_2CO_3, and **c** WJT+10% K_2CO_3 during CO_2 gasification at 3.5 MPa

weight decreased to less than 35 wt% during isothermal gasification, further proved that more K_2CO_3 would promote gasification. During isothermal gasification at 750, 800, or 900 °C, however, the differences in weight loss were not obvious. This is because gasification was nearly completed during heating stage, and the final temperature was thus insignificant for gasification.

13.3.1.3 Effect of Pressure

Figure 13.4 reveals the effect of pressure on the weight loss and conversion rate of WJT and WJT+10% K_2CO_3 as a function of time during CO_2 gasification at 750 °C. It is clear that at 2.0 MPa, the weight of WJT decreased to 70 wt% as temperature increased to 750 °C, and further decreased to 50 wt% after gasification for 350 min. This is much lower than that of WJT at 3.5 MPa of ca. 55% at 750 °C, and to 26 wt% after gasification (nearly completed) within the same time. These indicate that the increase in pressure promoted gasification of WJT. This was also applied for the weight loss of WJT+10% K_2CO_3 as illustrated in Fig. 13.4. The decrease in pressure promoted gasification of WJT+10% K_2CO_3. Moreover, the weight loss of WJT at 3.5 MPa was faster than that of WJT+10%K_2CO_3, suggesting that the effect of pressure on gasification might be superior to the effect of 10%K_2CO_3 addition.

It is also noted that t_{50} of these lignites decreased from 105 to 75, and 60 min as pressure increased to 3.5 MPa and K_2CO_3 addition to 10%, proving that the increase in pressure and pressure increased gasification of WJT. Moreover, the conversion

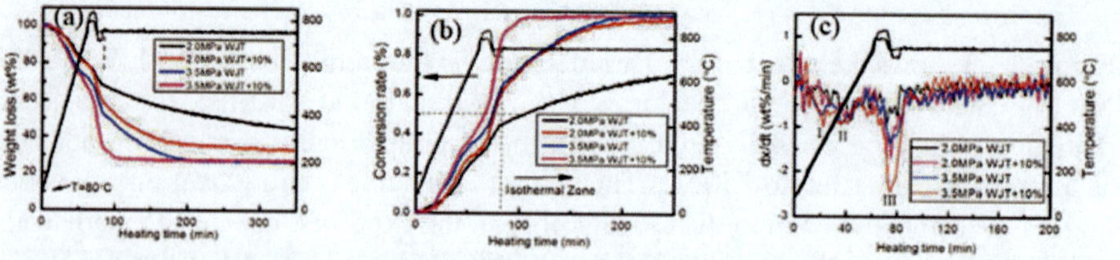

Fig. 13.4 Effect of pressure on the **a** weight loss and **b** conversion rate of WJT and WJT+10% K_2CO_3 during CO_2 gasification at 750 °C and pressures of 2.0 and 3.5 MPa

rate of WJT+10% K_2CO_3 at 2.0 MPa was less than that of WJT at 3.5 MPa when in non-isothermal stage, however its conversion rate was comparably to WJT at 3.5 MPa during isothermal gasification at 750 °C. This implies that the effect of 10% K_2CO_3 on gasification might be dependent on temperature. The catalytic effect of K_2CO_3 on gasification was low at lower temperatures, which however became significant at 750 °C. However, the effect of pressure on gasification, was significant during both non-isothermal and isothermal stages.

13.3.2 Ash Morphology

13.3.2.1 Effect of K_2CO_3

Figure 13.5 presents the morphological features and EDS analysis of the residue ashes of WJT with K_2CO_3 addition ratios of 0, 1, 5, and 10% during CO_2 gasification at 750 °C. It is clear that the sizes and shapes of the ash particles varied. The sizes of the ash particles ranged from sub microns to 200 μm as observed from SEM analysis. The shapes of these ash particles were irregular, and sintering between ash particles was seldom observed. This implies that most ash particles had relatively high melting-points and were thus not sintered. As 1% K_2CO_3 was added into the lignite, its ash particles (Fig. 13.5b) were also irregular in shape yet finer ash particles were less observed, implying that these finer particles might sinter together and therefore increase their sizes. As K_2CO_3 addition ratio further increased to 5% and 10%, the sizes of the ash particles were further increased as shown in Fig. 13.5c, d. Sintering between ash particles was observed, indicating that the addition of K_2CO_3 promoted ash sintering of the lignite ash at 750 °C.

EDS analysis on the ashes of WJT (Fig. 13.5a) reveals that the ash particles were heterogeneous in nature with varied contents of Si, Al, Ca, Fe, K, and Ca. As K_2CO_3 was added into the lignite, the contents of K in the ash increased as shown in Fig. 13.5b–d), suggesting that more K were presented in the ash, as expected due to K_2CO_3 addition. Looking into the ash chemistries of the sintered ashes, such as particles 2 and 1 in Fig. 13.5c and particles 1–5 in Fig. 13.5d, it was found that these particles were enriched in Si and K, and occasionally Ca, indicating the presence of K silicates, Ca-silicates and the eutectics between K-Si-Ca. The presence of these minerals or eutectics of low-melting points therefore promoted sintering of the ash particles.

13.3.2.2 Effect of Temperature

The morphological features of the ash of WJT, WJT+1% K_2CO_3, and WJT+10% K_2CO_3 during gasification at 3.5 MPa but different temperatures are presented in Figs. 13.6, 13.7 and 13.8. It is observable that the morphological features of WJT (Fig. 13.6) at 750, 800, 850, and 900 °C were largely the same, composing of the

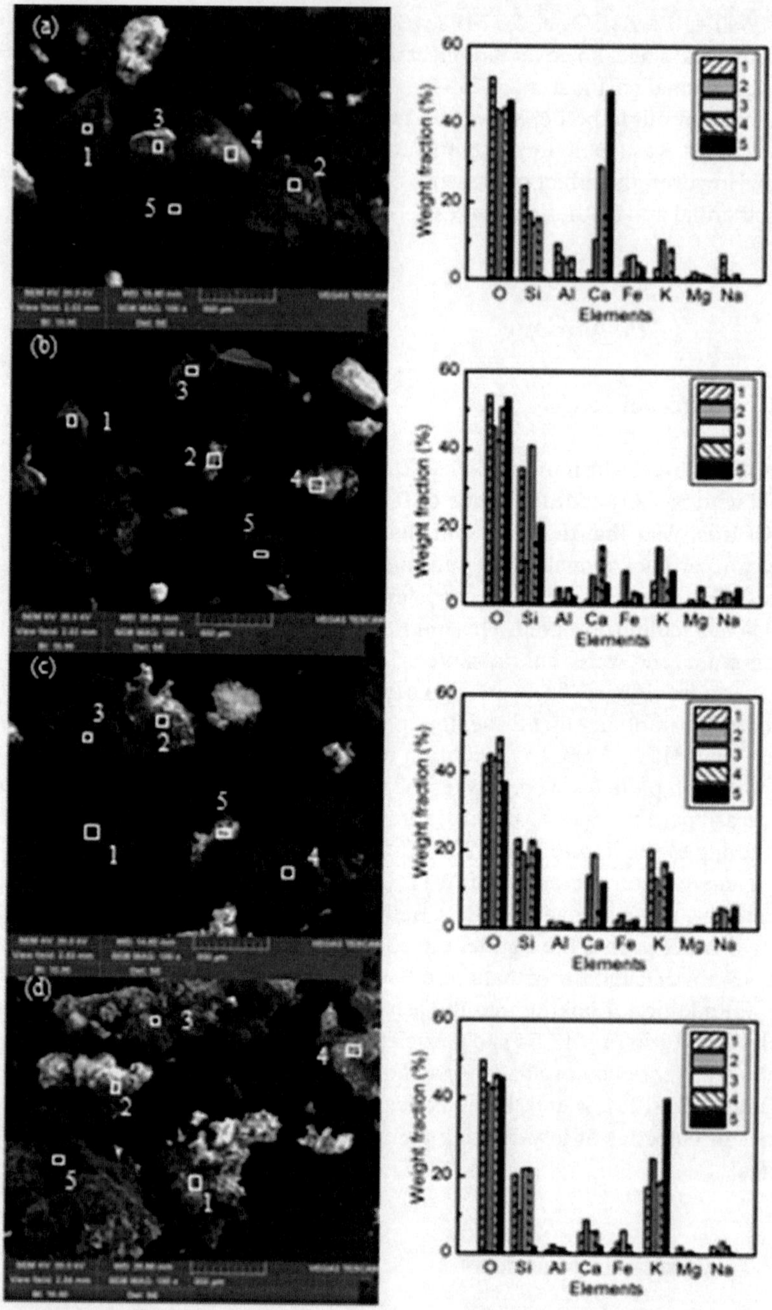

Fig. 13.5 Morphology of the obtained ashes during CO_2 gasification at 750 and 3.5 MPa of **a** WJT, **b** WJT+1% K_2CO_3, **c** WJT+5% K_2CO_3, and **d** WJT+10% K_2CO_3

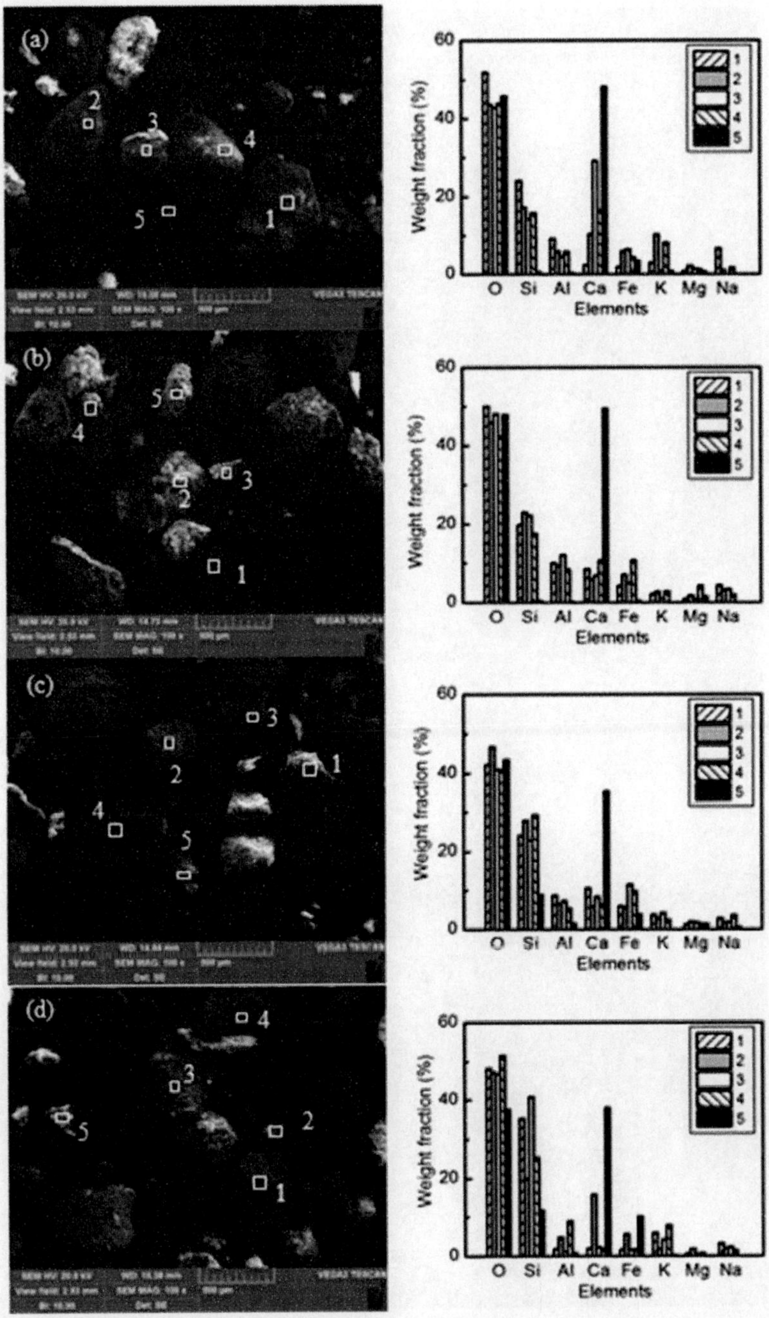

Fig. 13.6 Morphology of the obtained ashes during CO_2 gasification of WJT at 3.5 MPa, and temperatures of **a** 750 °C, **b** 800 °C, **c** 850 °C, and **d** 900 °C

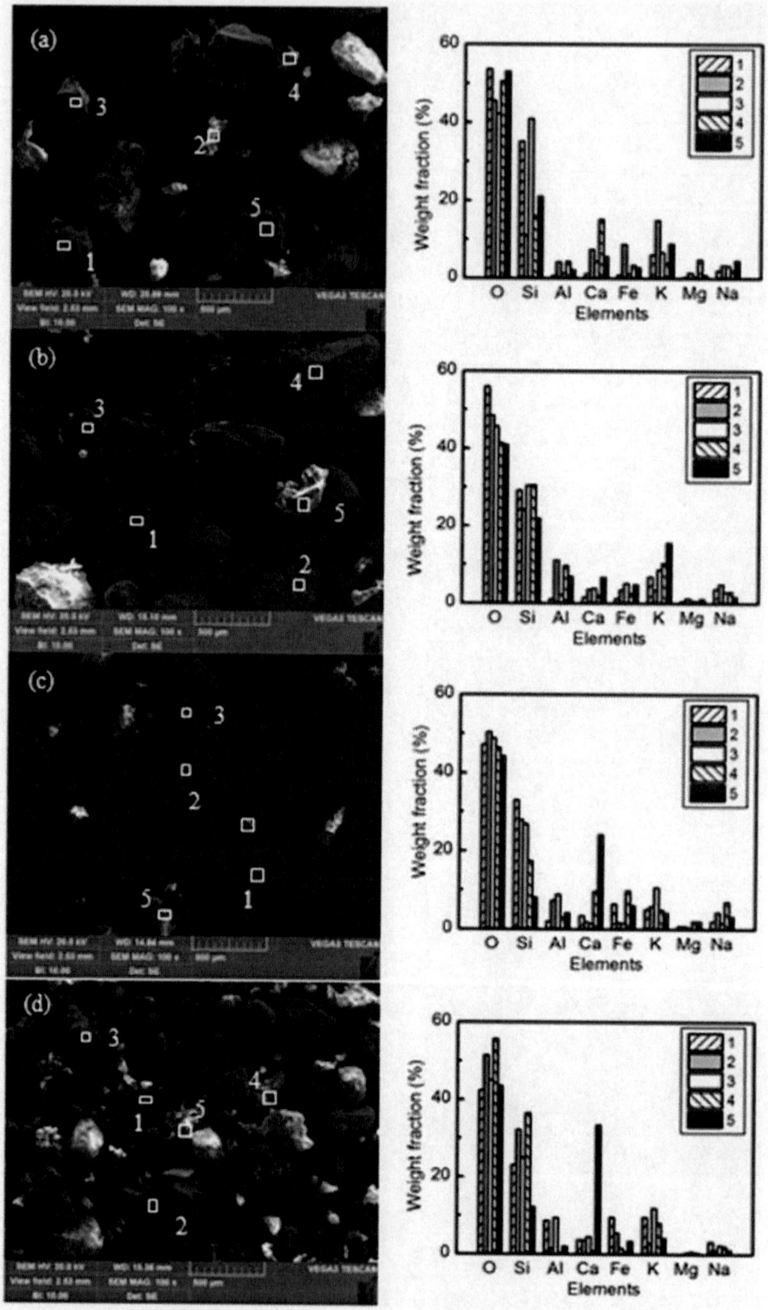

Fig. 13.7 Morphology of the obtained ashes during CO_2 gasification of WJT+1% K_2CO_3 at 3.5 MPa and temperatures of **a** 750 °C, **b** 800 °C, **c** 850 °C, and **d** 900 °C

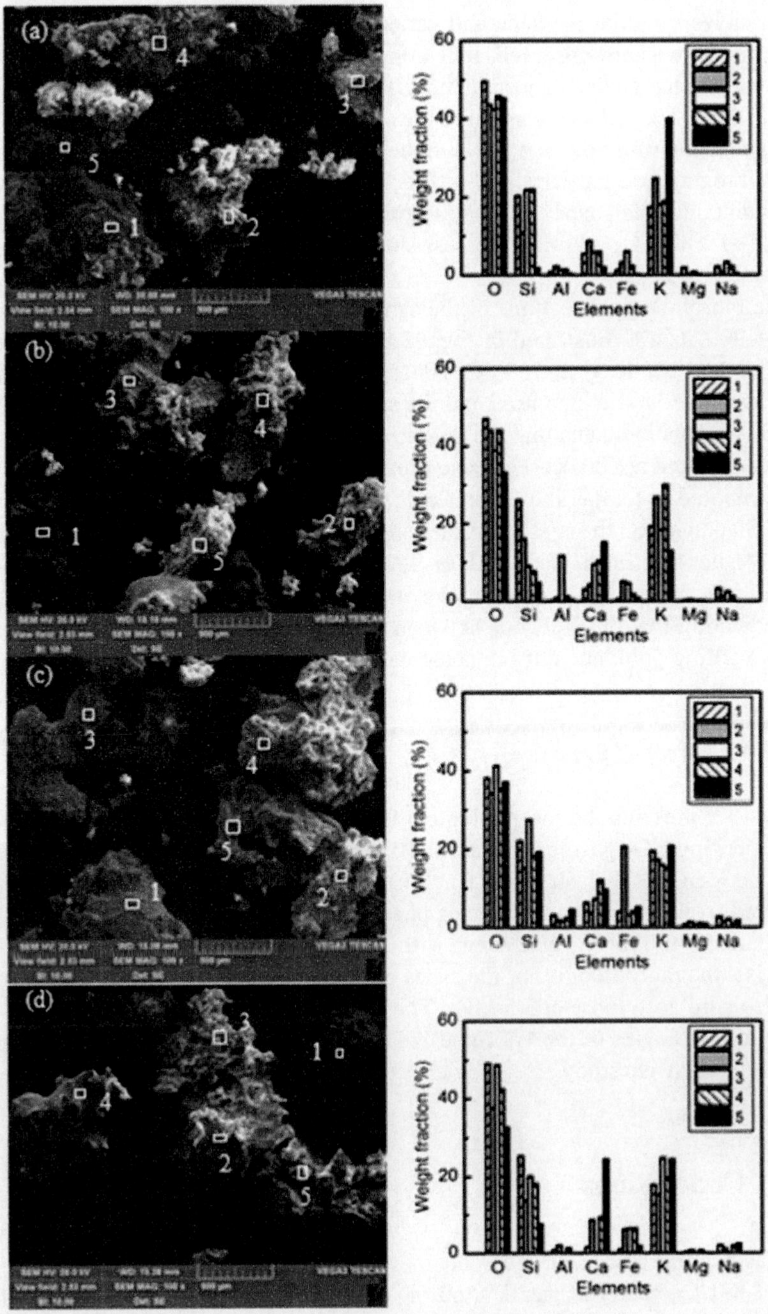

Fig. 13.8 Morphology of the obtained ashes during CO_2 gasification of WJT+10% K_2CO_3 at 3.5 MPa and temperatures of **a** 750 °C, **b** 800 °C, **c** 850 °C, and **d** 900 °C

ash particles irregular in shape and varied in sizes. In particular, sintering between ash particles was not observed, indicating that the ash of WJT was resistant to the temperature change during gasification at 750–900 °C. Similarly, the ash morphologies of WJT+1% K_2CO_3 were quite the same, without significant differences being observed. Sintering between ash particles might occur as certain ash particles had sintered together (e.g., particle 3 in Fig. 13.7b). However, the remaining ash particles maintained its shapes and sizes during gasification. These imply that the ash of WJT+1% K_2CO_3 addition was less sintered and also resistant during gasification temperatures.

The morphological features of the ash of WJT+10% K_2CO_3 during gasification at 750–900 °C are illustrated in Fig. 13.8. It is obvious that sintering between ash particles had occurred at 750 °C. As temperature increased, the sizes of the ash particles tended to be increased and the surface of the ash particles became smooth at 900 °C. This indicates that the degree of ash sintering increased and more ashes had been melted at 900 °C. This also proves that ash sintering of WJT+10% K_2CO_3 was promoted as temperature increased from 750 to 900 °C.

EDS analysis on these ash particles showed that the ashes of WJT and WJT+1% K_2CO_3 ether rich in Si, Ca, Si/Al, or Si/Al/K (Particles in Figs. 13.6 and 13.7). As 10% K_2CO_3 was added into the lignite, its ash particles were composed of K, Si, and Ca, indicating that the eutectics between K, Si, and Ca were formed. The presence of low melting point eutectics thus increased sintering between ash particles.

13.3.2.3 Effect of Pressure

Figure 13.9 presents the morphological features of the ash of WJT and WJT+10% K_2CO_3 during CO_2 gasification at 2.0 MPa and temperatures of 750 °C and 850 °C. It can be seen that the ash of WJT at 750 °C was composed of ashes with different sizes and irregular in shape. Similar phenomena was also observed for the lignite ash obtained at 850 °C, consistent with the morphologies of the ash at 3.5 MPa. Likewise, the morphologies of the ashes of WJT+10%K_2CO_3 at 750 °C and 850 °C was also similar to those at 3.5 MPa. These suggest that the effect of pressure alone on the morphologies of the WJT and WJT+10%K_2CO_3 ashes might require further and comprehensive study.

13.4 Conclusions

The gasification characteristics and morphology of the residue ash during pressurised K_2CO_3-catalytic gasification of a lignite at CO_2 atmosphere but different temperatures, pressures and K_2CO_3 addition ratio was investigated a High-Pressure Thermogravimetric Analyser. Results showed that the addition of K_2CO_3 promoted gasification of the lignite. The effect of K_2CO_3 on gasification was not evident when K_2CO_3 addition ratio was 1%, but became significant when K_2CO_3 addition ratio

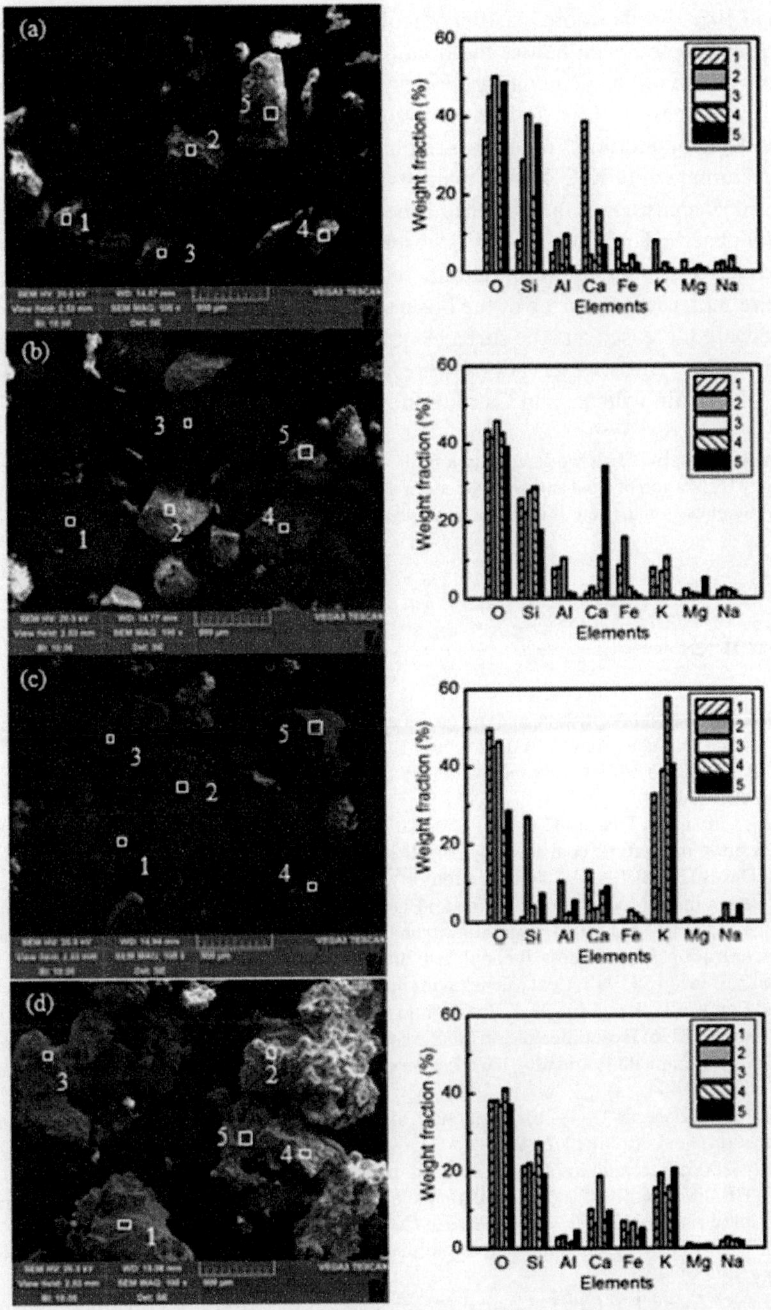

Fig. 13.9 Morphology of the ashes during CO_2 gasification at 2.0 MPa of **a** WJT at 750 °C **b** WJT at 850 °C, **c** WJT+10% K_2CO_3 at 750 °C, and **d** WJT+10% K_2CO_3 850 °C

reached 10%. Furthermore, gasification of the lignite with 10% K_2CO_3 addition was almost completed even before the final temperature 750 °C was reached, bringing the conversion index t_{50} much earlier than those without K_2CO_3 addition. Moreover, both the increase in final temperature and pressure benefited gasification, regardless of K_2CO_3 addition. Yet, pressure played a significant role in gasification at all temperatures, while K_2CO_3 was mainly active at temperatures no less than 750 °C. SEM–EDS analysis on the obtained ashes revealed that sintering of the lignite ash was not observed at 750 °C, but was observed and the degree of ash sintering further aggravated as K_2CO_3 addition ratio increased to 10%. Likewise, as temperature and pressure increased, the sizes of the observed ash particles of the 10% K_2CO_3-doped lignite were increased and the surfaces of the ashes became smooth, incurring severe ash sintering. These observations were attributed to the presence of low-melting points K-aluminosilicate and Ca-aluminosilicate in the ash as identified.

Acknowledgements This work was supported by the Foundation of State Key Laboratory of High-efficiency Utilization of Coal and Green Chemical Engineering (Grant No. 2017-K01), and National Natural Science Foundation of China (Grant No. 51706028).

References

Akyurtlu JF, Akyurtlu A (1995) Catalytic gasification of Pittsburgh coal char by potassium sulphate and ferrous sulphate mixtures. Fuel Process Technol 43(1):71–86

Bai J, Li W (2011) Effects of mineral matter and coal blending on gasification. Energy Fuels 25(3):1127–1131

Bruno G, Carvani L, Passoni G (1986) Correlation between potassium losses and mineral matter composition in catalytic coal steam gasification. Fuel 65(10):1473–1475

Chen J, Dai B (2012) XANES investigation on sulfur evolution during victorian brown coal char gasification in oxy-fuel combustion mode. Energy Fuels 26(8):4775–4782

Duchesne MA, Hall AD (2014) Fate of inorganic matter in entrained-flow slagging gasifiers: fuel characterization. Fuel Process Technol 118:208–217

Erickson T, Benson S (1992) Coal ash behavior in reducing environments. North Dakota University, Grand Forks, ND (United States). Energy and Environmental Research Center

Fan S, Yuan X (2016) Experimental and kinetic study of catalytic steam gasification of low rank coal with an environmentally friendly, inexpensive composite K_2CO_3–eggshell derived CaO catalyst. Fuel 165:397–404

Formella K, Leonhardt P (1986) Interaction of mineral matter in coal with potassium during gasification. Fuel 65(10):1470–1472

Gupta RP (2005) Research in Newcastle—past, present and future. Fuel 84(10):1176–1188

Hattingh BB, Everson RC (2011) Assessing the catalytic effect of coal ash constituents on the CO_2 gasification rate of high ash, South African Coal. Fuel Process Technol 92(10):2048–2054

Irfan MF (2011) Coal gasification in CO_2 atmosphere and its kinetics since 1948: a brief review. Energy 36(1):12–40

Jaffri G-e-R, Zhang J-Y (2008) Catalytic gasification characteristics of mixed black liquor and calcium catalyst in mixing (air/steam) atmosphere. J Fuel Chem Technol 36(4):406–414

Jing N, Wang Q (2011a) Effect of different reaction atmospheres on the sintering temperature of Jincheng coal ash under pressurized conditions. Fuel 90(8):2645–2651

Jing N, Wang Q (2011b) Effect of chemical composition on sintering behavior of Jingcheng coal ash under gasification atmosphere. Chem Eng Commun 199(2):189–202
Jing N, Wang Q (2013a) Effect of temperature and pressure on the mineralogical and fusion characteristics of Jincheng coal ash in simulated combustion and gasification environments. Fuel 104:647–655
Jing N, Wang Q (2013b) The sintering behavior of coal ash under pressurized conditions. Fuel 103:87–93
Jing N, Zhu M (2016) Effect of ash preparation method on the sintering characteristics of ashes from combustion of coal and biomass blends. Fuel 186:830–837
Kühn L, Plogmann H (1983) Reaction of catalysts with mineral matter during coal gasification. Fuel 62(2):205–208
Lee WJ, Kim SD (1995) Catalytic activity of alkali and transition metal salt mixtures for steam-char gasification. Fuel 74(9):1387–1393
Li F, Huang J (2010) Formation mechanism of slag during fluid-bed gasification of lignite. Energy Fuels 25(1):273–280
Li J, Zhu M (2017) Effect of coal blending and ashing temperature on ash sintering and fusion characteristics during combustion of Zhundong lignite. Fuel 195(Supplement C):131–142
Liang D, Xie Q (2018) Catalytic effect of alkali and alkaline earth metals in different occurrence modes in Zhundong coals. Asia-Pacific J Chem Eng 13(3):e2190
Liu Y, Guan Y-J (2018a) Gasification reactivity and morphology of coal chars formed in N_2 and CO_2 atmospheres. Chem Pap 7(8):2045–2054
Liu Y, Guan Y (2018b) CO_2 gasification performance and alkali/alkaline earth metals catalytic mechanism of Zhundong coal char. Korean J Chem Eng 35(4):859–866
Nzihou A, Stanmore B (2013) A review of catalysts for the gasification of biomass char, with some reference to coal. Energy 58:305–317
Sharma A, Takanohashi T (2008) Effect of catalyst addition on gasification reactivity of HyperCoal and coal with steam at 775–700°C. Fuel 87(12):2686–2690
Takematsu T, Maude C (1991) Coal gasification for igcc power generation. IEACR/37. IEA Coal Research, London
van Dyk JC, Waanders FB (2007) Manipulation of gasification coal feed in order to increase the ash fusion temperature of the coal enabling the gasifiers to operate at higher temperatures. Fuel 86(17–18):2728–2735
Wall TF, Liu GS (2002) The effects of pressure on coal reactions during pulverised coal combustion and gasification. Prog Energy Combust Sci 28(5):405–433
Wang J, Jiang M (2009) Steam gasification of coal char catalyzed by K_2CO_3 for enhanced production of hydrogen without formation of methane. Fuel 88(9):1572–1579
Wu Y, Wang J (2011) Potassium-catalyzed steam gasification of petroleum coke for H2 production: reactivity, selectivity and gas release. Fuel Process Technol 92(3):523–530
Ye DP, Agnew JB (1998) Gasification of a South Australian low-rank coal with carbon dioxide and steam: kinetics and reactivity studies. Fuel 77(11):1209–1219
Zhang L-X, Kudo S (2013) Catalytic effects of Na and Ca from inexpensive materials on in-situ steam gasification of char from rapid pyrolysis of low rank coal in a drop-tube reactor. Fuel Process Technol 113:1–7
Zhang J, Zhang L (2015a) Effect of bauxite additives on ash sintering characteristics during the K_2CO_3-catalyzed steam gasification of lignite. RSC Adv 5(9):6720–6727
Zhang J-G, Zhang L (2015b) Ash Sintering behavior of lignite in catalyzed steam gasification with kaolin as additive. Energy Technol 3(6):556–562
Zhao Y, Zhang W (2018) Kinetic characteristics of in-situ char-steam gasification following pyrolysis of a demineralized coal. Int J Hydrogen Energy 43(24):10991–11001
Zhao Y, Feng D, Zhang Y, Huang Y, Sun S (2016) Effect of pyrolysis temperature on char structure and chemical speciation of alkali and alkaline earth metallic species in biochar. Fuel Process Technol 141(Part 1):54–60

Chapter 14
Ash Deposition and Slagging Behavior of Xinjiang High-Alkali Coal in a 20MW$_{th}$ Cyclone Combustion Test Facility

Xiaojiang Wu, Yigong Zhou, Yezhu Sun, Zhongxiao Zhang, Mingqiang Li, Xiang Zhang, Kai Yan, Yuehua Li, Nan Chen, and Xinglei Hu

Abstract The high content of AAEMs (alkali and alkaline earth metals) in ash as well as its relating serious ash fouling slagging problems are the main reason for the unit boiler, which cannot absolutely fully use high-alkali coal at full generating capacity. In order to fully use those kinds of coal, the cyclone-fired boiler with slagging tap is induced in this paper. The ash melting behavior as well as the migration and transformation characteristics of AAEMs during cyclone-fired process were performed in a 20MWth cyclone combustion test facility. A number of slagging and deposit samples at different locations of boiler were selected and analyzed by using XRF, XRD and SEM–EDX analysis. The experimental results showed that the slag in cyclone burner contains higher content of Fe_2O_3, around 37.5–52.6 wt%, which is about 2.4 and 3.5 times than that of their parent coal ash. The content of Na_2O in slag is around 2.3–8.6 wt%, especially for the slag at location C, D and E, where can provide sufficient reaction time and contact area between flue gas and slag. The main existence form of sodium is Na–Al–Si eutectic compounds and being completely melted in cyclone with smooth surface. For the deposit ash on convective tubes, some Na–Al–Si and Ca–Mg–Si low melting eutectic compounds have a rounded boundary in the main particles due to the low melting temperature and low viscosity, which can adhere to the surface of other ash particles to agglomerate each other.

X. Wu · X. Zhang · K. Yan · Y. Li · N. Chen
Technical Department, Shanghai Boiler Works Co., Ltd, Shanghai 200245, People's Republic of China

Y. Zhou
Shangh Electric Power Generation Group, Shanghai 200240, People's Republic of China

Y. Sun
Huaneng International Power Co., Ltd, Beijing, People's Republic of China

X. Wu · Z. Zhang (✉) · M. Li · X. Hu
School of Mechanical Engineering, Shanghai Jiaotong University, Shanghai 200240, People's Republic of China
e-mail: zhzhx222@163.com; wxj9631@163.com

J. Lyu and S. Li (eds.), *Clean Coal and Sustainable Energy*, Environmental Science and Engineering,
https://doi.org/10.1007/978-981-16-1657-0_14

Keywords Alkali coal · Ash deposition · Slagging · Cyclone fired boiler

14.1 Introduction

The Xinjiang Zhundong high-alkai coal with a predicted reserve of 390 billion tons, which can be used around 100 years as a current coal consumption rate, can give a great support for an increasing energy demand of China (Zhang et al. 2013). However, the high content of AAEMs (alkali and alkaline earth metals) in ash as well as its relating serious ash fouling slagging problems is the main reason for the unit boiler, which cannot absolutely fully use high-alkali coal at full generating capacity (Wei et al. 2018). The most units have to be operated under a low generating capacity (70–80% BMCR) or use coal blends or silica-based additive which incurs an increased operating cost and limits the utilization of Xinjiang coal as a cheap fuel. The previous studies on zhundong coal ash fusion characteristics has shown that the high content of Na_2O, CaO and SO_3 in ash, and around 60% or above sodium was released during initial combustion stage, the rest of sodium are sodium aluminosilicates with lower melting temperature. The condensation of gaseous sodium salts will cause the serious ash deposition on the heat transfer surface and the molten sodium aluminosilicates will cause serious ash slagging problems in furnace (Wu et al. 2016; Wei et al. 2017). Furthermore, the deterioration of slagging can reduce the heat absorption in furnace and further improve the flue gas temperature at furnace outlet. The higher temperature of flue gas at convective zone of boiler can also aggravate the ash fouling behavior on the heat transfer tube surface due to the temperature is the most important factor affecting ash fouling behavior for high alkali coal (Kondratieva 2018). In order to fully use those kinds of coal, the cyclone-fired boiler with slagging tap were proposed in this paper. Since the high-alkali coal has strong slagging tendency, most of its ash, around 80% of ash content, can be melted in cyclone burner and tapped from the bottom of furnace. And then the ash concentration in flue gas can be reduced sharply as comparing with normal solid tap boiler, which is around 20–30% of ash was tapped from the bottom of furnace. On the other hand, the gas phase of AAEMs (Na, K, Ca, Fe etc.) can react with some aluminosilicates in slag to form some Na-, Fe- bearing minerals or some sodium sulfates in cyclone burner and be tapped from the bottom of furnace. The gas phase of AAEMs in flue gas can also be reduced and alleviate the ash fouling problems in boiler correspondingly. In order to verify the assumption above, the coal combustion and slagging producing process in cyclone burner were performed in a $20MW_{th}$ cyclone combustion test facility. The ash deposition/slagging behavior as well as the AAEMs migration and transformation mechanism were studied through testing a number of slagging and deposit samples at different location of boiler by using XRF, XRD and SEM–EDX etc.

14.2 Experimental

14.2.1 Coal Properties

One typical Xinjiang coal, namely JJM coal with high contents of sodium, magnesium and iron in ash was selected to use in this study. The properties of JJM coal are illustrated in Table 14.1. The contents Na_2O, MgO and Fe_2O_3 in ash are reached at 7.66%, 9.55% and 15.61% respectively. The B/A value, referring to the ratio of basic to acidic oxides in ash reaches 1.38. The higher B/A value is a direct sign of the strong fouling/slagging propensity of this ash sample that is prone to form low temperature melting compounds such as alkali silicates. The ash melting temperature was tested according to Chinese national standard (GB/T219-2008), the DT of ash is 1221 °C, the temperature of ST and FT are 38 and 43 °C higher than DT respectively. The short temperature range between DT and ST or FT means that the ash can be melted form initial melting to complete molten in a narrow temperature window (from 1221 to 1264 °C). Therefore, the serious slagging tendency can be reached at high temperature in the furnace. Furthermore, in order to make sure that the slag can be easily removed from the bottom of furnace, the slag flow characteristics play a critical role in stable operating of slag tap boiler. The critical viscosity and T_{cv} are most important factors for slagging taped boiler design and operating. Generally, if the T_{cv} is less than 1400 °C, it can provide steady operating conditions in boiler due to the continuously flows down to a tap hole at the bottom of furnace (Kondratieva 2018; Chen et al. 2017). The viscosity-temperature characteristic of JJM coal ash is shown in Fig. 14.1. As can be seen from Fig. 14.1 that the T_{cv} of JJM ash is 1258 °C,

Table 14.1 Coal properties of test coal

JJM coal			JJM coal		
Proximate analysis (wt%)			Ash composition (wt%)		
C_{ad}	%	62.37	SiO_2	%	26.52
H_{ad}	%	3.13	Al_2O_3	%	8.71
O_{ad}	%	10.64	Fe_2O_3	%	15.61
N_{ad}	%	0.65	CaO	%	15.95
$S_{t,ad}$	%	0.47	MgO	%	9.55
A_{ad}	%	6.18	K_2O	%	0.52
M_{ad}	%	16.74	Na_2O	%	7.66
V_{ad}	%	33.58	TiO_2	%	0.48
FC_{ad}	%	43.50	SO_3	%	14.43
$Q_{b,ad}$	MJ/kg	23.64	B/A	–	1.38
Ash fusion temperature (reducing atmosphere, °C)					
DT	°C	1221	ST	°C	1259
HT	°C	1260	FT	°C	1264

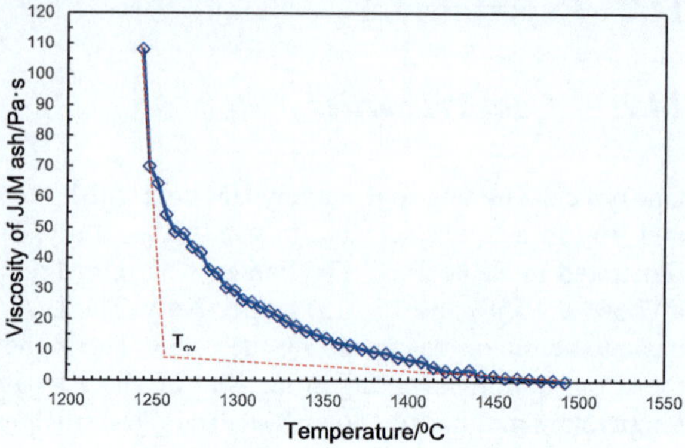

Fig. 14.1 The viscosity-temperature characteristics of JJM ash

which is lower than 1400 °C, and the slag of JJM coal can be easy to remove from the bottom of furnace.

14.2.2 20MWth Pilot Plant Test Facility

The experiments were performed in a 20MWth pilot plant facility, which mainly consists of coal milling system, cyclone-fired burner, furnace, fan system, water cooling system, flue gas on-line analyser, slaging quench treatment system and other auxiliary systems visualised in Fig. 14.2. The 20MWth cyclone-fired burner was designed with water-cooled wall lined with refractory lining. In the cyclone-fired burner, the coal ash will be melted due to the high combustion temperature (>1500 °C) in it. Furthermore, most melted ash particles can be thrown onto the refractory wall and further formed a slag layer on it under centrifugal force and high temperature on it. With the thickness of slag layer increasing, the slag will overflow from the cyclone and flow into the bottom of the furnace. The burned-out flue gas with some fine particles (un-burned carbon or some fine ash particles) were flow out the burner and continuing combust to maintain the the high temperature at bottom of furnace. After the molten ash particles were captured through the slag screen tubes at the bottom of the furnace, the flue gas with solid particles will enters the convection zone of furnace. The water-cooled panels were set up in the convective zone of furnace to simulate the ash deposition behavior on the tubes. During the test, the coal samples were pulverised with around 90% of mass fraction less than 138 and 45.7 μm average particle size (see Fig. 14.4). The pulverized coal flow with feeding rate at 4t/h were entrained in cyclone and rotated with rotating secondary air together. The pulverized coal mixed strongly with the secondary air and formed an intense swirling combustion flame. Some coal particles and melted or partial melted ashes were thrown on the wall and mixed with slag and air to form a strong slag film combustion. The excess oxygen in exhaust flue gas was controlled at around 3, 4 vol%. The detailed operation conditions are listed in Table 14.2. To mimic the

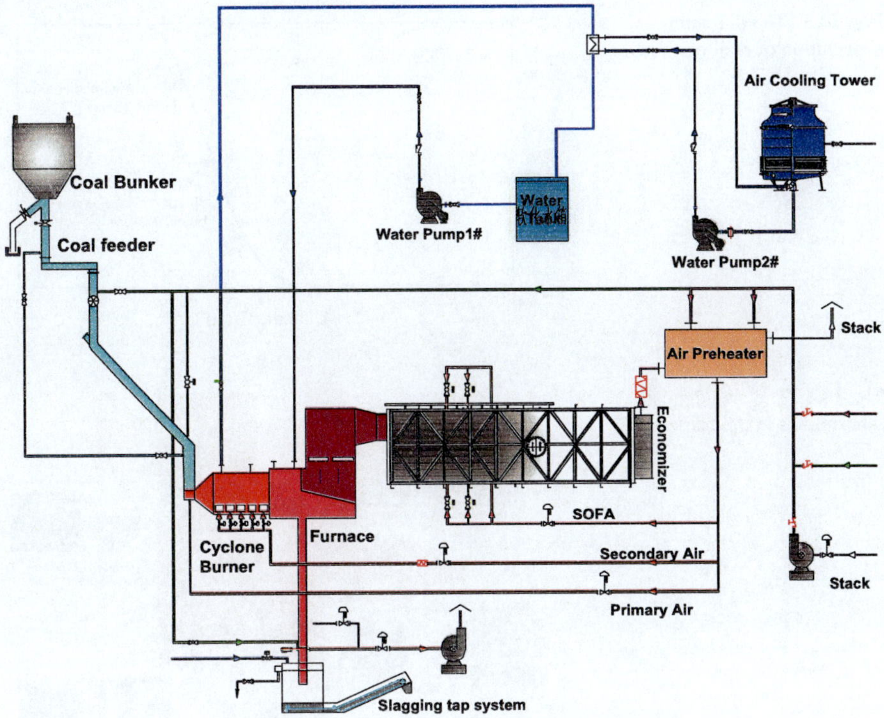

Fig. 14.2 Schematic diagram of 20MW$_{th}$ cyclone combustion test facility

ash deposition/slagging behavior in a real boiler, the heat transfer tubes with water cooling were installed at various locations in the furnace. After long time continue operating, the slag/deposit ash samples at various location were selected and analysed by using XRF, XRD and SEM–EDX test etc. (Fig. 14.3).

Table 14.2 Experimental operating conditions

Item	Unit	JJM coal
Heat input	MW	20.00
Coal feeding rate	kg/h	4000
Excess air ratio in cyclone	–	1.2
Oxygen in flue gas at exist of furnace	%	3, 4%
Flue gas temperature at exist of furnace	°C	~1000
Temperature of cyclone	°C	1350–1400
Operating time	h	~4 h

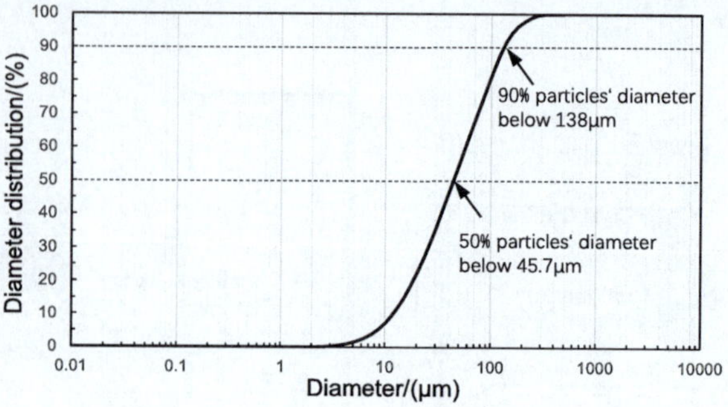

Fig. 14.3 The diameter distribution of coal particles

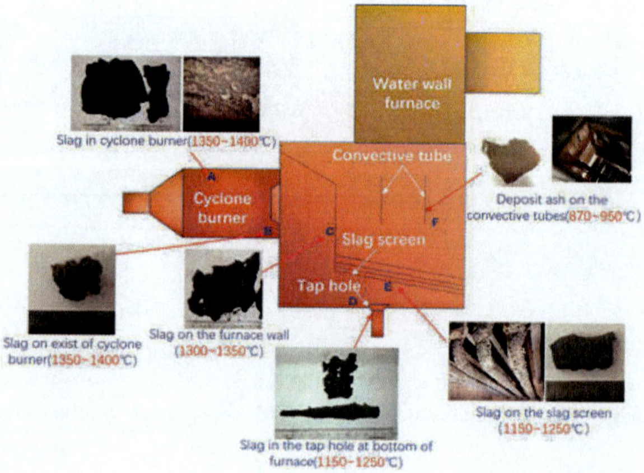

Fig. 14.4 The slag/deposit ash samples in the furnace

14.3 Results and Discussion

14.3.1 Slagging Behavior of Ash During Cyclone Combustion Process

The slag samples in cyclone burner and bottom of furnace were collected and tested by XRF, XRD and SEM–EDX analysis for exploring the migration and transformation characteristics of Na, K, Ca and Fe etc., during the slaging process. The photos and chemical composition of slag/deposit ash samples at different locations, which labeled as A, B, C, D, E, F were shown in Figs. 14.4 and 14.5 respectively. Furthermore, the enrichment factors (EF) of individual metals of all deposits is tabulated in Table 14.3. The EF value of an element was calculated by dividing its fraction in an ash deposit by the respective amount in the parent coal ash. As it can be seen from Figs. 14.4, 14.5 and Table 14.3, the Fe_2O_3 is highly enriched in the slag, which is accounted for 25.7–52.6 wt% in slag and the corresponding EF value varying from 1.65 to 3.16. The Na_2O in slag is varying for 1.57–8.6 wt%, especially for the slag at C, D and E, where can provide sufficient reaction time and contact area between

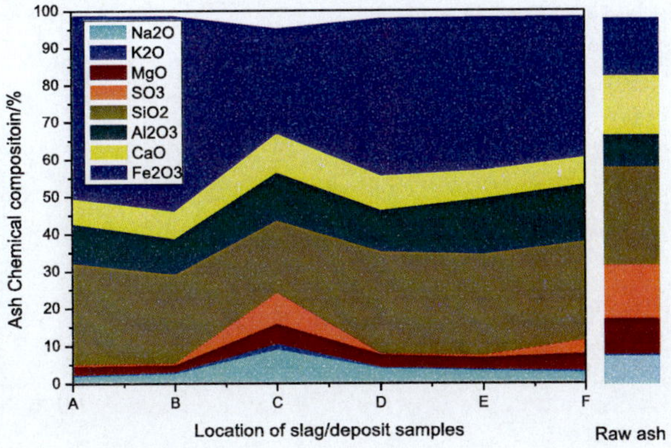

Fig. 14.5 Ash composition of slag/deposit samples

Table 14.3 EF of ash samples at different locations

Location	Na_2O	K_2O	MgO	SO_3	SiO_2	Al_2O_3	CaO	Fe_2O_3	Ti_2O
A	0.21	0.65	0.25	0.04	1.02	1.21	0.42	3.16	1.15
B	0.30	0.79	0.20	0.04	0.90	1.10	0.45	3.37	1.15
C	1.13	3.07	0.56	0.57	0.73	1.47	0.64	1.83	0.79
D	0.50	0.58	0.36	0.02	1.03	1.27	0.57	2.72	1.10
E	0.40	0.73	0.33	0.04	1.02	1.72	0.47	2.64	1.17
F	0.34	1.06	0.47	0.25	0.99	1.75	0.45	2.43	1.06

flue gas and slag. The gas phase sodium can react with liquid slag to form some low melting minerals such as Na–Al–Si minerals and sodium sulfates etc. and cause the relative high content of Na_2O in the slag at location of C, D and E in the furnace.

14.3.2 *Ash Deposition Behavior in the Convective Zone of Furnace*

The deposit ash on the convective tubes looks much looser and with high content of Al_2O_3 and low content of Na_2O in it. The B/A value of ash samples at at different locations were shown in Fig. 14.6. As it can be seen from Fig. 14.6, the B/A value of ash samples in cyclone and the bottom of furnace is higher than that of raw coal ash, the B/A value is ranged from 1.51 to 1.90, which has very strong slagging tendency, while the B/A value of samples E and F are relative low in the range from 1.25 to 1.30. The most alkali or alkali earth metals (Na, K, Fe, Ca) were formed as low melting minerals or salts and then made the deposits on convective tubes had relative high content of acidic oxides. The mineral composition of slag in cyclone (sample B) and deposit ash on convective tubes (sample F) was tested by XRD analysis. As it can be seen from Fig. 14.7, the main crystalline mineral phase of slag is augite,

Fig. 14.6 The B/A value of slag/deposit samples

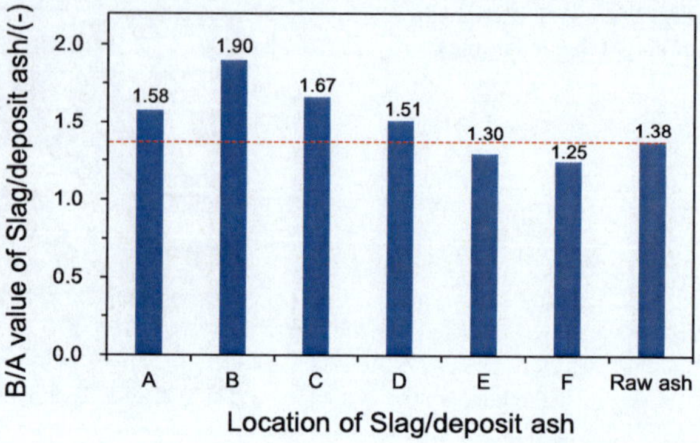

Fig. 14.7 XRD patterns of slag and deposit ash

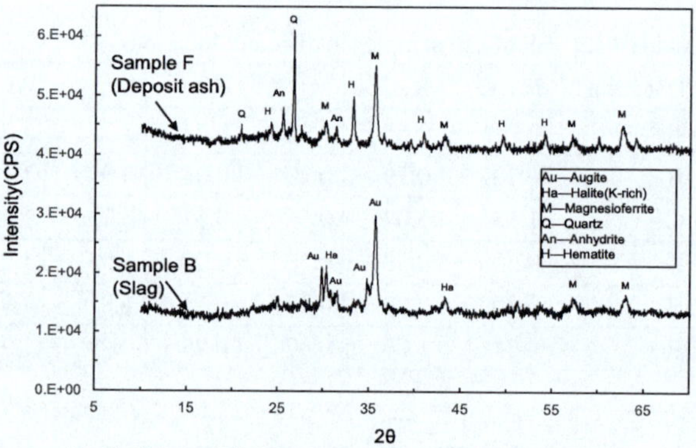

Au-Ca(Mg,Fe)Si_2O_6, Ha-$K_{0.4}Na_{0.6}Cl$, M-$MgFe_2$+$3O_4$
Q-SiO_2, An-$CaSO_4$, H-Fe_2O_3

magnesioferrite and halite, the other contents are mainly in an amorphous phase due to the totally melted. While for the the deposit ash, the main crystalline mineral phases are quartz, magnesioferrite, anhydrite and hematite. usually, the quartz is one of the main refractory minerals, which cause high melting temperature. The main low melting minerals were melted and go through with slag, the rest hard to melted ash will go with flue gas to the convective zone of boiler. This is one of the main reason for slagging tap boiler to use the high alkali coal.

14.3.3 Ash Slagging/Deposition Mechanism of Zhundong Coal Ash at Cyclone-Fired Condition

In order to identify the detail mechanism of ash melting behavior as well as the migration and transformation characteristics of AAEMs during cyclone-fired process, the slag in cyclone burner (sample B) and deposit ash at convective tubes (sample F)

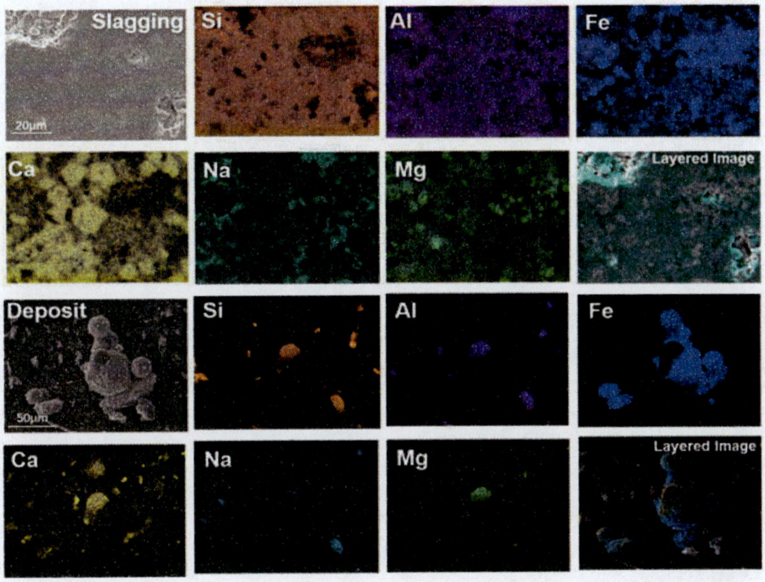

Fig. 14.8 SEM images and elemental mapping for slag and deposit ash

at different locations were tested by SEM–EDX for detail information on the slag and deposit ash structure and chemical distribution. The SEM images and elemental mapping of slag and deposit ash were shown in Fig. 14.8. For the slag in cyclone, the slag in cyclone has been completely melted, some Fe-bearing crystal minerals are dispersed in the slag and condensed to form iron-rich crystals during the slag cooling process. The elements Na, Al, Si looks like the similar distribution in the slag and mainly formed ash Na–Al–Si low melting eutectic compounds. For the deposit ash on convective tubes, Na–Al–Si and Ca–Mg–Si low melting eutectic compounds have a rounded boundary due to the low melting temperature and low viscosity, which are easy to adhere to the surface of large ash particles to agglomerate ash particles.

14.4 Conclusions

The ash melting behavior as well as the migration and transformation characteristics of AAEMs during cyclone-fired process were performed in a 20MW$_{th}$ cyclone combustion test facility. The major conclusions can be drawn as follows:

(1) The slag in cyclone burner contains higher content of Fe_2O_3, around 37.5–52.6wt%, which is about 2.4 and 3.5 times than that of their parent ash. The main Fe-bearing crystal minerals are dispersed in the slag and condensed to form iron-rich crystals during the slag cooling process.

(2) The content of Na_2O in slag is around 2.3–8.6wt%, especially for the slag at C, D and E, where can provide sufficient reaction time and contact area between flue gas and slag. The main existance form of sodium is Na–Al–Si eutectic compounds and being completely melted with smooth surface.

(3) The For the deposit ash on convective tubes, Na–Al–Si and Ca–Mg–Si low melting eutectic compounds have a rounded boundary due to the low melting temperature and low viscosity, which are easy to adhere to the surface of large ash particles to agglomerate ash particles. It is because the relative high content of SiO_2, Al_2O_3 in ash, most ash particles exist as solid and with relative weak ash slagging tendency.

Acknowledgements This work was supported by the National Key R&D Projects (2018YFB0604104), Program of Shanghai Technology Research Leader (19XD1431000) and Minhang District Leading Talents Project.

References

Chen XD, Kong LX, Bai J (2017) The key for sodium-rich coal utilization in entrained flow gasifier: the role of sodium on slag viscosity-temperature behavior at high temperatures. Appl Energy 206:1241–1249

Kondratieva A (2018) Ilyushechkinb A: flow behaviour of crystallising coal ash slags: shear viscosity, non-Newtonian flow and temperature of critical viscosity. Fuel 224:783–800

Wei B, Tan HZ, Wang YB et al (2017) Investigation of characteristics and formation mechanisms of deposits on different positions in full-scale boiler burning high alkali coal. Appl Therm Eng 119:449–458

Wei B, Tan HZ, Wang XB et al (2018) Investigation on ash deposition characteristics during Zhundong coal combustion. J Energy Inst 91:33–42

Wu XJ, Zhang X, Yan K et al (2016) Ash deposition and slagging behavior of Chinese Xinjiang high-alkali coal in 3 MW_{th} pilot-scale combustion test. Fuel 181:1191–1202

Zhang SY et al (2013) Situation of combustion utilization of high sodium coal. Proc CSEE 33:1–12

Chapter 15
Release Characteristics of Alkali Metals in Oxygen-Enriched Combustion of a Single Char Particle with Random Pore Model

Sibo Qu, Haiming Wang, and Changfu You

Abstract Alkali metals released from pulverized coal combustion and gasification are one of the important causes of ash-related problems, such as fouling, slagging, corrosion and atmospheric pollution. To investigate the release characteristics of alkali metals, a theoretical model was developed for the combustion of single char particle based on Random Pore Model (RPM). The particle burnout time, combustion temperature, and NaCl release characteristics of this model were verified with previous experimental results. The combustion of char particle with the size of 50–100 μm was then calculated using this model at the ambient temperature range of 1373 to 1573 K in two different gas atmospheres, i.e. O_2/N_2 and O_2/CO_2 atmospheres. The oxygen content in O_2/N_2 and O_2/CO_2 atmospheres varied from 21% to 100% and 50% to 100%, respectively. When comparing with char combustion in O_2/N_2 atmosphere, the particle burnout time and temperature both decreased in O_2/CO_2 atmosphere, which was also found to be beneficial to the control of alkali metals release. The simulation results showed that smaller particle size, higher temperature, and oxygen-enriched condition promoted char combustion, but in regard of alkali metals release control lower temperature was preferred.

Keywords Alkali metals release · Char combustion · Oxygen-enriched combustion · Random pore model (RPM)

15.1 Introduction

Alkali-rich coal, e.g., Zhundong Coal from Xinjiang, China, has a wide application prospect, because of its huge reserves and high quality characterized by low ash, aluminum, iron, sulfur, and trace element contents (Zhou et al. 2010). However, its high content of alkali metals is the main limitation. Previous research has shown that the Na in ash of Zhundong Coal is generally more than 5%, which is significantly

S. Qu · H. Wang · C. You (✉)
Key Laboratory for Thermal Science and Power Engineering of the Ministry of Education, Department of Energy and Power Engineering, Tsinghua University, Beijing 100084, China
e-mail: youcf@tsinghua.edu.cn

J. Lyu and S. Li (eds.), *Clean Coal and Sustainable Energy*, Environmental Science and Engineering,
https://doi.org/10.1007/978-981-16-1657-0_15

higher than the Na level in typical Chinese coal ranging from 1 to 2% (Yang et al. 2010). Alkali metals have the strongest sublimation tendency for the main mineral content in coal (Sugawara et al. 2002). The release of alkali metals to gas phase, especially Na (Li et al. 2015a), is generally accepted as the primary cause of serious ash-related problems, such as fouling, slagging and corrosion for many industrial systems (Takuwa and Naruse 2007). Further, Alkali metals emission contributes greatly to the formation of atmospheric pollutants, such as fine particulates (particle size less than 1 μm) due to homogeneous nucleation or heterogenous condensation (Takuwa and Naruse 2007; Takuwa et al. 2006; Naruse et al. 2005).

Meanwhile, for conventional utility boilers, oxygen-enriched combustion technology has attracted more attention in industrial application and scientific research. It has the advantages of high combustion efficiency, relatively low pollutants emission, and carbon capture benefits due to the concentrated carbon dioxide (Toftegaard et al. 2010). Char combustion is the main stage of pulverized coal combustion (Smith 1982). Under the condition of oxygen-enriched pulverized coal combustion, the time proportion for volatile matter release is further reduced. The char particle obtained after the coal pyrolysis is featured by its complicated porosity (Yu et al. 2007). Inner surface area is much larger than outer surface area of char particle, which leads to the important effect of inner surface on char combustion (Maffei et al. 2013). The random pore model (RPM) proposed by Bhatia and Perlmutter (1981) successfully described the effect of pore structure variation of solid reactants on the reaction rate by introducing pore structure parameter. Lu and Do (1992, 1994) applied the random pore, percolation, and grain model to predict activation rate of char in air, steam and CO. So far, RPM has been widely adopted and modified by researchers (Campbell et al. 2002). However, few researches were conducted on the coupling of char combustion with alkali metals release under oxygen-enriched condition using numerical methods.

To better understand the oxygen-enriched combustion process and the related alkali metal release characteristics, a theoretical model of small char particle based on chemical reaction kinetics, heat and mass transfer theory with RPM was proposed in this study. The calculation results with the proposed model were compared with previous experimental data for validation. The effects of particle size, ambient temperature, and oxygen ratio in O_2/N_2 and O_2/CO_2 atmospheres on char combustion and alkali metals release were studied. The release of a typical alkali metal, Na, during a single char particle combustion in O_2/N_2 and O_2/CO_2 atmospheres was then discussed.

15.2 Physical Model

A comprehensive theoretical model for temperature and components variation in a single char particle was developed based on heat and mass transfer theory, kinetics of chemical reaction, with the effect of pore structure described by RPM. The random pore sub-model with particle structure description were discussed in Sect. 2.1. The intrinsic kinetics sub-model for chemical reactions and the gasification of Na were discussed In Sect. 2.2.

Since the real combustion of char is very complex, the following assumptions are made to simplify the combustion process and improve the computational efficiency in the case that the particle size is small enough:

A. The ash skeleton exposed during combustion is not destroyed, and the particle size does not change significantly but the density decreases gradually;
B. The homogeneous reaction in the gas boundary layer is neglected under the assumption of single film theory;
C. The thermal conductivity is large enough to consider that the temperature across the whole particle is at the same level;
D. There is no gas concentration gradient between the surface and the internal pores of the particle.

15.2.1 Random Pore Sub-model

For spherical particle of given size (radius, R_p), the apparent surface area, A and volume, V can be described by

$$A = 4\pi {R_p}^2 \tag{15.1}$$

and

$$V = 4/3\pi {R_p}^3 \tag{15.2}$$

Thus, the initial mass of char particle, M_0 is

$$M_0 = \rho_0 V \tag{15.3}$$

where ρ_0 is initial density of char particle.

Assuming that the particle size remains unchanged, as the combustion proceeds, the mass and density of the particles decrease. At time = t, the carbon conversion rate, X_t, is

$$X_t = (M_{c,0} - M_{c,t})/M_{c,0} \tag{15.4}$$

where $M_{c,0}$ is initial mass of carbonaceous matter in char particle, and $M_{c,t}$ is the mass of carbonaceous matter at time = t. Considering that the char particle contains a certain amount of non-combustible matter, $M_{c,0}$ is calculated by

$$M_{c,0} = (1 - X_a)M_0 \tag{15.5}$$

where X_a is mass fraction of non-combustible substance in char particle. The total mass M_t of particle at time = t consists of $M_{c,t}$ and $M_{c,0}$.

The abundant pores of particle provide active surface for the gas–solid chemical reactions and make the gas-phase reactant more easily diffuse into the interior of char particle. Referring to the work of Bhatia and Perlmutter (1981), pore structure can be described by porosity, ε, tortuosity, τ, and specific surface area, S. According to Gil et al. (2011), at time = t, porosity, ε_t is calculate by

$$\varepsilon_t = \varepsilon_0 + (1 - \varepsilon_0)X_t \tag{15.6}$$

where ε_0 is initial porosity of char particle.

In agreement with the porosity evolution, at time = t, tortuosity, τ_t is calculate by

$$\tau_t = \tau_0 + (\tau_0 - 1)X_t \tag{15.7}$$

where τ_0 is initial porosity of char particle, which is taken as 1.414 (Wheeler 1951).

At time = t, specific surface area, S_t is calculated by

$$S_t = S_0 \cdot \frac{\rho_0}{\rho_t}(1 - X_t)\sqrt{1 - \psi \ln(1 - X_t)} \tag{15.8}$$

where S_0 is initial specific surface area of char particle, ρ_t is density of char particle at time = t, and ψ is structural parameter, which is considered to be 3 (Bhatia and Perlmutter 1981).

15.2.2 Chemical Reaction Sub-model

Only irreversible reactions were considered in this work. Assuming there are N reactions and M species are involved in the total chemical reaction system, the relation between reactants and products is:

$$\sum_{m \in M} v_{mn}{}^{(r)} = \sum_{m \in M} v_{mn}{}^{(p)} \tag{15.9}$$

where v is the stoichiometric number.

The production rate of species m ($\dot{\omega}_m$) is:

$$\dot{\omega}_m = \frac{W_m}{1000}\dot{\omega}_m^C = \frac{W_m}{1000}\sum_{n=1}^{N} v_{mn}{}^{(p)} Bk_n \prod_{m \in M} [m]^{v_{mn}{}^{(r)}} \tag{15.10}$$

where W_m is the molecular weight, g/mol; $\dot{\omega}_m{}^C$ is the molar production rate; *[m]* is the molar concentration of species *m*; coefficient *B* is $S_t M_t$ when $n = 1, 2$ or is $\varepsilon_t V$ when $n = 3$; k_n is the reaction rate constant of reaction *n*, which can be evaluated by the Arrhenius equation:

$$k_n = A_n \exp\left(-\frac{E_n}{RT}\right) \tag{15.11}$$

where A_n is pre-exponential factor of reaction k_n, E_n is activation energy of reaction *n*, and *R* is universal gas constant.

The heat production rate, $\dot{q}$ for total reactions is:

$$\dot{q} = -\sum_m^M h_m \dot{\omega}_m = \sum_n^N \dot{q}_n \tag{15.12}$$

where h_m is the enthalpy of species *m* and $\dot{q}_n$ is the heat production rate of reaction *n*.

15.2.2.1 Carbon Transformation

The carbon transformation can be summarized by the heterogeneous and homogeneous reactions showing as follows:

$$2C(s) + O_2(g) = 2CO(g) \quad R.1$$
$$C(s) + CO_2(g) = 2CO(g) \quad R.2$$
$$2CO(g) + O_2(g) = 2CO_2(g) \quad R.3$$

Table 15.1 lists the correlative kinetic parameters for the above three reactions.

As the mass of carbon in the particles decreases, its density variation can be described as:

$$\frac{\partial \rho}{\partial t} = -\sum_{n=1,2} \frac{W_c}{W_m} \frac{\upsilon_{cn}{}^{(r)}}{\upsilon_{mn}{}^{(p)}} \dot{\omega}_m = -\sum_{n=1,2} \frac{W_c}{1000} \upsilon_{cn}{}^{(r)} \dot{\omega}_n^C \tag{15.13}$$

where $\upsilon_{cn}{}^{(r)}$ is the stoichiometric number of carbon in reaction *n*, W_c is the molecular weight of carbon; and $\dot{\omega}_n^C$ is the molar rate of reaction *n*.

Table 15.1 Reaction rate and heat parameters

Reaction *n*	A_n	E_n	Reaction heat
R.1	1.97×10^7	1.98×10^5	2.21×10^5
R.2	1.29×10^5	1.91×10^5	-1.73×10^5
R.3	1.30×10^8	1.26×10^5	5.56×10^5

Table 15.2 Reaction rate and heat parameters

Reaction k	A_n	E_n	Reaction heat
R.4	6.32×10^9	2.81×10^5	-1.81×10^5

15.2.2.2 Alkali Metal Transformation

A series of experimental investigations confirmed that the water soluble alkali metals in the residue decreased with an increase in exposure time at high temperature (Li et al. 2015b; Wang et al. 2018; Kosminski et al. 2006; Song et al. 2016). Water soluble alkali metals such as alkali metal chlorides and organic alkali species are harmful components because of their proven harmfulness, high content and easy conversion to gas phase during the thermal conversion of coal. In this work, we took NaCl as an example to study the release characteristics of alkali metals. The gasification of NaCl solids was primarily determined by chemical reaction kinetics, the feasibility of which was confirmed by Yang (2017). Table 15.2 lists the correlative kinetic parameters that can be used to calculate the mass variation rate of NaCl for the following conversion:

$$\mathrm{NaCl(s)} = \mathrm{NaCl(g)} \quad \mathrm{R.4}$$

15.2.3 Diffusion Sub-model

Mass conservation can describe the variation of gas species in pores of char particle. The gas concentration of species m, C_m is

$$\varepsilon_t V \frac{\partial C_m}{\partial t} = \dot{\omega}_m - K_m A (C_m - C_{m,b}) \tag{15.14}$$

where K_m is mass transfer coefficient of species m, $C_{m,b}$ is ambient concentration of species m.

According to Lv and Liu (2018), K_m can be calculated by

$$K_m = (1 - \varepsilon_t) \cdot (1 - X_a) \cdot Sh D_{e,m} / (2R_p) \tag{15.15}$$

where Sh is Sherwood number, considered to be 2, $D_{e,m}$ is effective diffusion coefficient of species m, which is affected by both bulk diffusion and Knudsen diffusion in micropore of porous solid medium. The specific calculation of $D_{e,m}$ is

$$D_{e,m} = \left(\frac{1}{D_{b,m}} + \frac{1}{D_{K,m}} \right)^{-1} \tag{15.16}$$

where $D_{b,m}$ is the bulk diffusion coefficient, calculated by

$$D_{b,m} = \frac{435.7\, T^{1.5}}{p\left(V_m^{1/3} + V_{m'}^{1/3}\right)}\sqrt{\left(\frac{1}{W_m} + \frac{1}{W_{m'}}\right)} \tag{15.17}$$

where T is temperature, p is atmospheric pressure, V_m is liquid molar volume of species m at normal boiling point, while m' represents all gaseous components in space except species m. $D_{k,m}$ is Knudsen diffusion coefficient, calculated by

$$D_{K,m} = \frac{\varepsilon_t}{\tau_t} \cdot \frac{2}{3}\bar{r}_{pore}\sqrt{\left(\frac{8RT}{\pi W_m/1000}\right)} \tag{15.18}$$

where $\bar{r}_{pore}$ is average radius of pores.

15.2.4 Energy Sub-model

Energy conservation can describe the variation of temperature of char particle as follows:

$$M_t c_p \frac{\partial T}{\partial t} = hA(T - T_b) + \alpha\sigma A(T^4 - {T_b}^4) + \dot{q} \tag{15.19}$$

where c_p is heat capacity, h is convective heat transfer coefficient of ambient gas, T_b is ambient temperature, α is total thermal emissivity of particle surface and σ is Stefan-Boltzmann constant. h can be calculated by

$$h = Nu\lambda/R_p \tag{15.20}$$

where Nu is Nusselt number, considered to be 2. λ is thermal conductivity of ambient gas.

15.3 Model Validation

At first, calculation results of carbon consumption were compared with those from the Bejarano et al.'s experimental results (Bejarano and Levendis 2007). The combustion of PSOC-176 coal chars (pyrolyzed at 1600 K) with the size of 45–53 μm was conducted at 1300 K in O_2/N_2 atmosphere. The comparison of burn out times and the temperatures for the char combustion is shown in Figs. 15.1 and 15.2, respec-

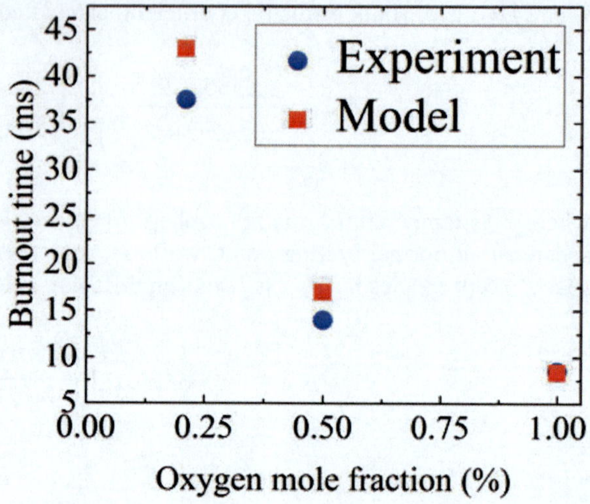

Fig. 15.1 Validation of burnout time

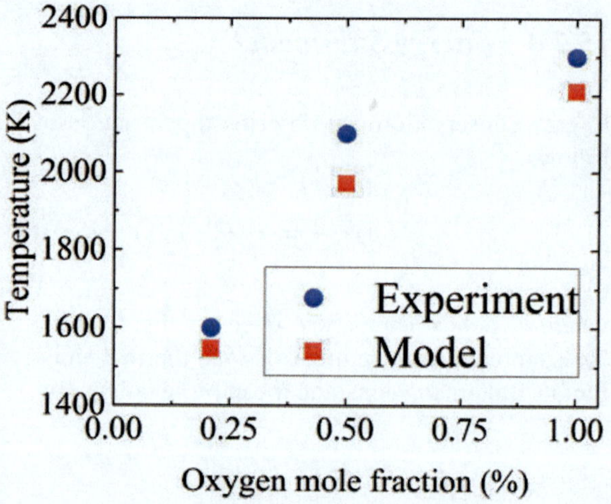

Fig. 15.2 Validation of temperature

tively. It can be seen that the errors from the current model are within the acceptable range. Then, NaCl release model from Yang (2017) was validated by calculating the variation of NaCl, accounting for 68% of Na content in Zhundong coal conversion rate as the temperature increased from 673 to 1473 K. Figure 15.3 shows that the model-based results are in agreement with the experimental data.

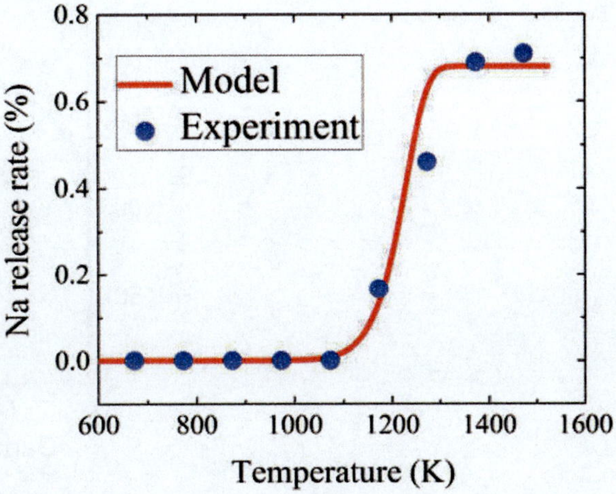

Fig. 15.3 Validation of Na conversion

15.4 Alkali Metal Release in O_2/N_2 and O_2/CO_2 Atmospheres

In the actual industrial application, coal gasification is another effective technology to promote its efficient and clean utilization. CO_2 is one of the most commonly used in coal gasification. Therefore, it is necessary to study coal consumption in different atmospheres.

In this section, the variations of carbon conversion rate, temperature, Na residual rate and Na in gas phase mass fraction during the char particle ($R = 50\ \mu m$) combustion at 1473 K ambient temperature were calculated and shown in Figs. 15.4, 15.5, 15.6 and 15.7, respectively.

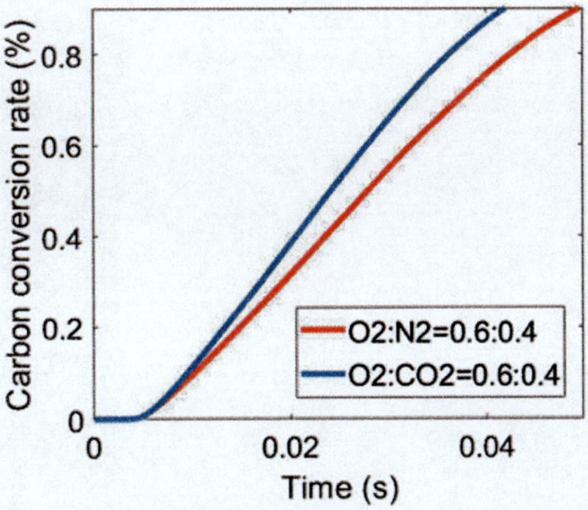

Fig. 15.4 Carbon conversion rate

Fig. 15.5 Temperature

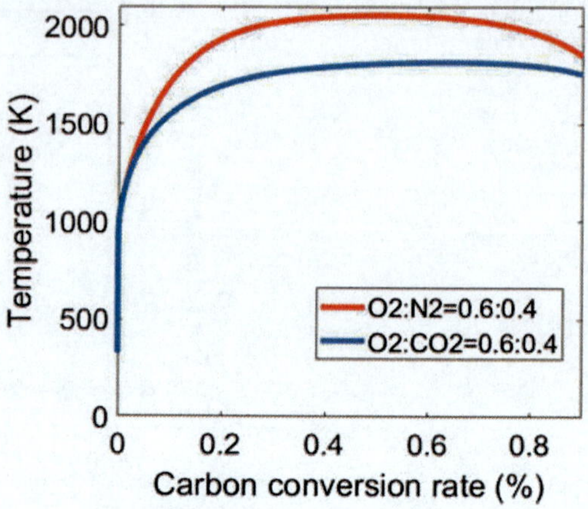

Fig. 15.6 NaCl residual rate

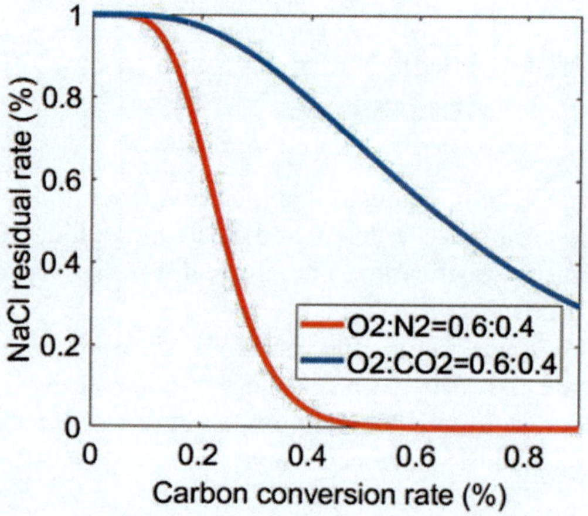

Fig. 15.7 NaCl-gas mass fraction

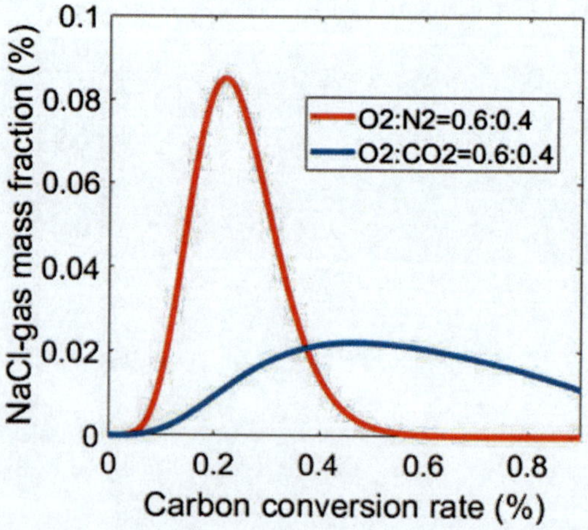

According to the results shown in Figs. 15.4 and 15.5, O_2/CO_2 atmosphere slightly shortened the burnout time and lowered the temperature of char combustion. From the calculation process, it was found that the proportion of the endothermic reaction between CO_2 and C to produce CO increased.

As can be seen in Fig. 15.6, O_2/CO_2 atmosphere made Na remain more in solid form in char particle. It can be inferred that it was the reduced burnout time and combustion temperature that limited the rapid release of Na in gas phase.

The mass fraction of Na in gas phase had an obvious peak value in O_2/N_2 atmosphere as shown in Fig. 15.7, when the carbon was consumed about 20%. It was found in calculation process that there was a competition between gasification of Na solids and gas diffusion. The former produced Na in gas phase and the latter transmitted it to environment. Thus, the chemical reaction (Na gasification) played a dominant role before the peak value, as initial stage of combustion, and with combustion proceeded, the effect of gas diffusion became increasingly important. Due to the smaller rate of Na release, the competition between chemical reaction and diffusion showed inapparent in O_2/CO_2 atmosphere.

15.5 Effects of Particle Size and Ambient Temperature

Particle size of char and ambient temperature are important parameters. Particle size directly changes the spatial scale of gas molecular diffusion and heat transfer. Ambient temperature determines the amount and rate of heat transfer to char. Both of the parameters can affect the rate of chemical reactions, and finally act on the combustion of char and alkali metals release.

In this section, Na release behavior of char particles with the size ranging from 50 to 100 μm at different ambient temperatures were studied. The results are shown in Figs. 15.8 and 15.9.

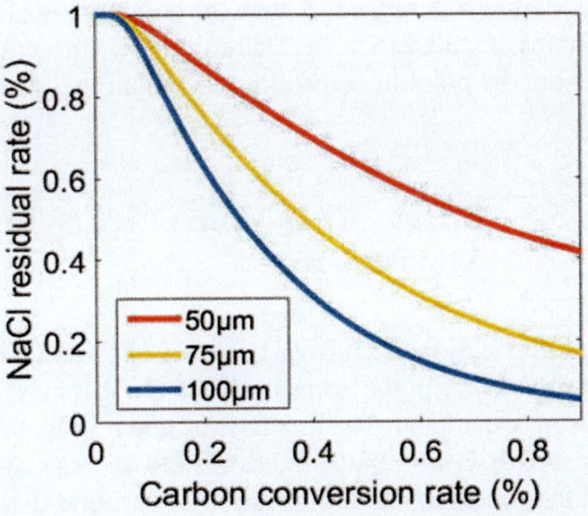

Fig. 15.8 Effect of particle size at the ambient temperature of 1473 K

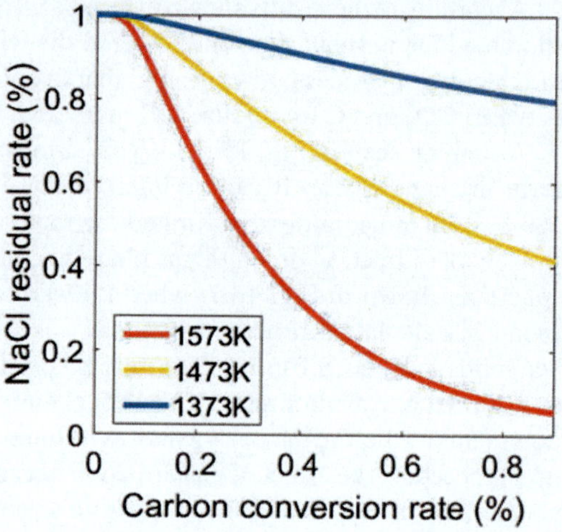

Fig. 15.9 Effect of ambient temperature on the Na release with the particle sized of 50 μm

According to the results shown in Fig. 15.8, with particle size increased, lesser Na remained as solid state before the carbon content of char particles was completely consumed. Minor difference in the combustion temperature was observed at different carbon conversion rate for the three different sizes of char particle, which indicates that the temperature is not significantly affected by particle size. It can be concluded that the main effect of char particle size is the time of carbon depletion. The carbon in small-sized char can be rapidly consumed before Na releasing in a large amount, which also terminates the combustion process in a short time. Thus, the rapid combustion of small-sized char particles is conducive to controlling the production of alkali metal gases.

According to the results shown in Fig. 15.9, with increasing the ambient temperature, more Na was released as gaseous phase before the carbon content of char particle was completely consumed. The rate increase of Na gasification reaction was obviously greater than that of carbon combustion reaction. Thus, Na release is more sensitive to temperature than the combustion of char particle. High temperature facilitates the release of Na. Therefore, low ambient temperature combustion is preferred from the point of controlling the alkali metals release.

15.6 Effects of O_2 Mole Fraction in O_2/N_2 and O_2/CO_2 Atmospheres

In this section, Na release behavior of 50 μm radius char particles at 1473 K ambient temperature in the atmosphere of O_2/N_2 and O_2/CO_2 with different oxygen content were calculated. The results are shown in Figs. 15.10 and 15.11.

With O_2/N_2 mole fraction ratio increased, Na release rate accelerated during the process of char combustion. It is found that the temperature of char combustion

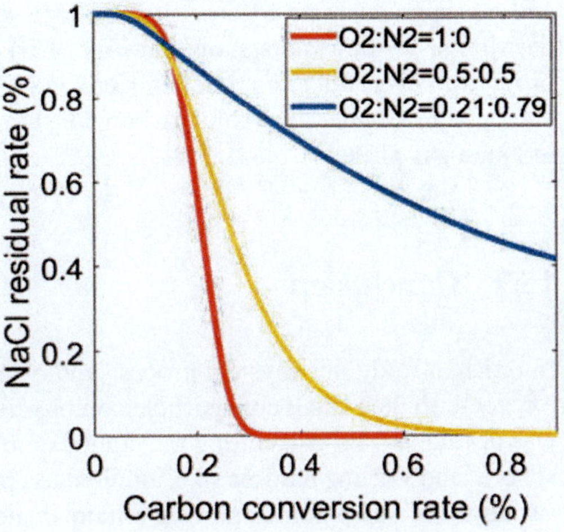

Fig. 15.10 Effect of O_2 ratio in O_2/N_2 atmosphere

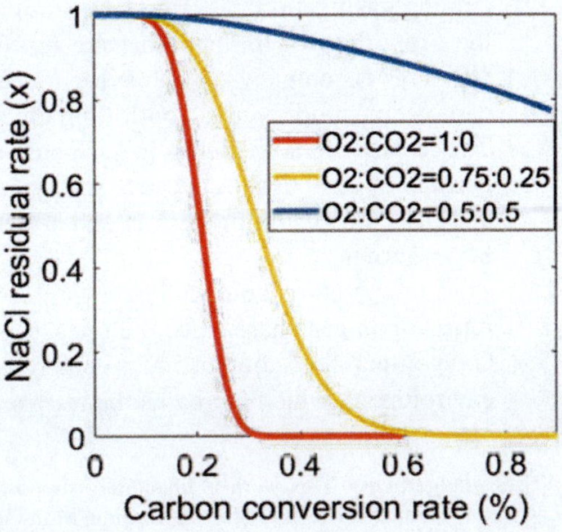

Fig. 15.11 Effect of O_2 ratio in O_2/CO_2 atmosphere

increased sharply when increasing the oxygen content. It is proved again that oxygen-enriched combustion is beneficial to the formation of high temperature environment, which increases the release of alkali metals in gas phase.

Similar to the char combustion in O_2/N_2 atmosphere, the release rate of Na in O_2/CO_2 also increased greatly as the oxygen content increased, as shown in Fig. 15.11. Comparing the results in Figs. 15.10 and 15.11, one can find that the effect of the combustion atmosphere on Na release is more obvious when the oxygen content is low. In the atmosphere of $O_2/CO_2 = 0.5{:}0.5$, only about 20% of Na released to gas phase, while all of the Na released when the char particle was completely consumed in the atmosphere of O_2/N_2 with the same ratio. It was found that the maximum combustion temperature of char particle was lower than that in O_2/N_2 atmosphere, and the burn out time of char particle in O_2/CO_2 atmosphere was shorter, which are similar to the results shown in Figs. 15.4 and 15.5. As a result, increasing the proportion of gasification reaction (R.2) is conducive to achieving low temperature combustion, accelerating solid carbon consumption and thus, controlling the release of Na in gas phase.

15.7 Conclusion

In order to study the physical process and characteristics of alkali metals release in oxygen-enriched small char particle combustion, a comprehensive simulation model was developed. By calculating the combustion process in O_2/N_2 and O_2/CO_2 atmospheres and varying particle size, ambient temperature and oxygen mole fraction, the variations of key parameters were obtained and discussed. The specific conclusions are as follows:

A. Adding gasification gas, e.g., CO_2, into char combustion system can increase the proportion of the endothermic reaction between CO_2 and C to produce CO. This is conducive to achieving low temperature combustion, accelerating carbon consumption and controlling the release of alkali metals in gas phase.
B. The release of alkali metals in gas phase can be reduced by using smaller char particle in combustion, because small-sized particle can be rapidly consumed and the combustion process is ended before alkali metals can be released in large amount.
C. Low temperature combustion is preferred for the control of alkali metals releasing in gas phase since the release rate is sensitive to high temperature.
D. Oxygen-enriched combustion is beneficial to the formation of high temperature environment, which increases the release of alkali metals in gas phase.

Acknowledgements This work is financially supported by the National Natural Science Foundation of China (51761125012) and Beijing Municipal Science and Technology Commission (Z191100002419005) and Huaneng Group science and technology research project (HNKJ20-H50) (U20GCZH02).

References

Bejarano PA, Levendis YA (2007) Combustion of coal chars in oxygen-enriched atmospheres. Combust Sci Technol 179(8):1569–1587

Bhatia SK, Perlmutter DD (1981) A random pore model for fluid-solid reactions: II. Diffusion and transport effects. AIChE J 27(2):247–254

Campbell PA, Mitchell RE, Ma L (2002) Characterization of coal char and biomass char reactivities to oxygen. Proc Combust Inst 29(1):519–526

Gil S, Mocek P, Bialik W (2011) Changes in total active centres on particle surfaces during coal pyrolysis, gasification and combustion. Chem Process Eng 32(2):155–169
Kosminski A, Ross DP, Agnew JB (2006) Influence of gas environment on reactions between sodium and silicon minerals during gasification of low-rank coal. Fuel Process Technol 490(87):953–962
Li GD, Li SQ, Qian H et al (2015a) Fine particulate formation and ash deposition during pulverized coal combustion of high-sodium lignite in a down-fired furnace. Fuel 143:430–437
Li WQ, Wang LY, Qiao Y et al (2015b) Effect of atmosphere on the release behavior of alkali and alkaline earth metals during coal oxy-fuel combustion. Fuel 139:164–170
Lu GQ, Do DD (1992) A kinetic study of coal reject-derived char activation with CO_2, H_2O, and air. Carbon 30(1):21–29
Lu GQ, Do DD (1994) Comparison of structural models for high-ash char gasification. Carbon 32(2):247–263
Lv T, Liu ZH (2018) Characteristics of single porous char particle oxy-fuel combustion with random pore model. Sci Technol Eng Chinese 18(7):12–17
Maffei T, Khatami R, Pierucci S et al (2013) Experimental and modeling study of single coal particle combustion in O_2/N_2 and Oxy-fuel (O_2/CO_2) atmospheres. Combust Flame 160(11):2559–2572
Naruse I, Kamihashira D, Miyauchi Y (2005) Fundamental ash deposition characteristics in pulverized coal reaction under high temperature conditions. Fuel 84:405–410
Smith IW (1982) The combustion rates of coal chars: a review. Sympos (Int) Combust 19(1):1045–1065
Song GL, Song WL, Qi XB et al (2016) Transformation characteristics of sodium of zhundong coal combustion/gasification in circulating fluidized bed. Energy Fuels 30:3473–3478
Sugawara K, Enda Y, Inoue H et al (2002) Dynamic behavior of trace elements during pyrolysis of coals. Fuel 81:1439–1443
Takuwa T, Naruse I (2007) Emission control of sodium compounds and their formation mechanisms during coal combustion. Proc Combust Inst 31:2863–2870
Takuwa T, Mkilaha ISN, Naruse I (2006) Mechanisms of fine particulates formation with alkali metal compounds during coal combustion. Fuel 85:671–678
Toftegaard MB, Brix J, Jensen PA et al (2010) Oxy-fuel combustion of solid fuels. Prog Energy Combust Sci 36(5):581–625
Wang ZS, Wang LY, Lin JY et al (2018) The influence of the addition of sodium on the transformation of alkali and alkaline-earth metals during oxy-fuel combustion. J Energy Inst 91:502–512
Wheeler A (1951) Reaction rates and selectivity in catalyst pores. Adv Catal 3(6):249–327
Yang YM (2017) Occurrence modes and release characteristics of alkali and alkaline earth metal species in zhundong coals. Tsinghua University, Beijing
Yang ZC, Liu JL, He HG (2010) Study on properties of Zhundong coal in Xinjiang region and type-selection for boilers burning this coal sort. Thermal Power Generat Chinese 39:38–44
Yu JL, Lucas JA, Wall TF (2007) Formation of the structure of chars during devolatilization of pulverized coal and its thermoproperties: a review. Prog Energy Combust Sci 33(2):135–170
Zhou J, Zhuang X, Andrés A et al (2010) Geochemistry and mineralogy of coal in the recently explored Zhundong large coal field in the Junggar basin, Xinjiang province, China. Int J Coal Geol 82(1–2):51–67

Chapter 16
Experimental Study on the Influence of Slagging and Fouling for Wall Temperature Distribution

Deli Li, Enlu Wang, Jinda Mao, Wei Wu, Kai Li, and Qi Wang

Abstract Coal is still the main fossil fuel for power plants in China, and heat resistance ratio of the fly ash slag and fouling layer is the key weight ratio for superheaters and re-heaters with coal fired boiler. Actually the slagging and fouling layers happened on the fireside surface of tubes are not uniform in the circumferential direction. To explore the influence level of the smoke flow on the slagging and fouling distribution, a tower type furnace facility was established, and the influence of slagging and fouling of a horizontal tube for the wall temperature distribution was studied. Meanwhile, the heat transfer model of clean tube and stained tube was established. A comparative study of clean tube surface and stained tube surface temperature distribution was conducted. The results showed that the wall temperature distribution with slagging and fouling was totally different with the clean condition. And the heat flux distribution of the clean tube was a single-humped curve, but the stained tube showed the double humped curve in the circumferential direction.

Keywords Wall temperature distribution · Slagging and fouling · Clean tube · Stained tube · Heat flux

16.1 Introduction

In the last decade, much attention has been drawn towards the burning of biomass or high-alkali lignite coal for power production. During the combustion process of such low-rank solid fuels, ash slagging and fouling are common problems that affect the operation and performance of heat transfer equipment (Robinson et al. 2002; Valmari et al. 1999). Slagging is defined as deposition of fly ash on heat-transfer surface and refractory in the furnace volume primarily subjected to radiant heat transfer (Dai et al. 2013). Although the name 'slag' suggests a fused or semi-fused ash, the term 'slagging' may also apply to sintered deposits and dry ash formed in liberally sized,

D. Li · E. Wang (✉) · J. Mao · W. Wu · K. Li · Q. Wang
Institute of Thermal Engineering, School of Mechanical Engineering, Shanghai Jiao Tong University, Shanghai 200240, China
e-mail: elwang@sjtu.edu.cn

J. Lyu and S. Li (eds.), *Clean Coal and Sustainable Energy*, Environmental Science and Engineering,
https://doi.org/10.1007/978-981-16-1657-0_16

low-pressure steam generator furnaces, or in furnaces fired with coals containing high moisture and alkaline earth ash. Fouling is defined as deposition in the heat-recovery section of the steam generator subject to convective heat exchange by fly ash quenched to a temperature below its predicted melting point, condensation by volatiles, or sulfidation by SO_3 (Bryers 1996). The deposits may vary from light sintering to complete fusion. The latter is due to the formation of lower melting sulfates.

With time away, the accumulated fireside deposits on the transfer surface result in the surface overheat (Bryers 1996). Otherwise, a desire to raise the temperature of the superheated steam to make the plants more efficient has also introduced slagging, and even corrosion problems. These issues of superheater tubes are closely related to the deposited material, which changes composition with fuel, boiler, and operational parameters. In order to control the surface temperature, several researchers pay attention to the ash formation and slagging deposition mechanism in basic (Han et al. 2019; Wu et al. 2018), and the others implore the relationship between slag/deposition and thermal resistance of heat transfer process based on engineering operation (Rezaei et al. 2000; Anderson et al. 1987). As known, in low-rank fuels combustion, the main part of the deposition (slagging or fouling) is formed by inertial impaction and thermophoresis. When the gas temperature is above melting point of impurities, melt and condensation happen around the tube outside surface, which is of great importance in view of deposition of potassium, chlorine, and sulphur.

It is apparent that the slagging and fouling vary the heat transfer characters of tube fireside. However, the influence level of the flue gas flow for the slagging and fouling and the distribution, are always not so clear. Thus, a tower type furnace facility was established, and the influence of slag and fouling of a horizontal tube for the wall temperature distribution was studied.

16.2 Experimental Setups and Methods

16.2.1 Experimental System and Setups

The experiments were performed in a vertical type of tower furnace, which was a pilot-scale, (60 kW), up-fired turbulent flow reactor that simulated the gas temperature and composition histories experienced by flue gas in a combustion system. The main body of furnace is a 6.2 m high and has a matched cooling system. The test sections are arranged along the furnace body height. And, in the test section, the cooling system is made up of several high-temperature steel pipes (Super 304H) and the corresponding high-pressure centrifugal fans. The K type thermocouples are arranged evenly outside the tube surface in the circumferential direction.

The air consumed, is controlled by a vertex flow meter with the range of 10–120 m^3/h and the precision of 1%. The combustion improver consumption (natural gas) is also controlled by a vortex flow meter, with the range of 4–20 m^3/h and the precision

Table 16.1 Chemical composition (as-received) of Jinggong coal used in the slagging experiments

Chemical analysis (wt%)	Values	Chemical analysis (wt%)	Values
C	65.55	Ash	11.38
H	5.1	Total water	5.7
O	16.92	Volatiles	31.28
N	0.73	LHV (MJ/kg)	25.07
S	0.74		
Al	0.2016		
Ca	1.6343		
K	0.0337		
Na	0.065		
Mg	0.596		

of 1%. The pulverized power coal is conveyed by a vibratory power feeder, with the feeder flow scale of 2 kg/h and precision of 5%. The coal used is a high- alkali coal in Zhundong, Xinjiang. The composition of the pulverized coals is shown in Table 16.1. Here, the C/N/H/O/S was detected by Vario EL CUBE element analyzer, with the C/N/H/O precision of 0.3%, and the S precision of 0.5%. The element of Al/Ca/Mg/K/Na, due to the small content in coal, was detected by a inductively coupled plasma emission spectrometer (ICP), iCAP6300 type of Thermo Fisher.

The flue gas composition is monitored by testo 340, produced by testo. Inc., Germany. The details of experimental system are demonstrated in Fig. 16.1.

Jinggong Coal, a high-alkali coal in Zhundong, Xinjiang was used for the experiments. The composition of the pulverized coal is shown in Table 16.1. Here, the C/N/H/O/S was detected by Vario EL CUBE element analyzer, with the C/N/H/O precision of 0.3%, and the S precision of 0.5%. The element of Al/Ca/Mg/K/Na, due to the small content in coal, was detected by a iductively coupled plasma emission spectrometer (ICP), iCAP6300 type of Thermo Fisher.

As presented in Table 16.1, the Jinggong coal is rich of high volatile and high alkali content. The promotion of volatiles to ignition and coal burnout determined that the coal is advanced for power generation, except for the slagging and fouling, even corrosion happened on the fireside. The effect of alkali metal content on the slagging and fouling characteristics of the heating surface also showes that the Jinggong coal is easy to raise the fouling and slagging (Li et al. 2016; Zhang et al. 2015; Wang et al. 2015).

16.2.2 Clean Tube Wall Temperature

In order to simulate the heating condition of super-heater or re-heater tube in high temperature flue gas at about 1050 °C, the test tube was horizontally arranged in the

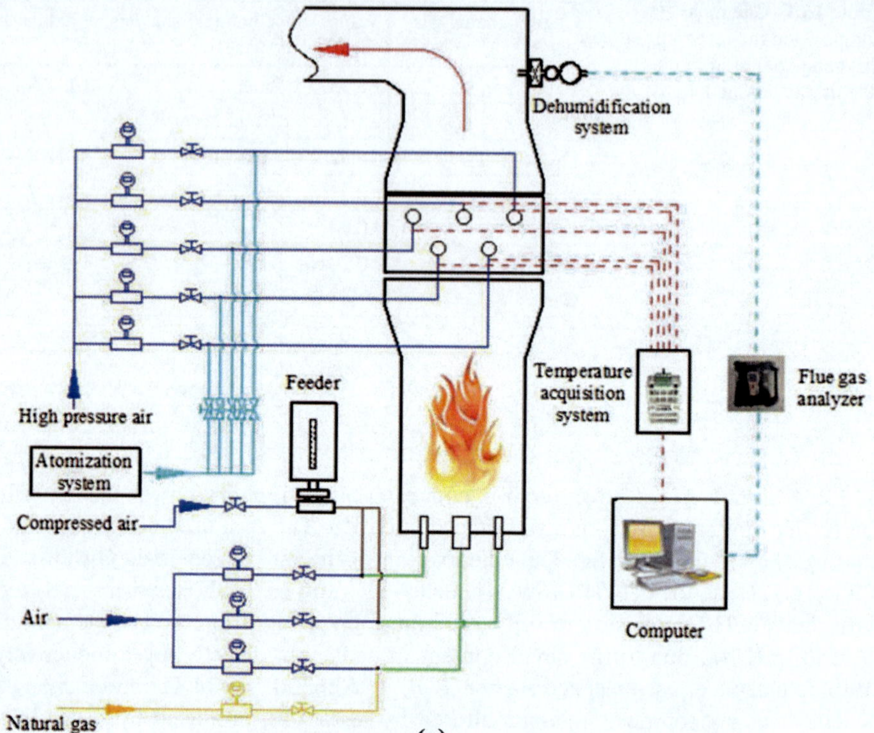

(a)

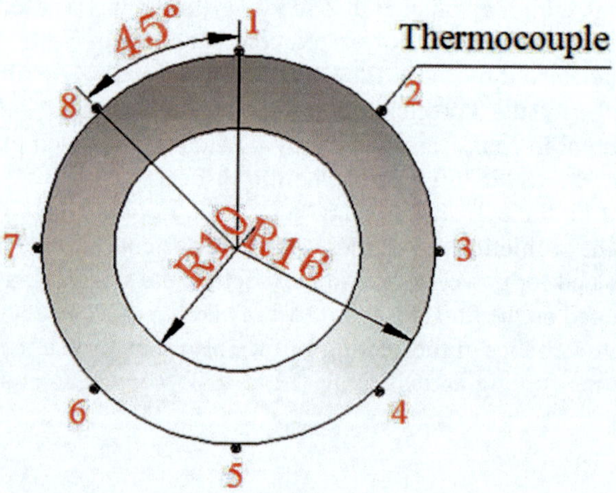

(b)

Fig. 16.1 Vertical tower test system and cooling tube. **a** Diagram of test system; **b** thermocouple distribution outside tube surface

Table 16.2 Natural gas components

Component	Volume fraction (%)
CH_4	78.9
C_2H_6	8.54
C_3H_8	5.73
C_4H_{10}	2.95
H_2O	0.37
N_2	3.15
CO_2	0.36

Table 16.3 Average value of component of flue gas generated by natural gas

Component	CO_2/%	O_2/%	CO/ppm	NOx/ppm	SO_2/ppm	Vapor/%
Value	12.2	6.8	400	220	60–100	19.18

chamber of combustion furnace, and the cooling medium flowed in each tube was controlled to adjust the each tube wall temperature. The atomized water/high-speed air was used as cooling medium.

For the high temperature heating surface arranged horizontally above the flame, the heat transfer on fireside usually contains both convection heat transfer and radiation of flue gas. Further, The un-uniformed distributed deposition then affects the heat transfer characteristics of the heating surface, in other word, it affects the temperature distribution of the heating tube wall temperature in the circumferential (Wang et al. 2015). So, in aims to explore the deposition effects on heat transfer resistance, the clean tube test and the polluted tube test were conducted.

In view of the natural gas generation no slagging and fouling on the surface of the test tube, the clean tube test was conducted in smoke atmosphere generated by natural gas combustion. And the natural gas component is listed in Table 16.2.

In order to explore the relationship between flue gas flow and the slagging and fouling and its influences, the smoke composition was kept consistent. The smoke composition in test was monitored by gas composition and temperature. The smoke component in details is shown in Table 16.3, which was close to that of the real boiler.

The non-uniform distribution of clean tube wall temperature in the circumferential direction is shown in Table 16.4.

16.2.3 Stained Tube Wall Temperature

In stained test, the temperature of flue gas was determined as 850–1150 °C, which was detected by the clean thermocouple. It should be noted that the temperature data derived from thermocouple for online monitoring is rather lower than the real

Table 16.4 Non-uniform distribution of clean tube wall temperature in the circumferential direction

Defined wall temperature/°C	1	2	3	4	5	6	7	8
650	355.56	541.17	619.69	672.06	654.75	602.43	502.29	409.83
700	455.49	604.28	679.69	719.11	706.03	666.98	584.04	510.76
750	550.55	673.30	748.63	767.64	753.07	734.41	662.11	593.44
800	610.29	721.62	798.11	811.36	804.25	783.11	717.23	646.40
850	655.91	757.30	837.57	849.19	842.06	827.82	752.38	685.77

Note 1–8 is matched with Fig. 16.1b

temperature of smoke. This is due to the accumulation of deposition around the thermocouple. In such condition, the temperature detected is the temperature of deposition layer, not the smoke.

In order to research the difference level of online temperature and real smoke temperature, two thermocouples were arranged at each measuring point. One thermocouple was placed in the flue gas for online monitoring the smoke temperature and the other was extended into the furnace intermittently to measure the smoke temperature. The intermittent monitoring temperature is larger than the online monitoring temperature, which was shown in Fig. 16.2. With the time away, the deposition around thermocouple has been formed and with the thickness kept the same. The online thermocouple with deposition layer, the temperature detected is the heat resistant shell of deposition. The deposition condition outside thermocouple is shown in Fig. 16.3. The other thermocouple will be cleaned after each measurement; the detected temperature can be viewed as the temperature of real flue gas. After the deposition layer formation, the difference value of the two remains 50 °C, in the case of 1100 °C flue gas temperature.

In consideration of slagging and fouling, the ash particles will hit the windward side of the tube wall and accumulate on the surface of heat transfer, which led the

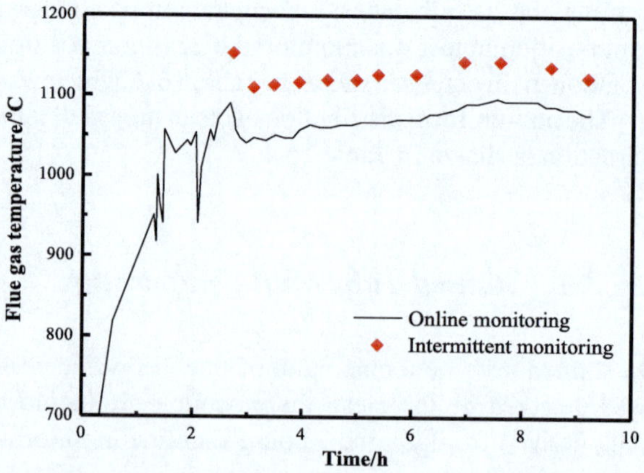

Fig. 16.2 Flue gas temperature derived from online monitoring or intermittent monitoring

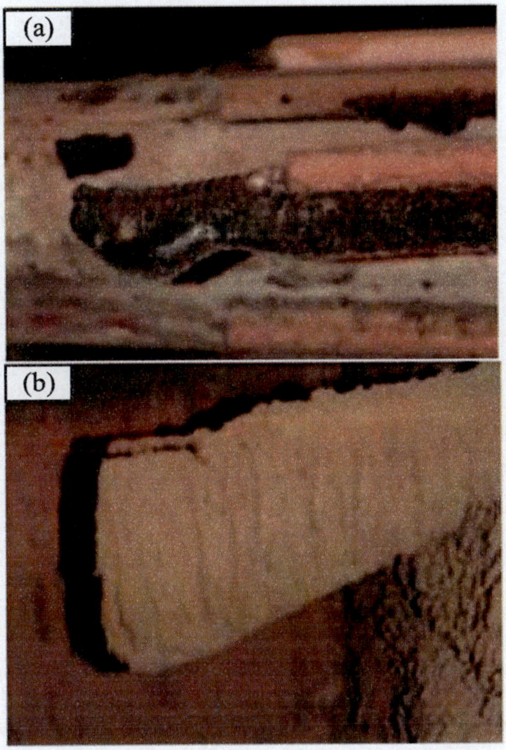

Fig. 16.3 The slagging and fouling outside of thermocouple in different temperature atmosphere. **a** the slagging outside the thermocouple; **b** the fouling outside the thermocouple

heat transfer characteristics to be different from that of the clean tube. Figure 16.4 shows the online monitoring temperature of flue gas stained tube in circumferential direction.

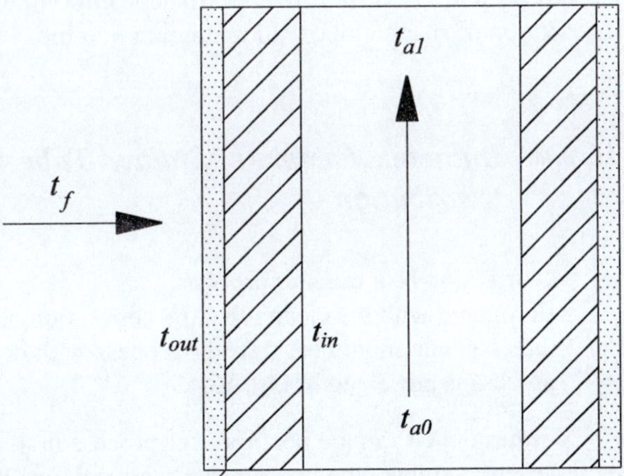

Fig. 16.4 Heat transfer process of stained tube in flue gas

Table 16.5 Flue gas component produced by Jinggong coal combustion

Component	O_2 (%)	CO (ppm)	CO_2 (%)	NO (ppm)	NOx (ppm)	SO_2 (ppm)
Value	5.39	217	13.66	348	365	123

Table 16.6 Non-uniform distribution of stained tube wall temperature in the circumferential direction

Position	T				
	650 °C	700 °C	750 °C	800 °C	850 °C
1	357.16	457.86	556.45	617.58	669.74
2	520.14	582.46	654.02	711.56	832.40
3	601.46	650.46	740.53	750.16	856.41
4	540.46	610.47	700.56	705.54	701.53
5	430.54	600.78	610.23	672.85	651.16
6	500.12	640.34	650.25	760.56	832.76
7	512.46	580.41	672.32	746.49	871.61
8	415.41	510.46	600.45	654.56	736.25

Note 1–8 is matched with Fig. 16.1b

The flue gas component produced in details is listed in Table 16.5. Comparing with the flue gas component in clean tube test (listed in Table 16.3), the corresponding component is similar, which can be considered as the same test condition in the two experiments.

Because the upper layer of deposition may be destroyed when operation, the temperature distribution here was used to characterize the contamination layer distribution. There are two theoretical bases as follows. On the one hand, the thermal resistance is much larger than that of steel; on the other hand, the thermal resistance is directly proportional to the thickness of deposition. Table 16.6 shows the non-uniform distribution of stained tube wall temperature in the circumferential direction.

16.2.4 *Influence Factors of Stained Tube Temperature Distribution*

(a) Convection heat transfer process
Compared with the clean tube, the convection heat transfer process of stained tube has one more heat transfer process with deposit layer. The heat transfer process is presented in Fig. 16.4.

Consider the clean pipe heat transfer process first. The formula for heat transfer by flow across single tube is given as (Yang and Tao 2006)

$$\begin{cases} Q_c = h_f A_f (t_f - t_{out}) \\ Q_a = h_a A_a \left(t_{in} - \dfrac{t_{a0} + t_{a1}}{2} \right) \\ Q = m_a C_p (t_{a1} - t_{a0}) \\ Q = \dfrac{t_{out} - t_{in}}{\frac{1}{2\pi\lambda_{steel} l} \ln \frac{d_1}{d_0}} \\ h_f = C \dfrac{\lambda_f}{d_1} \mathrm{Re}_f^n \, \mathrm{Pr}_f^{\frac{1}{3}} \\ h_a = 0.023 \dfrac{\lambda_a}{d_0} \mathrm{Re}_a^{0.8} \, \mathrm{Pr}_a^{0.4} \end{cases} \tag{16.1}$$

The conductive heat transfer process of stained tube is similar with that of clean tube, which is given directly

$$\begin{cases} Q = \dfrac{t_{out} - t_{in}}{\frac{1}{2\pi\lambda_{steel} l} \ln \frac{d_1}{d_0} + \frac{1}{2\pi\lambda_{stained} l} \ln \frac{d_2}{d_1}} \\ Q = \dfrac{t_{out} - t_w}{\frac{1}{2\pi\lambda_{stained} l} \ln \frac{d_2}{d_1}} = \dfrac{t_w - t_{in}}{\frac{1}{2\pi\lambda_{steel} l} \ln \frac{d_1}{d_0}} \end{cases} \tag{16.2}$$

When Eq. (16.2) introduced into Eq. (16.1), the tube surface temperature equation basis on cooling medium side is given as

$$t_{bi} = \frac{t_{a0} + t_{a1}}{2} + Q \left(\frac{\ln \frac{d_1}{d_0}}{2\pi\lambda_{steel} l} + \frac{1}{h_a A_a} \right) \tag{16.3}$$

Or the surface temperature equation basis on flue gas side is described as

$$t_{bi} = t_f - \frac{Q}{h_f A_f} - \frac{Q}{2\pi\lambda_{stained} l} \ln \frac{d_2}{d_1} \tag{16.4}$$

And the exhaust temperature of cooling medium is described as

$$t_{out} = \frac{t_{a0} + t_{a1}}{2} + Q \left(\frac{\ln \frac{d_1}{d_0}}{2\pi\lambda_{steel} l} + \frac{\ln \frac{d_2}{d_1}}{2\pi\lambda_{stained} l} + \frac{1}{h_a A_a} \right) \tag{16.5}$$

When the radiation process is considered, an radiation item need to be added in heat transfer equation, which is given as

$$Q_r = \varepsilon_{sys} \mathrm{A}_f \sigma \left(t_f^4 - t_{out}^4 \right) \tag{16.6}$$

where the system emissivity is

$$\varepsilon_{xt} = \frac{1}{\frac{1}{\varepsilon_{flame}} + \frac{1}{\varepsilon_w} - 1} \quad (16.7)$$

Adding of radiation item of Eq. (16.6) into system heat transfer flux Q, the model of wall surface temperature and cooling medium exhaust temperature are got.

16.3 Results and Discussions

16.3.1 *Circumferential Temperature Distribution of Clean Tube*

Table 16.4 shows the circumferential temperature distribution of clean tube in different cases. As shown, five tubes corresponding to five wall temperature level appears a similar pattern, that the wall temperature of the fire-facing side is obviously higher than that of the rear side. For each tube probe, the highest temperature appears at windward side, and the lowest temperature appears at the rear side. The max difference value between them can be reached 200 °C. To make the data clearly and directly, the relation curves between the wall surface temperature and the test position are given in Fig. 16.5.

In Fig. 16.5a, the same change rule of wall surface temperature with the increase of angle, which the temperature increases firstly until reaching the maximum at around 180° (windward center point) and then decrease rapidly. The temperature at the same test point increases with the heat load. In the view of thermal resistant performance of steel, the allowable temperature just need to be higher than the maximum temperature point, where appears around the windward center point. Figure 16.5b shows the same contents just in polar coordinate. As shown, the wall temperature distributes symmetrically, with the axis of 0–180° line. The surface temperature of windward is higher than that of rear side.

The high speed air/atomized water mist was used as cooling medium in cooling system. Besides, the cooling medium acts as a measurement of heat absorption. In this way, the volume flow of high speed air/water measured by vortex flow meter, and the input/exhaust temperature of cooling medium measured by K type thermocouple should be conducted. So the heat transfer rate of five defined wall temperature cases is presented in Table 16.7.

When the cooling medium is pure air, the absorption heat can be calculated as

$$Q_{absorp} = c_{air} m_{air} (t_{out} - t_{in}) \quad (16.8)$$

However, when the absorption heat load is large, the pure air is incompetent. In the case, mix of high speed air and atomized water mist is essential. Due to the temperature slippage of the moisture, the temperature of the mix of air and the water droplets is actually inconsistent. Then the Eq. (16.8) is not appropriate for

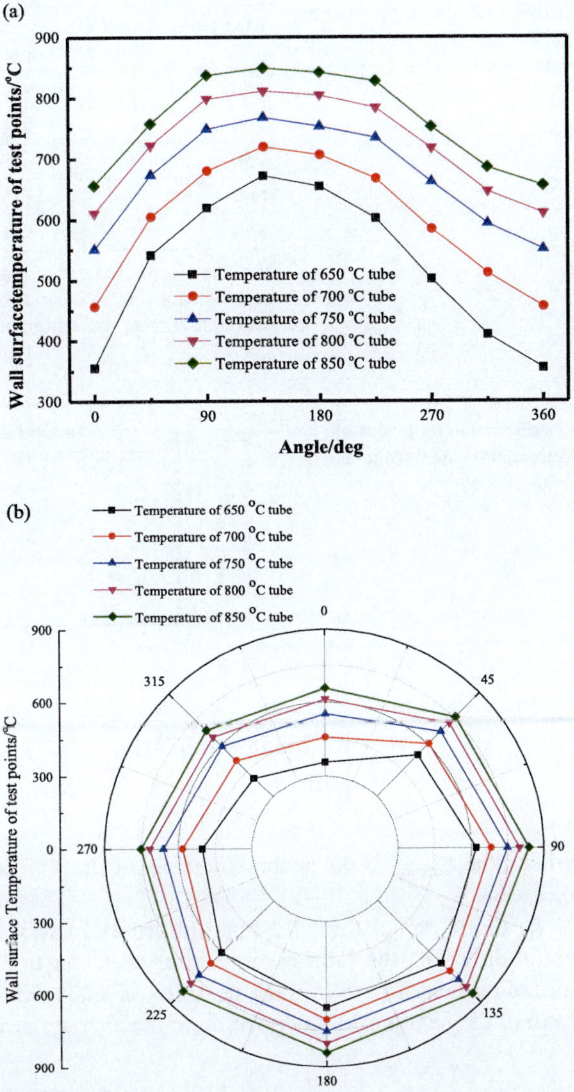

Fig. 16.5 Wall surface temperature of test points. **a** In rectangular coordinate; **b** in polar coordinate

the situation, and which is instead to use the method of thermal conductivity. The absorption heat here is defined as

$$Q_{absorp} = \frac{2\pi\lambda L(t_{outer} - t_{inner})}{\ln(r_{outer}/r_{inner})} \tag{16.9}$$

where, λ is the thermal conductivity coefficient of Super 304H steel, which shows in Fig. 16.6. The fitted λ equation is described as

Table 16.7 Heat transfer rate of five defined wall temperature cases

Max temperature/°C	Average heat transfer rate/W	R_1	R_2
650	844.31	1.25	0.6
700	911.50	1.22	0.74
750	1030.24	1.15	0.82
800	1115.67	1.13	0.83
850	1181.46	1.12	0.85

Note The R_1 is the ratio of maximum heat transfer rate to average heat transfer rate in clean tube test, and the R_2 is the minimum heat transfer rate to average heat transfer rate in clean tube test

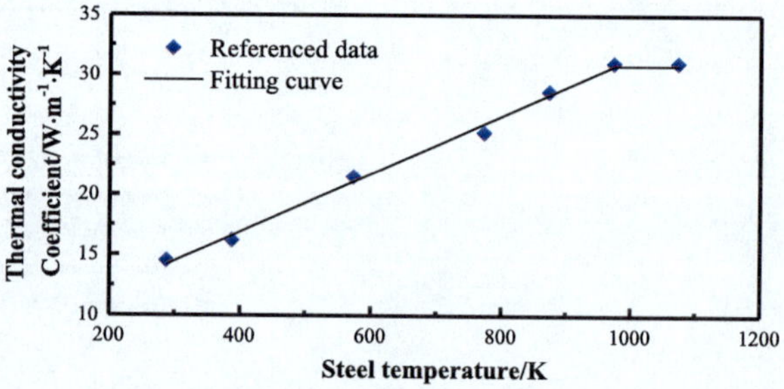

Fig. 16.6 Thermal conductivity coefficient of Super 304H steel (Mao and Wang 2006)

$$\lambda = \begin{cases} 7.23358 + 0.02419 t_{average} & t_{average} \leq 973K \\ 30.78 & t_{average} > 973K \end{cases} \tag{16.10}$$

where, the $t_{average}$ is the arithmetic mean of the wall temperature in circumferential direction.

In aim to demonstrate the non-uniformity of wall temperature in the circumferential direction, the ratio of maximum heat flux to average heat flux and ratio of minimum heat flux to average heat flux is analyzed. Here, we define the R as the ratio of the heat flux at any position to the average heat flux at same cross section.

$$R = \frac{heat\,transfer\,rate\,at\,any\,position}{average\,heat\,transfer\,rate} \tag{16.11}$$

Then, R is used as the criterion of the non-uniformity of wall temperature. The ratio can reflect the unevenness of the whole heat flux and convert the wall temperature into the heat flux value. In this way, a dimensionless coefficient is advanced in the future analysis.

According to the data of heat transfer rate, R changes between 0.6 and 1.25. That is to say, the heat flux always varies between 0.6 and 1.25 in the same cross section. The fitting equation of R is described as

$$C_R = 2e^{-10}\theta^4 - e^{-7}\theta^{-3} + e^{-5}\theta^2 + 0.00051\theta + 0.5914 \quad (16.12)$$

Through this method is derived from natural gas combustion, the scope of application is not limited to this. Firstly, the combustion fuel is natural gas, so that there is no additive thermal resistance except of that of steel tube wall, which the thermal resistance is too small to be considered. Secondly, the smoke component is at the same order of magnitude with the real smoke in boiler. Last but not the least, the equation is a non-dimension relationship, the variation such as flow velocity cannot raise an effect on the ratio R. So, the fitting equation is also appropriate to other case of heat transfer process of clean tube.

16.3.2 Circumferential Temperature Distribution of Stained Tube

In stained tube test, this work studied the heat transfer characteristics and thermal resistance variation of heat conduction process of tube wall. Table 16.6 shows the circumferential temperature distribution of stained tube in different cases. To make the data clearly and directly, the relation curves between the wall surface temperature and the test position are given in Fig. 16.7.

In Fig. 16.7, it is obvious that the rule is different from the clean tube experiment. In stained tube test, the wall surface temperature appears the bimodal distribution with the angle. The maximum appears at the angel of 90° and 270°, instead of the windward center (180°). On the other hand, the minimum appears at the windward center and rear side center. This can be explained by the non-uniform distribution of deposit layer. Literature found the deposit layer appeared bimodal distribution. At the angle of 45–90°, the thickness of deposit layer is large, and at the windward center and rear side center, the thickness is less. The thermal resistance is directly proportional to the thickness of deposition, so that the thermal resistance reaches maximum at 45–90°, and reduces to minimum at 0 and 180°. The heat transfer rate can be got using the same heat absorption method, which the results is listed in Table 16.8.

Seen from Table 16.8, the lower the wall temperature level is, the greater the decrease ratio γ. This phenomenon may be because of the fly ash condensation and thermophoresis. In one term, the hot fly ash, usually in melting state or half melting state, will condensate once hits the cold wall. The cold wall surface strengths the condensation process. In the other item, the temperature gradient of the cold wall accessory is large, that is to say, the thermal swimming force pushing the fly ash towards the wall becomes strong. Thus of these both improve the accumulation of fly ash. R_3 decreases but R_4 increases with the increase of wall temperature level. In other word, the distribution of wall temperature in circumferential direction trend to uniformity. This can also be indirectly explained the rules of decrease ratio γ.

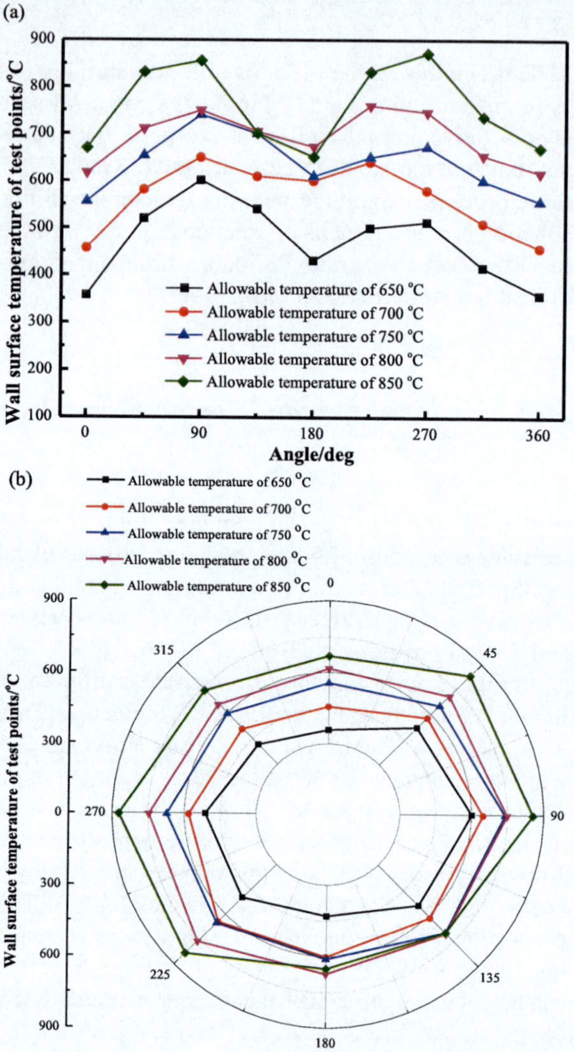

Fig. 16.7 Wall surface temperature of test points on stained tube. **a** In rectangular coordinate; **b** in polar coordinate

Due to the complex variation rules of deposit layer thickness, the uniformed equation for the heat transfer rate R just as in the clean tube cannot get. So here just give the fitting equation of stained tube wall temperature, which is demonstrated in Table 16.9. It is note that the fitting equation is only apply in the condition of the same fuel, same temperature level, and the short operation time.

Table 16.8 Heat transfer rate of five defined wall temperature cases in stained tube test

Defined wall temperature/°C	Average heat transfer rate/W	Decrease ratio γ (%)	R_3	R_4
650	725.81	14	1.26	0.72
700	858.53	4	1.16	0.79
750	976.96	5	1.17	0.86
800	1067.25	4	1.10	0.88
850	1173.76	0.6	1.13	0.87

Note The R_3 is the ratio of maximum heat transfer rate to average heat transfer rate in stained tube test, and the R_4 is the minimum heat transfer rate to average heat transfer rate in stained tube test; γ is the decrease ratio of heat transfer rate compared with the clean tube test

Table 16.9 Coefficient of fitting function described of stained tube temperature when Jinggong coal was used

Defined wall temperature/°C	Intercept	B1	B2	B3	B4	B5	B6
650	357.4	−0.62	0.20	−0.003	1.79E−5	−4.61E−8	4.34E−11
700	456.9	2.81	0.03	−7.95E−4	5.48E−6	−1.55E−8	1.56E−11
750	556.7	−2.96	0.22	−0.003	1.63E−5	−4.06E−8	3.74E−11
800	617.2	0.24	0.10	−0.002	1.10E−5	−3.00E−8	2.96E−11
850	668.8	1.05	0.17	−0.003	2.13E−5	−5.88E−8	5.82E−11

16.3.3 Influence Factors of Temperature Distribution of Stained Tube

Influence of deposit or not on wall temperature, deposit outer layer temperature and heat flux is presented in Fig. 16.8. As shown, the deposit does have a significant influence on tube wall temperature. The wall temperature decreases with the deposit

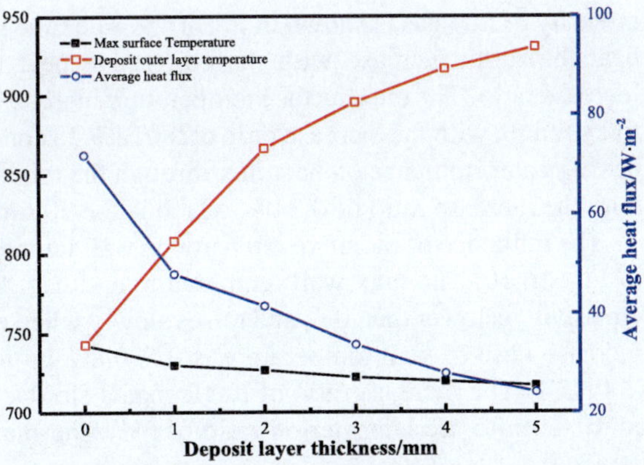

Fig. 16.8 Influence of deposit or not on wall temperature, deposit outer layer temperature and heat flux

layer thickness. Because the thickness of the deposit was able to reduce the surface temperature of the areas it covered (Phongphiphat et al. 2010). But it does not mean the deposit helps to protect the pipe. Once the tube covered by deposit, the thermal conductivity becomes large, which reduced the heat transfer. This can be illustrated by blue line in the figure. Due to the heat received by vapor in tube reduction, in order to reach the required temperature in super-heater or re-heater, the heat load has to raise a higher level, which resulted of higher flue gas temperature and more serious slagging or fouling. Eventually, it causes the heat exchange pipe overheated and burst.

In the initial stage of deposit, surface temperature and heat flux both decrease rapidly. After that, the two decrease with deposit thickness rather slightly and the decrease ratio keep constant. This can be explained by the deposit formation mechanism and its structure. The deposit consisted of two essentially different parts (Dai et al. 2013; Erickson et al. 1995). An inner uniform and smooth, white layer covers most of the surface area of the probe. This layer originated from condensation of vapor species and deposition of submicron particles. Formation of the initial deposit layer, consisting mainly of fine particles, occurs on the cool tube wall. Because of the low local temperatures, the initial ash deposit layer does not sinter to any significant extent. Partly for the above reason, the initial layer has a low thermal conductivity (Raask 1985). This explains that a little thickness increase results in great temperature drops in the initial stage. Outside the initial layer, bulk deposit growth can occur because of sticky ash particles adhering to themselves as well as by nonsticky ash particles adhering to the sticky part of the deposit surface. As additional deposition occurs, the thermal resistance increases, and the tube wall temperature goes to decrease. In this figure, it can also infer that the thermal resistance of bulk deposit is smaller than that of initial deposit layer.

Since slagging and fouling having a significant effect on heat transfer in heat exchange pipes, it is necessary to study the law of heat transfer tube. Based on the calculation of flue gas flowing across single tube, the main influence factors of surface temperature mainly contain the convection heat transfer of flue gas side, system emissivity, heat load of flue gas, and thermal conductivity of deposit.

The surface temperature and average heat transfer flux under different convective intensity of flue gas is shown in Fig. 16.9. The tube surface temperature and average heat flux both increase with the convective heat transfer strength with different increase ratio. The max surface temperature increases in direct proportion to convective strength with the increase ratio of 0.012%. In comparison, the convective strength has a greater influence on heat flux through the tube. The average heat flux increases with the increase ratio of 0.34%, which is the 30 times of surface temperature.

The influence of radiative intensity on wall temperature and heat flux is presented in Fig. 16.10. The max wall temperature and heat flux raises rapidly when system emissivity is lower than 0.6, and raises slowly when emissivity is larger than 0.6. The increase ratio of wall temperature is 0.5–0.8% in initial stage, but rapidly declines to 0.25%. The increase ratio of flux appears similar rules, that from the initial ratio of 10–20% to the later period ratio of 5%. The maximum of both appears near at

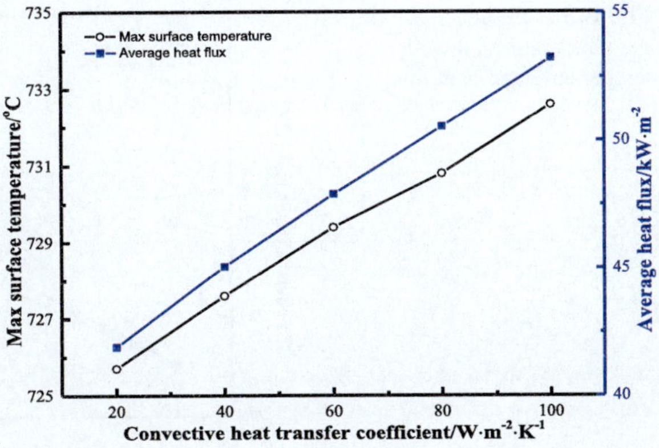

Fig. 16.9 Influence of flue gas heat transfer coefficient on wall temperature and heat flux

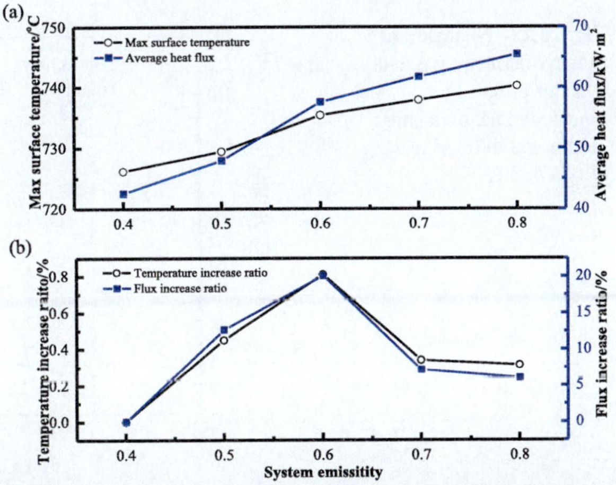

Fig. 16.10 Influence of system emissivity on wall temperature and heat flux

emissivity of 0.6. The influence on heat flux is two orders of magnitude larger than that of temperature.

The influence of flue gas temperature on wall temperature and heat flux is presented in Fig. 16.11. Max surface temperature increases with the flue gas with a ratio of 0.1 as flue gas temperature lower than 1000 °C, and increases with a ratio of 0.17 as flue gas temperature higher than 1000 °C. The average heat flux appears the similar rules. Heat flux increases with a ratio of 0.17 as flue temperature lower than 1000 °C, and increases with the ratio of 0.28.

In order to explore the weight ratio of influence of radiative/convective on tube surface temperature, the heat transfer model has been established, which is demonstrated in the Sect. 16.2 above. The convective heat transfer coefficient on fireside is at the range of 10–100 W/m^2 K (Yang and Tao 2006). Here, the maximum of convective heat transfer coefficient (100 W/m^2 K) is selected. And heat flux of the convection heat transfer and the total heat transfer appears in Fig. 16.12.

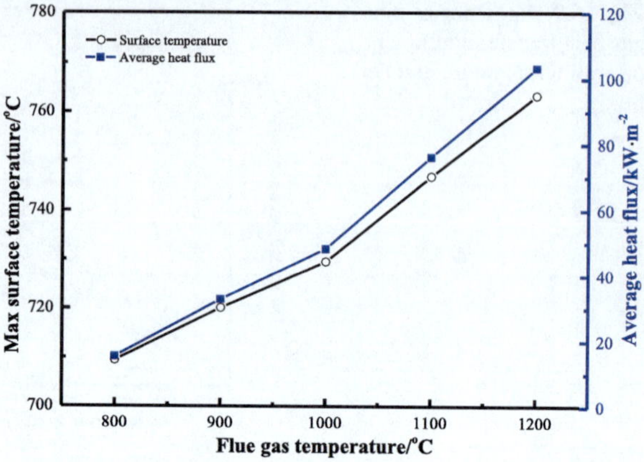

Fig. 16.11 Influence of flue gas temperature on wall temperature and heat flux

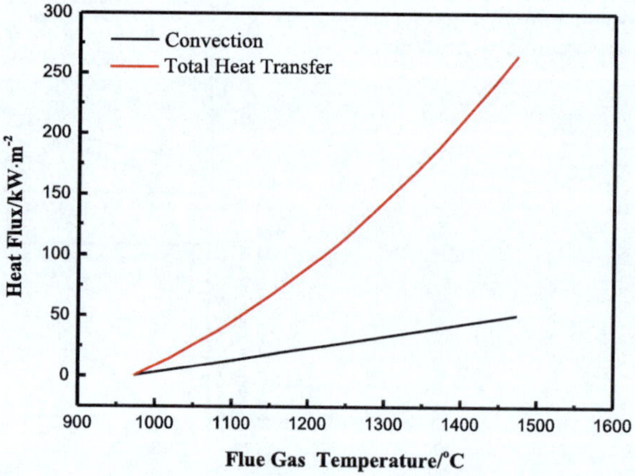

Fig. 16.12 Heat flux of radiative/convective heat transfer process when convective heat transfer coefficient defined as 100 W/m^2 K

As the flue gas temperature decreases, the radiation heat flux decreases rapidly and the total heat flux decreases rapidly. The radiation heat flux is three to four times of convective heat flux, the total maximum heat flux density of 124 kW/m^2, in which the radiation heat flux is 96 kW/m^2. Radiation and convection heat transfer of heat flux ratio is 1.5:1, the total heat flux decreased to 34.33 kW/m^2 with radiation heat flux of 24.33 kW/m^2. It can be concluded that the proportion of radiation heat transfer is larger than that of convective heat transfer as flue gas temperature above 800 °C, and vice versa. The tube surface temperature appears similar rules, because of the direct relation of heat flux to tube temperature. The influence of radiation heat transfer on the wall temperature is greater than that of convection heat transfer as flue gas temperature above 800 °C, and vice versa. For example, as calculated, the maximum of tube wall temperature is 744 °C under the 1000 °C smoke temperature; However, the temperature drops to 712 °C under 800 °C smoke temperature. The maximum of wall temperature decreases 4.4% when smoke temperature decreases from 1000 to 800 °C. When the convection heat transfer coefficient is less, the singular point

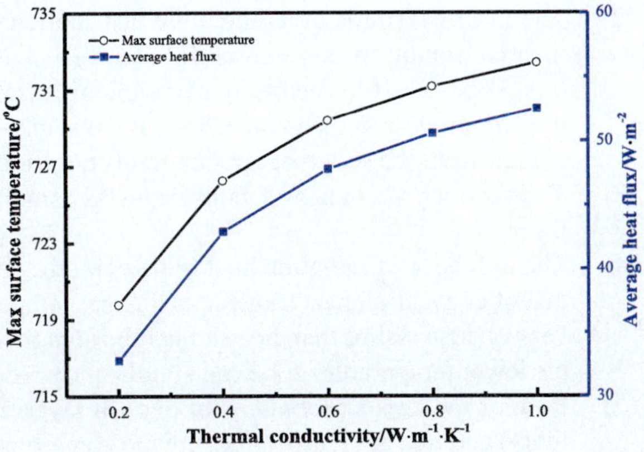

Fig. 16.13 Influence of deposit thermal conductivity on wall temperature and heat flux

of flue gas temperature will decrease. It also claims that the heating tube of high temperature zone arranged into the lower temperature zone can significantly reduce the tube wall temperature.

The influence of deposit thermal conductivity on wall temperature and heat flux is presented in Fig. 16.13. The max wall temperature and heat flux both grows with deposit thermal conductivity. The growth trend is slowing down with the conductivity.

In summarize of Figs. 16.8, 16.9, 16.10, 16.11, 16.12 and 16.13, we claim in qualitative that the deposit (contains of deposit layer thickness and thermal conductivity) and flue gas temperature are greatest influential on wall temperature; of secondary importance is the thermal resistance of the deposit layers; the least is the convective heat transfer on fireside. The flue gas temperature and deposit formation (deposit layer thickness) are mutual promotion and synergy for each other. A change in one necessarily leads to the change in the other. Moreover, the flue gas temperature is directly related to radiant heat transfer. So, in order to control the temperature of stained tube wall temperature, the most effective solution is to control the temperature of flue gas and to prevent slagging and fouling.

16.4 Conclusion

For understanding the influence of the fireside flow for the slag and fouling distribution, a tower type furnace facility was established. A comparative study of clean tube surface and stained tube surface temperature distribution was conducted. Meanwhile, the heat transfer model of clean tube and stained tube was established. The temperature circumferential distribution rule of clean tube was given and the influence of slagging and fouling on stained tube temperature distribution was also conducted. The main results are as follows:

(1) The temperature distribution of stained tube appears the bimodal distribution with the angle, but not the unimodal distribution of the clean tube.

(2) The Fitting formula of temperature distribution of clean tube under our experimental condition is given as follow. $C_R = 2e^{-10}\theta^4 - e^{-7}\theta^{-3} + e^{-5}\theta^2 + 0.0051\theta + 0.5914$. The fitting equation is also appropriate to other case of heat transfer process of clean tube. Through this method is derived from natural gas combustion, the scope of application is not limited to this.

(3) The thermal resistance of bulk deposit is smaller than that of initial deposit layer.

(4) The influence of radiation heat transfer on the wall temperature is greater than that of convection heat transfer as flue gas temperature above 800 °C, and vice versa. It also claims that the heating tube of high temperature zone arranged into the lower temperature zone can significantly reduce the tube wall temperature.

(5) Both of the deposit (contains of deposit layer thickness and thermal conductivity) and flue gas temperature are the greatest influential on wall temperature; of secondary importance is the thermal resistance of the deposit layers; the least is the convective heat transfer on fireside.

Acknowledgements This work was supported by the National Key R&D Program of China (2018YFB0604402).

References

Anderson DW, Viskanta R, Incropera FP (1987) Effective thermal conductivity of coal ash deposits at moderate to high temperatures. J Eng Gas Turbines Power 109(April):215–221

Bryers RW (1996) Fireside slagging, fouling, and high-temperature corrosion of heat-transfer surface due to impurities in steam-raising fuels. Prog Energy Combust Sci 22(1):29–120

Dai BQ, Low F, De Girolamo A, Wu X, Zhang L (2013) Characteristics of ash deposits in a pulverized lignite coal-fired boiler and the mass flow of major ash-forming inorganic elements. Energy Fuels 27(10):6198–6211

Erickson TA, Allan SE, McCollor DP et al (1995) Modelling of fouling and slagging in coal-fired utility boilers. Fuel Process Technol 44(1–3):155–171

Han J, Yu D, Wu J et al (2019) Fine ash formation and slagging deposition during combustion of silicon-rich biomasses and their blends with a low-rank coal. Energy & Fuels

Li G, Wang C, Yan Y, Jin X, Liu Y, Che D (2016) Release and transformation of sodium during combustion of Zhundong coals. J Energy Inst 89(1):48–56

Mao X, Wang P (2006) Evaluation of super 304H and XA704 tubes for ultra supercritical boiler. Dongfang Elect Rev 20(3):31–36

Phongphiphat A, Ryu C, Yang Y Bin et al (2010) Investigation into high-temperature corrosion in a large-scale municipal waste-to-energy plant. Corros Sci 52(12):3861–3874

Raask E (1985) mineral impurities in coal combustion: behavior, problems, and remedial measures. Taylor & Francis

Rezaei HR, Gupta RP, Bryant GW et al (2000) Thermal conductivity of coal ash and slags and models used. Fuel 79(13):1697–1710

Robinson AL, Junker H, Baxter LL (2002) Pilot-scale investigation of the influence of coal—biomass cofiring on ash deposition. Energy Fuels 16(2):343–355

Valmari T, Lind TM, Kauppinen EI, Sfiris G, Nilsson K, Maenhaut W (1999) Field study on ash behavior during circulating fluidized-bed combustion of biomass. 2. Ash deposition and alkali vapor condensation. Energy & Fuels 13(2):390–395

Wang X, Xu Z, Wei B et al (2015) The ash deposition mechanism in boilers burning Zhundong coal with high contents of sodium and calcium: a study from ash evaporating to condensing. Appl Therm Eng 80:150–159

Wu J, Yu D, Zeng X et al (2018) Ash formation and fouling during combustion of rice husk and its blends with a high alkali Xinjiang coal. Energy Fuels 32(1):416–424

Yang S, Tao W (2006) Heat transfer, 4th edn. Higher Education Press

Zhang X, Zhang H, Na Y (2015) Transformation of sodium during the ashing of Zhundong coal. Procedia Eng 102:305–314

Chapter 17
Numerical Investigation of Fly Ash Deposition onto Tube Bundles Inside Coal-Fired Boilers

Yipeng Li, Qian Huang, Lu Duan, and Shuiqing Li

Abstract The particle-laden flow across tube arrays is numerically investigated as a model problem of ash deposition onto the convection heat transfer surfaces inside practical boilers. The transverse pitch ratio is found to significantly affect the impaction efficiency onto each tube. For the frontmost tube, a decrease in the transverse pitch ratio enhances the impaction efficiency of fly ash particle of all sizes. Then, the downstream tubes are more frequently exposed to fine particles. Nevertheless, a small transverse pitch ratio makes it possible for coarse mode particles to reach the second- and third-row tubes in the adjacent column after rebound from the front tube. Finally, a comparison between experiments and simulation implies that the existing sticking/rebound model may overestimate the ash fouling rate.

Keywords Ash deposition · Tube bundle · CFD · Impaction efficiency · Sticking probability

17.1 Introduction

The emerging trends of flexible boiler loads and increasing economic requirements for burning low-rank coal in many countries like China have put higher demand to the availability, safety and efficiency of the unit operation. Among the several technical challenges, the ash-related problem is a major issue that hinders heat transfer, damages the tube surface, and even blocks the flow passenger (Bryers 1996). The current strategy to remove ash deposits in boiler operation is to deploy the sootblower (Valero and Cortes 1996). There is always a strong need to optimize the sootblowing procedure so that the cost can be minimized while the ash deposition problem is effectively mitigated. For this purpose, a prior prediction of ash deposition inside

Y. Li · Q. Huang (✉) · S. Li
Department of Energy and Power Engineering, Tsinghua University, Beijing 100084, China
e-mail: huangqian@tsinghua.edu.cn

L. Duan
China Coal Research Institute Company of Energy Conservation, Beijing 100013, China

J. Lyu and S. Li (eds.), *Clean Coal and Sustainable Energy*,
Environmental Science and Engineering,
https://doi.org/10.1007/978-981-16-1657-0_17

the boiler is helpful to the arrangement and commission of sootblowers at different locations.

As a long-standing research topic, extensive studies have been devoted to the fouling mechanisms and simulations (Bryers 1996; Valero and Cortes 1996; Huang et al. 2019). It is well known that the particles reach the surface of heat transfer tubes through fluid drag (inertia impaction), thermophoresis, and turbulence fluctuation (Bryers 1996). The impaction efficiency of particles in the flow around a single cylinder has been well understood with several useful correlations proposed, which is mainly determined by the particle Stokes number and the flow Reynolds number (Huang et al. 2019). However, whether the impacting particle can stick or rebound is a complex process. The viscosity model provides some most commonly-applied working formulas for the determination of the sticking probability (Walsh et al. 1990; Degereji et al. 2012; Kleinhans et al. 2017), but the validity and accuracy need further examination.

However, different from mechanistic studies of single-probe deposition in lab-scale furnaces, practical boilers are featured with inline tube bundles in the convection zone and generally more turbulent flows. The tube array introduces a blockage effect that may increase the impact efficiency of ash particles, as compared with the single probe cases (Zdravkovich and Bearman 1998). Moreover, the shielding effect of the frontmost tube can significantly reduce the impaction efficiency of incoming ash particles onto the downstream tubes. Some existing works simulated the ash deposit formation onto tube arrays and implied that the deposits are more prone to form on the lateral sides of the downstream tubes (Tomeczek and Waclawiak 2009). But there is still a lack of detailed investigation on the effects of the tube array configuration and flow condition on the impaction efficiency onto downstream tubes.

In this work, we numerically tackle the problem through computational fluid dynamics (CFD) simulations. The details of the simulation are given in the next section. In the section of Results, we quantify the impaction efficiency onto the tubes in different rows under various conditions. We then examine the performance of several sticking/rebound criteria.

17.2 Methodology

To reduce the computational cost, the inline tube bundle is simplified to 2D cases with the periodic boundary condition in the spanwise direction. As plotted in Fig. 1a, one column and three rows are involved in the computation domain. The geometrical structure is mainly characterized by the pitch ratios in the spanwise and longitude directions, namely, s_1/D and s_2/D. In this work, we fix $s_2/D = 2$ and vary s_1/D. The cases are listed in Table 17.1. Usually small values of s_1/D are encountered in the superheaters and reheaters (say, 1.5 or 2) (Tomeczek and Waclawiak 2009), while large values of s_1/D were used to reveal the single tube behaviors by eliminating the blockage effect (Zdravkovich and Bearman 1998). The tube diameter D is 40 mm and the surface is kept at 873 K. The flue gas is 1300 K and consists of 8 vol% of O_2,

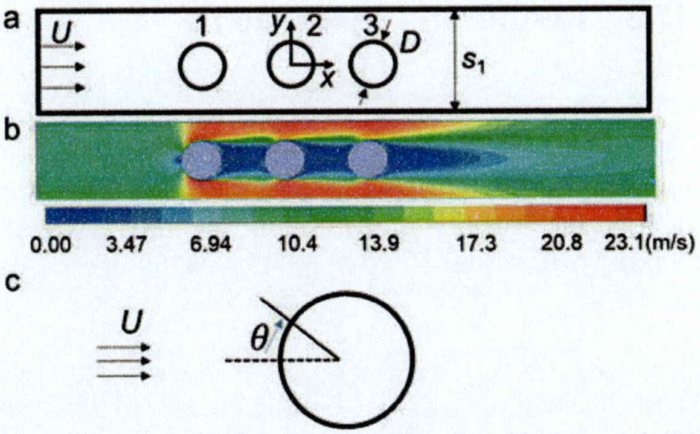

Fig. 17.1 **a** Schematic of the simulation case; **b** velocity magnitude for the case of $s_1/D = 1.5$ and $U = 10$ m/s; **c** azimuthal angle θ

Table 17.1 Simulation cases

Transverse pitch ratio	s_1/D	1.5, 2.0, 4.0, 6.0
Longitudinal pitch ratio	s_2/D	2.0
Tube diameter	D	40 mm
Inlet velocity	U	2 m/s, 10 m/s
Inlet temperature	T	1300 K
Numerical method	Steady state, k-ε model	
Ash particle density	ρ_p	2000 kg/m^3
DPM boundary condition	Stick, Rebound ($e = 1$)	

15 vol% of CO_2, 7 vol% of H_2O and balancing N_2. Two inlet gas velocities are used: 10 m/s for practical boilers (corresponding to $Re_D = 2400$), whereas 2 m/s ($Re_D = 480$) for our previous experiments in the 25 kW self-sustained combustor (Li et al. 2015; Huang et al. 2018).

The 2-D steady-state flow field is simulated with k-ε model using enhanced wall functions in ANSYS Fluent 15.0. The near wall mesh follows the requirements proposed in Ref. (Weber et al. 2013). Figure 1b shows a typical contour of the velocity magnitude in the case of $s_1/D = 1.5$ and $U = 10$ m/s. Clearly, the flow accelerates on the two sides of each tube, and a low velocity zone appears in the downstream region of each tube. The results have been validated as grid-free.

In the context of one-way coupling, ash particles injected into the known flow field are driven by the drag force and thermophoresis (Bryers 1996). The boundary condition for impacting particles is set as either "trap" (all stick) or "rebound" with the coefficient of restitution e being 1. The trajectories are tracked by the discrete phase model (DPM) so that the overall impaction efficiency and the azimuthal distribution of impacting particles can be determined. Figure 1c shows the definition of the azimuthal angle θ which ranges from $-180°$ to $180°$ and the zero value lies at the windward stagnation point.

17.3 Results and Discussion

We first investigate the impaction efficiency of ash particles onto the frontmost tube. The impaction efficiency η_i is defined as the ratio of impacting particles over those injected from the projected area of the tube. It is well known that η_i is heavily dependent on the Stokes number, a dimensionless parameter with the form (Friedlander 2000):

$$\mathrm{St} = C_c \frac{\rho_p d_p^2 U}{9 \mu D} \tag{17.1}$$

Here d_p is the particle diameter and μ is the fluid viscosity. (Pa·s). C_c is the slip correction coefficient with the detailed expression given in Ref. (Friedlander 2000). Figure 17.2 presents the impaction efficiency for different values of s_1/D and flow turbulence $\mathrm{Re_D}$. An isothermal DNS result with $\mathrm{Re_D} = 421$ (Haugen and Kragset 2010) is used for a comparison, which validates the accuracy of our simulation. Generally, a decrease in s_1/D enhances the impaction efficiency of inertial particles with St > 0.1 (see Fig. 2a), whereas the increase of flow Reynolds number suppresses the impaction efficiency of thermophoresis-driven fine particles with St < 0.1 (see Fig. 2b).

A clearer effect of s_1/D on the impaction efficiency is shown in Fig. 17.3 for ash particles with different values of St. Note that for different inlet velocity U, the same St corresponds to varied particle size d_p. We see that, with the increase of s_1/D, a

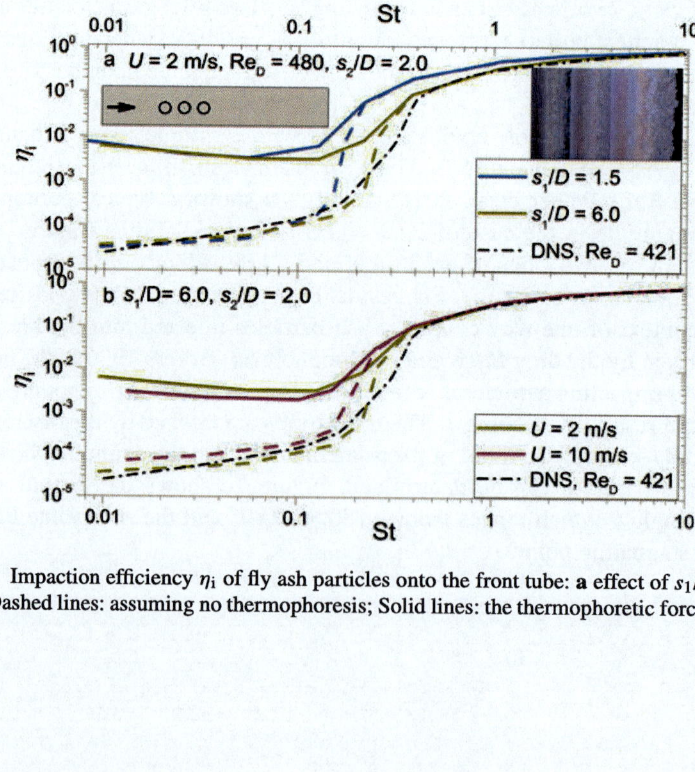

Fig. 17.2 Impaction efficiency η_i of fly ash particles onto the front tube: **a** effect of s_1/D; **b** effect of $\mathrm{Re_D}$. Dashed lines: assuming no thermophoresis; Solid lines: the thermophoretic force is present

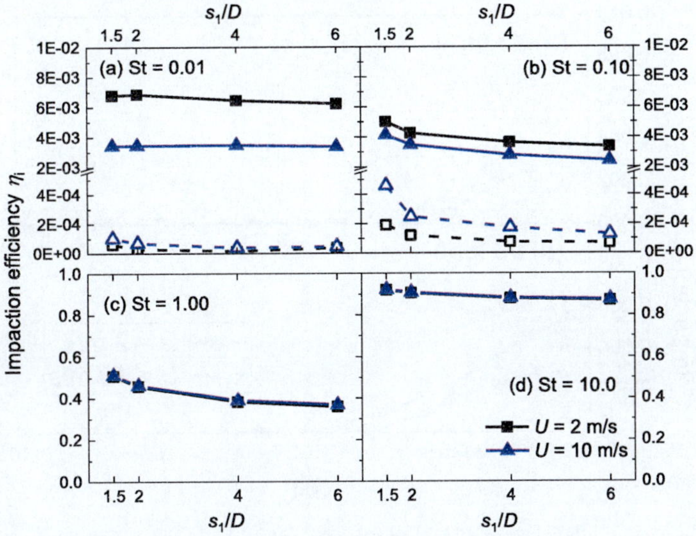

Fig. 17.3 Impaction efficiency η_i of fly ash particles onto the front tube. Solid lines: the thermophoretic force is present; Dashed lines: assuming no thermophoresis

decreasing trend can be observed for all particle sizes, though more apparent for larger particles. Meanwhile, the curve falls more slowly when s_1/D becomes as large as 4–6, during which the blockage effect weakens gradually. As for the effect of Re_D, it is seen that the impaction efficiency of smaller particles drops to a greater extent when the inlet gas velocity increases from 2 to 10 m/s. This implies that the formation of the inner deposited layer may become less efficient.

We then look into the impaction efficiency of ash particles onto the #2 and #3 tubes. In Fig. 17.4, it is assumed that all particles either stick or fully rebound after colliding with any tube for the cases of $s_1/D = 6$. Different from the #1 tube, the impaction efficiency onto the #2 or #3 tube decreases with St. In particular, the particles with St > 0.1 cannot reach the surfaces of #2 and #3 tubes regardless of the boundary conditions for incident particles. This is mainly attributed to the shielding effect of the upstream tubes that either trap or deviate incident particles. Fine particles are easier to follow the streamline so that they are more likely to impact the downstream tubes. For the case of $U = 2$ m/s, the #3 tube features higher impaction efficiencies than the #2 tube for fine particles, while this relation disappears for the case $U =$ 10 m/s.

Nevertheless, the fact that the sticking/rebound condition does not change the impaction efficiency onto downstream tubes becomes invalid when the tube bundle becomes much compact. Figure 17.5 shows the impaction efficiency of ash particles onto the #2 tube for the case of $s_1/D = 1.5$. It is found that the fine particles with St < 0.01 have the same impaction efficiency for the sticking and rebound conditions. On the contrary, larger particles with St > 1 exhibit a remarkable impaction efficiency which increases with St when the full rebound condition ($e = 1$) is adopted; otherwise

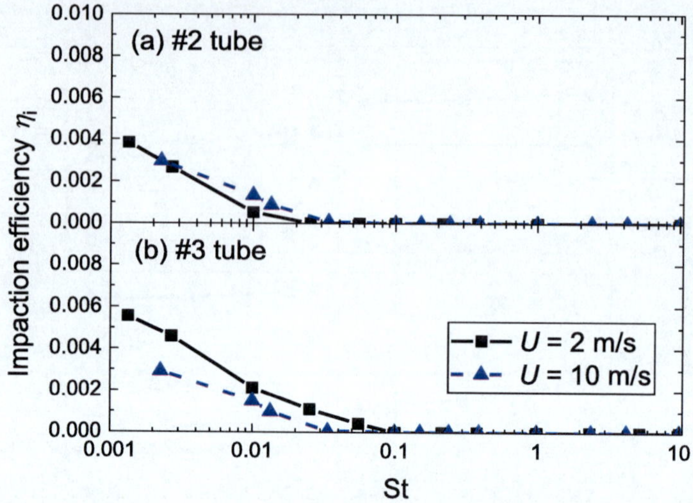

Fig. 17.4 Impaction efficiency η_i of fly ash particles onto the **a** #2 and **b** #3 tubes. Particles are assumed to either stick or fully rebound after impacting any tube. The cases of $s_1/D = 6$ are shown here

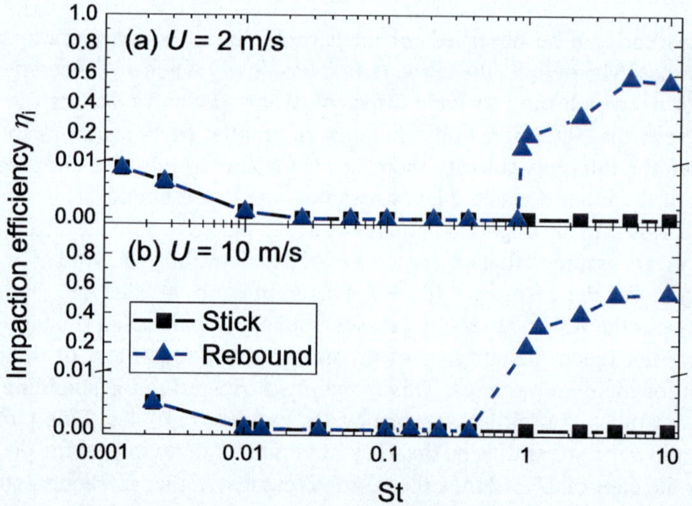

Fig. 17.5 Impaction efficiency η_i of fly ash particles onto the #2 tube for the cases of $s_1/D = 1.5$. **a** $U = 2$ m/s and **b** $U = 10$ m/s

the impaction efficiency keeps zero for the all sticking condition ($e = 0$). This significant difference can be better interpreted based on the particle trajectory. Indeed, many large particles that rebound from the first tube travel to the adjacent column which is dealt with by the periodic condition in our simulation. This interaction between multiple columns is impossible for large values of s_1/D.

Figure 17.6 further shows the azimuthal distribution density of the impacting particles onto the #2 tube for the case of $s_1/D = 1.5$ and $U = 10$ m/s. For fine particles with $d_p = 1$ μm (corresponding to St = 0.002), the distribution is roughly centered at ± 60° and has a large variance. This is caused by the great ability of 1 μm-particle to follow the streamline and spread in a wider range. Note that this distribution is the same for both sticking and rebound conditions of the first tube. For coarser particles with St = 1 and 10, though, only assuming full rebound condition ($e = 1$) can they reach the second tube. The particles with St = 1 also collide with the lateral sides of the tube with a much narrower variance, and can rarely reach the windward region between − 60° and + 60°. However, the coarse particle with St = 10 (corresponding to $d_p = 92.25$ μm) results in a nearly even impact distribution in the region of −60°~+60° and two tails in the regions of 60°~120° and −120°~−60°, respectively. Therefore, in the practical cases of superheaters and reheaters, the windward region of the second-row tube is expected to undergo intensive impaction of coarse mode fly ash.

We then examine the behaviors of sticking/rebound criterion on the ash deposition quantity based on our previous experiments on the 25 kW self-sustained pulverized coal combustor. The single-probe ash deposition rate under ~1300 K, particle size distribution of fly ash, and the elemental compositions of the size-segregated fly ash samples for a sodium-rich low-rank Zhundong lignite are reported in our previous works (Li et al. 2015; Huang et al. 2018). To simulate the ash deposition process, the particle impaction efficiency, particle temperature and impact velocity are obtained via Fluent DPM. As for the sticking probability η_s, unlike the abovementioned all sticking ($\eta_s = 1$) or full rebound ($\eta_s = 0$) assumption, a kind of viscosity model was widely used in the literature (Walsh et al. 1990):

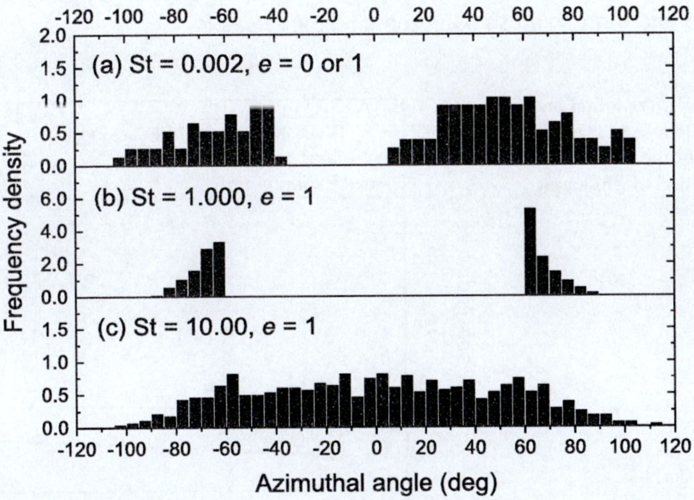

Fig. 17.6 Azimuthal distribution density of the impacting particles onto the #2 tube for the case of $s_1/D = 1.5$ and $U = 10$ m/s

$$\eta_s = \begin{cases} \frac{\mu_c}{\mu_p}, if\,\mu_p > \mu_c \\ 0, otherwise \end{cases} \tag{17.2}$$

Here μ_p denotes the particle viscosity and μ_c is a critical value of the particle viscosity. μ_p is assumed to be dependent on the particle temperature T_p and composition. The Watt-Faraday model is used in this work with the following form (Degereji et al. 2012):

$$\log(\mu_p) = \frac{10^7 M}{(T_p - 150)^2} + c \tag{17.3}$$

The free parameters M and c are determined by the weight percentages of SiO_2, Al_2O_3, equiv. Fe_2O_3, CaO and MgO of each ash particle. As for μ_c, a commonly used value is $\mu_c = 10^8$ Pa·s (Walsh et al. 1990; Degereji et al. 2012). Recently, a new formula was proposed that correlates μ_c with the kinetic energy of incident particle (Kleinhans et al. 2017):

$$\mu_c = 5 \times 10^{-12} \cdot E_{kin,p}^{-1.78} \tag{17.4}$$

The both criteria for μ_c are used in this work.

Figure 17.7 shows the such predicted and experimental deposited mass as a function of elapsed time for Zhundong lignite. Without deposit shedding, the fouling rates of the "$\mu_c = 10^8$ Pa·s" assumption, Eq. (17.4) assumption, and the experiment are 84.2 mg/min, 69.5 mg/min and 5.78 mg/min, respectively. Therefore, the use of kinetic-energy-dependent critical viscosity decreases, but still significantly overestimates, the fouling rate. This calls for more accurate quantification of the fly ash properties and more valid sticking/rebound models.

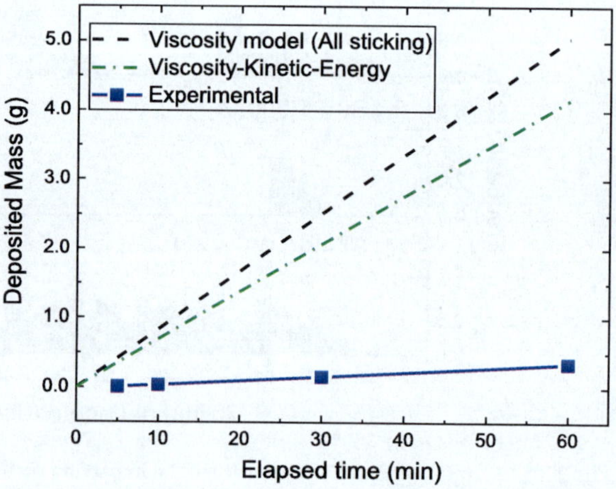

Fig. 17.7 Experimental and simulated deposited mass as a function of the elapsed time for a kind of Zhundong lignite

17.4 Conclusions

We numerically investigate the fly ash deposition behaviors onto the tube arrays with the periodic boundary condition. The impaction efficiency of particles onto the frontmost tube is quantified for different particle Stokes number, flow Reynolds number and transverse pitch ratio s_1/D. For the case of $s_1/D = 6$, the impaction efficiency of particles onto the second and third rows decreases with the particle Stokes number, and drops to zero when St > 0.1, regardless of the sticking/rebound condition. However, in the case of $s_1/D = 1.5$, coarse ash particles are able to impact the second-row tube in the adjacent column if the full-rebound condition is adopted. The resultant impaction density in the azimuthal direction is nearly evenly distributed in the range of −60°~+60°. Finally, we examine several viscosity models for the prediction of the fouling rate of Zhundong lignite, and find that the fouling rate is significantly overvalued. Hence, more valid sticking/rebound criteria are called for in future work.

Acknowledgements This work was supported by the National Key Research and Development Program of China (2017YFB0603203).

References

Bryers RW (1996) Fireside slagging, fouling, and high-temperature corrosion of heat-transfer surface due to impurities in steam-raising fuels. Prog Energy Combust Sci 22:29–120

Degereji MU, Ingham DB, Ma L et al (2012) Prediction of ash slagging propensity in a pulverized coal combustion furnace. Fuel 101:171–178

Friedlander SK (2000) Smoke, dust, and haze: fundamentals of aerosol dynamics. Oxford University Press, New York

Haugen NEL, Kragset S (2010) Particle impaction on a cylinder in a crossflow as function of stokes and reynolds numbers. J Fluid Mech 661(25):239–261

Huang Q, Zhang Y, Yao Q et al (2018) Mineral manipulation of Zhundong lignite towards fouling mitigation in a down-fired combustor. Fuel 232:519–529

Huang Q, Li SQ, Shao Y et al (2019) Dynamic evolution of impaction and sticking behaviors of fly ash particle in pulverized coal combustion. Proc Combust Inst 37:4419–4426

Kleinhans U, Wieland C, Babat S et al (2017) Ash particle sticking and rebound behavior: a mechanistic explanation and modeling approach. Proc Combust Inst 36(2):2341–2350

Li G, Li SQ, Huang Q et al (2015) Fine particulate formation and ash deposition during pulverized coal combustion of high-sodium lignite in a down-fired furnace. Fuel 143:430–437

Tomeczek J, Waclawiak K (2009) Two-dimensional modelling of deposits formation on platen superheaters in pulverized coal boilers. Fuel 88(8):1466–1471

Valero A, Cortes C (1996) Ash fouling in coal-fired utility boilers. Monitoring and optimization of on-load cleaning. Prog Energy Combust Sci 22:189–200

Walsh PM, Sayre AN, Loehden DO et al (1990) Deposition of bituminous coal ash on an isolated heat exchange tube: effects of coal properties on deposit growth. Prog Energy Combust Sci 16(4):327–345

Weber R, Schaffel-Mancini N, Mancini M et al (2013) Fly ash deposition modelling: requirements for accurate predictions of particle impaction on tubes using RANS-based computational fluid dynamics. Fuel 108:586–596

Zdravkovich MM, Bearman PW (1998) Flow around circular cylinders, vol 1: fundamentals. Oxford Science Publications

Chapter 18
Effect of Refractory Lining Thickness on Slag Layer Behavior in Cyclone Barrel

Chunli Tang, Yueyi Hu, Tao Zhu, Limin Wang, Yanhua Liu, and Defu Che

Abstract The cyclone-fired boiler has attracted much attention in recently yeares. The thickness of the refractory lining is closely related to the slag behavior which is of great importance for the operation of the cyclone-fired barrel. In this paper, the slag layer behavior in the cyclone barrel is investagated numerically when the refractory lining thickness is constant or varied. The results show that the refractory lining has a great effect on its surface temperature. Besides, the refractory lining should not be too thick to ensure the presence of the solid slag layer at the high temperature zone of the cyclone barrel. Therefore, in the cyclone barrel, the thickness of the refractory lining of the air inlets sections can be increased to reduce the heat dissipation, whereas the thickness of the refractory lining of the outlet section can be reduced to obtain a solid slag layer thick enough to prevent the erosion of the refractory lining. The proposed scheme with variable thickenss of the refractory lining is proved effective by the numerical simulation.

Keywords Slag layer behavior · Refractory lining · Thickness · Cyclone barrel · Numerical simulation

18.1 Introduction

The cyclone-fired boiler has been proposed by Babcock & Wilcox (B&W) Company in 1940s for the coal with low ash melt point (Kitto and Stultz 2005). With the development of the pollution control technology and combustion technology, the most serious problem with high NO_x emission for the tranditional cyclone-fired boiler can be overcome. Hence, the cyclone-fired boiler has attracted much attention due to its high combustion efficiency, high slag capture ratio and low NO_x emission

C. Tang · Y. Liu (✉)
School of Human Settlements and Civil Engineering, Xi'an Jiaotong University, Xi'an 710049, China
e-mail: yhliu@mail.xjtu.edu.cn

Y. Hu · T. Zhu · L. Wang · D. Che
School of Energy and Power Engineering, Xi'an Jiaotong University, Xi'an 710049, China

J. Lyu and S. Li (eds.), *Clean Coal and Sustainable Energy*, Environmental Science and Engineering,
https://doi.org/10.1007/978-981-16-1657-0_18

(Wu et al. 2017; Deng et al. 2016; Tang et al. 2017). Recently researches have also indicated that the cyclone-fired boiler is also suitable for the high alkali coal which is widely distributed in Xinjiang province of China (Deng et al. 2016; Wu et al. 2015). Hence, the Ministry of Science and Technology of China has encouraged the development of the cyclone-fired boiler for the high alkali coals (Ministry of Science and Technology of the People's Republic 2017).

The cyclone-fired boiler always composed of one or several cyclone barrels and the main furnace. The cyclone barrel is the key equipment where the burning mostly occurs. The cyclone barrel is enclosed with the water-cooled wall. Its internal wall is lined with refractory lining. In the cyclone barrel, most of the molten ash particles driven by the strong centrifugal force are trapped and then form the slag layer on the refractory lining. The refractory lining helps maintain high temperature level in the cyclone barrel and can protect the water-cooled wall from the erosion of the high speed particles and high temperature gas. The thickness of the refractory lining has a great effect on the behavior of the formed slag layer, which is closely related to the safe and efficient operation of the cyclone barrel.

Much research has been conducted on the slag layer behavior of the gasifier which is also using slag tapping by numerical simulation (Zhang et al. 2018a, b, c; Kim et al. 2017; Bi et al. 2015; Ye et al. 2015; Ye and Ryu 2015). Seggiani (Seggiani 1998) established a slag layer behavior model for describing the flow and heat transfer of the slag layer. Yong et al. (Yong and Ghoniem 2012; Yong et al. 2012) and Wang et al. (2007) have developed submodes based on the Seggiani's model for particle capture and wall burning more accurately. Bi et al. (2015) have predicted the slag layer thickness of the gasifiers enclosed with membrane wall or refractory wall by coupling the slag behavior model and 3-D gasifier model. The result indicated that the critical viscosity temperature and thermal resistance of the slag layer have a great influence on the slag behavior. The cyclone barrel is different from the gasifier in the operation condition, product compostion and cooling wall. In our previous work (Wu et al. 2017; Tang et al. 2017, 2019), a slag behavior model is established for the cyclone barrel. The effect of the operation condition and coal characteristics on the slag layer behavior were investigated numerically. The result indicated that the flow and heat transfer characteristics of the slag layer are closely releated to the coal characteristics.

In this paper, the slag layer behavior in the cyclone barrel is investagated numerically when the refractory lining thickness is constant or varies along the axial direction of the cyclone barrel. In addition, an optimized scheme of refractory lining thickness is seclected to achieve safer and more efficient operation of the cyclone barrel.

18.2 The Modeled Cyclone Barrel

The cyclone barrel with thermal power of 61 MW is selected for the study. The schematic diagram of the cyclone barrel is shown in Fig. 18.1. The primary air inlet is an annular passage at the top of the cyclone barrel. Two secondary air inlets are

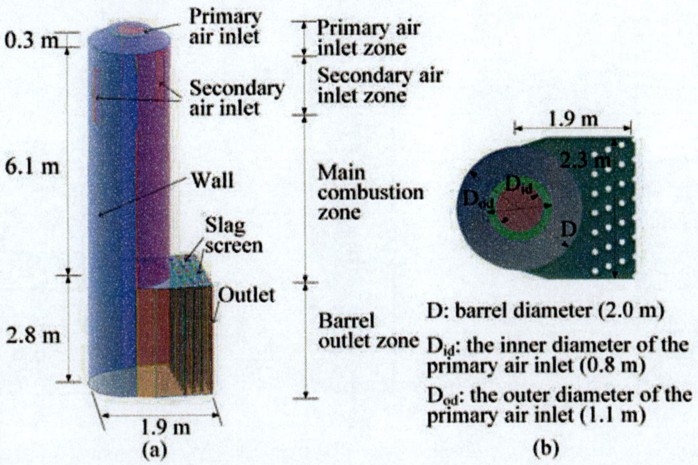

Fig. 18.1 The schematic diagram of the cyclone barrel

arranged in the upper part of the cyclone barrel. The diameter of the cyclone barrel is about 2.0 m and the height is about 9.2 m. Four rows slag tap tubes are arranged at the exit of the cyclone barrel. The generated high temperature flue gas flows into the main furnace through the slag screen. The wall of the cyclone barrel is divided to four zones of the primary air inlet zone, secondary air inlet zone, main combustion zone and barrel outlet zone for the convenicncc of analysis.

18.3 Numerical Models

The model is composed of two models: the model of the slag layer behavior and the model of the flow and combustion in the cyclone barrel. The models simulated the flow and combustion in the cyclone barrel are selected from the commercial software, as shown in Table 18.1. The flow and heat transfer characteristics of the slag layer are calculated by the slag behavior moel which is described by the User Defined Function. In orde to analyse the slag layer behavior, the cyclone barrel wall are divided to numerious control volumes. The mass, momentum and energy

Table 18.1 The selected models

Process	Model
Turbulent flow	Realizable *k*-ε
Gaseous combustion	Mixture-fraction/PDF
Devolatilization	Single rate
Char combustion	Kinetics/diffusion-limited
Radiation	Discrete Ordinates

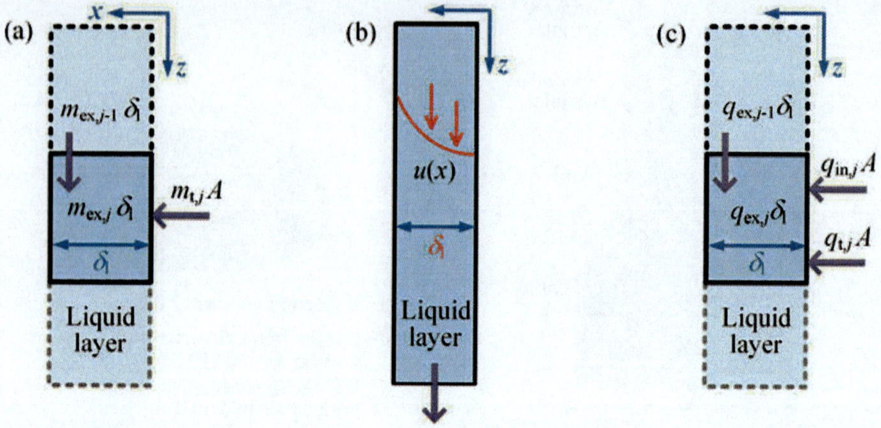

Fig. 18.2 Conservations: **a** mass conservation; **b** momentum conservation; and **c** energy conservation

conservations are established for each control volumes, as shown in Fig. 18.2. In the steady state, the mass increment of the slag in each control volum is zero, as shown in Fig. 18.2a. Hence, the mass conservation can be expressed by

$$m_{\text{ex},j}\delta_{\text{l}} = m_{\text{t},j}A + m_{\text{ex},j-1}\delta_{\text{l}} \tag{18.1}$$

where m_{ex} is the mass flux of the slag per unit length, j is the control volume index, m_{t} is the mass flux of the trapped particles per unit area, δ_{l} is the liquid slag layer thickness, A is the face area of each control volume on the surface of the slag layer.

The velocity and the thickness of the liquid slag layer can be calculated by the momentum conservation equation (as shown in Fig. 18.2b)

$$\begin{gathered} \frac{d}{dx}\left(\mu\frac{du}{dx}\right) = -\rho g, \\ \text{at } x = 0;\ \mu\frac{\partial u}{\partial x} = -\tau \\ \text{at } x = \delta_{\text{l}};\ u = 0 \end{gathered} \tag{18.2}$$

where μ, u and ρ are the viscosity, velocity and density of the liquid slag. The thickness of the liquid slag layer δ_{l} can be calculated by

$$\delta_{\text{l}} = \left(\sum_{j=1}^{j} m_{\text{ex},j}\mu \Big/ \left(\frac{1}{3}\rho^2 g + \frac{1}{4}\frac{u_{\text{p}}^2\rho^2 m_{\text{t},j}}{2\sum_{j=0}^{j} m_{\text{ex},j}}\right)\right)^{\frac{1}{3}} \tag{18.3}$$

The velocity of the liquid slag layer can be calculated by

$$u_l = \frac{1}{\delta_l}\int_0^{\delta_l} u(z)dr = \frac{\rho g \delta_l^2}{3\mu} + \frac{\tau\delta}{2\mu} \tag{18.4}$$

As shown in Fig. 18.2c, the energy conservation equation in each control volume can be expressed as

$$\begin{aligned} q_{ex,j}\delta_l &= q_{in,j}A + m_{t,j}C_p T_p A + q_{ex-1,j}\delta_l \\ &= \rho_{l,j}c_{l,j}\int_0^{\delta_{l,j}} u_j(z)T_j(z)dx \\ &\text{at } x = 0;\ T = T_s;\ \frac{\partial T}{\partial x} = -\frac{q_{in}}{k_l}, \\ &\text{at } x = \delta_l;\ T = T_{cv};\ \frac{\partial^2 T}{\partial x^2} = 0 \end{aligned} \tag{18.5}$$

q_{ex} is the heat flux of slag per unit length, q_{in} is the heat flux to the internal surface of cyclone barrel, c_p and T_p are the particle specific heat and particle temperature, respectively. Then the thickness of the solid slag layer and the surface temperatures of the refractory lining and the slag layers can be calculated by soving the equivations (1), (2) and (5) toghter. The behavior of the slag layer is used as the boundary condition for the model of the flow and combustion in the cyclone barrel.

18.4 Simulation Conditons

The material of the metal tube and the studs is the 21Cr1MoV. The refractory material of SiC is covered around the studs on the tube. The refractory material and the studs make up the refractory lining. The thermal conductivity of the refractory lining is 6.93 W/(m·K). Six cases with different thicknesses of the refractory lining are investigated in this study. The thicknesses of the refractory lining of Cases 1, 2 and 3 keep constants with 10, 20 and 30 mm along the axial direction of the cyclone barrel. The thicknesses of the refractory lining of Cases 4, 5 and 6 vary along the axial direction of the cyclone barrel. The simulation conditions are listed in Table 18.2. Case 2 is selected as the base case. Hence, the refractory lining thicknessses of the air inlets zones of Cases 4 and 6 are thicker than that of Case 2 by 10 mm. The refractory lining thicknessses of the outlet zone is thiner than that of Case 2 by 10 mm.

Table 18.2 Simulation conditions

Cases	Zones of the air inlets/mm	Main combustion zone/mm	Outlet zone/mm
Case 1	10	10	10
Case 2	20	20	20
Case 3	30	30	30
Case 4	30	20	20
Case 5	20	20	10
Case 6	30	20	10

18.5 Results and Discussion

18.5.1 Temperature and Heat Flux

Figures 18.3 and 18.4 show the surface temperatures of the liquid slag layer and the refractory lining under different cases, respectively. Figure 18.5 shows the heat flux transferred to the water-cooled wall. As shown in Fig. 18.3, the effect of the refractory lining thickness on the surface temperature of the liquid slag layer is almost negligible. The initial position of the slag layer formation is also hardly affected by the thickness of the refractory lining.

In the primary- and secondary-air inlets zones and the upper part of the main combustion zone, there is no slag layer is formed on the wall of the cyclone barrel.

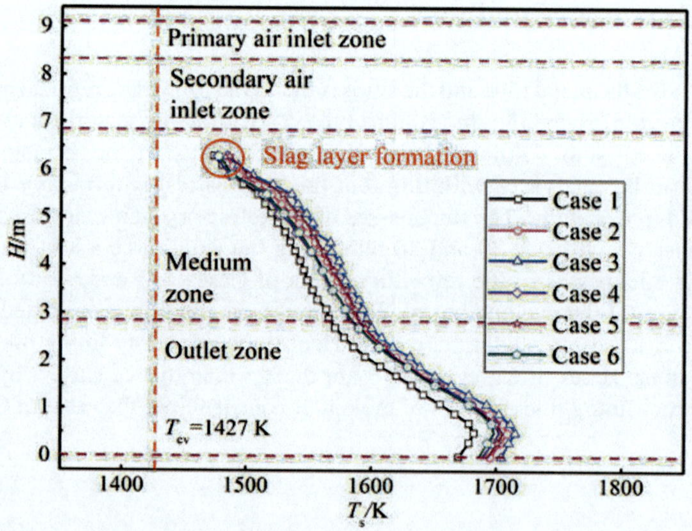

Fig. 18.3 Surface temperature of liquid slag

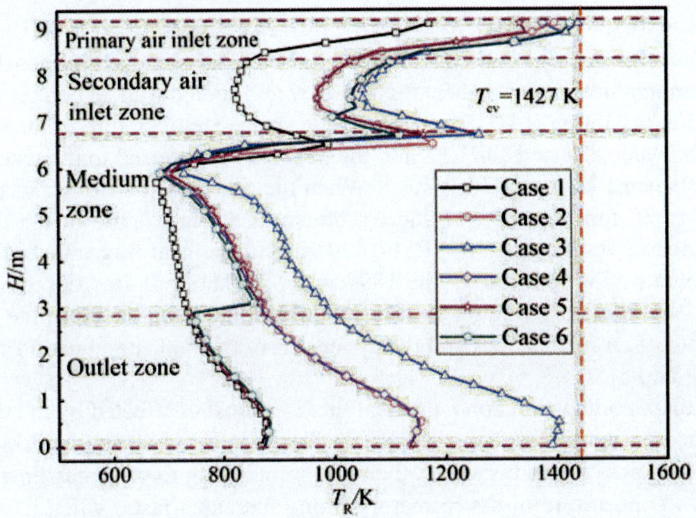

Fig. 18.4 The surface temperature of refractory material

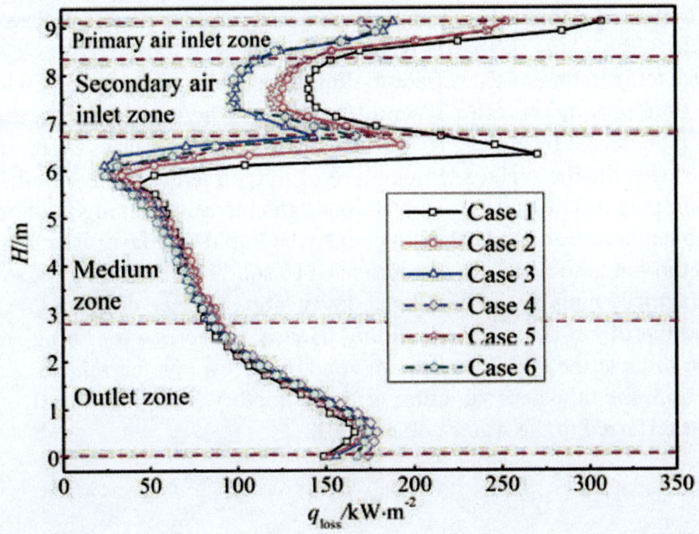

Fig. 18.5 Heat loss

As shown in Figs. 18.4 and 18.5, the thickness variation of the refractory lining can greatly affect both the surface temperature of the refractory lining and the heat flux to the water-cooled wall because the refractory lining acts as the insulating layer for the barrel. The surface temperature of the refractory lining increases with its thickness. The heat flux increase with the derease in the thickness of the refractory lining.

Compared with the cyclone barrel with the refractory lining of 10 mm (Case 1), the cyclone barrel with the refractory lining of 30 mm (Cases 3, 4 and 6) has the higher surface temperature and lower heat flux to the water-cooled wall. When the thickness of the refractory lining is 10 mm (Case 1), the surface temperature of the refractory lining is between 822 and 1171 K, and the heat flux transferred to the water-cooled wall is between 141 and 307 $kW{\cdot}m^{-2}$. When the thickness of the refractory lining increases to 30 mm (Cases 3, 4 and 6), the surface temperature of the refractory lining increases to the range of 1051-1439 K, and the heat flux transferred to the water-cooled wall is between 99 and 189 $kW{\cdot}m^{-2}$. Therefore, in order to reduce the heat loss, the refractory linging should be laid as thick as possible in the primary- and secondary-air inlets zones and the upper part of the main combustion zone (like Cases 3, 4 and 6).

In the main combustion zone, the heat flux is almost unaffected by the thickness. The surface temperature of the refractory lining is in the range of 700–1000 K when its thickness varies between 10 and 30 mm. Along the flue gas flow process, the surface temperature of the refractory lining increases faster when its thickness increases from 10 to 30 mm. In the junction of the main combustion zone and the outlet zone, the surface temperature increases by about 120 K when the thickness increases by 10 mm.

In the outlet zone, the surface temperature of the refractory lining increase slowly from 734 to 870 K along the barrel when the thickness is 10 mm (Cases 1, 5 and 6). The surface temperature of the refractory lining increases fastly from 980 to 1400 K when the thickness increases to 30 mm (Case 3). At the barrel bottom, the surface temperature increases by about 270 K when the thickness increases by 10 mm. It can be deduced that the the surface temperature of the refractory lining is might higher than the temperature of critical viscosity once the refractory lining is thick enough at the high temperure zone. At this time, only the liquid slag layer is covered on the surface of the water-cooled wall. The absence of a solid slag layer may cause abrasion of the refractory lining by the liquid slag layer. The presence of the solid slag layer can extend the life of the refractory lining. Hence, the refractory lining should not be laid too thick at the high temperature zone of the cyclone barrel. Besides, in the outlet section, the thickness variation of the refractory lining can hardly affect the heat flux transferred to the water-cooled wall.

18.5.2 Thickness and Velocity

Figures 18.6 and 18.7 show the effect of the refractory lining thickness on the thickness and velocity of the liquid slag layer, respectively. The thickness of the liquid slag layer is hardly affected by the thickness of the refractory lining. The thickness of the liquid slag layer is about 3–4 mm. However, the mean temperature of the liquid slag layer increases with the refractory lining, resulting in the decrease in the viscosity of the liquid slag layer. Hence, the liquid slag layer flow faster. When the thickness of the refractory lining is 10 mm (Case 1), the velocity of the liquid slag

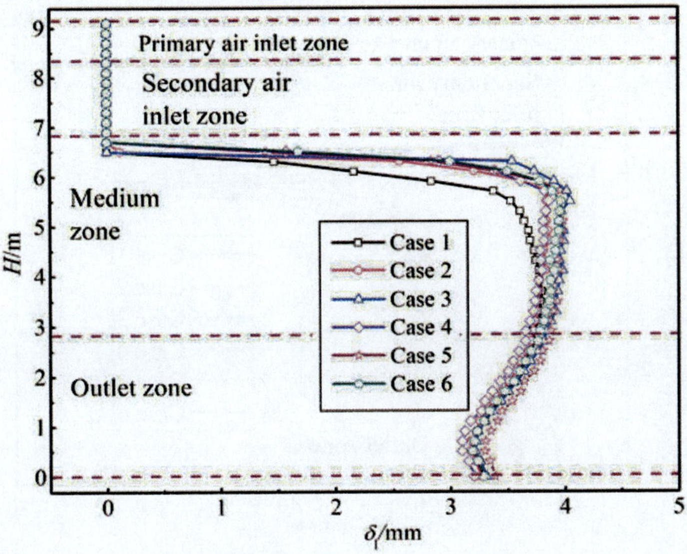

Fig. 18.6 The thickness of the liquid slag layer

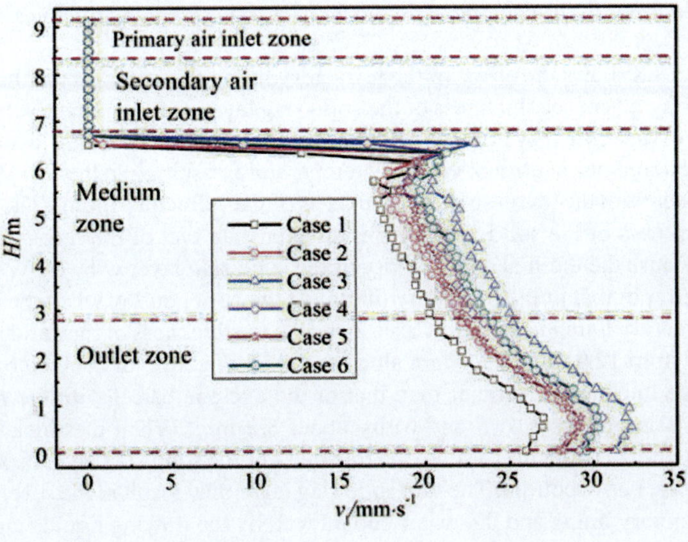

Fig. 18.7 The velocity of the liquid slag layer

layer increases from 19 to 27 mm·s^{-1} along the axial direction of the cyclone abrrel. When the thickness of the refractory lining is 30 mm (Case 3), the velocity of the liquid slag layer increases from 20 to 32 mm·s^{-1} along the axial direction of the cyclone abrrel. The velocity of the liquid slag layer increases by about 5 mm·s^{-1} at

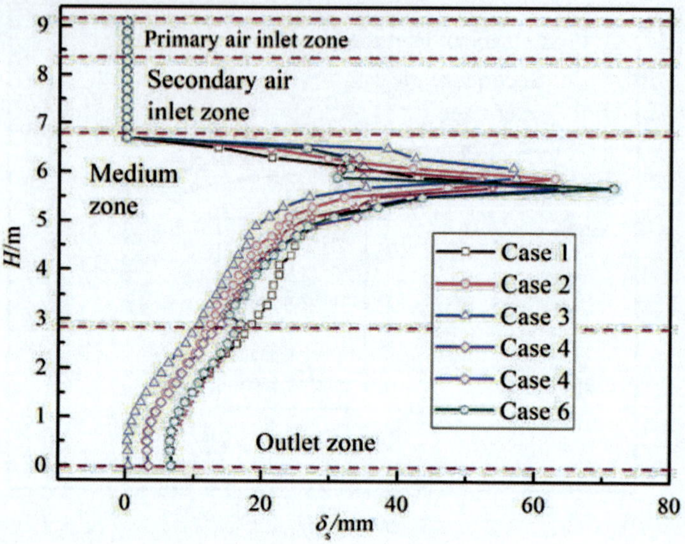

Fig. 18.8 The thickness of the solid slag layer

different loctions of the cyclone barrel when the refractory lining thickness increases 20 mm.

Figure 18.8 shows the effect of the refractory lining thickness on the thickness of the solid slag layer. The thickness of the solid slag layer can vary adaptively with the thickness of the refractory lining. The interface temperature of the slag layers of solid and liquid maintains at critical viscosity temperature, resulting in the solid slag layer turns thinner with the increase in the thickness of the refractory lining. The variation of the thickness of the solid slag layer is smaller than that of the refractory lining. This is because the thermal conductivity of the solid slag layer with 1.7 $W \cdot m^{-1} \cdot K^{-1}$ is one quarter of that of the refractory lining. At the barrel outlet, when the thickness of the refractory lining is 20 mm (Cases 2 and 4), the thickness of the solid slag layer decreases from 12.9 mm to 3.3 mm along the axial direction of cyclone barrel, and the average thickness is thinner than that of the cyclone barrel with the refractory lining of 10 mm (Cases 1, 5 and 6) by about 3.2 mm. When the thickess of the refractory lining is 30 mm (Case 3), the thickness of the solid slag layer is only about 0.3 mm at the barrel bottom. The thin solid slag layer may result in the severe erosion of the refractory lining and the water-cooled wall by the flowing liquid slag. Hence, the refractory lining should not be too thick to ensure the presence of the solid slag layer for proceting the wall-cooled wall (like Cases 1, 5 and 6). In addition, thinning the thickness of the refractory lining can reduce investment.

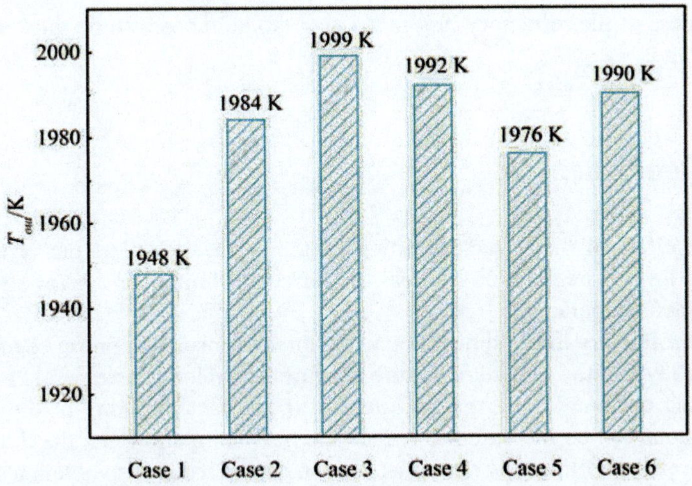

Fig. 18.9 The cross-section temperature of the barrel outlet

18.5.3 Outlet Temperature

Figure 18.9 shows the effect of refractory lining thickness on the cross-section temperature of the barrel outlet. Since the refractory lining can reduce the heat flux transfered to the working medium in the water-cooled wall without covered by the slag layer, the cross-section temperature at the outlet increases with the thickness of the refractory lining. The cross-section temperature is lowest (1948 K) among the six cases when the thickness of the refractory lining is 10 mm (Case 1), whereas the temperature increases to 1999 K when the thickness of refractory lining is 30 mm (Case 3). Compared Cases 2 and 5 and Cases 3, 4 and 6, it can be deduced that increase the thickness of the refractory lining at the air inlets of the primary and secondary would reduce the heat loss, resulting in the increase in the cross-section temperature of the outlet.

Therefore, among the six cases, case 6 presents best performance because different locations coveres refractory lining with different thickness. In the cyclone barrel, the thickness of the refractory lining of the air inlets sections can be increased to reduce the heat dissipation, but the thickness of the refractory lining of the outlet section can be reduced to obtain a solid slag layer thick enough to prevent the erosion of the refractory lining.

It should be noted that when the refractory lining thickness is variable along the axial direction of the cyclone barrel, it is also necessary to avoid the sudden change of the thickness of the refractory lining. This is because a sudden change in the thickness of the refractory lining may cause a dead corner in the cyclone barrel, resulting in the accumulation of the particles at the corners of the cyclone barrel. Hence, the refractory lining can be arranged by a gradual variation in thickness. Specifically, the thickness of the refractory ling at the slag tap hole needs to be thicker to avoid the blockage.

18.6 Conclusions

(1) The effect of the refractory lining on the surface templerature of the liquid slag layer is negligible. Whereas the refractory lining has a great effect on its surface temperature.
(2) The refractory lining should not be too thick to ensure the presence of the solid slag layer at the high temperature zone of the cyclone barrel.
(3) In the cyclone barrel, the thickness of the refractory lining of the air inlets sections can be increased to reduce the heat dissipation, but the thickness of the refractory lining of the outlet section can be reduced to obtain a solid slag layer thick enough to prevent the erosion of the refractory lining.

Acknowledgements This work has been financially supported by the National Key Research and Development Program of China under Grant No. 2018YFB0604101.

References

Bi D, Guan Q, Xuan W, Zhang J (2015) Combined slag flow model for entrained flow gasification. Fuel 150:565–572
Deng L, Tan X, Tang C, Che D (2016) A study on water-quenching waste heat recovery from molten slag of slag-tap boilers. Appl Therm Eng 108:538–545
Kim M, Ye I, Ryu C (2017) Numerical analysis on transient behaviors of slag layer in an entrained-flow coal gasifier. Fuel 196:532–542
Kitto JB, Stultz SC (2005) Steam: its generation and use. 41th edn. Babcock & Wilcox Company, Charlotte
Ministry of Science and Technology of the People's Republic (2017) Guidelines for the clean and efficient use of coal and new energy-saving technologies
Seggiani M (1998) Modelling and simulation of time varying slag flow in a prenflo entrained-flow gasifier. Fuel 77(14):1611–1621
Tang C, Deng L, Wu S, Bu Y, Che D (2017) Numerical simulation on the slag flow and heat transfer characteristics of the cyclone barrel for a cyclone-fired boiler. Numer Heat Transf Part A Appl 71(10):1052–1065
Tang C, Zhu T, Wang L, Deng L, Che D, Liu Y (2019) Effects of ash parameters and fluxing agent on slag layer behavior in cyclone barrel. Fuel 253:1140–1148
Wang XH, Zhao DQ, He LB, Jiang LQ, He Q, Chen Y (2007) Modeling of a coal-fired slagging combustor: development of a slag submodel. Combust Flame 149(3):249–260
Wu S, Deng L, Wang CA, Tang C, Bu Y, Che D (2017) A study on the adaptability of cyclone barrel to slag film in a cyclone-fired boiler. Appl Therm Eng 121:368–379
Wu S, Bai W, Tang C, Tan X, Wang CA, Che D (2015) A novel boiler design for high-sodium coal in power generation. American Society of Mechanical Engineers Power Division Power

Ye I, Ryu C (2015) Numerical modeling of slag flow and heat transfer on the wall of an entrained coal gasifier. Fuel 150:64–74

Ye I, Ryu C, Koo JH (2015) Influence of critical viscosity and its temperature on the slag behavior on the wall of an entrained coal gasifier. Appl Therm Eng 87:175–184

Yong SZ, Ghoniem A (2012) Modeling the slag layer in solid fuel gasification and combustion-two-way coupling with CFD. Fuel 97:457–466

Yong SZ, Gazzino M, Ghoniem A (2012) Modeling the slag layer in solid fuel gasification and combustion-formulation and sensitivity analysis. Fuel 92(1):162–170

Zhang BB, Shen ZJ, Liang QF, Xu JL, Liu HF (2018a) Modeling study of residence time of molten slag on the wall in an entrained flow gasifier. Fuel 212:437–447

Zhang BB, Shen ZJ, Liang QF, Xu JL, Liu HF (2018b) Numerical study of dynamic response analysis of slag behaviors in an entrained flow gasifier. Fuel 234:1071–1080

Zhang B, Shen Z, Liang Q, Xu J, Liu H (2018c) Modeling the slag flow and heat transfer on the bottom cone of a membrane wall entrained-flow gasifier. Fuel 226:1–9

Chapter 19
Characteristics of Alkali Metal Migration and Transformation During Pyrolysis of Naomaohu Coal

Junjie Fan, Jiaxiao Deng, Zhiyuan Ren, and Jianli Wang

Abstract The control of Na and K content plays important role on the quality of oil and gas from high alkali coal in Xinjiang. Effect of temperature on the migration and transformation characteristics of alkali metals within Naomaohu coal was investigated on a fixed-bed pyrolysis experimental platform under reducing atmosphere. The results show that Na in the raw coal mainly exists in H_2O-soluble form, and K mainly exists in insoluble form. When the pyrolysis temperature is lower than 600 °C, the H_2O-soluble Na changes to insoluble Na as the temperature increases. However, when the temperature is higher than 600 °C, the proportion of H_2O-soluble Na to insoluble Na decreases, and the content of H_2O-soluble Na increases with the, the relatively decreasing of insoluble Na. Under lower temperature conditions, the acid-soluble K content increases with the transformation of large numbers of insoluble K in the coal. When the temperature is higher than 600 °C, the decrease rate of insoluble K becomes slow down, and the content of H_2O-soluble K increases. In addition, the released Na and K mainly present in the tar, and the content in the gas is relatively small.

Keywords Migration and transformation · Occurrence form · Alkali metal · High alkali coal · Reducing atmosphere

19.1 Introduction

Naomaohu, located in the Hami Basin of Xinjiang, where natural resources and mineral resources are very rich, especial for coal resources. The coal bed of Naomaohu is one of the main mining areas of "West Coal East Transport", whose coalfield is relatively stable and mainly used as power coal or chemical coal, due to the high calorific value and relative environment friendly. However, Naomaohu coal contains many alkali metal substances, the release of alkali metal into the vapor

J. Fan (✉) · J. Deng · Z. Ren · J. Wang
School of Environment and Construction Engineering, University of Shanghai for Science and Technology, Shanghai 200093, China
e-mail: fjj2006jj@126.com

J. Lyu and S. Li (eds.), *Clean Coal and Sustainable Energy*,
Environmental Science and Engineering,
https://doi.org/10.1007/978-981-16-1657-0_19

phase has been known to cause serious problems, such as the erosion/corrosion of heating surface which affect the efficiency and safety of the boiler (Wei et al. 2008a). Faced on the current situation of oil and gas shortage in China, coal-to-liquid is one of the good ways for clean and efficient use of Naomaohu coal. As we know, coal conversion is a very complicated process, and the coal type, ash component, oxidation/ reduction condition, and temperature strongly affect the partitioning of these elements (Thompson and Agent 1999). So, how to solve the impact of alkali metal release and volatilization on the quality of coal-to-liquids would have great guiding value on improving the comprehensive utilization of Naomaohu coal resources.

Considerable studies have been carried out by scholars on the migration and transformation of alkali metals in coal during pyrolysis/gasification/combustion in China or in the world. Researches show that alkali metals in coals can be classified into four types by chemical extraction method, such as water-soluble form (H_2O-soluble), ammonium acetate-soluble form (NH_4Ac-soluble), hydrochloric acid-soluble form (HCl-soluble) or insoluble form (Wei et al. 2008b; Li et al. 2015; Wang et al. 2015a; Mourant et al. 2011), while H_2O-soluble and ammonium acetate-soluble form of minor vaporization elements Na and K are considered the most harmful (Zhang et al. 2001). Song et al. (2015) investigated the occurrence form and transformation of Na during during high-sodium Zhundong coal pyrolysis. It was found that H_2O-soluble Na is the dominant form in the coal, and Na is mainly released into the gas phase with the form of NaCl, and some part of H_2O-soluble alkali Na would convert into other forms (Wang et al. 2015a; Zhang et al. 2001). Wei et al. (2008a) studied the transformation law of alkali metals in lignite using a fixed bed heater during the pyrolysis processes. The H_2O-soluble Na and ammonium acetate-soluble Na gradually decreased with the increase of pyrolysis temperature. Na_2SO_4 plays an important role in the formation of stable forms such as $Na_2O{\cdot}Al_2O_3{\cdot}2SiO_2$. Li et al. (2000) summarized that bond break and formation are repeated between carbon matrix and alkali metal or alkaline earth metal during coal pyrolysis. The volatilization of alkali metal is determined by the bond strength and temperature is the main factors determining the volatilization of alkali metals. Luo et al. (2018) studied the Na migration of coal under N_2, CO_2 and H_2O atmosphere conditions. The results showed that sodium hydroxide vapor formed under water vapor can promote the volatilization of Na, and sodium carbonate would be found under CO_2 atmosphere, which inhibits the volatilization of Na. Quyn et al. (2002; Wu et al. 2002) studied the volatilization and catalysis of alkali metal in the pyrolysis and gasification process of Victoria brown coal using the thermogravimetric analyzer and fluidized bed/fixed bed reactor. It was concluded that NaCl would be easier to be released than carboxylate form in coal and mainly released by Na and Cl separately.

However, there are few studies on reducing atmosphere conditions and Naomaohu high alkali coal. In view of the important influence of reducing atmosphere and alkali metal content on coal conversion process and alkali metal transformation behavior, a fixed bed pyrolysis experiment platform was built to studied the occurrence and transformation law of alkali metals in Naomaohu coal, which can lay a theoretical foundation for exploring the controllable conversion pathway of alkali metals during the coal-to-liquid process of Naomaohu low-metabolization coal.

19.2 Experimental System

In this experiment, the experimental coal sample was pyrolyzed by a fixed rapid pyrolysis method. The whole experimental system consisted by gas flow control system, electric heating reaction device, tar condensation collection device and exhaust gas treatment device, which are shown in Fig. 19.1.

The quartz tube reactor has a quartz sand core in the middle, and the total length and inner diameter are about 750 and 21 mm respectively. The maximum heating temperature of the furnace is 1200 °C. While the electric heating reaches the predetermined temperature, N_2 will be introduced to the tube furnace for 10 min with the flow rate of 100 mL/min. After the air is removed, the flow gas would be adjusted to 20% H_2 + 80% N_2.When the gas flow become stable, 20 g of Naomaohu coal was poured into the reaction furnace with the particle sizes range from 0.5 to 1.0 mm. The pyrolysis product tar and the volatile alkali metal were collected by a condensing collection device and an exhaust gas treatment device respectively, and 1 mol/L HCl solution was used as the absorption liquid. After the end of the pyrolysis, N_2 would be continuously introduced to cool the char samples for the collection, although the quartz tube was taken out. The proximate analysis, ultimate analysis and ash analysis of Naomaohu coal are shown in Tables 19.1 and 19.2. The ash-making step in GB/T1574-2007 "Component Analysis Method of Coal Ash" was adopt to prepare coal ash, a certain amount of coal samples weight and placed in the muffle furnace, and the ashing temperature was 300 °C.

Fig. 19.1 Schematic diagram of the fixed bed pyrolysis system

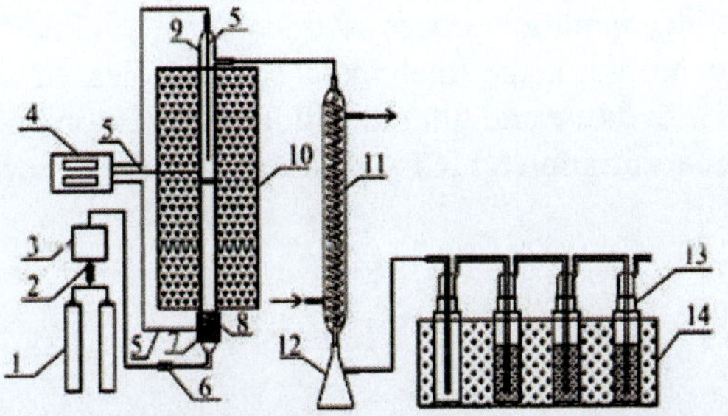

1.Cylinder gas, 2.flow meter, 3.premixer, 4.temperature controller, 5.thermocouple, 6.valve, 7.insulation cotton, 8.heating Belt, 9.quartz tube, 10.resisting furnace, 11.condensing tube, 12.tar collecting bottle, 13.washing bottle (diluted hydrochloric acid), 14.ice bath

Table 19.1 Proximate analysis and ultimate analysis of Naomaohu coal

Coal	Proximate analysis, %				Ultimate analysis, %				
	M_{ad}	A_{ad}	V_{ad}	FC_{ad}	N_{ad}	C_{ad}	H_{ad}	S_{ad}	O_{ad}
Naomaohu coal	3.25	6.93	46.22	43.6	0.84	67.16	5.09	0.25	16.48

Table 19.2 Ash composition of Naomaohu coal

Ash analysis, %											
CaO	Fe_2O_3	SiO_2	Al_2O_3	TiO_2	Na_2O	SrO	MgO	K_2O	SO_3	NiO	CuO
30.92	8.22	29.28	15.07	0.97	2.94	1.61	1.16	0.38	8.26	0.15	0.15

19.3 Measurement of Alkali Metals (Na, K)

Four kinds of occurrence forms of alkali metals in raw coal and semi-coke, namely H_2O-soluble state, ammonium acetate dissolved state, acid-soluble state and insoluble state, were measured by four-step chemical extraction method. Ultra-pure water can be used to extract the inorganic alkali metal form in the form of chloride crystals and hydrated forms, ammonium acetate solution can extract organic state in the form of carboxylate, the organic state of the form containing nitrogen or oxygen functional groups in the coal structure can only be obtained by dilute hydrochloric acid, and the residual alkali metal can be considered to be insoluble. Ultra-pure water (H_2O), ammonium acetate (CH_3COONH_4, 1 mol/L), hydrochloric acid (HCl, 1 mol/L) were used to extract the coal/coal coke sample and analysis their contents. 1 g of sample and 30 mL of extract solution added a erlenmeyer flask, constant temperature oscillated for 24 h with a constant temperature oscillator at 60 °C, and then the supernatant was extracted by centrifugation and dilute to 50 ml. A coupled plasma emission spectrometer (ICP-OES) was used to test the content of alkali metals. Digestion was used to determination the content of insoluble alkali metal and tar, the raw coal/semi-coke/tar sample weight is 0.1 g, the digestion solution was also tested by ICP-OES after the operation processes of constant volume (50 mL) and filtration. The method for determining the alkali metal content released into the gas phase during coal pyrolysis is to dilute and filter the dilute hydrochloric acid absorption liquid, and then test its concentration by ICP-OES to calculate the content.

19.4 Results and Discussion

19.4.1 Effect Of Temperature And Atmosphere On Semi-Coke And Tar Yield

Figure 19.2 shows the effect of temperature on the yield of semi-coke under different pyrolysis conditions. It can be seen from the figure that with the increase of pyrolysis temperature, the yield of semi-coke decreases rapidly at first and then gradually tends to be flat. When the pyrolysis temperature is 800 °C, the yield of pyrolysis char under H_2 atmosphere is the lowest, which is only 51%. Figure 19.3 shows the effect of temperature on tar yield. With the increase of temperature, tar yield increases first and then decreases, it reaches a peak value among the range of 500–600 °C.

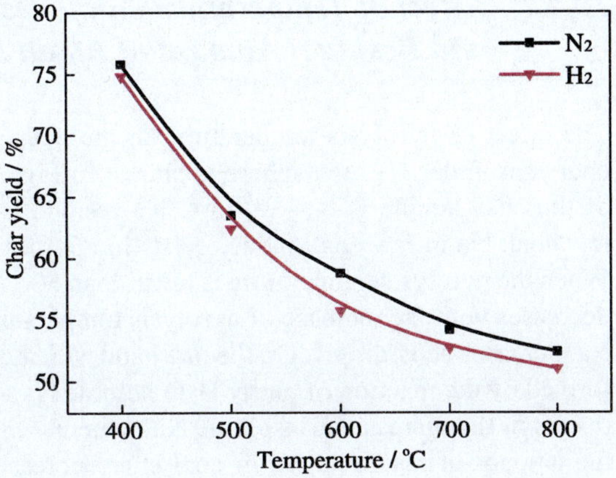

Fig. 19.2 Effects of temperature on char yield under different pyrolysis conditions

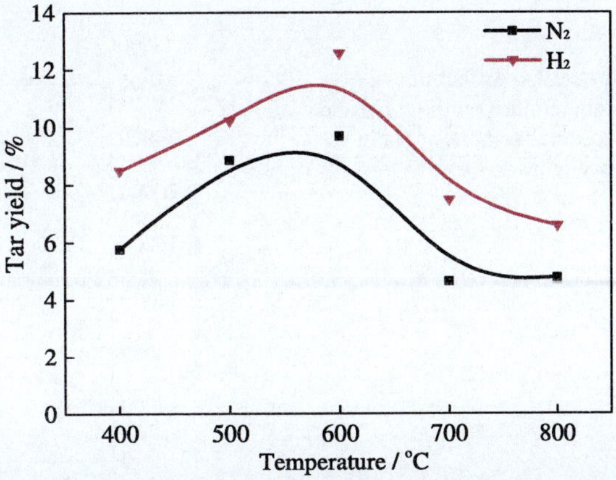

Fig. 19.3 Effect of temperature on tar yield under different pyrolysis conditions

The tar yield in H_2 atmosphere is significantly higher than that in N_2 atmosphere (Gao et al. 2015). This is mainly due to the cracking of aromatic rings, side chains, aliphatic chains and ether bonds induced by H_2 or H free radicals at high temperature, which promotes coal pyrolysis. In addition, the yield of tar is closely related to tar production and secondary cracking reaction. Under lower temperature conditions, the secondary cracking rate of tar is less than the production rate, and the yield of tar increases rapidly with the increase of temperature. However, when the temperature exceeds 600 °C, the secondary cracking rate of tar is faster than the production rate, and the tar yield decreases gradually under the high temperature thermodynamic environment. Therefore, the peak of tar yield occurs near 600 °C.

19.4.2 Effect of Temperature on the Occurrence Form and Residual Amount of Alkali Metal in Coal Char

The effect of pyrolysis temperature on the occurrence form and residue of Na in coal char under H_2 atmosphere is shown in Figs. 19.4 and 19.5. The particle size of the coal sample is 0.5–1.0 mm. Na residue is calculated by pyrolysis of 20 g raw coal. Na in raw coal mainly exists in the form of H_2O-soluble Na, about 62%. When the pyrolysis temperature is lower than 600 °C, the content of H_2O-soluble Na decreases with the increase of pyrolysis temperature, while the content of insoluble Na increases accordingly. On the one hand, silicate-aluminate (Wang et al. 2015b) is formed by the reaction of partly H_2O-soluble Na with SiO_2 during pyrolysis, which results in the increase of insoluble Na content. On the other hand, due to pyrolysis, the amount of insoluble Na in coal char decreases rapidly, and the concentration of insoluble Na in coal char keeps increasing, so the insoluble Na content tends to increase.

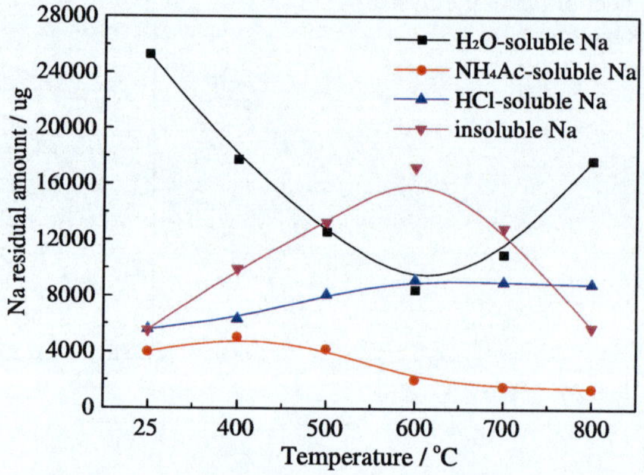

Fig. 19.4 Effect of temperature on the form and residual amount of Na in coal char

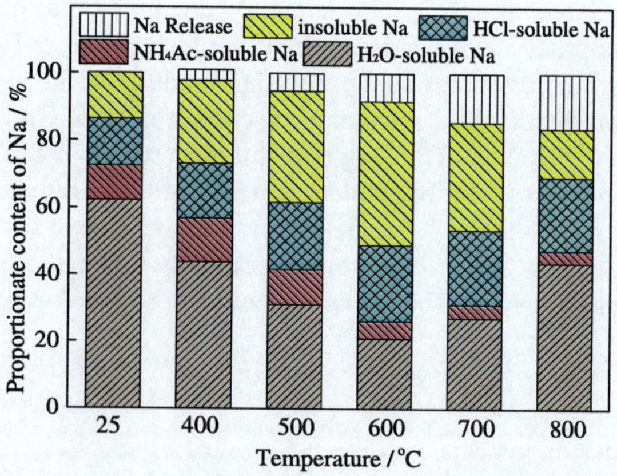

Fig. 19.5 Effect of temperature on the form and residual ratio of Na in coal char

However, when the pyrolysis temperature is higher than 600 °C, the H_2O-soluble Na content gradually increases, and the insoluble Na content decreases. Due to the rapid release of large amount of H_2O-soluble Na into the gas phase at high temperature, it does not react with SiO_2 and AlO_3. On the other hand, due to the high Ca content in Naomaohu coal, there is a competitive relationship between Ca, Na and SiO_2, Al_2O_3(Shen et al. 2015), resulting in the reduction of SiO_2 and AlO_3 which react with Na, so the insoluble Na content is gradually decreasing. In addition, the release of Na atoms combined with Cl^-, SO_4^{2-} in coal to form H_2O-soluble Na salts, which to some extent inhibited the reduction of H_2O-soluble Na content in coal char (Qi et al. 2016). With the increase of pyrolysis temperature, the content of acid-soluble Na increases slowly and that of ammonium acetate decreases correspondingly. This is because at lower temperature, H_2O-soluble Na in the form of NaCl first combines with carboxyl groups on coal surface to form ammonium acetate-soluble Na(–COONa). With the further increase of pyrolysis temperature, -COONa reacts with carbon to form –CNa or –CONa, resulting in the increase of acid-soluble Na content (Mckee et al. 1983). Therefore, the decrease of total Na content in coal pyrolysis process is mainly caused by the volatilization of H_2O-soluble Na.

$$2NaCl + 2SiO_2 + H_2O \rightarrow Na_2Si_2O_5 + 2HCl \tag{19.1}$$

$$Na_2SO_4 + 3SiO_2 \rightarrow Na_2O \cdot 3SiO_2 + SO_2 + 1/2O_2 \tag{19.2}$$

$$Na_2O \cdot 3SiO_2 + Al_2O_3 \rightarrow Na_2O \cdot Al_2O_3 \cdot 2SiO_2 + SiO_2 \tag{19.3}$$

The effect of pyrolysis temperature on the form and residual amount of K in coal char is shown in Figs. 19.6 and 19.7. The K in the raw coal mainly exists in the form of insoluble K, accounting for about 78% of the total K content. With the increase of pyrolysis temperature, the content of insoluble K in coal decreases sharply, and the amount of insoluble K in coal is released in large quantities. After 500 °C, the

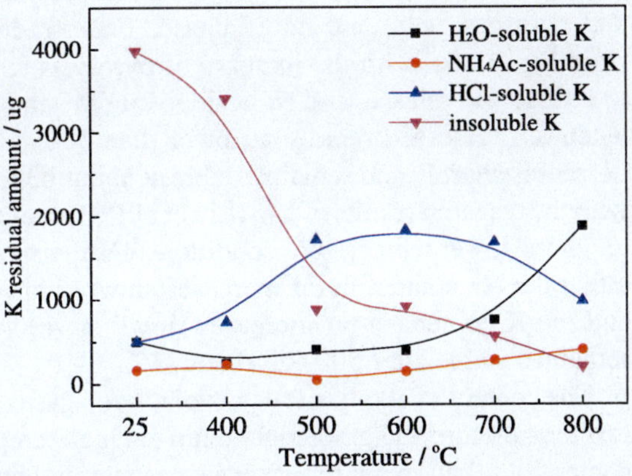

Fig. 19.6 Effect of temperature on the form and residual amount of K in coal char

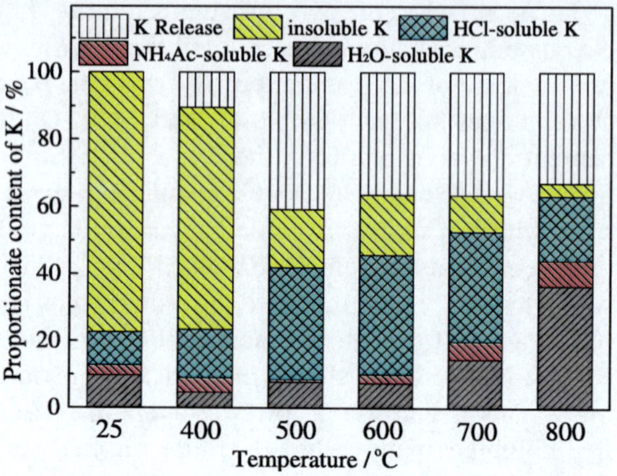

Fig. 19.7 Effect of temperature on the form and residual ratio of K in coal char

amount of insoluble K is basically stable (Liu et al. 2014). The content of acid-soluble K shows an upward trend, which is due to the transformation of insoluble K to acid-soluble K. When the temperature is higher than 600 °C, the rate of decline of insoluble K content slows down, acid soluble K content decreases, and water soluble K content increases. This is mainly due to the displacement of K from aluminosilicate by partial H_2O-soluble Na in the process of pyrolysis and volatilization (Raask 1985). The reaction pathway is:

$$mK_2O \cdot xSiO_2 \cdot yAl_2O_3 + 2mNaCl = mNa_2O \cdot xSiO_2 \cdot yAl_2O_3 + 2mKCl \quad (19.4)$$

19.4.3 *Effect of Temperature on Distribution Characteristics of Alkali Metals*

Figure 19.8 shows the effect of pyrolysis temperature on the content ratio of Na and K in gas, solid and liquid phases under reducing atmosphere. It can be seen from Fig. 19.8a, with the increase of pyrolysis temperature, alkali metals continue to diffuse and release, and Na residue rate in semi-coke decreases slowly. However, when K is released rapidly at lower than 500 °C, the K residue rate in coal char decreases sharply and remains stable at about 63% after 500 °C, which is consistent with the research results of Liu (Liu et al. 2014), alkali metal release rate is the fastest under the lower temperature conditions. Furthermore, it can be seen from Fig. 19.8b, c that the Na content in tar increases slowly with the pyrolysis temperature rising, and the K content in tar increases slowly at 400–500 °C. The content of K in tar remained stable after 500 °C, about 33%.

The content of alkali metals in volatilized gases showed a trend of rapid release at low temperature and gradual flattening at high temperature. At high temperature, the content of alkali metal in pyrolysis gas is only about 3.0% of the total. In addition,

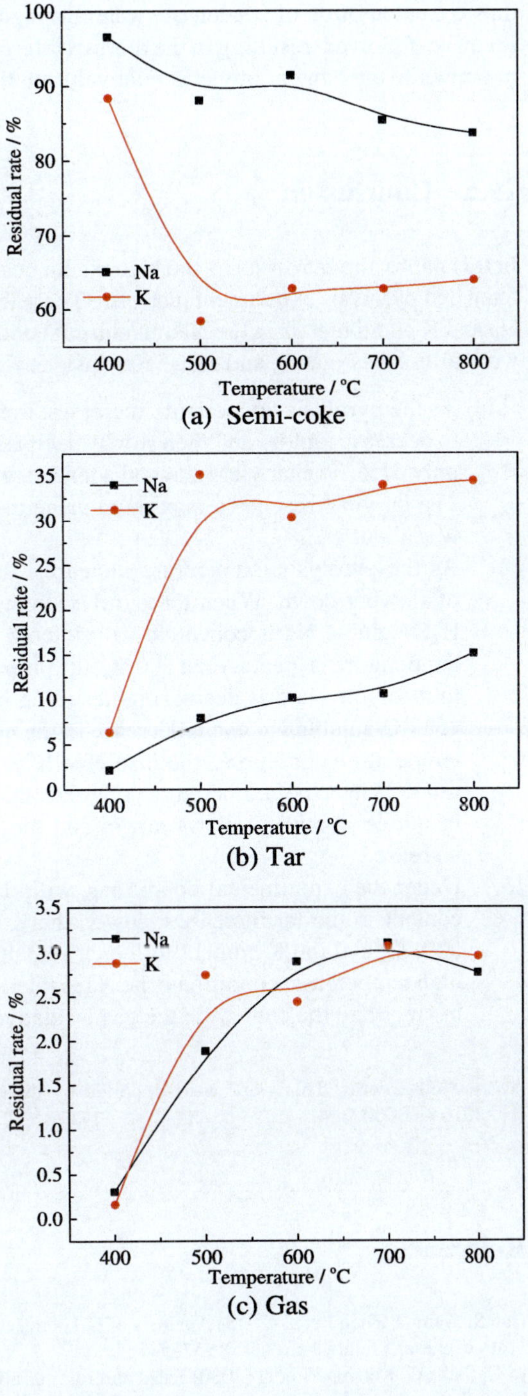

Fig. 19.8 Effect of temperature on the distribution of Na and K in coal char, tar and gas

due to condensation of condensate water during tar collection, some soluble Na and K can be dissolved, resulting in the thermal interpretation release of Na and K mainly in tar, while the content of gas is relatively small.

19.5 Conclusion

In this paper, the pyrolysis of the Naomaohu coal was carried out by constructing a fixed bed pyrolysis experiment platform. The release and transformation behavior of Na and K alkali metals in the Naomaohu coal under different pyrolysis temperatures were fully investigated, and some conclusions were obtained as follows:

(1) As the pyrolysis temperature increases, the semi-coke yield shows a tendency to decrease rapidly and then slowly decrease. The semi-coke yield at 800 °C is only 51%. The tar yield showed a tendency to increase first and then decrease. The tar yield reached a maximum value in the range of 500 to 600 °C, which was about 12.5%.

(2) As the pyrolysis temperature increases, the total Na content shows a trend of slowing down. When the pyrolysis temperature is lower than 600 °C, the H_2O-soluble Na is converted to insoluble Na. However, when the pyrolysis temperature is greater than 600 °C, the proportion of H_2O-soluble Na changing to insoluble Na was decreasing, resulting in a gradual increase in the content of H_2O-soluble Na and a decrease in the content of insoluble Na. Under low temperature conditions, the insoluble K in the coal is released in large quantities. When the temperature is higher than 600 °C, the decline rate of the insoluble K content slows down, and the H_2O-soluble K content is rapidly increased.

(3) Under the experimental conditions, with the increase of temperature, the Na content in the tar increased slowly, the K content increased rapidly at 400–500 °C, and the K content in the tar remained stable after 500 °C. Under the high temperature conditions, the Na K released by the pyrolysis mainly existed in tar, while the content in the gas is relatively small.

Acknowledgements This work was supported by the National Key R&D Program of China (No.2016YFB0600304).

References

Gao S, Wang J, Zhao J et al (2015) Analysis of CH_4 emission in the rapid pyrolysis of lignite in H_2 atmosphere. J Fuel Chem 43(5):537–545

Li C, Sathe C, Kershaw J et al (2000) Fates and roles of alkali and alkaline earth metals during the pyrolysis of a Victorian brown coal. Fuel 79(3):427–438

Li G, Wang C, Yan Y et al (2015) Release and transformation of sodium during combustion of Zhundong coals. J Energy Inst 89(1):48–56

Liu J, Wang Z, Xiang F et al (2014) Modes of occurrence and transformation of alkali metals in Zhundong coal during combustion. J Fuel Chem Technol 42(3):316–322

Luo A, Zhu P, Zhang J et al (2018) Effect of atmosphere on sodium migration during conversion of high sodium coals. J Fuel Chem Technol 46(5):513–520

Mckee D, Spiro C, Kosky P et al (1983) Catalysis of coal char gasification by alkali metal salts. Fuel 62(2):217–220

Mourant D, Wang ZH, He M et al (2011) Mallee wood fast pyrolysis: effects of alkali and alkaline earth metallic species on the yield and composition of bio-oil. Fuel 90(9):2915–2922

Qi X, Song G, Song W (2016) Transformation and migration of alkali metal with different occurrences of Zhundong high-alkali coal during gasification. J China Coal Soc 41(4):1011–1017

Quyn D, Wu H, Li C (2002) Volatilisation and catalytic effects of alkali and alkaline earth metallic species during the pyrolysis and gasification of Victorian brown coal. Part I. Volatilisation of Na and Cl from a Set of NaCI-loaded samples. Fuel 81(2):143–149

Raask E (1985) Mineral impurities in coal combustion, behavior, problems, and remedial measures. Hemisphere Publishing Corporation

Shen M, Qiu K, Huang Z et al (2015) Influence of kaolin on sodium retention and ash fusion characteristic during combustion of Zhundong coal. J Fuel Chem Technol 43(9):1044–1051

Song W, Song G, Zhang H et al (2015) Experimental study on migration characteristics in the process of pyrolysis of Zhundong high coal. J Fuel Chem 43(1):16–21

Thompson D, Agent B (1999) The mobilisation of sodium and potassium during coal combustion and gasification. Fuel 78:1679–1689

Wang C, Jin X, Wang Y et al (2015a) Release and transformation of sodium during pyrolysis of Zhundong coals. Energy Fuel 29(1):78–85

Wang W, Jia B, Yao H et al (2015b) Study on the migration law in the process of quality of Qingdong coal. J Eng Thermophys 36(12):2733–2737

Wei X, Huang J, Liu T et al (2008a) Transformation of alkali metals during pyrolysis and gasification of a lignite. Energy Fuels 22:1840–1844

Wei X, Huang J, Liu T et al (2008b) Transformation of alkali metals during pyrolysis and gasification of a lignite. Energy Fuel 22(3):1840–1844

Wu H, Quyn D, Li C (2002) Volatilisation and catalytic effects of alkali and alkaline earth metallic species during the pyrolysis and gasification of Victorian brown coal. Part III. The importance of the interactions between volatiles and char at high temperature. Fuel 81(8):1033–1039

Zhang J, Han C, Yan Z et al (2001) The varying characterization of alkali metals (Na, K) from coal during the initial stage of coal combustion. Energy Fuel 15(4):786–793

Part III
Pulverized Coal Combustion

Chapter 20
Numerical Investigation on Combustion Characteristics and NO_x Emission of Double-Reheat Tower Boiler at Different Loads

Yan Zhang, Shihao Ma, Jin Guo, Yuesheng Li, Zhengrong Zhu, Jiaqi He, Lei Deng, and Defu Che

Abstract In this study, a three-dimensional numerical investigation using a finite volume method is presented to obtain the velocity, temperature, and species distributions in a 660 MW double-reheat tower boiler fed with bituminous coal. The refined oxidation model of HCN and NO-char reaction model were used to substitute the default model by the user-defined functions (UDFs). The effects of different loads and excess air ratios on combustion characteristics and NO_x emission were studied numerically. The results show that temperature distribution is more uneven at the upper part of the furnace and the NO_x emission increases as the load decreases. The temperature distributions in the furnace have little differences under three different excess air ratios. That the oxygen concentration increases significantly as the increase of excess air ratios leads to an increase of NO_x emission at the furnace outlet. The minimum NO_x concentration at the furnace outlet, 308 ppm, appears under 100% THA with excess air ratio of 1.15. The simulation results in this study can provide reference for optimization of combustion characteristics in double-reheat tower boiler and illustrate a method to reduce NO_x emission.

Keywords Unit load · Excess air ratio · NO_x emission · Double-reheat · Numerical investigation

Y. Zhang · S. Ma · Z. Zhu · J. He · L. Deng (✉) · D. Che
State Key Laboratory of Multiphase Flow in Power Engineering, School of Energy and Power Engineering, Xi'an Jiaotong University, Xi'an 710049, China
e-mail: leideng@mail.xjtu.edu.cn

J. Guo
Guangdong Institue of Special Equipment and Research, Foshan 528251, China

Y. Li
Guangdong Institue of Special Equipment Inspection and Research, Shunde Institue of Inspection, Foshan 528300, China

J. Lyu and S. Li (eds.), *Clean Coal and Sustainable Energy*, Environmental Science and Engineering,
https://doi.org/10.1007/978-981-16-1657-0_20

20.1 Introduction

With the slowdown of economic growth and the rapid adjustment of the economic structure, China's overall electricity demand has declined. The contradiction between power supply and demand has shifted from electricity shortage to relative surplus. Meanwhile, to face the increasingly severe environmental regulations, China continues to promote energy structure adjustment and develop the clean energy rapidly. These all lead to the decline of our country's installed capacity of thermal power units while moving toward the development of large-capacity and high-parameter double-reheat unit (Tumanovskii and Shvarts 2017; Chang and Zhuo 2016). Besides, conditions of medium and low loads would become more universal during the unit operations due to the abundant hydropower or low power demand in late nights (Zhao and Fang 2017). However, most of the units lack experiences of long-term operation under medium and low loads, which makes the efficiency of the units decrease and the coal consumption increase. Therefore, it is necessary to carry out investigation on energy-saving operation optimization of large-capacity and high-parameter double-reheat units under conditions of wide loads to improve the efficiency and reduce pollutant emissions and power consumption.

Ultra-supercritical double-reheat units with high efficiency, high parameter and large capacity have become a main development tendency on the coal-fired power plants. But double-reheat units are still in the initial development stage in China. Several researches have analyzed the technical (Huo and Shi 2014) and economic characteristics (Jiang 2000; Yao and Zhu 2017), heat load distribution (Zhang and Yan 2016) and pollutant control (Guo and Zhang 2011) during the operation of double-reheat units. However, the sudies on ultra-supercritical double-reheat units are still limited.

Numerical simulation, which has been widely applied to investigate the combustion and pollutant formation process in pulverized coal fired boiler in the past few decades (Zhou and Mo 2011; Liu and Chen 2015), is more economical and convenient than experimental study. In this study, the effects of different loads and excess air ratios on combustion characteristics and NO_x emission of an ultra-supercritical double-reheat boiler were studied. In NO_x calculation, the revised NO_x model is used to replace the default model to accurately predict the NO_x emission. This work can provide reference for optimization of combustion organization for double-reheat tower boiler.

20.2 Boiler Decription

The simulations were based on a 660 MW double reheat tower boiler with bituminous coal, with the schematic of boiler geometry illustrates in Fig. 20.1a. The width, depth and height of the boiler are 18,150, 18,150 and 97,500 mm, respectively. Figure 20.1b shows the arrangement of nozzles in each corners. It includes the primary air (PA),

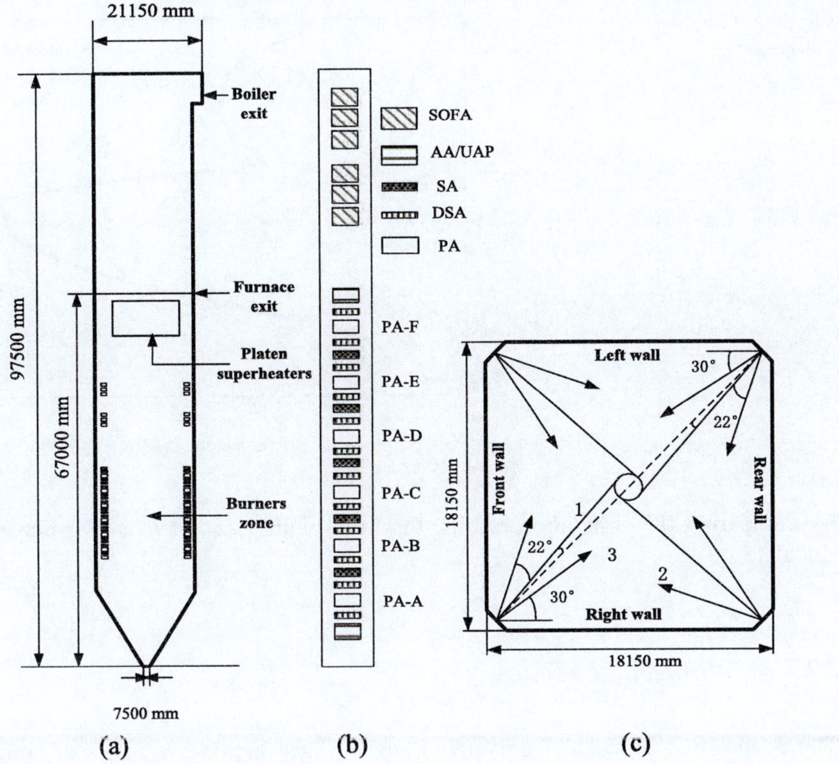

Fig. 20.1 Schematic configuration and burner nozzle distribution of the tangentially pulverized coal fired boiler

secondary air (SA) and separate over fire air (SOFA). In order to control the NO_x generation by air staging, two stage SOFA nozzles are arranged above the burners. Figure 20.1c presents the air injecting direction.

20.3 Modeling Methodology

20.3.1 Grid Independence

In this paper, a structured grid is used to simulate the coal combustion. Because of the intense combustion reaction in the burner area, the grid is refined locally to receive more accurate results in this area. The gird independence was verified with three different mesh systems, i.e., 1,565,078, 1,880,167 and 2,241,660 cells. Figure 20.2 decipts the temperature distribution in the three different mesh systems.

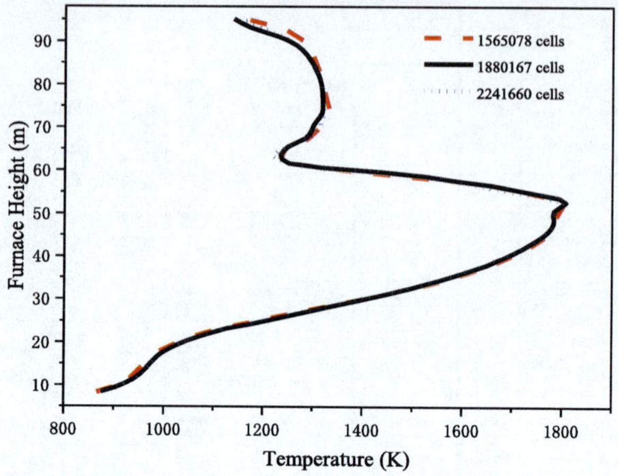

Fig. 20.2 Grid independence

By comparing the simulation results, the mesh with 1,880,167 cells is chosen to calculate.

20.3.2 Numerical Models

The combustion process of pulverized coal furnace includes a series of complex physical and chemical processes (Hu and Liu 2013). The standard $k - \varepsilon$ model was chosen in the present study. The radiation heat transfer was calculated by P-1 model (Drosatos and Nikolopoulos 2015). The WSGGM-domain based model was used to calculate the gas absorption coefficient. The volatile combustion is described by the two-step global reaction mechanism as follow. Coal particle trajectories were calculated using the Lagrangian approach.

$$CH_yO_x + (0.5 + 0.25y - 0.5x)O_2 \rightarrow CO + 0.5yH_2O \tag{20.1}$$

$$CO + 0.5O_2 \rightarrow CO_2 \tag{20.2}$$

The NO is generally discussed when investigating NO_x formation in coal-fired boilers. That's because the NO accounts for 95% of total NO_x. Three types of NO_x are generated, named as prompt NO_x, thermal NO_x and fuel NO_x. In large pulverized coal combustion system, the prompt NO_x production is little, which is generally neglecte (Panagiotis and Athanasios 2016; Álvarez and Yin 2013). Thus, only thermal NO_x and fuel NO_x are taken into consideration in the calculation process. NO_x generation is calculated by post-processing method (Park 2013).

The generation and reduction of thermal NO_x are determined by the extended Zeldovich mechanism as follows:

$$N_2 + O \underset{k_{-1}}{\overset{k_1}{\rightleftarrows}} N + NO \tag{20.3}$$

$$N + O_2 \underset{k_{-2}}{\overset{k_2}{\rightleftarrows}} O + NO \tag{20.4}$$

$$N + OH \underset{k_{-3}}{\overset{k_3}{\rightleftarrows}} H + NO \tag{20.5}$$

where k_1, k_2, k_3 present the forward reaction rate constant, while k_{-1}, k_{-2}, k_{-3} present the reverse reaction rate constant in the Eqs. (20.3–20.5). The six reaction rate constants are determined by the experimental results acquired by Hanson. Thus, the thermal NO formation rate can be described as the Eqs. (20.6). Besides, the concentrations of O, H, and OH in the Eqs. (20.3–20.6) are computed by the partial equilibrium approach.

$$\begin{aligned} \frac{d[NO]}{dt} = & k_1[O][N_2] + k_2[N][O_2] + k_3[N][OH] \\ & - k_{-1}[NO][N] - k_{-2}[NO][O] - k_{-3}[NO][H] \end{aligned} \tag{20.6}$$

Fuel NO_x is usually released in the form of volatile nitrogen and char nitrogen. Char nitrogen is converted to NO directly. During the transformation process from volatile nitrogen to the fuel NO_x, volatile nitrogen is first converted to HCN and NH_3, and then oxidation and reduction reactions occur between HCN, NH_3 and NO, O_2, depicted as the Eqs. (20.7–20.10).

$$HCN + O_2 \rightarrow NO + \cdots \tag{20.7}$$

$$HCN + NO \rightarrow N_2 + \cdots \tag{20.8}$$

$$NH_3 + O_2 \rightarrow NO + \cdots \tag{20.9}$$

$$NH_3 + NO \rightarrow N_2 + \cdots \tag{20.10}$$

In this study, the User Defined Functions (UDF) was used to substitute the default models for NO_x calculation, which has been verified by Zhang and Ito (2013) and Zha and Li (2017). Surface reaction of NO_x reduction by char nitrogen was calculated according to Levy and Chana (1981) mechanism.

The properties of the bituminous coal are presented in Table 20.1. The basic simulation case is based on the design case of 100% turbine heat acceptance (THA) operation. The effect of different unit loads and excess air ratios on the characteristics of NO_x emission are considered to study in the work. The three loads are 100% THA, 70% THA and 50% THA with the total excess air ratio of 1.15. The three excess air

Table 20.1 Basic information of the bituminous coal

Proximate analysis (ar, %)		Ultimate analysis (daf, %)	
w(*V*)	25.02	*w*(C)	78.86
w(*FC*)	43.62	*w*(H)	5.62
w(*A*)	14.96	*w*(O)	12.60
w(*M*)	16.40	*w*(N)	2.20
		w(S)	0.72
Q_{net} (MJ/kg)		20.36	

Table 20.2 Simulation cases of the 660 MW tower boiler

Case	Loads	Excess air ratios	Coal mills
Case 1	100% THA	1.15	A-E
Case 2	70% THA	1.15	B-E
Case 3	50% THA	1.15	C-E
Case 4	100% THA	1.20	A-E
Case 5	100% THA	1.25	A-E

ratios are 1.15, 1.20 and 1.25 with 100% THA load. All detailed simulation cases are listed in Table 20.2.

20.4 Results and Disscusion

20.4.1 Validation of the CFD simulation

In order to obtain credible and reasonable simulation results, the numerical calculation results were verified by the thermal calculation values with three different loads. Table 20.3 shows the comparison between the CFD results and thermal calculated value of furnace outlet temperature. The maximum deviation between the CFD results and thermal calculated value is 18 K under 100% THA operation.

Table 20.3 Comparison of furnace outlet temperature (K) between CFD results and thermal calculated value

Load	Excess air ratios	CFD	Thermal calculated
100% THA	1.15	1208	1226
70% THA	1.15	1146	1139
50% THA	1.20	1059	1074

20.4.2 Effects of Different Unit Loads on the Combustion Characteristics and NO_x Emission

Figure 20.3 shows the temperature distribution along furnace height under different unit loads. The coal quantity decreases with the load decrease. This results in reduction of the temperature level in the furnace. Besides, both the amount of pulverized coal and air are reduced, which leads to a decreasing gas fullness in the furnace. It can be seen from Fig. 20.3 that under the condition of 50% THA load, the temperature distribution at the upper part of the furnace is uneven. This can be explained by that the low gas fullness under the condition of 50% THA load results in the uneven velocity distribution in the furnace. Thus, heat exchanges between the gas and wall are different in different positions, which eventually leads to uneven temperature distribution at the upper part of the furnace. The temperature distribution showed a similar tendency with the results of Liu and Zhang (2019).

Figure 20.4 presents the average temperature on furnace cross-section along the height direction under different loads. The temperature distribution is similar under three boiler loads, while combustion details are different in the main combustion zone ranging from 24 to 41 m. In the main combustion zone, the temperature has a sharp decrease due to the injection of a large amount of primary air with the lower temperature. For the 50% THA load, that coal mills A-B in the lower position are closed causes that the temperature sharp decrease occurs in a higher position when compared with the 100% THA load and the 70% THA load. There is another sharp temperature decrease resulted from the injection of SOFA in the burnout zone. It is evident that the area-weighted average temperature at the furnace outlet T_{av} decreases as T_{av} (100% THA) > T_{av} (70% THA) > T_{av} (50% THA) with the highest temperature appearing at 100% THA load.

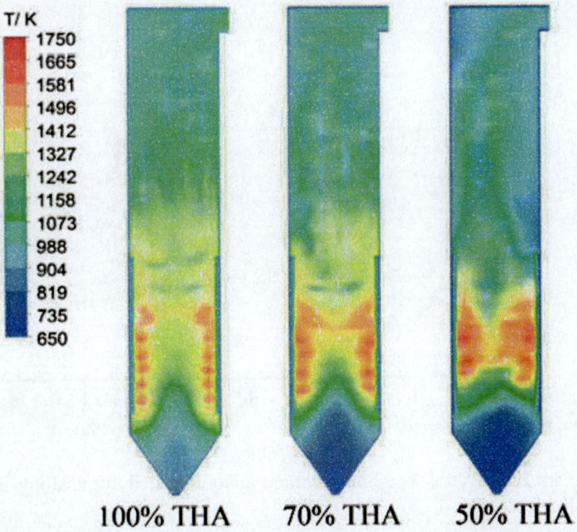

Fig. 20.3 Temperature distribution along the height direction under different loads

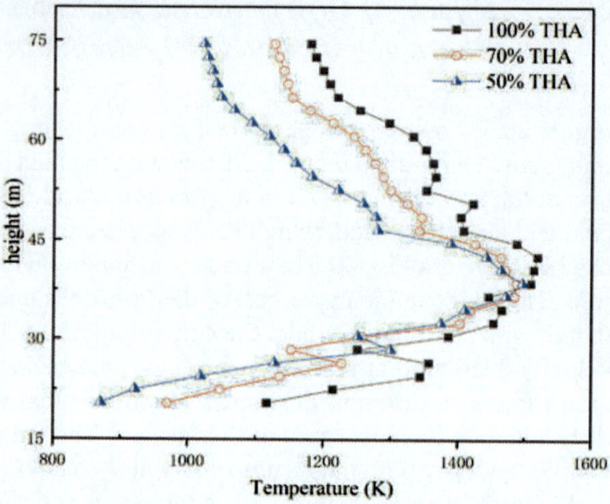

Fig. 20.4 Area-weighted average temperature along the height direction under different loads

Figure 20.5 shows area-weighted average species distribution along the height direction under different loads. With the injection of the air and the combustion reaction, the O_2 concentration C_{O2} changes drastically in the combustion zone. As shown in Fig. 20.5, the area-weighted average O_2 concentration at the furnace outlet C_{O2} descends as C_{O2} (100% THA) > C_{O2} (70% THA) > C_{O2} (50% THA). During the combustion of the furnace, most of O_2 is consumed to generate CO_2. The C_{CO2} decreases with the increasing loads as C_{CO2} (50% THA) > C_{CO2} (70% THA) > C_{CO2} (100% THA) with the highest C_{CO2} appearing at the 50-THA load. In the furnace

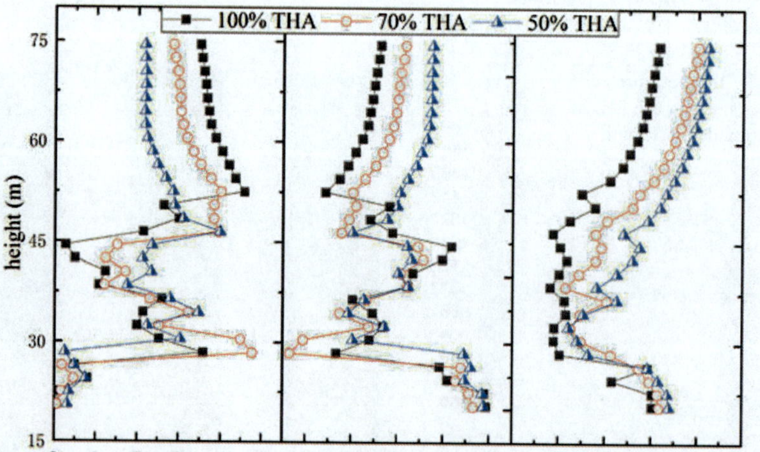

Fig. 20.5 Area-weighted average species distribution along the height direction under different loads

outlet, it is obviously seen that NO concentration decreases with increased loads as C_{NO} (50% THA) > C_{NO} (70% THA) > C_{NO} (100% THA). So it can be conclueded that NO_x concentrations in the flue gas often increases obviously with the decrease of the boiler operating load, which is consistent with the results of Zheng (Yang and Gao 2017).

20.4.3 Effects of Different Excess Air Ratios on the Combustion Characteristics and NO_x Emission

Temperature distributions along the height direction at different excess air ratios are demonstrated as Fig. 20.6. The combustion reaction near the nozzles in the main combustion zone is severe and the average temperature of the furnace gradually increases along the height of the furnace. After the overfire air with lower temperature is injected, the temperature in the furnace is reduced, while a small amount of heat is released as the char combustion. Above the main combustion zone, due to the heat exchange with water-cooled walls and superheater, the gas temperature begins to decrease. Under the three conditions of different excess air ratios, the temperature distributions in the furnace have little differences. This is because the oxygen concentration in the main combustion zone increases as the excess air ratio increases, which leads to that the combustion reaction in the main combustion zone is fully carried out and releases more heat. On the other hand, to heat the air with lower temperature causes the furnace temperature to decrease.

Figure 20.7 shows the O_2 distribution under different excess air ratios. As the excess air ratio increases, the oxygen concentration above the overfire air nozzles

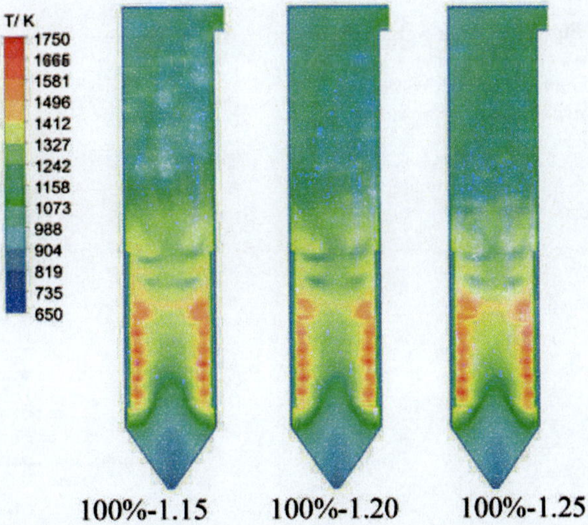

Fig. 20.6 Temperature distribution along the height direction at different excess air ratios

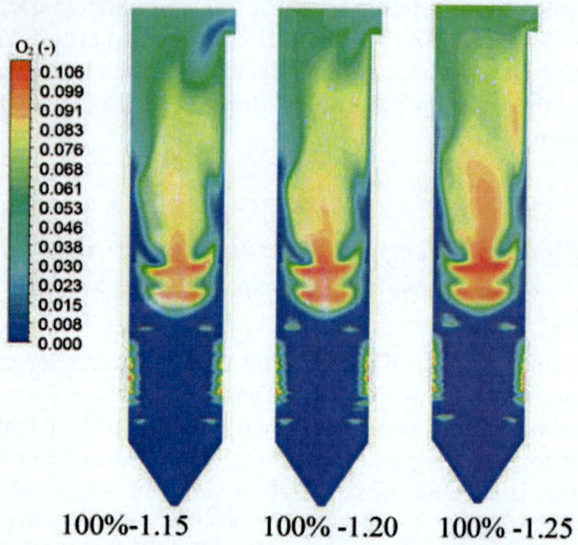

Fig. 20.7 O_2 distribution along the height direction under different excess air ratios

increases significantly. Figure 20.8 shows area-weighted average temperature and species distribution along the height direction under different air ratios. It can be seen from the Fig. 20.9 that the oxygen concentration is smaller in the main combustion zone. This is because the boiler adopts air-staging combustion technology, and the excess air ratio in the main combustion zone is less than 1.0, forming a strong reductive atmosphere, and the combustion reaction is insufficient. As the excess air ratio increases, the amount of oxygen concentration in the main combustion zone increases and combustion reaction in the furnace is more sufficient, so the high

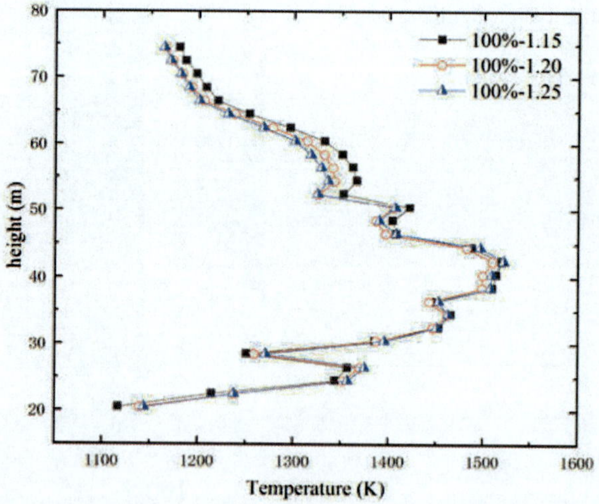

Fig. 20.8 Area-weighted average temperature along the height direction under different excess air ratios

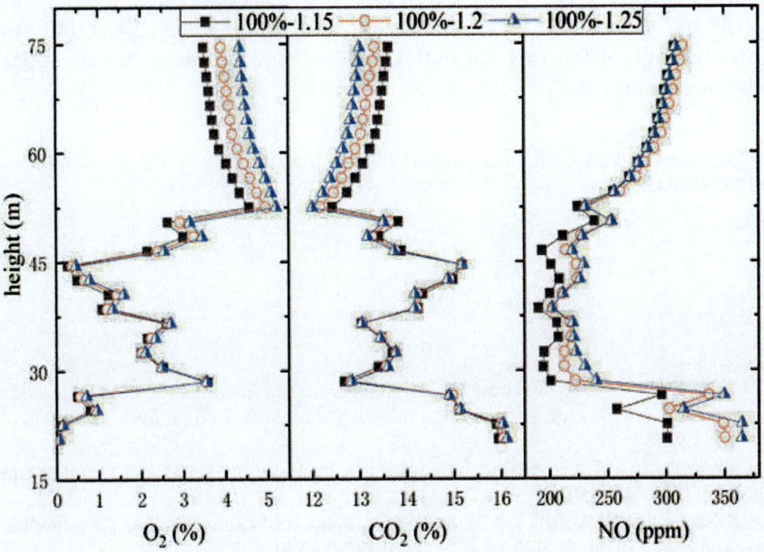

Fig. 20.9 Area-weighted average species distribution along the height direction under different excess air ratios

temperature region in the furnace gradually expands. With the injection of SOFA, the oxygen concentration increases significantly, and the char has almost burned out, so that the increase of excess air ratio will lead to an increase in NO_x emissions at the furnace outlet.

20.5 Conclusions

In this study, the effects of different loads and excess air ratios on combustion characteristics were studied by numerical simulation in an ultra-supercritical double-reheat unit. In NO_x calculation, the revised NO_x model is applied to replace the default model to accurately predict NO_x emission. The following conclusions are obtained:

1. As the load decreases, velocity distribution in the furnace are more uneven, which leads to more uneven temperature distribution at the upper part of the furnace. The NO_x emission increases while the boiler operates under the low and medium loads.
2. Under three conditions of different excess air ratios, the temperature distributions in the furnace have little differences. With the injection of SOFA, the oxygen concentration increases significantly, and the char has almost burned out, so that the increase of excess air ratios will lead to an increase in NO_x emission at the furnace outlet.

3. In the five simulation cases of the 660 MW tower boiler, the minimum NO_x concentration at the furnace outlet is 308 ppm, appearing under 100% THA with excess air ratio of 1.15.

Acknowledgements This work was supported by the National Key R&D Program of China (2017YFB0602102).

References

Álvarez L, Yin C (2013) Oxy-coal combustion in an entrained flow reactor: application of specific char and volatile combustion and radiation models for oxy-firing conditions. Energy 62(6):255–268

Chang S, Zhuo J (2016) Clean coal technologies in China: current status and future perspectives. Engineering 2(4):447–459

Drosatos P, Nikolopoulos N (2015) Numerical investigation of firing concepts for a flexible Greek lignite-fired power plant. Fuel Process Technol 142(2):370–395

Guo Z, Zhang S (2011) The pure heat conversion coefficient analysis method for thermal system in supercritical pressure power unit with double reheat cycles 2011

Hu L, Liu Y (2013) Effects of air staging conditions on the combustion and NO_x emission characteristics in a 600 MW wall fired utility boiler using lean coal. Energy Fuels 27(10):5831–5840

Huo H, Shi Q (2014) Influence of piping thermal efficiency on thermal economy of a 1000 MW double reheat power unit. J Chinese Soc Power Eng

Jiang D (2015) 1000 MW supercritial double reheat boiler technology. Value Eng

Levy JM, Chana LK (1981) NO/Char reactions at pulverized coal flame conditions. Symp Combust 18(1):111–120

Liu G, Chen Z (2015) Numerical simulations of flow, combustion characteristics, and NO_x emission for down-fired boiler with different arch-supplied over-fire air ratios. Appl Therm Eng 75(1):1034–1045

Liu H, Zhang L (2019) Effect of FGR position on the characteristics of combustion, emission and flue gas temperature deviation in a 1000 MW tower-type double-reheat boiler with deep-air-staging. Fuel 246:285–294

Panagiotis D, Athanasios N (2016) Numerical investigation of NO_x emissions for a flexible greek lignite–fired power plant. J Energy Eng

Park, Young H (2013) Numerical and experimental investigations on the gas temperature deviation in a large scale, advanced low NO_x, tangentially fired pulverized coal boiler. Fuel 104(2):641–646

Tumanovskii AG, Shvarts AL (2017) Review of the coal-fired, over-supercritical and ultra-supercritical steam power plants. Therm Eng 64(2):83–96

Yang Z, Gao X (2017) Impact of co-firing lean coal on NO_x emission of a large-scale pulverized coal-fired utility boiler during partial load operation. Korean J Chem Eng 34(4):1–8

Yao D, Zhu Y (2017) Design characteristics of 1000MW double reheat ultra-supercritical tower boiler. Boiler Technol

Zha Q, Li D (2017) Numerical evaluation of heat transfer and NO_x emissions under deep-air-staging conditions within a 600 MWe taentially fired pulverized-coal boiler. Appl Thermal Eng 116:170–181

Zhang J, Ito T (2013) Improvement of NO_x formation model for pulverized coal combustion by increasing oxidation rate of HCN. Fuel 113(2):697–706

Zhang W, Yan K (2016) Research on thermal hydrodynamic performance of water wall pipes for ultra-supercritical double reheat once-through boiler. J Eng Thermal Energy Power 31(8):75–80

Zhao S, Fang Q (2017) New fuel air control strategy for reducing NO_x emissions from corner-fired utility boilers at medium-low loads. Energy Fuels 31(7):6689–6699

Zhou H, Mo GY (2011) Numerical simulation of the NO_x emissions in a 1000 MW tangentially fired pulverized-coal boiler: influence of the multi-group arrangement of the separated over fire air. Energy Fuels 25(5):2004–2012

Chapter 21
Numerical Study on Influence of Platen Super-Heaters on Heat Deviation in a 600 MW Tangentially Fired Pulverized-Coal Boiler

Kai Chen, Yan Zhang, Lei Deng, and Defu Che

Abstract In order to investigate effects of platen super-heaters on the heat deviation in the cross-over pass, numerical simulation of a 600 MW tangentially fired pulverized-coal boiler has been studied. Two cases, building super-heaters or not, were established by FLUENT software in this study. Numerical results agree well with measured data of the utility boiler. Results show platen super-heaters break symmetry of velocity distribution at the furnace outlet. The position where peak temperature occurs is affected by platen super-heaters, while the position where peak velocity occurs is not. Platen super-heaters slightly reduce the temperature deviation at the furnace outlet, and enhances the velocity deviation at the furnace outlet. For this boiler, the velocity deviation is the main reason for the heat deviation in the cross-over pass. The simulation results in this study can provide reference for optimization of temperature deviation and velocity deviation in tangentially fired pulverized-coal boiler.

Keywords Platen super-heaters · Temperature deviation · Velocity deviation · Numerical simulation

21.1 Introduction

Tangentially fired boilers are commonly considered as steam generation facility in coal-fired power plant due to the advantage of stability and good fulfill-ness of flame in the furnace. However, the steam temperature distribution in the cross over pass of a tangentially fired boiler is quite uneven, thus may lead to local over-heating. Flue gas distributions (temperature and velocity distributions), structures of

K. Chen
School of Mechanic Engineering, Xijing University, Xi'an 710123, China

Y. Zhang · L. Deng · D. Che (✉)
State Key Laboratory of Multiphase Flow in Power Engineering, Xi'an Jiaotong University, Xi'an 710049, China
e-mail: dfche@mail.xjtu.edu.cn

J. Lyu and S. Li (eds.), *Clean Coal and Sustainable Energy*, Environmental Science and Engineering,
https://doi.org/10.1007/978-981-16-1657-0_21

heat exchangers and steam flow distributions jointly determine the degree of steam temperature deviation (Che 2008).

Flue gas distributions in the cross over pass are difficult to obtain from in-situ tests and numerical simulation is more economical and convenient than experimental study. Yin et al.(2003) studied factors that affect the temperature deviation, and proposed new boiler structures (a nose on front-wall and re-arranged platen super-heaters) for reducing the temperature deviation. Zhou et al. (2009) also proposed new boiler structures (furnace arch structures and arrangements of platen super-heaters) for alleviating the velocity deviation. Liu et al. (2014) analysed temperature and velocity deviations in a 1000 MW dual circle tangentially fired boiler, and found swing of over fire air (OFA) affected temperature and velocity deviations. Tian et al. (2015) investigated temperature and velocity deviations in a 700 MW tangentially fired boiler, and flue gas deviations can be reduced by tilting burners. Park et al. (2016) studied the steam temperature deviation in a 800 MW tangentially fired boiler, and found that changing the yaw and tilt angles of separated over fire air (SOFA) and closing coupled over fire air (CCOFA) can lead to the reduction of steam temperature deviation. In these studies, two methods for reducing temperature and velocity deviations are extracted. One is weakening the residual swirling intensity by optimizing injection angles of air (including primary air, secondary air, OFA, SOFA and CCOFA), and the other is optimizing boiler structures. Although Yin et al. (2003) and Zhou et al. (2009) re-arranged platen super-heaters for reducing temperature and velocity deviations, effects of platen super-heaters (with normal arrangement) on temperature and velocity deviations have not been studied.

In the present study, numerical simulation for a 600 MW tangentially coal-fired boiler is carried out. Two cases, building super-heaters or not, were simulated by FLUENT software. After validation of simulation results with measured data from coal-fired boiler, effects of platen super-heaters on temperature and velocity deviations are investigated.

21.2 Description of Boiler

Schematic configuration and burner nozzle distribution of the 600 MW tangentially coal-fired boiler is shown in Fig. 21.1 The height, width and depth of the boiler are 66,200, 19,558 and 15,400 mm, respectively. As shown in Fig. 21.1a, the furnace consists of the ash hopper zone, the combustion zone, the reduction zone, the SOFA zone, the burnout zone and the screen zone. In the screen zone, there are three groups of platen super-heaters, named as SH1, SH2 and SH3. SH1 and SH2 have six panels in Z-direction respectively and the spacing between adjacent panels is 3048 mm. SH3 has 25 panels in Z-direction and the spacing is 762 mm. Bundles following SH3 in the flue gas flow direction, arranged more compactly, are named as RH1, RH2 and SH4 in sequence, as shown in Fig. 21.1a, where RH represents the abbreviation for re-heater. SH5 and economizer (EZ) are placed in the rear pass.

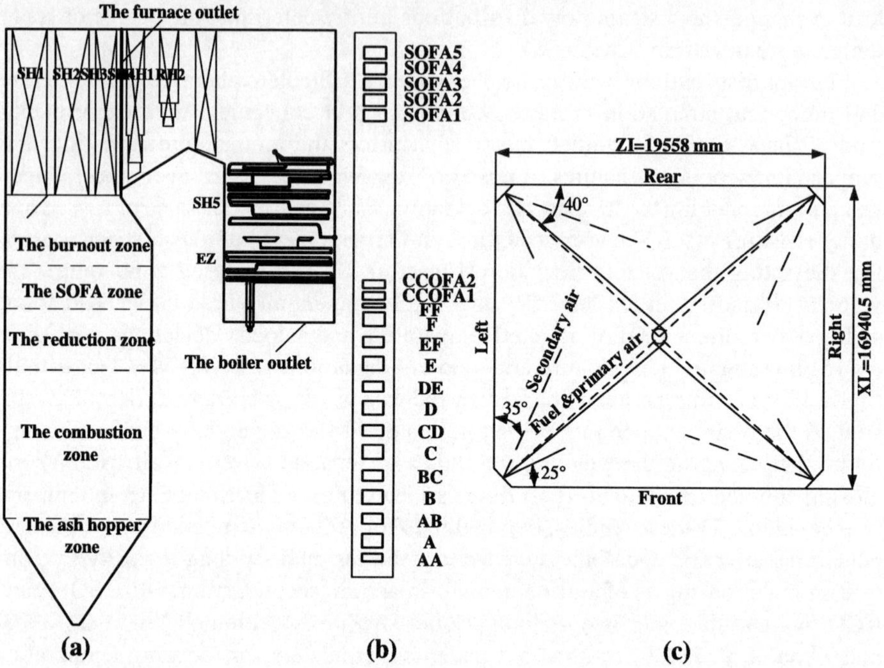

Fig. 21.1 Schematic configuration and burner nozzle distribution of the tangentially pulverized coal fired boiler

Vertical arrangements of burners at the corners are shown in Fig. 21.1b. A total of 6 primary air (PA) nozzles at one corner are divided into 6 rows, which are numbered from A to F. A secondary air (SA) nozzle is arranged between every two PA nozzles. Secondary air nozzle, named as AA, is installed below nozzle A. Two CCOFA nozzles and five SOFA nozzles are installed up the combustion zone.

Air from SA nozzles, CCOFA nozzles and SOFA nozzles is injected into furnace in horizontal direction. Air from PA nozzles is injected into furnace in both horizontal and vertical directions. Vertical incident angle of air from PA nozzles is 12.4 degree downward. Horizontal incident angles of air from nozzles are shown in Fig. 21.1c. Air from SA nozzles and SOFA nozzles forms contrary imaginary circles to reduce residual swirling intensity in the furnace.

The detailed operating information of the 600 MW boiler is shown in Table 21.1. During the boiler operation, five PA nozzles numbered from A to E are open and the PA nozzle F is standby. Besides, SOFA nozzles numbered from 2 to 5 are in operation and SOFA nozzle 1 is standby. Pulverized coal is carried into furnace with primary air equally. Coal mass flow rate is 62.01 kg·s^{-1} and excess air coefficient is chosen as 1.20.

A blended coal, comprised of Shenhua coal (80 wt%) and Baode coal (20 wt%), was used in the practical operation. Proximate and ultimate analyses of the blended coal is shown in Table 21.2. R90 of the coal is 18.0% and the uniformity exponent

Table 21.1 Operating information of the 600-MW boiler

Parameter	Location	Value
Air flow rates/kg·s^{-1}	PA (A, B, C, D, E)	4.84
	SA (AB, BC, CD, DE, EF)	10.26
	SA (FF)	2.43
	SA (AA)	5.68
	SA (A, B, C, D, E)	3.07
	CCOFA (CCOFA-1, CCOFA-2)	8.43
	SOFA (SOFA2, SOFA3, SOFA4, SOFA5)	9.36
Air inlet temperature /K	Primary air	350
	Secondary air	594
	CCOFA	594
	SOFA	594

Table 21.2 Proximate and ultimate analyses of the blended coal (as-received basis)

Coal properties	Shenhua coal	Baode coal	Blended Coal
Proximate analysis/wt%			
Moisture	12.60	16.50	13.38
Volatile matter	28.18	26.98	27.94
Fixed carbon	49.94	49.08	49.77
Ash	9.28	7.44	8.91
Ultimate analysis/wt%			
Hydrogen	3.81	3.64	3.78
Carbon	64.03	61.92	63.61
Sulfur	0.42	0.30	0.40
Nitrogen	0.83	0.75	0.81
Oxygen	9.03	9.45	9.11
Higher heating value/kJ·kg^{-1}	25,330	24,330	25,130

of the coal is 1.1. Average size of coal particles is 55.87 μm.

21.3 Numerical Simulation

21.3.1 Geometric Models

Figure 21.2 shows geometric models for the two cases in this study. The case modelling super-heaters is named as case 1 and the other as case 2. For case 1, super-heaters named as SH1, SH2 and SH3 in the upper furnace are set as double-sided walls with constant temperature (783 K). RH1, RH2, SH4, SH5, and EZ are set as porous mediums. Inertial resistance coefficients of heat-absorbing surfaces set as porous mediums are derived from measured pressure drop from the real boiler. However, no pressure drop was measured in this study. Therefore, inertial resistance coefficients from similar boilers (Yin et al. 2002, 2003; Diez et al. 2008) have been adopted in this study. Details of inertial resistance coefficients for these heat-absorbing surfaces are shown in Table 21.3. Temperature of water walls for both cases is constant (650 K).

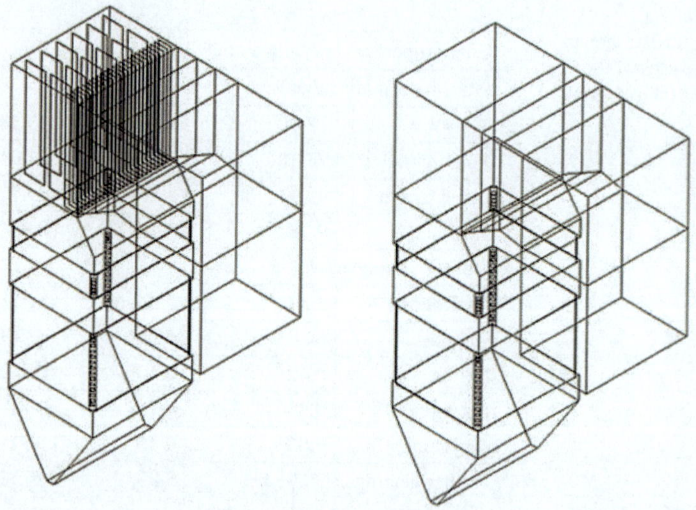

Fig. 21.2 Geometric models for the two cases

Table 21.3 Inertial resistance coefficients

	RH1	RH2	SH4	SH5 & EZ
X-direction	0.72	0.45	0.63	2
Y-direction	0.72	0.45	0.63	2
Z-direction	50	50	50	2

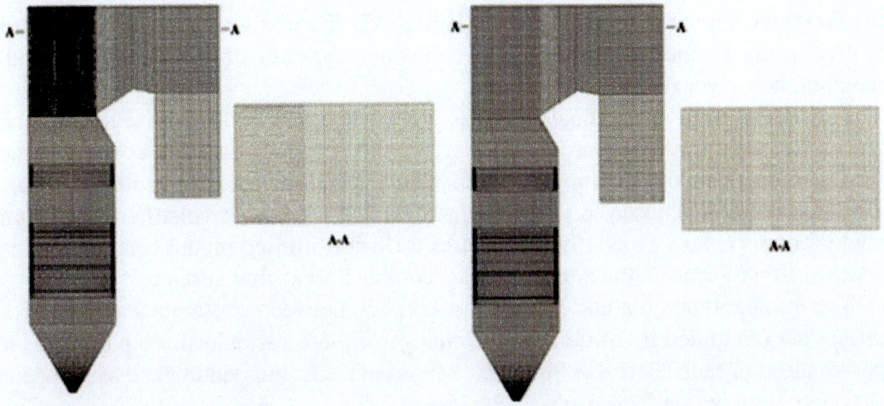

Fig. 21.3 Meshes for two cases

21.3.2 Meshes

Figure 21.3 shows meshes for two cases. For both cases, in the regions where a sudden change of flue gas velocity was expected, refined meshes were constructed. In the combustion zone and the SOFA zone, the grid lines are approximately along the flow direction. Difference between meshes lies in the mesh for the platen zone. Unstructured grids are used in case 1, while structured grids are used in case 2.

For grid independence test, three sets of meshes with different grids are built for each case. Cross-sectional average temperatures from different sets of meshes are compared for determining appropriate gird number for each case. Results show that meshes with 2,553,079 grids are accurate enough for case 1, and meshes with 2,346,582 grids are suitable for case 2.

21.3.3 Mathematical Models

When pulverized coal is injected into the furnace, coal particle is first heated by the flue gas, then consumes O_2 with the combustion process and releases heat to the flue gas. Two phases (continuous phase and discrete phase) are included in the combustion process. As the discrete phase occupies little space in pulverized coal combustion furnace, the Eulerian–Lagrangian approach is applied in this study. Sub-models, discretization and solving of equations are shown as follows.

For the continuous phase, realizable k–ε model is applied to close turbulent Reynolds equations. Probability density function (PDF) theory (Khalil 1980) is employed to simulate homogeneous combustion. The discrete ordinates (DO) model is used to simulate radiation heat transfer, due to its ability to calculate radiation in narrow space (Korytnyi et al. 2009). The absorption coefficient was calculated by the Weighted sum of gray gases model (WSGGM). Standard wall functions are used to account for viscous effects and resolve the rapid variation of flow variables within the boundary layer region.

For the discrete phase, coal particles are assumed to be spheres and obey the Rosin–Rammler distribution. Lagrangian method is used for tracing coal particles. Stochastic tracking model is applied to solve the turbulence intensity on coal particles. One-step model is chosen to simulate the releasing process of volatile matter from coal particles (Jones et al. 1999). Kinetics/diffusion-limited model (Field 1969) is adopted for calculating the heterogeneous combustion at char surface.

The mass, momentum and energy conservation between continuous and discrete phases are computed by examining the change of these variables for a particle as it passes through each control volume. Exchanges of mass, momentum and heat appear as source item in continuous phase equations.

Control volume method is used in the discretization of partial differential equations. First order upwind scheme is chosen for approximating the nonlinear terms in continuous phase equations. The semi-implicit method for pressure-linked equations (SIMPLE) algorithm is adopted to solve the relationship between velocity and pressure corrections and to obtain the pressure field. Exchanges of mass, momentum, and energy between the continuous and discrete phases are under-relaxed during the calculation.

21.4 Results and Discussion

21.4.1 Calculation Validation

As case 1 is more strictly built on structure of the 600 MW boiler, numerical results of case 1 are used for validation. Table 21.4 shows validation results with key design values. Table 21.4 shows the comparison between the numerical results and thermal design values. Biggest deviation of simulation results from design values is about 5%, which indicates that results from numerical simulations are in good agreement with key design values.

Table 21.4 Validation with key design values

	Design values	Numerical results
Temperature at the bottom of screen/K	1547	1552.9
O_2 mole fraction at the boiler outlet/%	3.5	3.5
Heat transfer of SH1 & SH2/MW	111.7	108.0
Heat transfer of SH3/MW	126.8	100.8

Besides, numerical results are validated by cross-sectional average temperature along the furnace height, as shown in Fig. 21.4. Variation trend of cross-sectional average temperature along the furnace height from numerical study is almost the same as that from experimental study. However, cross-sectional average temperature from experiments is 100~200 K lower than that from numerical simulations. This is because cross-sectional average temperature from experiments is the average temperature measured at limited points in the furnace. Since temperature at one cross-section is strongly non-uniform, experimental cross-sectional average temperature is not strictly equal to cross-sectional average temperature. Thus, it is understandable that cross-sectional average temperatures from experimental are generally lower than that from numerical simulation.

As the influence of platen super-heaters on the heat deviation is studied in this paper, temperature between platen super-heaters is validated by experimental data from infra-red thermometer (IRT-3000B), as shown in Fig. 21.5. The validation

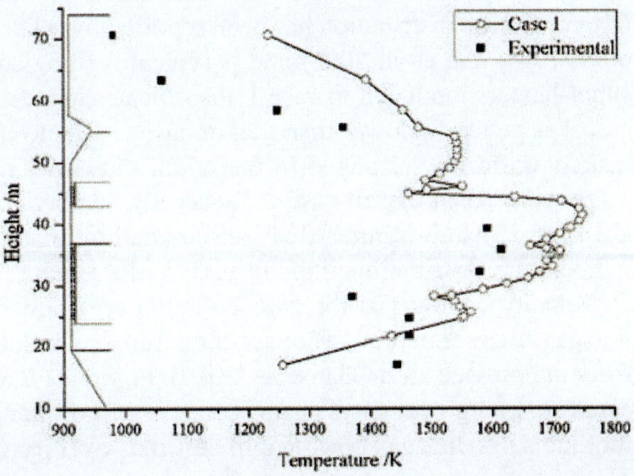

Fig. 21.4 Cross-sectional average temperature along the furnace height

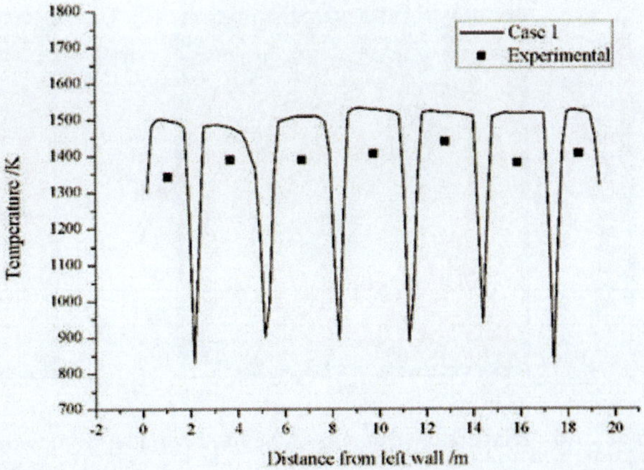

Fig. 21.5 Temperature between platen super-heaters

indicates numerical results generally agree well with experimental data. However, numerical results are 100~200 K higher than that from experimental data. In order to validate numerical results, heat balance based on reliable steam data from in-situ tests, is used for calculating temperature at the bottom of platen super-heaters. Calculation results show that temperature at the bottom of platen super-heaters from the simulation is more reliable than that from experimental data. Deviation between numerical and experimental results can be attributed to error caused by measurement environment.

In general, validations above indicate that simulation results are reliable.

21.4.2 Temperature and Velocity Contours

Figure 21.6 shows temperature contours at the furnace outlet for two cases. As shown in Fig. 21.6, lower center parts for both cases are in higher temperature. This kind of temperature distribution has been reported by other studies (Tian et al. 2015; Park et al. 2016; Yin et al. 2002) and is typical for tragically fired boilers. Since platen super-heaters modelled in case 1 absorb extra heat from flue gas, temperature from case 1 is generally lower than that of case 2. Due to heat absorbing of platen super-heaters uniformly arranged in the width direction, high temperature zone in case 1 is smaller than that in case 2. Generally, temperature distributions in both cases are approximately symmetrical, and modelling platen super-heaters will not affect temperature distribution at the furnace outlet much.

Velocity contours at the furnace outlet are shown in Fig. 21.7. For both cases, flue gas reverses in the lower part of the furnace outlet, as where flue gas is reflected from the furnace arch (Tian et al. 2015). Figure 21.7 also shows upper parts for two cases are in high velocity. That is because part of flue gas reflected from the furnace roof joins into flue gas flowing into the cross-over pass. Highest (lowest) velocity for

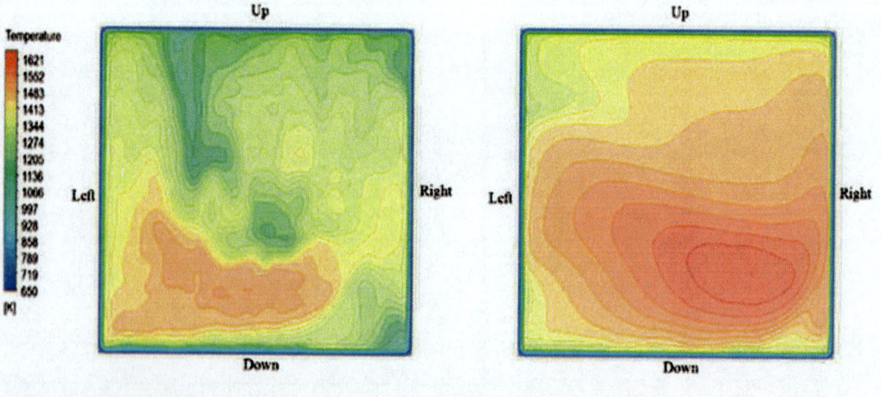

Fig. 21.6 Temperature contours at the furnace outlet from two cases

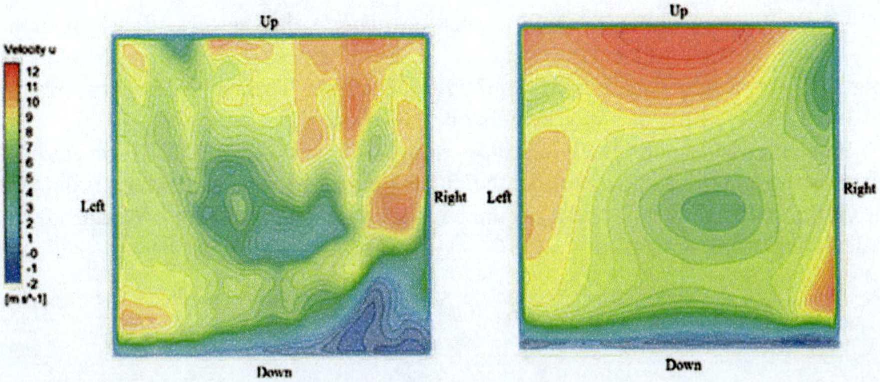

Fig. 21.7 Velocity contours at the furnace outlet for two cases

case 1 appears closer to right wall than that for case 2, since platen super-heaters cut flue gas apart and block exchange of flue gas in the width direction. Velocity contour from case 2 is more symmetrical than that from case 1.

21.4.3 Non-uniformity Analysis

In order to analysis the heat deviation quantitatively, non-uniformity coefficient M (Zhou et al. 2009; Liu et al. 2014; Tian et al. 2015) is introduced in this paper. The definition of M is as follows:

$$M_{\phi} = \left(\phi_{ave} + 3\sigma_{\phi}\right)/\phi_{ave} \tag{21.1}$$

$$\sigma_{\phi} = \sqrt{\frac{1}{n-1}\sum_{i=1}^{n}\left(\phi_i - \phi_{ave}\right)^2} \tag{21.2}$$

where ϕ is general physical parameter, subscript ave is the abbreviation of average, subscript i represents cell i. As ϕ in Eqs. (21.1) and (21.2) is substituted by temperature (T) or velocity (V), equations for calculating M_T or M_V are obtained, respectively.

Results of M_T and M_V at the furnace outlet are shown in Table 21.5. For both cases,

Table 21.5 M_T and M_V at the furnace outlet

Cases	M_T	M_V
Case 1	1.3618	2.3078
Case 2	1.4380	2.0964

M_T at the furnace outlet is small. Modelling platen super-heaters slightly reduces M_T at the furnace outlet. However, M_V at the furnace outlet is large. Modelling platen super-heaters enhances M_V at the furnace outlet. This enhancement is caused by block effect of platen super-heaters on flue gas, as mentioned above.

Parameter M quantitatively analyzes non-uniformity in a plane. To reveal heat deviation along the width direction, the dimensionless parameter ξ, which is defined as vertical average of a physical parameter, is introduced in this paper. Equations for calculating ξ are as follows:

$$\xi_\phi = \phi_{\text{ave}}^{\text{ver}} / \phi_{\text{ave}} \tag{21.3}$$

$$\phi_{\text{ave}}^{\text{ver}} = \frac{1}{n} \sum_{i=1}^{n} \phi_i \tag{21.4}$$

where ϕ is the general physical parameter, while superscript ver is the abbreviation of vertical, subscript ave is the abbreviation of average and subscript i represents cell i. As ϕ in Eqs. (21.3) and (21.4) is substituted by temperature (T) or velocity (V), equations for calculating ξ_T or ξ_V are obtained, respectively.

Figure 21.8 shows ξ_T and ξ_V distributions along the furnace width for the two cases. As a result of platen super-heaters modelled in case 1, ξ_T and ξ_V distributions from case 1 are more fluctuant than that from case 2. Max dimensionless velocity and dimensionless temperature from case 1 is larger than that from case 2. For both cases, the maximum of ξ_T is smaller than the maximum of ξ_V. This indicates velocity deviation is the main reason for the heat deviation. Furthermore, modelling platen super-heaters changes the position where peak dimensionless temperature occurs. As shown in Fig. 21.8a, dimensionless temperature reaches its maximum at left position in case 1, while in case 2 dimensionless temperature reaches its maximum at right

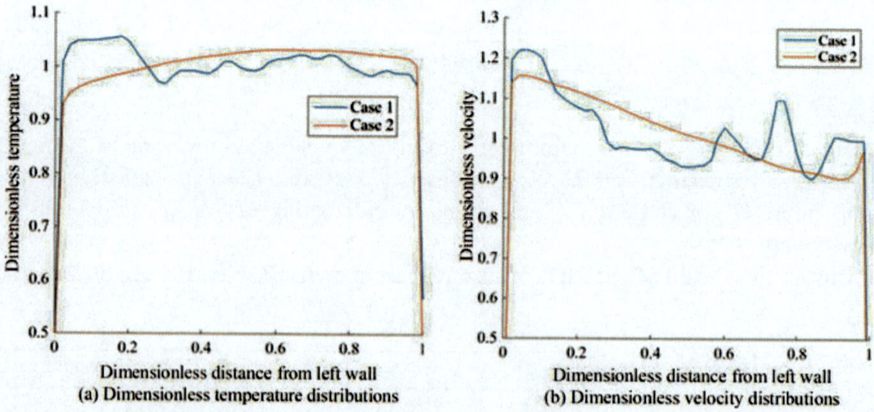

Fig. 21.8 Dimensionless temperature and velocity distributions for two cases

Table 21.6 E_T and E_V at the furnace outlet

Cases	E_T	E_V
Case 1	1.091	1.356
Case 2	1.109	1.283

position. However, modelling platen super-heaters will not change the position where peak dimensionless velocity occurs. As shown in Fig. 21.8b, peak dimensionless velocity in both cases occurs at left position.

For evaluating heat deviation in the width direction, the deviation factor E is introduced. The definition of E is as follows:

$$E_\phi = \xi_{\phi,\max}/\xi_{\phi,\min} \tag{21.5}$$

where ϕ is the general physical parameter, ξ the dimensionless parameter. As ϕ in Eqs. (21.5) is substituted by temperature (T) or velocity (V), equations for calculating E_T or E_V are obtained, respectively.

Results of E_T and E_V at the furnace outlet are shown in Table 21.6. For both cases, E_T at the furnace outlet is small. Modelling platen super-heaters slightly reduces E_T at the furnace outlet. However, E_V at the furnace outlet is large. Modelling platen super-heaters enhances E_V at the furnace outlet. Rules of E_T and E_V changing with cases are just the same as rules of M_T and E_V changing with cases. Analysis for both the non-uniformity coefficient M and the deviation factor E indicates platen super-heaters slightly reduces the temperature deviation, and enhances the velocity deviation at the furnace outlet. The velocity deviation is the main reason for the heat deviation. Optimizations on boiler structures and air injection angels will effectively reduce velocity deviation (Zhou et al. 2009; Liu et al. 2014; Tian et al. 2015), thus reduce heat deviation in the width direction.

21.5 Conclusions

Effects of platen super-heaters on temperature and velocity deviations has been numerically investigated in the present study. Main conclusions are as follows:

1. In the 600 MW tangentially coal-fired boiler, temperature and velocity contours indicate platen super-heaters has no effect on temperature distribution at the furnace outlet much. However, it breaks symmetry of velocity distribution at the furnace outlet.
2. In the 600 MW tangentially coal-fired boiler, analysis for dimensionless temperature and velocity distributions along the width direction indicates that platen super-heaters change the position where peak dimensionless temperature occurs. However, it does not change the position where peak dimensionless velocity occurs.

3. In the 600 MW tangentially coal-fired boiler, analysis for the non-uniformity coefficient and the deviation factor indicates platen super-heaters slightly reduces temperature deviation, and enhances velocity deviation at the furnace outlet. The velocity deviation is the main reason for the heat deviation.

Acknowledgements This work is financially supported by Shanghai Boiler Works Ltd under project title: Investigation on Combustion Systems for a 1000 MW Tangentially Fired Boiler Using Anthracite Coal. The authors gratefully acknowledge their support.

References

Che D (2008) Boilers—theory, design and operation. Xian Jiaotong University Press, Xi'an, China
Díez LI, Cortés C, Pallarés J (2008) Numerical investigation of NOx emissions from a tangentially-fired utility boiler under conventional and overfire air operation. Fuel 87(7):1259–1269
Field MA (1969) Rate of combustion of size-graded fractions of char from a low-rank coal between 1 200 K and 2 000 K. Combust Flame 13(3):237–252
Jones JM, Patterson PM, Pourkashanian M et al (1999) Modelling NOx formation in coal particle combustion at high temperature: an investigation of the devolatilisation kinetic factors. Fuel 78(10):1171–1179
Khalil E (1980) On the prediction of reaction rates in turbulent premixed confined flames. In: 18th aerospace sciences meeting. American institute of aeronautics and astronautics
Korytnyi E, Saveliev R, Perelman M et al (2009) Computational fluid dynamic simulations of coal-fired utility boilers: An engineering tool. Fuel 88(1):9–18
Liu H, Sha L, Xu L et al (2014) Modeling the occurrence and methods of reducing thermal deviations of upper furnace heating surfaces in a 1000 MW dual circle tangential firing single furnace ultra-supercritical boiler. Numer Heat Transf Part A-Appl 66(7):816–838
Park HY, Baek SH, Kim HH et al (2016) Reduction of main steam temperature deviation in a tangentially coal-fired, two pass boiler. Fuel 166:509–516
Tian D, Zhong L, Tan P, et al (2015) Influence of vertical burner tilt angle on the gas temperature deviation in a 700 MW low NOx tangentially fired pulverised-coal boiler. Fuel Process Technol
Yin C, Caillat S, Harion J-L et al (2002) Investigation of the flow, combustion, heat-transfer and emissions from a 609 MW utility tangentially fired pulverized-coal boiler. Fuel 81(8):997–1006
Yin C, Rosendahl L, Condra TJ (2003) Further study of the gas temperature deviation in large-scale tangentially coal-fired boilers. Fuel 82(9):1127–1137
Zhou Y, Xu T, Hui S et al (2009) Experimental and numerical study on the flow fields in upper furnace for large scale tangentially fired boilers. Appl Therm Eng 29(4):732–739

Chapter 22
Industrial Experimental and Numerical Simulation Study on the Combustion and NO_x Formation Characteristics in a 600MWe Utility Boiler with a Novel Swirl Burner Burning Bituminous Coal

Tao Shen, Zhengqi Li, Pisi Lu, Qiang Yu, Xin Song, and Jingyu guan

Abstract To minimize the NO_x formation, a novel low-NOx swirl pulverized-coal combustion technology characterized by the annulus recirculation was proposed by the Harbin Boiler Company Limited. By employing the industrial experiments to study the combustion and NO_x formation characteristics in a 600MWe utility boiler burning the bituminous coal, and simultaneously using the Reynolds Stress turbulence model and the Realizable k-ε model to reveal the in-burner flow characteristics and the full-scale combustion characteristics in the 600MWe utility boiler. The numerical results were in good consistent with the experimental results. The results show that an obvious annular recirculation zone is generated in the axial direction at the burner outlet zone. The recirculation zone brings the high-temperature flue gas back to the near burner zone and heats the primary coal/air flow, which promotes the pulverized coal ignition and maintains the flame stability. Additaionaly, the recirculating high temperature gas has the lowest oxygen content, is thus beneficial to the NO_x reduction. On reducing the blade angles of the inner and outer secondary air from 65° to 30°. The size of the recirculation zone is gradually reduced, the entraining capacity of the flue gas weakens, the coal ignition is simultaneously delayed. The oxygen contents in the main combustion zone increase, associated with the gradually weakened reducing atmosphere. The NO_x content is increased by 8.5%, while the CO content and carbon in fly ash are slightly reduced. While the blade angles becomes

T. Shen · Z. Li (✉) · J. guan
School of Energy Science and Engineering, Harbin Institute of Technology, Harbin 150001, Heilongjiang, China
e-mail: green@hit.edu.cn

T. Shen · P. Lu · Q. Yu · X. Song · J. guan
Harbin Boiler Company Limited, Harbin 150046, Heilongjiang, China

T. Shen · Q. Yu · X. Song · J. guan
State Key Laboratory of Efficient & Clean Coal-Fired Utility Boilers (Harbin Boiler Company Limited), Harbin 150046, Heilongjiang, China

P. Lu
SmartBurn LLC, 579 D'onotorio Dr, Madison, Wl 53719, USA

J. Lyu and S. Li (eds.), *Clean Coal and Sustainable Energy*,
Environmental Science and Engineering,
https://doi.org/10.1007/978-981-16-1657-0_22

larger, the outlet flow field of the burner is easy to fly, and the water wall is prone to coking、high temperature corrosion, etc. It is recommended that the blade angle of the inner and outer secondary air is set to 45° in the actual boiler operation. The numerical results show that due to the combined combustion effect of the secondary air and the pulverized coal, the combustion performance fluctuates. As the newly injected overfire air in the burnout zone, under the combined effects of the oxygen concentration and the high-temperature flue gas, the combustion performance under each case varies less, and the influence degree of adjusting the inner secondary air blade angle is greater than that of the outer secondary air.

Keywords Burner · Industrial experiment · Nmerical simulation · Recirculation zone · Pulverized coal combustion · Angle of inner/outer-secondary-air blade · NO_x

22.1 Introduction

China's energy mainly concentrates on the coal industry, and the power industry is one of the major aspects of the coal consumption. Therefore, ensuring the efficiency and stability of pulverized coal combustion is an important goal of the power industry. In the combustion equipment burning the low-volatile bituminous coal, the swirl burner has become the main technical means (National Bureau of Statistics of China 2015; Newby et al. 1994; Dang and Rochelle 2003; Beer 2000; Nainar and Veawab 2009).The swirl burner technology mainly relies on the recirculation zone bringing the high-temperature flue gas back to the near burner zone and heating the primary coal/air flow, which promotes the pulverized coal ignition and maintains the flame stability. When burning the low-volatile bituminous coal boiler, there are some problems in the operation of swirl burners introduced from abroad, such as high-temperature corrosion、unstable flame at low load, etc. Simultaneously, in order to reduce NO_x emissions, CO content and carbon in fly ash, the burner nozzle structure is more and more complicated to develop the swirl pulverized-coal combustion technology. Many swirl burners are unique two-phase flow composed of multi-swirl and DC concentric jets. In order to solve the above problems during boiler operation, many researchers have conducted the laboratory test and industrial experiments. Based on the radial dense-lean swirling pulverized coal concept, Li et al. (Rui 1998; Song et al. 2016) proposed a low NO_x technology application of centrally fuel rich swirl burner on the coal fired boiler in 2003. Chen et al. (2008) used three-dimensional PDA to study the gas–solid two-phase flow characteristics of the centrally fuel rich swirl burner, and obtained the influence of different structural parameters on the gas–solid flow characteristics. Wang et al. (Zhiqiang et al. 2007) employed the industrial experiments and numerical simulation to study the influence of the secondary air blade angle on the flow field in the furnace.

In view of the problems existing in the operation of the low-volatile bituminous coal boilers, Harbin Boiler Company Limited developed a novel low-NO_x swirl pulverized-coal combustion technology. In this paper the flow characteristics of a novel low-NO_x swirl burner are studied by the numerical simulation, including flow and distribution characteristics of pulverized coal concentration field. Simultaneously the industrial experimental and numerical simulation were studied on the combustion and NO_x formation characteristics in a 600 MWe supercritical utility boiler burning low-volatile bituminous coal. The results show that the the novel low-NO_x swirl burner technology achieves the desired results in practical application engineering, and can reduce NO_x formation effectively without increasing boiler efficiency.

22.2 Research Object

Figure 22.1 is the structure of a novel low-NO_x swirl burner. The swirl burner primary air is no swirl. A pulverized-coal feeder is arranged at the primary air duct, and a two-stage pulverized coal concentrator is installed in the primary air passage. A steady combustion ring and a steady-burning tooth are installed near the burner outlet. The main combustion air is divided into central air、DC secondary air、inner secondary air and outer secondary air. Central and DC secondary air has no swirl air. The inner and outer secondary air is axial swirling air. The inner and outer secondary air adopts 16 tangential blades, and the blade angle can be adjusted. The function of the central air is to provide an appropriate amount of air to the burner center to stabilize the combustion of the air and pulverized coal, and to prevent the burner nozzle from burning by adjusting the air ratio. The primary air carries the pulverized coal through the burner elbow to enter the primary airduct. By arranging the pulverized coal feeder at the inlet of the primary air pipet to ensure that the primary air/coal reaches a uniform distribution along the burner circumference. Then passing through the venture pulverized coal concentrator, the pulverized coal at the burner outlet zone respectively presents a high concentration and a low value inside and outside the burner center. The combustion air injects into the furnace at different combustion stages to enhance the mixing degree of air with the pulverized coal, together with the primary air forming an annular recirculation zone at the burner outlet. The recirculation zone brings the high-temperature flue gas back to

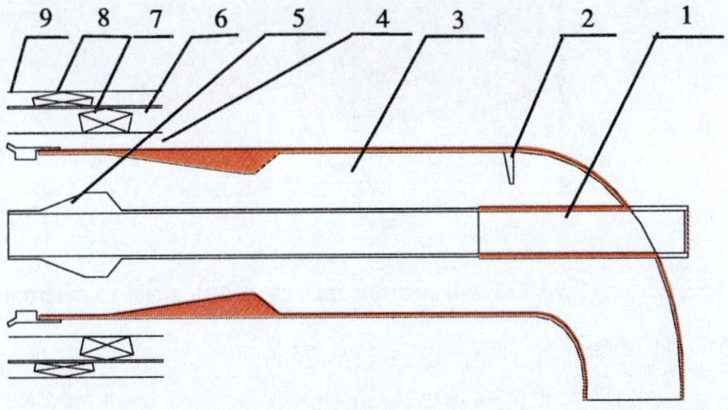

Fig. 22.1 Structure of low NO_x swirl burner: (1) Central air; (2) Pulverized-coal Feeder; (3) Primary air; (4) DC secondary air; (5) Concentrator; (6) Inner secondary air; (7) Inner secondary air blade; (8) Outer secondary air blade; (9) Outer secondary air

the near burner zone. Additaionaly, the recirculating high temperature gas has the lowest oxygen content, is thus beneficial to the NO_x reduction. Simultaneously the pulverized coal at the burner outlet respectively presents a high concentration and a low value inside and outside the burner center staging combustion with the main combustion air is also beneficial to the reduction of NO_x. The high-temperature flue gas in the recirculation zone can also heats the primary coal/air flow, which promotes the pulverized coal ignition and maintains the flame stability. The steady combustion tooth of the primary air outlet can increase the pulverized coal disturbance and promote the combustion of the pulverized coal. In summary, this burner can minimize the generation of NO_x emission while promoting stable combustion of pulverized coal without affecting boiler efficiency.

The furnance model of 600MWe bituminous coal is shown in Fig. 22.2. The boiler is all steel structural, full suspension structure, π-type boiler of a supercritical parameter, pump circulation, single reheat, single furnace balance draft, dry slagging and hypaethral placement, 6 sets of HP1003 middle storage pulverizing system. The coal burner developed a novel low-NO_x swirl burner for the Harbin Boiler Company Limited. Which are arranged on the front and rear walls. The front and rear walls are three rows of upper, middle and lower, each with 5 burners, a total of 30 burners. 16 OFA burners are located at the furnace upper part, with ten OFA burners distributed on the front and rear walls and six OFA burners distributed on the left and right walls.

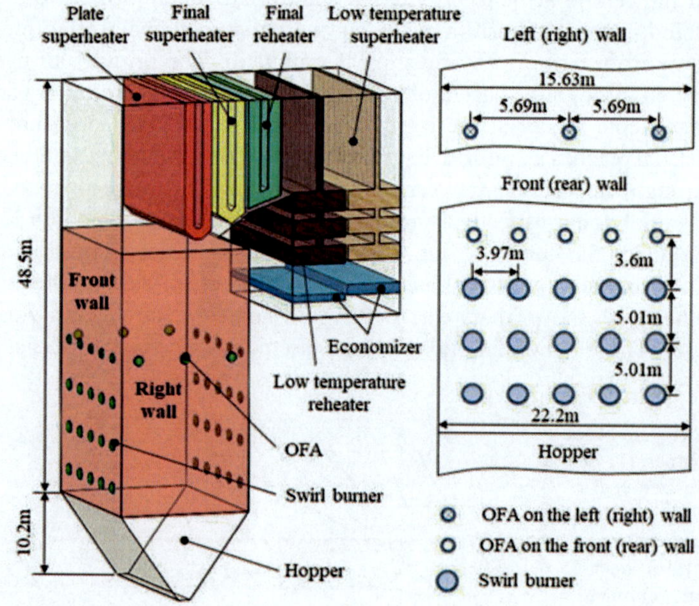

Fig. 22.2 The 600MWe furnace model of industrial experiment

22.3 Research Methods

22.3.1 Experimental Research Methods

22.3.1.1 Cold Experiment Test

The cold air aerodynamic field experiment and measurement system is shown in Fig. 22.3. The air passes through the air duct and enters the air chamber through the connecting pipe by the forced draft fan. The air distribution in the air chamber enters the experiment test system. By adjusting the valve opening of each branch, the air flow into each pipeline is adjusted. The pressure and flow measuring points are installed on each pipeline. The air velocity and direction of each burner nozzle are measured under the modeled air speed. In order to make the air flow characteristics of the furnace obtained by the cold modeling test approach to the flow characteristics of the furnace during the fired operation of the boiler, the cold modeling experiment should follow the following three modeling criteria: (1) The modeling burner and the actual burner are similar in geometry. In this paper, the molding experiment bench is reduced according to the actual furnace size by 0.8 times. (2) The air flows in the molded furnace is located in the second self stimulated domain. (3) The momentum ratio between the air flows entering the mold furnace is equal to the momentum ratio between the air flows in the actual furnace. The coordinate frame is placed directly in front of the burner nozzle. The test area is roughly long $\times$ width = 3000 mm $\times$ 3000 mm, the grid size of each test point is 80 mm $\times$ 80 mm. The red ribbon is placed at each grid point of the coordinate frame to trace the direction and inclination of the air at the burner outlet. Meanwhile using a single streamer frame to trace the fluid flow dip different areas of the upper, middle and lower parts of the burner outlet point by point and measure the angle. The experiment is carried out under normal temperature and pressure. The specific parameters of the single burner cold experiment are shown in Table 22.1.

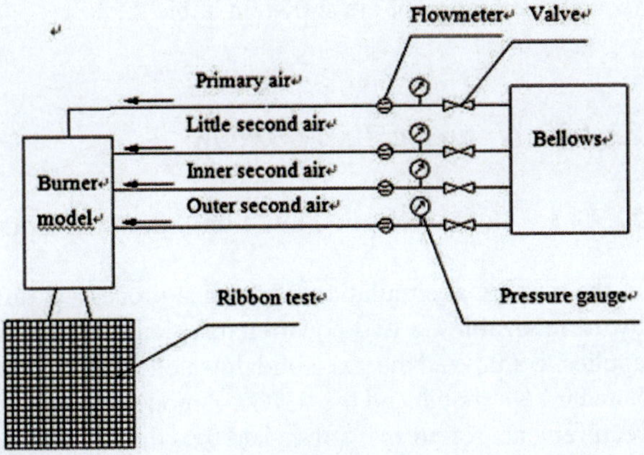

Fig. 22.3 The novel swirl burner cold air aerodynamic field experiment and measurement system

Table 22.1 Specific parameters of the single burner cold experiment

Parameter	Unit	Data
Central air speed	m/s	6.88
Primary air speed	m/s	20.74
DC secondary air speed	m/s	13.03
Inner secondary air speed	m/s	20.85
Outer secondary air speed	m/s	28.67
DC secondary air mass flow	kg/s	2.06
Inner secondary mass flow	kg/s	6.37
Outer secondary mass flow	kg/s	6.36

22.3.1.2 Industrial Experiment

In this paper, the experiment analysis of the low-volatile bituminous coal 600MWe boiler was carried out. The application effect of the HBC novel low-NO_x swirl burner in the actual project was studied. Simultaneously studying on the effect of changing the angle of the inner and outer secondary blades angle in the burner on the combustion characteristics of the furnace. The flue gas composition at the SCR inlet were measured under different working conditions.

The coal quality is kept stable during the test, and any interference test such as soot blowing is avoided. Each test condition was measured for 4 h. Measurements include: The flue gas composition at the SCR inlet were measured and analyzed under different working conditions. The CO, O_2, NO_x emission concentrations at the SCR inlet are measured by grid method. The arithmetic average of each measuring point is taken as the CO, O_2, NO_x emission concentration of the imported flue gas. The flue gas composition of the furnace horizontal outlet (SCR inlet) and the carbon in fly ash were measured using a Testo 350 flue gas analyzer. Before the measurement, the flue gas analyzer was calibrated using a standard concentration gas mixture, and the measurement errors of the NO_x and O_2 concentration were within 50 ppm and 1%, respectively. The specific working conditions and test parameters of the thermal industrial experiment are shown in Table 22.2.

22.3.2 *Numerical Simulation*

22.3.2.1 Cold Dynamic Field Simulation Methods

In the numerical simulation of furnace processes, turbulence is taken into account by the realizable k-ε two-equation turbulent model. The stochastic tracking model is applied to analyzed the gas–solid flow field. The realizable k-ε model differs from the Standard k-ε model and the RNG k-ε model, it can satisfy the mathematical model's requirements for normal stress, and thus the physical properties of turbulence can be

Table 22.2 Industrial experiment under variable inner and outer secondary air blade angle

Condition	Condition 1	Condition 2	Condition 3
Load (MW)	600	600	600
Operating layer damper Opening (%)	70	70	70
Outage layer damper opening (%)	20	20	20
Central damper opening (%)	10	10	10
OFA damper opening (%)	100	100	100
Burner damper opening (%)	100	100	100
Inner secondary air blade angle (°)	30 (45)	45 (45)	65 (45)
Outer secondary air blade angle (°)	45 (30)	45 (45)	45 (65)

more reasonably expressed. In order to simulate the effect of turbulent flow field on the motion of pulverized coal particles, we used the Lagrangian random orbit model to solve kinematic trajectory of pulverized coal particles. The basic idea of this model is to consider the motion of particles under the influence of the velocity of gas phase pulsation when calculating the random orbit of particles, that is, the instantaneous velocity of the gas is randomly given to account for the effect of fluid turbulence on the particles based on the instantaneous momentum equation of the particles. The solid phase flow calculated is a sparse suspension flow, and the particle phase volume fraction is small and negligible in this paper. Coupling calculation between gas phase and solid phase, the momentum, energy and component exchange between the particle phase and the gas phase flow field are calculated using the "Particle Source In Cell" algorithm.

In this paper, the novel low-NO_x swirl pulverized-coal burner of the flow characteristics was first carried out. The fluid flow area calculation grid in the burner uses a hexahedral calculation grid, and the total number of grids is 2.3 million. The turbulence calculation model is different from the whole furnace calculation. The Reynolds Stress model is adopted. The inlet adopts the mass flow boundary condition and the outlet is the pressure outlet boundary condition. Figure 22.4 shows the numerical simulation calculation grid model.

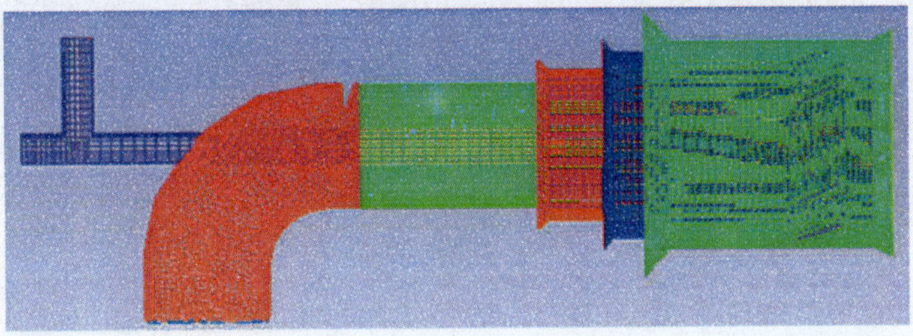

Fig. 22.4 Grid model

22.3.2.2 Thermal Furnace Simulation Methods

In this paper, the SIMPLE algorithm is used to consider the pressure–velocity coupling, and the turbulent flow field is simulated using the realizable k-ε model. Momentum equations, turbulent kinetic energy, turbulent energy dissipation rate, component concentration, energy and other control equations are all based on the second-order style. The simulation of the radiation heat transfer in the furnace uses the Discrete Ordinate method. The Lagrangian random orbit model is used to simulate the pulverized coal particle trajectory. Simulating pulverized coal process using a single equation model (Zhang et al. 1997). The simulates the surface combustion process of carbon particles with kinetic/diffusion-limited method (Zhang and Nieh 1997). Simulation of NO_x generation due to the extremely small amount of NO_x (Prompt NOx) produced by pulverized coal boilers. The calculations only consider the generation of thermal NO_x and fuel NO_x. Thermal NO_x adopts extended Zeldovich mechanism (Xu et al. 2000; Zhou et al. 2003). Fuel type NO_x is calculated using the De Soete model (Zhang et al. 2007; Thomas et al. 2007). Because all the basic equations are nonlinearcoupled differential equations, the Fluent separation solver wasused. The Semi-Implicit Method for Pressure Linked Equations (SIMPLE) algorithm of pressure correction was applied to consider the coupling of velocity and pressure. The second-order upwind scheme was used to iterate the turbulence intensity, dissipative rate, and Reynolds stress. The ga turbulence was specifically taken into account using the realizable k-ε model. The Lagrangian stochastic tracking model was used to analyze the pulverized coal flows, and gas/particle coupling was determined using the particle-sourcein-cell method. The P1 radiation model for the simulation of radiation heat transfer was used. The devolatilization process was modeled based on the two-competing-rate Kobayashi model, and the combustion of volatiles was modeled by employing the probability density function theory. Char combustion was modeled using a kinetic/diffusion-limited method. Little prompt NO_x was generated in the process of pulverized coal combustion; therefore, only thermal NO_x and fuel NO_x were considered in the numerical calculation. The fuel NO_x concentration was calculated using De Soete's model, and the thermal NO_x concentration was calculated using the extended Zeldovich mechanism. The generation of NO was calculated after the heat transfer, flow, and combustion calculations in the furnace were stable.

The working conditions arrangement of thermal numerical simulation is shown in Table 22.3. The main parameters are shown in Table 22.4. The calculated coal

Table 22.3 Thermal numerical simulation working condition parameters

Condition	Inner secondary air blade, °	Outer secondary air blade, °
1	65	45
2	45	45
3	30	45
4	30	30

Table 22.4 Thermal numerical simulation operating parameters

Parameter	Unit	Date
Total coal	t/h	226.8
Total air	t/h	2130.6
Excess air coeffient at the furnace outlet	/	1.19
Excess air coeffient at the burning domain	/	0.83
Pramary air temperature	°C	75
Secondary air temperature	°C	313
Pramary air mass flow	kg/s	4.18
Secondary air mass flow	kg/s	9.48
Central air mass flow	kg/s	0.123

Table 22.5 Thermal numerical simulation coal composition

Parameter	Unit	Data
M_{ar}	%	15.22
C_{ar}	%	60.14
H_{ar}	%	3.83
O_{ar}	%	7.53
N_{ar}	%	0.92
$S_{t,\ ar}$	%	0.49
A_{ar}	%	11.87
V_{daf}	%	37.34
$Q_{net.ar}$	kj/kg	22.194
R_{90}	%	23

quality is in accordance with the actual coal quality, and the coal composition is shown in Table 22.5.

22.4 Analysis of Results

22.4.1 Numerical Calculation and Experimental Comparison of Cold Aerodynamic Field

Figure 22.5 shows the comparison of the cold aerodynamic field between the experiment and the numerical simulation results. The inner and outer secondary air blade angle are both 45°. The numerical results were in good consistent with the experimental results. The results show that an obvious annular recirculation zone is generated in the axial direction at the burner outlet zone. The annular recirculation zone

Fig. 22.5 Numerical and experimental comparison of the recirculation zone

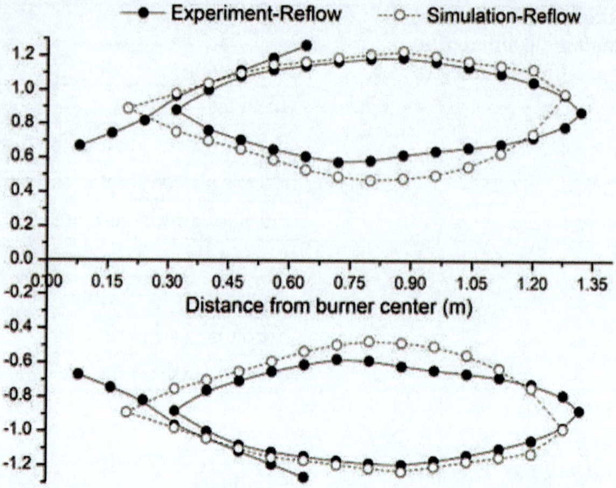

has the following effects on pulverized coal combustion: (1) Bringing the high-temperature flue gas back to the near burner zone and heats the primary coal/air flow, which promotes the pulverized coal ignition and maintains the flame stability. (2) The recirculating high temperature flue gas has the lowest oxygen content, is thus beneficial to the NO_x reduction.

22.4.2 *Full-scale Experimental and Numerical Simulation Results*

22.4.2.1 Numerical Simulation Calculation Results

Figure 22.6 shows the results of the velocity field distribution in the medium section of the furnace. As the angle of inner secondary air blade decreases from 65° to 30°, the annular recirculation zone of the burner outlet decreases gradually. And the flow field of the whole boiler is filled to a better extent. When the inner secondary air blade angle is about 65°, the burner outlet recirculation zone is the largest. But as shown it brings bad effects, the flow field of the burner outlet is unstable will easily generate flow field flash, which will easily cause the coking of the water wall and high-temperature corrosion. The whole flow field distribution of each burner outlet is different due to the swirl direction of the burner outlet, the interaction between adjacent burner jets and the influence of local flow field in the furnace. Therefore, it is necessary to study the flow field characteristics of the burner outlet in order to ensure the boiler can operate safely and achieve better combustion performance.

Figure 22.7 shows the distribution results of the furnace temperature under different working conditions. Figure 22.8 shows the comparison of the local temperature field at the outlet of the burner. As shown the recirculation zone of the burner

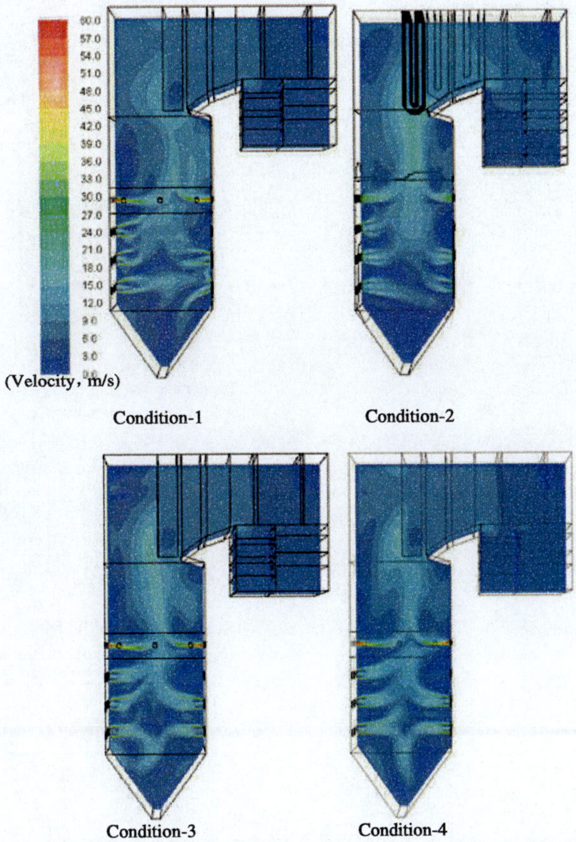

Fig. 22.6 Velocity distribution under different working conditions

outlet brings the high-temperature flue gas back to the near burner zone and heats the primary coal/air flow, which promotes the pulverized coal ignition and maintains the flame stability. As the angle of inner secondary air blade decreases from 65° to 30°, the expansion angle of the recirculation zone becomes smaller, the temperature in the recirculation zone is significantly reduced, and the entrainment ability of the surrounding high-temperature flue gas becomes weak, which is not conducive to the ignition and steady combustion of the pulverized coal. It can be seen from the working condition 1 that the temperature of the water wall region near the burner is higher than other working conditions. The reason is that When the inner secondary air blade angle is about 65°, the unstable flow field of the burner outlet causes the phenomenon of "flying edge" in the flow field. High temperature flue gas is sucked into the water wall area, which is prone to coking and slagging. This should be avoided in operation.

Figure 22.9 shows the temperature distributions along the direction of the burner centerline. The flow and temperature field of the lower burner is greatly affected by the air leakage at the bottom of the furnace, and the upper burner flow field is

Fig. 22.7 Temperature distribution under different working conditions

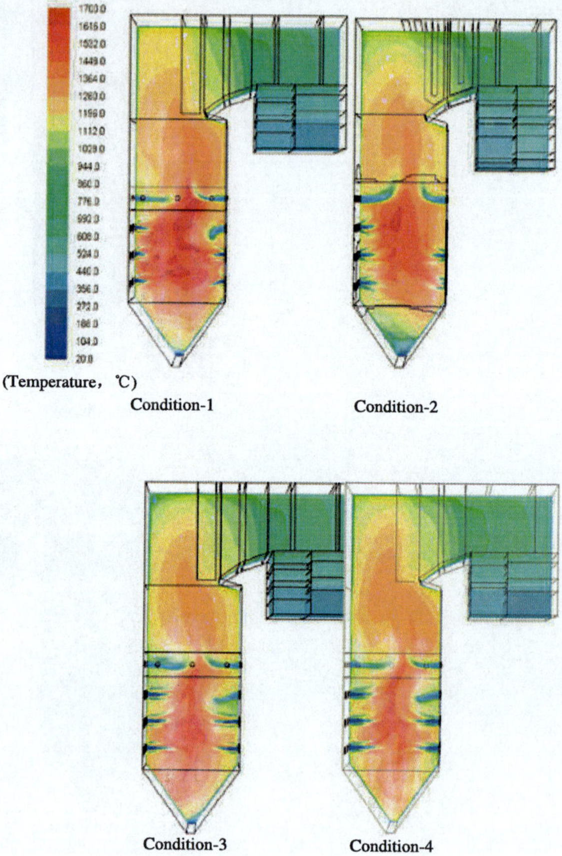

also affected by the ofa field. So the middle burner of front wall is chosen as the research object. The results show that when the angle of the inner secondary air blade decreases from 65° to 30°, the axial temperature field of the air flow in the central direction of the burner decreases obviously, and the pulverized coal ignition increases. The main reason is that as the recirculation zone is gradually reduced, the volume of the entrained high-temperature flue gas is reduced, and at the same pulverized coal concentration, the time for obtaining the heat required for ignition of pulverized coal increases; Compared with the influence of internal secondary air, the change of blade angle of outer secondary air has little effect on the air temperature field in the central direction of the burner.

Figure 22.10 shows the variation of average flue gas temperature along the height direction of the furnace with the cross-section of the lower burner as the reference. It can be seen that the introduction of low-temperature air and pulverized coal in each layer of the burner causes the average flue-gas temperature of the corresponding section of the furnace to decrease rapidly. However, as the combustion of pulverized coal continues, the temperature of the flue-gas rises rapidly, so that in the main

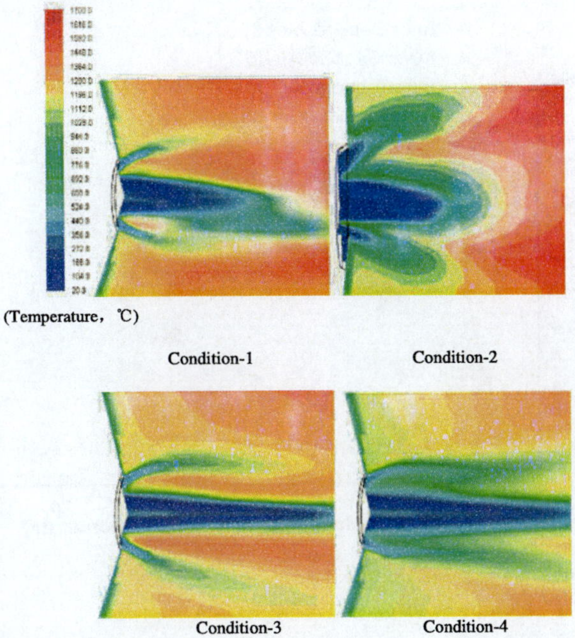

Fig. 22.8 Temperature distribution of the burner outlet

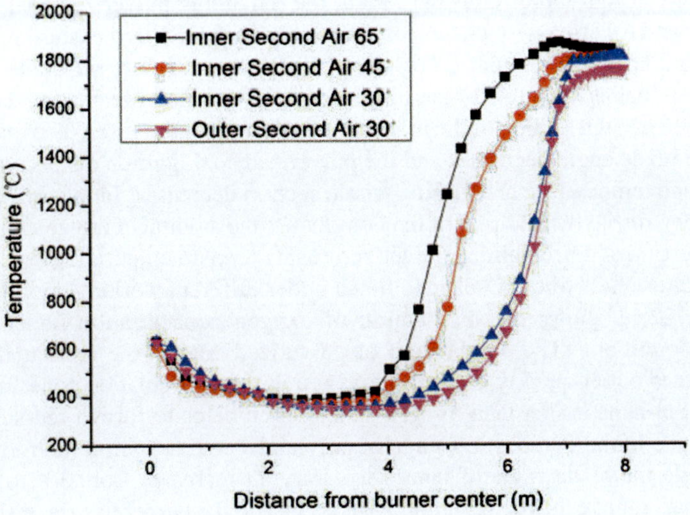

Fig. 22.9 Flue-gas temperature along the burner centerline direction

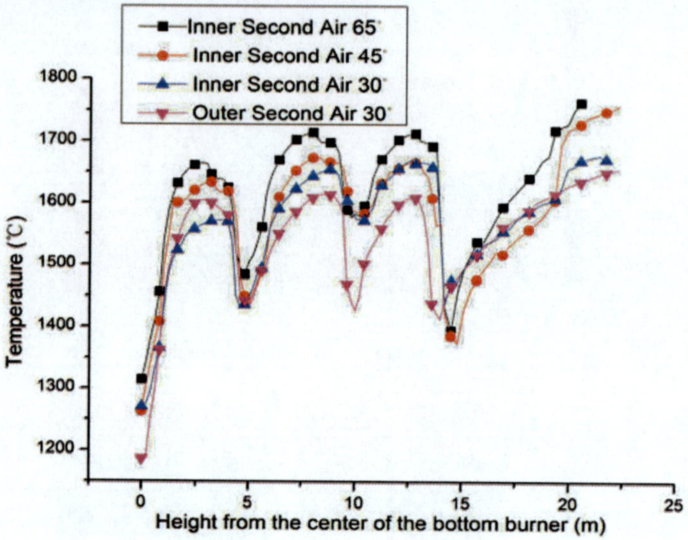

Fig. 22.10 Flue-gas temperature along the furnace height

combustion zone, the temperature of each section fluctuates along the direction of furnace height. The introduction of OFA burner has similar effect with that of burner jet. Due to the addition of OFA burner air, the remaining pulverized coal gradually burns out, and the cross-section temperature above each OFA burner shows an upward trend. Comparing the simulation results of different secondary air blade angles, it can be seen that as the angle of the inner secondary air blade decreases, the average temperature of each section of the main combustion zone decreases. The main reason is that the blade angle decreases and the pulverized coal ignition distance increases. The average temperature level of the whole section decreases. In the area above the OFA burner, due to the addition of the burnout air, the amount of oxygen required for the combustion of the remaining pulverized coal is supplemented, and the distribution of temperature field does not change much under different working conditions.

Figure 22.11 shows the distribution of oxygen concentration under different working conditions. The combustion of pulverized coal in the recirculation zone of the burner outlet quickly consumes oxygen in the recirculation zone, so that the recirculation zone is in a lean oxygen state, which helps to form a reducing atmosphere in the initial combustion stage of pulverized coal to inhibit the formation of NO_x. While the blade angle of inner secondary air increases from 30° to 65°. The oxygen-lean zone in the recirculation zone is close to the burner nozzle. It shows that the larger the angle of secondary air in the recirculation zone, the faster the pulverized coal is heated, and the earlier the volatile matter and its organic nitrogen content precipitate in the oxygen-poor zone, which is conducive to inhibiting the formation of NO_x.

Figure 22.12 shows the distribution of CO concentration under different working

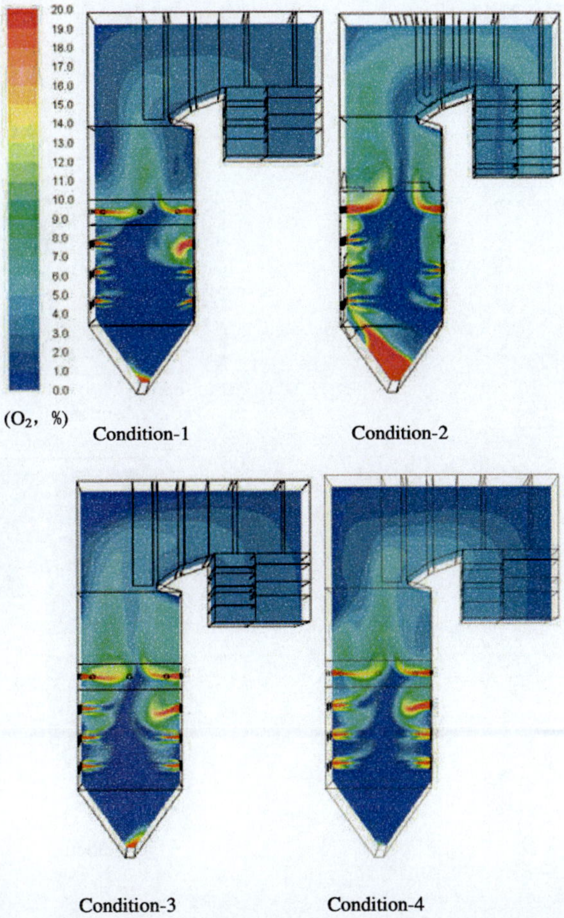

Fig. 22.11 Oxygen concentration under different working conditions

conditions. The model CO is an intermediate product of volatile combustion and carbon particle combustion. The CO concentration identifies the main area where the pulverized coal combustion occurrs. After CO is generated, it will be mixed with oxygen in the furnace, burned and generated CO_2. It can be seen from the results that the CO concentration in the recirculation zone of the burner outlet is high, indicating that the combustion of pulverized coal occurs in the corresponding region. Comparing the CO concentration distribution in the recirculation zone near the burner nozzle under different conditions, we can see the influence of different blade angles on entrainment and pulverized coal combustion in the recirculation zone: In working condition 1 and 2, because of the larger angle of the swirl blade, the larger recirculation zone is created. Therefore, more pulverized coal combustion occurs in the recirculation zone near the burner nozzle, which makes the CO concentration in the corresponding region higher. Further, since the amount of oxygen in the recirculation zone is insufficient to achieve complete combustion of the pulverized coal, the

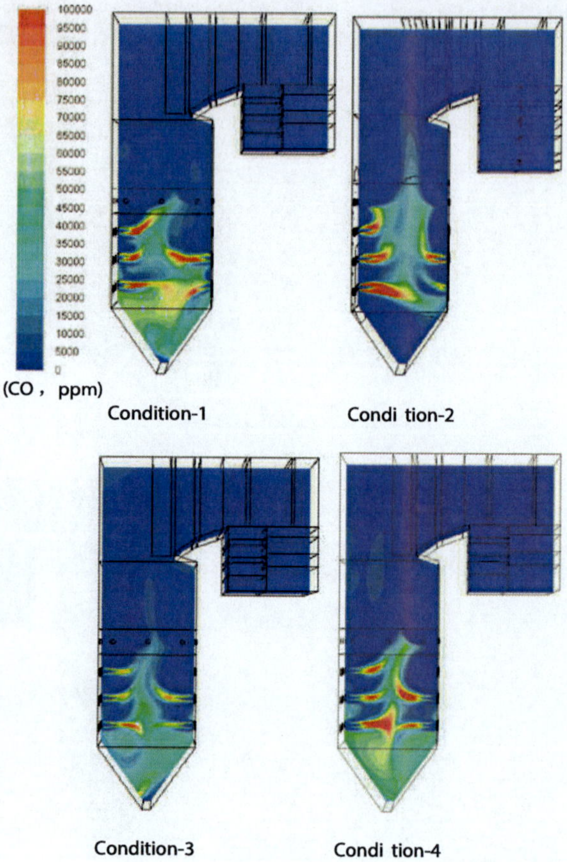

Fig. 22.12 CO concentration under different working conditions

combustion of the pulverized coal continues outside the recirculation zone, and the generated CO is thus distributed throughout the main combustion zone of the lower furnace. Since the staged combustion is used, the amount of oxygen in the lower furnace is not sufficient to convert the generated CO to CO_2, and therefore the CO concentration of the lower furnace is high. As the OFA burner air enters the furnace, the unburned CO of the lower furnace is completely converted to CO_2 downstream of the OFA burner. Therefore, the calculation results show that the novel low-NO_x swirl burner can reduce the NOx concentration while ensuring the efficiency and safety of the boiler operation.

Figure 22.13 shows the change of CO concentration in each section along the height of the furnace. The results show the similar to the trend shown in Fig. 22.10. In the main combustion zone, the CO concentration of each section fluctuates along the direction of furnace height due to the staged combustion of air and pulverized coal in the burner jet When it is in the area above the OFA burner, the CO in flue gas and the remaining pulverized coal is fully burned out due to the supplement of the oxygen from the OFA burner, and the CO concentration is gradually reduced. As the

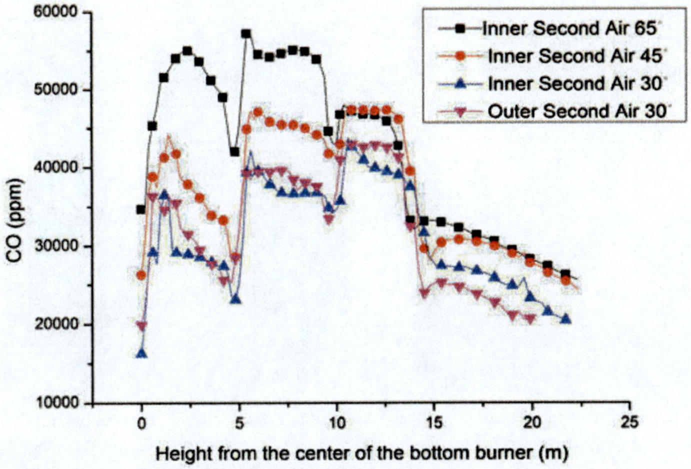

Fig. 22.13 CO concentration along the furnace height

angle of the inner secondary air blade decreases, the recirculation zone decreases, and the ignition and combustion degree of the pulverized coal lags behind, so that the CO concentration in the main combustion is relatively low. And in the later combustion stage, due to the addition of oxygen in the OFA burner, all of CO were oxidized, the CO concentration was significantly reduced. So the CO concentration did not change much under different working conditions.

Figure 22.14 shows the results of NO concentration distribution under different working conditions. Thc fucl nitrogcn in thc pulvcrizcd coal is mainly analyzcd with the volatilization in the form of HCN. In the oxygen-rich environment, it continues to react with oxygen to form NO. In an oxygen-poor environment, N_2 is formed by reduction reaction between HCN and NO. The main design criterion of the novel low NO_x swirl burner developed by HBC is to create a reducing (oxygen-poor) atmosphere in the initial stage of pulverized coal combustion to suppress NO formation. The oxidation reaction strengthens the reduction reaction of NO consumption. By optimizing the flow field near the burner nozzle, a reducing atmosphere is created while ensuring ignition and stable combustion of the pulverized coal, and NO generation is suppressed in the initial stage of coal combustion.

The results show that as the angle of the inner secondary air increases, the recirculation zone increases, the reducing atmosphere of the main combustion zone increases, and the NO_x emission decreases. The calculated data indicates that the novel low-NO_x swirl burner technology designed by the HBC makes the average emission level of NO_x at the furnace outlet lower than current advanced burner technology.

Figure 22.15 shows the results of NO concentration changes along the height of the furnace. Due to the "dilution effect" and staged combustion of the air introduced into each burner, the NO concentration in the main combustion zone fluctuates. In

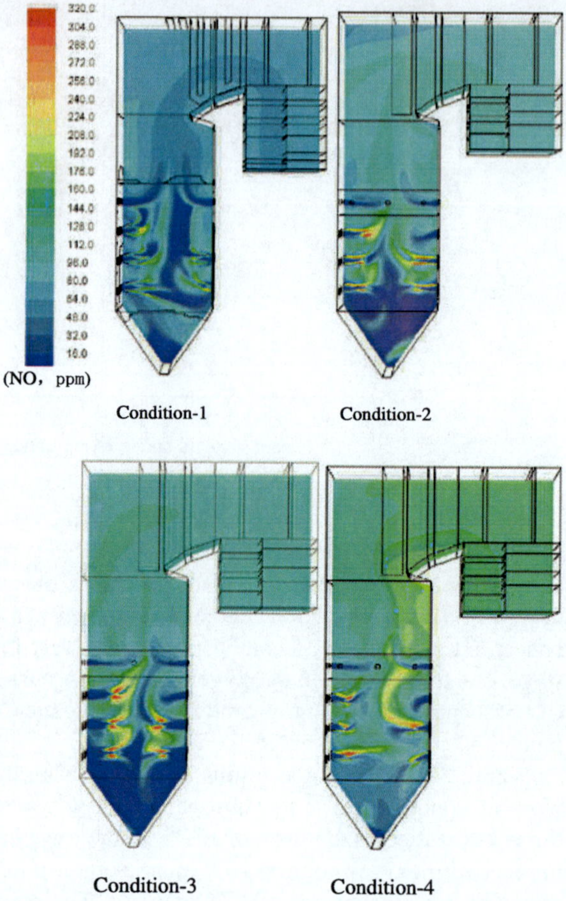

Fig. 22.14 NO concentration under different working conditions

the burnout zone, due to the fuel-nitrogen reacts with the oxygen in the burnout air to form NO, and the NO concentration gradually increases. When the reaction is completed, the NO concentration gradually decreases as the furnace temperature decreases. As the angle of inner secondary air blade decreases from 65° to 30°, the recirculation zone decreases, the staged combustion of the swirl burner is weakened, the fuel NO concentration increases.

Although the temperature decreases and the thermal NO increases, the fuel NO dominates the pulverized coal combustion, and the overall NO emission concentration increases gradually.

The statictics of numerical simulation results show that the NO_x emission level is 220–245 kg/Nm^3, which achieves the ultra-low emission standard. The CO emission concentration is below 100 ppm, and the carbon in fly ash is below 1%.

From the comparison of the calculation results, it shows that as the angle of the inner secondary air blade increases from 30° to 65°, the annular recirculation zone

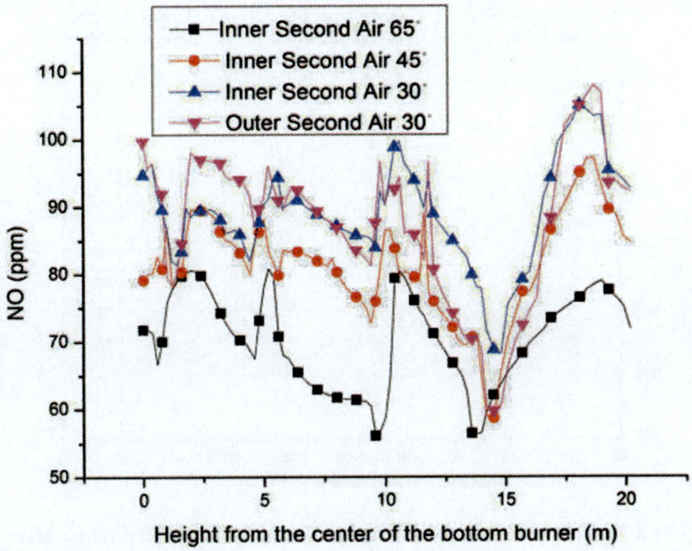

Fig. 22.15 NO concentration–Along the height of the furnace

gradually becomes larger, and the stronger the ability to entrain high-temperature flue gas, the more intense the pulverized coal combustion. The flue gas temperature of the outlet section of the heated surface becomes larger, and the reducing atmosphere reduces the NO_x emission level. On the contrary, the CO concentration and the carbon in fly ash combustibles increase slightly.

22.4.2.2 Analysis of Experiment Results

Figure 22.16 shows the variation of CO, O_2, NO_x emission concentration under different inner secondary air blade angles. The results show that the numerical results were in good consistent with the experimental results. As the angle of the inner secondary air blade increases from 30°to 65°, the NO_x content is reduced by 14%. Through analysing the cold experiment and the numerical simulation results, the influence of the inner secondary air swirl blade angle on NO_x should be considered from two aspects. On the one hand the action of the inner secondary air is recirculating high-temperature flue gas through the annular recirculation zone formed by the swirling flow. Reducing swirling angle will weaken the recirculation zone, increase the pulverized coal ignition, reduce the effect of burner staged combustion and increase the production of NO_x. On the other hand, the increase of the swirl angle will increase the inner secondary air resistance and reduce the inner secondary air volume. Intensification of the initial oxygen deficiency effect can also reduce the amount of NO_x production. While the effect of the inner secondary swirling blade angle on the ash flammable material is opposite to the NO_x concentration, and the

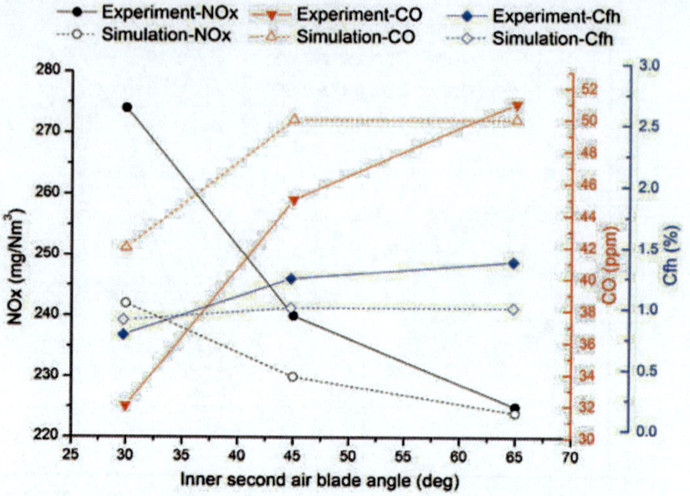

Fig. 22.16 Levels of CO and NO_x emissions and carbon in fly ash under different inner secondary air blade angle

inner secondary air angle is smaller, the oxygen concentration of initial combustion is sufficient and the burnout effect achieves better. From the experiment results, the NO_x emissions is below the 275 mg/Nm^3, CO concentration is below 50 ppm, the carbon in fly ash is below 1.5%, so the boiler efficiency does not change much.

Figure 22.17 shows the variation of CO, O_2, NO_x emission concentration under different outer secondary air blade angles. The outer secondary air acts as the outermost layer of swirling air of the swirl burner, which plays a role mainly in the radial expansion of the jet and the grading supply. The results show that as the angle of the outer secondary air blade increases from 30° to 65°, the NO_x concentration of the SCR inlet is reduced by 4%, which is consistent with the numerical simulation analysis that the increase of the outer secondary air swirling angle will delay the outer secondary air and coal mixing, enhance the degree of the air supply at the initial stage of combustion, which is easily forming a reducing atmosphere at the initial stage of combustion and reducing the generation of NO_x. Meanwhile the increase of swirl intensity of the outer secondary air will increase the resistance of burner nozzle, reduce the air flow in the main combustion zone, and strengthen the graded air supply in the whole furnace, which is also conducive to reducing the generation of NO_x. The results show that the increase of the angle of the outer secondary swirling blade increases the CO concentration and carbon in fly ash slightly, and has little effect on the boiler efficiency. From the variation of CO, O_2, NO_x emission concentration, It is analyzed that the influence of the change of the angle of the outer secondary air on the combustion performance is less than the influence of the inner secondary air.

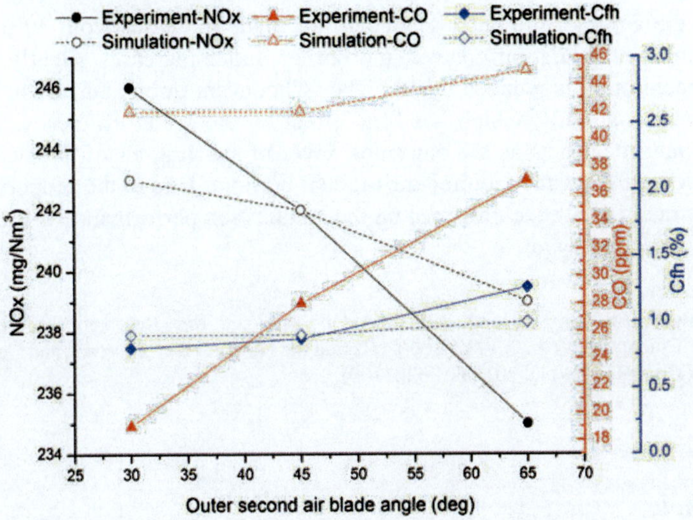

Fig. 22.17 Levels of CO and NOx emissions and carbon in fly ash under different outer secondary air blade angle

22.5 Conclusion

In this paper, the industrial experimental and numerical simulation were studied on the combustion and NO_x formation characteristics in a 600MWe supercritical utility boiler burning low-volatile bituminous coal. The conclusions are as follows:

(1) The results of cold aerodynamic field experiment were in good consistent with the numerical results. An obvious annular recirculation zone is generated in the axial direction at the burner outlet zone. The recirculation zone brings the high-temperature flue gas back to the near burner zone and heats the primary coal/air flow, which maintains the flame stability. Meanwhile increase the lower oxygen atmosphere and suppresses the formation of NO_x effectively. The distribution of NO_x emission, CO concentration and carbon in fly ash in the thermal numerical simulation calculation is consistent with the industrial experimental results.

(2) As the angle of the inner secondary air blade increases from 30° to 65°, the annular recirculation zone of the burner outlet increases gradually, pulverized coal ignition shorten. In the main combustion zone the temperature of the flue-gas increase, the NO_x emission concentration reduce and the CO concentration and carbon in fly ash increases slightly. In the above burnout zone, the remaining pulverized coal is fully burned out due to the supplement of the oxygen from the OFA burner, the CO concentration and carbon in fly ash does not change much. As the blade angle becomes larger the coking of the water wall and high-temperature corrosion should be avoided.

(3) As the angle of the outer secondary air blade increases from 30° to 65°, the annular recirculation zone of the burner outlet increases slightly. the NO_x concentration is reduced by 4%. The CO concentration and carbon in fly ash increases slightly, which has little effect on the boiler efficiency. The outer secondary air acts as the outermost layer of swirling air of the swirl burner, which provides the required air for coal burnout. Due to the proportion of air volume is small, the influence on the combustion performance is less than the inner secondary air.

Acknowledgements This research was funded by National Key Research and Development Program of China, Grant No. 2017YFB0602002 and the National Key Research and Development Program of China (Grant No. 2016YFC0203700).

References

Anke YZ, Qian S, Haizhu Z (2009) Effects of dual-channel velocity difference of pulverized coal burners on NOx emission. J Tongji Univ 37(4):537–539

Beer JM (2000) Combustion technology developments in power generation in re- sponse to environmental challenge. Prog Energy Combust Sci 26:301–327

Chen ZC, Li ZQ, Jing JP, Wang FQ, Chen LZ, Wu SH (2008) The influence of fuel bias in the primary air duct on the gas/particle flow characteristics near the swirl burner region. Fuel Process Technol 89:958–965

Dang H, Rochelle G (2003) CO_2 absorption rate and solubility in monoethanolamine/piperazine/water. Sep Sci Technol 38(2):337–357

Gong Z, Liu Z, Zhu Z (2014) Experimental study on semi-coke combustion and coal pyrolysisand combustion coupling. J China Coal Soc 39(2):519–525

Gushan TB, Lihong X, Guchang JZ (2005) Experimental study on aerodynamic field of dual-channel Swirling pulverized coal burner. Huazhong Electric Power 13(5):1–3

Li ZQ, Yang LB, Qiu PH, Sun R, Chen LZ, Sun SZ (2004) Experimental study of the combustion efficiency and formation of NO_x in an industrial pulverized coal combustor. Int J Energy Res 28:511–520

Nainar M, Veawab A (2009) Corrosion in CO_2 capture process using blended monoethanolamine and piperazine. Ind Eng Chem Res 48(20):9299–9306

National Bureau of Statistics of China (2015) China statistical year book-2015. China Statistics Press, Beijing, pp 179–193

Newby RA, Domeracki WF, McGuigan AW (1994) Integration of combustion turbine systems into pressurized fluidized bed combustion combined cycles. Am Soc Mech Eng 1–11

Rui S (1998) Experimental study and numerical simulation of flow characteristics of radial dense-lean Swirling Pulverized Coal burner. Ph.D. Dissertation, Harbin University of Technology, pp 87–108

Song M, Zeng L, Chen Z et al (2016) Industrial application of an improved multiple injection and multiple staging combustion technology in a 600 MWe supercritical down-fired boiler. Environ Sci Technol 50:1604–1610

Thomas LB, Francisco C, Sebastien C, Stanislas PC, Jacques B, Bernard B (2007) Coal combustion modelling of large power plant, for NOx abatement. Fuel 86:2213–2220

Xu M, Azevedo JLT, Carvalho MG (2000) Modeling of the combustion process and NOx emission in a utility boiler. Fuel 79:1611–1619

Xuefeng S, Renzhang Q, Xiaoxi M (1995) Experimental study on combustion characteristics and optimization of slotted blunt-body burners. Power Eng 15(5):12–17
Zhang J, Nieh S (1997) Simulation of gaseous combustion and heat transfer in a vortex combustor. Numer Heat Transfer A 32:697–713
Zhang J, Nieh S, Zhou LX, Lu HP (1997) Simulation of annular swirling turbulent flows with a new algebraic reynolds stress model. Numer Heat Transfer B 31:235–249
Zhang Y, Ding YJ, Wu ZS, Kong L, Chou T (2007) Modeling and coordinative optimization of NOx emission and efficiency of utility boilers with neural network. Korean J Chem Eng 24:1118–1123
Zhiqiang W, Shaozeng S, Xiaohui Z et al (2007) Effect of offset angles of mid-secondary air nozzles on flow sield in T-fired furnace. Chinese J Mech Eng 43(8):165–170
Zhou LX, Wang F, Zhang J (2003) Simulation of swirling combustion and NO formation using a USM turbulence-chemistry model. Fuel 82:1579–1586

Chapter 23
Distribution of Temperature and Characteristics of Soot Volume Fraction in MILD-OCC Flame

Jingwen Lu, Shuwei Zhang, Longhui Tan, and Jianyi Lu

Abstract Coal and methane were burnt in the MILD-OCC combustion flame under different combustion conditions to study the temperature distribution and characteristics of soot volume fraction. In this experiment, the flame temperature distribution is measured by thermocouple and the volume fraction (f_v) of soot at each sampling point is calculated, then combining with the mass concentration of soot by the filter weighing method to obtain the distribution characteristics of soot in the flame. The results of thermocouple particle densitometry (TPD) show that the formation and oxidation of soot at every point in the flame is related closely to its position and temperature in the flame. The f_v of soot in the center of the flame is negatively correlated with the temperature of the flame during the MILD-OCC combustion. The lower temperature region of the flame is conducive to the growth of soot, which leads to higher value of f_v. The higher temperature in the center of the flame is beneficial to the oxidation of the soot particles, resulting to lower value of f_v. The results of the filter membrane weighing method show that the formation of soot mainly occurs in the central region of the flame. When the radial distancc increases, the flame temperature rises, and the soot concentration decreases due to oxidation. On the whole, the distribution of soot measured by the filter membrane weighing method is roughly consistent with that calculated by the TPD, which provides a reference for further research on the MILD-OCC combustion.

Keywords Soot · MILD-OCC · Temperature · TPD

J. Lu · S. Zhang · L. Tan · J. Lu (✉)
Hebei Key Lab of Power Plant Flue Gas Multi-Pollutants Control, Department of Environmental Science and Engineering, North China Electric Power University, Baoding 071003, PR China
e-mail: lujianyi@tsinghua.org.cn

J. Lu
MOE Key Laboratory of Resources and Environmental Systems Optimization, College of Environmental Science and Engineering, North China Electric Power University, Beijing 102206, PR China

J. Lyu and S. Li (eds.), *Clean Coal and Sustainable Energy*, Environmental Science and Engineering,
https://doi.org/10.1007/978-981-16-1657-0_23

23.1 Introduction

Soot is produced during incomplete combustion of fuels (Na et al. 2009), which reduces energy efficiency (Lighty et al. 2000; David and Ravid 2011; Richter and Howard 2000). With the burning of fossil fuels, the emission of soot is increasing in the atmosphere. Due to the complex composition and strong adsorption capacity of soot particles, many pathogenic factors can be carried, which can lead to many diseases, especially respiratory diseases. Therefore, it has always been highly concerned in the field of international energy and environmental research. The formation of soot particles in the flame, the distribution of soot volume fraction and the flame temperature distribution are of great significance for the control of small particle pollutant emissions.

At present, the existing combustion technology still cannot meet the requirements both of high efficiency and low pollution (Pengfei et al. 2011). In recent years, the researches show that the oxy-combustion technology is the most widely used combustion technology and can reduce CO_2 emission, but it also faces the delay of fire and the reaction is not stable (Baojun et al. 2014). However, MILD combustion technology is regarded as one of the most promising and clean combustion technology (Cavaliere and Joannon 2004; Katsuki and Hasegawa 1998; Tsuji et al. 2003; Joannon et al. 2000; Sepmana, et al. 2013; Stadler et al. 2009). Therefore, considering the advantages and disadvantages of oxy-combustion and the unique characteristics of MILD combustion technology, some scholars have proposed innovative technology, which the combination of MILD combustion technology and OCC technology to form MILD-OCC technology (Liqi et al. 2014). It can achieve CO_2 enrichment and storage, and reduce the original amount of pollutants such as NO_X. At the same time, the technology has a uniform heat flux distribution characteristic, which can ensure the heat transfer safety of the heating surface of the boiler. Therefore, this topic mainly studies the temperature distribution in the combustion flame, the volume fraction and morphology of soot in the MILD-OCC.

23.2 Experimental

23.2.1 Experimental System

The present study uses a laboratory-scale modified Hencken burner (Xiong 2011) as shown in Fig. 23.1. The burner consists of 360 stainless steel fuel tubes arranged in a square 7 cm steel frame. The central coal tube is surrounded by eight fuel channels at the bottom of the burner. The inner and outer diameters of the fuel tubes were 1.2 mm and 1.8 mm respectively, and the tubes were uniformly distributed in the honeycomb ceramic piece. The oxidant is air which composed of oxygen and nitrogen, fed from the middle two pipes, and then rectified by the filter material, flowing out in the gap between the stainless steel capillary and the honeycomb ceramic piece. Finally,

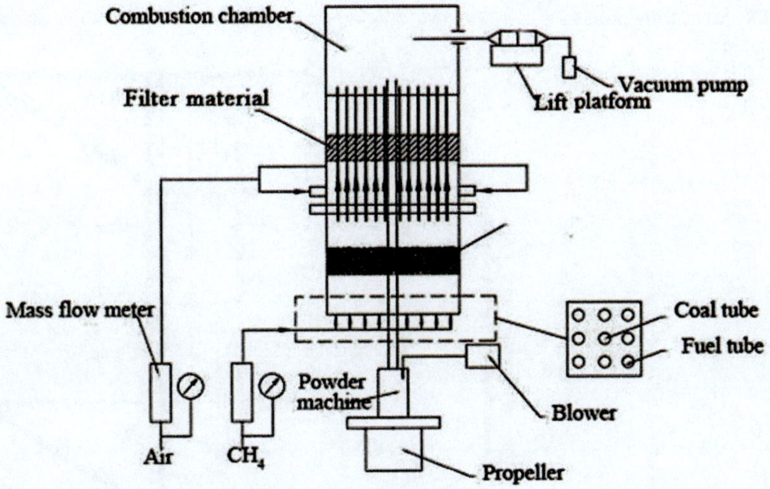

Fig. 23.1 A schematic diagram of the Hencken burner

the oxidizer flowed around the fuel tubes through the honeycomb structure and was combined with fuel to form a diffusion flame at the surface of the burner. In order to reduce external air interference, a quartz glass tube is placed outside the burner as a protective cover to separate the product of the burner from the outside, reducing the influence of external factors.

23.2.2 Measurement Method and Devices

Flame temperature was measured by an R-type thermocouple of 0.5 mm diameter wire and 1 mm bead diameter (Pt–Pt 13% Rh, with a measurement error of ±2%) in a rapid inserting way (Jianyi and Qinglong 2011), which was taken at the same positions as the soot sampling along the set points of the inverse diffusion flame (IDF). The position of the positive center of the flame combustion zone is zero point and is measured along the axial direction and the radial direction of the flame. The measurement point is as shown in Fig. 23.2. Soot samples were taken at different positions using a pinhole sampler. The sampling points were set 5, 15 and 25 mm away from the flame centerline along the flame edge and height above burner (HAB) at 10, 20, 30, 40 and 50 mm separately. The flow rates of air and methane were kept at 66.6 L/min and 4 L/min respectively.

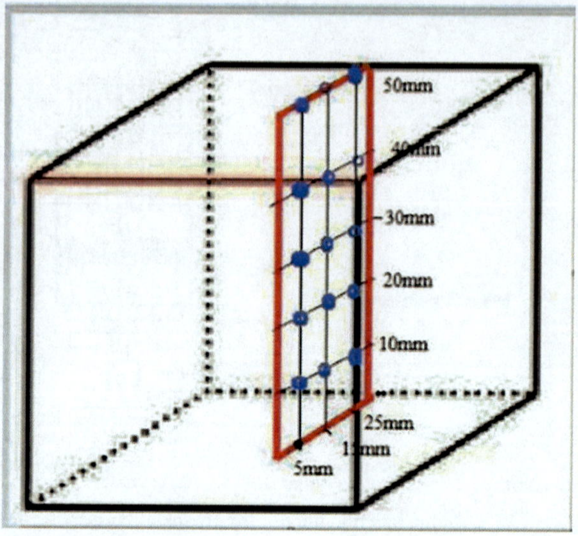

Fig. 23.2 Measuring points in the flame

23.2.3 *Thermocouple Particle Densitometry*

The volume fraction (f_v) of soot in the combustion flame is studied by the flame gas temperature distribution and the thermocouple deposition method. After the transient response phase of the thermocouple node, the node and the surrounding gas are in a thermal balance state, ignoring the heat conduction of the thermocouple wire. Then the local gas temperature (T_g) could be gained by the convective-radioactive energy balance shown as the following equations (Jianyi and Xiaobin 2013):

$$\varepsilon_j \delta T_j^4 = (k_{g0} Nu_j / 2d_j)(T_g^2 - T_j^2) \tag{23.1}$$

where ε_j = emissivity of the junction; σ = Stefan-Boltzmann constant; k_{g0} = thermal conductivity of the gas; Nu_j = Nusselt number and d_j = junction diameter of the thermocouple.

In the thermocouple deposition experiment, when the soot is deposited on the surface of the node, the emissivity of the node is increased. The node emissivity of a thermocouple can be calculated by the Bradley (1961) formula:

$$\varepsilon = 0.1083 \ln T_j - 0.5644 \tag{23.2}$$

Expression of thermal mobility D_T at thermocouple node (Cavaliere and Joannon 2004):

$$D_T = 3/4(1 + \pi\alpha/8)^{-1} \nu_g \tag{23.3}$$

where α = momentum adjustment factor (≈1); ν_g = aerodynamic viscosity.

Mass flow expression for particles moving to the surface of the joint due to heat transfer:

$$Q = (D_T Nu_j f_v \rho_p / 2d_j)(1 - (T_j/T_g)^2) \tag{23.4}$$

where Q = mass flow rate of soot particles per unit surface area and unit time; f_v = soot volume; ρ_p = soot particles density. When the diameter of the node changes, it can also be obtained from the rate of change of diameter with time:

$$Q = (\rho_d/2)d(d_j)/d_t \tag{23.5}$$

where ρ_d = density of soot deposited for the node.

Combine Eqs. 23.1 and 23.5 to get an expression of the change in node diameter over time:

$$\frac{d(d_j)}{d_t} = k_{g0} Nu_j \frac{T_j^{-3} - 2T_g^2 T_j^{-5}}{\varepsilon\sigma} \frac{d(T_j)}{d_t} \tag{23.6}$$

By connecting Eps. 23.1, 4~6 together, you can get the relationship of G with time.

$$G = mt + G(t = 0) \tag{23.7}$$

According to the $G = (1/4)(T_g/T_j)^8 - (1/6)(T_g/T_j)^6$, and draw the G value curve based on the experimental data. As shown in Fig. 23.3.

The change in the slope of the G value in the steady portion of the curve coincides with the gradual thickening of the soot layer over time, and a slope of m = 0.0011 can be obtained. Combine the following formulas to get f_v:

Fig. 23.3 Parameter G for each sample number

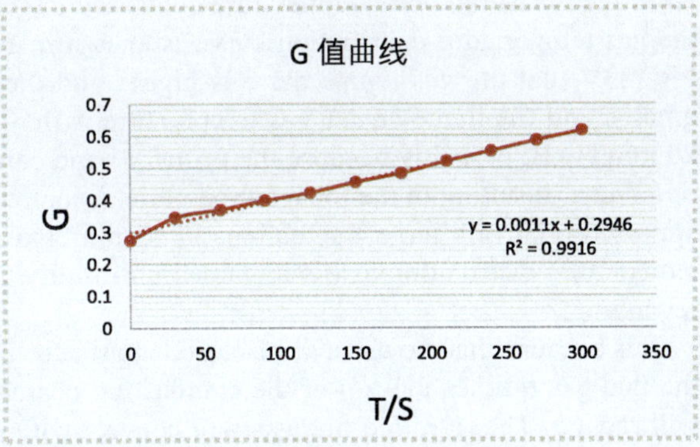

$$m = \beta f_v \tag{23.8}$$

$$\beta \equiv 2D_T \varepsilon_j \sigma^2 T_g^4 / (\varphi k_{g0}^2 Nu_j) \tag{23.9}$$

$$\nu_g = 1.29 \times 10^{-9} T_g^{1.65} \tag{23.10}$$

where Φ = Soot deposition share.

23.2.4 *The Filter Membrane Weighing Method*

This experiment used the filter weighing method (Zhe and Yongqiang 2018) to calculate and analyze the collected samples to study the distribution of the soot mass concentration in the MILD-OCC combustion flame with the axial and radial directions of the flame. A stainless steel capillary with a diameter of 0.1 mm is used as a pinhole sampler. The particulate matter at each sampling point is trapped on the polytetrafluoroetylene (PTFE) filter, and then the mass of the soot captured according to the difference in mass before and after the filter is sampled. Therefore, the ratio of the mass of the soot to the mass of the sampled air is the mass concentration of the soot.

23.3 Results and Discussion

23.3.1 *Measurement of Temperature Distribution in MILD-OCC Flame*

After the measured node temperature converted to gas temperature (Dexing 2000), the gas temperature distribution curve is shown in Fig. 23.4. It can be seen from Fig. 23.4 that the gas temperature is higher and the temperature curve fluctuates greatly, and the flame center gas temperature is the lowest at 10 mm, 15 mm and 20 mm HAB, probably because the primary wind carries the coal into the combustion flame, resulting in the local temperature drop, the temperature distribution also appeared relatively large fluctuations at 45 mm and 50 mm HAB, but overall the temperature distribution also maintained uniformity, which probably due to smoke rewinding.

It is because that we use a semi-closed approach above the combustion chamber, the flue gas reaches the top of the combustion chamber, and a part of the flue gas will rewind. The re-rolled flue gas will convect with the rising flue gas below, and

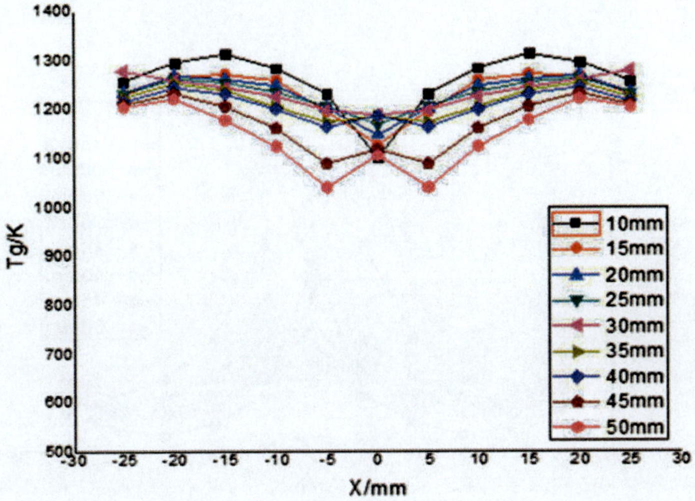

Fig. 23.4 Flame temperature distribution curves

most of the flue gas will flow to the combustion area, which has a reheating process in turn.

23.3.2 *Volume Fraction Distribution Characteristics of Soot in Combustion Flame*

The soot volume fraction distribution curve is plotted as shown in Fig. 23.5. As can be seen from the Fig. 23.5, the value of f_v is higher below 20 mm of the flame. Within 10 mm at the bottom of the flame, the f_v of soot in the center of the flame is higher, the f_v of the edge of the flame is almost zero, and the f_v of soot in the range of 10 to 20 mm is gradually decreased along the axial direction, and the peak of the f_v appears above the center of the flame, but the f_v of soot at the edge of the flame is still very low. Within at 25 to 40 mm HAB, the value of f_v did not change significantly with the increase of the flame height. In this range of HAB, the f_v of soot tends to be stable. However, the f_v of soot is increased in the range of 45 to 50 mm, and the f_v of soot in this range is relatively high compared to that of soot at the bottom of the flame.

It can be seen from the above discussion, in the axial direction, the f_v of soot rises first in the center of the flame, then slowly stabilizes in the region, finally rises. The f_v of soot in the edge region of the flame appears to be zero, and the range of variation is small. In the MILD-OCC, the f_v of soot in the positive center of the flame is inversely related to the temperature of the flame. The results show that the formation of soot has to undergo three stages of nucleation, growth and oxidation. Since the flame is in a lower temperature range from 10 to 20 mm HAB, the f_v of

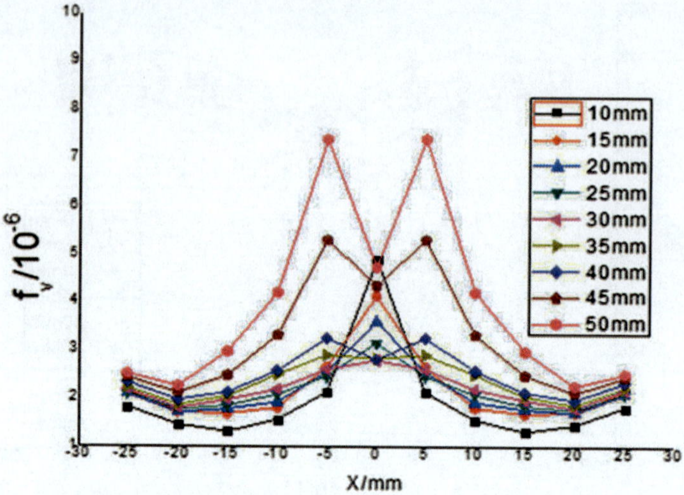

Fig. 23.5 Soot volume fraction distribution curves

soot is higher, indicating that low temperature is conducive to the growth of soot. Within 25 to 40 mm at HAB, the temperature in the center of the flame rises and tends to be stable, and the f_v of soot is low, indicating that the high temperature zone is beneficial to the oxidation of soot particles. It can be seen that the f_v of soot is a complex function of flame temperature and flame structure, which is closely related to both.

23.3.3 *Soot Mass Concentration Distribution in Combustion Flame*

From the analysis of the three soot concentration curves in Fig. 23.6, it can be seen that the concentration of soot at a distance of 5 mm from the center is reduced in the range of 10 to 20 mm HAB, and the elevated state is in the range of 20 to 30 mm HAB. The mass concentration of soot decreases in the range of 30 to 40 mm HAB because of soot oxidation, and in the range of 40 to 50 mm, the concentration of soot increases due to temperature drop. At a distance of 15 mm from the center, the concentration of soot is declining from 10 to 20 mm HAB. For the reason of analysis, it is less affected by the central low temperature area, resulting in the temperature to go up. At this time, the soot is in the oxidation stage, so the concentration of soot is decreased. However, in the range of 20 to 30 mm HAB, due to the increase of height, resulting in temperature drop, which is not conducive to soot oxidation, resulting in an increase in concentration. Within the scope of 30 to 50 mm, although the flame temperature declined, but volatility is very small, so the soot concentration is in a stable state. In the center of the distance of 25 mm soot concentration within the

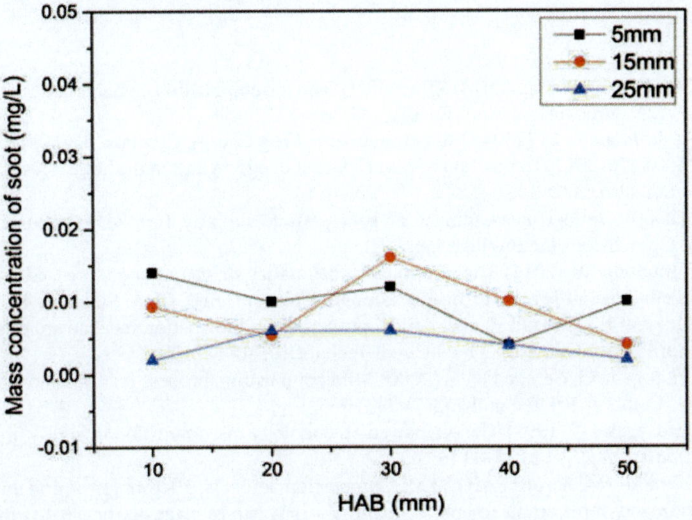

Fig. 23.6 Soot mass concentration distribution curves

scope of 10 to 50 mm, soot concentration is in a low state, which is in accordance with that of the results by using TPD method.

23.4 Conclusion

The temperature distribution of the combustion flame and the volume fraction and morphology of the soot were studied by MILD-OCC combustion method. The results show that the formation of soot in the flame is subject to different stages such as nucleation, growth and oxidation. The f_v of soot has a negative correlation with the temperature of the flame. In the HAB range of 10 to 20 mm, in a lower temperature range, the f_v of soot is higher, indicating that the low temperature is beneficial to the growth of soot. Within 25 to 40 mm HAB, the temperature in the center of the flame rises and tends to be stable, and the f_v of soot is low, indicating that the high temperature zone is beneficial to the oxidation of soot particles. This also provides references for further research on the use of MILD-OCC technology.

Acknowledgements This work was supported by the National Natural Science Foundation of China (Grant No.51761125011).

References

Baojun Y, Liqi Z, Fang H et al. (2014) Effect of H_2O on the combustion characteristics of pulverized coal in O_2/CO_2 atmosphere. Appl Energy 132:349–357

Cavaliere A, de Joannon M (2004) Mild combustion. Prog Energy Combust Sci 30(4):329–366

David MB, Ravid R (2011) Deposition of fractal-like soot aggregates in the human respiratory tract. J Aerosol Sci 42(6):372–386

Dexing D (2000) Optical measurement of soot particle initially formed in diffused flame [in Chinese]. Opto-Electronic Engineering. 27:1

Jianyi Lu, Qinglong W (2011) Distribution characteristics of gas temperature and soot fraction volume in ethylene/air inverse diffusion flame [in Chinese]. Acta Chim Sinica 69(8):1011–1016

Jianyi Lu, Xiaobin S (2013) Physicochemical properties and formation mechanism of soot during biomass burning [in Chinese]. J Fuel Chem Technol 10:1184–1190

Joannon M, Langella G, Beretta F et al (2000) Mild combustion: process features and technological constrains. Combust Sci Technol 153(1):33–50

Katsuki M, Hasegawa T (1998) The science and technology of combustion in highly preheated air. Proc Combust Inst 27(2):3135–3146

Lighty JS, Veranth JM, Sarofim AF (2000) Combustion aerosols: Factors governing their size and composition and implications to human health. J Air Waste Manage Assoc 50(9):1565–1611

Liqi Z, Zhihui M, Baojun Y et al. (2014) Coal powder oxygen-rich flameless combustion method and system: China, CN103615713A [in Chinese]

Na Y, Yongjie W, Min H, Limin Z, Yunhang Z (2009) Characterization and source identification of ambient organic carbon in $PM_{2.5}$ in urban and suburban sites of Beijing [in Chinese]. Acta Scientiae Circumstantiae 29(2):243–251

Pengfei L, Jianchun M, Dally BB et al (2011) Latest developments and trends in MILD combustion [in Chinese]. Sci China 41(2):135–149

Richter H, Howard JB (2000) Formation of polycyclic aromatic hydrocarbons and their growth to soot-a review of chemical reaction pathways. Prog Energy Combust Sci 26(4/6):565–608

Sepmana A, Mokhova AV et al. (2013) Numerical and experimental studies of the NO formation in laminar coflow diffusion flames on their transition to MILD combustion regime. Combust Flame 160(8):1364–1372

Stadler H, Ristic D, Forster M et al (2009) Nox-emissions from flame less coal combustion in air, Ar/O_2 and CO_2/O_2. Proc Combust Statute 32(2):3131–3138

Tsuji H, Gupta A, Hasegawa T et al (2003) High temperature air combustion: from energy conservation to pollution reduction. CRC Press, Florida

Xiong G (2011) Experimental study of soot formation during the combustion of coal and biomass [in Chinese]. Tsinghua University, pp 16–17

Zhe D, Yongqiang C (2018) Design of on-line smoke concentration monitoring system based on filter membrane weighing method [in Chinese]. Modern Electron Tech 41(11):145–146

Chapter 24
Experimental Investigation of Methane Assisted Pulverized Coal Flames Using an Optical Accessible Combustion Chamber

C. Axt, D. Zabrodiec, S. Pielsticker, T. Kreitzberg, O. Hatzfeld, and R. Kneer

Abstract The formation of inhalabe fine particles ($d_p \leq 10\mu m$) is an undesirable side effect of solid fuel combustion processes. These particles can accumulate in the human respiratory system and thus cause severe lung damage. Therefore, an understanding of the formation of these particles is of crucial importance to avoid or reduce the amount of fine particulate matter released into the atmosphere. For the investigation of particulate matter formation, a swirled methane assisted pulverized coal combustion test rig was developed, which allows intrusive and non-intrusive measurements to be performed during the combustion process. Within the scope of this work, the developed test rig is described and results of laser Doppler velocimetry (LDV) measurement are presented. The velocity measurements were performed at different height levels of the flame to investigate the flame structure. Within this study, it is shown that the experimental setup achieves stable and reproducible combustion conditions to allow detailed investigations in further works.

Keywords Swirl burner · Laser Doppler velocimetry (LDV) · Optical measurements

24.1 Introduction

Fine particulate matter (PM) in the size range of 0.1–2.5 μm produced by solid fuel combustion processes is a worldwide social- and health relevant problem which affects the breathing air quality. Compared with larger particles, fine particles are more minatory to human health because of their high specific surface area for toxic elements enrichment and long atmospheric residence time. Significant sources of anthropogenic fine particulate matter are biomass and coal fired power plants. In China, PM emissions from coal-fired power plants yields 3.81 million tons per year,

C. Axt (✉) · D. Zabrodiec · S. Pielsticker · T. Kreitzberg · O. Hatzfeld · R. Kneer
Institute of Heat and Mass Transfer, RWTH Aachen University, 52056 Aachen, Germany
e-mail: axt@wsa.rwth-aachen.de

J. Lyu and S. Li (eds.), *Clean Coal and Sustainable Energy*,
Environmental Science and Engineering,
https://doi.org/10.1007/978-981-16-1657-0_24

accounting for 44.6% of total mass of the released PM (Yao et al. 2009). Moreover, worldwide coal consumption is expected to rise due to a projected increase of electricity consumption, especially in developing countries (World energy outlook 2016).

In power stations, PM is captured by electrostatic precipitators or bag filters. However, the capturing efficiency for fine particulate matter of these methods is not very high. As a countermeasure, the formation of fine particulate matter in the combustion process should be minimized or avoided.

In order to predict fine particulate matter formation in solid fuel combustion, numerical models are developed which are then integrated into common Computational Fluid Dynamics (CFD) calculations. For the development of these models, the combustion process needs to be characterized in terms of temperature, velocity and local species concentration. However, the principle of fine particulate matter formation is far from being understood. Existing models are only valid for early combustion stages (pyrolysis) and were developed for laminar flows by now (Gao et al. 2017). Thus, so far, no model for the formation of fine particulate matter exists that takes the entire turbulent burnout process into account. Previous investigations within pulverized coal combustion systems rely mostly on invasive diagnostics. In comparison, non-intrusive optical diagnostics have the advantage of a high spatial and temporal resolution and have a much smaller influence on the flame. Therefore, a suitable combustion chamber for non-intrusive measurements has been developed and is presented in the following sections. In order to ensure that the test rig delivers reproducible results, LDV measurements have been carried out and are discussed in the results section.

24.2 Experimental Setup

For the investigations of coal combustion and the resulting particulate matter formation, an experimental test rig was developed. This setup allows the investigation of methane assisted pulverized coal flames. In Fig. 24.1 the scheme of the System is shown. The System consists of three major parts (I-III) and is described in the following sections.

24.2.1 Burner Design

The employed burner is a coal swirl burner initially proposed by Toporov et al. (Toporov et al. 2007) (I) and modified later on by Becker et al. (Becker et al. 2017). This kind of burner has the advantage, that it can generate aerodynamically stabilized flames. On the right hand side of Fig. 24.1 a cross-sectional view of the burner is shown. The central annular orifice (Ø30–34 mm, named primary), around a central bluff body, provides a premixed non-swirled fuel and air

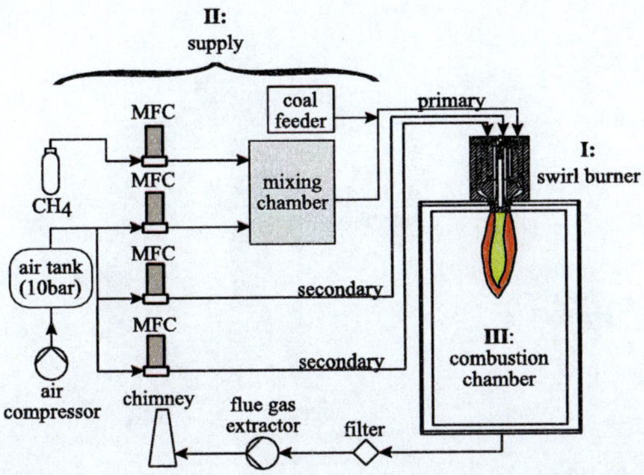

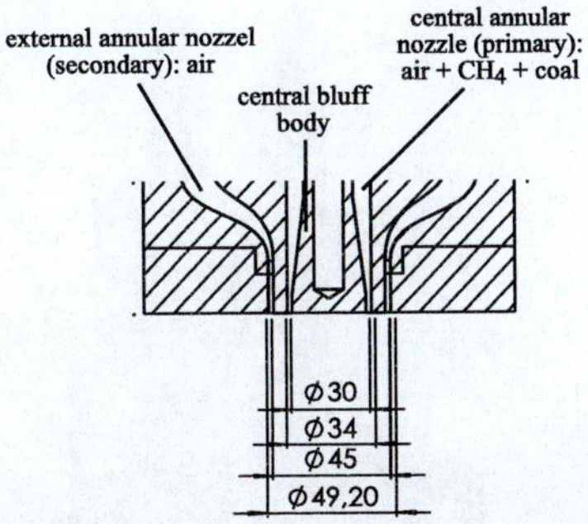

Fig. 24.1 Left: scheme of the experimental setup for investigations of methane assisted coal flames. The three major parts of this experimental setup are I: swirl burner, II: supply of educts and III: combustion chamber. Right: cross section of the swirl burner (dimensions are in mm)

mixture. The second annular orifice (Ø45–49.2 mm, named secondary) provides the swirled air flow, whereas the amount of swirl can be adjusted. The degree of swirl can be set from pure axial flow to strongly swirling jets. This burner has already been successfully employed in several previous experimental studies (Franchetti et al. 2016; Habermehl et al. 2012; Heil et al. 2009; Ikeda et al. 2012; Toporov et al. 2008).

24.2.2 *Combustion Chamber*

The frame of the combustion chamber (III) is made of stainless steel (UNS S31400) and can resist high temperatures (up to 1000 °C). The walls of the chamber are made of quartz glass (material EN08, company GVB Solutions GmbH). Between the quartz glass and the frame of the combustion chamber gaskets are used to avoid leakage (see Fig. 24.2) and also to reduce tension between quartz glass and frame. The used mica gasket can resist temperatures up to 1000 °C. The exhaust gases flow through the annular orifice (outlet combustion chamber) to the chimney. The inner dimensions inside of the chamber are 350 mm length, 350 mm width and 550 mm height.

24.2.3 *Gas and Particle Supply*

The adjustments of the supplied oxidizer and methane gases are controlled by four thermal mass flow controllers (MFC's left side of Fig. 24.1). The thermal MFC (Bronkhorst) are chosen because of their high accuracy (±0.5% percentage of reading

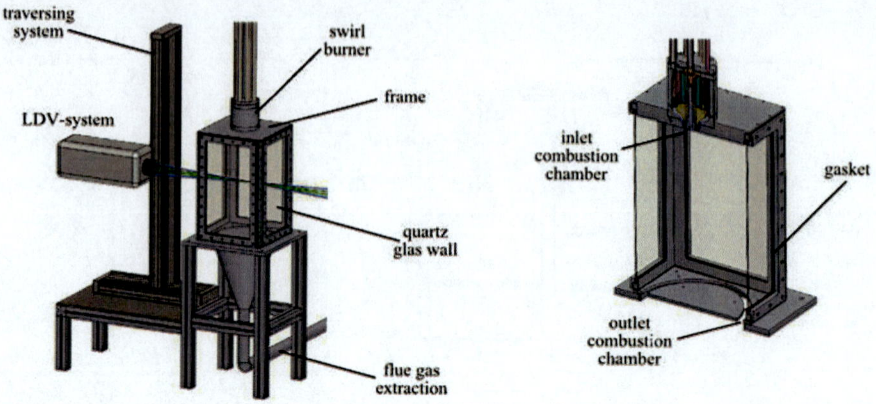

Fig. 24.2 Left: main parts of the experimental setup. Right: sectional representation of the combustion chamber

(RD) plus ± 0.1% percentage of full scale (FS)). For the generation of an aerosol of coal particles and primary stream, a solid aerosol generator (Palas BG 1000) is used.

24.2.4 Laser Doppler Velocimetry System

The used LDV system consists of two diode laser (Oxxius Laser Box: LCX-532S, LCX-561S) with the wavelengths 532 and 561 nm. The system can be moved to any position of the flame with the traversing system (Fig. 24.2). Each laser has a maximum power of 300 mW. The focal distance of the laser system is 300 mm in our experiments. The LDV measurements are carried out in backscattering mode (see also Mayinger 2011). The sending and receiving optics are integrated in a compact box that can be easily positioned to conduct measurements in the chamber. Doppler signal bursts were collected by means of photomultipliers and analyzed with a Dantec BSA processor.

24.3 Operation Parameters and Conditions

In this work, two operating conditions are defined. The first condition is the methane assisted pulverized coal flame (referred as coal-case). In this case, the coal particles are used as tracer particles for the LDV measurements. The second condition is the investigation of the flow field with aluminum oxide particles in an isothermal (referred as alu-cold-case) environment. In this case, the alumina particles act as tracers for LDV measurements.

24.3.1 Coal Case Measurements

The investigated flame in these experiments is a swirling down-fired methane assisted pulverized coal flame. The air for the experiments is dried, compressed and supplied by an air tank (pressure in the tank 10 bar). The used Zhundong Coal (lignite) is dried and ground. Figure 24.3 shows the coal particle size distribution of the Zhundong Coal. This lignite is characterized by its high volatile content (Table 24.1), which supports the formation of particulate matter and thus makes it easier to do

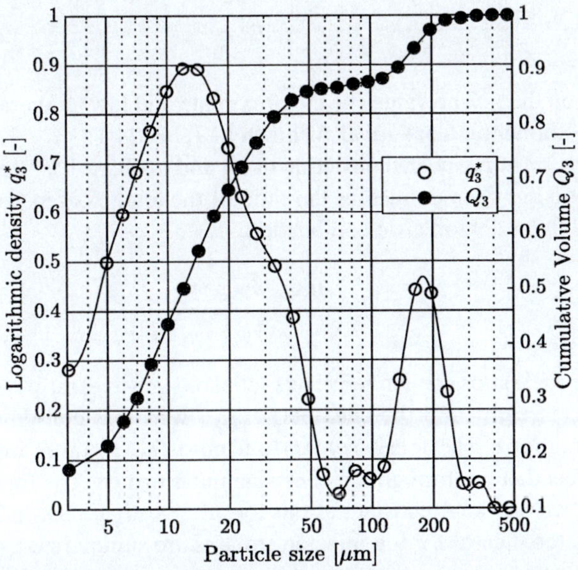

Fig. 24.3 Particle Size Distribution (PSD) of the Zhundong Coal (lignite). The coal was dried and ground before measurement

Table 24.1 Proximate and low temperature ash (LTA) analysis of Zhundong Coal. (values from Gao et al. 2015)

Proximate analysis (wt.%)				
Ash (dry)			Volatiles (dry)	
5.88			30.58	
LTA analysis (wt.%) (ash)				
SiO_2	Al_2O_3	Fe_2O_3	CaO	MgO
14.5	3.96	4.85	39.65	3.48
TiO_2	SO_3	K_2O	Na_2O	
0.36	25.97	0.69	7.48	

Table 24.2 Flow conditions for the educts. The air streams are given at a temperature of 293.15 K and a pressure at 4 bar. The methane stream is given at a temperature of 293.15 K and a pressure at 3 bar

Flow conditions				
CH_4	Air (primary)	Air swirled (secondary)	Al_2O_3	Coal
(m^3/h)	(m^3/h)	(m^3/h)	(g/min)	(g/min)
Coal (case)				
0.7	5.0	10.0	0.0	3.0
Alu-cold (case)				
0.0	5.52	10.0	3.0	0

measurements on them. The values of the proximate and low temperature ash (LTA) analysis of the Zhundong Coal are also shown in Table 24.1.

For this work, an atmosphere consisting of air and methane has been utilized. The oxygen-fuel ratio has been defined as the ratio of the total oxygen provided and the oxygen needed for stoichiometric combustion.

$$\lambda = \frac{\dot{m}_{O_2}/\dot{m}_{\text{fuel}}}{(\dot{m}_{O_2}/\dot{m}_{\text{fuel}})_{\text{stoichiometric}}} \tag{24.1}$$

Two oxygen-fuel ratios are defined: The global oxygen-fuel ratio (λ_{global}) includes the injected gas streams into the chamber; λ_{global} was kept constant at 1.8 for all the studied conditions. The local oxygen-fuel ratio (λ_{local}) takes into account the incoming streams that go through the central annular orifice. The local oxygen-fuel ratio at the burner was kept constant at 0.6 for all investigations. In Table 24.2 the flow conditions (coal case) for the injected streams are summarized.

24.3.2 *Alu-Cold Case Measurements*

To investigate the cold flow field, aluminum (tracer) particles were used for the velocity measurements. This measurement provides information about the flow field generated by the swirl burner, which remains unaffected by the changes through the flame. The particle size distribution of the alumina particles is defined by three diameters of the volume distribution, with the values given in Table 24.3.

Table 24.3 Particle size distribution of the alumina particles (MARTOXID® MR-70)

PSD (vol.%)		
D10% (μm)	D50% (μm)	D90% (μm)
0.2	0.6	2

For the cold measurements with alumina, the momentum of the injected streams was chosen the same as for the flame conditions. In Table 24.2, the flow conditions are summarized (alu-cold case).

24.4 Results and Discussion

In Fig. 24.4, the shape of the investigated flame is shown. For this, a photograph with an exposure time of 30 s was taken. The picture shows, that at this conditions a methane side flame was found (see also Becker et al. 2017).

In Fig. 24.5 the mean velocity is shown at the vertical axis of each sub figure in meter per seconds. The horizontal axis in each figure shows the radial coordinate in the combustion chamber wherein zero represents the center of the combustion chamber. On the left hand side of Fig. 24.5, the axial velocity of the particles is shown (U as the axial mean velocity). On the right hand side of Fig. 24.5, the tangential velocity is shown (V as the tangential mean velocity). The axial and tangential velocity is measured at different heights below the burner ($h = 1d$, $h = 2d$ and $h = 3d$). Where length d is related to the outside diameter of the secondary outlet (Ø49.2 mm). The grey dotted line represents the LDV measurements at the alu-cold condition with aluminum oxide as tracer particles. The black dotted line represents the LDV measurements with coal as tracer particles in a hot environment. The dots, rectangular and round, show the measured points that were considered.

24.4.1 Axial Mean Velocity

The curve profile of the axial mean velocities has two maxima. All axial profiles show a nearly symmetrical flame structure. This is especially important for a stable flame structure and thus also for a defined condition of numerical simulations.

Fig. 24.4 Photography of the flame with an exposure time of 30 s

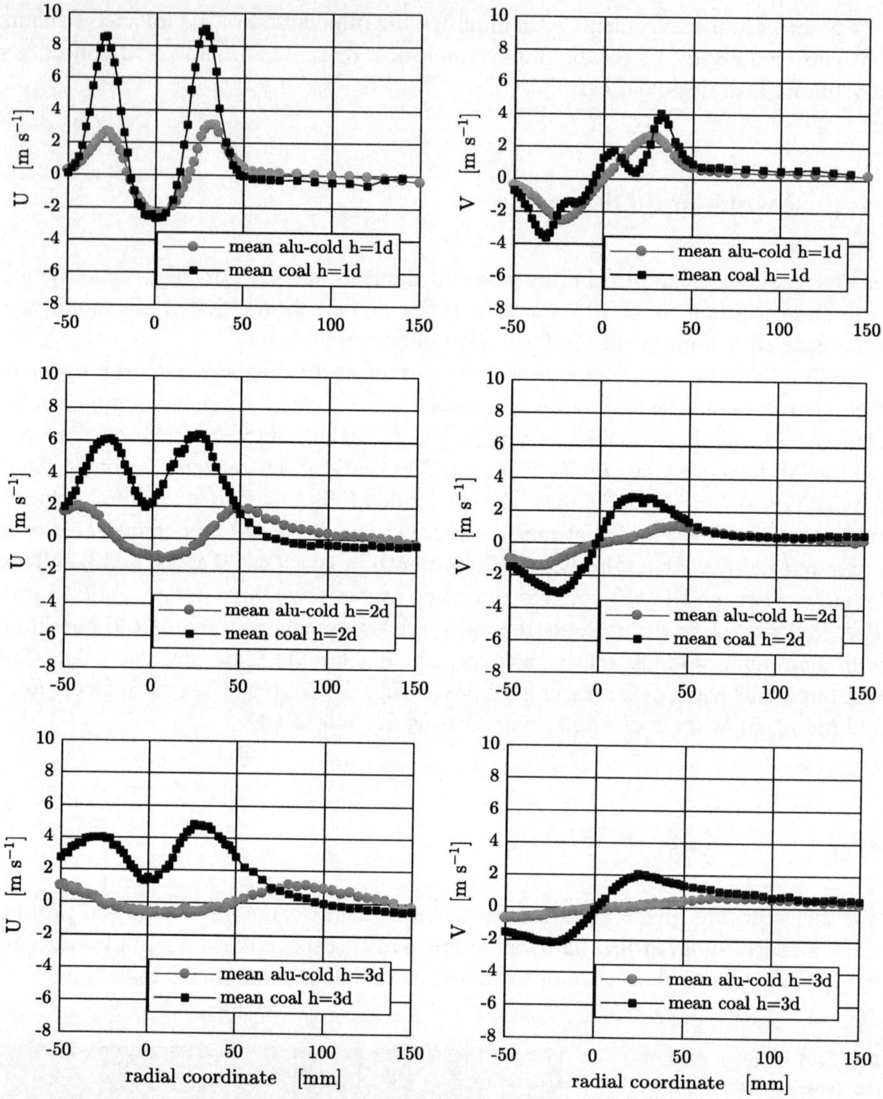

Fig. 24.5 LDV measurements of velocity. Left: axial velocity of the particles is shown (U as the axial mean velocity). Right: tangential velocity is shown (V as the tangential mean velocity). The h in the legend denotes the length below the burner

Further, it can be seen in the axial mean velocity figures, that the cold case magnitude velocity in comparison to the values from the hot measurements is much smaller. This can be attributed to the expansion of the gases at the higher temperatures and therefore higher velocities.

At h = 1d below the burner (49.2 mm), the flame and the cold measurements show negative axial velocities in the range of ± 15 mm radial coordinate. There the recirculation zone is formed, which contributes to flame stabilization. At h = 2d and thereafter no recirculation zone is perceived for the coal flame. The cold measurements, however, still have a recirculation zone, even at h = 3d.

24.4.2 *Tangential Mean Velocity*

The tangential velocity on the right hand side of Fig. 24.5 shows also a symmetrical profile around the centerline of the burner. This profile is point symmetrical (axial velocities are axis symmetrical) because the particles pass through the measurement volume in the opposite direction beyond the centerline. The centerline of the burner was determined at a tangential velocity of zero. Similar to the axial velocity, the tangential velocity of the flame is also higher in comparison to the cold conditions due to the expansion of the gases.

However, at h = 1d height below the burner, some differences between hot and cold appear. At cold conditions, the flow field shows only one negative and one positive maximum. In comparison, the flame shows two positive and negative local maxima. It is assumed that these effects are caused by vortices inside the flame (see also Taamallah et al. 2019).

24.4.3 *Root Mean Square (RMS) Intensities*

In Fig. 24.6, the RMS velocities are shown. The RMS velocity can be interpreted as the turbulence intensity of the flow. On the left hand side of the figure, the RMS of the axial velocities are shown (u_{rms}). On the right hand side, the tangential RMS velocities are shown (v_{rms}).

Around the centerline, all flame measurements show a higher RMS velocity (tangential and axial) compared to the cold flow field measurements. At outer radial coordinates, they become similar. However, at h = 1d, there are differences around the radial coordinate of 100 mm. At these points, the data rate of the particles drops.

The highest turbulence intensities occur in the vicinity outside of the recirculation zone at h = 1d. It can also be seen, that the maximum RMS values occur, where the highest velocity gradients are found.

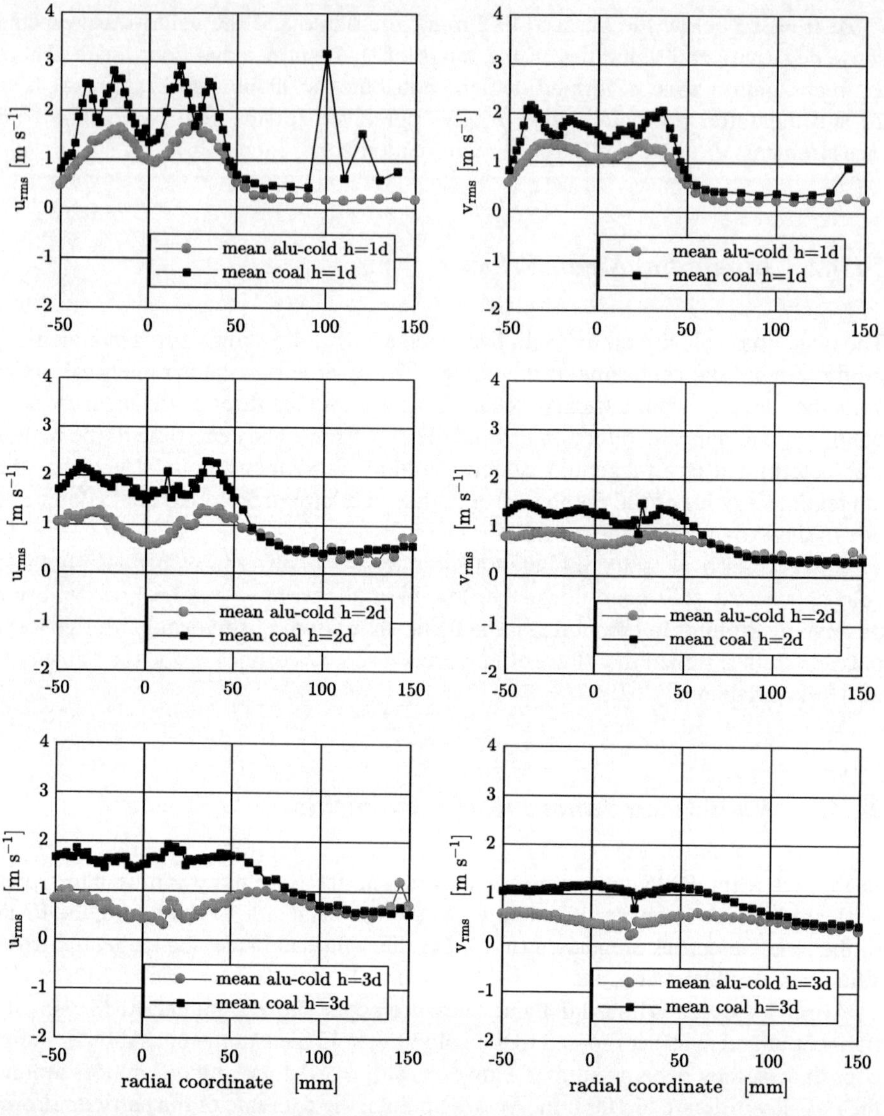

Fig. 24.6 LDV measurements of RMS velocity. Left: axial RMS velocity of the particles (u_{rms} as the axial RMS velocity). Right: tangential RMS velocity (v_{rms} as the tangential RMS velocity). The h in the legend denotes the length below the burner

24.5 Conclusion

In this work, a new experimental setup for optical measurements of methane assisted pulverized coal flames was presented. The combustion chamber was designed in such a way that high optical accessibility is possible (measurements can be carried out almost in the whole combustion chamber, see Fig. 24.2). In addition, the combustion chamber was equipped with high-temperature mica gaskets to prevent interferences from the outside.

For the investigations of the flame structure inside the test rig, LDV measurements were performed. The measurements show, that the flame has a symmetrical profile around the burner centerline and a recirculation zone can be found. The flame structure thus indicates that a stable flame structure can be achieved with the new test rig. It could also be shown that the maximum values of the RMS occur at the maximum gradients of the velocity. This indicates physical plausibility since increased velocity gradients lead to increased friction between the individual fluid layers and thus to increased turbulence.

The high optical accessibility in the chamber makes it possible to perform a variety of other optical measurements such as chemiluminescence, Phase-Selective Laser Induced Breakdown Spectroscopy (PS-LIBS), Phase Doppler Anemometry (PDA) and Laser Induced Incandescence (LII). Thus the new test rig provides the basis for further investigations on solid fuel combustion and therefore also on particulate matter formation.

Acknowledgements The authors would like to thank the German Research Foundation (DFG) and the National Natural Science Foundation of China (NSFC) for funding the joint Sino-German research project "Inorganic fine particulate matter formation during turbulent pulverized coal combustion" (Grant No. 392429716).

References

Becker LG, Kosaka H, Böhm B, Doost S, Knappstein R, Habermehl M et al (2017) Experimental investigation of flame stabilization inside the quarl of an oxyfuel swirl burner. Fuel 201:124–135

Chen L, Yong SZ, Ghoniem AF (2012) Oxy-fuel combustion of pulverized coal: characterization, fundamentals, stabilization and CFD modeling. Prog Energy Combust Sci 38(2):156–214

Franchetti BM, Cavallo Marincola F, Navarro-Martinez S, Kempf AM (2016) Large eddy simulation of a 100 kW th swirling oxy-coal furnace. Fuel 181:491–502

Gao Q, Li S, Yuan Y, Zhang Y, Yao Q (2015) Ultrafine particulate matter formation in the early stage of pulverized coal combustion of high-sodium lignite. Fuel 158:224–231

Gao Q, Li S, Yang M, Biswas P, Yao Q (2017) Measurement and numerical simulation of ultrafine particle size distribution in the early stage of high-sodium lignite combustion. Proc Combust Inst 36(2):2083–2090

Habermehl M, Erfurth J, Toporov D, Förster M, Kneer R (2012) Experimental and numerical investigations on a swirl oxycoal flame. Appl Therm Eng 49:161–169

Heil P, Toporov D, Stadler H, Tschunko S, Förster M, Kneer R (2009) Development of an oxycoal swirl burner operating at low O2 concentrations. Fuel 88(7):1269–1274

Ikeda M, Toporov D, Christ D, Stadler H, Förster M, Kneer R (2012) Trends in NOx emissions during pulverized fuel oxy-fuel combustion. Energy Fuels 26(6):3141–3149

Mayinger F (2011) Optical measurements: techniques and applications, 2nd edn. Springer, Berlin

Taamallah S, Dagan Y, Chakroun N, Shanbhogue SJ, Vogiatzaki K, Ghoniem AF (2019) Helical vortex core dynamics and flame interaction in turbulent premixed swirl combustion: a combined experimental and large eddy simulation investigation. Phys Fluids 31(2):25108

Toporov D, Forster M, Kneer R (2007) Burning pulverized coal in CO_2 atmosphere at low oxygen concentrations. Inter J Ener Clean Env 8(4):321–338
Toporov D, Bocian P, Heil P, Kellermann A, Stadler H, Tschunko S et al (2008) Detailed investigation of a pulverized fuel swirl flame in CO2/O2 atmosphere. Combust Flame 155(4):605–618
World energy outlook (2016) OECD/IEA, Paris
Yao Q, Li S-Q, Xu H-W, Zhuo J-K, Song Q (2009) Studies on formation and control of combustion particulate matter in China: a review. Energy 34(9):1296–1309

Chapter 25
Ignition Characteristics of Coal/Air Jets from Rectangular Nozzles

Lilin Hu, Pengyuan Liu, Yang Zhang, Lele Feng, Yuxin Wu, Junfu Lyu, and Hai Zhang

Abstract The ignition characteristics of pulverized coal-air jets from rectangular nozzles with different aspect ratios (ARs) were studied using a Hencken type entrained flow reactor. High-speed cameras and Mie scattering technique were employed to record the instantaneous flame images and particle motions under different exit velocities and coal concentrations of the primary air flow. The results revealed that similar to the gas flames, the flame length of the coal-air jet first increases and then decreases with increasing *Re* and finally barely changes when the flow is fully turbulent. For a coal/air flow from a rectangular nozzle, compared with the one from a round nozzle, the transient *Re* from laminar to turbulent for the rectangular jets is smaller. In the transient region, the flame length is obviously shorter at the same *Re*. The dissipation of large eddies is faster on the wide side, and the appearance of large eddy structure is delayed on both sides as the coal concentration increases. The simulation results show that for a rectangular nozzle the ignition of the coal jet begins on the two narrow sides, while for a round nozzle, coal jet ignition starts from the surrounding shear layers uniformly.

Keywords Pulverized coal · Combustion · Ignition · Slot nozzles · Free jet

25.1 Introduction

Burners are important for the coal-air jet ignition, flame stability, pollutant formation, and burnout of coal particles in a pulverized coal fired boiler (Pedel et al. 2013; Khatami and Levendis 2016; Obando et al. 2017; Fang et al. 2010; Steer et al. 2013). Though many burners have inner structures, the ones with free jet style are still of great interest, because they are not only popularly used, but also the fundamentals to study the fluid dynamic and combustion characteristics of other burners.

L. Hu · P. Liu · Y. Zhang (✉) · L. Feng · Y. Wu · J. Lyu · H. Zhang
Department of Energy and Power Engineering, Key Laboratory for Thermal Science and Power Engineering of Ministry Education, Tsinghua University, Beijing 100084, China
e-mail: zhangyang@tsinghua.edu.cn

J. Lyu and S. Li (eds.), *Clean Coal and Sustainable Energy*,
Environmental Science and Engineering,
https://doi.org/10.1007/978-981-16-1657-0_25

Besides the round ones (Beer and Chigier 1972), the rectangular burners are also widely used in the coal-fired boiler. Usually, these burners are nearly square-shaped in the cross section area with an aspect ratio (AR = length/width) close to unity. Previous studies found that the flow characteristics of a square jet are quite different from that of a round jet. For a rectangular jet, the presence of sharp corners deforms the formation of vortex structures, resulting in higher entrainment and stronger mixing (Gutmark and Grinstein 1999). At the same time, the growth rate of the shear layers along the narrow side is faster than that along the wide side, so there is a crossover point in the jet downstream after which the jet has an inverse shape against the nozzle exit. This phenomenon is called 'axis-switching' and is believed to be the mechanism for the increased entrainment (Gutmark and Grinstein 1999; Mi and Nathan 2010; Herzberg and Ho 1995).

In fact, some rectangular burners have a rather large AR (e.g., AR > 3), thereafter called slot burners. For example, in the arch-fired boilers made by Mitsui Babcock Company, the AR value of the primary air nozzle is ~6.5 (Kuang et al. 2013). Since the spreading rate is higher along the narrow side than along the wide side (Quinn 1992; Schadow et al. 1987), it is expected that a slot burner has even stronger entrainment capability and thereby more intensive heat and mass transfer than the burner with a smaller AR. The previous studies in the gaseous combustion found the nozzle configuration certainly affects the combustion characteristics, including flame length (Oh and Noh 2013), temperature (Branco et al. 2016), and turbulent burning velocities (Tamadonfar and Gülder 2016).

However, studies on ignition and combustion characteristics of the coal-air jets from slot burners are scarce. Coal combustion is much more complicated than gas combustion, as it includes more physicochemical processes such as devolatilization, ignition and subsequent char combustion (Shaddix and Molina 2009; Molina and Shaddix 2007; Hayashi et al. 2013). The existing results obtained from gaseous combustion could be insufficient or even invalid as AR is large.

Therefore, an experimental study on the ignition and combustion characteristics of coal-air free jets from slot nozzles was conducted using a well-established entrained flow reactor. A set of experiments with different ARs were performed. The instantaneous jet flame images were detected with a high-speed camera and Mie scattering technique, from which ignition distance and delay time were obtained. Effects of coal-air jet velocity (V_{pri}) and coal particle concentration (C_s) were also assessed. At the same time, CFD simulations were carried out under the experimental conditions.

25.2 Experimental Approach

25.2.1 Experimental System

As shown in Fig. 25.1, the experiments were conducted in a Hencken type entrained flow reactor, which consisted of a ceramic honeycomb with a cross-sectional area of

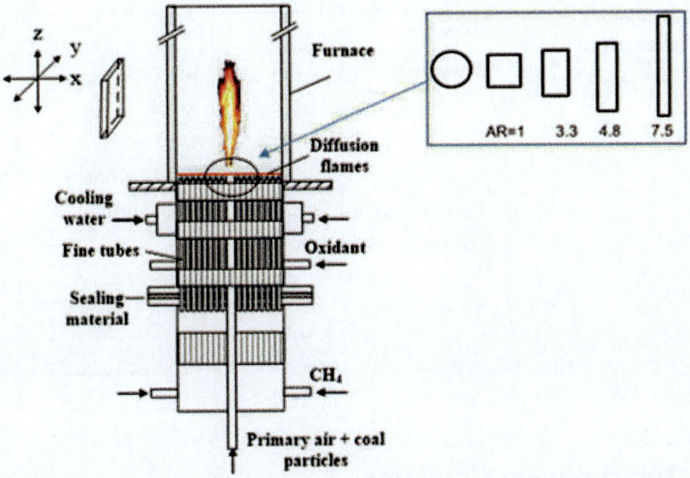

Fig. 25.1 Schematic diagram of the entrained flow reactor with Hencken burner and the configurations of nozzles

55 mm × 55 mm, through which 500 fine round steel tubes were inserted. CH_4 and oxidant (diluted by N_2) streams passed through the tubes alternatively to generate a matrix of small diffusion flames near the tube exits. The temperature and composition of the produced flue gas were adjusted by adjusting the mass flow rates of CH_4, O_2, and air streams. A quartz window was installed on one side of the furnace for optical assessment. Coal particles were injected into the furnace through a stainless-steel nozzle placed in the middle of the bottom plate. More details of the entrained flow reactor can be referred to our previous studies (Xu et al. 2016).

Four nozzles with the same area but different configuration were studied. Among them, the round nozzle had a diameter of 6.2 mm. Three slot nozzles roughly had the length (l) of 10 mm, 12 mm, 15 mm and the width (w) of 3 mm, 2.5 mm, 2 mm, respectively. Correspondingly, the ARs of the three slot nozzles were approximately 3.3, 4.8 and 7.5.

25.2.2 *Coal Properties*

The anthracite coal used in an arch-fired boiler with slot burners was tested. The properties of the tested coal are given in Table 25.1. The raw coal was ground and sieved before the experiments. The size distribution of tested coal was measured by a laser particle size analyzer and the mean diameter was 37 μm.

Table 25.1 Properties of tested coal (as received base)

Type	Item	Value
Proximate analysis (%)	Moisture	7.4
	Volatile	6.56
	Ash	29.79
	Fixed Carbon	56.25
Low heating value	20.89 (MJ/kg)	
Ultimate analysis (%)	C	56.23
	H	2.08
	N	0.72
	O	1.07
	S	2.71

25.2.3 *Measurement Conditions*

In the experiments, V_{pri} varied in the range of 1.42–7.96 m/s. Reynolds number based on V_{pri} and characteristic length d_0 are calculated by $\mathrm{Re} = V_{pri} \cdot d_0/\nu$, where d_0 is the equivalent hydraulic diameter of the nozzle, defined as $d_0 = 2lw/(l+w)$ for slot nozzles, and ν is the kinematic viscosity of air at 298 K. The maximum total flow rate through the nozzle was ~85 slpm (standard liters per minute). The temperature of the hot flue gas, close to the furnace temperature was controlled at ~1900 K. The O_2 concentration in the flue gas was maintained at ~13.3%. Coal feeding was supplied by a scraper coal feeder with a feeding rate varying in 2–8 g/min.

The combustion process was recorded by a high-speed CCD camera with a maximum frame rate of 2000 fps and a shutter speed of 30–1/8000 s. The height of the single image was ~22 cm and two images were simultaneously taken along the axis. Mie scattering technique was used to detect the particle motion. To illuminate the pulverized coal, a second-harmonic wave with 532 nm wavelength of Nd:YAG laser was used with a typical energy of 120 and 10 ns pulse width. The Nd:YAG laser and the CCD camera were synchronized using a pulse delay generator.

25.3 Numerical Approach

Numerical CFD simulations were conducted to assess the nozzle configuration effect on the ignition, especially on the horizontal sections, which was difficult to observe in the experiments. The time-averaged conservation equations for the gas phase were solved using Eulerian description and the particle phase was solved using Lagrangian description. The Realizable k-ε turbulence model and Discrete Ordinates (DO) model were adopted for turbulence and radiation description respectively. The Lagrangian stochastic tracking model was chosen to calculate the coal particle flows. The combustion of volatile was described by the probability density function (PDF) theory, while the char combustion was described by the diffusion/kinetics model. The properties of coal and the main boundary conditions used in the simulation were consistent with the experimental ones. The total mesh number was set to 328,000, at which the grid independency was satisfied. The simulation was done by FLUENT.

25.4 Results and Discussion

25.4.1 Ignition and Combustion Characteristics of Pulverized Coal-AIR Free Jets

Figure 25.2 illustrates the typical flame appearances of a laminar pulverized coal-air jet from a slot nozzle and a round nozzle. For both nozzles, the ignition of the coal jets occurs at a short distance from the nozzle exit. The flames develop along the flow centerline in a rather narrow region. The flame of the slot nozzle is slightly longer and less stable.

Figure 25.3 depicts the typical flame appearances of the coal-air jets in turbulent region (V_{pri} = 4.42 m/s). Compared with the laminar coal-air jet flame, for both nozzles, the turbulent one becomes much wider, shorter and brighter in the upstream region. The results indicate that turbulence enhances the entrainment and mixing with hot surroundings, accelerating the heating of coal particles and release of volatiles. At the same time, turbulence promotes the motion of particles and thus a number of discrete bright spots appear near the flame boundary.

Figure 25.4 presents the flame appearances ejecting from four different nozzles.

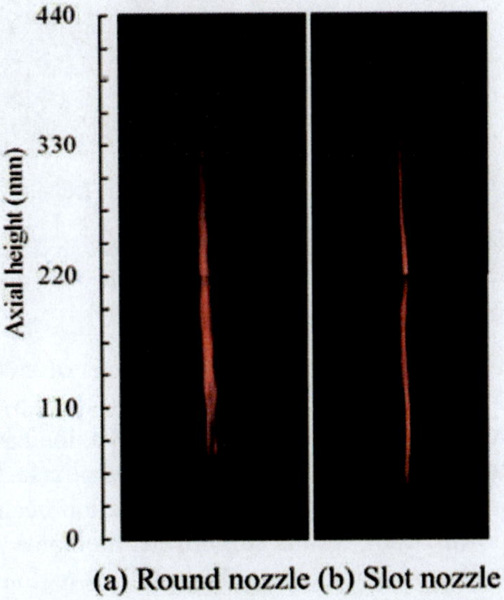

Fig. 25.2 Typical flame appearances of pulverized coal-air free jets in laminar region (AR = 7.5, viewed from the narrow side; V_{pri} = 1.42 m/s, $P_{O2,pri}$ = 21.0%, $P_{O2,sec}$ = 13.3%, C_s = 0.6 kg/Nm^3; Temperature of flue gas: 1900 K; Exposure time: 3 ms)

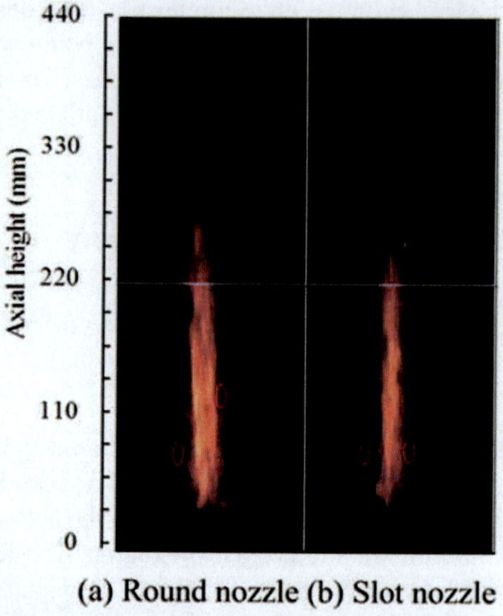

Fig. 25.3 Typical flame appearances of pulverized coal-air free jets in the turbulent region (AR = 7.5, viewed from narrow side; $P_{O2,pri} = 21.0\%$, $P_{O2,sec} = 13.3\%$, $C_s = 0.6$ kg/Nm^3; Temperature of flue gas: 1900 K; Exposure time: 3 ms)

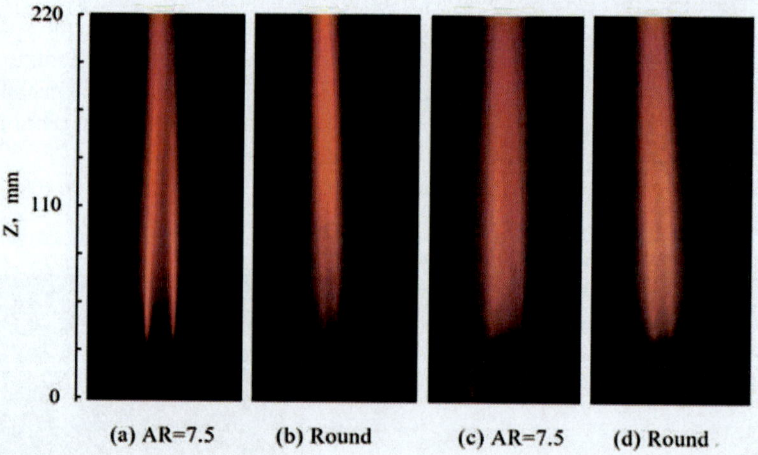

Fig. 25.4 Appearances of pulverized coal-air jets in the laminar **a**, **b** and turbulent **c**, **d** region (The images for AR = 7.5 are viewed from the wide side)

Each appearance was the average of 500 instantaneous images. As shown in Fig. 25.4a, b, for the slot nozzle (AR = 7.5), the laminar flame is brighter on the two narrow sides than in the middle area, indicating ignition of the flame jet begins from the two narrow sides. For the round nozzle, there is a narrow dark zone in the middle of the flame, indicating flame ignition occurs from the outer edge of the jet. In the downstream, for the slot nozzle, the flame needs noticeably longer time to become uniform luminance. In the turbulent region, for the slot nozzle, the flame intensity

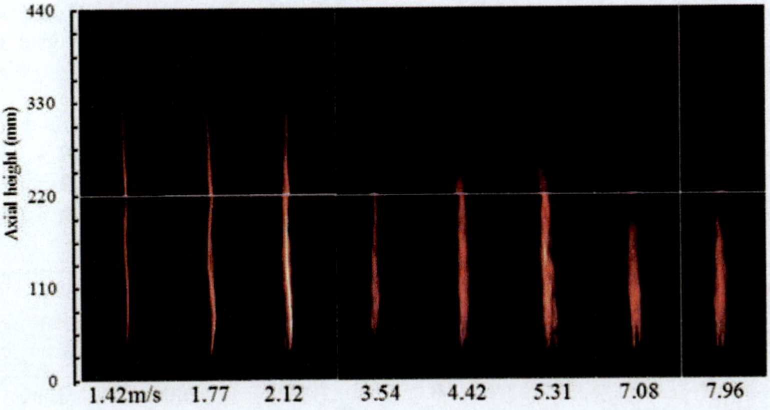

Fig. 25.5 Typical flame images of pulverized coal-air free jets at various V_{pri}'s (Viewed from narrow side, $P_{O2,pri} = 21.0\%$, $P_{O2,sec} = 13.3\%$, $C_s = 0.6$ kg/Nm^3; Temperature of flue gas: 1900 K; Exposure time: 3 ms)

at the edge is slightly greater than that in the middle area (Fig. 25.4c). The results indicate that the heat and mass transfer inside the jets increase. In the downstream, for the slot nozzle, the flame needs even shorter time to become uniform luminance.

25.4.2 *Effect of V_{pri} on the Flame Length of Coal-Air Free Jets from a Slot Nozzle*

Figure 25.5 illustrates typical images of pulverized coal-air free jet flames at different V_{pri} ejecting from a slot nozzle (AR = 7.5) at a given initial coal concentration C_s. In the laminar flow region (e.g., $V_{pri} < 2.12$ m/s), as V_{pri} increases, the jet flame becomes brighter and longer. When $V_{pri} = 3.54$ m/s, the flow becomes turbulent, and the flame height reduces rapidly and then keeps nearly constant as $V_{pri} > 4.42$ m/s. The effect of V_{pri} on mass input of coal particles and the turbulent intensity of the flow will be discussed in the later section.

25.4.3 *Flame Lengths of the Coal-Air Free Jets from the Slot and Round Nozzles*

Figure 25.6 presents the flame length L_f of the coal-air free jets from a round nozzle and three slot nozzles with different ARs. Similar to previous studies (Xu et al. 2016; Liu et al. 2011), the onset of flame ignition is defined at the location where the flame luminous intensity reached 50% of the peak value, while the flame tip is

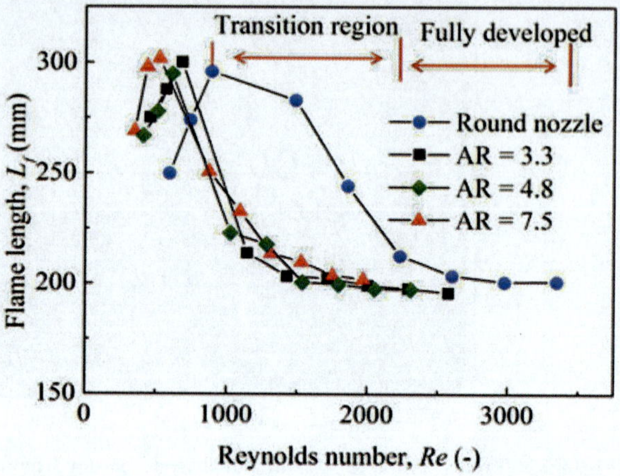

Fig. 25.6 Variations of the flame length from round and slot nozzles with *Re*

defined at the location where luminous intensity reached 20% of the peak value in the downstream. When the primary flow is laminar, L_f increases remarkably with V_{pri} for all nozzles. However, when the flow transits into turbulent, L_f first decreases when $V_{pri} = 2.12$–5.31 m/s and then remains nearly constant when $V_{pri} > 5.31$ m/s. The results show that the effect of flow pattern on L_f is significant.

When $V_{pri} < 2.12$ m/s, *Re*'s for the round and slot nozzles (AR = 3.3) are about 900 and 690, respectively. The flow is basically laminar and coal particles are concentrated along the jet axis. The slip velocity between the primary flow (1.4–2.1 m/s) and the co-flowing flue gas (~2 m/s) is small. Heat conduction and convective transfers between them are weak. Following the gas phase flame, the increase of V_{pri} or *Re* simply increases the axial length of the coal-air stream, and thereby L_f.

For the round nozzle, when $900 < Re < 2230$, the flow is in the transition region. Some eddies could appear in the jet flow to enhance coal particle dispersion from the primary air into the surrounding hot flue gas. As a result, heating of coal particles is accelerated, and so do the devolatilization and ignition of coal. Thus, the volatile and main coal particles burn faster and L_f becomes shorter. When $Re \geq 2230$, coal combustion occurs in a fully developed turbulent flow. Eddy dissipation dominates over the molecular transport. Also similar to a turbulent flame, V_{pri} barely affects L_f. For slot nozzles with AR = 3.3 and 4.8, the transition region spans over $680 < Re < 1720$ and subsequent fully developed region occurs when $Re > 1720$. For the slot nozzle with a larger AR of 7.5, the transition region exists even earlier.

Figure 25.7 shows the ignition distance L_I, defined as the distance between the nozzle exit and the ignition point, for the round and two slot nozzles at various V_{pri}'s. When $V_{pri} < 2.12$ m/s, L_I increases with V_{pri}, which is similar to the variation trend of the flame length. The increase of V_{pri} or *Re* simply reduces the reaction residence time and L_I increases. When $V_{pri} > 3.54$ m/s, strong interaction of the jet flow and hot flue gas induced turbulent eddies. As a result, a higher heating rate of coal particles is realized, which promotes the ignition and L_I becomes shorter. For an even higher

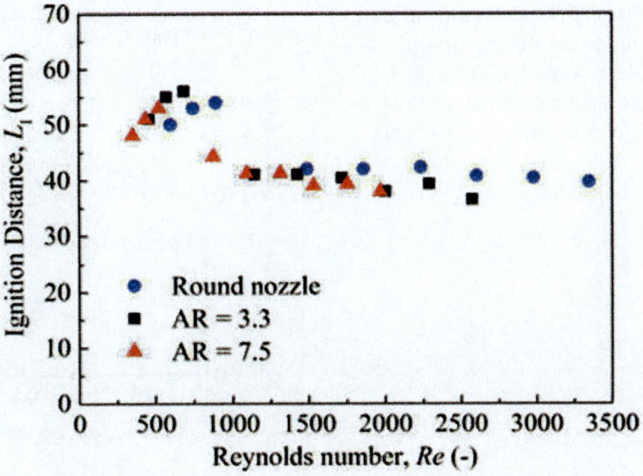

Fig. 25.7 Variations of the ignition distance with *Re* for the coal-air free jets from round and slot nozzles

V_{pri}, the variation trend is less obvious. L_I of the slot nozzles is slightly shorter than that of the round nozzle.

Figure 25.8 further describes the variations of the ignition delay time τ_i with *Re* of different nozzles. τ_i is calculated by $\tau_i = L_I/V_{pri}$, by assuming the coal particle velocity and gas velocity are the same in the ignition region. It can be seen that τ_i decreases rapidly with the increase of *Re* when $Re \leq 1500$ and the decrease rate becomes more gradual when $Re > 1500$. The τ_i in turbulent region could be 2–3 times shorter than that in laminar region. Turbulence greatly promotes the ignition of coal-air jets. τ_i of the slot nozzles is shorter than that of the round nozzles. In addition, with the increase of AR, τ_i becomes shorter. The results could be attributed to that with same exit area, slot nozzles have a larger circumferential surface area than the round nozzle, resulting in more intensive interaction between the hot surroundings with the coal-air jet. Thus, the heating rate is much faster for slot nozzles and τ_i is shorter.

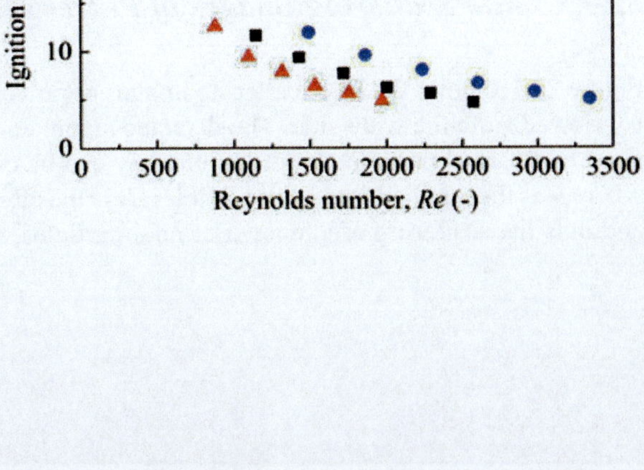

Fig. 25.8 Variations of the ignition delay time with *Re* for the coal-air free jets from different nozzles

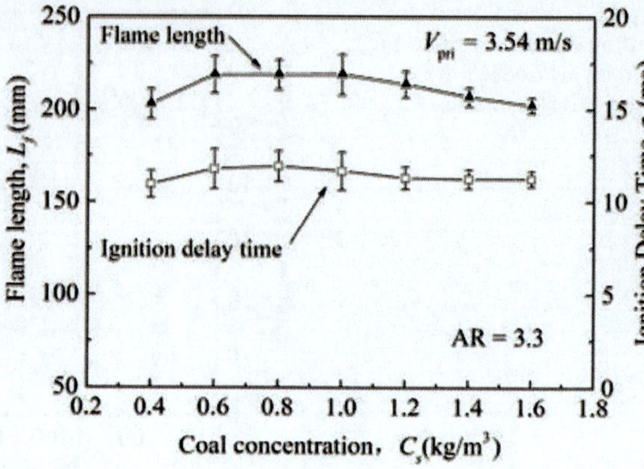

Fig. 25.9 Variations of the flame length and ignition delay time with coal concentration for the coal-air free jets from a slot nozzle

25.4.4 *Effect of Coal Concentration on the Flame Length and Ignition Delay Time*

Figure 25.9 shows L_f and τ_i of the coal-air free jet from a slot nozzle (AR = 3.3) when $C_s = 0.6$–1.6 kg/m^3 at a given $V_{pri} = 3.54$ m/s. It can be seen when $C_s <$ 0.6 kg/m^3, both L_f and τ_i increase with C_s. This could be attributed to that as C_s increases, the number density of coal particles increases, resulting in a lower heating rate of coal particles.

As C_s further increases, L_f is almost constant when $C_s = 0.6$–1 kg/m^3 and then gradually decreases with increasing C_s when $C_s > 1.0$ kg/m^3. At the same time, τ_i barely changes or even slightly decreases with increasing C_s when $C_s > 0.6$ kg/m^3. The phenomenon is due to the balance of increasing coal particle concentration in the flow and increasing the combustion intensity of volatile matter. On one hand, the increasing concentration of coal particles reduces the heating rate of coal particles, resulting in a longer L_f and τ_i. On the other hand, the total amount of released volatile matter increases with C_s, resulting in a high flame temperature and reaction rate, thereby a shorter L_f and τ_i. The results could be more clearly explained from Mie scattering images shown in next section.

For the round nozzle, when the primary flow is fully turbulent, the same characteristics were found by our previous studies (Xu et al. 2016).

25.4.5 *Mie Scattering Images of Pulverized Coal Particles*

Figure 25.10 shows the Mie scattering images of pulverized coal flame at the various C_s viewed from the wide side. The detected signal comes from the laser pulse scattered by the coal particle cloud. Its intensity can be calculated by $I_{sca} = N\sigma\pi D^2$, where N is the number density of particles, D is the effective diameter of the particles and σ is the scattering efficiency. For coal particles, σ has a weak dependence on

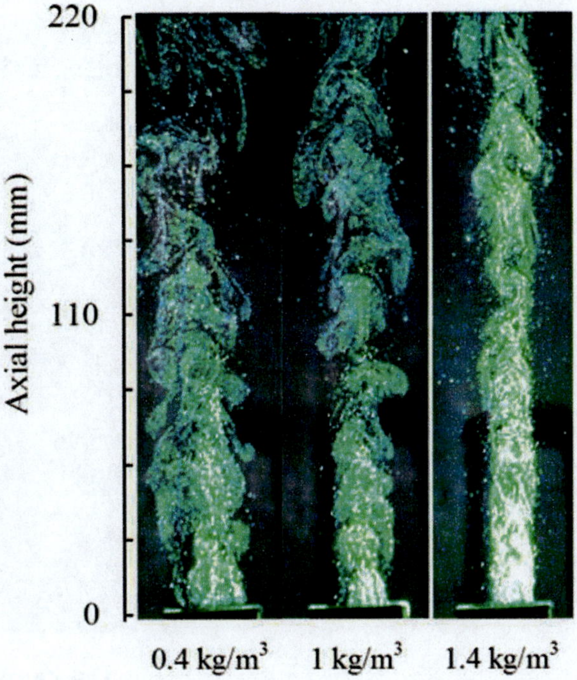

Fig. 25.10 Mie scattering of pulverized coal jet flames at different coal concentrations from a slot nozzle viewed from the wide side (AR = 3.3)

D, so I_{sca} is proportional to the surface areas of the particles per volume and can be used to represent the spatial distribution of the particles (Hayashi et al. 2013).

It can be seen that the number density of particles is very large near the nozzle exit and there are many large eddies along the jet. However, the location where the eddies commence is different at different C_s. When $C_s = 0.4$ kg/m^3, the particle-laden flow spreads quickly at the nozzle exit, with a number of large scale eddy structures appearing there. However, when $C_s = 1.4$ kg/m^3, most particles remain in the main flow at the exit zone and the large eddies appear in the further downstream. Most of the bright spots, which are regarded as the trajectories of large particles, appear in the jet center, while fine particles are observed near the boundary of the two-phase jet. Correspondingly, the scattering signal decays much slowly for the jet with a higher C_s.

Figure 25.11 shows the Mie scattering images of pulverized coal viewed from the narrow side. The lateral expansion of the jet becomes more evident. This is because the pressure field is non-uniformly distributed and pressure is higher on the narrow sides than the wide sides induced by shear-layer vortices (Quinn 1992). In combustion case, this effect leads to a higher spreading rate of particles along the narrow side.

Fig. 25.11 Mie scattering of pulverized coal jet flames at different coal concentrations from a slot nozzle viewed from the narrow side (AR = 3.3)

25.4.6 Simulation Results on Nozzle Effect

Figure 25.12 shows the simulated temperature distribution of cross section at different Z locations along the jet for slot and round nozzles when $V_{pri} = 3.54$ m/s. For the slot nozzle of AR = 7.5, due to the strong mixing with hot gas, the heating rate is faster on the narrow side than that in the middle area. As a result, the higher temperature air flow on both narrow sides squeezes the air flow in the middle area, and the axis-switching happens when Z = 40–60 mm. When Z = 100 mm, the high-temperature zone mainly locates near the two narrow sides, where the coal burns intensively. High-temperature region gradually transits into the middle area and forms an approximate ellipse shape when Z = 160 mm. For the slot nozzle of AR = 3.3, the development of flame is similar, but at the same location, the temperature is relatively lower. The shape of high-temperature region transits to circle more quickly. For the round nozzle, the coal jets have relatively uniform temperature in the outer shear layer, indicating that the coal jets ignite from the outer shear layer. The shape of high-temperature region is round.

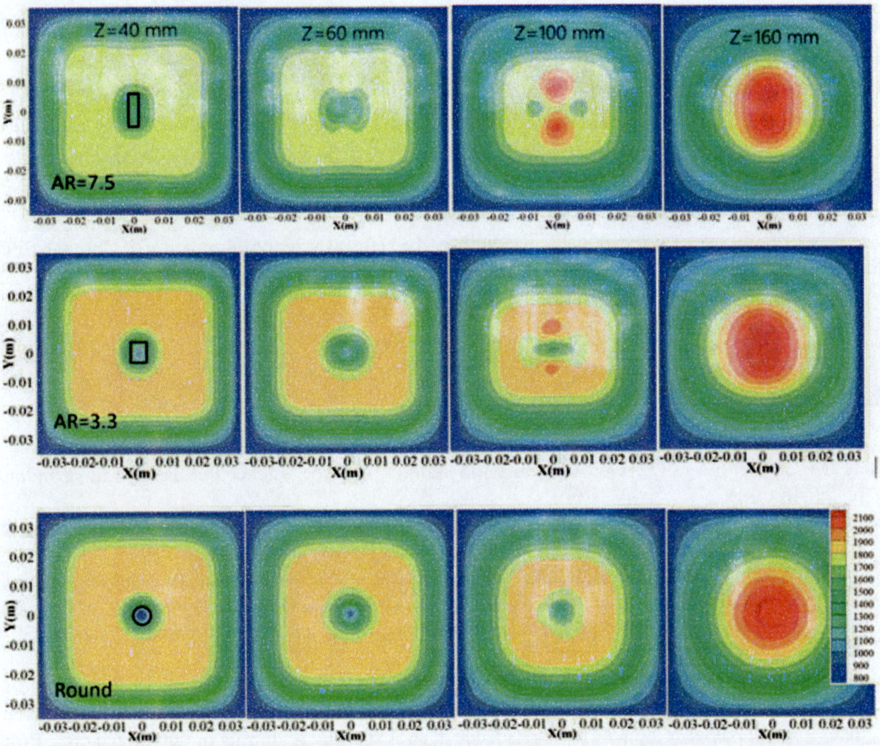

Fig. 25.12 Temperature distribution of coal jets with slot and round nozzles on different cross section along the jet (unit: K)

25.5 Conclusions

The ignition and combustion characteristics of pulverized coal-air free jets from slot nozzles with different aspect ratios were studied using a Hencken type entrained flow reactor. High-speed cameras and Mie scattering technique were employed to record the instantaneous flame image and particle motions under different primary air velocities (V_{pri}'s) and coal concentrations. The results reveal that for all nozzles, the flame length of the coal-air jet first increases and then decreases with increasing V_{pri} and finally keeps almost unchanged when the flow is fully turbulent. In the laminar region, the flame length from a slot nozzle is longer than that from the round nozzle with same exit area but much shorter in the transient region. In a turbulent jet flow, all jet flame become wider and brighter and have the same flame length in the fully turbulent region. The transient Reynolds number (*Re*) for the slot jets is smaller than the round jet. In the transient region between laminar and turbulent, the flame length is obvious shorter from the slot nozzles than that from the round nozzle at the same *Re*. For a jet flame from a slot nozzle, the spreading of large eddies is faster along the narrow side, and the appearance of large eddy structure is delayed for both sides as the coal concentration increases. The simulation results show that for the slot nozzles, the ignition of coal begins from the two narrow sides and then the high-temperature region expands to the middle of the wide side and the coal in the middle area is also ignited. For the round nozzle, coal ignition starts from the outer shear layers uniformly. The simulation results well explain the experimental observations.

Acknowledgements This work was supported by the Natural Science Foundation of China (NSFC #51761125011 and #51476088).

References

Beer JM, Chigier NA (1972) Combustion aerodynamics. Applied Science Publishers Ltd., London, pp 39–48

Branco J, Coelho PJ, Costa M (2016) Experimental and numerical investigation of turbulent diffusion flames in a laboratory combustor with a slot burner. Fuel 175:182–190

Fang QY, Wang HJ, Wei Y, Lei L, Duan XL, Zhou HC (2010) Numerical simulations of the slagging characteristics in a down-fired, pulverized-coal boiler furnace. Fuel Process Technol 91:88–96

Gutmark EJ, Grinstein FF (1999) Flow control with noncircular jets. Annu Rev Fluid Mech 31:239–272

Hayashi J, Hashimoto N, Nakatsuka N, Tsuji H, Watanabe H, Makino H, Akamatsu F (2013) Soot formation characteristics in a lab-scale turbulent pulverized coal flame with simultaneous planar measurements of laser induced incandescence of soot and Mie scattering of pulverized coal. Proc Combust Inst 34:2435–2443

Herzberg J, Ho CM (1995) Three-dimensional vortex dynamics in a rectangular sudden expansion. J Fluid Mech 289:1–27

Khatami R, Levendis YA (2016) An overview of coal rank influence on ignition and combustion phenomena at the particle level. Combust Flame 164:22–34

Kuang M, Li ZQ, Liu CL, Zhu QY (2013) Overall evaluation of combustion and NOx emissions for a down-fired 600 MWe supercritical boiler with multiple injection and multiple staging. Environ Sci Technol 47(9):4850–4858

Liu YH, Geier M, Molina A, Shaddix CR (2011) Pulverized coal stream ignition delay under conventional and oxy-fuel combustion conditions. Int J Greenh Gas Con 5:S36–S46

Mi J, Nathan GJ (2010) Statistical properties of turbulent free jets issuing from nine differently-shaped nozzles. Flow Turbul Combust 84:583–606

Molina A, Shaddix CR (2007) Ignition and devolatilization of pulverized bituminous coal particles during oxygen/carbon dioxide coal combustion. Proc Combust Inst 31:1905–1912

Obando J, Lezcano C, Amell A (2017) Experimental analysis of the addition and substitution of sub-bituminous pulverized coal in a natural gas premixed flame. Appl Therm Eng 125:232–239

Oh J, Noh D (2013) Lifted flame behavior of a non-premixed oxy-methane jet in a lab-scale slot burner. Fuel 103:862–868

Pedel J, Thornock JN, Smith PJ (2013) Ignition of co-axial turbulent diffusion oxy-coal jet flames: experiments and simulations collaboration. Combust Flame 160:1112–1128

Quinn WR (1992) Turbulent free jet flows issuing from sharp-edged rectangular slots: the influence of slot aspect ratio. Exp Therm Fluid Sci 5(2):203–215

Schadow KC, Wilson KJ, Lee MJ, Gutmark EJ (1987) Enhancement of mixing in reacting fuel-rich plumes issued from elliptical nozzles. J Propuls Power 3(2):145

Shaddix CR, Molina A (2009) Particle imaging of ignition and devolatilization of pulverized coal during oxy-fuel combustion. Proc Combust Inst 32(2):2091–2098

Steer J, Marsh R, Griffiths A, Malmgren A, Riley G (2013) Biomass co-firing trials on a down-fired utility boiler. Energy Convers Manage 66:285–294
Tamadonfar P, Gülder ÖL (2016) Effect of burner diameter on the burning velocity of premixed turbulent flames stabilized on Bunsen-type burners. Exp Therm Fluid Sci 73:42–48
Xu KL, Wu YX, Wang ZN, Yang YM, Zhang H (2016) Experimental study on ignition behavior of pulverized coal particle clouds in a turbulent jet. Fuel 167:218–225

Chapter 26
Numerical Simulation on Pulverized Coal Combustion Chamber with Air Cooling

Yongying Wang, Naiji Wang, and Shi Yang

Abstract To investigate the relationship between the temperature distribution in the air-cooled combustor, flame shape and gas distribution ratio, 3D modeling of 14 MW air-cooled combustor and its corresponding furnace was constructed and a series of numerical simulation cases were conducted in this article. The results show that: At the constant excess air ratio, the average temperature in the combustion chamber and the wall temperature both decrease with the increase of internal secondary air ratio. When the internal secondary air ratio is in the range of 0.3–0.4, the wall temperature is relatively high. The highest wall temperature is 930 K, which appears when the internal secondary air ratio is 0.4. When the internal/external secondary air ratio is 0.5/0.5, the flame fullness is at the highest value, which is an appropriate operating condition.

Keywords Pulverized coal combustion chamber · Air cooling · Flame shape

26.1 Introduction

Possessing the advantages of burning stability and low burn-out rate, the double-cone combustion chamber (Yongying et al. 2012, 2016; Jinhua and Naiji 2016) with independent space is suitable for industrial boilers which start-stop frequently. With the increasing of the combustion chamber volume and number, water cooling protection is no longer applicable to this kind of combustor. Air cooling technology has been widely applied in the combustor without independent space to protect the

---

Y. Wang (✉) · N. Wang · S. Yang
China Coal Research Institute Company of Energy Conservation CCRI, Beijing 100013, China
e-mail: yongying2006@163.com

State Key Laboratory of High Efficient Mining and Clean Utilization of Coal Resources, Beijing 100013, China

National Energy Technology & Equipment Laboratory of Coal Utilization and Emission Control, Beijing 100013, China

J. Lyu and S. Li (eds.), *Clean Coal and Sustainable Energy*,
Environmental Science and Engineering,
https://doi.org/10.1007/978-981-16-1657-0_26

combustor nozzle overheating (Yuejuan and Yongying 2018). Whether this technology can be used in the combustion chamber with independent space. To solve this problem, numerical simulation research on pulverized coal combustion chamber with independent space using air cooling was explored in this article.

In this article, temperature distribution and flame shape characteristics of air cooling combustion chamber were investigated by numerical simulation method (Jianquan et al. 2012; Liwei and Fengrui 2010), illustrated by the example of a 14 MW combustion system.

The conclusions of this article can provide reference for design and operation of combustion chamber with air cooling.

26.2 Model and Parameters

26.2.1 The Structure of Air-Cooled Combustor

The structure of 14 MW air-cooled combustor is displayed in Fig. 26.1. Pulverized coal is delivered into the combustion chamber through the primary air pipe, and reverse spraying as the blocking effect of reflow cap. Combustion-supporting wind in the chamber can be divided into two parts: external secondary air, internal secondary air. The internal secondary air rotates into the combustion chamber through the swirling vanes and forms a high temperature recirculation zone with the promotion of diffusion section in the combustion chamber. Reverse spraying pulverized coal is heated in the recirculation zone, and ignites at a high speed when encountering with the fresh oxygen in the internal secondary air at the blade roots. The burning coal particles, carried by the secondary air, spray out of the combustion chamber rotatingly. External secondary air, contacted with the flame in the contraction section and

Fig. 26.1 The structure of air-cooled combustor

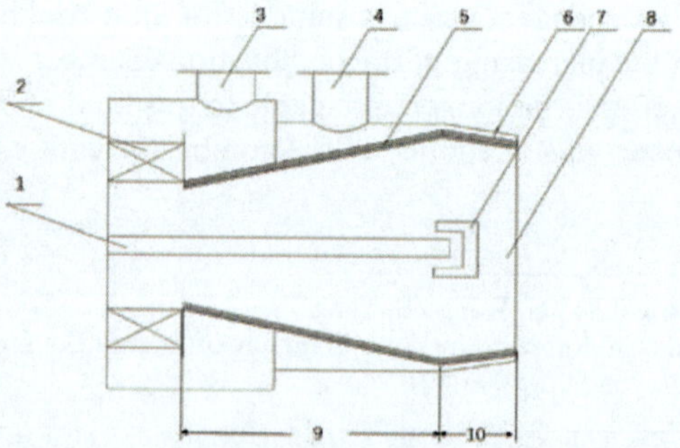

1-primary air pipe 2-swirling vanes 3-internal secondary air inlet 4-external secondary air inlet 5-combustion chamber wall 6-external secondary air outlet 7-reflow cap 8-combustion chamber outlet 9-diffusion section 10-contraction section

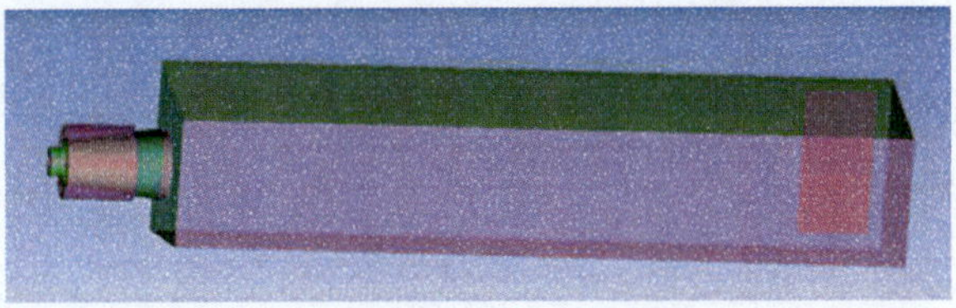

Fig. 26.2 Simulation object

Table 26.1 Internal/external secondary air ratio

Conditione	1	2	3	4	5	6	7
ratio	0.2/0.8	0.3/0.7	0.4/0.6	0.5/0.5	0.6/0.4	0.7/0.3	0.8/0.2

improve the combustion process, is cooling wind. The increase of external secondary air temperature is beneficial to improve the combustion efficiency.

26.2.2 Research Object and Operating Condition

As shown in Fig. 26.2, 14 MW air-cooled combustor and its corresponding furnace are select as the research object in this article. According to the actual situation of the site operation, the operating load of the boiler varies frequently with the change of customers' demands. In lower load condition, the temperature in the combustion chamber is relatively high, so the cooling effect in low load situation is a key object to investigate. In this article, the boiler load rate is controlled at 50%, excess air ratio is kept at 1.4. The combustion effect and the chamber wall temperature at different internal/external secondary air ratio with constant total air volume are investigated.Simulation conditiones are illustrated in Table 26.1.

26.2.3 Mathematical Model

Numerical simulation is conducted by CFD software. Realizable $K-\varepsilon$ turbulence model is utilized to simulate the swirling flow field in the combustor. Standard Wall Functions is used as Near-Wall Treatment. Discrete Ordinates Model is used as radiation model. Double Competition Mode is used to simulate the precipitation process of volatile. Eddy Dissipation Model is used to simulate the combustion process of volatile. Surface reaction rate model is used to simulate the combustion process of coke. The injection of coal particles is simulated by Discrete Phase Model. The number of continuous phase iterations per DPM iteration is 10,The accuracy of the simulation method mentioned above has been verified in literature (Renshan 2009; Siyuan et al. 2014). The material of the combustion chamber wall is set as 304 stainless steel (0Cr18Ni9). Total number of grids is about 2.4 million. Boundary

Table 26.2 Boundary conditions

Boundary conditions	Primary air inlet	Internal secondary air inlet	External secondary air inlet	Furnace outlet
Type	Velocity-inlet	Mass-flow-inlet	Mass-flow-inlet	Pressure-outlet

Table 26.3 Coal quality parameters

Proximate analysis (%)				Elemental analysis (%)					Heating value (MJ/kg)
A_{ar}	M_{ar}	V_{ar}	FC_{ar}	C_{ar}	H_{ar}	O_{ar}	N_{ar}	S_{ar}	$Q_{net.ar}$
8.62	5.00	30.97	55.41	69.43	3.25	13.02	0.85	0.27	27.81

conditions during calculation process are illustrated in Table 26.2. The coal quality parameters are illustrated in Table 26.3.

26.3 Results and Disscussion

26.3.1 *The Combustion Effect and the Chamber Wall Temperature*

The relationship between average temperature in the combustion chamber and internal/external secondary air ratio at constant excess air ratio is shown in Fig. 26.3. Internal secondary air provides the majority of the oxygen required for combustion and kinetic energy of the flame. So increasing the internal secondary air volume promotes the combusion heat release, but also shortens the coal particles' residence time in the combustion chamber. As a result, in theory, the relationship between chamber temperature and the internal secondary air ratio is non-monotonous.

However, from the simulation results in this article, the relationship between them is monotonous. The average temperature in the combustion chamber raises with decreasing of internal secondary air ratio. It means, when internal secondary air ratio is in the range of 0.2–0.8, the cooling effect of internal secondary air, caused by shorten the coal particles' residence time, is in main position.

As shown in Fig. 26.3, the wall temperature of combustion chamber also decreases as the internal secondary air ratio increases. Meanwhile, as shown in Fig. 26.4, the location of high temperature zone on the wall is different in different conditions. When the secondary air ratio is 0.2, the high temperature zone is in the middle of the wall. While it locates in the posterior part of the wall when secondary air ratio is 0.3. The location of the high temperature zone trends to move backwards as internal secondary air ratio increases. When the internal secondary air ratio is in the range

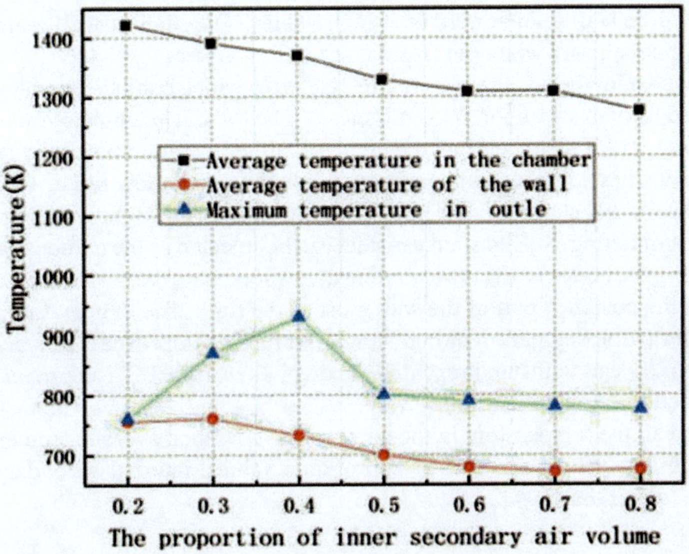

Fig. 26.3 Change of temperature via proportion of internal secondary air

Fig. 26.4 Wall temperature distribution at different internal/extenal secondary air ratios

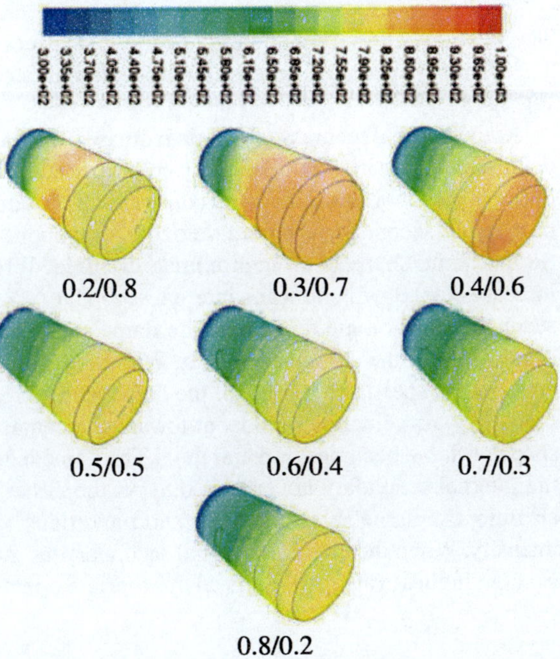

of 0.3–0.4, the wall temperature is at high values. The highest wall temperature is 930 K, which appears when the internal secondary air ratio is 0.4.

The value of highest wall temperature, related to which combustor material type to be selected, is of great important significance. The tolerance temperature of ordinary carbon steel is 693 K, lower than the highest wall temperature obtain by simulating, and it is not suitable for actual condition although it is at a low price. The tolerance temperature of 304 stainless steel is 1073 K, higher than the highest wall temperature obtain by simulating. So 304 stainless steel can be selected as the combustor material. Considering the boiler is always operating at low load rate, the high temperature zone locates in the posterior part of the wall most of the time. The temperature resistance of the contraction segment material should be further improved.310 stainless steel (06Cr25Ni20), can withstand repeated heating below 1000° C, is a proper choice of the contraction segment material.

To keep all the temperature in the combustion chamber, average temperature and highest temperature on the wall at appropriate values, internal secondary air ratio should be higher than 0.4.

26.3.2 *Flame Shape in the Furnace*

The internal/external secondary air ratio also affect the flame shape in the furnace. The flame shapes at different internal/external secondary air ratios are shown in Fig. 26.5.

As the internal secondary air ratio increases, the flame length first increase and then decreases. When internal secondary air ratio is 0.2, The flame speed at the combustor outlet is at a low value. Under the constraint of the constricted section, a great amount of external secondary air has a velocity component toward the center of the flame. So the flame shape is an approximate triangle. When internal secondary air ratio increases to 0.3, which means internal secondary air volume increase while external secondary air volume decreases, the flame shape is an approximate torch with the same width as the combustor outlet. With the further increase of internal secondary air ratio, the speed and rigidity of the flame both increases. At the same time, external secondary air velocity component toward the center of the flame decreases. So the shape of flame becomes thin and long. The flame length is at the highest value when the internal secondary air ratio is 0.5. As the father increase of internal secondary air ratio, the flame shape becomes short and thick, caused by the increase of swirl intensity. When the internal/external secondary air ratio is 0.5/0.5, the flame fullness is at the highest value, which is an appropriate operating condition.

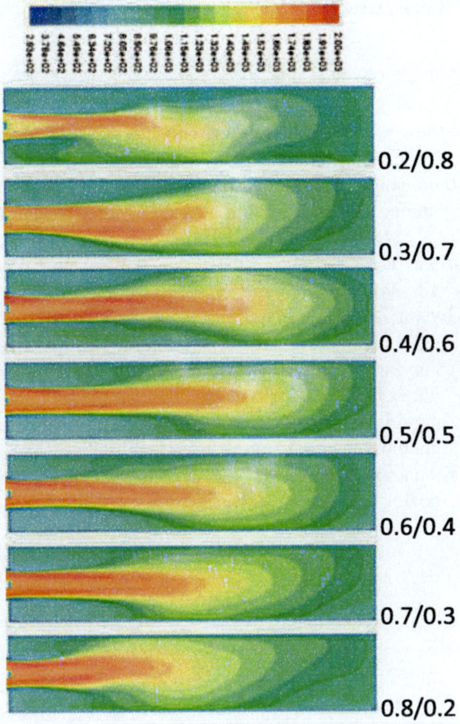

Fig. 26.5 Temperature distribution in the furnace at different operating condition

26.4 Conclusion

(1) At the constant excess air ratio, the average temperature in the combustion chamber and the wall temperature both decrease with the increase of internal secondary air ratio.
(2) When the internal secondary air ratio is in the range of 0.3–0.4, the wall temperature is relatively high. The highest wall temperature is 930 K, which appears when the internal secondary air ratio is 0.4.304 stainless steel can be selected as the wall material.
(3) As the internal secondary air ratio increases, the flame length first increase and then decreases.
(4) When the internal/external secondary air ratio is 0.5/0.5, the flame fullness is at the highest value, which is an appropriate operating condition.

Acknowledgements This work was supported by the Technology Innovation Foundation of China Coal Technology & Engineering Group (2018MA003) and the Technology Innovation Foundation of China Coal Research Institute (2016CX002、2017CX02).

References

Ji R (2009) Numerical simulation of combustion in the industrial pulverized-coal boiler. J China Coal Soc 34(12):1703–1706

Jinhua W, Naiji W (2016) Demonstration project of high efficiency pulverized coal fired industrial boiler for bulk coal clean utilization. Coal Eng 48(9):1–5, 10

Jianquan L, Baomin S, Guangcai Z et al (2012) Numerical simulation and optimization on stable combustion of a 1 000 MW ultra supercritical unit swirl combustion boiler. Proc CSEE 32(8):19–27, 144

Liwei D, Fengrui Li (2010) Mechanism and model realization of pulverized coal combustion simulation and calculation with Fluent. Zhejiang Electr Power 11:31–34

Siyuan J, Yongying W, Jianming Z et al (2014) Numerical simulation on middle volatile coal combustion in reversed injection burner. J China Coal Soc 39(6):1147–1153

Yongying W, Jianming Z, Jinfang Y (2012) Experiment study on cold-condition flow field in double cones fine coal combustion chambe. Clean Coal Technol 18(2):81–84

Yongying W, Shi Y, Xing L (2016) Experiment study of double-cone pulverized coal combustion chamber burning char. Clean Coal Technol 22(3):93–97

Yuejuan B, Yongying W (2018) Research development of typical low NO_X burners for pulverized coal. Coal Quality Technol 2:42–47

Chapter 27
Effect of Inner/Outer Secondary Air Mass Flow Rate on the Airflow Characteristics of the 14-MW Double Cone Burner

Nan Jia, Fang Niu, Pengzhong Liu, Pengtao Wang, and Jianming Zhou

Abstract The double cone burner is the only coal-fired burner in China that utilizes a annular reverse structure and has been industrialized. Cold airflow experiments on a 1:2 scaled model burner was preformed. Testo 425-hot wire anemometer was used to measure the cold aerodynamic characteristics of the double cone burner. Results show that with increasing the inner secondary air ratio, the length of the recirculation zone has not changed; The width of the recirculation zone does not change substantially without the cone, but it tends to generally increase with increaseing of the inner secondary air ratio with the cone; The shape of the recirculation zone is always a ring structure. When the cone is added, the axial dimensionless mean velocity becomes larger than that without the cone, but the jet profile becomes smaller. In both cases with and without the cone, the axial dimensionless mean velocity near the center of the burner is negative due to the reverse injection of the primary air into the cone.

Keywords Double cone burner · Annular reverse · Flow characteristics · Recirculation

27.1 Introduction

In the process of coal utilization, pollutants such as NO_x, SO_x, PM2.5 will be discharged. With the increasingly severe environmental situation in China, the control of pollutants is becoming more and more stringent. China first put forward the concept of "ultra-low emission" in 2012. In the "Government Work Report" of 2015, it is clearly required to "promote the transformation of ultra-low emission of coal-fired power plants", that is, the emission of SO_x not exceed 35 mg/m^3, the emission of

N. Jia · F. Niu (✉) · P. Liu · P. Wang · J. Zhou
China Coal Research Institute Company of Energy Conservation, Beijing 100013, China
e-mail: nf37@163.com

State Key Laboratory of High Efficient Mining and Clean Utilization of Coal Resources, Beijing 100013, China

J. Lyu and S. Li (eds.), *Clean Coal and Sustainable Energy*, Environmental Science and Engineering,
https://doi.org/10.1007/978-981-16-1657-0_27

NO_x should not exceed 50 mg/m^3, and the emission of PM2.5 should not exceed 10 mg/m^3. At present, China's power plant boilers have basically achieved ultra-low emissions, while industrial boilers still have high NO_x emissions, so it is necessary to further study industrial boilers.

Cold-state test is a method to study the flow field in burner. A lot of studies at home and abroad show that the three-dimensional velocity distribution, particle flow distribution and turbulence distribution obtained from the burner model established under the isothermal modelling criterion are very similar to the actual boiler combustion variation trend. Because of the advantages of time-saving, labor-saving, high efficiency and flexibility of cold-state test method, it is widely used.

Zhou et al. (2014) studied the cold flow field of an oxygen-rich burner with petroleum coke powder by means of low-temperature labeling tracer method. It was found that the mixing effect of the flow field of the high-speed primary air jet of the burner was significantly affected by the expander and the bluff body of the central tube. When the expanding angle was 45°, the expanding angle and range of the jet reached the maximum. Hao (2007) studied the cold flow field of multi-channel burner in rotary kiln by using temperature tracer method. It was found that the structure and operation parameters of multi-channel natural burner would affect the flow field distribution and mixing situation in the kiln. For example, with the increase of blade angle, the velocity attenuation accelerates and the jet area decreases, but the mixing intensity of the jet is greater. Ruan (2016) studied the cold flow field of a new oxygen-enriched pulverized coal burner by using temperature tracer method. It was found that with the decrease of primary air velocity, the reflux zone would increase, the initial position of the reflux zone would be moved back, and the gas mixing intensity in the reflux zone would be weakened. Shang (2016) studied the cold flow field of lean coal low-NOx swirl burner through the ribbon experiment. It was found that the main reflux zone of the prototype burner was annular reflux. When Jing et al. (2011) studied the flow field distribution of CFR burner, it was found by using three-dimensional hot film anemometer that with the decrease of primary air ratio, the peak values of axial, radial and tangential velocities increased, and the turbulence increased first and then decreased. The maximum length and diameter of the central reflux zone increased with the decrease of primary air rate. When Li et al. (2016) studied the influence of the angle of the secondary air blade on the flow field characteristics of CFR burner in a 300 MW low volatile coal-fired boiler, it was found that with the angle of the secondary air blade decreasing, the concentration of primary air in the direction of jet first decreased, then increased slightly, while the maximum axial mixing rate increased at the initial position of the burner outlet and then went down. By using the three-dimensional hot-film anemometer, it is found that the three-dimensional velocity increases as the angle of the outer secondary air blade decreases, and the front end of the maximum relative velocity field attenuates rapidly, and the relative reflux rate increases. When Bao (2015) studied the modification of low NO_x combustion system on 600 MW coal-fired thermal power boiler, it was found that the overall direction of airflow rotation was clockwise, the airflow distribution in the furnace was uniform, the track of primary air jet and tangential circle of rotating airflow were clear, there was no deviation and wall brushing phenomenon in the field boiler, which had good aerodynamic field.

At present, the dual-cone burner of the China Coal Research Institute Company of Energy Conservation has been industrialized in China and achieved good demonstration effect of energy saving and emission reduction, which has been widely recognized by the market. However, according to the "ultra-low emission standard", it is still difficult to achieve the emission concentration of NOx below 50 mg/m^3. This paper mainly studies the effects of different ISA/OSA (inner/outer secondary air) ratios on the aerodynamic characteristics of the 14-MW DC (Double cone) burner, and also discusses the influence of the cone on the flow field.

27.2 Experiment Section

27.2.1 *14-MW Double Cone Burner*

Figure 27.1 is the schematic diagram of the 14-MW DC burner with 10 radial vanes in the ISA duct. Pulverized coal carried by primary air flows into the primary air duct, then is reversed into the cone under the influence of reverse jet ring. a large annular recirculation zone of high temperature gases is formed by the swirling ISA jet and the reverse jet within the front cone, The radial vanes angle is 60°. The OSA cools the cone wall and reduces the incidence of coal/ash particle impaction on the

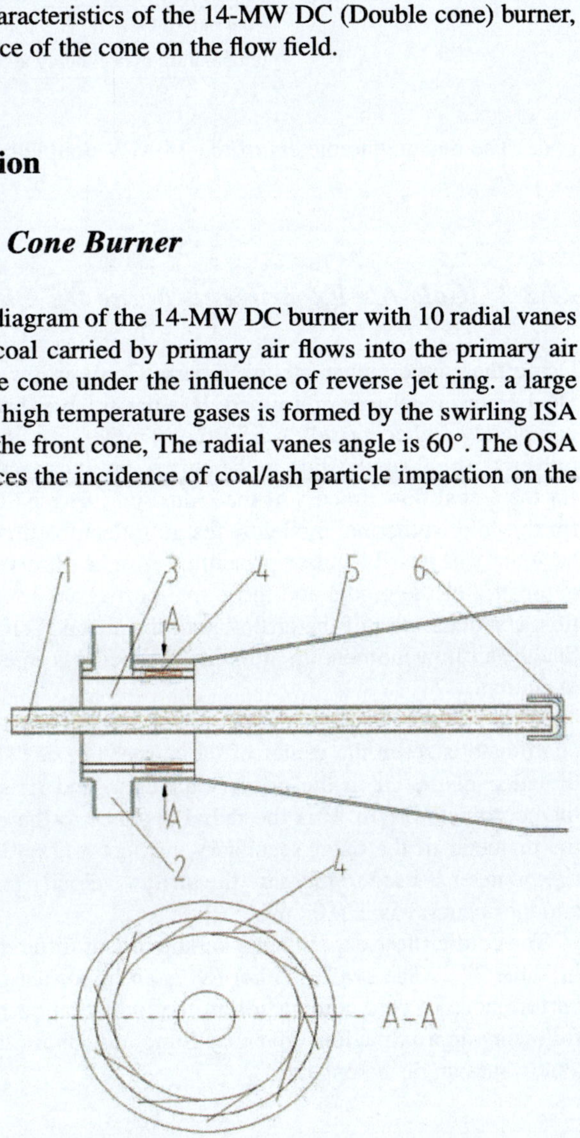

Fig. 27.1 Cross-section of the double cone burner: (1) primary air duct, (2) outer secondary air duct, (3) transitional duct, (4) inner secondary air duct, (5) radial vanes, (6) the cone,and (7) reverse jet ring

Table 27.1 Design parameters of the DC burner

Quantity	14-MW DC burner
Flow area of primary air (m^2)	0.0154
Flow rate of primary air (kg/s^{-1})	0.3432
Temperature of primary air (°C)	100
Flow area of ISA (m^2)	0.0578
Flow rate of ISA (kg/s^{-1})	1.0492
Flow area of OSA (m^2)	0.0694
Flow rate of OSA (kg/s^{-1})	2.098
Temperature of secondary air (°C)	20

cone. The design parameters of the 14-MW double cone burner are listed in Table 27.1.

27.2.2 *Cold Air Experiments of the DC Burner Model*

Using the isothermal modeling theory, Cold airflow experiments on a 1:2 scaled DC burner model was preformed. It is undeniable that cold air experiments cannot completely and accurately reflect the complex physical and chemical processes of pulverized coal combustion in the burner, but it can provide some qualitative results for the actual flow process of the industrial furnace (Ti et al. 2017, 2018). In order for the cold isothermal modeling test to reflect the flow field of the actual burner as accurately as possible, three principles must be observed during modeling. First, the geometric of the model and the actual burner are similar. Second, maintaining the flow characteristics of the airflow into the second self-modeling zone. Last, ensure that the air flow momentum ratio into the model is equal to those in the actual burner operation.

Figure 27.2 is a schematic view of the laboratory test facility. The origin of the coordinate is set in the center of the burner nozzle. The axial direction is the flow direction of the air in the primary air duct, and X is the axial distance from the the coordinate-origin. R is the radial distance to the centerline of the burner. D is the diameter of the outer secondary air ducts (D = 300 mm). Testo 425-hot wire anemometer is used to measure the airflow velocity under different test conditions, and the accuracy is ± 0.03 m/s.

The cold airflow experiments parameters of different ISA/OSA ratios illustrated in Table 27.2. The swirl number (S) is an important parameter to evaluate the the performance of coal combustion in the industrial burner. S is calculated using the following approximations given by Sung and Choi (2016) In the present tests, the axial vanes angle is kept at 60°.

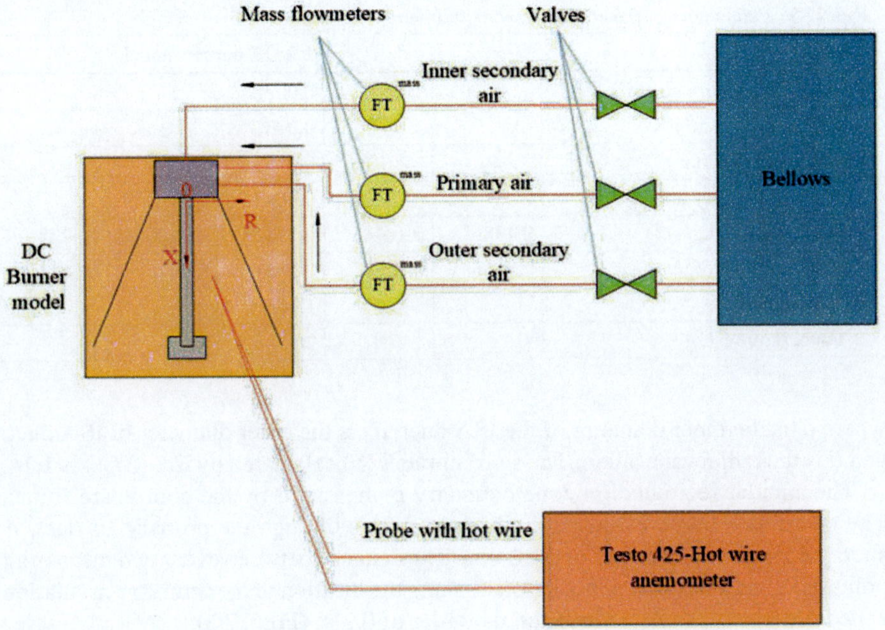

Fig. 27.2 Cold air experiment system

Fig. 27.3 Dimensionless axial velocity for different Reynolds numbers

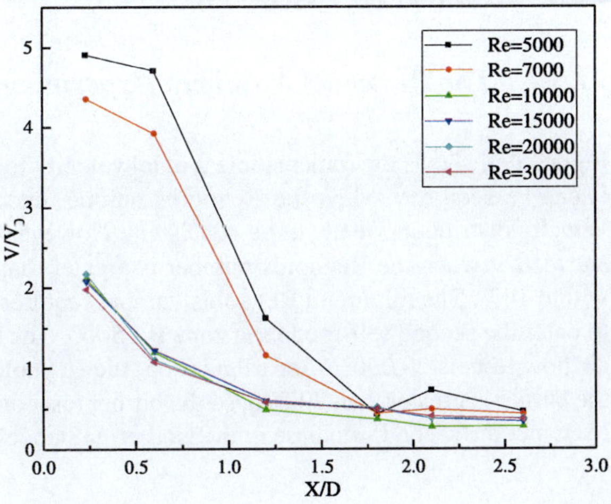

$$S = \frac{2}{3}\left[\frac{1 - (d_i/d_o)^3}{1 - (d_i/d_o)^2}\right]\tan\theta \tag{27.1}$$

Table 27.2 Parameters of the cold airflow experiments

Quantity			1:2 DC burner model		
Flow area of primary air (m^2)			0.0028		
Flow area of ISA (m^2)			0.0133		
Flow area of OSA (m^2)			0.0159		
Flow rate of primary air (kg/s^{-1})			0.0847		
Flow rate of ISA (kg/s^{-1})	0.1386	0.1614	0.1941	0.2426	0.3238
Flow rate of OSA (kg/s^{-1})	0.3466	0.3238	0.2911	0.2426	0.1614
Swirl number			1.114		
Air temperature (°C)			20		

where d_i is the inner diameter of the ISA duct, d_o is the outer diameter of ISA duct, and θ is the radial vane angle. The swirl number (S) calculated by Eq. (27.1) is 1.14.

The annular recirculation zone boundary is measured by the coordinate frame. The coordinate frame is placed on the burner axis through the primary air duct. A short yarn is tied to each grid of the frame. The distance between every two measuring points is 0.3 m. The uncertainty in establishing the location of the central recirculation zone border was estimated to be on the order of 0.3 m (Fig. 27.3).

27.3 Results and Discussion

27.3.1 *The Reynolds Numbers Experiments*

Figure 27.4 shows the dimensionless axial velocity for different Reynolds numbers. It can be seen that when the Reynolds number exceeds 7000, the dimensionless velocity distribution tends to be consistent. However, the error analysis of the data shows that when the Reynolds number is greater than 15,000, the velocity error is within 10%. The minimum Reynolds number required for the 1:2 DC burner model to enter the second self-modeling zone is 15000. The Reynolds number of this cold air flow tests is 20,000. After calculation, the Reynolds number in each air duct of the burner is greater than 20,000, so the burner test conditions can be considered has been met the second principle in the isothermal modeling theory.

27.3.2 *Dimensionless Axial Velocity Distribution*

Figures 27.5 and 27.6 show the flow field characteristics of the different ISA/OSA ratios without the cone. Figure 27.7 show the cold air characteristics of the different ISA/OSA ratios with the cone.

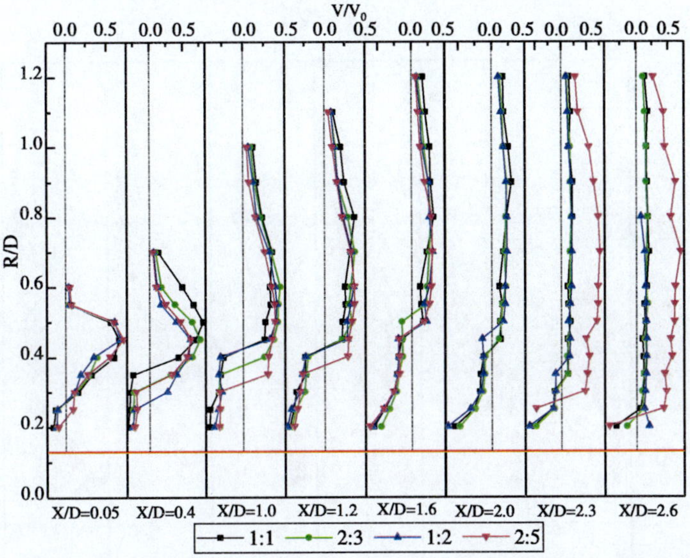

Fig. 27.4 Dimensionless axial mean velocities for different ISA/OSA ratios without the cone

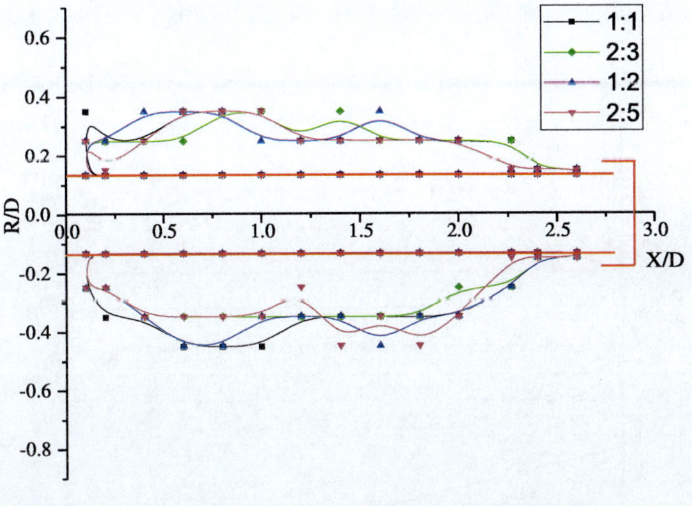

Fig. 27.5 Annular recirculation zone boundary of DC burner without the cone

As can be seen from Fig. 27.4, there is a peak at each cross section caused by the OSA, which increases with increasing OSA ratio. The average axial velocity of each cross section near the primary air duct of the burner is negative, which is caused by the reverse injection of primary air into the cone. The position of negative velocity increases with the increase of ISA ratio, and gradually moves away from the central axis. It also shows that there is a annular recirculation zone in this area. With the

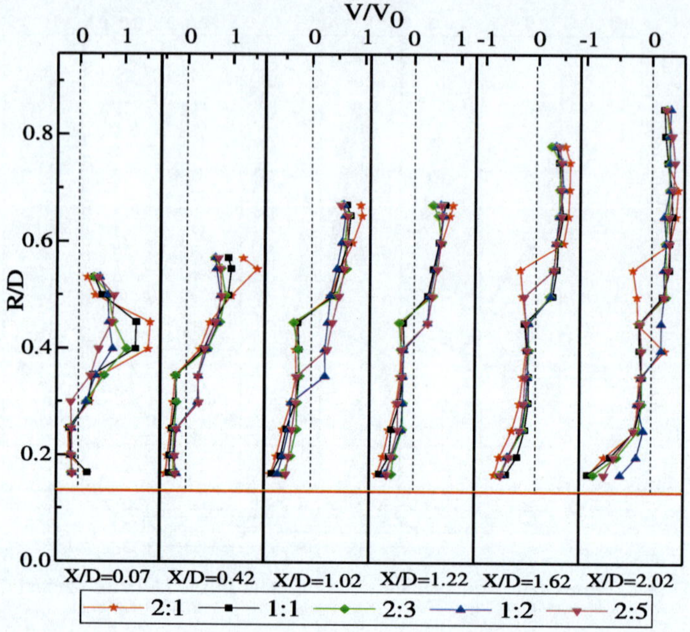

Fig. 27.6 Dimensionless axial mean velocities for different ISA/OSA ratios with the cone

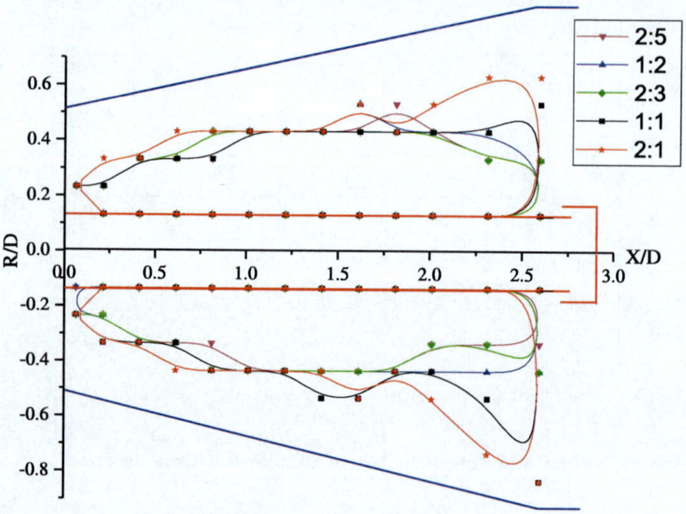

Fig. 27.7 Annular recirculation zone boundary of DC burner with the cone

increase of ISA ratio, the area of the recirculation zone becomes larger gradually. At section X/D = 1.6, the width of recirculation zone reaches the maximum.

Figure 27.5 shows annular recirculation zone boundary of DC burner without the cone. It can be seen that the shape of the recirculation zone does not change with the change of the ratio of the ISA/OSA. It is annular structure, which is caused by the special structure of the primary air duct of the DC burner. With the increase of ISA ratio from 0.1386 to 0.2426 kg/s, the length of the recirculation zone does not change and the width of the recirculation zone tends to increase slightly. This is because the recirculation zone of the DC burner is mainly formed by the Counter-jet flow. When the ratio of ISA/OSA is 1:2, the recirculation zone is symmetrical distribution, and the recirculation zone is asymmetrical under the other three conditions.

Figure 27.6 shows that after adding the cone, the jet profile becomes narrower and the velocity of secondary air increases. A mixing area of high-speed secondary air and low-speed reverse primary air near the burner outlet is formed in the cone. The mixing zone of high-speed reverse primary air and relatively low-speed secondary air formed near the reverse jet ring, so the shape of the recirculation zone in the cone becomes narrow in front and wide in back. When the ratio of ISA is 0.2426 and 0.3238 kg/s, the average axial velocity near the central line of the burner is positive at the cross section X/D = 0.07 near the burner outlet, because the ratio of ISA near the reverse primary air is too high. There is an obvious peak on the cross section X/D = 0.07, which is due to the rapid increase of the velocity in the region due to the addition of OSA, and the rapid decay of the velocity under the action of the cone wall. On cross sections X/D = 0.42, X/D = 1.02 and X/D = 1.22, the average axial dimensionless velocity $V/V0 > 0.5$ near the cone wall. The laminar air layer with a thickness of about 1 cm is formed near the conical wall of each section. It is speculated that the cold air layer will be formed in this area.

Figure 27.7 shows annular recirculation zone boundary of DC burner with the cone. It can be seen that the area of recirculation zone increases after adding cone, especially after section X/D = 1.0. At the same time, the shape of the recirculation zone becomes symmetrical after adding cones. When the ratio of ISA increases from 0.1386 to 0.1941 kg/s, the area of the recirculation zone near the reverse jet ring increases first and then decreases. When the ratio of ISA to OSA is 2:3, the shape of the recirculation zone is completely symmetrical, but when the ratio of ISA increases from 0.2426 to 0.3238 kg/s, the area of the recirculation zone increases rapidly. In the case of the cone and the case without the cone, the shape of the recirculation zone is always annular.

27.4 Conclusions

(1) In the case of cone, the velocity of the flow field in the cone increases as a whole and the flow field profile decreases, and an air layer 1 cm thick is formed on the cone wall. In both cases, there exists a peak value on the cross section of the axial average velocity distribution, and the peak value decreases with

the increase of the ISA ratio; the average axial velocity near the center of the burner is negative, which is caused by the reverse injection of the primary air of the DC burner.

(2) After adding cone, the area of recirculation zone becomes larger and more symmetrical distribution of recirculation zone; the area of recirculation zone first increases, then decreases, and then increases with the increase of ISA ratio, so there exists an ISA ratio which makes the minimum recirculation zone in cone, which is caused by the special structure of DC burner; without cone, the area of recirculation zone basically does not change with the ratio of ISA and OSA. This is because the recirculation zone is mainly formed by reverse jet. However, in both cases, the shape of the recirculation zone does not change with the ratio of the ISA and OSA, and the annular structure is always maintained.

Acknowledgements This work was supported by the China Coal Science and Technology Group Project (No. 2018-TD-MS007 and No. 2016ZYMS007).

References

Bao H (2015) The research on low NOx combustion of 600 MW coal-fired boiler. North China Electr Power Univ

Hao Y (2007)The research on cold model test of multi-channel rotary Kiln Burner and numerical calculation. Zhejiang University

Jing J, Li Z, Zhu Q et al (2011) Influence of primary air ratio on flow and combustion characteristics and NO emissions of a new swirl coal burner. Energy 36(2):1206–1213

Li S, Chen Z, Li X et al (2016) Effect of outer secondary-air vane angle on the flow and combustion characteristics and NO x, formation of the swirl burner in a 300-MW low-volatile coal-fired boiler with deep air staging. J Energy Inst 90(2):239–256

Ruan ZB (2016) Experimental studies of oxygen-fuel Swirl Burner and numerical simulation in the Furnace. Huazhong University of Science & Technology

Shang TK (2016) Experimental study and numerical simulation on a low-NOx semi-anthracite coal Swirl Burner. Tsinghua University

Sung Y, Choi G (2016) Non-intrusive optical diagnostics of co- and counter-swirling flames in a dual swirl pulverized coal combustion burner. Fuel 174:76–88

Ti S, Chen Z, Li Z et al (2017) Effect of outer secondary air vane angles on combustion characteristics and no x, emissions for centrally fuel rich swirl burner in a 600-MWe wall-fired pulverized-coal utility boiler. Appl Therm Eng 2017:S1359431117304696

Ti S, Chen Z, Li Z et al (2018) Influence of primary air cone length on combustion characteristics and NO x, emissions of a swirl burner from a 0.5 MW pulverized coal-fired furnace with air staging. Appl Energy 211:1179–1189

Zhou ZJ, Huang Y, Ji W et al (2014) Research on the cold flow field of petroleum coke powder oxygen-enriched burner by infrared image processing . Chin J Electr Eng 34(17):2794–2801

Chapter 28
Theoretical Analyses on Isolated Particle Ignition of Coal and Biomass

Ye Yuan, Hongpei Gao, Zhenghai Shi, Xianbin Sun, Ping Xiao, Shisen Xu, and Shuiqing Li

Abstract In this paper, an improved transient ignition model is used to study the ignition differences between coal and biomass particles, including ignition time and modes. Using the inflection condition for particle temperature, the heterogeneous ignition time can be determined. Meanwhile, the spatial gas phase temperature distribution can also indicate the homogeneous ignition time. The calculation has evaluated the particle ignition characteristics, with ambient temperature from 1200 to 1800 K, particle diameter from 2000 to 40 μm and oxygen mole fractions in the range of 10–30%. It can be found that heterogeneous ignition always occurs first for the range of coal particles from 40 to 100 μm with temperature increases from 1200 to 1800 K. Compared with small coal particles, 500–2000 μm biomass particle usually ignites homogeneously. The 500 μm biomass particles ignite almost at the same time with coal particles in pc boiler. The changing behavior of ignition temperature is also investigated in this work. The ignition temperatures of small coal particles are all around 1000 K, while ignition temperatures of big biomass particles vary from 650 to 500 K. In the end, the spatial gas temperature and oxygen species distribution are also used to characterize the particle ignition.

Keywords Coal · Biomass · Ignition time · Heterogeneous ignition · Homogeneous ignition

28.1 Introduction

Coal is the most abundant fossil fuel and accounts for about 27% of commercial primary energy usage in the world (Sami et al. 2001a). Due to the large reserves,

Y. Yuan (✉) · H. Gao · Z. Shi · X. Sun · P. Xiao · S. Xu
CFB Department, State Key Laboratory of Coal-Based Clean Energy, Huaneng Clean Energy Research Institute, Changping District, Beijing 102209, China
e-mail: yuanyethu@163.com

S. Li
Key Laboratory for Thermal Science and Power Engineering of Ministry of Education, Department of Energy and Power Engineering, Tsinghua University, Beijing 100084, China

J. Lyu and S. Li (eds.), *Clean Coal and Sustainable Energy*, Environmental Science and Engineering,
https://doi.org/10.1007/978-981-16-1657-0_28

the electricity generation in China is relied heavily on coal. Among the total power generation in China, 70% was provided by coal now and 55% is projected to be provided in 2020 (http://www.bp.com). Hence, coal will still be the dominant fuel in China continuously in the foreseeable future. However, the greenhouse gases (mainly including CO_2) and some air pollutants including SOx, NOx and particulate matter releasing during the coal combustion are significant. Considering the r public health and sustainable development, the Chinese government promulgated the energy policy to control the pollutant emission and reduce CO_2 emission by 60–65% from 2005 to 2030 (http://www.sdpc.gov.cn/).

Larges of technologies have been proposed and adopt for fewer gaseous and PM pollution emission from coal combustion. Among them, co-firing with biomass recognized as a less-expensive method has been widely used. Recent studies in Europe and the United States (Hein and Bemtgen 1998; Hughes and Tillman 1998; Easterly and Burnham 1996; Gold and Tillman 1996) have confirmed that burning biomass with fossil fuels can weaken the impact of power generation on the environment and the economics. The CO_2 net production of burning biomass with fossil fuels is lower than burning fossil fuels only, because the biomass is CO_2-neutral (Logistics et al. 2013). Due to the low contents of sulfur and low nitrogen in biomass, substituting biomass for coal reduces emissions of SO_x and NO_x (Savolainen 2003). Total fuel costs can also be reduced compared with building a biomass fired power plants. In addition, biomass has good ignition characteristics, so co-firing with biomass plays an important role in load peak regulation of thermal power plant. The ignition process is crucial in the boiler design because of its critical influence over flame stability (Simões et al. 2017). In order to obtain the better co-combustion of biomass and coal, this work presents the results of theoretical analyses on isolated particle ignition delay time of coal and biomass.

A brief discussion here is presented on the ignition of an isolated coal particle and the relation of the ignition residence time to the reaction parameters for coal (Yuan et al. 2014). Yuan et al. speculated on the ignition path of pulverized coal particles in Hencken flat flame burner (Yuan et al. 2014). It is directly observed from the cinematography that the high-rank bituminous coal particles ignited homogeneously called homogeneous gas ignition, while low-rank lignite coal particles experienced extensive bulk fragmentation and heterogeneous ignition (Du and Annamalai 1994; Howard and Essenhigh 1965,1967; Essenhigh et al. 1989; Annamalai and Durbetaki 1977; Khatami et al. 2012). Different types of coal have completely distinct ignition behaviors. Generally, ignition delay times decrease with increasing temperature, increasing oxygen concentration and decreasing particle size (Khatami et al. 2012; Levendis et al. 2011; Liu et al. 2011). In short, some works are intended to determine the ignition mode by some optic methods, whereas others applied to obtain the practical ignition time or temperature of pulverized coal particles. Besides that, the temperature or particle diameter effect on the coal ignition time is more important for the pc boiler design and these effects need to be studied systematically.

At present, most of the theoretical and experimental of researches ignition are focused on pulverized coal, while the researches of biomass ignition are rarely reported. Even though the ignition mode and time of coal has been relatively well characterized, some differences may arise when the fuel changing to biomass. Riaza et al. (2014) studied the combustion of four biomass fuels under different atmosphere conditions and found significant differences between biomass fuels and the coals. More than that, in the practical coal and biomass co-firing boiler, due to the limitation of the milling equipment, the diameter of the biomass fuel is much larger than that of coal. Compared with pure coal, the blending fuel of coal and biomass with the high volatile matter content leads to a lower ignition temperatures and more stable combustion. Therefore, there is an urgent need for the ignition characteristic research of big biomass particle, including the ignition time and ignition modes.

In particular, a theory on the transition of ignition between Heterogeneous and Homogeneous modes was proposed by Du and Annamalai (1994), well explaining the effects of ambient temperatures, coal particle sizes, volatile matter contents, kinetic parameters, etc. In this work, this model is used and developed to investigate the ignition differences between coal and biomass particles, including ignition time and modes. Information on the effect of temperature and particle diameter on the ignition of two fuel particles is significant both for understanding how to adjust burners and for computational fluid dynamic (CFD) modeling of the pc burner performance.

28.2 Methods

Consider a cold coal or biomass particle suddenly placed in a hot gas containing oxygen. The particle is heated and when the particle temperature reaches the pyrolysis level, thermal decomposition occurs and releases volatiles into the surrounding ambient. The calculation focuses on the solid phase particle and the gas phase species, which the grid number is 100. The following assumptions have been made for the ignition model.

(1) Spherical symmetry exists.
(2) $\rho D = \text{constant}$ and $Le = 1$.
(3) There is relative motion between the particle and the gas.
(4) The particle temperature is spatially uniform. The particle evolves volatiles uniformly throughout the particle swarm.
(5) The volatile consists of CH (30%), CH_4 (70%). Different contents of CH and CH_4 can meet the different ratio of H/C.
(6) The ideal gas law is applicable.

Particle Mass:

$$\frac{dm_P}{dt} = \frac{dm_V}{dt} + \frac{dm_C}{dt} \tag{28.1}$$

Particle density:

$$(\pi d_p^3)(\frac{d\rho_p}{dt}) = -\frac{dm_V}{dt} \tag{28.2}$$

Particle temperature: (F_B is blow effect coefficient, Fr is Knudsen coefficient)

$$\frac{dT_p}{dt}\rho_p V_p C_p = F_k F_B h A(T_{gas} - T_P) + \varepsilon\sigma A(T_w^4 - T_p^4) + \dot{M}_C h_C - \dot{M}_V h_V \tag{28.3}$$

$$F_B = \frac{(dm_p/dt 2\pi\rho_{gas} D d_p)}{(\exp(dm_p/dt 2\pi\rho_{gas} D d_p) - 1)} \tag{28.4}$$

$$F_r = 1 - \frac{(l_m/d_p)}{(l_m/d_p) + 0.17} \tag{28.5}$$

Pyrolysis kinetics:

$$\frac{dm_V}{dt} = A_V \exp(\frac{-E_V}{RT_P}) m_{Vu} \tag{28.6}$$

(Coal: 1800 K: $A_V = 1.586E15$ 1/s, $E_V = 263{,}000$ J/mol; 1500 K: $A_V = 1.461E8$ 1/s, $E_V = 144{,}500$ J/mol; 1200 K: $A_V = 8.087E4$ 1/s, $E_V = 79{,}400$ J/mol. Biomass: 1800 K: $A_V = 4.69E16$ 1/s, $E_V = 189{,}600$ J/mol; 1500 K: $A_V = 5.69E10$ 1/s, $E_V = 118{,}600$ J/mol; 1200 K: $A_V = 3.9E6$ 1/s, $E_V = 69{,}600$ J/mol (Sami et al. 2001b; Williams et al. 2001).

The volatile, fixed carbon and ash of coal in dry basis are 24%, 58% and 18% respectively. These are 72%, 19% and 9% for biomass fuel.

Char reaction kinetics:

$$\frac{dm_C}{dt} = M_C\rho_W/MW_{O_2} A_C \exp(\frac{-E_C}{RT_P}) Y_{O_2,w}\pi d_p^2 \tag{28.7}$$

(Coal: $A_C = 3020$ m/s, $E_C = 110$ kJ/mol; Biomass: $A_C = 31{,}000$ m/s, $E_C = 95$ kJ/mol (Sami et al. 2001b; Williams et al. 2001)).

Gas phase mass:

$$4\pi r^2\left(\frac{\partial\rho}{\partial t}\right) + \left(\frac{\partial\dot{m}}{\partial r}\right) = 0 \tag{28.8}$$

Gas phase momentum:

$$\left(\frac{\partial\rho U}{\partial t}\right) + U\left(\frac{\partial\rho U}{\partial r}\right) = \nabla P \tag{28.9}$$

Gas phase species:

$$4\pi r^2\rho\left(\frac{\partial Y_k}{\partial t}\right)+\left(\dot{m}\frac{\partial Y_k}{\partial r}\right)-[\frac{\partial}{\partial r}\left(4\pi r^2\rho D\left(\frac{\partial Y_k}{\partial r}\right)\right)]=\dot{w}_m'''4\pi r^2 \quad (28.10)$$

Gas phase energy:

$$4\pi r^2\rho\left(\frac{\partial h_T}{\partial t}\right)+\left(\dot{m}\frac{\partial h_T}{\partial r}\right)-[\frac{\partial}{\partial r}\left(4\pi r^2\rho D\left(\frac{\partial h_T}{\partial r}\right)\right)]=\dot{w}_h'''4\pi r^2 \quad (28.11)$$

Gas phase boundary conditions (The concentration of the gas component at the boundary is determined by the following equation, Sn is 1 when the corresponding reaction generates the component, Sn is −1 when the corresponding reaction consumes the component):

$$\frac{Y_{i,w}}{Y_i}=(\exp(X)-\sum_{k=1}^{K}\frac{\upsilon\rho_i Snd_p(\exp(X)-1)}{XSh\rho_i D})^{-1} \quad (28.12)$$

$$X=\frac{dm_p/dt}{(Sh\rho_{gas}D\pi d_p)} \quad (28.13)$$

$$Sh=2+0.6\mathrm{Re}^{1/2}Sc^{1/3} \quad (28.14)$$

A central difference explicit upwind scheme method is used to solve the above model. Using the inflection condition for particle temperature, the heterogeneous ignition time can be determined (Du and Annamalai 1994). The coal particle is ignited homogeneously, provided the gas-phase temperature of a shell (from r to (r + dr)) is greater than the temperature of both its adjacent shell. The overall ignition delay time and temperature represents the ignition mechanism occurs firstly.

28.3 Results and Discussions

28.3.1 Ignition Time of Coal and Biomass Particles

Fuel particles may experience homogeneous gas-phase ignition, heterogeneous ignition or hetero-homogeneous ignition under different operational conditions. Small coal particles (<100 μm) usually ignites heterogeneously and as the ambient temperature increases, a peak is observed in the gas phase indicating that secondary homogeneous ignition is occurring, which is termed as hetero-homogeneous joint ignition (Yuan et al. 2016). Figure 28.1 shows that the temperature and particle diameter effect on the ignition time of coal and biomass particles. The solid and open symbols denote

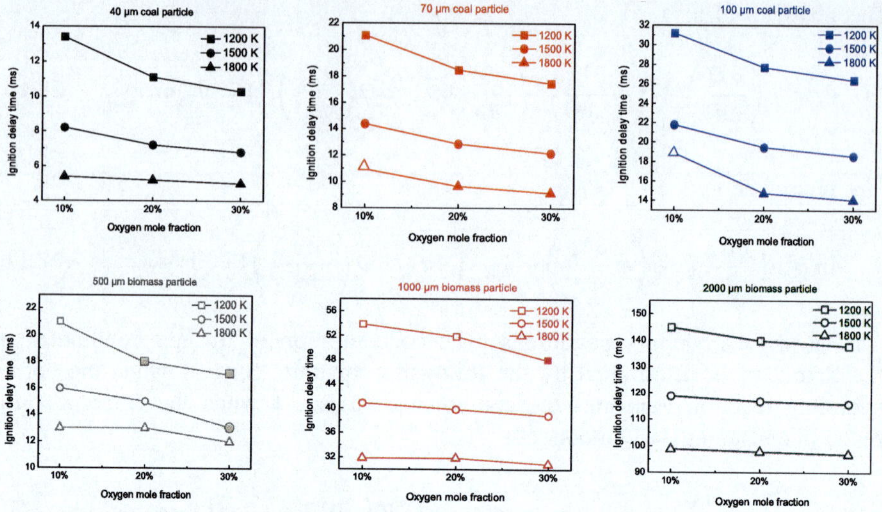

Fig. 28.1 The gas temperature, oxygen fraction and particle diameter effect on the ignition times of two fuel particles

primary heterogeneous and homogeneous ignition. Heterogeneous ignition always occurs first for the range of particles from 40 to 100 μm with temperature increases from 1200 to 1800 K. However, when the oxygen fraction becomes lower (10%) in 1500 and 1800 K, there is a transition from heterogeneous to primary homogenous ignition. It is related to the enhanced volatile releases when the ambient temperature is higher, accompanying the weak particle surface reaction in lower oxygen atmosphere. The industrial biomass particle diameter level (500–2000 μm) is adopted in this calculation. Compared with coal, biomass particle usually ignites homogeneously. It is because that the biomass volatile is higher and the particle-heating rate is apparently reduced as the particle diameter increases. Meanwhile, biomass fuel can also be ignited heterogeneously in small particle size and low ambient temperature (500 μm, 1200 K, 20% and 30% O_2; 500 μm, 1500 K, 30% O_2; 1000 μm, 1200 K, 30% O_2). It is the result of the competition between volatile evolution and heterogeneous surface reaction.

Within the heterogeneous ignition domain, the ignition time decreases as the temperature and oxygen fraction increases. For 40 μm coal particles, the coal ignition times for three O_2 mole fractions and three gas temperatures vary from 13.4 to 5 ms. As for 70 μm coal particles, the ignition times decrease from 21.8 to 6.8 ms. The ignition times of 100 μm coal are a little higher due to high particle heat capacity, which are from 31.2 to 10.3 ms. As the gas temperature increases, the reduction of ignition time becomes indistinct. However, the two cases, of which ignition mode transition occurs (70 μm, 10% O_2, 1800 K; 100 μm, 10% O_2, 1800 K), have a relative long ignition time. The heterogeneous reaction at coal surface is too weak due to the volatile barrier effect and the homogeneous ignition needs more time to trigger gas phase flame. Most biomass particles experience homogeneous ignition. The ignition time also decreases as the temperature and oxygen fraction increases mostly. The effect of oxygen mole fraction is weak to the ignition time. The differences between different oxygen conditions are negligible, especially in higher temperature ambient (1500 and 1800 K). The pyrolysis rate is mainly determined by the particle temperature, so it is easier to form gas phase flame as the ambient temperature increases. As can be found in Fig. 28.1, the ignition time of 500 μm biomass particle is decreased from around 21 to 12 ms. Meanwhile, ignition times of 1000 and 2000 μm biomass particle are separately reduced from about 52 to 31 and 140 to 98 ms. Similarly, when the ignition mode transition happens, the ignition time changing behavior will be affected. For the three conditions which heterogeneous ignition happens (500 μm, 20% O_2, 1200 K; 500 μm, 30% O_2, 1200 K; 500 μm, 30% O_2, 1500 K; 1000 μm, 30% O_2, 1200 K), the ignition times are all a little bit lower than that in the ambient temperature with homogeneous mode. It means that heterogeneous ignition occurs before the gas flame formation. Based on the calculation results, the 500 μm biomass particles ignite almost at the same time with coal particles in pc boiler. The ignition times of 1000 and 2000 μm biomass particles are relatively longer. Considering the short char burnout time of biomass particles, the 2000 μm biomass is applicable in pc boiler operation. What's more, the calculated ignition times are of great significance in pc boiler burner design and arrangement.

28.3.2 Ignition Temperatures of Coal and Biomass Particles

The ignition temperatures is termed as the particle temperature when the primary ignition happens. Figure 28.2 shows the gas temperature, oxygen fraction and particle diameter effect on the ignition temperatures of two fuel particles. For most coal particles with heterogeneous ignition domain, ignition temperature decreases as the oxygen mole fraction increases. The ambient temperature has a minor effect on the ignition temperature, except for 40 μm coal particle in 10% oxygen mole fraction. When the ignition transition occurs, the ignition temperature becomes a little higher, such as from 911 to 1085 K for 100 μm coal particles in 10% O_2 and 1800 K ambient. Being different with coal particles, most of the big biomass particles experience homogeneous ignition. Besides that, the changing behaviors of the two fuel are entirely different. The oxygen mole fraction doesn't change the ignition temperature too much. The ignition temperature increases as the ambient temperature increases. As also for the ignition mode transition cases, the ignition temperatures are a little lower, such as from 550 to 535 K for 500 μm biomass particle in 30% O_2 and 1500 K ambient. The ignition temperatures of small coal particles are all around 1000 K, while ignition temperatures of big biomass particles vary from 650 to 500 K. This is mainly attributed to the higher particle-heating rate for small particles. The particle heat capacity changes with the third power of particle diameter.

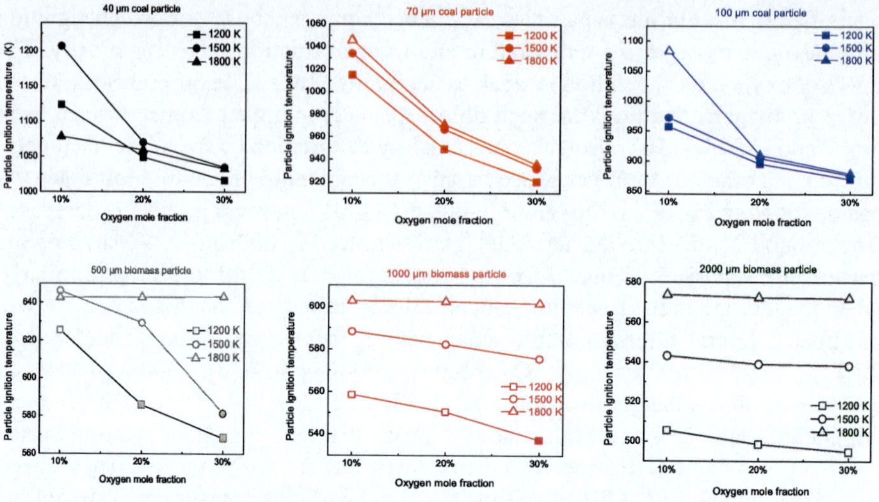

Fig. 28.2 The gas temperature, oxygen fraction and particle diameter effect on the ignition temperatures of two fuel particles

28.3.3 Spatial Temperature and Species Distribution

There are two main reasons for gas temperature variation: the heat transfer from the hot released volatile and the volatile exothermal oxidation. These two process makes different temperature distribution curves. If the temperature change is arose by the volatile matter heat transfer, the temperature distribution is a gradually increased or decreased curve, being equal to ambient temperature far away from the particle. If the change is caused by the reaction, there will be a bulge in the gas temperature distribution curve. The bulge position is a result of gas convection, diffusion and chemical reaction.

Figure 28.3 shows the spatial gas temperature distribution of coal and biomass particles in 1200, 1500 and 1800 K ambients (20% O_2). The diameter of coal and biomass particles are separately 70 and 1000 μm. There is no relevant bulge in 1200 K for coal particles, which verifies the heterogeneous ignition mode. As the temperature increases to 1500 and 1800 K, the bulge in the curve occurs at around 20 and 15 ms. It is bigger than the calculated particle ignition time mentioned above (12.9 and 9.7 ms), therefore it is corresponding to the secondary homogeneous ignition time. Besides that, the bulges almost disappear at 30 and 25 ms in 1500 and 1800 K ambients, which can be defined as the terminal time of volatile combustion. However, the judgment of volatilization combustion by the gas phase temperature is limited, because when the particle temperature is further increased, the reaction of the surrounding gas phase becomes difficult to distinguish from the temperature curve. For 1000 μm biomass particles, the bulge can be found in all the three above conditions. The bulge occurrence times are separately about 50, 40 and 30 ms, which is close to the

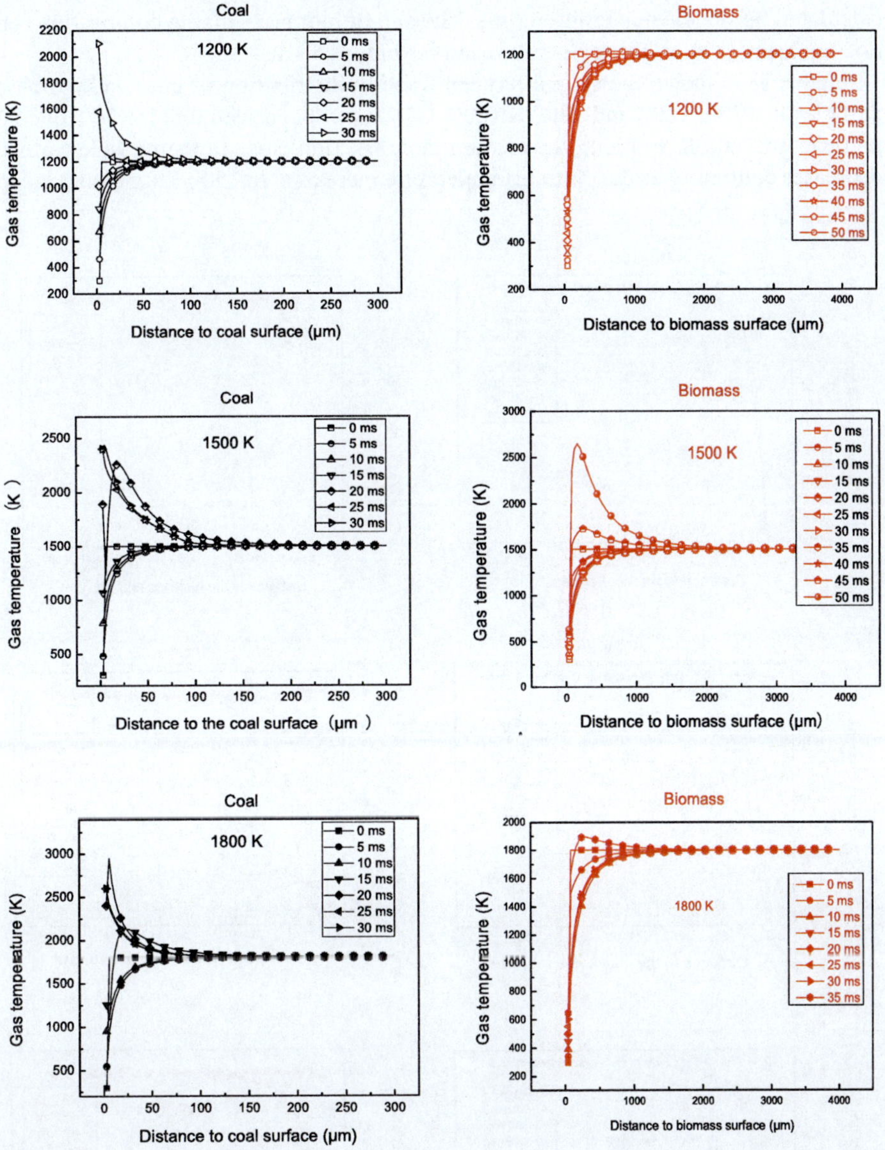

Fig. 28.3 Spatial gas temperature distribution of coal and biomass particles (1200, 1500 and 1800 K)

calculated homogeneous ignition time. The end of biomass volatile combustion can not be observed from the temperature curves before 50 ms.

Figure 28.4 shows the spatial oxygen fraction distribution of coal and biomass particles in 1200, 1500 and 1800 K (20% O_2). It can been seen that for 70 μm coal particles in 1200 K ambient, the oxygen mole fraction close to the particle surface decreases continuity as the particle temperature increases. In 1500 K ambient, due to

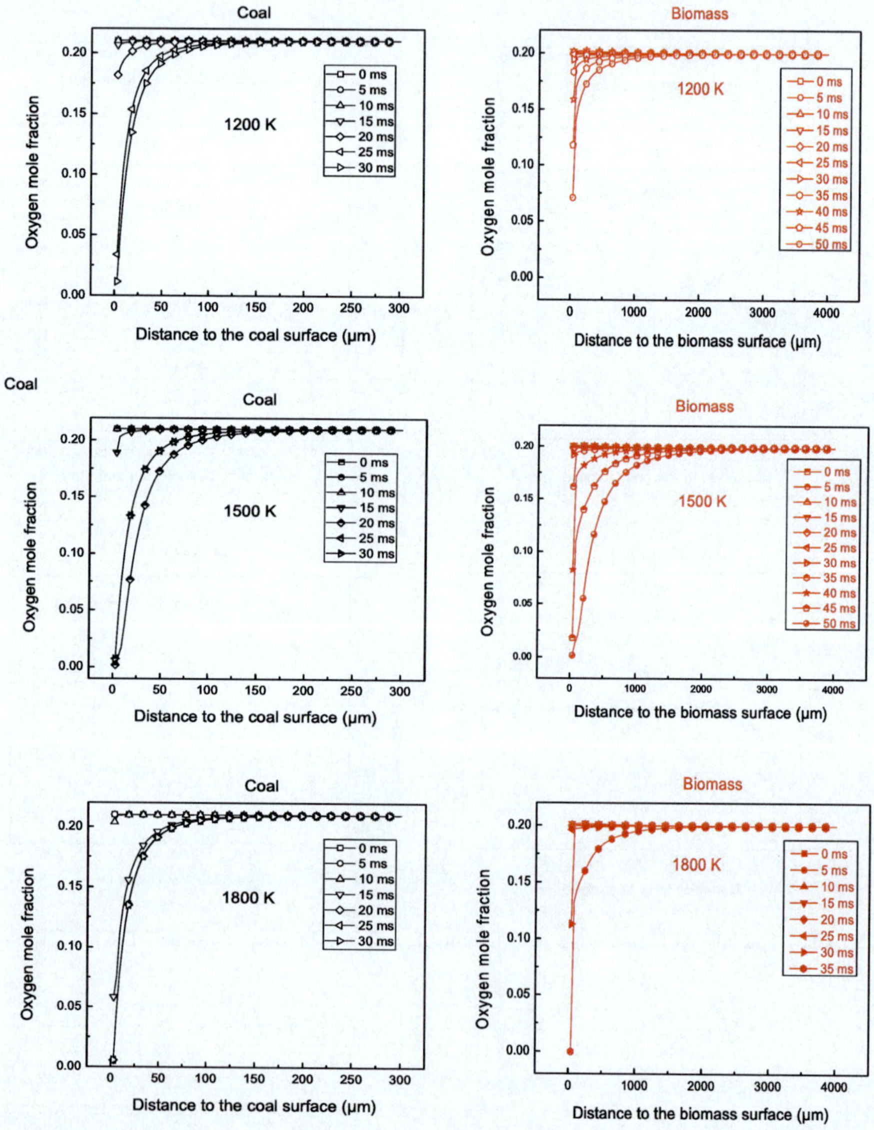

Fig. 28.4 Spatial oxygen fraction distribution of coal and biomass particles (1200, 1500 and 1800 K)

the volatile oxidation consumption, the oxygen fraction is lower in 20 ms, which is corresponding to the second homogeneous ignition. As for the 1800 K ambient, the oxygen is relatively lower in 15 ms. For 1000 μm biomass, the oxygen fractions all reach the lowest value in 50 ms, implying that the volatile oxidation doesn't finish before 50 ms.

28.4 Conclusion

By utilizing an improved transient ignition model, the ignition differences between coal and biomass particles, including ignition time and modes, are investigated in this work. Heterogeneous ignition always occurs first for the range of coal particles from 40 to 100 μm with temperature increases from 1200 to 1800 K. Compared with small coal particles, 500–2000 μm biomass particle usually ignites homogeneously. When the ignition mode transition occurs, the ignition time changing behavior will be altered to a certain degree. Based on the calculation results, the 500 μm biomass particles ignite almost at the same time with coal particles in pc boiler. For most coal particles with heterogeneous ignition domain, ignition temperature decreases as the oxygen mole fraction increases. The ambient temperature has a minor effect on the ignition temperature. On the contrary, the oxygen mole fraction doesn't change the biomass ignition temperature too much. The biomass ignition temperature increases as the ambient temperature increases. In the end, the spatial gas temperature and oxygen species distribution can also providing useful information on particle ignition.

Acknowledgements This work was mainly funded by Huaneng Clean Energy Research Institute Technology Development Fund (TB-16-CERI01). Special thanks are due to Prof. Marshall B. Long at Yale University for useful discussions and Prof. Thomas H. Fletcher at BYU for his coal short course at Tsinghua.

References

Annamalai K, Durbetaki P (1977) Combust Flame 29:193–208
B.P. http://www.bp.com
Du XY, Annamalai K (1994) Combust Flame 97:339–354
Essenhigh RH, Misra MK, Shaw DW (1989) Combust Flame 77:3–30
Easterly JL, Burnham M (1996) Biomass Bioenerg 10:79–92
Gold BA, Tillman DA (1996) Biomass Bioenerg 10:71–78
Hein KRG, Bemtgen JM (1998) Fuel Process Technol 54:159–169
Howard JB, Essenhigh RH (1965) Combust Flame 9:337–339
Howard JB, Essenhigh RH (1967) Proc Combust Inst 6:74–84
Hughes EE, Tillman DA (1998) Fuel Process Technol 54:127–142
Khatami R, Stivers C, Levendis YA (2012) Combust Flame 159:3554–3568
Levendis YA, Joshi K, Khatami R, Sarofim AF (2011) Combust Flame 158:452–465
Liu Y, Geier M, Molina A, Shaddix C (2011) Int J Greenh Gas Con 5:S36–S46

Logistics, Costs and GHG Impacts of Utility-Scale Cofiring with 20% Biomass (2013) Technical Report for U.S. Department of Energy Bioenergy Technology Office
Riaza J, Khatami R, Levendis YA, Álvarez L, Gil MV, Pevida C et al (2014) Biomass Bioenerg 64:162–174
Sami M, Annamalai K, Wooldridge M (2001a) Prog Energy Combust Sci 27:171–214
Sami M, Annamalai K, Wooldrige M (2001b) Prog Energy Combust Sci 27:171–214
Savolainen K (2003) Appl Energy 74:369–381
Simões G, Magalhães D, Rabaçal M et al (2017) Proc Combust Inst 36:2235–2242
The 13th Five-Year Plan for Economic and Social Development of the People's Republic of China. http://www.sdpc.gov.cn/
Williams A, Pourkashanian M, Jones JM (2001) Prog Energy Combust Sci 27:587–610
Yuan Y, Li SQ, Li GD et al (2014) Combust Flame 161:2458–2468
Yuan Y, Li SQ, Zhao FX et al (2016) Fuel 184:1000–1006

Chapter 29
Numerical Simulation on Combustion Characteristics of Co-Firing Biomass Syngas and Coal in a 660 MW Utility Tower Boiler

Shihao Ma, Yan Zhang, Zhengrong Zhu, Lei Deng, and Defu Che

Abstract Utilize the biomass energy and adopt carbon capture and storage technologies could reduce the CO_2 emission effectively. Biomass gasification is an effective ways to use the biomass. Co-firing biomass syngas with pulverized coal under oxy-fuel condition in utility boiler can combine the two advantages to reduce the CO_2 emission. In this paper, a new proposed combustion mechanism was used to investigate the combustion characteristics of co-firing different types biomass syngas and coal in a 660 MW double reheat tower boiler. The influence of syngas composition and oxy-fuel condition on temperature distributions and flue gas components was analyzed when total input heat of the boiler was steady. The results show that the temperature and the emission of the NO and CO_2 decrease in co-firing conditions, the NO mass concentration reduced up to 10.2% in co-firing cases. Biomass syngas with higher H_2 and less CO and CH_4 composition contributes to lower CO_2 emission and higher boiler temperature, which can improve plant performance. The oxy-fuel condition also affects the combustion characteristics to some extent. With the increase of O_2 volume fraction, the temperature increases and the O_2 consumption decreases. Moreover, the NO mass concentration increases in oxy-fuel combustion when co-firing biomass syngas and coal with a mixing ratio of 0.1. This study offers reference for boiler's operation and design optimization for biomass syngas co-firing.

Keywords Biomass syngas · Co-firing · Oxy-fuel combustion · NO emission · Numerical simulation

29.1 Introduction

Co-firing biomass with coal under oxy-fuel condition is regarded as an effective way to decrease the rate of fossil fuel use and reduce the emission of the carbon dioxide (Daood et al. 2013). Compared with biomass directly co-firing, the gasification based

S. Ma · Y. Zhang · Z. Zhu · L. Deng (✉) · D. Che
State Key Laboratory of Multiphase Flow in Power Engineering, School of Energy and Power Engineering, Xi'an Jiaotong University, Xi'an 710049, China
e-mail: leideng@mail.xjtu.edu.cn

J. Lyu and S. Li (eds.), *Clean Coal and Sustainable Energy*,
Environmental Science and Engineering,
https://doi.org/10.1007/978-981-16-1657-0_29

co-firing can not only achieve high-efficiency power generation (Basu et al. 2011), but also have less impact on existing coal-fired boilers and avoid most of the problems such as fouling, corrosion, and deactivation of SCR catalysts (Baxter 2005), which caused by the alkali and alkaline earth metals (AAEMs) contained in the biomass.

Over the last decades, both experimental and numerical studies have been conducted on biomass co-firing to elastrate the characteristics of different biomass co-firing with coal. Zakieh et al. (2014) investigated the impact of different biomass and the ratio of biomass to coal in co-firing plants. The results showed that increasing biomass quality improves plant performance, while NO emissions decrease as the biomass co-firing ratio increases. Dong et al. (2010) conducted a computational fluid dynamics (CFD) study of gasification based biomass co-firing in a 600 MW pulverized coal boiler. It was found that that NO emission was reduced by 50–70%.

But few researches investigated the influence of biomass syngas quality on temperature distribution and flue gas compusation. Moreover, co-firing biomass syngas and coal under oxy-fuel condition has been seldom investigated.

Carbon capture and storage (CCS) based on the oxygen combustion is another efficient way to reduce the emission of the carbon dioxide. Since the background gas changed from air to O_2/CO_2 atmosphere, the temperature, O_2 distributions and NO emissions are significantly different from air condition.

Black (2013) studyed the co-firing of wood and coal under air and oxy-fuel conditions in a 500 MW boiler, the result showed that the introduction of wood resulted in a higher furnace exit temperature. Bhuiyan and Naser (2015a) used a commercial CFD code to investigate the co-firing of biomass in a 550 MW tangentially fired furnace. They found the temperature decreased in oxy-fuel condition. Arias et al. (2008) analysed the effect of biomass co-firing with coal using recycled flue gas and suggested that better ignition could be achieved when O_2 fraction was 30% or higher. Unfortunately, only few researchers have studied the combustion characteristics of co-firing coal and biomass syngas under oxy-fuel condition.

In this study, a numerical study is conducted to simulate the co-firing of biomass syngas and coal in an existing 660 MW tower boiler. A gas phase combustion mechanism is implemented into the CFD code to simulate the co-firing of biomass syngas. Three types of biomass syngas and four oxy-fuel combustion cases are considered to investigate the influence of biomass syngas co-firing on the temperature, O_2 distributions and NO emissions of the boiler.

29.2 Methodology

29.2.1 Furnace and Burner Configuration

In this paper, a 660 MW double reheat tower boiler was considered for the numerical study. The height, width and depth of the boiler are 97,500, 18,150 and 18,150 mm,

Fig. 29.1 Schematic of the boiler

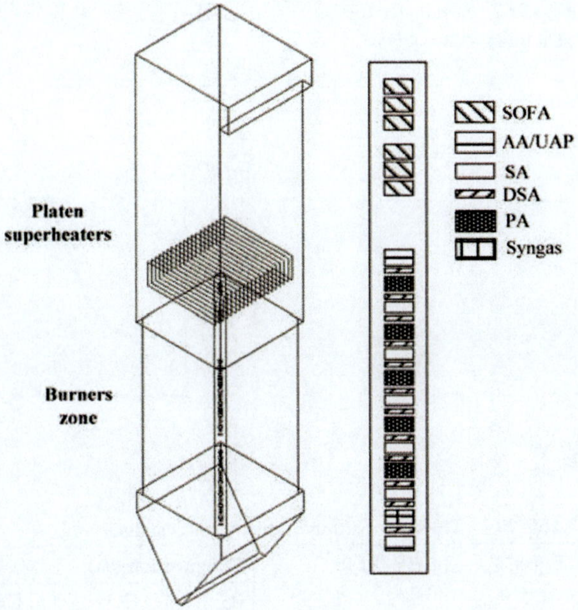

respectively. The boiler geometry, the positions of burners and separated overfire air (SOFA) nozzles are showed in the Fig. 29.1

29.2.2 Fuel Properties and Cases Conditions

Biomass syngas was injected through the lowest nozzle of the burner to substitute the use of the coal. Three different sets of mesh systems were selected for the grid independence test. The total number of cells for the system is 1565078, 1,880,167 and 2,241,660 respectively. Figure 29.2 shows the results for the grid independence test. 1,880,167 cells were used in the simulation zone.

Air is used as the gasification agent to gasify three different kinds of biomass wood, straw and palm, the obtained syngas composition showes in Table 29.1. The properties of the coal showes in Table 29.2.

To investigate the influence of different types of biomass syngas co-firing with coal on temperature, O_2 distribution and NO emission, the co-firing ratio kept at 10% when total input heat of the boiler was steady. In addition, the woody biomass syngas was choosed to investigate the co-combustion characteristics under the oxy-fuel condition, five working conditions were considered: one air-firing condition and four oxy-firing conditions (21, 24, 27, and 30% O_2 fraction). Case named 0.1–0.21 means the co-firing ratio is 10% and the O_2 fraction is 21%.

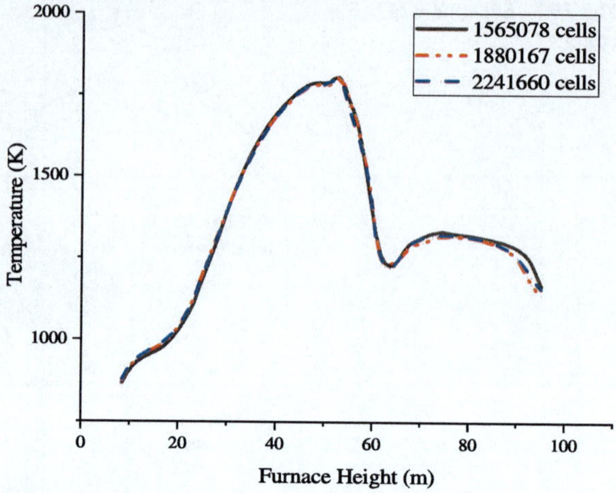

Fig. 29.2 Results for the grid independence test

Table 29.1 The composition of the biomass syngas

Types	LHV (kJ m^{-3})	Composition (%)				
		H_2	CO	CH_4	CO_2	N_2
Woody	5300	17.8	20.3	1.7	8.3	51.9
Straw	3906	5.6	17.89	10.05	10.86	55.6
Palm	4800	9.6	25.3	1.2	8.2	55.7

Table 29.2 Coal properties

Proximate analysis (wt %)	V_{ar}	A_{ar}	FC_{ar}	M_{ar}	
Ultimate analysis (wt %)	22.86	28.25	36.99	11.9	
	C_{ar}	H_{ar}	O_{ar}	N_{ar}	S_{ar}
	47.4	3.24	7.21	0.8	1.2

29.2.3 Computational Modeling

The standard k-ε model was applied to describe the gas-phase turbulence. The radiation heat transfer was calculated by P-1 model. Char combustion processes of coal were simulated by a diffusion/kinetics model. A two-competing-rates model was chosen to describe the devolatilization process of the coal.

Due to the introduction of the biomass syngas, the combustion processes of the blended fuel were different from the coal fired only. A gas phase combustion mechanisms is implemented into the CFD code to simulate the co-firing of biomass syngas and coal. The reaction equations can be written as follows:

$$CH_4 + 0.5O_2 = CO + 2H_2 \tag{29.1}$$

Table 29.3 Reaction mechanisms (units in cm, s, mol, and cal)

Reactions	*A*	β	*E*	Reaction orders
1	7.82×10^{12}	0	30.0×10^3	$[CH_4]^{0.5}[O_2]^{1.25}$
2	3.00×10^{11}	0	30.0×10^3	$[CH_4][H_2O]$
3	2.75×10^{12}	0	20.0×10^3	$[CO]\,[H_2O]$
4	1.21×10^{18}	−1	40.0×10^3	$[H_2]^{0.25}[O_2]^{1.5}$

$$CH_4 + H_2O = CO + 3H_2 \tag{29.2}$$

$$CO + H_2O = CO_2 + H_2 \tag{29.3}$$

$$H_2 + 0.5O_2 = H_2O \tag{29.4}$$

The kinetic rates of the reaction are showed in Table 29.3.

29.3 Results and Disscusion

29.3.1 Validation of the CFD Model

The numerical calculation for the coal fired case was first conducted with the purpose of validating the CFD model. The outlet temperature of the furnace obtained by numerical simulation is 1430 K, which is 4.6% lower than the given calculated temperature. This is acceptable for practical engineering. For oxy-fuel condition, the selected models had been widely applied in the previous oxy-firing investigations (Bhuiyan and Naser 2015b; Chen et al. 2012). These results confirm that the selected CFD model is suitable for the present work.

29.3.2 Effect of the Biomass Syngas Quality

Three cofiring cases and coal fired only case were modeled when the co-firing ratio keep at 10%. The excess air ratio at the outlet of furnace was kept at 1.15. Figure 29.3 shows the temperature distuibution of these cases.

As we can see from Fig. 29.3, there was a pick temperature in the primary zone since the reaction between the fuel and the oxidizer was very intense here, the temperature decreased in the upper furnace due to the heat transfer between the flue gas and the water wall. It is noteworthy that the average temperature in the co-firing case was lower than that for coal fired only. The reason for this phenomenon is that the

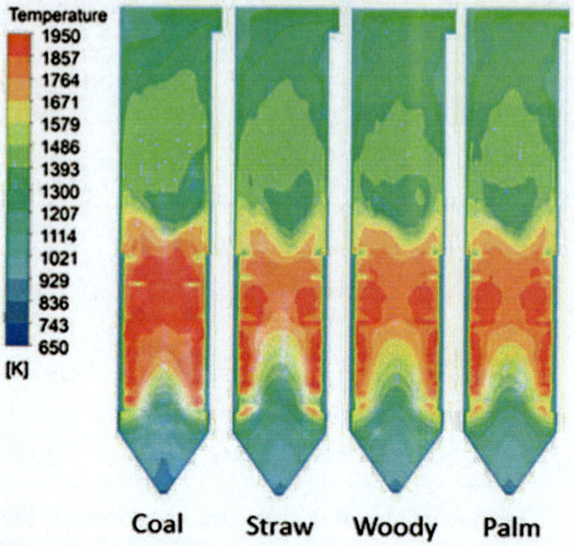

Fig. 29.3 Temperature distribution along the furnace height

calorific value of the biomass syngas is low and the biomass syngas contains a large amount of N_2 and CO_2, which will absorb the heat of furnace.

Compared with the other two kinds of co-firing cases, the co-firing of straw syngas caused a higher exhaust temperature. This may be explained by the lower calorific value of straw syngas, which results in a lower furnace temperature and lower heat transfer between the flue gas and the water wall. The reduction in temperature is also beneficial to control NO_x emissions (Hu et al. 2013).

The O_2 and CO_2 volume fraction along the furnace height in the co-firing cases are show in Fig. 29.4. With the injection of the oxidizer, there were two pick at the burner area and the OFA injection area separately. The O_2 volume fration decreased for fuel combustion will consumpt it in primary zone. For co-firing case, the average O_2 volume fraction in the primary zone is higher than that for coal fired only case. It was attributed to easier ignition and burnout for the biomass syngas. Dong et al. (2010) also find similar phenomenon.

The distribution of CO_2 showed the reverse trend of the O_2, which contributed to the combustion of the fuel. Moreover, compared to the coal fired only case, the CO_2 volume fraction is lower in co-firing cases and the woody syngas co-firing leads to lowest CO_2 volume fraction due to the woody syngas contain more H_2 and less CO and CH_4.

Figure 29.5 depicts the NO mass concentration under different co-firing cases. In the main combustion area, the NO concentration is highest, this caused by fuel combustion. In the OFA injection area, the NO mass concentration increased slightly, this is attribute to injection of the burn out air. Biomass syngas contain high reducing compounds such as hydrogen and carbon monoxide. These compounds will react

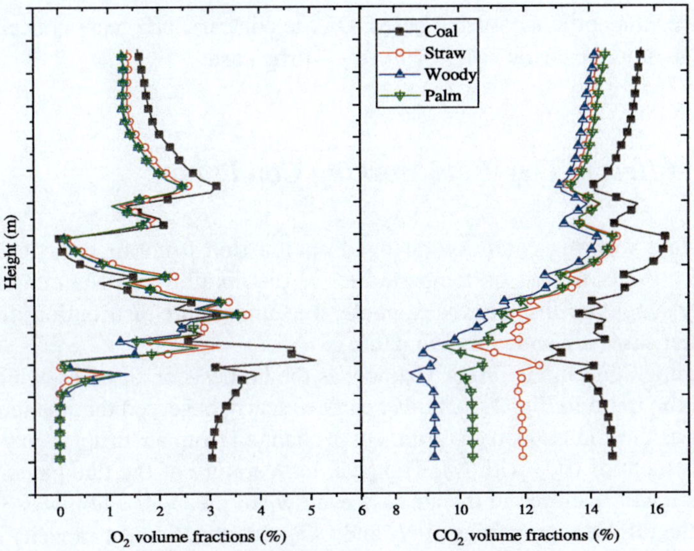

Fig. 29.4 CO_2 and O_2 volume fractions along the furnace height

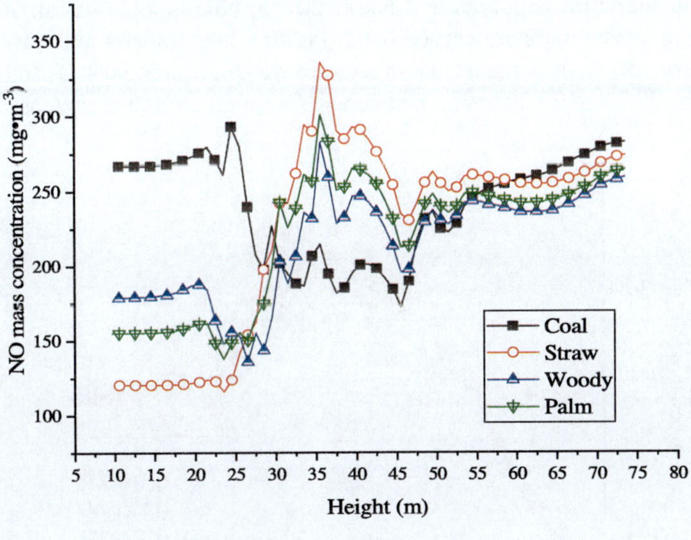

Fig. 29.5 NO mass concentration variations with furnace height

with NO. So, lower NO_x concentration obtained. This phenomenon have also been reported (Dong et al. 2010).

From this point of view, the syngas obtained from the gasification of wood has the highest calorific value and high reducing compounds. Therefore, the emission of

NO is lower than others co-firing cases. On the contrary, NO mass concentration is higher than other co-firing cases in straw co-firng case.

29.3.3 *Effect of Oxy-Fuel Working Condition*

The oxy-fuel working condition changed combustion from air to O_2/CO_2, which will have a direct impact on temperature, O_2 distribution and NO emission. Take the woody syngas co-firing as the example, the temperature distributions for various oxygen fuel cases are shown in Fig. 29.6.

The temperature in the primary zone was the highest for all cases, which agreed well with the trend in Fig. 29.3. It also can be clearly observed that the temperature decreases as the combustion environment is changed from air-firing to oxy-firing by the replacement of CO_2 with N_2, The peak temperature of the flue gas in oxy-fuel cases was lower than that in the air case even when the O_2 fraction was 30%. This is due to the different properties of N_2 and CO_2, the specific heat capacity of CO_2 is higher than N_2 and the gas temperature is harder to be risen by the combustion.

The increase of the O_2 volume fraction corresponds to a reduction of CO_2 amount, which contributed to a higher furnace temperature. Similar findings have been reported in literature (Bhuiyan and Naser 2015a; Nikolopoulos et al. 2011). The increased of the temperature caused better radition heat transfer and lower exhaust gas temperature, so the exhausted temperature was highest at 30% O_2 fraction.

As Figs. 29.7 and 29.8 shows, the average O_2 volume fraction increased with the

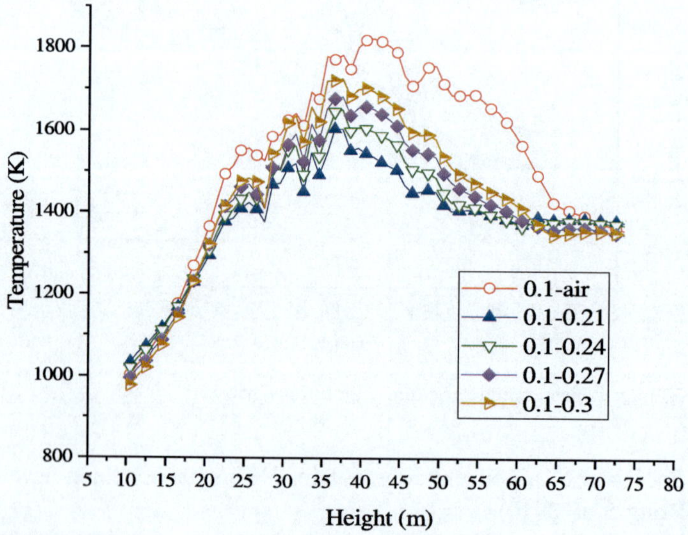

Fig. 29.6 Temperature along the furnace height for various oxy-fuel conditions with 10% woody syngas share

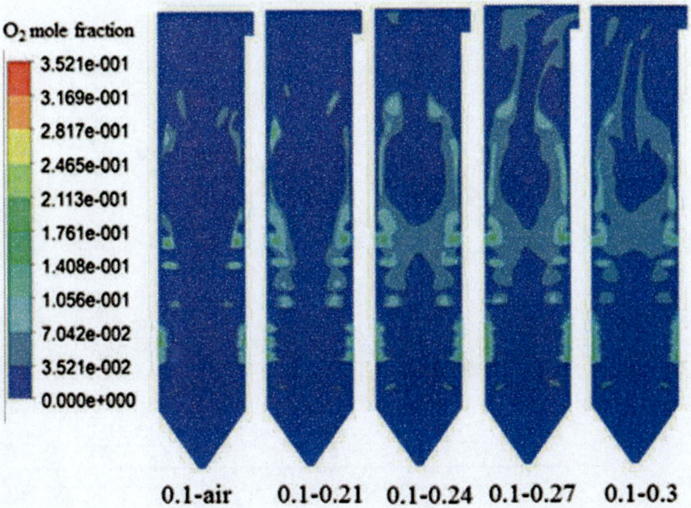

Fig. 29.7 O_2 distribution along the furnace height for various oxy-fuel conditions with 10% woody syngas share

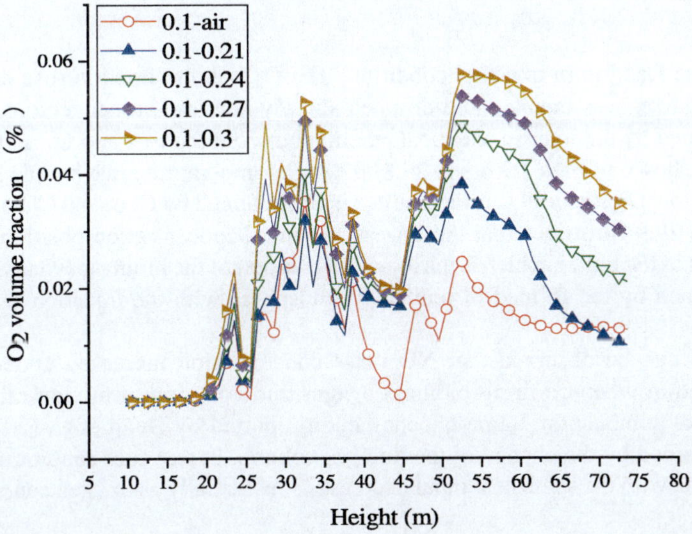

Fig. 29.8 O_2 volume fraction along the furnace height for various oxy-fuel conditions with 10% woody syngas share

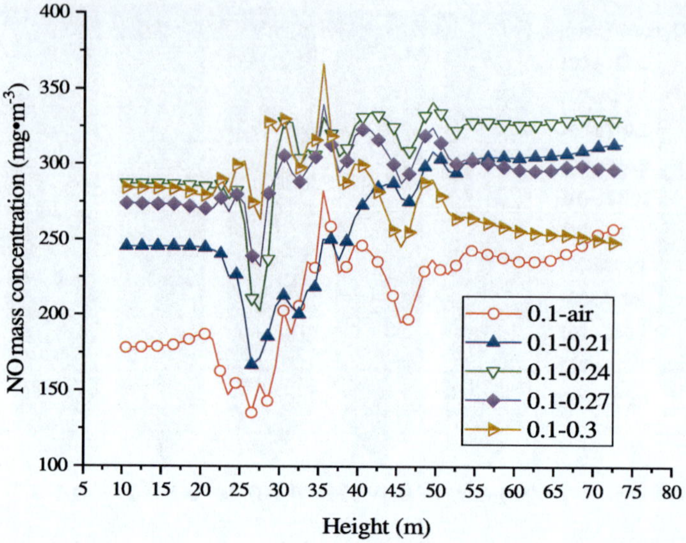

Fig. 29.9 NO mass concentration with furnace height for various oxy-fuel conditions with 10% woody syngas share

O_2 volume fraction in oxy-fuel condition. The O_2 volume fraction rose drastically as the oxidizer was injected and dropped sharply near the burner regions, this can be explained by the intense chemical reaction between the fuel and O_2 in the burner zone. As shown in Figs. 29.6 and 29.8, the high temperature zone coupled with the region of low O_2 fraction, similar results results obtained by Choi and Chang (2009).

Figure 29.9 illustrates that the lowest NO mass concentration position is at the 27 m high of the boiler, which is the injection position of the biomass syngas, this may be explained by the formed of reducing atmosphere with the injection of biomass syngas.

It also can be observed that NO mass concentration increased under oxygen concentration when co-firing biomass syngas and coal with a mixing ratio of 0.1 in oxy-fuel combustion. Same phoenomenon reported by Duan et al. (2015). This may explained by the change of the flue gas volume. In oxy-fuel condition, the flue gas volume with the same heat input decreases substantially as oxygen concentration increases.

In 0.1–0.3 case, the NO mass concentration showed a reverse trend to the O_2 volume fraction, which were in accordance with the studies reported by Wang et al. (2012). This may coused by the reaction between HNO, NH and NCO. The reaction (29.5) promoted with the temperature increased, which will caused the consumption of HNO in reaction (29.6). This reaction reduced the oxidization possibility of HNO and increased the probability of N_2 formation.

$$CO_2 + H = CO + OH \tag{29.5}$$

$$H + NO + M = HNO + M \tag{29.6}$$

29.4 Conclusions

In this study, a computational fluid dynamics (CFD) modelling has been conducted to simulate the performance of co-firing of biomass syngas and coal in a 660 MW tower boiler. The quality of the biomass syngas and oxy-fuel condition are considered. The following conclusions are obtained:

The average temperature in the co-firing cases under air condition was lower than that for coal fired only. The injection of biomass syngas can reduce the emission of the NO and CO_2. The NO mass concentration reduced 4.7%, 10.2% and 8.0% respectively in co-firing cases. Biomass syngas with higher calorific value contributed to lower CO_2 emisssion and higher boiler temperature, which can improve plant performance.

The NO mass concentration under oxy-fuel condition is higher than air combustion condition. With the increasing of O_2 volume fraction, the temperature increased to some extent and the O_2 volume fraction increased at furnace outlet.

Acknowledgements This work was supported by the Shaanxi Province Nature Science Foundation (2019JM-277).

References

Arias B, Pevida C, Rubiera F et al (2008) Effect of biomass blending on coal ignition and burnout during oxy-fuel combustion. Fuel 87(12):2753–2759

Basu P, Butler J, Leon MA (2011) Biomass co-firing options on the emission reduction and electricity generation costs in coal-fired power plants. Renew Energy 36(1):282–288

Baxter L (2005) Biomass-coal co-combustion: opportunity for affordable renewable energy. Fuel 84(10):1295–1302

Bhuiyan AA, Naser J (2015a) CFD modelling of co-firing of biomass with coal under oxy-fuel combustion in a large scale power plant. Fuel 159:150–168

Bhuiyan AA, Naser J (2015b) Computational modelling of co-firing of biomass with coal under oxy-fuel condition in a small scale furnace. Fuel 143:455–466

Black S, Pranzitelli A et al (2013) Effects of firing coal and biomass under oxy-fuel conditions in a power plant boiler using CFD modelling. Fuel 113(2):780–786

Chen L, Yong SZ, Ghoniem AF (2012) Oxy-fuel combustion of pulverized coal: characterization, fundamentals, stabilization and CFD modeling. Prog Energy Combust Sci 38(2):156–214

Choi CR, Chang NK (2009) Numerical investigation on the flow, combustion and NO_x emission characteristics in a 500 MW tangentially fired pulverized-coal boiler. Fuel 88(9):1720–1731

Daood SS, Nimmo et al (2013) NO_x control in coal combustion by combining biomass co-firing, oxygen; enrichment and SNCR. Fuel 105(1):283–292

Dong C, Yang Y, Rui Y et al (2010) Numerical modeling of the gasification based biomass co-firing in a 600 MW pulverized coal boiler. Appl Energy 87(9):2834–2838
Duan L, Duan Y, Zhao C et al (2015) NO emission during co-firing coal and biomass in an oxy-fuel circulating fluidized bed combustor. Fuel 150:8–13
Hu X, Dong C, Lu Q et al (2013) The influence of biomass gasification gas on the reduction of N_2O emissions in a fluidized bed. Energy Sour 35(15):1410–1417
Khorshidi Z, Ho MT, Wiley DE (2014) The impact of biomass quality and quantity on the performance and economics of co-firing plants with and without CO_2 capture. Int J Greenhouse Gas Control 21(2):191–202
Nikolopoulos N, Nikolopoulos A, Karampinis E et al (2011) Numerical investigation of the oxy-fuel combustion in large scale boilers adopting the ECO-Scrub technology. Fuel 90(1):198–214
Wang B, Sun LS, Su S et al (2012) A kinetic study of NO formation during oxy-fuel combustion of pyridine. Appl Energy 92(2):361–368

Chapter 30
Deactivation of V_2O_5–WO_3–TiO_2 SCR Catalyst by HF During Co-Firing Electrolytic Aluminum Waste with Coal

Zhengrong Zhu, Shihao Ma, Yikun Wang, Yan Zhang, Yuxiao Qiu, Lei Deng, and Defu Che

Abstract Fluorine-containing waste from the electrolytic aluminum industry would cause environmental pollutions if the waste is improperly disposed. It is important and urgent to develop a harmless treatment method for the waste. In this paper, co-firing electrolytic aluminum waste with coal was studied. And the deactivation of V_2O_5–WO_3–TiO_2 SCR catalyst by HF was investigated by a lab-scale experimental system. The results show that HF is harmful to the catalyst. The denitrification efficiency of catalyst sample which treated by flue gas with HF is reduced. The denitrification efficiency of sample increases with increasing work temperature and deactivation temperature. When the work temperature is 400 °C, the denitrification efficiency of catalyst sample is higher than 90%. Analysis by temperature programmed desorption of ammonia (NH_3-TPD), Brunauer–Emmett–Teller (BET), X-ray diffraction (XRD), and X-ray photoelectron spectroscopy (XPS) show the definite dispersion of F on the surface of SCR catalysts. HF would lead to decrease specific surface areas, surface acidity, and oxidation–reduction ability. Although the concentrations of O_β (surface adsorbed oxygen) increases owing to HF, which may interact with TiO_2 to destroy the carrier structure. Meanwhile, there is a chemical deactivation for the influence of HF on V_2O_5 and WO_3.

Keywords Deactivation · HF · Electrolytic aluminum waste · Co-firing · V_2O_5–WO_3–TiO_2

Z. Zhu · S. Ma · Y. Zhang · Y. Qiu · L. Deng (✉) · D. Che
State Key Laboratory of Multiphase Flow in Power Engineering, School of Energy and Power Engineering, Xi'an Jiaotong University, Xi'an 710049, China
e-mail: leideng@mail.xjtu.edu.cn

Y. Wang
Xi'an Thermal Power Research Institute Co., Ltd, Xi'an 710032, China

J. Lyu and S. Li (eds.), *Clean Coal and Sustainable Energy*, Environmental Science and Engineering,
https://doi.org/10.1007/978-981-16-1657-0_30

30.1 Introduction

At present, aluminum is an indispensable material for many industries. It is produced by the electrolytic aluminum technology. The electrolytic aluminum industry is with high energy consumption and high pollution emission (Ma and Shang 2016; Jun and Zeng 2014). In particular, the waste, which is produced by the lining of electrolytic aluminum tank, contains harmful components such as fluorides and cyanides. Most of fluorides and cyanides are water-soluble. These toxic substances would bring a great threat to human body and environment (Yaowu and Jianping 2018). The water-soluble F^- in waste discharged from China's electrolytic aluminum industry is about 2000 mg L^{-1} (Ma and Shang 2016). It greatly exceeding the standard requirements ($F^- < 50$ mg L^{-1}, $CN^- < 1$ mg L^{-1}) of the national "Hazardous Waste Identification Standard-Identification of Leaching Toxicity" (Thy and Jenkins 2010). Therefore, it is great important to achieve the harmless treatment of electrolytic aluminum waste.

Many methods have been applied for treatment of electrolytic aluminum waste, worldwide and in China. Most of them are still in the research and development stage. Fire treatment and wet treatment are more mature. And these two methods are widely used in industrial applications (Wang and Liu 2012). However, fluorine-containing exhaust gas produced during the application of fire treatment technology, which would cause serious corrosion on the equipments. During the wet treatment process, toxic and flammable gases such as HCN, H_2, NH_3, and CH_4 may be released, which may cause safety hazards. So it has been studied that the electrolytic aluminum waste could be co-fired with coal in the coal-fired power plant in this work. However, HF is also present in the coal-fired flue gas (Ukawa and Okino 1992). The amount of F contained in the electrolytic aluminum waste is much higher than that in the coal. Co-firing the waste with coal also produces much higher HF content than firing coal alone. The SCR catalyst is an important part of tail gas treatment and plays an important role in reduction of NO_x. Therefore, investigation the effect of HF on the performance of SCR catalyst is important basis for judging whether it is feasible to use co-firing technology to treat the electrolytic aluminum waste.

There are few reports on the effect of HF on SCR catalysts in recent years (Yang and Guo 2016). Several researchers (Yang and Guo 2016; Yu and Liu 2014; Jin and Shen 2017; Rui and Qin 2014; Deng and Zhong 2016; Xu and Ao 2008; Du and Qiu 2019; Zhang and Mao 2017; Wang and Wang 2015) have studied the effect of F on the catalyst. However, a consensus has not yet been reached. Rui and Qin (2014) and Deng and Zhong (2016) reported that the doping of F enhanced the activity of CeO_2-TiO_2 catalyst. It showed that the doping of F slightly increases the oxygen vacancies, specific surface area, and promotes the formation of surface hydroxyl groups. Yang and Guo (2016) pointed out that F heightened catalytic activity of CeO_2 catalyst. Xu and Ao (2008) and Du and Qiu (2019) also verified that F increased the photocatalytic activity of TiO_2. On the contrary, Wang and Wang (2015) found that the addition of F is adverse on the Mn/TiO_2 catalyst. Both F and Cl belong to halogen species. They have many similar properties. Studying influence of Cl on the catalyst acidity is of great significance for guiding how to study the influence of F on the catalyst acidity. Lisi and Lasorella (2004) have considered that HCl promoted the recovery of surface acidity of V_2O_5–WO_3–TiO_2 SCR catalyst. On the other hand, it also led to the reduction of V_2O_5. But Hou and Cai (2014) noted that HCl promoted the catalytic activity of V_2O_5/AC catalyst, which might be owing to the deposition of NH_4Cl on the surface of catalyst. It increased the surface acidity of the V_2O_5/AC catalyst, thereby increasing the SCR activity. However, the adverse effect of HCl on V_2O_5–WO_3–TiO_2 catalyst have been reported by Liu and Guo (2018). It may be caused by the chlorine deposition. In addition, Cl-containing compounds are also commonly used to modify catalysts. Chen and Jia (2018) have prepared $CuCl_2$–SCR catalyst by modified impregnation method. And they found the activity was significantly improved because the presence of HCl.

It is noteworthy that previous investigations have shown that the beneficial or harmful influence of F or Cl on the SCR catalyst acidity. It is affected by various factors such as the types of catalyst. However, the effect of HF on activity of V_2O_5–WO_3–TiO_2 catalyst has not been reported. At present, the catalyst is the most widely used as commercial denitration catalysts. So it is of great significance to explore the deactivation of HF on the catalyst. In this study, a lab-scale experimental system has been built, which is used to investigate the influence of HF on catalyst.

30.2 Experiment

30.2.1 Catalyst Preparation and Deactivation of Catalyst Sample

In this study, the honeycomb type V_2O_5–WO_3–TiO_2 SCR catalyst (Chengdu dongfang KWH Environmental Protection Catalysts Company, China) was studied. To facilitate experimentation with a quartz tube reactor, all catalyst samples were cut into cylinders with a diameter of 30 mm and a height of 60 mm.

The prepared catalyst samples were used in deactivation experiment, which was carried out in a fixed bed quartz tube reactor system. The system schematic is shown in Fig. 30.1. The experimental system is mainly composed of three parts: gas supply system, electric heating system and quartz tube reactor system. The inner diameter of quartz tube is 32 mm, the length is 1200 mm, and the maximum operating temperature is 1600 °C. In the experiment, in order to study only the influence of HF on the catalyst acidity, and to exclude the influence of other gases, only N_2 was used as the carrier gas. In other words, the catalyst would been exposed to the hot flue gas containing certain concentration of HF in order to accelerate deactivation of catalyst.

HF was derived from reaction of concentrated sulfuric acid and calcium fluoride at 250 °C, which is a common preparation method in a lab. In the experiment, 3.9 g of calcium fluoride and 4 mL of concentrated sulfuric acid are weighed, and the temperature was set. The position of quartz tube at 250 °C needs to be calibrated before the experiment.). Then the experimental process lasted for 2 h, the gas produced

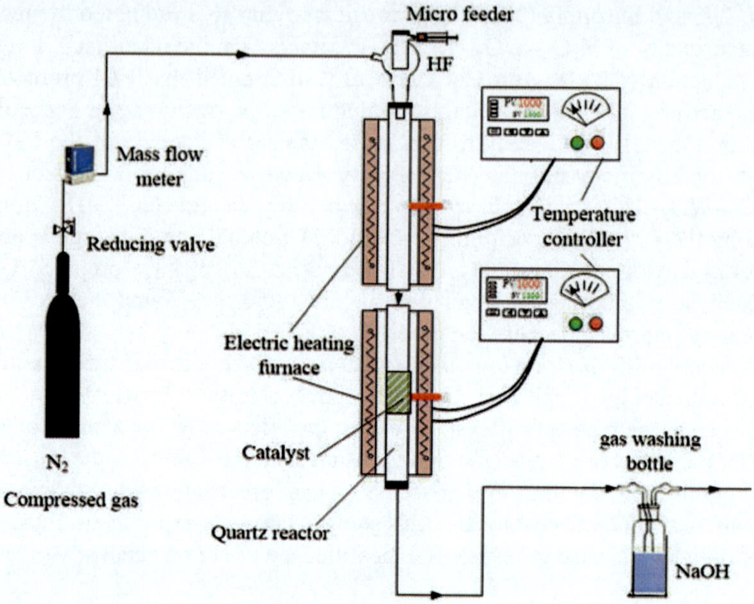

Fig. 30.1 Experiment system of catalyst deactivation

by the reaction through the catalyst, and finally the tail gas is absorbed by NaOH solution and then discharged outside. The samples were obtained by the experiment are respectively recorded as fresh, 300 °C-Used, 350 °C-Used, 400 °C-Used.

30.2.2 Catalytic Activity Tests

The catalytic activity tests can be carried out after the catalyst samples is prepared. The tests experimental system is shown in Fig. 30.2. It is mainly composed of four parts: gas supply system, electric heating system, quartz tube reaction system and flue gas analysis system. The gases in the simulated SCR reactor include NH_3, NO, O_2, and N_2. It is intended to mix the three gases NO, O_2 and N_2 in a certain volume ratio to simulate the flue gas produced during the actual boiler combustion, and then send the simulated flue gas and NH_3 into the quartz tube reactor. The reaction takes place under the catalysis of the SCR catalyst. The temperature controller is used to control the reaction temperature to simulate the operating conditions of actual boiler's SCR denitration system. The gas concentration of the quartz tube reactor inlet and outlet is measured by a Fourier infrared flue gas analyzer (Gasmet DX4000, Temet Instruments Oy, Finland).

For the catalytic activity tests, the gas ratio was presented in Table 30.1, then passed into a quartz tube reactor equipped with a SCR catalyst to measured concentration of

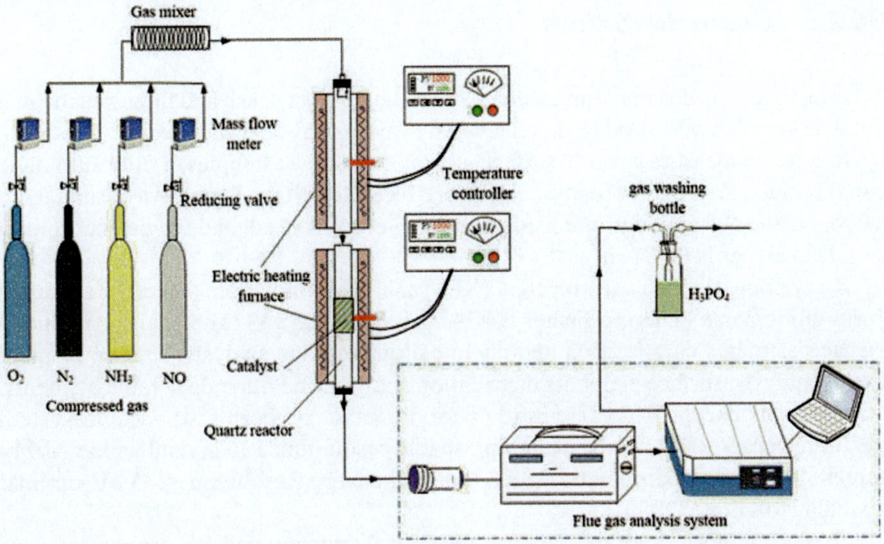

Fig. 30.2 Catalytic activity tests experimental system

Table 30.1 Measurement parameters of catalytic activity tests system

Treatment temperature (°C)	N_2 (L min^{-1})	NO (μL L^{-1})	NH_3 (μL L^{-1})	O_2 (%)	N_2
300	2.8	600	600	3	Balance gas
350					
400					

NO_x at different temperatures. In order to ensure the accuracy of the measurement, at least three times were measured under each set conditions, and average values were obtained as the concentration of NO_x in the outlet gas and that in the inlet gas to further calculate the efficiency of each catalyst sample.

The denitration efficiency of the SCR catalyst was calculated from concentration of NO_x at outlet and inlet of the reactor. And the change in the denitration activity of the catalyst samples was defined as X_{deNO_x}. It can be calculated by as:

$$X_{deNO_x} = 1 - \frac{C_{NO}^{out} + C_{NO_2}^{out}}{C_{NO}^{in} + C_{NO_2}^{in}} \tag{30.1}$$

Among of them,C_{NO}^{in}, $C_{NO_2}^{in}$, C_{NO}^{out} and $C_{NO_2}^{out}$ are the concentrations (μL L^{-1}) of NO and NO_2 in the inlet and outlet gases, respectively.

30.2.3 Characterizations

To investigate the deactivation mechanism of the catalyst, fresh and three deactivated catalyst samples were used to characterize by various physical and chemical methods.

In order to measure specific surface area of the catalyst samples, a fully automatic rapid surface area and porosity analyzer (BELSORP-Max, Japan) was used, and a 1000, 10 and 0.1 torr high-precision pressure sensor is used, and the detection limit of this analyzer is 0.001 $m^2\ g^{-1}$.

X-ray photoelectron spectroscopy (XPS) analyzer was used to detect the chemical form of elements (Thermo Fisher ESCALAB Xi + , USA). The analysis included surface elements composition and chemical quantitative analysis, surface element depth analysis, surface elements distribution analysis and other detectable elements. All elements except H and He in the periodic table can been test; vacuum system performance: 5×10^{-10} mbar, imaging spatial resolution: 1 μm; dual anode Al/Mg target: 400 W; monochromatic Al Kα, optimal energy Resolution ≤ 45 eV, optimal spatial resolution ≤20 ms.

Phases composition of the catalyst, the crystal structure and other parameters are measured by X-ray diffractometer (XRD-6100, Japan), the maximum power of X-ray output is 3 kW, the Cu target, the light tube power is 2.2 kW or more, and the vertical type θ/2θ is measured, an angle reproducibility is 0.0001°, and a minimum controllable step size is no more than 0.0001°.

The temperature-desorption of NH_3-TPD was carried out on a fully automatic chemical adsorption instrument (Auto Chem TM II 2920, USA), and the signal of NH_3 was detected by a thermal conductivity detector (TCD). Furnace: Temperature range: RT–1100 °C; Detector: Reference measurement dual gas path parallel heat conduction cell design, reserved interface for mass spectrometry; degassing temperature range: 40–450 °C, with 12 gas interfaces; Typical chemical adsorbate: hydrogen, oxygen, etc.

30.3 Resuits and Discussion

30.3.1 SCR Catalytic Activity

The dependence of denitration efficiency on V_2O_5–WO_3–TiO_2 SCR catalyst can been shown in Fig. 30.3. The fresh catalyst maintained over 92% at the work temperature (300–400 °C). At the same work temperature, the catalysts (300 °C-Used, 350 °C-Used, 400 °C-Used.), which were treated by exhaust gas with HF. The denitration efficiency of the above three cataly samples is lower than that of the fresh catalyst. So the three cataly samples are called deactivated catalysts. Whether deactivated catalysts or fresh catalysts, their denitration efficiency increased with the increase of work temperature. Moreover, the higher HF deactivation temperature, the lower the degree of deactivation. When the work temperature is 400 °C, it reaches more than

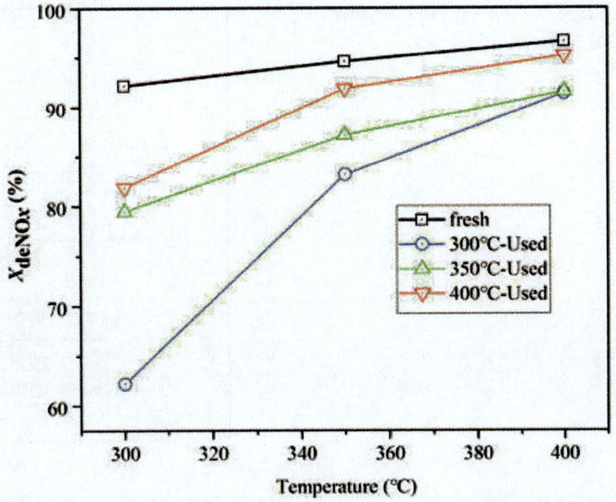

Fig. 30.3 Effect of catalysts and work temperature on denitration efficiency

90%. Therefore, in the tail gas treatment of the power plant, if work temperature of the catalyst is kept at 400 °C, then it will be greatly reduced for the effect of HF on the denitration efficiency of the catalyst.

30.3.2 NH_3-TPD Analysis

It has been proved to be crucial for the SCR reaction to adsorb NH_3 on the catalyst surface (Topsøe 1994), the adsorption of NH_3 is largely dependent on surface acidity. So the surface acidity of the catalyst samples would be analyzed. It can be observed in Fig. 30.4 that the NH_3 desorption laws of these four catalyst samples. Desorption curve of fresh catalyst is consistent basically with the results of Peña and Uphade (2004).

It is obvious that there is a desorption peak in the profile of each catalyst in the temperature range of 100–350 °C, which is attributed to the weak acid sites that is the Brønsted acid sites (Topsøe 1994; Mhamdi and Khaddar-Zine 2009). The desorption peak of fresh catalyst is the widest, so it illustrates the highest selective catalytic activity. The desorption peaks in the temperature range of 350–500 °C, which is the Lewis acid sites (Yang and Guo 2016; Guo and Zhou 2013), but its impact on the SCR reaction is not clear enough (Liu and Guo 2018). Thus the enhancement of NH_3 adsorption capacity of V_2O_5–WO_3–TiO_2 SCR catalyst by HF treatment should be an important factor for its high SCR activity.

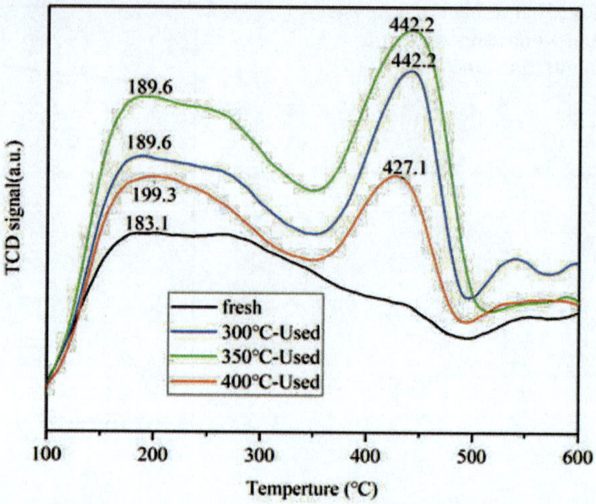

Fig. 30.4 NH_3-TPD profiles of different catalysts

30.3.3 BET and XRD Analysis

The specific surface area, pore volume and average pore diameter of the four catalyst samples are listed in Table 30.2. The specific surface area of deactivated catalysts are significantly lower than fresh one's (Yu and Liu 2014). And it follows the order of fresh > 400 °C-Used > 350 °C-Used > 300 °C-Used. It indicates the reduction of specific surface area leads to the adsorption amount of NH_3 decrease, and 300 °C-Used is reduced by 16.200 m^2 g^{-1}, which has the most reduction compared to others. Those results are similar as that of Yang and Guo (2016) and Gu and Liu (2010), it probably due to F inhibits in the surface micropores of TiO_2.

Figure 30.5 illustrates the XRD spectrograms of fresh and deactivated catalysts. The diffraction peak of V_2O_5 is not seen in Fig. 30.5. It means that V_2O_5 forms a dispersed or polymerized vanadium oxide species VO_x on the surface of catalyst carrier (TiO_2). If V_2O_5 is very low, V mainly exists in the form of isolated VO_x, and V_2O_5 is highly dispersed on the surface of TiO_2 (Bond and Tahir 1991). And WO_3 diffraction peaks is also not observed. That is to say, both WO_3 and V_2O_5 are highly dispersed on the TiO_2. And there is strong interaction between the dispersed active component and the carrier (Reddy and Rao 2007). Moreover, F-containing peaks are

Table 30.2 Bet test results of four catalyst samples

Samples	BET surface area A/(m^2 g^{-1})	Pore volume v/(cm^3 g^{-1})	Average pore diameter d/nm
Fresh	50.996 ± 1.698	0.124	9.754
300°C-Used	34.796 ± 2.289	0.058	6.724
350°C-Used	45.685 ± 1.305	0.066	5.687
400°C-Used	49.106 ± 0.210	0.101	8.345

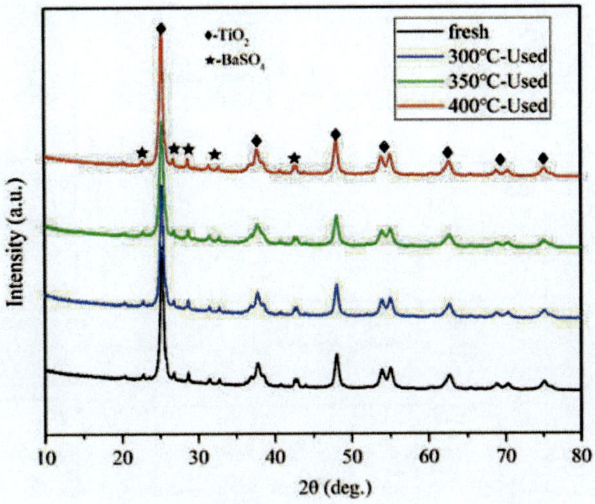

Fig. 30.5 XRD patterns of the four catalyst samples

detected, indicating that F is high dispersibility on catalyst, which is consistent with the F1s pattern of XPS analysis (Fig. 30.6 (IV) F1s).

30.3.4 XPS Analysis

In the standard SCR reaction, which is based on the Eley-Ridel mechanism (Eigenmann and Maciejewski 2006). NH_3 is adsorbed primitively on the active sites of the catalyst, and then reacts with gaseous nitrogen oxides, thereby the reduced species in the catalyst are reoxidized by oxygen. In the fast SCR reaction mechanism, the reoxidation is replaced by NO_2, which is a more effective oxidant (Koebel and Madia 2002; Ramis and Busca 1990). The standard SCR follows the Eley-Ridal mechanism, but it would become the Langmuir–Hinshelwood mechanism if there is NO_2 in the feed (Blanco and Avila 1993). Irfan et al. (2008) reported that adsorbing oxygen on the catalyst is short inr the Langmuir–Hinshelwood mechanism, so it is not as effective as the fast SCR.

The characterization results of XPS are shown in Fig. 30.6 and Table 30.3. The positional shift of the V2p peak is small can be seen from Fig. 30.6 (I), which seems that there is an effect of HF on V. In Fig. 30.6 (II), the spectrum of O1s shows two peaks could be identified. The first appears lattice oxygen O_α (529.0–530.0 eV), and the second one is surface adsorption oxygen O_β (531.1–531.9 eV). O_β as the most active oxygen in the SCR reaction, it is easy to interact with oxygen in the gas or molecules adsorbed on the surface of the catalyst. O_β could promote NO oxidation to NO_2 and accelerate the "fast SCR" reaction (Wang and Wang 2015; Irfan and Goo 2008). The O1s concentration of the deactivated catalysts are significantly reduced is presented in Table 30.3. Combined with the spectrogram of Fig. 30.6 (II), the peak

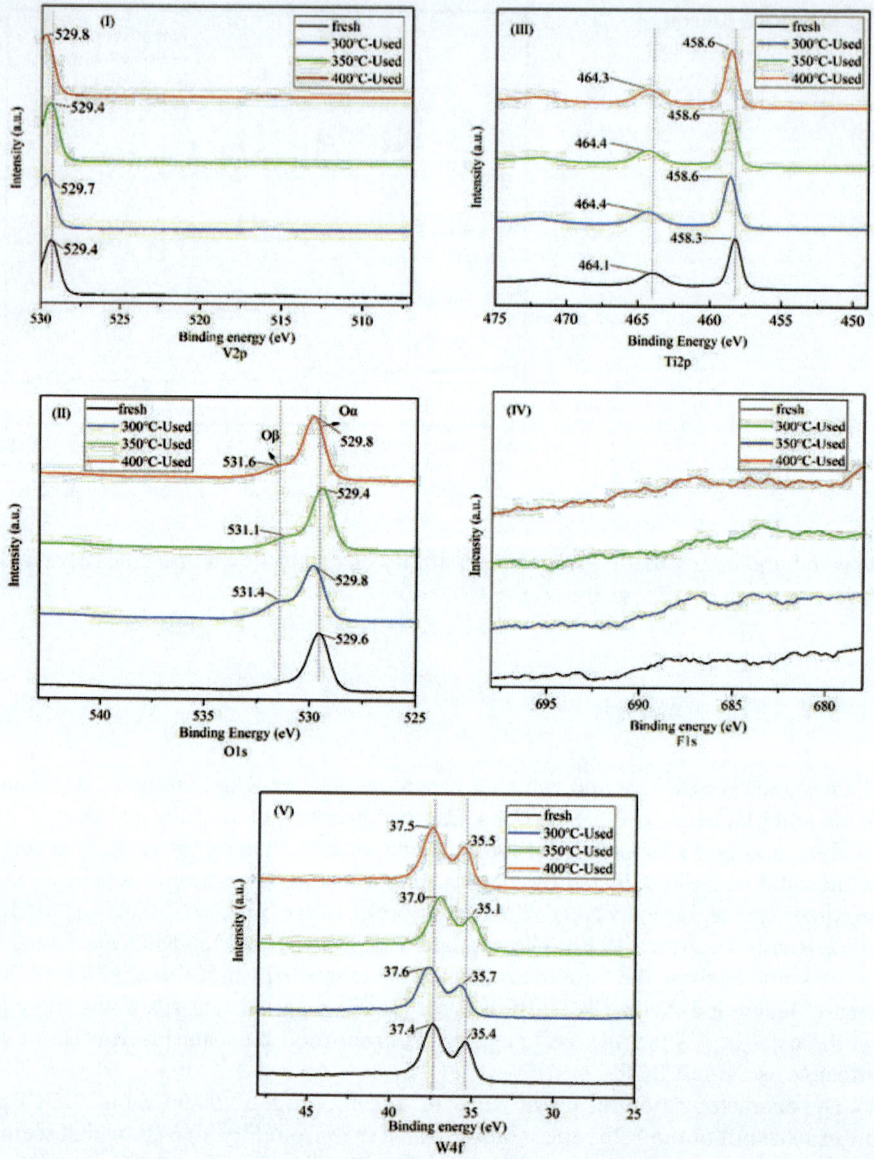

Fig. 30.6 XPS spectra of four catalyst samples: (I) V2p, (II) O1s, (III) Ti2p, (IV) F1s, and (V) W4f

of O_α moved to a higher binding energy compared with fresh except for the sample 350 °C-Used. It is possible that there is greater interaction between F and O_α in TiO_2 (Nicosia and Czekaj 2008).

To evaluate the stability of the TiO_2, Fig. 30.6 (III) shows the Ti2p spectrogram of different catalysts. There are two peaks at 458.3 and 464.1 eV, while the two peaks

Table 30.3 Surface element composition of four catalyst samples measured by XPS

Sample	Ti2p (at. %)	O1s (at. %)	N1s (at. %)	S2p (at. %)	Others (at. %)
Fresh	20.05	61.33	15.09	1.45	2.08
300 °C-Used	19.56	57.40	16.23	3.93	2.88
350 °C-Used	19.20	60.99	14.72	2.66	2.43
400 °C-Used	20.04	58.57	13.82	2.36	5.21

of deactivated catalysts at 458.6 and 464.4 eV (464.3 eV for 400 °C-Used), $Ti^{4+}2p_{3/2}$ and $Ti^{4+}2p_{1/2}$ (Zhang and Zhong 2013). The Ti2p peak position of the deactivated catalysts is shifted a little in the direction of binding energy increase. It may be due to the interaction of TiO_2 with V or W species, and the interaction of HF with the carrier (Balikdjian and Davidson 2000) (Fig. 30.6 (I) V2p, (V) W4f). There is no peak in the F1s spectrogram (Fig. 30.6 (IV) F1s), which implies that there is dispersibility of F on the catalyst.

30.4 Conclusions

In this paper, the effect of V_2O_5–WO_3–TiO_2 SCR catalyst by HF was investigated in a lab-scale experimental system. There are conclusions of the study, which can be drawn as follows:

(1) The activity tests results show that HF is adverse to catalyst. The activity of fresh catalyst and deactivated catalyst samples increases with increasing the work temperature and deactivation temperature. When the work temperature is 400 °C, the denitrification efficiency of the catalysts is higher than 90%.
(2) HF could lead to surface Brønsted acid sites of catalysts decrease, and the adsorption property to NH_3 decrease. So the redox ability is lower.
(3) HF causes the specific surface area of catalysts decreas. It seems that HF is likely interact with the TiO_2 carrier, which destroys the carrier structure of the catalys.
(4) Although HF increases the concentration of adsorbed oxygen O_β, there is interaction between HF and the components (V_2O_5 and WO_3) of catalyst. It can be attributed to chemical deactivation, which is greater harm to catalyst.

Acknowledgements This work was supported by the National Key R&D Program of China (2017YFB0602102).

References

Balikdjian JP, Davidson A (2000) Sintering and phase transformation of V-loaded anatase materials containing bulk and surface V species. J Phys Chem B 104(38):8931–8939

Blanco J, Avila P (1993) Influence of nitrogen dioxide on the selective reduction of NO_x with a catalyst of copper and nickel oxides. Appl Catal A 96(2):331–343

Bond GC, Tahir SF (1991) Vanadium oxide monolayer catalysts preparation, characterization and catalytic activity. Appl Catal 71(1):1–31

Chen C, Jia W (2018) Mechanism of Hg0 Oxidation in the presence of HCl over a $CuCl_2$-modified SCR Catalyst. J Mater Sci 53(14):10001–10012

Deng Z, Zhong Q (2016) Effect of fluorine additives on the performance of amorphous Ce–Ti catalyst and its promotional progress on ozone for NO_x $(x = 1, 2)$ removal at low temperature. J Fluorine Chem 191:120–128

Du M, Qiu B (2019) Fluorine doped TiO_2/mesocellular foams with an efficient photocatalytic activity. Catal Today 327:340–346

Eigenmann F, Maciejewski M (2006) Selective reduction of NO by NH_3 over manganese–cerium mixed oxides: relation between adsorption, redox and catalytic behavior. Appl Catal B 62(3):311–318

Gu T, Liu Y (2010) The enhanced performance of ceria with surface sulfation for selective catalytic reduction of NO by NH3. Catal Commun 12(4):310–313

Guo RT, Zhou Y (2013) Effect of preparation methods on the performance of CeO_2/Al_2O_3 catalysts for selective catalytic reduction of NO with NH_3. J Ind Eng Chem 19(6):2022–2025

Hou Y, Cai G (2014) Effect of HCl on V_2O_5/AC catalyst for NO reduction by NH_3 at low temperatures. Chem Eng J 247(6):59–65

Irfan MF, Goo JH (2008) Co_3O_4 based catalysts for no oxidation and NOx reduction in fast SCR process. Appl Catal B 78(3):267–274

Jin Q, Shen Y (2017) Effect of fluorine additive on $CeO_2(ZrO_2)/TiO_2$ for selective catalytic reduction of NO by NH_3. J Colloid Interface Sci 487:401–409

Jun Du J, Zeng P (2014) The study on the new technology of exhaust gas cascade utilization from electrolytic aluminum industry

Koebel M, Madia G (2002) Enhanced reoxidation of vanadia by NO_2 in the fast SCR reaction. J Catal 209(1):159–165

Liu SW, Guo RT (2018) The deactivation effect of Cl on V/TiO_2 catalyst for NH_3-SCR process: a drift study. J Energy Inst

Lisi L, Lasorella G (2004) Single and combined deactivating effect of alkali metals and HCl on commercial SCR catalysts. Appl Catal B 50(4):251–258

Ma J, Shang X (2016) Directions for development of hazardous waste treatment technologies in electrolytic aluminum industry. Environ Prot Chem Ind (in Chinese)

Mhamdi M, Khaddar-Zine S (2009) Influence of the cobalt salt precursors on the cobalt speciation and catalytic properties of H-ZSM-5 modified with cobalt by solid-state ion exchange reaction. Appl Catal A 357(1):42–50

Nicosia D, Czekaj I (2008) Chemical deactivation of $V_2O_5/WO_3–TiO_2$ SCR catalysts by additives and impurities from fuels, iubrication oils and urea solution: Part II. characterization study of the effect of alkali and alkaline earth metals. Appl Catal B: Environ 77 (3):228–236

Peña DA, Uphade BS (2004) TiO_2-supported metal oxide catalysts for low-temperature selective catalytic reduction of NO with NH_3: I. evaluation and characterization of first row transition metals. J Catal 221(2):421–431

Ramis G, Busca G (1990) Fourier transform-infrared study of the adsorption and coadsorption of nitric oxide, nitrogen dioxide and ammonia on vanadia-titania and mechanism of selective catalytic reduction. Appl Catal 64(C):259–278

Reddy B, Rao K (2007) Structural characterization and oxidehydrogenation activity of CeO_2/Al_2O_3 and $V_2O_5/CeO_2/Al_2O_3$ catalysts

Rui Z, Qin Z (2014) Promotional effect of fluorine on the selective catalytic reduction of NO with NH_3 over CeO_2–TiO_2 catalyst at low temperature. Appl Surf Sci 289(289):237–244
Topsøe NY (1994) Mechanism of the selective catalytic reduction of nitric oxide by ammonia elucidated by in situ on-line fourier transform infrared spectroscopy. Science 265(5176):1217–1219
Thy P, Jenkins BM (2010) Bed agglomeration in fluidized combustor fueled by wood and rice straw blends. Fuel Process Technol 91(11):1464–1485
Ukawa N, Okino S (1992) The effects of fluoride complexes in wet limestone flue gas desulfurization. J Chem Eng Jpn 25(2):146–152
Wang J, Liu H (2012) Study on harmless and resources recovery treatment technology of waste cathode carbon blocks from electrolytic aluminum
Wang P, Wang QS (2015) The influence of F and Cl on Mn/TiO_2 catalyst for selective catalytic reduction of NO with NH_3: a comparative study. Catal Commun 71:84–87
Xu J, Ao Y (2008) Low-temperature preparation of f-doped TiO_2 film and its photocatalytic activity under solar light. Appl Surf Sci 254(10):3033–3038
Yang NZ, Guo RT (2016) The enhanced performance of ceria by HF treatment for selective catalytic reduction of NO with NH_3. Fuel 179:305–311
Yaowu W, Jianping P (2018) Separation and recycling of spent carbon cathode blocks in the aluminum industry by the vacuum distillation process. Jom 70(9):1877–1882
Yu W, Liu X (2014) Enhanced visible light photocatalytic degradation of methylene blue by F-doped TiO_2. Appl Surf Sci 319:107–112
Zhang N, Mao D (2017) Selective conversion of bio-ethanol to propene over nano-HZSM-5 zeolite: remarkably enhanced catalytic performance by fluorine modification. Fuel Process Technol 167:50–60
Zhang S, Zhong Q (2013) Promotional effect of WO_3 on O_2- over V_2O_5/TiO_2 catalyst for selective catalytic reduction of NO with NH_3. J Mol Catal: Chem 373:108–113

Chapter 31
Numerical Study on Combustion and NO_x Emission Characteristics of Co-Firing Semi-Coke and Coal in a Tangentially-Fired Utility Boiler

Qinqin Feng, Chang'an Wang, Pengqian Wang, Zhichao Wang, Wei Yao, Lei Zhao, Yongbo Du, and Defu Che

Abstract With the booming coal industry and the increasing production of semi-coke, a large number of small particles and powder pyrolysis semi-coke are urgent to be exploited. Co-firing semi-coke with coal is a potential approach to achieve clean and efficient utilization of such low-volatile fuel. In this paper, the co-combustion performance of semi-coke and coal in a 135 MW tangentially-fired boiler was investigated by numerical simulation. The influences of semi-coke blending ratio, blending mode and injection position on the combustion efficiency and NO_x generation characteristics of the utility boiler were extensively analyzed. The simulation results indicated that the NO_x emission at the furnace outlet was elevated and the burn-out ratio declined with an increase in semi-coke blending ratio. Different heights of semi-coke injected position could lead to different combustion efficiency and NO_x emissions. The semi-coke was recommended to be injected from the middle layer of burners to obtain low carbon content in fly ash. The blending methods (in-furnace versus out-furnace) had certain impacts on the NO_x emission and carbon content in fly ash. High burn-out ratio could be obtained when the semi-coke and coal were injected from different burners.

Keywords Semi-coke · Co-firing · Combustion efficiency · NO_x emission

---

Q. Feng · C. Wang (✉) · P. Wang · L. Zhao · Y. Du · D. Che
State Key Laboratory of Multiphase Flow in Power Engineering, School of Energy and Power Engineering, Xi'an Jiaotong University, Xi'an 710049, PR China
e-mail: changanwang@mail.xjtu.edu.cn

Z. Wang · W. Yao
Xi'an Thermal Power Research Institute Co., LTD, Xi'an 710054, PR China

J. Lyu and S. Li (eds.), *Clean Coal and Sustainable Energy*, Environmental Science and Engineering,
https://doi.org/10.1007/978-981-16-1657-0_31

31.1 Introduction

Coal chemical technology based on coal gasification and pyrolysis technology has become an important development direction for efficient, clean and economic utilization of coal (Pan and Qu 2017; Zhou and Zhong 2018). Semi-coke is a solid product with a feature of low volatile content and high calorific value obtained by the pyrolysis of lignite or high-volatile bituminous coal at moderate and low temperatures, which is widely used in metallurgy, chemical industry, and adsorbent fields (Zhang and He 2017; Gong and Wenfei 2018). However, a large amount of semi-coke needs to be solved due to the insufficient demand in the traditional semi-coke market. The efficient and safe utilization of semi-coke as power fuel may be a feasible solution to expand the application range of semi-coke. The co-combustion performance of semi-coke and coal in circulating fluidized bed (CFB) boiler has been extensively investigated in previous studies (Yao and Zhu 2015; Yang and Wang 2018; Yang and Lu 2016). Considering the current situation that majority of existing utility boilers in China are pulverized coal boilers, it is a sensible option to consume semi-coke as power fuel in pulverized coal boilers.

Numerous experimental and numerical studies have been carried out about the combustion characteristics of blended fuels (Yao and He 2019; Moon and Sung 2015; Lu and Wang 2017). Moon and Sung (2015) performed experiments in the pulverized coal-fired furnace and investigated the co-combustion performance of lignite and bituminous, and the results indicated that the NO_x concentration is minimum under the 10% blending ratio of lignite. Baek and Park (2014) conducted simulations in a tangentially coal-fired boiler on the influence of the blending method and revealled that the carbon in ash was greatly affected but the NO_x emission was slightly influenced. However, the co-firing of semi-coke and coal could bring about different influences on the combustion characteristics and NO_x emissions due to the marked differences of semi-coke in fuel properties.

In the present study, the numerical approach was employed to explore the optimal condition of co-firing bituminous coal and semi-coke in a 135 MW tangentially fired boiler. Different ratios, positions and methods of blending semi-coke have different influences on the co-combustion performance and NO_x emissions. The present study can provide a guidance for the further development of semi-coke as clean and efficient power fuel.

31.2 Boiler Description and Mesh Generation

The 135 MW utility boiler in this study is a π-shaped solid-state slag pulverized coal one with natural circulation. The furnace has a section of 10,380 × 10,380 mm and a height of 32,187 mm. Figure 31.1 shows the schematic structure of the burners and boiler. The boiler adopted four-corner tangential combustion mode (cutting diameter Φ700 mm, pulverized coal jet and angle between the two sides of the wall were 42°

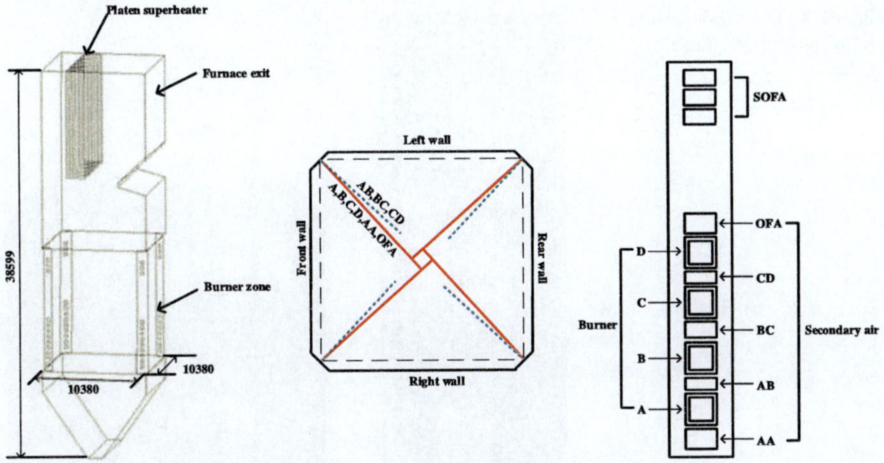

Fig. 31.1 The schematic configuration and burner arrangement of the boiler

and 48°, respectively), and four layers of burners were symmetrically arranged at the four corners. As shown in Fig. 31.1b, there are 5° deviation between the primary air and secondary air, except for the secondary air in the bottom layer (AA) and over fire air (OFA), to eliminate the residual swirl in the horizontal flue gas pass. Three layers of separate over fire air (SOFA) were placed 6000 mm above the main combustion zone to achieve deep-air-staging combustion.

The boiler was divided into the structured mesh system by the Integrated Computer Engineering (ICEM). As shown in Fig. 31.2, the mesh in the main combustion zone was refined to improve the accuracy of the combustion simulations, and the O-grid method was applied to enhance the efficiency of the grids near the surface. The number of meshes had an influence on the computational accuracy and speed. The average temperature distributions along the furnace under the mesh systems of 927 544, 1 223 280 and 1 608 336 cells were demonstrated in Fig. 31.3, and the mesh system with 1 223 280 cells was finally adopted to achieve the accuracy of engineering requirements and the purpose of economic efficiency.

31.3 Numerical Models and Operating Conditions

The simulations were carried out by the software Fluent 14.5 in the present study. The combustion of pulverized coal in the furnace is a complex physical and chemical process, including the flow, combustion and heat transfer of pulverized coal flame, to predict the gas phase flow field distribution, pulverized coal particle movement trajectory, composition concentration distribution of combustion products, flue gas temperature field and so on. The realizable k-ε model was adopted to simulate the turbulent flow in the gas phase and the P1 model was applied to calculate the radiation

Fig. 31.2 The mesh system of the tangentially fired boiler

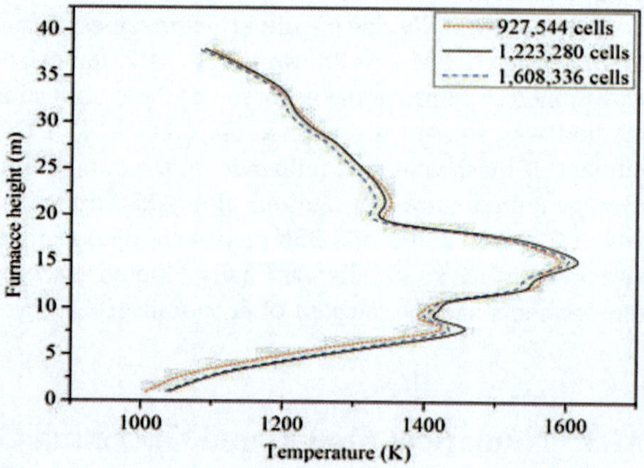

Fig. 31.3 The average temperature along the furnace for grid independence validation

heat transfer. The species transport model was adopted to describe the interaction between gases and particles, and the particle movement was predicted under the Lagrange model. The finite-rate/eddy-dissipation model was applied to depict the turbulence-chemistry interaction. The two-competing-reactions model was used to describe the volatile pyrolysis, and the combustion of char was represented by the kinetics/diffusion-limited model. It is generally believed that the NO_x concentration

Table 31.1 The properties of fuels

Sample		Semi-coke	HY
Proximate analysis (wt%)	M_{ar}	11.20	10.00
	A_{ar}	8.98	6.24
	V_{ar}	8.75	32.15
	FC_{ar}	71.07	54.61
Ultimate analysis (wt%)	C_{ar}	72.80	69.07
	H_{ar}	1.60	4.20
	O_{ar}	4.21	8.92
	N_{ar}	0.81	1.00
	S_{ar}	0.40	0.57
$Q_{net,ar}$	(MJ/kg)	25.84	26.37

is not enough to influence the flow field and the main combustion process in numerical simulations. Hence, the calculation of NO_x adopts the method of the post-processing, which is based on the stability of combustion temperature field and flow field. It was believed that the reaction process of nitrogen in coal was that fuel-N was first mainly converted to HCN and NH_3, and then NH_3 and HCN were reduced to N_2 or oxidized to NO through competitive reactions (Du and Wang 2018; Wang and Feng 2019).

Eight cases were carried out based on the same fuel heat to explore the combustion performance and NO_x emissions under various blending ratios and blending methods in the 135 MW utility boiler. The properties of semi-coke and bituminous coal considered in the in-situ tests and simulations are illustrated in Table 31.1 and the detailed simulation cases with specific parameters are depicted in Table 31.2. The acronym "Out-furnace" represents the conditions that semi-coke and coal are premixed before the furnace, and the acronym "In-furnace" represents the conditions that semi-coke and coal are entered into the furnace from different burners, and the acronym Case T1 is the numerical simulation under in-situ data as the basis conditions. The temperatures of the primary air and secondary air are 343 K and 583 K, respectively. The primary air contains the moisture in fuels, and the total excess air

Table 31.2 Detailed operation conditions of simulation cases

Case	Blending ratio (%)	Running mills	Burners for semi-coke	Blending method
T1	0	ACD	/	/
1	17	ACD	ACD	In-furnace
2	33	ACD	ACD	In-furnace
3	45	ACD	ACD	In-furnace
4	67	ACD	ACD	In-furnace
5	33	ACD	A	Out-furnace
6	33	ACD	C	Out-furnace
7	33	ACD	D	Out-furnace

coefficient remains unchanged at 1.2. When the load is 100%, the three-layer burners of A, C and D are put into use, and the B-layer burner is closed for backup.

31.4 Results and Discussion

The simulation results can be influenced by the variation of models, and therefore the model validation is necessary by comparing the simulation results with the in-situ test data. The comparison of some indicators at the furnace outlet is shown in Table 31.3. The data deviation is within the acceptable scope of the applications and the selected models can be considered to be reasonable.

31.4.1 Influence of Semi-Coke Blending Ratio

The average temperature and species concentrations along the furnace under various blending ratios are shown in Fig. 31.4. It can be seen from Fig. 31.4 that the average temperature in the main combustion zone is the highest, while the O_2 concentration is relatively low due to the injection of fuels. The variation of average temperature along the furnace height is slight with the increase of semi-coke blending ratio because the fuel heat remains unchanged. The NO mole fraction increases when more semi-coke is sent into the furnace.

Some indicators at the furnace outlet under various blending ratios are demonstrated in Table 31.4. The average temperature at the furnace outlet fluctuates somewhat, but the overall fluctuation range does not exceed 10 K, which has little effect on the performance of the superheater and reheater. The simulation results indicate that the ignition and combustion of fuels are postponed with the increase of semi-coke blending ratio. Considering that semi-coke is relatively difficult to ignite and burn compared with bituminous coal, the carbon content in fly ash increases and the burn-out ratio decreases. The NO_x concentration shows an increasing trend. When the blending ratio of semi-coke is as high as 67%, the NO_x concentration and carbon content in fly ash are up to 357.5 mg/m^3 (@6% O_2) and 8.63%, respectively. Compared with the single combustion of bituminous coal, the increase of 97.9 mg/m^3

Table 31.3 The comparision of some indicators at the furnace outlet

	Temperature (K)	NO_x concentration (mg/m^3, 6%O_2)	Carbon content in fly ash (%)
Experiment	1124.0	229.0	4.80
Simulation	1119.1	259.6	4.57

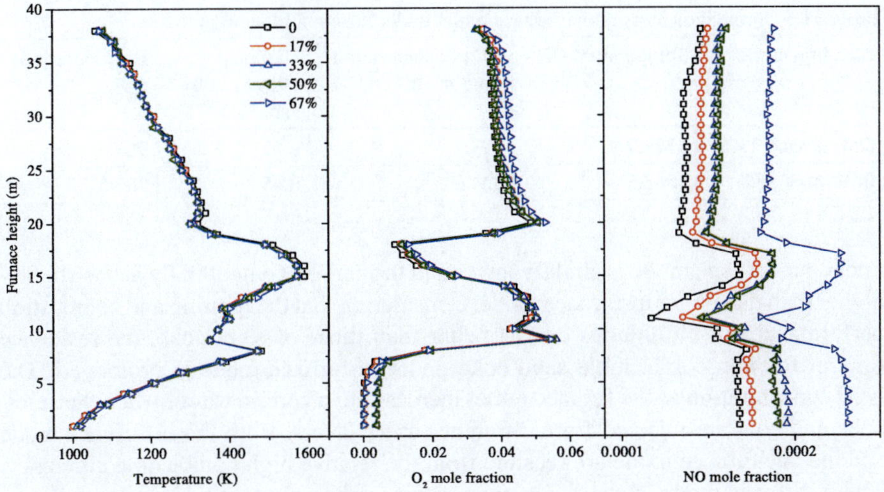

Fig. 31.4 The average temperature and species concentrations along the furnace under various blending ratios

Table 31.4 Some indicators at the furnace outlet under various blending ratios

Semi-coke blending ratio	Temperature (K)	NO_x concentration (mg/m^3, 6%O_2)	Carbon content in fly ash (%)	Burn-out ratio (%)
0	1119.1	259.6	4.57	96.43
17%	1123.5	278.5	5.79	96.27
33%	1118.7	293.2	6.82	96.06
50%	1117.5	295.9	8.18	95.48
67%	1113.5	357.5	8.63	95.17

(@6% O_2) and one time is obvious. The semi-coke blending ratio of 33% is recommended in this utility boiler to achieve low NO_x concentration and high combustion efficiency.

31.4.2 Influence of Blending Mode

Different blending modes can bring ahout the influence on the co-combustion and NO_x emissions characteristics. The comparison of several indicators at the furnace outlet under two blending modes is illustrated in Table 31.5. Semi-coke is sprayed into the furnace from the C-layer burner under 33% blending ratio. It can be found that the average temperature under out-furnace mode is 30 K higher than that under in-furnace mode when the blending ratio of semi-coke is 33%. The burn-out ratio

Table 31.5 Some indicators at the furnace outlet under different blending modes

Blending mode	Temperature (K)	NO_x concentration (mg/m^3, 6%O_2)	Carbon content in fly ash (%)	Burn-out ratio (%)
Out-furnace 33%	1118.7	293.2	6.82	96.06
In-furnace 33%	1088.5	324.6	6.45	96.45

under out-furnace mode is slightly lower and the carbon content in fly ash is slightly higher than that under in-furnace mode. Considering that the ignition and combustion performances of bituminous coal is better than those of semi-coke, the residence time of the non-combustible semi-coke under in-furnace mode is prolonged. The NO_x concentration at the furnace outlet increases to a certain extent when the semi-coke and coal are injected from different burner layers. Both the in-furnace mode and the out-furnace mode are feasible from the relative high combustion efficiency. The in-furnace mode is conducive to reducing carbon content in fly ash, while the NO_x concentration at the furnace outlet under in-furnace mode of such a semi-coke burner configuration is higher than that under out-furnace mode. Combined with the previous esearch, it is possible that the two indicators of combustion efficiency and pollutant discharge under an optimal in-furnace condition are better than that under out-furnace mode, which needs further exploration.

31.4.3 Influence of Semi-Coke Injection Position

There are different injection positions of semi-coke when the in-furnace mode is adopted and the semi-coke blending ratio is 33%. Here, three conditions that semi-coke is respectively sprayed from the layer-A, layer-C and layer-D are investigated. The cross-sections temperature distributions at three burner layers under different semi-coke injection positions are shown in Fig. 31.5. It can be seen from Fig. 31.5 that the temperature distribution of the burner section forms a clear annular area. The primary air carried the pulverized coal is sprayed into the furnace from four corners of the furnace, which is tangent to the imaginary circle. The airflow strongly rolates in the furnace. The pulverized coal spirals up continuous under the action of strong swirl after on fire, resulting in that the temperature at the annular outer ring corresponding to the burner is the highest, while the temperature at the center of the furnace is the lowest. In addition, the annular high temperature zone corresponding to the lowest injection position of pulverized coal is the smallest. As the airflow develops upward, the temperature difference between the center and the side of the furnace decreases gradually, which indicates that the flame center is reasonable and the simulation results are reasonable. For the same layer of the burner, the temperature of semi-coke injected slightly raises compared with bituminous coal, mainly due to the difference in calorific value. The high temperature center in the furnace moves up gradually with the increase in the injection position of semi-coke.

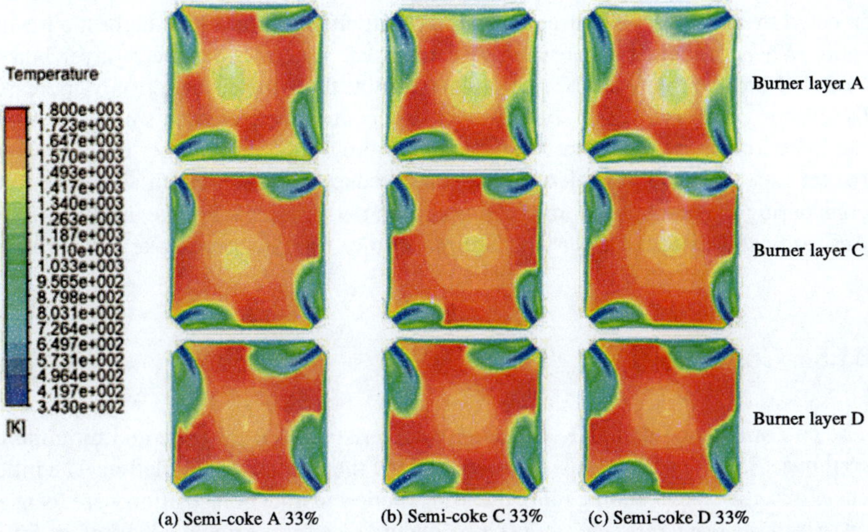

Fig. 31.5 The cross-sections temperature distribution at three burner layers under different semi-coke injection positions

Table 31.6 shows the comparison of furnace outlet indicators under different semi-coke injection positions. As can be seen from Table 31.6, the average temperature at the furnace outlet is defined within limited difference when the injection position of semi-coke varies, and the average temperature at which the topmost layer is injected into semi-coke has a modest growth. By comparing the three working conditions, it can be found that the burn-out ratio significantly decreases when the semi-coke is injected from the A-layer burner. The injection of semi-coke with high fixed carbon content from the low-temperature bottom burner carries out the deterioration of combustion performance and the obvious increase of the bottom slag and carbon content in bottom slag. Moreover, the injection of semi-coke from the bottom burner into the low-temperature zone of the furnace leads to difficult ignition and stable combustion due to the feature of low volatile. The carbon content in fly ash shows an increase by raising the injection position of semi-coke, while the NO_x concentration at the furnace outlet rises accordingly. The combustion performance and residence time of semi-coke is promoted and prolonged when the condition that semi-coke is

Table 31.6 Some indicators at the furnace outlet under different semi-coke injection positions

Semi-coke injection position	Temperature (K)	NO_x concentration (mg/m^3,6%O_2)	Carbon content in fly ash (%)	Burn-out ratio (%)
A	1088.9	306.5	4.81	95.28
C	1088.5	324.6	6.45	96.45
D	1094.8	411.9	8.01	95.98

injected from the C-layer burner is adopted, which gives rise to the highest burn-out ratio. When the injection position of semi-coke changes from the A-layer burner to the C-layer burner, the NO_x concentration at the furnace outlet only increases by 18.1 mg/m^3 (@6% O_2), but the NO_x concentration increases by 26.9% when the injection position of semi-coke changes from the C-layer burner to the D-layer burner. It is not recommended to spray semi-coke from the bottom and top burner considering the combustion and NO_x emission performance. For the utility boiler in this study, the C-layer burner is an optimal choice to inject semi-coke into furnace.

31.5 Conclusion

The co-combustion and NO_x emission charateristics of semi-coke and bituminous coal in a 135 MW utility boiler was studied via the numerical simulation. The influences of semi-coke blending ratio, blending mode and injection position were focused on to provide some reference and technical guidance for the practical test of co-firing semi-coke and coal in power plants. The simulation results indicated that the average temperature at the furnace outlet fluctuated slightly within 10 K with an increase in the blending ratio of semi-coke. The NO_x emission and carbon content in fly ash were affected significantly when the blending ratio of semi-coke exceeded 33%. Hence, the blending ratio of 33% semi-coke was recommended in this utility boiler to achieve low NO_x concentration and high combustion efficiency. For the combustion performance of the utility boiler, both in-furnace mode and out-furnace mode were feasible. Under the conditions of semi-coke burner configuration selected the present study, the in-furnace mode was conducive to reducing carbon content in fly ash, but caused an increase in the NO_x concentration compared with the out-furnace mode. The injection position of semi-coke from the middle layer burner was optimal to achieve a balance between the combustion efficiency and NO_x emission.

Acknowledgements The authors acknowledge financial support from the National Key R&D Program of China (2017YFB0602003) and the China Postdoctoral Science Foundation (2018M641885).

References

Baek S, Park H (2014) The effect of the coal blending method in a coal fired boiler on carbon in ash and nox emission. Fuel 128:62–70

Du Y, Wang C (2018) Computational fluid dynamics investigation on the effect of co-firing semi-coke and bituminous coal in a 300 MW tangentially fired boiler. Proc Inst Mech Eng, Part a: J Power Energy 233(2):221–231

Gong Z, Wenfei W (2018) Combination of catalytic combustion and catalytic denitration on semi-coke with Fe_2O_3 and CeO_2. Catal Today 318:59–65

Lu Y, Wang Y (2017) Investigation of ash fusion characteristics and migration of sodium during co-combustion of zhundong coal and oil shale. Therm Eng 121:224–233
Moon C, Sung Y (2015) NO_x emissions and burnout characteristics of bituminous coal, lignite, and their blends in a pulverized coal-fired furnace. Exp Thermal Fluid Sci 62:99–108
Pan D, Qu X (2017) Effect of gasified semi-coke on coal pyrolysis in the poly-generation of CFB gasification combined with coal pyrolysis. J Anal Appl Pyrol 127:461–467
Wang C, Feng Q (2019) Numerical investigation on co-firing characteristics of semi-coke and lean coal in a 600 MW supercritical wall-fired boiler. Appl Sci 9(5):889–905
Yang Y, Lu X (2016) Experimental study on combustion of low calorific oil shale semicoke in fluidized bed system. Energy Fuels 30(11):9882–9890
Yang Y, Wang Q (2018) Combustion behaviors and pollutant emission characteristics of low calorific oil shale and its semi-coke in a lab-scale fluidized bed combustor. Appl Energy 211:631–638
Yao Y, Zhu J (2015) Experimental study on nitrogen transformation in combustion of pulverized semi-coke preheated in a circulating fluidized bed. Energy Fuels 29(6):3985–3991
Yao H, He B (2019) Thermogravimetric analyses of oxy-fuel co-combustion of semi-coke and bituminous coal. Appl Therm Eng 156:708–721
Zhang K, He Y (2017) Multi-stage semi-coke activation for the removal of SO_2 and NO. Fuel 210:738–747
Zhou G, Zhong W (2018) Experimental study on characteristics of pressurized grade conversion of coal. Fuel 234:965–973

Chapter 32
Investigation on NO_x Formation Characteristics During Semi-coke Air-Staged Combustion

Hui Li, Naiji Wang, Shi Yang, Xin Zhang, Yuhong Cui, and Jianming Zhou

Abstract Semi-coke is the product of low-rank coal by pyrolysis at low temperature. If semi-coke could be used as fuel of industrial pulverized coal boilers, it will widen the fuel range of the industrial pulverized coal boilers and effectively promote the coal staged utilization. As the semi-coke need higher temperature than bituminous coal for its ignition and combustion process, the NO_x emission will rapidly increase with the rising of temperature. So, decreasing the NO_x is an important task in its utilization. In this paper, the NO_x emission rules at the higher fuel-rich zone temperature and properties of semi-coke air-staged combustion were explored by two-stage drop-tube furnace. In the air-staged combustion experiments, the influence of fuel-rich zone temperature and the ratio of air on NO_x emission and combustion behavior were investigated. The results indicate that the NO_xemission concentration of non-staged combustion rises with fuel-rich zone temperature and the excess air coefficient in its combustion process. The air-staged combustion could visibly reduce the NO_x emission in the combustion process. As the Fig. 32.1 shows, the optimum ratio of secondary air is 0.56, at which NO_x emission concentration is under 120 mg/m3 and the burn-out rates were above 90%. The conclusions of the burn-out rate and the decrease rate of NO_x can be used to guide the industry enlargement experiment in running conditions.

Keywords Semi-coke · Low-NO_x combustion · Ratio of secondary air · Air-staged combustion

---

H. Li (✉) · N. Wang · S. Yang · X. Zhang · Y. Cui · J. Zhou
China Coal Research Institute Company of Energy Conservation Corporation Ltd, Beijing 100013, China
e-mail: 447110302@qq.com

State Key Laboratory of Coal Mining and Environmental Protection (Coal Research Institute), Beijing 100013, China

National Energy Technology and Equipment Laboratory of Coal Utilization and Emission Control (Coal Research Institute), Beijing 100013, China

S. Yang
Coal Science and Technology Research Institute, Beijing 100013, China

J. Lyu and S. Li (eds.), *Clean Coal and Sustainable Energy*, Environmental Science and Engineering,
https://doi.org/10.1007/978-981-16-1657-0_32

32.1 Introduction

Oil and gas products, produced by low-rank coal pyrolysis at low temperature, is a kind of high value added raw material. And the semi-coke which is produced after pyrolysis process is used as fuel usually (Qinlong and Jia 2013; Gang 2013; Xin 2015). The market of the pulverized coal industrial boiler will be widen and the coal grading process will be promoted, if semi-coke can be used as the fuel of pulverized coal industrial boiler (Fang 2015; Yongying et al. 2016; Zhiqiang et al. 2014). As semi-coke has a low content of volatile matter, it needs higher temperature, which may increase difficulty in controlling NO_x, to insure its ignition and stable combustion process. So, multiple technical measures are needed to logically organize combustion and decrease the formation of NO_x. Research shows that in higher heating rate, the volatile devolatilization of semi-coke will be faster. Besides, there are more reductive micromolecular substances in the volatile devolatilize and it results in easier ignition (Taniguchi et al. 2011, 2010; Bai et al. 2014; Suda et al. 2002; Tamura et al. 2015; Yongfeng et al. 2009). Meanwhile, reducing atmosphere is more easily formed in the fuel-rich zone when air-staged combustion is used to control NO_x (Fan et al. 2009; Qin 2013). Above all, exploring the NO_x formation and combustion characteristics at higher fuel-rich zone temperature when air-staged combustion is used is the key to use semi-coke as the fuel of pulverized coal industrial boiler. In this paper, NO_x formation and combustion characteristics was researched in air-staged and non-staged condition on self-building drop-tube funace system. SEM and nitrogen adsorption methods were used to characterize the ash and explain the characteristics concluded from the air-staged combustion experiment .

32.2 Experimental Methods and Equipment

Air-staged and non-staged combustion experiment of semi-coke was performanced on self-building drop-tube furnace system on which we can investigate combustion of semi-coke particle abstracted from pulverized semi-coke combustion in industrial boiler. The drop-tube furnace system is shown as Fig. 32.1.

In this paper, the experimental temperature of the fuel-rich zone is among 1000–1400 °C, and the temperature of the fuel-lean zone is 1400 °C invariably. The excess air coefficient is among 1.0–1.3 in non-staged combustion experiment. In air-staged combustion experiment, we study the influence of different ratio of the secondary air and burn-out air to NO_x formation and combustion characteristics and the total excess air coefficient is 1.20. We adjust the volume of the secondary air and burn-out air to change the ratio of the two stages of air. The secondary air ratio is set as a research indicator. The ratio of the secondary air is among 0.4–0.8 and the ratio of the first air is 5% invariably.

The micro-feeder was calibrated before the experiment, and the result is shown in Fig. 32.2. The experimental feed rate was chosen to be 3.5 g/min for 22 r/s. The

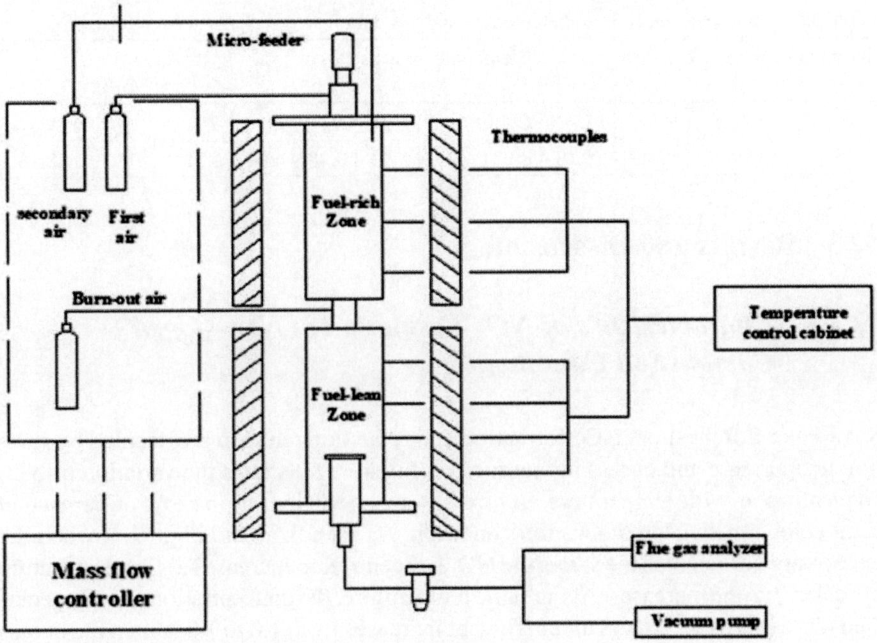

Fig. 32.1 Schematic diagram of drop-tube furnace system

Fig. 32.2 Curve graph of the rating test of the micro feeder

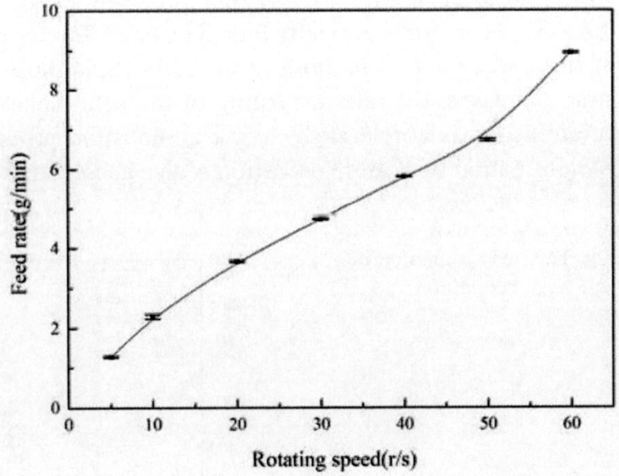

flue gas data was measured at the exit of the furnace and converted to O_2 6%. The experimental semi-coke was made into a 200-mesh semi-coke powder. The industrial analysis and elemental analysis data are shown in Table 32.1.

Table 32.1 Industrial analysis and elemental analysis data

Industrial analysis/%				Elemental analysis/%					Q / MJ/kg
M_{ad}	A_d	V_{daf}	FC_d	C_d	H_d	N_d	O_d	$S_{t,d}$	$Q_{net,ar}$
1.45	10.96	12.19	78.18	80.24	1.53	0.88	6.07	0.31	28.18

32.3 Results and Disscusion

32.3.1 Characteristics of NO_x Formation at Non-staged Combustion Condition

Semi-coke Combustion, NO_x formation and reduction reaction are decided by reaction temperature and excess air coefficient. Figure 32.3 shows the variation of NO_x concentration with temperature and excess air coefficient at the exit of furnace at semi-coke non-staged combustion condition. As seen from the Fig. 32.3, when the excess air coefficient is the same, the NO_x concentration increases as the temperature of fuel-rich zone increases. At the same temperature, the concentration of NO_x generated increases as the excess air coefficient increases from 1.0 to 1.3. When the excess air coefficient is increased from 1.0 to 1.15, the concentration of NOx is increased significantly. And when the air excess coefficient is increased from 1.15 to 1.30, the amount of NOx formation increases slowly. The main reason is that the combustion air is sent to the combustion system at one time. At the experimental temperature, the NO_x generated is basically fuel-type NOx. Fuel-type NO_x is oxidized by volatile matter and char N (Glarborg et al. 2003; Spliethoff et al. 1996). As the temperature increases, the rate of heating of the semi-coke is accelerated, the release of volatilization is accelerated. So the combustion process is intensified. The rate of NO_xformation by N atom oxidation is also accelerated (He et al. 2004; Chunbo et al.

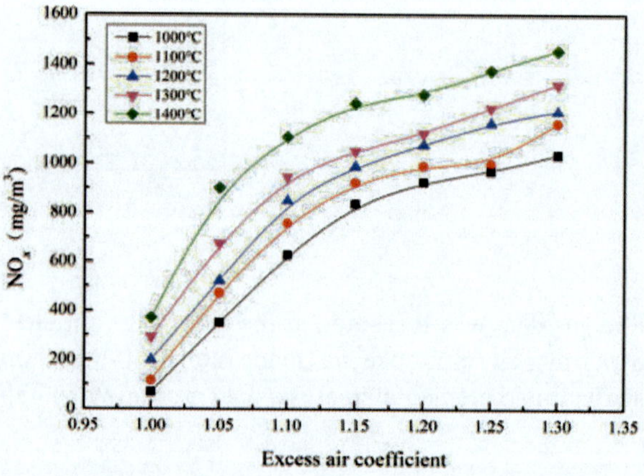

Fig. 32.3 NO_x emission in non-staged condition

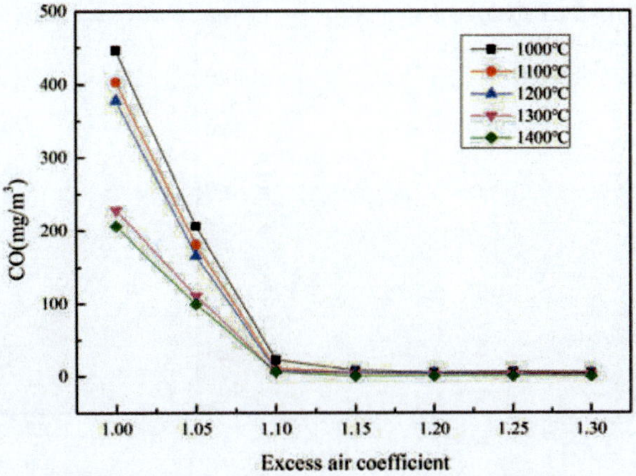

Fig. 32.4 CO emission in non-staged condition

2015; Huan et al. 2015). However, when the excess air system is increased to a certain value, the burning rate of the semi-coke reaches a limit, and therefore, even if the air excess coefficient is continuously increased, the rate of generation of NO_x is not significantly increased.

The law of CO formation under non-staged combustion conditions is shown in Fig. 32.4. It can be seen from the figure that as the temperature of the fuel-rich zone increases and the excess air coefficient increases, the CO concentration of the outlet flue gas decreases, indicating that the combustion rate is accelerated and the combustion is more sufficient. When the air excess coefficient is in the range of 1.0–1.15, the CO rises obviously. When the air excess coefficient is in the range of 1.15–1.30, the CO change is not obvious, indicating that the combustion process has reached a steady state. This law is basically the same as the NO_x generation law.

32.3.2 *Characteristics of NO_x Formation at Air-Staged Combustion Condition*

32.3.2.1 The Influence of Fuel-Rich Zone Temperature

The characteristics of NO_x formation during the semi-coke air staged combustion process is shown in Fig. 32.5. It can be seen from the figure that as the temperature of the fuel-rich zone increases, the concentration of NO_x increases first and then decreases. When the secondary air ratio is greater than 0.56, the concentration of NO_x increases as the temperature increases. And when the secondary air ratio is less than 0.56, the concentration of NOx generated decreases as the temperature increases. The reasons for the above phenomenon can be summarized as the following three points.

Fig. 32.5 NO_x emission in semi-coke air-staged combustion

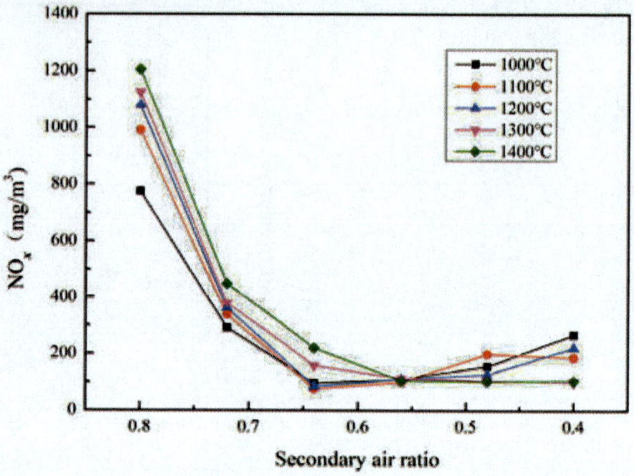

(1) After the semi-coke particles enter the combustion zone, they are rapidly heat up, and the N in the fuel rapidly releases with the volatiles, forming nitrogen-containing intermediates such as HCN and NHi. These nitrogen-containing intermediates are oxidized to form NOx or reduce the produced NO_x to N_2. The conversion pathway of fuel N depends on temperature and reaction atmosphere.

(2) Due to the low volatiles of semi-coke, the NOx generated during combustion is mainly derived from char, and the oxidation of N to NOx and heterogeneous reduction of NO_x by char are all enhanced at high temperature.

(3) When the secondary air ratio is greater than 0.56, due to the higher oxygen concentration in the fuel-rich zone, the rate of oxidation of nitrogen-containing intermediates such as HCN and NHi to NOx is higher than that of the NOx reduction reaction rate. Therefore, the total amount of NO_x generated is increased. When the ratio of secondary air is less than 0.56, the reducing atmosphere in the fuel-rich zone is stronger. With the increase of the temperature of the fuel-rich zone, the rate of NO_x reduction reaction increases greatly, which is greater than oxidation rate due to the increase of temperature, so the total amount of NOx generated is reduced.

Figure 32.6 shows the law of NOx reduction rate compared with air non-staged combustion. It is similar to the law of NO_x formation concentration. As the temperature of the fuel-rich zone increases, when the secondary air ratio is greater than 0.56, the NO_x reduction rate decrease, when the secondary air ratio is less than 0.56, the NO_x concentration reduction rate increases. The law of the burn-out rate of the semi-coke air—staged combustion is shown in Fig. 32.7. As the temperature of the fuel-rich zone increases, the burnout rate generally increases.

From the above analysis, we can know increasing the temperature of the fuel-rich zone can reduce the amount of NO_x generated under the premise of ensuring the burn-out rate.

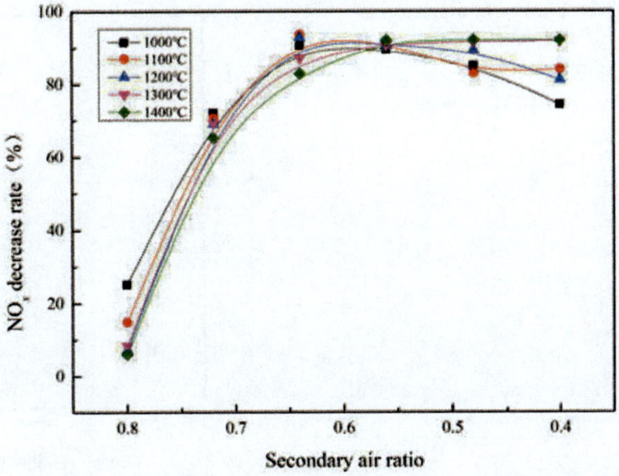

Fig. 32.6 Decrease rate of NO_x in air-staged condition

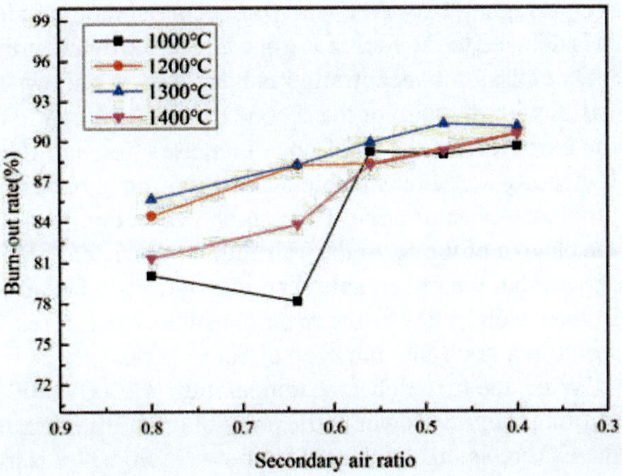

Fig. 32.7 Burn-out rate in semi-coke air-staged combustion

32.3.2.2 The Influence of the Secondary Air Ratio

At all experimental fuel-rich zone temperature, the NO_x concentration decreases first and then increases or then stabilizes as the ratio of secondary air decreases. When the temperature of the fuel-rich zone is 1000–1200 °C, the secondary air ratio turning point of NO_x concentration is 0.64. When the secondary air ratio is greater than 0.64, the NO_x concentration decreases as the secondary air ratio decreases. When the secondary air ratio is less than 0.64, the concentration of NO_x rises slightly as the secondary air ratio decreases. When the temperature of the fuel-rich zone is 1300–1400 °C, the turning point of the secondary air ratio is 0.56. When the ratio of the secondary air is large, the oxygen concentration in the vicinity of the semi-coke particles is high, the concentration of the reducing gas such as CO is low, the amount of NO_x generated is larger than the amount of reduction, and the concentration of

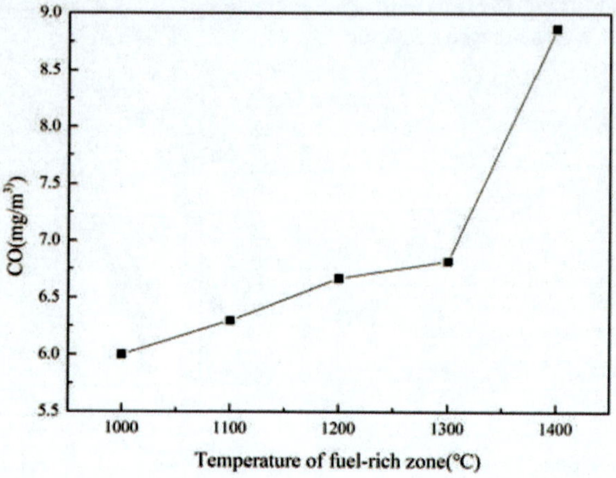

Fig. 32.8 CO emission in different fuel-rich zone temperature

NO_x is higher. However, when the secondary air ratio is too low, the release of the fuel N is delayed to the fuel-lean zone due to the delay of the combustion process, and the NOx emission concentration is increased. As shown in Fig. 32.7, the burn-out rate varies with the ratio of the secondary air. As the ratio of the secondary air decreases, the burnt rate of the semi-coke increases first and then decreases or then stabilizes, indicating that a reasonable air distribution is beneficial to improve the combustion efficiency of semi-coke. Then, as shown in Fig. 32.6, the rate of NOx reduction with the change of the secondary air ratio is consistent with the NOx emission law, which proves that the reasonable secondary air ratio for air-staged combustion has a great influence on its low nitrogen combustion effect. A reasonable secondary air ratio will result in a good low nitrogen effect.

When the fuel-rich zone temperature is 1000–1200 °C, the NO_x emission concentration is slightly lower at the point of 0.64. But the burnout rate is higher at 0.56. As the CO emission concentration shown in Fig. 32.8 is integrated, under the five experimental fuel-rich zone temperature, the CO emission concentration is lower when the secondary air ratio is 0.56, so 0.56 is selected as the optimal secondary air ratio among the experimental five secondary air ratios of 0.4–0.8 under the experimental conditions.

32.3.3 *Microscopic Pore Structure and Morphology of Air-Graded Combustion Ash*

Figure 32.9 shows the pore size distribution of the ash sample obtained from the pulverized semi-coke air-staged combustion. The results in the figure are the pore size distribution of the ash samples obtained when the secondary combustion zone temperature is 1300 °C and the secondary air ratios are 0.4, 0.56 and 0.8, respectively. When the secondary air ratio is 0.56, the probability of the pores appearing at the

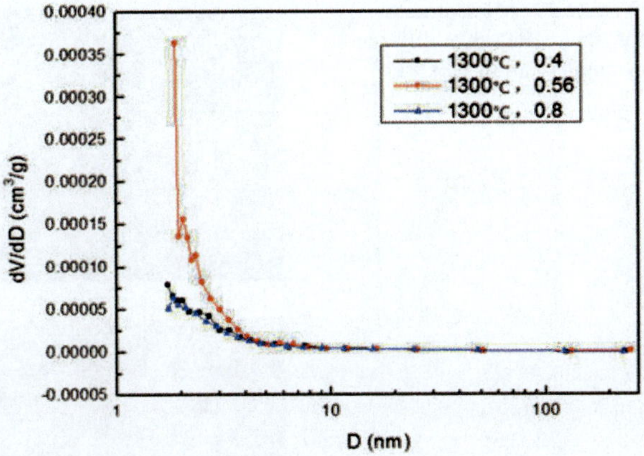

Fig. 32.9 Pore distribution of semi-coke ash at different ratio of secondary air

smaller pore size is much greater than when the secondary air ratio is 0.4 and 0.8. When the secondary air ratio is 0.4 and 0.8, the pore size distribution is not much different. As shown in Table 32.2, when the secondary air ratio is 0.56, the specific surface area of the ash sample is the largest, indicating that at this secondary air ratio (0.56), the pores on the surface of the ash have good formation and development (Thommes 2015; Lorenz and Tamurac 2000). So volatile matter devolatilization is eaier, and the combustion of semi-coke and the reduction reaction of NO_x progress well. The NO_x emission law (Fig. 32.5) also reaches the optimum value at this point (0.56). When the secondary air ratio is 0.8, the total pore volume is the smallest, (0.005914 cm3/g), and the specific surface area of the ash sample is also the smallest. When the secondary air ratio is 0.4, the total pore volume is the largest (0.017286 cm^3/g,).But the specific surface area is also small (2.28 m^2/g). It is indicated that under above two secondary air ratios, the volatile matter devolatilization is not as good as the secondary air ratio of 0.56, which is not conducive to the formation and development of pores. The combustion reaction is not as good as the secondary air ratio of 0.56, which proves once again. the importance of a reasonable air distribution. In general, among these three secondary air ratio, 0.56 is the optimal secondary air ratio, at which sufficient combustion of the semi-coke and low NO_x emission can ba ensured.

Figure 32.10 shows the microscopic morphology of the ash sample obtained by the air-staged combustion of the semi-coke at the ratio of the secondary air of 0.4,

Table 32.2 Characteristic parameters of semi-coke and ash at ratio of different secondary air

Conditions	BET surface m^2/g	Total pore volume cm^3/g	Pore volume ratio/surface ratio (%/%)		
			<2 nm	2–50 nm	>50 nm
1300,0.4	2.28	0.0128	1.5/19.04	17.28/70.04	81.22/10.92
1300,0.56	3.221	0.0127	2.17/16.52	21.27/76.53	76.56/6.95
1300,0.8	1.908	0.0059	2.71/18.29	33.21/74.42	64.08/7.09

Fig. 32.10 Microstructure of semi-coke ash at different secondary air at 1300 °C

0.56 and 0.8 respectively, when the temperature of the fuel-rich zone is 1300 °C. It can be seen from the figure that when the secondary air ratio is 0.56, compared with the case where the secondary air ratio is 0.4 and 0.8, the ash-like particles are smaller and the inter-particle adhesion is more, indicating that the combustion is performed better under the secondary air ratio and the reaction is more adequate. It can also be seen from Fig. 32.10 that when the secondary air ratio is 0.56, the surface of the ash sample is the most abundant, which proves the conclusion of the analysis of the mesoporous structure more intuitively. It shows when the secondary air ratio is 0.56, it is most conducive to the combustion of semi-coke and the precipitation of volatiles, and it is conducive to the formation and development of pores.

32.4 Conclusions

(1) In semi-coke non-staged combustion, the NO_x emission concentration increases as the temperature of the fuel-rich zone increases. As the air coefficient increases, the NOx emission concentration increases; when the air coefficient is greater than 1.15, the NO_x emission concentration increases to a smaller extent, and the combustion reaction and NOx formation reaction reach a certain limit.

(2) The results of semi-coke air-staged combustion experiments show that the comprehensive NOx emission concentration, burn-out rate and NO_x abatement efficiency change law, under the reasonable air distribution conditions, improve the fuel-rich zone temperature is conducive to semi-coke combustion and NO_x reduction.

(3) From the semi-coke air-staged combustion experiment, it can also be concluded that within the experimental conditions, the optimal secondary air ratio is 0.56, and the NO_x emission concentration is below 120 mg/m^3 under the secondary air ratio. Besides, both the burn-out rate and NO_x reduction ratio are higher.

Acknowledgements This work was supported by the China Coal Science and Technology Group Innovation Fund Project (No. 2018MS003).

References

Bai W, Li H, Deng L (2014) Air-staged combustion characteristics of pulverized coal nder high temperature and strong reducing atmosphere conditions. Energy and Fuels 28:1820–1828

Chunbo W, Tao F, Ming L (2015) Experimental study on coupling and NO emission of volatiles and coal char during the isothermal combustion of pulverized coal. J China Coal Soc 03:665–670

Fang N (2015) Feasibility of semi coke combustion in industrial pulverized coal boiler. Clean Coal Technol 21(2):106–168

Fan W, Lin Z, Li Y, Kuang J, Zhang M (2009) Effect of air-staging on anthracite combustion and NO_Xformation. Energy Fuels 23(2):111–120

Gang D (2013) The research of the properties of blue coke used as a partial substitution for coal in Blast Furnace injection. Xi'an University of Architecture and Technology

Glarborg P, Jensen AD, Johnsson JE (2003) Fuel nitrogen conversion in solid fuel fired systems. Progress Energy Combust Sci 29(2):89–113

He R, Suda T, Takafuji M (2004) Analysis of low NO emission in high temperature air combustion for pulverized coal. Fuel 83(9):1133–1141

Huan S, Ruiqiong L, Xingzhi H (2015) Pulverized coal combustion characteristics at constant high temperatures and model simulation. Electr Sci Eng (09):1–6

Lorenz E, Carreab M, Tamurac N (2000) The role of char surface structure development in pulverized fuel combustion. Fuel 79(10):1161–1172

Qi YF, Zhang MC, Zhang J, Tian FG (2009) Experimental investigation on fast pyrolysis kinetic characteristics of micro-pulverized coal. Chem Eng 03:62–65+74

Qin Z (2013) Numerical simulation on low NO_xcombustion of 600 MW W-shape flame boiler. Zhe Jiang University

Qinlong Z, Jia Y (2013) Development of Yulin semi-coke industry. Ind Sci Tribune 08:61–62

Spliethoff H, Greul U, Rudiger H (1996) Basic effects on NO_x emissions in air staging and reburning at a bench-scale test facility. Fuel 75(5):560–540

Suda T, Takafuji M, Hirata T (2002) A study of combustion behavior of pulverized coal in high-temperature air. Proc Combust Inst, Japan

Taniguchi M, Kamikawa Y, Okazaki T (2010) A role of hydrocarbon reaction for NO_X formation and reduction in fuel-rich pulverized coal combustion. Combust Flame 157(8):1456–1466

Taniguchi M, Kamikawa Y, Okazaki T (2011) Staged combustion properties for pulverized coals at high temperature. Combust Flame 2261–2271

Thommes M (2015) Physisorption of gases, with special reference to the evaluation of surface area and pore size distribution (IUPAC Technical Report). Pure Appl Chem. 87(9/10), 1051–1069. ISSN: 00334545.

Tamura M, Watanabe S, Komaba K (2015) etc Combustion behaviour of pulverised coal in high temperature air condition for utility boilers. Appl Therm Eng 75:445–450

Xin Z (2015) Feasibility on semicoke substitute for anthracite in energy conservation and emissions reduction. Clean Coal Technol 03:103–106

Yongying W, Shi Y, Xing L (2016) Experiment study of double-cone pulverized coal combustion chamber burning char. Clean Coal Technol 22(3):93–97

Zhiqiang G, Zhicheng L, Yeping Z (2014) Experimental study on semi-coke combustion and coal pyrolysis and combustion coupling. J China Coal Soc (S2):519–525

Chapter 33
Numerical Investigation on Combustion Stability of Coal Slurry in the Double Cone Burner

Mo RiGen, Yang Shi, and Liu JianHang

Abstract Fluent reaction flow has been widely used in the simulation of uniform or non-uniform reaction flow including boilers, gas turbines and rocket engines. Flow field characteristics, material concentration and pollutant generation can be obtained (Travin et al. in Advances in LES of complex flows, vol 65, pp 239–254, 2002). In order to study the combustion stability of coal slurry in the double cone burner, numerical investigation based on Reynolds-Averaged Navier Stokes (RANS) equations (Chen and Ghoniem in Energy Fuels 26:4783–4798, 2012) was explored in this paper. Besides, a reverse jet device for atomization is set up in the pulverized coal double cone reverse jet burner. and a test platform to verify the numerical simulation results is obtained. The simulated data are verified by the experimental data. The results show that a numerical simulation platform for coal slurry combustion process has been established. Based on the test and calculation results obtained, the pulverized coal double cone burner is optimized for the coal slurry combustion. The technical measures include that the diameter of the burner outlet is reduced from 500 to 450 mm, the angle of the front cone is increased from 7° to 9°, the swirling flow strength is increased by 20%, and the preheating temperature of the combustion air is increased from 20 to 150 °C. The simulation results show that after optimization, the temperature in the burner increases by 120 °C, the combustion process is effectively strengthened, and the front cone pressure is reduced. the temperature in the burner can be increased, and the combustion of coal slurry can be more efficient. Therefore, the adoption of technical means can effectively solve the problem of coal slurry combustion organization.

Keywords Numerical simulation · Coal slurry · Combustion · Burner

M. RiGen (✉) · Y. Shi · L. JianHang
China Coal Research Institute Company of Energy Conservation Corporation Ltd., Beijing 100013, China
e-mail: ccripeking@foxmail.com

State Key Laboratory of Coal Mining and Environmental Protection (China Coal Research Institute), Beijing 100013, China

National Energy Technology and Equipment Laboratory of Coal Utilization and Emission Control (China Coal Research Institute), Beijing 100013, China

J. Lyu and S. Li (eds.), *Clean Coal and Sustainable Energy*, Environmental Science and Engineering,
https://doi.org/10.1007/978-981-16-1657-0_33

33.1 Introduction

Coal water slurry (Cheng et al. 2008), which approximately consists of 70% coal, 30% moisture, and a small amount of additives (Mishra et al. 2002), is a coal-based clean fuel that can be used in industrial boiler. The preparation technology of coal slurry can utilize wasteliquid generated during various industrial processes, which provides a feasible path for wasteliquid treatment. However, the existing coal slurry combustion technology generally has such problems as difficulty in ignition, low burnout rate, unreasonable combustion organization, and can not ensure stable operation. The coal powder double cone burner developed by the China Coal Research Institute is an advanced technology system for coal combustion and utilization. It is verified by engineering practice that the double cone inverse injection burner has the characteristics such as small ignition power, short starting and stopping time, high thermal efficiency, stable combustion and wide load regulation. It is of great significance to study the adaptability of the combustion mechanism and the combustion process of coal slurry to the development of coal slurry dual cone reverse burner.

The purpose of this study is to apply pulverized coal combustion technologies to coal slurry combustion. The industry scale test and further simulation are carried out in this paper. An attempt is made to identify the optimum design of burner.

33.2 Experimental Set up

The working flow of the burner is as follows. The primary air which carries pulverized coal is fed into the primary air pipe. The pulverized coal is initially heated at the end of the primary air pipe where some of the pulverized coal releases volatiles. Pulverized coal is ejected through the gap between the primary air pipe and the reflow cap. Under the combined action of secondary air swirl and forced reverse jet, a strong swirl region and the primary flame is formed in front of the reflow cap. The jet flow of pulverized coal is heated by the flame at the outlet of the reflow cap. The combustion process of volatile consumes oxygen, which promotes the pulverized coal to release volatiles. Hot volatiles and coke are fully mixed with the secondary air at the joint of the cones where the main flame is formed. And then the airflow carrying pulverized coke enters the furnace and continue to burn while about half of the burning air flows back to the front cone of the burner.

For the same kind of coal, the main difference between the combustion process of pulverized coal and coal slurry is the influence of evaporation and reaction of water. Therefore, this paper investigates the combustion of coal slurry in the burner through a water injection test in the pulverized coal burner, which provides basis for calculation (Fig. 33.1).

Test system process is shown in Fig. 33.2. An atomizing nozzle is arranged in the primary air pipe. Water atomized by compressed air is sprayed out through the gap between the reflow cap and the primary air pipe. The branches for water are all

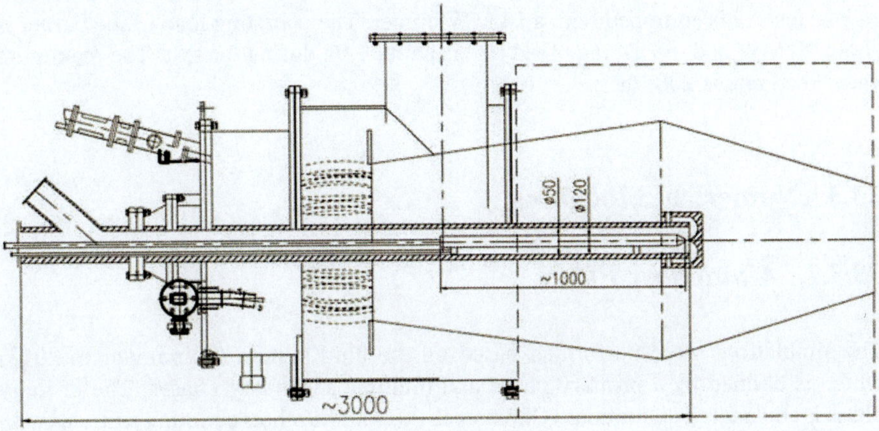

Fig. 33.1 Coal water slurry burner

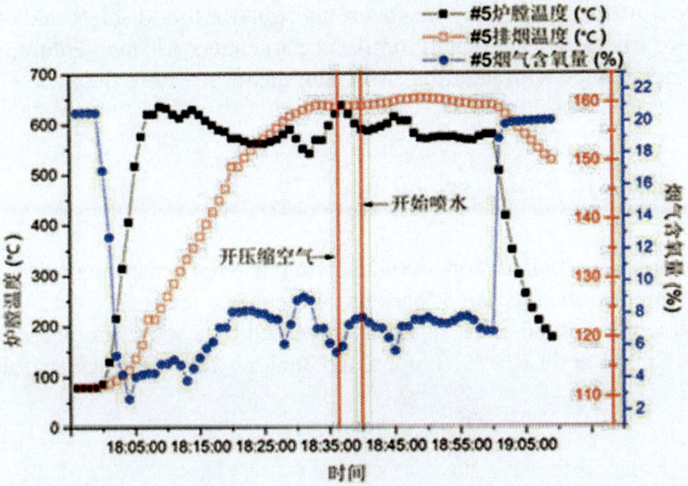

Fig. 33.2 Combustion under 200 L\h spraying water supply

of pressure-resistant flexible pipe, which is convenient for installation, disassembly and replacement. Firstly, the test rig is operated according to normal pulverized coal ignition program. After the system is stable, compressed air and water from the constant pressure pump are introduced through the atomizing nozzle.

Before the experiment, the atomization effect of the atomizing gun was tested. The effective atomization operation parameters such as pump pressure and compressed air pressured were determined.

The test rig is equipped with a 14 MW burner. The operating load of the burner is about 10 MW and the average feed rate is about 1 t/h during the test. The maximum water feed rate is 200 l/h.

33.3 Numerical Modelling

33.3.1 Multiphase Flows

The simulations were performed based on the fluid system (Subramaniam 2013) which is defined by a primary phase and multiple secondary phases. The primary phase is considered continuous (Ghose et al. 2014). The others denoting particles with different sizes are dispersed within the continuous phase. A Eulerian framework is used to solve the continuous phase, while the discrete phase is solved in a Lagrangian framework. Particles can exchange heat (Ghose et al. 2014), mass and momentum with the continuous gas phase. The governing equations for discrete and continuous phases were calculated alternately until the parameters for momentum, mass and energy do not change with iteration. And then the results were output.

33.3.2 Discrete Phase Model

During the process of calculation, the water content in proximate analysis is governed by the decomposition equation of particle surface (Fuertes et al. 1994). Other water content, which is defined as jet liquid, is governed by a set of droplet vaporization equations (Crespo and Li 1975; Tamura and Tanasawa 1958) shown as follows.

$$\frac{dm_\rho}{dt} = k_c A_\rho \rho \ln(1 + B_m) \tag{33.1}$$

$$B_m = \frac{Y_{i,s} - Y_{i,\infty}}{1 - Y_{i,s}} \tag{33.2}$$

where, k_{c} represents mass transfer rate, $Y_{\mathrm{i,s}}$ represents generation rate of component i.

When the temperature of particles is lower than the evaporation temperature, there is only the heat transfer process of particles calculated based on the heating equation of inert component. And the particles do not participate in chemical reaction and mass exchanging with continuous phase.

When the temperature of particles reaches the boiling point, therer are energy and mass exchanging between the particles and the continuous phase, which is expressed as a source in the equations of the continuous phase as shown below:

$$m_p c_p \frac{dT_p}{dt} = hA_p(T_\infty - T_p) + \varepsilon_p A_p \sigma_b(\theta_R^4 - T_p^4) \quad (33.3)$$

where, m_p is the mass of the paticle, c_p refers to specific heat, A_p is the surface area of the paticle, T_∞ is the local temperature of the continuous phase, h is the heat transfer coefficient of the surface area of particles, ε_p is the emissivity of particles, σ_b represents the Boltzmann constant and θ_R^4 is the radiation temperature of particles.

33.3.3 Combustion Model

For the continuous phase turbulent combustion (Yavuzkurt and Ha 1991), the eddy dissipation model (Liu et al. 2012; Raman et al. 2006; Menon et al. 2003) is used to calculate the reaction rate. It is assumed that the chemical reaction rate is much faster than the turbulent mixing rate. The rate of chemical reaction is controlled by the rate at which the unburned gas parcels break up into smaller ones under turbulence. Fuel and oxidant enter the reaction zone through turbulent slowly and burn rapidly. In short, the turbulent combustion process is controlled by the mixing process of fuel and oxidant.

The devolatilisation (Jones et al. 2005) and combustion of the volatile gases from the coal particle are represented using a two-step overall reaction model (Arenillas et al. 2001; Gadiou et al. 2002). The first step is devolatilisation. The second step is the oxidation of CO.

The rate of heterogeneous surface reaction of coke is controlled by the kinetic equation and the diffusion (Heynderickx et al. 2005; Zhang et al. 2010). The model equations representing the relationship above are as follows.

$$\frac{dm_p}{dt} = -A_p pox \frac{D_0 \Re}{D_0 + \Re} \quad (33.4)$$

$$\Re = C_Z e^{-\left(E/RT_p\right)} \quad (33.5)$$

$$D_0 = C_1 \frac{[(T_P + T_\infty)/2]^{0.75}}{d_p} \quad (33.6)$$

where, D_0 is diffusion coefficient, C_1 is density, T_p is temperature of coal slurry droplets, d_p is diameter of coal slurry droplets; dynamic reaction rate; C_Z is pre-exponential factor; E is activation energy; Ap is surface area of coal slurry droplets; pox is partial pressure of oxidant in atmosphere; M_P is mass of coal slurry.

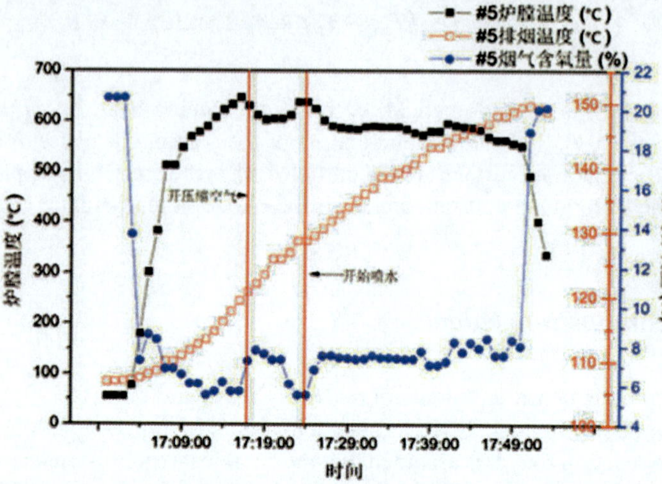

Fig. 33.3 Combustion under 200 L\h spraying water supply equipped with the bluff-body

33.4 Results

33.4.1 Experimental Results

Figure 33.2 is the combustion under the conditions of 200 L\h spraying water supply and 1 t\h pulverized coal supply. The outlet temperature of the burner is 120 °C lower than that of the pulverized coal combustion. The temperature of the end cap of the burner is 241 °C. The possible reason is that compressed air and atomized water increase the pulverized coal ejection speed. At the same time, the presence of water delays the combustion process and leds to the pressure increasing.

In order to reduce the temperature of the end cap of the burner, a bluff-body flame stabilizer was installed on the primary air pipe. Compared with the one without the bluff-body, the end cap temperature of the burner decreases by 100 °C. The maximum water supply is 200 L\h. The exhaust gas measurement of the boiler shows that before and after spraying water, the CO content increased by 40%, and NO_X decreased by 30% (Fig. 33.3).

33.4.2 Comparison Between Simulation and Experiment

The combustion simulation of pulverized coal and coal water slurry was carried out under the experimental conditions using the model mentioned above. Comparisons between calculated and experimental data are shown in Table 33.1.

Table 33.1 Compare the calculated data with the experimental data

	Temperature		Pressure	
	Coal/°C	Slurry/°C	Coal/Pa	Slurry/Pa
Experimental data	910	800		
Simulated data	957	838	3100	3800

Temperature is an effective data for the combustion process. Since the distance between the outlet and measurement point, the outlet temperature obtained by simulation is higher than the experimental data. But It is still in a reasonable range of errors. From the temperature comparison in Table 33.1, the simulated data closely match the experimental results, confirming that the overall temperature field is correctly represented.

The pressure in the burner can not be measured during the test process. However, in the process of coal slurry combustion test, the burner had the situation of the exhaust of burning particles, which indicates that the pressure in the burner under the condition of coal slurry combustion is higer than that of pulverized coal combustion. The simulation results also showed the change.

The emission of NOx is also very sensitive to the change of combustion process. So the change of NOx emission is also an important reference data. The NOx emission measured by experiment is from the gas through heat exchanger. The calculated NOx value is form the burner outlet. So there is a difference between the two sets of data. And it is acceptable. Both of them show that the NO_X emission of coal slurry combustion is lower than that of pulverized coal combustion.

Based on the comparison and analysis of the above data, it can be seen that the numerical model established in this paper has a good fitting between the simulation results and the test results of the combustion process of the double cone burner.

33.4.3 Simulation Results

In order to solve the problem of excessive temperature of the end cap of burner, the analysis based on the calculation results is as follows.

As showen in Fig. 33.4, The flow area of pulverized coal in the front cone of the burner can be divided into the following parts:

1. The first area ‘a’, shorter in length, near the reflux channel, did not ignite.. The second area ‘b’ is the location where the initial flame begins to generate.
2. The third area ‘c’, high temperature reflux zone, a large amount of volatilization, intense combustion, high temperature.
3. The fourth region ‘d’, high temperature, reflux region, lack of oxygen.
4. The fifth region ‘e’, high temperature, strong mixing region.
5. The sixth region ‘f’, the high temperature pulverized coal meets the secondary air and forms a tubular flame, and the coke begins to burn.

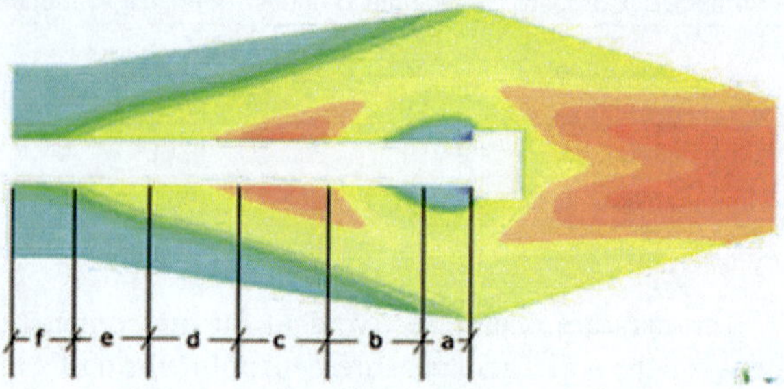

Fig. 33.4 The simulated temperature distribution of coal combustion

According to the above zoning, it can be concluded that the combustion process of pulverized coal is ejected from the reflux cap and heated by the initial flame. Then, through the high temperature reflux zone, the devolatilization occured rapidly due to the insufficient oxygen concentration. In the sixth region, the combustible gas carries coke and meets the secondary air. The combustible gas is ignited first, and then the temperature rises to ignite the coke, forming the main flame in the burner.

Figure 33.5 is the simulation temperature distribution of coal slurry combustion under the operating conditions of the experimental process. Compared with Fig. 33.4, it is not difficult to see that the main flame of coal slurry combustion process is stronger than that of pulverized coal. At the same time, the initial flame of coal slurry combustion is stronger than that of pulverized coal combustion. The simulation results show that the velocity field of the front cone is 28 m/s under the condition of coal slurry and 20 m/s under the condition of pulverized coal. With the increase of reverse injection speed, each region of coal slurry combustion condition extends

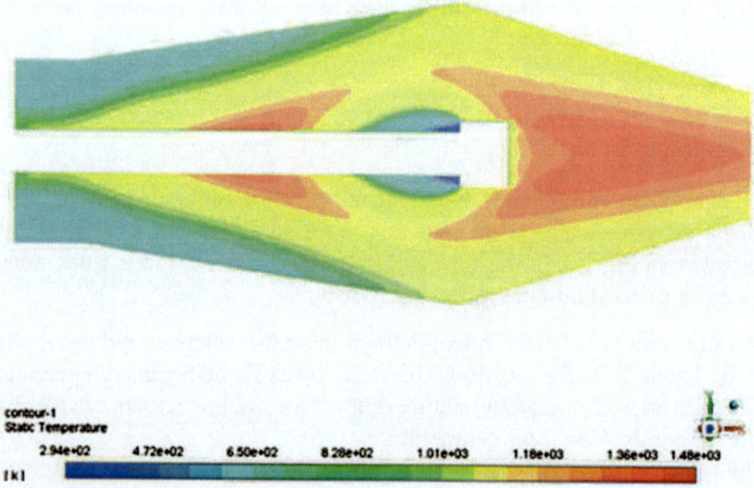

Fig. 33.5 The simulated temperature distribution of coal slurry combustion under the test process 2

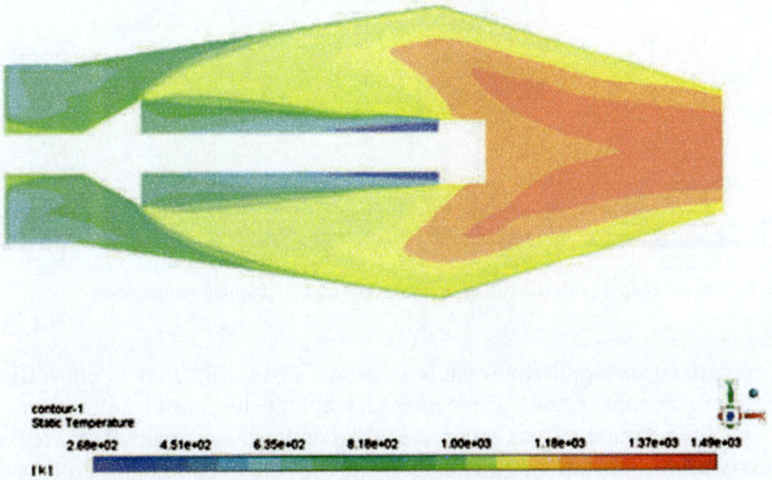

Fig. 33.6 The simulated temperature distribution of coal slurry combustion with bluff-body

forward cone direction compared with that of pulverized coal combustion condition. This causes the primary flame to extend to the end cover of the burner and causes the over-temperature phenomenon of the end cover. In the actual test, the velocity may be higher than the calculated data, which is determined by the following two factors: firstly, the effective flow cross-section area in the primary air pipe is reduced by adding atomizing gun in it; secondly, the outlet velocity of the compressed air of the atomizing gun is much higher than that of the primary air, so it can accelerate the coal slurry droplets.

In order to reduce the overtemperature of the end cap during the combustion of coal slurry, a bluff-body was added to the front cone of the burner in the test. The simulation calculation under the corresponding working conditions is shown in Fig. 33.6.

The bluff-body is designed to adjust the flue gas reflow rate of front cone to adapt to the combustion characteristics of different kinds of fuel. During the experiment, the bluff-body blocked the ejection material from the exit of the reflux cap, and extended the first area mentioned above. The test results show that the temperature of the end cap with bluff-body decreases to 143 °C, and the temperature of the end cap without blunt body reaches 241 °C. The calculation results show that the average temperature of the sixth region decreases from about 800–640 °C after installing bluff- body. Bluff-body may be an effective technical method to solve the contradiction between high-speed jet and residence time of coal slurry in the process of developing coal slurry burner.

Another important phenomenon in the test process is the overpressure of the burner. The specific manifestation is that the gas carrying the burning coke particles is ejected from the positions connected by flanges. The reason for the termination of the test is the damage of the equipment caused by the overpressure of the burner.

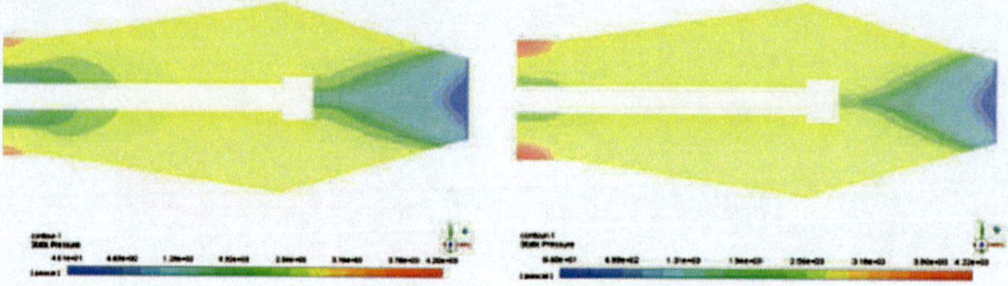

Fig. 33.7 The simulated pressure distribution of coal and coal slurry combustion

The comparison of simulation results is shown in Fig. 33.7. It is not difficult to see that the pressure in the burner increases obviously due to the evaporation of water and the increase of gas volume in the combustion condition of coal slurry. The simulation results are in good agreement with the experimental process. It can also be seen from the observation of the simulation results that the increase of water vapor partial pressure mainly increases the pressure in the area 'f' and the main flame area of the rear cone. These two positions are very important for the stable combustion in the burner. The increase of steam pressure means the decrease of oxygen concentration. The limit of stable combustion in the burner is restricted by this phenomenon. Although there is no fire break due to insufficient oxygen concentration in the test process, the relevant tests are necessary for the development of coal slurry burners.

Based on the obtaining temperature field, the transport equation of NOx pollutant concentration was calculated. And the effects of turbulence and temperature on the formation of NOx were simulated. The distribution of NOx obtained is shown in Fig. 33.8. Because coal slurry contains a lot of water, the vaporization of the water can reduce the core temperature of the flame area. In addition, the presence of water vapor also reduces the oxygen concentration inside the torch. These two aspects have inhibited the NOx emission from coal slurry combustion process. As can be seen from the figure, the NOx emission of coal slurry combustion process is significantly lower than that of pulverized coal combustion process. Volume integration of NOx concentration in the calculation domain shows that the emission of NOx from coal slurry combustion process is 20.23% lower than that from pulverized coal combustion process.

33.4.4 Design of Burner

Based on the above calculation and test results, the structural parameters of 14 MW pulverized coal burner are optimized. The specific measures are as follows:

(1) Increase swirl intensity, increase high temperature flue gas reflowing, and promote ignition;

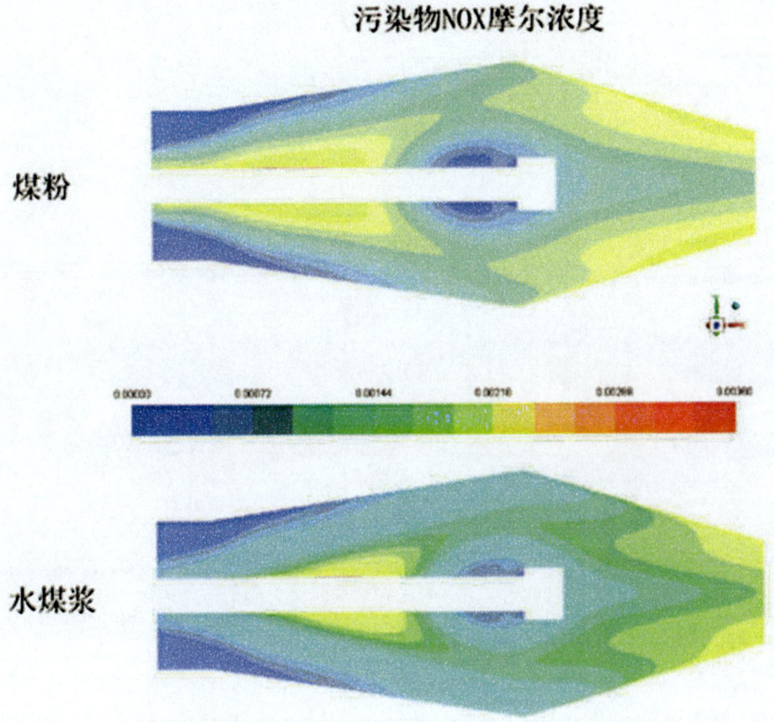

Fig. 33.8 NOx concentration distribution

(2) The front cone of the burner is equipped with a guard belt to ensure the combustion stability.
(3) Increase properly the angle of burner front cone to increase the flue gas reflow;
(4) Appropriately reducing the diameter of the burner outlet to promote the combustion process in the burner;
(5) Increase the air preheater, reduce the exhaust gas temperature, and increase the temperature of combustion-supporting air to facilitate ignition and combustion.

The optimum parameters were obtained by burner process calculation. The diameter of burner outlet was reduced from 500 to 450 mm, the angle of front cone was increased from 7 to 9°, and the swirl intensity was increased by 20%. The preheating temperature of combustion-supporting air was increased from 20 to 150 °C.

A computational model is established and simulated. The results are as follows:

From Fig. 33.9, the combustion process of the optimized burner front cone is enhanced and the ignition is strengthened. The fluctuation of coal slurry temperature decreases, and the average temperature is 120 °C higher than that before optimization. The pressure of the front cone is obviously lower than that before optimization, which is basically close to pulverized coal stituation. The distribution of CO and oxygen

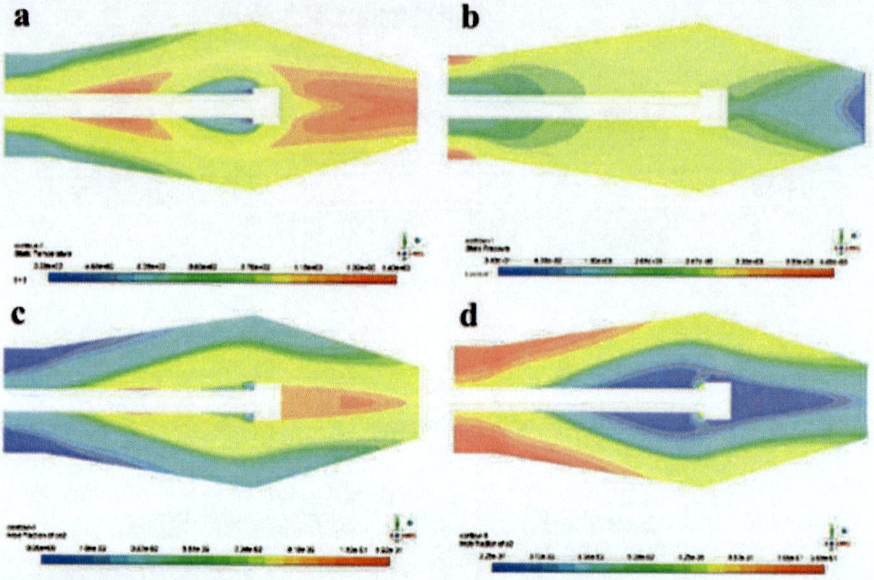

Fig. 33.9 Simulation results of optimized double cone burners (**a**: static temperature; **b**: static pressure distribution; **c**: mole fraction of CO_2; **d**: mole fraction of O_2)

indicates that the combustion process of coal slurry in the burner is more sufficient than before.

The results of numerical simulation show that the combustion structure of coal slurry can be solved by optimizing the burner structure. Since the difference of coal slurry properties, in order to ensure the reliable operation of the equipment, on the basis of the above technology, the combustion stabilization technologies are proposed for the reference of the further research and development of coal slurry burner.

Combustion stabilization 1: natural gas accompanying combustion. Calculation shows that the ratio of natural gas accompanying combustion is 10–20%, that is, the maximum mixing amount of natural gas in a single boiler is 750 Nm^3/h.

Combustion stabilization 2: oxygen-enriched combustion. An oxygen nozzle is arranged in the atomizing spray gun to form a local high temperature stable torch in the ignition area of the burner, which plays a role in intensifying ignition and combustion. Oxygen supply can be prepared by mature technologies such as adsorption and membrane separation. Calculation shows that the oxygen price is calculated at 0.9 yuan/kg, and the cost per GJ is increased by 0.11 yuan.

33.5 Conclusion

The test process is to install a atomizing spray gun in the primary air pipe. Experiments show that the combustion in the burner can be stabilized under the condition of 200 L/h water injection supply. The phenomenon of overtemperature and overpressure of burner appeared during the test. After the bluff-body is installed in the burner, the phenomenon is alleviated.

The operation conditions of the test process were simulated. The results show that the simulation of the combustion process of coal slurry in a double cone burner is in good agreement with the actual situation, and the conclusions obtained in this paper have reference value. On this basis, the pulverized coal double-cone burner is optimized and simulated again. The results show that after optimization, the ignition and combustion process of coal slurry can be effectively strengthened, the pressure of front cone can be reduced, the temperature in the burner can be increased, and the combustion of coal slurry can be more efficient.

Acknowledgements This work was supported by the Coal Science and Technology Development Fund (NO. 2018CX02).

References

Arenillas A, Rubiera F, Pevida C et al (2001) A comparison of different methods for predicting coal devolatilisation kinetics. J Anal Appl Pyrol 58(2):685–701

Chen L, Ghoniem AF (2012) Simulation of oxy-coal combustion in a 100 kWth test facility using RANS and LES: a validation study. Energy Fuels 26(8):4783–4798

Cheng J, Zhou J, Li Y et al (2008) Improvement of coal water slurry property through coal physicochemical modifications by microwave irradiation and thermal heat. Energy Fuels 22(4):2422–2428

Crespo A, Li AA (1975) Unsteady effects in droplet evaporation and combustion. Combust Sci Technol 11(1–2):9–18

Fuertes AB, Marbán G, Pis J (1994) Combustion kinetics of coke particles in a fluidized bed reactor. Fuel Process Technol 38(3):193–210

Gadiou R, Bouzidi Y, Prado G (2002) The devolatilisation of millimetre sized coal particles at high heating rate: The influence of pressure on the structure and reactivity of the char. Fuel 81(16):2121–2130

Ghose P, Patra J, Datta A et al (2014) Effect of air flow distribution on soot formation and radiative heat transfer in a model liquid fuel spray combustor firing kerosene. Int J Heat Mass Transf 74(5):143–155

Heynderickx GJ, Schools EM, Marin GB (2005) Coke combustion and gasification kinetics in ethane steam crackers. AIChE J 51(5):1415–1428

Jones JM, Kubacki M, Kubica K et al (2005) Devolatilisation characteristics of coal and biomass blends. J Anal Appl Pyrol 74(1–2):502–511

Liu J, Chen S, Liu Z et al (2012) Mathematical modeling of air– and oxy-coal confined swirling flames on two extended eddy-dissipation models. Ind Eng Chem Res 51(2):691–703

Menon S, Genin F, Chernyavsky B (2003) Large eddy simulation of scramjet combustion using a subgrid mixing/combustion model. In: 12th AIAA international space planes and hypersonic systems and technologies

Mishra SK, Senapati PK, Panda D (2002) Rheological behavior of coal-water slurry. Energy Sources 24(2):159–167

Raman V, Pitsch H, Fox RO (2006) Eulerian transported probability density function sub-filter model for large-eddy simulations of turbulent combustion. Combust Theor Model 10(3):439–458

Subramaniam S (2013) Lagrangian-Eulerian methods for multiphase flows. Prog Energy Combust Sci 39(2–3):215–245

Tamarin AI, Galershtein DM, Shuklina VM (1982) Heat transfer and the combustion temperature of coke particles in a fluidized bed. J Eng Phys 42(1):14–19

Tamura Z, Tanasawa Y (1958) Evaporation and combustion of a drop contacting with a hot surface. Symp Combust 7(1):509–522

Travin A, Shur M, Strelets M et al (2002) Physical and numerical upgrades in the detached-eddy simulation of complex turbulent flows. Advances in LES of complex flows, vol 65, pp 239–254

Yavuzkurt S, Ha MY (1991) A model of the enhancement of combustion of coal-water slurry fuels using high-intensity acoustic fields. J Energy Res Technol 113(4):268

Zhang X, Sui Z, Zhou X et al (2010) Modeling and simulation of coke combustion regeneration for coked Cr2O3/Al2O3 propane dehydrogenation catalyst. Chin J Chem Eng 18(4):618–625

Chapter 34
Experimental Study of NO_x Emission and Char Characteristics

Jialun Wang, Dongfang Wang, Mingming Wang, Mingyan Gu, and Yuyu Lin

Abstract The NOx emission and char characteristics during coal combustion process in a one-dimension furnace for different air ratios were studied combined with BET, SEM and FTIR analysis. Nitrogen-containing organic compounds in the rapid pyrolysis products of char were determined by pyrolysis chromatography/mass spectrometry (Py-GC/MS). The results obtained shown that when air ratios in the main combustion zone decreased from 0.85 to 0.65, the exit NOx concentration decreased from 315 to 125 ppm. The variation trends of the oxygenic functional group content were contrary to the NOx. The Py-GC/MS experimental results showed that the benzonitrile (BN) released in large amounts from char taken in the primary 400 mm distance. The char specific surface areas (SSA) changed with the coal volatile matters releasing and char burning process along the furnace. With the progress of coal combustion, the numbers of surface micropore first increased and then decreased, following that the macropores appeared, and finally most pores collapsed and chars broke into small particles.

Keywords Coal · NOx · BET · Py-GC/MS · FTIR

34.1 Introduction

At present, air-staged combustion is a "low-NOx emission" combustion technique commonly adopted in coal-fired power plants. In this technique, air is supplied in two stages, the carbon monoxide (CO) will be produced through the combustion of pulverized coal in oxygen-lean atmospheres in the main combustion zone and NHi can be produced through the decomposition of fuel nitrogen, thus reducing existing NOx and inhibiting its further production (Shen et al. 2011, 2013; Cheng

---

J. Wang · D. Wang · M. Wang · M. Gu (✉) · Y. Lin (✉)
School of Energy and Environment, Anhui University of Technology, Ma'anshan 243002, China
e-mail: gumy@ahut.edu.cn

Y. Lin
e-mail: linyuyu88@126.com

J. Lyu and S. Li (eds.), *Clean Coal and Sustainable Energy*,
Environmental Science and Engineering,
https://doi.org/10.1007/978-981-16-1657-0_34

2010; Wen et al. 2019; Li 2017). According to the results of the drop tube furnace experiment conducted by Chen et al., when stoichiometric ratio declined from 1.0 to 0.7, NOx emission concentration decreased from 661.89 to 169.99 mg Nm^{-3} (Chen et al. 2014). Yang et al. studied the influence of stoichiometric ratio of the reduction zone on the emission of NOx during air-staged combustion, and the results revealed a great influence on the production of NOx in the reduction zone (Yang et al. 2014).

According to existing studies, the surface characteristics of coal char particles have a great influence on the on the formation of NOx (Xu et al. 2018). Aarna and Suuberg discovered that changing specific surface area (SSA) under experimental conditions could modify the reactivity of NO-coal coke (Aarna and Suuberg 1999). Jia founded that SSA changed constantly during the combustion of coal char; and with the increase of the combustion reaction rate of coal char, the increasing amplitude of its SSA declined (Jia 2014). Xiao et al. showed that at temperature below 1000 °C, addition of an appropriate amount of oxygen in the reduction reaction of NO-coal coke could significantly promote the reduction of NOx, mainly due to an increase in the surface oxygenic functional groups of coal char (Xiao et al. 2012). According to other related researchers, treating coal char in oxygen atmospheres can improve its reducing capacity, mainly because the oxygen-containing stable complex C(O) produced on particle surface enhances the reduction reactivity of coal char. In this case, the complex mainly exists in the form of phenol, carboxyl, anhydride, and ether bonds, which rapidly and continuously consume NOx and thus reduce NOx during coal combustion (Wang et al. 2017, 2018).

Utilization of Fourier transform infrared (FTIR) spectroscopy and pyrolysis–gas chromatography/mass spectrometry (Py-GC/MS) to study the migration characteristics of the nitrogenous substances in coal char helps in in-depth analysis of the producing characteristics of NOx (Luo et al. 2018; Jiang et al. 2019). Hou et al. adopted FTIR spectroscopy to compare the yields of NO_2 during the combustion of SM coal and SM char, and founded that NO_2 began to escape at 300 °C in the entire combustion process, escaped NO_2 reached the maximum value at 1000 °C, and its content produced from SM char was lower than that produced from SM coal (Hou et al. 2018). Arenillas et al. conducted an experimental study on the influencing factors of coal char NO reduction by TG-MS-FTIR, indicating that heterogeneous reduction depends largely on temperature (Arenillas et al. 2002). Wei et al. adopted Py-GC/MS to study the pyrolysis charaterists of charcoal. According to their findings, the pyrolysis products of char from Pingdingshan coal is dominated by aromatic compounds, and also contains nitrogenous compounds and other compounds (Wei et al. 2017).

During air-staged combustion, the morphological characteristics and nitrogenous substances of the coal char in the primary combustion zone varied significantly with the combustion of pulverized coal, and exerted a great influence on the production characteristics of NOx. This study adopted a one-dimensional (1D) furnace to study pulverized coal combustion characteristics under main combustion zone air ratios (f_M) of 0.65, 0.75, and 0.85. Further, distributions of the surface oxygenic functional groups, nitrogenous groups, SSA, and morphological characteristics of coal char along the furnace height were investigated, and a theoretical foundation was laid for deep staged combustion.

34.2 Experimental Equipment

The test bench of the 1D furnace consists of furnace body, temperature control system, feeding system, air distribution system, cooling water system, sampling gun, and flue gas analyzer system, as shown in Fig. 34.1.

The 1D furnace is a vertical-type tubular electric heating furnace length of 1500 mm, and adopts PID intelligent temperature-controlled heating system. There is also an airtight high-temperature quartz glass tube with a length of 1800 mm, an inner diameter of 80 mm, and a constant-temperature interval length of 1200 mm. Shenhua coal was used in the experiments performed in this study. Composition of coal is listed in Table 34.1, and the experimental conditions are summarized in Table 34.2.

Particle size of the pulverized coal used in the experiment was analyzed using an OMEC LS-C (IIA) laser particle size analyzer. Char analysis was carried out using a Nicolet 6700 FTIR spectrometer (Thermo Nicolet Corporation, the US). Coal char pyrolysis was carried out using a CDS5200HP-R pyrolyzer (CDS, the US), and product was analyzed using a QP2020 gas chromatography/mass spectrometer (SHIMADZU, Japan), configured with Rtx-5 ms capillary column with size of 30 m

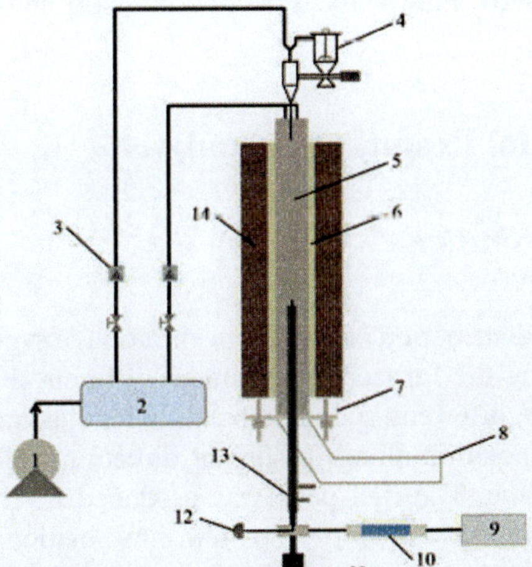

1-air compressor; 2-pressure stabilizer; 3-mass flow meter; 4-coal screw feeder; 5-high-temperature quartz tube; 6-resistive heater; 7-bottom fixed pallet; 8-exhaust gas tube; 9-flue gas analyzer; 10-drying filter; 11-coal char collector; 12-thermocouple; 13-water-cooled sampling gun; 14-temperature control thermocouple; 15-thermal-protective coating.

Fig. 34.1 Schematic illustration of experimental system of the one-dimensional furnace

Table 34.1 Characteristics of SH coal

Ultimate analysis, wt% (as air-dried)				
C	H	O	N	S
71.64	4.3	18.29	0.82	0.14
Proximate analysis, wt% (as air-dried)				
Volatile matter	Ash	Moisture	Fixed carbon	Low heating value (MJ/kg)
33.33	4.25	0.56	61.86	21.72

Table 34.2 Experimental parameters

Coal mass rate (g/min)	Atmosphere	Excess air ratio	Furnace temperature (°C)	Coal diameter (μm)
7.27	21%O_2 + 79%N_2	1.2	1000	78
Main combustion zone air ratio		Main combustion zone air stream (L/min)		
0.65		39.18		
0.75		45.21		
0.85		51.23		

× 0.25 mm × 0.25 μm (Restek, the US). Surface pore characteristics of char were analyzed using a micro active ASAP 2460 Version 2.01 low-temperature nitrogen adsorption apparatus (Micromeritics Instrument Corporation, the US). The scanning electron microscope used for the analysis of coal char morphology was JSM-6510LV.

34.3 Experimental Results and Analysis

34.3.1 NOx Distribution

Figure 34.2 shows the distribution of NOx concentration, oxygen concentration, and CO concentration along the furnace height under different main combustion zone air ratios. For the three different main combustion zone air ratios, NOx shows the first peak at 600 mm position along the height direction, followed by the second one after a slight decline. The first peak was produced by the release of volatile components; NOx content was low at 200–400 mm position, mainly because the heated coal particles released less volatile components and produced less NOx. The second peak was produced by the combustion of coal char. When $f_M = 0.65$ and 0.75, the peak of NOx emerged at 1000 mm position; when $f_M = 0.85$, its emergence position moved slightly backward (1200 mm).

When main combustion zone air ratios increased from 0.65 to 0.85, the first peak of NOx increased from 277 to 401 ppm, the second peak increased from 185 to 359 ppm,

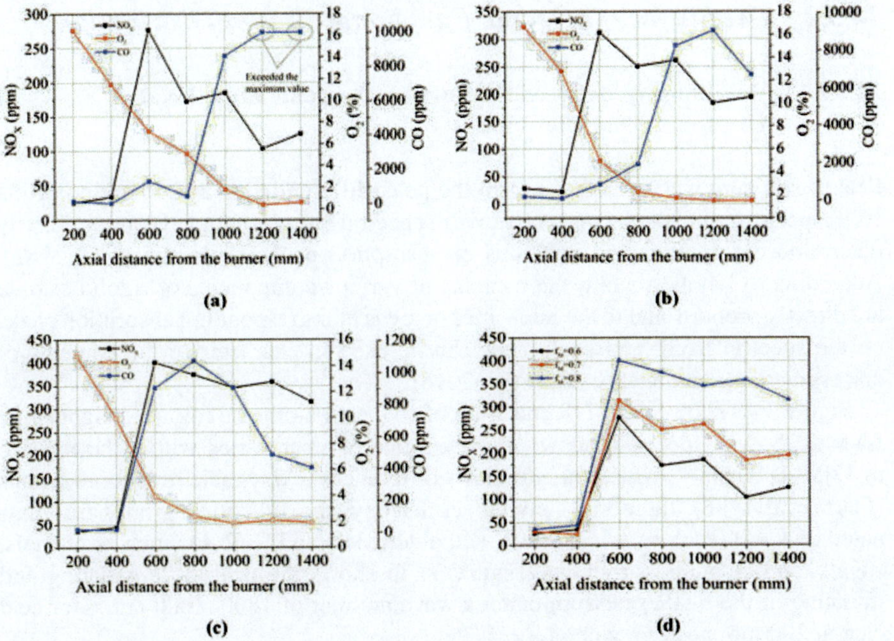

Fig. 34.2 NOx, O_2, and CO concentrations as functions of axial distance from the burner with different burner combustion zone air ratios: **a** $f_M = 0.65$; **b** $f_M = 0.75$; **c** $f_M = 0.85$

and the exit concentration of NOx increased from 125 to 315 ppm. CO concentration presented a single peak along the furnace height, but peak position varied under different main combustion zone air ratios. When $f_M = 0.75$, its peak emerged at 1200 mm position; and when $f_M = 0.85$, its peak emerged at 800 mm position. When $f_M = 0.65$, CO concentration exceeded the maximum range (10,000 ppm) of the flue gas analyzer after 1200 mm position. This indicated that, when f_M was 0.65, the incomplete combustion degree of coal char increased after 1200 mm position, and the CO concentration in the products increased. For the three different main combustion zone air ratios, oxygen concentration gradually declined along the furnace height, and exit concentration was uniformly lower than 2%.

The above mentioned analysis showed that for the primary combustion zone of air-staged combustion, when main combustion zone air ratios decreased from 0.85 to 0.65, the exit concentration of NOx declined from 315 to 125 ppm.

34.3.2 *The Surface Oxygenic Functional Groups Distribution*

34.3.2.1 Relationship Between the Surface Oxygenic Functional Groups of Coal Char and NO_X

Peakfit software was used to perform the peak-differentiating and imitating of the FTIR spectrum, and the second derivative of spectral line was adopted to preliminarily determine the number and positions of absorption peaks (Solomon et al. 1982). According to Lambert's law, the contents of various components of a solid sample are directly proportional to the intensities or areas of corresponding absorption peaks on the spectral band, and peak areas can be used to characterize the contents of corresponding components (Yan et al. 2016).

Figure 34.3 shows the FTIR spectrum of coal char along the furnace height when $f_M = 0.75$. The spectrum shows that the wave number varies within the interval of 1800–1000 cm^{-1}, which mainly shows differences in oxygenic functional groups (Ibarra et al. 1996). Table 34.3 summarizes that oxygenic functional groups are dominated by C=O (such as acids, esters, and aldehydes) and C–R–O (such as phenols, sugars, and alcohols) structures. Figure 34.3b shows the peak-differentiating and imitating of the FTIR spectrum under a wave number of 1800–1000 cm^{-1} for coal char at 200 mm position when $f_M = 0.75$.

Figure 34.4 shows the absorption peak intensities of the surface oxygenic functional groups of coal char along the furnace height under different main combustion zone air ratios and the changes in NOx. Clearly, when $f_M = 0.65$, the absorption peak intensity of oxygenic functional groups first gradually declined along the furnace height, then reached the first valley 2.83 $m^2\ g^{-1}$ at 600 mm position, and finally reached the peak 8.86 $m^2\ g^{-1}$ at 800 mm position. The corresponding NOx concentration reached the maximum value 277 ppm at 600 mm position. With the progress of combustion, the absorption peak intensity reached the second valley 0.98 $m^2\ g^{-1}$ at 1200 mm position, while the corresponding NOx concentration increased from

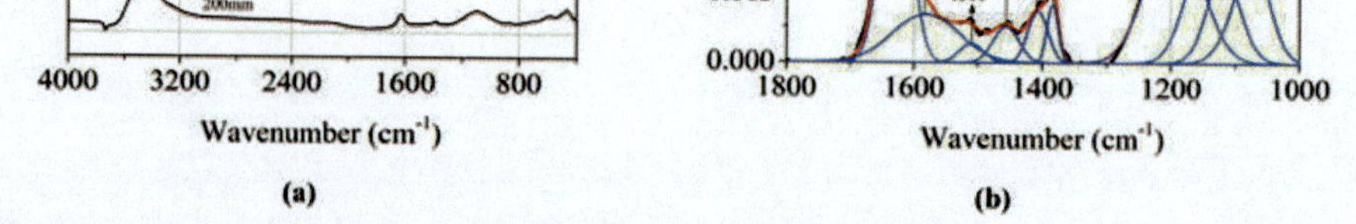

Fig. 34.3 The FTIR analysis of chars (**a**) with fM = 0.75 and infrared spectrum fitting results of coal char with fM = 0.75 at 200 mm position

Table 34.3 Band assignments of FTIR spectra in the 1800–1000 cm^{-1} region (Ibarra et al. 1996)

Absorption frequency (cm^{-1})	Functional group structure
1772	Aryl esters
1703	Carboxyl acids
1650	Conjugated C=O
1618	Aromatic C=C
1586	Aromatic C=C
1502	Aromatic C=C
1458	δ as.CH3-, CH2-
1437	Aromatic C=C
1410	δ as.CH-(CH3); δ OH
1377	δ s.CH3-Ar, R
1350	δ s.CH2-C=O
1274	ν C-O in aryl ethers
1222	ν C-O and δ OH, phenoxy structures, ethers
1195	ν C-O phenols, ethers
1168	ν C-O phenols, ethers
1138	ν C-O, tert. alcohols, ethers
1094	ν C-O sec, alcohols
1036	Alkyl ethers, Si–O

Note ν stretching vibration; δ deformation vibration; *s* symmetry; *as* asymmetry

173 ppm at 800 mm position to the second peak 185 ppm at 1000 mm position. When $f_M = 0.75$, the absorption peak intensity of oxygenic functional groups first declined to the first valley 3.29 $m^2\ g^{-1}$ at 600 mm position and then increased to the first peak 4.79 $m^2\ g^{-1}$ at 800 mm position, while the corresponding NOx concentration declined from the maximum value 313 ppm at 600 mm position to 251 ppm at 800 mm position. Along the furnace height, the absorption peak intensity of oxygenic functional groups again declined to the second valley 4.03 $m^2\ g^{-1}$ at 1000 mm position, while NOx concentration increased up to 262 ppm at 1000 mm position. With the progress of combustion, the absorption peak intensity of oxygenic functional groups increased to the second peak 4.87 $m^2\ g^{-1}$ at 1200 mm position. When $f_M = 0.85$, the absorption peak intensity of oxygenic functional groups first increased slightly at 400 mm position based on that at 200 mm position, and then reached the valley 5.03 $m^2\ g^{-1}$ at 800 mm position. With the progress of combustion, it reached the peak 7.66 $m^2\ g^{-1}$ at 1000 mm position. In contrast, NOx concentration first declined from the maximum value 401 ppm at 600 mm to 346 ppm at 1000 mm position, and then gradually declined to 4.67 $m^2\ g^{-1}$ at 1400 mm position. NOx concentration first increased slightly, and then declined within the interval of 1000–1400 mm position.

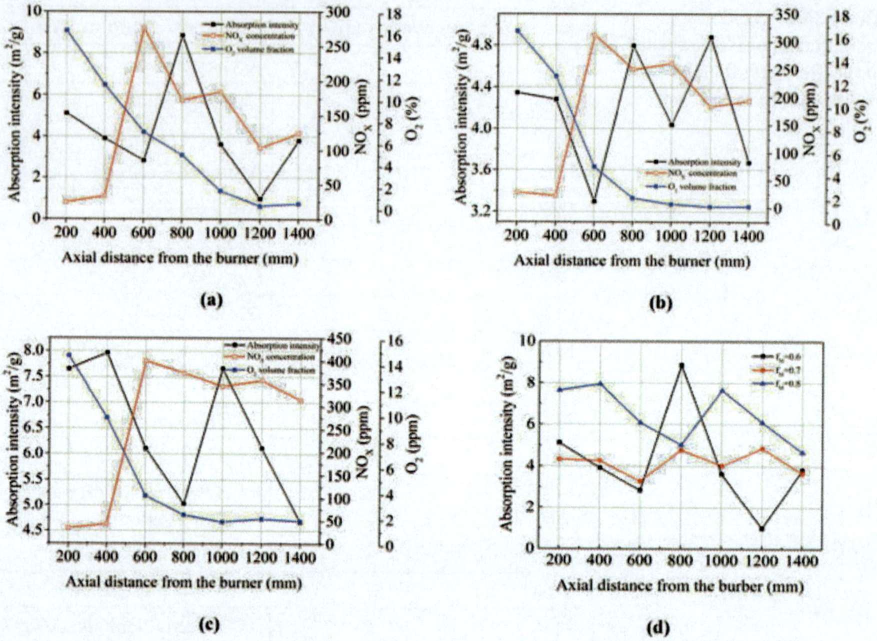

Fig. 34.4 The absorption intensity of oxygen-containing functional group, NO_X, and O_2 concentrations as functions of axial distance from the burner with different main combustion zone air ratios: **a** $f_M = 0.65$; **b** $f_M = 0.75$; **c** $f_M = 0.85$

According to the above mentioned analysis, the variation trends of the oxygenic functional group content were mainly contrary to the NOx.

34.3.2.2 Relationship Between the Nitrogenous Groups Among the Pyrolysis Products of Coal Char and NOx

Figure 34.5 shows the changes in the contents of benzonitrile and phenylethylamine among the pyrolysis products of coal char under different main combustion zone air ratios along the furnace height. Clearly, under different fs values, the contents of BN and BNEP presented basically the same variation trend. At 200 mm position, the HCN precursors, namely, BN and BNEP in coal char were released in large amounts, and the nitrogenous functional groups in coal changed before 200 mm position. At 400 mm position, the contents of the two HCN precursors were low, indicating that the two nitrogenous precursors were released in large amounts in coal char before 400 mm position. At 600 mm position, NOx concentration reached the maximum value, mainly because the two HCN precursors were oxidized in large amounts into NOx. At a high temperature, BNe was released in small amount while BNEP were released in large amount, indicating that it was easier for char nitrogen to produce the latter product through thermal decomposition. When the coke caught fire, the release

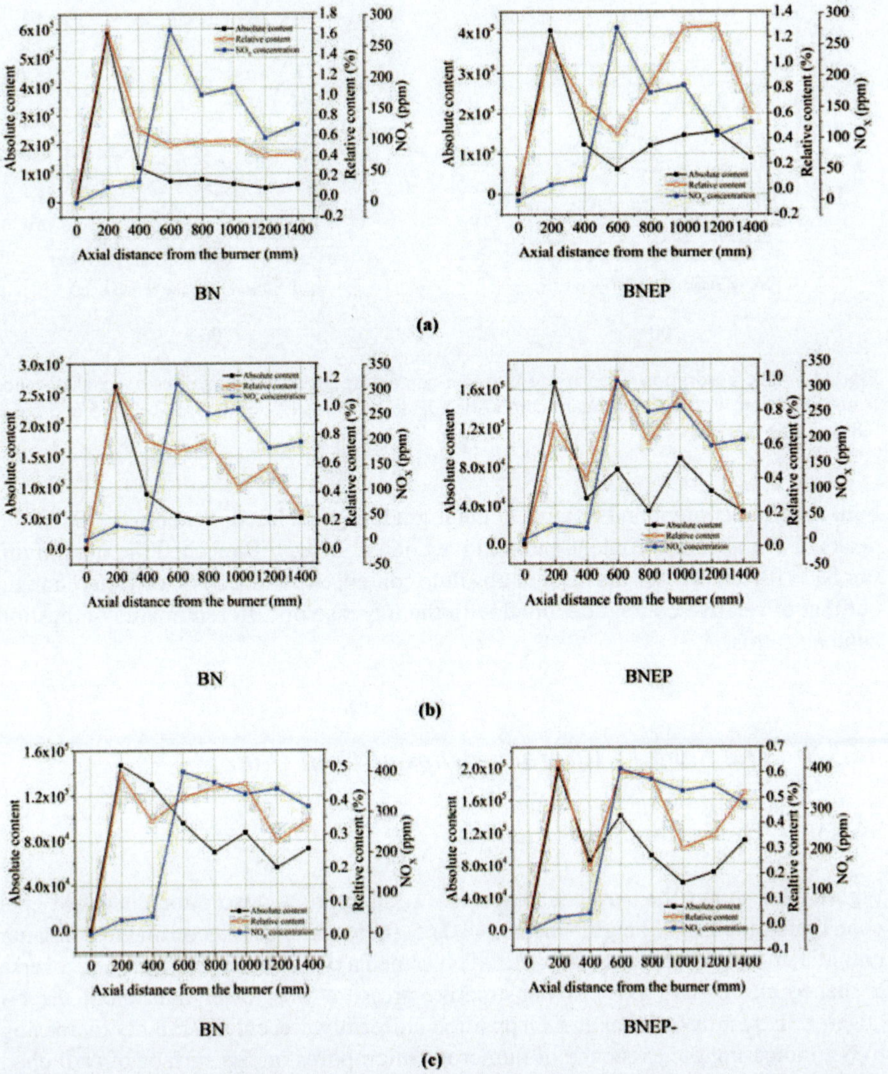

Fig. 34.5 The absorption intensity of nitrogen-containing groups and NO_X concentrations as functions of axial distance from the burner with different main combustion zone air ratios: **a** $f_M = 0.65$; **b** $f_M = 0.75$; **c** $f_M = 0.85$

of the two precursors still continued, resulting in the increase of the corresponding NOx concentration again at 1000 mm position.

Figure 34.6 demonstrates the changes in the contents of the nitrogenous groups among the pyrolysis products of coal char under different f_M. Clearly, under different main combustion zone air ratios, the absolute content and relative content of BN presented basically the same variation trend. With the increase of f_M, the peaks of

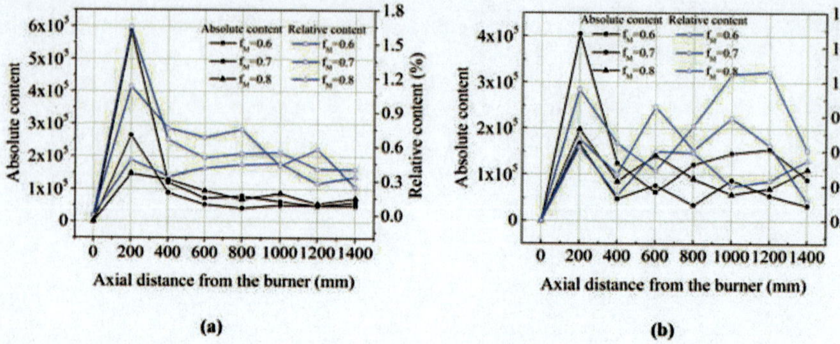

Fig. 34.6 The absorption intensity of nitrogen-containing groups as functions of axial distance from the burner with different main combustion zone air ratios ($f_M = 0.65$; $f_M = 0.75$ and $f_M = 0.85$): **a** BN; **b** BNEP

both absolute content and relative content gradually declined. When $f_M = 0.65$, the peaks of the absolute content and relative content of BNEP reached the maximum. For $f_M = 0.75$ and 0.85, the peak of absolute content experienced no obvious change, but that of relative content declined with the increase of different main combustion zone air ratios.

34.3.3 The Surface Characteristics of Coal Char

34.3.3.1 The Surface Pores Analysis

Figure 34.7 shows the surface isothermal N_2 adsorption–desorption curves of coal char along the furnace height when $f_M = 0.65$, 0.75, and 0.85. According to Brunauer classification, coal char at each position presented a typical II-type curve, i.e., reverse S-shaped curve (Lv 2009). When relative pressure was lower than 0.05, the N_2 adsorption quantity of char at each position uniformly presented a rapidly increasing trend, indicating the existence of numerous micropores on the surface of coal char. When relative pressure was lower than 0.5, mainly micropore filling and monolayer adsorption occurred; and when it was above 0.5, mainly multilayer adsorption and capillary condensation took place. When relative pressure was about 0.5, there was no overlapping between adsorption–desorption isothermal curves, and the desorption curve was above the adsorption curve, resulting in hysteresis of adsorption. When relative pressure was close to 1, adsorption quantity increased rapidly again, mainly because adsorption occurred on the surface of large pores and adsorption–desorption isothermal curves were basically identical (Li et al. 2009; Howaniec 2016; Sadhukhan et al. 2009). Under the same relative pressure, for coal char at each position, the greater the degree of separation between adsorption–desorption isothermal curves, the higher the corresponding pore content. When $f_M = 0.65$, with the increase of distance along

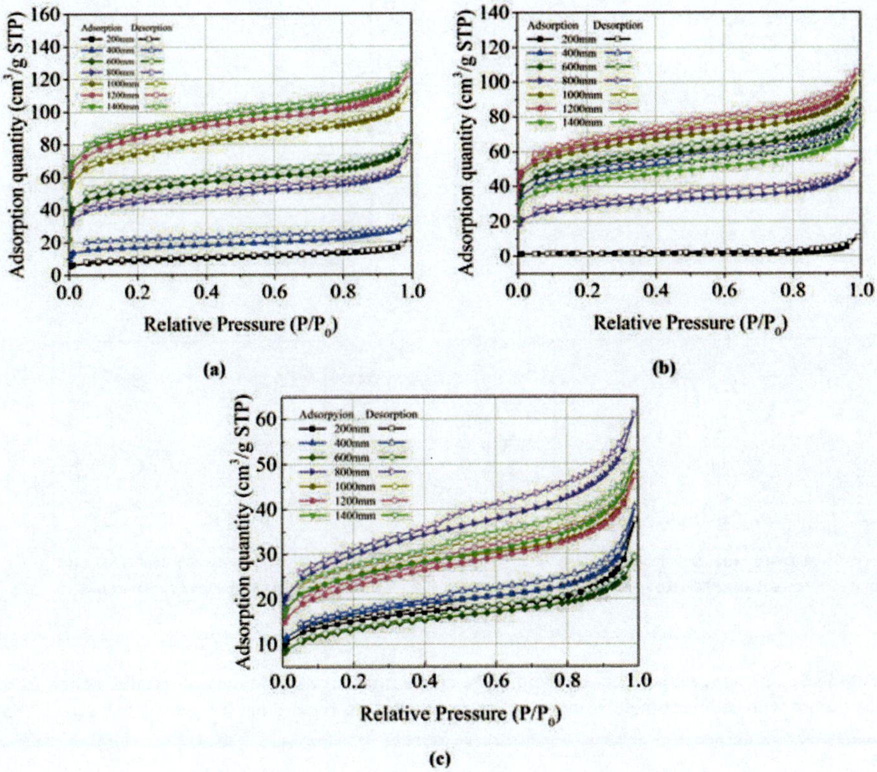

Fig. 34.7 Adsorption/desorption isotherms of char particles as functions of axial distance from the burner with different main combustion zone air ratios: **a** $f_M = 0.65$; **b** $f_M = 0.75$; **c** $f_M = 0.85$

the furnace height, the N_2 adsorption quantity of coal char increased as well. When $f_M = 0.75$ and 0.85, with the progress of combustion, adsorption quantity first increased, and then decreased. With the increase of f_M, adsorption–desorption curves presented different reverse S-shaped trends. When relative pressure was close to 1, the greater the fs value, the more obvious the changes in the reverse S-shaped curves.

34.3.3.2 The Specific Surface Area Analysis

Figure 34.8 demonstrates that when main combustion zone air ratio $f_M = 0.65$, the particles were in a stage of absorption of heat and release of volatile components at 200–400 mm; the SSA increased to the peak 179 $m^2\ g^{-1}$ at 400 mm position, accompanied by the production of nitrogenous small molecules. With the progress of combustion, the SSA first gradually declined to the valley 139 $m^2\ g^{-1}$ at 800 mm position, and then increased again to 277 $m^2\ g^{-1}$ at 1400 mm position. The corresponding nitrogenous small molecules were oxidized into NOx at 600 mm position.

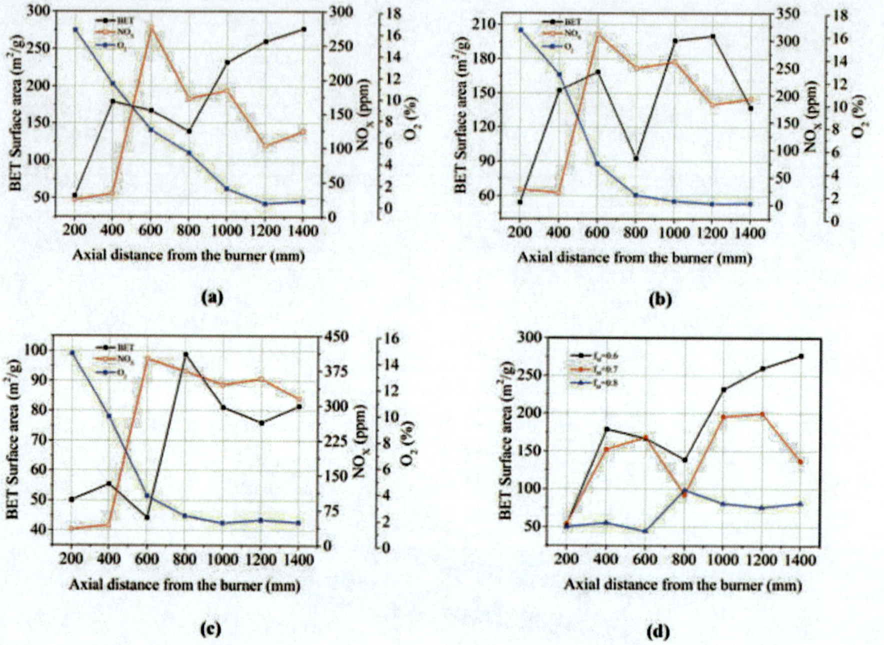

Fig. 34.8 The specific surface area and NOx concentrations as functions of axial distance from the burner with different main combustion zone air ratios: **a** $f_M = 0.65$; **b** $f_M = 0.75$; **c** $f_M = 0.85$

The entry into the stage of combustion of coal char caused NOx concentration to first increase slightly and then gradually decline. When $f_M = 0.75$, the SSA at 600 mm position reached the first peak 167 $m^2\ g^{-1}$, which was slightly lower when $f_M = 0.65$. At 800 mm position, the SSA of particles reached the valley 92 $m^2\ g^{-1}$; later it reached the second peak 200 $m^2\ g^{-1}$ at 1200 mm position. The corresponding NOx concentration first reached the peak at 600 mm, and then began to decline. After entering the stage of combustion of coal char at 1000 mm, NOx concentration first increased slightly, and then continued to decline. When $f_M = 0.85$, SSA reached the peak 98 $m^2\ g^{-1}$ at 800 mm position. Before 800 mm position, SSA was small, and rarely increased; the corresponding NOx concentration first reached the peak 401 ppm at 600 mm position, and then gradually declined to 315 ppm at 1400 mm position.

34.3.3.3 Scanning Electron Microscopy-Based Surface Morphology Analysis

Figure 34.10 shows the SEM images of char along the furnace height under different main combustion zone air ratios. In 200–400 mm distance, the coal particles were in a stage of release of internal moisture and volatile components. In this period,

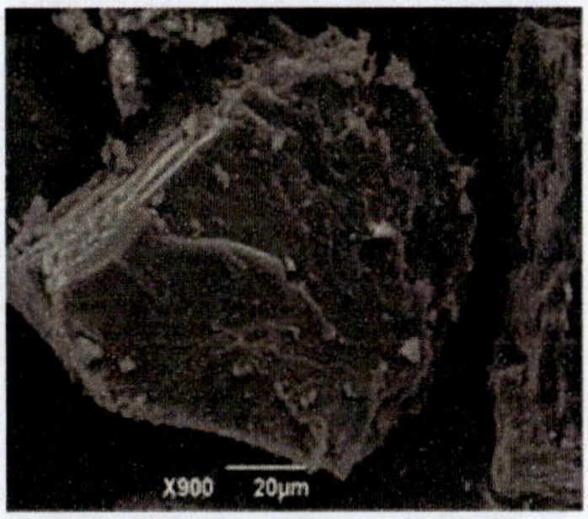

Fig. 34.9 SEM images of raw coal

pore channels could not meet the demand of the release of volatile components in large amount, thus the volatile components gathered inside char and resulted in the expansion of particles, causing the surface pores of coal char to increase. This can also be confirmed from Fig. 34.6. With the release and combustion of volatile components, a lot of heat was released, resulting in the plastic deformation of coal char; and the morphology of surface of particles also changed from angular (Fig. 34.9) to smooth and round (Sadhukhan et al. 2009). At 600–1400 mm position, the small surface pores of char began to decrease, accompanied by the appearance of large pores. This was possibly attributed to the fact that the char had almost completed the stage of release of volatile components, and begun to enter a stage of coke combustion, where small pores collapsed and merged into larger ones (Hu et al. 2001). With the progress of the combustion of char, the entire structure of char collapsed, formed irregular shapes, and eventually caused the crushing and fragmentation of char. Figure 34.10 exhibits that under different main combustion zone air ratios, the changes in the surface pore characteristics of char were basically the same. With the increase of f_M, the surface pore characteristics of char became more advanced, combustion became more complete, and the crushing of pore structures became more intense.

34.4 Conclusions

Coal combustion NOx emission and char surface characteristics along a one-dimension furnace at different air ratios were studied by gas analyzer, FTIR, Py-GC/MS, BET and SEM analysis. Conclusions were as follows:

(1) With the decrease of f_M, the NOx concentrations at the exit and the peak of NOx curve along the furnace decreased. When $f_M = 0.65$, the exit concentration of NOx was less than half of that when $f_M = 0.85$, and the exit concentration of NOx when $f_M = 0.75$ was between the two values.

(2) Analysis of the surface oxygenic functional groups of char indicated that the variation trends of the oxygenic functional group content were mainly contrary to the NOx distributions at different air ratios. BN and BNEP were released in

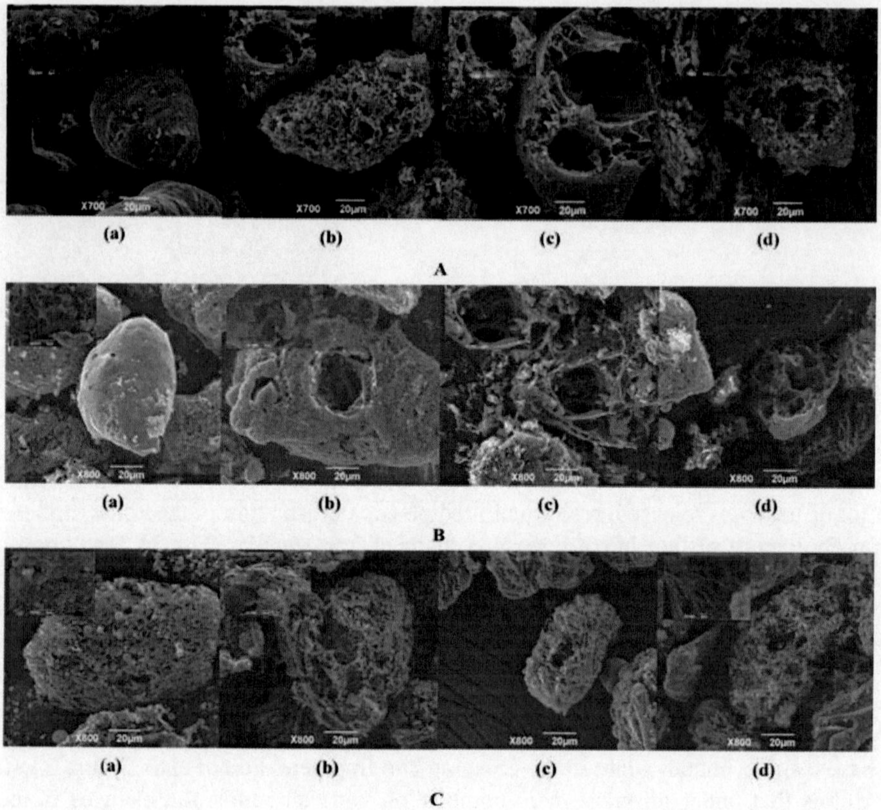

Fig. 34.10 SEM images of char particles with different main combustion zone air ratios: **A** ($f_M = 0.65$); **B** ($f_M = 0.75$); **C** ($f_M = 0.85$); **a** (200 mm); **b** (600 mm); **c** (1000 mm); **d** (1400 mm)

large amounts in char before 400 mm position, and oxidized in large amounts into NOx after 600 mm position.

(3) The adsorption–desorption curves of coal char showed a typical reverse S shape. With the increase of main combustion zone air ratios, adsorption–desorption curves presented different reverse S-shaped trends. In the coal char combustion zone, the higher SSA of char is more selective for NOx reduction.

(4) There was a shift from the dominance by micropores to the gradual increase of large pores on the surface of pulverized char particles along the furnace height. Large pores began to merge and decrease on the surface of char, and the particles began to collapse into irregular shapes.

Acknowledgements This work was supported by the National key basic R & D project of China (2017YFB0601805) and the National Natural Science Foundation of China (51776001, 51376008, 51506128).

References

Aarna I, Suuberg EM (1999) The role of carbon monoxide in the NO-carbon reaction. Energy Fuels 13(6):1145–1153

Arenillas A, Rubiera F, Pis JJ (2002) Nitric oxide reduction in coal combustion: role of char surface complexes in heterogeneous reactions. Environ Sci Technol 36(24):5498–5503

Chen Z, Wang Z, Li Z et al (2014) Experimental investigation into pulverized-coal combustion performance and NO formation using sub-stoichiometric ratios. Energy 73:844–855

Cheng J (2010) Application of air-staged combustion for energy-conservation and pollutant-reduction in utility boilers. Energy Conserv Technol 28(3):236–240

Hou J, Ma Y, Li S et al (2018) A comparative study on characteristics of sulfur and nitrogen transformation and gaseous emission for combustion of bituminous coal and char. Carbon Resour Convers 1(1):86–93

Howaniec N (2016) The effects of pressure on coal chars porous structure development. Fuel 172:118–123

Hu S, Sun XX, Li M et al (2001) Study on surface structure of coal during combusion with SEM analysis. J Fuel Chem Technol 29(5):463–467

Ibarra J, Munoz E, Moliner R (1996) FTIR study of the evolution of coal structure during the coalification process. Org Geochem 24(6):725–735

Jia LC (2014) Experimental study on the influence of Char pore structure on Char nitrogen release characteristics. Harbin Institute of Technology

Jiang Y, Zong P, Tian B et al (2019) Pyrolysis behaviors and product distribution of Shenmu coal at high heating rate: a study using TG-FTIR and Py-GC/MS. Energy Convers Manag 179:72–80

Li XL, Xu SQ, Zhou ZJ et al (2009) Effect of prolysis condition on the pore structure of Huainan coal char. Coal Convers 32(4):8–12

Li PX (2017) Principle and application of air staged combustion technology. Boiler Technol 2017(4)

Luo L, Yao W, Liu J et al (2018) Evolution of NO_X precursors of superfine pulverized coal with a fixed bed in N_2 and CO2. Fuel 234:263–275

Lv YC (2009) Study on pore structure and functional group changes during mixed coal combustion. Huazhong University of Science and Technology

Sadhukhan AK, Gupta P, Saha RK (2009) Characterization of porous structure of coal char from a single devolatilized coal particle: Coal combustion in a fluidized bed. Fuel Process Technol 90(5):692–700

Shen J, Jiang XM, Liu JX et al (2011) Investigation of NOx emissions for superfine pulverized coal in air-staging combustion

Shen J, Liu JX, Zhang H et al (2013) NOx emission characteristics of superfine pulverized anthracite coal in air-staged combustion. Energy Convers Manag 74:454–461

Solomon PR, Hamblen DG, Carangelo RM (1982) Applications of Fourier transform IR spectroscopy in fuel science

Wang Z, Sun R, Ismail TM et al (2017) Characterization of coal char surface behavior after a heterogeneous oxidative treatment. Fuel 210:154–164

Wang Z, Sun R, Zhao Y et al (2018) Investigation of demineralized coal char surface behaviour and reducing characteristics after partial oxidative treatment under an O2 atmosphere. Fuel 233:658–668

Wei SY, Song JZ, Peng PA et al (2017) Py-GC/MS study on the characteristics of soot and charcoal derived from biomass materials and coal. Geochimica 2017(3)

Wen L, Ziqu O, Xiaoyang C et al (2019) The influence of air-stage method on flameless combustion of coal gasification fly ash with coal self-preheating technology. Fuel 235:1368–1376

Xiao M, Wang JC, Li Y et al (2012) Influence of the oxygen-contained functional groups on the surface of coal coke at a high temperature on the NO-coal coke reduction reaction. J Eng Therm Energy Power 27(2):227–231

Xu J, Sun R, Ismail TM et al (2018) Nitrogen/NO conversion characteristics of coal chars prepared using different pyrolysis procedures under combustion conditions. Fuel 211:484–491

Yan XZ, Qiu J, Yin Y et al (2016) Effects of functional groups in lignite on its combustion characteristics. Coal Sci Technol 4:169–174

Yang J, Sun R, Sun S et al (2014) Experimental study on NOx reduction from staging combustion of high volatile pulverized coals. Part 1. Air staging. Fuel Process Technol 126:266–275

Part IV
Fluidized Bed Combustion

Chapter 35
Study on Penetrability of Central Secondary Air Jet in CFB and Its Influences on Fluidization Characteristics

Chen Lin, Juan Yu, Yao Zhang, Fan Feng, and Zhong-xiao Zhang

Abstract In this paper, cold-state experiments were conducted in a self-built circulating fluidized bed to study the influences of total air volume, ratio of the volume of secondary air to total air (SAR), velocity of secondary air and the secondary injection height on the penetrability of secondary air, the chamber's pressure drop, the particle circulating flow rate and the distribution of particle concentration. Results suggest that when the total air volume rises or SAR decreases, the chamber's overall pressure drop and circulating flow rate increase. Apparent particle concentration increases substantially when total air volume rises and the increase of SAR is capable of promoting particle concentration around the area where the secondary air is injected. Penetration depth of secondary air jet increases with the increase of SAR, jet velocity and injection height while the total air volume alone has little effect on jet's penetration depth.

Keywords CFB · Secondary air · Penetrability · Pressure drop

35.1 Introduction

The function of secondary air (SA) in a circulating fluidized bed is primarily providing oxygen for fuel combustion and enhancing the turbulence of materials in the furnace. Ren (2011) proposed the problem of poor penetration of SA results in lack of combustion air in the central area of the furnace. In addition, chamber's pressure drop, particle circulating flow rate and particle's concentration distribution are critical for the operation of CFB. Too much pressure drop increases energy consumption and load of fans, which would reduce the boiler's efficiency and stability. The particle circulating flow rate, to some extent, represents fluidization states. The bubbling fluidized bed has zero circulating flow rate while fast fluidized bed has a wide range

C. Lin (✉) · J. Yu · Y. Zhang · F. Feng · Z. Zhang
School of Mechanical Engineering, Shanghai Jiao Tong University, Shanghai 200240, China
e-mail: lynch0923@sjtu.edu.cn

J. Lyu and S. Li (eds.), *Clean Coal and Sustainable Energy*,
Environmental Science and Engineering,
https://doi.org/10.1007/978-981-16-1657-0_35

of circulating flow rate, depending on different circumstances. However, high circulating flow rate makes great demands on cyclone and bed materials are bound to be continuously lost. Reasonable distribution of particles along the height direction of chamber satisfies boiler's load and at the same time reduces abrasion of heated side.

Yang et al. (2008), through experiments on the penetrability of secondary air, proposed three schemes to increase penetrability by increasing the secondary air velocity, nozzle diameter and raising the height of injection. Chen et al. (2007), who carried out experimental research and numerical simulation, concluded that the secondary air significantly changed the distribution of material concentration in the furnace and the penetration depth of SA increased approximately with the increase of wind speed as a power function. Cao et al. (2011), through calculation, found that the penetration depth of SA jet is proportional to the nozzle diameter and jet velocity, inversely proportional to the particle concentration, particle density and diameter. Also, the jet temperature has influences on its penetrability. Sun et al. (2010) compared the secondary air of both central and wall layout. He found the particle concentration under the area of SA injection height increases with the increase of SAR for both layouts.

Nevertheless, previous investigations mainly inject secondary air through wall layout. Research focusing on the penetrability of central layout secondary air and its influences on fluidization characteristics is rare. This work, upon a self-built CFB cold-state test platform, studied the influences of total air volume, ratio of the volume of secondary air to total air (SAR), velocity of secondary air and the secondary injection height on the penetrability of secondary air, the chamber's pressure drop, the particle circulating flow rate and the distribution of particle concentration.

35.2 Experiment

Figure 35.1 is the CFB cold-state test system. To observe the state of fluidization, the chamber is made of organic glass of 1.5 m high, 50 mm in inner diameter and 5 mm in thickness. The maximum wind amount for the fan is 21 m^3/h, displayed and regulated by a rotary flowmeter and a valve. Heated secondary air is injected from the center of the furnace tube. The nozzle has four circular outlets of 3 mm diameter, distributed horizontally and evenly. The injecting direction is opposed to the radial direction of the furnace tube. The plane where air distribution plate is located is set as 0 cm in chamber height. The pressure gauging points are distributed on the side for every 5 cm below 40 cm of the chamber and every 10 cm above it. Pressure is measured by testo510, a differential manometer. Arenaceous quartz is adopted as bed materials with true density of 2650 kg/m^3 and bulk density of 1480 kg/m^3 respectively.

According to the similarity of heat and mass transfer (Yang et al. 2008; Liu et al. 2007), heated air is used as tracer gas. The non-dimensional residual temperature $\theta = \frac{T-T_1}{T_2-T_1}$ is adopted to represent the penetrability of secondary air jet. In the equation, θ is non-dimensional residual temperature; T is the temperature of gauging

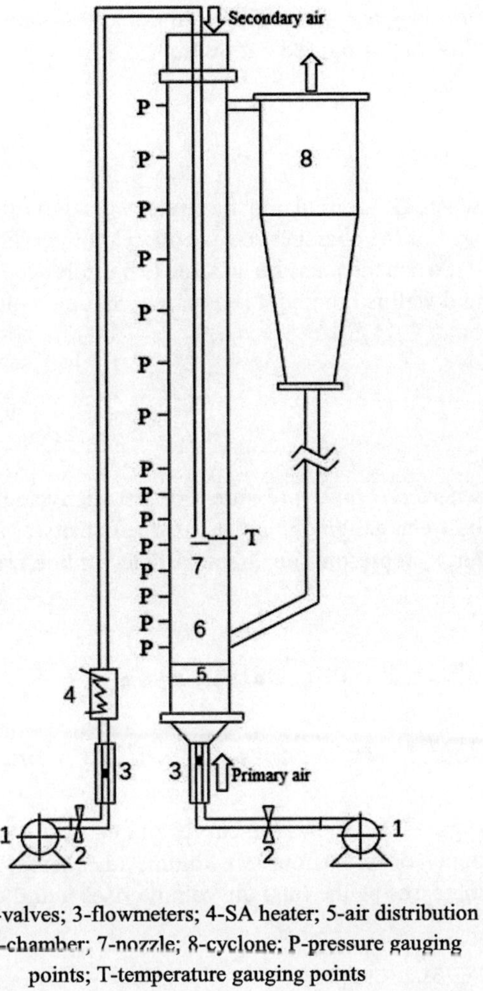

Fig. 35.1 CFB cold-state test platform. 1-fans; 2-valves; 3-flowmeters; 4-SA heater; 5-air distribution plate; 6-chamber; 7-nozzle; 8-cyclone; P-pressure gauging points; T-temperature gauging points

points; T_1 is the temperature of primary air, i.e. the room temperature; T_2 is the temperature of secondary air, heated to around 50 °C and kept stable under the same working conditions. The variation of air volume with temperature is taken into consideration. Through pre-experiment, the position where θ declines to 0.5 is defined as penetration depth x. The ratio of penetration depth and inner diameter of the furnace tube is defined as penetration rate: $r_p = \frac{x}{d_0}$. It is noteworthy that $\theta = 0.5$ is merely an artificial hypothesis that it is applicable to this paper's experiments.

When conducting the experiment, the bed material of 3 cm in height is placed upon the air distribution plate. Later the fan is turned on and adjusted to the set wind amount. The pressure of different chamber height and the temperature of secondary air jet are then measured in turn. The return tube of cyclone is connected to a collecting vessel to gather particles lifted off of the chamber over a period of time. The gathered particles are weighed. According to the mass of gathered materials, the particle circulating flow rate is figured out by Eq. (35.1).

$$G_s = \frac{m_c}{S\Delta t} \tag{35.1}$$

where G_s is circulating flow rate; m_c is the collected particle mass; S is cross-sectional area of the chamber; Δt is collection interval.

When the pressure loss due to particle acceleration and friction among air, particle and wall is ignored, the apparent particle volume fraction can be figured out (Wang et al. 2015):

$$\frac{\Delta P}{\Delta Z} \approx \left[\rho_p \varepsilon_s + \rho_g (1 - \varepsilon_s)\right] g \tag{35.2}$$

where ΔP is the pressure drop between gauging point; ΔZ is the distance in height between gauging point; ρ_p is true density of arenaceous quartz; ρ_g is the density of air; ε_s represents apparent particle volume fraction.

35.3 Results and Analysis

35.3.1 Influences of Total Air Volume

Figure 35.2 shows the curves of chamber's pressure drop and circulating flow rate under different total air volume and SAR. The chamber's overall pressure drop increases as the total air volume rises and the growth rate has an increasing trend.

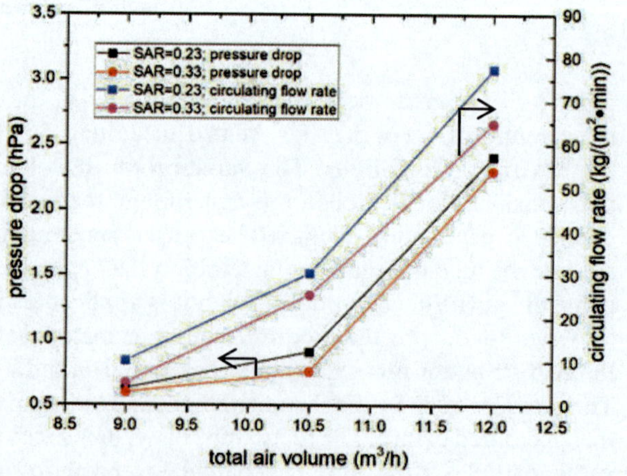

Fig. 35.2 Chamber's pressure drop and circulating flow rate under different total air volume and SAR

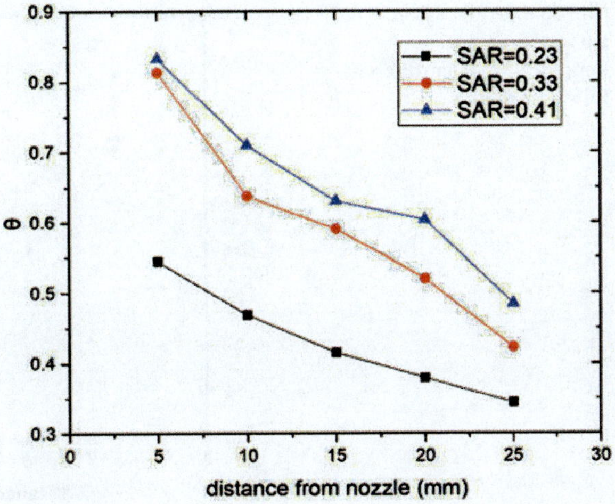

Fig. 35.6 Attenuation curves of residual temperature under different SARs

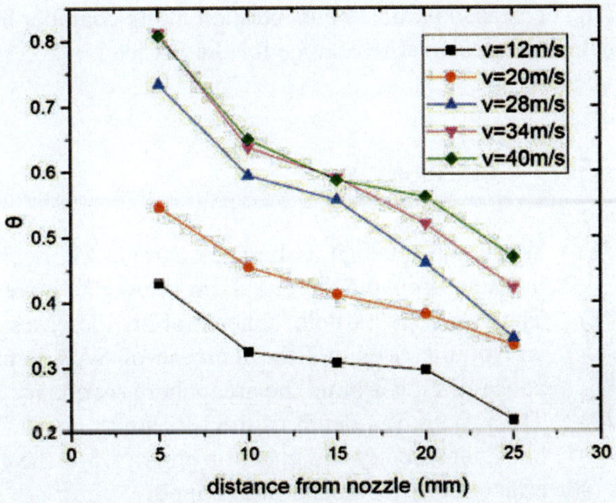

Fig. 35.7 Attenuation curves of residual temperature under different secondary air velocities

attenuation pattern and their penetration rate r_p all lie within 0.3–0.36. This means simply altering the total air volume when SAR is fixed has little effect on the jet penetration depth. When the total air volume is fixed at 10.5 m^3/h, the penetration rate increases with the rise of SAR but the growth rate recedes. The penetration rate first soars from 0.3 (SAR = 0.23) to 0.84 (SAR = 0.33), and then rises to 0.97 (SAR = 0.41). With the elevation of secondary air velocity, the penetration rate increases. However, the increase slows down evidently when the velocity reaches 28 m/s. The penetration rate increases by 0.44 as the velocity elevates from 20 to 28 ms/s while it merely increases by 0.28 when air velocity elevates from 28 to 40 m/s. The penetration also deepens significantly with the rise of injection height.

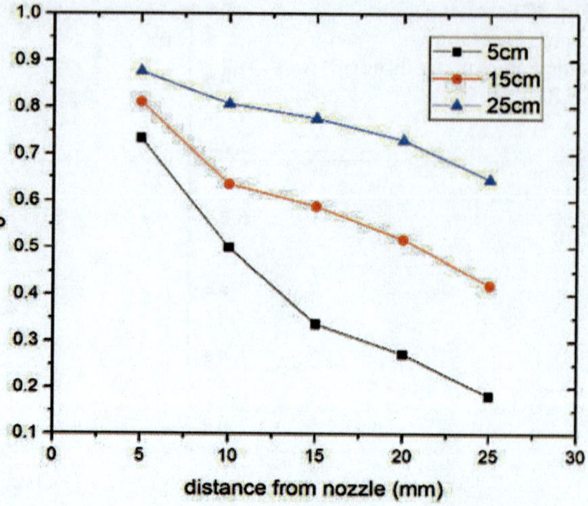

Fig. 35.8 Attenuation curves of residual temperature under different injection heights

The decreased particle concentration along chamber height contributes to this effect which causes great resistance for the SA jet.

35.4 Conclusions

(1) When the total air volume rises or SAR decreases, the chamber's overall pressure drop and the circulating flow rate increase.
(2) The apparent particle concentration increases substantially when the total air volume rises and the increase of SAR is capable of promoting particle concentration around the area where secondary air is injected.
(3) The penetration depth of the secondary air jet increases with the increase of SAR, jet velocity and injection height while the total air volume alone has little effect on the jet's penetration depth.

Acknowledgements The authors gratefully acknowledge the financial support from National Key R&D Program of China, Grant No. 2016YFB0600202.

References

Cao H, Miao Z-Q, Xiao F (2011) Numerical simulation of influencing factors for secondary air penetration depth in circulating fluidized bed boiler. Boiler Technol

Chen J-H, Lu X-F, Liu H-Z et al (2007) Experimental study and numerical simulation of secondary air's jet length in circulating fluidized beds. J Power Eng 27(6):895–898

Liu J, Chen J-H, Lu X-F et al (2007) Cold model experiments on the characteristic of secondary air injection in CFB boiler. Power Syst Eng

Marzocchella A, Arena U (1996) Hydrodynamics of a circulating fluidized bed operated with different secondary air injection devices. Powder Technol 87(3):185–191

Ren F-Q (2011) Status research of the circulating fluidized bed secondary air jet. Energy Conserv Technol

Sun S-Z, Wang Z-Y, Du M-K et al (2010) Design and experimental study of the centrally-located secondary air of a circulating fluidized bed. Reneng Dongli Gongcheng/j Eng Therm Energy Power 25(1):51–56

Wang X-F, Shuai D-P, LV Q-G et al (2015) Experimental study on circulating fluidized bed with annular furnace and six cyclones. Clean Coal Technol

Yang J-H, Yang H-R, Yue G-X (2008) Experimental study on secondary air jet penetration in circulating fluidized bed. J Power Eng

Chapter 36
The Fundamental Studies on Combustion of Sintering Flue Gas in Circulating Fluidized Bed

Xin Tao, Lujian Chen, Yiqun Huang, Runxia Cai, Hairui Yang, Shouyu Zhang, and Junfu Lyu

Abstract The development of sintering flue gas circulating fluidized bed (CFB) treatment technology is very important because there are many disadvantages in the existing treatment technology. In order to understand this technology, it is necessary to study the combustion of sintering flue gas in a circulating fluidized bed. Experiments were performed to study the combustion characteristics of Zuoyun bituminous coal in this oxygen-deficient atmosphere by a large-capacity TGA (LC-TGA). By changing reaction parameters, the influence of combustion temperature and gas atmosphere on coal combustion were analyzed. The experimental results confirm that the change of oxygen concentration and combustion temperature will greatly affect the combustion of coal in the low oxygen atmosphere. With the increase of oxygen mole fraction and combustion temperature, the combustion rate of coal will gradually accelerate. The presence of CO in the combustion atmosphere will not change this trend, but it does greatly affect the combustion characteristics of coal, which is particularly evident at 800 and 850 °C. CO has obvious oxygen-grabbing behavior at high temperature, and the influence on coal combustion seems to be closely related to the temperature. The present experimental results will provide the reference for the further development of sintering flue gas circulating fluidized bed technology.

Keywords CFB · TGA · Coal · CO · Low-oxygen atmosphere

X. Tao · L. Chen · S. Zhang (✉)
Department of Thermal Engineering, University of Shanghai for Science and Technology, Shanghai, China
e-mail: 37487418@qq.com

X. Tao · Y. Huang · R. Cai · H. Yang · J. Lyu (✉)
Key Laboratory for Thermal Science and Power Engineering of Ministry of Education, Department of Energy and Power Engineering, Tsinghua University, Beijing, China
e-mail: lvjf@mail.tsinghua.edu.cn

J. Lyu and S. Li (eds.), *Clean Coal and Sustainable Energy*,
Environmental Science and Engineering,
https://doi.org/10.1007/978-981-16-1657-0_36

36.1 Introduction

The iron and steel industry has made tremendous contributions to the rapid development of the economy, but it also brought a lot of pollution. Sintering flue gas, one of the pollutants, which is produced by sintering machine in the process of steelmaking, has the characteristics of large amount of smoke and pollution (Wang et al. 2017). As its emission will cause great damage to the environment, the treatment of sintering flue gas has important significance.

Nowadays, the treatment methods of sintering flue gas mainly include limestone method and activated carbon adsorption method (He and Yao 2018; Wang and Zhong 2016). However, these methods often have many disadvantages, such as high operating cost and easy blockage of equipment. In order to better control sintering flue gas, new treatment methods need to be developed urgently. Among the many proposed methods, treating sintering flue gas by blast furnace combustion (Wei et al. 2016) has proved the possibility. As circulating fluidized bed (CFB) has many excellent characteristics in combustion, burning sintering flue gas in CFB furnace has attract much attention.

The research of sintering flue gas in circulating fluidized bed combustor has just started, which can provide some reference for the further development and application of this technology. The fundamental study of coal combustion characteristics is often carried out in thermogravimetric analysis (TGA). Qu and Liu (2017) used thermogravimetric experiments to study the effect of different oxygen concentration on coal combustion under hypoxic conditions and found that the effects of oxygen concentration change on each characteristic temperature point were quite different. Cong et al. (2019) studied the influence of particle size on the combustion performance of coal by TGA, and found that if the heating rate was high, the influence of particle size on the combustion performance of coal was small. However, in actual working conditions, coal is often instantaneously added to a high temperature environment, which has obviously difference from the conditions in TGA (Liu et al. 2012; Wang and Che 2010).

As sintering flue has CO as high as almost 1% and low oxygen content, to better understand the combustion characteristics of coal under this atmosphere, this paper studies the combustion characteristics of carbon monoxide with different oxygen content and at different temperatures.

The combustion characteristics of coal in oxygen-deficient atmosphere of sintering flue gas were studied by a large-capacity TGA (LC-TGA) which is able to heat coal quickly to the set temperature. And the effect of CO on combustion was also preliminary studied.

Table 36.1 Proximate and elemental analyses of the coal sample

Name	ZY bituminous coal
Proximate analysis (%), air-dried basis	
Moisture	4.42
Ash	33.23
Volatile	25.65
Fixed carbon	36.69
Elemental analysis (%), dry ash-free basis	
Carbon	43.86
Hydrogen	3.33
Oxygen	13.43
Nitrogen	1.00
Sulfur	0.71

36.2 Methodology

36.2.1 Materials

In this experiment, Zuoyun (ZY) bituminous coal was selected as the coal type. The coal was crushed and screened firstly, and then particle with size of 0–300 μm was stored in sample bags for the experiments. The results of proximate analysis and elemental analysis of coal samples are listed in Table 36.1.

The combustion characteristics of coal samples were studied by weight loss balance mass fraction α and combustion weight loss rate ν. Among them, the weight loss balance mass fraction is the ratio of residual mass of coal sample to initial mass at t time. The expression is as follows:

$$\alpha = \frac{mt}{m0} \times 100\% \tag{36.1}$$

m0 is the initial mass, mt is the mass at t time.

The rate of combustion weight loss ν is the derivative of the weight loss balance mass fraction. The expression is as follows:

$$\nu = \frac{d\alpha}{dt} \tag{36.2}$$

36.2.2 Experimental

The schematic of the LC-TGA is shown in Fig. 36.1, which specific parameters have

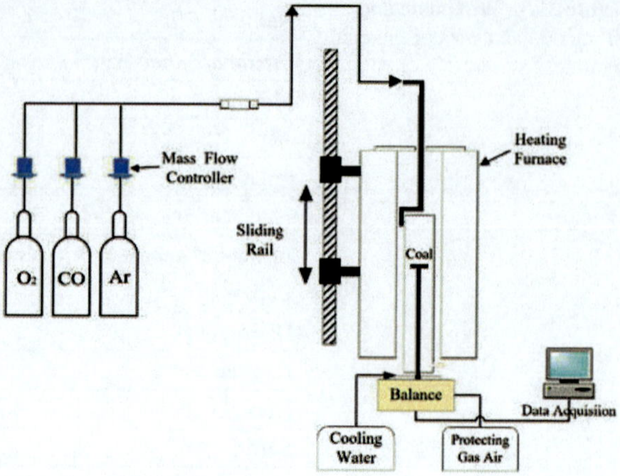

Fig. 36.1 The schematic of the large-capacity thermogravimetric analyzer (LC-TGA) system

been described in detail by previous work (Cai et al. 2019). The system is composed of an intelligent temperature control tubular furnace, data acquisition system, gas circuit and so on. The maximum temperature of the TGA is 1200 °C and the heating rate of coal can reach 20 °C/s. The maximum weight of the TGA is over 100 g and the sensitivity of its balance is ±0.1 mg.

At the beginning of the experiment, the gases needed were mixed by the mass flow controllers. Then the mixing gas was added to the TGA to react with the coal at the rate of 1 L/min. As the flow rate of the gas and heating rate are both high, the experimental results obtained from the LC-TGA seem to be better to reflect the combustion condition of coal in CFB. The mass changes of the coal were measured by the balance and were recorded by the computer at a frequency of about 60 times per minute.

Thermogravimetric experiments at constant temperature have been carried out. In order to simulate the actual combustion atmosphere, oxygen, CO and argon were mixed in a certain proportion in the experiment to get a low-oxygen gas containing 1% CO. The oxygen mole fractions were 8%, 10%, 13%, 16%, 18% and 21%, respectively. In this experiment, TGA was heated to 750, 800, 850 and 900 °C, and then the argon was injected into TGA for a period of time at the rate of 1 L/min to remove the air in it. When the TGA was filled with argon, the coal samples were placed on the quartz crucible. At the same moment, the heating furnace was fallen to the bottom at the speed of 5 mm/s through the sliding rail. At last, the combustion atmosphere was changed to achieve the target working condition. The gas flow rate was always controlled at 1 L/min by the mass flow controller during the whole process to reduce the possible impact of air flow changes. The quality of each sample is about 100 mg. In addition, in order to study the influence of CO containing in the gas, the comparative experiments were carried out. In these experiments, oxygen concentration was controlled at 7.5 and 8%.

Before each test, a blank TG run was performed with an empty crucible and the corresponding blank TG curve was deducted from the TG curve measured with the presence of a sample in the crucible.

36.3 Results and Discussion

36.3.1 Comparison of Combustion Behavior of Coal Under Different Oxygen Contents

The CFB usually operates at a relatively stable temperature, so the influence of oxygen concentration on coal combustion at a fixed temperature was evaluated first. The combustion atmosphere is the oxygen-deficient gas consisting of CO, O_2 and Ar. In this experiment, six different oxygen mole fractions (8, 10, 13, 16, 1 and 21%) were selected to study the role of oxygen in coal combustion at different temperatures. After the experiments, the general trend of the coal mass changing seemed to be similar. So, the experimental results at 750 and 850 °C were selected to describe the effects of oxygen concentration change on coal combustion characteristics, which is shown in Fig. 36.2. Both figs are composed of the curves of weight loss balance mass fraction α and combustion weight loss rate ν.

Different from the usual TGA heating experiment, the coal in the experiment was directly placed in the high temperature environment, so volatile combustion and coke combustion will occur quickly after the short dehydration. In order to understand the combustion characteristics of coal more directly, the time required for coal mass to be reduced to the minimum was simplified as the burnout time of coal in this paper, which is also the time when the combustion weight loss rate returns to zero.

As shown in Fig. 36.2, with the increasing of oxygen concentration, the burnout time of coal decreases gradually. The burnout time of coal at 8% oxygen mole fraction is even several times that of coal at 21% oxygen mole fraction. It means that the oxygen concentration of gas has a great influence on the burnout time of coal, which is similar to the experiment results of other researchers (Hurt et al. 1996; Zeng et al. 2019).

At the beginning of the reaction, with oxygen mole fraction of 21%, it has the highest weight loss rate. It indicates that the increase of oxygen concentration is beneficial to the release and combustion of volatiles. And the rapid decrease of weight loss rate also represents the rapid decrease of volatiles.

The burnout time of coal decreases greatly when the oxygen mole fraction changes from 8 to 10%, which is almost the same as the decreasing of time when the oxygen mole fraction changes from 10 to 16%. It means that in the case of low oxygen, although the increase of oxygen concentration can improve the combustion of coal, it is not a linear increase. The possible factors contributing to results are as follows: (1) The increased oxygen concentration may lead to faster destruction of carbon structure in coal and the destruction of pore structure will make it difficult for the

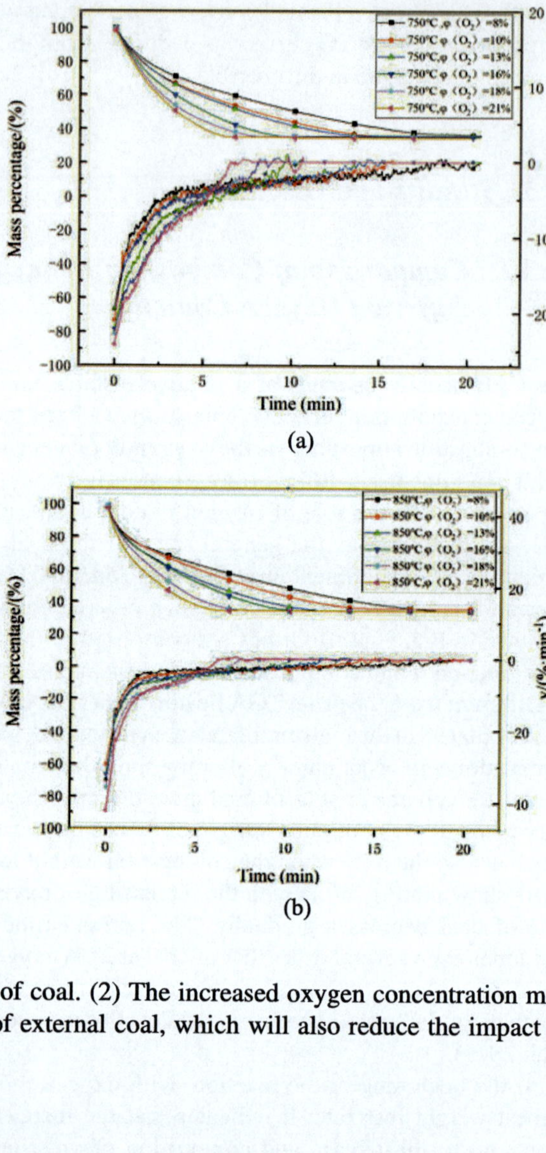

Fig. 36.2 Influence of concentration of oxygen on combustion. **a** 750 °C; **b** 850 °C

oxygen to attach the internal of coal. (2) The increased oxygen concentration may lead to faster ash formation of external coal, which will also reduce the impact of oxygen.

36.3.2 Comparison of Combustion Behavior of Coal Under Different Temperatures

One of the characteristics of CFB combustion technology of sintering flue gas is that the oxygen mole fraction at the exit can reach about 8% and the oxygen mole fraction of sintering flue gas is generally 13–21%. Therefore, 8 and 13% oxygen mole fractions were selected to describe the influence of temperature change on coal combustion. Figure 36.3 shows the results of thermogravimetric experiments at four

Fig. 36.3 Influence of combustion temperature on combustion. (1) 8% oxygen; (2) 13% oxygen

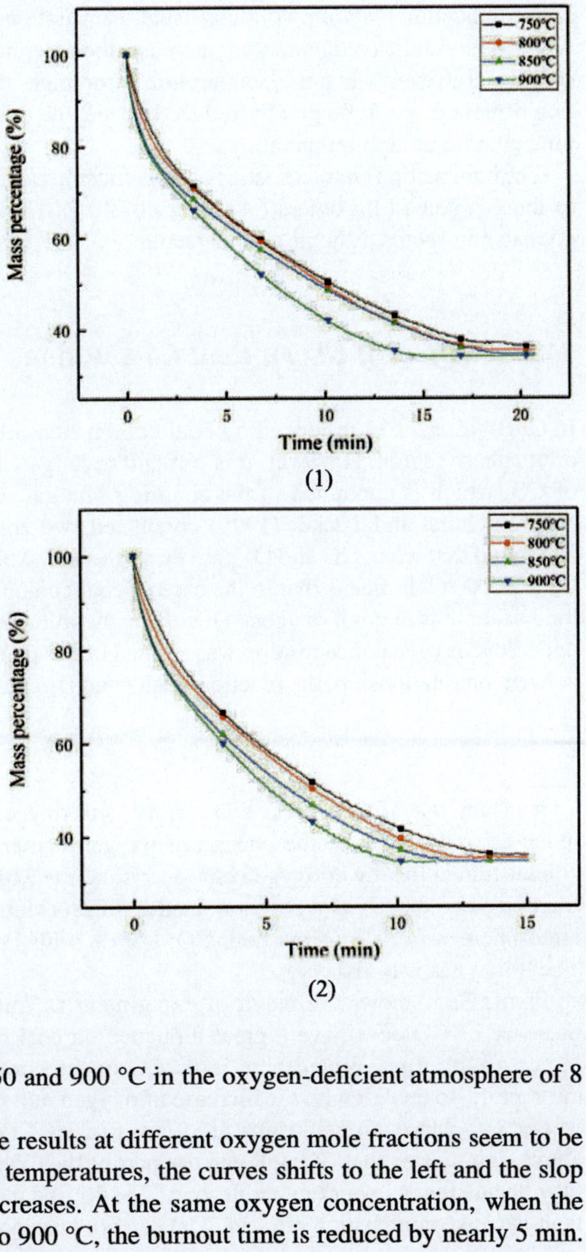

temperatures of 750, 800, 850 and 900 °C in the oxygen-deficient atmosphere of 8 and 13% oxygen.

As shown in Fig. 36.3, the results at different oxygen mole fractions seem to be similar. With the increase of temperatures, the curves shifts to the left and the slop of the weight loss curves increases. At the same oxygen concentration, when the temperature rises from 750 to 900 °C, the burnout time is reduced by nearly 5 min. It indicates that when the oxygen concentration is the same, raising the temperature can improve the combustion of coal.

As shown in Fig. 36.3, when the temperature rises from 750 to 800 °C, the initial section of the curve drifts a lot. However, the initial section of the curves almost coincide at 850 and 900 °C. It is well known that in the initial stage of coal combustion reaction, volatiles usually separate out quickly. So it indicates that when the combustion temperature increase, the precipitation of volatiles will be more quickly. However, when the temperature is too high, the sensitivity of the precipitation of volatiles will be greatly reduced. It may be due to the fast speed of volatiles precipitation at high temperature.

The combustion characteristics of coal under hypoxic conditions above are similar to those obtained by others (Wang et al. 2015). However, the role of CO in coal combustion seems difficult to understand.

36.3.3 *Effect of CO on Coal Combustion*

In CFB boilers, CO produced by coal combustion often reacts with oxygen in the dilute phase region. However, it is difficult to judge whether the high concentration of CO, which is contained in the sintering flue gas, will affect the combustion of coal. Hayhurst and Tucker (1990) conducted two-zone fluidized bed experiments and found that when CO and O_2 pass through sand, 0.5% CO can be burned with 5% O_2 at 1000 °C. It means that in the case of coal combustion, CO is bound to burn as long as there is enough oxygen. Therefore, the influence of CO on coal combustion under low oxygen concentration was studied in this paper. Regardless of the complex conversion relationship, the reaction of CO and O_2 at high temperature is as follows:

$$2CO + O \rightarrow 2CO_2$$

It means that if CO reacts completely with oxygen, CO is consumed twice as much as oxygen. Since the change of oxygen content has bigger impact on coal combustion at the lower oxygen concentration, the experiments at 7.5 and 8% mole fraction oxygen w/o CO were conducted at different temperatures. The three reaction atmosphere were: 8% O_2 without CO; 8% O_2 with 1% CO; 7.5% O_2 without CO. The filling gas was always Ar.

Figure 36.4 shows the result of experiment at 750 °C. It can be seen that the presence of CO does have a great influence on coal combustion and these effects increase with time. When there is no CO in the combustion gas, the weight loss curve shifts to the left with the increase of oxygen mole fraction from 7.5 to 8%. The weight loss curve of combustion at 8% O_2 and 1% CO lies between the two curves above. It indicates that CO will capture part of the oxygen and burn at 750 °C. This may be due to the lower temperature. At 750 °C, the gas reaction rate is much lower than the coke combustion rate, and CO can not burn completely .

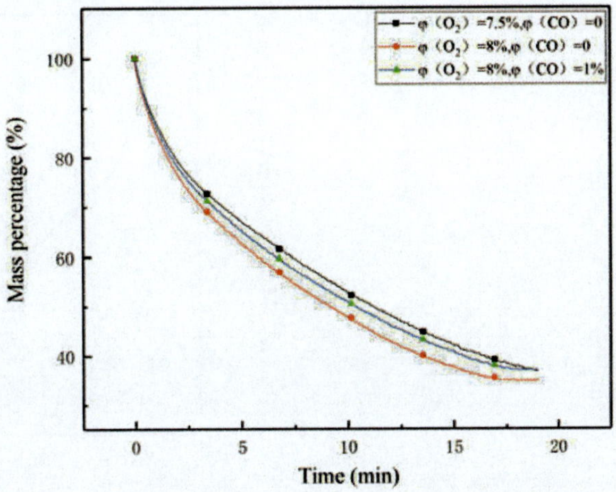

Fig. 36.4 Influence of CO on coal combustion at 750 °C

With the increase of the temperature, the experimental results change greatly. Figure 36.5 shows the experimental results at 800 and 850 °C. It can be seen that the presence or absence of CO has little effect on the reaction rate at the beginning of reaction. It indicates that the oxygen capture capability of coal is much stronger than that of CO at the stage of the release of volatiles. However, with the decrease of volatiles, the weight loss curve at the condition of 8% oxygen and 1% CO coincides with curve at the condition of 7.5% oxygen gradually. It indicates that CO reacts completely with oxygen at 800 and 850 °C. It means that the reaction of CO with oxygen takes precedence over that of coke with oxygen at these range of temperature.

Figure 36.6 shows the results at 900 °C, and there are some differences from the results shown in Fig. 36.5. After a short period of volatiles release, the weight loss curves do not coincide with each other. That is to say, CO still have an influence on combustion at this time, but it can not burn out completely anymore. It may be due to the increased oxygen demand and the faster combustion of coke at high temperature.

To sum up, the the oxygen capture capability of CO increases with the increase of temperature, but there seems to be a peak point corresponding to the combustion temperature.

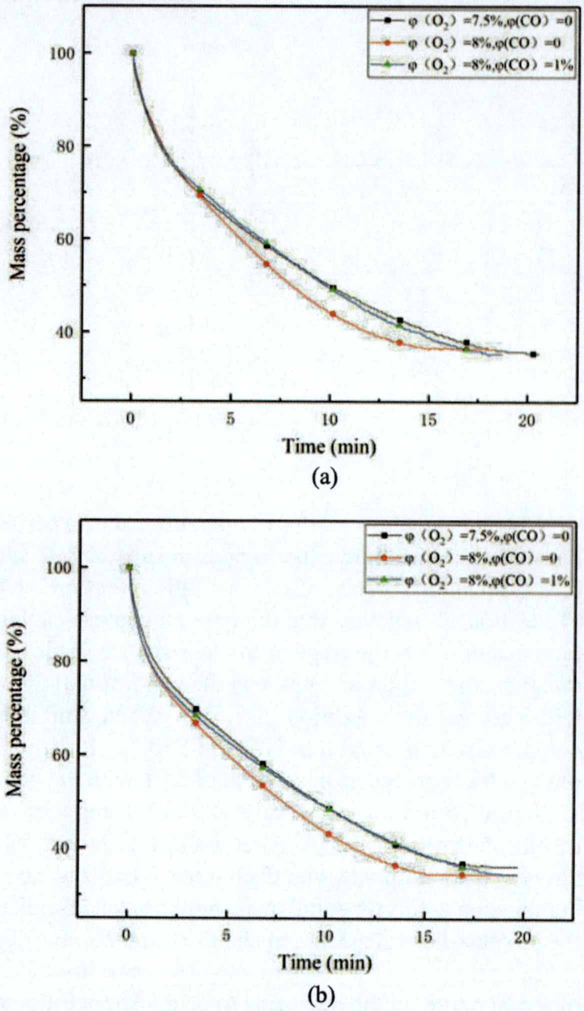

Fig. 36.5 Influence of CO on coal combustion. **a** 800 °C; **b** 900 °C

36.4 Conclusion

The combustion characteristics of ZY bituminous coal in the atmosphere of sintering flue gas were studied by a large-capacity TGA (LC-TGA). Different oxygen concentrations and combustion temperatures were selected for comparative experiments. The influence of reaction temperature and reaction atmosphere on coal combustion were analyzed at different conditions. Following conclusions can be drawn:

(1) The experimental results confirm that the change of oxygen concentration has a great influence on coal combustion. With the increase of oxygen concentration, the combustion rate of coal increases remarkably.

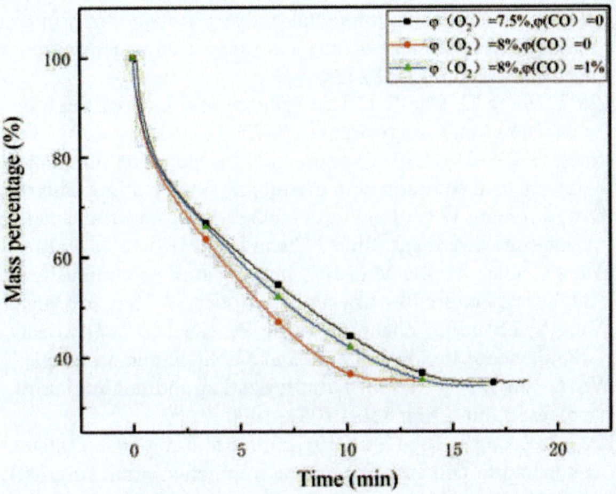

Fig. 36.6 Influence of CO on coal combustion at 900 °C

(2) Under the condition of low oxygen, the lower the oxygen mole fraction is, the greater influence of the change of oxygen concentration on coal combustion is.
(3) When coal burns in the low-oxygen atmosphere of sintering flue gas, increasing temperature will reduce the burnout time.
(4) The existence of CO in combustion gas will not affect the overall trend of coal combustion, but it does have a great impact on coal combustion, which is closely related to the change of combustion temperature.
(5) At the normal combustion temperature of circulating fluidized bed, the combustion priority of oxygen and CO seems to be higher than that of oxygen and coal in a certain range. This conclusion needs to be demonstrated by further study.

Acknowledgements The support by National Natural Science Foundation of China (U1710251) is greatly appreciated.

References

Cai R, Huang Y, Li Y, Wu Y, Zhang H, Yang H, Lyu J (2019) Effects of the limestone particle size on the sulfation reactivity at low SO_2 concentrations using a LC-TGA. Materials. https://doi.org/10.3390/ma12091496

Cong K, Zhang Y, Han F, Li Q (2019) Influence of particles sizes on combustion characteristics of coal particles in oxygen-deficient atmosphere. Energy 170(1):840–848

Hayhurst AN, Tucker RF (1990) The combustion of carbon monoxide in a two-zone fluidized bed. Combust Flame 79(3):175–189

He T, Yao Y (2018) Treatment of sintering flue gas discharge. Shandong Metallurgy 40(4):44–46

Hurt RH, Lunden MM, Brehob EG (1996) Symp Combust 26(2):3169–3177

Qu L, Liu Q (2017) Experimental study on influence of different oxygen concentrations on reaction energy level of coal spontaneous combustion under thermogravimetric experimental conditions. J Saf Sci Technol 13(8):134–139

Liu Y, Wang C, Che D (2012) Ignition and kinetics analysis of coal combustion in low oxygen concentration. Energy Sources 34(9):810–819

Wang C, Che D (2010) Experimental investigation on combustion characteristics of coals in low oxygen concentration with thermogravimetry. J Eng Thermophys 31(10):1785–1788

Wang J, Zhong W (2016) Simultaneous desulfurization and denitrification of sintering flue gas via composite absorbent. Chin J Chem Eng 24(8):1104–1111

Wang C, Qiao M, Lei M (2015) Investigation on combustion characteristics of pulverized coal in high-temperature low-oxygen atmosphere. J Chin Soc Power Eng 35(8):619–631

Wang S, Zhang Q, Zhang G, Wang Z, Zhu P (2017) Effects of sintering flue gas properties on simultaneous removal of SO2 and NO by ammonia-Fe (II). J Energy Inst 90(4):522–527

Wei F, Liu J, Ji Y (2016) Desulfurization and denitration of sintering flue gas by blast furnace process. Chin J Eng 38(8):1082–1090

Zeng Z, Zhang T, Zhou Y (2019) Ignition and combustion characteristics of coal particles under high temperature and low-oxygen environments mimicking MILD oxygen combustion conditions. 253(1):1104–1113

Chapter 37
Emission Characteristic of NO_x in CFB Boiler at Low Load

Lujian Chen, Xin Tao, Shouyu Zhang, Hairui Yang, and Junfu Lyu

Abstract In order to study the NOx emission of CFB boiler at low load, a bubbling fluidized bed test was established to investigate the effects of bed temperature, excess air ratio and particle size of ZY coal on NOx emission. A continuous feeder was used to ensure the feeding rate steady, so that the significant parameter excess air ratio can be confirmed. By comparing the NOx emissions of different conditions, the influence law of three parameters can be obtained. The results showed that the bed temperature had a continuous promoting effect on the formation of NOx when bed temperature increased from 700 to 850 °C. At a given temperature, the NOx emission depended mainly on the excess air ratio. With the increasing of excess air ratio, the NOx emission increased first and then decreased. In the same condition, the NOx emission generated in the combustion process of larger particle coal was lower than the value of small particle coal.

Keywords CFB combustion · NO_x emission · Bed temperature · Excess air ratio · Particle size

37.1 Introduction

In recent years, with the increasing installed capacity of coal-fired units and the grid connection of new energy power generation in China, the peak-to-valley difference of the power grid has become larger (Chen 2009). The regulation peak load is facing tremendous pressure. Circulating fluidized bed (CFB) boiler has the advantages of wide fuel adaptability, good load regulation performance, high combustion efficiency and low pollutant emission (Cai et al. 2017). There is much hot bed material for heat

L. Chen · X. Tao · S. Zhang (✉)
School of Energy and Power Engineering, University of Shanghai for Science and Technology, Shanghai 200093, China
e-mail: zhangsy-guo@163.com

L. Chen · X. Tao · H. Yang · J. Lyu (✉)
Department of Energy and Power Engineering, Tsinghua University, Beijing 100084, China
e-mail: lvjf@mail.tsinghua.edu.cn

J. Lyu and S. Li (eds.), *Clean Coal and Sustainable Energy*,
Environmental Science and Engineering,
https://doi.org/10.1007/978-981-16-1657-0_37

storage in CFB boiler when it is in operation, so CFB boiler has a good performance at low load, which means that it has enormous potential in regulation peak load (Feng et al. 2003).

Because of the increasingly prominent environmental problems, the pollutant emission requirement of NO_x on CFB power plants becomes more stringent, especially at low load, which caused few attentions so far. Previous studies have shown that the NOx emission is influenced by many factors, including bed temperature, bed pressure, excess air ratio, ratio of primary and secondary air, position of upper-level secondary air, coal particle distribution, coal combustion characteristic and so on (Varol et al. 2014; Gungor 2009; Li et al. 2017). In addition, the impacts of these factors are not independent. Generally, the flow pattern in the furnace is composed of the bubbling bed formed by large particles at the bottom and the fast bed formed by fine particles in the freeboard. Li et al. (2016a, 2013) found that the dense phase zone of CFB is an important area for the formation and reduction of NOx. In operational boilers, it may have difficulties in studying the effect of single factor on NOx emission, so a laboratory-scale fluidized bed was built to simulate the NO_x emission in the dense zone of CFB boiler at low load with low bed temperature and relatively high excess air ratio. The experimental results are significant and will be used to guide the actual CFB boiler operation.

37.2 Experimental Work

37.2.1 NOx Emission Characteristic in a Bubbling Fluidized Bed

37.2.1.1 Bubbling Fluidized Bed Test Rig

The schematic of the bubbling fluidized bed system was shown in Fig. 37.1. The picture of continuous feeder was shown in Fig. 37.2. Two particle size of ZY coals were added from the top of the quartz tube by a continuous feeder and heated by the hot bed material quickly. At same time, air entered through the hot cullet at the bottom of the quartz tube and reacted with ZY coal. The air mass flow rate was controlled by a mass flow controller. A K-type thermocouple, whose measurement accuracy is ±0.4%, was placed above the sand core to monitor the bed temperature. The top of the quartz tube was installed inside a water-cooled jacket, which was fixed on the furnace to cool the high temperature flue gas. Cooled flue gas entered the gas analyzer and the data was recorded by data acquisition.

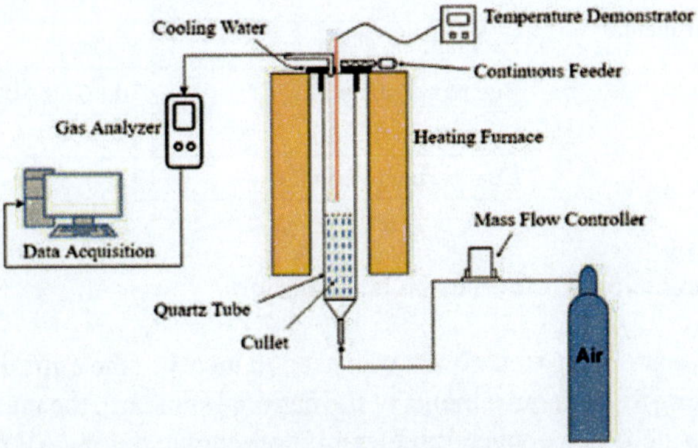

Fig. 37.1 The schematic of the bubbling fluidized bed (BFB) test

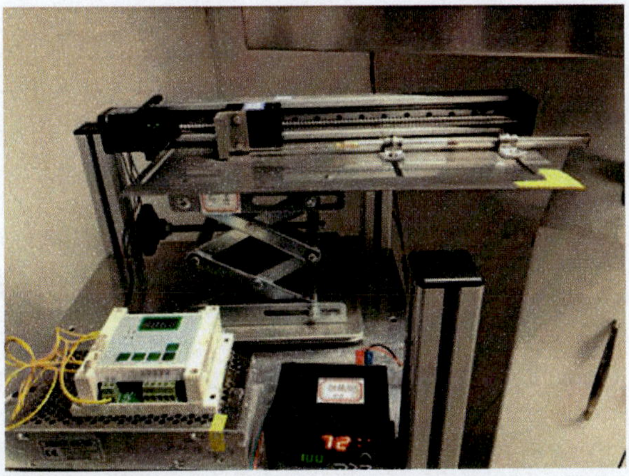

Fig. 37.2 The picture of continuous feeder

37.2.1.2 Coal Analysis

The coal for experiment was Zuoyun bituminous coal (ZY coal). The proximate analysis, elemental analysis and heat value are listed in Table 37.1.

Table 37.1 Proximate analysis, elemental analysis and heat value

Sample	C_{ar}	H_{ar}	O_{ar}	N_{ar}	S_{ar}
ZY coal	43.86	3.33	13.43	1.00	0.71
A_{ar}	M_{ar}	V_{ar}	FC_{ar}	Q_{ar} kJ/kg	
33.23	4.42	25.65	36.70	16,162	

Table 37.2 Experimental conditions

Parameters	Value
Bed temperature (T)/°C	700, 750, 800, 850
Excess air ratio (a)	1.2, 1.4, 1.6, 1.8, 2.0
Coal particle size/mm	1.0–1.2, 2.0–2.2

37.2.1.3 Calculation Method of NOx Emission

In this paper, a Testo 340 gas analyzer was used to measure the emission of NO, CO, and O_2. According to the requirements of the national standard, the measured value of NO was transformed. The conversion formula was shown below. All NOx emissions were calculated with this conversion formula based on 6% O_2 concentration.

$$NOx(\text{mg/Nm}^3) = \frac{NO(\text{ppm})}{0.95} \times \frac{\frac{46}{22.4}}{\frac{273}{273+20}} \times \frac{21-6}{21-O_2} \tag{37.1}$$

37.2.1.4 Experimental Conditions

According to the field experience, the bed temperature of CFB boiler was lower than 850 °C when the load was at low condition, so the experimental bed temperature was selected at 700, 750, 800 and 850 °C. The bed material was quartz sand of 212–425 μm. The feedstock was decreased when CFB boiler was at low load, in order to ensure the minimum fluidization velocity, the fluidization airflow cannot be reduced in proportion, so the excess air ratio was selected at 1.2, 1.4, 1.6, 1.8 and 2.0, respectively. Two different particle sizes ZY coal were sieved by standard sieves into two groups with narrow cuts, including 1.0–1.2 mm and 2.0–2.2 mm. All the experimental conditions were listed in Table 37.2.

37.3 Results and Discussion

37.3.1 Experimental Stability Test

The excess air ratio was a significant parameter in this paper. In order to maintain the excess air ratio stable, a continuous feeder and mass flow controller were introduced. The continuous feeder was composed of a stepping motor, a controller, a driver and a power supply. Coal feed rate and air rate were controlled respectively. The coal feed rate was 1 g/min. Necessary tests were done to confirm the stability of continuous feeder (Table 37.3).

Table 37.3 The stability test of the continuous feeder

Number	1.0–1.2 mm g/min	2.0–2.2 mm g/min
1	0.961	0.969
2	0.952	1.069
3	1.007	0.942
4	0.962	1.046
5	0.979	1.034
Average error/%	3.06	4.76

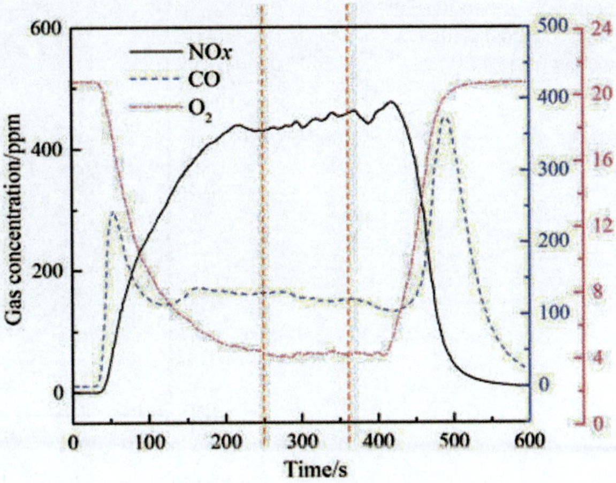

Fig. 37.3 NOx concentration of ZY coal combustion at 700 °C with 1.2 excess air ratio

The results showed that the average errors of the continuous feeder with two particle size coals were less than 5% and under the error range.

As shown in Fig. 37.3, the NOx emission was zero at the beginning of the measurement, because the response time of gas analyzer was about 40 s. After this period, the NOx emission increased rather fast and then turned into a gentle changing area until the coal feeding stopped at 360 s. The NOx emission decreased in a short time after coal feeding stopped. Because of the burning of coal, the O_2 concentration gradually decreased and then tended to stabilize between 250 and 360 s. Besides, there was similar rule for change in CO concentration. It was obvious that the fluctuations of three gas concentrations between 250 and 360 s were small. This meant that the BFB system achieved dynamic balance in the period. Thus, the steady-state concentration of NOx also can be obtained. According to this result, the coal feeding time in following experiments was selected at 360 s.

37.3.2 *Comparison of NOx Emission in Different Bed Temperature*

As shown in Fig. 37.4a, the NOx emission of 1.0–1.2 mm ZY coal increased when bed temperature increased from 700 to 850 °C, but the promoting effect was weakened when bed temperature increased. The similar phenomenon was found in Fig. 37.4b.

Fig. 37.4 Effect of bed temperature on NOx emission of two particle sizes ZY coal with 1.2 excess air ratio

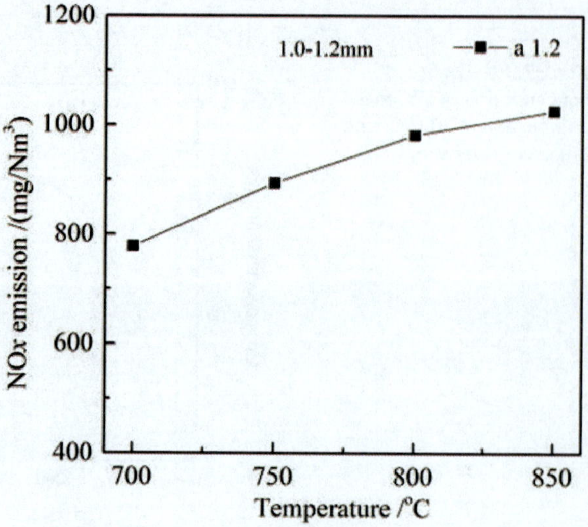

(a)

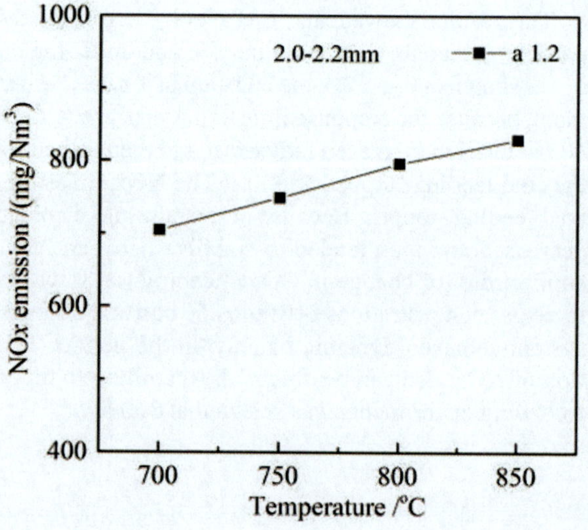

(b)

Because the release rate and reaction rate of fuel nitrogen were slow at low temperature, fuel nitrogen cannot be released quickly and participate in the reaction to generate NOx. As the temperature improved, the release rate and reaction rate of fuel nitrogen improved. Meantime, with the increase of char reactivity and CO reactivity, the reduction rate of NOx over char increased (Lyu et al. 2018). But the amount of char and CO were decreased when bed temperature increased. Some scholars demonstrate that some metal oxides can catalyze the reduction of NOx (Li et al. 2016b; Zhao et al. 2003a, b). Sørensen et al. (2001), Sun et al. (2009) found that the higher temperature will be beneficial to the reduction of NOx. The measurement value of NOx emission depended on the reduction reaction of NOx and the generating reaction of NOx.

37.3.3 Comparison of NOx Emission in Different Excess Air Ratio

As shown in Fig. 37.5a, b, with the increasing of excess air ratio, the NOx emission also increased. Because oxidizing atmosphere is beneficial for conversion reaction of fuel nitrogen. An increase in the excess air ratio meant that the fuel has more opportunities to participate in the reaction with oxygen to generate NOx. When the excess air ratio was more than 1.6, the promotion of excess air ratio to NOx formation will weaken. Because the amount of oxygen was excessive at this time, increasing the amount of air will dilute NOx. Therefore, reducing the value of the excess air ratio will facilitates control of NOx emissions while ensuring a minimum fluidization velocity.

37.3.4 Effects of Coal Particle Size on the NOx Emission

As shown in Fig. 37.6, in different conditions, the larger particle coal eventually released less NOx than small particle coal when the bed material size was determined. After small particle coal was added into the quartz tube and heated by hot bed material, it burnt at a high rate and produced a lot of NOx, but the NOx generation rate of large particle coal was slow relatively. Lager specific surface area and faster heating rate of small particle coal were beneficial for it to release much fuel nitrogen.

Generally, the dense zone of CFB boiler is a reducing atmosphere, complex reduction reactions on the surface of char can reduce NOx emission. So it is advised to control the coal particle size distribution for a lower NOx emission.

Fig. 37.5 Effect of excess air ratio on NOx emission of two particle sizes ZY coal at different temperature

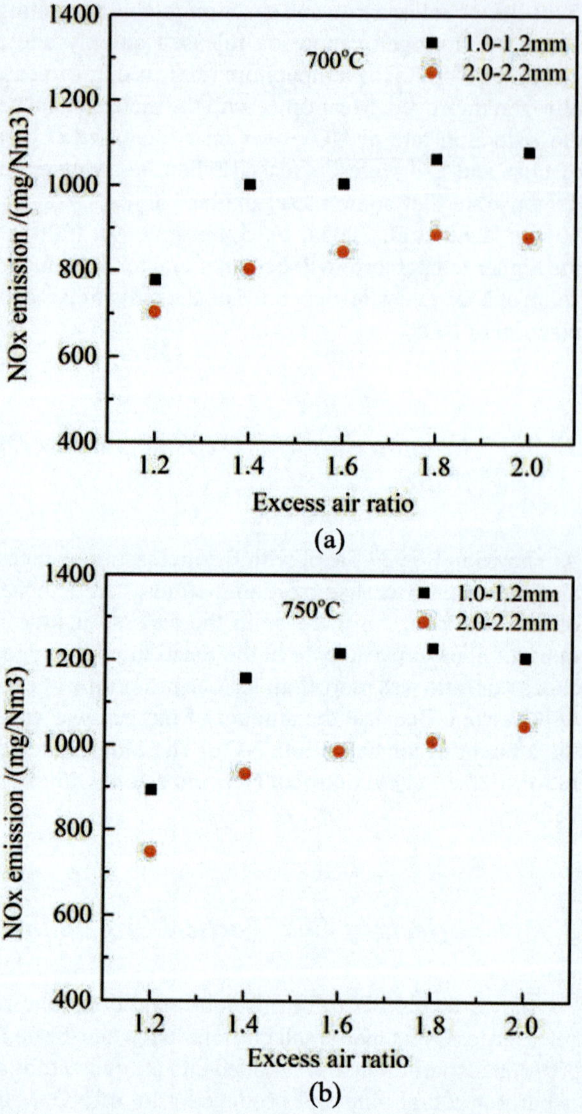

Fig. 37.6 Effects of coal particle size on the NOx emission at 700 °C with several excess air ratio

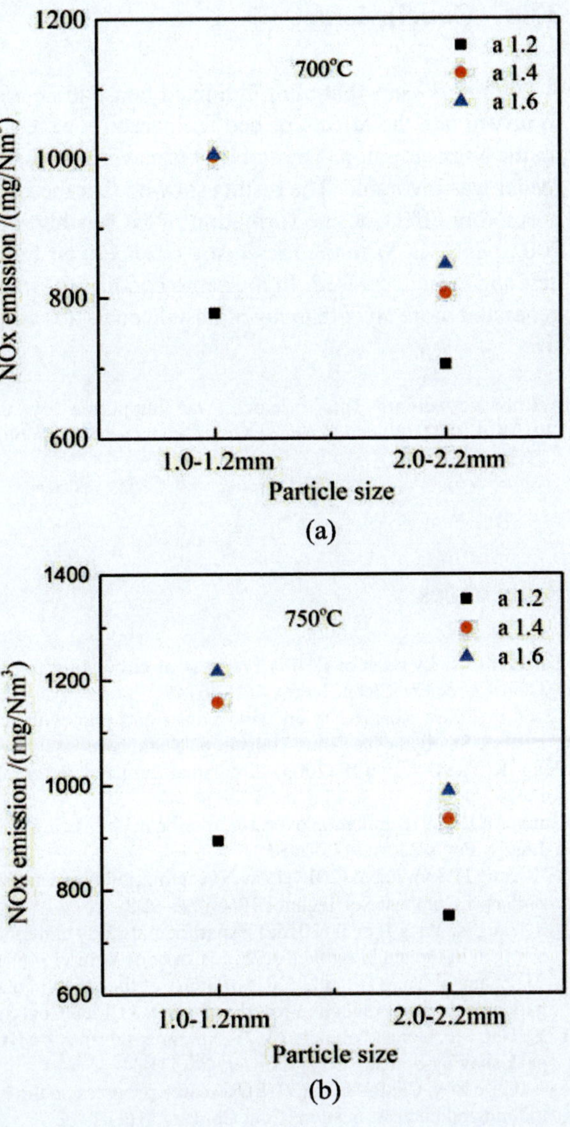

37.4 Conclusions

A laboratory-scale bubbling fluidized bed with a continuous feeder was developed to investigate the effects of bed temperature, excess air ratio and coal particle size on the NOx emission. The stability tests verified that the performance of continuous feeder was favorable. The results showed that the bed temperature had a continuous promoting effect on the formation of NOx when bed temperature increased from 700 to 850 °C. With the increasing of excess air ratio, the NOx emission increased first and then decreased. In the same conditions, small particle coal combusted and generated more NOx than lager particle coal. It is advised to control the coal particle size.

Acknowledgements This research was supported by the National Key Research Plan (2019YFE0102100) and Huaneng Group science and technology research project (HNKJ20-H50) KTHT-U20GJJS06.

References

Cai R, Ke X, Lyu J et al (2017) Progress of circulating fluidized bed combustion technology in China: a review. Clean Energy 1(1):36–49

Chen Y (2009) Researches on wind power and photovoltaic generation integration into power system. Shanghai Jiao Tong University, Shanghai

Feng JK, Yue GX, Lyu JF (2003) Circulating fluidized bed combustion boiler. China Electric Power Press

Gungor A (2009) Simulation of NOx emission in circulating fluidized beds burning low-grade fuels. Energy Fuels 23(5):2475–2481

Li J, Yang H, Lyu J et al (2013) Low NOx emission characteristic of low energy consumption CFB boilers. J Combust Sci Technol 19(4):293–298

Li J, Yang X, Yang H et al (2016a) Experimental study and modeling of NOx generation from char nitrogen in the bubbling bed. J China Coal Soc 41(6):1546–1553

Li JJ, Zhang Y, Yang HR et al (2016b) Study of the effect of ash composition in coal on the kinetic parameters of NO reduction reaction by CO. J China Coal Soc 41(10):2448–2453

Li Q, Gao MF, Dong CW et al (2017) Experimental study on NOx emissions characteristics of CFB peak shaving boiler. Energy Conserv 36(11):34–36

Lyu JF, Ke XW, Cai RX et al (2018) Research progress on the kinetics of NOx reduction over chars in fluidized bed combustion. Coal Convers 41(1):1–12

Sørensen CO, Johnsson JE, Jensen A (2001) Reduction of NO over wheat straw char. Energy Fuels 15(6):1359–1368

Sun S, Zhang J, Hu X et al (2009) Studies of NO—char reaction kinetics obtained from drop-tube furnace and thermogravimetric experiments. Energy Fuels 23(1):74–80

Varol M, Atimtay AT, Olgun H et al (2014) Emission characteristics of co-combustion of a low calorie and high sulfur–lignite coal and woodchips in a circulating fluidized bed combustor: part 1. Effect excess air ratio. Fuel 117(6):792–800

Zhao Z, Qiu J, Li W et al (2003a) Influence of mineral matter in coal on decomposition of NO over coal chars and emission of NO during char combustion. Fuel 82(8):949–957

Zhao Z, Wen L, Qiu J et al (2003b) Effect of Na, Ca and Fe on the evolution of nitrogen species during pyrolysis and combustion of model chars. Fuel 82(15):1839–1844

Chapter 38
A Method to Measure the Solid Circulation Rate in CFB Boilers

Yangxin Zhang, Lu Cheng, Yang Zhang, Baoguo Fan, Hai Zhang, and Junfu Lyu

Abstract Circulating fluidized bed (CFB) combustion technology is attracting recently due to its low cost in pollution control. Researches show solid circulation rate (G_s) is an important parameter to CFB boiler design and of great significance in pollution control during the operation of CFB boilers. In view of the complex and rough measuring environments such as high temperature and severe abrasion, the measurement of G_s in a CFB boiler remains an unsolved problem in the literature. Thus, this paper proposes a method to quantitatively measure G_s in CFB boilers based on the heat transfer mechanism. In this method, the relation of G_s to the heat transfer coefficient between hot solid particles and heat exchange device was established, and thereby this method can overcome the high temperature and the abrasion problems. Experiments in a hot test rig and numerical simulation using the Barracuda package were conducted to obtain quantitative relation between G_s and heat transfer coefficient, as well as the distribution of the solid particle concentration. Finally, a linear relationship between G_s and the heat transfer coefficient/solid concentration of the cross-section in the standpipe were found and fitting functional formulas were presented as well, proving that the heat transfer method was a feasible and potential way to be applied in actual CFB boilers.

Keywords Circulating fluidized bed · Gas–solid flow · Heat transfer · Solid circulation rate

Y. Zhang · L. Cheng · B. Fan
Department of Thermal Engineering, Taiyuan University of Technology, Taiyuan 030024, China

Y. Zhang · L. Cheng · Y. Zhang (✉) · H. Zhang · J. Lyu
Key Laboratory for Thermal Science and Power Engineering of Ministry of Education, Department of Energy and Power Engineering, Tsinghua University, Beijing 100084, China
e-mail: yang-zhang@tsinghua.edu.cn

J. Lyu and S. Li (eds.), *Clean Coal and Sustainable Energy*, Environmental Science and Engineering,
https://doi.org/10.1007/978-981-16-1657-0_38

38.1 Introduction

As an efficient and clean coal combustion technology, the circulating fluidized bed (CFB) boiler technology has been developed rapidly during the past decades in China. In CFB boilers, circulating solids including pulverized coal, ash particles, and bed material are the media of the heat and mass transfer, maintaining a uniform temperature in the furnace. The solid circulation rate is directly related to the material balance and the pressure balance (Yao et al. 2010). Cai et al. (2018), Yue et al. (2017) demonstrated the fluidization state specification principle and showed that the solid circulation rate (G_s) is a momentous parameter to affect the fluidization state, once G_s and the gas velocity u_f are fixed, the state is determined. Consequently, acquiring the specific value of G_s is of great significance and it will be very useful for practical operation and design. Thus, an accurate and sustainable method to achieve on-line measuring G_s in CFB boilers is required.

The existing G_s measurement methods in the literature can be classified as invasive and non-invasive methods based on whether the measuring devices are inserted into gas–solid flow in pipelines or not. Non-invasive methods mainly include electrical sensors, attenuation and scattering methods, resonance sensors, tomography methods, digital imaging, etc. (Zheng and Liu 2010, 2011; Qian et al. 2015; Pang et al. 2018; Seraj et al. 2013). In addition, a magnetic tracer method was also reported recently (Guío-Pérez et al. 2017). Although these non-invasive methods can avoid destructing the flow, they still have plenty of problems like the high cost, the breakdown at high temperature and the sensitivity to the measuring environment. Considering the actual measuring condition, non-invasive methods are obviously not good options. On the contrary, invasive methods are relatively cheaper and more practical. Liu et al. (1995) designed a turbine meter to measure G_s which relates the solid flow rate to the velocity of the turbine. Hu et al. (2011) proposed an impact flow meter which determines G_s using the magnitude of the impact force. Wang et al. (2016) used a vibration probe sensor to measure a solid flow rate in a gas–solid flow. However, all of the mentioned invasive methods are only applied to the cold state, which is of great difference compared with the high temperature environment and the abrasion is a considerable issue as well.

Based on the aforementioned consideration, a new idea to measure G_s originated from the heat transfer between the hot gas–solid flow and the cold wall. This method relates G_s to the heat transfer coefficient which can be more conveniently experimentally determined. In addition, this method can be applied to high-temperature environments. Due to the cooling effect, the abrasion problem, commonly seen in the invasive method, is also weakened. Thus, it is more appropriate and has a greater possibility to be applied in the actual measuring environment in a CFB boiler. This method is denoted by "heat transfer method" hereafter in the present paper.

The core of the heat transfer method is the relation between the heat transfer coefficient and solid mass flow rate on the cross-section as shown in Fig. 38.1. This requires the heat transfer from the solid to the surface is approximately a monotonic function of the solid material that instantaneously flushes through the heat exchange

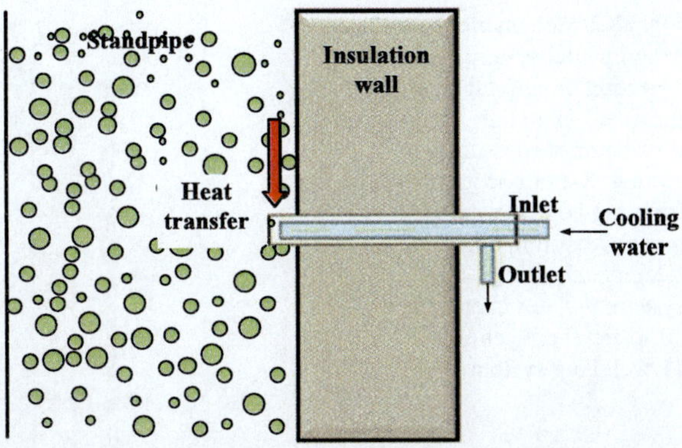

Fig. 38.1 Description of the method

surface. The standpipe between the cyclone and the loop seal is a good place to apply the heat transfer method due to its one-way downstream in the upper portion of the standpipe rather than the complex back mixing in risers. The heat transfer coefficient is convenient to acquire through temperature measurements. Once the monotonic relation between G_s and the heat transfer coefficient is given, the value of G_s can be therefore obtained. In addition, since the measuring device is inserted into the standpipe for a small distance to avoid destroying the flow state, the measuring value only represented one point on the whole cross-section. In order to obtain the exact value of G_s, it is necessary to study the mass flux distribution of the solid material in the standpipe.

This paper is organized as follows. In Sect. 38.2, the basic methods of experimentation and numerical simulation are generally introduced. Section 38.3 gives the experimental study results which form the relation between G_s and the heat release coefficient, and present the numerical results of the solid material distribution in the standpipe. The findings in the present study give a firm fundamental basis to develop this heat transfer method.

38.2 Methods

38.2.1 Experimental Setup

Experiments were conducted in a hot test rig as shown in Fig. 38.2. It primarily consisted of a tube furnace, a measuring section, a temperature acquisition system, a water pump, and a weighing system. The overall height was 4.2 m. The heating tube had a height of 2.3 m and a diameter of 0.06 m with a maximum holding load being 6.5 kg. The bed material used was quartz sand, whose physical properties are listed in Table 38.1. The solid flowed through the central square tube which had a cross-section area of 0.04 × 0.06 m^2. Between the wall and inner part, there were insulating layers to minimize the heat loss.

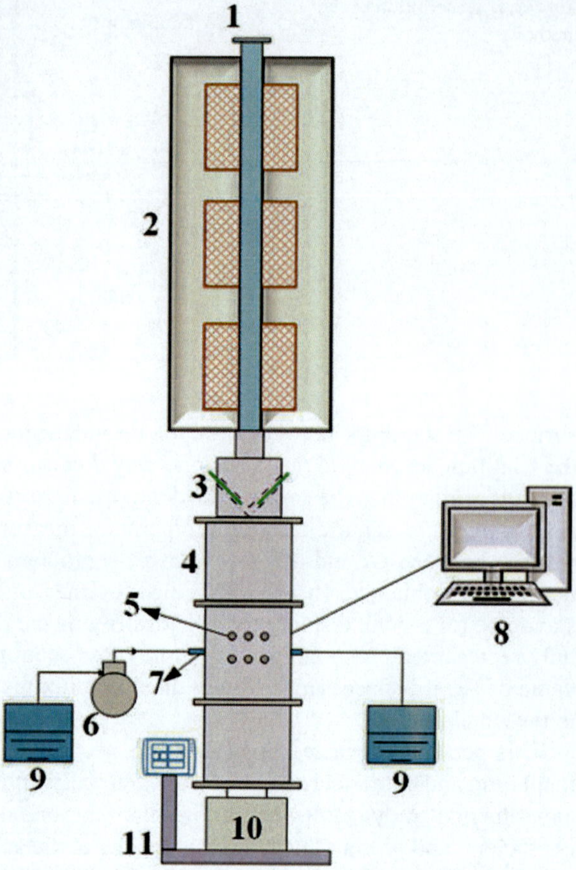

Fig. 38.2 Schematics of the experimental system. 1-heating tube; 2-tube furnace; 3-material distributor; 4-measuring section; 5-thermocouple metering hole; 6-water pump; 7-cooling tube; 8-temperature acquisition system; 9-water bucket; 10-material collection; 11-weighing system

Table 38.1 Physical properties of quartz sand

ρ/kg m^{-3}	Size range/μm	d_{50}/μm
2650	106–212	171

During experiments, quartz sand was heated to 300 °C first. When the sand reached the target temperature, the horizontal baffle, acting as a valve for the sand was removed so that sand fell down into the measuring section along the central square tube and the heat transfer process from the hot sand to the water-cooled tube occurred simultaneously. The solid flow rates varied by controlling the opening degree of baffles. The solid flow rate was proved to be very stable as shown in Fig. 38.3. Round 1, 2 and 3 mean that the calibration experiments were repeated three times at each opening degree and $G_{s,ave}$ was the average of the three results. The temperature acquisition system included eight thermocouples (level II degree of accuracy and 0.2 s response time) and an automatic recorder (1 s collecting frequency). Six of the thermocouples were set to detect the sand temperature near the cooling tube, and the

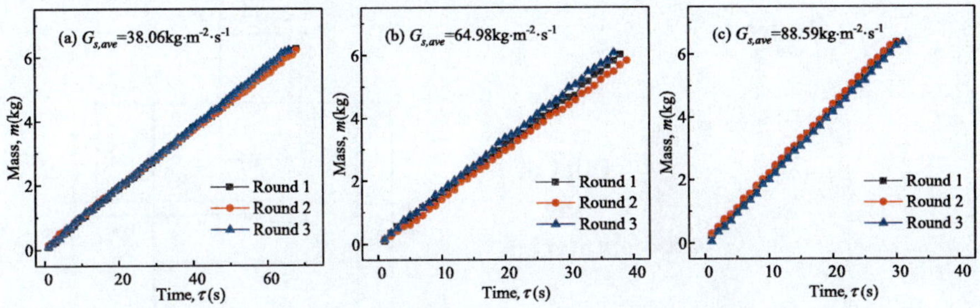

Fig. 38.3 Calibration of solid flow rate at three different opening degree

other two were placed at the inlet and the outlet of the water tube to acquire the water temperature difference. The cooling water was provided through a metering pump which has an operational range of 0.01–50 mL/min. The uncertainty of the water flow rate was less than 1%. The weighting system was used to record the mass of quartz sand falling during the experiments. The gradient of sand mass as a function of time was regarded as the mass flow rate. G_s was then the ratio of mass flow rate over the sectional area of the central square tube.

38.2.2 Numerical Simulation

Standpipe is an important component connecting cyclone and loop seal. The solid flow inside the standpipe includes two regimes: dilute phase in the upper part and the dense phase at the bottom. A number of studies have discussed its mechanism and impacts (Yao et al. 2011; Dieter et al. 2014; Wang et al. 2013). However, the flow pattern of the dilute phase in the standpipe remains unclear. Based on some studies on the numerical simulation of CFB boilers (Huang et al. 2018; Zhang et al. 2013; Wang et al. 2014a, b; Qiu et al. 2016), it is found the software Barracuda VR (CPFD Software, LLC, NM, USA) could be used to numerically study the solid concentration distribution. Barracuda is a relatively mature commercial software specializing in large-scale gas–solid flow simulation in the industrial field. Based on the concept of MP-PIC (multiphase-particle-in-cell) method, it can effectively solve the coupling calculation problem of fluid and a large number of particles in the three-dimension problem.

In the present study, a numerical model was constructed using Barracuda as shown in Fig. 38.4. Since the proposed method was designed for in-situ measurement, the size of the model was mainly based on an actual 220t/h CFB boiler. Besides, since the measurements take place in the dilute phase of the standpipe, thus the model merely contained a cyclone with the standpipe. In this case, the interphase function was defined by Wenyu-Ergun model. Detailed parameter settings were shown in Table 38.2, and the results of the simulation will be discussed in the next section.

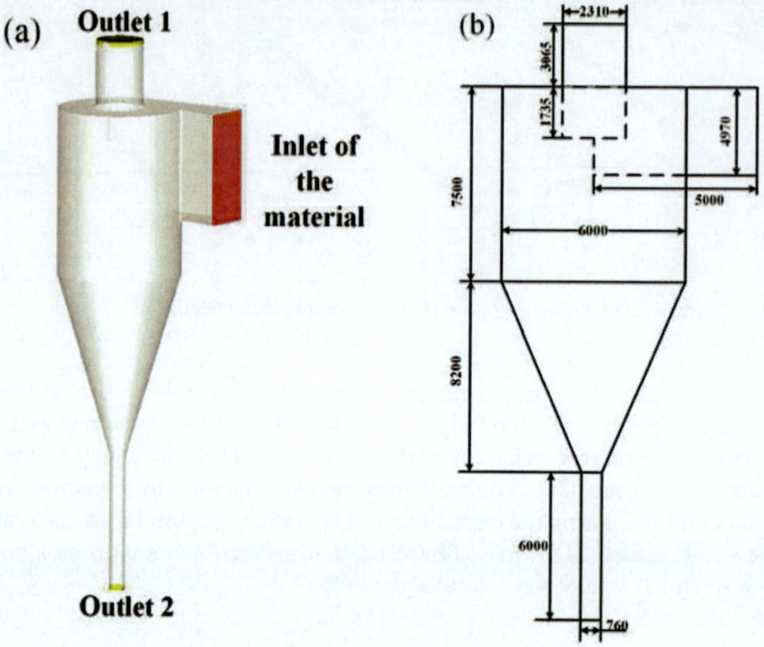

Fig. 38.4 **a** The structure of the 3D model; **b** geometry dimension of the model

Table 38.2 Detailed parameters setting

Parameter	Value
Gas velocity (m/s)	20
Gas density (kg/m^3)	1.17
Solid density (kg/m^3)	2650
Solid mass flow (kg/m^2 s)	18.4
Total time (s)	15
Time step (s)	3×10^{-3}
Close pack fraction limit	0.55
Normal-to-wall momentum retention	0.3
Particle–wall interaction	0.99

38.3 Result and Discussion

38.3.1 Experimental Results

The experiments were conducted under different solid flow rates and the case of each opening degree of the baffle board was repeated 3 times. The actual measured solid flow rates are shown in Fig. 38.5.

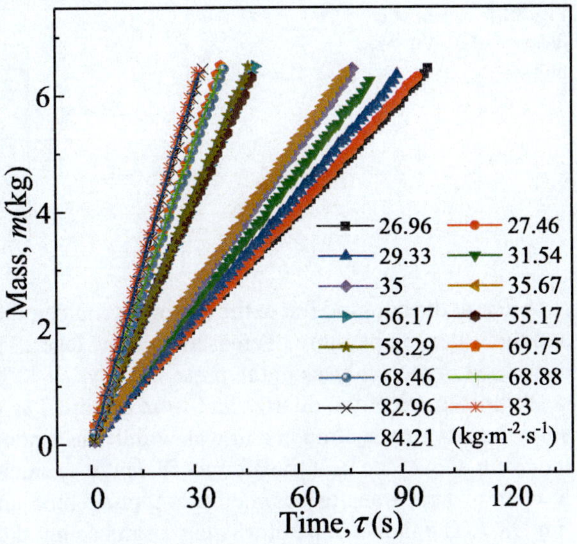

Fig. 38.5 The actual measured solid flow rates during the experiments

The flow rate of cooling water was fixed at 50 mL/min, and the temperature changes of water of every case are presented in Fig. 38.6. As demonstrated in Fig. 38.6, the water temperature rapidly increased at the very beginning and then reached a relatively steady state.

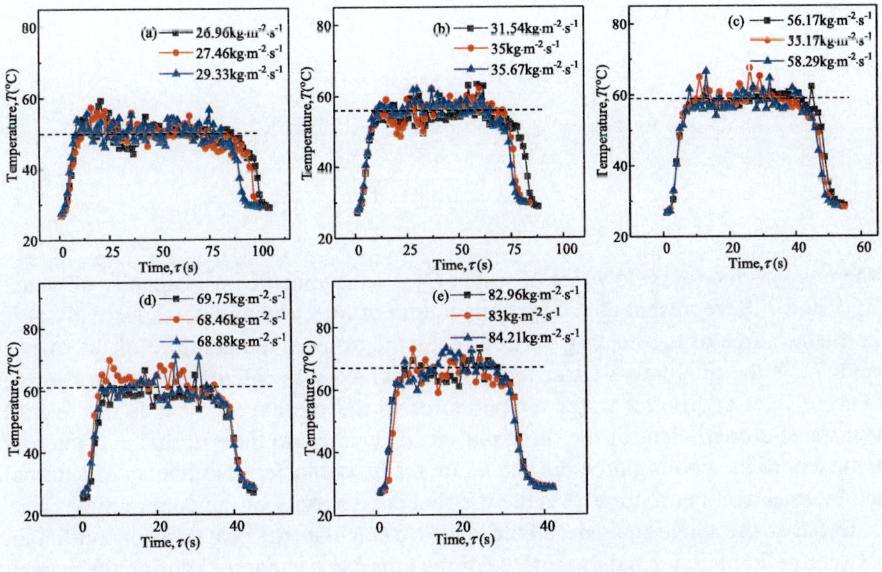

Fig. 38.6 Variations of outlet water temperature at different solid flow rates

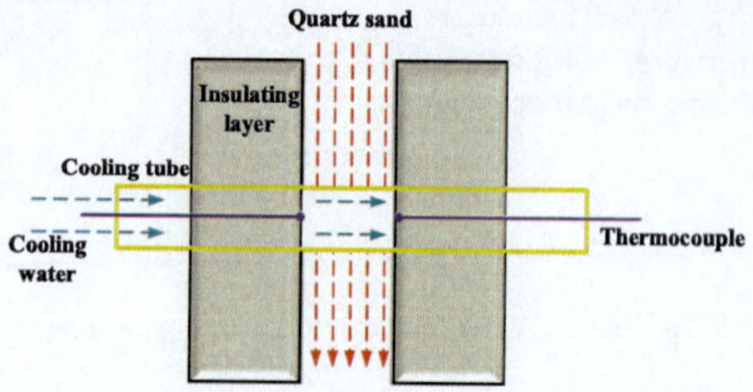

Fig. 38.7 Schematic diagram of heat transfer process

After all the hot sand fell to the bottom container, the heat transfer process stopped and the water temperature decreased to a low level. Thus, only the steady-state data were used in the analyses in the present study.

Temperatures of the quartz sand were obtained as well. The quartz sand temperature did not vary too much and was within the range of 260–290 °C. According to the result above, the heat transfer coefficient was calculated. First of all, the detailed process of heat transfer between the cooling tube and solid particles is shown in Fig. 38.7. The high-temperature quartz sand scoured the surface of the cooling tube made of copper and lead to an increase of the cooling water temperature. The actual length l of the heating section was 6 cm. The thermocouples were set exactly at each end of the heating section. When this process reached a steady state, the temperatures of water of inlet and outlet and quartz sand were nearly constant. The whole heat transfer process was divided into three phases: (1) heat convection between hot sand and tube surface, (2) heat conduction inside the tube wall, and (3) heat convection between water and tube inner surface (Yang and Tao 2006), which were described by Eqs. (38.1) and (38.2).

$$k\frac{q_{m,w}C_{p,w}T_{w,out}T_{w,in}}{AT_sT_f} \tag{38.1}$$

$$k\frac{1}{\frac{1}{h_i}\frac{d_o}{d_i}\frac{d_o}{2}\ln\frac{d_o}{d_i}\frac{1}{h_o}} \tag{38.2}$$

where $q_{m,w}$ is the mass flow rate of water. $C_{p,w}$ is the specific heat capacity of water. $T_{w,out}$ and $T_{w,in}$ represent the water temperature of inlet and outlet, respectively. A is the surface area of the heating section. T_s is the average temperature of the quartz sand. T_f is the qualitative water temperature which was calculated as the average value of inlet and outlet water temperatures in the present study. k is the overall heat transfer coefficient of the three phases. d_o and d_i are the external and internal diameters of the cooling tube. h_o and h_i are the heat transfer coefficients of external and internal heat convection. λ is the thermal conductivity of the cooling tube.

Based on the above analysis, all the values to calculate the heat transfer coefficient between solid material and the surface of the tube can be acquired during the experiments. In the present study, h was calculated at every instantaneous steady state. The results are shown in Fig. 38.8.

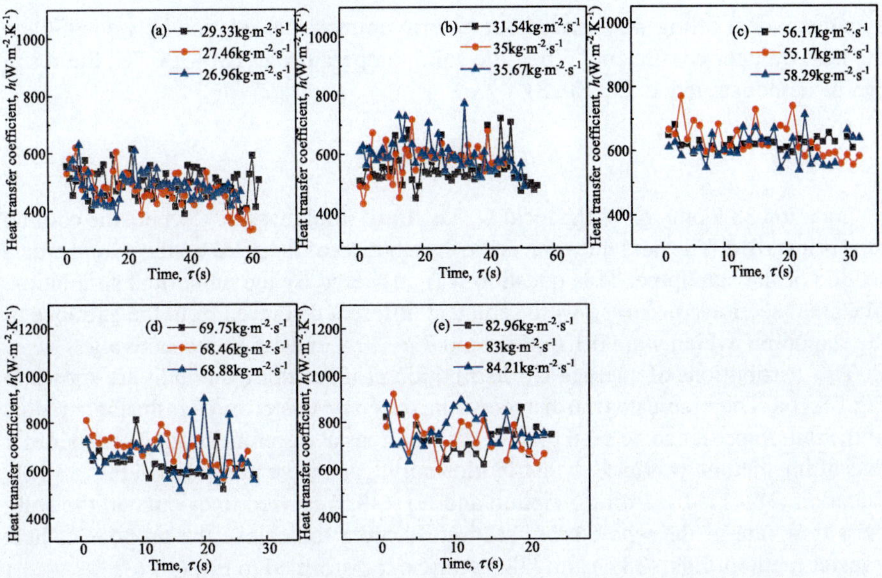

Fig. 38.8 Transient heat transfer coefficient change

As shown Fig. 38.8, at approximately the same G_s, the h value fluctuated around a certain value, and h increased as G_s increased. Finally, the average of all the transient h values at a certain G_s is regarded as the heat transfer coefficient at the specific G_s.

The obtained heat transfer coefficients are summarized in Fig. 38.9. The heat transfer coefficient increases as the solid flow rate increased. This is straightforward that the increase of the solid concentration resulted in an enhanced heat transfer. The trend of the heat transfer coefficient as a function of the solid flow rate approximately agreed to the linear law, which proved the feasibility of the proposed method. Thus,

Fig. 38.9 Variation of heat transfer coefficient with G_s

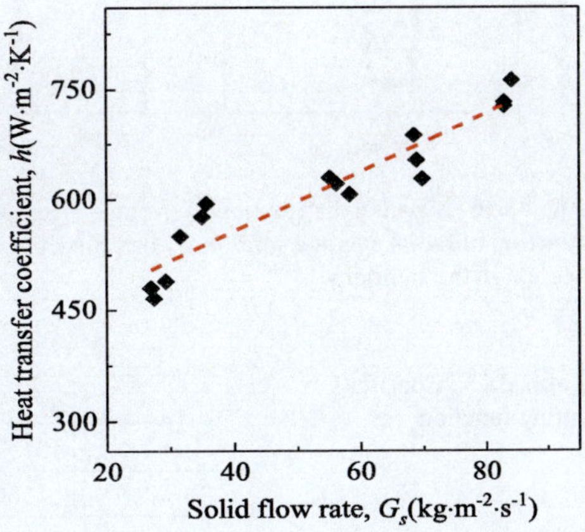

a mathematical fitting was carried out to form a functional relation between G_s and the heat transfer coefficient. When the solid temperature is 280–300 °C, the result can be demonstrated as Eq. (38.3).

$$h\ 3.99G_s\ 396.99 \tag{38.3}$$

Equation 38.3 only gives the local G_s, i.e., local solid mass flux, where the cooling tube is installed. The next question is the distribution of the solid on the whole cross-section of the standpipe. This question was answered by the numerical simulation. The cross sections discussed were settled at different distances from the entrance of the standpipe, which were 0.1 m, 0.4 m, 0.7 m, 1 m, and 1.5 m respectively.

The distributions of average G_s distribution profiles along the radii are shown in Fig. 38.10a. The r indicates the distance from the circle center and R is the inner radius of the standpipe. It can be seen that the distributions at different heights were similar thus fitting method was used in distribution curves and were presented in Fig. 38.10b. Equation (38.4) is the fitting function and Eq. (38.5) is used to calculated the solid mass flow rate of the whole cross section. In order to achieve the propose of heat transfer method, Eqs. (38.4) and (38.5) can be transformed to Eq. (38.6).

$$G_s^*(r) = \begin{cases} a_1r^2 + b_1r + c_1 & -R \le r < -0.468R;\ 0.468R < r \le R \\ a_2r^2 + b_2r + c_2 & -0.468R \le r \le 0.468R \end{cases} \tag{38.4}$$

where $G_s{}^*$ is the local solid flow rate. a_i, b_i*and* c_i ($i = 1$ and 2) are constants which are presented in Table 38.3.

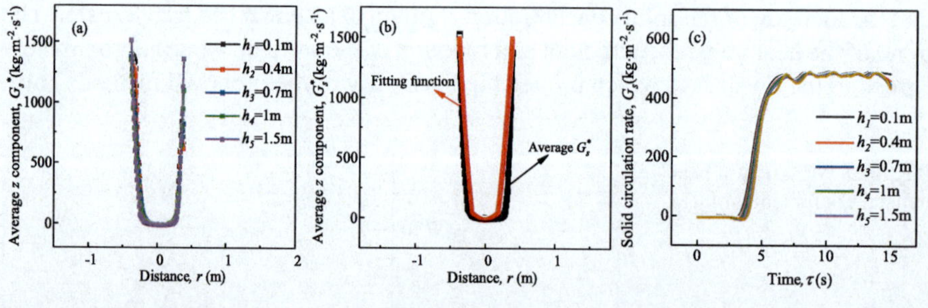

Fig. 38.10 **a** Radial distribution of average *z* component of mass flux at different heights; **b** fitting function of radial average solid mass flux distribution; **c** average solid mass flow rate on the cross-section in the standpipe

Table 38.3 Constants of fitting function

i	a_i	b_i	c_i
1	13157.58	0	−396.09
2	1566.83	0	0

$$Gs \cdot A = -\int_{-R}^{-0.486R} G_s^*(r)\pi r dr - \int_{-0.486R}^{0} G_s^*(r)\pi r dr + \int_{0}^{0.486R} G_s^*(r)\pi r dr + \int_{0.486R}^{R} G_s^*(r)\pi r dr \quad (38.5)$$

$$Gs \cdot A = \pi f_1 (-\int_{G_s^*(-R)}^{G_s^*(-0.486R)} G_s^* dG_s^* + \int_{G_s^*(0.468R)}^{G_s^*(R)} G_s^* dG_s^*) - \frac{\pi}{3133.66}(\int_{G_s^*(-0.486R)}^{G_s^*(0)} G_s^* dG_s^* - \int_{G_s^*(0)}^{G_s^*(0.468R)} G_s^* dG_s^*) \quad (38.6)$$

where A is the area of standpipe cross section. f1 is

$$f_1 = \frac{\sqrt{-4a_1c_1 + 4a_1G_s^*}}{2a_1\sqrt{-4a_1c_1 + 4a_1G_s^*}}.$$

The mass flux of the cross section of different heights was also acquired from Fig. 38.10c. The average mass flux at various heights reached a steady state after 5 s, and almost was stabilized at ~500 kg m^{-2} s^{-1}, which was also corresponding with the calculation results using Eqs. (38.4) and (38.5).

From the computation, the distribution of the solid on a cross section was fitted to Eq. (38.6). Using both Eqs. (38.3) and (38.6), the overall G_s in the standpipe can be obtained. One should note that actually the experimental condition of the present study is not exactly the same as the real CFB boiler. This study still gives the pioneering work to confirm the feasibility of the heat transfer method to measure G_s of the real CFB boilers. Experiments at higher temperatures will be a future study.

38.4 Conclusion

In this work, a heat transfer method for measuring the solid circulating rate in CFB boilers was proposed and studied in the laboratory scale. A linear relation between heat transfer coefficient of the solid flow to the cold surface and solid flow rate was established through hot-test experiments, which indicated the feasibility and possibility of the proposed method. Numerical simulation of the solid flow inside the standpipe was also conducted to discover the solid material concentration distribution inside the standpipe of an actual CFB boiler. An empirical correlation of the distribution of solid material mass flux on the cross-section of the dilute phase area in standpipe was given so that the overall cross-section flow rate could be derived from the solid flow rate at one point during the practical measuring process. Combining the experimental and numerical results, a relatively complete method was given. Higher temperature experiments will be conducted in the future to meet the needs of the real CFB boilers.

Acknowledgements This work was supported by the National Key Research and Development Program of China (2018YFF0216002).

References

Cai R, Hai Z, Man Z (2018) Development and application of the design principle of fluidization state specification in CFB coal combustion. Fuel Process Technol 174:41–52

Dieter H, Charitos A, Bidwe AR, Wei A (2014) Study of standpipe and loop seal behavior in a circulating fluidized bed for Geldart B particles. In: International conference on circulating fluidized beds & fluidization technology-CFB, Oregon, USA

Guío-Pérez DC, Dietrich F, Cala JNF (2017) Estimation of solids circulation rate through magnetic tracer tests. Powder Technol 316:650–657

Hu N, Yao X, Chao J (2011) Impacting method for on-line measurement of solid circulating rate in CFB. J Chin Soc Power Eng 31:485–490

Huang Y, Zhang M, Lyu J (2018) Effects of gas leakage on the separation performance of a cyclone, part 2: simulation. Chem Eng Res Des 36:906–915

Liu J, Huan B, Su Y (1995) Measurement on solids recirculating flowrate of circulating fluidized bed. Eng Chem Metallurgy 16:257–262

Pang L, Shao Y, Geng C (2018) Measurement of solid mass flow rate by a non-intrusive microwave method. Powder Technol 323:525–532

Qian X, Yong Y, Wang L (2015) An integrated multi-channel electrostatic sensing and digital imaging system for the on-line measurement of biomass–coal particles in fuel injection pipelines. Fuel 151:2–10

Qiu G, Ye J, Wang H (2016) CPFD simulation of gas-solids flow in annular combustion chamber of large-scale circulating fluidized bed. J Univ Chin Acad Sci 33:218–222

Seraj H, Rahmat MF, Khaled M (2013) Measurement of velocity of solid/air two phase fluid using electrostatic sensors and cross correlation technique. Scientia Iranica 20:786–792

Wang X, Ma Z, Lei J (2013) Numerical simulation for the loop seal in the circulating fluidized bed and experimental validation. Appl Therm Eng 52:141–149

Wang Q, Yang H, Wang P (2014a) Application of CPFD method in the simulation of a circulating fluidized bed with a loop seal, part I—determination of modeling parameters. Powder Technol 253:814–821

Wang Q, Yang H, Wang P (2014b) Application of CPFD method in the simulation of a circulating fluidized bed with a loop seal, part II-investigation of solids circulation. Powder Technol 253:822–828

Wang C, Yu H, Zhan N (2016) A vibration probe sensor for mass flow rate measurement of gas-solid two-phase flow. Sens Rev 36:200–206

Yang S, Tao W (2006) Heat transfer, 4th edn. Beijing (chapter 6)

Yao X, Yang S, Chao J (2010) Experiment study of solids circulating rate's effect on the pressure loop in circulating fluidized bed. Proceeding of the CSEE 20:1–6

Yao X, Yang H, Zhang H (2011) Gas-solid flow behavior in the standpipe of a circulating fluidized bed with a loop seal. Energy Fuels 25:246–250

Yue G, Cai R, Lu J (2017) From a CFB reactor to a CFB boiler-the review of R&D progress of CFB coal combustion technology in China. Powder Technol 316:18–28
Zhang R, Yang H, Lyu J (2013) Application of CPFD approach on gas-solid flow and combustion in industrial CFB boilers. Proceeding of the CSEE 33:75–83
Zheng Y, Liu Q (2010) Review of certain key issues in indirect measurements of the mass flow rate of solids in pneumatic conveying pipelines. Measurement 43:727–773
Zheng Y, Liu Q (2011) Review of techniques for the mass flow rate measurement of pneumatically conveyed solids. Measurement 44:589–604

Chapter 39
Reliability Analysis of 100 MW and Above Circulating Fluidized Bed Boiler Units in China in 2017

Xia Zhou, Jianfeng Li, Yuge Yao, Shujie Chen, Hong Zhou, Qing Liu, and Geng Chen

Abstract In order to demonstrate the overall operation reliability level of circulating fluidized bed (CFB) boiler units in China and provide a useful reference for their operation and development, the statistics of the operation reliability of more than 130 CFB boiler units in China were collected and analyzed. The results show that the 200 MW class units had the longest utilization hours which exceeds 4500 h. The 135 MW class units were most reliable, each of which met 0.65 unplanned outages on average in 2017. The reliability level of the 300 MW class units was comparatively low due to some new units. They met 1.1 and 0.96 planned outages and unplanned outages on average respectively. The unplanned outages occurred twice in the 600 MW supercritical CFB boiler unit in 2017 and had occurred five times in total since its commissioning, three of which were caused by its boiler system. It could be concluded that the boiler systems have the greatest impact on the reliability of the units. The high-temperature heating surface and the combustion system are the main parts of a boiler causing the unplanned outages. The wear and crack are the main kinds of technical faults. In addition, the product quality, maintenance and operation management of the power plants are also very important for the reliability of the CFB boiler units.

Keywords Circulating fluidized bed boiler · Operation reliability · Unplanned outage · Boiler system · Technical fault

---

X. Zhou · J. Li · H. Zhou · G. Chen
China Electricity Council, Beijing 100761, China

Y. Yao (✉) · S. Chen · Q. Liu
Key Laboratory for Thermal Science and Power Engineering of Ministry of Education, Department of Energy and Power Engineering, Tsinghua University, Beijing 100084, China
e-mail: yyg18@mails.tsinghua.edu.cn

J. Lyu and S. Li (eds.), *Clean Coal and Sustainable Energy*, Environmental Science and Engineering,
https://doi.org/10.1007/978-981-16-1657-0_39

39.1 Introduction

China is rich in coal resources but poor in oil and gas. The current situation that coal is the major resource and the coal-fired power is the major power supply form in China will not change in the near future (Song et al. 2015). By the end of 2017, the capacity of coal-fired power plants was 989.1 million kW, accounting for 55.22% of the total capacity in China. And the coal-fired power generation was 4.15 trillion kWh, accounting for 64.67% of the total generation in China.

Circulating fluidized bed (CFB) boiler units are indispensable in China's coal-fired generating units. Due to better fuel flexibility and lower emission of SO_2 and NO_X, CFB combustion technology can be broadly applied to use low-quality fuels (Yue et al. 2017; Cai et al. 2018; Arjunwadkar et al. 2016; Yao et al. 2019). CFB boiler units also have a good ability of load regulation (Gao et al. 2016). Up to now, more than a hundred 300 MW class CFB boiler units, about twenty 350 MW class supercritical CFB boiler units and the only 600 MW supercritical CFB boiler unit in the world has been put into operation in China.

With the wide application of CFB boiler units in China, its operation reliability becomes a key parameter to the safety and stability of power grids and benefit of the power enterprises. Therefore, the statistics of the operation reliability of more than 130 CFB boiler units in China in 2017 were collected and analyzed. On the one hand, the statistics could demonstrate the overall operation reliability level of CFB boiler units in China. On the other hand, statistical analysis could provide a useful reference for their operation and development.

39.2 Statistics of Basic Reliability Indexes

Figure 39.1 shows the statistics of the numbers and total capacities of more than 130 CFB boiler units in 2017 divided into four different capacity classes. The 135

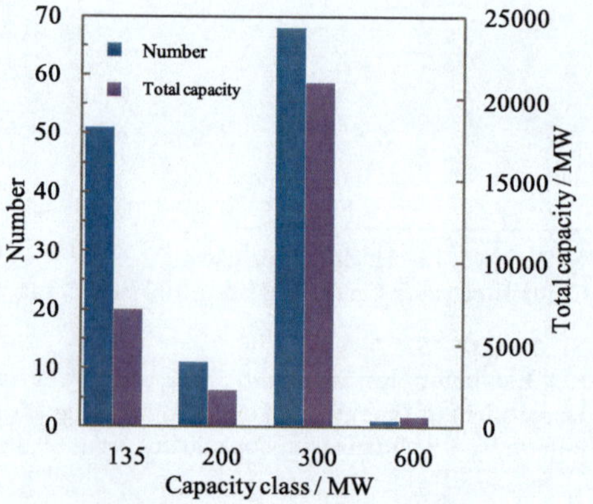

Fig. 39.1 The numbers and total capacities of CFB boiler units in 2017

and 300 MW classes units were the main units in China. However, the number of 135 MW class units decreased in recent years due to the development and promotion of high capacity units in China. The 200 MW class only had a small portion of all the units. The No. 61 unit in Baima Power Plant was the only 600 MW class unit in the world.

The utilization hours, defined as the equivalent full-load operation hours, is an important reliability index of a unit, which also, as for power enterprises, directly links to the benefit of a power plant. As shown in Fig. 39.2, the 200 MW class units had the longest utilization hours among four different capacity classes in 2017, which exceeds 4500 h. The utilization hours of 135 and 300 MW classes could be regarded as equal. The 600 MW class unit's utilization hours was merely about 2000 h.

Figure 39.3 shows the operation hours and standby hours of four different capacity class CFB boiler units. The sum of the operation hours and standby hours is defined as available hours. The statistics shown in Fig. 39.2 are quite similar to those in Fig. 39.3. Most 200 MW class units are enterprise-owned in China and have to supply power to enterprises production so that they operated longer compared with

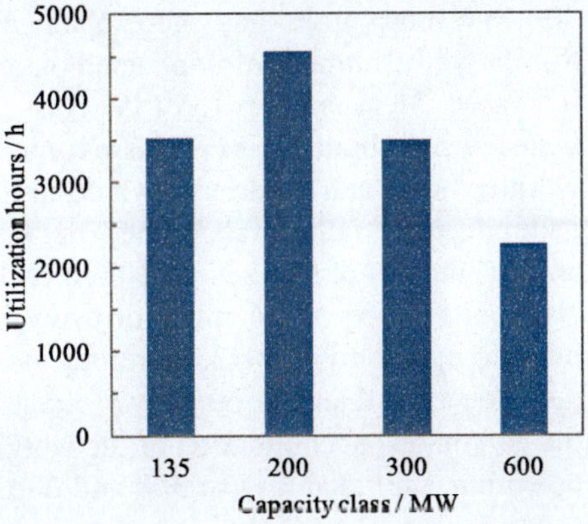

Fig. 39.2 The utilization hours of CFB boiler units in 2017

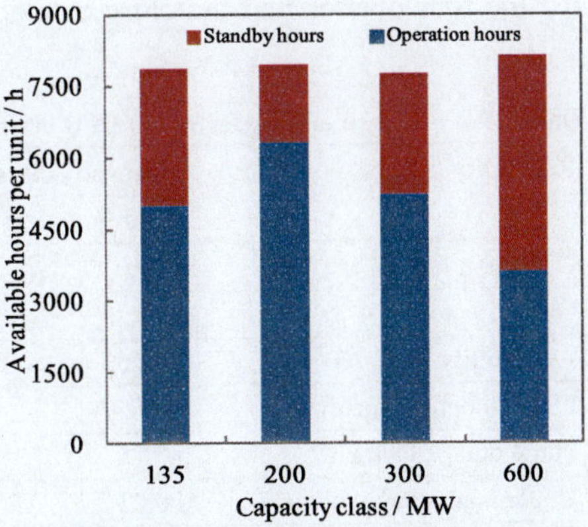

Fig. 39.3 The available hours of CFB boiler units in 2017

Table 39.1 The statistics of outages occurred in CFB boiler units in 2017

The number and the average of different kinds of outages (per unit)	Capacity class/MW			
	135	200	300	600
The number of planned outages	0.84	0.82	1.10	1
Average planned outage hours/h	673.55	858.38	677.14	606.20
The number of force outages	0.57	0.64	0.79	1
Average force outage hours/h	111.88	111.38	132.14	4.98
The number of unplanned outages	0.65	0.91	0.96	2
Average unplanned outage hours/h	108.32	119.08	136.83	6.89

135 and 300 MW classes in 2017. Therefore, they had the longest operation hours and also utilization hours. The No. 61 unit in Baima Power Plant is located in Sichuan Province, one of the major hydropower provinces in China. This unit operated under relatively low load and had the most standby hours and the least operation hours due to the operation of the hydropower units. Therefore, it had the least utilization hours.

Table 39.1 shows the statistics of outages occurred in CFB boiler units. The 300 MW class units had more planned outages than 135 and 200 MW classes. Similarly, the number of unplanned outages per unit increased as the capacity class increased. Most units in 135 MW classes were put into operation earlier and there were few new-built 135 MW classes units. Thus, the operators had accumulated more working hours and experience which could provide the units with better maintenance and reduce the failure rates. They met 0.65 unplanned outages on average in 2017, almost same as pulverized coal-fired boiler units in China. However, new 300 MW class units were put into operation every recent year. And the 600 MW class unit was put into operation in 2013, therefore. Both of these two classes units went through more outages than the other two classes in 2017. It can be speculated that with the continuous accumulation of the working experience and the optimization of the operation, the reliability of 300 and 600 MW classes will be enhanced in the near future.

Table 39.2 shows the statistics of main reliability factors of CFB boiler units. All the four classes had the close availability factors around 90% in 2017, which

Table 39.2 Main reliability factors of CFB boiler units in 2017

Reliability factor	Capacity class/MW			
	135	200	300	600
Operation factor/%	59.01	72.33	60.59	41.28
Output factor/%	70.22	71.95	66.97	62.70
Availability factor/%	92.52	90.78	89.86	92.92
Planned outage factor/%	6.65	8.04	8.62	6.92
Force outage factor/%	0.75	0.81	1.21	0.06
Unplanned outage factor/%	0.83	1.24	1.52	0.16

Table 39.3 Main reliability factors of the No. 61 unit in Baima Power Plant from 2013 to 2017

Reliability factor	Year				
	2013	2014	2015	2016	2017
Operation factor/%	72.59	74.82	44.60	46.84	41.28
Output factor/%	66.77	71.81	69.55	60.31	62.70
Availability factor/%	85.04	93.50	82.80	100.00	92.92
Planned outage factor/%	14.94	6.50	16.28	0.00	6.92
Force outage factor/%	0.00	0.00	0.92	0.00	0.06
Unplanned outage factor/%	0.02	0.00	0.92	0.00	0.16

means that the overall reliability of the CFB boiler units in China is high. Although the 600 MW class unit went through two unplanned outages in 2017, each of them lasted only 6.89 h on average shown in Table 39.1. Therefore, it had the lowest unplanned outage factor.

The No. 61 unit in Baima Power Plant is the largest capacity CFB boiler unit in the world. As the only 600 MW class CFB boiler unit, its reliability is very important and has been attracting great attention in the industry since it was put into operation in April 2013. The unplanned outages occurred 5 times in total from April 2013 to the end of 2017, three of which belongs to the force outages and two belongs to the unplanned outages. The longest outage lasted 67.07 h while the shortest lasted only 1.18 h. Only three outages had occurred in the boiler system, caused by the crack of the water wall tube, slag discharge fault, and hot air leakage, respectively. Table 39.3 shows the complete statistical data of reliability factors of the No. 61 unit in Baima Power Plant from 2013 to 2017. In these five years, this unit had kept on a considerable operation reliability level.

39.3 Analysis of Unplanned Outages

The unplanned outages of units bring great economic losses to the power enterprises. The indirect economic loss including the power loss, power grid assessment cost and reputation loss, etc. cost even more than the direct economic loss such as the repair and restart costs of the units. In addition, the unplanned deratings of a unit also reduce the load and cause the power loss so that reduce the unit economy.

According to *Reliability evaluation code for generating equipment* (DL/T 793.2-2017), the unplanned outages of the coal-fired units can be divided into five types (UO1–UO5). The UO1 means the unit encounters the most emergency events which cause or require the immediate shutdown. On the contrary, UO5 is with the least urgency. The UO1, UO2 and UO3 are collectively called the force outages. Similarly, the unplanned deratings of the coal-fired units can also be divided into four different types. Statistics of different types of unplanned outages and deratings of the CFB boiler units in 2017 are summarized in Table 39.4. None of the CFB boiler units went

Table 39.4 Statistics of unplanned outages and deratings of the CFB boiler units in 2017

Reliability index	Type			
	1	2	3	4
The number of unplanned outages	66	29	7	19
Unplanned outage hours/h	10,016.57	2980.72	496.72	2605.46
The number of unplanned deratings	18	11	5	1
Equivalent outage hours/h	16.69	11.28	69.00	10.02

through the UO5 in 2017. The UO1 occurred most frequently, which accounted for 54.55% of the total unplanned outages. It also accounted for 60% of the unplanned outages hours. These indicate that most faults in CFB boiler units brought about the immediate shutdown and then would cause great economic losses.

As shown in Table 39.4, there were only 35 unplanned deratings and 106.99 h in total. The negative impact of unplanned deratings on CFB boiler units was very small and even could be neglected compared to the unplanned outages. Therefore, this paper will not make further analysis of the unplanned derating.

In order to explore the occurrence regularity of unplanned outages, where the unplanned outages occurred need to be known. Table 39.5 shows the systems in which the unplanned outages occurred. There were 84 unplanned outages occurred in boiler systems, accounting for 69.42% of the total and 77.82% of the unplanned outage hours. Therefore, the boilers system is the least reliable part of the entire CFB boiler unit which needs better operation, maintenance, and also technical management. The turbine systems caused the second most unplanned outages, accounting for 17.36% of the total and 11.17% of the unplanned outage hours. The average unplanned outage hours of different systems are also summarized in Table 39.5. A shorter average unplanned outage hours implies the system is easier to be maintained and has better ability in recovery from an unplanned outage. The transformer systems had the longest average unplanned outage hours which is near 300 h and much longer than all other systems. In spite of only three unplanned outages in 2017, the faults of power transformer systems can cause greater trouble in units operation.

Table 39.5 Statistics of the systems in which the unplanned outages occurred

System	Unplanned outages		
	Number	Hours/h	Average hours//h
Boiler	84	12528.93	149.15
Turbine	21	1798.08	85.62
Transformer	3	824.38	274.79
Protection	4	391.63	97.91
Auxiliary power	2	298.65	149.33
Generator	5	183.80	36.76
Other	2	74.00	37.00

To further analyze the reliability of the units, the statistics of boiler components and technical faults causing unplanned outages in boiler systems were collected and summarized in Tables 39.6 and 39.7 respectively. As shown in Table 39.6, the water walls were the major parts which contributed to 29.76% of the total unplanned outages in boiler systems and 43.88% of the unplanned outage hours in 2017. The heating surfaces, including water walls, superheaters, and reheaters, caused 44 unplanned outages in total, accounting for 52.38% of the total and 68.05% of the unplanned outage hours. The heating surfaces in CFB boilers are arranged in the furnaces or the fluidized bed heat exchangers, where high concentration solid particles move up and down and can easily cause severe wear and even crack on tubes of the heating surface. Therefore, it can be speculated that the wear and crack should be the main kinds of technical faults, which is proved in Table 39.7, As shown in Table 39.7, the wear and crack caused more than half of unplanned outages in boilers systems, which also accounted for more than 65% of the unplanned outage hours.

In addition, near one in five unplanned outages in boilers systems were attributed to combustion systems due to its various kinds of faults such as coke, blockage (slag discharge fault), burnout and ash leakage.

The causes of the faults summarized in Table 39.7 are classified in Table 39.8. As shown in Table 39.8, the product quality, maintenance and operation management of the power plants are very important for the operation reliability of the CFB boiler units. The causes listed in Table 39.8 are linked to different departments or organizations. The manufacturers are in charge of product quality. The construction companies are in charge of construction and installation. And the design institutes are in charge of the planning and design of a power plant unit. Fuel procurement is handled by power enterprises. In fact, restricted by the fuel market in China, often the power enterprises cannot purchase the design coal considering the benefit, which

Table 39.6 Statistics of the boiler components causing unplanned outages

Components	Unplanned outages in boiler systems		
	Number	Hours/h	Average hours/h
Water wall	25	5497.63	219.91
Superheater	11	1863.53	169.41
Combustion system	16	1720.62	107.54
Reheater	8	1164.58	145.57
Fuel system	5	580.87	116.17
Economizer	3	538.67	179.56
Flue	4	507.15	126.79
Desulfurization system	4	389.93	97.48
Dust removal system	1	109.08	109.08
Fan	5	72.35	14.47
External Heat Exchanger	1	57.95	57.95
Other	1	26.57	26.57

Table 39.7 Statistics of the technical faults causing unplanned outages in boiler systems

Technical faults	Unplanned outages in boiler systems		
	Number	Hours/h	Average hours/h
Wear	25	4122.03	164.88
Crack	20	4054.16	202.71
Open welding	4	943.92	235.98
Coke	4	676.76	169.19
Blockage	8	590.92	73.87
Fracture	3	476.54	158.85
Burnout	2	319.77	159.89
Ash leakage	3	292.54	97.51
Other	15	1052.29	70.15

Table 39.8 Classification of the technical faults by cause

Causes	Unplanned outages in boiler systems		
	Number	Hours/h	Average hours/h
Product quality	23	3904.48	169.76
Maintenance	20	3077.46	153.87
Operation management	15	2598.02	173.20
Aging	12	1871.81	155.98
Fuel	9	688.95	76.55
Construction and installation	3	231.42	77.14
Planning and design	2	156.79	78.40

may reduce the operation reliability. In addition, the maintenance, operation management, and equipment aging are directly linked to the power plants. The promotion of the operation reliability needs the efforts of not only one of the ahead mentioned departments or organizations but all of them.

39.4 Conclusion

Statistics of the reliability of more than 130 CFB boiler units in China in 2017 were collected and analyzed. It can be concluded that CFB boiler units in China had a high overall operation reliability level. Boiler systems are regarded as the major system affecting the reliability which contributed to most of the unplanned outages. Further analysis into unplanned outages showed that the high-temperature heating surfaces and combustion systems are two major boiler components causing unplanned outages. The wear and crack are two main kinds of technical faults. The product quality, maintenance and operation management of the power plants are very important for the reliability of the CFB boiler units. What's more, the promotion of the overall reliability of CFB boilers units needs improvement in many aspects, which needs the efforts of various departments or organizations including the manufacturers, construction companies, design institutes, power enterprises and power plants.

Acknowledgements This work was supported by Key Project of the National Thirteen-Five Year Research Program of China (Project Number: 2016YFB0600201).

References

Arjunwadkar A, Basu P, Acharya B (2016) A review of some operation and maintenance issues of CFBC boilers. Appl Therm Eng 102:672–694

Cai R, Zhang M, Mo X et al (2018) Operation characteristics of external heat exchangers in the 600MW supercritical CFB boiler. Fuel Process Technol 172:65–71

DL/T 793.2-2017. Reliability evaluation code for generating equipment. Part 2: coal-fired units

Gao M, Hong F, Liu J (2016) Investigation on energy storage and quick load change control of subcritical circulating fluidized bed boiler units. Appl Energy 185:463–471

Song C, Li M, Zhang F et al (2015) A data envelopment analysis for energy efficiency of coal-fired power units in China. Energy Convers Manage 102:121–130

Yao Y, Cai R, Zhang Y et al (2019) A method to measure the tube-wall temperature in CFB boilers. Appl Therm Eng 153:493–500

Yue G, Cai R, Lu J et al (2017) From a CFB reactor to a CFB boiler—the review of R&D progress of CFB coal combustion technology in China. Powder Technol 316:18–28

Chapter 40
Development of Advanced Ultra Super Critical Circulating Fluidized Bed Technology by DongFang Boiler

Nie Li, Gong Liming, Xue Dayong, Lu Jiayi, Wei Lixiao, and Huang Min

Abstract In this paper, the research and development of circulating fluidized bed (CFB) technology by Dongfang Boiler Group Co., Ltd (Dongfang Boiler) in recent years and the understanding of its future development trend are summarized. Based on several decades' accumulation, Dongfang Boiler has developed two types of supercritical CFB boilers (350 MW and 600 MW, 25.4 MPa/571 °C/569 °C), and both types have been in operation. Additionally, supported by major national research and development program (2016YFB0600204), the scheme of 660 MW advanced ultra supercritical (29.4 MPa/605 °C/623 °C) CFB boiler has also been completed and is shown in the present work. In order to use the same material as that of pulverized coal fired boiler with similar parameters, one of the most important researches is to reduce the heat deviation in external heat exchanger (EHE). According to the field test on a 600 MW supercritical CFB boiler and simulation of the flow characteristics of EHE, the deviation characteristic is reappeared, and solution to reduce the heat deviation have been proposed. Additionally, with the stringent requirement of pollutant emission from coal-fired units in China, a series of flue gas cleaning technologies can be used. The recommended flue gas cleaning system for a CFB boiler is discussed.

Keywords Circulating fluidized bed · Advanced ultra super-critical · Heat deviation

N. Li (✉)
Institute for Thermal Power Engineering, State Key Laboratory of Clean Energy Utilization, Zhejiang University, Hangzhou 310027, China
e-mail: niel@dbc.com.cn

N. Li · G. Liming · X. Dayong · L. Jiayi (✉) · W. Lixiao · H. Min
Dongfang Boiler Group Co. Ltd., Chengdu 611731, China
e-mail: dbcjszx-cfb@dbc.com.cn

J. Lyu and S. Li (eds.), *Clean Coal and Sustainable Energy*, Environmental Science and Engineering,
https://doi.org/10.1007/978-981-16-1657-0_40

40.1 Introduction

As a highly efficient clean coal combustion technology, the circulating fluidized bed (CFB) boiler, which plays a significant role in environmental protection and energy saving, has been remarkably developed and widely popularized (Yue et al. 2009; Feng et al. 2003; Cen et al. 1997; Nie et al. 2008).

Dongfang Boiler Group Co., Ltd. (Dongfang Boiler), which is a state-run enterprise and one of the largest bases for manufacturing large electro-mechanical products in China, is dedicated to providing advanced equipments and excellent services for the world's energy and environmental protection industries.

Dongfang Boiler is also one of the largest companies who research and develop CFB technologies around the world.

In the late 1990s, 135–150 MW grade CFB boilers with reheat system were developed by us and the products have been in operation in succession from 2003. Also in 2003, the development of 300 MW grade subcritical CFB boiler started, and the first two units had been in operation in 2008 in HeShuYuan power plant in China (Nie et al. 2007), and at that time, they were also the largest CFB boilers all over the world.

For further improve the unit efficiency, the research and development of supercritical CFB boiler has been started in 2005. In 2013, the 600 MW supercritical CFB boiler was put into commercial operation in Sichuan Baima CFB Demonstration Power Plant Co. Ltd. (Wang et al. 2013). And at present, it is the only CFB boiler with capacity up to 600 MW.

Based on the consumers' requirement, DongFang boiler developed another type of supercritical CFB boiler, which capacity is more suitable for actual condition and the world's first batch of 350 MW supercritical CFB boilers has been in operation successively from 2015 (Su et al. 2010).

So far, there are dozens of supercritical CFB boilers already in operation in China, and therefore, the time is ripe to research and develop more economical, more efficient ultra-supercritical circulating fluidized bed boiler (http://www.sasac.gov.cn/n86114/n326638/c2440160/content.html).

On the other hand, with the continuous improvement of the emission reduction targets for coal-fired generating units, new coal-fired units' emission shall meet the emission limits of gas turbines standard in China, that is under 6% O_2 condition, the concentration of SO_2, NO_x and particle matter shall be lower than 35 mg/Nm^3, 50 mg/Nm^3 and 10 mg/Nm^3, respectively. Therefore, how to reduce the emission of pollutants in coal combustion process is also an urgent problem to be solved.

In this back ground and also supported by major national research and development program (2016YFB0600204), the scheme of 660 MW advanced ultra supercritical (29.4 MPa/605 °C/623 °C) CFB boiler has been completed and is shown in the present work along with the low emission technologies.

40.2 660 MW Advanced Ultra Supercritical CFB Boiler Scheme and Key Technical Problem Analysis

40.2.1 Boiler Scheme

Previous study show that the unit efficiency can be additional improved by 0.2–0.25% for every 1 MPa increase in initial steam pressure and 0.15–0.20% for every 10 °C increase in reheated steam temperature (Maoliang and Guangzhou 2014). For this reason, the advanced ultra-supercritical parameter is chosen for boiler development. The main steam pressure is 29.4 MPa and the main steam and reheat steam temperatures are 605 and 623 °C. The steam parameters of the boiler are shown in Table 40.1. The coal specification of the boiler is shown in Table 40.2.

From Table 40.2, the coal is a kind of anthracite, with total sulfur about 3%, calorific value nearly 4300 kcal/kg, and dry ash-free base volatile matter less than 10%, which is difficult to ignite and burn out, and also will generate much SO_2 emission.

Table 40.1 Steam parameters of 660 MW advanced ultra supercritical CFB boiler

Item	Unit	Parameter
Final stage super heater		
Steam flow	t/h	2002
Steam pressure	MPa (g)	29.4
Steam temperature	°C	605
Final stage reheater		
Steam flow	t/h	1685
Steam pressure	MPa (g)	6.17
Steam temperature	°C	623
Feed water temperature	°C	304

Table 40.2 Coal specifications of 660 MW advanced ultra supercritical CFB boiler

Symbol	Unit	Parameter
Mt	%	7.5
Aar	%	37.77
Vdaf	%	10.32
Car	%	47.66
Har	%	2.14
Nar	%	1.12
Oar	%	0.83
St,ar	%	2.98
Qnet,ar	kJ/kg	18,370

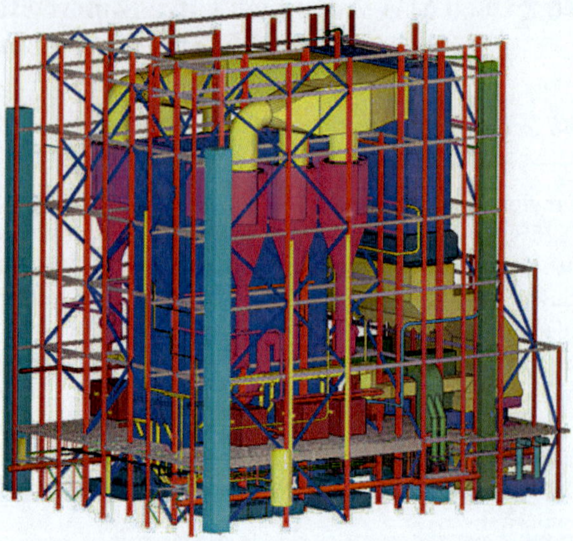

Fig. 40.1 Schematic diagram of the 660 MW advanced ultra-super critical CFB boiler

For large-scale circulating fluidized bed boilers with up to 660 MW capacity, the overall layout needs to consider comprehensively the factors such as heating surface arrangement, anti-erosion, secondary air penetration, and scheme maturity and so on. Based on the comparison of different types of furnaces, the boiler scheme adopts an H-type furnace and external circulation loop arrangement of single furnace and double air distributor structure. The overall view of the boiler is shown in Fig. 40.1.

In the furnace, there are platen super heaters. Coal feeders are placed at the ash duct between loop seal and furnace. The primary air boxes are arranged under air distributor which is made of membrane water wall of the furnace. During the start up process, under bed igniter is mainly adopted. Primary air and the hot gas generated in the ignition ducts are introduced into both of the air boxes. Secondary air is introduced into the furnace in 2 layers mainly at inner side and outer side wall. Ten bottom ash discharge ports are arranged in the inner side wall above air distributor, corresponding to ten roller ash coolers. Six high efficient steam cooled cyclone separators are located at both sides of the furnace. The auto-balance loop seals, each having one input and two outputs to improve the uniformity of recycling material, locate under the cyclones, and each equipped with an external heat exchanger (EHE).

The reheat steam temperature is controlled by adjusting the ash flow rate into the EHE or directly returns to furnace. In the rear pass locate low-temperature reheater and low temperature super heater. The economizer is arranged in the lower part of rear pass and just above two four-section regenerative air preheaters.

Three stages of spraying water desuperheaters are arranged in the superheat system. There are also spraying water desuperheaters in reheat system for emergency condition.

40.2.2 Key Technical Problem Analysis During Boiler Development

In order to use the same material as that of pulverized coal fired boiler with similar parameter, for this boiler, by using SA-335P92 as final steam pipe and final heating surfaces headers, the main steam and reheat steam temperatures can be maintained at 605 and 623 °C. The 1000 MW pulverized coal boiler with same steam pressure and temperature has been put into operation (Maoliang and Guangzhou 2014). However, the allowable temperature of SA-335P92 material is 650 °C (ASME BPVC II D-2013), and considering the margin of 10 °C, the maximum temperature of steam should not exceed 640 °C, which means the deviation temperature of last reheat steam must be controlled within 17 °C.

As we know, CFB boilers have lower combustion temperature. In order to reach the steam temperature above 600 °C and meanwhile have good steam temperature characteristics at low load, the final super heater and reheater should be arranged in the furnace or EHE, while the high temperature reheater of pulverized coal boilers is arranged in the horizontal flue gas pass. Normally, the heat flux of CFB boiler's furnace and EHE is higher than that of pulverized coal boiler's horizontal flue gas pass, which makes the thermal deviation of heating surface in CFB boiler's furnace and EHE is higher, and the deviation is usually about dozens of degrees. Therefore, it is difficult to control the steam temperature deviation within 17 °C for a CFB boiler.

In this boiler scheme, the final stage reheater is located in EHE, so one of the most important researches is to reduce the heat deviation in it; hence the following studies are carried out.

40.3 Deviation Characteristics and Its Solution

40.3.1 Gas Solid Flow and Deviation Characteristics in EHE

Figure 40.2 shows the structure of an EHE, it can be seen that the ash enters into the EHE's empty section from one side, and then it would be entrained by the air to the upper part of the empty section and turn over the partition wall into the second section where the steam can be heated.

After that, the ash temperature would be cooled to about 700 °C and returned to furnace. The steam inlet and outlet are both from one side.

In order to take an insight view of the gas–solid two phase flow in the EHE, Barracuda software is used to simulate it.

Barracuda is numerical calculation software based on Computational Particle Fluid Dynamics (CPFD) concept. It still uses Eulerian method to deal with fluid phase. The momentum equation is expressed by Navier–Stokes equation. For particle phase, Barracuda uses Lagrangian method to describe and couple with fluid phase equation. It can consider the interaction between particles; and several researchers

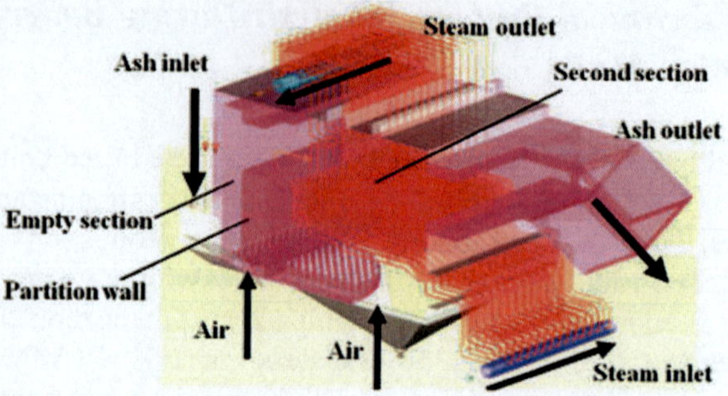

Fig. 40.2 A schematic of an EHE

have used this software to simulate the gas–solid flow characteristics in boilers and chemical reactors, and obtained reasonable prediction results (Zhao et al. 2006; Snider et al. 2010; Parker 2011). For the drag force model, EMMS model is selected (Li and Kwauk 1994). Additionally, sampled ash from the real CFB boiler's EHE is used for the particle size distribution in the simulation (Jiayi 2012). And in this simulation, the ash particle size is divided into 40 grades (Fig. 40.3).

In order to shorten the simulation period and meanwhile show the flow characteristics, only 1/16 of the real EHE geometry size is chosen after several preliminary simulations and the grids number is about 600,000. The heating surface tubes are also considered, as shown in Fig. 40.4.

Figure 40.5 shows the flow characteristics in EHE. It can be seen that fresh ash particles mainly enter two flow channels after turning over the partition wall. One part of particles is entrained directly to the outlet by the momentum of existing particle surge and the upper gas on the bed surface. This part of particles is almost not involved in heat transfer. While the other part of particles enters to the bottom of the section and entrained by the gas and gradually accelerated and flow upward and forward to the outlet at the end. The later part of particles passes through the whole heating surface tube bundle, transferring the heat from the particles to steam.

Fig. 40.3 Particle size distributions for the simulation (Jiayi 2012)

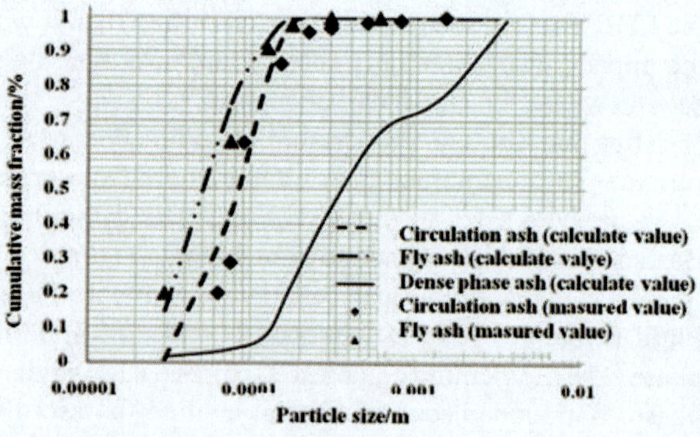

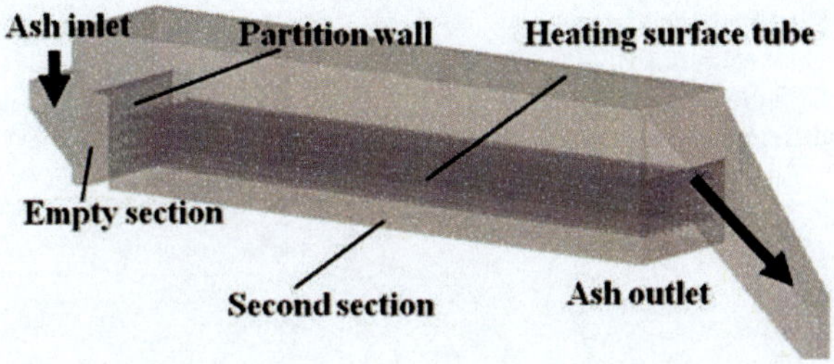

Fig. 40.4 Simulation model of the EHE

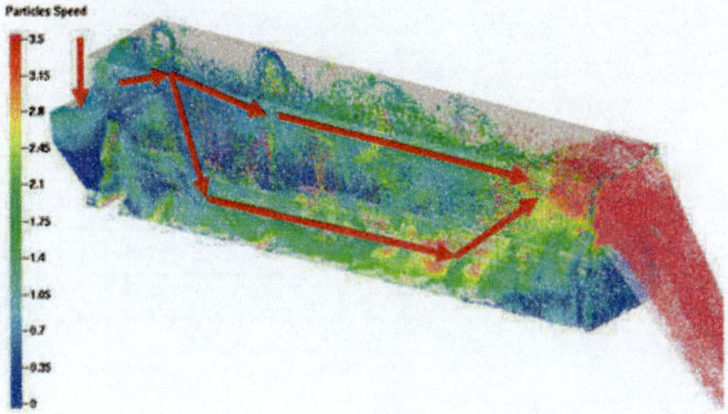

Fig. 40.5 Flow characteristics in EHE

Since the height and width in the simulation are both only 1/4 of the real EHE, the deviation characteristics is not obvious. Therefore, field test is carried on to make it reappear.

In the 600 MW CFB boiler's final reheater EHE, there are totally 40 tube panels, and in each panel, there are 7 tubes and hang tube in parallel connection, and wall temperature test points are arranged.

Fields test results are shown in Fig. 40.6, it can be seen that the heating surface of the final reheater has thermal deviation. And the outlet wall temperatures of the reheaters along the width of the EHE show a trend of higher temperature at the middle and lower temperature at side. Previous studies showed that there are some deviations in an EHE (Zheng et al. 2016; Wang et al. 2001; Yang lei Experimental study on external heat exchanger of circulating fluidized bed boiler 2007; Wenqing 2017), but did not give enough suggestions.

From Fig. 40.6, the outlet wall temperatures of 1–3 tube panels at both sides are lower, while the temperature difference among the other tube panels are small, the maximum deviation is less than 30 °C. The deviation mainly occurs in the 3 panels on both sides. The wall temperature is lower than the average about 30–50 °C, and the distribution is symmetrical.

As mentioned above, during operation, the ash enters the EHE from one side, and the steam comes in and out of the EHE also from one side. These are all not axisymmetric of the EHE. However, from the more symmetrical distribution of the thermal deviation in Fig. 40.6, it is analyzed that the thermal deviation may not cause by the uneven structure, but probably by the gas–solid two phase flow.

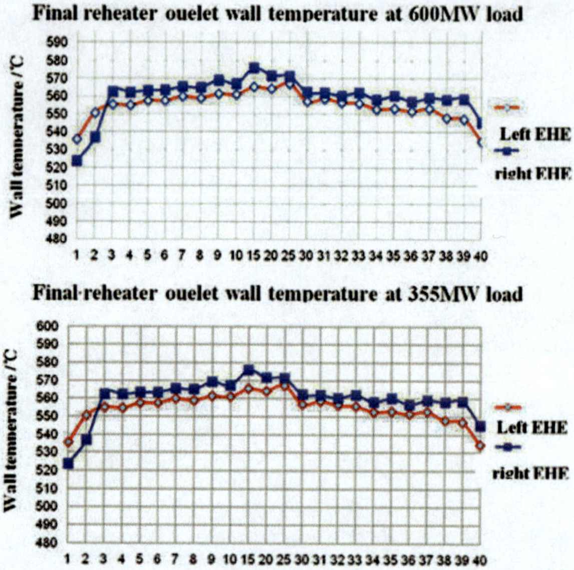

Fig. 40.6 Deviation characteristics of 600 MW CFB boiler EHE under different loads

Additionally, it should be noted that although these deviations exist in subcritical or supercritical CFB boilers, they are all within the allowable range of materials, not affecting safe operation. However, the above deviations will exceed the allowable temperature of SA-335P92 material under the condition of advanced ultra supercritical parameters, so the deviation needs to be reduced or eliminated.

40.3.2 *Solution on Thermal Deviation*

For boiler manufacture, there are several methods to reduce the deviation; the following solution based on adjusting the steam flow.

From the above analysis, the temperature deviation is mainly due to the gas–solid two phase flow, and from the view of heat transfer, the deviation can be reflected on the heat flux. Since the steam temperature and wall temperature distribution at the heat exchanger outlet is measured, then the heat flux in the EHE can be calculated.

That is:

$$\Delta H \times q = A \times h \times \Delta T \tag{40.1}$$

where:

H enthalpy of steam;
q steam flow in the EHE's each panel;
A heating surface area of EHE's each panel;

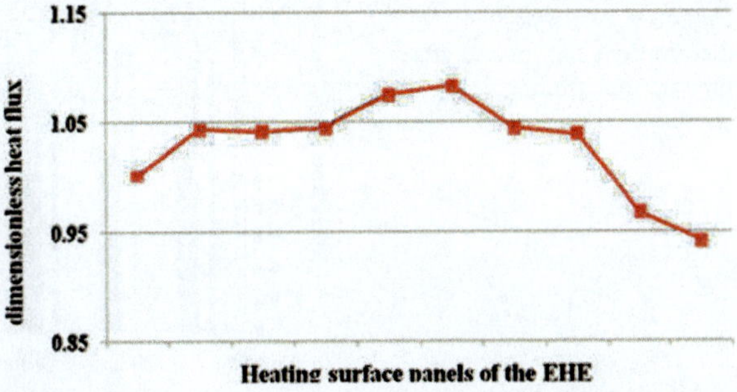

Fig. 40.7 Dimensionless heat flux distribution in the EHE

h heat flux;

T average temperature of steam/ash.

At the same time, the average heat flux can also be calculated, and then the deviation coefficient can be obtained.

Figure 40.7 shows the 10 panels' dimensionless heat flux distribution in the EHE. In Fig. 40.7, the heat flux of each panel is compared with the first panel's heat flux. It can be seen that the heat flux distribution has a certain degree of non-uniformity, lower on both sides and higher in the middle, which is consistent with the measured temperature distribution.

According to the above analysis, Fig. 40.8 shows the comparison between the calculated and measured wall temperatures. It can be seen that the calculated values match good with the measured values. It shows that the calculation model can correctly reflect the actual deviation characteristics of the EHE.

For the advanced ultra super critical CFB boiler, since its EHE's structure, size, heating surface arrangement and operation condition are all the same or very close, so the deviation coefficient can be used as an input for boiler design. Then, the heat deviation can be reduced by adjusting the steam flow in each panel.

By changing the resistance characteristics of each panel, the mass flow rate of each panel can be adjusted.

The flow distribution of one panel's tubes and hanging tube before and after optimization is shown in Fig. 40.9.

It can be seen that before optimization, the mass flow which near the side wall is larger. The calculation results show that under the same resistance conditions, in

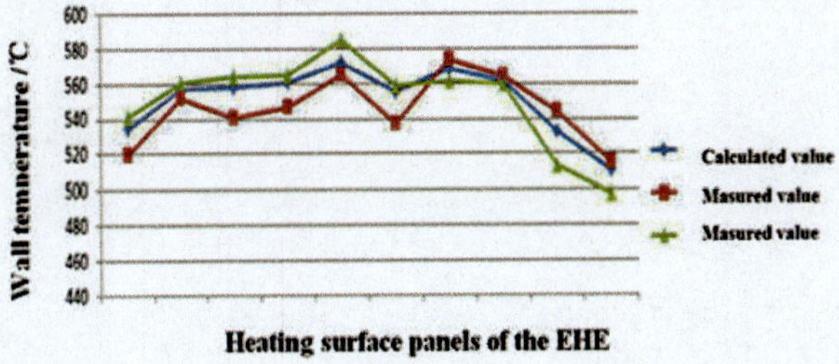

Fig. 40.8 Comparison between calculated values and the measured value of tube

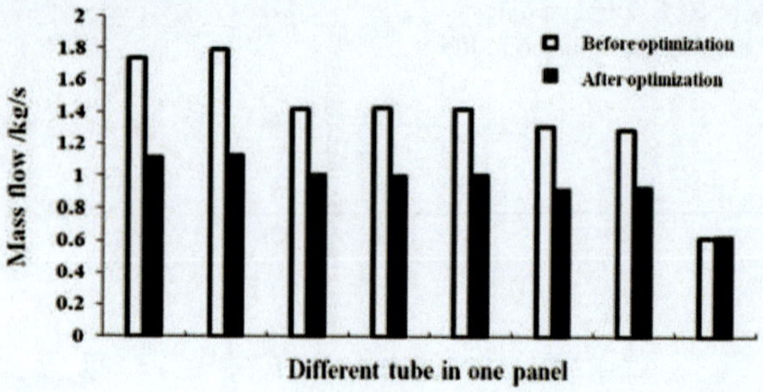

Fig. 40.9 Mass flow distributions before and after pipe optimization

order to reduce the temperature deviation, the flow distribution of each pipe shall be more uniform by means of throttling measures.

Thermodynamic calculation parameters of high temperature reheater under boiler maximum continuous rating (BMCR) condition after adjusting flow distribution are shown in Fig. 40.10.

From the calculation results, it can be seen that the wall temperature deviation of the EHE has been significantly improved. The maximum steam temperature is less than the allowable temperature of SA-335P92 pipe, and there is still a certain margin. That means the scheme is safe and reliable.

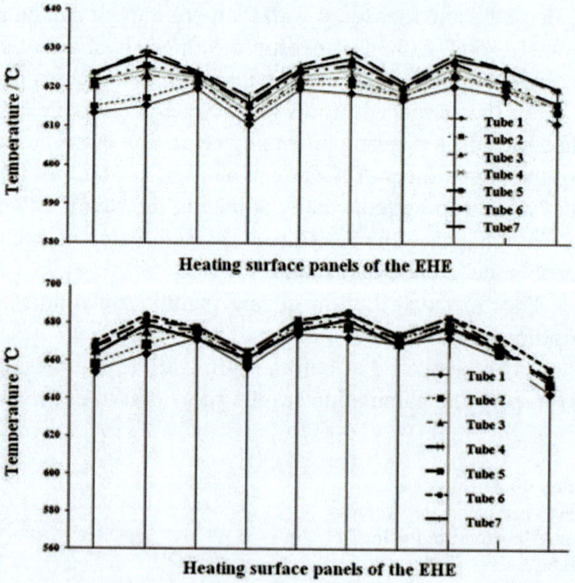

Fig. 40.10 Results of optimal fitting

40.4 Environmental Protection Schemes for CFB Boilers

At present, the thermal power plant emission limit of NOx, SO_2 and PM are respectively 50, 35 and 10 mg/Nm^3 in China (Development and Reform Energy 2014), which is the most stringent international environmental standard (see Fig. 40.11).

For a CFB boiler, there are mainly two schemes to achieve extremely low emissions. One is CFB boiler with selective non-catalytic reduction (SNCR) De-NOx system, circulating fluidized bed semi-dry flue gas desulfurization (CFB-FGD) and electrostatic precipitator (ESP), and the other is CFB boiler with SNCR and wet FGD, both as shown in Fig. 40.12.

In order to match the new emission requirements, CFB boilers also need to add FGD devices. Therefore, it is necessary to coordinate the share between desulfurization in furnace and desulfurization in FGD, and take into account the interaction between the generation and removal of a series of pollutants.

In the aspect of desulfurization, we can consider the technical route of combining desulfurization in furnace with wet or semi-dry process at rear pass; in the aspect of denitrification, SNCR can realize extremely low emission of NOx and reserve space for SCR catalyst; in the aspect of dust removal, we can determine it according to the choice of technical route of desulfurization. With the progress of combustion technology and according to the recent operated CFB boilers actual emissions, it is recommended to choose the former scheme for normal CFB boilers, and choose the later one for high sulfur coal fired CFB boilers.

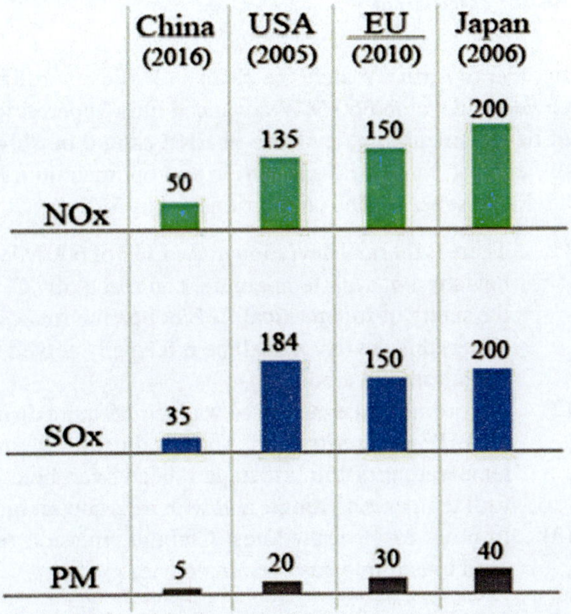

Fig. 40.11 Emission requirements in different country and region

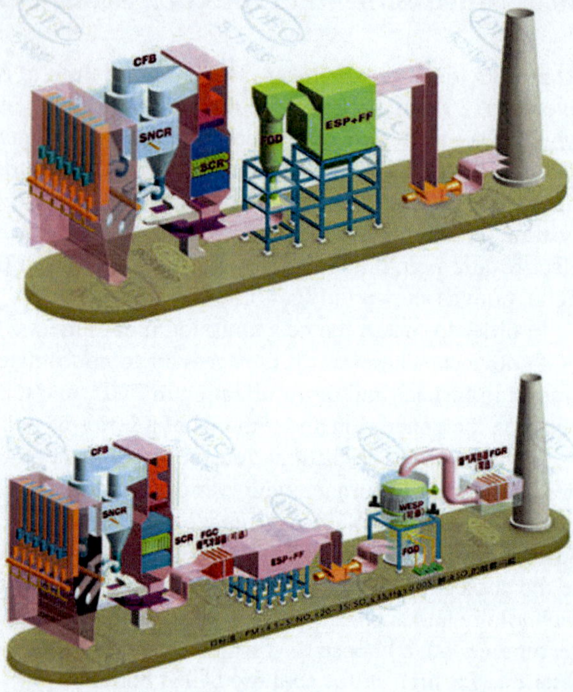

Fig. 40.12 Two schemes of extremely low emission for a CFB boiler

40.5 Summary

In order to further realize the clean and efficient utilization of low grade coal, it is a choice to develop 660 MW advanced ultra-supercritical CFB boiler. In the process of development, the deviation of EHE should be solved. Through the combination of field test, numerical simulation and optimization, Dongfang Boiler puts forward a reliable scheme. The conclusions are as follows:

(1) There is thermal deviation in the EHE of 600 MW supercritical CFB boiler. The deviation of wall temperature can reach 50 °C. The deviation does not affect the safety of supercritical CFB boiler, but for advanced ultra supercritical CFB boiler, this deviation shall be effectively solved to ensure the safe and reliable operation of the boiler.

(2) According to the measured wall temperature distribution, the actual heat flux in the EHE is reconstructed. And by adjusting the steam flow distribution, the wall temperature of the last stage reheater can be controlled within the allowable wall temperature range, and with a certain margin to achieve safe and reliable.

(3) In order to meet the latest Chinese emission requirement, CFB boilers also need to add flue gas cleaning devices.

Acknowledgements This work was supported by major national research and development program (2016YFB0600204) and Sichuan Science and Technology Program (2018GZ0157).

References

ASME BPVC II D-2013. ASME boiler and pressure vessel code materials part D properties (Metric)

Cen K, Ni M, Luo Z (1997) Theory, design and operation of circulating fluidized bed boiler. China Electric Power Press, Beijing

Development and Reform Energy (2014) 2093. Action plan for energy conservation and emission reduction and transformation of coal and electricity (2014–2020)

Feng J, Yue G, Lv J (2003) Circulating fluidized bed combustion boiler. China Electric Power Press, Beijing

http://www.sasac.gov.cn/n86114/n326638/c2440160/content.html

Li JH, Kwauk MS (1994) Particle fluid two phase flow, the energy minimization multiscale method. Metallurgical Industry Press, Beijing

Lu J (2012) Study on particle population balance and heat balance of large-scale circulating fluidized bed boiler. Ph.D. dissertation of Chongqing University, Chongqing

Maoliang Li, Guangzhou Yi (2014) Design scheme of Dongfang 1000MW advanced ultra supercritical boiler. Dongfang Electric Rev 28(2):24–28

Nie L, Wang P, Huo S, Yao B (2007) Development and design of oriental 300 MW circulating fluidized bed boiler. Dongfang Electric Rev 21(2):33–42

Nie L, Wang P, Peng L (2008) Design of 600MW supercritical circulating fluidized bed boiler. Power Eng 28(5):701–706

Parker J (2011) Validation of CFD model for polysilicon deposition and production of silicon fines in a silane deposition FBR. Int J Chem Reactor Eng 9(1):A40

Snider D, Guenther C, Dalton J et al (2010) CPFD Eulerian-Lagrangian numerical scheme applied to the NETL bench-top chemical looping experiment. In: 1st international conference on chemical looping, Lyon. 2010 circulating fluidized bed boiler technology conference proceedings, Changsha, China (in Chinese)

Su H, Nie L, Yang X et al (2010) Development and design of oriental 350 MW supercritical circulating fluidized bed boiler. Dongfang Boiler 2010(1):1–6

Wang H, Lu X, Wang Q (2013) Analysis of operation characteristics of 600 MW supercritical CFB boiler. Chin Soc Eng Thermophys

Wang Q, Shi Z, Cheng L (2001) Experimental study on external mechanical heat exchanger controlled by non mechanical valves. Power Eng 21(4):1324–1330

Yang lei experimental study on external heat exchanger of circulating fluidized bed boiler. Master's thesis, Chongqing University, Chongqing, Power Engineering College (2007)

Yue G, Yang H, Lu J et al (2009) Latest development of CFB boilers in China. In: 20th international conference on fluidized bed combustion, Xi'an, China

Zhang W, Sun S, Lu X (2017) Comparative study on operating characteristics of external heat exchangers for large CFB boilers. J Eng Thermal Energy Power 32(6):97–101

Zhao P, Snider D, Williams K (2006) Computational particle-fluid dynamics simulations of a commercial-scale turbulent fluidized bed reactor. In: AIChE annual meeting, San Francisco

Zheng X, Xu P, Hu X (2016) Experimental study on operation adjustment of 600MW high critical CFB boiler high external external heat exchanger. Dongfang Electric Rev 30:23–26

Chapter 41
Cold-Experimental Study About Pressure Resistance of CFB Wind Caps

Boheng Tong, Hongyu Zeng, Qingfeng Zhang, and Junfu Lyu

Abstract Air distributor is one of the most important devices to maintain stable and high-efficient operation of circulating fluidized bed boilers. And the pressure drop of the air distributor determine the effect of air distribution and efficiency. The wind caps are the essential parts of air distributors, whose structures determine the pressure drop. Therefore, predicting the resistance of wind caps is one of the most important issues in CFB boiler design. At present, there is little study and theoretical inference about the pressure drop prediction when the diameters of internal structure change. In this study, cold-modelling experiments were made to study the flow characteristics and the resistance regulation as geometry parameters changes qualitatively.

Keywords Cold-experiment · Wind cap · Pressure resistance

41.1 Introduction

The development trends of circulated fluidized bed boiler (CFB boiler) is larger generation load, thus the area of furnance and air distribution plate is larger. Realizing effective and uniform air distribution is becoming more and more difficult.

Air distributor is one of the most important devices to maintain stable and high-efficient operation of circulating fluidized bed boilers (Yue et al. 2005). And the pressure drops of the air distributor determine the effect of air distribution and efficiency (Yang et al. 2009).

The wind caps are the essential parts of air distributors, whose structures determine the pressure drop. At present, the geometry parameters in wind caps design result from empirical generalizations, which may cause deviations. Larger resistance coefficient than designed value result in more energy consumption, while smaller ones

B. Tong · Q. Zhang
North China Electric Power Research Institute, Beijing 100037, China

B. Tong · H. Zeng · J. Lyu (✉)
Institute of Thermal Engineering, Tsinghua University, Beijing 100084, China
e-mail: lvjf@mail.tsinghua.edu.cn

J. Lyu and S. Li (eds.), *Clean Coal and Sustainable Energy*,
Environmental Science and Engineering,
https://doi.org/10.1007/978-981-16-1657-0_41

result in air distribution nonuniformity, material leakage and even furnance shut down. Therefore, predicting the resistance of wind caps is one of the most important issues in CFB boiler design.

The major research direction of wind caps includes three parts: measuring resistance coefficient, inventing novel design, CFD calculating. Yang et al. (2010) studied about the critical characteristics of slag leakage. Lyu (2015) performed an experiment on bell type wind caps in a 300 MW CFB Boiler. Feng et al. (2007) did a cold-state experiment comparing with three types of wind caps. Niu et al. (2009), analysed the abrasion and materials about wind caps. In a word, at present, there is little study and theoretical inference about the resistance prediction and internal pressure whilst the diameters of structure change. Therefore, it's difficult to make accurate design according to preset pressure.

In this study, cold-modelling experiments were made to study the flow characteristics, internal pressure distribution and the resistance regulation as geometry parameters changes qualitatively. The experimental results showed the regulation of resistance coefficient via inner structure changes. By qualitatively analyzing experiment results, this paper indicates the changing trend of pressure resistance by the variation of geometric construction. Moreover, pressure inside the wind cap is measured and pressure distribution was studied. Also, the internal flow characteristics and resistance distribution is studied. In conclusion, pressure resistance characteristics were studied, and experiment data procide a reference for structure design and further pressure resistance prediction.

41.2 Experimental Methods

Research object is bell-cap style wind caps which are most commonly used in CFB boilers. Model experiments were made in cold state. Basic information is shown in Table 41.1.

Previous studies shown that pressure drop of wind caps are quadratic proportional to inlet velocity. Wind cap resistance coefficiency is usually designed as formula (41.1):

Table 41.1 Basic information

Object	Units	Value
Primary air flow	m3/h	315,000.0
Primary air temperature	K	300
Air distribution area	m^2	130
Wind caps amount	One	480
Wind chamber pressure	kPa	11.5
Bed pressure drop	kPa	8

$$\Delta P = \frac{1}{2}\xi \rho v^2 \quad (41.1)$$

where ΔP is the pressure drop from inlet to outlet and ξ is the resistance coefficiency.

By cold-experimental study, pressure drop and velocity can be measured, therefore resistance coefficiency can be calculated. It's worth noting that the overall and partial coefficiency are both suitable for this formula.

In this study, the wind caps are divided into five parts: inner tube, outlet of the inner tube, cap, slit and the outlet holes. As shown in Fig. 41.1, there are five measure points inside the wind caps, which measure inlet and outlet pressure of each five parts. D_1 defines the diameter of inner tube. D_2 refers to the parameter of holes on the pore plate. D3 refers to width of slit. D_4 refers to outlet diameters.

Pitot tubes were used to measure inside the wind cap. The outlet gauge pressure of the wind cap is considered zero. In order to study different geometry parameter wind caps, the experimental model can be divided into several detachable parts. By assembling different components, changes of resistance coefficiency via sizes can be studied.

As formula (41.2) shown, each pressure drop ΔP_1, ΔP_2, ΔP_3, ΔP_4, ΔP_5 is measured. Also the relations between pressure drop coefficient and velocity, geometry parameters are qualitatively analyzed.

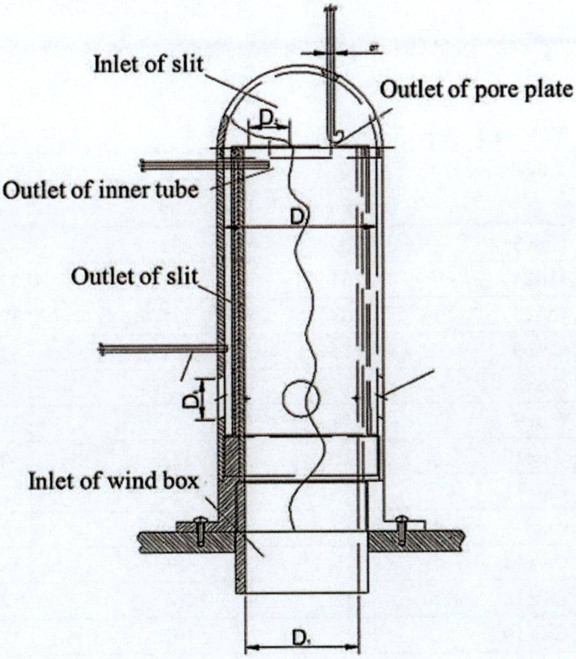

Fig. 41.1 Measure points of wind caps

$$\Delta P = \Delta P_1 + \Delta P_2 + \Delta P_3 + \Delta P_4 + \Delta P_5 = \sum_{i=1}^{5} \frac{1}{2} \xi_i \rho v_i^2 \quad (41.2)$$

$$\xi_i = f(D_i, L_i, ...)$$

The system diagram is shown in Fig. 41.2. It consists of a 1000 m^3/h Roots blower, a model wind cap and a wind box sealed up with a flange. There are pitot tubes joint with U-type tubes measuring pressure, a rotor flow meter and a thermometer.

Experimental cases are shown in Table 41.2. One size change while the others

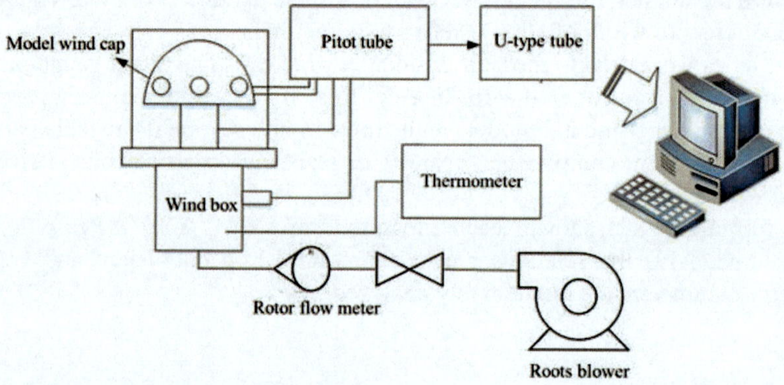

Fig. 41.2 Cold-state experiment system diagram

Table 41.2 Experiment cases

Subject	D_1	D_2	D_3	D_4
	mm	mm	mm	mm
Case3	100	131	32	34
Case4	104	131	32	34
Case5	108	131	32	34
Case6	115	131	32	34
Case8	115	131	36	34
Case9	115	131	38	34
Case10	115	131	40	34
Case12	115	132	32	34
Case13	115	134	32	34
Case14	115	136	32	34
Case18	115	131	32	35
Case19	115	131	32	36
Case20	115	131	32	37
Case21	115	131	32	38

maintain in every group of cases. 300, 350, 400, 450, 500 Nm^3/h flow rate are measured respectively in every case.

41.3 Results and Analysis

In this study, a length-minification model experiment was made. According to self-model theory, when Re > 10^4, inertia force plays a leading role, and at that time Eu has nothing to do with Re.

$$Eu = \frac{\Delta P}{\rho v^2} = 2\xi \tag{41.3}$$

As shown in formula 41.3, Eu becomes a constant, therefore coefficiency ξ is as same as in practical condition. In this experiment, the smallest Re number is calculated as shown in formula 4.

$$\text{Re} = \frac{\rho v D}{\mu} = 70163.62 > 10^4 \tag{41.4}$$

where ρ is air density, kg/m^3, and μ is air coefficient of kinetic viscosity, Pa s. Therefore, self-modeling conditions are satisfied in this experiment.

Inner pressure distribution is shown in Fig. 41.3. This figure shows the pressure drop trend inside the wind cap. The total pressure resistance of the wind cap is about 3000 Pa. The pore plate, cap and slit make up a large proportion of total resistance. And as flow rate grows, these three resistances grow more than other parts, which

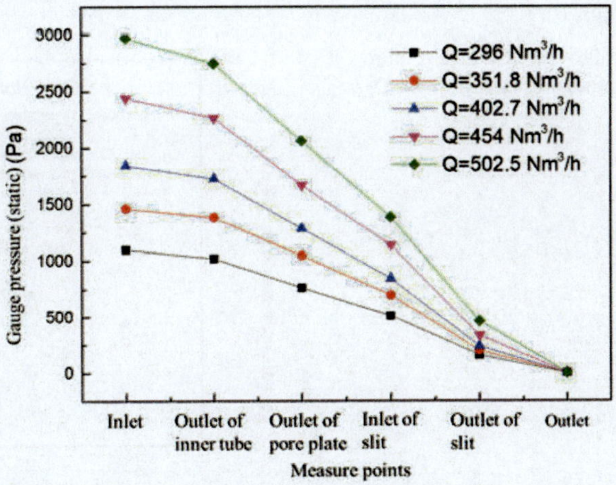

Fig. 41.3 Change of gauge pressure via measure points

means that they have bigger pressure coefficiency. This is concerned with their narrow size. In the discussion below, this paper will focus on these three parts.

Setting up with different sizes of inner tube, the diagram of inner tube, D_1 is 100, 104, 108, 115 mm respectively. The pressure drop of inner tube, ΔP_1 is less than 200 Pa when flow rate $Q = 500$ m^3/h. Also, as D_1 changes, ΔP_1 changes a little. This is because inner tube has a simple structure and relatively large size. Therefore, resistance coefficiency hardly changes.

As for pore plate pressure drop ΔP_2, the determine size includes inner tube diameter D_1 and holes diameter D_2, as shown in Fig. 41.4. The independent parameter is the velocity of holes, v_2.

The experiment results are shown in Fig. 41.5. At the same velocity, the less diameters of holes, the more pressure drop. This is because as the holes become smaller, the throttling phenomenon become more obvious, which makes the energy consumption higher and resistance coefficiency larger.

Maintaining holes diameters D_2, the change of pressure drop via D_1 changing is shown in Fig. 41.6. At the same velocity, the more D_1, the more pressure drop is, which means the more coefficiency is. Compared to Fig. 41.5, the influence to the pressure drop is smaller.

As for cap pressure drop ΔP_3, this paper defined outlet velocity v_3 as independent parameter. The structure includes two important geometry parameters, D_2 and width of slit d, as shown in formula 41.5.

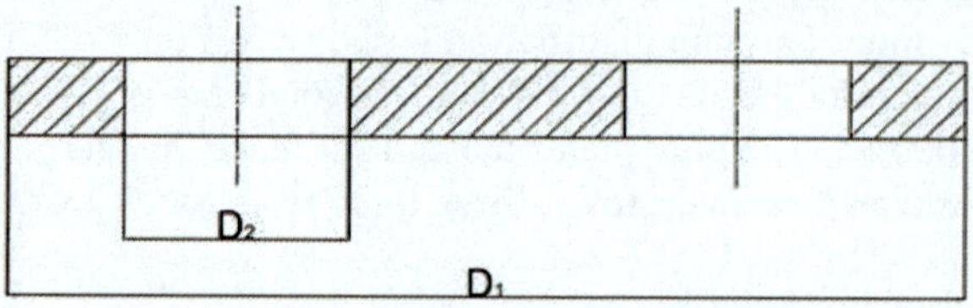

Fig. 41.4 Geometry structure of pore plate

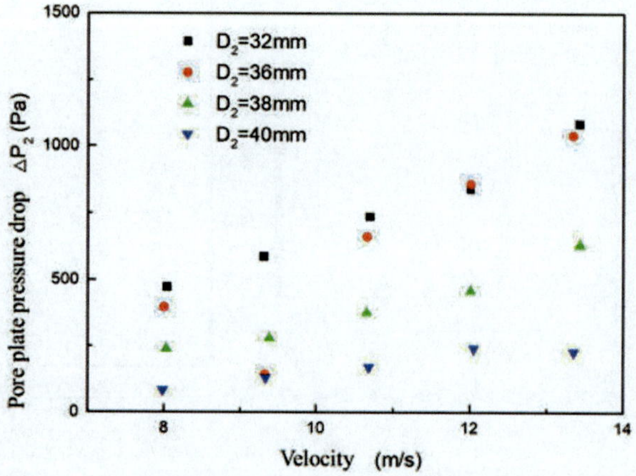

Fig. 41.5 Change of pore plate pressure drop via D_2

Fig. 41.6 Change of pore plate pressure drop via D_1

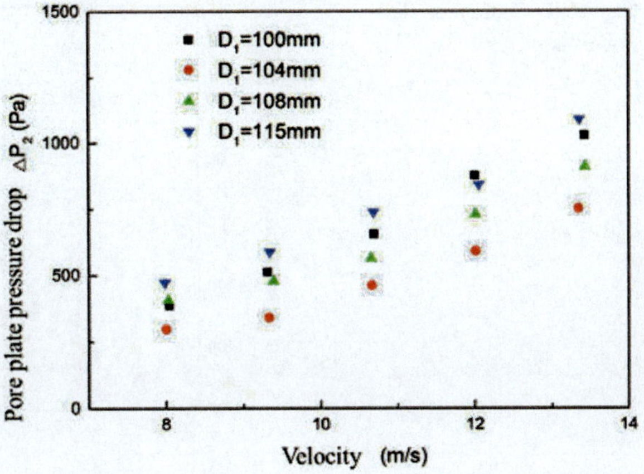

$$d = D_3 - D_2$$
$$v_3 = \frac{Q}{\frac{\pi}{4}\left(D_3^2 - D_2^2\right)} \quad (41.5)$$

As shown in Fig. 41.7, at the same velocity v_3, the slit width has little influence on ΔP_3. And the more D_2, the more pressure drop is. When D_2 increases, the flow rate increases as velocity maintains. This will cause more eddies and energy dissipation. Thus, resistance coefficiency increases.

The change of slit pressure drop via slit width d_1 is shown in Fig. 41.8. The resistance is relatively as high as 1400 Pa, making up large percentages. However, as the width changes, the pressure drop hardly changes. Therefore, the resistance coefficiency can be estimated as a constant, and the determining factor is velocity.

The structure of wind cap outlet is relatively simple. As shown in Fig. 41.9, the

Fig. 41.7 Change of cap pressure drop via D_2

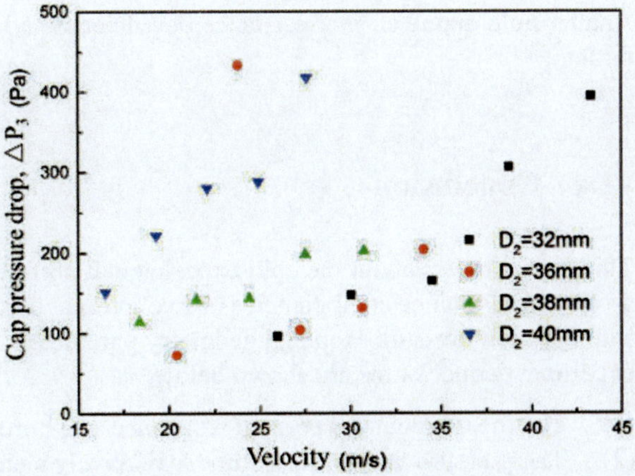

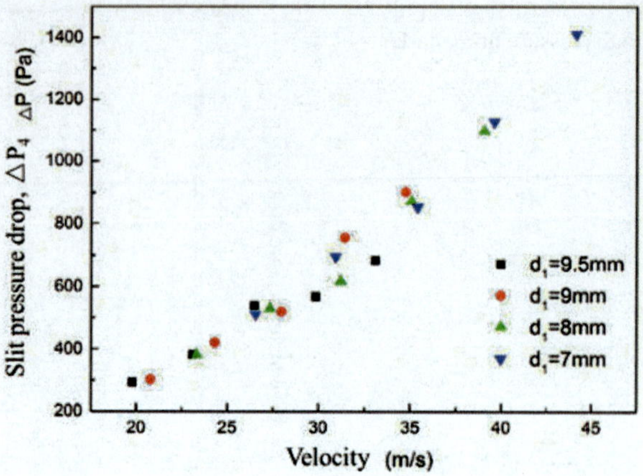

Fig. 41.8 Change of slit pressure drop via d_1

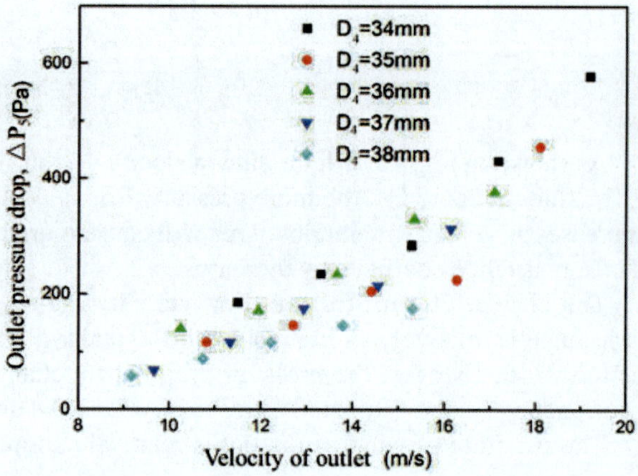

Fig. 41.9 Change of slit pressure drop via D_4

smaller hole diameter, the resistance coefficiency is bigger at the same velocity of outlet.

41.4 Conclusion

This paper studies about the cold-experimental performance of wind caps. Pressure drops and resistance coefficiencies of five internal parts are measured. The relationship between pressure drop and geometry parameters are qualitatively studied. The experiment conclusions are shown below.

(1) The most influential parts of resistance are: pore plate, cap and slit.
(2) The pressure drop of inner tube is relatively small.

(3) The resistance coefficiency relates to inner tube diameter, D_1 and hole diameter D_2. The smaller D_1, the bigger D_2, the larger coefficiency is. D_1 has more effects than D_2.
(4) The cap pressure drop is relate to entrance injection. The bigger D_2, the larger coefficiency is.
(5) The width of slit has less influence on resistance coefficiency.
(6) The smaller D_4, the larger coefficiency of outlet is.

In conclusion, cold-experimental studies about the fluent characteristics and provide references to wind cap design. Furthermore, experimental data is collected for establishing resistance coefficiency mathematical model.

References

Feng B (2007) Research on the nozzle character on the distributor plate in the CFB boiler and heat-transfer in the furnance. Shanghai Jiao Tong University

Lye J et al (2015) Numerical and experimental research on the distributor plate nozzles in a 1 069 t/h boiler. Electr Power Sci Eng 05:66–73

Niu J et al (2009) Comparative analysis on the structures and performances of wind caps of circulating fludized bed boilers. Tsinghua University

Yang H, Zhang H, Yang S et al (2009) Effect of bed pressure drop on performance of a CFB boiler. Energy Fuels 23(6):2886–2890

Yang HR, Zhang H, Lu JF et al (2010) Novel CFB boiler technology with reconstruction of its fluidization state. In: Proceedings of the 20th international conference on fluidized bed combustion. Springer, Berlin, Heidelberg, pp 195–199

Yue G, Lu J, Zhang H et al (2005) Design theory of circulating fluidized bed boilers. In: International conference on fluidized bed combustion, pp 135–146

Chapter 42
Progress and Main Technical Characteristics of Ultra-supercritical Circulating Fluidized Bed Boiler

Shengwei Xin, Yingping Li, Peng Zhang, Changhua Hu, Man Zhang, and Hu Wang

Abstract The research and development progress of ultra-supercritical circulating fluidized bed boiler in major scientific research institutions and boiler manufacturing plants at home and abroad is described, and the furnace type schemes and technical characteristics of different boilers types are discussed in detail. To research and develop ultra-supercritical CFB boiler technology, main technological difficulties were analyzed from heating surface safety, hydrodynamics safety, maintaining reheat steam temperature at low boiler load and low-cost ultra-low emission technology, and then the solutions to these difficulties were put forward. It guarantees the research and development of ultra-supercritical CFB boiler and provides technical support to continue to maintain the leading position of China's CFB power technology.

Keywords Ultra-supercritical · Circulating fluidized bed boiler · Ultra-low emission · Pulverized coal boiler

Coal is still the main energy source in China at present and even in the future decades. Washing, selecting, and upgrading of coal are thus necessary to maintain its clean and effective utilization (Zhou 2015). A large number of coal slime, gangue, washed coal and other by-products with low calorific value are inevitably produced as a consequence. It is therefore essential to utilize these by-products with improved energy efficiency and ecological benefits. Circulating fluidized bed technology for power generation has been developed rapidly by virtue of its wide fuel adaptability (Yu et al. 2004), high efficiency in desulphurization and nitrogen suppression with low cost in furnace, and is considered to be the best choice for clean and efficient utilization of low calorific value coal.

S. Xin (✉) · P. Zhang · C. Hu · H. Wang
China Energy Group CFB Technology R&D Center, Xi'an 710065, China
e-mail: xinswei@163.com

Y. Li
China Energy Group GUO NENG Energy Group, Beijing 100033, China

M. Zhang
Tsinghua University, Beijing 100084, China

J. Lyu and S. Li (eds.), *Clean Coal and Sustainable Energy*,
Environmental Science and Engineering,
https://doi.org/10.1007/978-981-16-1657-0_42

The supercritical CFB technology in China has reached a leading level in the world and has been widely popularized and applied since the successful operation of the world's first 600 MW Supercritical CFB demonstration unit with completely independent intellectual property rights in April 2013 (Xin et al. 2015). By June 2019, 31 supercritical CFB units have been put into production in China. This promotes the development of CFB technology in the world, and plays a positive role in upgrading of the relevant technology and equipment level (Xin et al. 2017; Yue et al. 2016; Henan 2016). The parameters and capacity of steam turbines and generators have also reached a high level according to the production and supply of major main engine plants in China. Despite this, even higher parameters are the pursuit to further improve efficiency and reduce emissions, so as to lead the worldwide development of CFB technology for power generation (Lingwen et al. 2019). Scientific research institutions at home and abroad and major boiler manufacturers including Tsinghua University, China Huaneng Group Clean Energy Technology Research Institute, foster wheeler Company of the United States, Alstom company of France, Dongfang Boiler Factory, Harbin Boiler Factory, Shanghai Boiler Factory etc. have therefore carried out the research and development of ultra-supercritical CFB boilers.

Based on a comprehensive summary of the existing boiler type and technical characteristics of the super/ultra-super critical CFB boiler both at home and abroad, the main technical difficulties of the ultra-supercritical CFB boiler are highlighted, with their solutions given in this paper. This provides significant technical supports for the research and development of ultra-supercritical CFB boiler.

42.1 Research and Development of Ultra-supercritical CFB Technology

42.1.1 Tsinghua University

Tsinghua University has designed a simple ultra-supercritical CFB boiler. The steam parameters are 26.15 MPa/603 °C/605 °C (Lu et al. 2014). This boiler adopts a M-shape arrangement with single air distributor. This could effectively avoid the "turn over problem of bed materials" when underpants and legs are designed and avoid over-high pressure head of the primary fan. The power supply is thus reduced and the operation is simplified (Olaleye et al. 2015). Without the use of external heat exchanger, four cyclone separators with diameter of 10.85 m are set up to ensure high separation efficiency of the separator and compact boiler structure which is beneficial to reduce the expenditure of boiler manufacture (Fig. 42.1).

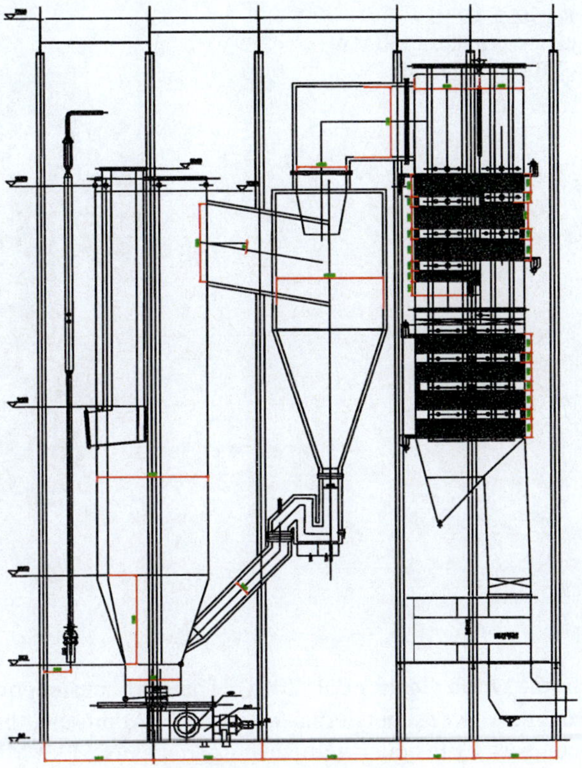

Fig. 42.1 Scheme design of simple ultra-supercritical CFB boiler of Tsinghua University

42.1.2 Clean Energy Technology Research Institute of Huaneng Group

CERI has carried out the research work of 350 MW advanced ultra-supercritical CFB boiler with steam parameter 36.8 MPa/705 °C/737 °C, based on the design, manufacture and engineering experience of 330 and 600 MW CFB boilers (Sun 2014). A technical scheme of 350 MW ultra-supercritical CFB boilers with steam temperature of 700 °C has been formed. The boiler adopts compact pneumatic heat exchanger, with a total of four superheaters. The reheater system is arranged in three stages (Fig. 42.2).

42.1.3 American Foster Wheeler Company

The South Korean Sanchok 4 × 550 MW ultra-supercritical CFB boiler units designed and manufactured by FW in the United States has been put into operation at present, which has the highest steam parameter of 25.7 MPa/603 °C/603 °C)

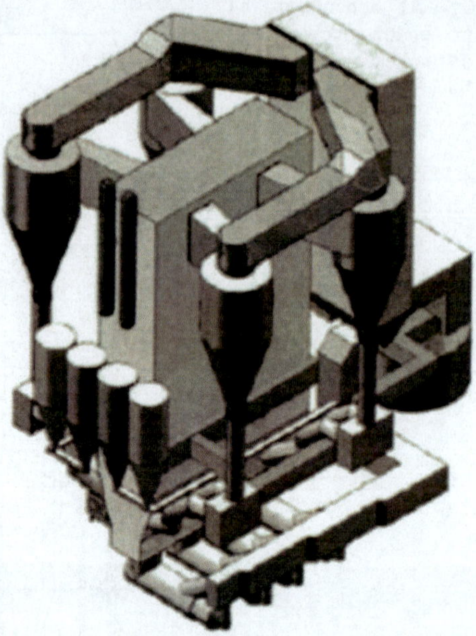

Fig. 42.2 CERI ultra-supercritical 350 MW CFB boiler scheme

in the world (Jantti et al. 2012). The coal consumption of the unit is 290 g/kWh, and the power supply efficiency is 43%. At present, the 800 MW ultra-supercritical compact CFB boiler with steam parameters (30.9 MPa/604 °C/621 °C) has been conceptually designed by FW (Jantti and Rasanen 2011), whose unit efficiency can reach 45% (Fig. 42.3).

42.1.4 ALSTOM Company of French

Alstom has designed a 600 MW (27.6 MPa/602 °C/602 °C) ultra-supercritical CFB boiler for French electric power company (EDF) (Olaleye et al. 2015). The optimized high-temperature cyclone separator and external heat exchanger are used. Compared with traditional CFB power stations of the same size, fuel consumption and CO_2 emissions can be reduced by 6%.

42.1.5 Three Major Boiler Manufacturers in China

With the promotion of the national key R & D program "Research and Development and demonstration of Ultra-Supercritical circulating Fluidized bed Technology", the National Energy Group led the domestic well-known scientific research institutes and

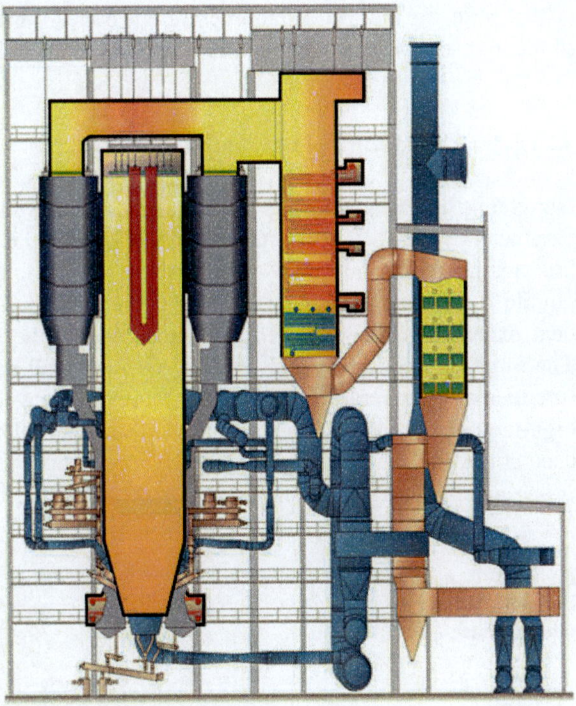

Fig. 42.3 The South Korean Sanchok ultra-supercritical CFB boiler

three major boiler manufacturers to jointly carry out the research and development of ultra-supercritical CFB power generation technology. The project is intended to build a 660 MW high efficiency ultra-supercritical CFB demonstration unit, and meanwhile achieve low-cost ultra-low emissions, whose energy consumption is close to or reach that of the pulverized coal-fired boiler unit of the same capacity (Expected indexes are shown in Table 42.1). At present, the three boiler manufacturersb in China have carried out certain research and development for conventional and high-efficiency ultra-supercritical CFB boilers, and a preliminary high-efficiency ultra-supercritical 660MWCFB boiler scheme has been formed. The main steam pressure of the boiler

Table 42.1 Expected index of 660 MW ultra-supercritical CFB boiler (Cai et al. 2016)

Index	Unit	Numerical value
Unit capacity	MW	660
Efficiency of boiler	%	>92
Net coal consumption	g/kWh	<290
SO_2 emission concentration	mg/Nm^3	<35
NO_x emission concentration	mg/Nm^3	<50
Dust emission concentration	mg/Nm^3	<10

is 29.4 MPa, the main steam temperature is 605 °C, and the outlet steam temperature of the high-temperature reheater is 623 °C (Cai et al. 2016).

42.1.5.1 Dongfang Boiler Manufacturer

Dongfang Boiler Factory has raised a scheme of high efficiency ultra-supercritical parameters (29.3 MPa/605 °C/623 °C) (Dongfang Boiler Group Co., Ltd. 2015). Following Baima 600 MW supercritical CFB boiler with single furnace and double distributor layout, six steam-cooled cyclone separators and six external heat exchangers are arranged on both sides of the boiler (shown in Fig. 42.4). The furnace is equipped with middle partition wall and high-temperature reheater. Intermediate-temperature superheater I, intermediate-temperature superheater II and high-temperature reheater are arranged in six external heat exchangers. Single flue is used at the tail, where low-temperature superheater, low-temperature reheaters and economizer are arranged (shown in Fig. 42.5). At the same time, in order to achieve

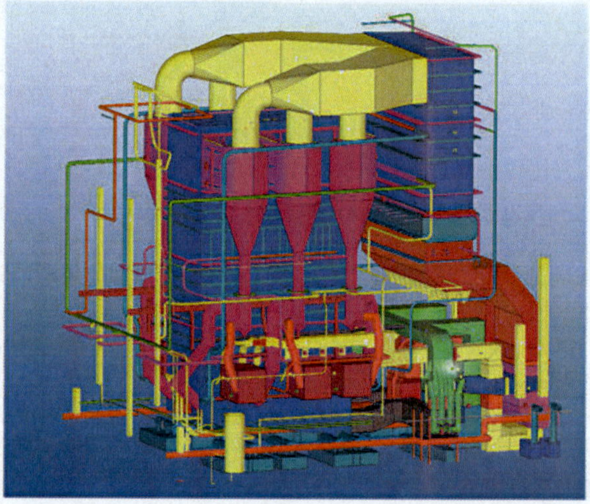

Fig. 42.4 General layout of Dongfang boiler type scheme boiler

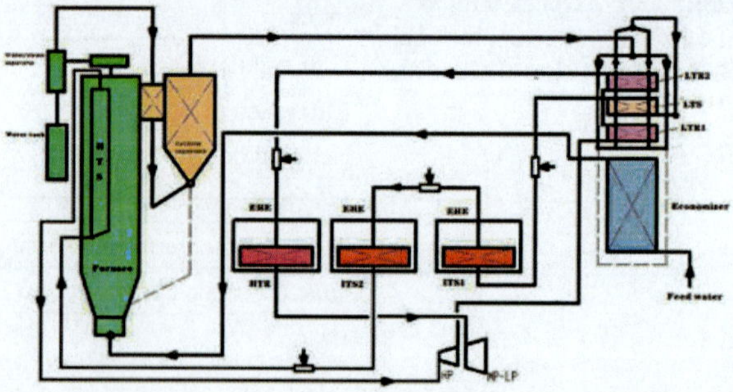

Fig. 42.5 Steam-water system diagram of Dongfang boiler type scheme

ultra-low emission, Dongfang Boiler Factory has developed an efficient secondary air technology (Dongfang boiler co., Ltd. 2014), which can achieve the original NO_x emission to no more than 50 mg/Nm^3 and has been verified on the existing 350 MW supercritical CFB boilers in operation. It will provide a reference for the development and manufacture of ultra-supercritical CFB boilers.

42.1.5.2 Harbin Boiler Manufacturer

The 660 MW ultra-supercritical CFB boiler of Harbin Boiler Factory adopts the M-shape arrangement, with a single furnace and a single air distributor, four steam-cooled cyclones with a diameter of 10.5 m, and four external heat exchangers (Xin et al. 2017) (shown in Fig. 42.6). The external heat exchangers are all equipped with medium temperature superheaters to avoid the thermal deviation and safety of the heating surface caused by the increase of steam parameters; the tail is double flue, and the reheat steam temperature is adjusted by the flue gas baffle. In the boiler design, low level operation, reasonable fan selection, separator optimization and waste heat depth utilization technology are used to further reduce energy consumption, and the use of fluid state reconstruction ultra-low emission technology, selection of suitable bed temperature, staged combustion etc. technological innovation to achieve ultra-low emission of NO_x.

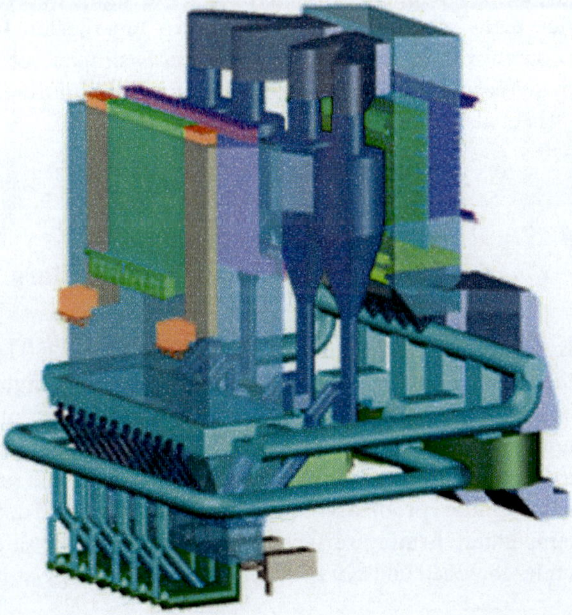

Fig. 42.6 General layout of Harbin boiler type scheme boiler

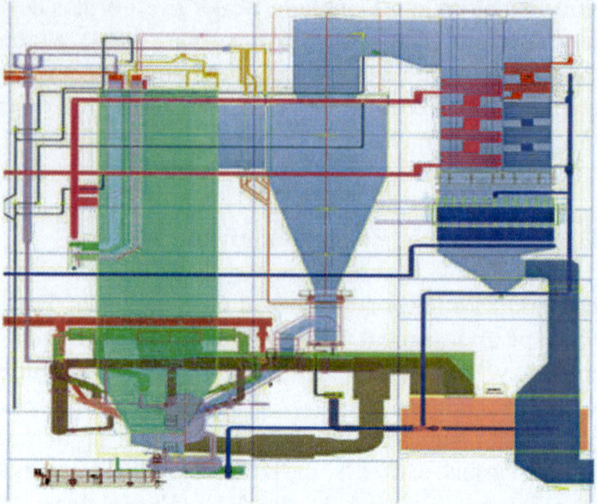

Fig. 42.7 General layout of boiler type scheme boiler

42.1.5.3 Shanghai Boiler Manufacturer

The Shanghai Boiler adopts a single furnace type, single air distributor and M-type arrangement of four steam-cooled separators without external heat exchanger (shown in Fig. 42.7). The furnace is equipped with secondary rising water cooling screen, low-temperature superheater, low-temperature reheaters. The tail has double flue paths where the low temperature superheater, low temperature reheater and economizer are equipped. The system is simple, yet the furnace size is large with many heating surfaces within. The flow field in the furnace requires further study (Xin et al. 2017).

42.2 Research and Development of Key Technologies for Ultra-supercritical CFB Boilers

Based on the design and manufacture of 600 and 350 MW supercritical CFB boilers, the technical difficulties and risks of developing high efficiency supercritical CFB boilers are relatively small. However, as steam parameters including pressure and temperature are much higher than those of the existing supercritical CFB units particularly the increase of superheater outlet pressure and reheat steam temperature, some new technical problems require further research and development. This includes high temperature heating surface safety, hydrodynamic safety, reheat steam temperature under low load and ultra-low emission technology at low cost.

42.2.1 *Safety of High Temperature Heating Surface*

The reheated steam temperature of ultra-supercritical CFB boilers is raised from 603 to 623 °C, but the total increase in enthalpy of reheated steam decreases due to the large increase of the inlet working fluid temperature at the cold end of reheat steam. The design of reheater heating surface mainly considers the wall temperature safety of high temperature reheater after temperature rising and the steam temperature characteristics of reheater system under low load. At present, SUPER304 and HR3C (Zhang et al. 2012b) are widely used as the heating surface materials for high temperature reheaters of high-efficiency ultra-supercritical boilers. The maximum allowable wall temperatures of these two materials are 705 and 730 °C (Zhang et al. 2012b). The combustion temperature of CFB boiler is low and in order to ensure the steam temperature above 600 °C, the high temperature reheater must be arranged in the furnace or external heat exchanger, while the high temperature reheater of pulverized coal boiler should be arranged in the horizontal flue. Because the heat loads of the furnace and the external bed are higher than that of the horizontal flue, the thermal deviation of the heating surface in the furnace and the external bed is large and difficult to control at a small range. Sun et al. (2014) analyzed the wall temperature characteristics of high temperature reheater of 600 MW supercritical CFB boiler, which showed that there is a certain thermal deviation of the external high temperature reheater, but it is still within the temperature allowable range; Xingsheng et al. (2016) carried out experimental study on operation adjustment of external high temperature reheater for 600 MW supercritical CFB boiler; Xu Peng et al. (Dongfang Boiler Group Co., Ltd. 2016) carried out simulation study on particle gas–solid flow characteristics of high temperature heating surface of external heat exchanger. Based on the thermal deviation control technology of 600 MW supercritical CFB boiler, the gas–solid two-phase flow is adjusted, and the arrangement of the heating surface is optimized. The influence mechanism of gas–solid two-phase flow on thermal deviation is deeply understood, and the gas–solid two-phase flow is reduced by adjusting the gas–solid two-phase flow. The thermal deviation fundamentally solves the problem of thermal deviation. At the same time, the position of the reheater is reasonably arranged, the structure of the reheater is optimized, and the number of stages and the increase of the heating surface of each level are reasonably distributed, and the temperature deviation can be controlled within a reasonable range.

42.2.2 *Hydrodynamic Safety of Water Wall*

The water wall of 600 MW supercritical CFB boiler adopts low mass flow rate hydrodynamic technology (Li et al. 2008; Harbin Boiler Company Limited 2017). Compared with 600 MW supercritical CFB boilers, 660 MW supercritical CFB boilers have the same main steam flow rate, higher steam pressure and temperature. The average mass flow rate of water wall under BMCR condition was calculated by

Table 42.2 Average mass flow rate of the water wall obtained from three boiler manufacturers

Boiler Manufacturer	Mass flow rate (kg/m^2·s)	Hydrodynamic calculation program
Dongfang boiler manufacturer	~730	Siemens
Harbin boiler mmanufacturer	930	Xi'an Jiaotong University
Shanghai boiler manufacturer	1140	Xi'an Jiaotong University

using different hydrodynamic calculation program software under three boiler plants (shown in Table 42.2) (Harbin Boiler Company Limited 2017). The numerical calculation and experimental study on the hydrodynamic characteristics of supercritical (ultra-supercritical) water wall have been carried out successively. The results show that the design and operation of supercritical (ultra-supercritical) water wall are safe and reliable (Jianxiong 2010; Cheng et al. 2015; Wang et al. 2017). The water wall of high-efficiency ultra-supercritical CFB boiler adopts low mass flow rate hydrodynamic technology and the general trend of hydrodynamic characteristics is basically consistent with the measured data of Baima 600 MW supercritical CFB boiler. It conforms to the characteristics of positive flow response. The temperature deviation of outlet working fluid, wall temperature deviation and dynamic stability are within the safe range, and the hydrodynamic characteristics are safe and reliable.

42.2.3 *Low Load Reheating Steam Temperature*

When the CFB boiler is running at low load, the upper temperature of the furnace is lower and the circulating ash volume is less. Therefore, it is difficult to ensure the reheat steam temperature reaches 623 °C at low load. It is necessary to study and optimize the design of the thermal system to improve the low-load unit. Thermal efficiency. Zhang et al. (2012a) compared the distribution of furnace temperature under different loads of large CFB boilers with or without external heat exchangers. The research shows that there is still a large heat transfer temperature difference in the boiler with external bed under low load, and the steam temperature characteristics are also Better; the boiler manufacturer has proposed a detailed thermal calculation and refinement of the heated surface layout scheme to ensure that the reheat steam temperature of the high-efficiency ultra-supercritical CFB boiler at low load still reaches the rated value. Dongfang Boiler Plant (Dongfang Boiler Group Co., Ltd. 2015) and Harbin Boiler Plant (Xin et al. 2017) all proposed to adopt an external heat exchanger scheme to adjust the bed temperature through the return of the external heat exchanger. In addition, the Eastern Boiler Plant arranges the final reheater in an external heat exchanger and ensures the reheat steam temperature by optimizing the arrangement of the heated surface of the external heat exchanger.

42.2.4 *Low Cost Ultra-low Emission Technology*

At present, CFB boilers cannot meet the national ultra-low emission requirements solely depending on their own pollutant control characteristics. The potential of high efficiency desulfurization and nitrogen suppression in CFB boilers needs to be further tapped. The cost of combined removal inside and outside of the boiler is high. The existing ultra-low emission technology route has a high pressure to remove many pollutants, and its environmental protection advantages are gradually weakened. However, if the ultra-low emission technology of pulverized coal boilers is advanced in CFB boilers. Transplantation can achieve ultra-low emission, but the cost is high and the inherent advantages of CFB are discarded. Existing research shows that in the technology of ultra-low emission of NO_x, through bed temperature control and reasonable air distribution, the reduction atmosphere in the furnace will be further created, in addition, the flow field, temperature field in the furnace, the quality of the bed and the secondary air will be also optimized, so as to realize further high-efficiency staged combustion and reduction denitrification, and tap the potential of denitrification in the furnace. It is expected that the ultra-low emission of NO_x will be realized directly in the furnace (Yue et al. 2016; Dongfang Boiler Group Co., Ltd. 2014). The deep removal of SO_2 can be achieved by controlling the temperature of the bed, the particle size of the desulfurized limestone, and the desulfurization reaction time (Hu et al. 2012), and developing a new high-efficiency cyclone separator, returning system, air distribution device, coal feeding and limestone preparation and joining system. Auxiliary equipment, to achieve higher desulfurization efficiency in the CFB furnace at a lower Ca/S. Therefore, the innovative research on ultra-supercritical CFB boilers, based on the source generation control and the ultra-low emission technology of efficient synergistic removal in the furnace, synergistically realize the research and development of NO_x and SO_2 deep decontamination technology, fully exploit the CFB boiler low pollutants. With the advantage of cost control, we strive to achieve ultra-low emission of NO_x and SO_2 in the furnace, and a simple desulfurization and denitration device outside the furnace as a hot standby.

42.3 Conclusion

(1) Under the current energy structure of our country, the development of ultra-supercritical CFB power generation technology is the only way to further improve unit efficiency and reduce energy consumption. It is also an effective way to clean and efficient use of low calorific value fuel. It is of great significance to maintain the leading position of CFB power generation technology in our country.

(2) The ultra-supercritical CFB boiler scheme proposed by various scientific research institutions and boiler manufacturers at home and abroad is technically feasible. Under the latest national energy conservation and emission

reduction policy, ultra-supercritical CFB power generation technology must be developed on a large scale and must be developed. Low-cost ultra-low emission technology, and close to or reach the energy consumption level of the same capacity pulverized coal boiler unit in terms of energy consumption.

(3) High-efficiency ultra-supercritical CFB power generation technology has been included in the national "13th Five-Year Plan" and "Made in China 2025", and the key issues such as the development of high temperature heating surface and hydrodynamic safety are all developed. Can be further overcome through technological development. With the development and commissioning of ultra-supercritical CFB boilers, CFB combustion technology will reach a new height and become a power station boiler combustion technology that goes hand in hand with the combustion technology of pulverized coal fired boilers.

Acknowledgements This work was supported by the National Key Research & Development Program of China (NO. 2016YFB0600201).

References

Cai R, Lu J, Ling W et al (2016) Progress of supercritical and ultra-supercritical circulating fluidized bed boiler technology. Electr Power 49(12):1–7

Cheng L, Xu L, Xia Y et al (2015) Key issues and solutions in development of the 600 MW CFB boiler. Proc CSEE 35(21):5520–5532

Dongfang Boiler Co., Ltd. (2014) Secondary air distribution method for circulating fluidized bed boiler. Chinese patent 201410516049.2

Dongfang Boiler Group Co., Ltd. (2015) Specification for scheme design and evaluation of self-developed 660MW ultra-supercritical CFB boiler. Dongfang Boiler Group Co., Ltd, Chengdu

Dongfang Boiler Group Co., Ltd. (2016) The method of eliminating the thermal deviation of working fluid in external fluidized bed. Chinese patent 201610102256.2

Dongfang Boiler Group Co., Ltd. (2014) Secondary air distribution method for CFB boiler. Chinese patent 201410516049.2

Harbin Boiler Company Limited (2017) Specification for design evaluation of 660 MW high efficiency ultra supercritical CFB boiler. Harbin Boiler Company Limited, Harbin

Henan (2016) Energy efficiency alignment standard and competition data of CFB units in 2015. China Unified Power Evaluation Consultation Institute, Technical Exchange Service Collaboration Network for CFB Units

Hu C, Lu X (2012) Equipment and operation for 600MW supercritical CFB boiler. China Electric Power Press, Beijing

Jantti T, Rasanen K (2011) Circulating fluidized bed technology towards 800MW scale-lagisza 460 MW super-critical CFB operation experience. In: Power Gen Europe. Milan, Italy

Jantti T, Nuortimo K, Ruuskanen M et al (2012) Samcheok green power 4×550 MWe super-critical circulating fluidized bed steam generation in South Korea. In: Power Gen Europe. Colon, Germany

Jianxiong M (2010) Development of supercritical/ultra-supercritical CFB boiler technology. Electr Power Constr 31(1):1–6

Lingwen W, Lyu J et al (2019) Research and development progress of the 660 MW ultra-supercritical circulating fluidized bed boiler. Proc CSEE 39(9):2515–2513

Lu J, Man Z, Yang H et al (2014) Conceptual design of a simplified 660MW ultra-supercritical circulating fluidized bed boiler. Proc CSEE 34(5):741–747

Nie L, Wang P, Peng L et al (2008) Design of 600MW supercritical circulating fluidized bed boiler. J Power Eng 28(5):701–706

Olaleye AK, Wang M, Xu C et al (2015) Dynamic modelling, validation and analysis of coal-fired super-critical once-through boiler-turbine-generator systems under stringent UK grid requirement. In: Clean coal technology and sustainable development. Beijing, China

Sun X (2014) Plan design and research of 700°C ultra-supercritical circulating fluidized bed boiler. Proc CSEE 34(23):3977–3982

Sun X, Hu C, Li X et al (2014) Tube wall temperature characteristic of external heat exchanger in 600 MW supercritical CFB boiler. Electr Power Constr 35(4):6–9

Wang S, Wang W, Shen Z et al (2017) Heat-transfer performance of vertical water wall in an ultra-supercritical pressure boiler with high efficiency and wide regulation load at low mass flux. J Chin Soc Power Eng 37(2):85–90

Xin S, Li X, Zhang W et al (2015) The operation of the 600MW super-critical CFB boiler unit. In: Clean coal technology and sustainable development. Beijing, China

Xin S-W, Han P, Nie L et al (2017) Research on furnace types and key technologies of ultra-supercritical CFB boiler. Power Syst Eng 33(4):1–4+8

Xingsheng Z, Peng X, Xiukui H et al (2016) Experimental research on adjustment of operation of high temperature external heat exchanger for the 600 MW super-critical CFB boiler. Dongfang Electr Rev 30:23–26

Yu L, Lu J, Wang Z et al (2004) Prospective research progress of combustion technology for circulating fluidized beds. J Eng Therm Energy Power 19(4):336–342

Yue G, Lu J, Xu P et al (2016) The up-to-date development and future of circulating fluidized bed combustion technology. Electr Power 49(1):1–13

Zhang M, Wu H, Sun Y et al (2012a) Operation characteristics of fluidized bed heat exchanger of large-scale circulating fluidized bed boiler. Proc CSEE 32(14):42–48

Zhang Y, Cai X, Huang S (2012b) Materials in 700°C advanced ultra-supercritical coal-fired units. Electr Power 45(2):16–21

Zhou C (2015) Supply-demand prediction of coal by area in 2020 and 2025. China Coal 41(7):5–8

Chapter 43
Experimental Study on Peaking Performance of 145 MW Circulating Fluidized Bed Boiler Unit with Retrofitting for Low-Vacuum Heat Supply

Xuhui Zhang, Zhonghua Zhao, Gui Long Xiong, Fuxing Cui, Xinguang Dong, Qingchuan Zhao, Ke Liu, Haichao Wang, and Jinglong Liu

Abstract The peaking performance of 145 MW circulating fluidized bed (CFB) boiler unit with retrofitting for low-vacuum heat supply were studied in this article. Peaking depth, peaking rate, tube wall temperature of superheater and reheater and bed temperature were measured. Unit load curve analysis shows that the peak load capacity of the test unit is significantly reduced after low-vacuum heating reform with the minimum output of 41.4%. The load reduction rate of 0.99 MW/min and the load increasing rate of 0.71 MW/min is relatively small because of the large thermal inertia of CFB boiler. The wall temperature of the superheater and reheater of low load and variable load stages could be higher than that of high load stages, which is easy to cause overheating and tubc bursting. The results of bed temperature and furnace negative pressure indicate the excellent combustion stability of CFB boiler at low load.

Keywords Peaking performance · CFB boiler · Low-vacuum heat supply

43.1 Introduction

The national "13th Five-Year Energy Plan" clearly put forward that renewable power integration and consumption should be strengthened, and the large-scale consumption of wind power and photovoltaics forced the thermal power units gradually shift from power production to undertake peaking tasks and provide support services (Zhang 2017). However, the installed capacity of renewable energy increases rapidly,

X. Zhang (✉) · Z. Zhao · F. Cui · X. Dong · Q. Zhao · K. Liu · H. Wang · J. Liu
State Grid Shandong Electric Power Research Institute, Jinan 250002, China
e-mail: 15098758582@163.com

G. L. Xiong
School of Resources Environment & Chemical Engineering, Nanchang University, Nanchang 330000, China

J. Lyu and S. Li (eds.), *Clean Coal and Sustainable Energy*,
Environmental Science and Engineering,
https://doi.org/10.1007/978-981-16-1657-0_43

especially the distributed photovoltaics, causing a serious shortage of peaking space in the power grid under frequent peaking of thermal power units in some areas (Gu et al. 2016; Wang et al. 2010; Shen et al. 2014). During heating period in winter, the power of some units is fixed by heat, making the adjustment performance of area grid dropping dramatically (Lv et al. 2019; Wang et al. 2018). In order to meet the growing demand for renewable energy consumption, it is necessary to encourage thermal power units to participate in peak shaving.

The circulating fluidized bed boiler is operated by the circulating combustion of high-temperature bed material in the fluidized furnace, keeping the temperature distribution inside the furnace uniform. CFB unit is always regarded as one of the important peaking sources due to its strong combustion stability under low load. Meanwhile, some CFB units with relatively low capacity are often used as heating units, some of which are retrofitted for low-vacuum heat supply. There peaking performance is reduced, as a result. Therefore, the peak shaving performance of CFB boiler units during heating period needs further study (Jun et al. 2018; Jia 2014).

At present, there are many researches on peaking performance of CFB heating units. The main methods used to study the peak-regulation performance of heating units include steam turbine thermal test method, condition diagram analysis method and thermodynamic calculation method. The economic analysis of 350 MW super-critical low vacuum heating unit showed that the heat economy of low vacuum heating unit is greatly improved compared with that of conventional pumping and condensing unit. EBSILON software was used to establish the prediction model of peak load regulation performance of heating units, and the relationship between heat extraction volume and unit load was obtained (Lv 2019; Wang et al. 2018). Multiple theoretical analysis and experimental research showed the peak-regulation performance of the unit would deteriorate under the condition of high steam extraction quantity of heating (Xing et al. 2018).

However, the current studies are only focused on theoretical analysis, few experimental studies were carried out at present. In order to study the peak-regulation characteristics of the CFB boiler unit with retrofitting for low-vacuum heat supply, a peaking performance test were carried out in a 145 MW natural circulation fluidized bed boiler unit in a power plant.

43.2 Experimental Apparatus and Procedure

43.2.1 Experimental Apparatus

The boiler of the test unit is an ultra-high pressure natural CFB drum furnace with rated evaporation of 465 t/h, adopting the technical route of single chamber, membrane type water wall, balanced ventilation, primary intermediate reheat and solid slag discharge. High temperature adiabatic cyclone separator is adopted for the

Table 43.1 Coal quality parameters

Item	Design coal	Checked coal	Burned coal
Car (%)	59.36	55.45	46.65
Sar (%)	1.5	2.47	1.22
Aar (%)	28.4	32.58	28.23
Mar (%)	5.8	4.9	9.4
Vdaf (%)	12.81	14.81	25.2
Qnet.ar (kJ/kg)	22,316	20,055	20,478

separation of circulating materials. The boiler heat exchange surfaces is designed as suspension structure. Both the designed coal and the checked coal are lean coal, and the actual coal burned in the experiment is bituminous coal, which has a large deviation from the designed coal. The specific coal quality parameters are shown in Table 43.1.

The steam turbine of the unit was originally the extraction condensing steam turbine with the structure of ultra-high pressure primary intermediate reheat, double cylinder, double exhaust steam and single shaft. In order to further improve the heating capacity of the unit, the unit has been reformed with low vacuum heating. The original low-pressure rotor was reformed by installing 2 inoperative blades root and exhaust steam guide ring after removing the blades and baffles of last two stages. The condenser was transformed by changing the inlet and outlet water chamber to volute circular arc water chamber.

43.2.2 Experimental Procedure

The test process were divided into 4 stages:

(1) operating stably at initial load for a period of time;
(2) reducing the load from initial load to the lowest at the maximum rate it can reach;
(3) operating steadily for 2 h at the lowest output;
(4) increasing the output to the maximum from the lowest load it can reach at the maximum rate.

43.3 Results and Discussion

43.3.1 Depth of Peak Regulation

In order to compare the difference between the test unit and some other non-heating units with CFB boiler, peak regulation capacity tests were also carried out in three

Table 43.2 Condition of the units

Unit	Type	Rated capacity (MW)	Minimum output (MW)
Test unit	High backpressure heating	145	60
C1	Straight condensing	145	55
C2	Straight condensing	220	84
C3	Straight condensing	700	243

other CFB units. Table 43.2 shows the brief introduction of the test unit and 3 pure condensing units. Figure 43.1 shows their comparison of the depth of peak regulation. Compared with other 3 pure condensing units, the peak shaving capacity of test unit is clearly insufficient, which minimum output could only achieve 41.4%. This is because the unit follows the operation mode of "fixing electricity by heat". When the electrical load of the unit drops, the back pressure of condenser and outlet temperature of circulating water decrease, then the unit's external heating capacity decreases. In order to meet the heating requirements, the unit's electric power can not be reduced to the lowest output under the normal back pressure. The maximum electric load of test unit can reach 120 MW, while the minimum electric load can only be reduced to 60 MW, indicating the unit load interval narrowed. After heat supply retrofitting, the capacity of unit with high load and low load is both weakened.

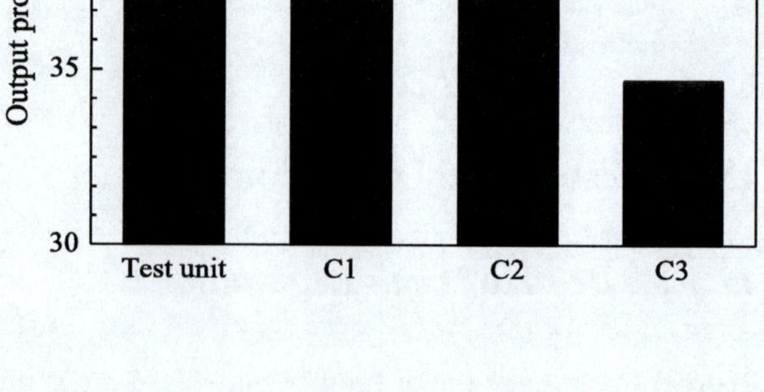

Fig. 43.1 Depth of peak regulation of 4 units

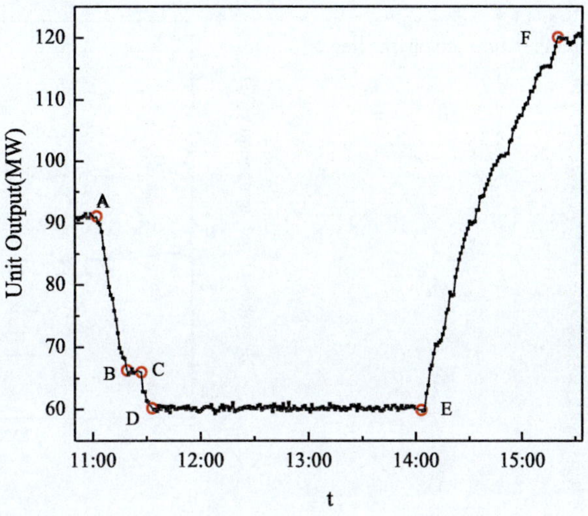

Fig. 43.2 Load change curve of the test unit

43.3.2 Rate of Peak Regulation

The rate of peak regulation is calculated by

$$v = |(L_2 - L_1)/(t_2 - t_1)|$$

where L_1 and L_2 is the initial load and the end load respectively, t_1 and t_2 is the start time and end time respectively.

Figure 43.2 shows the load change curve of the test unit. During the test, the unit load is reduced from point A to point B. After that, the unit operates stably, until point C. Then the load is reduced from point C to point D, and the stable operation is maintained to point E. Finally, the load is increased from point E to point F of maximum output. The load decreasing rate of AB section and CD section is 1.44 and 0.97 MW/min, respectively. It can be seen obviously that the rate of load reduction under lower load is much lower. The overall speed of load decreasing is 1.07 MW/min, and the load rise rate of EF segment is 0.71 MW/min. Compared with other coal-fired units, some obvious deficiency is existed in the variable load performance of circulating fluidized bed boiler units. A large number of circulating materials with high temperature enable the CFB boiler to burn stably at low load, but limit the rate of load adjustment on the other hand.

43.3.3 Tube Wall Temperature of Superheater and Reheater

Figures 43.3 and 43.4 show the wall temperature of superheater and reheater during

Fig. 43.3 Tube wall temperature of superheater

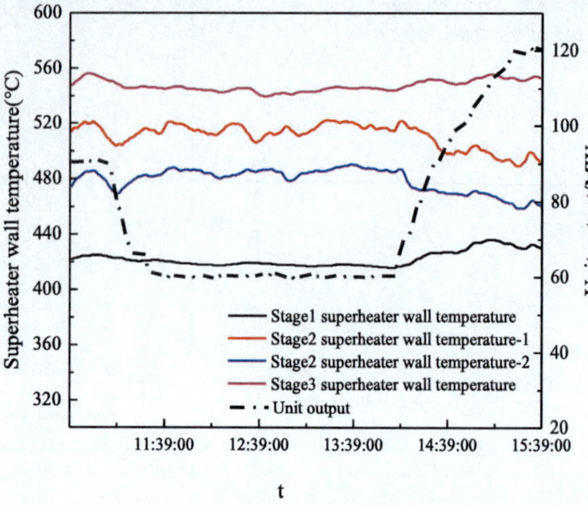

Fig. 43.4 Tube wall temperature of reheater

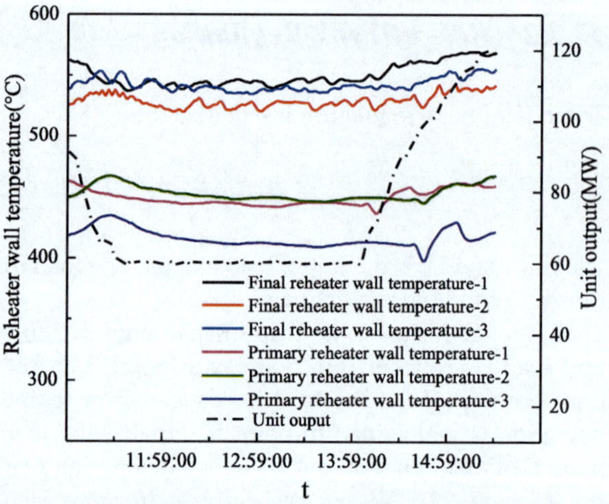

the test, respectively. It can be seen from the figure that as the load decreases, the wall temperature of the superheater shows a downward trend. This is because the coal put into the furnace is reduced when the unit is running down load, the wall of the tube is reduced by heat as well. When operating in the low load section, the wall temperature of the primary superheater and the tertiary superheater is smaller than that of the high load section, while the wall temperature of the secondary superheater increases, which means that the superheater overheating does not only occur at high load. There is also a large risk of overheating in the high load section and low load section. When the load rises, the wall temperature of the first-stage superheater and the third-stage superheater rises significantly, while the wall temperature of the secondary superheater drops significantly.

It can be seen from Fig. 43.4 that the maximum value of the reheater wall temperature occurs when the load changes, showing a sharp fluctuation of the wall temperature at the stage of variable load. When the load is stable, the wall temperature gradually stabilizes.

43.3.4 Bed Temperature and Furnace Negative Pressure

The bed temperature of each layer and furnace negative pressure were measured to study combustion stability in the experiment. Figure 43.5 shows variation of bed temperature of CFB boiler under different loads. It can be seen that the bed temperature increases with the increase of the load and decreases with the decrease of the load. During the test, the upper, middle and lower bed temperature is relatively stable. To a certain extent, it reflects the strong stable combustion capacity of the circulating fluidized bed boiler under low load, which is suitable for peak shaving.

Figure 43.6 shows the change of furnace negative pressure. Furnace negative pressure is an important parameter that reflects the stability of combustion condition and it is one of the important parameters to control and monitor in operation (Juhua and shengqi 2011). Once the combustion condition of the furnace changes, the negative pressure of the furnace will change accordingly. When operating in the depth peak load state, the negative pressure of the test boiler basically maintains a stable negative pressure state with small fluctuation, which can indicate the stable combustion of the boiler. In the stage of variable load, especially when the load is raised, the negative pressure of the furnace fluctuates violently due to the increase of flue gas volume. At this time, the phenomenon of furnace negative pressure turning to positive pressure is obvious, need to closely monitor the operation of the unit.

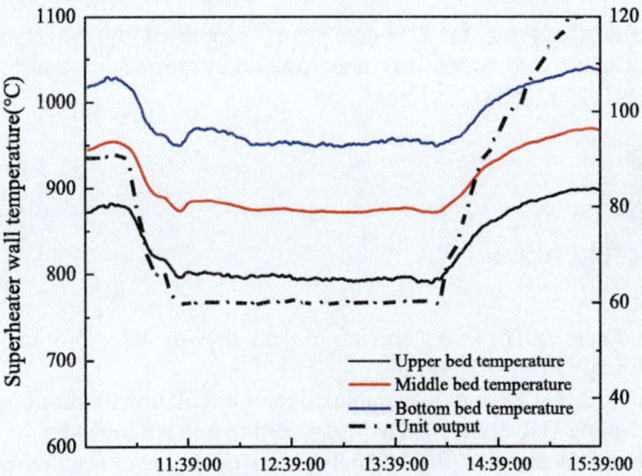

Fig. 43.5 Variation of bed temperature of boilers

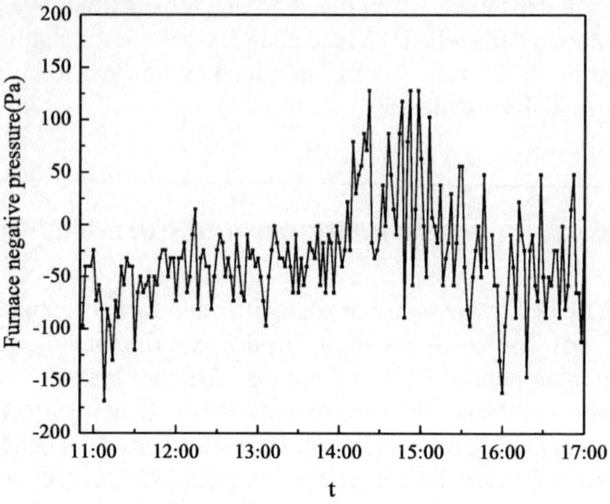

Fig. 43.6 Furnace negative pressure

43.4 Conclusions

The peak load capacity of the test unit is significantly reduced after low-vacuum heating reform with the minimum output of 41.4%, while other 3 units without heating can reach peaking depth of about 35%. The load reduction rate of 0.99 MW/min and the load increasing rate of 0.71 MW/min is relatively small because of the large thermal inertia of CFB boiler. The wall temperature of the superheater and reheater of low load and variable load stages could be higher than that of high load stages, which is easy to cause overheating and tube bursting. The results of bed temperature and furnace negative pressure indicate the excellent combustion stability of CFB boiler at low load.

Acknowledgements This work was supported by the National Natural Science Foundation of China (No. 51666011) and Youth key project of jiangxi natural science foundation (No. 20171ACB21008).

References

Gu Y et al (2016) Overall review of peak shaving for coal-fired power units in China. Renew Sustain Energy Rev 54:723–731

Jia W (2014) Research on application of circulating fluidized bed boiler in industrial plant heating. Beijing University of Civil Engineering and Architecture

Juhua Z, Shengqi H (2011) Analysis and treatment of negative pressure anomaly to positive furnace ash and dust fault. J Shandong Electr Power Tech Coll 14(01):46–48

Jun C, Suxia M, Zhongyuan L et al (2018) Study on the pollutant emission characteristic under the dynamic compound combustion of low calorific coal-high calorific pulverized coal. J Eng Therm Energy Power 33(10):60–66

Lv K, Li J, An J et al (2019) Thermodynamic Characteristics of a 330mw unit with high back-pressure circulating water heating. Turbine Technol 61(01):59–62+66

Shen J, Cheng C, Wu X et al (2014) Optimization of peak load among multiple provincial power grids under a certain dispatching authority. Energy 74:494–505

Wang Z, Su A, Lu S (2010) Analysis on capacity of wind power integrated into Liaoning power grid based on power balance. Autom Electr Power Syst 3:019

Wang W, Guo L, Lu S (2018) Analysis of peak shaving performance and coal consumption of cogeneration unit. Turbine Technol 60(06):457–460

Xing C, Jingfeng Z, Lihua C et al (2018) Study on the peaking load adjustment ability of thermal power unit. Turbine Technol 60(04):299–302

Zhang Y (2017) Evolutionary game theory on thermal power peaking under large scale of wind power integration. North China Electric Power University, Beijing

Chapter 44
The Technology and Application of a New Type of Coal Water Slurry-Fired CFB with High Efficiency and Clean Combustion

Yuping Hao, Ruiguo Li, and Yujiao Chen

Abstract This paper introduces the core technology of coal water slurry-fired CFB with high efficiency and clean combustion, based on flow pattern reconstruction. The design of key boiler components and characteristics of boiler and system are also introduced. This technology realizes high desulfurization efficiency as well as less NO_x emission at the furnace exit that sharply reduce investment and operating cost. In the last part of the paper, it expounds the application references and development prospects.

Keywords Coal water slurry (CWS) boiler · Fluidization state specification · Flow pattern reconstruction · Circulating fluidization · High efficiency and clean combustion

44.1 Introduction

For high efficient and clean utilization of coal and to reduce requirement on water slurry particle size, a new type of coal water slurry (CWS)-fired CFB boiler technology was jointly developed by Qingdao Clear, Tsinghua University and Taiyuan Boiler Group on the basis of FSS design principle and flow pattern reconstruction theory. Owing to this, investment and operation cost have been reduced while realizing ultra-low emission. The CWS consist of 65% coal (dry basis), 34% water and 1% additives, or to be changed as required. The particle mass fraction of CWS is over 50%, size range from 100 to 300 μm, larger than the conventional atomized CWS as average size of 74 μm.

Y. Hao · R. Li · Y. Chen (✉)
Qingdao Clear Environmental Group Co., Ltd., Shandong 266071, China
e-mail: 820473950@qq.com

J. Lyu and S. Li (eds.), *Clean Coal and Sustainable Energy*, Environmental Science and Engineering,
https://doi.org/10.1007/978-981-16-1657-0_44

44.2 Core Technology of Boiler Design

The key point of this technology is when the efficiency of separator improves, fine ash increases in circulating material, the lower bed inventory can be reduced properly while still ensuring boiler normal operation. Material concentration in the secondary air area decreases and penetrating disturbance enhances optimizing the gas–solid mixing effects in the upper part of furnace, thus improve the boiler combustion efficiency. The demand for material fluidization reduces, dynamic pressure of primary air and secondary air decreases that lower the power consuming of fans. When inventory reduces, concentration in the lower part of furnace decreases, the wear in the lower concentrated phase of furnace especially at the junction of anti-wear layer and membrane wall recedes.

44.2.1 FSS Design Principle

In May 2005, Academician Guangxi YUE from the Department of Energy and Power Engineering of Tsinghua University introduced FSS design principle for the first time at the 18th CFB Combustion International Conference held in Canada. He put forward the concept of "fast bed + bubble bed", determining the random drift state in the design. FSS design theory assumes a certain state, that is, the material concentration distribution in the combustion chamber is quantitative at full capacity while heat transfer coefficient distribution along the bed height is also determined. The theoretical basis is that there is an artificial adjustable parameter in the CFB, namely the bed inventory, which can adjust the fluidization state back to the design state by adjusting the bed inventory when circulation quantity or material concentration drift.

Tsinghua University established a circulating bed flow pattern according to fluidization velocity and ash circulating flow rate, provided the guidance diagram for the selection of FSS design parameters (Fig. 44.1). This chart indicates the optional area of flow pattern in circulating bed boiler and the position of circulating bed combustion technology A-I domestic or abroad in the chart (Yue 2015).

For fast bed which lay in top of CFB is polymorphic at the same flow rate, a CFB state should be determined as the reference for full load calculation when designing the CFB. That is to say, the fluidization velocity and ash circulating flow rate under full load should be determined in quantities: the fluidization velocity of furnace is less than 5 m/s; the ash circulating flow rate in furnace is 2–2.5 kg/m^3 (standard state).

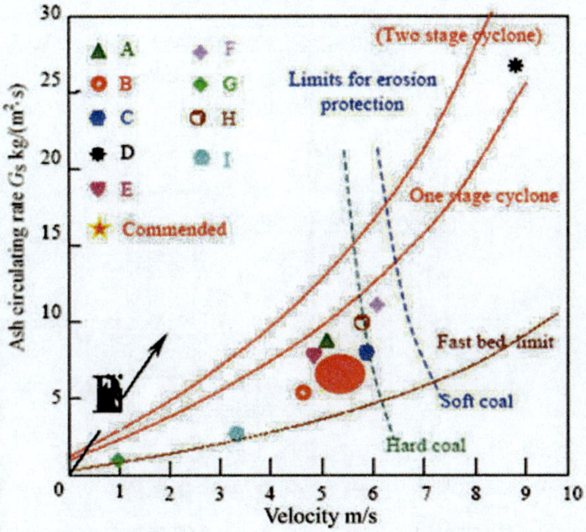

Fig. 44.1 Circulating bed pattern (Yue 2012)

44.2.2 Flow Pattern Reconstruction

Flow pattern reconstruction refers to maintaining fast bed fast fluidization status in the upper part of the furnace, meanwhile, maintaining the fluidization velocity and ash circulating flow rate. By optimizing the quality of bed inventory and changing the concentration distribution in the lower part of the furnace, the wear on the heating surface and energy consumption of the fans are reduced. The change of flow pattern in the lower part of furnace and its influence on combustion, heat transfer, mixing and pollutant generation is the key of the technology.

Flow pattern reconstruction makes the ratio of coarse and fine particles and the ratio of primary and secondary air in the furnace to be arranged and combined in the best state.

The core of flow pattern reconstruction is to improve the quality of bed material by fine particle circulation. The material circulation system consists of furnace, high temperature adiabatic cyclone separator, return valve, air supply system, should be capable of extremely high separation efficiency. Finer particle size of circulating material is the basis of improving combustion efficiency, thus we chose thinner bed inventory layer for operation. When separation efficiency improves, the particle size of fly ash reduce to $d_{50} = 10–12\ \mu m$, $d_{99} = 90–110\ \mu m$, comparing to that of traditional circulating fluidized bed $d_{50} = 20\ \mu m$ (Zhu and Yang 2004).

44.3 Furnace Structure and Design Key Points

This CWS-fired CFB boiler is a single drum, two pass type arrangement (Fig. 44.2).

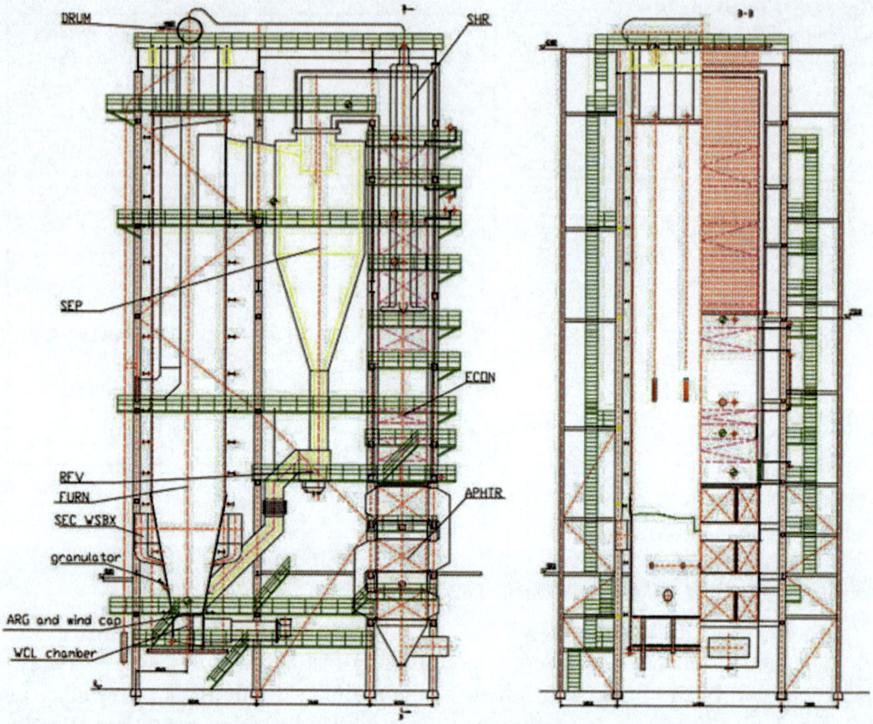

Fig. 44.2 Coal-water slurry boiler body structure diagram

The boil proper mainly consists of furnace, adiabatic cyclone separator, self-balance return valve and convection flue gas duct. Membrane walls are applied in the furnace, adiabatic separator locates in the middle of the boiler, super-heater, economizer and air pre-heater are arranged from top to bottom along the duct from top to bottom.

In order to realize the flow pattern reconstruction of material circulation, the design of key components is optimized in the aspect of boiler structure design.

44.3.1 Determine Reasonable Furnace Structure According to CWS

According to the principle of FSS theory, the fluidization velocity, the structure and layout of furnace are designed for different kinds of fuel. The designed fluidization velocity in furnace is less than 5.0 m/s. Furnace consists of whole sealing membrane walls, the cross section aspect ratio is about 2:1 rectangular structure ensuring uniform flow between gas and solid phases. The heating surface is about 15% larger than that of the conventional coal-fired CFB with the same capacity.

The height of the furnace is increased to ensure the residence time of material in the furnace to be over 4 s. Combustion efficiency is improved by the well mixing of materials. The pipes in the dense phase at the bottom of the membrane walls shrink with the vertical direction at a certain angle. The bottom is the water-cooled distributor.

44.3.2 Application of High Efficiency Adiabatic Separator

Large diameter cylinder is adopt in adiabatic separator for the low rising flue gas velocity. The acceleration section design of separator was optimized: the outlet structure of flue gas is adopted on both sides. Because the acceleration rate of gas is faster than that of solid, the longer inlet acceleration section ensures the uniform flow of gas–solid two phases. Separator inlet is inclined downward with a narrow width and sufficient height that ensure the flow rate of flue gas for the material to be concentrated and avoiding short circuit. Center cylinder bias, reducing the wear of the center cylinder.

44.3.3 Application of High Flow Rate and Low Energy Consumption Self-balance Loop Seal

Key design points of self-balance loop seal: roots blower is applied for air supply system to the return valve rather than common header system. It is the key to improve the separation efficiency of circulating material system to solve secondary elutriation problem of separator by optimizing the design of return valve. Return air amount, pressure and velocity of circulating material in the riser are determined by the minimum flow conditions, maximum return pressure conditions and return air non-channeling conditions respectively.

44.3.4 Application of Changeable Jacketed Bell-Type Air Nozzle

The blast cap consists of blast cap head, jacket and inner core (Fig. 44.3). Key design points: labyrinth structure is adopted by solving slag pour-out problem; Smaller volume and hole diameter are designed to reduce the wear of blast cap; Jacket system is designed to avoid wear of inner core; Typical slot is designed facilitating the change of blast cap; Wearing parts are jacket and blast cap while inner core has a long service life; Precision casting process ensures the uniform size of blast cap with lower primary air resistance.

Fig. 44.3 Structure of combined blast cap

44.3.5 Application of Front Air Chamber with Special Structure

Key design point: layout of front air chamber along the width of furnace; the design is similar to the isobaric air chamber of bubbling bed in order to reduce the depth of bed inventory layer; Adopt low resistance blast cap to uniform distribution of air.

44.3.6 Application of Reasonable Air Distribution Device

Design overall horizontal layout of water-cooled air distributor, precision cast bell type cap; The motor power of primary air fan is sharply reduced by adopting the thin running bed inventory layer and low resistance air distribution system.

44.3.7 Application of Low Resistance, Large Momentum with Powerful Penetrating Power Bellows Structure

Key design point: single-layer layout of secondary air, most of the primary air consumption is below 2.5 m height, secondary air can enhance the gas–solid mixing effect on the upper part of the furnace. Select the proper concentration of circulating materials in the furnace to lower the back pressure at the entrance of secondary air. Although the increase of circulating material can improve the separation efficiency of separator, high concentration affect the impenetrability of secondary air results in the drop of combustion efficiency.

44.4 Application of Reasonable Combustion Cycle and Gas/Air System

In the combustion cycle and flue gas/air system, CWS drops are put into the lower part of the combustion chamber by special granulator. The hot bed inventory consist of quartz sand with temperature range from 830 to 870 °C. CWS rapidly precipitates moisture, volatile and combusts with the heating of hot bed materials. In fluidization status, CWS disintegrated into fine particles which are taken out of dense phase zone by hot flue gas and burn in the suspension chamber (Jiang et al. 2006). Primary and secondary fans supply the air what boiler combustion needs. The air sent by primary fan is heated by primary air pre-heater then introduced into the water-cooled air chamber via wind tunnel and into the combustion chamber through blast cap on distribution plate. Air sent by secondary fan is heated by the secondary air pre-heater sprayed into the furnace by the nozzles distributed on the furnace wall to strengthen the disturbance and mixing. Fuel and air are mixed combustion in fluidized furnace and heat transfer between heating surfaces. Flue gas in the furnace (carrying large quantity of unburned carbon particles) continue to burn in the upper part of furnace. Flue gas with large amount of materials enters adiabatic cyclone separator through the outlet of furnace, most of the materials are separated and sent back to the furnace via return feeder. After separation, the flue gas travels through steering chamber, super-heater, economizer, primary and secondary air heater to flue gas treatment equipment.

44.5 Environmental Performance

44.5.1 Basic Reaction Principle and Correlation of Desulfurization in Furnace and NO_x

The desulfurization reaction: $S + O_2 \rightarrow SO_2$, $2SO_2 + 2CaO + O_2 \rightarrow 2CaSO_4$, requires oxygen;

NO_x generation: $2N + O_2 \rightarrow 2NO$, reducing atmosphere can reduce or avoid the generation of NO;

The circulation fluidized bed fuel is entered into the bottom of furnace while the oxygen distribution is well controlled. The lower part is reducing atmosphere to prevent the generation of NO while the upper part is oxidizing atmosphere to facilitate desulfurization.

44.5.2 High Desulfurization Efficiency in the Furnace

The high separation efficiency of circulating material system means the high separation efficiency of limestone injected for desulfurization. The surface of limestone powder is constantly worn and formed in the furnace to ensure high desulfurization efficiency.

Design temperature of furnace considering boiler combustion efficiency and environmental emission performance in conditions of bed temperature 830–870 °C (at maximum continuous evaporation rate). This temperature range is the best for limestone desulfurization (Zhu and Yang 2004).

44.5.3 Low NO_x Initial Emission

(1) The core of low NO_x original emission determined by several components below: higher separation efficiency of circulating material, finer the particle size of circulating material, better quality of bed and denser of lower part of furnace with reducing atmosphere. Because of the high separation efficiency of circulating material system, circulating quantity is guaranteed that boiler can operate at a low oxygen atmosphere.

(2) Determine the ratio of primary and secondary air according to the quality of coal. It is better to enlarge the proportion of secondary air, increase the height of the secondary air-distribution plate in order to delay its entrance, strengthening the high reducing atmosphere in the bottom of the furnace.

(3) Design temperature of furnace reducing the generation of NO_x. Original emission of NO_x is less than 50 mg/m^3(at maximum continuous evaporation rate).

(4) CWS boiler has advantages in low NO_x original emission performance. Operation data proofs part of CWS burn in the hypoxia status that can generate CO and H_2 in the lower part of the furnace that enhance the reducing atmosphere helps to suppress NO_x production.

44.5.4 Inhibition of PM2.5 Emission

The cluster effect occurs of fine particles when adding additives to CWS; Cluster of fine particles is easier to be removed by dust collector that effectively suppressed the generation and emission of $PM_{2.5}$ fine particles.

Additionally, CWS is transported into the ash storage by air after combustion. Ash has good activity and high comprehensive utilization value.

44.6 Advantages of the Boiler

Achievement: 90% higher boiler efficiency and saving 20–30% fan energy; clean and eco-friendly, high desulfurization efficiency in furnace, the original NO_x emission <50 mg/m^3; save 30% of environmental protection investment and 80% of operating costs; safe and reliable, with almost zero wear in the heating surface. The continuous running time of the boiler reaches over 6500 h; CWS is non-flammable and non-explosive fluid fuel with high safety in transportation, storage and usage.

The advantages of the boiler are obvious: low combustion temperature 830–870 °C, it operates stable and seldom coke; The cycle combustion of CWS is realized with high combustion efficiency; High efficiency adiabatic cyclone separator and returning-material device are adopted that improves the utilization rate of material hugely reducing the temperature of bed; Low temperature combustion is assisted by spraying limestone powder in the furnace makes desulfurization during combustion also; Low temperature combustion and optimized ratio of primary and secondary air inhibit NO_x formation. Boiler can stably operate under 30–100% load with typical coal; Low pressure bed is adopted with lower primary air pressure compare to coal-fired circulating fluidized bed boiler, which greatly reduces the power consumption; The boiler has better adaptability to CWS fuel which reduces the requirement of CWS quality and fuel cost; High degree of automation, safety and stable; The process of fuel in the furnace is totally enclosed which has a clean environment; CWS boiler is less land occupation.

44.7 Applications

Qingdao Clear Environmental Group Co., Ltd., Tsinghua University and Taiyuan Boiler Group jointly developed and promoted such new type of CWS-fired CFB. There are 34 sets of the orders till now about a total boiler output 3125 tons/h. They are applied to district heating or CHP in industrial park. Among them, 25 sets have been in service, about a total output of 2080 tons/h. References are in Shandong Jinan, Qingdao, Dongying, and in Shanxi Heshun etc.

44.7.1 Awards

In 2017, the new type of CWS-fired CFB with high efficiency and clean combustion boiler technology was awarded the major energy-saving achievement of Shandong Province [Shandong Government Office (2017) No. 169]; In 2017, confirmed by the Ministry of Science and Technology, this technology has reached the domestic advanced level; In Nov. 2018, the technology was listed by Ministry of Industry and Information Technology of China in "National Directory Of Industrial Energy-Saving Technology And Equipment" as a coal high efficiency and clean utilization technology; In Dec. 2018, the technology was selected as "Advanced Technology for Clean Production in Shandong Province" by Shandong Department of Industry and Information Technology [Shandong Department of Industry and Information Technology (2018) No. 78]. The technology is a major technological breakthrough in the field of clean coal and environmental protection in recent years. The application of the technology (central heating, combined heat and power generation) has become the important development direction of China.

44.7.2 References

44.7.2.1 2 × 70 MW New CWS Hot Water Boiler in Lingxiucheng in Jinan (Kang et al. 2018)

(1) Basic information of the project

Jinan Lingxiucheng project locates in the south of Jinan second-ring South Road. Till 2017, the heating area is about 3.5 million square meters which is expected to reach 6.34 million square meters in 2020. Jinan Thermal Power Co., Ltd. Lingxiucheng is responsible for supplying heat during winter. Phase I 2 × 58 MW coal-fired chain-type hot water boilers were built in 2008 and 2012 respectively; In the beginning of 2016, 2 × 70 MW new CWS-fired boilers started construction and were put into production on Dec. 23, 2016, replacing the two former chain boilers as heat sources.

(2) Project operation

The project is a district heating project with a heating period of 120 days in a year. The boilers run continuously and the indoor temperature can reach above 20 °C. Boiler performance test shows that the boiler thermal efficiency is over 90%. New CWS fuel is transported by closed tank lorries with zero dust spilling which greatly improved the environment of the heat supply plant (Fig. 44.4).

(3) Environmental analysis

Compared with the original coal fired chain boiler, 7530 tons of standard coal is saved per furnace in each heating season, Reduce emissions by 20,418 tons of CO_2, 636.9 tons of SO_2 and 311.3 tons of NO_x.

(4) Economic analysis

The operating cost (Last period of heating season data).

Consumption of CWS: 68,464 tons; Consumption of electricity: 7.02 million kwh; Consumption of water: 94372 m^3; Unit heating area equivalents to standard coal amount: 12.27 kgce/m^2.

Fig. 44.4 Comparison of heat plant environment before and after the retrofit

44.7.2.2 Haiyang Longfeng Thermal Power Co., Ltd. Retrofit Project

Haiyang Longfeng locates in the east of Haiyang economic and technological development zone, in the center of Haiyang city, Fengcheng industrial park and Bicheng industrial park. It covers an area of 80,000 m^2, with total investment about 474 million yuan, officially put into operation in Oct. 2004.

There are three 75/5.29-M5 coal-fired CFB boilers before reconstruction (rated pressure: 5.29 MPa, steam temperature: 450 °C), 2 × 15 MW extraction condensing turbo-generator unit, ESP bag dust removal + limestone-gypsum method + wet electric dust removal as gas treatment. The fuel is low LHV coal and gangue mined in Longkuang coal mine (The annual average LHV is about 3000 kcal/kg). Two boilers are operated normally, steam is supplied to the enterprises in industrial park and their cement production. It can meet 700,000 square meters heating demand in winter. Because the efficiency of original boiler is only 83% with high energy consumption, low operation efficiency, dust and nitrogen oxide emissions substandard. In 2017, for improving quality and increasing efficiency, the company decided to retrofit the #1 boiler into a CWS-fired boiler with high efficiency.

Project retrofit plan: Supporting construction of CWS system with annual output of about 300,000 tons/year. Part of the former boiler is retrofitted into a CWS boiler. The main tasks are retrofitting the separator, loop seal, secondary air system, fluidized bed system, water-cooled surface in dense and dilute phase area, slurry feeding granulator in front of furnace, bed inventory supplementary system, castable and thermal insulation materials. Other heating surfaces and structures are maintained, in order to make the technique target meet the new CWS boiler standard, and improve the boiler thermal efficiency.

After retrofitting, the innovation of new CWS boiler is mainly reflected in: (1) Boiler thermal efficiency increased by more than 7% (from 83% to above 90%) that save 6000 tons of coal per year. (2) The original emission concentration of NO_x is 46.65 mg/Nm^3 ($O_2 = 6\%$). (3) Compared with the former boiler, the fan can save about 30% electricity.

Haiyang Longfeng retrofit was based on EMC model that shares the benefits in 5 years for the energy saving, which builds a good example for the conventional CFB boiler to be retrofitted to CWS-fired boiler.

44.8 Development Prospects

Nowadays, coal-fired industrial boilers dominated in many parts of China. The utilization of scattered coal (transportation, storage, combustion, emission technologies) is out of date which bring about pollution to the environment. The new CWS-fired boiler technology is regarded as a clean combustion technology to replace coal burning which can solve scattering problem during transportation, meanwhile, make energy conservation and emission reduction to improve our environment.

Sustainable development strategy of China provides the great opportunity for CWS-fired boiler technology popularization. New CWS-fired boiler technology will play an important role in clean energy supply in cities, towns and new rural areas, make contributions to the new and old kinetic energy conversion and national air pollution control.

References

Jiang X, Ma Y et al (2006) Coal water slurry fluidization and suspension high efficiency and clean combustion technology research and application. Chem Eng 34(1):62–65

Kang H, Li R et al (2018) New type of coal water slurry-fired CFB district heating references, problems and suggestions. Wechat «Coal to Clean Coal»

Yue G (2012) The energy-saving CFB boiler technology based on flow pattern reconstruction. Tsinghua University

Yue G (2015) Development review on CFB combustion technology and outlook of industrial boiler ultra-low emission technology. Tsinghua University

Zhu G, Yang Xu (2004) CFB boiler design and calculation. Tsinghua University Press, Beijing

Chapter 45
Study on Pollutants Control of Circulating Fluidized Bed Boiler Burning Low Calorific Value Coal

Wanzhu Wu, Ruixin Li, Hu Wang, and Congyang Gu

Abstract In recent years, more and more low calorific value coal such as slime and gangue are used in circulating fluidized bed (CFB) boilers. Under this condition, it is difficult to control pollutants such as SO_2 and NO_X due to the change of ash content and particle size of coal. Results implied that when low calorific value coal was burnt in a CFB boiler, the change of circulating ash quantity and particle size distribution of CFB boiler cause the fluctuation of furnace bed pressure, as well as the returning blockage of loop seal, and the deterioration of combustion conditions inside furnace leads to a large fluctuation of pollutants emission. The fluctuation of furnace bed pressure is reduced by adjusting the relationship between air and coal volume and introducing bed pressure to modify the control strategy of primary air flow on both sides of the furnace. Effective measures are put forward to deal with the returning blockage of loop seal in the CFB boiler. The influence of operating parameters such as oxygen on pollutants emission and the coupling relationship between SO_2 and NO_X emission are both studied through experiments. On this basis, the optimization proposal of control strategy of limestone adding system is put forward, which is applied in a 300 MW CFB boiler, and good results are achieved. The results can provide a reference for controlling pollutants of CFB units burning low calorific value coal.

Keywords CFB boiler · Low calorific value coal · Pollutants control

W. Wu (✉) · H. Wang · C. Gu
CHN Energy CFB Research and Development Center, Xi'an 710065, China
e-mail: wuwanzhu@126.com

R. Li
CHN Energy Guoneng Group Co., Ltd., Beijing 100033, China

J. Lyu and S. Li (eds.), *Clean Coal and Sustainable Energy*,
Environmental Science and Engineering,
https://doi.org/10.1007/978-981-16-1657-0_45

45.1 Introduction

Coal washing and processing and utilization according to its quality are the primary way of clean and efficient use of coal (Liu et al. 2018). Coal washing and processing will produce a lot of low-calorific coal by-products, such as coal slime and gangue (Li 2018; Zhao 2011). By the year 2020, the selection rate of raw coal in China will be over 80%, and the output of low calorific value coal such as slime and gangue will be greatly increased. It has been proved that circulating fluidized bed (CFB) power generation technology is the best way for large-scale clean utilization of low calorific value coal (Yue et al. 2016).

With the increase of the national requirements on the emissions of thermal power units, the emission reduction for the CFB unit is becoming more and more serious (Xin 2017; Duan 2017; Ye 2018). In China government requires the emissions of the newly built coal-fired units to approach or basically reach the emission limit of the gas turbine, that means, the emissions of the dust, SO_2 and NO_X are less than 10, 35 and 50 mg/Nm^3 respectively (Zheng and Zhou 2017).

Due to the change of coal market, in recent years, the power plants buy more low-calorific coal as boiler fuel or mix the slime and gangue with the coal to improve the operation economy of the power plants (Wu 2017). However, when some CFB power plants use low calorific value coal, because of the change of ash content and particle size of coal, the control of SO_2, NO_X and other pollutants is faced with the problems of large fluctuation and high instantaneous concentration. It is difficult to meet the demand of ultra-low emission, and some units even run down the load of units due to the difficulty in controlling environmental parameters.

45.2 Effect of Burning Low Calorific Value Coal on Pollutant Emission of CFB Boilers

The SO_2 emission of CFB power plant is generally controlled by adding limestone and control the mole ratio of calcium and sulfur (Wang 2018; Gao et al. 2014). The effect of desulfurization reaction is also greatly affected by the bed temperature, operating oxygen, upper differential pressure and bed temperature uniformity of the boiler (Hong 2011). Under the condition of burning low calorific value coal, because of more ash and moisture, the coal particle size also changes, which results in the change of combustion conditions of CFB units and the increase of circulating ash quantity of CFB boilers. The rise of upper differential pressure and bed pressure leads to the lateral fluctuation of bed pressure on both sides of the furnace and the vibration of loop seal. The combustion and fluidization of CFB furnace are deteriorated. The parameters such as bed temperature, bed pressure and oxygen content deviate from the original working conditions, resulting in the reduction of limestone desulfurization reaction efficiency in CFB boiler, which leads to the problems of high SO_2 emission, large fluctuation range and high Ca/S ratio in the furnace.

The NO_X emission of CFB boiler is mainly affected by the coal quality, combustion temperature, combustion uniformity and the strength of reduction atmosphere in the furnace (Li et al. 2013). The results show that the special reductive atmosphere caused by gas–solid two-phase flow at low temperature is the main reason for the lower NO_X emission of CFB boiler (Cai et al. 2016). By reasonably controlling the bed temperature of CFB units, adjusting the ratio of primary and secondary air, maintaining sufficient circulating ash, and maintaining the appropriate combustion temperature and reducing atmosphere in the furnace, most CFB units can control the original NO_X emission concentration below 200 mg/Nm^3 (Liu et al. 2014). However, in the case of a larger proportion of low calorific value coal such as slime and gangue, due to its low calorific value, high ash content and high moisture, the volatile matter of coal increases, the content of gangue increases and the amount of circulating ash decreases. This may lead to the increase of furnace bed temperature, the uneven distribution of temperature field and the destruction of the original gas–solid two-phase flow combustion conditions of CFB boilers, which all lead to the increase of NO_X emission.

In order to control the fluctuation of SO_2 emission, excessive addition of limestone often occurs in some CFB power plants, which further promotes the formation of NO_X in the catalytic furnace. At the same time, in order to inhibit the increase of NO_X emission, excessive reduction of unit operating oxygen often occurs, which leads to the reduction of limestone desulfurization reaction efficiency and the increase of SO_2 emission.

Generally speaking, the influencing factors and control methods of SO_2 and NO_X emission of CFB units are coupled with each other, which leads to many difficulties in pollutant control under the condition of burning low calorific value coal. In view of the above difficulties, the problems and solutions are analyzed based on the practical operation test experience.

45.3 Bed Pressure and Emission Fluctuation of Boilers Burning Low-Calorific-Value Coal

To study coal characteristics on lateral fluctuation of bed pressure inside furnace, operational parameters were analyzed in a 300 MW CFB boiler. Table 45.1 shows coal characteristics of the boiler. It can be seen that when the boiler burnt different coal types, low-calorific-value coal had lower calorific value and higher ash content than that of the common coal, and low calorific value of the coal could be reduced to 13.84 MJ/kg. Meanwhile, the boiler burnt low-calorific-value coal, fraction of fine particles inside furnace after crushing and screening increased obviously, and coal particle share below 1 mm increased from 16.46 to 49.96%.

The combustion of low calorific value coal causes the increase of ash concentration, as well as bed pressure and upper differential pressure inside furnace, and it results in a left–right-side pressure imbalance and large fluctuations in the boiler.

Table 45.1 Coal characteristics of the boiler

Common coal		Low-calorific-value coal	
Item	Value	Item	Value
M_{ar}	2.80	M_{ar}	4.80
A_{ar}	41.02	A_{ar}	47.17
FC_{ad}	43.7	FC_{ad}	33.03
V_{daf}	24.04	V_{daf}	33.58
$Q_{ar,net}$	17.58	$Q_{ar,net}$	13.84
$S_{t,d}$	0.56	$S_{t,d}$	0.51
H_{ad}	2.49	H_{ad}	2.39

Note weight unit wt%; $Q_{ar,net}$ unit MJ/kg

The fluctuation of the material inside furnace leads to the instability of combustion, the fluctuation of boiler operation parameters and the increase of pollutant emission concentration, which threatens the safety and environmental stability of the CFB units.

45.3.1 Adjusting the Function Relationship Between Air and Coal Volume

It is considered that with the increase of bed pressure, the resistance of the distributor will decreases relatively, and the uneven distribution of air in the furnace becomes more prone (Wu et al. 2014). The resistance of distributor varies in the same direction with primary air volume. During the boiler load reduces, the resistance of air distributor decreases rapidly with the primary air flow rapidly falling off, and then the mismatch of primary air pressure, bed pressure and distributor resistance occurs in the boiler, which results in a large fluctuation of bed pressure on both the left and right sides, as well as the fluctuation of bed temperature and pollutant emission.

Through modifying the slope of the function curve of coal and air volume, the change of air volume is relatively slow in the stage of load reduction, which has a better effect on avoiding the fluctuation of bed pressure under the condition of burning low calorific value coal. After that, the fluctuation of bed pressure on both the left side (side A) and right side (side B) of the furnace has been significantly improved when the boiler load changes, shown in Fig. 45.1.

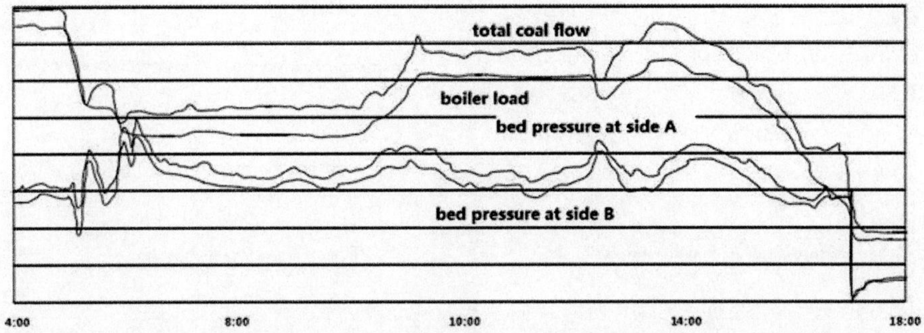

Fig. 45.1 Bed pressure profile after modifying function of coal and air volume

45.3.2 Modification of Primary Air Volume by Bed Pressure Function

The air-bed system of CFB boilers includes fan, fluidized air, bed material and all related pipelines and resistance components. The instability of gas–solid flow will produce one-dimensional pressure wave propagating along the direction of air flow (Hu et al. 2013). Through the analysis of the automatic control logic strategy of primary air, it is found that the feed-forward control of bed pressure deviation is not introduced in primary air volume regulation, and the response rate of primary air volume deviation control is slow. The fluctuation of bed pressure caused by the above reasons is hard to eliminate, thus the combustion conditions continue to fluctuate, and the SO_2 and NO_X emissions are difficult to control. Therefore, a scheme to adjust the deviation of primary air volume on both the left and right sides by increasing the bed pressure deviation is proposed. At the same time, the integral time of the primary air volume control PID module in DCS system is reduced and the coeffcients of proportionality are increased, so as to improve the response speed and variation range of the primary air volume.

After the modification is completed, the artificial disturbance test is carried out. Pressure difference between the left and right sides of the furnace is about 3 kPa caused by manual operation, and pressure fluctuation of the left and right sides of the primary wind plenum and the bed pressure are observed. It shows that the primary air volume and pressure of primary wind plenum at both sides of the furnace tend to be stable and convergent after adjustment, as shown in Fig. 45.2. By optimizing the logical parameters of air flow, the bed pressure can adjust the feed-forward air flow, and the response rate of air flow is faster. The cross-fluctuation of primary air and bed pressure is effectively improved. The combustion condition inside furnace is more stable, and the fluctuation of SO_2 and NO_X emissions are significantly reduced.

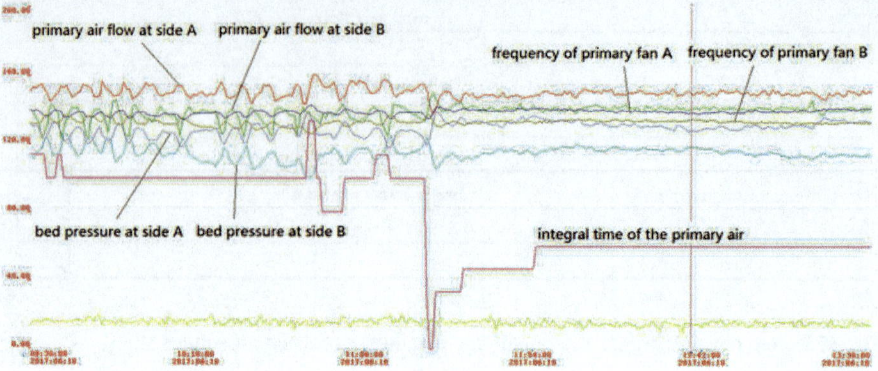

Fig. 45.2 Comparison of parameters before and after optimization of air volume control strategy

45.3.3 Circulating Ash Flow Blockage in a Loop Seal

When CFB boiler burns low calorific value coal, the phenomenon of circulating ash flow blockage in loop seal occurs from time to time (Guo 2016; Wang et al. 2015). This is due to the continuous improvement of cyclone separator efficiency and the change of coal quality, the particle size of circulating ash in standpipe is getting finer and finer, and the viscosity of materials increases. Under the fluidization of standpipe aeration air, it is easier to generate slugging phenomenon. With the increase of ash content of coal, external circulating ash flow of CFB boiler exceeds the maximum flow rate of loop seal, which is also one of the reasons for the circulating ash flow blockage in standpipe. To solve these problems, the following improvements are proposed.

(1) Put fluidized air pressure and air volume into automatic control in order to ensure the stability of air pressure and air volume in the loop seal. The fluidized air system of CFB units in service mostly shares common main pipe (Xing and Liu 2017). Some power plants are not put into automatic control of fluidized air pressure and flow rate of fluidized air branch pipe. When burning low calorific value coal, the increase of circulating ash will cause the pressure fluctuation of the standpipe, and the fluctuation of the overall air pressure and air volume of the fluidized air system will be further caused, leading to the flow blockage in loop seal.

(2) Change the ratio of returning air to aeration air. Due to the change of coal quality, the agglomeration characteristics of fine particles are enhanced, and the fine particles of bed material accumulate in the standpipe, resulting in slugging. Excessive aeration air flow in standpipe will cause material layer to be blown through, and then the pressure balance between furnace and loop seal to be destroyed. There is too little aeration air flow in standpipe to change in the status of moving bed of materials, resulting in insufficient power for fine ash transportation and material flow blockage in loop seal. According to the

different coal quality, the optimum ratio of aeration air is 20–40% of returning air in a loop seal.

(3) Maintain the stability of furnace bed pressure. The furnace and loop seal of CFB boiler are parallel pressure systems. If the combustion condition fluctuates, bed pressure fluctuation in the furnace follows, and the back pressure at outlet of the loop seal fluctuates, which leads to the overall fluctuation of the loop seal system. Therefore, by modifying control functions and parameters of the primary air system, good combustion conditon, stable material circulation system and bed pressure inside furnace could be ensured to maintain the smooth flow in the loop seal.

45.4 Optimization of Pollutant Control Parameters for Burning Low-Calorific-Value Coal

Oxygen consumption is needed for limestone desulfurization reaction in CFB boiler, so there is a quantitative relationship between SO_2 emission concentration and oxygen content. On the other hand, the formation of NO_X has a strong relationship with the oxygen content of the boiler, and the addition of excessive limestone can promote the formation of NO_X (Liu and Gibbs 1998). Therefore, when burning low calorific value coal, the parameters of controlling pollutant emission concentration in CFB boilers are coupled with each other, SO_2 emission is difficult to predict efficiently (Lyngfelt and Leckner 1999),which makes it difficult to control pollutants. Through experiments, the appropriate operating parameters and the control strategy of limestone are optimized, which can effectively reduce the fluctuation of pollutant emission caused by burning low calorific value coal.

45.4.1 Experiments on the Coupling Relationship of Pollutant Emission

Test are carried out on a 300 MW subcritical CFB boiler, and the oxygen was adjusted under the stable boiler load. Operating parameters of the boiler and the emission of SO_2 and NO_X were obtained after the oxygen was changed, and the influence of oxygen on the pollutant emission of the CFB boiler was studied. At the same time, the coupling relationship between SO_2 and NO_X emission concentration is also studied and analyzed. During the test, the oxygen varied from 2.3 to 4.0%, and boiler load and coal flow were almost the same, as well as limestone flow.

The change of NO_X and SO_2 emission with oxygen is shown in Fig. 45.3. The test results show that with the increase of oxygen content, SO_2 emission decreases and NO_X emission increases exponentially. The analysis shows that there is an optimum range of operating oxygen in CFB boilers. In this range, SO_2 emission can be maintained at a low level, and good NO_X suppression effect can be achieved. Under the

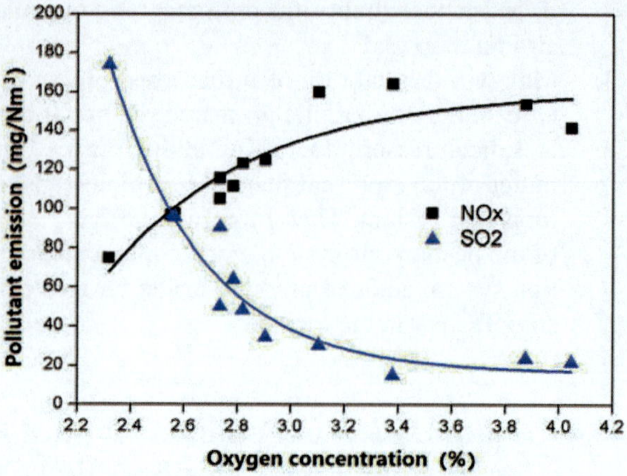

Fig. 45.3 Change of NO_X and SO_2 emission with oxygen

current operating conditions of the unit, the optimal pollutant control effect can be achieved by maintaining the oxygen content in the range of 2.8–3.2%. Other CFB boilers can adopt the same scheme and carry out tests to find out the optimal oxygen control range under different loads.

The coupling relationship between SO_2 and NO_X emission is also analyzed through experiments, as shown in Fig. 45.4. Under stable load conditions of 150 and 190 MW, there is a proportional coupling relationship between SO_2 and NO_X emission. Based on this coupling relationship, the influence of SO_2 emission concentration on NO_X emission can be quantitatively described by fitting the measured data of SO_2 and NO_X emission concentration in different CFB units during the period of load stability, which provides a reference for pollutant emission control strategy under the condition of burning low calorific value coal. The target SO_2 emission concentration is kept in a reasonable range, so that both SO_2 and NO_X emission concentration can reach an ideal value.

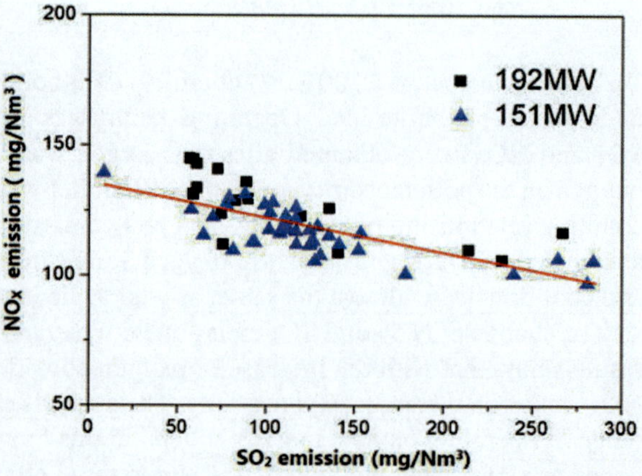

Fig. 45.4 Change of NO_X emission with SO_2 emission

45.4.2 Control Strategy Optimization of Limestone Feeding System

When the CFB boiler burning low calorific value coal, SO_2 emission at outlet of some CFB units are prone to short-term fluctuation and instantaneous over-standard due to the reduction of circulating ash and the rise of bed temperature. Based on the analysis of desulfurization control strategy of limestone adding system in the CFB boiler, optimization and improvement measures are put forward and applied in a 300 MW CFB boiler.

In a certain period of time, when SO_2 emission concentration is stable, the ratio of limestone addition to coal feed volume tend to a fixed value, which shows that molar ratio of calcium to sulfur of desulfurization system inside furnace is basically unchanged during this period of time. By using the above proportional relationship, the feed-forward of coal feed volume on limestone control is introduced. The limestone addition amount and coal feed amount in a period of time are integrated, and the proportional relationship is calculated. The results are introduced into the feed-forward of the limestone control system. It can increase the amount of limestone before the start of the increase of SO_2 while increasing the amount of coal in the loading stage, which can effectively control the rapid increase of SO_2 emission concentration in advance.

In order to restrain the rapid fluctuation of SO_2 emission concentration, the differential term of SO_2 change rate is added to the limestone control strategy, and the correction of SO_2 change rate to the amount of limestone addition is strengthened.

It can be seen from the experiment result shown in Fig. 45.3 that the operating oxygen has a great influence on SO_2 emission concentration. However, in the conventional limestone control strategy, the variation of operating oxygen is not taken into account. Therefore, it is proposed to quantify the relationship between the operating oxygen content and SO_2 concentration by using the results of the actual test, and to introduce the correction of the change of oxygen content in the limestone control strategy. When the oxygen content changes, the amount of limestone addition inside furnace can be changed in time to restrain the fluctuation of SO_2 emission.

Through the optimization of the limestone control strategy mentioned above, it has been applied to a 300 MW CFB boiler and achieved good results. The effective linkage among the amount of limestone addition, coal feed volume represented by power generation and SO_2 emission concentration is realized. After then, Both fluctuation range of SO_2 emission concentration and molar ratio of calcium to sulfur of desulfurization system inside furnace are reduced, and the hourly mean value of SO_2 emission is more stable, as shown in Fig. 45.5, Thus the environmental protection stability and economy of the unit are obviously improved.

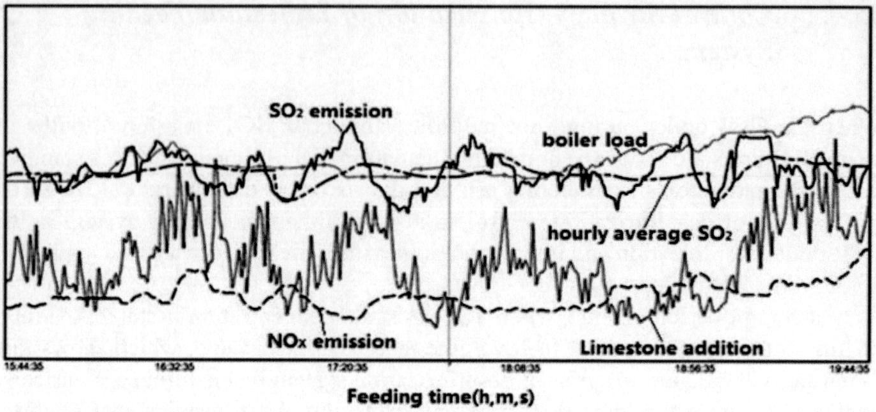

Fig. 45.5 SO_2 emission after optimization of control strategy

45.5 Conclusion

Under the condition of burning low calorific value coal in CFB power plants, the changes of ash content and particle size of coal inside furnace lead to the fluctuation of bed pressure and the returning blockage of loop seal, and then the deterioration of combustion conditions in the furnace and the difficulty of controlling pollutant emission happen.

By means of adjusting the function relationship between air supply volume and coal feed volume, which reduced the change rate of air volume during load reduction and introducing bed pressure to modify primary air volume on both sides of the furnace, the fluctuation of bed pressure of the boiler effectively reduce. Meanwhile, By automatic control of fluidization air pressure and flow rate, adjusting the ratio of returning air to aeration air inside loop seal and maintaining bed pressure stability inside furnace, the fluctuation of pollutant emissions caused by the returning blockage of loop seal could be effectively controlled.

Field test demonstrates the coupling relationship between SO_2 and NO_X emission inside furnace of CFB boilers, and the optimum operation range of oxygen at outlet of the boiler was obtained. On this basis, lots of measures such as optimization of limestone addition control system, introducing feed forward control of function of coal feed and limestone volume, introducing oxygen fuction to modify limestone volume, and so on, have been applied to 300 MW CFB boilers and pollutant emissions of the boilers are effectively reduced, and the operation stability and economy of power plants are also improved.

Acknowledgements This work is financially supported by the National Key Research & Development Program of China (No. 2016YFB0600205).

References

Cai RX, Lu JF, Ling W (2016) Progress of supercritical and ultra-supercritical circulating fluidized bed boiler technology. Electr Power 49(12):1–7

Duan SB (2017) Technical route of air pollutants ultra-low emission modification for 300 MW CFB boilers. Clean Coal Technol 26(6):103–107

Gao MM, Yue GX, Lei XJ (2014) Research on limestone control of circulating fluidized bed boiler. J Chin Soc Power Eng 34(10):760–777

Guo ZT (2016) Analysis on measures to prevent recycle interruption of circulating fluidized bed boiler. Ind Boil 2:49–51

Hong FM (2011) Desulfurization experiment with limestone in CFB boiler burning Fujian anthracite. Gas Heat 31(7):4–9

Hu N, Li JJ, Liu XM (2013) Mechanism study of lateral bed pressure wave of large scale CFB boilers. Proc CSEE 33(20):1–7

Li MH (2018) A review on 60 years of coal preparation in China. Coal Eng 46(10):24–29

Li JJ, Yang HR, Lu JF (2013) Low NO_X emission characteristic of low energy consumption CFB boilers. J Combust Sci Technol 19(4):293–298

Liu H, Gibbs BM (1998) The influence of limestone addition at different positions on gaseous emissions from a coal-fired circulating fluidized bed combustor. Fuel 77(14):1569–1577

Liu RZ, Fan ZG, Liu PY (2018) Discussion on efficient and clean utilization mode of coal resources. China Resour Compr Util 36(12):61–63

Liu ZQ, Ma H, Zhang GL (2014) Integrated technology of desulfurization and denitration in CFB boilers. Therm Power Gener 43(5):121–127

Lyngfelt A, Leckner B (1999) Sulphur capture in circulating fluidized-bed boilers: can the effciency be predicted. Chem Eng Sci 54(11):5573–5584

Wang XM (2018) Technology about dry desulfurization and semi-dry desulfurization. Electr Power Technol Environ Prot 34(1):45–48

Wang YZ, Qi YJ, Li J (2015) Analysis and preventive adjustment of recycle material abnormal in CFB boiler. Power Syst Eng 33(3):33–34

Wu WZ (2017) Technical economic study on type selection of slag cooler for a 660 MW ultra-supercritical CFB boiler. Clean Coal Technol 23(2):103–107

Wu HH, Sun JN, Wang Z (2014) Impacts of bed inventory and air distributor resistance on bed overturn in circulating fluidized bed with pant-legs. J Chin Soc Power Eng 34(1):7–18

Xin SW (2017) Study on modification of SO_2 ultra-low emission in CFB boiler. Electr Power Technol Environ Prot 33(4):10–13

Xing XF, Liu M, Li KY (2017) Study on fan operation mode of circulating fluidized bed boiler material return system. Shanxi Elect Power 205(4):60–62

Ye XL (2018) Flow field optimization study of SNCR+SCR combined denitrification on CFB boiler. Electr Power Technol Environ Prot 33(3):15–18

Yue GX, Lv JF, Xu P (2016) The up-to-date development and future of circulating fluidized bed combustion technology. Electr Power 49(1):1–13

Zhao ZG (2011) Present development status of coal preparation technology and its influence on coal clean utilization in China. Sci Pap Online 6(3):215–219

Zheng TT, Zhou YG, Jin QY (2017) Integrated removal and ultra-low emission of multiple pollutants for coal-fired power plants. Therm Power Gener 46(4):11–15

Part V
Low Carbon Energy

Chapter 46
CO_2 Adsorption Performance of Na/K-Impregnated MgO

Jie Zou, Ning Ding, and Cong Luo

Abstract As a carbon dioxide solid sorbents, magnesium oxide has been widely studied due to its wide distribution and low price. But pure MgO under intermediate temperature (250–500 °C) showed a poor CO_2 capture capacity. In our study, an effective method for improving the cyclic adsorption stability is to prepare the MgO-based sorbents promoted by sodium/potassium nitrate and citric acid through the mixing-impregnation method. The sorbents were determined by powder X-ray diffraction analyzer (XRD), thermogravimetric analyzer (TGA), and N_2 adsorption and desorption were measured at 77 K. The morphology of the sorbents was investigated by scanning electron microscopy (SEM). The results showed that as-prepared composites promoted by sodium nitrate and citric acid exhibited a better adsorption capacity than those modified by potassium nitrate and citric acid, which was 3.9158 mmol/g during the cyclic CO_2 adsorption reaction of calcination temperature of 450 °C and carbonation temperature of 350 °C. After 30 cycles of adsorption and desorption, the capture capacity of this sorbents was stable at 3.769 mmol/g. The SEM indicated that the pore morphology of these modified sorbents becomes abundant. Overall, the as-prepared Na/K impregnated MgO sorbent was promising for cyclic intermediate temperature CO_2 capture.

Keywords CO_2 capture · Intermediate temperature · MgO sorbents · Citric acid · Mixing-impregnation method

---

J. Zou · C. Luo (✉)
China-EU Institute for Clean and Renewable Energy, Huazhong University of Science and Technology, Wuhan 430074, China
e-mail: cluo@hust.edu.cn

State Key Laboratory of Coal Combustion, Huazhong University of Science and Technology, Wuhan 430074, China

N. Ding
Hebei Ji-Yan Energy Science and Technology Research Institute CO. LTD., 9 Yuhua West Road, Shijiazhuang 050051, China

J. Lyu and S. Li (eds.), *Clean Coal and Sustainable Energy*, Environmental Science and Engineering,
https://doi.org/10.1007/978-981-16-1657-0_46

46.1 Introduction

Since the beginning of the industrial revolution, people have been using fossil fuels more and more frequently, resulting in a large number of greenhouse gas emissions (Intergovernmental Panel on Climate Change 2005). Moreover, account for about 75% of CO_2 emissions (Ochoa-Fernández et al. 2009). GHGs emissions have led to global warming and a series of extreme climate changes. Although renewable energy has been developed for the current stage, the primary source of energy supply is still fossil fuels. The inevitable GHGs emissions have become a significant challenge for the world.

Carbon capture and storage (CCS) technology is the most attractive and potential way to mitigate the greenhouse effect. The targets of CCS are significant sources of greenhouse gas emissions, such as fossil fuel power plants. Currently, different substances for CO_2 capture have been studied, such as lithium silicate (Amorim et al. 2016), magnesium oxide (Lee et al. 2010), sodium or potassium bicarbonate, calcium oxide (Armutlulu et al. 2017), ionic liquids (Li et al. 2017). Metal oxide-based sorbents have shown great interest in effective sorption of CO_2 from synthesis gas and flue gas because they can be applied at ambient, intermediate, and elevated temperatures and pressures (Vu et al. 2014a). MgO-based CO_2 sorbents have been widely studied for its sizeable CO_2 adsorption capacity (24.8 mmol/g in theory) at intermediate temperatures (Hassanzadeh and Abbasian 2010; Lee and Park 2015; Xiao et al. 2011). Generally, one mole of MgO can absorb one mole of CO_2 per stoichiometry. However, the actual adsorption capacity of natural MgO is deficient (<1 mmol/g) for its poor CO_2 capture kinetics (Yang et al. 2013).

MgO-based sorbents that are doping and modifying with porous supporters or promoters have an excellent potential for CO_2 adsorption. Mayorga et al. first suggested MgO-based with double-salt promoted containing magnesium oxide and alkali metal carbonates and nitrates, which express excellent adsorption capacity for CO_2 capture at intermediate temperatures, with a highest regenerable capacity of 11 mmol/g demonstrated using pressure swing regeneration at 375 °C (Mayorga et al. 2001).

In recent years, researchers have studied magnesium oxide modified by nitrate and found that it can improve the adsorption performance. MgO-based sorbents modified by $NaNO_3$ and KNO_3 were prepared by coprecipitation to have a CO_2 uptake capacity of about 30 wt% (Lee et al. 2014a), and prepared by $LiNO_3$ with aerogel method showed a capacity of 5.14 wt% (Vu et al. 2014b). The problem is that the MgO-based sorbents promoted by nitrate appear poor cyclic stability. In the meantime, researchers began to study the performance of the sorbents after the addition of carbonate. The Na–Mg double salt, prepared by a precipitation method, showed unique CO_2 sorption characteristics in the temperature range of 300–500 °C and a remarkably high CO_2 sorption uptake of 3.48 mmol/g (15.3 wt%) was measured at 375 °C and was maintained during repeated sorption/desorption cycles (Lee et al. 2014b). In general, the way to increase the stability of cyclic adsorption of modified sorbents is still a hot topic.

Our research group proposed some effective way to enhance the stability of CO_2 sorbents and oxygen carriers (Ding et al. 2017, 2016; Luo et al. 2013, 2010; Xu et al. 2018a, b, c, d, e). In this paper, The modified MgO-based sorbents were prepared by the conventional impregnation method combined with the direct mixing method. Instead of synthesis, we use pure MgO mixed with nitrate solution. To make citric acid the leading variable in the experiment, We chose a low-concentration salt solution for impregnation. Furthermore, for comparison, the adsorption performance, MgO-based sorbents promoted by $NaNO_3$ and Na_2CO_3, were prepared to observe the adsorption capacity of modified sorbents with a nitrate solution a low mass fraction. Higher CO_2 adsorption capacity was measured at 350 °C with sodium salt modified sorbents by the impregnation method with the addition of citric acid. Moreover, composites modified by KNO_3 and citric acid also showed a favorable CO_2 capture. The pore morphology of sodium and potassium salt modified sorbents becomes abundant. In the cyclic CO_2 adsorption test by TGA, the sodium salt-modified sorbents' thermal stability was also improved. Overall, our preparation and further systematic evaluation of the sorbents demonstrated that modified MgO sorbents were promising for cyclic intermediate temperature CO_2 capture.

46.2 Experimental Section

Nitrate-modified MgO-based sorbents were prepared by the impregnation method. First, 8 g MgO and 2 g $NaNO_3$ were added to 100 ml deionized water and stirred at 50 °C for 1 h in an electromagnetic stirrer. Then, a 20 wt% citric acid solution was added by dropper. To compare the influence of different amounts of citric acid on the adsorption performance, add 25 ml (about 5 g citric acid) and 50 ml (10 g) citric acid solutions of different volumes into the stirred slurry, and then the slurry was stirred for 2 h. Finally, it was placed in the oven to dry at 110 °C for 10 h and calcined at 450 °C for 4 h in the muffle furnace. For MgO modified by potassium nitrate, 8 g MgO and 2 g KNO_3 were mixed with 100 ml deionized water and stirred for 1 h at 50 °C. Then 50 ml citric acid solution with a mass fraction of 20% was added. The mixture was further stirred for 2 h and dried for 10 h at 110 °C in the oven, calcined at 450 °C for 4 h. $NaNO_3$ modified MgO, and KNO_3 was prepared similarly without citric acid as a contrast sample. The powder of dried $NaNO_3$ and MgO was impregnated with 6 and 2 g Na_2CO_3, and the subsequent steps were consistent with obtaining modified MgO with both $NaNO_3$ and Na_2CO_3. The powder modified by citric acid and $NaNO_3$ prepared was directly mixed with 0.5 g $NaNO_3$ at 1.5 g, and then artificially mixed for 20 min to obtain citric acid and directly mixed with $NaNO_3$ modified MgO-based sorbents.

The cyclic adsorption experiment was carried out under a thermogravimetric analyzer (STA 2500 Regulus, NETZSCH). The sample was first pre-calcined at 450 °C for 60 min and carbonated at 350 °C for 60 min, then desorbed at 450 °C for 40 min. The calcination atmosphere during the process was 150 ml/min of N_2, and the condition of carbonation was under an atmosphere mixed 120 ml/min CO_2

with 30 ml/min N_2. The quality of all tested samples in the cyclic adsorption test was around 20 mg.

The crystalline parameters of as-prepared sorbents were investigated by powder X-ray diffraction analyzer (XRD, Empyrean, PANalytical B.V.) with Cu-Kα radiation with the 2θ range of 10°–80° at a scanning speed of 0.05°/s. The sorbents' microscopic morphology was determined with field emission scanning electron microscopy (FSEM, Sirion 200, FEIC) with 20 kV of accelerating voltage. The specific surface area and pore volume were calculated by the Brunauer–Emmett–Teller (BET) method. Nitrogen adsorption/desorption isotherms were measured at the boiling point of liquid nitrogen with an analyzer (ASAP 2020-M).

46.3 Results and Discussion

For the MgO-based sorbents prepared by impregnation without citric acid, Fig. 46.1a shows that the mass change in the 12 cycles was low, only 0.366% (adsorption capacity was 0.08325 mmol/g). When the 50 ml citric acid was added to prepare

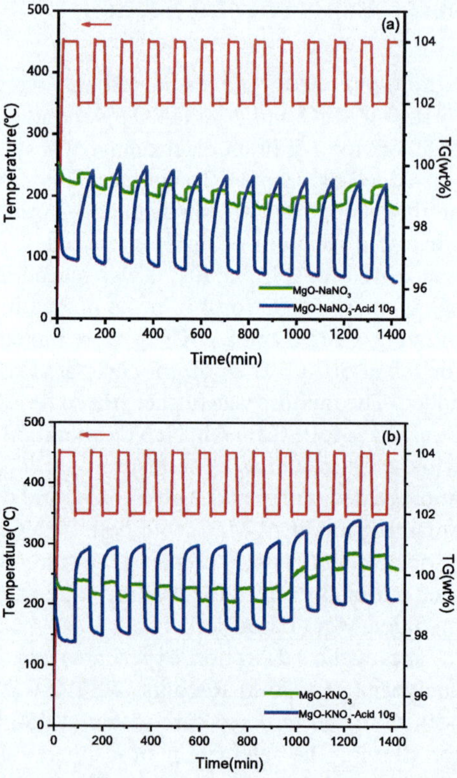

Fig. 46.1 Comparison of cyclic adsorption performance between **a** $NaNO_3$–MgO and citric acid (10 g)–$NaNO_3$–MgO, **b** KNO_3–MgO and citric acid (10 g)–KNO_3–MgO

the MgO-based sorbents, the average mass change was 2.971% (0.675 mmol/g). Although the overall uptake capacity was still low, for the ordinary sample that was impregnated by $NaNO_3$, the adsorption performance has improved by contrast.

For MgO-based sorbents modified by KNO_3, Fig. 46.1b presents that its performance in 12 cyclic adsorption experiments was similar to that of sorbents modified by $NaNO_3$, with an average mass change of 0.4032% (adsorption capacity of 0.09163 mmol/g). After the addition of citric acid, its adsorption curve improved, and the average mass change increased to 2.7674% (0.62896 mmol/g). For both $NaNO_3$ and KNO_3 modified MgO-based sorbents, the addition of citric acid tends to improve their adsorption performance.

The effect of the change of citric acid addition amount on the sorbents only discussed in the modified MgO-based sorbents prepared by $NaNO_3$, so the citric acid solution with a mass fraction of 20 wt% in different volumes of 25 ml (5 g) and 50 ml (10 g) was added in the preparation to make the difference. The data obtained in the cycling experiment showed that the average mass change of the 12 cycles was 3.537 and 2.971% for the addition of citric acid of 5 and 10 g, respectively. That is to say, the performance of modified magnesium oxide by $NaNO_3$ prepared with 5 g citric acid became better, as shown in the Fig. 46.2, although the increase was not much, it indicated the influence of citric acid on its performance, getting better.

Figure 46.3 (N_2 adsorption/desorption isotherms) shows that the sorbents impregnated by $NaNO_3$ and citric acid have a hysteresis loop at the relative pressure between 0.9 and 1, indicating the existence of ordered mesopores, The increase in adsorbed volume also shows the richness of its internal pore structure. It can be seen from Table 46.1 that not only the nitrate impregnation can improve the specific surface area of the sample, but also the surface area of the sample modified by citric acid is increased (about 48.3 m^2/g), and with the increase of citric acid content, it has an upward trend. However, the addition of citric acid harms the average pore size of the sorbents. With the citric acid adding, the BJH average pore size decreased from 22.51 to 16.9 nm with the pore volume increased from 0.1397 to 0.2648 cm^3/g,

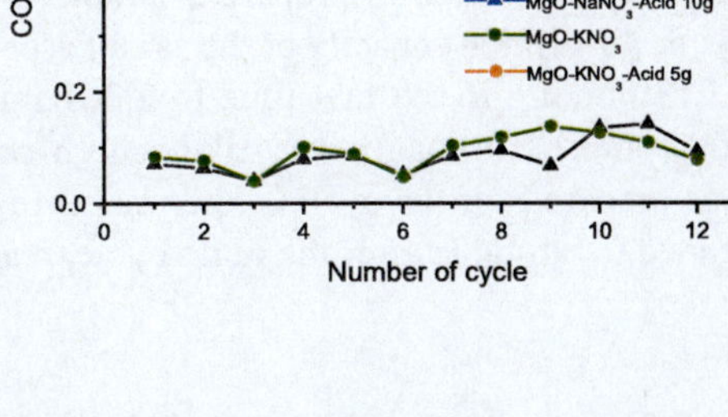

Fig. 46.2 Adsorption capacity of different impregnation composites

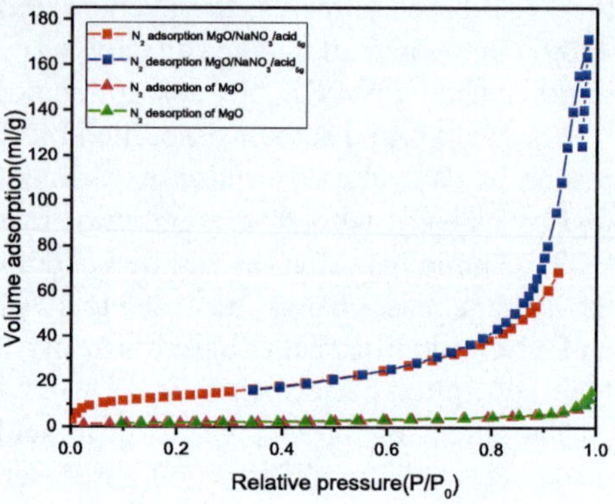

Fig. 46.3 N_2 adsorption–desorption isotherms for the sorbents samples and pure MgO

Table 46.1 Textural properties of the MgO-based sample

Sample	BET surface area (m^2/g)	Pore volume (cm^3/g)	BJH average pore size (nm)
Pure MgO	6.0380	0.0238	13.17
$MgO/NaNO_3$	18.7284	0.1397	22.51
$MgO/NaNO_3/acid_{5\ g}$	48.2842	0.2648	16.90
$MgO/NaNO_3/acid_{10\ g}$	62.5706	0.2494	11.53
MgO/KNO_3	19.7229	0.1457	21.68
$MgO/KNO_3/acid_{10\ g}$	28.9864	0.1884	18.76

indicating the crystals of the modified sample may have occupied the mesoporous pores, and the internal structure may have become abundant.

Citric acid is an organic acid with a melting point of 156 °C and a boiling point of 310 °C. During its thermal decomposition, carbon dioxide, carbon monoxide, water vapor, and anhydride ketones will produce. At high temperatures, these products will further be burned. Due to the gas's escape and combustion, the pore structure of the sorbents will change. The external surface morphology can be seen in the images obtained by FSEM. Figure 46.4 indicates as-prepared sorbents have a rough and porous appearance, which means the reaction of carbonation will become faster and better due to the gas-phase diffusion on the surface area.

Due to the low adsorption capacity of the samples prepared before, the samples were impregnated with $NaNO_3$ in the same method and content. And on this basis, Na_2CO_3 was added to prepare a sample for comparison. Figure 46.5 indicates the cyclic adsorption capacity of the samples obtained showed an increasing trend, from 0.918 mmol/g in the first time to 2.361 mmol/g in the last time. At this time, the adsorption performance was still not high, so a group of directly mixed sodium nitrate was prepared for the cyclic experiment. In the results, the adsorption performance tended to be stable after the ninth cycle, reaching 3.93 mmol/g.

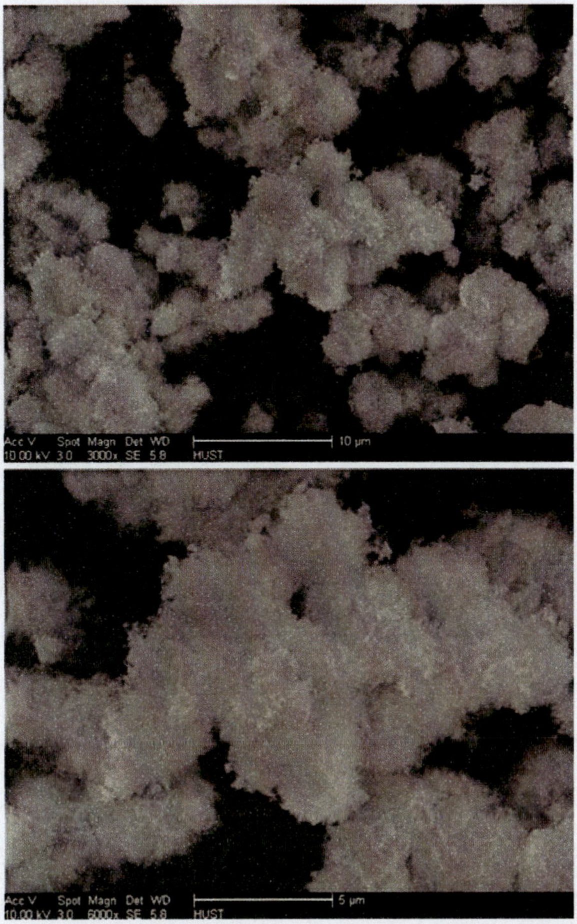

Fig. 46.4 FSEM photos of MgO–$NaNO_3$–Acid$_{5\ g}$

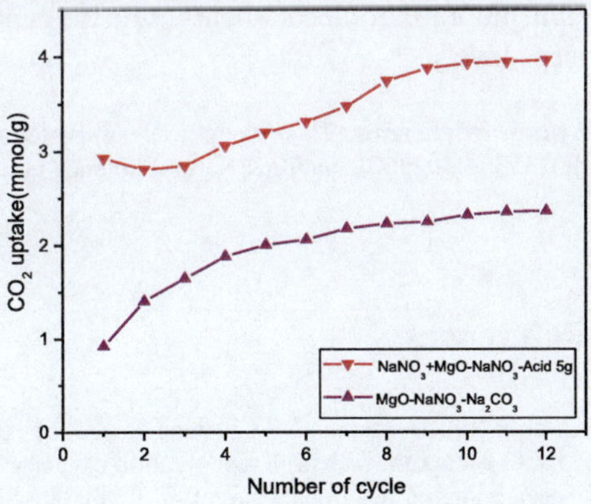

Fig. 46.5 Performance comparison between direct mixing $NaNO_3$ and MgO–$NaNO_3$–Na_2CO_3

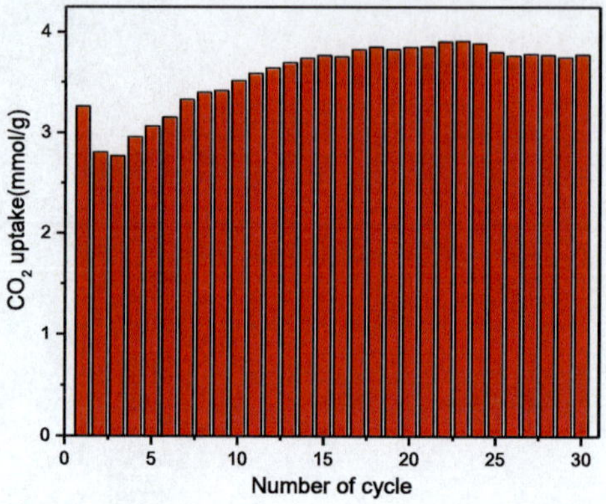

Fig. 46.6 CO_2 uptake capacity of direct mixing $NaNO_3$ with MgO–$NaNO_3$–acid 5 g

On account of the stable adsorption performance obtained by directly mixing sodium nitrate, the cycle number increases to 30 times, Fig. 46.6 shows after the first appearance of high adsorption capacity, the performance gradually declines. However, after the third time, the uptake capacity was on the rise, peaked at 23 times (3.9158 mmol/g), and decreased slightly. Finally, the capture capacity tended to be stable in the last six cycles (3.769 mmol/g).

46.4 Conclusion

The MgO-based sorbents, modified by the mixing-impregnation method using citric acid and sodium nitrate simultaneously, have a high CO_2 adsorption capacity (0.675 mmol/g without mixing $NaNO_3$ directly, 3.93 mmol/g with the $NaNO_3$ mixing), and can maintain excellent resistance for sintering. The sorbents promoted by potassium nitrate and citric acid also showed a better performance than the potassium nitrate-modified sorbents, with the capture capacity of 0.092 and 0.629 mmol/g, respectively.

Acknowledgements This work was supported by National Key R&D Program of China (2016YFE0102500), National Natural Science Foundation of China (No. 51606076)

References

Amorim SM, Domenico MD, Dantas TLP, José HJ, Moreira RFPM (2016) Lithium orthosilicate for CO_2 capture with high regeneration capacity: kinetic study and modeling of carbonation and decarbonation reactions. Chem Eng J 283:388–396

Armutlulu A, Naeem M, Liu HJ, Kim SM, Kierzkowska A, Fedorov A, Müller CR (2017) Multi-shelled CaO microspheres stabilized by atomic layer deposition of Al_2O_3 for enhanced CO_2 capture performance. Adv Mater 29(41):1702896

Ding H, Xu Y, Luo C, Zheng Y, Shen Q, Liu Z, Zhang L (2016) Synthesis and characteristics of BaSrCoFe-based perovskite as a functional material for chemical looping gasification of coal. Int J Hydrog Energy 41(48):22846–22855

Ding H, Xu Y, Luo C, Wang Q, Li S, Cai G, Zhang L, Zheng Y, Shen Q (2017) Oxygen desorption behavior of sol-gel derived perovskite-type oxides in a pressurized fixed bed reactor. Chem Eng J 323:340–346

Hassanzadeh A, Abbasian J (2010) Regenerable MgO-based sorbents for high-temperature CO_2 removal from syngas: 1. Sorbent development, evaluation, and reaction modeling. Fuel 89(6):1287–1297

Intergovernmental Panel on Climate Change (IPCC) (2005) IPCC special report on carbon dioxide capture and storage. http://www.ipcc.ch

Lee S-Y, Park S-L (2015) A review of solid adsorbents for carbon dioxide capture. J Ind Eng Chem 23:1–11

Lee JM, Min YJ, Lee KB, Jeon SG, Na JG, Ryu HJ (2010) Enhancement of CO_2 sorption uptake on hydrotalcite by impregnation with K_2CO_3. Langmuir 26(24):18788–18797

Lee HJ, Kim JH, Kim JW, Cho SJ (2014a) Structure transformation of Na-Mg based salts for CO_2 capture and storage at high temperature probed with variable temperature X-ray powder diffraction. Energy Procedia 63:253–265

Lee CH, Mun SY, Lee KB (2014b) Characteristics of Na–Mg double salt for high-temperature CO_2 sorption. Chem Eng J 258:367–373

Li X, Zhang L, Zhou D, Liu W, Zhu X, Xu Y, Zheng Y, Zheng C (2017) Elemental mercury capture from flue gas by a supported ionic liquid phase adsorbent. Energy Fuels 31(1):714–723

Luo C, Zheng Y, Ding N, Wu QL, Bian GA, Zheng CG (2010) Development and performance of CaO/La_2O_3 sorbents during calcium looping cycles for CO_2 capture. Ind Eng Chem Res 49(22):11778–11784

Luo C, Zheng Y, Yin JJ, Qin CL, Ding N, Zheng CG, Feng B (2013) Effect of sulfation during oxy-fuel calcination stage in calcium looping on CO_2 capture performance of CaO-based sorbents. Energy Fuels 27(2):1008–1014

Mayorga SG, Weigel SJ, Gaffney TR, Brzozowski JR (2001) Carbon dioxide adsorbents containing magnesium oxide suitable for use at high temperatures. US Patent

Ochoa Fernández E, Zhao T, Rønning M, Chen D (2009) Effects of steam addition on the properties of high temperature ceramic CO_2 acceptors. J Environ Eng 135(9):397–403

Vu A, Park Y, Jeon PR (2014a) Mesoporous MgO sorbent promoted with KNO_3 for CO_2 capture at intermediate temperatures. Chem Eng J 258:254–264

Vu AT, Park Y, Jeon PR, Lee CH (2014b) Mesoporous MgO sorbent promoted with KNO_3 for CO_2 capture at intermediate temperatures. Chem Eng J 258:254–264

Xiao G, Singh R, Chaffee A, Webley P (2011) Advanced adsorbents based on MgO and K_2CO_3 for capture of CO_2 at elevated temperatures. Int J Greenhouse Gas Control 5(4):634–639

Xu Y, Ding H, Luo C, Zheng Y, Zhang Q, Li X, Sun J, Zhang L (2018a) Potential synergy of chlorine and potassium and sodium elements in carbonation enhancement of CaO-based sorbents. ACS Sustain Chem Eng 6(9):11677–11684

Xu Y, Ding H, Luo C, Zheng Y, Li X, Xu Y, Zhang Z, Zhao W, Zhang L (2018b) Increasing porosity of molded calcium-based sorbents by glucose templating for cyclic CO_2 capture. Chem Eng Technol 41(5):956–963

Xu Y, Ding H, Luo C, Zhang Q, Zheng Y, Li X, Hu Y, Zhang L (2018c) NaBr-enhanced CaO-based sorbents with a macropore-stabilized microstructure for CO_2 capture. Energy Fuels 32(8):8571–8578

Xu Y, Ding H, Luo C, Zheng Y, Xu Y, Li X, Zhang Z, Shen C, Zhang L (2018) Porous spherical calcium-based sorbents prepared by a bamboo templating method for cyclic CO_2 capture. Fuel 219:94–102

Xu Y, Ding H, Luo C, Zheng Y, Xu Y, Li X, Zhang Z, Shen C, Zhang L (2018) Effect of lignin, cellulose and hemicellulose on calcium looping behavior of CaO-based sorbents derived from extrusion-spherization method. Chem Eng J 334:2520–2529

Yang X, Zhao L, Xiao L (2013) Effect of $NaNO_3$ on MgO–$CaCO_3$ absorbent for CO_2 capture at warm temperature. Energy Fuels 27(13):7645–7653

Chapter 47
Comprehensive Analyses on Activation Agents of Amines and Nanoparticles for TETA-Based CO_2 Capture Absorbents

Jiang Yanchi, Zhang Zhongxiao, Yu Juan, Zhao Ziqi, Fan Junjie, Li Boyang, Jia Mengchuan, and Mu Aiwei

Abstract The issue of climate warming and emissions to the core of global public focus. Chemical absorption is the most possible CCS technology for large-scale commercial application, in which the absorbents were the most critical factor to determine the CO_2 capture performance. In this study, TETA-based high-performance absorbents by adding activation agents of nanoparticles and other amines. In this study, TETA-based high-performance absorbents were prepared by adding activation agents with nanoparticles and amines. The ab/desorption performance was tested by a bubble reactor. The critical parameters, such as ab/desorption rate, equilibrium time and CO_2-loading, were comprehensively summarized and learned by AHP method. The results proves that Ab/desorption CO_2-loading occupies the most important part with the value of 30.77 and 24.36%, respectively. The sequences of amine activation agents is 2.5%PZ > 2.5%DEA > 2.5%MDEA > 2.5%AMP > 2.5%TEA and 5.0%PZ > 5.0%AMP > 5.0%DEA > 5.0%MDEA > 5.0%TEA. For amine activation agents, the addition of PZ shows the optimal overall CO_2 capture performance with the value of 0.593 and 0.580. The activation agent of DEA also shows impressive overall CO_2 capture performance. The cheap price of 8 k RMB/t gives it the potential for developing high-performance absorbents with the increase of mass fraction, the overall scores of AL_2O_3-TETA nanofluids increase first and then decrease. The Brown motion of nanoparticles can effectively promote the heat/mass transfer performance of the absorbents.

Keywords Carbon capture · Absorbents · Nanofluids · Analytic hierarchy process

---

J. Yanchi · Z. Zhongxiao (✉) · Y. Juan · L. Boyang
School of Mechanical Engineering, Shanghai Jiao Tong University, Shanghai 200240, China
e-mail: zhangzhongxiao@sjtu.edu.cn

Z. Ziqi · F. Junjie · J. Mengchuan · M. Aiwei
College of Architecture and Environment, Shanghai University of Technology, Shanghai 200093, China

J. Lyu and S. Li (eds.), *Clean Coal and Sustainable Energy*, Environmental Science and Engineering,
https://doi.org/10.1007/978-981-16-1657-0_47

47.1 Introduction

Following the signing of Kyoto Protocol in the twentieth century, the Copenhagen World Climate Summit once again pushed the issue of climate warming and emissions to the core of global public focus (Teir et al. 2007). Based on the differences in public perception of the correlation between climate change and greenhouse effect, and the constraints of the maturity of existing carbon dioxide capture and storage technologies on economic development (Sun et al. 2016), the conference finally failed to reach a restrictive treaty (Asif et al. 2018).

Dust, heavy metals and unsaturated water vapor in coal-fired flue gas make it very difficult to manufacture adsorption separation method and membrane separation method (Li et al. 2017). The characteristics of high flue gas flow rate, low partial pressure and high outlet temperature also bring relatively difficult problems to solve for low temperature separation method and physical absorption method. Based on improved inclusion, technical maturity, commercial feasibility and wide application, chemical absorption is considered to be one of the most feasible post-combustion capture methods. For post-combustion method, the performance of absorbents determined the overall application of CCS system (Oexmann et al. 2012).

Since the 1930s, amines have been used to remove CO_2. These compounds have been widely used because of their advantages of large absorption, high absorption rate, low reagent cost and recyclability. Chemical absorption is the most possible decarbonization technology for large-scale commercial application (Theo et al. 2016). Alcohol-amine decarbonization process is mature and has been applied in the field of natural gas industry for many years. Common reagents include MEA (monoethanolamine), DEA (diethanolamine), DIPA (diisopropanolamine) and MDEA (methyl diethanolamine). Enylamine (DETA, TETA, etc.) has attracted the attention of researchers at home and abroad for its higher absorption rate and capacity, easier regeneration and more amino groups per molar molecule than traditional absorbents (Yuan et al. 2018).

Thus, it is potential to establish novel high-performance absorbents by addition activation agents into TETA solution. However, comprehensive analysis model should be adopted to measure the detailed parameters of the absorbents. Analytic Hierarchy Process (AHP) is a multi-criteria decision-making method that combines qualitative and quantitative analysis, which was put forward by Satty et al. in 1970s (Tapia et al. 2017). The characteristic of this method is to construct a hierarchical structure model after thorough analysis of the essence, influencing factors and internal relations of complex decision-making problems, and then use less quantitative information to mathematicise the thinking process of decision-making. It is especially suitable for situations where people's qualitative judgment plays a major role and decision results are difficult to measure directly and accurately (Teir et al. 2007).

Accordingly, it is necessary to prepare TETA-based high-performance absorbents by adding activation agents of nanoparticles and other amines. The ab/desorption performance was tested by a bubble reactor. Then the parameters, such as ab/desorption rate, equilibrium time and CO_2-loading, were comprehensively summarized and learned by AHP method, while the overall and group rankings were listed in this work.

47.2 Materials and Methods

47.2.1 *Materials*

In this experiment, the base solution used as absorbent is triethylenetetramine (TETA), with purity over 99.9%. Shanghai Titan Hi-tech Co., Ltd., China; the additives of mixed amine are ethanolamine (MEA), diethanolamine (DEA), triethanolamine (TEA), diethylenetriamine (DETA), steric hindrance amine (AMP) and piperazine (PZ), with purity over 99.9%. Shanghai Aladditive of Aladdin Biochemical Technology Co., Ltd., China; using nanoparticles of AL_2O_3, with purity of 99.9%, particle size of 10–50 nm, Shanghai McLean Biochemical Technology Co., Ltd., China.

47.2.2 *Experimental Process*

As shown in Fig. 47.1, bubble reaction system is used in absorption performance experiment. The test system mainly includes electric heating jacket, gas path, mass flowmeter, condensation system, load analysis device, etc. The temperature of solution in five flasks can be measured and displayed by the thermocouple carried by the

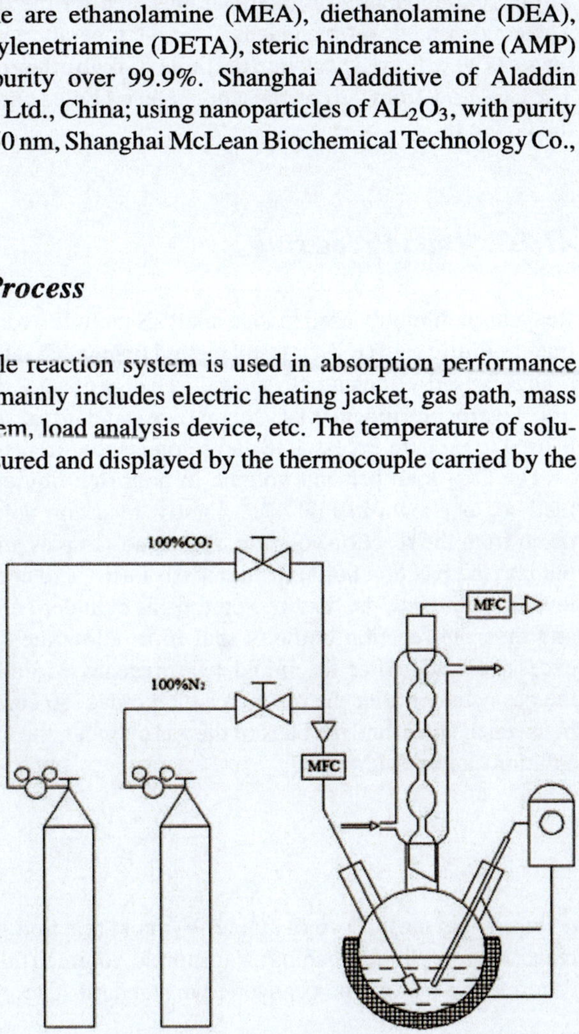

Fig. 47.1 Experimental system for the ab/desorption performance of TETA-based solution

heating sleeve itself. The solution can be heated and maintained at the set value. In order to avoid the loss of amine reagents and vapor evaporation caused by excessive temperature in the absorption reaction process, a spherical condensation tube was installed on the upper part of the three flasks. The condensated water condensed high temperature water vapor and some volatile reagents were flowed back into the three flasks by the condensated water and solution. The CO_2 flow from the condensation tube is cooled and the mass flow is used to show the flow rate.

The temperature of solution in three flasks can be measured and displayed by the thermocouple carried by the heating sleeve itself. The solution can be heated and maintained at the set value. In order to avoid the loss of amine reagents and vapor evaporation caused by excessive temperature in the absorption reaction process, a spherical condensation tube was installed on the upper part of the three flasks. The condensated water condensed high temperature water vapor and some volatile reagents were flowed back into the three flasks by the condensated water and solution. The CO_2 flow from the condensation tube is cooled and the mass flow is used to show the flow rate.

47.2.3 Data Processing

Reagents commonly used in load analysis include hydrochloric acid (HCL), concentrated sulfuric acid (H_2SO_4), and methyl orange ($C_{14}H_{14}N_3NaO_3S$), in which methyl orange is used as indicator to show the change of acid–base content in acid–base reaction. Hydrochloric acid is used to prepare standard titration solution, and sulfuric acid is used to prepare excess acid and sample solution reaction.

The CO_2 load per unit volume in poor/rich liquids was measured by acid leak method, as shown in Fig. 47.2. Firstly, a certain volume of absorbent solution is taken from the reaction solution, and these samples are put into the extraction tank, put into the reaction bottle with excessive acid, and cover the piston. After the liquid level is stabilized, the readings on the gas cylinder before the reaction are recorded, and then the reaction bottle is shaken to allow the sample solution to react with excessive acid. After the liquid level reaches equilibrium again, the readings on the gas cylinder after the reaction are recorded. According to the reading difference between the front and the back of the gas cylinder, the CO_2 content in the unit volume solution can be determined.

$$\alpha = \frac{(V_2 - V_1)}{22.4 \times V_0 \times M} \times f \tag{47.1}$$

α Represents mol CO_2/mol amine; V_2: post-reaction gas cylinder reading; V_1: pre-reaction gas cylinder reading; V_0: sample volume (ml), M: solution concentration; f: correction factor when converted to standard state, $f = 273/(273 + t)$.

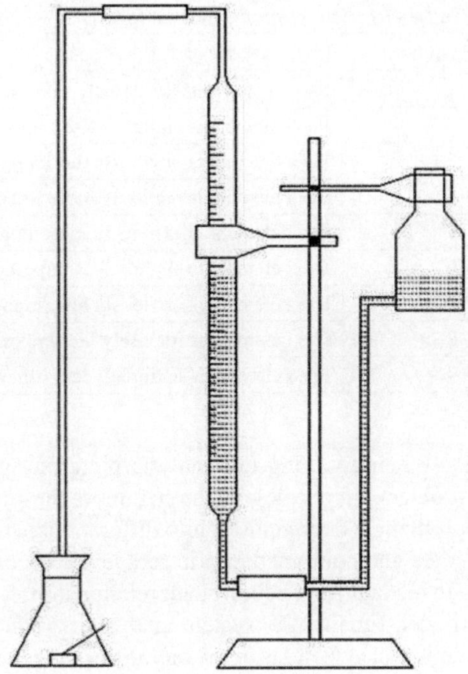

Fig. 47.2 CO_2-loading measuring system

Absorption rate N_A. The absorption rate of the bubble absorption system is expressed as:

$$N_{ab} = Q_i - Q_o \tag{47.2}$$

Q_i and Q_o are inlet and outlet air flow, ml/min respectively.

Energy consumption. The energy consumption of the whole desorption process can be described as follows.

47.3 AHP Analysis

47.3.1 Dividing Levels

The evaluation method is a very important part in the study of amine formula screening. The quality of the method directly affects the performance of the final selected amine formula. Based on the investigation of the existing evaluation methods and the merits and demerits of each method, it is decided to adopt the method of multi-index comprehensive evaluation to evaluate the formula of amine solution, and to determine the weight of each index by analytic hierarchy process.

Table 47.1 The 1–9 scale method for AHP

Number	Importance level	Assignment
1	The i, j elements are equally important	1
2	The i element is slightly more important than the j element	3
3	The i element is obviously more important than the j element	5
4	The i element is intensely more important than the j element	7
5	The i element is extremely more important than the j element	9
6	The i element is slightly less important than the j element	1/3
7	The i element is obviously less important than the j element	1/5
8	The i element is intensely less important than the j element	1/7
9	The i element is extremely less important than the j element	1/9

When applying the analytic hierarchy process, we should first hierarchize the problem. According to the nature of the problem and the goal to be achieved, the problem is decomposed into different constituent factors, and the factors are aggregated and combined in different levels according to the interrelated influence of the factors and their subordinate relationship, forming a multi-level analytical structure model. Finally, the system analysis is reduced to the determination of the relative importance weights or the ranking of the relative advantages and disadvantages of the lowest level, relative to the highest level. In ranking calculation, the single ranking problem of factors at each level can be simplified to a series of paired factors. In order to quantify the comparative judgment, the analytic hierarchy process (AHP) introduces the 1–9 scale method (Table. 47.1) and writes it in the form of judgment matrix. After forming the judgment matrix, we can calculate the maximum eigenvalue of the judgment matrix and its corresponding eigenvector, and calculate the relative importance weight of a certain layer for a certain element of the upper layer.

After calculating the single ranking weights of a certain level relative to each factor of the upper level, the total ranking weights of the upper level can be calculated by weighting synthesis of the weights of the factors themselves. In a word, the relative importance weights or relative ranking values of the lowest level factors relative to the highest level can be calculated from top to bottom.

47.3.2 Establishment of Judgment Matrix

As shown in Fig. 47.3, according to the logical relationship of the graph, it is necessary to determine the relative importance of each index between the target layer and the criterion layer. As far as this paper is concerned, Criteria 1 has four indicators: absorption capacity, desorption capacity, economy and safety for chemical absorbents. Absorption capacity and desorption capacity of absorbents are the most fundamental factors that directly determine the CO_2 capture capacity of chemical

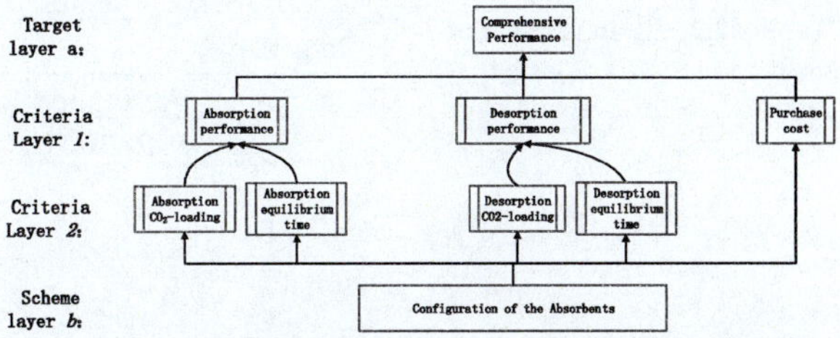

Fig. 47.3 Logical relationship of the graph

absorbents, and they are placed in the most important degree of approximate equality. The economy directly determines the current application value and development prospects of the absorbent, and is also a very important factor, so this paper regards it as a medium-important evaluation criterion. Safety involves the toxicity and volatility of absorbents, which affect the harmfulness to human and environment when using reagents, volatilization and explosion in use and storage. It is also an important criterion. However, some of the risks can be controlled and avoided through strict regulations and certain means. The noise is smaller than the previous factors. Therefore, the judgment matrix of criterion 1 is obtained.

The target layer in the figure is the desired goal, which is the only requirement. In this paper, the amine solution formula with the best comprehensive performance is presented. The criterion layer is the next layer of the target layer, including all kinds of factors affecting the realization of the goal. In the usual evaluation problems, the realization of the target level is often constrained by various factors, and there is an interactive relationship among them. Therefore, we should determine the main criteria and sub-criteria, and distinguish the subordinate relationship among the elements. The upper elements represent a set composed of several lower elements, and play a dominant role in the elements of the set. The properties of the elements in the same set are similar.

From Table 47.2 can be seen that the criterion layer is divided into two levels according to the characteristics of the evaluation index. The first level is absorption performance, desorption performance and purchase cost. The second level includes absorption CO_2-loading and absorption equilibrium time, desorption rate and regeneration time, and the purchase cost. The elements in the second layer are assigned to the subordinates of each element in the first layer, in which the absorption rate is expressed by the characteristic absorption rate, and the purchase cost of the first layer has only one subordinate element. In order to achieve the best comprehensive performance of the amine solution formula, the different amine solution formulations under different experimental ratios were put into the scheme layer and placed at the lowest level of hierarchical division.

Table 47.2 Establishment of evaluation matrix at target layer

1. Target layer

Target A	Absorption capacity B_1	Desorption capacity B_2	Purchase cost B_3
Absorption capacity B_1	1	1	2
Desorption capacity B_2	1	1	2
Purchase cost B_3	1/2	1/2	1

2. Criteria layer B_1-absorption capacity

Absorption capacity B_1	Absorption CO_2-loading C_1	Absorption equilibrium time C_2
Absorption CO_2-loading C_1	1	3
Absorption equilibrium time C_2	1/3	1

3. Criteria layer B2-desorption capacity

Desorption capacity B_2	Desorption CO_2-loading C_3	Desorption equilibrium time C_4
Desorption CO_2-loading C_3	1	2
Desorption equilibrium time C_4	1/2	1

47.3.3 Consistency Test

The consistency of judgement matrix refers to the consistency of judgements among experts when judging the importance of indicators, so as to avoid conflicting results. The inconsistency is very easy to occur under the condition of multi-level judgment, but the degree of inconsistency is different under different conditions.

According to matrix theory, if λ satisfies:

$$Ax = \lambda x \tag{47.3}$$

Then λ is the eigenvalue of A, and for all $a_{ii} = 1$, there is

$$\sum_{i=1}^{n} \lambda_i = n \tag{47.4}$$

When the matrix has complete consistency,

$$\lambda_1 = \lambda_{max} = n \tag{47.5}$$

The rest of the eigenvalues are all 0, and when matrix A does not have complete consistency,

$$\lambda_1 = \lambda_{max} > n \tag{47.6}$$

The rest of the eigenvalues λ_2, λ_3, λ_n have the following relations:

$$\sum_{i=2}^{n} \lambda_i = n - \lambda_{max} \tag{47.7}$$

When the judgment matrix cannot guarantee complete consistency, the characteristic roots of the corresponding judgment matrix will also change, so the consistency of judgment can be tested by the change of the characteristic roots of the judgment matrix. Therefore, in the analytic hierarchy process (AHP), the negative average value of the remaining eigenvalues besides the maximum eigenvalue of the judgment matrix is introduced as an index to measure the consistency of the deviation of the judgment matrix.

$$CI = \frac{\lambda_{max} - n}{n - 1} \tag{47.8}$$

Check the consistency of decision makers' thinking. The larger the *CI* value, the greater the deviation of the judgment matrix from the complete consistency; the smaller the *CI* value (close to 0), the better the consistency of the judgment matrix. When the judgment matrix has complete consistency, $CI = 0$; When the judgment matrix has satisfactory consistency, the average random consistency index *RI* value of the judgment matrix should be introduced. For judgment matrices of order 1–9, the *RI* values are as Table 47.3.

When the order is greater than 2, the ratio of *CI* to *RI* is called *CR*. When $CR = CI/RI < 0.10$, it can be considered that the judgment matrix has satisfactory consistency. Otherwise, the judgment matrix needs to be adjusted.

Table 47.3 *RI* values for AHP method

1	2	3	4	5
0.00	0.00	0.58	0.90	1.12
6	7	8	9	
1.24	1.32	1.41	1.45	

47.3.4 Signal Index Weight

The problem of hierarchical single ranking can be reduced to the problem of calculating the maximum eigenvalue and its eigenvector of the judgment matrix. Generally speaking, calculating the maximum eigenvalue and its corresponding eigenvector of the judgment matrix does not require high accuracy, because the judgment matrix itself has a considerable error range. Moreover, the priority weights of various factors in the hierarchy given by analytic hierarchy process essentially express some qualitative concepts. Therefore, the approximate maximum eigenvalues and their corresponding eigenvectors are usually obtained by iterative method on computer. In this paper, the calculating steps of the square root method for calculating the maximum eigenvalue of a matrix and its corresponding eigenvectors are given.

W is the required eigenvector.

$$W = [W_1, W_2, \ldots, W_n] \tag{47.9}$$

Computing the Maximum Eigenroot of Judgment Matrix

$$\lambda_{\max} = \sum_{i=1}^{n} \frac{(AW)_i}{nW_i} \tag{47.10}$$

Among them, $(AW)_i$ represents the first element of vector AW.

47.4 Results and Discussion

47.4.1 Weight Calculation

The characteristic parameters of the matrix have been calculated in Table 47.4. It can be seen that the CR value of matrix B_1 and B_2 are 0.00, while the CR value of A is $0.01 < 0.1$, which indicates that these matrixes meet the consistency requirements. This proves the construction of matrix is reasonable and the logical contradictions among various elements can be effectively avoided.

Then, the weight value can be calculated from Eq. (47.9) and shown in Fig. 47.4. The sequence is Absorption CO_2-loading > Absorption equilibrium time > Desorption CO_2-loading > Desorption equilibrium time > Purchase cost. It can be seen that the Ab/desorption CO_2-loading occupies the most important part with the value of 30.77 and 24.36%, respectively. And the rest factors possess the weight of almost 15%. Thus the CO_2-loading is considered as the vital factor of the overall CO_2 capture performance of the absorbents.

Table. 47.4 Key Parameters of characteristic matrix

	M_i	W_i		$W_{i,ov}$	W_{i1}
B_1	2.00	1.41	2.00	2.12	0.67
	0.50	0.71	2.00	2.12	0.33
	λ	λ_{max}	*CI*	*RI*	*CR*
	1.33	2.00	0.00	0.00	0.00
	0.67	2.00	0.00	0.00	0.00
B_2	M_i	W_i		$W_{i,ov}$	W_{i1}
	0.33	0.58	2.00	2.31	0.25
	3.00	1.73	2.00	2.31	0.75
	λ	λ_{max}	*CI*	*RI*	*CR*
	0.50	2.00	0.00	0.00	0.00
	1.50	2.00	0.00	0.00	0.00
A	M_i	W_i		$W_{i,ov}$	W_{i1}
	21.00	2.76	3.00	4.12	0.67
	1.00	1.00	3.00	4.12	0.24
	0.05	0.36	3.00	4.12	0.09
	λ	λ_{max}	*CI*	*RI*	*CR*
	2.01	3.01	0.00	0.58	0.01
	0.73	3.01	0.00	0.58	0.01
	0.26	3.01	0.00	0.58	0.01

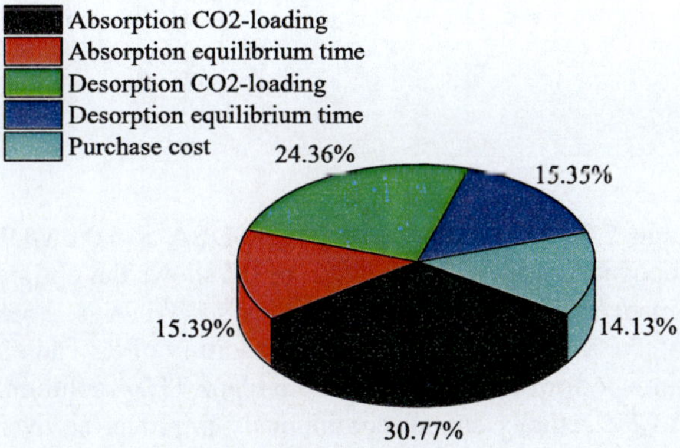

Fig. 47.4 Calculation results for the weight value

47.4.2 Comprehensive Performance

According to the weight of calculation, the index is calculated by linear weighting method and the results are shown in Figs. 47.5 and 47.6. The sequences of amine activation agents is 2.5%PZ > 2.5%DEA > 2.5%MDEA > 2.5%AMP > 2.5%TEA

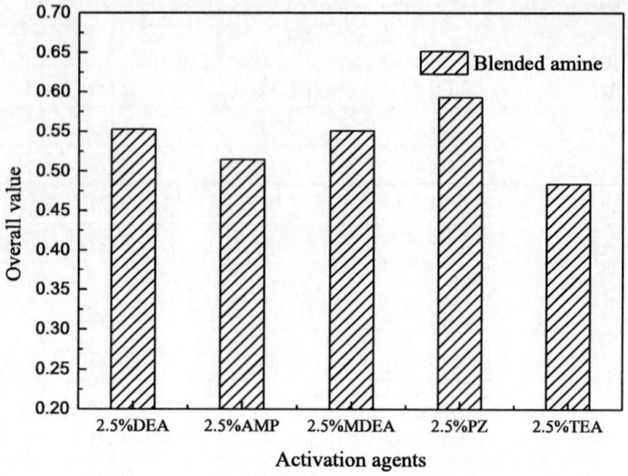

Fig. 47.5 Overall scores for amine activation agents with the mass fraction of 2.5%

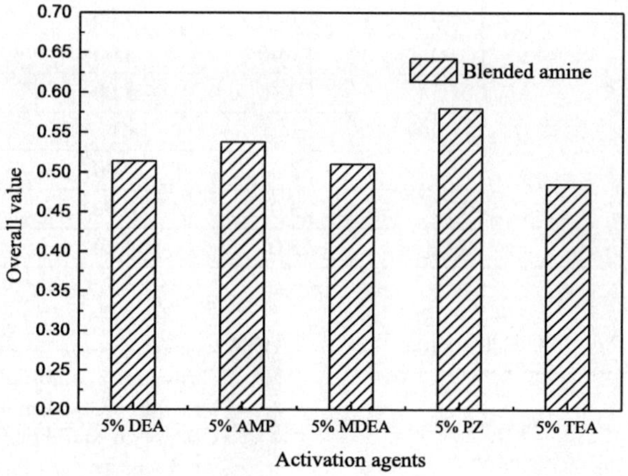

Fig. 47.6 Overall scores for amine activation agents with the mass fraction of 5%

and 5.0%PZ > 5.0%AMP > 5.0%DEA > 5.0%MDEA > 5.0% TEA. For amine activation agents, the addition of PZ shows the optimal overall CO_2 capture performance with the value of 0.593 (12.5%TETA + 2.5%PZ) and 0.580 (15.0%TETA + 5.0%PZ). This is because the addition of PZ can effectively promote the absorption performance compared with blank TETA solution. However, the undue addition of PZ actually cannot continuously improve the overall performance. Though PZ has higher absorption rate than traditional primary and secondary amines, thus the absorption performance of 5%PZ is 2.0% higher than 2.5%PZ. However, the carbamate generated by PZ is hard for desorption, which will reduce the desorption rate and energy, so the desorption performance get slightly weakened. Moreover, the purchase cost of PZ is 43.1 k RMB/t, which is 69.1% higher than TETA, thus the absorbent's overall performance is clipped.

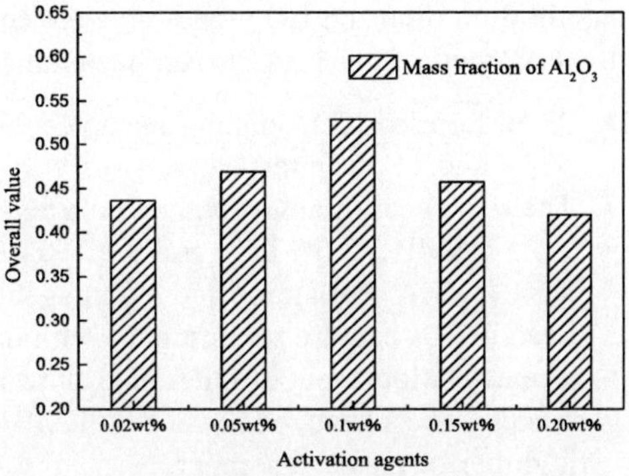

Fig. 47.7 Overall scores for AL_2O_3 nanofluids

Besides that, the activation agent of DEA also shows impressive overall CO_2 capture performance. Compared with enamine TETA, the secondary amine DEA doesn't prevail in ab/desorption performance, but the cheap price of 8 k RMB/t gives it the potential for developing high-performance absorbent. Nevertheless, the addition of 2.5%TEA has not achieved satisfactory effect, which shows a decline of overall CO_2 capture performance even lower than blank TETA solution. In addition, the scores for the ab/desorption are respectively 0.186 and 0.183, which is 16.81 and 18.63 lower than the absorbent with activation agent of DEA.

Figure 47.7 shows the effect of adding AL_2O_3 nanoparticles into the TETA absorbent on CO_2 capture performance. It can be seen that with the increase of mass fraction, the overall scores of AL_2O_3-TETA nanofluids increase first and then decrease. This is because the Brown motion of nanoparticles can effectively promote the heat/mass transfer performance of the absorbents. Kars etc. proved that the solid particles on the gas–liquid interface can penetrate into the boundary layer of gas–liquid mass transfer because their particle size is smaller than the thickness of the mass transfer interface, and can stay in the boundary layer (Kars et al. 1979). After adsorbing a certain amount of gas molecules, the fluid micro aggregates pass through the interface layer and return to the main body of the liquid phase. However, the too much addition of nanoparticles damps down the enhancement effect, because when the concentration of nanofluids increases, the collision between particles intensifies, which increases the instability of nanofluids and make them easy to be aggregated. Under this condition, the Brown motion weakens to cause the mass/heat transfer performance decline of the absorbents.

47.5 Conclusion

In this study, TETA-based high-performance absorbents were prepared by adding activation agents with nanoparticles and amines. The ab/desorption performance was tested by a bubble reactor. The critical parameters, such as ab/desorption rate, equilibrium time and CO_2-loading, were comprehensively summarized and learned by AHP method. The results can be summarized as follows:

- Ab/desorption CO_2-loading occupies the most important part with the value of 30.77 and 24.36%, respectively.
- The sequences of amine activation agents is 2.5%PZ > 2.5%DEA > 2.5%MDEA > 2.5%AMP > 2.5%TEA and 5.0%PZ > 5.0%AMP > 5.0%DEA > 5.0%MDEA > 5.0% TEA. For amine activation agents, the addition of PZ shows the optimal overall CO_2 capture performance with the value of 0.593 and 0.580.
- The activation agent of DEA also shows impressive overall CO_2 capture performance. The cheap price of 8 k RMB/t gives it the potential for developing high-performance absorbent.
- With the increase of mass fraction, the overall scores of AL_2O_3-TETA nanofluids increase first and then decrease. The Brown motion of nanoparticles can effectively promote the heat/mass transfer performance of the absorbents.

Acknowledgements Research and Development Program of China (Grant 2016YFE0102500) funded this work.

References

Asif M, Suleman M, Haq I, Jamal SA (2018) Post-combustion CO_2 capture with chemical absorption and hybrid system: current status and challenges. Greenh Gases Sci Technol 8(6):998–1031

Kars RL, Best RJ, Drinkenburg AAH (1979) The sorption of propane in slurries of active carbon in water. Chem Eng J 17(3):201–210

Li W, Landon J, Irvin B, Zheng L, Ruh K, Kong L, Pelgen J, Link D, Figueroa JD, Thompson J, Nikolic H, Liu K (2017) Use of carbon steel for construction of post-combustion CO_2 capture facilities: a pilot-scale corrosion study. Ind Eng Chem Res 56(16):4792–4803

Oexmann J, Kather A, Linnenberg S, Liebenthal U (2012) Post-combustion CO_2 capture: chemical absorption processes in coal-fired steam power plants. Greenh Gases Sci Technol 2(2):80–98

Sun J, Su C-W, Shao G-L (2016) Is carbon dioxide emission convergence in the ten largest economies? Int J Green Energy 13(5):454–461

Tapia JFD, Promentilla MAB, Tseng M-L, Tan RR (2017) Screening of carbon dioxide utilization options using hybrid analytic hierarchy process-data envelopment analysis method. J Clean Prod 165:1361–1370

Teir S, Kuusik R, Fogelholm C-J, Zevenhoven R (2007) Production of magnesium carbonates from serpentinite for long-term storage of CO_2. Int J Miner Process 85(1–3):1–15

Theo WL, Lim JS, Hashim H, Mustaffa AA, Ho WS (2016) Review of pre-combustion capture and ionic liquid in carbon capture and storage. Appl Energy 183:1633–1663

Yuan M, Gao G, Hu X, Luo X, Huang Y, Jin B, Liang Z (2018) Premodified sepiolite functionalized with triethylenetetramine as an effective and inexpensive adsorbent for CO_2 capture. Ind Eng Chem Res 57(18):6189–6200

Chapter 48
Experimental Study on Carbon Capture Performance of Polyimide Hollow Fiber Membrane in Post-combustion Process

Liu Yang, Yanchi Jiang, Juan Yu, Ziqi Zhao, Mengchuan Jia, and Aiwei Mu

Abstract Carbon capture and storage (CCS) technology is effective in mitigating the negative effect of the greenhouse gases. At the same time, membrane-based technology for the capture of CO_2 has attracted worldwide attention. In this paper, an experimental study on the carbon capture performance of polyimide (PI) hollow fiber membrane in post-combustion process is carried out to analyze the PI membrane system separation performance of single-stage and two-stage membrane separation methods under different pressure, temperature and gas flow rate conditions. For single-stage membrane separation system, increasing feed gas pressure, feed gas temperature and gas flow rate is beneficial to enhance membrane CO_2 permeability. For the two-stage membrane separation system, the feed gas pressure of the first/the second stage pressure, as well as the gas flow rate, can effectively influence the CO_2 concentration on permeate side, which can reach up to 91.01%. Furthermore, the two-stage method has 40–60% higher CO_2 concentration than single-stage method due to the additional membrane stage.

Keywords Membrane separation · PI membrane · Two-stage method · Post-combustion capture

48.1 Introduction

Carbon dioxide (CO_2), a major component of greenhouse gases in the atmosphere, accounts for a large part of global warming (Kvamsdal et al. 2009). According to the 5th assessment report from the Intergovernmental Panel on Climate Change (IPCC), human activities, especially energy consumptions, play a predominant factor of CO_2 emissions (IPCC 2005; International Energy Agency 2015; Wang et al. 2017). The

L. Yang · Y. Jiang · J. Yu (✉)
School of Mechanical Engineering, Shanghai Jiao Tong University, Shanghai 200240, China
e-mail: yujuan@sjtu.edu.cn

Z. Zhao · M. Jia · A. Mu
School of Environment and Architectural Engineering, University of Shanghai for Science and Technologyent, Shanghai 200093, China

J. Lyu and S. Li (eds.), *Clean Coal and Sustainable Energy*,
Environmental Science and Engineering,
https://doi.org/10.1007/978-981-16-1657-0_48

CO_2 emissions from fossil fuel combustion account for about 91% of the total CO_2 emissions from human activities. The International Energy Outlook (IEO) report in 2016 states that global power generation will grow 69% in 2040 (Man et al. 2016). Notably, the state of relying on traditional fuels (especially coal (Merkel et al. 2009)) to provide energy for most countries in the world, to some extent, cannot be changed in a short time. That means we will increase the CO_2 emissions in order to meet our needs. However, the rising level of CO_2 concentration leads to not only the rise of the average temperature of the global climate (Li et al. 2010), unexpectedly, but also the higher number of flood disasters and drought problems. How to effectively control and reduce carbon dioxide emissions and slow down the greenhouse effect while maintaining sustained economic growth and social needs has become a serious problem facing the whole world. The technology of carbon dioxide capture and storage (CCS), considered as the key to mitigating environmental temperature changes, provides a promising way to solve this dilemma (Wang et al. 2017).

CCS realizes the safe storage of CO_2 by collecting, compressing and sealing carbon dioxide in the air, thus preventing CO_2 from returning to the air. Basically, it can be divided into three main aspects: post-combustion, pre-combustion, and oxy-fuel combustion (Figueroa et al. 2008). Compared with pre-combustion and oxy-fuel combustion, post-combustion technology has a high degree of compatibility with the transformation of existing power plants (so that generic power generation technologies, such as coal-fired power plants, can continue to be used). Meanwhile, it is high matured and has a large number of practical applications (Farjami et al. 2015), which attracts worldwide attentions.

For post-combustion carbon capture, there are four specific methods: chemical absorption, adsorption, membrane separation, and cryogenic distillation. But as the "end-of-pipe" technology, the low partial pressure of CO_2 of the exhaust gas (about 13–15%) can be the biggest barrier to the commercial application (Wang et al. 2017). On the one hand, chemical absorption has become the most widely used carbon capture technology in industry due to its good separation effect for flue gas with low partial pressure of CO_2. On the other hand, it is more mature than other three methods. However, the large use of absorbents contaminating the environment and the energy-intensive character of chemical absorption verify that it is not a long-term solution (Gong et al. 2006). In the 1980s, membrane separation technology began to be used for gas separation.

The membrane separation method is economic and environmentally friendly, and thus can be considered as a promising post-combustion technical method, which attracted the attention of many researchers. Additionally, its advantages of compact equipment and small area occupation makes it a good competitor. In recent years, many studies have been conducted. Whereas, most of the existing work jut focuses on finding out the best membrane materials referring (Hussain et al. 2015; He and Hägg 2010). Meanwhile, the process intensification research is mostly realized by numerical calculation presuming the permeability and selectivity of membranes, examples are: Hussain et al. 2015, Kotowicz et al. 2010. Nevertheless, the experimental data about membrane modules is relatively less.

Membrane-based carbon separation method for post-combustion mainly relies on the pressure difference between the feed side and the permeate side, as well as the selectivity of the gas to the membrane material. Therefore, its carbon capture performance is not only related to membrane materials, but also to the design of the separation systems (Giordano et al. 2017). However, experimental studies both on membrane materials and membrane separation systems are still rare. This paper aims to analyze the carbon capture performance of polyimide hollow fiber membranes in post-combustion, discussing and comparing single-stage and two-stage membrane separation methods, through specific experiments. A 3–5 m^3/h experimental platform was established. The goal of this paper is to provide the experimental data about membrane modules filling the blank in membrane technology researches and lay the foundation for subsequent optimization researches.

48.2 Experiment

48.2.1 Material

The membrane material with glass polymer as the main component has strong brittleness and good molecular structure stability, and the requirements for permeable membrane materials of gas molecules of different sizes are strict. Therefore, the membrane has a low permeability coefficient and a high separation factor. Polyimide membrane is the most widely applied. The molecular formula of polyimide (Shi et al. 2002) is shown in Fig. 48.1.

There are three common types of gas membrane separators: flat plate type, spiral coil type and hollow fiber type. Membrane separators require larger packing area per unit volume and good contact between gas and membrane surface. Among the three types, hollow fiber membrane has high packing density, can provide more membrane area, and possesses good economical effect. These experiments take PI hollow fiber membrane module as the research object considering the influence of membrane performance parameters, such as permeability and selectivity.

48.2.2 Basic Principle of Membrane Separation

The solution-diffusion theory, as shown in Fig. 48.2, can be the basic principle to govern the membrane-based separation process (Stern 1994; Koros and Fleming 1993; Koros et al. 1988; Wijmans and Baker 1995).

$$F = \frac{\Delta X}{\delta} \tag{48.1}$$

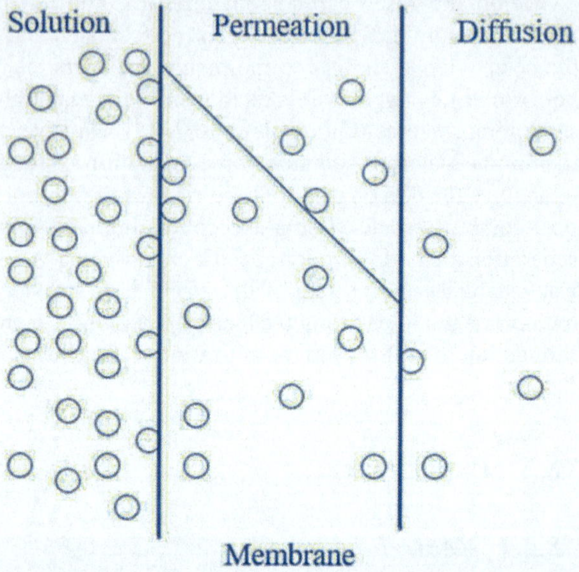

Fig. 48.1 The molecular formula of polyimide, Where N was methylpyrrolidone

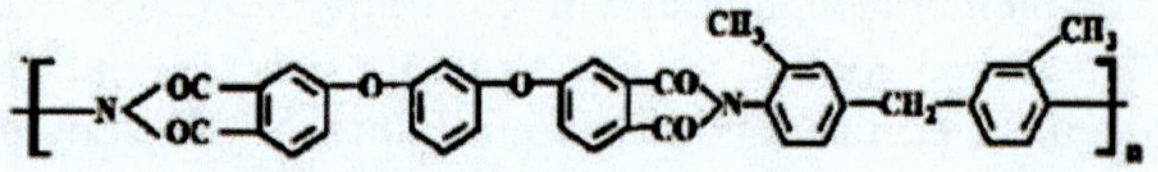

Fig. 48.2 Schematic diagram of solution-diffusion mechanism

where F was the driving force on both sides of the membrane; ΔX was a chemical difference, mainly including pressure, concentration, temperature, etc. δ was the membrane thickness.

According to the solution-diffusion theory, four steps as follows can be used to explain the process of CO_2 passing through membrane modules:

a. CO_2 reaches the membrane surface, means contact.
b. CO_2 dissolves on the membrane surface, means solution.
c. CO_2 passes through the membrane by means of the concentration gradient, unstable process, means permeation.
d. CO_2 reaches the other side of the membrane and the desorption rate is stable, means diffusion.

There are two well-known conservation equations. The first one is material conservation equation:

$$q_f = q_r + q_p \tag{48.2}$$

The second one is the component conservation equation of CO_2:

$$\alpha_{CO_2/N_2} = \frac{y_{p,CO_2}/(1 - y_{p,CO_2})}{x_{f,CO_2}/(1 - x_{f,CO_2})}$$

$$x_{f,CO_2}q_f = x_{r,CO_2}q_r + y_{p,CO_2}q_p \tag{48.3}$$

where q_f, q_r, and q_p were respectively the gas flow of the feed side, the retentate side, and the permeate side; x_{f,CO_2} was the CO_2 fraction of the feed gas; x_{r,CO_2} was the CO_2 fraction of the retentate gas; y_{p,CO_2} was the CO_2 fraction of the permeate gas.

It is worth mentioning that the component conservation equation is available for every constituent of flue gas.

The separation characteristic of the PI membrane module was defined by permeability and selectivity factors (P and α). For the process, CO_2 removal efficiency (η) and CO_2 concentration of permeate gas (y_{p,CO_2}) were adopted to evaluate the overall performance of different processes.

$$\alpha_{CO_2/N_2} = \frac{y_{p,CO_2}/\left(1 - y_{p,CO_2}\right)}{x_{f,CO_2}/\left(1 - x_{f,CO_2}\right)} \tag{48.4}$$

$$P_{CO_2} = \frac{Q \times \delta}{A \times \Delta p} \tag{48.5}$$

$$\eta = \frac{q_p y_{p,CO_2}}{q_f x_{f,CO_2}} \cdot 100\% \tag{48.6}$$

where Δp was pressure difference; A was effective area; δ was membranc thickness, Q was the CO_2 gas flow of the permeate side.

48.2.3 Operating Conditions

As shown in Fig. 48.3, a 3–5 m^3/h experimental platform was established to study the carbon capture characteristics of polyimide (PI) membrane module. The parameters, such as pressure, temperature and gas flow rate, were considered to measure the one/two stage CO_2 membrane separation methods.

Feed gas consists of 15% carbon dioxide and 85% nitrogen. For the two-stage membrane separation system, the permeate gas of the first membrane enters the second membrane as a feed gas, while the retentate gas of the two membranes gets together to the gas-terminal section. The single stage membrane separation experimental conditions are shown in Table 48.1. Meanwhile, Table 48.2 demonstrates the experimental conditions of the two-stage membrane separation system.

Based on the single stage membrane separation system, the two membrane modules are coupled in series to form a two-stage membrane separation system as shown in the Fig. 48.3. And the separation characteristics are explored by conducting the experiments on the basis of the experimental conditions in Table 48.2.

Fig. 48.3 System diagram of carbon capture with two-stage membrane separation method

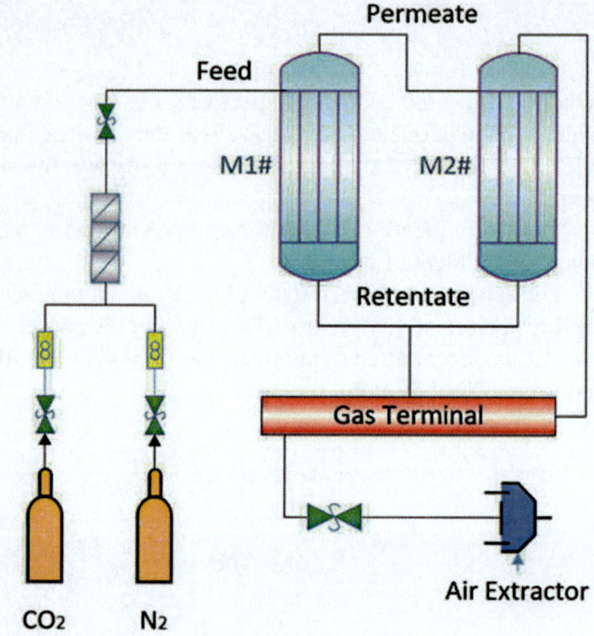

Table 48.1 Single stage membrane separation experimental conditions

Feed pressure (MPa)	Feed gas flow rate (L/min)	Temperature (°C)
0.3 0.5 0.7 0.9 1	35	45
0.7	30 35 40 45 50 55	45
0.7	35	25 30 35 40 45

Table 48.2 Two stage membrane separation experimental conditions

Feed gas pressure of the first stage membrane module (MPa)	Feed gas pressure of the second stage membrane module (MPa)	Feed gas flow rate (L/min)	Temperature (°C)
0.3 0.5 0.7 0.9 1	0.9	45	25
0.9	0.3 0.5 0.7 0.9 1	45	45
0.9	0.9	25 30 35 40 45	45

48.3 Results and Discussion

In this work, PI hollow fiber membrane modules were used to test the single-stage and two-stage membrane separation methods under different pressure, different temperature and different flow rate conditions. The experimental results of the two methods were analyzed and compared.

48.3.1 Experimental Results of Single-Stage Membrane

It can be seen from Fig. 48.4 that as the pressure of the feed gas increases, the selectivity and the permeability factors of the PI membrane have a significant upward trend. When the pressure of the feed gas changes from 0.3 to 0.8 Mpa, the selectivity and the permeability factors increase monotonically. The slope is large, and at the same time, when the pressure of the feed gas is changed from 0.8 to 1.0 Mpa, the rise is gentle. Because, under high pressure conditions, for an ideal gas, the condensable nature of CO_2 can lead the polyimide molecule of the separation membrane to have a swelling effect, which can greatly improve the performance of CO_2 separation. It should also be noted that an increase in pressure causes a decrease in the free volume inside the polyimide membrane and the volume of the microcavity in the membrane, which will inhibit the gas permeation to some extent. However, the polymeric chain structure, as shown in Fig. 48.1, possessed by the PI material has relatively good rigidity and can weaken the negative influence of the reduction of the free volume.

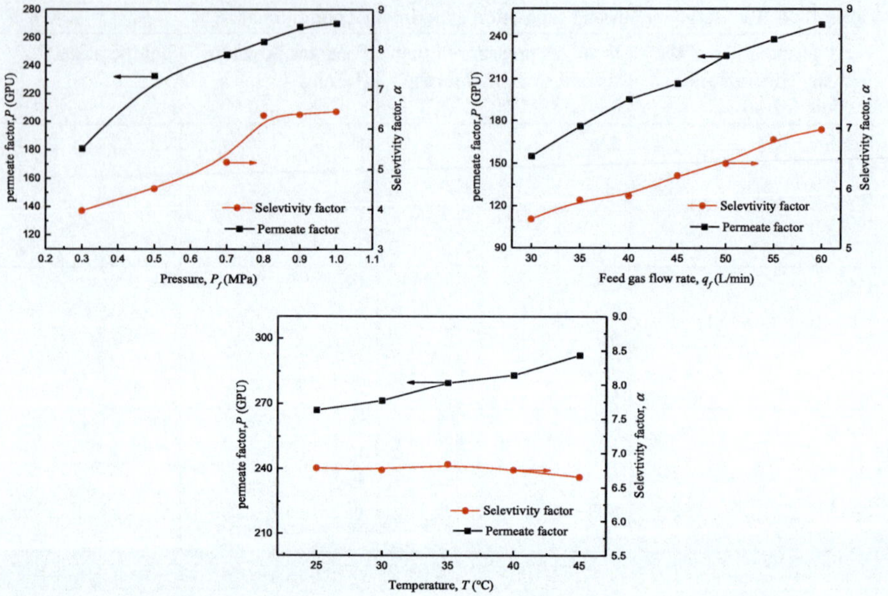

Fig. 48.4 CO_2 separation performance of single stage membrane

Therefore, when the pressure is considerably high, the CO_2/N_2 separation coefficient shows a trend of slowing down, but there is no decline. When the feed gas pressure is 1.0 Mpa, the PI membrane selectivity factor can reach about 6.3, at the same time, the permeability factor can realize maximum value of 269 GPU (GPU = $10^{-6}cm^3(STP)/(cm^2 \cdot s \cdot cmHg)$).

Similarly, as the feed gas flow rate increases from 30 to 60 L/min, the permeability and selectivity of the PI membrane increase steadily. This indicates that, to a certain extent, the increase of the feed gas flow rate promotes the permeability and selectivity performance of the polyimide hollow fiber membrane. During the process of the mixed gas passing through the separation membrane, concentration polarization occurs near the surface of the separation membrane. This phenomenon is due to the "aggregation and concentration" of the solute, which will weaken the mass transfer process. The increase of the intake gas amount can effectively reduce the influence of the concentration polarization phenomenon and promote mass transfer.

It is worth noting that the separation performance of the membrane is related to the temperature of the feed gas. Theoretically, the membrane penetration process follows the Arrhenius rule (Shi et al. 2002). The relationship between diffusion coefficient (D) and temperature (T) is as follows:

$$D = D_0 Exp\left(-\frac{E_d}{RT}\right) \tag{48.7}$$

where E_d was the apparent activation energy of diffusion.

Increasing the temperature will undermine the solubility in the membrane. As can be seen from the above figures, with the increase of temperature, the permeability factor increased while the selectivity factor almost unchanged. The permeability coefficient increased from 266.97 to 292.19 GPU and improved by 9.44% with the increase of temperature. The separation coefficient remained unchanged at about 6.77. For the actual polyimide membrane module, the gas permeation process is mainly affected by the gas diffusion process in the membrane. But the increase of temperature will improve the permeability of CO_2 in the separation membrane to a certain extent. Possibly, it is because the competition between CO_2 and N_2 existing in the PI hollow fiber membrane mass process leading to the unchanged selectivity factor. Hence, it is of vital importance to choose the appropriate temperature for membrane-based CO_2 separation in post-combustion to obtain the best separation performance.

48.3.2 *Experimental Results of Two-Stage Membrane*

Experiments of two-stage membrane separation system were carried out. The effects of the feed gas pressure of the first stage membrane, the feed gas pressure of the second stage membrane and the total feed gas flow rate on the separation performance of the PI hollow fiber two-stage membrane system were analyzed. It can be seen from Fig. 48.5 that as the inlet pressure of the first stage membrane increases, apparently, the CO_2 concentration of the permeate gas and the removal efficiency of CO_2 increases accordingly. However, as the intake pressure of the second-stage membrane increases, the CO_2 concentration of the permeate gas rises steadily and stabilizes at about 91.01%. The principle of this results is similar to the pressure growth process of single-stage membrane separation. It is achieved by the swelling effect of polyimide to enforce the infiltration process of CO_2 in the membrane due to the increases of pressure. The second-stage membrane intake CO_2 concentration is much higher than the first-stage membrane intake CO_2 concentration due to the separation of the first-stage membrane. Correspondingly, the inlet flow rate of the second membrane is smaller. In the case of a change in the inlet pressure of the second stage membrane, the separation efficiency of the two-stage membrane separation method is mainly affected by the permeation and separation performance of the second-stage membrane. The CO_2 removal efficiency increases as the intake pressure of the second-stage membrane increases.

Simultaneously, the feed gas flow rate is also an important factor. It can be seen from the figures that as the feed gas flow rate increases, the carbon dioxide concentration of the permeate gas increases from 83.82 to 91.01%, but the carbon dioxide removal rate decreases from 73.51 to 53.33%. Kim et al. found out similar results (Kim et al. 2016). This is because the increase in the flow rate of the first membrane feed gas can effectively increases the gas flow rate and CO_2 concentration at the permeate end of the first stage. Accordingly, however, the proportion of CO_2 loss at

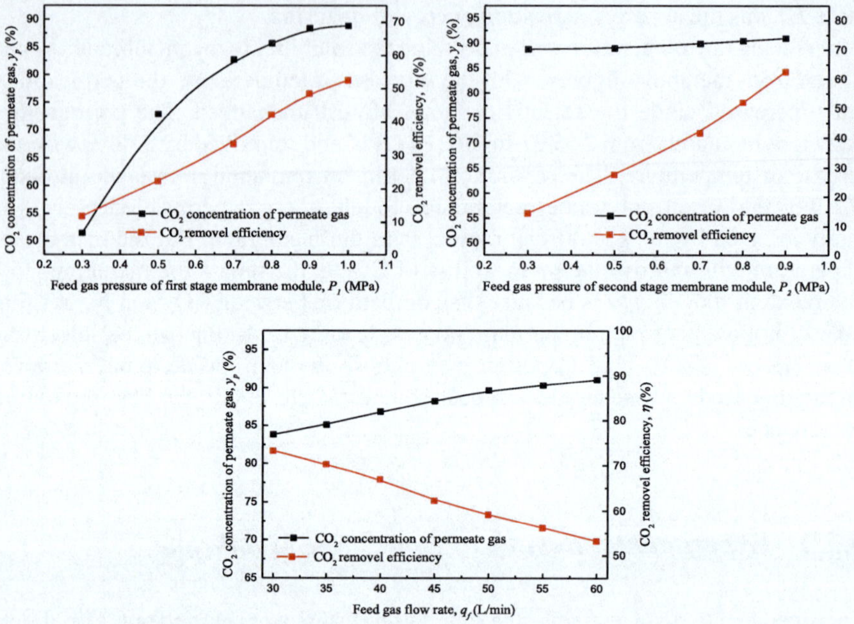

Fig. 48.5 CO_2 separation performance of two stage membrane

the retentate end is also increased, thereby impairing the carbon capture efficiency of the entire process to some extent so that the CO_2 removal efficiency decreases.

48.4 Conclusions

In this paper, through a series of experiments, by analyzing the effects of different temperatures, different pressures and different feed gas flow rates on the CO_2 capture performance of polyimide hollow fiber membrane in post-combustion process, the following conclusions can be drawn:

(1) For the single-stage membrane separation of PI hollow fiber membrane module, increasing gas flow rate is conducive to reducing polarization effect, thus improving the CO_2 concentration at the permeation side. Meanwhile, rising the feed gas pressure and gas temperature is conducive to improving the membrane CO_2 permeability, thereby improving the CO_2 removal efficiency (η) of the membrane module.

(2) Based on the experimental results of single-stage membrane, for the two-stage method, the feed gas pressure of one/two-stage pressure, as well as the flow rate, can effectively influence the CO_2 concentration, which can reach up to 91.01%.

(3) The two-stage method has 40–60% higher CO_2 concentration than single-stage method, because of the additional membrane stage.

Experiments were carried out to analyze the separation performance of PI hollow fiber membrane modules that filled the gap in membrane separation technology study. We hope to provide the experimental data about membrane modules for future membrane system researches.

Acknowledgements Research and Development Program of China (Grant 2017YFB0603403) funded this work.

References

Farjami M, Moghadassi A, Vatanpour V (2015) Modeling and simulation of CO_2 removal in a polyvinylidene fluoride hollow fiber membrane contactor with computational fluid dynamics. Chem Eng Process 98:41–51

Figueroa JD et al (2008) Advances in CO_2 capture technology—The U.S. Department of Energy's Carbon Sequestration Program. Int J Greenh Gas Control 2(1):9–20

Giordano L, Roizard D, Bounaceur R, Favre E (2017) Energy penalty of a single stage gas permeation process for CO_2 capture in post-combustion: a rigorous parametric analysis of temperature, humidity and membrane performances. Energy Procedia 114:636–641

Gong Y, Wang Z, Wang S (2006) Experiments and simulation of CO_2 removal by mixed amines in a hollow fiber membrane module. Chem Eng Process 45:652–660

He X, Hägg M-B (2010) Hollow fiber carbon membranes: Investigations for CO_2 capture. J Membr Sci 378(1–2):1–9

Hussain A, Farrukh S, Minhas FT (2015) Two-stage membrane system for post-combustion CO_2 capture application. Energy Fuels 29(10):6664–6669

International Energy Agency (2015) Key trends in CO_2 emissions from fuel combustion. International Energy Agency, Paris

IPCC (2005) IPCC Special Report on Carbon Dioxide Capture and Storage. United Kingdom & New York, USA, 2005.

Kim SH et al (2016) Comparative feasibility study of CO_2 capture in hollow fiber membrane processes based on process models and heat exchanger analysis. Chem Eng Res Des 117:659–669

Koros WJ, Fleming GK (1993) Membrane-based gas separation (Review). J Membr Sci 83:1–80

Koros WJ, Fleming GK, Jordan SM, Kim TH, Hoehn HH (1988) Polymeric membrane materials for solution-diffusion based permeation separations. Prog Polym Sci 13:339–401

Kotowicz J, Chmielniak T, Katarzyna Janusz-Szymańska (2010) The influence of membrane CO_2 separation on the efficiency of a coal-fired power plant. Energy 35(2):841–850

Kvamsdal HM, Jakobsen JP, Hoff KA (2009) Dynamic modeling and simulation of a CO_2 absorber column for post-combustion CO_2 capture. Chem Eng Process 48:135–144

Li L, Wen X, Fu X, Wang F, Zhao N, Xiao F, Wei W, Sun Y (2010) MgO/Al_2O_3 sorbent for CO_2 capture. Energy Fuels 24(10):5773–5780

Man Y, Yang S, Qian Y (2016) Integrated process for synthetic natural gas production from coal and coke-oven gas with high energy efficiency and low emission. Energy Convers Manage 117:162–170

Merkel TC, Lin H, Wei H, Baker R (2009) Power plant post-combustion carbon dioxide capture: an opportunity for membranes. J Membr Sci 359(1)

Shi B-L, Wu Y-L, Li G-M (2002) Studies of the separation of methanol-MTBE mixtures with hollow fibermembranes spun with the polyimide. J Funct Polym (02):167–170 (Chinese)

Stern SA (1994) Polymers for gas separation: the next decade (Review). J Membr Sci 94:1–65
Wang X, Wang D, Song M, Xin C, Zeng W (2017) Tetraethylenepentamine-modified activated semicoke for CO_2 capture from flue gas. Energy Fuels 31:3055–3061
Wang Y, Zhao L, Otto A, Robinius M, Stolten D (2017) A review of post-combustion CO_2 capture technologies from coal-fired power plants. Energy Procedia 114
Wijmans JG, Baker RW (1995) The solution-diffution model: a review. J Membr Sci 107:1–21

Chapter 49
Simulation Study on Separation of CO_2 from Flue Gas in Coal-Fired Power Plant by Membrane Method

Boyang Li, Juan Yu, Fan Feng, Zhongxiao Zhang, and Xinwei Guo

Abstract A simplified model of membrane separation was improved for predicting the CO_2 separation from the flue gas of coal-fired power plant. Simultaneously, a single-stage membrane separation system was built under laboratory conditions that uses a hollow fiber membrane as the separation core module. Different factors affecting the separation effect of the system were experimented and the simulated and experimental values were compared. The comparison results show that the variation trend of the simulated values is basically consistent with the experimental values, and within the allowable error range of the engineering, the simulated values can approximately reflect the experimental values. This model is suitable for the engineering calculation of CO_2 membrane separation of flue gas in coal-fired power plants.

Keywords CO_2 membrane separation · Simulation · Flue gas · Coal-fired power plant

49.1 Introduction

Nowadays, the greenhouse effect caused by human activities is becoming more and more significant. By 2017, global carbon dioxide emissions had reached 33.444 billion tons, an increase of 12.55% compared with 2009 (British Petroleum Corporation 2018). Faced with the increasingly serious problem of excessive carbon emissions, carbon capture and utilization technology has increasingly become the focus of attention. Among them, the research on carbon dioxide capture in flue gas of coal-fired power plants has become a hot issue in this field.

At present, the research on carbon dioxide capture in flue gas of coal-fired power plants mainly focuses on post-combustion capture technologies that can be divided

B. Li · J. Yu (✉) · F. Feng · Z. Zhang
School of Mechanical Engineering, Shanghai Jiao Tong University, Shanghai 200240, China
e-mail: yujuan@sjtu.edu.cn

X. Guo
School of Energy and Power Engineering, University of Shanghai for Science and Technology, Shanghai 200093, China

J. Lyu and S. Li (eds.), *Clean Coal and Sustainable Energy*, Environmental Science and Engineering,
https://doi.org/10.1007/978-981-16-1657-0_49

into the following four types: chemical absorption, physical adsorption, low temperature fractionation and membrane separation (Zhongyang 2012). In recent years, with the continuous development of membrane technology, more and more scholars focus on membrane separation. Among them, some scholars have studied the application prospects of membrane separation through simulation. For example, Binns et al. (2015) compared the energy consumption of the separation system at a given CO_2 recovery rate when the membrane separation or the chemical absorption method was used alone or when the membrane separation and the chemical absorption method were used together. The results show that compared with other methods, the purity of CO_2 obtained by membrane separation alone is lower, but the energy consumption of the system is the lowest. Shao et al. (2013) also conducted simulation studies, pointing out that membrane separation is more economical than amine absorption in capturing carbon dioxide in flue gas from coal-fired power plants. In addition, Arias et al. (2016) systematically designed a multi-stage membrane system for carbon capture using mathematical programming and optimization method based on superstructure, with a given product CO_2 concentration (90–98%) and CO_2 recovery rate (90–98%) as the goal of minimizing total cost. Kim et al. (2017) modeled the separation membrane based on differential theory and compared the simulation results with the experimental results. It is found that the simulation results are in good agreement with the experimental results within the allowable range of errors and can be used for more in-depth study. The achievements of the above scholars have played a positive role in further promoting the research of CO_2 membrane separation technology. However, the model based on mathematical programming, superstructure and differential theory is not convenient for engineering application. To solve this problem, Luhang (2017) put forward a simple model, which is more suitable for engineering application. However, the assumption that the permeation rate and membrane separation coefficient are fixed values, which still has room for improvement compared with the actual situation.

Based on the experimental results, a model of separation membrane was built in this paper. The permeation rate of the model is variable. Under laboratory conditions, different factors affecting the separation effect of single-stage membrane separation system were experimented in this work and the simulated values were compared with the experimental values.

49.2 Simulation and Experiment

49.2.1 Model

According to the characteristics of hollow fiber membranes, this work simpifies the hollow fiber membrane into a cylinder as shown in Fig. 49.1. The wall of the cylinder is a separating membrane that the mixture can pass through. The upper surface is sealed so that the gas can not pass through. The lower surface is connected with

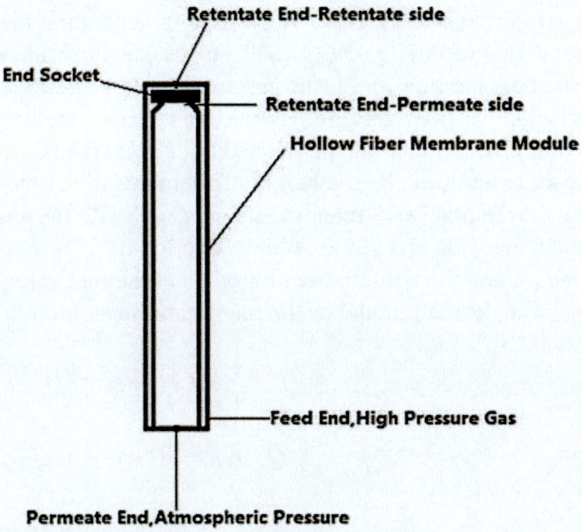

Fig. 49.1 Schematic diagram of hollow fiber membrane

the outlet of the permeate end, and the gas passing through the membrane leaves the membrane assembly accordingly. Near the upper surface is the retentate end. Among them, the retentate end–retentate side is outside the membrane, and the retentate end–permeate side is inside the membrane. Near the lower surface are the feed end and the permeate end, in which the feed end is outside the membrane and the permeate end is inside the membrane.

Considering the gas flow in the membrane module, the following assumptions were made: the flow in the hollow fiber membrane is approximately countercurrent and the pressure drop on the permeate side is neglected (Li et al. 2019).

The CO_2 recovery rate of the membrane separation system $\left(\varphi_{CO_2}\right)$, N_2 recovery rate of the membrane separation system $\left(\varphi_{N_2}\right)$, CO_2 permeation rate $\left(J_{CO_2}\right)$ and N_2 permeation rate $\left(J_{N_2}\right)$ used in the modeling process are defined as follows (Zhang et al. 2017):

$$\varphi_{CO_2} = \frac{Q_P b_{PC}}{Q_F a_{FC}} \tag{49.1}$$

$$\varphi_{N_2} = \frac{Q_P b_{PN}}{Q_F a_{FN}} \tag{49.2}$$

$$J_{CO_2} = \frac{Q_P b_{PC}}{A_{mem}(p_F a_{FC} - p_P b_{PC})} \tag{49.3}$$

$$J_{N_2} = \frac{Q_P b_{PN}}{A_{mem}(p_F a_{FN} - p_P b_{PN})} \tag{49.4}$$

In the formulas, A_{mem} is the area of separation membrane, Q_F(STP) is the total flow rate of inlet gas, Q_P (STP) is the gas flow rate at the permeate end, p_F is the inlet gas pressure, p_P is the gas pressure at the permeate end, a_{FC} is the CO_2 volume fraction of inlet gas, a_{FN} is the N_2 volume fraction of inlet gas, b_{PC} is the CO_2 volume fraction at the permeate end, b_{PN} is the N_2 volume fraction at the permeate end. In addition, let the height of filaments in the module be H H, the gas flow rate at the retentate end–retentate side is Q_R (STP), the gas pressure at the retentate end–retentate side is p_R, the volume fraction of CO_2 at the retentate end–retentate side isa_{RC}, and the volume fraction of N_2 at the retentate end–retentate side is a_{RN}.

The detailed model of the membrane separation is as follows:

$$Q_F = Q_P + Q_R \tag{49.5}$$

$$Q_F a_{FC} = Q_P b_{PC} + Q_R a_{RC} \tag{49.6}$$

$$Q_F a_{FN} = Q_P b_{PN} + Q_R a_{RN} \tag{49.7}$$

where $a_{FC} + a_{FN} = 1$, $b_{PC} + b_{PN} = 1$, $a_{RC} + a_{RN} = 1$.

Set the volume fraction of CO_2 at the height H of the filaments (retentate end–permeate side) be $\mathrm{y_{HC}b_{HC}}$, then according to the definition of the volume fraction, the following equation can be obtained:

$$b_{HC} = \frac{J_{CO2}(p_R a_{RC} - p_P b_{HC})}{J_{CO2}(p_R a_{RC} - p_P b_{HC}) + J_{N2}[p_R a_{RN} - p_P(1 - b_{HC})]} \tag{49.8}$$

By substituting J_{CO_2}, J_{N_2}, p_R and p_P into Formula (49.8) and combining the results obtained from the above steps, Formula (49.8) can be solved to obtain b_{HC}.

The flux of CO_2 through the membrane (Q_C) can be obtained from the following formula:

$$Q_C = J_{CO_2} A_{mem} \overline{\Delta P_{CO_2}} \tag{49.9}$$

where, the average partial pressure difference of CO_2 on both sides of the module membrane can be obtained by using logarithmic average method, that is,

$$\overline{\Delta P_{CO_2}} = \frac{(p_F a_{FC} - p_P b_{PC}) - (p_R a_{RC} - p_P b_{HC})}{\ln(\frac{p_F a_{FC} - p_P b_{PC}}{p_R a_{RC} - p_P b_{HC}})} \tag{49.10}$$

Thus, the calculated CO_2 recovery rate of the system (φ'_{CO_2}) is

$$\varphi'_{CO_2} = \frac{Q_C}{Q_F a_{FC}} = \frac{J_{CO2} A_{mem} \overline{\Delta P_{CO_2}}}{Q_F a_{FC}} \tag{49.11}$$

Similarly, the calculated N_2 recovery rate of the system ($\varphi'_{CO_2}\varphi'_{N_2}$) can be written.

49.2.2 *Procedure*

Firstly, the initial values of CO_2 recovery (φ_{CO_2}) and N_2 recovery (φ_{N_2}) are assumed, and then the parameters of feed end, permeate end and retentate end–retentate side are obtained by using the conservation Eqs. (49.5)–(49.7) combined with Eqs. (49.1)–(49.2) with the known quantities of Q_F anda_{FC}. Then, solving the Eqs. (49.8)–(49.11), the new recovery rates of CO_2 and N_2 (φ'_{CO_2} and $\varphi'_{CO_2}\varphi'_{N_2}$) can be obtained. The convergence conditions of the model are as follows: $(\varphi_{CO_2} - \varphi'_{CO_2})^2$ and $(\varphi_{N_2} - \varphi'_{N_2})^2$ is less than the given error limit (ε). That is to say, when the model does not satisfy the convergence conditions, we change the φ_{CO_2} and φ_{N_2} to iterate again until the convergence conditions are satisfied.

The flow chart of the simulation calculation is summarized in Fig. 49.2.

49.2.3 *Experiment*

In this paper, a single-stage membrane separation system was built under laboratory conditions. The schematic diagram of the system is shown in Fig. 49.3.

The system mainly consists of the following parts: CO_2 cylinder gas, N_2 cylinder gas, mass flowmeter, filter, heating belt, thermocouple, hollow fiber membrane, gas flow meter and gas analyzer.

CO_2 and N_2 were used to simulate the flue gas of coal-fired power plants. CO_2 and N_2 were supplied by cylinder gas. In the experiment, the volume fractions of different inlet gas components were proportioned by mass flowmeter. The experiment was carried out at 25–45 °C. The required temperatures were realized by heating belt and thermocouple. The feed pressure required for gas separation was provided by cylinder gas, and pressure regulating valves were used to achieve a specific feed pressure (Li et al. 2019).

49.3 Results and Discussion

In this paper, the related parameters under different working conditions were calculated by the above simulation method. The results are shown in Figs. 49.4 and 49.5.

In Figs. 49.4 and 49.5, the data points represent the experimental results, and the curves are the simulation results. It is not difficult to see that the simulation results are basically consistent with the trend of the experimental results.

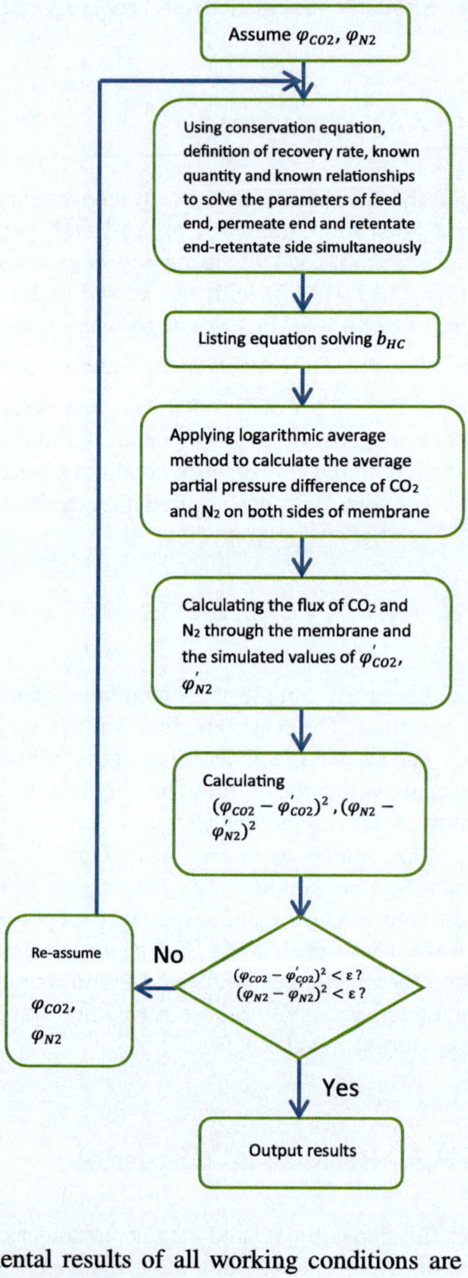

Fig. 49.2 Flow chart of simulation calculation

Further, the simulation and experimental results of all working conditions are plotted in scatter plots, as shown in Figs. 49.6 and 49.7.

Figure 49.6 shows that the relative error between the calculated CO_2 volume fraction at the permeate end and the experimental results is within or near +10% for most conditions.

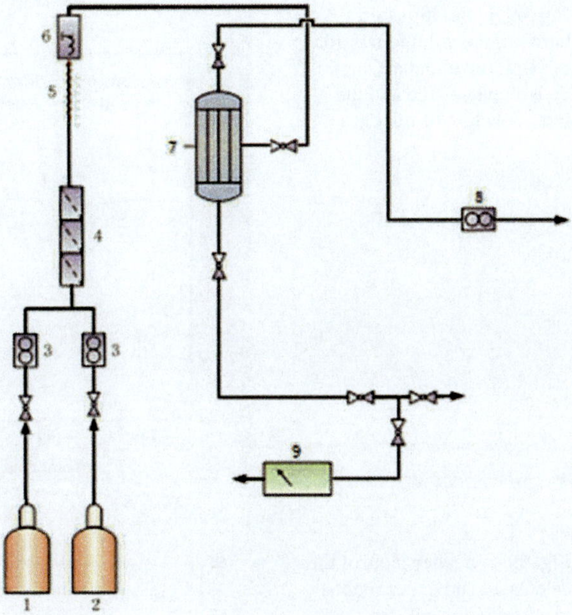

Fig. 49.3 Diagram of membrane separation system. 1—CO_2 cylinder gas 2—N_2 cylinder gas 3—mass flowmeter 4—filter 5—heating belt 6—thermocouple 7—hollow fiber membrane 8—gas flow meter 9—gas analyzer

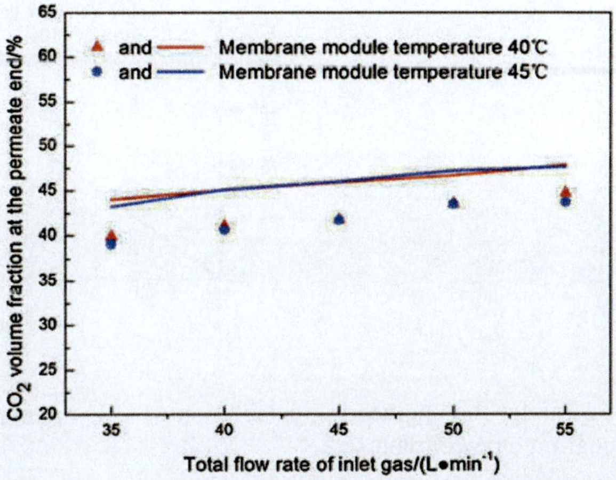

Fig. 49.4 Relationship between the CO_2 volume fraction at the permeate end and the total flow rate of inlet gas

Figure 49.7 shows that the relative error between the calculated volume fraction of N_2 at the retentate end–retentate side and the experimental results is within −5%. Combining the results of Figs. 49.6 and 49.7, it can be seen that the relative error between the simulation results and the experimental results is within or near ±10% for most working conditions. In the allowable error range, the simulation results can approximately reflect the experimental results. Therefore, this model can provide

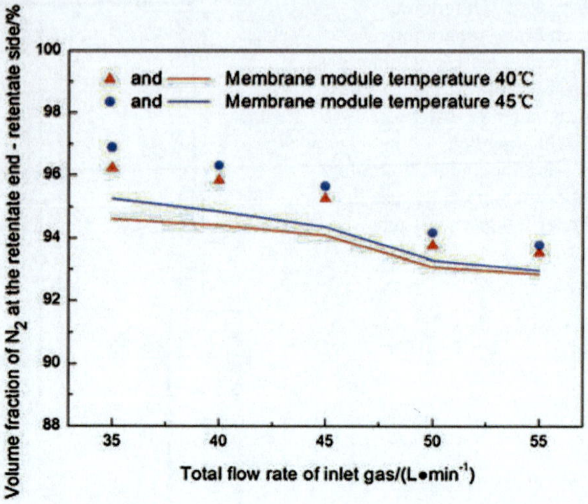

Fig. 49.5 Relationship between the volume fraction of N_2 at the retentate end–retentate side and the total flow rate of inlet gas

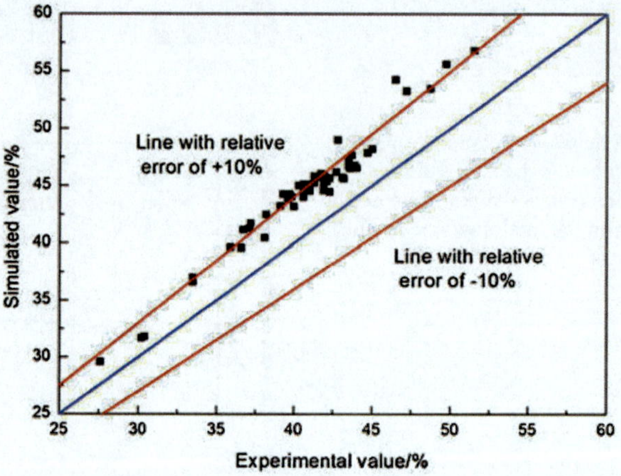

Fig. 49.6 Scatter plots of the calculated and experimental results of CO_2 volume fraction at the permeate end

a basis for the engineering calculation of CO_2 membrane separation of flue gas in coal-fired power plants.

49.4 Conclusion

Based on the experimental results, a model of separation membrane was established. The permeation rate of the model is variable. In addition, a single-stage membrane separation system was built in this paper using hollow fiber membranes under laboratory conditions. Different factors affecting single-stage membrane separation were experimented and the simulated and experimental values were compared.

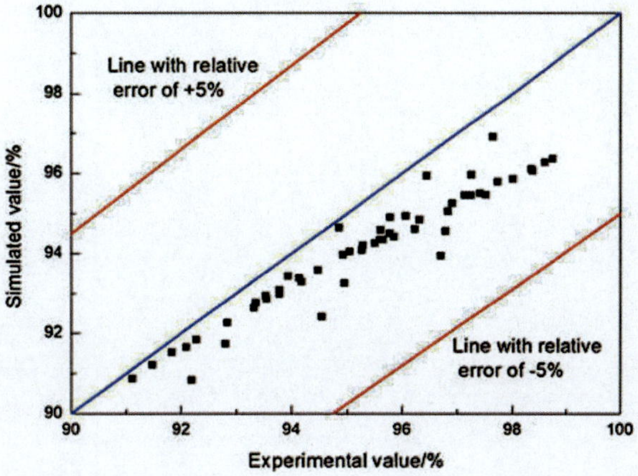

Fig. 49.7 Scatter plots of the calculated and experimental results of the volume fraction of N_2 at the retentate end–retentate side

The comparison results show that within the allowable error range, the simulated values can approximately reflect the experimental values. Therefore, this model has certain reference significance for the engineering application of CO_2 membrane separation of flue gas in coal-fired power plants.

Acknowledgements The authors of this paper gratefully acknowledge financial support from National Key R&D Program of China, Grant No. 2017YFB0603403.

References

Arias AM, Mussati MC, Mores PL, Scenna NJ, Caballero JA, Mussati SF (2016) Optimization of multi-stage membrane systems for CO_2 capture from flue gas. Int J Greenh Gas Control 53:371–390

Binns M, Se-Young Oh, Kwak D-H, Kim J-K (2015) Analysis of hybrid membrane and chemical absorption systems for CO_2 capture. Korean J Chem Eng 32(3):383–389

British Petroleum Corporation (2018) BP World Energy Statistics Yearbook. Beijing Grammy Digital Graphic Production Co., Ltd, Beijing, p 49

Kim SH, Kim J-K, Yeo J-G, Yeo Y-K (2017) Comparative feasibility study of CO_2 capture in hollow fiber membrane processes based on process models and heat exchanger analysis. Chem Eng Res Des 117:659–669

Li B, Yu J, Zhang Z, Jiang Y (2019) Experimental study and simulation of CO_2 membrane separation of coal-fired flue gas. In: Xiamen, The 11th Asia-Pacific power and energy engineering conference (APPEEC 2019)

Luhang W (2017) Simulation for CO_2 capture by using membrane gas separation from flue gas in power plant. Qingdao University of Science and Technology, Qingdao, pp 64–70

Shao P, Dal-Cin MM, Guiver MD, Kumar A (2013) Simulation of membrane-based CO_2 capture in a coal-fired power plant. J Membr Sci 427:451–459

Zhang L, Hu B, Wang X, Li J, Zhou X, Yang L (2017) CO_2 separation by PI hollow fiber membrane from desulfurized flue gas. Proc CSEE 37(9):2637–2643

Zhongyang L et al (2012) Carbon dioxide capture, storage and utilization technology. China Electric Power Press, Beijing, p 16

Chapter 50
Experimental Study on Slagging and Fouling Behaviors During Oxy-Fuel Combustion of Zhundong Coal Using a Drop Tube Furnace

Lei Zhao, Chang'an Wang, Yueyi Hu, Ruijin Sun, Guantao Tang, Jin Guo, Zhiming Jiang, Yongbo Du, and Defu Che

Abstract Zhundong coal featuring high content of alkali metals induces severe slagging and fouling problems during its utilization in utility boilers. Oxy-fuel combustion technology can achieve a large amount of CO_2 capture and reduce the NO_x emission. Unfortunately, in-depth investigation has been seldom carried out on the process of ash deposition and slagging of high-alkali coal during oxy-fuel combustion. In the present study, the oxy-fuel combustion of high-alkali coals were experimentally studied using a non-cooled sampling probe in a drop tube furnace. The fouling and slagging mechanisms were investigated in various atmospheres, temperatures and different O_2 concentrations. In addition, the slag samples were characterized by X-ray diffraction (XRD) and Scanning Electron Microscopy (SEM) coupled with Energy Dispersive X-ray Spectrometer (EDS). Experimental results indicated that under air and oxy-fuel conditions with the same O_2 content, the types of elements and minerals were almost the same. With the temperature increased, the adhesion between the slag samples was strengthened and the pore structure slag was increased. The degree of slag melting and sintering was aggravated at the temperature of 1200 °C and the slag structure was complicated. As the oxygen concentration was increased, some of the spherical particles had completely melted on the surface of the massive ash, resulting in an uneven surface. The present work can provide new insights into the slagging and ash deposition of Zhundong high-alkali coal during oxy-fuel combustion.

L. Zhao · C. Wang (✉) · Y. Hu · R. Sun · G. Tang · Y. Du · D. Che
State Key Laboratory of Multiphase Flow in Power Engineering, School of Energy and Power Engineering, Xi'an Jiaotong University, Xi'an 710049, China
e-mail: changanwang@mail.xjtu.edu.cn

J. Guo
Guangdong Institute of Special Equipment Inspection and Research, Foshan 528251, China

Z. Jiang · D. Che
Shunde Institute of Inspection, Guangdong Institute of Special Equipment Inspection and Research, Foshan 528251, China

J. Lyu and S. Li (eds.), *Clean Coal and Sustainable Energy*, Environmental Science and Engineering,
https://doi.org/10.1007/978-981-16-1657-0_50

Keywords Zhundong coal · Oxy-fuel combustion · High-alkali · Slagging · Fouling

50.1 Introduction

Zhundong coalfield with abundant coal resources is the largest integrated coalfield in China with a predicted reserve of 390 billion tons (Li and Li 2018; Wang and Li 2018). However, due to the large amount of alkali and alkaline earth metals (AAEMs) within Zhundong coal, it is easy to cause ash-related problems such as contamination, ash deposition and slagging on the heating surface during combustion, which seriously restricts the large-scale utilization of Zhundong coal (Li and Zhu 2016; Liu and Wang 2018). Oxy-fuel combustion technology is a CO_2 emission reduction control technology that can better undertake the existing coal-fired technologies (Toftegaard and Brix 2010; Li and Xu 2019). Compared to air combustion, oxy-fuel combustion can not only achieve CO_2 emission reduction, but also effectively reduce NO_x and SO_x emissions (Wall and Stanger 2011). However, the problems of ash deposition and slagging issues during the oxy-fuel combustion process are still likely to occur. Therefore, research and development of oxy-combustion technology of high-alkali coal has important scientific significance for promoting the efficient development and utilization of Zhundong coal and environmental protection.

Undoubtedly, alkali and alkaline earth metals in coal play a crucial role in the slagging characteristics and numerous studies have been conducted on AAEMs. Su and Pohl (2001) pointed that the Fe_2O_3/CaO molar ratio correlated the slagging propensity and the worst slagging occured when the ratio approached 1.0. Li and Li (2015) found that due to the forming of low melting-temperature wollastonite and other eutectic mixtures, Zhundong lignite brought apparent slagging on the wall near burner inlet around 1150 °C. Similarly, Wu and Zhang (2016) also found that abundant magnetite and wollastonite that were initially melted and agglomerated to form particles at high temperature contributed to the severe slagging behaviors in the radiative section. Wei and Tan (2018) found that sodium (Na), calcium (Ca), and iron (Fe) were the main elements leading to slagging when the gas temperature was about 1000 °C during Zhundong coal combustion, but the corresponding mechanisms were quite different. The difference in combustion atmosphere also affects coal combustion characteristics and ash deposition and slagging characteristics, but there are still no uniform conclusions. Engin and Atakül (2018) reported that insignificant difference was observed in the combustion behavior of coal between air and oxy-fuel condition with 21% O_2. Qi and Ji (2018) found that the slagging characteristics of different atmospheres were different. In the O_2/CO_2 atmosphere, the surface was eroded into a groove due to gasification, and Fe_2O_3 was reduced to FeO which enhanced slagging. Nevertheless, the effect of O_2/N_2 atmosphere was weak. Fryda and Sobrino (2012) found that the chemical and mineralogical compositions did not seem to differ significantly between air and oxy-fuel conditions, but ash deposition tended to be higher under oxy-fuel conditon. However, the results of other scholars showed that the ash deposition rate in the air combustion mode was larger than that in oxy-fuel mode, and there was only tiny change in the types of minerals in different atmospheres at the same temperature but a slight difference in the content (Li and Li 2013; Han and Wang 2019). Therefore, although much research on the mechanisms of ash deposition and slagging in air and oxy-fuel atmospheres, these conclusions might be contradictory to each other due to different experimental methods or different samples. The vast majority of studies only focus on the slagging characteristics at a single temperature under air condition, while rare literature on how the evolution of slagging with temperature changed under oxy-fuel atmosphere have been mentioned. In addition, the evolution of elemental composition and minerals in ash has yet to be fully understood.

The present study aims to reveal the fouling and slagging behaviors of Zhundong coal in oxy-fuel combustion using a drop tube furnace. The ashes were collected at various experimental conditions, including different gas temperatures, atmospheres and oxygen concentrations. The microstructure, elemental composition and crystalline minerals of ash were further analyzed using scanning electron microscope coupled with energy dispersive spectrometer (SEM-EDS) and X-ray diffractometer (XRD). The experimental results could provide new insights into solving the slagging and ash deposition issues of Zhundong high-alkali coal.

50.2 Experimental

The coal sample selected in the present experiment was Wucaiwan (WCW) bituminous coal from Xinjiang Zhundong coalfield and the particle size of the coal sample was sieved to 75–105 μm. The ash composition analysis of WCW coal was shown in Table 50.1. The main components in the ash are CaO, SO_3, SiO_2, and Fe_2O_3. The contents of SO_3 and SiO_2 account for 21.65 and 17.08%, respectively. Therefore, the combustion products of WCW coal are likey to be principally sulfate and silicate.

The experiments were carried out in a high temperature drop tube furnace (DTF) system, as shown in Fig. 50.1. The system includes particle feeder, gas-flow controller, combustion reaction system, sampling probe and temperature controller. The sampling probe was made of silicon carbon material and its ability to withstand high temperatures can reach 1500 °C or more. At the same time, the sampling site was detachable. In order to simulate the real boiler furnace combustion, the temperature of the combustion zone (length 1.2 m) was controlled to 1400 °C, and the sampling zone temperature (length 0.8 m) was set as 1300 °C. The pulverized coal sample was sent out by the spiral micro-powder feeding machine and the powder feeding

Table 50.1 Analysis of ash components of experimental coal

SiO_2	Al_2O_3	Fe_2O_3	CaO	MgO	Na_2O	SO_3
17.08	6.99	11.60	27.53	7.42	6.08	21.65

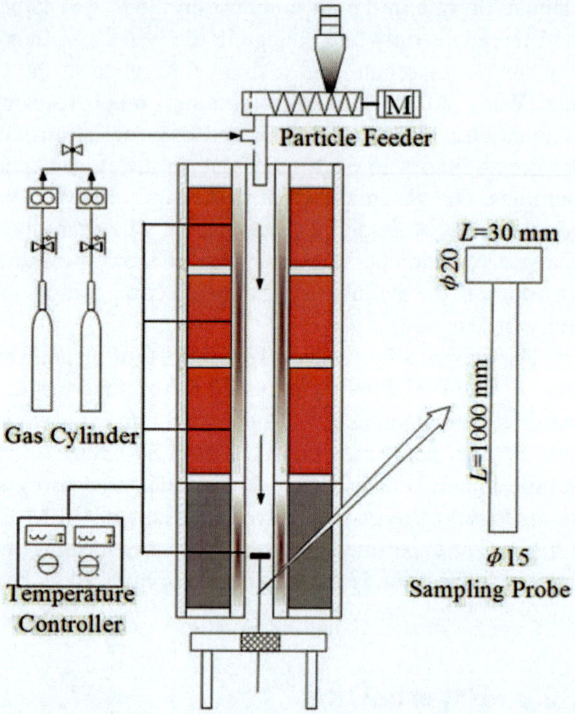

Fig. 50.1 The schematic diagram of experimental system

rate was controlled at 300 mg min^{-1}. The pulverized coal was put into the furnace for combustion, and the combustion products were collected by the sampling probe. The sampling probe was inserted from the bottom of the drop tube furnace and was sampled at different flue gas temperatures by adjusting the height of the sampling probe. The micro-structure and surface element composition of the combustion products on the probe were tested using a scanning electron microscopy (SEM) equipped with energy dispersive X-ray spectroscopy (SU3500, Tianmei). The distributions of the mineral phases in Zhundong coal slag were identified by an X-ray diffraction instrument (D8 ADVANCE, Bruker AXS).

50.3 Results and Discussion

50.3.1 Slagging Characteristics in Air and Oxy-Fuel Atmospheres

Except for the flue gas temperature, the atmosphere was also one of the causes of ash deposition and slagging and the change of atmosphere could affect the existence of minerals in ash (Fryda and Sobrino 2010). The evolution of crystal phase minerals

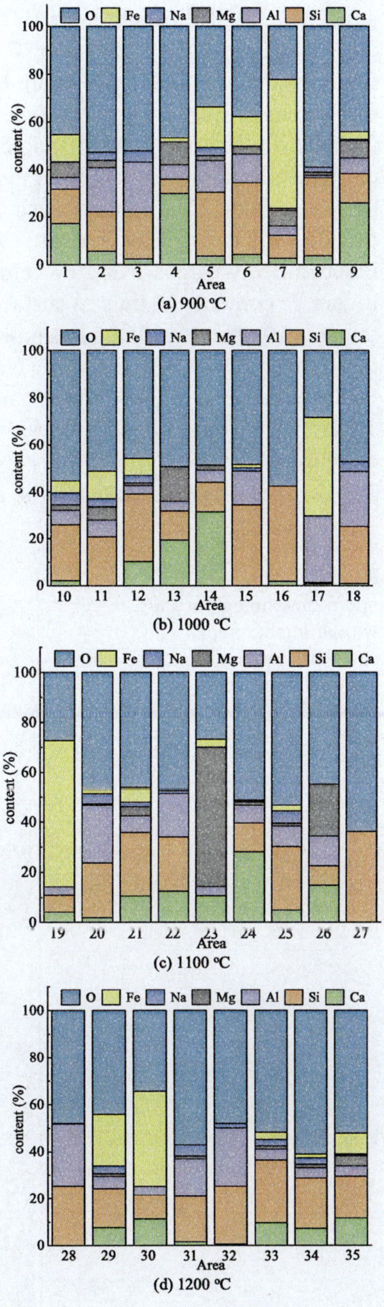

Fig. 50.4 The elemental compositions of ash surface in various areas

spherical ash of different sizes in the slag, such as areas 2, 4, 8, and 9 at 900 °C, and areas 13, 14 and 17 at 1000 °C. The main elements of the smaller spherical ash (such as area 4, 9, 13, and 14) are Ca, Si, Al and Mg, and there are a small amount of Fe in individual areas. The content of alkaline earth metals (Ca and Mg) is much higher than the content of Si and Al, which implies that the areas with high alkaline earth metals are not prone to slagging, probably due to their high melting point. The main elements of the spherical ash of the area 8 are Si and O. Hence, it is inferred to be SiO_2, indicating that there are still many SiO_2 in the ash which have not reacted independently. Areas 2 and 17 belong to the larger spherical ash, but area 17 has a higher Fe content and thus its surface is rough. The same phenomenon also occurs in the area 19 when the flue gas temperature is 1100 °C. This is mainly because under oxy-fuel combustion, the slag surface is eroded into grooves due to gasification, and Fe_2O_3 is reduced to FeO, which can further exacerbate slagging. The three types of 1200 °C ash were analyzed in the previous section and Fig. 50.5 shows the elemental composition of each types. The elemental composition of the area 28 and the areas 31, 32 are relatively close, and the main elements are O, Si and Al. However, areas

Fig. 50.5 The micro-morphology of slag with different oxygen concentrations under oxy-fuel condition

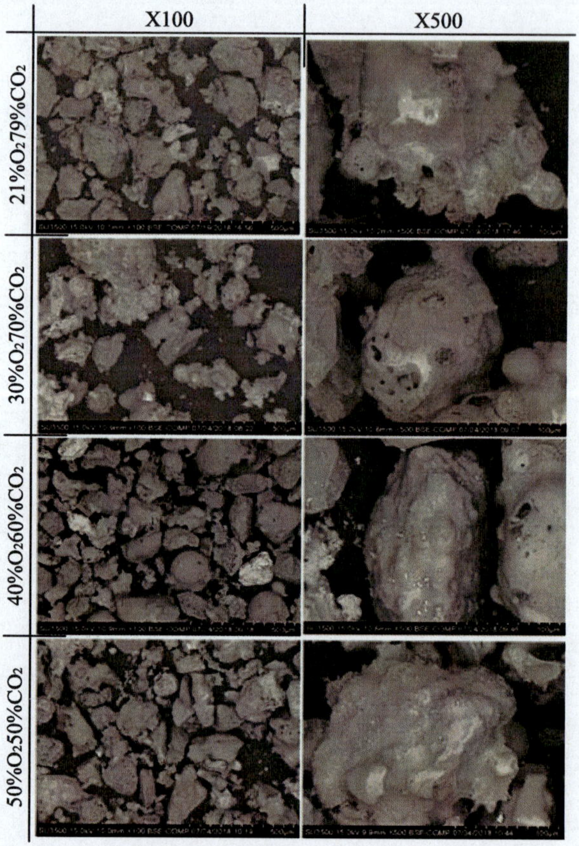

31 and 32 have a small amount of Na and Ca element, accordingly the integration of the area is more serious. The elemental compositions of the areas 33, 34 and 35 are similar, but there is a slight difference in conten.

50.3.3 Slagging Characteristics at Different Oxygen Concentrations

The effects of O_2 concentrations on the fouling and slagging characteristics of WCW coal at a flue gas temperature of 1100 °C were elucidated. In order to ensure the complete burnout of pulverized coal, the O_2 flow provided in the experiment was 1.2 times the O_2 flow required for the theoretical complete combustion. The flow rate of O_2 was maintained in each condition, and the O_2 concentration was adjusted by adjusting the flow rate of CO_2. The oxygen concentration is an important factor causing the volatilization of minerals in ash. According to the existing research results, as the oxygen concentration was increased, the slagging tendency was more serious. The types of minerals in the ash were almost unchanged, but the content of each component had a significant variation.

Figure 50.5 shows the effect of different oxygen concentrations on the microstructure of WCW coal slag samples. It can be seen from Fig. 50.5 that the slag samples have substantially completely melted at this temperature. The slag sample under different O_2 concentrations is composed of irregular block ash and the independent ash is rare. The difference in the micro-morphology of different oxygen concentrations is unobvious at 100 times, but it can still be seen that there is an increase in pore-like ash, which indicates that the slagging situation is slightly aggravated with the increase of oxygen concentration. There is a significant difference when the scanning electron microscope is zoomed to 500 times. At low O_2 concentrations, significant spherical particles embedded in the surface of the massive ash can still be observed. As the O_2 concentration increases, a portion of the spherical particles has completely melted on the surface of the massive ash, leading to a rough surface.

50.4 Conclusions

In this paper, the fouling and slagging behaviors of Zhundong coal in oxy-fuel combustion was experimentally studied using a high temperature drop tube furnace. The influences of combustion atmosphere (air and oxy-fuel), gas temperature (900–1200 °C) and oxygen concentration (21–50%) on the ash depositon and slagging were investigated. The major conclusions can be drawn as follows:

With the same oxygen concentration, the characteristic peaks correspond to the same mineral species during air and oxy-fuel combustion, but the content of each mineral component has changed to some extent.

As the temperature increases, the ash slagging degree is aggravated. The adhesion between the ash particle increases and the pore structure slag is raised. The independent spherical ash particle reduces and disappears completely above 1100 °C. The lesser amount of sodium is mainly present on the surface of massive ash or large spherical ash, forming the low temperature eutectic with other elements to capture ash particles. With the oxygen concentration increased, some of the spherical particles have completely melted on the surface of the massive ash, leading to a rough surface.

Acknowledgements The authors acknowledge financial support from the National Key R&D Program of China (2017YFB0602003) and the National Natural Science Foundation of China (51506163).

References

Engin B, Ataküll H (2018) Air and oxy-fuel combustion kinetics of low rank lignites. J Energy Inst 91(2):311–322

Fryda L, Sobrino C (2010) Study on ash deposition under oxy-fuel combustion of coal/biomass blends. Fuel 89(8):1889–1902

Fryda L, Sobrino C (2012) Study of ash deposition during coal combustion under oxy-fuel conditions. Fuel 92(1):308–317

Han T, Wang C (2019) Experimental study on ash deposition of zhundong coal in oxy-fuel combustion. J Energy Inst. https://doi.org/10.1016/j.joei.2019.01.004

Li G, Li S (2013) Comparison of particulate formation and ash deposition under oxy-fuel and conventional pulverized coal combustions. Fuel 106:544–551

Li G, Li S (2015) Fine particulate formation and ash deposition during pulverized coal combustion of high-sodium lignite in a down-fired furnace. Fuel 143:430–437

Li X, Li J (2018) Clean and efficient utilization of sodium-rich Zhundong coals in China: behaviors of sodium species during thermal conversion processes. Fuel 218:162–173

Li S, Xu Y (2019) Measurements and modelling of oxy-fuel coal combustion. Proc Combust Inst 37(3):2643–2661

Li J, Zhu M (2016) Characterisation of ash deposits on a probe at different temperatures during combustion of a Zhundong lignite in a drop tube furnace. Fuel Process Technol 144:155–163

Liu Y, Wang Z (2018) Inhibition of sodium release from zhundong coal via the addition of mineral additives: a combination of online multi-point LIBS and offline experimental measurements. Fuel 212:498–505

Qi Y, Ji S (2018) Experimental study of the high-temperature slagging characteristics of coal ash-biomass ash blends under different atmospheres. J Energy Inst. https://doi.org/10.1016/j.joei.2018.11.003

Su S, Pohl J (2001) Slagging propensities of blended coals. Fuel 80(9):1351–1360

Toftegaard M, Brix J (2010) Oxy-fuel combustion of solid fuels. Prog Energy Combust Sci 36(5):581–625

Wall T, Stanger R (2011) Demonstrations of coal-fired oxy-fuel technology for carbon capture and storage and issues with commercial deployment. Int J Greenh Gas Control 5:S5–S15

Wang C, Li G (2018) Ash deposition and sodium migration behaviors during combustion of zhundong coals in a drop tube furnace. J Energy Inst 91(2):251–261
Wei B, Tan H (2018) Investigation on ash deposition characteristics during zhundong coal combustion. J Energy Inst 91(1):33–42
Wu X, Zhang X (2016) Ash deposition and slagging behavior of Chinese Xinjiang high-alkali coal in 3 MW_{th} pilot-scale combustion test. Fuel 181:1191–1202

Chapter 51
Thermodynamic Study on the Utility Oxy-Fuel Boiler with Different Oxygen Volume Fractions

Kai Li, Enlu Wang, Deli Li, Lifen Wang, Naveed Husnain, and Qi Wang

Abstract Oxy-fuel combustion technology is one of the most promising technologies for capturing CO_2 emitted from coal-fired power plants. For the operation of an oxy-fuel combustion utility boiler, its thermodynamic features should be clarified. For this purpose, a calculation model for oxy-fuel combustion boiler was established, and the thermodynamic characteristics of a 2102 t/h oxy-fuel utility boiler were explored. The results show that at a given oxygen volume fraction, the start-up process of an oxy-fuel boiler can be completed in a short time. Along with the increasing of the oxygen volume fraction, from 22 to 29%, it showed a strong effect on furnace radiant surfaces and furnace volumes. And that variation of CO_2 enthalpy played a vital role in the flue gas enthalpy, the oxygen volume fraction was shown to affect the various parts of the furnace differently. The calculations in present research could contribute to the oxy-fuel boiler design and operation.

Keywords Oxy-fuel boiler · Thermodynamic study · Oxygen volume fraction · Furnace dimensions

51.1 Introduction

As known, the combustion of coal is always along with the emission of greenhouse gas. Especially the emission of CO_2 has gained great focus in recent years in connection with the discussions of global warming (Holloway 2005; Pipitone and Bolland 2009; González-Salazar 2015). In order to reduce the CO_2 emissions from power plants, carbon capture and storage (CCS) technology have been brought up (Baciocchi et al. 2006; Ahmed et al. 2016, 2017). To capture the CO_2 from a power plant, the main purpose of using this technology is to generate a flue gas with a high concentration of CO_2 and water vapor (Johansson et al. 2010; Stanger et al. 2015;

K. Li · E. Wang · D. Li · L. Wang (✉) · N. Husnain · Q. Wang
Institute of Theraml Engineering, School of Mechanical Engineering, Shanghai Jiao Tong University, Shanghai 200240, China
e-mail: wlwang@sjtu.edu.cn

J. Lyu and S. Li (eds.), *Clean Coal and Sustainable Energy*,
Environmental Science and Engineering,
https://doi.org/10.1007/978-981-16-1657-0_51

Yin and Yan 2016); and then separate the CO_2 from the flue gas by dehydration and low-temperature purification processes.

For achieving the concentration purpose of CO_2, one of the high-potential methods is the oxy-fuel (O_2/CO_2 recycle) combustion, where instead of using air as the oxidizer, pure oxygen or a mixture of O_2 and recycled flue gas is used to generate high CO_2 concentration product gas. Next, captured CO_2 can be stored or utilized in subsequent additional processes. In term of controlling CO_2 emission by concentration from the fossil fuel and biomass-fired boiler, oxy-fuel combustion in power plant is confirmed to be one of the most prospective technologies (Chen et al. 2012).

Although there are several researches on oxy-fuel combustion (Yin and Yan 2016; Chen et al. 2012; Hong et al. 2015; Fujimori and Yamada 2013; Toftegaard et al. 2010; Normann et al. 2008; Scheffknecht et al. 2011; Wang et al. 2012, 2017), most focus on the efficiency analysis of oxy-fuel boiler and effects of O_2 concentration or the bench-scale/pilot-scale test, the researches on start-up stability and effect of the oxygen volume fraction on the large-scale boiler are deficient. As a pioneer, Toftegaard et al. (2010) made an extensive literature study on oxy-fuel combustion with a special focus on combustion fundamentals. They highlighted the lack of research in a pilot and plant scale and testing in areas related to the application of the oxy-fuel technique in combustion research. Under this condition, the efforts on the design of oxy-fuel boiler have been accompanied by ambitious experimental campaigns of Foster Wheeler 30 MW_{th} oxy-fuel CFB (Circulating fluidized bed) boiler, which is a pilot test scale (Hotta 2012; Lupion et al. 2013). A bench-scale (3 MW_{th}) oxy-fuel CFB boiler (Nsakala and Liljedahl 2007; Turek, et al. 2006) has been validated by Alstom. Meanwhile, Metso has made efforts on the design of oxy-fuel boiler including a 4 MW_{th} bench scale test (Turek, et al. 2006; Combustion 2014; Seddighi Khavidak et al. 2015) and a 100 kW_{th} pilot-scale oxy-fuel CFB boiler (Sadegh Seddighi et al. 2013).

Since the scale-up process largely imposes risks to the investor and manufacturer, the thermodynamics of oxy-fuel boilers requires studies. Further, it is also known from previous studies that the efforts on the thermodynamics of plant scale oxy-fuel boiler and its applications are poor, so that more attentions should be paid for the purpose of promotion and utilization of the advanced oxy-fuel technology on plant scale.

By previous experimental data (Anthony 2012; Kuivalainen et al. 2010; Stewart et al. 2012; Tan et al. 2012; Wang et al. 2008; Wu et al. 2011) on the 0.8 MW_{th} and 100 kW_{th} oxy-CFB boilers, it is presented that the enriched oxygen in furnace has large influence on heat release and heat transfer, mainly due to the differences in radiation capability of flue gas. Then, the determined furnace diameter and heat transfer surface is variable corresponding to the variable O_2 concentration.

In addition, the plant scale oxy-fuel boiler is based on the smooth and efficient operation of the start-up period with the flue gas recycle (FGR) (Wu et al. 2018; Sung et al. 2018; Becher et al. 2011; Zhou and Moyeda 2010; Hu et al. 2012). In studies involving of start-up period, Starkloff et al. (2015) and Oko and Wang (2014) both modeled the entire boiler process with validations and viewed the start-up process is vital in boiler operation. In particular, Li et al. (2005) focused on the studies of the start-up process and stated the important role in boiler operation of that. Actually, there are two start-up schemes for the oxy-fuel boiler, such as the conventional air start-up scheme and the O_2/CO_2 start-up scheme (Hu and Jinyue 2012). Thus, the first important issue is to confirm the start-up characteristics of the oxy-fuel boiler on plant scale, especially the time arrived in a steady state. That is because the start-up success or not, or start-up performance determines whether the boiler can operate safely or not.

Whereas, concentrated oxygen (normally between 22–97% O_2) is used instead of atmospheric air in an oxy-fuel boiler, and it is mixed with recycled flue gas (RFG) as oxidant (Conti et al. 2016; Hindiyarti et al. 2007; Toporov 2014; Yuzbasi and Selçuk 2011). This results in a high CO_2 concentration (normally between 65–85% dry bases) in the flue gas. As well, the furnace flue gas composition and effective heat transfer (Anthony 2012; Kuivalainen et al. 2010; Stewart et al. 2012; Tan et al. 2012; Wang et al. 2008; Wu et al. 2011) are obviously influenced by the oxygen concentration (oxygen volume fraction is used in present paper), thus the furnace flue gas theoretical temperature, the furnace flue gas radiant capacity, and the dimensions of the furnace in the oxy-fuel combustion mode would also be influenced by the oxygen volume fraction. But the extent of furnace dimensional change caused by oxygen volume fraction is still unknown for the oxy-fuel boiler on plant scale.

In the present work, the flue gas circulating system of a 2102 t/h pulverized coal (PC) boiler is used as the object. Calculation model over time in the start-up period with FGR system is established under the two start-up schemes. Furthermore, the effects of oxygen volume fraction on furnace dimensions and flue gas characteristics are discussed.

51.2 Calculation Models

51.2.1 Models of Start-up Schemes

Start-up schemes for oxy-fuel boiler are presented in Fig. 51.1. The O_2/CO_2 start-up scheme is shown in Fig. 51.1a and the conventional air start-up scheme is shown in Fig. 51.1b. The properties of coals used in the simulations are summarized in Table 51.1, which is analyzed by Elementar™ Elemental Analyzer, with the accuracy of 0.1% in C/H/N/S detection and 0.2% in O detection. And the boiler system parameters are presented in Table 51.2.

The oxidant in the oxy-fuel boiler is the mixture of separated oxygen and circulating flue gas, the ratio of O_2/CO_2 in the mixture is proposed in view of clearly and concisely explanation. Firstly, a certain quantity of separated oxygen and circulating flue gas from boiler back-pass mixes, in which the quantities of oxygen and circulating flue gas are determined by the constituent of flue gas, the fuel flow quantity, the FGR (Hu and Jinyue 2012) and the excess oxygen factor of the oxidant. Then the oxidant gas is injected into the furnace with coal powder for continuous combustion

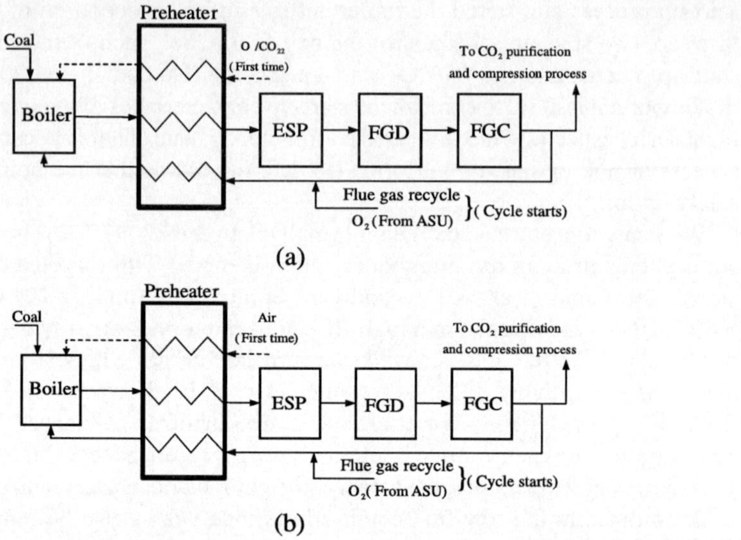

Fig. 51.1 Start-up schemes for oxy-fuel boiler. **a** O_2/CO_2 start-up; **b** conventional air start-up

Table 51.1 Ultimate analysis of the bituminous coal (as received basis)

Item	Value (%)	Item	Value (%)
C_{ar}	61.54	S_{ar}	0.62
H_{ar}	3.65	A_{ar}	17.98
O_{ar}	8.63	M_{ar}	6.6
N_{ar}	0.98	$Q_{ar,net,p}$ (kJ/kg)	24,010

Table 51.2 Boiler system parameters

Item	Value	Item	Value
Superheated steam flow D_{gr} (t/h)	2102	Reheated steam pressure p_{zr} (MPa)	4.72/4.52
Superheated steam pressure p_{gr} (MPa)	25.4	Reheated steam temperature t_{zr} (°C)	322/569
Superheated steam temperature t_{gr} (°C)	571	Feed water Pressure p_{gs} (MPa)	29.33
Reheated steam flow D_{zr} (t/h)	1761	Feed water temperature t_{gs} (°C)	282

and new flue gas with different proportion of constituents is produced. At last, heat release and radiation ability of flue gas and heat transfer between flue gas and heat exchange are calculated. After this, another circulation starts until the calculation is convergent and start-up period reaches a steady state.

Another one is so-called the conventional air start-up scheme as shown in Fig. 51.1b, which is the most common start-up type in the conventional air–fuel boiler and is verified to be true by the application. The only difference from the O_2/CO_2 start-up scheme is that the air, instead of the O_2/CO_2 mixture, is utilized as an oxidant for burning coal powder at the first stage of the start-up process in the oxy-fuel boiler. When coal powder is burning continuously in the air, the oxidant begins to switch to the O_2/CO_2 mixture for the follow-up combustion process. The circulation process, which is the same as the former scheme, repeats for a series of circulations, then reaches a steady state in oxy-fuel combustion.

In the oxy-fuel combustion, the different oxidant (air or an O_2/CO_2 mixture) resulted that the time spent to reach a steady state of the two start-up schemes is different. Therefore, it is necessary to deal with the two schemes in different ways during the calculations.

Meanwhile, two FGR systems are involved in the oxy-fuel combustion boiler: dry FGR system (mainly used for the pulverized coal drying and pneumatic transportation) and wet FGR system (mainly used for oxidant supply) (Hu et al. 2012; Hu and Jinyue 2012).

In dry flue gas circulation system, both primary and secondary circulating gas first flow through the dust catcher in back-pass, thereafter pass through the dehydrators where the flue gas loses its majority of moisture. In comparison, only the primary circulating gas flows through the dehydrator after dust catcher and loses its moisture in wet FGR system. The primary circulating gas enters into primary gas circulating fan. And the secondary circulating gas bypasses the dust catcher and enters the secondary gas circulating fan. Then the primary and secondary gas in both circulation systems carry the pulverized coal to the furnace and ensure the coal powder burned out completely. The circulation processing diagram is shown in Fig. 51.2.

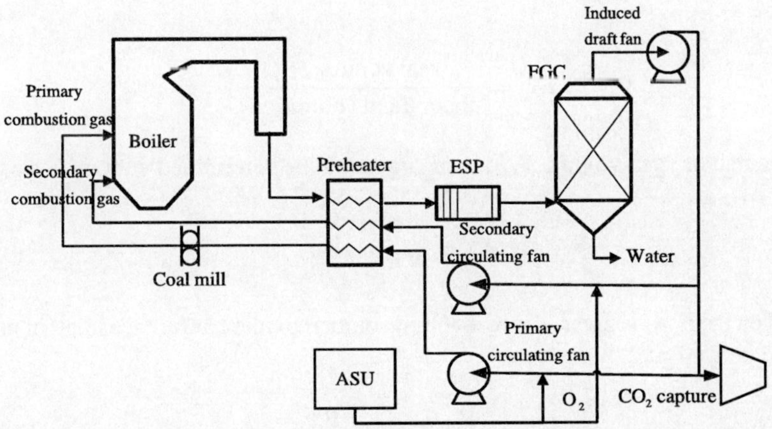

Fig. 51.2 Scheme of the flue gas recycle system in oxy-fuel boiler

51.2.2 Models of Furnace Heat Transfer

Furnace heat transfer model has been studied by Johansson et al. (2010); Kakaras et al. (2007); Taler et al. (2009). Until now, the zero-dimension model is still the main method used for furnace heat transfer calculations, which is called Gurvitch Method (Che 2008). There are two branches of Gurvitch method. In one branch the platen heated surfaces are included in the whole furnace; whereas, the other branch divides the furnace into two parts: the upper furnace and the lower furnace, and the furnace and platen heated surfaces are treated as an entirety. For a better accuracy, the latter separating method is more popular now when modern utility boilers are of huge dimensions. Therefore, the zero-dimensional model with separating furnace calculation method is employed to do thermodynamic calculations of the upper and lower furnaces in present work, respectively.

51.2.3 Models of Oxidant Gas

As O_2 is an oxidant in all combustion, the volume fraction of O_2 exiting in total oxidant is expressed with $r_{\mathrm{r,\,O2}}$, in %. When the coal rank is specified, theoretical oxygen volume (m^3/kg), which is the stoichiometric oxygen amount needed for fully burning out of per kilogram of coal, should be as follows (Che 2008):

$$V_{O_2}^0 = 0.01886C_{\mathrm{ar}} + 0.0556H_{\mathrm{ar}} + 0.007S_{\mathrm{ar}} - 0.007O_{\mathrm{ar}} \quad (51.1)$$

In order to ensure fully burning out the fuel in practice, the real oxygen consumption is larger than the theoretical oxygen volume, which is defined as excess oxygen factor α_{O2}:

$$\alpha_{O_2} = \frac{\text{real volume of } O_2}{\text{theoratical volume of } O_2} \quad (51.2)$$

The real oxygen volume V_{O2}, in m^3/kg, could be determined with excess oxygen factor α_{O2} as:

$$V_{O_2} = \alpha_{O_2} / V_{O_2}^0 \quad (51.3)$$

When the $r_{\mathrm{r,\,O2}}$ is given, the real volume of total oxidant at furnace inlet, in m3/kg, should be:

$$V_{O_2} = \alpha_{O_2} / V_{O_2}^0 \quad (51.4)$$

Then, the boiler begins for the flue gas circulation. Therefore, it is necessary to re-calculate the oxidant amount. As the oxidant amount consists of separated oxygen volume and circulating flue gas volume, the volume of separated oxygen from the air separated unit (ASU) and circulating flue gas should be analyzed. The nth gas circulation volumes satisfy the following relations.

$$\left(V_{O_2}\right)_n = \left(V_{O_2}^{ASU}\right)_n + \left(r_{O_2}\right)_{n-1}\left(V_y^{rec}\right)_n \tag{51.5}$$

$$(V_r)_n = \left(V_{O_2}^{ASU}\right)_n + \left(V_y^{rec}\right)_n \tag{51.6}$$

According to Eqs. (51.3) to (51.6), the total circulating flue gas volume $(V^{rec}{}_y)_n$ during the nth gas circulation volume of the oxy-fuel boiler and the separated oxygen volume $(V^{ASU}{}_{O2})_n$ are obtained respectively, as follows:

$$\left(V_y^{rec}\right)_n = \frac{V_r - V_{O_2}}{1-\left(r_{O_2}\right)_{n-1}} = \frac{\alpha_{O_2}(0.01886C_{ar} + 0.0556H_{ar} + 0.007S_{ar} - 0.007O_{ar})(1 - r_{r,O_2})}{r_{r,O_2}\left[1-\left(r_{O_2}\right)_{n-1}\right]} \tag{51.7}$$

$$\left(V_{O_2}^{ASU}\right)_n = \frac{V_{O_2} - \left(r_{O_2}\right)_{n-1}V_r}{1-\left(r_{O_2}\right)_{n-1}} = \frac{\alpha_{O_2}(0.01886C_{ar} + 0.0556H_{ar} + 0.007S_{ar} - 0.007O_{ar})\left(r_{r,O_2} - \left(r_{O_2}\right)_{n-1}\right)}{r_{r,O_2}\left[1-\left(r_{O_2}\right)_{n-1}\right]} \tag{51.8}$$

where, $(V^{rec}{}_y)_n$ represents circulating gas volume from behind the induced fans during the nth circulation, in m^3/kg; $(V^{ASU}{}_{O2})_n$ represents separated oxygen from ASU during the nth circulation, in m^3/kg; $(r_{O2})_{n-1}$ represents the oxygen volume fraction in total circulating flue gas during nth gas circulation, i.e. the volumetric O_2 fraction in flue gas volume during (n-1)$_{th}$ circulation at boiler back-pass after induced fans, in %. Its value could be obtained by iterating the calculation of boiler gas feature (n = 1, 2, 3 ......).

Therefore, when boiler fuel and excess oxygen α_{O2} factor are determined, and the oxygen volume fraction $r_{r,O2}$ in total oxidant keeps invariant, the Eqs. (51.7) and (51.8) provide formulas to determine the relationship between circulating flue gas volumes, separated oxygen volume and the circulation sequential number in the oxy-fuel boiler.

51.2.4 Models of Flue Gas Recycle (FGR)

Since the oxy-fuel combustion technology is closely associated with flue gas recirculation system and few studies on the flue gas have been done to investigate the FGR characterizations in an oxy-fuel combustion as well as the arrangement of flue gas, especially the allocation of primary and secondary circulating flue gas (PRFG and SRFG), which are equal to primary and secondary air in the conventional boiler here. The PRFG amount is vital for the quality of the combustion process in a furnace for a stable ignition (Che 2008; Namkhat and Jugjai 2010). An optimized ratio of PRFG ought to be chosen during boiler design stage based on its designed fuel rank and the type of associated pulverizing system. Generally, the optimized ratio of PRFG should be chosen according to the boiler design guidebook, and then the ratio of SRFG would be determined as the corresponding that of PRFG.

The combustion process in an oxy-fuel boiler is similar to the conventional air-blown pulverized coal combustion boiler. As known, the ratio of PRFG r_1 in the oxy-fuel boiler may take a value of 25–40% (Che 2008) and that of SRFG is $r_2 = 1\text{-}r_1$ generally when a bituminous coal with a volatile component $V_{\text{daf}} \leq 30\%$.

The calculation diagram for primary and secondary oxidant volumes is shown in Fig. 51.3. In order to prevent against the self-ignition of coal powder in the pulverizing system, the ambient air is taken as an analogy so that the oxygen concentration in PRFG maintains at 21% in the modeling. As well, the oxygen concentration in SRFG is determined by oxygen volume available from ASU and the secondary circulating flue gas volume.

Presuming that the total circulating flue gas volume $(V^{\text{rec}}{}_{\text{y}})_{\text{n}}$ after the induced draft fan and the total oxygen volume $(V^{\text{ASU}}{}_{\text{O2}})_{\text{n}}$ drawn from the ASU is given during nth

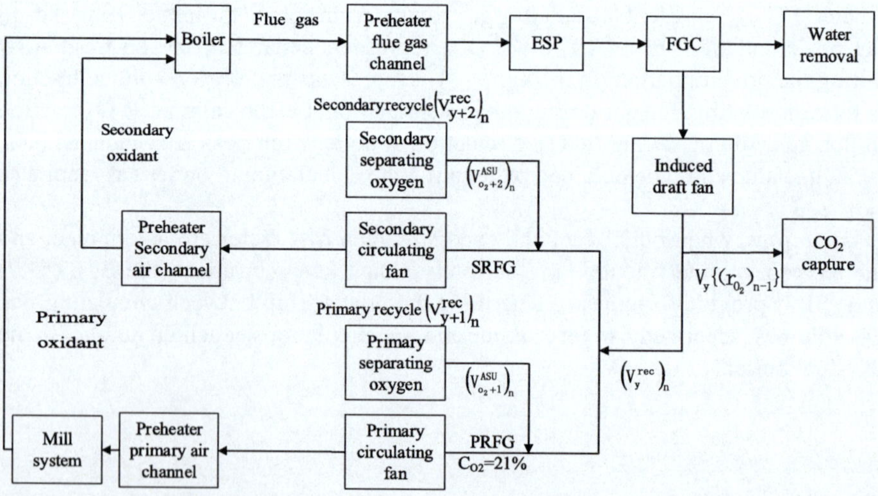

Fig. 51.3 Diagram on volumes calculation of PRFG and SRFG in oxy-fuel boiler

circulation, the total circulating flue gas volume and total oxygen volume drawn from ASU can get as follows:

$$\left(V_{y+1}^{rec}\right)_n+\left(V_{y+2}^{rec}\right)_n=\left(V_{y}^{rec}\right)_n \tag{51.9}$$

$$\left(V_{O_2+1}^{ASU}\right)_n+\left(V_{O_2+2}^{ASU}\right)_n=\left(V_{O_2}^{ASU}\right)_n \tag{51.10}$$

Presuming that the ratio of PRFG to SRFG is r_1/r_2, there is:

$$\frac{r_1}{r_2}=\frac{\left(V_{y+1}^{rec}\right)_n+\left(V_{O_2+1}^{ASU}\right)_n}{\left(V_{y+2}^{rec}\right)_n+\left(V_{O_2+2}^{ASU}\right)_n} \tag{51.11}$$

In order to prevent against self-ignition of coal powder in the pulverized system, it is necessary to maintain the oxygen concentration at $C_1 = 21\%$. Therefore, Eq. (51.11) can also be described as follows:

$$\left(r_{O_2}\right)_{n-1}\left(V_{y+1}^{rec}\right)_n+\left(V_{O_2+1}^{ASU}\right)_n=0.21\left[\left(V_{y+1}^{rec}\right)_n+\left(V_{O_2+1}^{ASU}\right)_n\right] \tag{51.12}$$

where, $(V^{rec}{}_{y+1})_n$ and $(V^{rec}{}_{y+2})_n$ represent primary and secondary circulating flue gas volumes drawn from boiler back-pass during the nth gas circulation respectively, in m^3/kg; $(V^{ASU}{}_{O2+1})_n$ and $(V^{ASU}{}_{O2+2})_n$ represent the primary and secondary separated oxygen volumes drawn from ASU during the nth gas circulation, in m^3/kg, ($n = 1$, 2, 3 ......).

Combining Eqs. (51.7) to (51.12), the primary and secondary circulating flue gas volumes during the nth circulation can be described as follows:

$$\left(V_{y+1}^{rec}\right)_n=\frac{0.79r_1}{\left[1-\left(r_{O_2}\right)_{n-1}\right]r_{r,O_2}}\alpha_{O_2}V_{O_2}^0 \tag{51.13}$$

$$\left(V_{y+2}^{rec}\right)_n=\frac{1-r_{r,O_2}-0.79r_1}{\left[1-\left(r_{O_2}\right)_{n-1}\right]r_{r,O_2}}\alpha_{O_2}V_{O_2}^0 \tag{51.14}$$

And the amount of PRFG and SRFG during the nth circulation can be obtained as follows:

$$\left(V_{O_2+1}^{ASU}\right)_n=\frac{\left[0.21-\left(r_{O_2}\right)_{n-1}\right]r_1}{\left[1-\left(r_{O_2}\right)_{n-1}\right]r_{r,O_2}}\alpha_{O_2}V_{O_2}^0 \tag{51.15}$$

$$\left(V_{O_2+2}^{ASU}\right)_n = \frac{\left[r_{r,O_2} - \left(r_{O_2}\right)_{n-1}\right] - \left[0.21 - \left(r_{O_2}\right)_{n-1}\right] r_1}{\left[1 - \left(r_{O_2}\right)_{n-1}\right] r_{r,O_2}} \alpha_{O_2} V_{O_2}^0 \tag{51.16}$$

As well, the primary $(V^1{}_r)_n$ and $(V^2{}_r)_n$ secondary oxidant volumes during the nth circulation respectively as follows:

$$\left(V_r^1\right)_n = \frac{r_1 \alpha_{O_2} (0.01886 C_{ar} + 0.0556 H_{ar} + 0.007 S_{ar} - 0.007 O_{ar})}{r_{r,O_2}} \tag{51.17}$$

$$\left(V_r^2\right)_n = \frac{r_2 \alpha_{O_2} (0.01886 C_{ar} + 0.0556 H_{ar} + 0.007 S_{ar} - 0.007 O_{ar})}{r_{r,O_2}} \tag{51.18}$$

When the oxygen concentration in the primary oxidant is maintained at $C_1 =$ 21%, the oxygen concentration in secondary oxidant is:

$$C_2 = \frac{r_{r,O_2} - 0.21 r_1}{1 - r_1} \tag{51.19}$$

Therefore, presuming that the coal rank is determined for an oxy-fuel combustion boiler and the pulverizing system configuration is also determined, an optimized first oxidant rate r_1 can be chosen based on the design guidelines which can ensure the stable, safe and effective ignition and burning of the coal powder.

If the excess oxygen factor α_{O2} and the volume percentage $r_{r,\,O2}$ are known at the furnace inlet, the primary and secondary circulating flue gas volumes, the primary and secondary separated oxygen volumes, the primary and secondary oxidant volumes, and the oxygen volume fractions in primary and secondary oxidants can be calculated by solving Eqs. (51.13) to (51.19).

The flue gas constituents and the volume fractions are quite different depending on different start-up schemes. Table 51.3 shows the flue gas volume calculations during the start-up period with the two start-up schemes, respectively.

Presuming that the total air leakage factors of the furnace and all the heated components are$\sum \Delta\alpha_i$ ($\Delta\alpha_i$ is air leakage factor at the ith heated component of the boiler), then the volume of air leaked is $\Delta V = \sum \Delta\alpha_i V0$. The calculations for flue gas volumes and volume fractions before and after the dehydrators are shown in Table 51.4.

The influence of start-up schemes to the furnace flue gas characteristics appeared only in the start-ups. As the start-up period reaches steady state, the flue gas characteristics are the same, no matter which scheme is employed. The calculation of the furnace flue gas volume after reaching steady circulation state is shown in Table 51.5.

Table 51.3 Calculation of furnace gas volumes during start-up period

Parameter	Calculation formation	
	O_2/CO_2 start-up scheme	Conventional air start-up scheme
Volume of CO_2 in flue gas	$(V_{CO_2})_0^{lt} = 0.01886C_{ar} + 0.007S_{ar} + (1 - r_{r,O_2})V_r$	$(V_{CO_2})_0^{lt} = 0.01886C_{ar} + 0.007S_{ar}$
Volume of N_2 in flue gas	$(V_{N_2})_0^{lt} = 0.008N_{ar}$	$(V_{N_2})_0^{lt} = 0.008N_{ar} + 0.79(V^0 + \Delta V^0)$
Volume of H_2O in flue gas	$(V_{H_2O})_0^{lt} = 0.1118H_{ar} + 0.0124M_{ar}$	$(V_{H_2O})_0^{lt} = 0.1118H_{ar} + 0.0124M_{ar} + 0.0161(V^0 + \Delta V^0)$
Volume of O_2 in flue gas	$(V_{O_2})_0^{lt} = (\alpha_{O_2} - 1)V_{O_2}^0$	$(V_{O_2})_0^{lt} = (\alpha_{O_2} - 1)V_{O_2}^0$
Total volume of flue gas	$(V_y)_0^{lt} = (V_{CO_2})_0^{lt} + (V_{N_2})_0^{lt} + (V_{H_2O})_0^{lt} + (V_{O_2})_0^{lt}$	$V^0 = (0.01886C_{ar} + 0.0556H_{ar} + 0.007S_{ar} - 0.007O_{ar})/0.21$ $\Delta V^0 = (r_{r,O_2} - 1)(0.01886C_{ar} + 0.0556H_{ar} + 0.007S_{ar} - 0.007O_{ar})/0.21$ $(V_y)_0^{1t} = V^0 + \Delta V^0$

51.3 Results and Discussions

As the sample calculation case, the bituminous coal is used in the 2102 t/h supercritical oxy-fuel pulverized-coal combustion boiler, the ratio of O_2/CO_2 in oxidant at the furnace inlet remains constant at 27/73, and the excess oxygen factor α_{O2} is 1.20. Furthermore, the oxygen volume fraction, the structure calculation and heat transfer calculation are shown in Tables 51.1 and 51.2, respectively.

51.3.1 Stability Analyses of the Start-up Period

The conventional air start-up scheme and O_2/CO_2 start-up scheme are both studied and discussed with different air leakage ratio and dehydrate rate, which are presented in Figs. 51.4 and 51.5. It is shown that there are similar variation rules for circulating flue gas volumes versus circulation numbers (time) in both start-up schemes.

For stability analyses of the start-up period, it is essential to establish a criterion for judging whether the flue gas circulation reaches a steady state or not. It is provided that the oxy-fuel boiler has reached a steady state if the relative difference ratio of

Table 51.4 Calculation of the flue gas volume during dehydration

Parameter	Calculation formation	
	Before dehydration	After dehydration
Volume and volume fraction of CO_2 in flue gas	$(V_{CO_2})'_n = (V_{CO_2})'^{lt}_n$ $(r_{CO_2})'_0 = (V_{CO_2})'_0/(V_y)'_0$	$(V_{CO_2})_0 = (V_{CO_2})'_0$ $(r_{CO_2})_0 = (V_{CO_2})_0/(V_y)_0$
Volume and volume fraction of N_2 in flue gas	$(V_{N_2})'_n = (V_{N_2})'^{lt}_n + 0.79\Delta V$ $(r_{N_2})'_0 = (V_{N_2})'_0/(V_y)'_0$	$(V_{N_2})_0 = (V_{N_2})'_0$ $(r_{N_2})_0 = (V_{N_2})_0/(V_y)_0$
Volume and volume fraction of O_2 in flue gas	$(V_{O_2})'_0 = (V_{O_2})'^{lt}_0 + 0.21\Delta V$ $(r_{O_2})'_0 = (V_{O_2})'_0/(V_y)'_0$	$(V_{O_2})_0 = (V_{O_2})'_0$ $(r_{O_2})_0 = (V_{O_2})_0/(V_y)_0$
Volume and volume fraction of O_2 in flue gas	$(V_{H_2O})'_0 = (V_{H_2O})'^{lt}_0 + 0.0161\Delta V (r_{H_2O})'_0 = (V_{H_2O})'_0 / (V_Y)'_0$	$(V_{H_2O})_0 = (r_{H_2O})_0 (V_y)_0$ $(r_{H_2O})_0 = k_0 (r_{H_2O})'_0$
Total volume of flue gas	$(V_y)'_0 = (V_{CO2})'_0 + (V_{N2})'_0 + (V_{H2O})'_0 + (V_{O2})'_0$	$(V_y)_0 = \frac{(V_{CO2})_0 + (V_{N2})_0 + (V_{O2})_0}{1-(r_{H2O})_0}$
Dehydration rate	$k_n = \frac{volume fraction of H_2O behind dehydration at the nth circulation}{volume fraction of H_2O before dehydration at the n^{th} circulation}$	

volume fraction between two successive circulations is less than 1×10^{-6}. Before reaching a steady state, the volumes of circulating flue gas in the latter calculation are higher than that in the former calculation, which can be explained by Eq. (51.7).

According to the determined criterion, the oxy-fuel boiler in the O_2/CO_2 start-up scheme reaches a steady state until 15 circulations while it reaches a steady state until 25 circulations in the conventional air start-up scheme. The time spent to reach a steady state of the O_2/CO_2 start-up scheme is 40% less than the other one. That is, the oxy-fuel boiler in the O_2/CO_2 start-up scheme reaches a steady state faster than the conventional air start-up scheme.

In the O_2/CO_2 start-up scheme, the volumes of circulating flue gas increase more gently as shown in Fig. 51.4, because the constituents of inlet oxidant gas of the O_2/CO_2 start-up in the first circulation is similar to circulating flue gas (oxidant gas in the O_2/CO_2 start-up is the mixture of separated oxygen and circulating flue gas, while the that in the conventional air start-up is fresh air instead). And this also leads to the higher oscillation of gas volume in the conventional air start-up, which is un-beneficial to boiler starts smoothly.

Table 51.5 Calculation of the flue gas volume on steady state

Parameter	Calculation formation
Volume and volume fraction of CO_2 in flue gas	$(V_{CO2})_{\rm n}^{\rm lt} =$ $0.01886C_{\rm ar} + 0.007S_{\rm ar} + (r_{CO2})_{\rm n\text{-}1}\left(V_{\rm y}^{\rm rec}\right)_{\rm n}$ $\left(r_{CO_2}\right)_{\rm n}^{\rm lt}=\left(V_{CO_2}\right)_{\rm n}^{\rm lt}/\left(V_{\rm y}\right)_{\rm n}^{\rm lt}$
Volume and volume fraction of N_2 in flue gas	$\left(V_{N_2}\right)_{\rm n}^{\rm lt} = 0.008N_{\rm ar} + \left(r_{N_2}\right)_{\rm n\text{-}1}\left(V_{\rm y}^{\rm rec}\right)_{\rm n}$ $\left(r_{N_2}\right)_{\rm n}^{\rm lt}=\left(V_{N_2}\right)_{\rm n}^{\rm lt}/\left(V_{\rm y}\right)_{\rm n}^{\rm lt}$
Volume and volume fraction of H_2O in flue gas	$\left(V_{H_2O}\right)_{\rm n}^{\rm lt} =$ $0.1118H_{\rm ar} + 0.0124M_{\rm ar} + \left(r_{H_2O}\right)_{n-1}\left(V_{\rm y}^{\rm rec}\right)_{\rm n}$ $\left(r_{H_2O}\right)_{\rm n}^{\rm lt}=\left(V_{H_2O}\right)_{\rm n}^{\rm lt}/\left(V_{\rm y}\right)_{\rm n}^{\rm lt}$
Volume and volume fraction of O_2 in flue gas	$\left(V_{O_2}\right)_{\rm n}^{\rm lt} = \left(\alpha_{O_2} - 1\right)V_{O_2}^{0}$ $\left(r_{O_2}\right)_{\rm n}^{\rm lt}=\left(V_{O_2}\right)_{\rm n}^{\rm lt}/\left(V_{\rm y}\right)_{\rm n}^{\rm lt}$
Total volume of the flue gas	$\left(V_{\rm y}\right)_{\rm n}^{\rm lt} =$ $\left(V_{CO_2}\right)_{\rm n}^{\rm lt} + \left(V_{N_2}\right)_{\rm n}^{\rm lt} + \left(V_{H_2O}\right)_{\rm n}^{\rm lt} + \left(V_{O_2}\right)_{\rm n}^{\rm lt}$

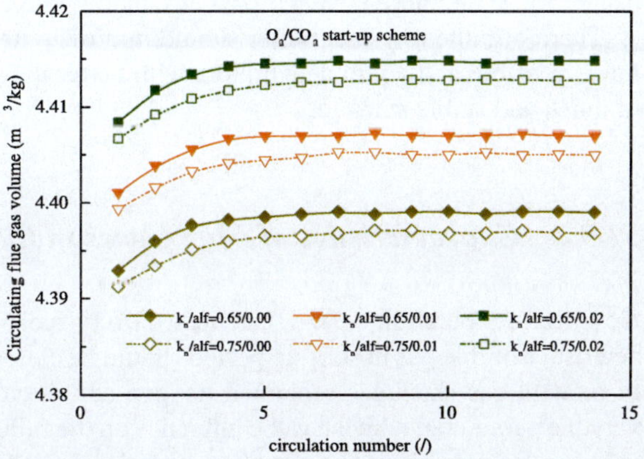

Fig. 51.4 Circulating flue gas volumes with respect to circulation number under O_2/CO_2 start-up scheme

Based on the states of Che (2008), dehydration has a vital influence on the constituent proportion of circulating flue gas, which is closely related to the start-up process using circulating flue gas. Meanwhile, air leakage will change the oxygen amount in flue gas and oxidant gas. Lupion et al. (2013) conducted the initial operational experiences of 30MWth oxy-fuel CFB boiler including the fixed air leakage case of 3%, but no research on its effects was given.

Hence, it is necessary to consider the influence of dehydration rates (the ratio of the volume fraction of H_2O after dehydration to that before dehydration) and

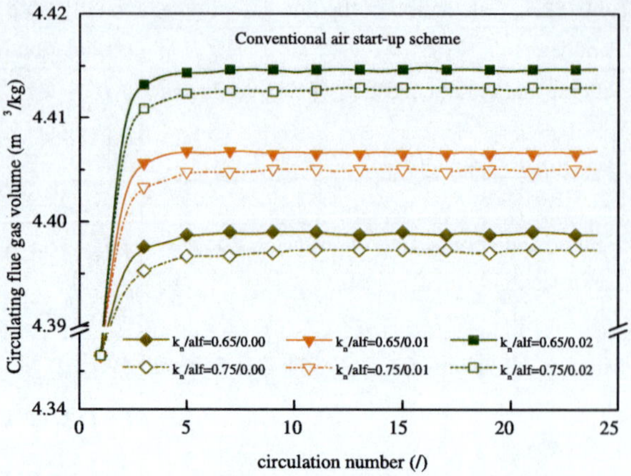

Fig. 51.5 Circulating flue gas volumes with respect to circulation number under conventional air start-up scheme

furnace air leakage factors to the flue gas circulating process. The circulating flue gas volume increases with the increase in air leakage factors; whereas, the circulating flue gas volume increases with the decrease in dehydration rate, which is shown correspondingly by real and broken lines in Figs. 51.4 and 51.5.

By comparing groups of real and broken lines of dehydration rates 0.65 and 0.75, it is found that the less the dehydration rate k_n is, and the stronger the dehydrating capability is, the more the back-pass circulating flue gas volume becomes.

Therefore, the oxy-fuel boiler should minimize its furnace air leakage factor as far as possible and equip dehydrators with better dehydration rates for the purpose of quick and stable start-up.

51.3.2 *Characteristics of Flue Gases in the Start-up Period*

Now that the operability of the start-up in the oxy-fuel boiler has been identified, characteristics of flue gas in start-up period should be cleared. In this period, constituents of oxidant gas (includes separated oxygen and circulating flue gas) is a transient variable parameter, which has the influence on the followed thermodynamic analysis in heat release and heat transfer characteristics. In order to quantify the process, two parameters, volumes of separated oxygen and CO_2 volume fraction in circulating flue gas are selected as the study object. The details are as follows:

Figures 51.6 and 51.7 show the volumes of separated oxygen with respect to circulation number in the oxy-fuel boiler for the O_2/CO_2 and the conventional air start-up schemes, respectively. It is found that the trends of the volumes of separated oxygen with respect to the circulation numbers are similar for both start-up schemes.

The influence of furnace air leakage to separated oxygen volume could be expressed as that the more the furnace air leakage factor is, the less the separated oxygen volume will be; that means, more air leakage supplies to more oxygen and

Fig. 51.6 Separated oxygen volumes with respect to circulation number under O_2/CO_2 start-up scheme

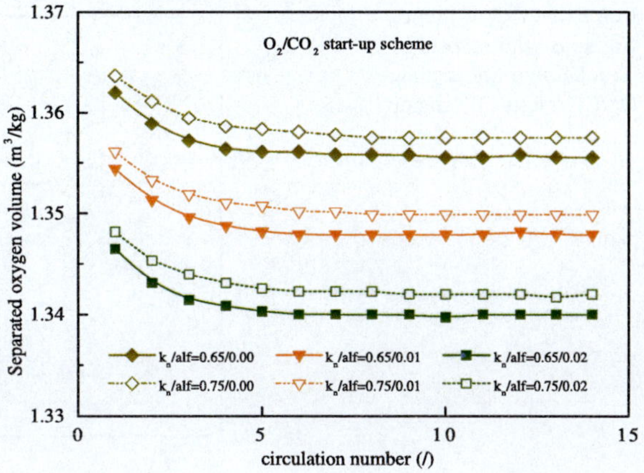

Fig. 51.7 Separated oxygen volumes with respect to circulation number under conventional air start-up scheme

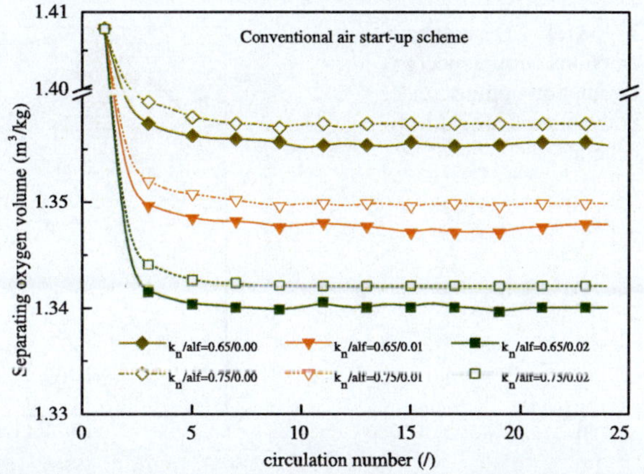

less separated oxygen volume is needed for full combustion. But air leakage means the additional impurities (mostly nitrogen) is put into flue gas in the furnace, and it will lead to increase of circulating flue gas amount, which verified the results of Figs. 51.4 and 51.5. Furthermore, the separated oxygen volume increases with the decrease in dehydration rate which states by comparison of the curves with different dehydration rates under the same air leakage.

Figure 51.8 shows the CO_2 volume fraction in furnace flue gas with respect to circulation number under the O_2/CO_2 start-up scheme. It is found that the CO_2 volume fraction decreases smoothly until reaching certain values after about 15 circulations when boiler operation switch from start-up state to a steady state.

Figure 51.9 shows that the CO_2 volume fraction in furnace flue gas with respect to circulation number under the conventional air start-up scheme in the same oxy-fuel combustion boiler. It can be observed that the CO_2 volume fraction in the furnace flue gas increases continuously with the increase in the circulation number value and

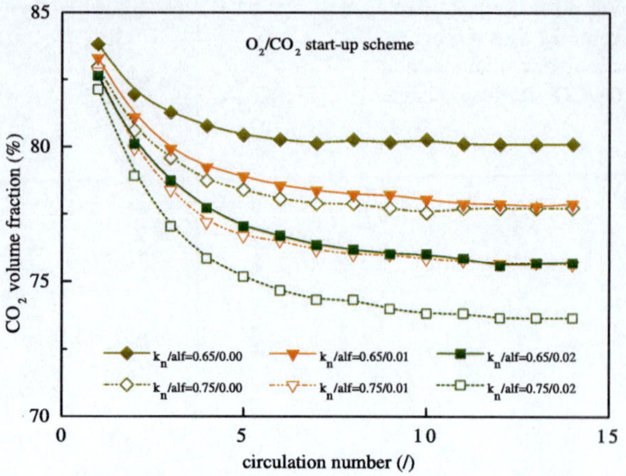

Fig. 51.8 CO_2 volume fractions with respect to circulation number under O^2/CO^2 start-up scheme

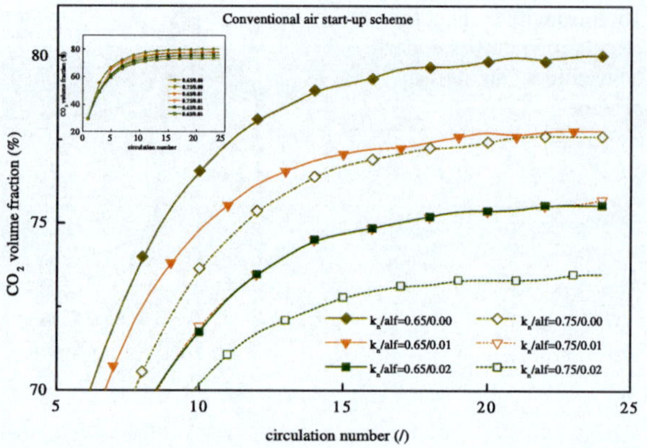

Fig. 51.9 CO_2 volume fractions with respect to circulation number under conventional air start-up scheme

reaches a constant value at the range from 30 to 80% at steady state, with a sharp rise in start circulations.

Comparing the CO_2 volume fractions shown in the two figures, it is found that the influences of furnace air leakage factor and the dehydration rate on the volume fraction of CO_2 in furnace flue gas are just the same under both start-up schemes. The furnace air leakage would decrease CO_2 volume fraction in the furnace. The more the air leakage factor is, the less the furnace CO_2 will be; after the boiler reaches a steady state. Meanwhile, the less the value of dehydration ssrate k_n and the higher the value of dehydrating capability, the larger will be the value of furnace volume fraction of CO_2. Therefore, in order to increase the volume fraction of tri-atom gas in the furnace, and to improve furnace flue gas radiation capability, it is suggested to avoid air leakage into the furnace and to utilize the high-efficiency dehydrators in the oxy-fuel boiler for the concentration of CO_2.

Figure 51.10 shows that the N_2 volume fraction in furnace flue gas becomes larger with the increase in circulation number value under the O_2/CO_2 start-up scheme. Even

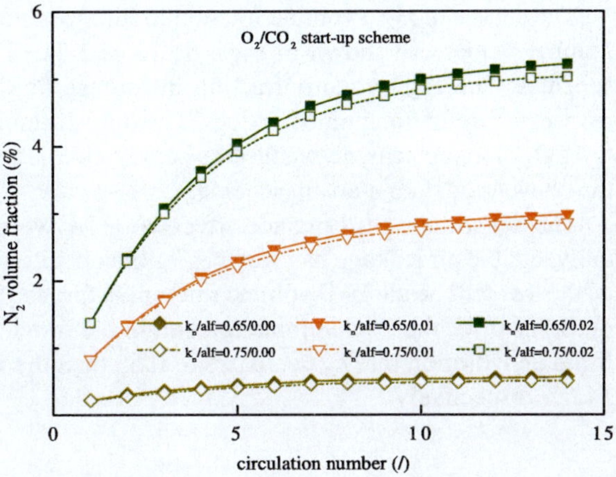

Fig. 51.10 N_2 volume fractions with respect to circulation number under O_2/CO_2 start-up scheme

a very small air leakage could make a drastic increase in N_2 volume fraction. For example, the N_2 volume fraction in furnace flue gas is 0.545% without air leakage when boiler operation reaches a steady state. If the furnace air leakage factor increases to 0.02, the N_2 volume fraction becomes 5.306%.

Analyzing N_2 volume fraction curves under the constant air leakage factor, it can be found that the dehydration capability affects the N_2 volume fraction slightly. The difference of N_2 volume fractions under two dehydration rates is quite small. But the difference would increase with the increase in the air leakage amount.

As shown in Fig. 51.11, N_2 volume fraction under the conventional air start-up scheme decreases drastically from 54% in the first circulation to 12% in the seventh circulation. After the eighth circulation, N_2 volume fractions would slow down its decreasing. N_2 volume fraction is 0.545% at steady state without air leakage. Whereas, the fraction values are 3% and 5%, corresponding to the air leakage factor of 0.01 and 0.02, respectively.

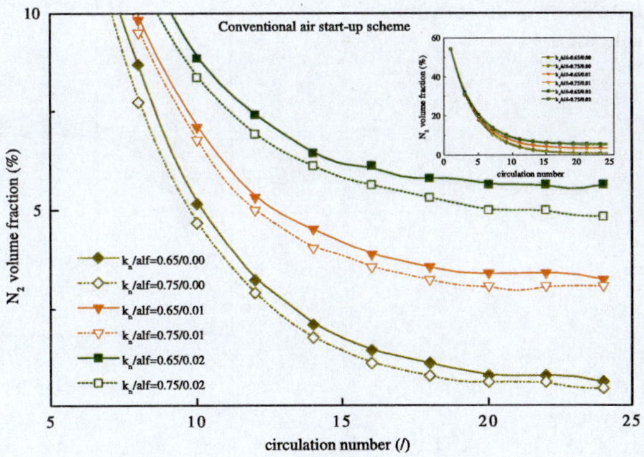

Fig. 51.11 N_2 volume fractions with respect to circulation number under conventional air start-up scheme

The curves of H_2O volume fraction in furnace flue gas with respect to circulation number values are shown in Figs. 51.12 and 51.13 for the two different start-up schemes. The H_2O volume fraction in furnace flue gas both would increase with respect to circulation number values. The only difference is that the system under the O_2/CO_2 start-up scheme reaches its steady circulation state earlier than that under the conventional air start-up scheme.

The dehydrator performance affects the H_2O volume fraction in flue gas essentially, but the air leakage has very negligible effects. The less the dehydration rate k_n is, the less will be the H_2O volume fraction in furnace flue gas. When the dehydration rate k_n is 0.75, the H_2O volume fraction in the furnace is about 17% at steady state. If the dehydration rate k_n decreases to 0.65 then the H_2O volume fraction becomes 14.6% respectively.

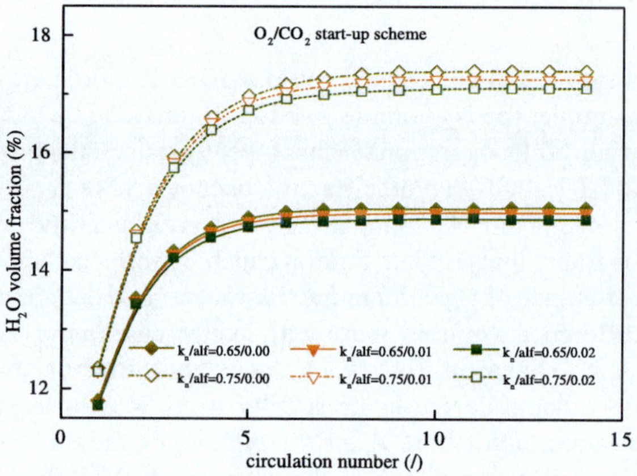

Fig. 51.12 H_2O volume fractions with respect to circulation number under O_2/CO_2 start-up scheme

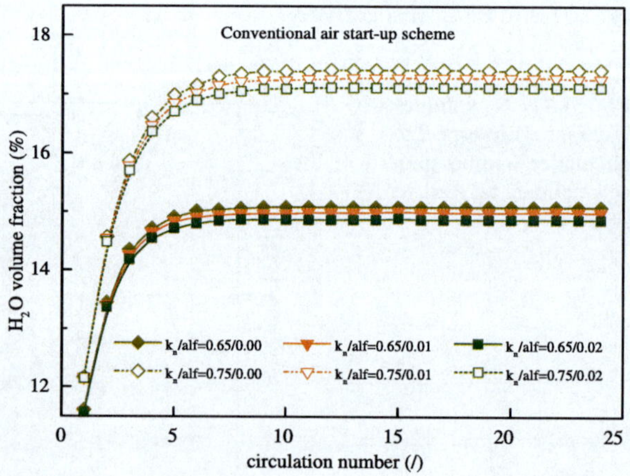

Fig. 51.13 H_2O volume fractions with respect to circulation number under conventional air start-up scheme

51.3.3 Effects of Oxygen Volume Fraction

Until now, it is confirmed that the start-up periods of the oxy-fuel boiler with two schemes reach a steady period in a short time. Furthermore, the attention turns to pay on the effects of oxygen volume fraction on the oxy-fuel boiler, especially the furnace dimension, theoretical flame temperature and enthalpy of furnace flue gas.

The furnace structure calculation is carried out at the following geometry unchanged: the cross-sectional dimensions of the lower furnace, the trapezoid shapes and dimensions of furnace nose and hopper. And the principle for determining the height of the rectangular belt F_7 in the upper furnace is that the flue gas velocity between the rear panels of the platen super-heater should keep constant at 8.12 m/s, which is the same as the conventional air combustion mode.

The given dimensions of lower furnace are as follows for all oxy-gen volume fractions: the furnace width B as 18.816 m, furnace depth C as 18.144 m, the depth at hopper lowest bottom C_{hd} as 9.72 m, the half-height of hopper h_{hd} as 6.02 m, its slope angle as 50°, the trapezoid height of the furnace nose h_{tx} as 5.172 m, its lower slope angle θ_1 as 50°, the depth at exit window c_{ck} as 13.804 m. The structure of upper furnace includes mainly: the rectangular belt at furnace ceiling F_7, the trapezoid belt at furnace ceiling F_8. The height of the furnace nose remains at 0.568 m, its upper slope angle θ_2 at 30°. The corresponding diagram of the boiler structure is shown in Fig. 51.14.

The change in the vertical dimensions of the furnace sections with respect to the oxygen volume fraction is shown in Fig. 51.15. The total height of the furnace increases from 46.692 m to 57.262 m with the increase in oxygen volume fraction from 22 to 29%, and the growth rate is 2.162 m per oxygen volume fraction of 1%. In comparison, the height of the upper furnace is getting smaller with the increase in oxygen volume fraction, and the growth rate is −0.654 m per oxygen volume fraction of 1%. It can be explained that the total flue gas volume decreases with the increase in the oxygen volume fraction, so it is necessary to reduce the height of

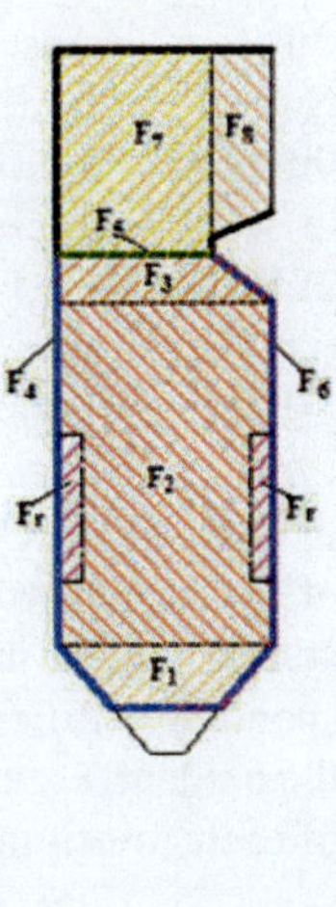

Fig. 51.14 Scheme of 2-D divisions of oxy-fuel boiler furnace

Fig. 51.15 Vertical furnace dimensions with oxygen volume fraction

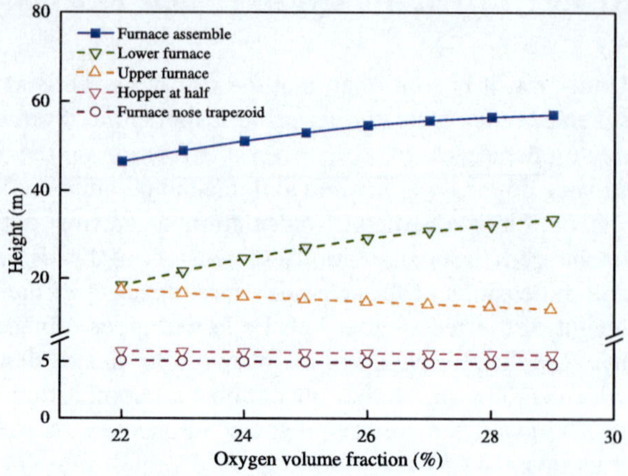

rectangular belt for the required flue gas velocity at 8.12 m/s and increase the height of lower furnace to receive more heat from furnace flue gas.

Furnace volume and the radiant surface area of the lower furnace with oxygen volume fraction are shown in Fig. 51.16. Increasing oxygen volume fraction leads to a higher height of the lower furnace, thus the total radiant surface area in lower furnace increases with the increase in oxygen volume fraction. As all the aforementioned factors have an influence on the radiation performance of the furnace, therefore it is necessary to observe the effects of oxygen volume fraction on the furnace radiant surfaces and the furnace volumes.

In Fig. 51.16a, along with the increase in the oxygen volume fraction within the range from 22 to 29%, furnace radiant surface increases sharply from 2612.79 m^2 to 3726.90 m^2 and the growth rate is 160 m^2 per oxygen volume fraction of 1%. Meanwhile, the furnace radiant surface at an O_2 concentration of 29% would be 41.25% more than that at an O_2 concentration of 22%.

In Fig. 51.16b, the volume of lower furnace increases but the volume of the upper furnace decreases with the increase in the oxygen volume fraction, which is due to decrease of upper furnace height with oxygen volume fraction. The total volume of the entire furnace at an O_2 volume fraction of 29% is 23.89% more than that at an O_2 volume fraction of 22%. The lower furnace volume at an O_2 volume fraction of 29% is 54.61% larger than that at an O_2 volume fraction of 22%. However, the upper furnace volume at an O_2 volume fraction of 29% is 25.97% smaller than that at an O_2 concentration of 22%.

The oxidant at the inlet of the oxy-fuel combustion furnace is a gaseous mixture of CO_2, N_2, H_2O and O_2 with a given oxygen volume fraction, which is used instead of the oxygen volume fraction of 21% as in the ambient air of the conventional air combustion furnace. Therefore, different oxygen volume fractions would be corresponding to different compositions of the oxidant, hence also to different enthalpies of the oxidant. Figures 51.17 and 51.18 show the curves of volume and volume fraction of each constituent of the oxidant with oxygen volume fraction, respectively.

Fig. 51.16 Radiant surface and furnace volumes with oxygen volume fraction. **a** Radiant surface; **b** Furnace volumes

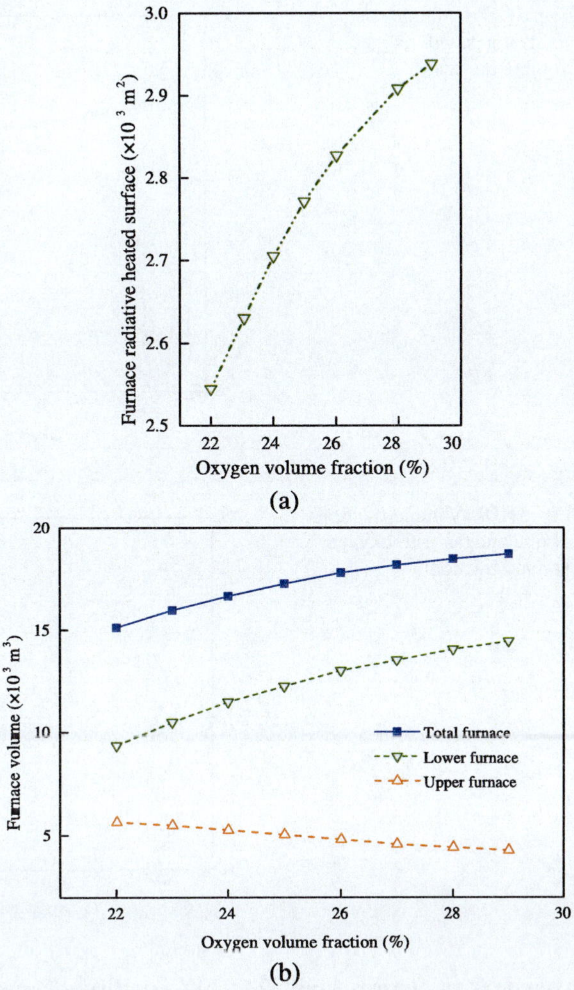

It can be seen from Fig. 51.17 that the volumes of CO_2, N_2 and H_2O decrease with the increase in oxygen volume fraction, as the O_2 volume remains constant at 1.554 m^3 per kilogram. Meanwhile, the total volume of the oxidant at the furnace inlet decrease also in the similar trend, and CO_2 volume decreases more quickly than N_2 and H_2O. This is also stated by Hu and Jinyue (2012) that FGR ratio decreased approximately linear with the increase of oxygen concentration in oxidant. That is to say, the higher the volume fraction of O_2, the less the volume fraction of CO_2. The volume fractions of CO_2 and N_2 decrease linear with the increase in the oxygen volume fraction (Fig. 51.18a) as shown in Fig. 51.18; whereas, the volume fractions of H_2O increase near linear with the increase in the oxygen volume fraction (Fig. 51.18b). By calculation, the volume fraction of O_2 rises by 31.8% as oxygen

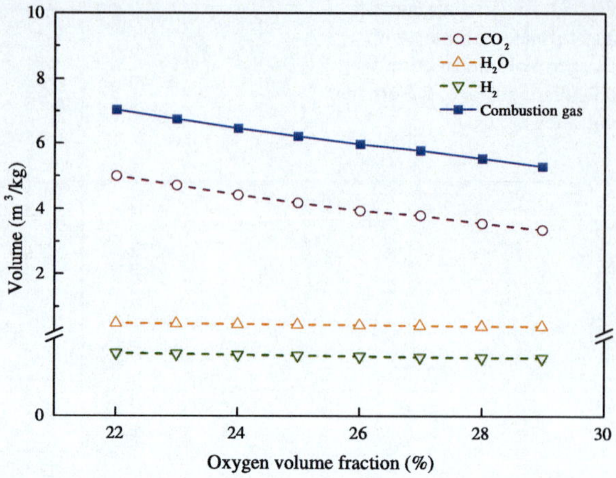

Fig. 51.17 Volumes of oxidant gas with oxygen volume fraction

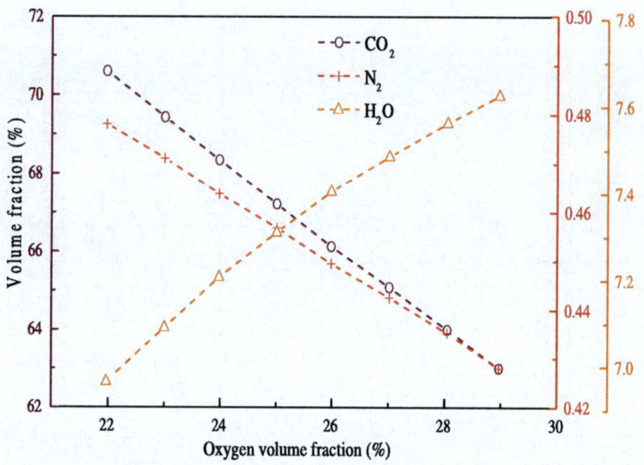

Fig. 51.18 Volume fractions of oxidant gas with oxygen volume fraction

volume fraction rises from 22 to 29%, but the volume fraction of CO_2 decreases by 12.7%.

A comparison between the enthalpies of oxidant in oxy-fuel mode and combustion air in the conventional mode with respect to the oxygen volume fraction is shown in Fig. 51.19. It is found that the enthalpy of the oxidant in oxy-fuel mode decreases gradually from 4122.77 kJ/kg to 3058.51 kJ/kg, with the increase in oxygen volume fraction from 22 to 29%. Whereas, the enthalpies of combustion air in the conventional air combustion boiler remain constant at 2995.39 kJ/kg. With oxygen volume fraction of 22%, the combustion air enthalpy is 27.35% less than oxidant enthalpy in oxy-fuel mode, whereas, with oxygen volume fraction of 29%, the combustion air enthalpy is 2.06% less than oxidant enthalpy in oxy-fuel mode.

As a matter of fact, the tri-atomic gas (mainly CO_2 and water vapor) is of strong emittance and absorbance of the heat radiation. Therefore, the volume fraction of tri-atomic gas (which has been discussed in Fig. 51.17) is of important influence to

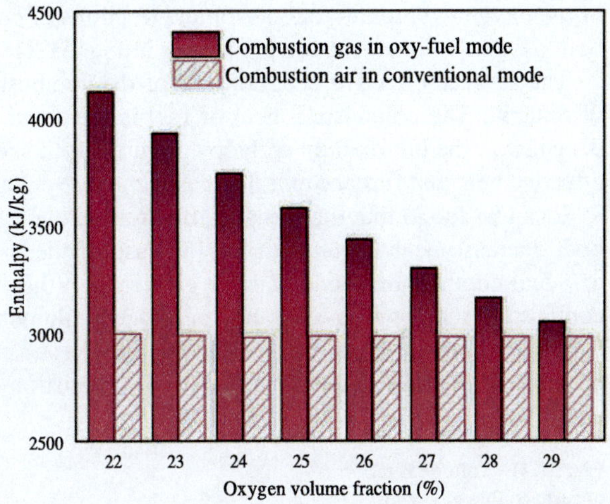

Fig. 51.19 Enthalpies of oxidant gas and fresh air with oxygen volume fraction

furnace flue gas radiative heat transfer due to the much tri-atomic gas within in the atmosphere.

In Fig. 51.20, O_2 volume in furnace flue gas is independent of oxygen volume fraction in the oxidant and remains constant at 0.259m^3/kg. But volumes of the total furnace flue gas and its CO_2 volume decrease at nearly the same trend with the increase of the oxygen volume fraction. The volume of furnace flue gas decreases by 33.0% from 7.4 to 5.7m^3/kg, while the volume of CO_2 decreases by 26.2% from 6.1 to 4.5m^3/kg. The volume fractions of CO_2 with 22% and 29% oxygen volume fraction are 82.9% and 79.6% respectively. The concentration of CO_2 is a little diluted.

That is to say, the influence of the volume of furnace flue gas on the CO_2 emission is dominant, while the volume of furnace flue gas is mainly affected by oxygen volume fraction. Volumes of N_2 and H_2O both decrease just a little with the increase

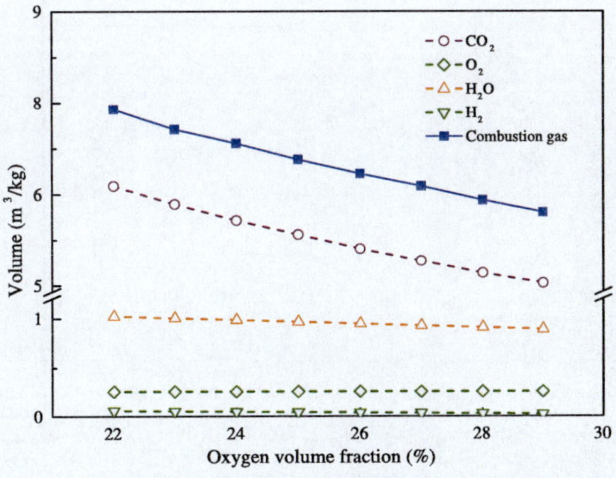

Fig. 51.20 Volumes of furnace flue gas with oxygen volume fraction

of the oxygen volume fraction. Volume fractions of constituents of furnace flue gas with oxygen volume fraction are shown in Fig. 51.21.

The furnace effective heat consists of the combustion heat of fuel and the heat of oxidant. The combustion heat of fuel is the same. Thus, furnace effective heat depends on the hot oxidant enthalpy. Figure 51.22 shows the curves of the furnace effective heat and furnace exit flue gas enthalpy versus oxygen volume fraction.

It can be found that the furnace effective heat and furnace exit flue gas enthalpy both decrease near linear with the increase of the oxygen volume fraction in the oxy-fuel combustion mode and are greater than the values in the conventional air combustion mode within the range of oxygen volume fraction from 22 to 29%.

If the temperatures at the furnace exit are kept as same in the two combustion modes, the flue gas enthalpy at oxy-fuel combustion furnace is higher than in the

Fig. 51.21 Volume fractions of furnace flue gas with oxygen volume fraction

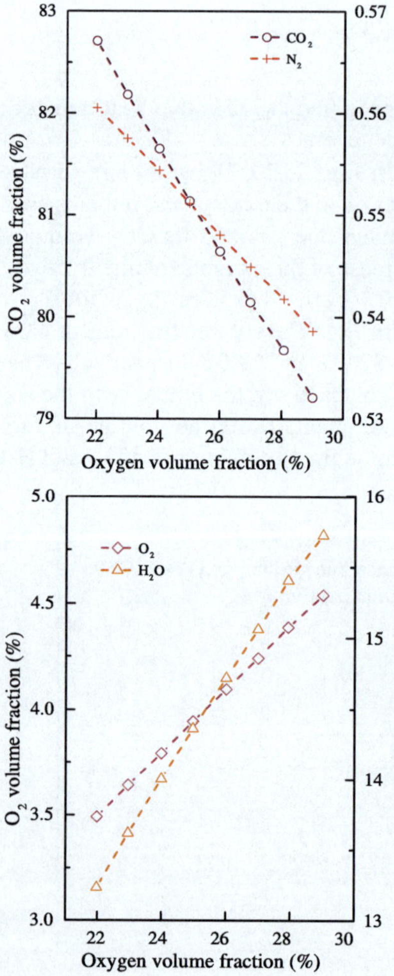

Fig. 51.22 Furnace effective heat and flue gas enthalpy at furnace outlet with oxygen volume fraction in both oxy-fuel combustion and conventional air combustion **a** Furnace effective heat; **b** Flue gas enthalpy at furnace outlet

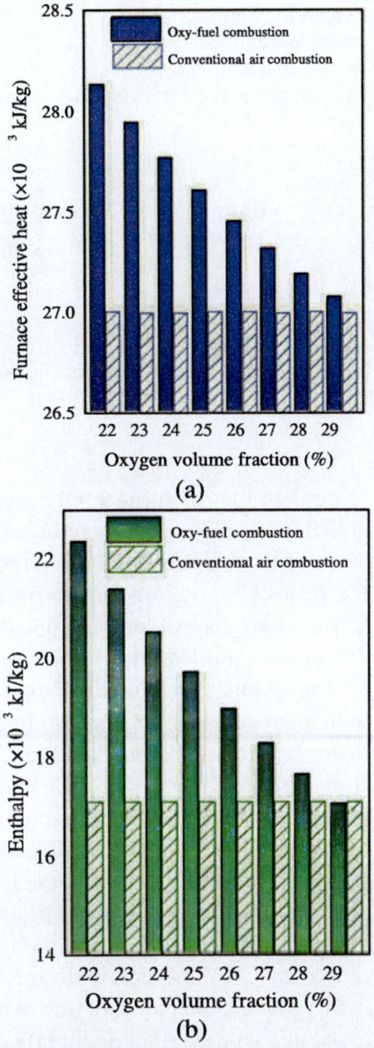

conventional air combustion furnace. For the volume of CO_2 in oxidant is much more than that in combustion air, the CO_2 volume in the furnace exit flue gas is also far more than that at the conventional air combustion mode in the same manner. As well, the specific heat capacity of CO_2 is the highest among the four gaseous products of CO_2, H_2O, N_2, and O_2.

The exit temperature of the lower furnace is kept as 1365 °C. The flue gas enthalpy at the furnace exit is the sum of the enthalpies of the four gaseous constituents. Increase in the oxygen volume fraction makes the total volume of the four constituents decrease (see Fig. 51.20). Therefore, the hot flue gas enthalpy at furnace exit is getting smaller (see Fig. 51.22b). The decrease in flue gas enthalpy at the furnace exit is much

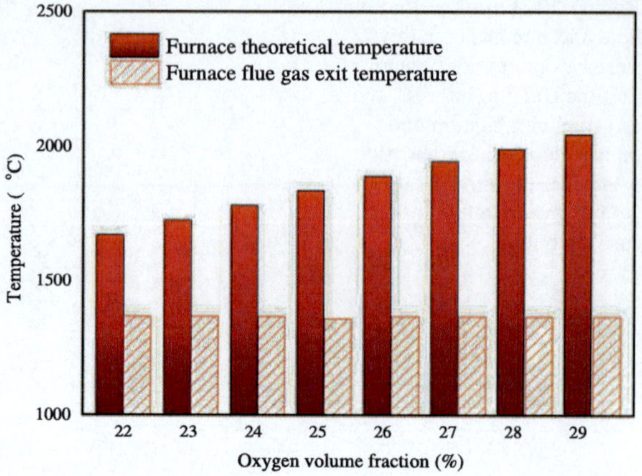

Fig. 51.23 Furnace theoretical temperature and flue gas exit temperature with oxygen volume fraction in oxy-fuel boiler

more than that of furnace effective heat, which the reduced proportion in flue gas enthalpy is 23.7% and the reduced proportion in furnace effective heat is 3.8%.

In order to understand the importance of theoretical combustion temperature in the furnace, variations in the furnace theoretical combustion temperature as well furnace flue gas exit temperature with respect to oxygen volume fraction in oxy-fuel combustion mode is shown in Fig. 51.23.

The results show that the theoretical combustion temperature increases gradually with the increase in oxygen volume fraction, and the flue gas exit temperature of the lower furnace maintains at 1365°C. The theoretical combustion temperature raises from 1670.5 °C to 2046.2 °C, which the percentage of increase is 22.5% with the increase in oxygen volume fraction from 22 to 29%. Because of the emissive power of flame follows the Stenfan-Boltzmann Law, which the power is proportional to the fourth power of temperature, the talent of radiation heat transfer increases by 125% with the increase in oxygen volume fraction from 22 to 29%. It can be concluded that the oxy-fuel combustion mode is an effective way to increase heat transfer and reduce the CO_2 emission with constant thermal power. Combined with the discussion in Fig. 51.16a, the radiative power and radiant surface both increase with the increase in oxygen volume fraction, in other words, the higher the oxygen volume fraction is, the stronger the radiation ability is.

In a comparison of Fig. 51.19 and Fig. 51.23, it states that the volume fractions of the furnace flue gas affect the enthalpy of flue gas a little more than temperature. This may be explained that the variations of oxygen volume fraction have a direct effect on volume of flue gas and enthalpies of flue gas, but a relative week effect of temperature on inherent physical properties of flue gas such as enthalpy. Furthermore, the flue gas enthalpy variation is mainly affected by CO_2 enthalpy, which is explained in comparison of Figs. 51.21 and 51.22. For the volume of CO_2 in oxidant is much more than the other constituents in combustion air, the CO_2 volume in the furnace exit flue gas is also far more than that in the same manner. So, the variation of CO_2 enthalpy plays a vital role in the effects of flue gas enthalpy. Although the furnace effective heat decreases with the increase in oxygen volume fraction, the theoretical combustion temperature increases with the rise in oxygen volume fraction, which is shown in Figs. 51.22 and 51.23.

51.4 Conclusions

For understanding the start-up period and the thermodynamic characteristics of an oxy-fuel boiler on plant scale, the start-up model under O_2/CO_2 and conventional air schemes were established. Meanwhile, the thermodynamic calculation on the oxy-fuel boiler on plant scale was given. The results are as follows:

(1) Start-up period of O_2/CO_2 scheme reaches a steady state within 15 circulations, which is faster than that of the conventional air schemes within 25 circulations.
(2) Even a very small air leakage could make a drastic increase of N_2 volume fraction. The N_2 volume fraction in furnace flue gas could increase sharply from 0.545% to 5.306%, if the furnace air leakage factor increases from 0.00 to 0.02.
(3) The air leakage rate and dehydration rate of the flue gas both have a sensitive effect on the flue gas volume fractions in start-up period. As the dehydration rate changes from 0.65 to 0.75 with the air leakage factor of 0.00, the volume fraction of H_2O increases from 15% to 17.5%. As the air leakage factor just increase from 0.01 to 0.02 with the dehydrate ratio of 0.65, the volume fraction of CO_2 decreases from 81 to 76% under O_2/CO_2 start-up scheme, but increase from 76 to 81% under the conventional air schemes.
(4) Oxygen volume fraction has a strong effect on the furnace radiant heat surfaces and the furnace volumes. In the calculation case, furnace radiant heat surface increases sharply in the growth rate of 160 m^2 per oxygen volume fraction of 1%; the volume of lower furnace increases but the volume of the upper furnace decreases with the increase in the oxygen volume fraction, and the total volume of the entire furnace at O_2 volume fraction of 29% is 23.89% more than that at O_2 volume fraction of 22%.
(5) The variations of oxygen volume fraction have a linear effect on volume of flue gas and enthalpies of flue gas, and the variation of CO_2 enthalpy plays a vital role in the effects of flue gas enthalpy, the theoretical combustion temperature raises in the percentage of 22.5% as the oxygen volume fraction increases from 22 to 29%.
(6) The oxygen volume fraction has different effects on the different part of the furnace. As the oxygen volume fraction increases from 22 to 29%, the height of the lower furnace increases with the rate of 2.162 m per oxygen volume fraction of 1%, but the upper furnace decrease with the rate of 0.654 m per oxygen volume fraction of 1%.

References

Ahmed U, Zahid U, Jeong YS, Lee CJ, Han C (2016) IGCC process intensification for simultaneous power generation and CO_2 capture. Chem Eng Process 101:72–86

Ahmed U, Kim C, Zahid U, Lee CJ, Han C (2017) Integration of IGCC and methane reforming process for power generation with CO_2 capture. Chem Eng Process 111:14–24

Anthony EJB (2012) Oxy-fuel firing technology for power generation. In: Handbook of climate change mitigation. Springer, New York, US

Baciocchi R, Storti G, Mazzotti M (2006) Process design and energy requirements for the capture of carbon dioxide from air. Chem Eng Process 45:1047–1058

Becher V, Bohn JP, Goanta A, Spliethoff H (2011) A combustion concept for oxyfuel processes with low recirculation rate - Experimental validation. Combust Flame 158:1542–1552

Che D (2008) Boilers—theory, design and operation. Xian Jiao Tong University Press, Xian

Chen L, Yong SZ, Ghoniem AF (2012) Oxy-fuel combustion of pulverized coal: characterization, fundamentals, stabilization and CFD modeling. Prog Energy Combust Sci 38:156–214

Conti J, Holtberg P, Diefenderfer J, LaRose A, Turnure JT, Westfall L (2016) International Energy Outlook 2016 With Projections to 2040

Fujimori T, Yamada T (2013) Realization of oxyfuel combustion for near zero emission power generation. Proc Combust Inst 34:2111–2130

González-Salazar MA (2015) Recent developments in carbon dioxide capture technologies for gas turbine power generation. Int J Greenhouse Gas Control 34:106–116

Hindiyarti L, Glarborg P, Marshall P (2007) Reactions of so3 with the O/H radical pool under combustion conditions. J Phys Chem A 111:3984–3991

Holloway S (2005) Underground sequestration of carbon dioxide—a viable greenhouse gas mitigation option. Energy 30:2318–2333

Hong SK, Noh DS, Lee EK (2015) Improvement in thermal efficiency of regenerator system by using oxy-fuel combustion. Appl Therm Eng 87:648–654

Hotta A et al (2012) Development and demonstration of oxy-fuel CFB technology. J South Afr Inst Min Metallurgy 112:430–436

Hu Y, Jinyue Y (2012) Characterization of flue gas in oxy-coal combustion processes for CO2 capture. Appl Energy 90:113–121

Hu Y, Yan J, Li H (2012) Effects of flue gas recycle on oxy-coal power generation systems. Appl Energy 97:255–263

Johansson R, Andersson K, Leckner B, Thunman H (2010) Models for gaseous radiative heat transfer applied to oxy-fuel conditions in boilers. Int J Heat Mass Transf 53:220–230

Kakaras E, Koumanakos A, Doukelis A, Giannakopoulos D, Vorrias I (2007) Oxyfuel boiler design in a lignite-fired power plant. Fuel 86:2144–2150

Kuivalainen R, Eriksson T, Hotta A, Sacristán ASB, Jubitero JM, Ballesteros JC, McCalden D (2010) Development and demonstration of oxy-fuel CFB technology. Coal Gen Conference, 2010

Li B, Chen T, Yang D (2005) DBSSP—A computer program for simulation of controlled circulation boiler and natural circulation boiler start up behavior. Energy Convers Manag 46:533–549

Lupion M, Alvarez I, Otero P, Kuivalainen R, Lantto J, Hotta A et al (2013) 30 MWth CIUDEN Oxy-CFB boiler—first experiences. Energy Proc 37:6179–6188

Namkhat A, Jugjai S (2010) Primary air entrainment characteristics for a self-aspirating burner: model and experiments. Energy 35:1701–1708

Normann F, Andersson K, Leckner B, Johnsson F (2008) High-temperature reduction of nitrogen oxides in oxy-fuel combustion. Fuel 87:3579–3585

Nsakala N, Liljedahl GN, Turek DG (2007) Commercialization development of oxygen fired CFB for greenhouse gas control

Oko E, Wang M (2014) Dynamic modelling, validation and analysis of coal-fired subcritical power plant. Fuel 135:292–300

Pipitone G, Bolland O (2009) Power generation with CO_2 capture: technology for CO_2 purification. Int J Greenhouse Gas Control 3:528–534

Sadegh Seddighi K, Pallarès D, Normann F, Johnsson F (2013) Carbon monoxide formation during oxy-fuel-fired fluidized-bed combustion. Energy Fuels 27:2275–2282

Scheffknecht G, Al-Makhadmeh L, Schnell U, Maier J (2011) Oxy-fuel coal combustion-a review of the current state-of-the-art. Int J Greenhouse Gas Control 5:16–35

Seddighi S (2014) Combustion, fluid dynamics and heat transfer in oxy-fuel and air-fired CFB boilers. Chalmers University of Technology

Seddighi Khavidak S, Pallarès D, Normann F, Johnsson F, Ylä-Outinen V (2015) Heat transfer in a 4-MWth circulating fluidized bed furnace operated under oxy-fired and air-fired conditions: Modeling and measurements. Int J Greenhouse Gas Control 37:264–273

Stanger R, Wall T, Spörl R, Paneru M, Grathwohl S, Weidmann M et al (2015) Oxyfuel combustion for CO2capture in power plants. Int J Greenhouse Gas Control 40:55–125

Starkloff R, Alobaid F, Karner K, Epple B, Schmitz M, Boehm F (2015) Development and validation of a dynamic simulation model for a large coal-fired power plant. Appl Therm Eng 91:496–506

Stewart MC, Symonds RT, Manovic V, MacChi A, Anthony EJ (2012) Effects of steam on the sulfation of limestone and NOxformation in an air- and oxy-fired pilot-scale circulating fluidized bed combustor. Fuel 92:107–115

Sung JH, Back SK, Jeong BM, Kim JH, Choi HS, Jang HN et al (2018) Oxy-fuel co-combustion of sewage sludge and wood pellets with flue gas recirculation in a circulating fluidized bed. Fuel Process Technol 172:79–85

Taler J, Duda P, Weglowski B, Zima W, Gradziel S, Sobota T et al (2009) Identification of local heat flux to membrane water-walls in steam boilers. Fuel 88:305–311

Tan Y, Jia L, Wu Y (2012) Anthony EJ. Experiences and results on a 0.8MWth oxy-fuel operation pilot-scale circulating fluidized bed. Appl Energy 92:343–347

Toftegaard MB, Brix J, Jensen PA, Glarborg P, Jensen AD (2010) Oxy-fuel combustion of solid fuels. Prog Energy Combust Sci 36:581–625

Toporov D (2014) Combustion of pulverised coal in a mixture of oxygen and recycled flue gas. Disputatio 6:45–66

Turek DG, et al (2006) Expanding the clean coal portfolio: oxyfiring to enhance CO_2 capture. Pittsburgh Coal Conference

Wang C, Jia L, Tan Y, Anthony EJ (2008) Carbonation of fly ash in oxy-fuel CFB combustion. Fuel 87:1108–1114

Wang C, Zhang X, Liu Y, Che D (2012) Pyrolysis and combustion characteristics of coals in oxyfuel combustion. Appl Energy 97:264–273

Wang Z, Liu M, Cheng X, He Y, Hu Y, Ma C (2017) Experimental study on oxy-fuel combustion of heavy oil. Int J Hydrogen Energy 42:20306–20315

Wu Y, Wang C, Tan Y, Jia L, Anthony EJ (2011) Characterization of ashes from a 100kWthpilot-scale circulating fluidized bed with oxy-fuel combustion. Appl Energy 88:2940–2948

Wu Y, Liu D, Duan L, Ma J, Xiong J, Chen X (2018) Three-dimensional CFD simulation of oxy-fuel combustion in a circulating fluidized bed with warm flue gas recycle. Fuel 216:596–611

Yin C, Yan J (2016) Oxy-fuel combustion of pulverized fuels: combustion fundamentals and modeling. Appl Energy 162:742–762

Yuzbasi NS, Selçuk N (2011) Air and oxy-fuel combustion characteristics of biomass/lignite blends in TGA-FTIR. Fuel Process Technol 92:1101–1108

Zhou W, Moyeda D (2010) Process evaluation of oxy-fuel combustion with flue gas recycle in a conventional utility boiler. Energy Fuels 24:2162–2169

Part VI
Emissions Control

Chapter 52
Formation of SO_3 in Flue Gas Under SNCR Conditions

Kang Wang, Wenfeng Shen, Yang Zhang, Yu Peng, Hai Zhang, Hairui Yang, and Junfu Lyu

Abstract Experimental and numerical studies were conducted to investigate the formation of sulfur trioxide (SO^3) during the selective non-catalytic reduction (SNCR) process. The effects of the inlet NH_3/NO ratio (R_{AN}), reaction temperature, inlet mole fractions of SO_2, O_2, CO and H_2O were assessed. The experiments were conducted using a perfectly stirred reactor (PSR) and SO_3 mole fraction was determined using the sulfur balance method. Corresponding numerical simulation was performed using detailed chemistry developed by Mueller and coworkers. Both experimental and numerical results revealed that the SO_3 formation was considerably affected by R_{AN}, SO_2 and O_2 mole fractions. The experimental results demonstrated that under typical SNCR conditions, 0.5 ~ 1.0% of SO_2 was converted into SO_3, and SO_3 mole fraction was 5–10 ppm. The SO_3 formation was noticeably enhanced by the addition of NH_3 when $R_{AN} < 0.5$. The conversion rate decreased as the initial SO_2 increased. A small amount of O_2 could promote the SO_3 formation remarkably, but this effect became much weaker as inlet O_2 mole fraction $\geq 1\%$. The numerical simulation indicated that the increase of the reaction temperature significantly promoted the SO_3 formation when the temperature was above 1173 K. A small amount of CO could significantly enhance the SO_3 formation. The H_2O addition could inhibit SO_3 formation. The detailed chemical kinetic analyses showed that the main reaction paths of the SO_3 formation were the oxidization reaction of SO_2 with O radical via $SO_2 + O\ (+M) \rightarrow SO_3\ (+ M)$ (52.3) and the one of SO_2 with NO_2 via $SO_2 + NO_2 \rightarrow SO_3 + NO$ (52.4). The effect of the operational parameters, i.e., R_{AN}, reaction temperature, and SO_2, O_2, CO, H_2O mole fractions, could be well explained by the variation of the reaction rates of Eqs. 52.3) and (52.4).

Keywords Selective non-catalytic reduction · Ammonia · Sulfur dioxide · Sulfur trioxide

---

K. Wang · W. Shen · Y. Zhang (✉) · Y. Peng · H. Zhang · H. Yang · J. Lyu
Key Laboratory for Thermal Science and Power Engineering of Ministry of Education, Department of Energy and Power Engineering, Tsinghua University, Beijing 100084, China
e-mail: yang-zhang@tsinghua.edu.cn

J. Lyu and S. Li (eds.), *Clean Coal and Sustainable Energy*, Environmental Science and Engineering,
https://doi.org/10.1007/978-981-16-1657-0_52

In a coal-fired boiler, the sulfur content in coal is mostly converted into sulfur dioxide (SO_2), the precursor of acid rain and aerosol formation, being a hazardous source for human health. The other type of sulfur oxide is sulfur trioxide (SO_3). Despite SO_3 formed during the combustion process could cause corrosion problems of the membrane water wall of the furnace (Dahl 1992; Stein-Brzozowska et al. 2014), its amount is trivial, only 1–2% of that of SO_2. Therefore, traditionally, SO_3 is not seriously considered as a pollutant in the emission control regulation yet (MEP 2011).

Recently, the SO_3 released from a coal-fired power plant, even with a small amount, becomes a great concern. There are two main reasons for such great concern. One is that SO_3 could significantly enhance the ash deposition on the air preheater. To stringently control NO_x emission, the selective catalytic reduction (SCR) catalyst is popularly used in a coal-fired power plant and a rather large amount of ammonia solution is sprayed over the catalyst. Some SO_2 in the flue gas is catalyzed by SCR catalyst and converted to SO_3. SO_3 then reacts with the residue ammonia, forming ammonium sulfate. Ammonium sulfate is the main cause of the severe plugging in the air preheater (Anthony and Granatstein 2001; Matsuda et al. 1982). The other reason is that once SO_3 escapes from the flue gas desulfurization (FGD) device, it could promote the formation of aerosol particles or sulfuric acid by interacting dust particles (Moser 2006; Kouravand and Kermani 2018; Adams and Senior 2006) or moisture (Wang 2008). Though SO_3 is water-soluble and most of the gaseous SO_3 is absorbed by the droplets, due to the insufficient absorption of the sorbent and the limited residence time, more than half of SO_3, in form of the sulfuric acid aerosol, can still escape from FGD to the air (Kouravand and Kermani 2018). The gaseous sulfuric acid aerosols facilitate the formation of blue plume (Adams and Senior 2006; Wang 2008; Gao et al. 2013).

The SO_3 formation during the coal combustion and the additional SO_3 formation caused by the use of SCR devices have been studied rather widely (Kouravand and Kermani 2018; Adams and Senior 2006; Wang 2008; Gao et al. 2013; Xiao et al. 2016; Marier and Dibbs 1974). However, in some coal-fired power plants, especially in the circulating fluidized bed (CFB) ones, the selective non-catalytic reduction (SNCR) technology is alternatively used for flue gas denitrification (Marier and Dibbs 1974; Li et al. 2013; Shen et al. 2017). The effect of the SNCR process on the SO_2/SO_3 conversion has not been concluded. On the one hand, our previous numerical study found the presence of NH_3 changes the reaction pathway of the SO_2 oxidation and enhances the formation of SO_3 (Shen et al. 2017). On the other hand, according to Hu et al. (2018), the presence of NH_3 increases the amount of free radicals H and HO_2 and then enhances the destruction of SO_3. So far, no experimental studies have been conducted on the conversion of SO_2 into SO_3 and its influencing factors under SNCR conditions remains unclear.

Consequently, in this paper, experimental and numerical studies on the SO_3 formation under SNCR conditions were conducted. The influencing factors, including inlet NH_3/NO ration (R_{AN}), inlet mole fractions of SO_2, O_2, CO and H_2O, and the reaction temperature were accessed. Numerical simulation with detailed chemistry was performed to further shed light on the mechanisms of SO_3 formation in a homogenous environment with the presence of NH_3. The results of this study are of special significance for the design and operation of the SNCR system.

52.1 Experimental and Numerical Approaches

52.1.1 Experiment Approach

An experimental system, as schematically shown in Fig. 52.1, was built to study the conversion of SO_2 into SO_3 under SNCR conditions. During the experiments, certain amounts of purely deionized H_2O (atomized by a nebulizer) and NO was sent into a mixing chamber and mixed with the simulated flue gases with preset mole fractions of CO_2, SO_2, N_2, O_2 and NH_3. The mass flow rate of H_2O was controlled by a micro-syringe pump and the simulated flue gas streams were controlled by a six-way mass flow meter. The mixed gases flowed through a perfectly stirred reactor (PSR), which was placed in a tubular furnace. As shown in Fig. 52.1, the PSR contains three main parts: two circular tubes (one for the reactant inlet and the other the product outlet) and a reaction chamber with four quadruple circular-spiral nozzles. The nozzles enhanced the mixing of the gases and the species inside the reaction chamber was regarded as a homogeneous mixture (Park et al. 2009). The PSR is made of quartz glass, and it operated in the temperature range of 800 ~ 1300 K.

The SO_2 mole fraction was measured using a flue gas analyzer, made by Madur Inc., with the resolution of 1 ppm. Direct measurement of SO_3 mole fraction is challenging as SO_3 is corrosive. When SO_2 is oxidized, it can be converted into SO_3 and some other S-containing species. Based on the numerical studies by Mueller et al.

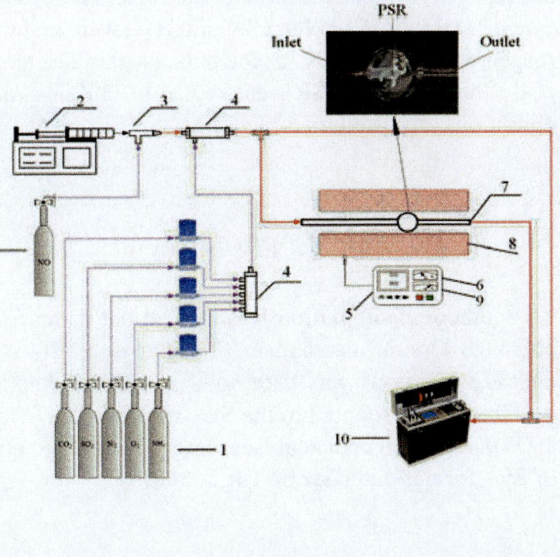

Fig. 52.1 Schematics of the experimental system (1-standard gas; 2-syringe pump; 3-nebulizer; 4-mixing chamber; 5-temperature controller; 6-ammeter; 7-perfectly stirred reactor; 8-tubular furnace; 9-voltmeter; 10- flue gas analyzer)

Table 52.1 Flue gas conditions of the experimental cases

No	SO_2	NO	NH_3	O_2	CO_2	H_2O	N_2
	ppm	Ppm	ppm	%	%	%	%
1	2000	300	0	3.5	15	5	76.27
2	2000	300	150	3.5	15	5	76.26
3	2000	300	300	3.5	15	5	76.24
4	2000	300	450	3.5	15	5	76.23
5	2000	300	600	3.5	15	5	76.21
6	1500	300	450	3.5	15	5	76.28
7	1000	300	450	3.5	15	5	76.33
8	500	300	450	3.5	15	5	76.38
9	2000	300	450	0	15	5	89.73
10	2000	300	450	1	15	5	78.73
11	2000	300	450	2	15	5	77.73
12	2000	300	450	5	15	5	74.73

(2000) and Fleig et al. (2011), among the oxidation products, SO_3 is approximately an order of magnitude higher than any other S-containing species. Thus, in the present study, the formation of SO_3 was determined by the sulfur balance method by assuming all consumed SO_2 was converted to SO_3. Namely, the amount of SO_3 formation was determined by the subtraction of the SO_2 mole fractions measured at the inlet and outlet of PSR. The detailed introduction and validation of this sulfur balance method are provided in Xiang et al. 2017a; Xiang et al. 2017b). This method is simple and of acceptable accuracy for the homogeneous reaction system of the present study.

The experiments were conducted at different R_{AN}'s, varying from 0 to 2. The mole fractions of SO_2 and O_2 were set within the ranges of 500–2000 ppm and 0–5% respectively, consistent with the typical flue gas conditions of a coal-fired boiler. The mole fractions of NO, CO_2 and H_2O were maintained at 300 ppm, 15%, and 5%, respectively by using N_2 as the balance gas. The temperature and the residence time of the flue gas in the PSR were kept at 1173 K and 2 s, respectively. The experimental conditions are listed in Table 52.1.

52.1.2 Numerical Approach

The numerical simulation was carried out using Chemkin-Pro software. Mueller's chemical kinetic mechanism (Mueller et al. 2000), which consists of 42 species and 222 reactions was used to describe the detailed chemistry. This mechanism was frequently adopted in the SO_3 studies (e.g., (Mueller et al. 2000; Glarborg et al. 1995)). Reaction path analyses were performed to obtain an insightful understanding of SO_3 formation under SNCR condition.

The inlet and outlet tubes of the PSR reactor was also partially placed inside the tubular furnace. When the reactants flowed through the inlet tube, the reactants were heated and the reactions could begin before the reactants entered the PSR chamber. Similarly, the reaction could continue inside the outlet tube as the products were still hot. Thus, to be consistent with the experiments, a series-connected reactor group was built in the modeling. The inlet and outlet tubes were simulated using two individual plug-flow reactors (PFRs), connected to a PSR module. The temperature distribution of the PFRs was set to be consistent with the thermocouple measured value. Actually, the simulation results exhibited that the > 90% of the conversion occurred inside the reactor chamber, as the inlet and outlet tubes were rather thin allowing very short residence time.

As shown in Table 52.2, R_{AN} and the mole fraction variations of SO_2 and O_2 were consistent with those used in the experiments. The reaction temperature varied from 1123 to 1323 K, with an interval of 50 K.

Table 52.2 Flue gas conditions used in the simulation

No	SO_2	NO	NH_3	O_2	CO_2	H_2O	CO	N_2
	ppm	ppm	ppm	%	%	%	ppm	%
1	2000	300	0	3.5	15	5	0	76.27
2	2000	300	150	3.5	15	5	0	76.26
3	2000	300	300	3.5	15	5	0	76.24
4	2000	300	450	3.5	15	5	0	76.23
5	2000	300	600	3.5	15	5	0	76.21
6	1500	300	450	3.5	15	5	0	76.28
7	1000	300	450	3.5	15	5	0	76.33
8	500	300	450	3.5	15	5	0	76.38
9	2000	300	450	0	15	5	0	79.73
10	2000	300	450	1	15	5	0	78.73
11	2000	300	450	2	15	5	0	77.73
12	2000	300	450	5	15	5	0	74.73
13	2000	300	450	3.5	15	5	100	76.22
14	2000	300	450	3.5	15	5	250	76.20
15	2000	300	450	3.5	15	5	500	76.18
16	2000	300	450	3.5	15	5	750	76.15
17	2000	300	450	3.5	15	5	1000	76.13
18	2000	300	450	3.5	15	2.5	0	78.73
19	2000	300	450	3.5	15	7.5	0	73.73
20	2000	300	450	3.5	15	0	0	81.23

52.2 Results and Discussion

52.2.1 The SO_3 Formation Under Typical SNCR Condition

Figure 52.2 shows the SO_3 formation under different R_{AN}'s at a typical SNCR condition of the inlet SO_2 mole fraction $[SO_2]_{in} = 2000$ ppm, inlet O_2 mole fraction $[O_2]_{in} = 3.5\%$, reaction temperature $T = 1173$ K and residence time $\tau = 2.0$ s. When no NH_3 was added, ~ 5 ppm SO_3 was produced. When NH_3 was present, the outlet SO_3 mole fraction $[SO_3]_{out}$ kept at ~ 10 ppm, corresponding to the conversion rate of ~ 0.5%, being only 1/4 ~ 1/3 of that under the SCR condition (Wu et al. 2006). Figure 52.2 also exhibits that the prediction using the Mueller's mechanism trendwisely agreed well with the experimental data.

The experimental and numerical results show that $[SO_3]_{out}$ increased much faster as R_{AN} increased when $R_{AN} < 0.5$, while it increased much more gradually as R_{AN} increased when $R_{AN} \geq 0.5$. To figure out the cause of the different variation rates of $[SO_3]_{out}$ under different R_{AN}'s, reaction pathway analyses were conducted and the results are shown in Fig. 52.3. Four main pathways were found to contribute to the SO_3 formation under SNCR conditions:

$$HOSO_2 + O_2 \rightarrow SO_3 + HO_2 \tag{52.1}$$

$$SO_2 + O_2 \rightarrow SO_3 + O \tag{52.2}$$

$$SO_2 + O(+M) \rightarrow SO_3(+M) \tag{52.3}$$

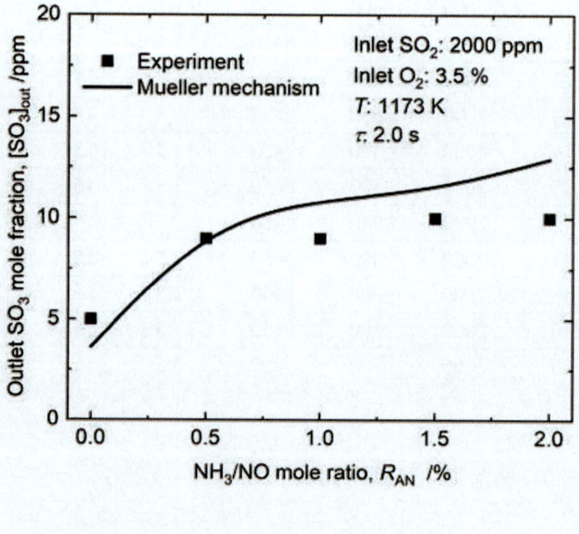

Fig. 52.2 Effect of inlet NH_3/NO mole ratio on the SO_3 formation

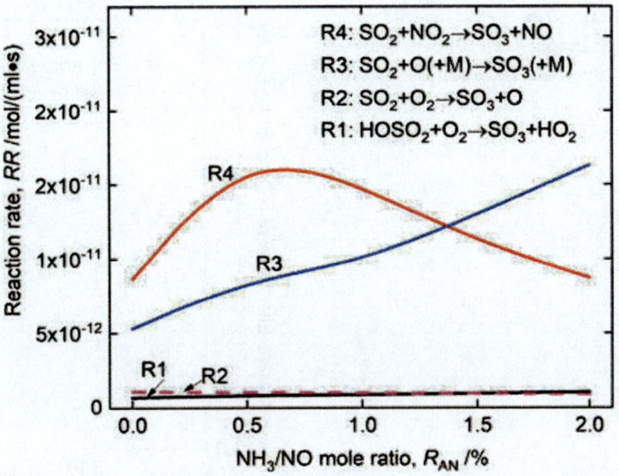

Fig. 52.3 Reaction rates of the SO_3 formation pathways at different inlet NH_3/NO ratios

$$SO_2 + NO_2 \rightarrow SO_3 + NO \tag{52.4}$$

Since reactions with the O_2 molecule, i.e., (52.1) and (52.2) were found to be rather slow, thus, (52.3) and (52.4) dominated the SO_3 formation under the tested SNCR conditions. The reaction rates of (52.3) and (52.4) were actually very sensitive to R_{AN}, as shown in Fig. 52.3. Moreover, the reaction rate of (52.4) first increased and then decreased as R_{AN} increased, demonstrating a non-monotonic variation trend. When $R_{AN} \leq 1$, (52.4) dominated the overall SO_3 formation. When $R_{AN} > 1$, (52.3) became important and it could play the leading role on the SO_3 formation as $R_{AN} >$ 1.5.

Figure 52.4 shows the numerically simulated mole fractions of NO, NO_2 and NO_x (NO + NO_2) at different R_{AN}'s. Clearly, the addition of NH_3 promoted NO_x removal. A peak emerged on the NO_2 mole fraction curve at $R_{AN} \approx 0.7$, consistent with the results shown in Fig. 52.3, The result indicates that during SNCR when a

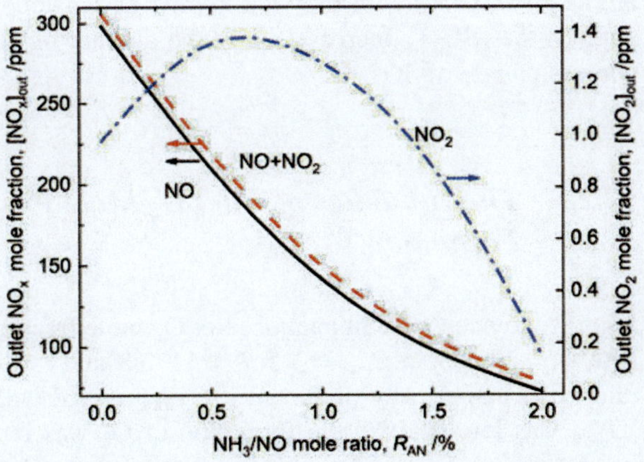

Fig. 52.4 Variation of NO_x and NO_2 mole fraction with respect to the inlet NH_3/NO ratio

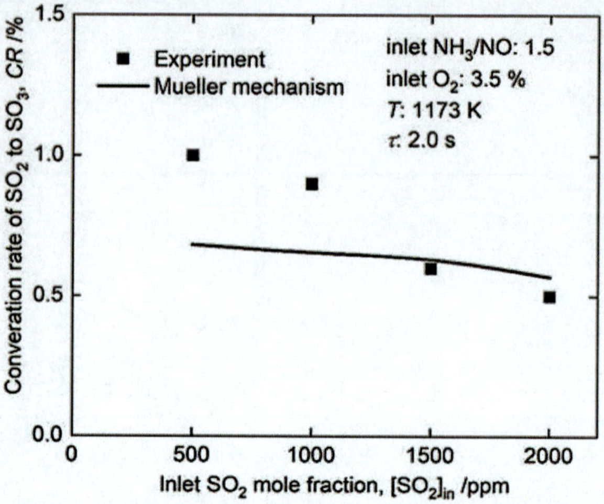

Fig. 52.5 Effect of inlet SO_2 mole fraction on the SO_2 conversion rate

small amount of NH_3 is added, NO_2 mole fraction increases, promoting the formation of SO_3 through (52.4). Nevertheless, the mole fraction of NO_2 was 2–3 orders of magnitude less than that of NO in the present SNCR conditions.

52.2.2 *The Influence of Inlet SO_2 Mole Fraction on SO_3 Formation*

Figure 52.5 shows the variation of conversion rate of SO_2 with respect to $[SO_2]_{in}$ at $R_{AN} = 1.5$, $T = 1173$ K and$\tau = 2.0$ s. As $[SO_2]_{in}$ increased, the conversion rate slightly decreased, from 1.0% to 0.5%. The predicted values at lower $[SO_2]_{in}$ were lower than the measured data, but, again, they exhibited a rather good trendwise agreement, consistent with the finding of Fleig et al. (Fleig et al. 2009). The results are also consistent with the well-known Le Chatelier principle (Liu et al. 1996), namely, the $[SO_2]_{in}$ increase raises the amount of SO_3 formation, but lowers the conversion rate of SO_2.

52.2.3 *The Influence of Inlet O_2 Mole Fraction on SO_3 Formation*

Figure 52.6 shows the impact of inlet O_2 mole fraction $[O_2]_{in}$ on SO_3 formation at $[SO_2]_{in} = 2000$ ppm, $R_{AN} = 1.5$, $T = 1173$ K and $\tau = 2.0$ s. Within the experimental range, the numerically predicted $[SO_3]_{out}$ agreed with the measured value. When $[O_2]_{in}$ was low (< 1%), the formation of O_3 was remarkably promoted as $[O_2]_{in}$

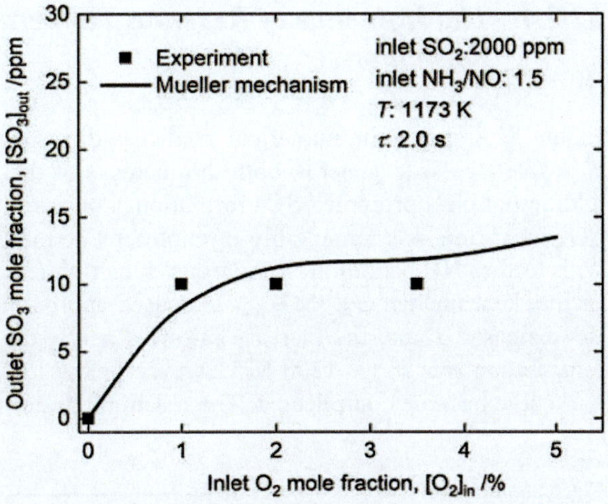

Fig. 52.6 Effect of inlet O_2 mole fraction on the SO_3 formation

increased. When $[O_2]_{in} \geq 1\%$, the increase rate of the SO_3 formation with respect to the increased $[O_2]_{in}$ was relatively slow and $[SO_3]_{out}$ kept at ~ 10 ppm. Figure 52.7 shows that the rapid increase of $[SO_3]_{out}$ with respect to the increased $[O_2]_{in}$ when $[O_2]_{in} < 1\%$ was contributed by (52.3). However, the reaction rate of (52.3) varied non-monotonically with the increased $[O_2]_{in}$. It increased to a peak value at $[O_2]_{in} \approx 1\%$ then decreased with the further increase of $[O_2]_{in}$. The reason for this phenomenon is that the branching reaction rate of (52.5) increased rapidly as $[O_2]_{in}$ increased from 0 to 1%, producing abundant O radicals. When $[O_2]_{in} > 1\%$, the reaction rate of (52.5) decreased and (52.3) thereby slowed down as $[O_2]_{in}$ further increased.

$$O_2 + H \rightarrow O + OH \text{ (R6)} \tag{52.5}$$

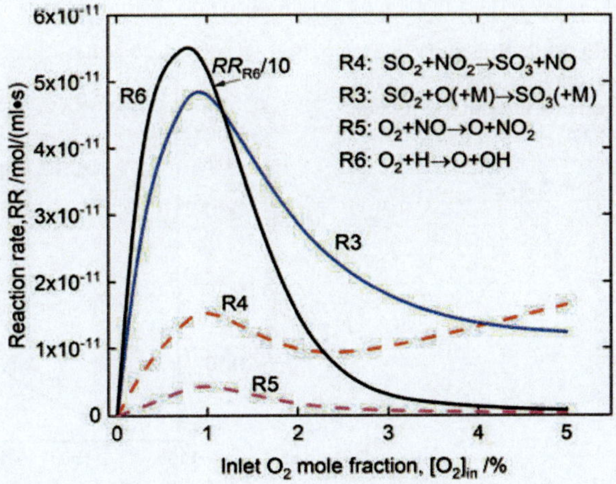

Fig. 52.7 Effect of inlet O_2 mole fraction on the reaction rates of (52.3) and (52.4)

52.2.4 The Influence of Reaction Temperature on SO_3 Formation

Figure 52.8 shows the numerical predict with respect to the reaction temperature T at five R_{AN}'s. In general, both the increase of the reaction temperature and the addition of NH_3 promoted SO_3 formation. Consistent with the finding in Fig. 52.3, SO_3 formation was remarkably promoted by a small amount of NH_3 addition but with further NH_3 addition, it increased much more slowly when other influencing factors kept unchanged. $[SO_3]_{out}$ increased approximately linearly with respect to the increased T and the increase rate was nearly the same at each R_{AN} when the temperature was above 1200 K. However, below 1200 K, the slope of the $[SO_3]_{out}$ ~ T curve became complicated. The reaction rate for (52.3) and (52.4) in Fig. 52.9

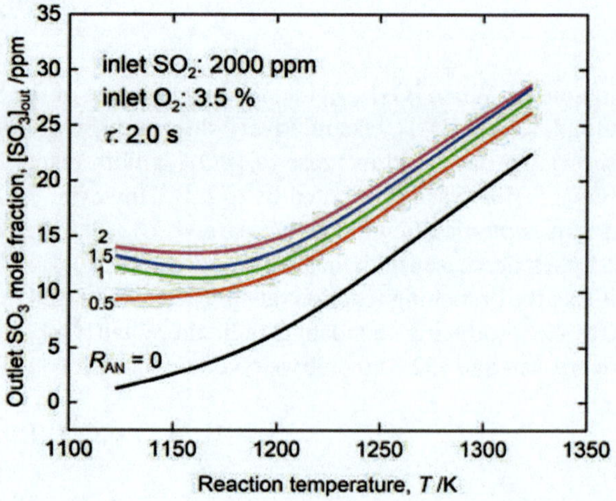

Fig. 52.8 Effect of the reaction temperature on the SO_3 formation

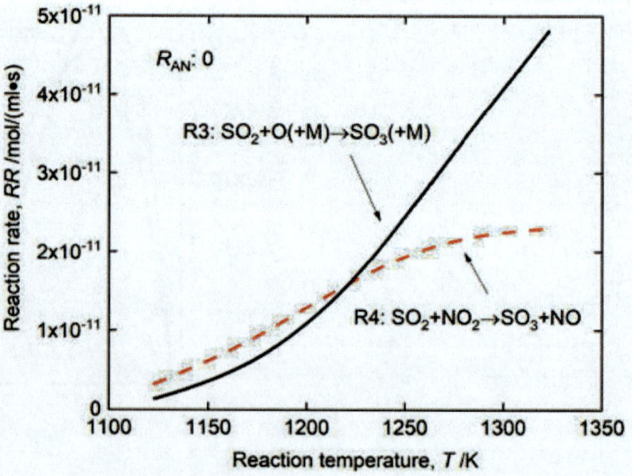

Fig. 52.9 Effect of the reaction temperature on the reaction rates of (52.3) and (52.4)

increased with T at $R_{AN} = 0$.

52.2.5 The Influence of Inlet CO Mole Fraction on SO_3 Formation

Figure 52.10 depicts the variation of $[SO_3]_{out}$ with respect to the inlet CO mole fraction $[CO]_{in}$ at various T's and $R_{AN} = 1.5$. The increase of $[CO]_{in}$ promoted the SO_3 formation at $T = 1123$, 1173, 1223 and 1273 K, and the promotion was more effective at lower T's and lower $[CO]_{in}$. For the case of $T = 1323$ K, $[SO_3]_{out}$ first increased as $[CO]_{in}$ increased from 0 to ~ 700 ppm, and then moderately decreased as $[CO]_{in}$ further increased from 700 ppm to ~ 1000 ppm. When $[CO]_{in} = 0$, the difference among $[SO_3]_{out}$'s at different T's were rather large, being ~ 16 ppm. As $[CO]_{in}$ approached 1000 ppm, $[SO_3]_{out}$'s at different T's converged to a narrow range of 33–35 ppm. The results indicate that it is important to keep the CO mole fraction lowest possible to retard SO_3 formation. When a small amount of CO exists, the reaction temperature should be lowered down to suppress SO_3. Numerical simulation shows that the reaction rate of (52.4) increased but the reaction rate of (52.3) decreased as $[CO]_{in}$ increased at $T = 1173$ K, as depicted in Fig. 52.11. The variation of the reaction rate of (52.3) and (52.4) explains the $[SO_3]_{out}$ variation with respect to $[CO]_{in}$.

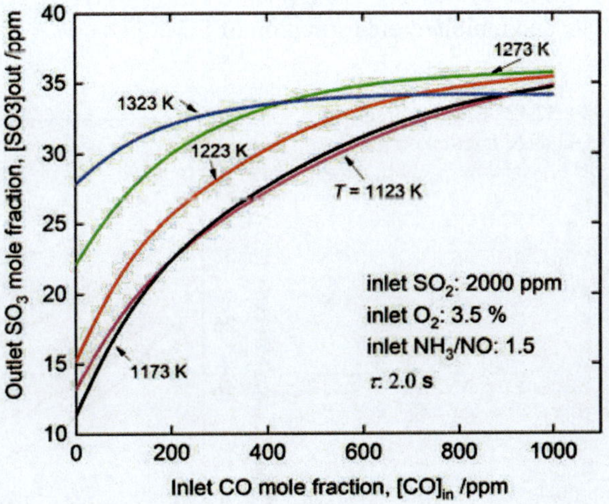

Fig. 52.10 Effect of inlet CO mole fraction on the SO_3 formation

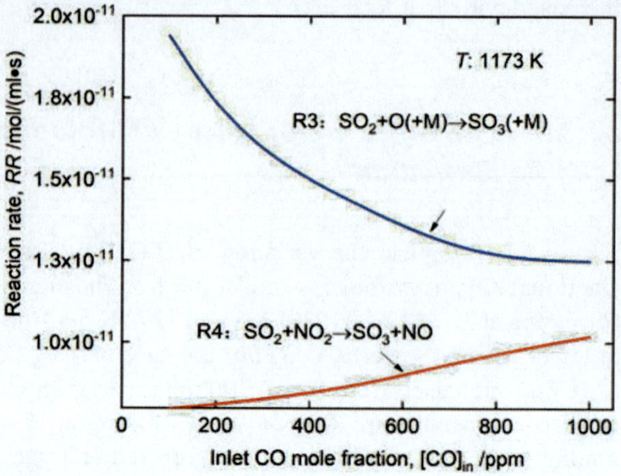

Fig. 52.11 Effect of inlet CO mole fraction on the reaction rates of (52.3) and (52.4)

52.2.6 The Influence of Inlet H_2O Mole Fraction on SO_3 Formation

Shown in Fig. 52.12, as H_2O inlet mole fraction $[H_2O]_{in}$ increased, the presence of H_2O inhibited the SO_3 formation at all the tested temperatures. Figure 52.13 demonstrated that H_2O consumed O atom through the reaction $H_2O + O \rightarrow OH + OH$ (R7), which slowed down the reaction (52.3) at $T = 1173$ K. Moreover, H_2O also reacted with N_2 through $H_2O + N_2 \rightarrow NH_2 + NO$ (R8), which produced more NO and inhibited the formation of SO_3 (52.4).

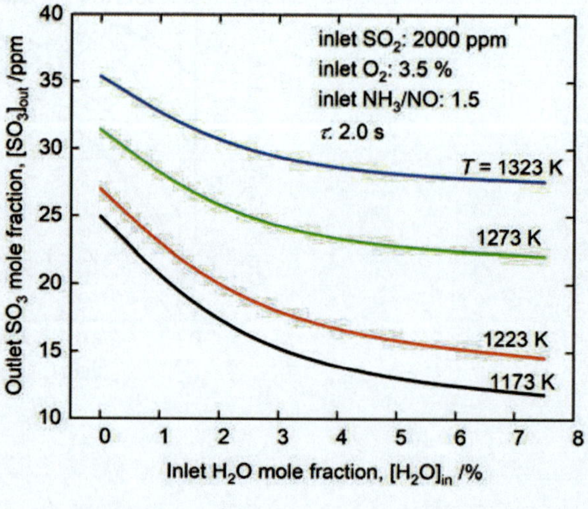

Fig. 52.12 Effect of inlet H_2O mole fraction on the SO_3 formation

Fig. 52.13 Effect of inlet H_2O mole fraction on the reaction rates of (52.3) and (52.4)

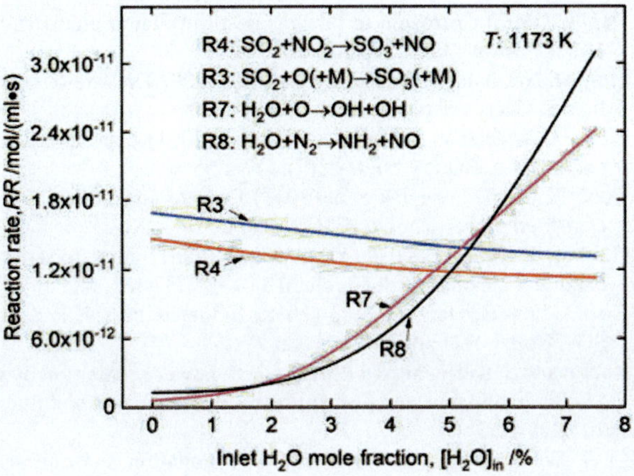

52.3 Conclusions

Experimental and numerical studies were conducted to investigate the SO_3 formation under SNCR conditions. Both experimental and numerical results revealed that SO_3 formation was apparently affected by R_{AN}, SO_2 and O_2 mole fractions, and at typical SNCR conditions, ~ 0.5% -1.0% SO_2 was converted into SO_3, and formed SO_3 mole fraction was 5–10 ppm. The SO_3 formation increased rapidly as R_{AN} increased when $R_{AN} < 0.5$. The conversion rate decreased with initial SO_2. A small amount of O_2 promoted SO_3 formation remarkably, but the effect became much weaker as the initial O_2 mole fraction $\geq$ 1%. The numerical simulation further revealed that the increase of reaction temperature significantly promoted SO_3 formation. A small amount of CO remarkably enhanced SO_3 formation at a lower SNCR temperature. In a rather wide range, the H_2O addition could inhibit SO_3 formation. Using detailed chemical kinetic calculation, it can be found that reactions via $SO_2 + O\ (+M) \rightarrow SO_3\ (+M)$ (52.3) and $SO_2 + NO_2 \rightarrow SO_3 + NO$ (52.4) were the dominant paths for SO_3 formation. All the effects of the considered operational parameters can be well explained by the variation of reaction rates of (52.3) and (52.4).

Acknowledgements This work was supported by the National Natural Science Foundation of China (U1710251).

References

Adams B, Senior C (2006) Curbing the blue plume: SO_3 formation and mitigation. Power 150(4):39–41

Anthony EJ, Granatstein DL (2001) Sulfation phenomena in fluidized bed combustion systems. Prog Energy Combust Sci 27(2):215–236

Dahl L (1992) Corrosion in flue gas desulfurization plants and other low temperature equipment. Mater Corros 43(6):298–304

Fleig D, Normann F, Andersson K et al (2009) The fate of sulphur during oxy-fuel combustion of lignite. Energy Procedia 1(1):383–390

Fleig D, Andersson K, Johnsson F et al (2011) conversion of sulfur during pulverized oxy-coal combustion. Energy Fuels 25(2):647–655

Gao Y, Luan T, Peng JW et al (2013) $DeNO_x$ performance of SCR catalyst for exhaust gas from coal-fired power plant. CIESC J 64(7):2611–2618

Glarborg P, Kubel D, Kristensen PG et al (1995) Interactions of CO, NO_x and H_2O under post-flame conditions. Combust Sci Technol 110–111(1):461–485

Hu JG, Li WH, Han GY et al (2018) Influence of SNCR denitration reform on the formation of SO_3. Boiler Technol 49(2):75–80

Kouravand S, Kermani AM (2018) Clean power production by simultaneous reduction of NO_x and SO_x contaminants using Mazut Nano-Emulsion and wet flue gas desulfurization. J Cleaner Prod 201:229–235

Li Q, Wu YX, Yang HR et al (2013) Simulation and optimization of SNCR process. CIESC J 64(5):1789–1796

Liu Z-K, Ågren J, Hillert M (1996) Application of the Le Chatelier principle on gas reactions. Fluid Phase Equilib 121(1):167–177

Marier P, Dibbs HP (1974) The catalytic conversion of SO_2 to SO_3 by fly ash and the capture of SO_2 and SO_3 by CaO and MgO. Thermochim Acta 8(1):155–165

Matsuda S, Kamo T, Kato A et al (1982) Deposition of ammonium bisulfate in the selective catalytic reduction of nitrogen oxides with ammonia. Ind Eng Chem Prod Res Dev 21(1):48–52

MEP (2011) Emission standards of air pollutants for thermal power plants, GB13223–2011. Ministry of Environmental Protection of China

Moser RE (2006) SO_3's impacts on plant O&M: part I. Power 150(8):40–40

Mueller MA, Yetter RA, Dryer FL (2000) Kinetic modeling of the $CO/H_2O/O_2/NO/SO_2$ system: implications for high-pressure fall-off in the $SO_2 + O(+M) = SO_3(+M)$ reaction. Int J Chem Kinet 32(6):317–339

Park S-H, Lee K-M, Hwang C-H (2009) Influences of heat loss on NO_x formation in a premixed CH_4/air-fueled combustor. Energy Fuels 23(9):4378–4384

Shen WF, Xiang BX, Zhang H et al (2017) Numerical simulation on formation of SO_3 during SNCR process in pulverized coal-fired boiler. CIESC J 68(8):3225–3231

Stein-Brzozowska G, Norling R, Viklund P et al (2014) Fireside corrosion during oxyfuel combustion considering various SO_2 contents. Energy Procedia 51:135–147

Wang HZ (2008) Analysis of SCR impact on denitration efficiency and SO_2 translation. Elect Power Sci Eng 24(5):17–21

Wu N, Song Q, Li SQ et al (2006) Measurement of SO_2 and SO_3 in SCR flue gas denitrification. Coal Conversion 29(02):84–87

Xiang BX, Shen WF, Zhang M et al (2017a) Effects of different factors on sulfur trioxide formations in a coal-fired circulating fluidized bed boiler. Chem Eng Sci 172:262–277

Xiang BX, Zhang M, Wu YX et al (2017b) experimental and modeling studies on sulfur trioxide of flue gas in a coal-fired boiler. Energy Fuel 31(6):6284–6297

Xiao HP, Dong L, Ning X (2016) Heterogeneous catalytic mechanism of SO_2 oxidation with Fe_2O_3. In: Proceedings of the CSEE 2016, vol 21, p 36

Chapter 53
The Effect of Oxygen-Coal Equivalent Ratio on the Rapid Preparation of Powdered Activated Coke in Low Oxygen Atmosphere

Binxuan Zhou, Yuan Zhao, Cheng Li, Tao Wang, Ping Zhou, Zhanlong Song, and Chunyuan Ma

Abstract Oxygen is a good activator for the rapid preparation of the powdered activated coke (PAC) in the low oxygen atmosphere, and the oxygen-coal equivalent ratio is an important parameter in the preparation process. In order to obtain the optimal range of oxygen-coal equivalent ratio, it was adjusted by changing the amount of coal supplied under a certain gas distribution (6% O_2 and 94% N_2).The changes of the SO_2 adsorption capacity of the PACs, the burn-off rate of raw coal and the production rate of gas phase products with the equivalent ratio were studied. In addition, the BET and SEM were used to characterize the pore structure of the PACs. The results showed that with the increase of the equivalent ratio, the SO_2 adsorption capacity of the PACs increases first and then decreases, reaching the maximum at 0.5, which is 102.2 mg/g; the burn-off rate increases as a whole, but the increment is not obvious, which is in the range of 0.2 ~ 0.4. Combined with the analysis of gas phase products, BET and SEM, it is concluded that the PACs prepared when the oxygen-coal equivalent ratio is in the range of 0.2 ~ 0.5 have relatively developed pore structure. To sum up, the suitable oxygen-coal equivalent ratio range about the preparation of PAC is 0.3 ~ 0.5. This conclusion provides reference for future industrial application.

Keywords Oxygen-coal equivalent ratio · Low oxygen atmosphere · Powdered activated coke · SO_2 adsorption

---

B. Zhou · Y. Zhao · C. Li · T. Wang (✉) · P. Zhou · Z. Song · C. Ma
National Engineering Laboratory of Coal-Fired Pollution Reduction, Shandong University, Jinan 250061, China
e-mail: wtao@sdu.edu.cn

T. Wang
Key Laboratory of Coal Resources Exploration and Comprehensive Utilization, Ministry of Land and Resources, Xian 710021, China

J. Lyu and S. Li (eds.), *Clean Coal and Sustainable Energy*,
Environmental Science and Engineering,
https://doi.org/10.1007/978-981-16-1657-0_53

53.1 Introduction

As is known to all, SO_2 emission has caused great harm to the environment and human health (Liu et al. 2010; Sage and Ford 2016; Gao et al. 2011). Although at present, the most commonly used desulfurization technology is the wet flue gas desulphurization, it has been faced with high operating costs, large water consumption, secondary pollution and other problems (Xu et al. 2000; Zhang et al. 2017). In comparison, activated coke desulfurization technology has attracted more and more attention due to its advantages of water saving, environmental protection, sulfur resource recovery and recyclable adsorption materials (Pi et al. 2017; Sun et al. 2015; Yan et al. 2013). As an effective SO_2 adsorbent in this technology, carbon-based adsorbents such as activated carbon (Lisovskii et al. 1997; Rubio and Izquierdo 1998; Ma et al. 2003), coke (Tsuji and Shiraishi 1997a,b; Li et al. 2008), semi-coke (Liu et al. 2003a,b; Shangguan et al. 2008) and activated carbon fibers (Mochida et al. 2000; Gaur et al. 2006; Brasquet and Cloirec 1997) have been studied by many scientific research institutions. However, the most concerned problem is to prepare activated carbon/coke with optimal SO_2 adsorption capacity at low production cost.

In general, the preparation methods of activated carbon/coke include physical activation method and chemical activation method. Chemical activation method is limited due to its high preparation cost and environmental impact caused by the use of chemical reagents (hydroxides, carbonates, H_3PO_4 or $ZnCl_2$). Physical activation is considered to be more environmentally friendly and widely used (Yuan et al. 2018; Heidari et al. 2014). The traditional physical activation method is divided into one-step method and two-step method. For the two-step method, firstly, the materials are carbonized in the inert atmosphere, of whose temperature is usually below 800 °C; and then under the activation of components to complete the activation, the activation temperature is usually at 700–1000 °C. For the one-step method, carbonization and activation are no longer independent of each other (Şentorun-Shalaby et al. 2006; Plaza et al. 2014). Comparatively speaking, one-step method has the advantages of low energy consumption, short time period and low cost.

Physical activation method often uses water vapor, CO_2, O_2 (Air), or a mixture of them as the activating agent. Among them, when oxygen is involved in activation, the reaction is too intense, the activation process is not easy to control, and the oxygen concentration is not very high, generally 2–11% is ideal (Plaza et al. 2014; Inagaki et al. 2006; Zhen et al. 2014). In order to realize rapid preparation and reduce the cost as much as possible, a new technology for the rapid preparation of powdered active coke from pulverized coal was proposed. The reason why pulverized coal is chosen is that when the pulverized coal (< 100 μm) is heated, the heating rate can reach 10^3 ~ 10^4 k/s (Hayashi et al. 2000). And studies have shown that the pore structure for the preparation of activated coke under the high heating rate (10^3 ~ 10^4 k/s) is more developed than that prepared under low heating rate (Cai et al. 1996; Cetin et al. 2005).

For this process, we have done some preliminary research and draw the corresponding conclusions. Firstly, the selection of coal is closely related to the performance of the PAC prepared. The lignite based PAC has a large SO_2 adsorption capacity and low yield; the bituminous coal based PAC has a relatively low SO_2 adsorption capacity and a relatively high yield; and the anthracite based PAC prepared by the process has poor adsorption capacity (Zhen 2016). In addition, using the classical optimization method, that is, optimizing one parameter at a time, the parameters suitable for the preparation of PAC are as follows: the temperature is 850 ~ 950 °C, the oxygen concentration is 6–8%, the water vapor concentration is 20–35%, and the CO_2 concentration has little effect (Zhang et al. 2016; Zhen et al. 2015). For the oxygen-coal equivalent ratio (similar to the excess oxygen coefficient in coal combustion), as its increases, the burn-off rate increases and the SO_2 adsorption capacity also changes. At present, the lowest oxygen-coal equivalent ratio is only tested to 0.3 (Ke 2017). In this process, when the gas distribution is fixed, the equivalent ratio is adjusted by adjusting the coal feed, that is, when the coal feed is large, the equivalent ratio is low and the yield of PAC per unit time is high; when the coal feed is small the equivalent ratio is high and the yield of PAC per unit time is low. The magnitude of the equivalent ratio is directly related to the output of the device in the future. Therefore, it is necessary to study the effect of the oxygen-coal equivalent ratio on the preparation of PAC.

Therefore, in this paper, we prepared the PACs at different oxygen-coal equivalent ratio using a drop-tube reactor system. Under the condition of fixed gas distribution, it is to achieve different equivalent conditions by adjusting the coal feed amount. Two key indicators about the preparation of PAC, performance index (SO_2 adsorption capacity) and economic index (burn-off rate), were analyzed. The activation of oxygen and the evolution of pores at different equivalent ratios were deduced. Finally, the optimum equivalent ratio range is obtained. And the conclusion provides reference for the future industrial application.

53.2 Experiment and Characterization

53.2.1 Preparation of Experimental Materials

In this study, the ShengLi lignite coal (SL-coal) was used as the experimental materials to prepare the PAC. The ultimate and proximate analysis of the SL-coal are given in Table 53.1. The experimental materials were dried at 105°C for 8 h, then crushed, ground, and sieved to obtain a particle size of 60 ~ 90 μm.

Table 53.1 Proximate and ultimate analysis of the SL-coal

Proximate (wt %)	M_{ad}	V_d	A_d	FC_d	
	2.42	40.59	14.48	44.93	
Ultimate (wt %)	C_d	H_d	O_d	N_d	S_d
	63.77	3.47	16.50	1.20	0.58

53.2.2 Preparation of PAC

A drop-tube reactor (DTR) system was used to prepare the PAC, as shown in Fig. 53.1. The system consists of five parts: gas system, micro feeder, DTR, PAC collector and vent treatment system. The micro feeder was used to continuously feed the pulverized coal into the DTR, and nitrogen was used as carrier gas with a flow rate of 6 L/min (standard condition). The gas system consists of pure oxygen, pure nitrogen and each flow is controlled by mass flow meter. The DTR is a stainless steel tube with an inner diameter of 80 mm and a length of 2000 mm. The maximum temperature that DTR can reach is 1200 °C, and the length of constant temperature zone is 1200 mm. PAC was collected in a container at the bottom of the DTR through a water-cooled sampling tube. After filtering, the final gas was released into the atmosphere. And

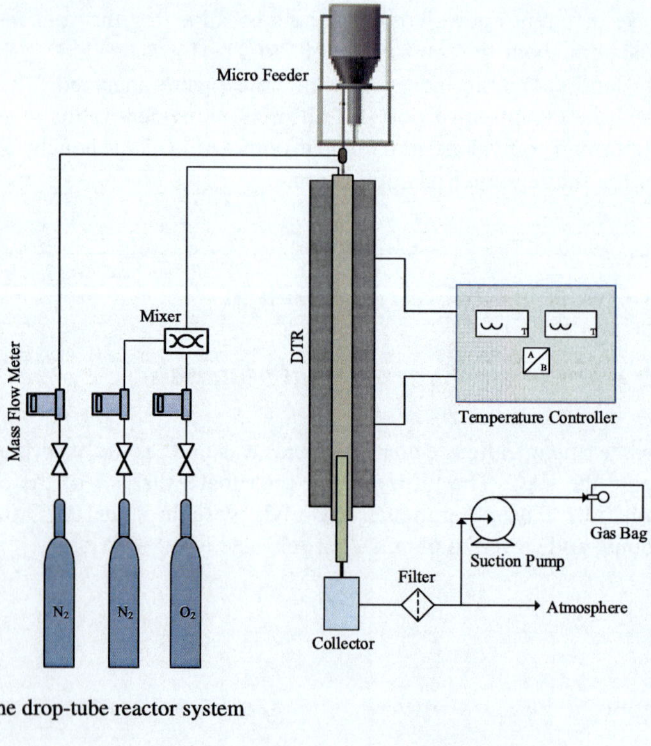

Fig. 53.1 The drop-tube reactor system

in the exhaust line, some gas is extracted by a suction pump and stored in the sealed air bag for gas samples analysis and test.

In these experiments, the reaction temperature was 950°C, and the reaction atmosphere was a low oxygen atmosphere with a volume concentration of 6% O_2 and 94% N_2. According to the calculation of settling velocity about the particles in the literature (Zhen 2016), the total gas flow under standard condition was 15.89 L/min, and the residence time of the particles in the DTR is 4 s.

The oxygen-coal equivalent ratio (α) is similar to the excess oxygen coefficient in coal combustion, as calculated by Eq. (53.1).

$$\alpha = \frac{V_{O_2}}{m \times (1.866\frac{C_{ar}}{100} + 5.55\frac{H_{ar}}{100} + 0.7\frac{S_{ar}}{100} - 0.7\frac{O_{ar}}{100})} \tag{53.1}$$

where, α is the oxygen-coal equivalent ratio, V_{O_2} is the volume flow rate of oxygen in the gas distribution (L/min), m is the mass flow rate of the coal (g/min). C_{ar}, H_{ar}, S_{ar} and O_{ar} are the mass fraction of C, H, S and O elements in raw coal respectively (%).

According to Eq. (53.1), 8 sets of experiments were designed by adjusting the amount of coal to change the α, and the α were 0.05, 0.1, 0.2, 0.3, 0.4, 0.5, 0.7, and 1.0, respectively.

53.2.3 SO_2 Adsorption Capacity

The SO_2 adsorption experiments were carried out in a fixed bed reactor using 1 g of sample at 75°C. The experimental system consists of a tubular reactor (inner diameter is 17 mm, wall thickness is 1.5 mm), placed in a vertical furnace, with a system of valves and mass flow controllers in order to select the flow and the composition of the inlet gas. The gas volumetric composition used in experiments was: SO_2, 1500 ppm; O_2, 6%; water vapor, 8%; N_2, balance. The SO_2 concentration was continuously monitored with an on-line Fourier transform infrared gas analyzer (GASMET-DX4000, Finland) until 2 h. The amount of SO_2 removed expressed in mg SO_2 g^{-1} PAC was calculated by integration of SO_2 conversion versus time curves (Izquierdo et al. 2003; Zhu et al. 2011).

53.2.4 Characterization

The moisture, ash and volatile content of each sample were measured by SDTGA5000 industrial analyzer, and the fixed carbon content was calculated by differential subtraction.

The pore structure characteristics of samples were determined by N_2 adsorption at 77 K using Autosorb-IQ fully automatic surface and pore size distribution analyzer. The BET surface area was calculated from the isotherm using the Srunauer Emmett Teller equation. The micropore surface was calculated by t-plot method. The micropore volume was obtained by DR method. The total pore volume was calculated by the analysis software at relative pressure larger than 0.97 (Hu et al. 2008). The surface area of mesopore and macropore, volume of mesopore and macropore were calculated by differential subtraction.

The surface morphology of all samples were observed using a Thermal Field Emission Scanning Electron Microscope (SUPRATM55). And the magnification are 2000 and 10,000 times.

The collected gas samples were analyzed by the refinery gas analyzer (CLARUS 580 GC, Perkinelmer, America) for their composition and concentration. The instrument can simultaneously measure the contents of H_2, CO, CO_2, O_2, N_2 and some hydrocarbons (C_1 ~ C_5) in one sample injection.

53.3 Results and Discussion

53.3.1 SO_2 Adsorption Capacity

The SO_2 adsorption capacity is the main performance index to evaluate PAC. Figure 53.2 shows the SO_2 breakthrough curves of the PACs prepared in these experiments. It can be seen that the adsorption performance of the PACs are poor when α is too small or too large. For example, when α is 0.05, SO_2 penetrates the adsorption layer at the very beginning; about 0.1, SO_2 is completely adsorbed in the first 3 ~ 5 min, and then penetrates the adsorption layer; about 1.0, SO_2 can be completely adsorbed for nearly 10 min at the initial stage, but the SO_2 concentration increases rapidly after breakthrough. When α is 0.2 ~ 0.7, the PACs show a similar SO_2

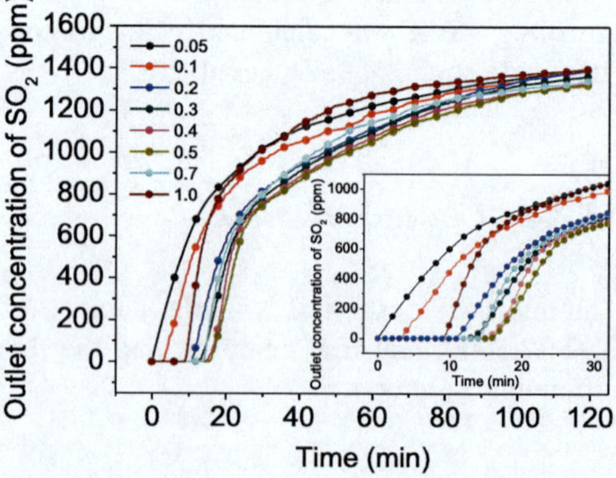

Fig. 53.2 The SO_2 breakthrough curve of the PACs

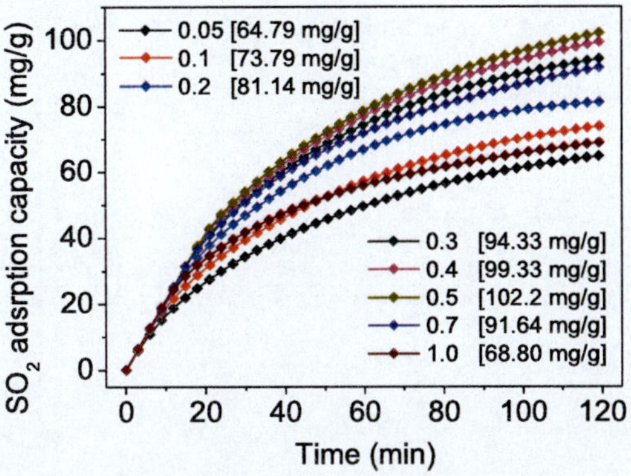

Fig. 53.3 The SO_2 adsorption capacity of the PACs

adsorption dynamic behavior. However, the time that SO_2 is completely adsorbed and the increase rate of SO_2 concentration after breakthrough are different. It is these differences that lead to the differences in SO_2 adsorption capacity.

Figure 53.3 shows the corresponding SO_2 adsorption capacity. It can be seen that with the increase of the equivalent ratio, the SO_2 adsorption capacity of the PACs increases first and then decreases. In the range of 0.3 to 0.7, the SO_2 adsorption capacity of PACs are relatively high. When α is 0.5, the SO_2 adsorption capacity is the highest, reaching 102.2 mg/g.

53.3.2 *Burn-Off Rate*

The burn-off rate (R) is the main economic index to evaluate the preparation of PAC, which is directly related to the yield of PAC. It is calculated according to the ash balance method, that is, assuming that the absolute amount of ash does not change before and after the reaction, as shown in the Eq. (53.2).

$$R = 1 - \frac{A_{coal}}{A_{PAC}} \times 100\% \tag{53.2}$$

where, A_{coal} and A_{PAC} are respectively the dry base ash mass fraction (%) in raw coal and PAC.

As known from Eq. (53.2), the burn-off rate of raw coal mainly includes two parts, the burn-off of volatiles and fixed carbon. So after R is known, the burn-off rate of volatiles ($R_{[V]}$) and fixed carbon ($R_{[FC]}$) can be calculated according to Eqs. (53.3) and (53.4).

$$R_{[V]} = \frac{V_{coal} - (1 - R) \times V_{PAC}}{V_{coal}} \tag{53.3}$$

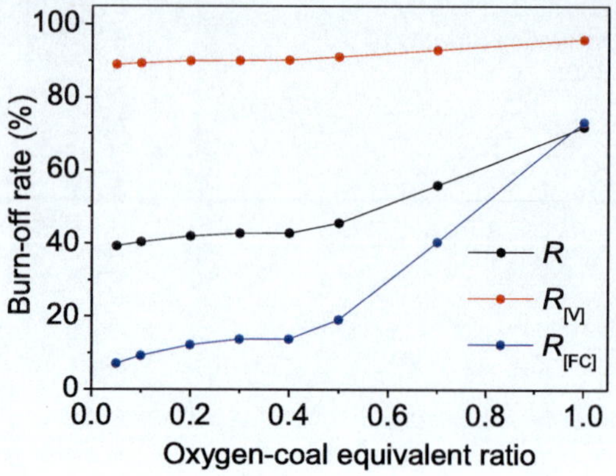

Fig. 53.4 The burn-off rate at different oxygen-coal equivalent ratio

$$R_{[FC]} = \frac{FC_{coal} - (1 - R) \times FC_{PAC}}{FC_{coal}} \tag{53.4}$$

where, V_{coal} and V_{PAC} are respectively the dry base volatiles mass fraction in raw coal and powdered coke industrial analysis. And FC_{coal} and FC_{PAC} are the fixed carbon.

Figure 53.4 shows the change of burn-off rate via the oxygen-coal equivalent ratio. It can be seen that the burn-off rate increases as a whole with the oxygen-coal equivalent ratio increases, and the main reason for the increase of the burn-off rate is that the rate of fixed carbon loss increases significantly. The precipitation rate of volatile is less affected by the equivalent ratio. In the range of 0.05 to 0.2, the main reason of the fixed carbon loss is the reaction between volatile decomposition products and fixed carbon. In the range of 0.2 to 0.4, the ignition loss rate tends to be balanced, and the reason is the homogeneous reaction between oxygen and volatile decomposition products. When the equivalence ratio is greater than 0.4, the fixed carbon loss increases significantly, that is excessive oxygen is added, and the heterogeneous reaction between oxygen and fixed carbon occurs.

53.3.3 Gas Products Analysis

In the preparation of PAC, the yield of each component in gas products are calculated according to the N_2 balance method, that is, assuming that the absolute amount of N_2 does not change before and after the reaction, as shown in the Eq. (53.5).

$$y_i = \frac{\frac{\varphi_i}{\varphi_{N_2}} \times V_{N_2}^0 - V_i^0}{m} \tag{53.5}$$

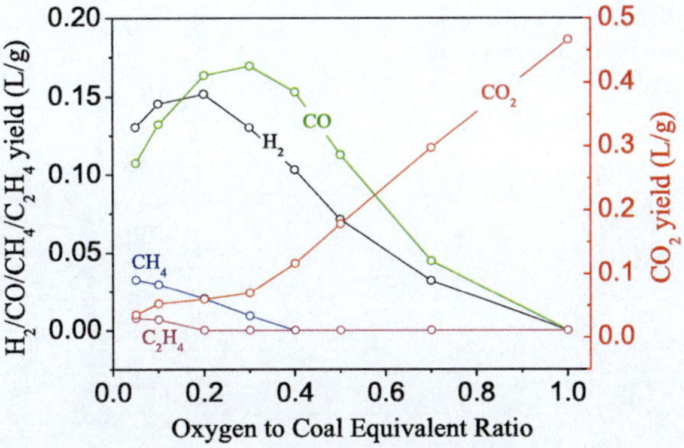

Fig. 53.5 The yield of gas products at different oxygen-coal equivalent ratio

where, y_i is the yield of the i gas (L/g), V_i^0, $V_{N_2}^0$ are the volume flow rate of the i gas and the nitrogen in the carrier gas (L/min), φ_i, φ_{N_2} are the volume fraction of the i gas and the nitrogen detected by the refinery gas analyzer (%), and m is the mass flow rate of the coal (g/min).

Figure 53.5 shows the the yield of each component at diffenert equivalent ratio. The role of O_2 in the rapid preparation of the PAC can be inferred from this figure. When α goes from 0.05 to 0.2, the yield of H_2 and CO are increasing, while the yield of CH_4 and C_2H_4 are decreasing, what mainly happens is the release and decomposition of volatiles in this range. When α goes from 0.2 to 0.4, the yield of H_2 begins to decline, indicating that excessive O_2 begins to consume H_2. While the yield of CO increases first and then declines, and the yield of CO_2 increases rapidly, according to Fig. 53.4, indicating that partial O_2 reacts with fixed carbon to form CO and CO_2. The activation of O_2 begins to occur. When α is greater than 0.4, the yield of H_2 and CO decline rapidly, the yield of CO_2 increase rapidly, and the $R_{[FV]}$ also increased sharply, indicating that excessive O_2 begins to consume H_2, CO and fixed carbon vastly.

53.3.4 Pore Structure and Surface Morphology

The nitrogen adsorption isotherms of all PACs are shown in Fig. 53.6, and the corresponding pore structure parameters are given in Table 53.2. From these information, we can infer the law of pore evolution during the preparation of PACs with the change of the equivalent ratio. And the surface morphology of the PACs as shown in Fig. 53.7 can help to explain these inferences.

According to the International Union of Pure and Applied Chemistry (IUPAC) classification, these isotherm patterns behave a combined type I and type IV adsorption–desorption features indicating that micropores, mesopores and/or macropores coexist in these samples(Sun et al. 2015; Hu et al. 2008). However, depending on different equivalent ratios, the isotherms of PACs display clear differences. When α

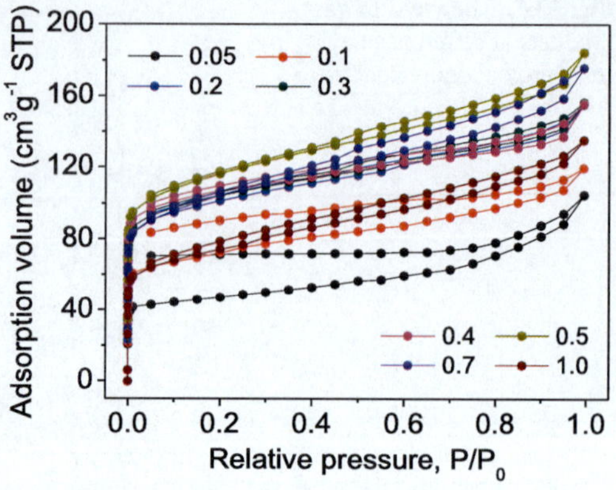

Fig. 53.6 The nitrogen sorption isotherms of the PACs

is 0.05 or 0.1, the adsorption capacity of N_2 are lower than that of other samples, showing that the number of pores is relatively small. In addition, the desorption isotherm do not coincide with the adsorption isotherm, indicating that at low equivalence ratio, the pore structure frame of the PACs are unstable and change during the BET test. At low equivalent ratio, the pore structure of PACs is mainly the channel formed after volatiles releasing. As shown in Fig. 53.7 (a) and (b), there are some slit-shaped pores and cylindrical pores on the surface of PACs, and there are also a large number of small pores inside these pores. These pores are mainly formed by volatiles releasing. When α is from 0.2 to 0.4, there is no obvious difference between the adsorption isotherms, and the desorption isotherms and the adsorption isotherms basically coincide in the end. Besides, the hysteresis loop of these PACs confirm the hierarchically porous structure. That is to say, the activation of oxygen is very effective in this range. In Fig. 53.7 (c), (d) and (e), there are abundant pore structures on the surface of the PACs, and inside these pores there are also abundant pore structures, that is to say, these PACs have layered porous structures. According to the data in Table 53.2, as the equivalence ratio increases in this range, the total number of pores increases, the number of micropores increases, and the number of mesopores/macropores decreases, indicating that O_2 is starting to ablate the mesopores/macropores surface, exposing the previously hidden micropore structure. When α increases to 0.5, the total surface area, total pore volume and other pore parameters increase further except for the number of micropores, that is, O_2 begins to ablate micropores, and some micropores evolve towards the mesopores/macropores. When α increases to 0.7, the ablation of micropore is more obvious, and the adsorption–desorption isotherm shows obvious hysteresis loop, indicating that there is a greater number of mesopores/macropores. Figure 53.7 (g) shows that the wall of pores inside the PAC is smooth and the layered porous structure gradually disappears. When α increases to 1.0, the pore frame was seriously burned and the pore structure was seriously damaged as shown in Fig. 53.7 (h).

Table 53.2 The pore structure information of the PACs

Sample	Pore structure parameters					
	S_{BET} (m^2/g)	S_{mi} (m^2/g)	S_{me+ma} (m^2/g)	V_t (cm^3/g)	V_{mi} (cm^3/g)	V_{me+ma} (cm^3/g)
PAC-0.05	200	177	23	0.141	0.097	0.044
PAC-0.1	268	188	80	0.185	0.107	0.078
PAC-0.2	376	264	112	0.240	0.147	0.093
PAC-0.3	387	279	108	0.239	0.151	0.088
PAC-0.4	404	312	91	0.241	0.165	0.076
PAC-0.5	426	287	139	0.285	0.160	0.125
PAC-0.7	376	225	151	0.209	0.144	0.065
PAC-1.0	267	127	140	0.196	0.106	0.090
S_{BET}: total surface area				V_t: total pore volume		
S_{mi}: surface area of micropores				V_{mi}: pore volume of micropores		
S_{me+ma}: surface area of mesopores and macropores				V_{me+ma}: pore volume of mesopores and macropores		

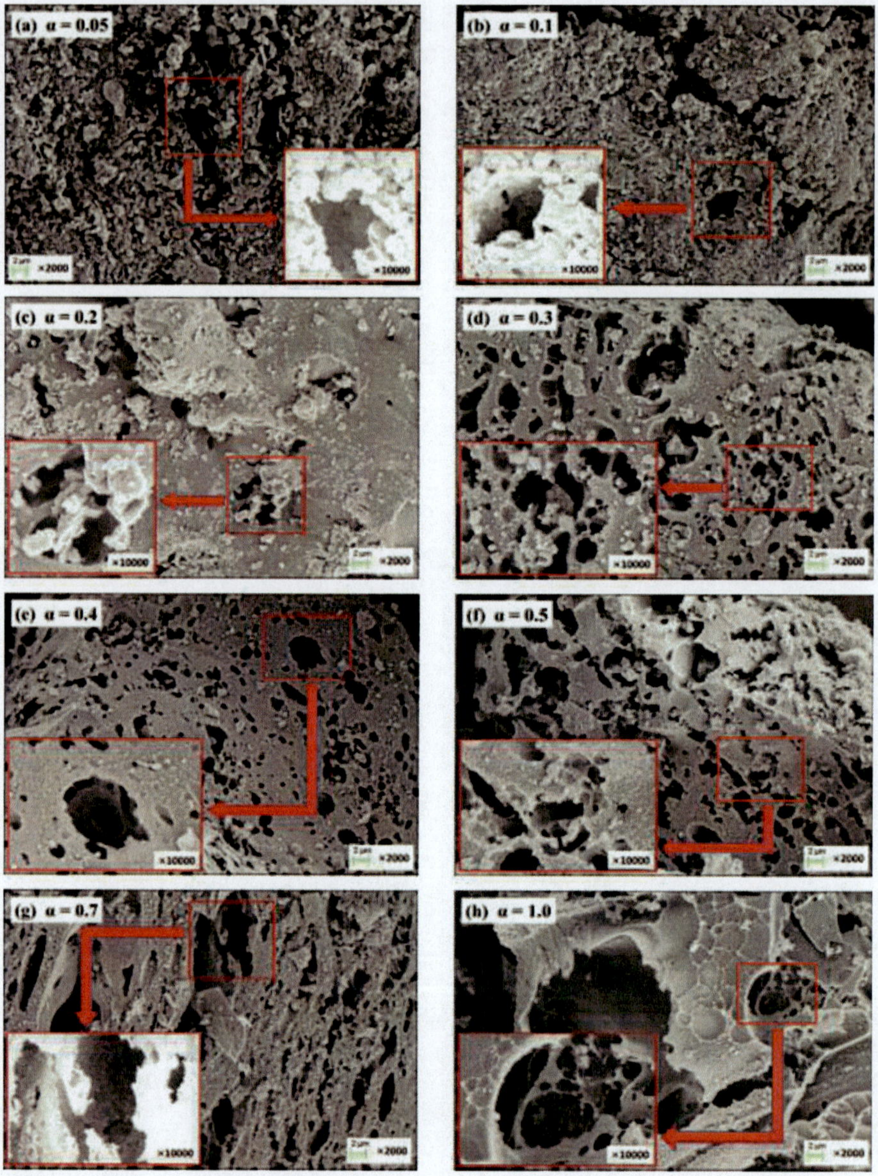

Fig. 53.7 The surface morphology of the PACs

53.4 Conclusions

The oxygen-coal equivalent ratio is an important factor to be considered in the process of rapid preparation of PAC under low oxygen atmosphere. In this paper, the PAC with different equivalent ratios is prepared by adjusting the coal feed amount under the condition of fixed gas distribution. Through the analysis of SO_2 adsorption capacity, burn-off rate, gas products, BET and SEM, the following conclusions are obtained:

1. With the increase of the equivalent ratio, the SO_2 adsorption capacity of the PACs increases first and then decreases, reaching the maximum at 0.5, which is 102.2 mg/g; the burn-off rate increases as a whole, and it increases slowly before 0.5 and increases rapidly after 0.5.
2. At low equivalent ratio ($\alpha \leq 0.1$), the number of pores is small, and the surface area of PACs is less than 300 m^2/g. Besides, the pore frame is unstable. When α is from 0.2 to 0.4, O_2 is starting to ablate the mesopores/macropores surface, exposing the previously hidden micropore structure. When $\alpha \geq 0.5$, O_2 begins to ablate micropores, and some micropores evolve towards the mesopores/ macropores. As α continues to increase, the ablation of micropores becomes more and more obvious. When it increases to 1.0, the pore frame is seriously burned and the pore structure is seriously damaged.
3. The optimal range of oxygen-coal equivalent ratio is 0.3 ~ 0.5. In this range, the PACs prepared have high SO_2 adsorption capacity, well developed pore structure, and the burn-off is relatively low.

Acknowledgements This work was supported by the National Key R&D Program of China (No.2017YFB0602902).

References

Brasquet C, Le Cloirec P (1997) Adsorption onto activated carbon fibers: application to water and air treatments. Carbon 35(9):1307–1313

Cai HY, Ell AJG, Chatzakis IN, Lim JY, Dugwell DR, Kandiyoti AR (1996) Combustion reactivity and morphological change in coal chars: effect of pyrolysis temperature, heating rate and pressure. Fuel 75(1):15–24

Cetin E, Gupta R, Moghtaderi B (2005) Effect of pyrolysis pressure and heating rate on radiata pine char structure and apparent gasification reactivity. Fuel 84(10):1328–1334

Gao H, Li C, Zeng G, Zhang W, Shi L, Li S, Zeng Y, Fan X, Wen Q, Shu X (2011) Flue gas desulphurization based on limestone-gypsum with a novel wet-type PCF device. Sep Purif Technol 76(3):253–260

Gaur V, Asthana R, Verma N (2006) Removal of SO2 by activated carbon fibers in the presence of O2 and H2O. Carbon 44(1):46–60

Hayashi JI, Takahashi H, Iwatsuki M, Essaki K, Tsutsumi A, Chiba T (2000) Rapid conversion of tar and char from pyrolysis of a brown coal by reactions with steam in a drop-tube reactor. Fuel 79(3–4):439–447

Heidari A, Younesi H, Rashidi A, Ghoreyshi A (2014) Adsorptive removal of CO 2 on highly microporous activated carbons prepared from eucalyptus camaldulensis wood: effect of chemical activation. J Taiwan Inst Chem Eng 45(2):579–588

Hu Q, Lu Y, Meisner GP (2008) Preparation of nanoporous carbon particles and their cryogenic hydrogen storage capacities. J Phy Chem C 112(5):1516–1523

Inagaki M, Nishikawa T, Oshida K, Fukuyama K, Hatakeyama Y, Nishikawa K (2006) Air oxidation of carbon spheres. II micropore development. Adsorp Sci Technol 24(1): 55–64

Izquierdo MT, Rubio B, Mayoral C, Andres JM (2003) Low cost coal-based carbons for combined SO2 and NO removal from exhaust gas. Fuel 82:147–151

Ke L (2017) Experimental study on preparation of powder activated coke and by-product pyrolysis gas by rapid pyrolysis of pulverized coal, Shandong University, University in Jinan, China

Li J, Kobayashi N, Hu Y (2008) The activated coke preparation for SO2 adsorption by using flue gas from coal power plant. Chem Eng Process 47(1):118–127

Lisovskii A, Semiat R, Aharoni C (1997) Adsorption of sulfur dioxide by active carbon treated by nitric acid: I. effect of the treatment on adsorption of SO2 and extractability of the acid formed. Carbon 35: 10–11 (1639–1643)

Liu Q, Li C, Li Y (2003a) SO2 removal from flue gas by activated semi-cokes: 1 the preparation of catalysts and determination of operating conditions. Carbon 41(12): 2217–2223

Liu Q, Guan JS, Li J, Li C (2003b) SO_2 removal from flue gas by activated semi-cokes: 2 effects of physical structures and chemical properties on SO_2 removal activity. Carbon 41(12): 2225–2230

Liu Y, Bisson TM, Yang H, Xu Z (2010) Recent developments in novel sorbents for flue gas clean up. Fuel Process Technol 91(10):1175–1197

Ma J, Liu Z, Liu S, Zhu Z (2003) A regenerable Fe/AC desulfurizer for SO2 adsorption at low temperatures. Appl Catal B 45(4):301–309

Mochida I, Korai Y, Shirahama M, Kawano S, Hada T, Seo Y, Yoshikawa M, Yasutake A (2000) Removal of SOx and NOx over activated carbon fibers. Carbon 38(2):227–239

Pi X, Sun F, Gao J, Zhu Y, Wang L, Qu Z, Liu H, Zhao G (2017) Microwave irradiation induced high-efficiency regeneration for desulfurized activated coke: a comparative study with conventional thermal regeneration. Energy Fuels 31(9):9693–9702

Plaza MG, González AS, Pis JJ, Rubiera F, Pevida C (2014) Production of microporous biochars by single-step oxidation: effect of activation conditions on CO2 capture. Appl Energy 114:551–562

Rubio B, Izquierdo MT (1998) Low cost adsorbents for low temperature cleaning of flue gases. Fuel 77(6):631–637

Sage PW, Ford NWJ (2016) Review of sorbent injection processes for low-cost sulphur dioxide control. Proc Inst Mech Eng J Power Energy 210(3):183–190

Şentorun-Shalaby Ç, Uçak-Astarlıogˇlu MG, Artok L, Sarıcı Ç (2006) Preparation and characterization of activated carbons by one-step steam pyrolysis/activation from apricot stones. Microporous Mesoporous Mater 88(1–3):126–134

Shangguan J, Li C, Miao M, Yang Z (2008) Surface characterization and SO2 removal activity of activated semi-coke with heat treatment. New Carbon Mater 23(1):37–43

Sun F, Gao J, Liu X, Tang X, Wu S (2015) A systematic investigation of SO 2 removal dynamics by coal-based activated cokes: the synergic enhancement effect of hierarchical pore configuration and gas components. Appl Surf Sci 357:1895–1901

Tsuji K, Shiraishi I (1997a) Combined desulfurization, denitrification and reduction of air toxics using activated coke: 1 activity of activated coke. Fuel 76(6):549–553

Tsuji K, Shiraishi I (1997b) Combined desulfurization, denitrification and reduction of air toxics using activated coke: 2 process applications and performance of activated coke. Fuel 76(6): 555–560

Xu G, Guo Q, Kaneko T, Kato K (2000) A new semi-dry desulfurization process using a powder-particle spouted bed. Adv Environ Res 4(1):9–18

Yan Z, Liu L, Zhang Y, Liang J, Wang J, Zhang Z, Wang X (2013) Activated semi-coke in SO_2 removal from flue gas: selection of activation methodology and desulfurization mechanism study. Energy Fuels 27(6):3080–3089

Yuan X, Choi SW, Jang E, Lee KB (2018) Chemically activated microporous carbons derived from petroleum coke: performance evaluation for CF4 adsorption. Chem Eng J 336:297–305

Zhang Z, Wang T, Ke L, Zhao X, Ma C, Powder-activated semicokes prepared from coal fast pyrolysis: influence of oxygen and steam atmosphere on pore structure. Energy Fuels

Zhang K, He Y, Wang Z, Huang T, Li Q, Kumar S, Cen K (2017) Multi-stage semi-coke activation for the removal of SO2 and NO. Fuel 210:738–747

Zhen Z (2016) Study on fast preparation of powder activated semi-coke and characteristics of SO2 adsorption. Shandong University, JINAN

Zhen Z, Tao W, Chun-yuan M, Tai F, Tian-xing M (2014) Effect of oxygen concentration on activated char pore structure during low oxygen fast pyrolysis. J China Coal Soc 39(10):2107–2113

Zhen Z, Tao W, Tai F, Chunyuan M (2015) Influence of Steam and Oxygen on Pore Structure Evolution of Active Semi-Char. J Combust Sci Technol 21(4):318–323

Zhu Y, Gao J, Li Y, Sun F, Gao J, Wu S, Qin Y (2011) Preparation of activated carbons for SO2 adsorption by CO2 and steam activation. J Taiwan Inst Chem Eng

Chapter 54
Experimental Study on Synergistic Dust Removal of Desulfurization Tower of Coal-Fired Thermal Power Unit

Shen Zhen, Meng Lei, and Lyu Junfu

Abstract Dust is one of the flue gas pollutants in coal-fired power plants. In coal-fired power plants without wet ESP installed, the synergistic dust removal effect of wet desulfurization tower directly affects the final dust emission concentration. In order to explore the synergistic dedusting effect of wet desulfurization tower and improve the synergistic dedusting performance of wet desulfurization tower, tests were carried out on five 300 MW coal-fired units in actual operation. The reliability and advantages of low pressure impactor (ELPI) measurement method were verified by analyzing and comparing the test results of different measurement methods for particulate matter. Through testing the dust concentration at the entrance and exit of the desulfurization tower, obtained the influence rule of the structure of the wet desulfurization tower on its synergistic dedusting effect, revealed the irregular change of the efficiency of WFGD for PM10 and PM2.5, and expounded that the desulfurization tower measures adopted for ultra-low SO2 emission are helpful to improve the synergistic dust removal effect.

Keywords Coal-fired power plant · Wet desulfurization tower · Synergistic dedusting · Ultra-low emission

54.1 Introduction

Due to the characteristics of energy endowment and the limitation of environmental capacity in China, China has continuously improved the emission standards of flue gas pollutants from coal-fired power plants in the past ten years, and coal-fired power plants have basically implemented ultra-low emission transformation. Among them, the dust emission requirement is lower than that of 10 mg/m^3. According to the

---

S. Zhen (✉) · L. Junfu
Institute of Theraml Engineering, Tsinghua University, Beijing 100084, China
e-mail: shenz@dteg.com.cn

S. Zhen · M. Lei
Datang Environmental Industry Group CO., LTD, Beijing 100097, China

J. Lyu and S. Li (eds.), *Clean Coal and Sustainable Energy*,
Environmental Science and Engineering,
https://doi.org/10.1007/978-981-16-1657-0_54

various ways of dust control, ultra-low emission transformation can be divided into two technical routes, one is to install wet ESP at the outlet of desulphurization tower, the other is to improve the performance of synergistic dust removal through desulphurization tower transformation (Hua et al. 2013; Qiliang et al. 2013; Xiang et al. 2009). The former has larger investment and higher operating cost and produces waste water. The latter has lower investment and operating cost, and is easy to transform (Xuehai and Zhaopeng 2015; Xu and Jianjun 2015). Therefore, the technical route of ultra-low emission transformation based on synergistic dedusting effect of desulphurization tower transformation has a good application prospect, and has become one of the hotspots of scholars' research.

The basic process of wet desulphurization is flue gas scrubbing in which process SO_2 is dissolved and captured by slurry. At the same time of flue gas desulfurization, the dust in flue gas is partly scrubbed by desulfurization slurry, thus realizing the dust removal in the process of desulfurization, which is called the synergistic dust removal of wet flue gas desulfurization.

The synergistic dust removal of desulfurization tower mainly includes two aspects. First, most of the dust in flue gas is absorbed by spray slurry collision, and then part of the dust is captured by high efficiency demister and the droplet content at the outlet of desulphurization tower is reduced, so as to realize the ultra-low emission of dust (Wenzheng et al. 2016). It is worth noting that the dust in flue gas entering desulfurization tower is partly trapped by desulfurization tower, which is conducive to reducing dust emission; At the same time, a small amount of desulphurization slurry droplets escape from the desulphurization tower, in which the desulphurizer particles or desulphurization product particles and the dissolved soluble salt increase the dust concentration in flue gas, which is not conducive to reducing the dust emission. There has been controversy about the synergistic dust removal effect of desulfurization tower, and there is a lack of filed measurement data (Sheng et al. 2011). This paper makes an in-depth study on it.

In order to study the synergistic dust removal characteristics of desulfurization tower, the test was carried out on 300 MW coal-fired units in actual operation, analyzed and compared the different testing methods of particulate matter, tested the particle size distribution and concentration at the entrance and outlet of desulfurization tower, obtained the dust concentration data at the entrance and outlet of desulfurization tower under different conditions of different units, explored the influence of desulfurization tower structure on its synergistic dust removal effect, and provided a basis for improving the synergistic dust removal efficiency of desulfurization tower.

54.2 Particular Matter Concentration Measurement Method

The dust concentration at the outlet of desulphurization tower is low and the moisture content is higher, so it is difficult to measure accurately. In this paper, the traditional weighing method and sampling method of electrical low pressure impactor (ELPI) were combined to test, and each working condition was tested by different methods, which could verify each other and improve the accuracy of test results.

Figure 54.1 shows the diagram of ELPI dust test system. The main sampling and testing instruments needed for testing are sampling gun, electric heater, cyclone separator, diffusion dryer, electrical low pressure impactor (ELPI), etc. The principle and structure of ELPI are shown in (Fig. 54.2; Hanxiao et al. 2014).

ELPI can measure the particle size distribution, classification number concentration and mass concentration of particles in the range of 30 to 10 μm in real time. The sampling frequency is 10 Hz and the flow rate is 10 L/min. The operation principle of ELPI can be divided into three parts: charge of particles, screening according to particle size classification in particle impactor, and detection of current at all levels by multi-channel sensitive electrometer. The charge efficiency of different particle sizes in the charger is known, so the induced current signal is directly proportional to the concentration and size of the particles. ELPI can measure the concentration of PM10 and PM2.5 online in real time, and it can also be used as a sampling and collecting device. Under the condition that ELPI is not charged, the impact plates at all levels act as filter cartridges. The classification quality of PM10 and PM2.5 collected on ELPI filter membranes can be tested by traditional weighing method.

In practical testing, ELPI method is used to test first, and then traditional weighing method is used to test. Through the above operation, the classification mass of PM10 (especially PM2.5) particles collected on the filter membrane of ELPI, as well as the whole range ash mass and PM2.5 mass obtained by weighing method are obtained. When using ELPI method, flue gas is sampled continuously in the sampling hole of the desulfurization tower for 120–180 min. The classified charges, number concentration and particle size distribution of PM10 (especially PM2.5) with different particle sizes in the desulfurization tower could be obtained in real time. When using traditional weighing method, flue gas is sampled continuously for 120 min to 180 min through sampling holes in the desulfurization tower. The full range ash samples and PM2.5 ash samples are collected by cyclone separator respectively. After sampling, the preset filter membrane is taken out and dried in the oven for 2 h, in which the oven temperature is about 105 °C, and then it is taken out and cooled to room temperature

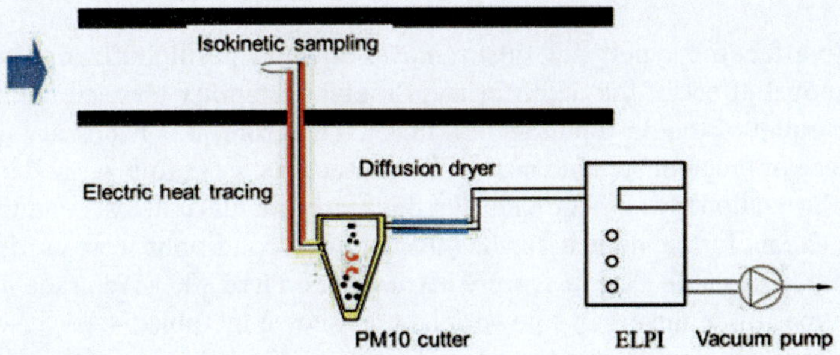

Fig. 54.1 Diagram of ELPI test system (Hanxiao et al. 2014)

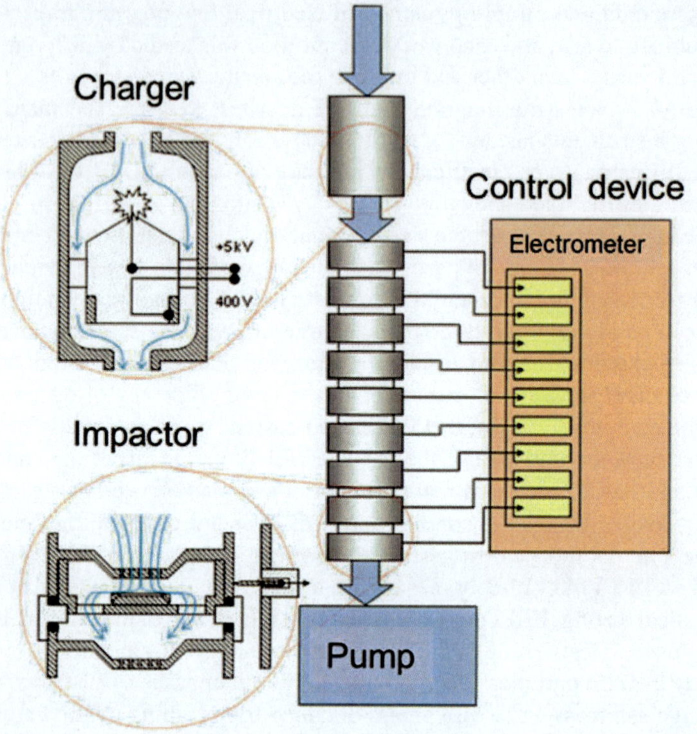

Fig. 54.2 ELPI structure diagram (Hanxiao et al. 2014)

in a drying bottle; Then the mass of particles on the filter membrane with different particle sizes is measured by weighing with 1/100000 (precision 0.01 mg) balance (Zheng et al. 2013).

54.3 Measurement Object

The key to affect the synergistic dust removal effect of desulphurization tower lies in fog removal effect of the demister and droplet portability characteristics of flue gas at the outlet of the desulphurization tower. Therefore, it is necessary to change performance of the demister and portability characteristics of flue gas at the outlet of the desulphurization tower to droplets. For this reason, actual unit test conditions were designed, except for the demisters, the other external conditions were as identical or comparable as possible. The tests were carried out on five 300 MW grade units with different types of demisters in a power plant, as shown in Table 54.1.

According to load, different test conditions were divided to study the variation of synergistic dust removal efficiency of desulphurization tower under different working

Table 54.1 General situation of test power plant

Unit number	Unit capacity	Type of desulfurization tower	Type of demister	Export control indicators
1	320 MW	Spray scrubber	Imported three-level ridge type demister	$SO_2 < 35$ mg/m^3 Dust concentration < 5 mg/m^3
2	320 MW	Spray scrubber	Domestic three-level ridge type demister	$SO_2 < 35$ mg/m^3 Dust concentration < 5 mg/m^3
3	320 MW	Spray scrubber	Tube Bundle demister	$SO_2 < 35$ mg/Nm^3 Dust concentration < 5 mg/m^3
4	300 MW	Spray scrubber with high efficiency desulfurization equipment	Tube Bundle demister	$SO_2 < 35$ mg/m^3 Dust concentration < 10 mg/m^3
5	300 MW	Spray scrubber with highefficiency desulfurization equipment	Domestic three-level ridge type demister	$SO_2 < 35$ mg/m^3 Dust concentration < 10 mg/m^3

conditions and configurations. In order to ensure the reliability of the tests, repeated tests have been carried out. Under each test condition, operation parameters were stable as far as possible, and coal quality was maintained in the whole test process.

54.4 Result and Discussion

54.4.1 Comparison of the Two Measurement Methods

Two different methods, membrane weighing method and ELPI method, were used to test the concentration of PM10, PM2.5 in flue gas of desulphurization tower of five 300 MW grade units in parallel. The results were summarized in Fig. 54.3. It can be seen that for most working conditions, the results measured by the two methods were basically the same. The measurement results of PM10, membrane weighing method were about 25% lower than those of ELPI method, and the measurement results of ELPI method tend to be more conservative, as shown in Fig. 54.3a. However, the reliability of PM2.5 measurement by ELPI method was obviously improved, only 15% of the measurement point error was relatively large, most of which were within 20%, as shown in Fig. 54.3b.

According to the Chinese national standard, this paper analyzes the data based on the weighing method, and the advantage of ELPI method is that it can identify the

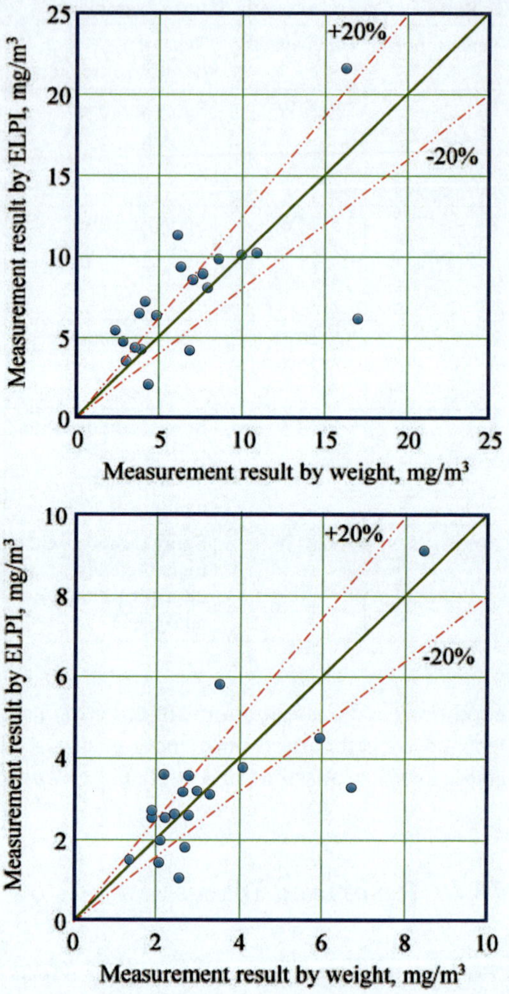

Fig. 54.3 Measurement results comparison of the two measurement methods

wrong working conditions in the weighing method according to the detailed particle size distribution characteristics. The combination of the two is very important for accurate testing.

54.4.2 *Effect of Demister on Synergistic Dust Removal of Desulfurization Tower*

As mentioned earlier, the synergistic dust removal effect of desulphurization tower is first shown as the separation effect of demister on droplets, so the synergistic dust removal efficiency of desulphurization tower with different demister performance is different, and the measurement results are shown in Fig. 54.4.

From the test results of five units, the synergistic dust removal efficiency of desulphurization tower for PM10 is generally about 50%, which is shown in Fig. 54.3a, but the synergistic dust removal efficiency for PM2.5 is low, at about 30%,see Fig. 54.3b. The test result of Fig. 54.3 also shows that different demisters have a significant difference to the synergistic dust removal effect of desulfurization tower. The tube bundle demister is more favorable for synergistic dust removal, and the synergistic removal effect of the imported three-level ridge type demister on the PM10 is better than that of the domestic three-level ridge type demister, but there is no advantage of imported equipment on the efficiency of synergistic removal of PM2.5.

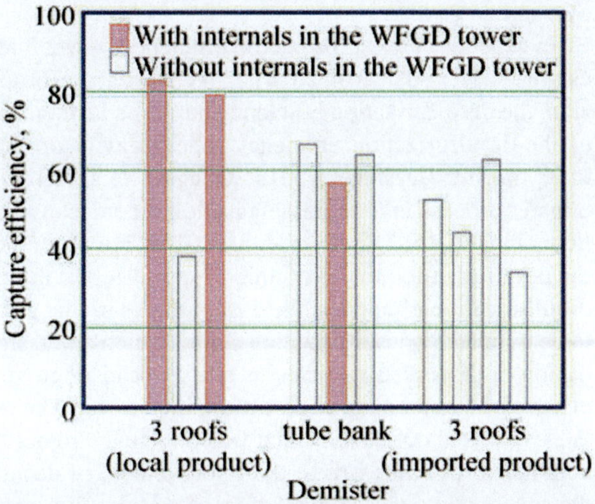

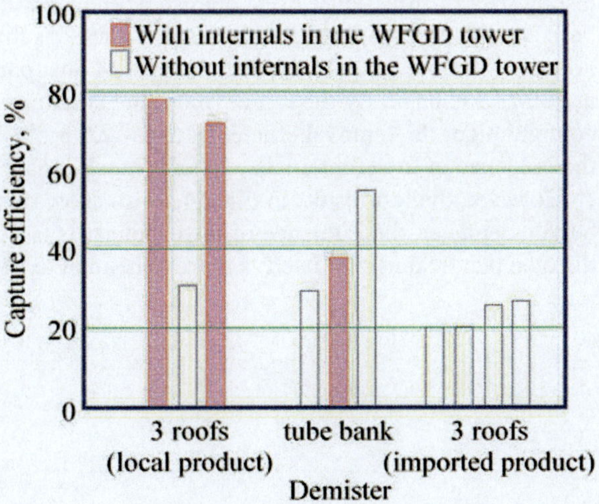

Fig. 54.4 The effect of the demister and tower internals on the dust removal efficiency of WFGD

54.4.3 Effect of Tower Internals of Desulfurization Tower on Cooperative Dust Removal

From the test results, it can be seen that not only the influence of demister is very important, but also the influence of desulphurization tower internals seems to be more obvious in the case of three-level ridge type demister. The capture of dust in flue gas by desulfurization tower, one part of the particles coexists with droplets captured by demister, including hard particles in escape droplet suspension and soluble salts, and the other part of dispersed particles exists independently. These particles are not combined with slurry in desulfurization tower, and the number of these particles depends on the touch conditions between gas and slurry. Fortunately, the contact between droplets and particles is in the same direction as that between droplets and SO_2 in flue gas.

In general, the desulfurization efficiency of wet flue gas desulfurization (WFGD) can reach over 98% (Min 2015). In recent years, in order to realize the ultra-low emission, the desulfurization efficiency has to be further improved, and the improvement of the desulfurization efficiency relies heavily on the enhancement of gas–liquid mass transfer (Deshuang 2014; Donghui et al. 2015). Enhancing gas–liquid mass transfer process is to increase gas–liquid contact area and turbulence of multiphase flow (Ping 2015; Qingshan et al. 2018). For example, the inner parts installed in the desulfurization tower of unit $4^{\#}$ and $5^{\#}$ tested in this paper, while improving the desulfurization efficiency, the dust particles in flue gas were undoubtedly intensified to be trapped by desulfurization slurry, which could significantly reduce the concentration of dispersed particles in flue gas and improve the synergistic dust removal effect of the inner parts of desulfurization tower. The removal efficiency of PM10 by WFGD was about 80%, which was obviously higher than about 30% under normal conditions. In other words, the improvement of desulfurization effect can improve the synergistic dust removal effect of wet desulfurization, as shown in Fig. 54.4 for testing data of three-level ridge demister produced in China. This improvement is based on the contact between dust particles and desulphurization slurry, which does not seem to have much to do with the size of dust particles, so the effect on PM10 and PM2.5 is basically the same. Under the condition of three-level ridge demister configuration, the removal efficiency of PM2.5 by WFGD is as high as 70% due to the influence of tower internals, which is much higher than that of 40% in general.

However, the test results in Fig. 54.4 also show that for the configuration of tube bundle demister, the influence of tower internals is very small, and it is likely that the tube bundle demister itself is more beneficial to the synergistic dust removal.

54.4.4 *Effect of Load on Synergistic Dust Removal of Desulfurization Tower*

The synergistic dust removal effect of desulphurization tower is related to the operating conditions, especially the load rate, and the test results are shown in Fig. 54.5. Under different load conditions, the inlet flue gas conditions of the desulphurization tower change, such as the lower the load rate, the lower the inlet flue gas temperature, and the smaller the flue gas flow rate, but the dust concentration does not change much. The synergistic dust removal efficiency of desulphurization tower is related to the inlet dust concentration to a certain extent. The changes of flue gas properties at the inlet of desulphurization tower have an impact on the synergistic dust removal

Fig. 54.5 The effect of load on the dust capture efficiency of WFGD

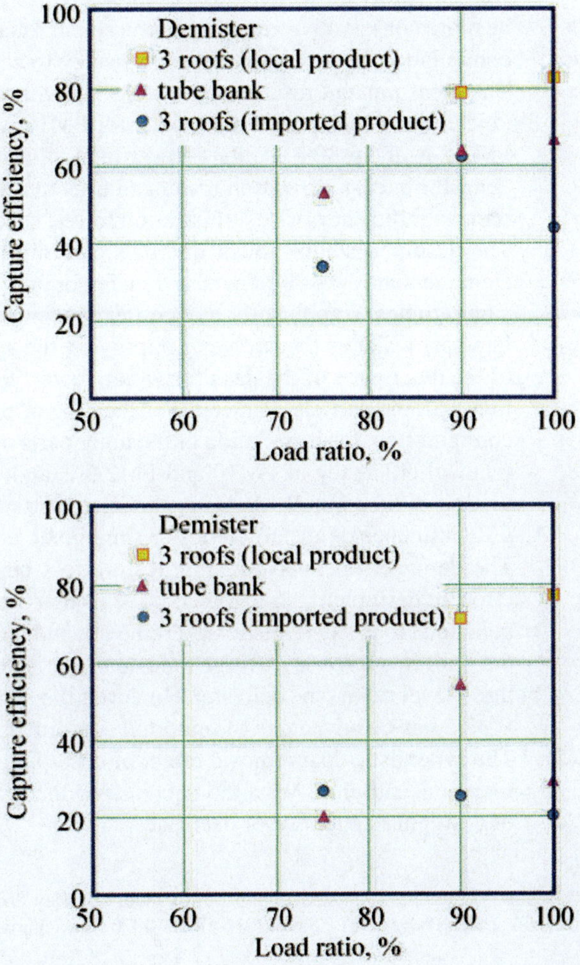

effect. The overall trend is that with the decrease of load rate, the synergistic dust removal efficiency of desulphurization tower decreases.

54.5 Conclusion

Based on the research of ultra-low emission dust test methods of thermal units, the synergistic dust removal efficiency of PM10 and PM2.5 in desulphurization tower system of 5 units was tested in this paper. The conclusions are as follows:

(1) The membrane weighing method and ELPI method were used to test the PM10 and PM2.5 emission of 5 units in parallel, and the results of the two methods were basically the same. The measurement of PM10 by ELPI method tends to be more conservative, and the measurement reliability of PM2.5 is higher. The combination of the two methods is the key to accurate measurement.
(2) The measurement results of five units show that the synergistic dust removal efficiency of desulphurization tower for PM10 is generally about 50%, and for PM2.5 is about 30%. Adding tower internals and installing three-stage ridge demister in desulfurization tower can effectively improve the synergistic dust removal efficiency of desulfurization tower, especially for PM2.5.
(3) The desulphurization tower internals can improve the desulphurization efficiency, enhance gas–liquid mass transfer, improve the contact between droplets and particles, significantly reduce the concentration of dispersed particles in flue gas, which is conducive to improving the synergistic dust removal effect of the inner parts of the desulfurization tower, and the effect has little correlation with the particle diameter. For the case of three-level ridge type demister configuration, the installation of the inner parts of desulfurization tower makes removal efficiency of PM10 and PM2.5 is above 70%, but under the configuration of tube bundle demister the inner parts of desulfurization tower have little influence on the efficiency of synergistic dust removal.
(4) The demister does have a significant effect on the synergistic dust removal effect of desulphurization tower. It seems that the tube bundle demister is more beneficial to the synergistic dust removal, and the synergistic removal effect of the imported three-level ridge type demister is better than that of the domestic three-level ridge type demister. However, the synergistic removal efficiency of PM2.5 has no advantage of imported equipment.
(5) The synergistic dust removal effect of desulphurization tower is related to the inlet concentration. With the decrease of load rate, the synergistic efficiency of desulphurization tower decreases.

Acknowledgements Financial support of this work by Key Project of the National Thirteen-Five Year Research Program of China (2016YFB0600201) is acknowledged.

References

Deshuang ZHAI (2014) Boiler super-clean emission technological renovation for coal-fired power plants. East China Electric Power 42(10):2218–2221

Donghui Z, Ye Z, Runru Z, et al (2015) Ultra-low air pollutant control technologies for coal-fired flue gas and its economic analysis. Electric Power Construct 36(5): 125–130

Hanxiao L, Shaoping Z, Yuping Y, et al (2014) Experimental research on the PM2.5 in inlet and outlet flue gas using elpi for wesp of ultra-clean emission power plants. Electric Power 47(12): 37–41

Min Z (2015) Ultralow emission transform technology and application for one 600 MW coal-fired unit. Jiangsu Elect Eng 34(3): 78–80

Mo H, Zhu FH, Wang S, et al (2013) Application of WESP in coal-fired power plants and its effect on emission reduction of PM2.5. Electric Power 46(11): 62–65

Niemel V, Lamminen E, Laitinen A (2008) A novel method for particle sampling and size-classified electrical charge measurement at power plant environment. In: Proceeding of the 11th international conference on electrostatic precipitation. Zhejiang University Press, Hangzhou, pp 228–233

Ping XU (2015) Analysis and selection of the ultraclean emissions technology route for coal fired power units. Huadian Technol 37(9):55–58

Qiliang LI, Yuan BAI, Zhonghua LI et al (2013) Study on technology of multi-pollutant co-benefit control for coal-fired gas to meet new standards. Electric Power Technol Environ Protect 29(3):6–9

Qingshan Y, Lingling M, Hubin Z, et al (2015) Research on reformation strategy of thermal power plant for "untra clean emissions. Huadian Technol

Sheng WANG, Fahua ZHU, Huimin WANG et al (2011) Fine particle emission characteristics from coal-fired power plants based on field tests. Acta Sci Circum 31(3):630–635

Wenzheng S, Mengmeng Y, Xuhui Z, et al (2016) Ultra-low emission technical route of coal-fired power plants and the cooperative removal. In: Proceedings of the CSEE, 36(16), pp 4308–4318

Xiang GAO, Zuliang WU, Zhen DU et al (2009) Study on the collaborative removal of multi-pollutants from flue gas. Environ Pollut Cont 31(12):84–90

XiannaW, Lin Z, Yanliang J, et al (2015) Measures of ultra-low emission technology of coal-fired power plant smoke. Electric Power Technol Environ Protect 31(4): 47–49

Xu Z, Jianjun L (2015) The study of ultra-low emission technology in coal-fired power plant. Sichuan Chem Ind 18(5): 55–58

Xuehai Z, Zhaopeng J (2015) Modification scheme comparison of ultra low flue gas emission in 600MW coal fired generating unit. Huadian Technol 37(10): 65–68+78

Yafeng W. Comparison about particles samples for coal-fired pollution sources and its application in power plants. PhD thesis, Chinese Research Academy of Environmental Science, Beijing

Zheng W, Jianming X, Yueyang X, et al (2013) Research on testing methods for PM2. 5 emission in coal-fired power plants. J Environ Eng Technol 3(2): 133–137

Zhengda Y, Qianyun C, Tao Y, et al (2015) Experimental study on simultaneous control of SO2 and PM by wet electrostatic precipitator. J Eng Thermophys 6: 1365–1370

Chapter 55
Analysis of Desulphurization Process in Circulating Fluidized Bed Boiler by Calcium and Sulfur Conservation

Zhang Xin, Wa Naijin, Yang Shi, and Li Ting

Abstract During the project of the technical transformation of a 75t/h CFB boiler, the desulphurization rate, the utilization ratio of CaO, the sintered ratio of CaO, etc. can be got, by the calcium conservation and sulfur conservation according to the analysis of coal and ash. In this project, the desulphurization rate in furnace is 14.34% before the technical transformation, and the utilization ratio of CaO is 21.1%, the sintered ratio of CaO is 39.2%. After the technical transformation, the desulphurization rate in furnace is 76.34%, the utilization ratio of CaO is 52.27%, and the sintered ratio of CaO is 23.74%.

Keywords Circulating fluidized bed (CFB) · Desulfurization · Calcium conservation · Sulfur conservation · f-CaO · Desulfurization ratio

55.1 Introduction

Furnace desulfuration in circulating fluidized bed boilers (CFBB) is an effective desulfurization technology with low cost. In inideal conditions, the desulfurization efficiency of furnace desulfuration technology can achieve to 90%. However, in practical operations, the desulfurization effect is far inferior to that in inideal conditions.

As a major and harmful flue gas pollutant, SOx has always been under strict control in China. And to the year of 2020, the emission concentration of SOx will be controlled to no more than 35 mg/m^3 (Senwang and Bin 2017; Youfu, 2017; Debo et al. 2017; Shibing et al. 2018; Houyong 2017; Yanfeng et al. 2009).Whether the actual situation of furnace desulfuration technology in circulating CFBB can meet

Z. Xin · W. Naijin · Y. Shi (✉) · L. Ting
China Coal Research Institute Company of Energy Conservation, Beijing 102606, China

National Key Lab of Coal Resources High Efficient Mining and Clean Utilization, Coal Research Institute, Beijing 100013, China

National Key Lab of Technical Equipment for Efficient Coal Utilization Energy Conservation & Environmental Protection, Beijing 100013, China

J. Lyu and S. Li (eds.), *Clean Coal and Sustainable Energy*, Environmental Science and Engineering,
https://doi.org/10.1007/978-981-16-1657-0_55

the demand of future policy criterion? Whether this desulfurization technology can be further optimized? These two questions are realistic problems deserves attention greatly.

To answer these two questions, the measurement of desulfurization efficiency of furnace desulfuration technology in existing CFB boilers is absolutely necessary method.

Traditionally, desulfurization efficiency can be obtained by measuring the molar quantities of SO2 in inlet and outlet of the reactor. Although, this method is simple, intermediate process of desulfurization is neglected and desulfurizer utilization potential can't be evaluate. So, this method is helpless for the optimization of furnace desulfuration technology. Baoguo and Yun-han 2007 has proposed the calcium conservation method to obtain the desulfurization efficiency of furnace desulfuration technology. Based on mass conservation principle, the conversion rate of desulfurizer and desulfurization efficiency of furnace desulfuration can be obtained by component analysis of desulfurization product. At the same time, influence factors of desulfurization efficiency and modification schemes can be evaluated by the calcium conservation method. However, the operation of this method is complicated, high cost and needs several precision equipments. So the calcium conservation method is difficult to be widely used in industrial field.

Based on mass conservation principle, the simplified calcium/sulfur conservation methed for calculating desulfurization efficiency of furnace desulfuration technology is proposed in this article. In the method, quantitative analysis of desulfurization product and f-Cao content in fly ash (Ting 2016) is simplified. Desulfurization efficiency and desulfurizer utilization rate can be calculated rapidly.

Using this method, desulfurization efficiency and desulfurizer utilization rate in a 75 t/h CFB boiler with furnace desulfuration are calculated in this article. Actual effect of furnace desulfuration is also evaluated.

55.2 Simplified Calcium/Sulfur Conservation Methed

Based on mass conservation principle, the mass of calcium before combustion equals to that after combustion. Meanwhile, the mass of sulfur before combustion equals to that after combustion.

55.2.1 Sulfur Conservation

Base on sulfur conservation shown in Fig. 55.1, including direct sulfur conservation and counter sulfur conservation method, desulfurization efficiency can be calculated. By direct sulfur conservation method, desulfurization efficiency equals to:

flue gas volume × SO2 Concentration in flue gas/(fuel quantity × sulfur content in coal).

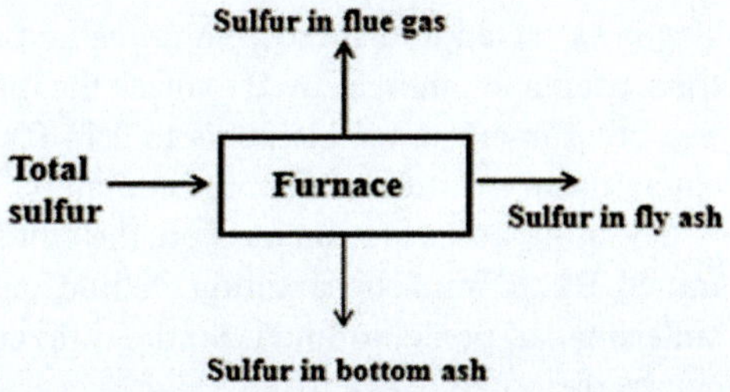

Fig. 55.1 The schema of sulfur conservation

By counter sulfur conservation method desulfurization efficiency equals to:

fuel quantity × ash content in coal × sulfur content in ash and slag/(fuel quantity × sulfur content in coal).

As Fig. 55.1 illustrated, sulfur conservation is only used to calculate the desulfurization efficiency, but can not be used to analyse economical efficiency and other interior problems of the desulfurization process.

55.2.2 *Calcium Conservation*

The "calcium" concerned in calcium conservation method means the calcium which can react with SO2,such as CaO, Ca(OH)2. By calcium conservation method, utilization ratio of desulfurizer can be obtained.So calcium conservation method can evalued the economical efficiency of the desulfurization process.

The schema of calcium conservation is shown in Fig. 55.2. As Fig. 55.2 shows, calcium before combustion can be divided into two forms: one is calcium in coal as fired, the other is calcium in the applied desulfurizer. And calcium after combustion includes calcium in desulfurization product such as CaSO3, unreacted calcium in ash and slag, calcium with no reaction activity caused by high temperature sintering. Among them, the first and third form of calcium can be measured by conventional methods, but how to determine the amount of unreacted calcium in ash and slag needs to be considered.

In our previous research, a new method to measure unreacted calcium amount has been proposed (Ting 2016). In this method, the unreacted calcium in ash and

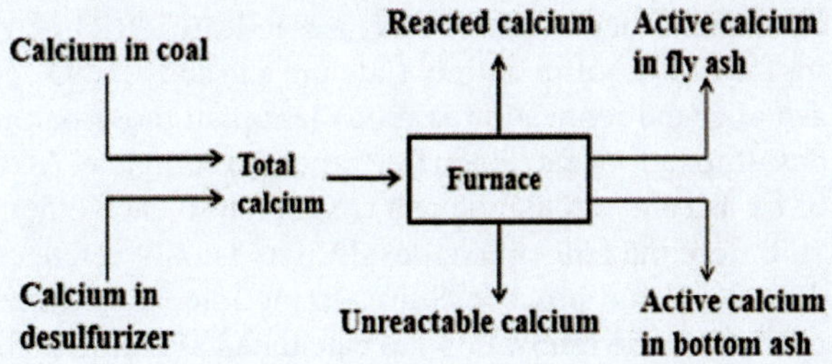

Fig. 55.2 The schema of calcium conservation

slag is extracted into sucrose solution, and titrated with EDTA reagent. At the same time, triethanolamine is used to mask the interference of iron, aluminum, manganese and other interference elements in ash. By this method, real quantity of unreacted calcium can be measured more accurately.

By sulfur conservation method, the amount of calcium reacted SO2 can be determined. By calcium conservation method, calcium with no reaction activity caused by sintering can be determined. So the utilization efficiency and status of desulfurizer can be quantitative evaluated.

55.3 Application Example

Simplified calcium/sulfur conservation method has already been used in desulphurization technical reformation of a 75 t/h medium temperature-medium pressure CFB boiler. Initially, double alkali method is merely applicated for desulfurization. So the initial emission of SO2 is in the range of 1,500–3,000 mg/m^3, far beyond the limit promulgated by government.

Considering the user's actual situation, technical transformation plan, adding proper amount of limestone into the coal but without improvement on desulfurization device at the furnace tail, was proposed. The concrete measures are that blending the fine limestone particles into the coal at the coal feeding belt to maintain the calcium and sulfur molar ratio in the range of 1.5–2.5 (Qinhjie and Yongjun 2016; Jianheng 2011; Bin et al. 2017; Xuebing 2006; Bin et al. 2017).

Comprehensive quantitative assessment results of desulfurization effect after transformation are obtained by sample examination and balance calculation. The sampling objects are the coal, bottom ashes, circulating ashes and fly ashes before and after the renovation. The testing projects include Proximate analysis, elemental analysis, ash composition analysis and f-CaO content analysis.

Detailed results are illustrated in Tables. 55.1, 55.2, 55.3 f-CaO content in the ash is measured by sucrose leaching EDTA titration method.

As illustrated in Table. 55.1, sulfur content of raw coal is higher than 1%, so the initial emission of SO2 is high as well. Whether adding limestone or not, the coal property is relatively stable.

As illustrated in Table. 55.2, calcium content in raw coal is not high, initial calcium/sulfur molar ratio is only 0.68:1.By adding limestone into raw coal, calcium/sulfur molar ratio increases to 1.46:1. Rich active CaO environment is benefit for furnace desulfurization. Calcium and active CaO content in fly ash and circulating ash after the renovation is much less than those before the renovation. That shows desulfurization reaction in the furnace is complete. At the same time, the composition of fly ash and circulating ash are similar to each other.

Before the renovation, desulfurizer is only active calcium in the raw coal during desulfurization process. Sulfur-fixing rate, CaO utilization rate and CaO sintering rate before the renovation are calculated as follows:

Table 55.1 Industrial analysis of coal and ash

Sample	Proximate analysis					
	$M_{t,\ ar}$ (%)	A_{ar} (%)	V_{daf} (%)	FC_{ar} (%)	$Q_{gr,\ ar}$ MJ/kg	$Q_{net,\ ar}$ MJ/kg
Raw coal	10.1	40.75	20.85	28.3	15.55	14.76
Circulating ash	0.49	94.76	2.18	2.57	1.3	-
Fly ash	0.54	94.46	2.46	2.54	1.44	-
Raw coal*	11.4	35 0.12	24.12	29.36	16.83	15.99
Bottom ash*	0	98.15	0	1.85	0.17	-
Circulating ash*	0.14	93.68	3.58	2.6	1.58	-
Fly ash*	0.68	91.22	3.96	4.14	2.09	-

*adding 5% limestone

Table 55.2 Elements analysis of coal and ash

Sample	Elemental analysis				
	$S_{t,\ ar}$ (%)	C_{ar} (%)	H_{ar} (%)	O_{ar} (%)	N_{ar} (%)
Raw coal	1.14	38.12	2.68	0.8	6.41
Circulating ash	0.36	-	-	-	-
Fly ash	0.38	-	-	-	-
Raw coal*	1.15	42.42	2.8	0.81	6.3
Bottom ash*	0.49	-	-	0.06	-
Circulating ash*	0.35	-	-	0.12	-
Fly ash*	0.72	-	-	0.3	-

*adding 5% limestone

Table 55.3 Composition and F-CaO analysis of coal and ash

Sample	SiO_2 (%)	Al_2O_3 (%)	Fe_2O_3 (%)	CaO (%)	MgO (%)	f-CaO (%)
Raw coal	56.08	25.82	5.71	3.32	1.28	-
Circulating ash	-	-	-	-	-	4.08
Fly ash	-	-	-	-	-	1.25
Raw coal*	53.03	24.46	5.51	8.37	1.5	-
Bottom ash*	54.9	22.2	4.02	10.78	1.74	6.67
Circulating ash*	53.24	22.55	5	11.09	1.44	5.03
Fly ash*	53.28	24.71	5.3	7.77	1.44	1.94

*adding 5% limestone

Balance calculation 1

According to sulfur balance principle shown in Fig. 55.1, a part of gross sulfur burns into SO2, another gross sulfur reacts and becomes the component of ash in the form of sulfate and sulfite. Sulfur-fixing rate can be calculated as formula 55.1:

$$1 \times S_{t1} = 1 \times A_1 \times \frac{(1-R_A)}{A_{UA1}} \times S_{UA1} + 1 \times A_1 \times \frac{R_A}{A_{CA1}} \times S_{CA1} + 1 \times S_{t1} \times R_s \tag{55.1}$$

According to the statistics data, Ra is set as 95%. So Rs equals to 85.7% and Sulfur-fixing rate 1-Rs equals to 14.3%。

Balance calculation 2:

According to calcium balance principle shown in Fig. 55.2, the desulfurizer is the majority of CaO in raw coal. Another part of CaO in raw coal, in the form of sinter, becomes the component of ash. Calcium balance during the desulfurization process can be expressed as formula 55.2:

$$\begin{aligned} 1 \times \frac{Ca_{o1}}{100} = m_{ACa} + m_{DCa} + 1 \times R_A \times \frac{A_{CA1}}{100} \times \frac{Ca_{CA1}}{100} \\ + 1 \times (1-R_A) \times \frac{A_{UA1}}{100} \times \frac{Ca_{UA1}}{100} \end{aligned} \tag{55.2}$$

As the composition of fly ash and circulating ash are similar to each other, the circulating ash can be considered as fly ash during calculation process.

$$m_{ACa} + m_{DCa} = 2.005 \times 10^{-2} kg \tag{55.3}$$

The quality of CaO, reacts with SOs, can be calculated as formula (55.4):

$$m_{ACa} = \frac{1}{A_{c1}} \times 100 \times \frac{S_t}{100} \times (1-R_s) \times \frac{56}{32} \tag{55.4}$$

According to formula (55.3) and (55.4), the quality sintered CaO, mDCa, can be calculated.

$$m_{DCa} = 1.303 \times 10^{-2} kg \tag{55.5}$$

Sintering rate of CaO can be calculated as formula (55.6):

$$R_{DCa} = \frac{m_{DCa}}{\left(\frac{Ca_{o1}}{100}\right)} = 39.2\% \tag{55.6}$$

Utilization rate of CaO can be calculated as formula (55.7):

$$R_{ACa} = \frac{m_{ACa}}{\left(\frac{A_c}{100} \times \frac{Ca_{o1}}{100}\right)} = 21.1\% \tag{55.7}$$

Three important parameters, sulfur-fixing rate, CaO utilization rate, CaO sintering rate, have been calculated. Before the renovation, sulfur-fixing rate equals to 14.3%, CaO utilization rate equals to 21.1%, CaO sintering rate equals to 39.2%.From the results, it seems that, the sulfur-fixing rate and CaO utilization rate are at low values, while the CaO sintering rate is relatively high before the renovation. The reason for that is the incorrect temperature distribution and relative high temperature in the CFBB furnace.

As a result, the renovation plan is proposed: a part of flue gas behind the induced draft fan is introduced as fluidizing air into the bottom of the furnace, which can eliminate the local high temperature zone and decrease the temperature gradient along the height in the furnace.

The renovation plan includes two sides: (55.1) adding proper amount of limestone to increase the desulfurizer content in the fuel. (55.2) optimizing the temperature distribution in the furnace by increasing the fluidizing airflow. Effect of the renovation is calculated as follows:

Balance calculation 3

CaO content of the fuel, composed of CaO in the raw coal and CAO in limestone, is calculated as formula (55.8):

$$1 \times \frac{Ca_{o2}}{100} = 1 \times R_{Ca} \times \frac{56}{100} + 1 \times (1 - R_{Ca}) \times \frac{Ca_{o1}}{100} \tag{55.8}$$

Actual addition ratio of CaO, Rca equals to 9.59%, which can satisfy the demand of desulfurization.

Balance calculation 4

According to sulfur balance principle, sulfur-fixing rate after renovation is calculated as formula (55.9):

$$1 \times S_{t2} = 1 \times A_2 \times \frac{(1 - R_A)}{A_{UA2}} \times S_{UA2} + 1 \times A_2 \times \frac{R_A}{A_{CA2}} \times S_{CA2} + 1 \times S_{t2} \times R_s \tag{55.9}$$

sulfur-fixing rate,1-Rs equals to 76.3%.

Calcium balance during the desulfurization process can be expressed as formula 55.10:

$$1 \times \frac{Ca_{o2}}{100} = m_{ACa} + m_{DCa} + 1 \times R_A \times \frac{A_{CA2}}{100} \times \frac{Ca_{CA2}}{100} + 1 \times (1 - R_A) \times \frac{A_{UA2}}{100} \times \frac{Ca_{UA2}}{100} \tag{55.10}$$

Total mass of CaO is

$$m_{ACa} + m_{DCa} = 6.36 \times 10^{-2} kg \tag{55.11}$$

The quality of CaO, reacts with SOs, can be calculated as formula (55.12):

$$m_{ACa} = \frac{1}{A_2} \times 100 \times \frac{S_{t2}}{100} \times (1 - R_s) \times \frac{56}{32} \tag{55.12}$$

The quality of CaO equals to 4.375$\times$10^{-2}kg as formula:

$$m_{DCa} = 1.987 \times 10^{-2} kg \tag{55.13}$$

Sintering rate of CaO can be calculated out and equals to 23.7%.CaO utilization rate equals to 52.3%.

Three important parameters, sulfur-fixing rate, CaO utilization rate, CaO sintering rate, have been calculated. After the renovation, sulfur-fixing rate equals to 76.3%, CaO utilization rate equals to 52.3%,CaO sintering rate equals to 23.7%.

From the results, it seems that, the sulfur-fixing rate and CaO utilization rate are obviously improved. And CaO sintering rate is also decreased by the optimization of furnace temperature distribution.

55.4 Conclusion

In this article, effect of furnace desulfurization process was evaluated by calcium-sulfur balance method. Sulfur-fixing rate, CaO utilization rate, CaO sintering rate after the renovation were calculated by sampling test data analysis. The conclusions are inducted as follows:

(1) Before the renovation, initial calcium/sulfur molar ratio is 0.68:1.Sulfur-fixing rate is 14.3%,CaO utilization rate is only 21.1%, CaO sintering rate is 39.2%.And 39.7% desulfurization material is unutilized;
(2) To improved desulfurization effect, limestone was added into raw coal, the mass mixing rate is 5%.Initial calcium/sulfur molar ratio is 1.46:1.Sulfur-fixing rate is 76.3%,CaO utilization rate is 52.3%,CaO sintering rate is 23.7%.And 24% desulfurization material is unutilized;
(3) Generally, Optimum reaction temperature of desulfurization is in the range of 850–860°C.If the furnace temperature exceed 900°C, sulfur-fixing rate decreases significantly (Fuqiang et al. 2017; Zhensen et al. 2017).Flue gas recirculation technology was used in the renovation to optimize the furnace temperature distribution and to decrease the furnace temperature.

Acknowledgements This work was supported by the China Coal Science and Technology Group Innovation Fund Project (No.2018-TD-ZD001).

References

Baoguo FAN, Yun-han XIAO et al (2007) An analytic method featuring calcium conservation during a gas-solid calcium-based desulfurization reaction. J Eng Thermal Energy Power 22(1):101–104

Bin IANG, Zhongpan LIU, Qiong WU et al (2017) Experimental research on the effect of limestone characteristics on SO2 emission from 1MW CFB boiler. Coal Process Compr Utiliz 8:71–80

Bin L, Zhongpan L, Qiong W et al (2017) Pollutant emission performance of additive-upgraded high-sulfur coal in CFB combustion. Clean Coal Technol 23(6): 94–100

Debo LI, Tinghua ZENG, Yongjiang CAI et al (2017) Research on ultra-low emission key technology for circulating fluidized bed boiler and its engineering application. Guangdong Electric Power 30(3):1–6

Fuqiang Z, Yang L, Yan et al Z (2017) Development of circulating fluidized bed hot water boiler using in-furnace desulfurization and low nitrogen combustion technology. Thermoelectric Technol (3): 10–14

Houyong Z (2017) Operational discussion on ultra-low emission of pollutants from CFB boilers. Metallurgical power (11): 46–49

Jianheng W (2013) Influence of combustion desulfurization on SO2, NOX, CO pollutant emissions of fujian anthracite-fired circulating fluidized bed boiler. Industrial Boiler (3): 5–10

Qinhjie REN, Yongjun NIU (2016) Circulating fluidized bed boiler blended combusting high calcium coal gangue desulfurization performance test. Huadian Technol 38(10):17–19

Senwang JIN, Bin LIU et al (2017) Experimental study on improving the SO2 ultra low emission economy of 300 MWCFB boiler. Clean Coal Technol 23(4):77–83

Shibing WANG, Rui ZHOU, Hongjie HOU et al (2018) Study on ultra low emission technology route of low calorific value slime coal gangue power plant. Coal Mine Modern 1:86–91

Ting LI (2016) Determination of f-Cao content in fly ash for NGD desulfuration. Clean Coal Technol 22(4):42–44

Xuebing LIU (2006) Research and application of calcium-adding desulfurization technology for circulating fluidized bed boiler. Clean Coal Technol 12(3):88–90

Yanfeng Y, Yanqiu Y, Guichang C (2009) Application of desulfurization technology in circulating fluidized bed boiler. Clean coal technology 15(3): 111–112

Youfu W (2017) Study on SO2 ultra low emission technology on circulating fluidized bed boiler. J Clean Coal Technol 23(2): 108–113

Zhensen YANG, Bin LIU, Ningwu CHEN et al (2017) Experimental research on desulfuration efficiency improvement of CFB boiler. Clean Coal Technol 23(6):94–100

Chapter 56
Simulation Investigation on NO_x Emission Characteristics and Mechanisms During Co-combustion of Fossil Fuels with Different Fuel-Nitrogen Distributions via CHEMKIN

Chaowei Wang, Chang'an Wang, Lin Zhao, Maobo Yuan, Pengqian Wang, Yongbo Du, and Defu Che

Abstract There are over 100 million tons of semi-coke produced in China every year, and use of the semi-coke in utility coal-fired boilers is a promising approach. In this paper, a blending model of semi-coke and bituminous coal was established via CHEMKIN. The NO_x emission was investigated by analyses of sensitivity and production rate, while the effects of stoichiometric ratio of main combustion zone, blending ratio, and combustion temperature on migration of fuel nitrogen were further elucidated. The simulation results showed that the conversion ratio of NO_x increased monotonically with an increase in the stoichiometric ratio of main combustion zone. The temperature of plug flow reactor (PFR) I exhibited a significant effect on NO_x emission, but other reactors had limited influences. The NO_x conversion ratio was enlarged with the increase of semi-coke blending ratio. The influence of blending proportion on NO_x emission was significant when the stoichiometric ratio of main combustion zone was small. Compared with high blending ratio, the main reactions with low blending ratio had stronger inhibitory effect on the formation of NO. The present study was of guiding significance for reducing NO_x emission in the utility boilers co-firing semi-coke.

Keywords NO_x emission · Co-combustion · Fuel-nitrogen · Semi-coke · CHEMKIN

C. Wang · C. Wang (✉) · L. Zhao · M. Yuan · P. Wang · Y. Du · D. Che
State Key Laboratory of Multiphase Flow in Power Engineering, School of Energy and Power Engineering, Xi'an Jiaotong University, Xi'an 710049, China
e-mail: changanwang@mail.xjtu.edu.cn

J. Lyu and S. Li (eds.), *Clean Coal and Sustainable Energy*, Environmental Science and Engineering,
https://doi.org/10.1007/978-981-16-1657-0_56

56.1 Introduction

Semi-coke is a kind of low-volatile solid carbon product, which features low content of volatile and high specific resistance and extensive porosity (Yang and Wang 2017; Liu and Yao 2017; Hou and Shang 2016). Compared with bituminous coal, combustion of semi-coke produces more NO_x. The grading transformation of low rank coal in China is developing continuously, therefore the semi-coke production every year would reach hundreds of millions of tons. Some scholars studied the co-combustion of semi-coke and bituminous coal to realize the conversion of coal into different grades via experiment (Yang and Wang 2017; Liu and Yao 2017; Hou and Shang 2016; Li and Li 2001; Fang 2007; Huang and Jiang 2013; Zhang and Li 2017; Fang and Gao 2003; Yan and Chen 2006; Liu and He 2018; Zhang and Jia 2019; Wang and Wu 2017), while few scholars have studied the NO_x emission of co-combustion of semi-coke and bituminous coal by CHEMKIN. The migration mechanisms of fuel-nitrogen in the process of co-combustion of semi-coke and bituminous coal have yet to be fully understood. In this paper, a model about co-combustion of semi-coke and bituminous coal via CHEMKIN was established, to study NO_x emission during co-combustion process and revealed the migration mechanisms of fuel-nitrogen from various solid fuels.

CHEMKIN, a chemical reaction kinetics software, can explore the chemical reaction mechanism in the reaction process. From CHEMKIN, the conversion from fuel-nitrogen into NO_x and the variation of fuel-nitrogen conversion path can be observed. Niu and Shang (2017) studied the NO_x emission using CHEMKIN combined with mature NO_x reduction mechanism to study the various influencing factors. Hashemi and Hansen (2011) used CHEMKIN to establish the gas reaction model when the nitrogen chemical reaction of pulverized coal in fuel-lean combustion was studied. CHEMKIN is an effective tool for studying NO_x emission characteristics and mechanisms during combustion of fossil fuels.

Yang and Wang (2017) conducted a series of studies on semi-coke combustion using experimental methods. But Yang and Wang (2017) focused on the efficiency of the boiler not the NO_x emission of semi-coke combustion. Liu and Yao (2017) studied the adaptability and grinding characteristics of the powder system in the process of semi-coke blending, and proposed that semi-coke should be blended in power plants with large grinding output margin in order to ensure safety and economy. Niu (2015) conducted an experimental study on an industrial boiler and analyzed the ignition, steady combustion and burnout of semi-coke. Li and Ma(2016) took semi-coke as a heat carrier, which played a catalytic role in the pyrolysis process of coal coke in the pyrolysis of shendong long-flame coal. The results showed that the four types of semi-coke had catalytic cracking effects on volatile components, but thermal cracking had more products than catalytic cracking. Qin and Wang (2017) studied the thermal behaviour of semi-coke, and indicated that the activation energy of the mixture first decreased and then increased with the reaction degree. Many scholars (Li and Li 2001; Fang 2007; Liu and He 2018; Zhang and Jia 2019; Ma and Guo 2018; Xie 2004; Baek and Park 2014) have studied the co-combustion of blended coal, and analyzed the pollutant emission and combustion characteristics. However, the existing studies have not revealed the migration of fuel-nitrogen and the transformation mechanisms of NO_x in the co-combustion of semi-coke and bituminous coal, while the research in this aspect is not yet clear.

In the present, the NO_x emission of co-combustion of semi-coke and bituminous coal was simulated using the plug flow reactors (as PFR) in CHEMKIN, and the effects of blending ratio (as x), stoichiometric ratio of main combustion zone (as α_1), combustion temperature in main combustion zone (as T_1) were further discussed. The rate of production (as ROP) and sensitivity analyses were applied for the interpretation of the migration mechanisms of fuel-nitrogen from various solid fuels. The migration mechanism of fuel-nitrogen and the influence of different factors on NO_x emission in the co-combustion process were elucidated. The simulation results provided guidance for the co-combustion of semi-coke and bituminous coal in the utility boilers.

56.2 Models

The reaction of homogeneous phase in CHEMKIN has been extensively employed (Ping and Zhou 2007), and the reactions of homogeneous phase in CHEMKIN have been used during study coal combustion, and the present study also applies the homogeneous phase reactions between volatiles of semi-coke and bituminous coal. The volatiles of bituminous coal are mostly methane (Fuxian 1992), therefore the combustion process of bituminous coal is simulated by methane combustion. According to the previous study (Zhong and Xu 1995), the precursors of NO_x are mainly NH_3 and HCN, and the ratio of HCN/ (HCN + NH_3) in bituminous coal is usually about 0.7. Some scholars (Zhou and Zou 2006) proposed that the proportion of fuel-nitrogen in coal is generally between 0.2%–3.4%. Here, the proportion of fuel-nitrogen was set as 1%. The combustion characteristics of semi-coke are similar to that of anthracite. The proportion of HCN/ (HCN + NH_3) in anthracite coal is about 0.9, while the HCN/ (HCN + NH_3) of semi-coke was assumed as 0.9 in the present simulation.

Table 56.1 shows the basic case conditions. In the present simulation, the stoichiometric ratio of main combustion zone was 0.9, and the stoichiometric ratio

Table 56.1 Basic case conditions

	Main combustion zone	Burnout combustion zone
Stoichiometric ratio	$\alpha_1 = 0.9$	$\alpha_2 = 0.3$
Temperature (°C)	1600	1100
Length (mm)	500	500
Position of blending (mm)	100	

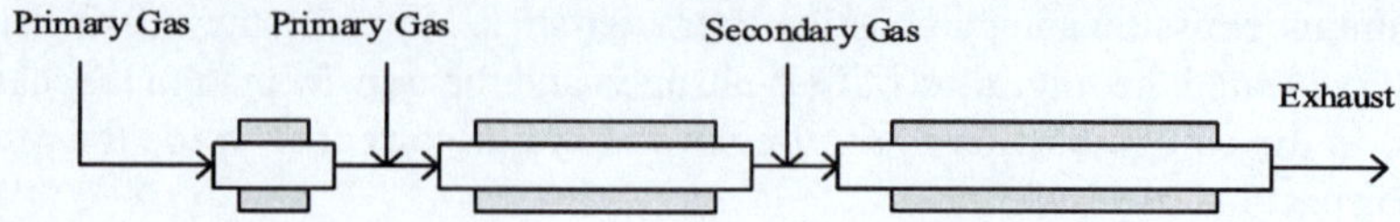

Fig. 56.1 The physical model

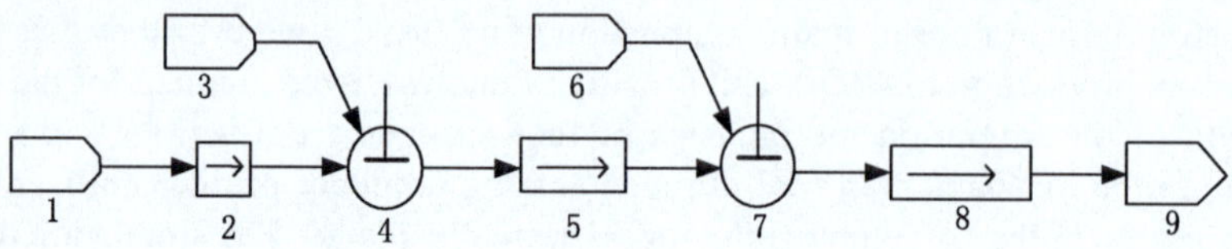

Fig. 56.2 Reaction flow chart of the co-combustion model: 1-external source of inlet gas I, 2-plug flow reactor I, 3-external source of inlet gas II, 4-non-reactive mixer I, 5-plug flow reactor II, 6-external source of inlet gas III, 7-non-reactive mixer II, 8-plug flow reactor III, 9-outlet flow of reactor

of burnout zone was 0.3. The combustion temperature of main combustion zone was 1600 °C, and the combustion temperature of burnout combustion zone was 1100°C. Fig. 56.1 depicts the physical model with the PFR I and PFR II comprised of the main combustion zone, and PFR III being burnout zone. As shown in Fig. 56.2, the reaction flow chart of the co-combustion model was established in CHEMKIN.

56.3 Results and Discussion

56.3.1 The Influence of Combustion Temperature

There are three PFR reactors within the model of co-combustion of semi-coke and bituminous coal. As the combustion temperature changes, the emission characteristics of NO_x are shown in Fig. 56.3. The combustion temperature range of main combustion zone covers combustion of cyclone furnace and pulverized coal furnace, and is set as 1300–1700°C. It can be found from Fig. 56.3 that the conversion rates of NO_x vary considerably with the combustion temperature of PFR I reactor. As the PFR I reactor temperature rises, the NO_x conversion shows ascendant trend, but the extents of rise are different. The combustion temperature of PFR I reactor mainly affects the combustion process of bituminous coal. Hence, when the blending ratio of semi-coke is low, the change of NO_x conversion ratio is the most obvious. The NO_x conversion ratio increases with the rise of combustion temperature in various blending ratios owing to more thermodynamic NO_x generating with the rise of temperature. Wang and Wang (2018) experimentally studied the co-combustion of bituminous coal and anthracite when the blending ratio was 50% at 1500 °C, and the results showed that

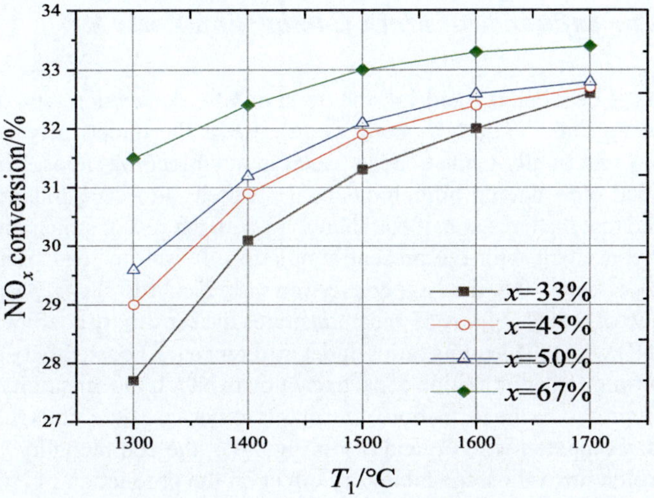

Fig. 56.3 Change of NO_xconversion via PFR I combustion temperature ($\alpha_1 = 0.9$)

the NO_x emissions were promoted with the temperature, which is consistent with the present simulation results.

The main conversion paths of fuel-nitrogen to NO in the process of co-combustion are shown in Fig. 56.4. The main reactions involved are similar at various conditions. As shown in the red box in Fig. 56.4, the precursors of NO are mostly NCO and NNH. While in air-staging combustion, the precursors of NO are mostly NCO and HNO. Another difference is that a little NO_2 and N_2O would be produced in the co-combustion process, while this phenomenon does not exist in the air-staging combustion. In addition, the paths for NO_x transforming into N_2 are more abundant due to the relatively large number of substances involved in the reaction in the process of co combustion. Hence, it is also difficult to reduce the NO_x conversion ratio significantly due to the diversification of reactions and reactants.

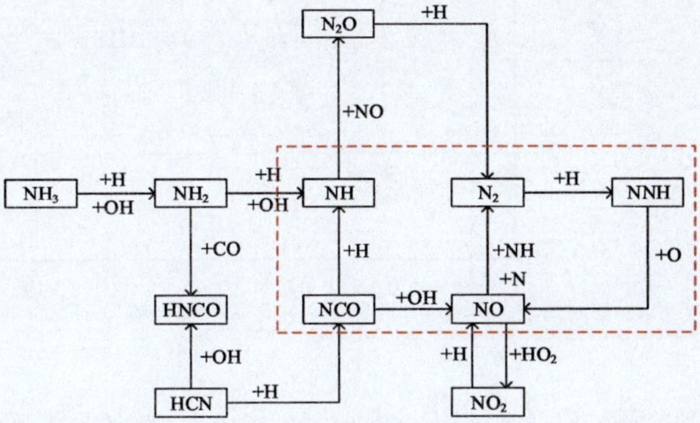

Fig. 56.4 Migration path of fuel-nitrogen in the co-combustion of semi-coke and bituminous coal ($\alpha_1 = 0.9$, $T_1 = 1600$ °C, $T_2 = 1100$°C)

56.3.2 The Influence of Main Combustion Zone A

When the α_1 is between 0.7 and 1.4, the trend of NO_x conversion ratio at the outlet is shown in Fig. 56.5. The NO_x conversion ratio at the outlet increases with the rise of the α_1 and finally tends to be constant in any blending ratios. Liu and Chen (2017) studied 60% datong bituminous coal blended with 40% Indonesia lignite, and indicated that the emission of NO increased with the rise of stoichiometric ratio, which was consistent with the present simulation. It can be seen from Fig. 56.5 that when α_1 is below 0.8, the NO_x conversion ratio has little change, especially the blending ratio of < 50%. Figure 56.6 demonstrates the sensitivity coefficients of each reaction to NO with the blending ratios under different α_1. The sensitivity coefficients reflect the promotion or inhibition of each reaction to NO. If the sensitivity coefficient is positive, it promotes the formation of reactants, otherwise, it inhibits the formation of reactants. Compared with (a) and (b) in Fig. 56.6, the commonality is that, even at different blending ratios, the inhibitory effect on the production of NO decreases with the rise of α_1. The main reactions are as follows:

$$H + NO + M \rightarrow HNO + M.$$

$$NH + NO \rightarrow N_2O + H.$$

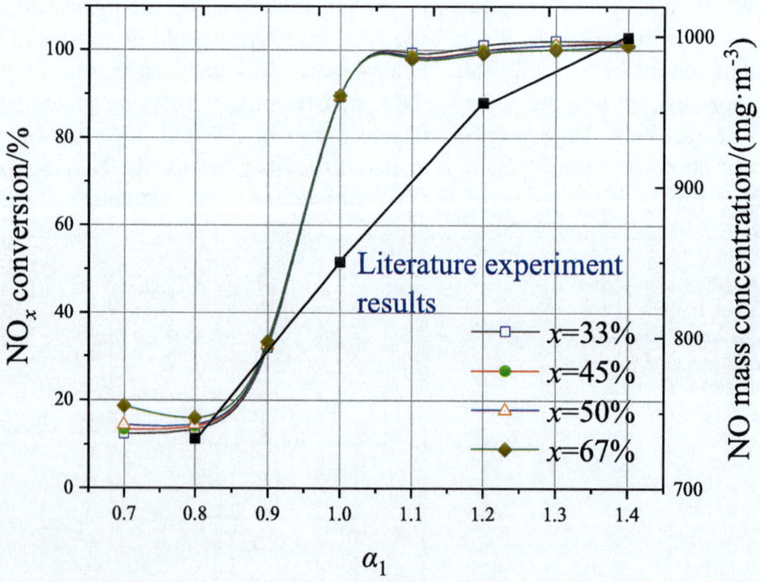

Fig. 56.5 Change of NO_x conversion via α_1 ($T_1 = 1600$ °C, $T_2 = 1100$ °C)

Fig. 56.6 Change of the sensitivity coefficient of each main reaction to NO via α_1 (**a**) $x = 33\%$, (**b**) $x = 67\%$ ($T_1 = 1600$ °C, $T_2 = 1100$ °C)

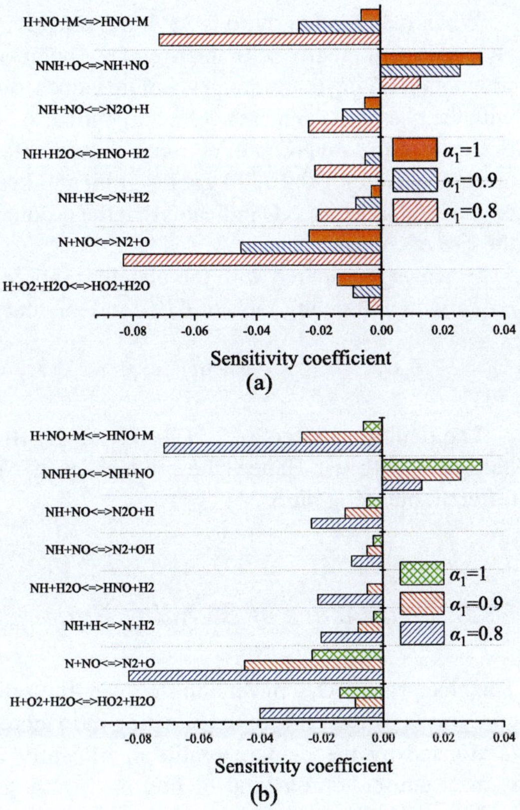

$$NH + H_2O \rightarrow HNO + H_2.$$

$$NH + H \rightarrow N + H_2.$$

$$N + NO \rightarrow N_2 + O.$$

The following reaction shows that the promotion of NO production gradually increased with the increase of α_1:

$$NNH + O \rightarrow NH + NO.$$

However, sensitivity coefficients of the following reaction are different with diverse blending ratios:

$$H + O_2 + H_2O \rightarrow HO_2 + H_2O.$$

When the blending ratio is 33%, the sensitivity coefficient of $H + O_2 + H_2O \rightarrow HO_2 + H_2O$ increases with the rise of α_1, but it could not change the trend of the NO production owing to the small absolute values, thus the overall trend of NO increases with the rise of α_1. When the blending ratio is 67% and α_1 is 0.8, the inhibitory effect of this reaction on NO production is remarkable, while the differences between the effects on α_1 being 0.9 and 1 are insignificant. In addition, the overall inhibitory trend on the production of NO indicates that the production of NO increases gradually with the rise of α_1.

However, compared with the blending ratio of 33%, the sensitivity coefficient is larger with a blending ratio of 67%, and the reaction is following:

$$NH + NO \rightarrow N_2 + OH.$$

The inhibitory effect of $NH + NO \rightarrow N_2 + OH$ on the production of NO also decreases with α_1. Hence, the production of NO increases with the rise of α_1 at different blending ratios.

56.3.3 Influence of Blending Ratio

The change of NO_x conversion via blending ratio is shown in Fig. 56.7. The NO_x conversion increases with the blending ratio increases at any α_1.The NO_x conversion is affected by the blending ratio significantly when the α_1 is small. The reason is that semi-coke is a kind of fuel having the characteristics of low burnout rate and low volatiles. The NO_x emission increases with the proportion of semi-coke increases. However, the influence of reducing atmosphere on the NO_x generation in the combustion of bituminous coal is remarkable when the α_1 is small. Therefore, the NO_x emission is affected by the proportion of bituminous coal greatly. According to experiments on the combustion of semi-coke in power station pulverized coal boilers of Yang and Wang (2017), the emission concentration of NO_x rises slightly with the increase of the proportion of semi-coke, which is consistent with the present simulation results. Besides, Wang (2015) studied the co-combustion of anthracite and bituminous coal adopting experimental means to ensure $\alpha = 1.2$, and concluded that the emission concentration of NO_x increased with the rise of the blending ratio of anthracite.

The Fig. 56.8 below shows the sensitivity analysis of the blending ratios of 33% and 50%. It can be found that the sensitivity coefficients of these two blending ratios are not significantly different. The sensitivity coefficients of the following three reactions are positive, which promotes the generation of NO:

$$NNH + O \rightarrow NH + NO.$$

$$N + OH \rightarrow NO + H.$$

Fig. 56.7 Change of NO_x conversion via blending ratio

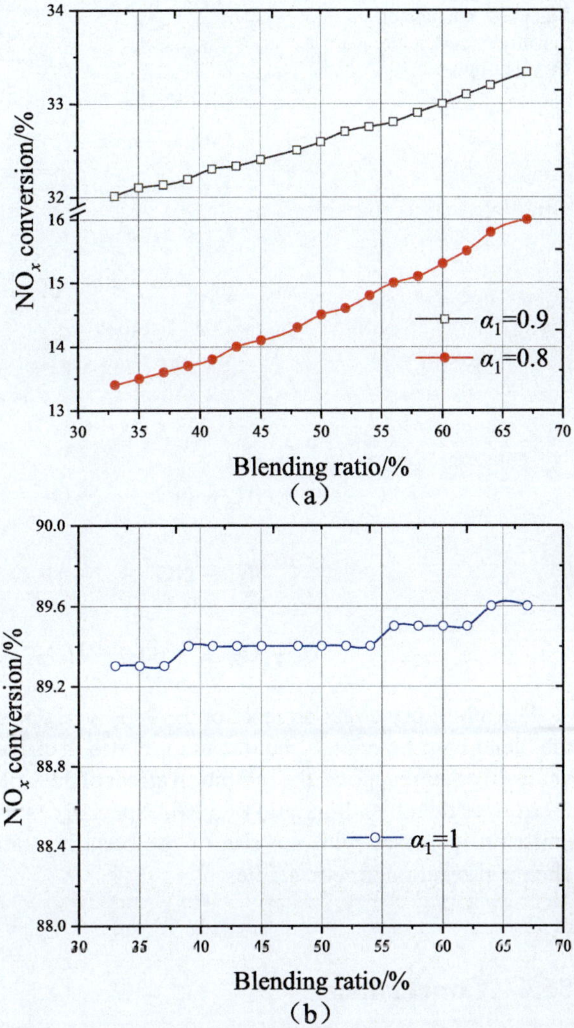

$$HNO + H \rightarrow H_2 + NO.$$

The differences between sensitivity coefficients of $NNH + O \rightarrow NH + NO$ and $N + OH \rightarrow NO + H$ are small, while the differences between the sensitivity coefficients of different blending ratios from $HNO + H \rightarrow H_2 + NO$ are relatively large. However, the absolute value of sensitivity coefficient on $HNO + H \rightarrow H_2 + NO$ is small, therefore it would not play a decisive role in the NO_x emission. The decisive roles are the reactions that inhibit the generation of NO as follows:

$$H + NO + M \rightarrow HNO + M.$$

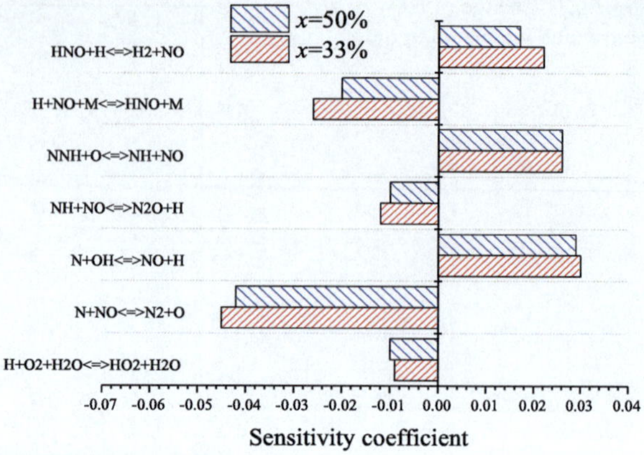

Fig. 56.8 Change of Sensitivity coefficient via blending ratios

$$NH + NO \rightarrow N_2O + H.$$

$$N + NO \rightarrow N_2 + O.$$

$$H + O_2 + H_2O \rightarrow HO_2 + H_2O.$$

The inhibitory effect on reaction $H + O_2 + H_2O \rightarrow HO_2 + H_2O$ is significant at the blending ratio of 50%, but it cannot play a decisive role due to its small absolute value. In other reactions, the inhibition effect of 33% blending ratio on NO generation is stronger than blending ratio of 50%, especially $N + NO \rightarrow N_2 + O$. Hence, NO_x emission increases with the rise of the blending ratios both in experimental and chemical reaction kinetic studies.

56.4 Conclusions

The model of co-combustion of semi-coke and bituminous coal was established by three plug flow reactors (PFR) via CHEMKIN, and the NO_x emission characteristics in the co-combustion process were simulated. The migration mechanisms of fuel-nitrogen from various solid fuels were revealed, and the conclusions were as follows:

(1) The combustion temperature of PFR I had obvious influence on NO_x conversion. When the blending ratio was below 50%, the NO_x emission could rise significantly with the combustion temperature, because the bituminous coal in the main combustion zone was mainly affected. In addition, the main path of fuel-nitrogen converting into NO_x in the process of co-combustion was obtained. Compared with air-staging combustion, there was more reactant NNH in the co-combustion reaction system.

(2) The NO_x conversion increased with the rise of stoichiometric ratio of main combustion zone (α_1). It was found that increasing α_1, partial reactors inhibiting the NO_x formation were reduced.

(3) The NO_x conversion ratio increased with the rise of the blending ratios. Moreover, the influence of NO_x emission on the blending ratio was remarkable when the α_1 was small. Compared with high blending ratio, the main reactions with low blending ratios had stronger inhibitory effect on the formation of NO.

Acknowledgements This work was financially supported by the National Key R&D Program of China (2018YFB0604103).

References

Baek S, Park H (2014) The effect of the coal blending method in a coal fired boiler on carbon in Ash and NO_x emission. Fuel 128(14):62–70

Fang Q (2007) Numerical simulation and experimental study of low volatile coal and its mixed coal combustion. Huazhong University of Science and Technology

Fang L, Gao Z (2003) Experimental study on performance of NO_x emission for low volatilization coals. Proc CSEE 23(8):211–214

Fuxian Wu (1992) Preliminary study on the gas composition in the volatile product of coal. Coal Geol China 3:27–29

Hashemi H, Hansen S (2011) A model for nitrogen chemistry in oxy-fuel combustion of pulverized coal. Energy Fuels 25(10):4280–4289

Hou J, Shang W (2016) Comparison of emission from Lantan (semi-coke) instead of raw coal for clean and efficient combustion. Coal Technol 35(8):287–289

Huang Z, Jiang J (2013) Research on CFB boiler large proportion coal slime co-combustion test. Proc CSEE 33(s1):112–116

Li Y, Li S (2001) Experimental study on the formation of NO_x of brown and brown-blending coal combustion. Chinese Soc Electr Eng 21(8):34–36

Li X, Ma J (2016) Semi-coke as solid heat carrier for low-temperature coal tar upgrading. Fuel Process Technol 143:79–85

Liu Y, Chen Q (2017) Experimental investigation on NO emission characteristics during pulverized coal combustion in O_2/CO_2 and O_2/N_2 atmospheres. Thermal Power Generat 46(4):39–45

Liu Y, He Y (2018) Characteristics of Alkali species release from a burning coal/biomass blend. Appl Energy 215:523–531

Liu J, Yao W (2017) Adaptability analysis for semi-coke in different mill systems. Thermal Power Generat 46(2):7–14

Ma L, Guo A (2018) Combustion interactions of blended coals in an O_2/CO_2 mixture in a drop-tube furnace: experimental investigation and numerical simulation. Appl Therm Eng 145:184–200

Niu F (2015) Feasibility of semi coke combustion in industrial pulverized coal boiler. Clean Coal Technol 2:106–108

Niu Y, Shang T (2017) Effect of pulverized coal preheating on NO_x reduction during combustion. Energy Fuels 31(4)

Ping C, Zhou J (2007) Research on the pyrolysis kinetics of blended coals. Proc CSEE 27(17):6–10

Qin H, Wang W (2017) Thermal behavior research for co-combustion of furfural residue and oil shale semi-coke. Appl Therm Eng 120:19–25

Wang T (2015) Experimental study on the influence of coal blending method on the burnout rate and NO_x emission of mixed coal. Huazhong University of Science and Technology

Wang C, Wang P (2018) Experimental study on NO_x reduction in oxy-fuel combustion using synthetic coals with pyridinic or pyrrolic nitrogen. Appl Sci-Basel 8:2499–2510

Wang C, Wu S (2017) Study on correlations of coal chemical properties based on database of real-time data. Appl Energy 204:1115–1123

Xie J (2004) Study on test of combustion optimization for 350 MW boiler firing blended coals in Hejing power plant. Shanghai Jiaotong University

Yan W, Chen Y (2006) Performances of Pulverized-coal boilers burning heavy slagging blending coals. Proc CSEE 26(14):93–97

Yang Z, Wang Z (2017) Experimental research on pulverized coal power station blending with semi-coke. Boiler Technol 23(3):80–85

Zhang J, Jia X (2019) Experimental investigation on combustion and NO formation characteristics of semi-coke and bituminous coal blends. Fuel 247:87–96

Zhang J, Li X (2017) Experimental study on Shenhua coal blending in million tower furnace unit. Clean Coal Technol 23(1):65–68

Zhong B, Xu X (1995) Numerical simulation of NO_x formation in combustion processes. J Combust Sci Technol 02:120–128

Zhou Y, Zou P (2006) Experimental study of effects of pulverized coal characteristics on the conversion of fuel nitrogen to NO_x. Proc CSEE 15:63–67

Chapter 57
Experimental Study on Combustion of Pulverized Char Preheated by a Circulating Fluidized Bed

F. Pan, J. G. Zhu, J. Z. Liu, Y. Zhang, and S. J. Zhu

Abstract The preheating combustion of pulverized char was conducted in a 30 kW test platform, which consists of a circulating fluidized bed (CFB) and a down-fired combustor (DFC). Properties of high-temperature coal gas and preheated char particles were studied under higher air equivalence ratio in circulating fluidized bed, and the fuel gas compositions and temperature distribution along the down-fired combustor were analyzed. It was also found that parameter optimization helped to reduce the NO_x emission, the NO_x emission was reduced to 109.9 mg/m^3(@6%O_2) and the combustion efficiency was improved to 97.73% by changing the tertiary gas distribution, which achieved clean and efficient combustion of ultra-low volatile fuel.

Keywords Circulating fluidized bed · Preheating · Pulverized char · Combustion · NO_x emission

57.1 Introduction

Coal is the main primary energy in China, and plays a strategic role in China's economic growth. By the end of 2017, coal combustion accounted for 68.6% of the country's primary energy utilization. Coal combustion is the main source of nitrogen oxide emissions, and has caused severe air pollution problems. How to effectively control the emission of nitrogen oxides is an urgent problem to be solved. In recent years, low NOx combustion technologies such as air-staged combustion technology, fuel-staged combustion technology, high temperature air combustion technology, MILD combustion technology have developed rapidly. These technologies decrease the emission of nitrogen oxides by inhibiting the process of nitrogen transformation. Selective catalytic reduction (SCR) and selective non-catalytic reduction (SNCR)

F. Pan · J. G. Zhu (✉) · J. Z. Liu · Y. Zhang · S. J. Zhu
Institute of Engineering Thermophysics, Chinese Academy of Sciences, Beijing 100190, China
e-mail: zhujianguo@iet.cn

F. Pan · J. G. Zhu · Y. Zhang · S. J. Zhu
University of Chinese Academy of Sciences, Beijing 100049, China

J. Lyu and S. Li (eds.), *Clean Coal and Sustainable Energy*,
Environmental Science and Engineering,
https://doi.org/10.1007/978-981-16-1657-0_57

reduce the emission of nitrogen oxides after combustion, but the economic cost is high and may lead to escape of ammonia. With the requirement of nitrogen oxides emission is further improved, more effective emission reduction methods are required controling the emission of nitrogen oxides.

Low-rank coal accounts for a large proportion of coal reserves in China. Its main clean and efficient utilization way is to convert coal into gas or liquid fuel. In the actual pyrolysis and gasification process, a large number of char will be produced, and the energy content is also high. In view of the characteristics of low moisture and volatile content of semi-coke, the ignition point is high, it is difficult in burning out and the NO_x emission is far beyond the national standard, achieving clean and efficient combustion of char is an urgent problem to be solved.

In order to achieve clean and efficient combustion of the ultra-low volatile fuels, the Institute of Engineering Thermophysics of the Chinese Academy of Sciences (IET, CAS) proposed a new technology based on preheating pulverized coal in a circulating fluidized bed (CFB). The pulverized coal is first preheated at a temperature higher than 800°C in a CFB, then char particles and high-temperature coal gas are burned in a down-fired combustor (DFC), achieving high efficiency combustion and low emission of NO_x. Previous studies focused on the preheating and combustion characteristics of fuel and the mechanisms of NO_x emission control.

In order to further decrease NOx emission, a relative high air equivalence ratio is used in the CFB with the hope of a high conversion of fuel-N to N2. In the support of this thinking, some experiments were carried out in a 30 kW test rig.

57.2 Experimental Section

57.2.1 Apparatus and Method

The schematic illustration of the test platform is shown in Fig. 57.1. It consisted of three parts: CFB, DFC and auxiliary system. The inner diameter and height of the riser of the CFB are 78 mm and 1500 mm. Primary air provideed oxygen for partially gasification and combustion of pulverized coal. Gas produced during preheating was defined as high-temperature coal gas and solid particles were defined as preheated char particles. A horizontal pipe was installed between CFB and DFC to transport preheated fuel. The DFC had a diameter of 260 mm and a height of 3000 mm. Three tertiary air nozzles were arranged at 200 mm, 600 mm and 1200 mm from the top of the DFC.

In order to ensure the stable operation of the experiment, temperature and pressure measurement points were set on the CFB and DFC. Four temperature measuring points (Ni–Cr/Ni-Si thermocouple) were arranged at distance of 100 mm, 500 mm, 1450 mm from the bottom of the riser and arranged at the return leg. Five temperature measuring points (Pt/Pt–Rh thermocouple) were organized at the distance of 100 mm, 400 mm, 900 mm, 1400 mm and 2400 mm from the top of the DFC. The pressure

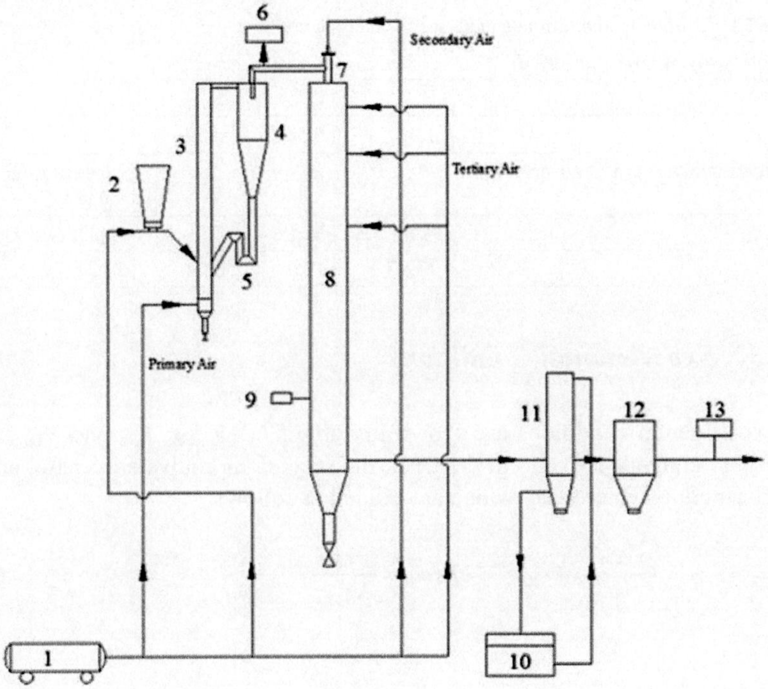

Fig. 57.1 Schematic diagram of test system. 1-Air Compressor, 2-Screw Feeder, 3-Riser, 4-Cyclone, 5-U-valve, 6-Sampling Port, 7-Secondary Air Nozzle, 8- Down-fired Combustor, 9-Sampling Port, 10-Water Tank, 11-Water Cooler, 12-Bag Filter, 13-Gas Analyzer

measuring points were located at the three temperature measuring points in the riser and the outlet of the cyclone.

Preheated char particles were collected by filter cartridges for proximate analysis and ultimate analysis. A Testo-310 analyzer and a gas chromatographic analyzer were used to analyze the high-temperature coal gas. What's more, Gasmet FTIR DX-4000 and KM9106 gas analyzer were used to measure and analyze gas in the DFC.

57.2.2 Fuel Characteristic

The experimental coal is Shenmu char with particle size ranging from 0.1 to 0.355 mm. The fuel characteristics are listed in Table 57.1. The CFB bed material is quartz sand with particle size ranging from 0.1 to 0.5 mm.

Table 57.1 Proximate and ultimate analyses of Shenmu char

Ultimate analysis (wt%, air-dried)				
C	H	O	S	N
73.73	1.92	5.54	0.29	0.78
Proximate analysis (wt%, air-dried)				Low calorific value
M	A	VM	FC	$Q_{net,ar}$ (MJ/kg)
3.73	13.97	13.9	68.4	23.32

57.2.3 Experimental Conditions

The experimental conditions are shown in Table 57.2. λ_{CFB}, λ_{Se} and λ_{Te} are the primary air equivalence ratio in the CFB, the second air equivalence ratio, and the tertiary air equivalence ratio, which are difined as follows:

$$\lambda_{CFB} = \frac{A_{CFB}}{A_{S\,toic}} \tag{57.1}$$

$$\lambda_{S\,e} = \frac{A_{S\,e}}{A_{S\,toic}} \tag{57.2}$$

$$\lambda_{Te} = \frac{A_{Te}}{A_{S\,toic}} \tag{57.3}$$

where A_{CFB}, A_{Se}, A_{Te} and Atoic are air flow rate in the CFB, the secondary air flow rate, the tertiary air flow rate and the theoretical air requirement for pulverized char combustion. During the experiment, the temperature of the CFB was kept at about 880°C, and the λ_{Stoic} is the excess air ratio, fixed at about 1.05. In case 1, the gas along the DFC is sampled to analyze combustion and emission characteristics. On the basis of case 1, case 2 and 3 change the distribution of tertiary air to study the influence characteristics, "200 + 600 + 1200" means that the tertiary gas is injected evenly at 200 mm, 600 mm and 1200 mm below the top of the DFC.

Table 57.2 Experimental conditions

Items	Case1	Case2	Case3
λ_{CFB}	0.52	0.52	0.52
λ_{Se}	0.23	0.23	0.23
Λ_{Te}	0.31	0.31	0.31
Tertiary gas distribution	200 + 600 + 1200	600 + 1200	1200

57.3 Results and Discussion

57.3.1 Preheating Characteristics in CFB

Figures 57.2 and 57.3 are the curves of temperature and pressure drop of CFB. It can be found that the temperature and pressure drop of CFB remained stable during the experiment, showing that the preheating process in the CFB was stable. The outlet temperature of cyclone was maintained at about 820 C, which was much higher than ignition temperature of the Shenmu char. When the preheated fuel enters the DFC, it will ignite immediately and burn in the DFC. It shows that this combustion technology can realize ignition and stable combustion of low volatile coal such as Shenmu char.

Table 57.3 shows the results of proximate and ultimate analyses of preheated char particles. The conversion rate of each component can be calculated by ash balance method, the carbon conversion ratio in the preheating process is 45.2%, meaning that 45.2% of the carbon has been transformed into high-temperature coal gas in the CFB. The nitrogen conversion ratio in the preheating process is 50.2%, meaning that 49.8% of the fuel-N would be converted to NO_x or N_2 in the DFC.

Table 57.4 shows that the main components of high-temperature coal gas at the outlet of CFB are CO, H_2 and CH_4, with average volume fractions of 8.83%, 2.20% and 0.29%, respectively. The low heating value (LHV) of high-temperature coal gas is 1.50 MJ/m^3 on a cool dry basis. Because of the reducing atmosphere in the CFB, the concentration of NO, NO_2, N_2O and O_2 at the outlet of CFB is 0. The nitrogen mainly exists in the form of N_2, NH_3 and HCN at the outlet of CFB. According to

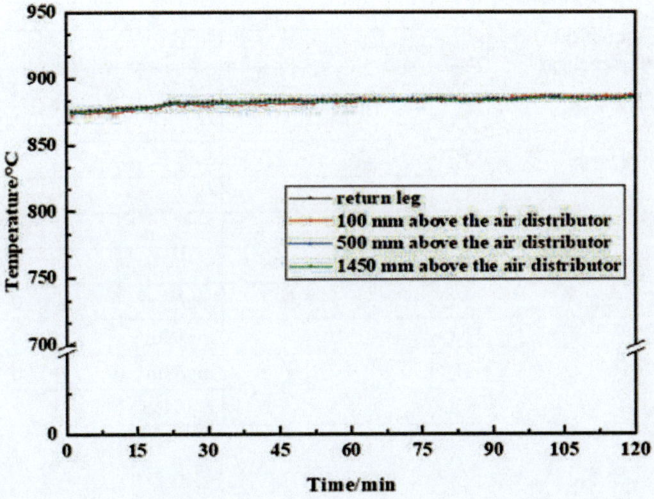

Fig. 57.2 Temperature distribution in the circulating fluidized bed (CFB)

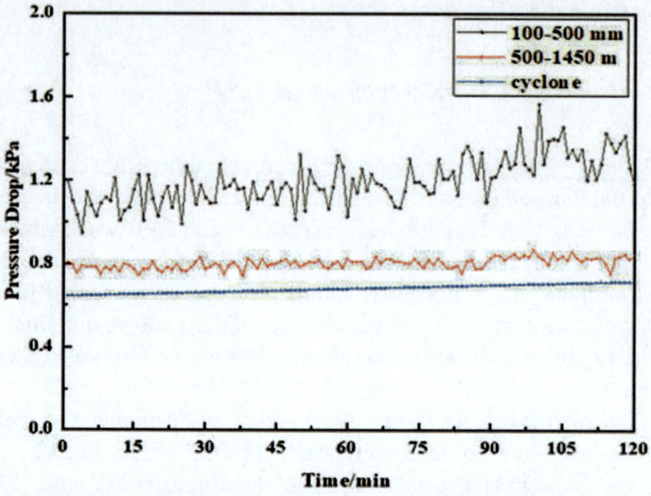

Fig. 57.3 Pressure drop distribution in the circulating fluidized bed (CFB)

Table 57.3 Proximate and ultimate analyses of preheated char particles

Ultimate analysis (wt%, air-dried)				
C	H	O	S	N
70.26	0.65	1.69	0.82	0.68

Proximate analysis (wt%, air-dried)			
M	A	VM	FC
1.64	24.26	6.24	67.86

Table 57.4 Components of high-temperature coal gas

Gas composition	Unit	Value
N_2	%	72.73
CO	%	8.83
H_2	%	2.20
CO_2	%	15.95
CH_4	%	0.29
O_2	%	0
NO	mg/Nm^3	0
NO_2	mg/Nm^3	0
N_2O	mg/Nm^3	0
QL	MJ/Nm^3	1.50

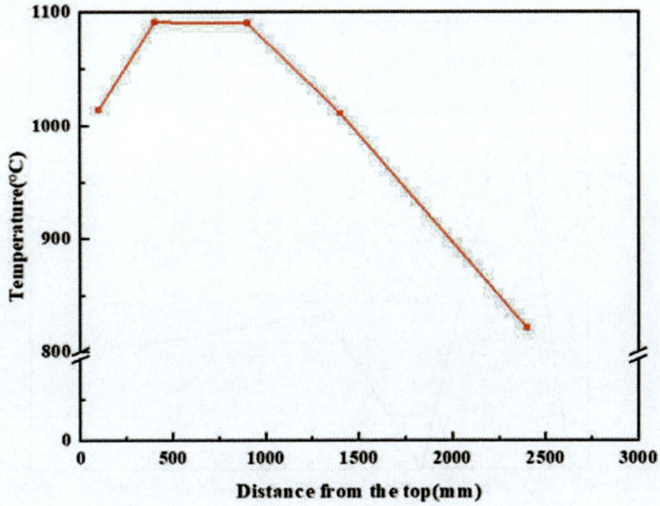

Fig. 57.4 Temperature distribution in the down-fired combustor (DFC)

the previous research, about 30–45% of fuel-N is converted to N_2, which is one of the important reasons why preheating in CFB can reduce the emission of NO_x.

57.3.2 Combustion Characteristics in DFC

In this experimental study, case 1 was taken as the basic condition. Figure 57.4 shows temperature distributions along the DFC. The temperature reaches the maximum at 100 mm below the top of the DFC, which is 1093°C. The lowest temperature is 821°C at the outlet of DFC. At this time, the combustion efficiency is 97.1%, and the CO emission is 113.8 mg/m^3, which indicates that efficient and stable combustion of char can be achieved in the DFC.

The temperature along the DFC is lower than 1200°C, so the thermodynamic NO_x can be neglected. As we can see 49.8% of fuel-N still exists in preheated char particles, which will be released during combustion. At the same time, NH_3 and HCN in high-temperature coal gas will react and eventually convert into N_2 or NO_x.

57.3.2.1 NO Emission Characteristics in DFC

Figure 57.5 shows NO concentrations along the down-fired combustor (DFC). The NO concentration at 100 mm below the top of the DFC was 0 mg/m^3, which indicateed that oxygen was rapidly consumed after the preheated fuel and secondary air were injected into the DFC, so this area was at a strong reducing atmosphere. At

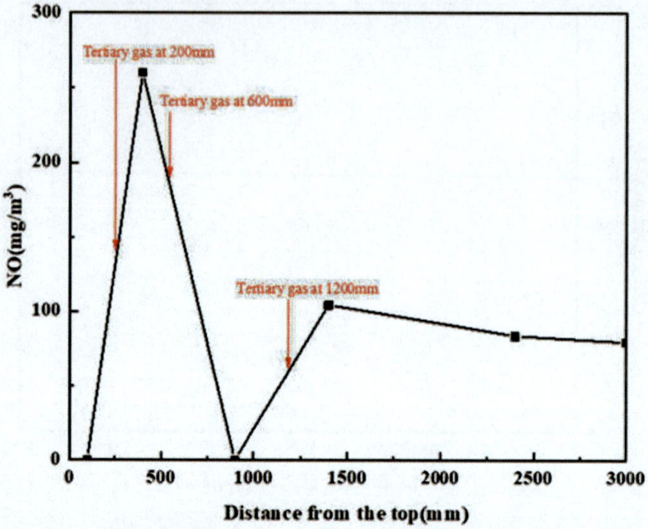

Fig. 57.5 NO concentrations along the down-fired combustor (DFC)

400 mm below the top of the DFC, NO concentration reached a peak of 256 mg/m^3, which indicates that a high oxygen concentration area was formed in this area due to the injection of tertiary air at 200 mm. At 900 mm from the top of the DFC, NO concentration was 0 mg/m^3, and CO concentration was also high at this time, indicating that the reduction reaction in this region was still dominant. After the tertiary air were injected at 1200 mm from the top of the DFC, the oxidation reaction dominated, and the NO concentration at 1400 mm was 104.5 mg/m^3. Different distribution positions of tertiary air have great influence on NO emission characteristics along DFC. In addition, the migration and transformation of nitrogen in DFC is significant for the emission of NO_x.

57.3.2.2 Effects of Tertiary Air Nozzle Position on Combustion Characteristics and NO_x Emissions

Three positions of the tertiary air were used to study the influence of the air distribution on combustion characteristics and NOx emission. As shown in Fig. 57.6, when the position of tertiary air distribution moved down, the maximum temperature of DFC moved down. Under case 2, combustion efficiency was the highest, reaching 97.73%.

Figure 57.7 shows the NO_x and CO emissions at the outlet of DFC. The results show that the lowest NO_x emission was 109.9 mg/m^3 when the tertiary air was injected from 600 and 1200 mm below the top of the DFC. When the tertiary air changed from single-layer to multi-layer, the emission of NO_x decreases. This is

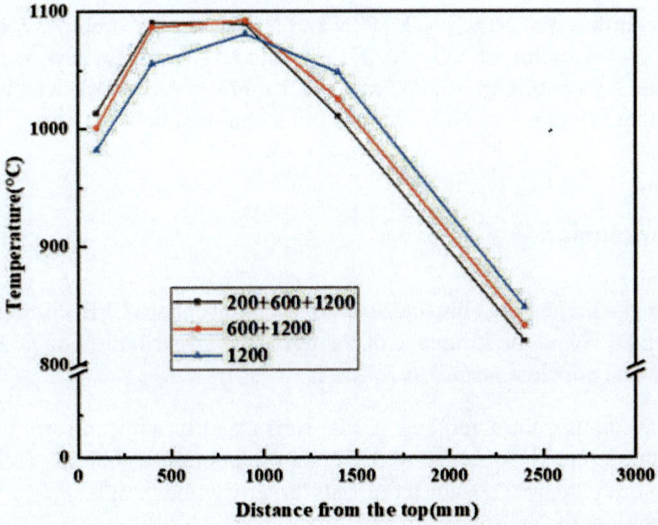

Fig. 57.6 Temperature distribution along the down-fired combustor (DFC) with different tertiary air distributions

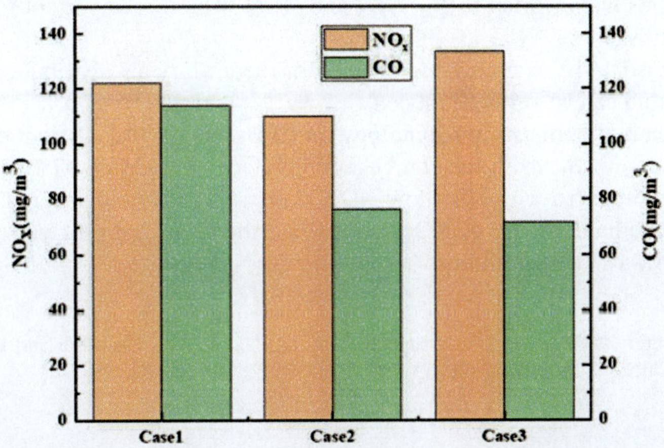

Fig. 57.7 NO_x and CO emissions at the outlet of DFC with different tertiary air distributions

due to the high oxygen concentration in the DFC under the single-layer air distribution condition, which will promote the formation of NO_x. When the air distribution changed from three-layer air distribution to double-layer air distribution, the NO_x emission decreased obviously. This is because the reduction area was prolonged and more fuel-N was converted into N_2. Changing the tertiary air distribution helps to reduce the NO_x emission.

The concentration of NO_2, N_2O, HCN and NH_3 was measured. It was found that the relative concentration of NO_2, N_2O, HCN and NH_3 was very low, almost 0. And the calculated concentration of NO_x was 122.1 mg/m^3 (@6% O_2), which was much lower than that of other low NO_x combustion technologies.

57.4 Conclusion

In this paper, the preheating characteristics of Shenmu char in CFB, combustion characteristics in DFC and the influence of the tertiary air distribution on NO_x emission are studied. The conclusions are as follows:

(1) The circulating fluidized bed (CFB) runs steadily with uniform temperature distribution and is stable at 880°C. High-temperature coal gas and preheated char particles enter DFC and combusts directly with secondary air, which solves the problem of ignition.
(2) The high-temperature coal gas at the outlet of the CFB mainly consists of CO_2, N_2, CO, H_2, and CH_4 and the concentration of NO, NO_2 and N_2O is 0, because of the strong reductive atmosphere. 50.2% of the fuel N in the preheated char particles is converted to N_2, NH_3 and HCN, which is important for reducing the emission of NO_x.
(3) By reasonably arranging the position of the tertiary air distribution, the NO_x emission can be further reduced.
(4) Preheating combustion technology on CFB realizes the efficient combustion of Shenmu char with the combustion efficiency exceeding 97%. At the same time, compared with other low NO_x combustion technologies, it significantly reduces the emission of NO_x and realizes the clean and efficient combustion of ultra-low volatile fuel.

Acknowledgements This work was supported by the Nation Key Research and Development Program of China (2018YFB0604201).

References

Coda B, Kluger F, FöRtsch D et al (1998) Coal-Nitrogen release and NO\r, \r, *x*\r, evolution in air-staged combustion. J Energy Fuels 12(6):1322–1327

Cullinane J, Rochelle G (2004) Carbon dioxide absorption with aqueous potassium carbonate promoted by piperazine. Chem Eng Sci 59(1):3619–3630

Hardison L C, Nagl G J, Addison G E. NO_x, reduction by the Econ-Nox™ SCR process[M]// Environmental Progress. 2006.

Jødal M, Lauridsen T L, Dam-Johansen K (2010) NOx removal on a coal-fired utility boiler by selective non-catalytic reduction. J Environ Prog Sustaine Energy 11(4):296–301

Liu H, Hampartsoumian E, Gibbs BM (1997) Evaluation of the optimal fuel characteristics for efficient NO reduction by coal reburning. J Fuel 76(11):985–993

Liu C, Hui S, Pan S et al (2015) The influence of air distribution on gas-fired coal preheating method for NO emissions reduction. J Fuel 139:206–212

Liu C, Hui S, Zhang X, Wang D, Zhuang H, Wang X (2015) Influence of type of burner on NO emissions for pulverized coal preheating method. J Appl Thermal Eng 85:278–286

Muzio LJ, Quartucy GC (1997) Implementing NOx control: research to application. J Prog Energy Combust Sci 23(3):233–266

Ouyang Z, Lu Q, Zhu J (2013) An experimental study on NOx emissions in combustion of pulverized coal preheated in a circulating fluidized bed. J Energy Fuels 27(12):7724–7729

Ristic D, Schuster A, Scheffknecht G (2010) On the potential of flameless oxidation to reduce NOx emissions from pulverized coal combustion. Int Flame Res Found Combust J

Suda T, Takafuji M, Hirata T et al (2002) A study of combustion behavior of pulverized coal in high-temperature air. J Proc Combust Inst 29(1):503–509

Wang A, Cai J, Xie G (2007) Numerical study of flame properties and nitrogen oxide formation in high temperature air combustion. Challenges of power engineering and environment

Weber R, Smart JP (2005) On the (MILD) combustion of gaseous, liquid, and solid fuels in high temperature preheated air. Proc Combust Inst 30:2623–2629

Yao Y, Jianguo Z, Qinggang L et al (2015) Experimental study on preheated combustion of pulverized semi-coke. J Therm Sci 24(4):370–377

You CF, Xu XC (2010) Coal combustion and its pollution control in China. J Energy 35(11):4467–4472

Zhong BJ, Shi WW, Fu WB (2002a) Effects of fuel characteristics on the NO reduction during the reburning with coals. J Fuel Process Technol 79(2):93–106

Zhong BJ, Shi WW, Fu W (2002b) Effects of fuel characteristics on the NO reduction during the reburning with coals. J Fuel Process Technol 79(2):93–106

Zhu S, Lyu Q, Zhu J, Liang C (2019) Experimental study on NOx emissions of pulverized bituminous coal combustion preheated by a circulating fluidized bed. J Energy Inst 92(2):247–256

Chapter 58
Experimental Study on Optimal Adjustment of SCR Ammonia Injection for 1000 MW Coal-Fired Unit Based on Multi-field Cooperative Diagnosis

Jinglong Liu, Fanjun Hou, Limeng Zhang, Chuanjun Duan, Haojie Liu, Xudong Zhang, Zhihong Hu, and Xingsen Yang

Abstract Taking the SCR flue gas denitrification system of a 1000MW coal-fired boiler in a power plant as the research object, through the coordinated test of the temperature field, flue gas field, NO_x concentration field and NH_3 concentration field at the entrance and exit of SCR, taking into account the influence of temperature, flow field and concentration field on denitrification reaction comprehensively, the ammonia injection optimization adjustment test was carried out under 900MW load, and the ammonia nitrogen molar ratio distribution at the entrance of SCR was optimized to improve the denitrification reaction. The uniformity of NO_x distribution at SCR outlet was improved, and the ammonia escape concentration was reduced. The validation tests were carried out under 750 MW and 500 MW loads, and the NO_x concentration distribution and ammonia escape concentration at SCR outlet were tested.

Keywords Optimization of ammonia injection · Multi-field cooperative diagnosis · Experimental study

Nitrogen oxide (NO_x) is one of the main air pollutants, and its large discharge will cause a series of environmental problems such as photochemical smoke, acid rain, stratospheric ozone layer destruction and global warming [13]. Coal-fired power plants are an important source of nitrogen oxide emissions. In recent years, the national emission reduction index of NO_x is becoming more and more strict. In 2014 and 2015, the state has successively issued the Action Plan for Upgrading and Reforming Energy Saving and Emission Reduction of Coal Power Plants (2014–2020) and the Work Plan for Overall Implementation of Ultra-low Emissions and Energy-saving Transformation of Coal-fired Power Plants [1, 2]. By 2020, ultra-low emission transformation will be realized in coal-fired power stations above 300MW (NO_x emission concentration is not higher than $50mg/Nm^3$), and all new coal-fired

---

J. Liu (✉) · F. Hou · L. Zhang · C. Duan · H. Liu · X. Zhang · Z. Hu · X. Yang
State Grid Shandong Electric Power Research Institute, Jinan 250000, China
e-mail: 470901183@qq.com

J. Lyu and S. Li (eds.), *Clean Coal and Sustainable Energy*, Environmental Science and Engineering,
https://doi.org/10.1007/978-981-16-1657-0_58

power stations must meet ultra-low emission levels. Selective catalytic reduction (SCR) of flue gas denitrification technology has been widely used in domestic coal-fired power stations because of its mature technology, high denitrification efficiency and reliable operation [15].

The efficiency of SCR flue gas denitration is affected by various factors such as inlet flue gas flow field, NO_x concentration field, ammonia concentration field, flue gas temperature field and catalyst activity [14]. According to the actual operation, the flue gas flow field and the flue gas temperature are related to the boiler load and the combustion in the furnace, and the catalyst can only be replaced after failure, so that the ammonia injection flow becomes one of the key factors that can be adjusted daily[3]. If the amount of ammonia spray is too small, it is difficult to ensure that the NO_x emission standard is met; if the amount of the ammonia spray is too large, not only the waste of NH_3 is caused, but also the ammonia escape can react with SO_3 and H_2O in the flue gas to generate the hydrogen sulfate (ABS) [11], resulting in boiler tail equipment ash blockage and reduction of the safety and economy of the unit.

For the existing coal-fired units, the optimization and adjustment test of ammonia injection system is an effective means to reduce the deviation of NO_x mass concentration distribution and ammonia escape at the outlet of denitrification system. Through a large number of field ammonia injection optimization tests, the domestic scholars have realized the uniform distribution of NO_x at the outlet of denitrification system and the reduction of ammonia escape rate [4–17].But most of them are based on the test of NO_x concentration field at SCR outlet. The adjustment of ammonia injection valve is based on experience, and there are few optimal adjustments considering multi-field collaborative diagnosis.

During the operation of a 1000 MW unit, the pressure difference of air preheater increases, which leads to the fluctuation of furnace negative pressure. It is considered that the main reason is that the ammonia escape rate is high due to the non-uniform spraying of ammonia, and the amount of hydrogen sulfate generated is high. Therefore, under 900 MW load, considering the factors such as concentration field, temperature field, velocity field and so on, the ammonia injection optimization adjustment test of SCR denitrification system of the unit was carried out, and the influence of load change on the adjustment was also taken into account. The NOx concentration distribution and NH_3 concentration distribution at the outlet of SCR reactor were re-measure under 750 MW and 500 MW conditions to verify the effect of the field optimization test.

58.1 Object of Study

58.1.1 Overview of Units

The object of this paper is the SCR denitrification device of a 1000 MW unit in a power plant. The denitration device is a high-temperature high-dust arrangement, and is arranged between the economizer and the air preheater, the absorbent is urea. Two SCR reactors are arranged on the boiler, the catalyst in the reactor is arranged according to the '2 + 1' layer, the first two-layer catalyst is first installed, the uppermost layer is a reserved layer. The catalyst is a honeycomb, and the main active components are TiO_2, V_2O_5 and WO_3. The arrangement mode of the denitration system is shown in Fig. 58.1, and the main design parameters are shown in Table 58.1.

The injection grille is arranged on the vertical flue at the entrance of SCR, and 4×18 vents are arranged in each reactor to ensure the uniform mixing of ammonia

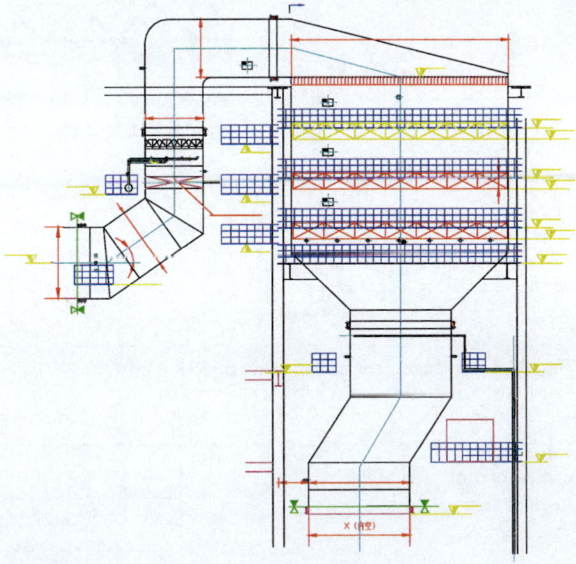

Fig. 58.1 Arrangement of the SCR reactor

Table 58.1 Main design parameters of SCR reactor

Item	Value
Design service temperature/°C	372.2
Allowable temperature/°C	295–430
Denitrification efficiency/%	85
Ammonia escape rate/(μL/L)	3
Transform rate of SO_2/SO_3/%	0.75

and flue gas. The injection system adopts rake arrangement, that is, every four vents share a main pipe. A total of 18 manual valves are installed in each reactor. The arrangement of ammonia injection grille is shown in Fig. 58.2. The corresponding area of each manual valve is shown in Fig. 58.3.

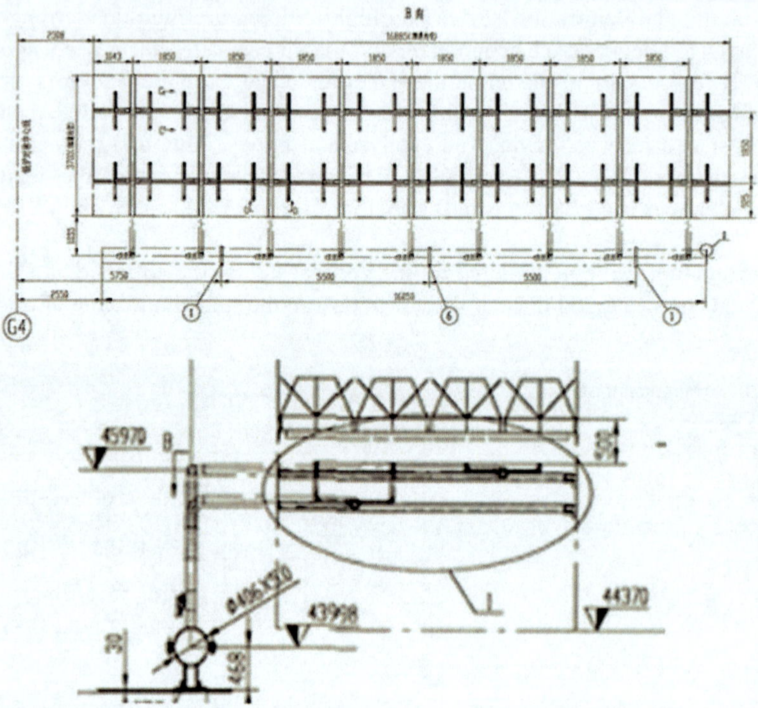

Fig. 58.2 Arrangement of ammonia injection grille

Fig. 58.3 The corresponding area of each manual valve

Boiler tail									A
17	15	13	11	9	7	5	3	1	
18	16	14	12	10	8	6	4	2	
Boiler tail									B
1	3	5	7	9	11	13	15	17	
2	4	6	8	10	12	14	16	18	

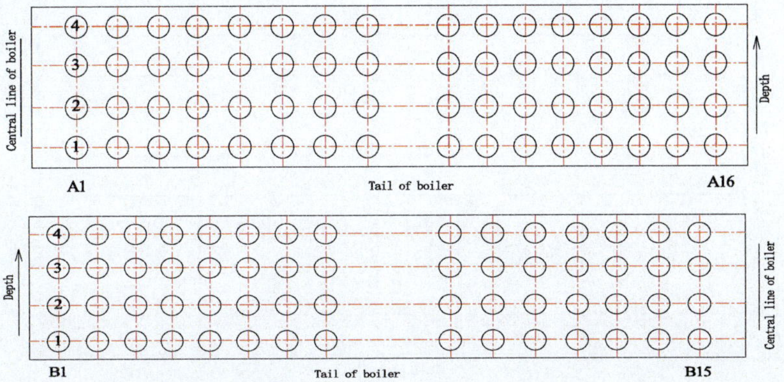

Fig. 58.4 Layout of entrance measuring points

58.1.2 Test Methods

58.1.2.1 Layout of Measuring Points

The existing measuring points of the inlet and outlet of the SCR reactor are not strictly arranged according to the grid method, and the replacement of the measuring point needs to wait for the opportunity of shutdown and maintenance, so only the existing measuring points can be used for testing. Since the width of the outlet flue is up to 6.4 m, the existing measuring holes on both sides are used for measurement, and the measuring points are arranged according to the grid section method. The measuring points are arranged as shown in Figs. 58.4 and 58.5.

The flue gas temperature field, flue gas flow field, NO_x mass concentration field at inlet and outlet of the two SCR reactors were tested by the existing measuring points and then the contour map of flue gas temperature field, flow field and NO_x concentration field was drawn by using Sufer software and Kriging method [7]. According to the contour map, the flue gas temperature, velocity and NO_x concentration of each partition of ammonia injection grid were obtained, and then the corresponding adjustment was made.

58.1.2.2 Optimization and Adjustment Method

Firstly, the temperature field, velocity field and flue gas composition distribution at the inlet and outlet of SCR reactor were measured under 900MW load, and the ammonia injection was optimized accordingly. Then the verification test were carried out under 750 MW and 500 MW load by measuring the NO_x concentration distribution and ammonia escape at outlet of the SCR.

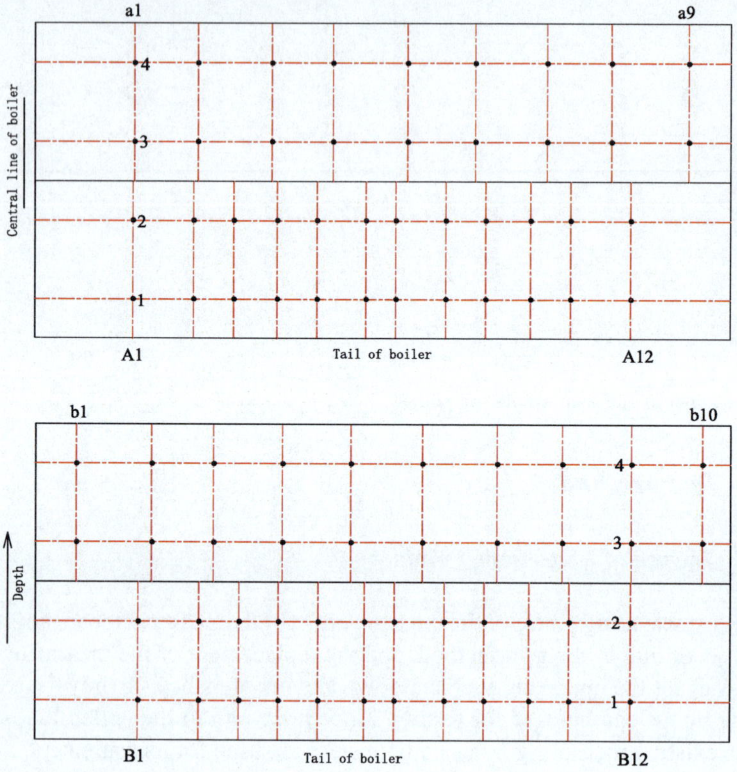

Fig. 58.5 Layout of outlet measuring points

The proportion of the ammonia injection amount in each partition to the total amount was calculated according to the flue gas velocity and NO_x mass concentration corresponding to each partition of the AIG at inlet of the SCR. Then the weight-K of regulating valve of each ammonia injection branch was obtained, and the corresponding opening adjustment was carried out. The regulating valve was fine adjusted according to the distribution of NO_x mass concentration at outlet ofz the SCR after adjustment until the coefficient of variation-Cv of NO_x mass concentration at outlet of the SCR was less than 20%. The flow graph of the ammonia injection optimization test is shown in Fig. 58.6.

The weight K of ammonia injection valve is calculated as follows:

$$K_i = \frac{v_i c_i}{\sum_{i=1}^{n} v_i c_i} \tag{58.1}$$

K_i-weight of ammonia injection valve of i partition, that is the proportion of ammonia injection in this zone to the total ammonia injection; v_i-flue gas velocity of i partition, m/s; c_i-mass concentration of NO_x of i partition, mg/Nm^3.

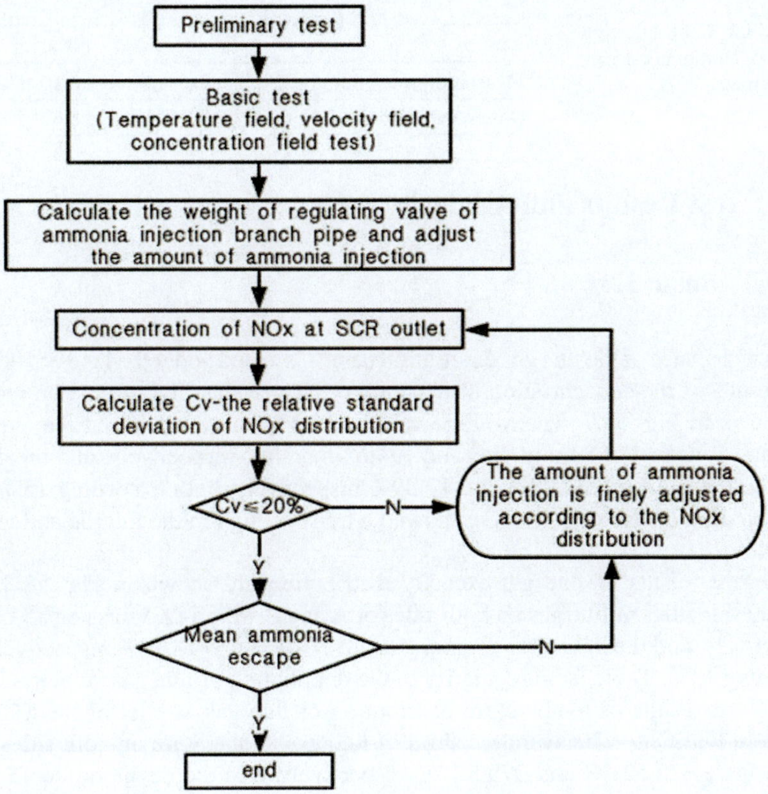

Fig. 58.6 Flow graph of ammonia injection optimization experiment for SCR denitration system

Cv is calculated as follows:

$$Cv = \frac{\sigma}{\overline{x}} \times 100\% \tag{58.2}$$

$$\sigma = \sqrt{\frac{1}{n-1}\sum_{i-1}^{n}(x_i - \overline{x})^2} \tag{58.3}$$

$$\overline{x} = \frac{1}{n}\sum_{i}^{n} x_i \tag{58.4}$$

Cv-coefficient of variation; σ-standard deviation; $\overline{x}$-average value; x_i-value of measuring point i; n-number of measuring points.

The average concentration of ammonia escape was measured by the on-line measuring device of full section installed in the power plant. The device was verified before the test, and the calibration results are shown in Table 58.2.

Table 58.2 Calibration results of ammonia escape on-line meter

Item	A	B
Measured value/ppm	8.91	8.10
Online value/ppm	8.98	8.62

58.2 Test Results and Analysis

58.2.1 Basic Test

The performance of the flue gas denitration device is tested under the load of 900MW. The results of the concentration distribution of NO_x at the inlet of the two reactors are shown in Fig. 58.7. The average values of NO_x concentration at the entrance of A and B were 179.53 mg/Nm^3 and 194.63 mg/Nm^3 respectively, and the deviation coefficients were 21.08% and 17.39% respectively, both exceeding 15%. The concentration distribution of NOx showed a trend of high in the middle and low on both sides.

The test results of flue gas velocity distribution are shown in Fig. 58.8. The average velocities of flue gas on both sides of A and B were 18.65 m/s and 15.69 m/s respectively, and the deviation coefficients are 16.44% and 18.79% respectively, all exceeding 15%. There is no regularity in the distribution of flue gas velocity.

The test results of temperature distribution of flue gas at inlet of the SCR are shown in Fig. 58.9. The average values of flue gas temperature on both sides of A and B were 372.52 °C and 379.52 °C respectively, and the deviation coefficients were 2.89% and 3.67% respectively. The distribution of flue gas temperature at inlet of the SCR was respectively uniform, and the distribution of flue gas temperature was generally higher in the middle and lower on both sides.

Test results of NO_x concentration distribution at outlet of the SCR are shown in Fig. 58.10. The average concentrations of NO_x on both sides of A and B were 9.83

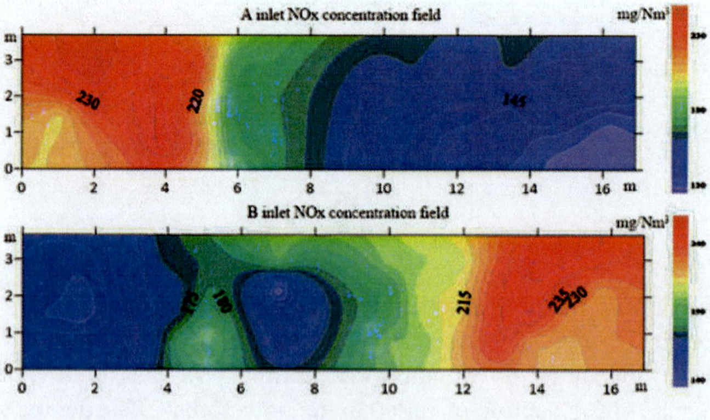

Fig. 58.7 NO_x concentration distribution map at inlet of the SCR under 900 MW operation conditions

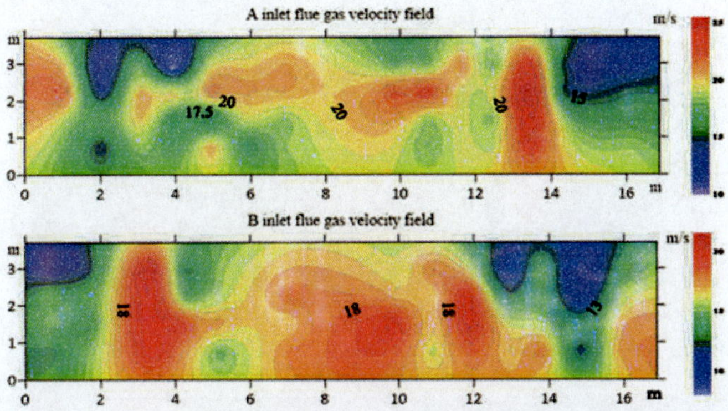

Fig. 58.8 Flue gas velocity distribution map at inlet of the SCR under 900 MW conditions

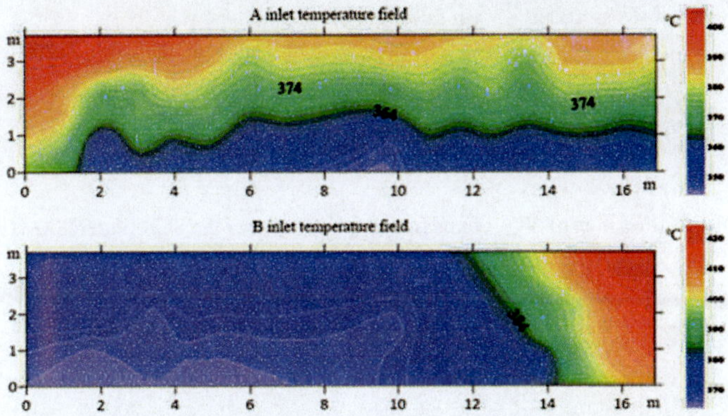

Fig. 58.9 Flue gas temperature distribution map at inlet of the SCR under 900 MW Operation Conditions

mg/Nm^3 and 12.38 mg/Nm^3 respectively, and the deviation coefficients were 28.58% and 64.37% respectively, which were over 20%. The denitration efficiency on both sides reached 93.5% or more, exceeding the design value of 85%.

The test results of temperature distribution of flue gas at outlet of the SCR are shown in Fig. 58.11. The average values of flue gas temperature on both sides of A and B were 368.06 °C and 375.10 °C respectively, and the deviation coefficients were 1.70% and 2.05% respectively, which were less than 3%. The flue gas temperature at outlet of the SCR was relatively uniform, and the distribution of flue gas temperature was approximately the same as that at the entrance, generally showing a trend of high in the middle and low on both sides. The temperature drops of flue gas on both sides of A and B were 4.46 °C and 4.42 °C respectively.

The average values of ammonia slip on both sides A and B at outlet of the SCR were 8.91 ppm and 8.10 ppm respectively, and there was a significant excess of ammonia injection.

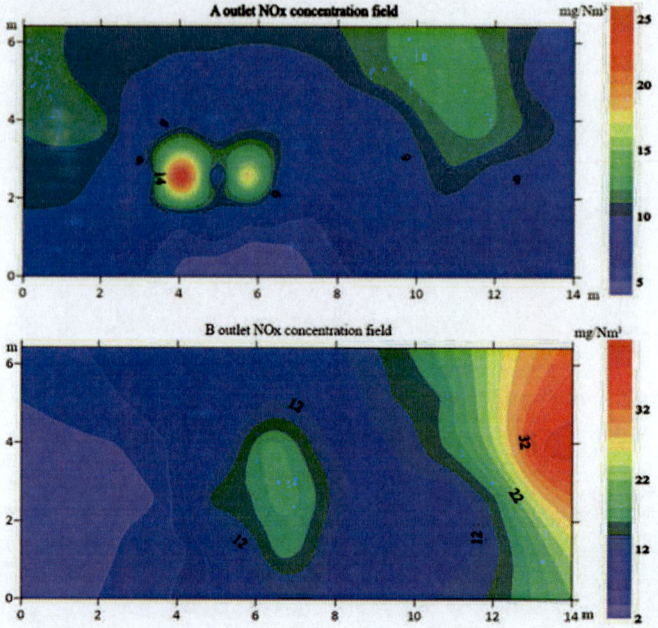

Fig. 58.10 Distribution map of NO_x concentration at outlet at of the SCR under 900 MW conditions

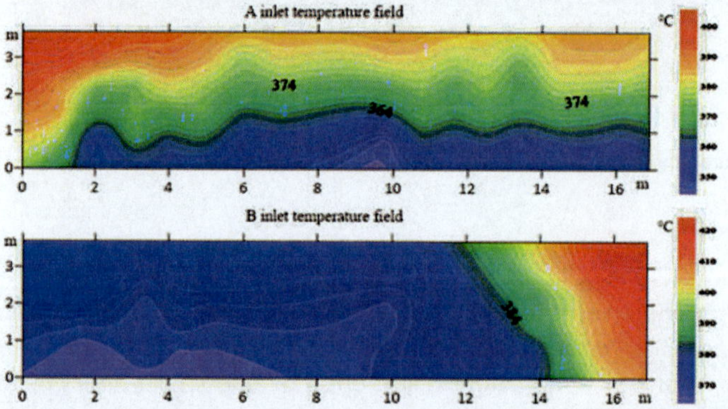

Fig. 58.11 Flue gas temperature distribution map at outlet of the SCR under 900 MW Conditions

The basic test shows that the distribution of NO_x concentration field and velocity field at inlet of the SCR is not uniform, and the deviation coefficient is over 15%. This indicates that the uniformity of flue gas distribution at inlet of the SCR has decreased greatly due to ash deposit, abrasion and the change of combustion in furnace after long-term operation. The flue gas temperature distribution in SCR reactor is uniform and it is in the range of catalyst operation temperature. Distribution deviation of NO_x concentration field at outlet of the SCR exceeds the allowable value (20%), and ammonia escape concentration seriously exceeds the allowable value. Therefore, the distribution of NO_x concentration and velocity field at inlet of the SCR should be considered comprehensively, and the ammonia injection amount should be optimized and adjusted to match the ammonia-nitrogen molar ratio in each region so as to reduce the deviation of NO_x concentration distribution and ammonia escape concentration at outlet of the SCR.

58.2.2 *Test Results After Optimization*

According to the test results of NO_x concentration field and flue gas velocity field at SCR inlet, the weight of each valve was calculated according to formula (58.1) as shown in Table 58.3.

According to Table 58.3, the opening degrees of the valves on both sides of A and B were adjusted separately, and the valves were finely adjusted according to the distribution deviation coefficient of the NO_x concentration field at outlet of the SCR. The final adjusted test results are shown in Fig. 58.12. The adjusted valve opening degree basically corresponds to the calculated valve weight, and some valves have deviations, which was related to the ash blockage and activity reduction of part of the catalyst.

After optimum adjustment of ammonia injection, the average NO_x concentration at outlet of reactor A was 22.74 mg/Nm^3, and the deviation coefficient was 19.53%. The average concentration of NO_x at the outlet of reactor B was 10.72 mg/Nm^3, and the deviation coefficient was 17.97%. The deviation coefficients of NO_x concentration distribution at the outlet of SCR denitrification unit were both controlled within 20%, and the uniformity was obviously improved. The average concentration of ammonia escape at the outlet of reactor A and B were 4.28 ppm and 4.03 ppm respectively, which decreased significantly.

Table 58.3 Weight of each value

A				B			
No.	K	No.	K	No.	K	No.	K
1	0.063	10	0.054	1	0.069	10	0.056
2	0.083	11	0.046	2	0.061	11	0.051
3	0.066	12	0.050	3	0.066	12	0.053
4	0.073	13	0.044	4	0.059	13	0.056
5	0.069	14	0.048	5	0.074	14	0.043
6	0.071	15	0.047	6	0.058	15	0.052
7	0.052	16	0.045	7	0.063	16	0.053
8	0.060	17	0.039	8	0.060	17	0.037
9	0.053	18	0.036	9	0.057	18	0.031

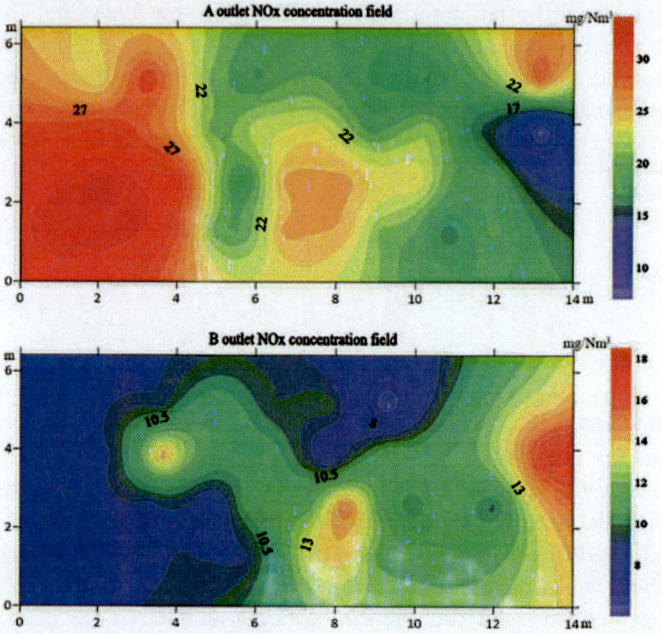

Fig. 58.12 Distribution map of NO_x concentration at outlet at of the SCR under 900 MW conditions after optimization

The average value of ammonia escape is still higher than 3 ppm, because the plant requires that the concentration of NOx outlet be controlled below 20 mg/Nm^3. Under the condition of only two layers of catalyst installed, the denitrification efficiency can only be improved by increasing the molar ratio of ammonia to nitrogen. At this time, the denitrification efficiency of SCR operation exceeds the design value, and there is an overall ammonia injection excess.

58.2.3 Verification Test

The uniformity of NO_x distribution and the concentration of ammonia escape at outlet of the SCR were measured under 750 MW and 500 MW loads respectively. The test results of 750 MW load are shown in Fig. 58.13. The deviations of NO_x concentration distribution at the outlet of reactor A and B are 32.87% and 61.75% respectively. The average values of ammonia escape on both sides of reactor A and B are 3.49 ppm and 2.92 ppm respectively.

The test results of 500 MW load are shown in Fig. 58.14. The deviations of NOx concentration distribution at the outlet of reactor A and B were 38.49% and 63.28% respectively. The average values of ammonia escape on both sides of reactor A and B were 2.30 ppm and 2.08 ppm respectively.

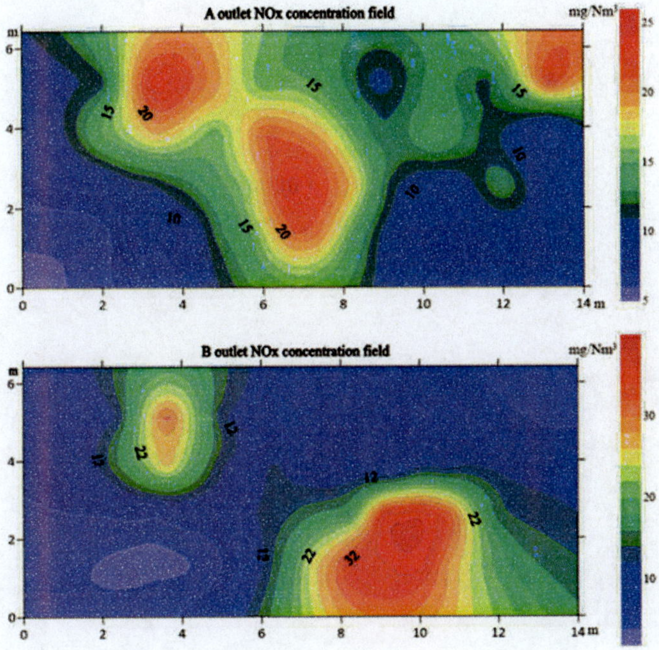

Fig. 58.13 Distribution map of NO_x concentration at outlet at of the SCR under 750 MW conditions after optimization

The verification test results show that the average value of ammonia slip under 750 MW and 500 MW conditions is below 3.5 ppm, which is lower than that of 900 MW. This is because under the partial load, the overall flow velocity of flue gas decreases, the contact time of NO_x and NH_3 increases, and denitrification efficiency increased. However, due to the change of combination mode of the grinding, the flue gas field and NO_x concentration distribution at inlet of the SCR change, the NO_x distribution uniformity decreases and the deviation coefficient is more than 30%. Therefore, after the high-load ammonia injection optimization, the unit may still have uneven ammonia spray in partial load conditions. The influence of the flow field change caused by the load change on the ammonia injection optimization result should be considered, and the corresponding ammonia injection amount adjustment should be carried out.

58.3 Conclusion

Aiming at the existing measuring points of the unit, an optimized method of ammonia injection based on multi-field collaborative diagnosis was proposed. Through the on-site ammonia-smelting optimization adjustment test, the NO_x mass concentration

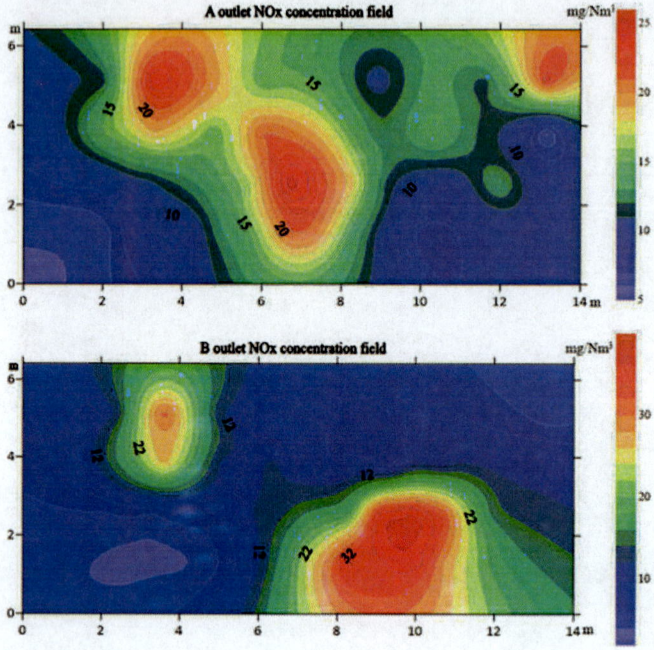

Fig. 58.14 Distribution map of NO_x concentration at outlet at of the SCR under 500 MW conditions after optimization

distribution of the uniform denitration system was obtained, and the ammonia escape rate was reduced. However, when the unit load changes, the NO_x mass concentration distribution at outlet of the SCR is still uneven, and the deviation coefficient increases.

After long-term operation, the uniformity of flue gas field and NO_x concentration field at SCR inlet decreases. The uniformity of flow field and NO_x concentration distribution should be guaranteed by improving or optimizing the diversion device of flue at SCR inlet to improve denitrification efficiency.

In order to ensure the ultra-low emission of the unit, the third layer catalyst should be installed to increase the residence time of flue gas and improve the denitrification efficiency. Under the existing conditions, it is suggested that the concentration of NO_x at SCR outlet should be increased appropriately and the denitrification efficiency should be reduced in order to reduce the ammonia escape rate.

The experimental results show that the flow field distribution at SCR inlet varies with load, and different flue gas distribution corresponds to different weight distribution of ammonia injection valve. Deep peak shaving of coal-fired units has become normal. Load fluctuation of coal-fired units is large, and it is impossible to operate at high load for a long time. Therefore, the study of on-line ammonia injection optimization technology based on variable load conditions is of great significance for the safe and economic operation of SCR.

References

Cheng M, Zhong J, Liao Y, Fan J, Gao Z (2016) Optimization experiments of ammonia-injection in SCR denitrification system based on changeable flow field. Thermal Power Gener 45(12):136–142

Debo Li, Yongjin Liao, Tinghua Zeng, Jianhua Deng, Kai Xu, Fangyong Li (2016) Discussion on key problems in adjustment experiment for ammonia injectiongrid of SCR denitration system. Guangdong Electr Power 29(7):17–21

Ministry of Environmental Protection, National Development and Reform Commission, National Energy Administration: Notice on Printing and Distributing the Work Plan for Overall Implementation of Ultra-low Emissions and Energy-saving Transformation of Coal-fired Power Plants. 2015-12-11

Ji Weng, Zheng Wang, Xiaohai Li, Xingxing Xie (2017) Optimization of partition controlled ammonia-injection grid in SCR denitrification system. Chin J Environ Eng 11(5):289–293

Jie Wang, Chunhong Liu, Xiaozhong Tong, Biao Chen (2018) Application of Kriging method for flow analysis of denitrification reactor. Proce CSEE 38(6):1751–1759

Liang C, Shen Y (2012) Optimal operation of selective catalytic reduction (SCR) flue gas denitrification system in 1000 MW unit. Electr Power 45(1):41–44

National Development and Reform Commission, Ministry of Environmental Protection, National Energy Administration: Notice on Printing and Distributing the Action Plan for Upgrading and Reforming Energy Saving and Emission Reduction of Coal Power Plants (2014–2020). 2014-09-12

Sheng Sun, Jianle He, Fazhi Liu (2018) Optimization and adjustment experiment study on ammonia injection grid of SCR denitration system. Environ Develop 30(2):124–125

Shuangchen MA, Yue Deng, Wenlong WU (2016) Experimental research on ABS formation characteristics in SCR denitrification process. J Chin Soc Power Eng 36(2):143–150

Xin-gang Zhou, Xiao Lin, Qing-chuan Zhao (2016) Optimal experiment of selective catalytic reduction in 300MW coal-fired units. Power Syst Eng 32(2):43–46

Xingxing Cheng, Xiaotao Bi (2014) A review of recent advances in selective catalytic NOx reduction reactor technologies. Particuology 16:1–18

Yan-bin Ma, Guo-an Tang, Guang-wei Chen, Qian-wei Feng (2017) Optimal Experiment of selective cataly reduction in 1000MW power station. Power Gener Air Condit 38(4):16–20

Yongjin Liao (2017) Ammonia spraying optimization of an SCR denitrificaion system based on RBF neural network. J Chin Soc Power Eng 37(11):76–82

Yu-guang NIU, Yan PAN, Xiao-Bin LI (2019) Review and prospect of automatic control system for flue gas SCR denitration in steam power station. J Eng Thermal Energy Power 34(4):1–9

Yuming Zhao (2017) Study on ammonia injection optimization of SCR system based on flue gas field uniformity diagnosis. Chem Eng Equip 7:302–304

Zhaojun Fang, Lipeng Jin, Yubao Song, Junwu Zhao, Lele Wang (2014) Performance optimization and maximum denitrationefficiency analysis for SCR-DeNOxpower plants. Thermal Power Gener 43(7):163–166

Zhou H, Zhang Z, Zhang J (2011) Mixed structure-radial basis function neural network optimal control on spraying ammonia flow for supercritical boiler flue gas denitrification. Proc CSEE 31(5):108–113

Chapter 59
Effects of High Temperature on NH_3/NO Reactions in the Absence of Oxygen

Zhixiang Zhu, Degui Bi, Juan Yu, Zhongxiao Zhang, and Chen Lin

Abstract The influence of industrial processing parameters, especially high temperature, on NO_x abatement in the absence of oxygen has been experimentally. NO_x reduction efficiency is significantly promoted with increasing residence time and NSR and optimal residence time and NSR are 0.7 s and 1.5, respectively, when temperature exceeds 1400 °C. NO_x reduction is strongly dependent on temperature. When temperature is lower than 1000 °C, NO consumption is hindered due to lack of O radicals. The denitrification efficiency is significantly promoted with the increase of temperature because thermal decomposition of CO_2 and NO is quite sensitive to temperature. However, NO formation from pyrolysis of HNO begins plays an important role since temperature exceeds 1400 °C, which results in decline in NO_x reduction efficiency. And the peak value of NO reduction efficiency can reach almost 100% at temperature range of 1300–1400 °C with NSR of 1.5. Four chemistry mechanisms have been adopted to simulate NO_x reduction by ammonia. Validation shows that results calculated by POLIMI chemistry mechanism agrees better with experimental data than other 3 mechanisms.

Keywords Oxygen · Temperature · Rate of production · Denitrification efficiency

59.1 Introduction

NO_x reduction by nitrogen reagent has been investigated widely and deeply by domestic and foreign scholars. Selective non- catalytic reduction (SNCR) technology is the most cost-effective NO_x emission control technology (Hwang et al. 2009). However, optimal NO_x reduction achieved so far in practical utility units is only about 40% (Ljungdahl and Larfeldt 2001; Zhu and Li 2010) due to large temperature gradient, short residence time, dynamic boiler conditions, poor mixing, and so on (Zhou et al. 2010). And the fact also limits NO_x reduction efficiency that NO can be reduced into N_2 in a narrow temperature range of 850–1100 °C (Zhang

Z. Zhu · D. Bi · J. Yu (✉) · Z. Zhang · C. Lin
School of Mechanical Engineering, Shanghai Jiao Tong University, Shanghai 200240, China
e-mail: yujuan@sjtu.edu.cn

J. Lyu and S. Li (eds.), *Clean Coal and Sustainable Energy*,
Environmental Science and Engineering,
https://doi.org/10.1007/978-981-16-1657-0_59

et al. 2008; Mahmoudi et al. 2010) in the presence of oxygen. When temperature exceeds optimal temperature, nitrogen reagent is prone to be oxidized in oxidizing atmosphere (Skreiberg et al. 2004; Moo and Chin 1997; Zandaryaa et al. 2001).

Therefore, nitrogen reagent injection into highly reducing atmosphere is considered to be a solution, in which circumstances nitrogen reagent only reacts with NO rather than oxygen. EPRI and REI names this technology Rich Reagent Injection (RRI) and application of RRI in two existing cyclone fired utility boilers equipped with overfire air results in a 30% increase of NO_x reduction with temperature of 1873–2073 K and excess air ratio in primary combustion zone of 0.9 (Cremer et al. 2001, 2003). Experiment is conducted in the drop tube furnace to investigate the influence of excess air ratio in primary zone and temperature on NO_x reduction by urea solution (Bi et al. 2019; Yue et al. 2017). And results show that highly reducing atmosphere and high temperature is benefitable for NO abatement. Lu et al. find that optimal temperature is shifted to a much higher temperature and NO_x reduction efficiency is promoted under extremely low O_2 oxygen, i.e. about 0.01% (Lu and Lu 2009). However, oxygen still exist even in high reducing atmosphere. And to the author's knowledge, NO_x reduction by nitrogen reagent without the presence of oxygen is seldom studied.

The objective of this paper is to investigate the influence of extremely high temperature on NO_x reduction by ammonia injection without the presence of oxygen and find out chemistry mechanism.

59.2 Methods

59.2.1 Experimental Set-up

A bench-scale testing facility has been established to investigate the influence of high temperature on NO reduction by ammonia in the absence of oxygen. Schematic diagram of experimental system is shown in Fig. 59.1. Different gases flow successively into mass flow controller (MFC), check valve and then enter the mixing chamber where mixing process happens. MFC accurately measures and controls mass flow rates despites gas volume fluctuated due to pressure or temperature changes and the possibility of backflow due to uneven pressure distributions in parallel tubes is significantly reduced via the utilization of check valve. Premixed gases enter the reactor, a quartz tube with the length of 2373 mm and the diameter of 36 mm. The reactor is electric heated and the maximum temperature can be reached inside of it is 1600 °C. The flue gas can be pumped into the Optima 7 gas analysis and the gas component of flue gas along axial direction can be obtained via shifting the position of the water-cooled probe made of stainless steel. The measurement accuracy of Optima 7 for O_2 and NO are 0.1% and 0.0001%, respectively. Furthermore, the rest of flue gas is released outdoors by induced draft fan.

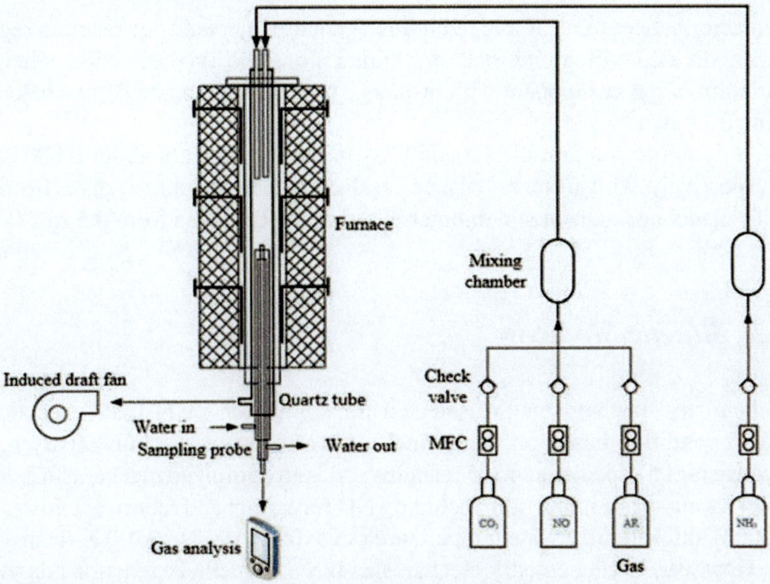

Fig. 59.1 Schematic diagram of the bench-scale testing facility

It is essential to obtain the temperature distribution along axial direction of the reactor in order to determine the region where homogeneous reactions take place. As shown in Fig. 59.2, the reaction region characterized with constant temperature is about 900 mm in length and the beginning of it is located 300 mm from the inlet

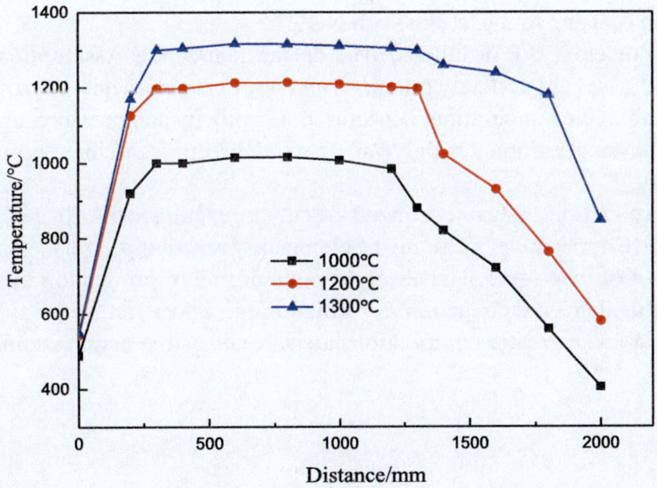

Fig. 59.2 Temperature distribution along axial direction

of the reactor, where flue gas and reduction agent are injected. The reaction region is uniformly divided with an interval of 100 mm along the axial direction. Therefore, the variation of gas component with residence time can be obtained once velocity of gas flow is known.

The volumetric fraction of NO and CO_2 in the flue gas are about 0.0005% and 16% respectively, with argon as balance gas. Reaction temperature ranges from 1000 to 1450 °C and normalized stoichiometric ratio (NSR) ranges from 0.5 to 2.0.

59.2.2 Kinetic Modeling

Four chemistry mechanisms are selected for simulation of NH_3/NO reactions in anaerobic condition based on Chemkin 17.0 Pro software. GRI chemistry mechanism consists of 53 species and 325 reactons and is an optimized mechanism designed to model natural gas combustion, including NO formation and reburn chemistry (Han et al. 2003) and valid in a wide temperature range from 1000 to 2500 K (Han and Lu 2007). However, GRI chemistry mechanisms lacks elementary reactions describing NO_X reduction through selective non-catalytic reduction (SNCR) reactions. Therefore, GRI-SNCR chemistry mechanism adds 26 elementary reactions (Caton and Xia 2004) on the basis of GRI chemistry mechanism. Chemistry mechanism constructed by Cao et al. (2009), abbreviated as CAO in this paper, consists of 66 species and 461 reactions. It is a detailed mechanism constructed to model NO_X reduction, oxidization of gaseous additives like H_2, CO and CH_4, and interaction between gaseous additives and NO_X. Mechanism concluded by scholars from Politecnico di Milano, abbreviated as POLIMI in this paper, consists of 115 species and 2141 reactions. And it is a detailed mechanism of the pyrolysis, partial oxidation and combustion of hydrocarbon fuels up to 3 C atoms with NO_X.

Plug flow reactor (PFR) is utilized to model the quartz tube. Assumptions are made that different gases are perfectly premixed and physical parameters of particles in the radial direction, like temperature, concentration, velocity and pressure, are the same. Therefore, physical parameters only vary in axial direction and there is no back flow in the reactor.

Rate of production analysis is utilized to analyze reaction paths. Rate of production can be positive or negative, meaning the elementary reaction promotes the formation or conversion of the specie. The absolute value of rate of production means contributions of the elementary reaction to formation or conversion of the specie. Thus, reactions that make greater contributions are selected as the main reaction paths.

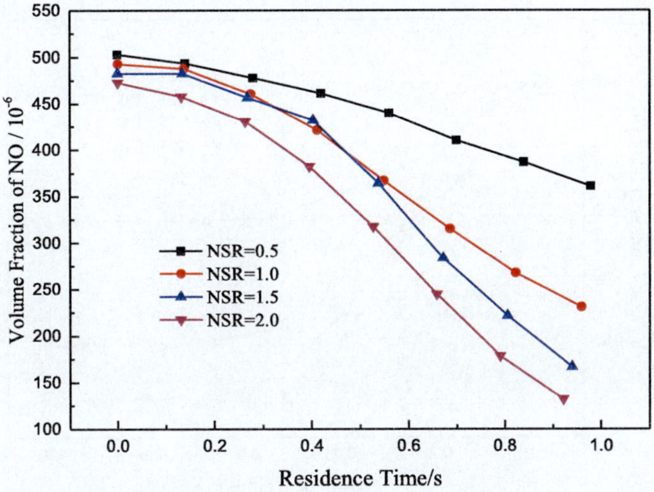

Fig. 59.3 The influence of residence time on NO reduction at 1200 °C

59.3 Results and Discussion

59.3.1 The Influence of Residence Time

Experiments are conducted to investigate the influence of residence time on NO reduction and results obtained at 1200 °C and 1400 °C are shown in Fig. 59.3 and Fig. 59.4 respectively. It is apparent that NO reduction is promoted as residence time increases. However, the trends of NO concentration variation at 1200 and 1400 °C are quite different. NO concentration decreases slowly at first and then sharply at 1200 °C while the trend is opposite at 1400 °C, which is caused by different reaction rates. The reaction rate is relatively slow at 1200 °C and finite mixing rate is of great significance to NO consumption. As residence time increases, gases mix more sufficiently and rate of NO consumption increases. When temperature is 1400 °C, the influence of temperature on reaction rate is much stronger than that of mixing process. NO concentration decreases sharply due to high temperature and reaction rate decreases with the consumption of reactants as residence time increases.

59.3.2 The Influence of NSR

The influence of NSR on denitrification efficiency at different temperatures is shown in Fig. 59.5 and the residence time is 1.0 s. When temperature exceeds 1300 °C, temperature has little influence on NO reduction since there is little difference between results obtained at 1300–1450 °C. Furthermore, 1.5 is considered to be

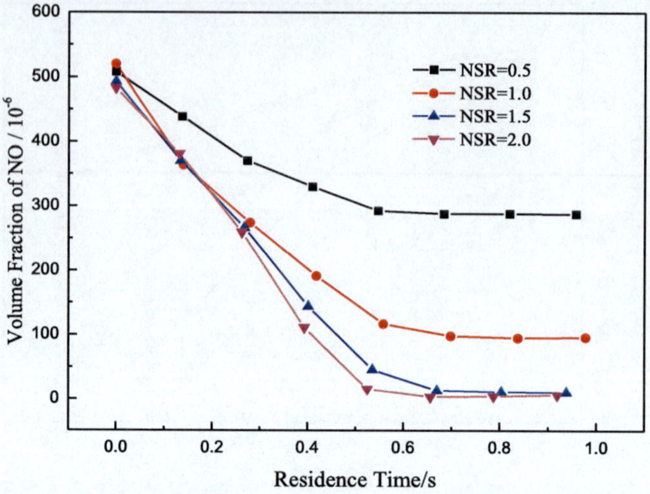

Fig. 59.4 The influence of residence time on NO reduction at 1400 °C

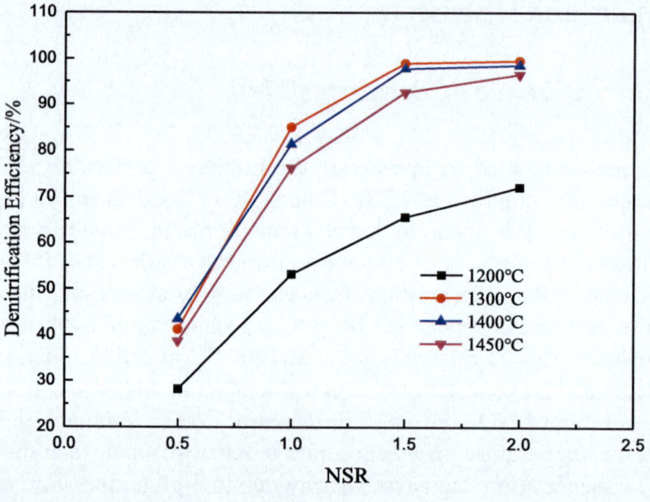

Fig. 59.5 The influence of NSR on denitrification efficiency

the optimal NSR since increase of NSR will no longer promote NO reduction and may lead to ammonia slip in the outlet of the reactor. However, the denitrification efficiency keeps rising with the increase of NSR when temperature is below 1200 °C. Because homogeneous reaction is comparatively slow at lower temperature and the residence time is not sufficient for it to reach equilibrium. More ammonia injection help increase reactant concentration and thus reaction rate is slightly improved.

59.3.3 The Influence of Temperature

The influence of temperature on NO reduction by ammonia with residence time of about 0.8 s in anaerobic condition is investigated experimentally. As shown in Fig. 59.6, NO reduction is evidently promoted as the temperature grows but the denitrification efficiency declines when temperature exceed 1400 °C. Therefore, optimal temperature is within 1300 and 1400 °C. When temperature is 1000 °C, the denitrification efficiency is lower than 10% with NSR of 2.0. So, it can be considered that the homogeneous reactions between NH_3 and NO in anerobic condition cannot occur when temperature is lower than 1000 °C, which is consistent with conclusions of references (Kasuya et al. 1995; Fan et al. 2014). It is widely acknowledged that NO can be reduced into N_2 only when NH_3 is converted into NH_2. NH_2 radicals can be generated in the presence of O or OH radicals via following reactions:

$$NH_3 + O = NH_2 + OH$$
$$NH_3 + OH = NH_2 + H_2O$$

The O and OH radicals in above reactions initially result from the thermal decomposition of O_2 and H_2O. As a result, NO can't be reduced into N_2 by ammonia without oxygen when temperature is not high enough.

And because NH_3 won't be oxidized into NO without the presence of oxygen, the maximum denitrification efficiency with NSR greater than 1.5 can reach almost 100%, which is greater than that of SNCR.

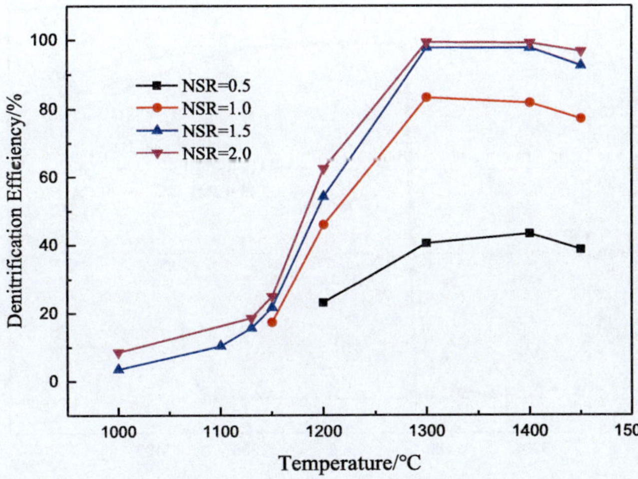

Fig. 59.6 The influence of temperature on denitrification efficiency

59.3.4 Validation of Kinetic Model

Experimental data obtained with residence time of about 0.8 s and NSR of 1.5 are compared with results calculated by Chemkin software. 4 chemistry mechanism are adopted to simulate NO reduction by NH_3 without oxygen and comparisons are shown in Fig. 59.7. Results calculated by GRI chemistry mechanism don't agree well with experimental data due to lack of elementary reactions describing NO consumption by SNCR. And POLIMI chemistry mechanism fit the best with experimental data, especial at extremely high temperature, since it is applicable for high temperature condition.

As temperature is lower than 1100 °C, experimental data are slightly larger than calculated results. Because it is inevitable that pure industrial gases contain gas impurity in practice, NO reduction can be provoked due to the presence of trace oxygen at relatively low temperature. However, calculated denitrification efficiency is greater than experimental data as temperature rises. Denitrification efficiency calculated reaches almost 100% at 1200 °C while it obtained in experiment is just about 50%. Assumption that gases are perfectly premixed is made in kinetic modeling while it takes time for gas mixing process. When reaction temperature is low, there is not sufficient time for homogeneous reactions to reach equilibrium due to finite mixing rate, which has been illustrated in Fig. 59.3 already. When temperature is higher enough, mixing process has less influence on reaction rate compared to temperature and reaction time is sufficient to reach reaction equilibrium. So calculated results agree well with experimental data.

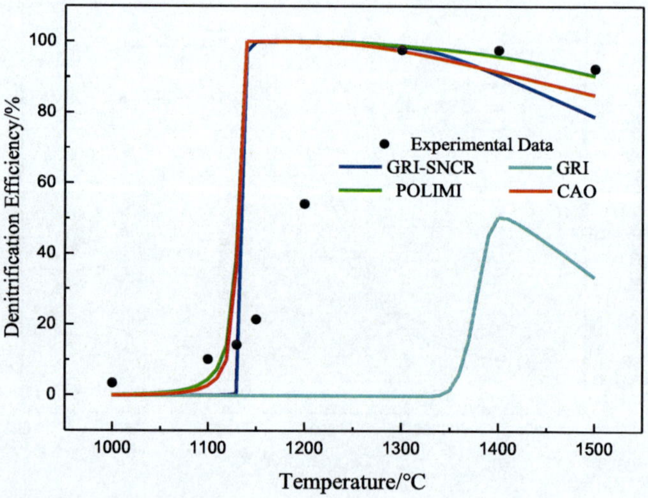

Fig. 59.7 Comparison between experimental data and calculated results

59.3.5 Analysis of Reaction Mechanism

Rate of production analysis is adopted to investigate the reaction mechanism for NH_3/NO reactions in the absence of oxygen.

It is widely known that NO cannot be reduced into N_2 by ammonia injection without oxygen within temperature range of SNCR. But with the increase of temperature, thermal decomposition of CO_2 and NO in high temperature results in generation of O radicals via following reactions:

$$CO_2(+M) = O + CO\ (+M)$$
$$2NO = O + N_2O$$

Then NH_3 is converted into NH_2 with the presence of oxygen via following reaction:

$$NH_3 + O = NH_2 + OH$$

Although above reactions are not dominant, generation of NH_2 is quite essential for supplies of OH radicals through following reactions:

$$NH_2 + NO = NNH + OH$$
$$NNH + M = N_2 + H + M$$
$$CO_2 + H = CO + OH$$

Only with the presence of OH radicals, NH_3 can be converted into NH_2:

$$NH_3 + OH = NH_2 + H_2O$$

NO can be reduced by NH_2 into nitrogen intermediates and finally be converted into N_2 via reactions as follow:

$$NH_2 + NO = N_2 + H_2O$$
$$NH_2 + NO = NNH + OH$$
$$NNH + M = N_2 + H + M$$

Temperature requirement for NH_3/NO reactions in the absence of oxygen is higher than that for SNCR because O radicals are generated from thermal decomposition of CO_2 and NO without the presence of oxygen while are resulted from hydrolysis of oxygen in SNCR.

When temperature exceeds 1400 °C, NO reduction is retreated with increasing temperature. It is due to generation of HNCO and NH radicals at high temperature via following reactions:

$$NH_2 + CO_2 = HNCO + OH$$
$$NH_2 + OH = NH + H_2O$$

HNCO radicals are then converted into NH or NCO radicals and NCO radicals are subsequently converted into N_2:

$$NCO + NO = N_2O + CO$$
$$N_2O\ (+M) = N_2 + O\ (+M)$$
$$NCO + NO = N_2 + CO_2$$

Although part of NH radicals converted into N_2O and finally converted into N_2, NH radicals can also be oxidized to NO:

$$NH + CO_2 = HNO + CO$$
$$HNO + M = H + NO + M$$

Thermal decomposition of HNO leads to increase of NO concentration in the reactor, which is strongly temperature independent. Therefore, final NO concentration is determined by competition of NO consumption and NO formation from pyrolysis of HNO.

59.4 Conclusion

The influence of process operation parameters on NO reduction by NH_3 injection in the absence of oxygen has been experimentally and theoretically investigated. When temperature is lower than 1200 °C, NO reduction is significantly promoted with the increase of residence time and NSR. However, as temperature exceeds 1300 °C, there is an optimal residence of 0.7 s and an optimal NSR of 1.5 because reaction rate is significantly promoted and residence time of 1.0 s is sufficient for homogeneous reactions to reach equilibrium.

Four chemistry mechanisms are adopted to simulate NO reduction by NH_3 without oxygen and results calculated by POLIMI agrees quite well with experimental data. Rate of production analysis is utilized to investigate the reaction path. NO can be reduced only when temperature exceeds 1100 °C because high temperature is essential for the thermal decomposition of CO_2 and NO. NO can be reduced into N_2 via a sequence of reactions with the presence of O radicals. And NO consumption increases as temperature rises because increasing temperature leads to more generation of O radicals. However, intermediate nitrogen specie HNO are generated and successively converted into NO at high temperature. Thermal decomposition of HNO is promoted with increasing temperature, which cause decreasing denitrification efficiency when temperature exceeds 1400 °C.

Acknowledgements This work was supported by the National Key R&D Program of China (No. 2018YFB0604202).

References

Bi D, Zhang Z, Dong J, Zhu Z, Yu J (2019) Effect of stoichiometry and temperature on NOx reduction by reagent injection in the fuel-rich zone of pulverized coal combustion. Energ Fuel 33:1501–1508

Cao Q, Wu S, Lui H, Liu D, Qiu P (2009) Experimental and modeling study of the effects of multicomponent gas additives on selective non-catalytic reduction process. Chemosphere 76:1199–1205

Caton J, Xia Z (2004) The selective non-catalytic removal (SNCR) of nitric oxides from engine exhaust streams: comparison of three processes. Trans ASME 126(2):234–240

Cremer M, Chiodo A et al (2001) Sub 0.15 LB/MBTU NOx emissions achieved with ALTA on a 500 MW cyclone-fired boiler. In: EPRI-DOE-EPA combined utility air pollution control symposium: the MEGA symposium

Cremer M, Wang H, Boll D, Schindler E, Vasquez E (2003) Improved rich reagent injection (RRI) performance for NOx control in coal fired utility boilers. In: U.S. DOE conference on SCR and SNCR for NOx control

Cremer M, Adams B et al. Design and demonstration of rich reagent injection (RRI) for NOx reduction at Conectiv's B.L. England Station

Fan W, Zhu T, Sun Y, Lv D (2014) Effects of gas compositions on NOx reduction by selective non-catalytic reduction with ammonia in a simulated cement precalciner atmosphere. Chemosphere 113:182–187

Han K, Lu C (2007) Kinetic model and simulation of promoted selective non-catalytic reduction by sodium carbonate. Chin J Chem Eng 15:512–519

Han X, Rückert F, Schnell U, Hein K, Koger S, Bockhorn H (2003) Computational modeling of NOx reburning by hydrocarbons in a coal furnace with reduced kinetics. Combust Sci Technol 175:523–544

Hwang I, Minoya H, Matsuto T, Matsuo T, Matsumoto A, Sameshima R (2009) Removal of ammonium chloride generated by ammonia slip from the SNCR process in municipal solid waste incinerators. Chemosphere 74:1379–1384

Kasuya F, Glarborg P, Johnsson J, Dam-Johansen K (1995) The thermal DeNOx process: influence of partial pressures and temperature. Chem Eng Sci 50:1455–1466

Ljungdahl B, Larfeldt J (2001) Optimised NH_3 injection in CFB boilers. Powder Technol 120:55–62

Lu Z, Lu J (2009) Influences of O_2 concentration on NO reduction and N_2O formation in thermal DeNOx process. Combust Flame 156:1303–1315

Mahmoudi S, Baeyens J, Seville J (2010) NOx formation and selective non-catalytic reduction (SNCR) in a fluidized bed combustor of biomass. Biomass Bioenerg 34:1393–1409

Moo B, Chin F (1997) Low temperature SNCR process for NO control. Sci Total Environ 198:73–78

Skreiberg Ø, Kilpinen P, Glarborg P (2004) Ammonia chemistry below 1400 K under fuel-rich conditions in a flow reactor. Combust Flame 136:501–518

Yue P, Zhang Z, Zhang J, Bi D (2017) NOx reduction by urea solution in fuel-rich pulverized coal combustion. Energy Sour Part a: Recov Util Environ Effects 39:2090–2097

Zandaryaa S, Gavasci R, Lombardi F, Fiore A (2001) Nitrogen oxides from waste incineration: control by selective non-catalytic reduction. Chemosphere 42:491–497

Zhang Y, Cai N, Yang J, Xu B (2008) Experimental and modeling study of the effect of CH_4 and pulverized coal on selective non-catalytic reduction process. Chemosphere 73:650–656

Zhou W, Marquez A, Moyeda D, Nareddy S, Frato J, Yu G et al (2010) Design and test of a selective noncatalytic reduction (SNCR) system for full-scale refinery CO boilers to achieve high NOx removal. Energy Fuel 24:3936–3941

Zhu L, Li W (2010) Application of selective non-catalytic reduction denitrification. Guangdong Electr Power 23(8):69–71+90

Chapter 60
Study on Kinetic Model of NO_x Reduction Overall Reaction of Fluidized-Bed Flue Gas

Yao Zhang, Chen Lin, Juan Yu, Jindong Jiang, Fan Feng, and Zhongxiao Zhang

Abstract There have been a great number of investigations on kinetic model of NOx reduction reaction. However, it was found that it is difficult to apply the reaction model conveniently in the comprehensive numerical simulation of fluidized bed combustion due to the complexity of the kinetic model. Therefore, it is necessary to establish a simplified overall reaction kinetic model to adapt to the complex simulation computation of fluidized bed. In this paper, the Arrhenius coefficients of the overall reaction of NOx reduction were obtained by comparing with detailed mechanism model and optimized by using a genetic algorithm. The result showed that the overall reaction model can predict the change of NOx reduction reaction temperature window, the denitrification efficiency and the ammonia slip accurately.

Keywords Overall reaction · Kinetic model · Denitrification · NOx reduction · Fluidized-bed

60.1 Introduction

In the practical engineering application, the NOx reduction reaction model mainly involves two influencing factors, chemical reaction kinetics and fluid dynamics. Therefore, in the simulation calculation of the reaction model, the coupling calculation of chemical reaction kinetics and fluid dynamics is needed to make more reliable prediction of the reaction (Jepsen 2018). However, due to the highly computational difficulty brought by coupling calculation, it is necessary to adopt certain means to simplify one of the chemical reaction kinetic models or fluid dynamic models (Mitsos et al. 2008).

At present, there are two main types of feasible simplification methods. The first method is to simplify the reaction kinetics (Vujanović et al. 2005), while the flue gas flow is described and calculated in detail. The second kind of method is to simplify the model of the flue gas flow and to calculate the chemical reaction

Y. Zhang · C. Lin · J. Yu (✉) · J. Jiang · F. Feng · Z. Zhang
School of Mechanical Engineering, Shanghai Jiao Tong University, Shanghai 200240, China
e-mail: yujuan@sjtu.edu.cn

J. Lyu and S. Li (eds.), *Clean Coal and Sustainable Energy*,
Environmental Science and Engineering,
https://doi.org/10.1007/978-981-16-1657-0_60

in detail. According to the previous research results (Cao 2009), the second method cannot well describe the denitrification process of NOx reduction reaction in practical engineering. Therefore, the first simplified method is of higher research value.

From the actual research process, it can be clearly found that the coupling calculation between the simplified reaction mechanism model and the detailed fluid dynamics model is more widely used. Liang and Zhong (2009) and others, to study the simulation of selective non-catalytic reduction (SNCR) process by the model of laminar finite rate with FLUNT, firstly introduced simplified chemical reaction mechanism of SNCR process, and avoided the problem of the large amount of calculation caused by the simulation of a large number of elementary reactions, which makes the SNCR process of large boiler NOx removal simulation become feasible. Similarly, the simplified denitrification reaction model with ammonia as reducing agent and the eddy dissipation concept (EDC) model in FLUENT software were used to simulate the industrial application of SNCR process in Taicang port at 1025t/h (Liang 2010). Sun (2016) constructed two simplified mechanism models. Based on it, the distribution of NO concentration and denitrification in CFB boiler cyclone separator were studied by using Reynolds stress turbulence model (RSM) and discrete phase model (DPM).

In general, in the process of building NOx reduction reaction model, the construction of simplified chemical kinetics reaction is an important step to reduce the calculation amount coupled with the detailed fluid dynamics model. This is an important reason for studying on the kinetic model of NOx reduction overall reaction of fluidized-bed flue gas in this paper.

60.2 Model Description

The basic idea is to use the two-step competitive reaction as the NOx reduction overall reactions:

$$NH_3 + NO + 1/4O_2 \rightarrow N_2 + 3/2H_2O \tag{60.1}$$

$$NH_3 + 5/4O_2 \rightarrow NO + 3/2H_2O \tag{60.2}$$

The following formula is used to calculate the chemical reaction rate of the overall reaction model.

$$\mathrm{r1} = K_r C_{NO} C_{NH3} \tag{60.3}$$

$$\mathrm{r2} = K_{ox} C_{O2} C_{NH3} \tag{60.4}$$

where *Kr* and *Kox* are the reaction rate constants of reaction (60.1) and reaction (60.2) respectively. On this basis, the reaction rate of NO and NH_3 in the reaction system can be expressed by the overall reaction rate constant, as shown in the following formulas:

$$R_{\mathrm{NO}} = -K_r C_{NO} C_{NH3} + K_{ox} C_{O2} C_{NH3} \tag{60.5}$$

$$R_{\mathrm{NH3}} = -K_r C_{NO} C_{NH3} - K_{ox} C_{O2} C_{NH3} \tag{60.6}$$

According to the Arrhenius chemical reaction rate formula (modified by Adams), *Kr* and *Kox* can be expressed as follow:

$$\mathrm{K_{fi}} = A_i \mathrm{T}^{\beta_i} \exp(-E_i/\mathrm{R_c T}) \tag{60.7}$$

A_i is the pre-exponential factors of the overall reaction, β_i is the temperature factor and E_i is the activation energy required for the reaction. In order to build the kinetic model of NOx reduction overall reactions, we need to find these three parameters of each reaction.

Due to the wide range of Arrhenius parameters of the overall reaction, the time complexity of most solution methods will be too large, so it is a very appropriate choice to adopt evolutionary algorithm. In this paper, genetic algorithm is adopted to optimize the problem to find the best overall reaction parameters.

Evaluation criteria need to be developed to assess the feasibility of Arrhenius parameters. The evaluation criteria used in this paper is that the sum of the second power of the difference between the concentration of NO and NH_3 calculated by the detailed mechanism reaction and the overall reaction at each time in a range of temperature.

$$Err_{mean} = \frac{1}{MN} \sum_{j=0}^{M} \sum_{i=0}^{N} \left[\frac{2\left(C'_{\mathrm{NO}\_i\_j} - C_{\mathrm{NO}\_i\_j}\right)^2}{C^2_{\mathrm{NO}_0}} + \frac{\left(C'_{\mathrm{NH3}\_i\_j} - C_{\mathrm{NH3}\_i\_j}\right)^2}{C^2_{\mathrm{NH3}_0}} \right] \tag{60.8}$$

$C'_{\mathrm{NO}\_i\_j}$ and $C'_{\mathrm{NH3}\_i\_j}$ are the concentration of NO and NH_3 based on the overall reaction at the temperature of j and the moment of i. $C_{\mathrm{NO}\_i\_j}$ and $C_{\mathrm{NH3}\_i\_j}$ are the concentration of NO and NH_3 based on the detailed mechanism reaction at the same temperature of j and the same moment of i. N represents the number of samples for the reaction residence time, that is, the entire reaction time is discretized into N moments for investigation. M represents the number of temperature samples for the reaction temperature segment, that is, the selected temperature interval is discretized into M temperatures for investigation.

The concentrations of NO and NH_3 at each time are solved by using the overall reaction model, and compared with the calculation results based on detailed mechanism reaction, so as to obtain the optimal combination of *Kr* and *Kox*.

Table 60.1 Arrhenius parameters of NOx reduction overall reaction

Temperature (K)	*Ar* (cgs)	βr	*Er* (cal/mol)	*Aox* (cgs)	βox	*Eox* (cal/mol)
1100–1400	1.266×10^8	6.150	8.730×10^4	5.843×10^7	6.973	1.187×10^5

Through the iterative operation of genetic algorithm, a set of optimal overall reaction parameters were found. The Arrhenius coefficient for the NOx reduction overall reaction is shown in Table 60.1 in the temperature interval from 1100 to 1400 K, and the relative sum of squares error from the genetic algorithm is 0.003238591.

It can be found from the calculation results of evaluation criteria that the overall reaction model can accurately describe the process of NOx reduction reaction in the temperature range of 1100–1400 K. In the next part of this paper, the overall reaction model is tested.

60.3 Results and Discussion

In this section, the calculation results of the overall reaction and the detailed mechanism reaction are compared in detail, under the temperature range of 1000–1500 K and different residence times.

Moreover, to guarantee the wide applicability of the overall reaction model under different residence time, the results at three residence times, 0.15, 0.3 and 0.6 s, were also compared.

60.3.1 Denitrification

In the reaction process, it is important to study the denitrification rule and explore the change of NO concentration. It can be clearly seen from Figs. 60.1, 60.2 and 60.3 that under different residence times and determined temperature intervals, the overall reaction parameters solved by genetic algorithm can describe the progress of reaction, and the calculation results can well coincide with the numerical simulation results of detailed mechanism reaction.

The concentration of NO begins to change at 1100 K. There is a slight error in the calculation of the overall reaction compared with the detailed mechanism reaction, but the error is not conspicuous. In addition, the optimal reaction temperature and denitration efficiency are accurately reflected under different residence times. For example, as can be seen from Fig. 60.1, the overall reaction model does reflect the optimal denitrification efficiency at the temperature of 1250 K, which is consistent with the calculated results of the detailed mechanism reaction. The reason why the concentration of NO at the outlet is greater than the concentration at the entrance in the temperature range of 1450–1500 K is that in the case of high temperature,

Fig. 60.1 Simulation results of NO with the overall reaction and detail reaction at residence time of 0.15 s and different temperatures

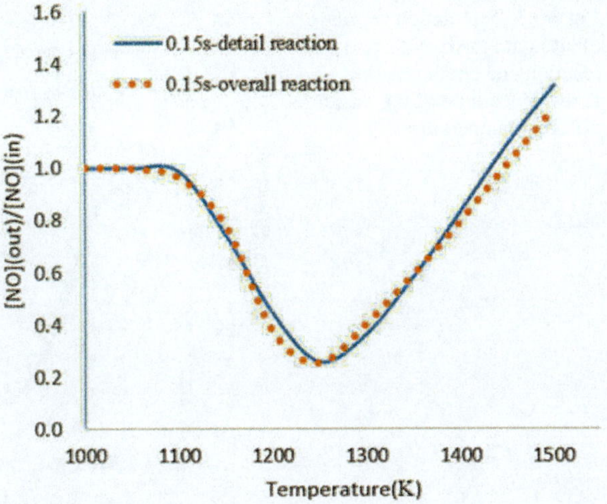

Fig. 60.2 Simulation results of NO with the overall reaction and detail reaction at residence time of 0.3 s and different temperatures

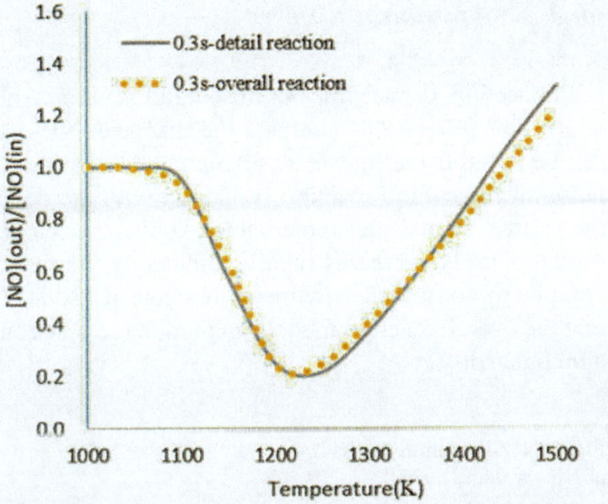

the reaction (60.2) plays a leading role in the NOx reduction overall reaction, which oxidizes NH_3 into more NO. It highlights the overall reaction model can reflect the best denitrification efficiency and the impact of reaction temperature window, from 1100 to 1400 K.

When the residence time is 0.3 s or 0.6 s, the calculation results of the overall reaction are most consistent with those of the detailed mechanism reaction. When the residence time is 0.15 s, there is some little error. Moreover, with the increase of reaction residence time, the optimal reaction temperature of the overall reaction model moves to the left, and the denitration efficiency increases, which is consistent with the calculated results of the detailed mechanism reaction. It highlights the overall reaction model can also reflect the impact of reaction residence time.

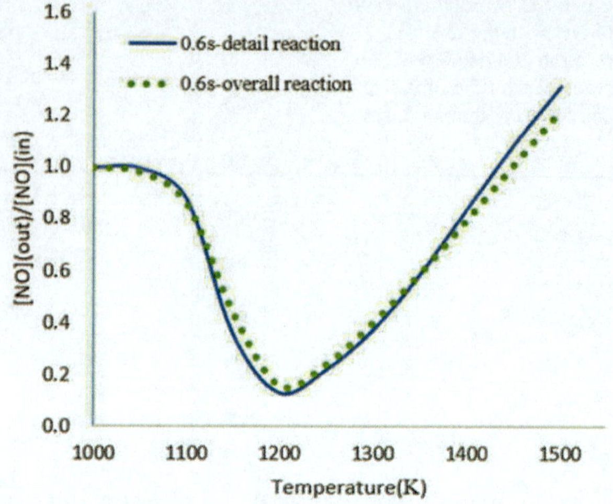

Fig. 60.3 Simulation results of NO with the overall reaction and detail reaction at residence time of 0.6 s and different temperatures

60.3.2 Ammonia Slip

In this section, the accuracy of the overall reaction model in predicting the ammonia slip will be reflected by studying the change of NH_3 concentration in the reaction. It can be found from Figs. 60.4, 60.5 and 60.6 that there is a small error in the calculation of the overall reaction compared with the detailed mechanism reaction, but the relative error is not remarkable. And as the temperature increases, the concentration of NH_3 decreases rapidly, indicating the fast consumption of NH_3. NH_3 is completely consumed in temperature range of 1200–1300 K. Once again, it is proved that the overall reaction model can predict the reaction temperature window and the ammonia slip.

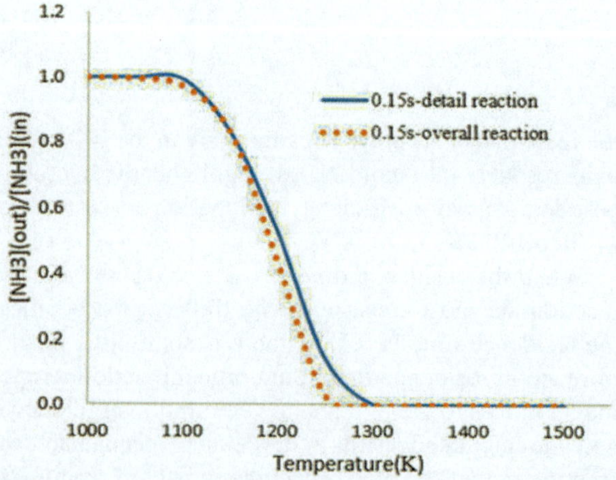

Fig. 60.4 Simulation results of NH_3 with the overall reaction and detail reaction at residence time of 0.15 s and different temperatures

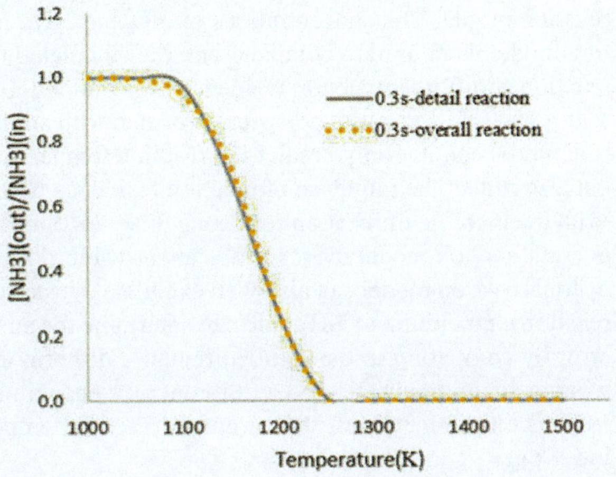

Fig. 60.5 Simulation results of NH_3 with the overall reaction and detail reaction at residence time of 0.3 s and different temperatures

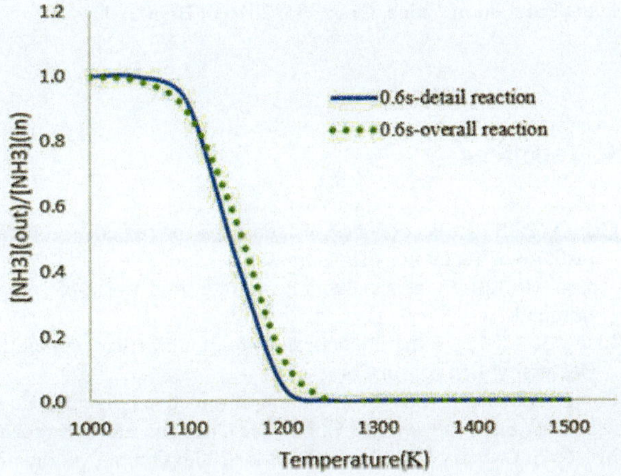

Fig. 60.6 Simulation results of NH_3 with the overall reaction and detail reaction at residence time of 0.6 s and different temperatures

Similarly, as the reaction residence time increases, the reaction temperature of NH_3 moves to the lower value. By comparing NH_3 concentration at different residence times, it can be found that the overall reaction can predict the influence of residence time on the ammonia slip in NOx reduction reaction.

60.4 Conclusion

In this paper, the genetic algorithm was used to compare the calculation results of the overall reaction and the detailed mechanism reaction, and the optimal Arrhenius parameters were obtained to calculate the reaction constant of the NOx reduction overall reaction. The gas component concentration can be solved by using the overall reaction model. The concentrations of NO and NH_3 calculated by the overall reaction model were used to compare with the calculated results of detailed mechanism reaction at different reaction residence time and within a certain temperature range, 1000–1500 K. From the perspective of denitrification efficiency, the overall reaction model can not only predict the reaction temperature window of denitrification, but also reflect the influence of reaction residence time on denitrification efficiency, With the increase of reaction residence time, the optimal reaction temperature of the overall reaction model moves to the lower value. In terms of the ammonia slip, the overall reaction model can also well calculate the starting temperature and complete reaction temperature of NH_3, and can determine the ammonia slip with an acceptable error by comparing to the detailed model. Furthermore, the change of temperature window of ammonia slip under different residence times showed that as the reaction residence time increases, the complete reaction temperature of NH_3 moves to the lower one.

Acknowledgements The authors gratefully acknowledge the financial support from National Key R&D Program of China, Grant No. 2016YFB0600202.

References

Cao Q (2009) Effects of gaseous additives for selective non-catalytic reduction of NOx. Harbin Institute of Technology (in Chinese)

Jepsen MS (2018) NOx reduction in grate-fired waste-to-energy plants. Technical University of Denmark

Liang X (2010) Experimental study and numerical simulation of SNCR process. Southeast University, Jiangsu (in Chinese)

Liang X, Zhong Z (2009) Simulation of selective non-catalytic reduction process by the model of laminar finite rate with FLUNT. Proc Chin Soc Electr Eng 29(35):96–101 (in Chinese)

Mitsos A, Oxberry GM, Barton PI et al (2008) Optimal automatic reaction and species elimination in kinetic mechanisms. Combust Flame 155(1–2):118–132

Sun T (2016) Numerical simulation of SNCR denitrification and NO reduction kinetics of a circulating fluidized bed boiler. Nanjing Normal University, Jiangsu (in Chinese)

Vujanović M, Schneider D, Baburić M et al (2005) A comprehensive modelling of NOx formation in combustion systems based on reduced chemical reaction mechanisms. In: CD Dubrovnik conference on sustainable development of energy

Chapter 61
The Mechanisms and Applications of NO_x Reduction by Low-NO_x Burner Coupling Deep Air-Staging Technology in Pulverized Coal

Xiaolei Cheng, Naiji Wang, Xin Zhang, Yongying Wang, and Long Chen

Abstract In order to reduce the original NOx emission from coal combustion, and lighten the excessive redeucer used in the De-NOx post-treatment which will lead to the corrosion of steam heat exchanger tube and economical efficiency, a new De-NOx technology combined with the De-NOx combustor and the deep air stage combustion was developed, and the technology was testified by theory, numerical simulation, pilot experiment, and industrial experiment. FR/ED (Finite Rate/Eddy Dissipation) model which considering the chemical reactions involved in the gasification process is used in the simulation, and it showed a sufficient accuracy with the experiment results. A strong reductive atmosphere including CO and CH_4 was detected with a low stoichiometric ratio both by numerical simulation and experiment in a 7 MW double-cone De-NOx burner, the CO and H_2 contents could be 15.457% and 0.992 for SR of 0.456, respectively, while no NOx content was detected at the burner outlet. Method Oxygen-enriched air used in the burner could intensify the reductive atmosphere, and the CO content at the center burner increased form 9.54 to 20.258%, when the oxygen content increased from 21 to 28.3%. For the 7 MW pilot boiler system, the initial NOx emission of the boiler can reach to 159 mg/m^3, and the total reduction of NOx generation was more than 70.9%, when the oxygen content in secondary air was 28.3% and the proportion of staged air in tertiary air was 41.2%. Utilization of De-NOx technology in long-term industrial experiment for 40t/h steam boiler showed near no influence on the boiler operation, original NOx of less than 200 mg could be reached at the extreme condition. When the original NOx emission reduced from 697 to 300 mg/m^3, the heat loss was less than 0.5%, and more than 50% of ammonia water was saved for SCR, which leading about 1 yuan per ton steam saving of the cost.

Keywords NOx emission control · Deep air-staging · Low-NOx combustor · Numerical simulation

X. Cheng (✉) · N. Wang · X. Zhang · Y. Wang · L. Chen
China Coal Research Institute Company of Energy Conservation, Beijing 100013, China
e-mail: chengxl328@126.com

China Coal Science and Technology Research Institute, Beijing 100013, China

J. Lyu and S. Li (eds.), *Clean Coal and Sustainable Energy*, Environmental Science and Engineering,
https://doi.org/10.1007/978-981-16-1657-0_61

61.1 Introduction

Due to serious environment pollution problem, severe restriction of air pollutants especially for NOx was formulated in China during the past years. As for coal combustion, which account for more than 66% of total NOx emission, the NOx emission standard is 50 mg/m^3 in some key areas. Compared to power plant, the coal industrial boilers of heat power range from 7 to 80 MW, which consume 640 million tons coal per year by 600,000 sets of boiler, are facing more difficult technical problems for NOx emission control, since the shorter reaction time and smaller capacity.

For both power plant and industrial boiler, NOx post-treatment technologies of flue gas such as Selective Catalytic Reduction (SCR) and Selective Non-catalytic Reduction (SNCR) must be applied to meet the NOx emission demand. However, excessive reducers used during de-NOx process discharged into the atmosphere would course ammonia slip problems which will lead to corrosion of steam heat exchanger tube (Wilburn 2004; Burke and Johnson 1980; Ye et al. 2016). De-NOx combustion technology was an efficient, low-cost and clean way to reduce the NOx emission through coal combustion. Generally, the original NOx emission for coal-fire power plant could be reduced from 600 mg/m^3 (@6 vol% O_2 dry) to as low as 200 mg/m^3 by utilizing de-NOx combustion technology such as de-NOx burner, air-stage combustion, flue gas recirculation, fuel stag combustion (Chui and Gao 2010; Man et al. 2005). However, since the different operation conditions compared to coal-fire power plant, the original NOx emission was about 600 mg/m^3 (@9 vol% O_2 dry) for pulverized coal industrial boiler, new De-NOx technology suitable for them should be developed.

Among all the de-NOx combustion technologies, air stage combustion is the basic and widely-used method. The coal combustion process is divided into reductive zone, where partial combustion and gasification reaction happened to form a reductive atmosphere, and the burnout zone where unreacted char and gas was burnout. The reduced substances such as CO, HCN and NH_3 can react with NO to inhibit the formation of fuel-type NOx. De-NOx burner is the key factor to generate reductive atmosphere. The combination of air stage combustion and de-NOx burner was studied in this article.

Many researchers have conducted numerical and experimental study on the de-NOx combustors. Zhou et al. (2014) found an annular recirculation zone between the primary and secondary air jets which was advantageous to lower NOx formation in the HT-NR3 burner. Researches of Tsumura et al. (2003) showed that the NOx reduction load extension(NR-LE) type burner could achieve a NOx initial emission to 200 mg/m^3 with minimum load of 50%. Zhou et al. (2018) found that the annular recirculation zone generated between the primary and secondary air jets to promote the rapid ignition and combustion of pulverized coal particles to consume oxygen, and then a reducing region is formed as fuel-rich environment to contribute to in-flame NOx reduction, and when the burner stoichiometric ratio of 0.75 is adopted, a proper low-NOx efficiency and burn-out ratio is achieved. A new coal preheating technology which is based on the CFB boiler was proposed by Institute of Engineering Thermophysics, Chinese Academy of Science (Ouyang et al. 2018; Liu et al. 2019), theory and experiments were carried out, and a CO content of 9% in the preheating burner. Also, oxy-fuel combustion technology is conducted to low-NOx burner by some researchers to coal combustion and sufficient low NOx emission was observed by coupling with air-staged combustion (Scheffknecht et al. 2011; Watanabe et al. 2011).

Low NOx effects of air stage combustion were observed by many researchers, and the key factors are mainly the stoichiometric ratio (SR) of the burner or the primary combustion area, and the burnout stream injection position. Shao et al. (2017) analyzed the effects of excess air coefficient in main combustion zone, residence time of flue gas, temperature, coal type and particle size of pulverized coal on low-NOx efficiency for air-staged combustion. Li (2017) studied the effects of oxygen content, proportion and distribution of exhaust air, location of exhaust air and residence time of reduction zone in the main combustion area, and verified the low nitrogen effect of deep tangential graded air distribution in power plants. Zhou (2016) used air-staged combustion technology in small and medium-sized pulverized coal boilers, and the emission of NOx was lower to 387 mg/m^3. Sun et al. (2013) and Li et al. (2016) mainly studied the low-NOx combustion effect of air fractionation from the reaction mechanism.

Among all the De-NOx technologies, the key points are to maintain the reductive atmosphere during the early combustion stage, which can restrain the production of NOx and reduce NOx already produced, and even the burn-out stage to avoid local high temperature region. In this article, a new Low-NOx combustion technology was conducted by coupling the Low-NOx combustor and deep air staged combustion. The combustor with reverse primary air flow and double-cone structure could intensify the combustion, thus more than 50% of the coal would be reacted in the combustor while only about 60% of the air inject into it, so there would be an extremely high content of reductive gas mainly CO and CH_4, which could reduce the original NOx emission during the combustion process; Low-NOx combustion in the boiler furnace would be realized by utilizing the deep air-staging method while 40% of the air will be injected at the middle of the furnace, by which a strong reductive atmosphere was provided. Numerical simulations, pilot system experiments, and long-term industrial experiments were conducted in the article.

61.2 Low-NOx Burner Simulation and Experiments

The main goal for low-NOx burner is to produce a strong reductive atmosphere and high carbon conversion ratio. The effect of high oxygen content of air on the reductive atmosphere produced by the combustion process has been proved by the utilization of oxy-fuel combustion or oxy-rich combustion (Stanger et al. 2015; Fan et al. 2017; Shaddix and Molina 2011). Efforts should be made to determine how the excess ratio

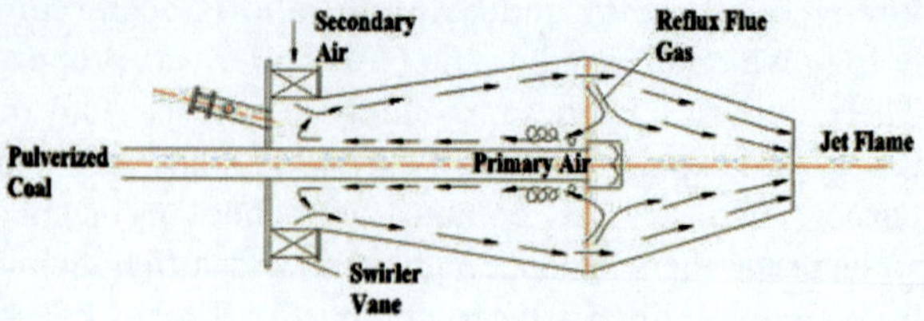

Fig. 61.1 Double cone combustor with centre reverse spraying

and oxygen content will affect the reaction in low-NOx burner, and how deep can the reductive atmosphere and carbon conversion could be reached.

61.2.1 *Structure of Double Cone Burner*

Double cone burner is a low-NOx burner developed by China Coal Research Institute Company of Energy Conservation (Mo et al. 2018; Jiang et al. 2014), as shown in Fig. 61.1. Pulverized coal is carried to the middle of the burner by primary air, and flows reversely towards the front of the burner, where it is mixed with strong swirl secondary air. Coal is preheated, ignited and steady combusted during the mixing process. The large scale of burner provides sufficient capacity for coal combustion, the carbon conversion rate can be up to 60%, if for proper air distribution mode.

61.2.2 *Numerical Simulation of Double Cone Burner*

CFD simulation method has been widely used in the study of pulverized coal combustion characteristics. But for the strong reductive atmosphere reaction in the burner, not only combustion reaction occurred, the partial gasification reaction between coal particles, oxygen and steam water also should be considered. Since the difference from that of conventional pulverized coal combustion, an appropriate reaction model should be chosen. In the simulation of pulverized coal combustion, the most commonly used chemical turbulence interaction models are the mixed fraction PDF (Probability Density Function) method and the ED Model (Eddy Dissipation Model) in the component transport model. However, neither of the two reaction models takes into account only the char combustion reaction process, and the simulation of gasification reaction under strong reducing atmosphere is not considered, which leads to poor accuracy of simulation results (Zhang et al. 2017; Duan et al. 2017). In this paper, the FR/ED (Finite Rate/Eddy Dissipation) model is used in the gas phase reaction model. Considering the chemical reactions involved in the gasification process, the reaction of pulverized coal in strong reducing atmosphere can be accurately simulated. The reaction equation and the reaction mechanism parameters are shown in Table 61.1.

Table 61.1 Reaction and kinetic mechanisms for oxygen rich combustion

	Reactions	Pre-exponential factor	Activation energy j/kgmol
1	vol + 1.3422O_2 → 1.2893CO + 1.9997H_2O + 0.0288N_2	2.119×10^{11}	2.027×10^{8}
2	CO + 0.5O_2 → CO_2	2.239×10^{12}	1.7×10^{8}
3	$C_{(s)}$ + 0.5O_2 → CO	267	1.26×10^{8}
4	$C_{(s)}$ + CO_2 → 2CO	2100	2.98×10^{8}
5	$C_{(s)}$ + H_2O → CO + H_2	36	1.54×10^{8}
6	H_2 + 0.5O_2 → H_2O	8.83×10^{8}	1×10^{8}
7	CO + H_2O → CO_2 + H_2	2.98×10^{12}	3.69×10^{8}

In this paper, Ansys Fluent is used for numerical simulation. The other main model parameters involved in the simulation are as follows: the turbulence model uses Realizable k-ε model, which can simulate the rotating jet in turbulent flow relatively accurately; the burner meets the condition of optical depth greater than 1, and the radiation model uses P1 model, which has high calculation efficiency; the char volatilization reaction uses double competing model, two reactions act on different temperature, and the reaction kinetics parameters are shown in Table 61.2. The Lagrangian DPM discrete phase model is used for gas–solid phase flow, which is suitable for the model with volume fraction less than 10%, and the DPM model uses surface source injection.

The structure of the 7 MW double-cone burner is simplified, which is divided into primary air inlet, secondary air inlet, outlet and burner boundary. Ansys ICEM is used to mesh the 7 MW double-cone burner. The number of meshes is 465,000 (Fig. 61.2).

The numerical simulation uses Shenfu Dongsheng bituminous coal as fuel. The proximate analysis and ultimate analysis are shown in Table 61.3.

Table 61.2 Reaction kinetics parameters of two competing rates model

	Pre-exponential rate	Activation energy J/kg mol	Weighting factor
First rate	2×10^{5}	1.046×10^{8}	0.3
Second rate	1.3×10^{7}	1.674×10^{8}	1

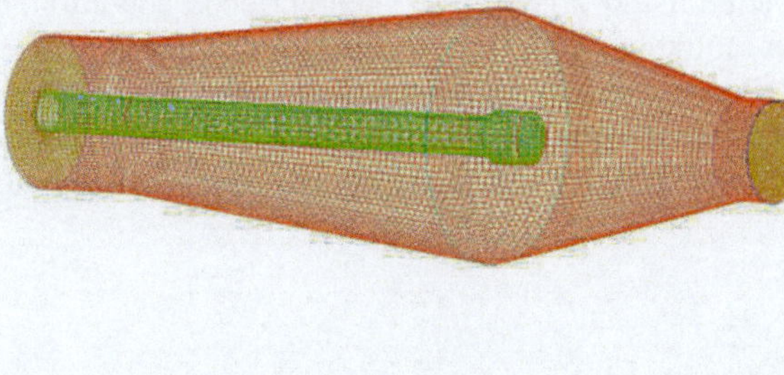

Fig. 61.2 Meshing of double cone combustor

Table 61.3 Proximate and ultimate analysis of utilized coal %

Ultimate analysis					Proximate analysis			
C_{ad}	H_{ad}	O_{ad}	S_{ad}	N_{ad}	M_{ad}	A_{ad}	V_{ad}	FC_{ad}
68.98	3.92	11.22	0.50	0.83	5.13	9.42	32.49	52.96

The combustion characteristics of the burner under typical air condition and oxygen enrichment condition were compared. For both conditions, the excess air ratio of secondary air (the ratio of oxygen required for combustion to theoretical oxygen) is 0.61, and the oxygen concentration for air and oxygen-rich is 21 and 25.9%, separately. The combustion temperature of the burner under oxygen-enriched condition is higher than that under air combustion condition, and the ignition area is earlier and the exit temperature of the burner is more uniform. The average temperatures in the burner under air and oxygen enrichment conditions are 897 °C and 1007 °C, respectively. The average temperatures at the burner outlet are 1255 and 1356 °C, respectively. In the case of oxygen enrichment, the reaction rate with pulverized coal increases with the increase of oxygen concentration in the air, while the amount of N_2 and other unreactive gases decreases, the absorption and combustion heat release decreases, and the reaction combustion in the burner increases. With the increase of oxygen concentration, the total amount of air decreases, the velocity of flue gas in the burner decreases, and the residence time increases, which is also conducive to the combustion reaction of pulverized coal and the improvement of combustion. Because of the decrease of secondary air velocity, the mixing position of primary air and secondary air in reverse injection is closer to the front of burner, which further improves the residence time of pulverized coal combustion. The outlet velocity of the burner decreases, the swirl intensity decreases, the air near the wall burner outlet is stratified, and the temperature distribution at the burner outlet is more uniform.

Secondary air propagates forward and primary air propagates backward in the burner. The intersection of the two air is the starting point of the reaction of pulverized coal combustion to produce CO_2 and partial gasification to produce CO, and the CO content rises rapidly. Similar to the analysis of reaction temperature, oxygen-enriched primary air carries pulverized coal farther back and has higher CO content in front of the burner. At the same time, due to the increase of partial gasification reaction and the decrease of N_2 and other gas components, the overall CO content in the burner is significantly higher than that in the air combustion mode by using oxygen-enriched partial gasification. The average CO concentration in the burner was 5.48% in air condition and 7.17% in oxygen-enriched condition, respectively. The average CO concentration at the burner outlet was 5.23 and 6.35%.

Because of the oxygen-enriched atmosphere, the reaction degree of pulverized coal in the burner has also been greatly improved. The carbon conversion rates under air and oxygen-enriched conditions are 61.54% and 86.27%, respectively (Fig. 61.3).

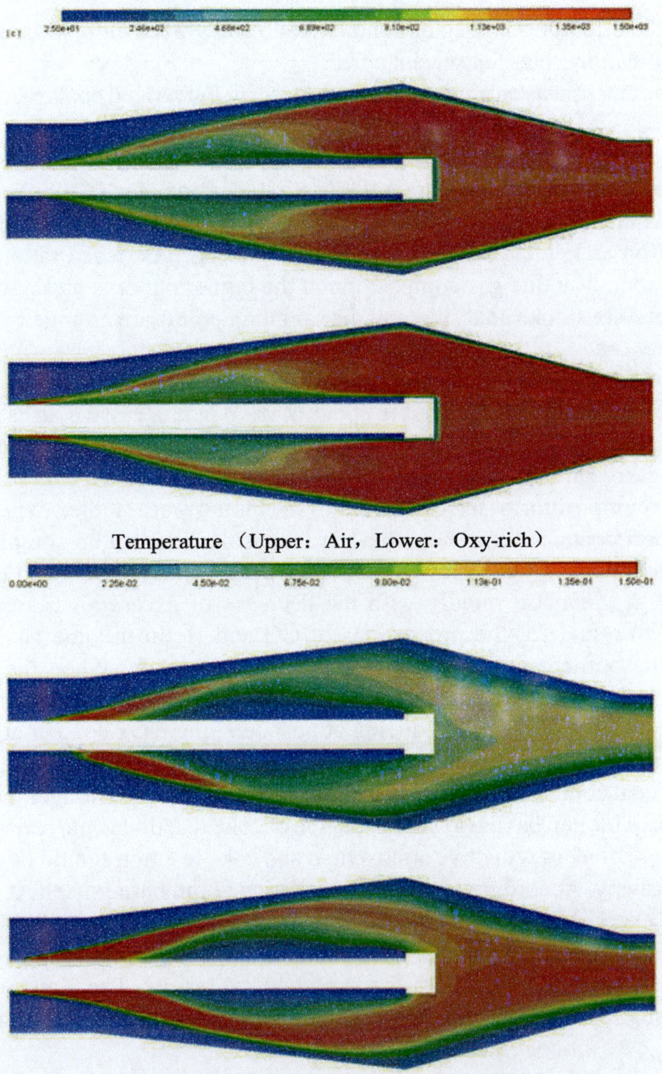

Fig. 61.3 Simulation results of typical condition

61.2.3 Pilot Scale Experiments for Burner and the Combustion System

The effect of oxygen-enriched conditions on partial gasification reaction in burner is preliminarily discussed through numerical simulation. It also needs to be verified by field experiments. In the experiment of partial gasification with oxygen enrichment in burner, the effects of oxygen enrichment ratio and graded air distribution ratio on partial gasification in burner were studied.

The experiment was carried out on a 10t/h steam pulverized coal industrial boiler in Erdos, Inner Mongolia. The burner of the boiler is the same as that used in the previous section. The pulverized coal is Shenfu Dongsheng bituminous coal, the same as the simulation, as shown in Table 61.1. Pure oxygen is supplied by liquid oxygen tank after vaporization, and mixed with secondary air in accordance with the set proportion and enters the burner to form the oxygen-enriched ratio required in the experiment. The flue gas composition at the burner outlet is measured by Vario high temperature smoke analyzer, and the sampling points are arranged at the center of the burner outlet.

Due to the limited reaction time of pulverized coal in the burner, the maximum reaction degree of pulverized coal is about 60%. When excessive air is introduced into the burner, an oxidizing atmosphere will be formed, which is not conducive to the partial gasification of pulverized coal. The effects of different excess air ratio on the flue gas composition at the exit center of the burner were studied experimentally.

The experimental results show that partial gasification can be completed in the burner under the condition of low excess air coefficient, and the CO and H_2 contents in the flue gas increase rapidly with the decrease of excess air ratio. When the excess air ratio is 1.25, the proportions of CO and H_2 in the flue gas at the exit center of the burner are 2.512% and 0.184%, respectively. When the excess air coefficient is 0.456, the proportions of CO and H_2 increase to 15.457% and 0.992%, respectively. Under the strong reducing atmosphere, the NOx content at the burner outlet is 0, which greatly reduces the influence of NOx formation at the initial stage of pulverized coal combustion. The lower the excess air ratio, the stronger the reductive atmosphere of burner, but the lower the degree of coke reaction in pulverized coal, the contradiction effect of reductive atmosphere and coke reaction should be considered comprehensively. According to the characteristics of the burner itself, it is better to control the excess air coefficient in the burner at about 0.5.

The accuracy of the numerical simulation results is verified by comparing the numerical simulation results with the experimental results. The simulation of CO content is slightly higher when the excess air coefficient is high, and slightly lower when the excess air coefficient is low (Fig. 61.4).

Compared with air combustion, the oxygen-enriched combustion method can produce a more reductive atmosphere and a higher CO content in the flue gas. Under the condition that the total oxygen content (excess air coefficient) in the air of the burner is unchanged, the experiments of oxygen-enriched combustion of the burner under different oxygen-enriched concentration are carried out. Because of the presence of oxygen-enriched atmosphere, the content of N_2 and other unreacted components during the combustion process decreases, and the combustion reaction temperature increases. Under the same conditions, the degree of char reaction increases significantly. When the oxygen content in the air increases from 21 to 28.3%, the CO content at the outlet center of the burner increases from 9.54 to 20.258%. Even excluding the factors of increasing the oxygen content, the CO production increases by 49.3%, and the CH_4 production decreases slightly (Fig. 61.5).

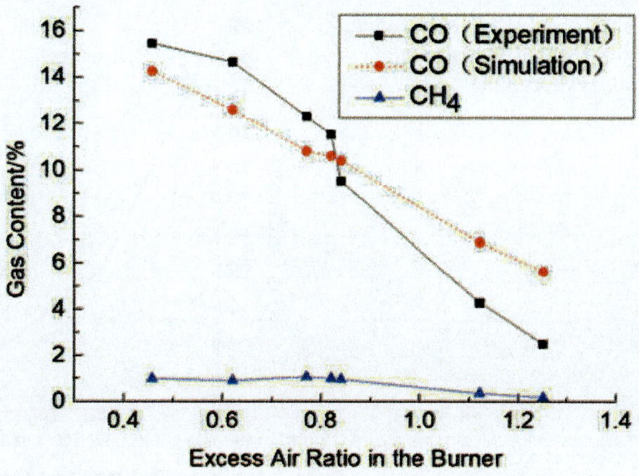

Fig. 61.4 Effects of excess air ratio on the air composition from combustor outlet

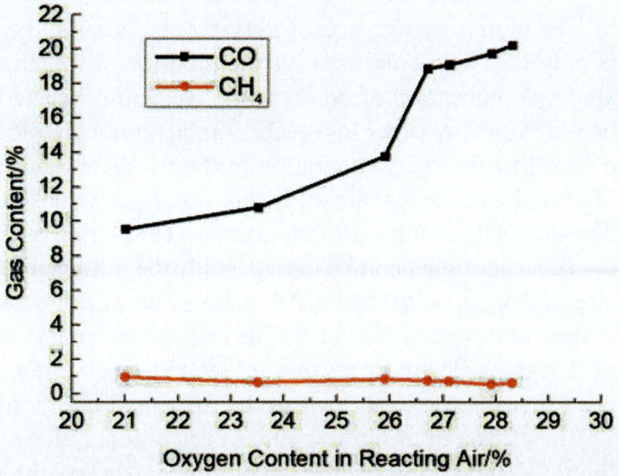

Fig. 61.5 Effects of oxygen enrichment ratio on the air composition from combustor outlet

However, because the burner itself belongs to the strong swirl burner, the exhaust gas composition at different radial position of the burning exit is also different. The secondary air of the burner is a strong swirling air, which is close to the wall in the burner and forms an air layer near the wall to cooling the boundary. The central position of the burner, especially near the exit, has a relatively low air content and a stronger reductive atmosphere. The reductive gas content at the burner outlet is high in the middle and low in the edge.

The distribution of reductive gases measured under oxygen-enriched conditions is shown in Fig. 61.6. Under this condition, the CO content at the outlet center of the burner is 20.025%, the CO content at 30 mm away from the center is rapidly reduced to 8.262%, and the CO content at the edge of the burner (150 mm) is 1.26%. The average CO content in burner cross section is 3.15%.

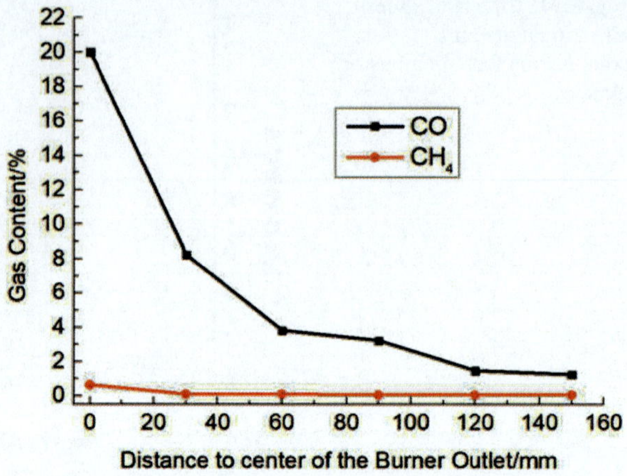

Fig. 61.6 Air distribution at different radial distance of combustor outlet

The partial gasification of pulverized coal with oxygen enrichment in the burner is only the basic condition for the final low-nitrogen combustion. The NOx generated by combustion of combustible gas and unreacted char in the furnace needs to be controlled by other low-NOx combustion technologies to achieve the purpose of controlling the NOx generation in the whole combustion process. In this section, a 10t/h pulverized coal steam boiler system in Ordos is taken as the research object. The low-NOx combustion experiments are carried out by partial oxygen-enrich gasification and staged combustion to study the influence of oxygen enrichment ratio and staged degree on the initial NOx emission of the boiler. The arrangement of boiler system is shown in Fig. 61.7. The burner is arranged downward at an inclined angle of 8 degrees. The primary air and secondary air enter the burner, and the tertiary air is arranged at the lower part of the front wall burner of the boiler.

The effects of different oxygen enrichment ratios on low-NOx combustion under the same air staging degree were studied. The results are shown in Fig. 61.8. When all the air required for combustion enters the furnace from the burner, the initial NOx emission of the boiler is 546 mg/m^3, and the initial NOx emission decreases

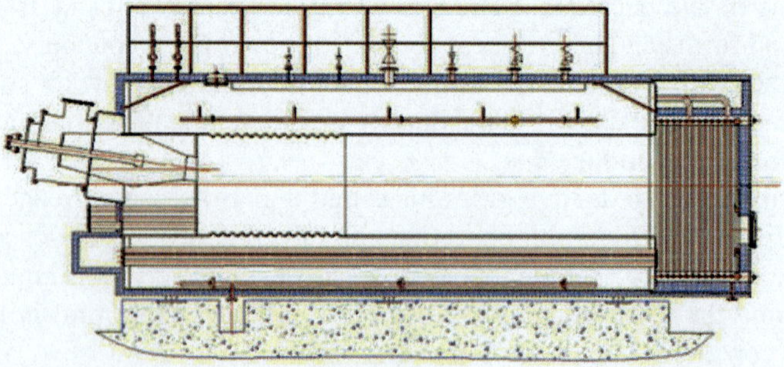

Fig. 61.7 Structure of the 10t/h steam boiler system

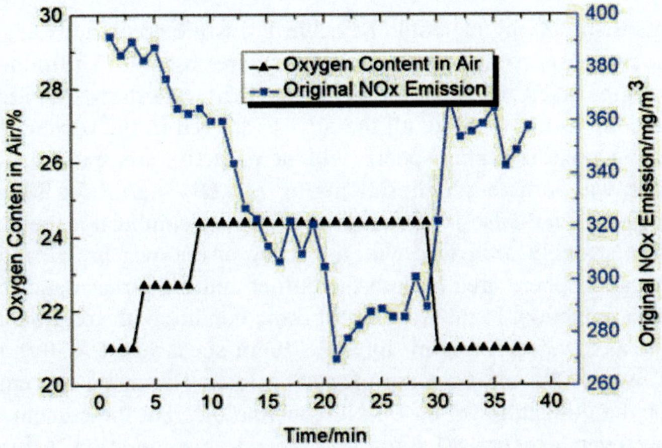

Fig. 61.8 Effects of oxygen enrichment ratio on original NOx emission for boiler

to 390 mg/m^3 when the proportion of tertiary air is 35%. When the oxygen content in the air rises from 21 to 22.7%, the initial emission of NOx decreases from 390 to 358 mg/m^3 with a decrease of 8.2%, while for an increase of oxygen content to 24.4%, the initial emission of NO_x decreases to 285 mg/m^3 with a decrease of 26.9%.

The experimental results show that both oxygen-enriched partial gasification and air-staging combustion can effectively reduce the initial emission of NOx. The staged air mainly ensures that the burner and the front of the furnace are in a reductive atmosphere, and hardly produces NOx. The rest air injected to middle of the furnace to combust the rest char while produce a small amount of NOx. The oxygen-enriched way in the burner improves the carbon conversion rate and reductive gas in the burner. Volume concentration enlarges the action area of reducing atmosphere in combustion process, while the amount of char and char NOx produced in burnout zone decreases, which further reduces the initial emission of NOx. When the oxygen content in secondary air is 28.3% and the proportion of staged air in tertiary air is 41.2%, the initial NOx emission of the boiler can reach 159 mg/m^3, and the total reduction of NOx generation can reach 70.9%.

61.3 Industrial Experiments

The combination of De-NOx burner and deep air-stage combustion method to reduce the original NOx emission was carried out in pilot scale experiments. The industrial device applications of the theory should be focused on the constancy and the comprehensive influence of the system. Experiments in the pulverized coal industrial boiler were carried out on a 40t/h steam boiler. The results showed that the original NOx emission less than 200 mg/m^3 could be achieved, while no obvious negative influences for the combustion efficiency of the boiler were observed. The influences of the air-staging degree on NOx emission for different load were shown in Fig. 61.9. With no air stage combustion method, all the air was flowed in the burner, the reaction in which was in a oxidized atmosphere, with no reductive area existed. The original NOx emission was sufficient high, 621 mg/m^3 and 694 mg/m^3 for 30t/h and 40t/h load, respectively. For the De-NOx burner and air stage combustion method, original NOx emission strongly decreased with the rising of the over fire air ratio, showed the reductive atmosphere area in both the burner and the furnace and the effect of NOx emission reduction. In the extreme working condition, the original NOx emission could be as low as 172 mg/m^3 by a 36.7% air stage ratio for 30t/h load, while 195 mg/m^3 by a 47.8% air stage ratio for 40t/h load. The rising CO emission was observed for the utilization of the De-NOx technology, but the content of CO was still very low, even less than 80 mg/m^3 for the extreme condition. All the working condition was taken for a long term running, and no obvious negative influence was observed.

The heat loss of exhaust gas and solid incomplete combustion may be affected by the utilization of the De-NOx technology. The temperature distribution in the furnace will be different, leading to the change of exhaust gas temperature, which will affect the heat loss of exhaust gas and boiler combustion efficiency. It will also affect the burnout of pulverized coal, the carbon content in fly ash and slag, and then affect the heat loss of solid incomplete combustion and boiler combustion efficiency.

The flue gas temperature of the boiler is relatively stable, and the difference between the highest and lowest temperatures is less than 3 °C. Under the suitable working condition, the exhaust gas temperature of the boiler decreases. The results show that the heat efficiency of the boiler is ±0.2% affected by the heat loss of exhaust gas.

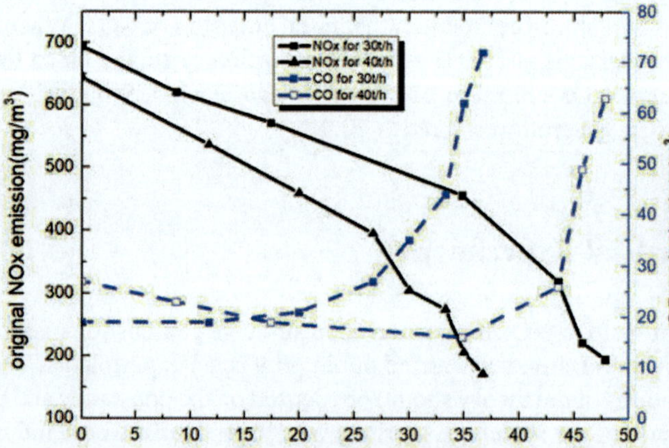

Fig. 61.9 The influence over-fire-air ratio on the original NOx emission in 40t/h steam boiler

Table 61.4 Effect of air-staging on combustion efficiency

	Fly ash/%			Slag from bottom furnace/%		
	V	FC	Loss ratio	Vad	FC	Loss ratio
Unstaged	3.44	3.94	7.38	7.84	74.54	80.52
Staged	3.19	8.20	11.34	9.44	48.91	58.35

The fly ash and slag under the condition of normal combustion and De-NOx combustion with 300 mg/m^3 NOx emission were compared. The results show that the De-NOx combustion method has a certain impact on the burnout of pulverized coal, and the carbon content in fly ash increases, as shown in Table 61.4. The heat loss of fly ash is 0.39% and 0.87% for normal condition and De-NOx condition, respectively, and 0.48% more loss in boiler overall thermal efficiency. The slag yield is 15 and 45 kg/h for the two condition, 0.33% loss under normal condition, 0.98% under De-NOx condition, and 0.62% more heat loss in De-NOx condition. Generally speaking, when the boiler used the De-NOx combustion method, a heat loss of less than 0.5% was detected, while more than 50% of ammonia water was saved for SCR, which leading about 1yuan/t steam saving of the cost.

61.4 Conclusions

A new De-NOx technology combined with the De-NOx combustor and the deep air stage combustion was introduced, and theory and experiment were conducted in this work. The results are as follow:

1. By using the double-cone De-NOx burner with a low stoichiometric ratio, a strong reductive atmosphere including CO and CH_4 was formed both by numerical simulation and experiment in a 7 MW burner. The CO and H_2 content could be 15.457% and 0.992 for SR of 0.456, respectively, while no NOx content was detected at the burner outlet.
2. Oxygen-enriched air used in the burner could intensify the reductive atmosphere formed. The CO content at the center burner increased form 9.54 to 20.258%, when the oxygen content increased from 21 to 28.3%.
3. For the 7 MW pilot boiler system, the oxygen content in secondary air is 28.3% and the proportion of staged air in tertiary air is 41.2%, the initial NOx emission of the boiler can reach to 159 mg/m^3, and the total reduction of NOx generation was more than 70.9%.
4. Utilization of De-NOx technology in long-term industrial experiment for 40t/h steam boiler showed near no influence on the boiler operation, original NOx of less than 200 mg/m^3 could be reached at the extreme condition. When the original NOx emission reduced from 694 to 300 mg/m^3, the heat loss was less than 0.5%, and more than 50% of ammonia water was saved for SCR, which leading about 1yuan/t steam saving of the cost.

Acknowledgements This work was supported by the China Coal Research Institute Innovation Fund project (No. 2017CX02), and the China Coal Science and Technology Group Innovation Fund Project (No. 2018MS003).

References

Burke JM, Johnson KL (1980) An investigation of ammonium sulfate/bisulfate formation and deposition in air preheater. Radian Corporation, Nashville, Tennesses, USA

Chui EH, Gao H (2010) Estimation of NOx emissions from coal-fired utility boilers. Fuel 89:2977–2984

Duan Y, Zhang L, Mao Z et al (2017) Numerical simulation of MILD combustion on coal-water slurry. Clean Coal Technol 23(4):36–41+46 (in Chinese)

Fan W, Li Y, Guo Q et al (2017) Coal-nitrogen release and NOx evolution in the oxidant-staged combustion of coal. Energy. https://doi.org/10.1016/j.energy.2017.02.130

Jiang S, Wang Y, Zhou J et al (2014) Numerical simulation on middle volatile coal combustion in reversed injection burner. J China Coal Soc 39(6):1147–1153 (in Chinese)

Li P (2017) Principle and application of air staged combustion technology. Boiler Technol 48(4):45–50 (in Chinese)

Li H, Zhou J, Liu G et al (2016) Experiment study on NOx emission features during staged-air combustion process of pulverized coal. Coal Sci Technol 44(12):194–201 (in Chinese)

Liu W, Ouyang Z, Cao X, Na Y (2019) The influence of air-stage method on flameless combustion of coal gasification fly ash with coal self-preheating technology. Fuel 235:1368–1376

Man CK, Gibbins JR, Witkamp JG, Zhang J (2005) Coal characterisation for NOx prediction in air-staged combustion of pulverised coals. Fuel 84:2190–2195

Mo R, Wang N, Cheng X (2018) Numerical analysis of coal water mixture combustion in double cone reversed injection burner. Clean Coal Technol 24(2):83–89 (in Chinese)

Ouyang Z, Liu W, Man C, Zhu J, Liu J (2018) Experimental study on combustion, flame and NOx emission of pulverized coal preheated by a preheating burner. Fuel Process Technol 179:197–202

Scheffknecht G, Al-Makhadmeh L, Schnell U (2011) Oxy-fuel coal combustion—a review of the current state-of-the-art. Int J Greenh Gas Control 5:16–35

Shaddix CR, Molina A (2011) Fundamental investigation of NOx formation during oxy-fuel combustion of pulverized coal. Proc Combust Inst 33:1723–1730

Shao W, Wang W, Pan W (2017) Mechanism of NOx emission reduction by air-staged combustion and the influencing factors. Power Equip 31(1):23–25 (in Chinese)

Stanger R, Wall T, Sporl R (2015) Oxy-fuel combustion for CO_2 capture in power plants. Int J Greenh Gas Control 40:55–125

Sun B, Wang D, Duan E et al (2013) Investigation on NOx formation characteristics under air-staged combustion. J Chin Soc Power Eng 33(4):261–266 (in Chinese)

Tsumuraa T, Okazaki H, Dernjatin P, Savolainen K (2003) Reducing the minimum load and NOx emissions for lignite-fired boiler by applying a stable-flame concept. Appl Energy 74:415–424

Watanabe H, Hichiro Y, Okazaki K (2011) NOx formation and reduction mechanisms in staged O_2/CO_2 combustion. Combust Flame 158(7):1255–1263

Wilburn RT (2004) Wright TL (2004) SCR ammonia slip distribution in coal plant effluents and dependence upon SO_3. Power Plant Chem 6(5):295–304

Ye D, Qu R, Song H et al (2016) New insights into the various decomposition and reactivity behaviors of NH_4HSO_4 with NO on V_2O_5/TiO_2 catalyst surfaces. Chem Eng J 283:846–854

Zhang Y, Zhang Y, Yang T et al (2017) Numerical simulation on combustion of lignite semi-char in cyclone furnace. Therm Power Gener 46(2):42–48+54 (in Chinese)

Zhou X (2016) Application of low NOx retrofit technology in small and medium pulverized coal fired boiler. Ind Furn 38(3):43–48 (in Chinese)

Zhou H, Yang Y, Dong K, Liu HZ, Shen YL, Cen KF (2014) Influence of the gas particle flow characteristics of a low-NOx swirl burner on the formation of high temperature corrosion. Fuel 134:595–602

Zhou CY, Wang YQ, Jin QY, Zhou YG (2018) Mechanisms analysis on the pulverized coal combustion flame stability and NOx emission in a swirl burner with deep air staging. J Energy Inst 1–13

Chapter 62
Experimental Research on NO_x Emission Characteristics Based on Combined Removal Technology of Multi-pollutant with Ash Calcium Recycling

Ting Li, Naji Wang, and Shi Yang

Abstract As one of the feasible technologies suitable for industrial boilers and kilns to reduce pollutant emissions to ultra low emission standards, combined removal technology of multi-pollutant is attracting more attention in recent years. As a combined removal technology of multi-pollutant, a low temperature flue gas deep purification technology based on semi-dry method was proposed by China Coal Research Institute. In this article, NO_2 absorption characteristics of $Ca(OH)_2$ was investigated in fixed bed and bubbling bed reactor. The composition of reaction products was detected by FTIR method. The result shows that: when the temperature is among 70–80 °C, oxygen content is 5%, the relative humidity is in the range of 40–60%, the absorption rate in fixed bed test is in the range of 20–30%, nitrate and nitrite exist in the reaction products; while in bubbling bed test, the absorption rate is above 90%. Industrial test of this technology was implemented on a 30 t/h boiler. The result shows that: at the proper operation condition, the absorption rate is in the range of 81–90%. It is found that the absorption rate of NO_2 by calcium-based absorbent is as follows: bubbling bed (wet method) $\geq$ airflow bed (semi-dry method) > fixed bed (close to dry method). The results are mainly affected by the humidification mode.

Keywords NOx removal technology · Combined removal technology · Ultra low emission

T. Li (✉) · N. Wang · S. Yang
Coal Science and Technology Research Institute, Beijing 100013, China
e-mail: tingli_2008@126.com

State Key Laboratory of High Efficient Mining and Clean Utilization of Coal Resources, Beijing 100013, China

National Energy Technology & Equipment Laboratory of Coal Utilization and Emission Control, Beijing 100013, China

J. Lyu and S. Li (eds.), *Clean Coal and Sustainable Energy*, Environmental Science and Engineering,
https://doi.org/10.1007/978-981-16-1657-0_62

62.1 Introduction

As the haze weather in many cities of China has been increasing in recent years, the environmental protection pressure has risen sharply, and the government's policy on the atmospheric pollutants control has also become increasingly strict. As one of the atmospheric pollutants sources, pollutants from the flue gas of industrial boilers must be controlled within ultra-low emissions imperatively. Compared with large-scale power station boilers, industrial boilers, with smaller capacity and dispersed location distribution, have many difficulties to achieve ultra-low emissions of flue gas. According to the previous practical experience, intensive and integrated flue gas multi-contaminant synergistic removal technology (Zhang et al. 2006; Mok and Lee 2006; Hutson et al. 2008; Pourmohammadbagher et al. 2011) is the development direction to achieve ultra-low emission of such boilers.

By analyzing the characteristics of flue gas in the tail of industrial boilers, a low temperature flue gas deep purification technology based on semi-dry method was proposed by China Coal Research Institute. In this technology system, high-rate calcium-based ash desulfurization technology (Li and Zhang 2015) is combined with low-temperature denitrification technology, and SOx, NOx and dust can be removed to ultra-low emissions in a integrated device.

Simultaneous removal of sulfur dioxide and nitrogen oxides using the calcium-based absorbent in a fluidized bed reactor is the key technology in this technology system. At present, the semi-dry integrated desulfurization and dust removal in circulating fluidized bed/flowing bed using calcium-based absorbent is a mature technology, and it has been utilized in many engineering cases (Wang et al. 2000, 2011). While, researches and engineering cases on the integrated removal technology with semi-dry desulfurization and oxidative denitrification are quite insufficient. Investigated in laboratory and verified on industrial equipment, in this article.

62.2 Combined Removal Technology of Multi-pollutant with Circulating Calcium-Based Ash

As shown in Fig. 62.1, the principle of Combined Removal Technology of Multi-Pollutant is illustrated as follows. NO in flue gas is oxidized to NO_2 by using strong oxidants, and then NO_2 with high reactive activity is introduced into the tubular dense phase flow bed reactor. Under the humid and cooling condition in the reactor, SO_3, SO_2, NO_2 and other pollution are removed by the absorption of active calcium-based material. Combined NO pre-oxidation with semi-dry removal technology, desulfurization rate can reach to 90%, denitration rate can reach to 60%.

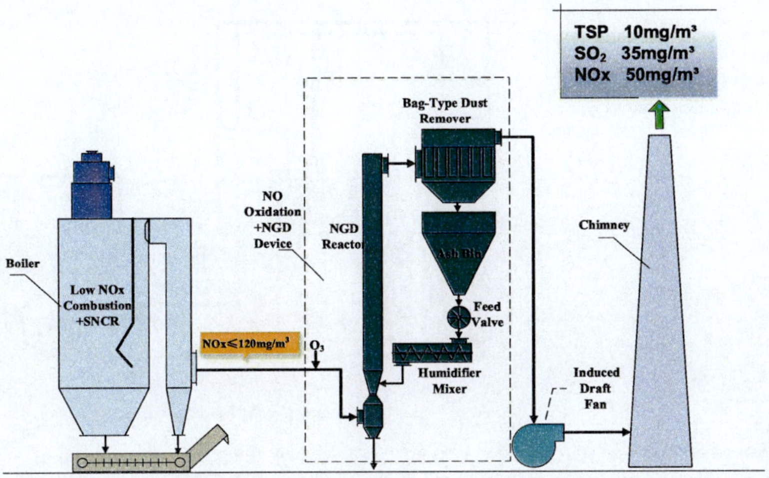

Fig. 62.1 Flow chart of combined removal technology of ash and calcium circulating pollutants

62.3 Experiment

Chemical reaction of $Ca(OH)_2$ and NO_2 is expressed as follows:

$$3NO_2 + H_2O = 2HNO_3 + NO \tag{62.1}$$

$$4NO_2 + O_2 + 2H_2O = 4HNO_3 \tag{62.2}$$

$$2NO_2 + H_2O = HNO_3 + HNO_2 \tag{62.3}$$

$$2HNO_3 + Ca(OH)_2 = Ca(NO_3)_2 + 2H_2O \tag{62.4}$$

$$2HNO_2 + Ca(OH)_2 = Ca(NO_2)_2 + 2H_2O \tag{62.5}$$

In this article, the adsorption characteristics of $Ca(OH)_2$ to NO_2 in fixed bed and bubbling bed reactors under low temperature conditions were respectively investigated. The structure of fixed bed absorption device is shown in Fig. 62.2a. The device is consist of a gas mixing preheater, an absorption reactor and a steam generator. The simulated flue gas is humidified by bubbling method. Calcium hydroxide powder, with more than 90% active content, is used as the absorbent. Quartz sand with 1 mm average particle size is used as inert dispersant. The structure of the bubble bed absorption device is shown in Fig. 62.2b. Calcium hydroxide suspension is used as absorbent. The gas used in this experiment is high-purity N_2, high-purity O_2, and

Fig. 62.2 Schematic diagram of NO_2 calcium based absorption experimental device

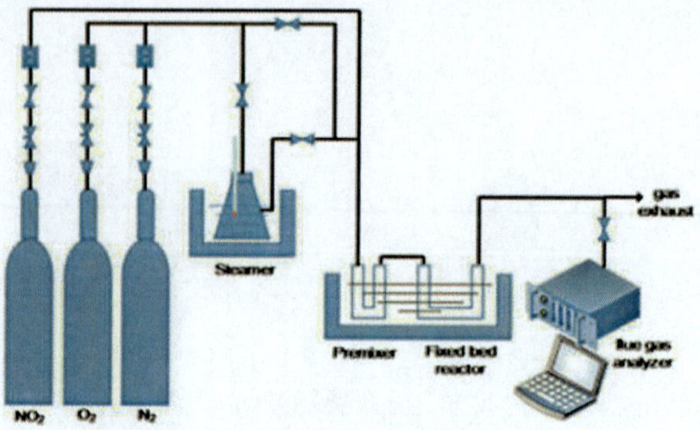

(a) fixed bed device

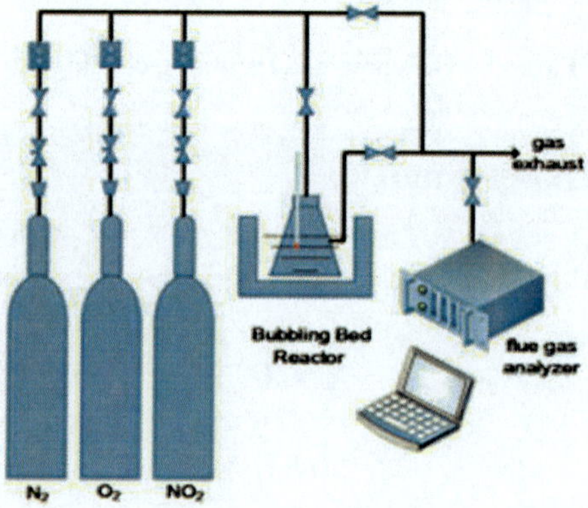

(b) bubble bed device

1% NO_2 (N_2 is a balance gas). The pollutant gas composition is measured by Detu 350 flue gas analyzer.

In order to verify the actual absorption effect of the calcium-based absorbent to NO_2 under the semi-dry method condition, industrial tests were carried out on a flue gas purification device matching pulverized coal boiler. Limited by site conditions, sodium chlorite solution is selected as the oxidant instead of ozone. Fly ash blended with a portion of $Ca(OH)_2$ was used as absorbent. Figure 62.3 is the schematic view of the oxidative denitration test device. The NOx concentrations at the inlet and outlet of the reactor were simultaneously measured by Detu flue gas analyzers, and converted to 9% of the reference oxygen. Since the sodium chlorite solution is easily decomposed in the pre-exposure, it was directly injected into the reactor for reaction. NO oxidation and absorption reactions were almost simultaneously performed. In this article, the approximate absorption rate and total denitration rate of NO_x were calculated according to the NO_2 conversion concentration at the inlet and outlet, shown as follows:

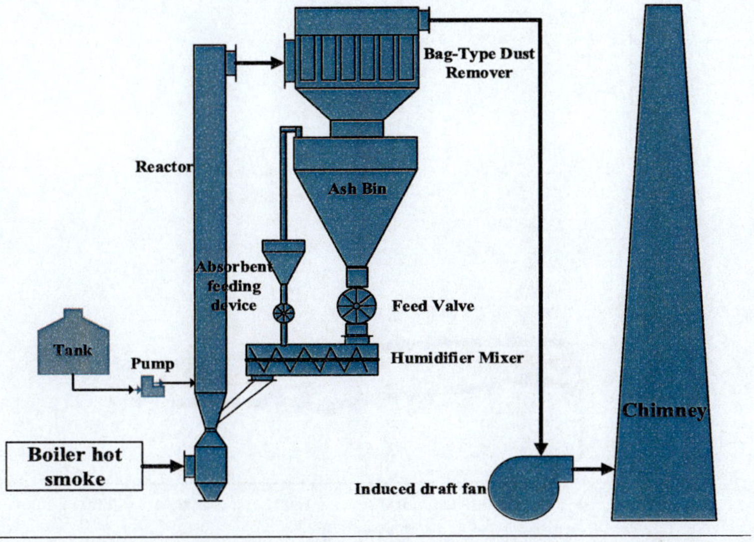

Fig. 62.3 Schematic diagram of oxidation denitrification test device

$$\eta_a = 100 \times \frac{1 - C_{outNO2}}{C_{inNO} - C_{outNO}} \quad (62.6)$$

$$\eta = 100 \times \frac{C_{inNOx} - C_{outNOx}}{C_{inNOx}} \quad (62.7)$$

η_a—NO_2 absorption rate, %; η—total denitration rate of NO_x, %; C_{inNO}/C_{inNOx}—reactor inlet conversion concentration of NO/NOx, mg/m^3; $C_{outNO}/C_{outNO2}/C_{outNOx}$—reactor outlet conversion concentration of NO/NOx, mg/m^3.

62.4 Results and Discussion

62.4.1 The Fixed Bed

The effects of relative humidity, reaction temperature, initial concentration of NO_2 on the absorption of NO_2 by calcium-based absorbent were investigated in fixed bed reactor. The composition of denitration product is measured by infrared spectrum. The experimental results are shown in Figs. 62.4, 62.5 and 62.6, and the infrared characterization results are shown in Fig. 62.7.

In the fixed bed reactor, water vapor condenses on the solid absorbent particles' surface to form a liquid film (Nelli and Rochelle 1996, 1998). The amount of

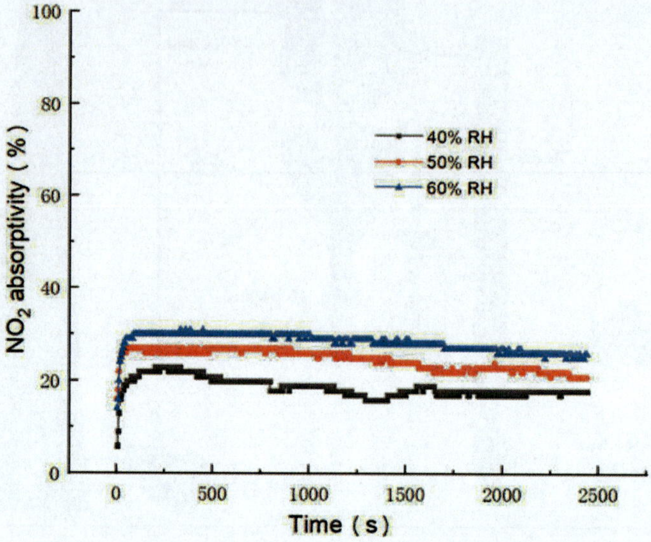

70 ℃, NO_2 200ppm, O_2 5%, N_2 1500ml, $Ca(OH)_2$ 1g

Fig. 62.4 Comparison of absorption effect of $Ca(OH)_2$ on NO_2 at different relative humidity

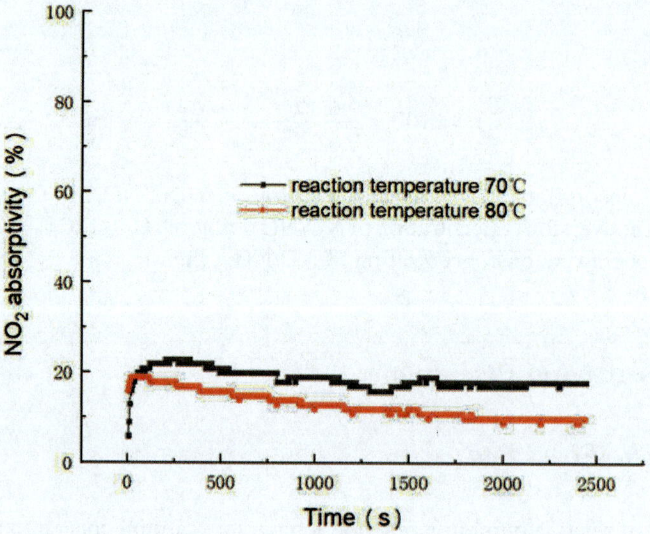

40% RH, NO_2 200ppm, O_2 5%, N_2 1500ml, $Ca(OH)_2$ 1g

Fig. 62.5 Comparison of absorption effects of $Ca(OH)_2$ on NO_2 at different temperatures

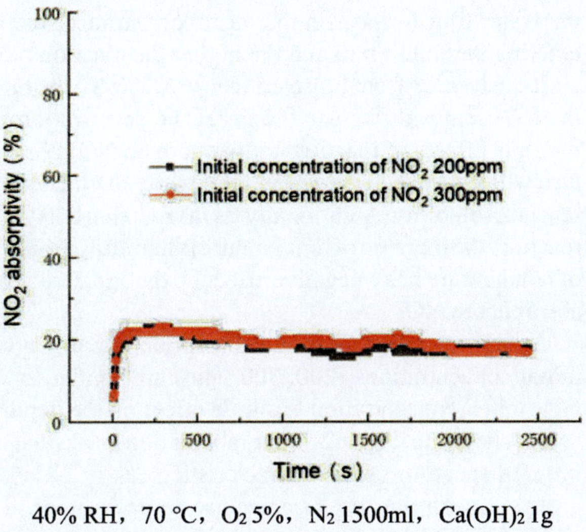

Fig. 62.6 Comparison of absorption effect of $Ca(OH)_2$ on NO_2 at different initial concentrations of NO_2

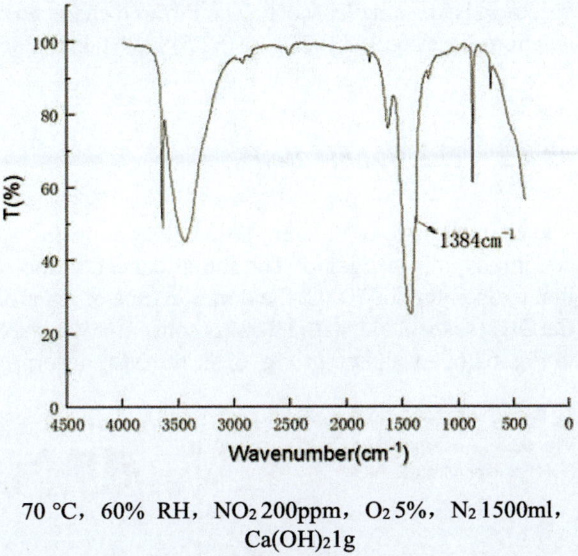

Fig. 62.7 IR results of denitration products

condensed water increases with increase of the relative humidity of the introduced gas. As NO_2 is soluble in water, the reaction form between NO_2 and $Ca(OH)_2$ can be converted from gas–solid heterogeneous reaction to ionic reaction in the liquid membrane. As shown in Fig. 62.4, when the relative humidity is in the range of 50–60%, the absorption rate of $Ca(OH)_2$ to NO_2 increases with the increase of relative humidity. The reason for that is the larger the relative humidity and the area of the water film formed on the absorbent surface, the more the amount of reactants entering the liquid film and the higher the reaction rate.

It can be seen from Fig. 62.5 that at 70–80 °C, the higher the temperature, the lower the NO_2 removal rate and the faster the deactivation rate. The reason for that is the multiple effects of reaction temperature on NO_2 absorption reaction. High temperature will improve the reaction rate, while the increase in temperature decreases the reactant solubility. Additionally, as the reaction of NO_2 and $Ca(OH)_2$ is an exothermic reaction, the increase of temperature is harmful to the reaction. In general, the increase of temperature has a negative effect on the apparent absorption rate of calcium-based absorbent to NO_2.

As shown in Fig. 62.6, the absorption rate of $Ca(OH)_2$ to NO_2 at different NO_2 initial concentrations (200, 300 ppm) are similar to each other. It is indicated that NO_2 initial concentration has little effect on the denitration effect.

As shown in Fig. 62.7, the absorption peak that appears at 1384 cm^{-1} in the infrared spectrum is the characteristic peak of NO^{3-}/NO^{2-}. As the two functional groups' absorption peaks are close to each other, it is impossible to distinguish which functional group is the specific one. It is indicated that nitrate or nitrite exists in the denitration product. As the duration reaction is relatively short, the peak in the spectrum is at a smaller value. The infrared characterization results confirm that the reactions described as (62.4) or (62.5) have taken place.

62.4.2 The Bubble Bed

The characteristic of calcium-based absorption to NO_2 in bubbling bed under wet conditions is investigated. The initial concentration of NO_2 is 200 ppm, the reaction temperature of 70 °C, the duration time of reaction is 4000 s. The absorption of $Ca(OH)_2$ suspension, with 1% mass concentration, to NO_2 is ploted second by second in Fig. 62.6. As shown in Fig. 62.8, the NO_2 absorption rate at the reactor outlet is

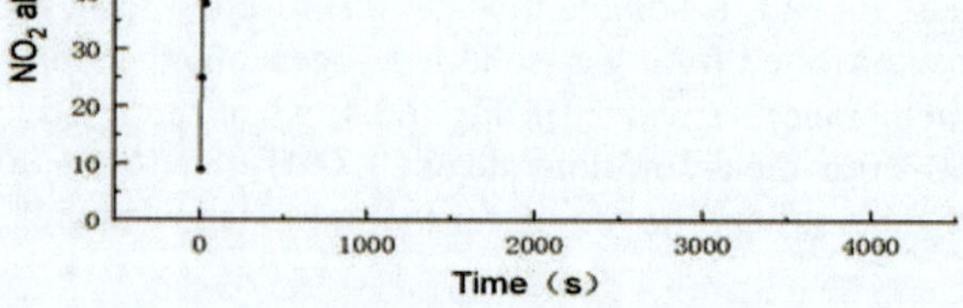

Fig. 62.8 Absorption of $Ca(OH)_2$ suspension to NO_2

nearly 100% firstly, and is continuously higher than 85% during the entire experimental process. The average removal rate of NO_2 is higher than 90%. Compared with the fixed bed reactor, the the absorption rate of calcium to NO_2 in the bubbling bed reactor is much higher. The reason for that is liquid water in the bubbling bed providers a contact medium, where the reaction between gaseous NO_2 and solid $Ca(OH)_2$ converts from gas–solid reaction with low reaction rate to ionization reaction with high reaction rate (Suchak et al. 1990). In a certain range, the increase of liquid water volume condensed at the solid $Ca(OH)_2$ partile surface can improve the absorption effected. Meanwhile, in the fixed bed reactor, the absorbent is in a stacked state which leads the poor mass transfer and a low absorption rate. NO has not been detected at the reactor outlet, which indicates that the alkaline conditions where the recation reacts effectively inhibits the occurrence of side reaction as follow:

$$3NO_2 + H_2O = NO + 2HNO_3 \tag{62.8}$$

62.4.3 *Industrial Test*

In the stable condition, calcium-based absorption to NO_2 and total NOx removal rate were obtained. The specific results are shown in Fig. 62.9. The test conditions are as follows: boiler load is 20 t/h, sodium chlorite solution concentration is 20% and its' flow rate is 135 L/h, import NO average conversion concentration is 226 mg/m^3, import NO_2 average conversion concentration is 11 mg/m^3, imported SO_2 average conversion concentration is 556 mg/m^3, the temperatures of the reactor inlet is 149 °C, the temperatures of the reactor outlet is 83 °C.

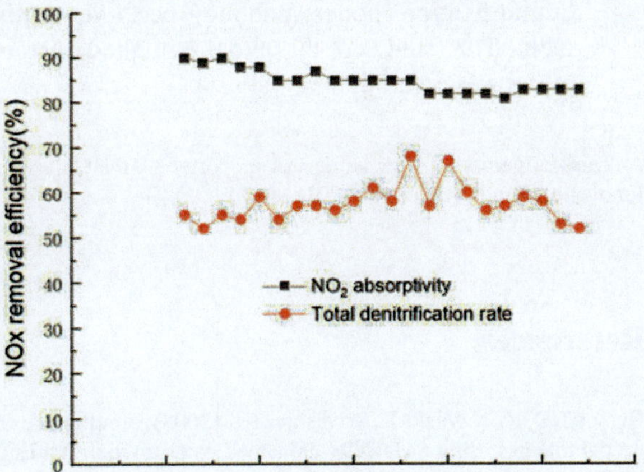

Fig. 62.9 Experimental results of denitrification of calcium based absorbent

As shown in Fig. 62.9, during the 20 min test period, the NO_2 absorption rate is maintained in the range of 81–90% and the total denitration rate is 52–68%. Combined with the results of basic experiments, it is found that the sequence of absorption rate to NO_2 by calcium-based absorbents is as follows: bubbling bed (wet method) $\geq$ flow bed (semi-dry method) > fixed bed (close to dry method).

This result is mainly affected by the way of humidification, that is, the presence and relative amount of liquid water are the key factors affecting the ionized absorption reaction.

62.5 Conclusion

The synergistic removal of multiple pollutants is a feasible technical path to achieve ultra-low emission of industrial boilers.

The main conclusions of this article are as follows:

(1) In fixed-bed reactor condition, the absorption rate of pure $Ca(OH)_2$ to NO_2 is 20–30% when the relative humidity is in the range of 40–60%, O_2 content 5%, the recation temperature is 70–80 °C. By infrared characterization results, nitrate or nitrite exists in the denitration product.

(2) In the bubbling bed reactor condition, the average denitration rate is above 90% when the inlet NO_2 concentration is 200 ppm, $Ca(OH)_2$ suspension concentration is 1%, O_2 content is 5%, and the recation temperature is 70 °C.

(3) Industrial test results show that when the concentration of sodium chlorite solution is 20%, the NO concentration is 226 ppm, and the temperature of the reactor outlet is 80°C, the NO_2 absorption rate is in the range of 81–90%, and the NOx removal rate is 52–68%.

(4) The sequence of absorption rate to NO_2 by calcium-based absorbent in different humidification modes: bubbling bed (wet method) $\geq$ airflow bed (semi-dry method) > fixed bed (close to dry method), semi-dry denitration effect is close to the wet method.

Acknowledgements This work was supported by the China Coal Science and Technology Group Innovation Fund Project (No. 2018MS003).

References

Hutson ND, Krzyzynska R, Srivastava RK (2008) Simultaneous removal of SO_2, NOx, and Hg from coal flue gas using a $NaClO_2$-enhanced wet scrubber. Ind Eng Chem Res 47(16):5825–5831

Li T, Zhang X (2015) Integrated technology of circulating flue gas desulfuration and dust removal by active calcium in fly ash. Clean Coal Technol 21(2):69–71

Mok YS, Lee H-J (2006) Removal of sulfur dioxide and nitrogen oxides by using ozone injection and absorption-reduction technique. Fuel Process Technol 87(7):591–597

Nelli CH, Rochelle GT (1996) Nitrogen dioxide reaction with alkaline solids. Ind Eng Chem Res 35:999–1005

Nelli CH, Rochelle GT (1998) Simultaneous sulfur dioxide and nitrogen dioxide removal by calcium hydroxide and calcium silicate solids. Air Waste Manage Assoc 48:819–828

Pourmohammadbagher A, Jamshidi E, Ale-Ebrahim H et al (2011) Study on simultaneous removal of NOx and SO_2 with $NaClO_2$ in a novel swirl wet system. Ind Eng Chem Res 50(13):8278–8284

Suchak NJ, Jethani KR, Joshi JB (1990) Absorption of nitrogen oxides in alkaline solutions: selective manufacture of sodium nitrite. Ind Eng Chem Res 29:1492–1502

Wang A, Qi H, You C et al (2000) Experimental study of FGD technology by CFB at 400–800 °C. J Combust Sci Technol 64(4):351–355

Wang Y, Wang H, Duan Y (2011) Experimental research of NID semi-dry desulphurization of flue gas. Boiler Technol 42(3):72–75

Zhang H, Tong H, Wang S et al (2006) Simultaneous removal of SO_2 and NO from flue gas with calcium-based sorbent at low temperature. Ind Eng Chem Res 45(18):6099–6103

Chapter 63
Selective Catalytic Reduction of NO_x with NH_3 Using Coal Ash Catalyst

Shagufta Fareed, Enlu Wang, Naveed Husnain, Kai Li, and Deli Li

Abstract The selective catalytic reduction (SCR) of NO_x with NH_3 has been investigated with coal ash. Coal samples from different regions of the People Republic of China are taken and processed to be used as the SCR catalysts. The influences of different coal ash catalysts, as well as the influences of calcination temperatures on the NO_x conversion of catalysts, are analyzed. The methods of XRF, XRD, and BET are used to characterize the materials. It is found that the calcined sample at 800 °C with Ca: Fe: Al: Si ratio of 0.051: 0.123: 0.511: 1, exhibited good NH_3-SCR activity in a broad temperature range of 200–500 °C. As well, the influences of $[NH_3]/[NO]$ molar ratio and oxygen concentration on the SCR activity are also studied.

Keywords Coal ash · NO_x · Selective catalytic reduction

63.1 Introduction

NO_x emissions produced by the power plants, vehicles and diesel engines, are one of the major causes of air pollution (Liu et al. 2011). These emissions cause acidification of rain, ozone depletion and photochemical smog, which are hazardous for environment and the human health (Zhao et al. 2009). Up till now, a number of technologies have been developed to control the NO_x emissions, but SCR of NO_x with NH_3, is the most effective technology to remove NO_x emissions from the environment (Wang and Gui 2013). The removal efficiency of this technology mainly depends on the choice of the catalyst used in SCR (Chmielarz et al. 2005). Several types of catalysts namely zeolites (Long and Yang 2000; Li et al. 2010), noble metal oxides (Lee et al. 2008; Itoh et al. 2009) and transition metal oxide (Shi et al. 2013; Schneider et al. 1995), have been discovered for this purpose. But V_2O_5/TiO_2 or V_2O_5–WO_3/TiO_2 catalysts are commonly used in SCR, which are efficient in SCR of NO_x with NH_3, and are good resistant to SO_2 poisoning. The operating temperature range for these catalysts is 350–450 °C. Due to high temperature range, the SCR reactor should be

S. Fareed · E. Wang (✉) · N. Husnain · K. Li · D. Li
School of Mechanical Engineering, Shanghai Jiao Tong University, Shanghai 200240, China
e-mail: elwang@sjtu.edu.cn

J. Lyu and S. Li (eds.), *Clean Coal and Sustainable Energy*,
Environmental Science and Engineering,
https://doi.org/10.1007/978-981-16-1657-0_63

placed upstream of the dust removal device and desulfurizing unit, to avoid reheating of flue gas. But this arrangement of SCR reactor causes catalyst deactivation due to high concentration of SO_2 and dust (Gao et al. 2013). Moreover, flue gas contains coal ash and sulphur dioxide which is dangerous for SCR catalysts. Therefore, the SCR catalyst is changed frequently to sustain the catalytic activity, which causes high operational cost (Ahonen 1997). So, the development of low cost catalyst is much needed.

Coal fly ash is a waste material released from coal fired power plants, which causes numerous problems like water, soil and environment pollution (Yao et al. 2015). So the dumping of coal fly ash is a major issue. Coal fly ash has good pore structure and large pore volume so, it is used in concrete production (Yao et al. 2015), and is also used to make zeolites because of its high alumina and silica contents (Steenbruggen and Hollman 1998).

Previously, a study has been reported on the utilization of fly ash as a catalyst support in SCR (Xuan et al. 2003). But coal ash catalyst is still not used directly in SCR technology. Inspired by this need, some kinds of coal ash as a catalyst in SCR have been investigated in this study. The catalytic activity of SCR catalyst was also examined in fixed bed quartz reactor. The properties of catalyst and crystalline structures were characterized by X-ray fluorescent spectrometry (XRF), specific surface area Brunauer–Emmett–Teller (BET) analysis and X-ray powder diffraction (XRD). In addition, the effects of calcination temperature, $[NH_3]/[NO]$ molar ratio, and oxygen concentration on the SCR activity has been also studied.

63.2 Experimental Section

63.2.1 Catalyst Preparation

Coal samples from different regions (Anhui, Xinhua, Mangolia, Shanxi, and Xinjiang) of People Republic of China were taken and used as catalysts. The treatment of samples was as follows: initially, the samples were ground by using ZM-200 Ultra Centrifugal Mill (Retsch, Germany) to get 150-mesh powders. After getting this powder, these ground samples were calcined at 800 °C for 2 h in electric heated furnace for making ash. Coal samples from Anhui region were also calcined at 600, 1000, 1200 and 1800 °C for 2 h in heated furnance to investigate the effect of calcination temperature on catalysts.

63.2.2 Catalyst Characterization

XRF-1800 sequential X-ray fluorescence spectrometer (Shimadzu, Japan) was used to obtain elemental composition of coal ash catalyst.

Brunauer–Emmett–Teller (BET) specific surface area was measured from the N_2 adsorption at 77 k by using Autosorb-IQ3 (Quantachrome; Anton Paar, Austria) analyzer. After the analysis 0.09–0.1 g of catalyst was heated at 105 °C for 2 h below N_2 atmosphere.

X-ray powder diffraction (XRD) measurements were carried out on D8 Advance X-ray diffractometer (Bruker, Germany) with CuKα radiation at room temperature. The scan rate of diffraction pattern was 1° min^{-1}, with a resolution of 0.02° and the diffraction pattern was taken in a 2θ range of 10–90°.

63.2.3 Catalysts Activity Measurement

Catalytic performance test was evaluated in fixed-bed quartz reactor with 6 mm i.d × 500 mm length at atmospheric pressure. The schematic diagram of the experimental setup is shown in Fig. 63.1.

The experimental system is consisted of four units: a simulated flue gas system, an electrically heated system, a fixed-bed quartz reactor, and a flue gas analyzer system. For each test, 3 g of catalyst was loaded in vertically quartz reactor and was heated at the required reaction temperature by an electric heated furnace. NO_x concentrations at inlet and outlet were measured by an FT-IR flue gas analyzer (DX-4000, Gasmet, Finland). The gas mixture was consisted of 600 ppm NO, 600 ppm NH_3, 3 vol% O_2, and N_2 balance (with a corresponding gas hourly space velocity (GHSV) of 3,000 h^{-1}). Mass flow controllers (MFCs) (CS200A, CS200D, Sevenstar, China) were used to control the flow of simulated flue gas.

Catalytic performance test was calculated by NO_x conversions, which was expressed as following equation:

$$\eta = \frac{[NO_x]_{in} - [NO_x]_{out}}{[NO_x]_{in}} \times 100\% \quad (63.1)$$

where $[NO_x] = [NO] + [NO_2]$, η represents the NO_x conversion (%), $[NO_x]_{in}$ and $[NO_x]_{out}$ denotes the concentrations at inlet and outlet of the reactor.

Fig. 63.1 Schematic diagram of experimental system

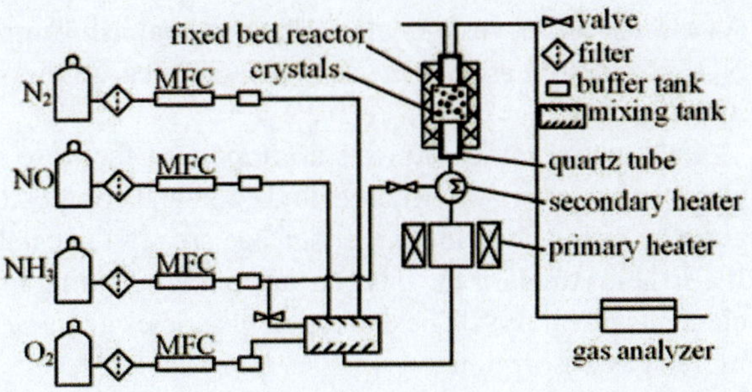

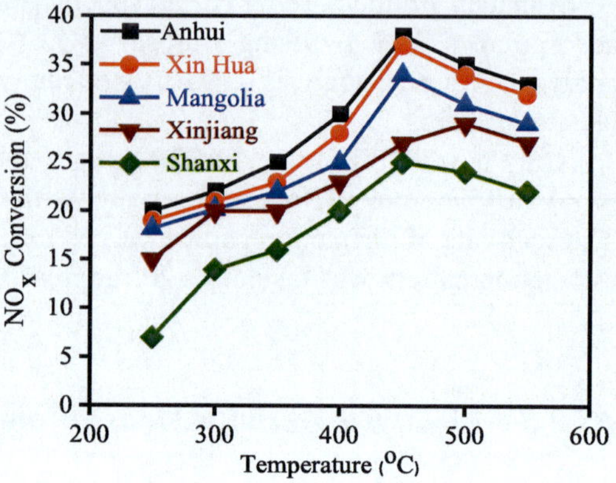

Fig. 63.2 Catalytic activity of thermally treated different coal samples at 800 °C

63.3 Results and Discussions

63.3.1 Catalyst Activity Test

Figure 63.2 shows the NO_x conversion of different coal samples calcined at 800 °C. It is clear from Fig. 63.2 that the NO_x conversion has been increased by increasing the temperature. At 250 °C, the coal ash sample from Anhui region gives 20% NO_x conversion, while Xinhua and Mangolia coal ash samples show approximately 19%. Other coal ash samples show NO_x conversion below than 20% at the temperature of 250 °C. As the temperature reaches 450 °C, the NO_x conversion of Anhui coal ash sample increases significantly and reaches at 38%. Beyond 450 °C, the NO_x conversion decreases gradually and gives 33% NO_x at 550 °C. Other coal ash samples have different NO_x conversion due to different elemental composition, structure and surface properties. It can be concluded that catalytic activity of coal ash sample from Anhui region is best among the selected samples.

63.3.2 Effect of Calcination Temperature

As discussed above, it is clear that the coal ash sample of Anhui region gives best NO_x conversion among the selected samples, so this catalyst is selected for further investigation.

Calcination temperature is an important factor to enhance the catalytic activity. Coal ash has complex structure and chemical composition. Thus, calcination temperature helps in achieving larger surface area of the catalyst for SCR catalytic activity. It also helps to make the oxygen vacancies by aiding in losing oxygen from the active elements of catalyst. These oxygen vacancies generate active sites which are helpful in flue gas adsorption.

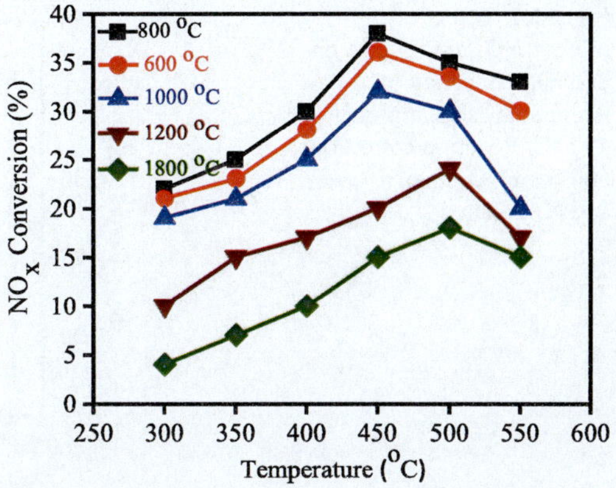

Fig. 63.3 Effect of calcination temperature on de-NO_x efficiency of Anhui coal ash sample. Reaction conditions: 600 ppm NO, 600 ppm NH_3, 3 vol% O_2, and N_2 balance

Figure 63.3 illustrates the NO_x conversion of Anhuicoal ash samples calcined at different calcination temperatures. Coal ash sample calcined at 800 °C has better NO_x conversion than the samples calcined at 600, 1000, 1200 and 1800 °C. At 300 °C, the coal ash sample (800 °C) gives approximately 22% NO_x and increases gradually up till 450 °C. At 450 °C it shows the maximum NO_x conversions, which are about 38% and it decreases for temperatures beyond this point. If the sample is calcined at low temperature it will not be helpful to get good surface performance. However, the higher calcination temperature can change the crystalline shape of active elements due to sintering of the catalyst and is also harmful for oxygen vacancies (Ruan 2001).

63.3.3 *Effect of [NH_3]/[NO] Molar Ratio on SCR Activity*

[NH_3]/[NO] molar ratio is an essential factor for the NO_x elimination in SCR. To obtain the influence of [NH_3]/[NO] molar ratio on NO_x elimination, several experiments are completed at different [NH_3]/[NO] molar ratios at 450 °C. The influence of [NH_3]/[NO] molar ratio on catalytic activity is presented in Fig. 63.4. It is clear from the figure that NO_x conversion of catalyst increases suddenly when the molar ratio of [NH_3]/[NO] is less than 1 and changes slowly at higher molar ratio of [NH_3]/[NO]. So, some increment in molar ratio is used for catalyst safety. 1.2 molar ratio of [NH_3]/[NO] is appropriate for good NO removal and there is no need to increase the molar ratio beyond 1.2.

From previous research (Lázaro et al. 2006; Mochida et al. 2000; Boyano et al. 2008), it can be deduced that the chemical reaction of NO and NH_3 can be divided into two steps. In first step, the reaction was performed in gas phase; while in the second step, the reaction was performed on the catalyst surface. The chemical reaction of NH_3 and NO is slow in the gase phase so the main focus of the reaction is on the catalyst surface. Firstly, the NH_3 was absorbed on the catalyst surface and then the reaction of NO and NH_3 took place on the surface. When NH_3 concentration is

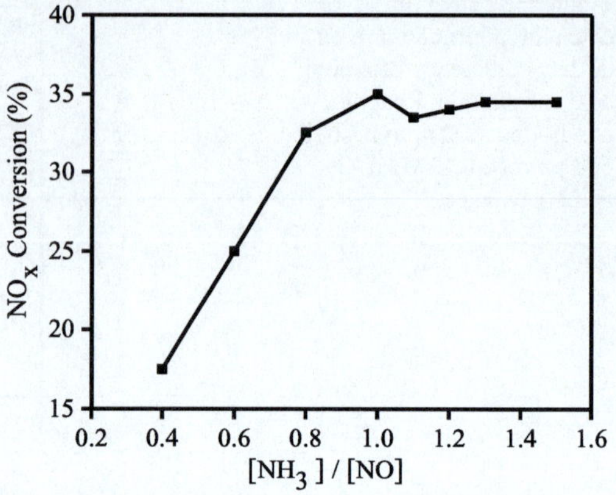

Fig. 63.4 Effect of $[NH_3]/[NO]$ molar ratio on de-NO_x efficiency of Anhuicoal ash sample at 450 °C. Reaction conditions: 600 ppm NO, 3 vol% O_2, and N_2 balance

increased, the more amount of NH_3 is adsorbed on the catalyst surface and more NH_3 reacts with NO. Hence, the adsorption positions on the catalyst surface are limited so the concentration of NH_3 is also limited. Thus, more concentration of NH_3 cannot increase the NO_x conversion.

63.3.4 Effect of Oxygen Concentration on SCR Activity

Oxygen concentration has a significant effect on the catalyst activity in SCR. Figure 63.5 shows the effect of O_2 concentration on de-NO_x efficiency of Anhui coal ash sample. It is clear from the figure that the NO_x conversion of coal ash catalyst increases with the increasing O_2 concentration, particularly when the O_2 concentration is below 3%. At 450 °C, 29% NO_x conversion is achieved by using 0.5% oxygen concertation. At 3% O_2 concentration, it gives approximately 35%

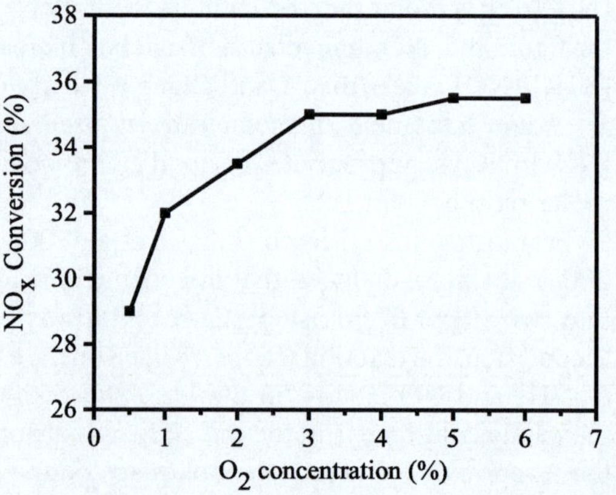

Fig. 63.5 Effect of O_2 concentration on de-NO_x efficiency of Anhui coal ash sample at 450 °C. Reaction conditions: 600 ppm NO, 600 ppm NH_3, and N_2 balance

Table 63.1 Main components of the different coal samples

Coal	SiO_2	Al_2O_3	Fe_2O_3	CaO
Xin Hua	32.10	10.34	11.65	25.52
Mangolia	55.69	4.60	9.79	3.80
Anhui	59.54	30.46	7.35	3.05
Shanxi	52.21	28.94	4.19	5.95
Xinjiang	43.24	18.11	7.39	11.18

NO_x conversion. After 3% O_2 concentration, there is no substantial further increase in the NO_x conversion. It is proved that the O_2 concentration improves the catalytic activity in SCR and the suitable range of O_2 concentration for catalyst is 3% under the given conditions.

63.3.5 Catalyst Characterization

63.3.5.1 XRF

Elemental analysis of different coal samples which are thermally treated at 800 °C, is shown in Table 63.1. The main components of coal are SiO_2, Al_2O_3, Fe_2O_3 and CaO. The NO_x conversion of all the coal ash samples, are different from each other due to the different elemental composition. According to the above results, the coal ash sample from Anhui region shows the best catalytic activity in SCR. Anhui coal ash sample exhibits good results of NO_x conversion due to the higher ratio of SiO_2 and Al_2O_3 and the lower ratio of CaO. The higher ratio of CaO breaks the chemical structure of catalyst and reduces the activity of a catalyst (Li et al. 2016, 2017). So the structure of this catalyst is decided by SiO_2 and Al_2O_3. While other coal ash samples follow the rule of higher the Fe contents, better will be the SCR performance (Yao et al. 2015).

63.3.5.2 BET Analysis

BET surface area has strong effect on catalytic activity (Wan et al. 2014), and calcination temperature greatly affects the surface area of the sample (Tang et al. 2007). Table 63.2 shows the BET surface analysis of the coal ash samples. It shows the BET surface area of not calcined sample and calcined samples at particular temperatures. From the table, it has been observed that the calcined samples show higher specific surface area than the samples which are not calcined. Furthermore, specific surface area decreases with the increasing of calcination temperatures. The coal samples at 600 and 800 °C, show large BET surface areas with a small difference among them. While, the samples calcined at higher temperature has lower BET surface

Table 63.2 Anhui coal ash sample's BET surface area at different calcination temperatures

Calcination temp (°C)	BET surface area (m^2/g)
Not calcined	6.45
600	13.58
800	13.15
1000	10.32
1200	1.94
1800	1.12

area. According to Zhang et al. (2015), the higher calcination temperature decreases the BET surface area due to particle growth. In other words, the higher calcination temperature decreases the porosity of a catalyst. In (Singh et al. 2018), the powder coal is used in pulverized coal fired boiler and temperature range for coal combustion is taken 1300–1700 °C, and the porosity of fly ash obtained from the pulverized coal fired boiler is very low. While the temperature range for combustion of coal in circulating fluidized bed combustion boiler (CFBC) is 800–900 °C. Fly ash formed at this temperature gets separated from the flue gas in the cyclone. Most of the large particles of fly ash are recycled back to the furnace and about 50% of its small particles are collected by the precipitator. This fly ash has good surface area and porosity. So, it can be used as a catalyst in SCR and can give good NO_x conversion. Thus, it can be concluded that an extremely high calcination temperature decreases the SCR catalytic activity.

63.3.5.3 XRD

Calcination temperature has strong effect on the structure of catalyst (Liu et al. 2011). XRD pattern for calcined sample at 800 °C and not calcined sample, are shown in Fig. 63.6. Coal sample has a complex chemical composition and structure. It can be seen that coal sample contains different minerals and elements. Figure 63.6 shows the diffraction peaks of calcined and not calcined samples. The calcined sample has diffraction peaks at 20.82°, 26.62°, 31.43°, 33.15°, 36.51°, 39.43°, 42.32°, 45.02°, 50.04°, 54.78°, 69.88°, 68.13°, 75.48°, 81.34° and 83.31°. According to JCPDS-033-1161, the diffraction peaks of 20.82°, 26.62°, 45.02°, 54.74° show the presence of Quartz (SiO_2), while the diffraction peaks of 31.43°, 39.43° and 75.48° indicate the existence of Mullite ($Al_6O_{13}Si_2$) (JCPDS-001-0613). Other diffraction peaks of 50.04°, 81.34° and 83.31° display the appearance of Hematite (Fe_2O_3) (JCPDS-033-0664).

The sample without calcination shows the diffraction peaks at 12.1°, 19.8°, 20.30°, 20.8°, 25.2°, 26.6°, 30.6°, 34.9°, 35.8°, 39.4°, 42.29°, 45.68°, 50.04°, 54.96°, 59.83°, 62.13°, 68.17°, 73.28°, 75.4°, 79.6°, 81.31° and 83.60°. According to JCPDS-0331161, the diffraction peaks of 20.30°, 26.6°, 42.29°, 45.68°, 54.96° and 75.4°, show the presence of Quartz (SiO_2). The diffraction peaks of 20.8°, 25.2°, 35.8°,

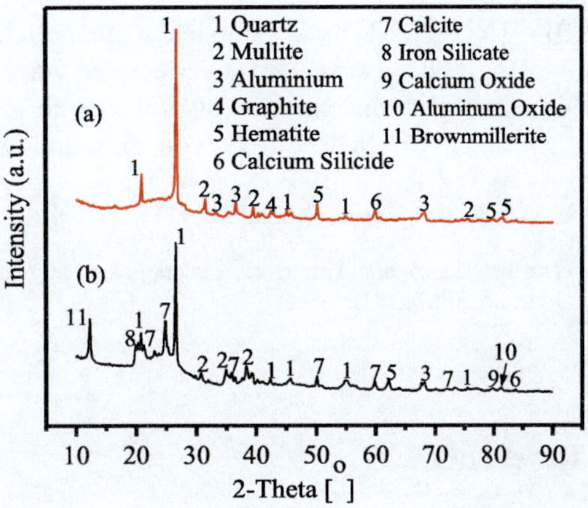

Fig. 63.6 XRD patterns of coal ash samples: **a** calcined at 800 °C and **b** not calcined

50.04° and 73.28°, show the Calcite ($CaCO_3$) (JCPDS-005-0586). The diffraction peak of 79.6° indicates the presence of Calcium Oxide (CaO) (JCPDS-001-1160).

XRD pattern of the not calcined sample shows that it contains most of the quartz (SiO_2), Calcite ($CaCO_3$) and Calcium oxide (CaO) peaks. From previous studies (Li et al. 2016, 2017), it has been reported that the calcium species are not good for the catalytic performance. It harms the catalyst and decreases its activity in SCR. So, the NO_x conversion of the not calcined sample is not good. After calcination, the main diffraction peaks are Mullite ($Al_6O_{13}Si_2$), Graphite and Hematite (Fe_2O_3). Hematite contents enhance the catalytic activity in the SCR, so it shows good NO_x conversion than the not calcined sample.

63.4 Conclusions

In the present study, coal ash samples from different regions of the People Republic of China are taken and processed to be used as a catalyst for the SCR of NO_x with NH_3. After investigating the results achieved from chemical and physical characterization of the coal samples, the conclusions were made as follow:

(1) Coal ash which is a solid waste and creates many environmental problems can be used as a catalyst in the SCR technology. Our experimental results indicate that the coal ash has a potential to give good NO_x conversion in SCR. The thermally treated coal sample from Anhui region with Ca: Fe: Al: Si ratio of 0.051: 0.123: 0.511: 1, exhibited good NO_x conversion in a broad temperature range of 200–500 °C.

(2) Calcination temperature had a strong effect in NH_3-SCR activity test. Higher calcination temperature decreases the BET surface area and reduces the activity of catalyst in the SCR.

(3) $[NH_3]/[NO]$ molar ratio had a great effect on the NO_x conversion in SCR. The best NO_x conversion was obtained when the $[NH_3]/[NO]$ molar ratio was 1.
(4) Oxygen concentration also played an important role to improve the catalytic activity in SCR. The O_2 concentration was 3 vol% for best NO_x conversion in the SCR.

Acknowledgements This work was supported by National Natural Science Foundation of China (Grant no. 50676057).

References

Ahonen M (1997) Long-time experience in catalytic flue gas cleaning and catalytic NO_x reduction in biofueled boilers. Fuel Energy Abstr 38(3):164

Boyano A, Lombardo N, Gálvez ME, Lázaro MJ, Moliner R (2008) Vanadium-loaded carbon-based monoliths for the on-board NO reduction: experimental study of operating conditions. Chem Eng J 144(3):343–351

Chmielarz L, Kustrowski P, Lasocha AR, Dziembaj R (2005) Selective oxidation of ammonia to nitrogen on transition metal containing mixed metal oxides. Appl Catal B 58:235–244

Gao R, Zhang D, Liu X, Shi L, Maitarad P et al (2013) Enhanced catalytic performance of V_2O_5–$WO_3/Fe_2O_3/TiO_2$ microspheres for selective catalytic reduction of NO by NH_3. Catal Sci Technol 3:191

Itoh M, Saito M, Takehara M, Motoki K, Iwamoto J et al (2009) Influence of supported-metal characteristics on deNO_x catalytic activity over Pt/CeO_2. J Mol Catal A Chem 304:159–165

Lázaro MJ, Gálvez ME, Ruiz C, Juan R, Moliner R (2006) Vanadium loaded carbon-based catalysts for the reduction of nitric oxide. Appl Catal B Environ 68(3–4):130–138

Lee JH, Schmieg SJ, Oh SH (2008) Improved NO_x reduction over the staged Ag/Al_2O_3 catalyst system. Appl Catal A Gen 342:78–86

Li J, Zhu R, Cheng Y, Lambert CK, Yang RT (2010) Mechanism of propene poisoning on Fe-ZSM-5 for selective catalytic reduction of NO(x) with ammonia. Environ Sci Technol 44:1799–1805

Li X, Li X, Chen J, Li J, Hao J (2016) An efficient novel regeneration method for Ca-poisoning V_2O_5-WO_3/TiO_2 catalyst. Catal Commun 87:45–48

Li X, Li X, Yang R, Mo J, Li J, Hao J (2017) The poisoning effects of calcium on V_2O_5-WO_3/TiO_2 catalyst for the SCR reaction: comparison of different forms of calcium. Molecular Catal 434:16–24

Liu F, Asakura k, He H, Liu Y, Shan W (2011) Influence of calcination temperature on iron titanate catalyst for the selective catalytic reduction of NO_x with NH_3. Catal Today 164:520–527

Liu F, He H, Zhang C, Shan W, Shi X (2011) Mechanism of the selective catalytic reduction of NO_x with NH_3 over environmental-friendly iron titanate catalyst. Catal Today 175:18–25

Long RQ, Yang RT (2000) Characterization of Fe-ZSM-5 catalyst for selective catalytic reduction of nitric oxide by ammonia. J Catal 194:80–90

Mochida I, Shirahama N, Kawano S, Korai Y, Yasutake A et al (2000) NO oxidation over activated carbon fiber (ACF). Part 1. Extended kinetics over a pitch based ACF of very large surface area. Fuel 79(14):1713–1723

Ruan HD, Frost RL, Kloprogge JT (2001) The behavior of hydroxyl units of synthetic goethite and its dehydroxylated product hematite. Spectrochim Acta Part A Mol Biomol Spectrosc 57(13):2575–2586

Schneider H, Maciejewski M, Kohler K, Wokaun A, Baiker A (1995) Chromia supported on Titania: VI. Properties of different chromium oxide phases in the catalytic reduction of NO by NH_3 studied by in Situ diffuse reflectance FTIR spectroscopy. J Catal 157(2):312–320

Shi YN, Chen S, Sun H, Shu Y, Quan X (2013) Low-temperature selective catalytic reduction of NO_x with NH_3 over hierarchically macro-mesoporous Mn/ TiO_2. Catal Commun 42:10–13

Singh A, Sharma V, Mittal S, Pandey G, Mudgal D, Gupta P (2018) An overview of problems and solutions for components subjected to fireside of boilers. Int J Indus Chem 9:1–15

Steenbruggen G, Hollman GG (1998) The synthesis of zeolites from fly ash and the properties of the zeolite products. J Geochem Explor 62(1–3):305–309

Tang X, Hao J, Xu W, Li J (2007) Low temperature selective catalytic reduction of NO_x with NH_3 over amorphous MnO_x catalysts prepared by three methods. Catal Commun 8:329–334

Wan YP, ZhaoWR, Tang Y, Li L, Wang HJ, Cui YL et al (2014) Ni-Mn bi-metal oxide catalysts for the low temperature SCR removal of NO with NH_3. Appl Catal B Environ 148:114–122

Wang X, Gui K (2013) Fe_2O_3 particles as superior catalysts for low temperature selective catalytic reduction of NO with NH3. J Environ Sci 25(12):2469–2475

Xuan X, Yue C, Li S, Yao Q (2003) Selective catalytic reduction of NO by ammonia with fly ash catalyst. Fuel 82:575–579

Yao Z, Ji X, Sarker P, Tang J, Ge L et al (2015) A comprehensive review on the applications of coal fly ash. Earth Sci Rev 141:105–121

Zhang X, Lv S, Jia X, Wu X, Meng F (2015) Low-temperature selective catalytic reduction of NO with NH_3 over manganese oxides supported on fly ash-palygorskite. J Thermodyn Catal 6:3

Zhao Q, Xiang J, Sun L, Su S, Hu S (2009) Adsorption and oxidation of NH_3 and NO over sol-gel-derived CuO-CeO_2-MnOx/y-Al_2O_3 catalysts. Energy Fuels 23(3):1539–1544

Chapter 64
Effects of Different Precipitants on the De–NO Efficiency of the Fe_2O_3 Catalyst Synthesized by Co-precipitation Method

Naveed Husnain, Enlu Wang, Shagufta Fareed, Kai Li, Deli Li, and Qi Wang

Abstract Fe_2O_3 catalysts synthesized by the co-precipitation method with two different precipitants (NH_4OH/Na_2CO_3) were experimentally investigated in the selective catalytic reduction (SCR) of NO with NH_3. It was found that the catalyst in which NH_4OH was used as precipitant exhibited high NO conversion (above 80% from 250–400 °C). XRD, BET, EDS, and FT-IR characterizations were done to examine the catalysts. The catalyst prepared by using NH_4OH precipitant exhibited a lower crystallization degree together with better pore structure, and an increase in the O/Fe ratio, which was helpful for the NH_3-SCR reaction. Also, the SO_2 tolerance of the catalyst was investigated. The results exhibited that the addition of SO_2 gradually decreased the SCR activity. FT-IR analysis of the catalyst exhibited that after the addition of SO_2 in the De–NO reaction the development of ammonium sulfate species on the surface of the catalyst caused pore plugging and was responsible for a reduction in the NH_3-SCR activity.

Keywords Selective catalytic reduction · NO · Co-precipitation · Fe_2O_3 · Precipitants

64.1 Introduction

(NO_x, x = 1, 2) has been credited to many serious environmental issues like fine particle pollution, ozone depletion, smog, and rain, etc. (Husnain et al. 2019a). Because of these hazardous characteristics, countries around the globe have shown keen interest in minimizing NO_x production by making strict legislations about NO_x (Busca et al. 1998; Heck 1999; Liu and Woo Ihl 2006). To implement these strict legislations, different technologies have been implemented for NO_x reduction. For the application in power plants powered by coal-burning, NH_3-SCR of NO_x has been

N. Husnain · E. Wang (✉) · S. Fareed · K. Li · D. Li · Q. Wang
Institute of Thermal Energy Engineering, Shanghai Jiao Tong University, Shanghai 200240, China
e-mail: elwang@sjtu.edu.cn

J. Lyu and S. Li (eds.), *Clean Coal and Sustainable Energy*, Environmental Science and Engineering,
https://doi.org/10.1007/978-981-16-1657-0_64

used worldwide due to its better efficiency as compared to other NO_x reduction techniques (Yang et al. 2012a; Yao et al. 2010; Gui et al. 2015). In NH_3-SCR different catalysts have been used. Among these, noble metal catalysts exhibit high NO reduction, but they have their drawbacks such as the high price, less SO_2 tolerance, and narrow effective temperature range (Kang et al. 2006; Roy et al. 2008; Busca and Larrubia 2005; Yadav and Prasad 2016). Although, the V_2O_5-WO_3/TiO_2 catalyst based SCR technology is commercially operational, yet it has some issues, such as the thin temperature range (300–400 °C), SO_3 formation due to the oxidation of SO_2 (Dunn et al. 1998), N_2O formation at higher temperatures (Husnain et al. 2019b), and V is poisonousness to the atmosphere as well (Ballen et al. 2009). This has resulted in a lot of focus to develop V-free NH_3-SCR catalysts (Husnain et al. 2019c; Wang et al. 2016; Yang et al. 2012b; Liu et al. 2016). Due to its prominent thermal stability and environmentally benign characteristics, a great interest is shown in the iron oxide-based catalysts in recent years for the reduction of NO_x with NH_3. Numerous synthesis methods have been reported for the Fe_2O_3 particles, but the most common method to synthesize the Fe_2O_3 particles is the co-precipitation method (Singh et al. 2008; Wu and Gao 2012; Lin et al. 2012).

Motivated by the idea to examine the influence of different precipitants on the De–NO efficiency of Fe_2O_3 catalysts, we prepared Fe_2O_3 catalysts by co-precipitation method by using different precipitants. One amino precipitant (NH_4OH) and one sodium metal precipitant (Na_2CO_3) was used to facilitate the precipitation process. XRD, BET, EDS, and FT-IR characterizations were done to examine the samples. It was found that the catalyst in which NH_4OH was used as precipitant exhibited lower crystallization degree, better pore structure, an increase in O/Fe ratio, and more sites of lattice oxygen were exposed to the crystal surface of the catalyst, which was helpful for the NH_3-SCR reaction. Also, the SO_2 tolerance of the catalyst was examined. The results displayed that the addition of SO_2 gradually decreased the SCR activity. FT-IR analysis of a fresh sample, as well as a sample after deactivation of the catalyst, was conducted. It was found that after introducing SO_2 in the reaction chamber proceeding De–NO reaction, the development of ammonium sulfate species on the surface of the catalyst caused pore plugging and was responsible for a reduction in the NH_3-SCR activity.

64.2 Catalyst Preparation

In this study, the catalysts were synthesized by the co-precipitation method. A detailed procedure is described as follows: 2.0 M aqueous solutions of ferric chloride ($FeCl_3$) and 1.0 M ferrous chloride tetrahydrate ($FeCl_2 \cdot 4H_2O$) were taken and pre-mixed. This resulting solution is then stirred at 25 °C for 60 min. To maintain the *p*H of the solution at 10–11, a definite quantity of the precipitator (NH_4OH/Na_2CO_3) was added to the solution along with continuous stirring (Singh et al. 2008). Then deionized water was used to wash the resulting precipitate several times and then it was filtrated under suction and then evaporated to dryness at 120 °C. At last, the particles were calcined at 300 °C in the presence of air for 4 h, and 60 mesh powders were made after crushing.

64.3 Characterization

X-ray powder diffraction (XRD) characterizations were done on a D8 Discover X-ray diffractometer (manufactured by Bruker, Germany) to find out the crystalline structures of the materials.

Brunauer–Emmett–Teller (BET) surface properties of the materials were observed by an Autosorb-IQ3 (Quantachrome; Anton Paar, Austria).

Energy dispersive spectrum analysis (EDS) of the materials was obtained by a Scanning electron microscope (SEM) machine model Sirion 200 manufactured by Thermo Fisher Scientific, USA.

Fourier transform spectrometer (Nicolet Nexus 470) was used to investigate the root cause of the SO_2 poisoning on the catalyst's surface with a recording range of 4000 cm^{-1}.

64.4 Experimental Setup

Our experimental system has a simulated flue gas setup with a horizontal fixed-bed reactor made of quartz, and a test rig that is electrically heated. The schematic of our experimental setup is shown in Fig. 64.1.

The equation used to measure the NO conversion of the catalysts is as follow:

$$\eta = \frac{[NO]_{in} - [NO]_{out}}{[NO]_{in}} \times 100\% \tag{64.1}$$

where η represents the NO conversion (%), $[NO]_{in}$ denotes the inlet concentrations of NO (ppm), and $[NO]_{out}$ denotes the outlet concentrations of NO (ppm).

64.5 Results and Discussion

64.5.1 NH_3-SCR Activity

Figure 64.2 shows the results of NO conversion of the Fe_2O_3 catalysts prepared by the co-precipitation method using different precipitants. It demonstrated that the NO conversion of the catalysts gradually increased with the rise in temperature from 100–175 °C. After that, the NO conversion increased sharply from 175–250 °C. It is

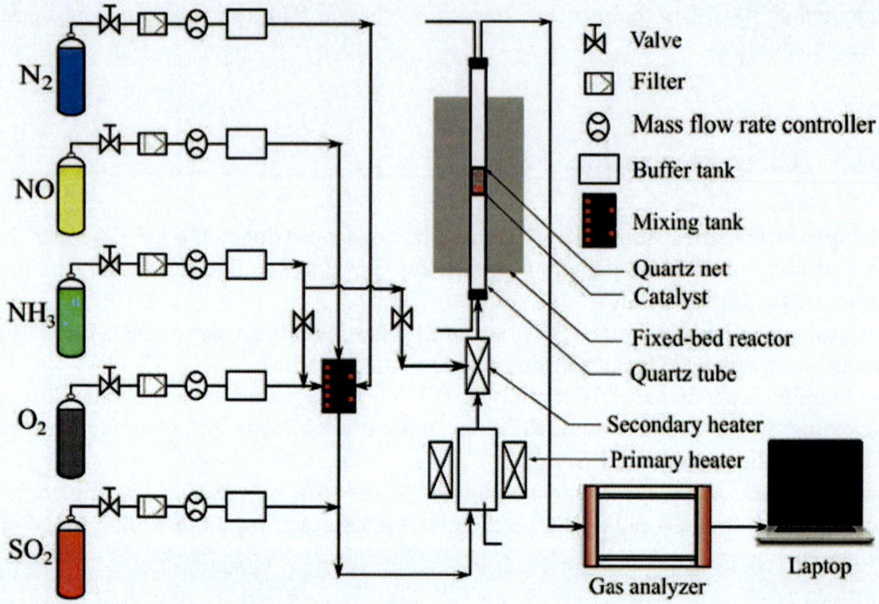

Fig. 64.1 Diagram of the experimental system

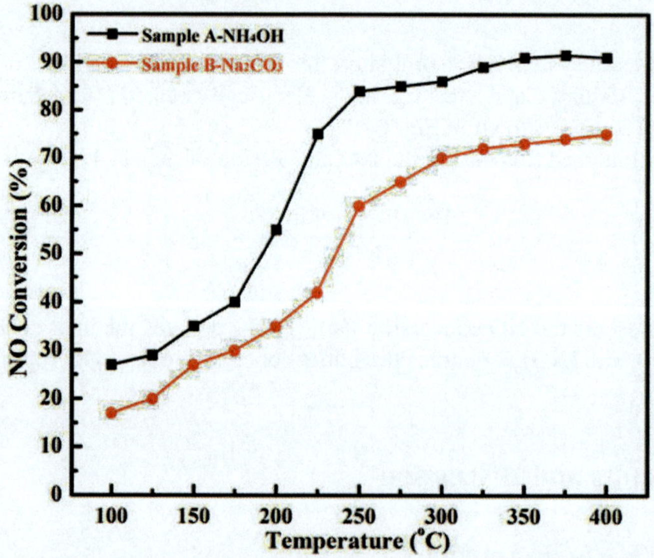

Fig. 64.2 NO conversion of the Fe_2O_3 catalysts. Reaction conditions: 500 ppm NO, 500 ppm NH_3, 3 vol% O_2, N_2 balance, and GHSV = 20,000 h^{-1}

known that the maghemite phase exhibits better NH_3-SCR activity than the hematite at lower temperatures. So, the presence of the maghemite phase might be the reason behind this sharp increase in NO conversion at low temperatures (Liu et al. 2013). Above 250 °C, the rise in the NO conversion is gradual, and the overall activity remained above 80% from 250–400 °C.

The influence of the precipitants on the De–NO efficiency of the Fe_2O_3 catalysts was also investigated in Fig. 64.2. By analyzing and comparing the results, it can be concluded that the characteristics of the synthesized Fe_2O_3 catalysts depend upon the cations and anions in the precipitants used, which affects the NO conversion of the catalysts. The sample A-NH_4OH exhibited better NO conversion than the sample B-Na_2CO_3. At 250 °C, the NO conversion by the sample A-NH_4OH (amino precipitant) with OH^- anions was approximately 28% higher than that of sample B-Na_2CO_3 (sodium metal precipitant) containing $CO_3{}^{2-}$ anions.

64.5.2 *Crystalline Phase Analysis (XRD)*

Figure 64.3 illustrates the X-ray diffractions of the samples manufactured by different precipitants (NH_4OH, and Na_2CO_3). The results demonstrated that the main phase in both the sample A-NH_4OH and sample B-Na_2CO_3 were α-Fe_2O_3 (hematite) and γ-Fe_2O_3 (maghemite). The diffraction peaks of the catalysts, where 2θ equals

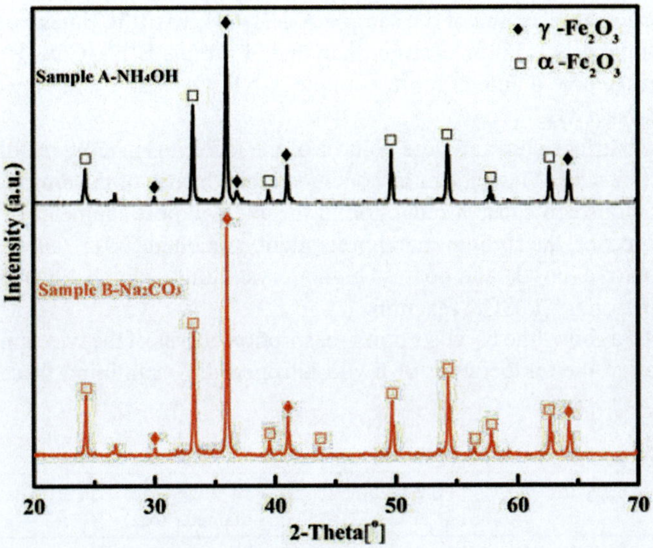

Fig. 64.3 XRD spectra of the Fe_2O_3 catalysts

30.2°, 35.7°, 37.25°, 40.2°, and 64.2° showed the presence of γ-Fe_2O_3 (JCPDS 39-1346), and 2θ equals 24.2°, 33.2°, 39.4°, 43.6°, 49.6°, 54.2°, 56.3°, 57.6°, and 62.6° correspond to the diffraction planes of α-Fe_2O_3 (JCPDS 89-8104).

The Scherrer equation is shown in the following.

$$D = 0.94 \times \lambda/(\beta \times \cos\theta) \quad (64.2)$$

where D is the average diameter of particles, β represents the half-width of diffraction peak, λ is the wavelength of incident X-ray, and θ represents the diffraction angle.

It can be seen that both the catalysts contain sharp crystalline peaks which indicate the good crystallinity in the catalysts. However, the intensities of the sample B-Na_2CO_3 were found higher than the sample A-NH_4OH, as well it contained sharper peaks. It is evident from Eq. (64.2) that the crystallite size of a catalyst is inversely related to the half-width of the diffraction peak. Thus, the crystallinity of sample A-NH_4OH is less than that of the sample B-Na_2CO_3. This could be due to the development of α-Fe_2O_3. The afore-mentioned factors resulted in the decrease in the catalytic activity of the sample B-Na_2CO_3.

64.5.3 Microstructure Analysis (BET)

Table 64.1 illustrates the textural properties of the Fe_2O_3 catalysts. It could be seen that the specific surface area of the sample A-NH_4OH was 1.36 times larger, as well the pore volume was 1.54 times larger than that of sample B-Na_2CO_3. However, the value of average pore diameter for the sample A-NH_4OH was 1.4 nm lower than that of sample B-Na_2CO_3.

The larger surface area and pore volume of the materials usually resulted in more available active sites. Thus, it can be concluded that the use of the amino precipitant with OH^- anions can cause a reduction in the average pore diameter of the Fe_2O_3 catalyst. Moreover, the sodium metal precipitant contained ${CO_3}^{2-}$ anions promotes the formation of α-Fe_2O_3 and possess higher crystallinity, which adversely affected the NO reduction in the SCR reaction.

Figure 64.4a shows the N_2 adsorption–desorption curves of the two samples which are measured at the temperature of liquid nitrogen. By examining these isotherms

Table 64.1 Surface properties of the Fe_2O_3 samples

Samples	S_{BET} ($m^2\ g^{-1}$)	Pore volume ($cm^3\ g^{-1}$)	Average pore diameter (nm)	Particles size (nm)
Sample A-NH_4OH	29.4	0.17	19.8	102.0
Sample B-Na_2CO_3	21.6	0.11	21.2	127.2

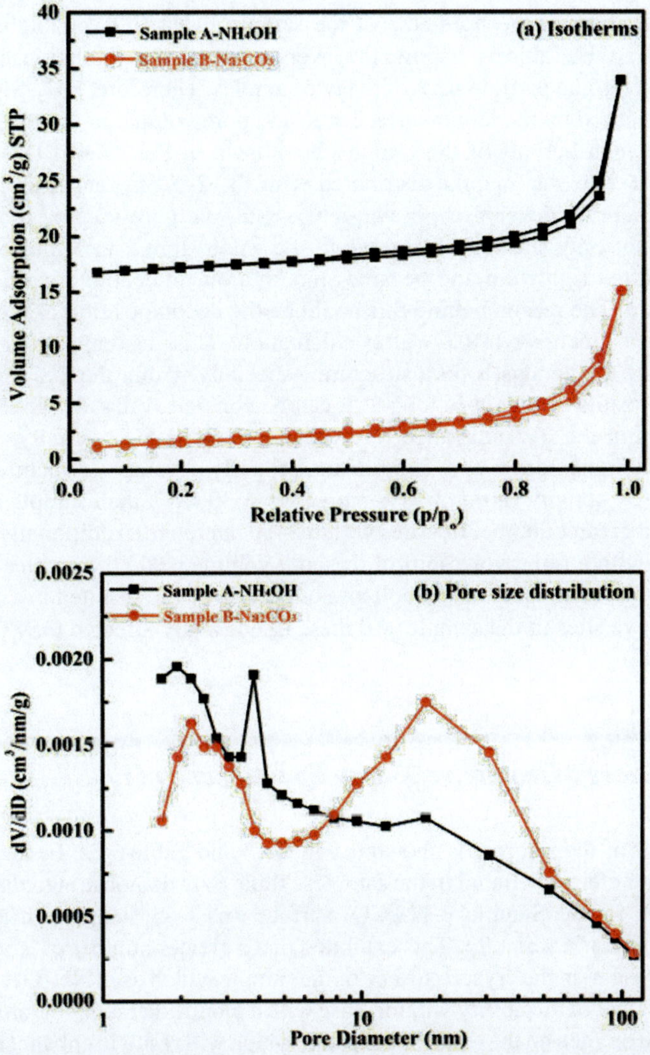

Fig. 64.4 **a** N_2 adsorption–desorption and **b** distribution of pore size of the Fe_2O_3 catalysts

concerning the IUPAC classification, it was found that both the isotherms resembled type II isotherms (Xiong et al. 2016; Li et al. 2016). This type of isotherm represents the microporous materials. As well, the hysteresis loop was similar to the H3 type loop, which verified the presence of a slit-shaped pore structure within the catalyst (Xiong et al. 2016; Sing et al. 1985). The comparison of the closure points of the hysteresis loop concluded that those of the sample A-NH_4OH are at lower P/P_o than that of sample B-Na_2CO_3. This illustrated that a smaller number of micropores are present in the sample B-Na_2CO_3. Besides, higher P/P_o for the sample B-Na_2CO_3

depicted that the average pore size of the sample B-$Na_2CO_3$was higher than the sample A-NH_4OH. Table 64.1 shows the inversely proportional relationship between the surface areas and particle sizes of the two samples. Therefore, NH_3-SCR reaction could be benefited by the larger surface area and pore volume of the material.

Pore size distributions of the samples are shown in Fig. 64.4b. The sample A-NH_4OH pore size was mainly distributed with the 2–5 nm range, as well it also contains a properly developed pore structure. But, the pore volume of the sample B-Na_2CO_3 not only showed a decrease in the value from 2 to 5 nm, but also the peak has shifted rightward and became smooth a bit, indicating a reduction in its pore structure. The reason behind this could be the decomposition of the carbonate species in the precursor to CO_2 after calcination. This decomposition has a role in eliminating the catalyst's pore structure, especially within the 2–5 nm range. In light of the results shown in Fig. 64.2, it can be concluded that the greater number of pores within the 2–5 nm range resulted in enhanced SCR activity. Sample A-NH_4OH containing pore within 2–5 nm and properly developed structure displayed improved SCR activity (around 18% increased at 300 °C) than sample B-Na_2CO_3 comprising a greater number of pores within 2–50 nm range. Additionally, in sample B-Na_2CO_3, with a right word shift of the pore volume peak, the greater number of pores of 20–70 nm range, and the collapse of pore structure resulted in a decrease in available active sites in the sample. All these issues badly affected the SCR activity of the sample.

64.5.4 Energy Dispersive X-Ray Spectroscopy (EDS)

EDS analysis of the samples is shown in Fig. 64.5 and Table 64.2. Fe and O are the obvious major elements found in the samples. Table 64.2 demonstrated the atomicity ratio of O/Fe for the Sample B-Na_2CO_3 surface was 1.45, whereas, in the sample A-NH_4OH, the ratio was 1.73. This exhibited that a greater number of lattice oxygen sites were present on the crystal surface of the sample which used NH_4OH as precipitant. The catalyst oxidizability will increase with a plentiful amount of lattice oxygen and adsorbed oxygen on the catalyst's surface, which will result in enhancement of the SCR activity of the catalyst. The NH_4OH precipitant based Fe_2O_3 catalyst acquired plentiful lattice oxygen, which favorably affected the NH_3-SCR reaction.

64.5.5 SO_2 Tolerance

To find the tolerance of the Fe_2O_3 catalyst, we have taken the Sample A-NH_4OH and tested it against sulfur dioxide poisoning. Experiments were conducted at 250 °C by introducing a 100 ppm sulfur dioxide supply in the feed gas after 30 min of a stable reaction. After 100 min a minor decrease in NO conversion could be seen with turning on the SO_2 supply, but after that, a sharp decline in NO conversion could be seen in

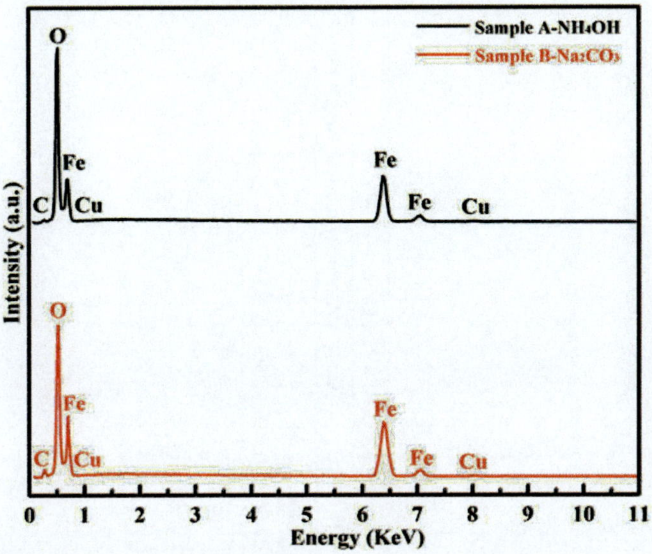

Fig. 64.5 EDS spectra of the Fe_2O_3 catalysts

the next 50 min. After 150 min, the NO conversion has almost reached a stable level of 51%. It is illustrated from the figure that the catalyst exhibited a good tolerance towards SO_2 poisoning at 250 °C within the first 50 min of the reaction. However, after that point, substantial deactivation could be seen with a sharp decrease in the NO conversion of the catalysts. It can be concluded from these experiments that the Fe_2O_3 catalyst was substantially deactivated by the presence of sulfur dioxide, and this could have resulted from the development of sulfate species on the catalyst's surface (Wang et al. 2018; Liu et al. 2011; Li et al. 2012) (Fig. 64.6).

64.5.6 FT-IT Analysis of Fresh and Deactivated Catalyst

To further investigate the statement above, FT-IR analysis of the catalysts was conducted (Fig. 64.7). It can be seen from the spectra that the deactivated catalysts spectra exhibited some new bands at 3252, 1398, and 1122 cm^{-1}. The band at 3252 cm^{-1} can be ascribed to the N–H stretching vibration of NH^{4+} ions (Xie et al. 2004). Whereas, the bands at 1398, and 1122 cm^{-1} can be linked to the adsorption of $SO_4{}^{2-}$ (Pietrogiacomi et al. 2009). This FT-IR analysis confirms that during the NH_3-SCR reaction, the ammonium sulfate $(NH_4)_2SO_4$ and ammonium bisulfate (NH_4HSO_4) was formed on the surface of the catalyst. This, in turn, will lead the way to the catalyst's deactivation.

Table 64.2 EDS investigation of the samples

Catalyst sample	Atomicity % (at%)				Weight % (wt%)				O/Fe	
	Fe	O	C	Cu	Fe	O	C	Cu	Atomicity ratio	Weight ratio
Sample A-NH_4OH	35.42	61.58	0.93	1.21	62.53	32.19	0.34	4.94	1.73	0.52
Sample B-Na_2CO_3	39.79	57.68	1.39	0.84	66.69	27.56	0.52	2.56	1.45	0.42

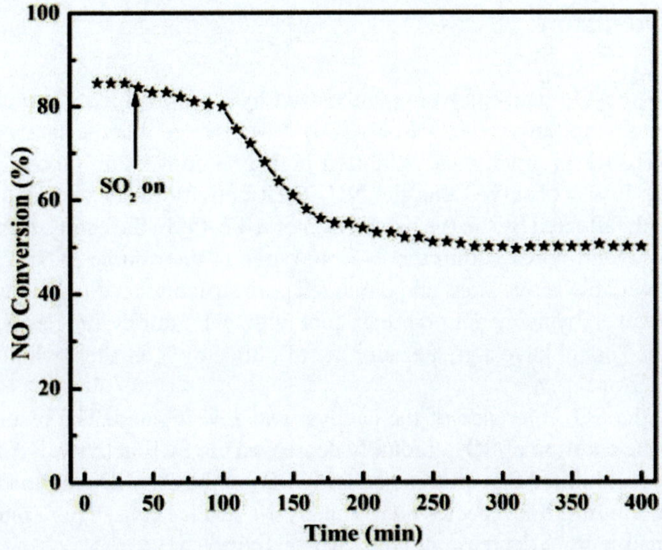

Fig. 64.6 SO_2 tolerance of the Fe_2O_3 catalyst at 250 °C: Reaction conditions: 500 ppm NO, 500 ppm NH_3, 3 vol% O_2, 100 ppm SO_2, GHSV = 20,000 h^{-1}, and N_2 balance

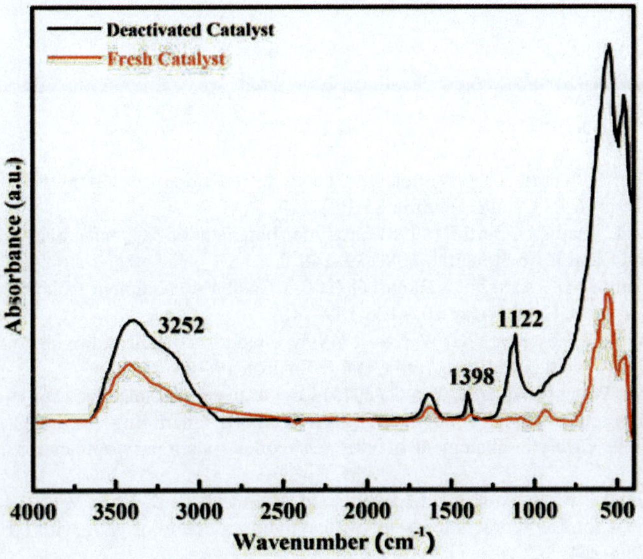

Fig. 64.7 FT-IR spectra of the Fe_2O_3 catalyst

64.6 Conclusion

In this study, Fe_2O_3 catalysts were synthesized by a co-precipitation method using two different precipitants (NH_4OH/Na_2CO_3). It was found that the catalyst prepared by using NH_4OH as precipitant exhibited high NO conversion (above 80% from 250–400 °C. It was observed that the NH_3-SCR activity of the sample B-Na_2CO_3 was negatively affected by the the production of α-Fe_2O_3 in the catalyst. As well, the decrease in several pores within the 2–5 nm range in the sample B-Na_2CO_3, lesser number of available active sites, and damaged pore structure resulted in low catalytic activity. However, by using amino precipitant with OH^- anions, the Fe_2O_3 catalyst's crystal surface could have a greater number of lattice oxygen sites, which benefited the SCR reaction.

As well, the SO_2 tolerance of the catalyst was investigated. The results demonstrated that the addition of SO_2 gradually decreased the SCR activity. FT-IR analysis of the catalyst exhibited that after introducing SO_2 in the De–NO reaction the formation of ammonium sulfate species on the catalyst's surface caused pore plugging and were responsible for a decrease in NH_3-SCR activity.

Acknowledgements We are thankful to the National Natural Science Foundation of China (Grant no. 50676057), which has provided the necessary support for this work.

References

Ballen P, Geiger B, Kureti S (2009) Selective catalytic reduction of NO_x by NH_3 on Fe/HBEA zeolite catalysts. Appl Catal B Environ 85:109–119

Busca G, Lietti L, Ramis G, Berti F (1998) Catalytic abatement of NO_x: chemical and mechanistic aspects. Appl Catal B Environ 107–108:139–148

Busca G, Larrubia MA, Arrighi L, Ramis G (2005) Catalytic abatement of NO_x: chemical and mechanical aspects. Catal Today 107–108:139–148

Dunn PJ, Koppula RP, Stenger GH, Wachs EI (1998) Oxidation of sulfur dioxide to sulfur trioxide over supported vanadia catalysts. Appl Catal B Environ 19:103–117

Gui K, Liang H, Wang F, Wang X, Yao G (2015) Low-temperature selective catalytic reduction of NO on an iron ore catalyst in a magnetically fluidized bed. Chem Eng Technol 38:1537–1542

Heck MR (1999) Catalytic abatement of nitrogen oxides—stationary applications. Catal Today 53:519–523

Husnain N, Wang E, Li K, Anwar T. M, Mehmood A, Gul M, Li D, Jinda M (2019a) Iron oxide-based catalysts for low-temperature selective catalytic reduction of NO_x with NH_3. Rev Chem Eng 35(2):239–264

Husnain N, Wang E, Fareed S (2019b) Low temperature selective catalytic reduction of NO with NH_3 over natural iron ore catalyst. Catalysts 9:956

Husnain N, Wang E, Fareed S, Anwar TM (2019c) Comparision of the low-temperature NH_3-SCR performance of γ-Fe_2O_3 catalysts prepared by two different methods. Catalysts 9:1018

Kang M, Kim JD, Park DE, Kim MJ, Yie EJ, Kim HS, Hopeweeks L, Erying ME (2006) Two-stage catalyst system for selective catalytic reduction of NO_x by NH_3. Appl Catal B Environ 68:21–27

Li Q, Yang H, Ma Z, Zhang X (2012) Selective catalytic reduction of NO with NH_3 over CuO X-carbonaceous materials. Catal Commun 17:8–12

Li Y, Wan Y, Li Y, Zhan S, Guan Q, Tian Y (2016) Low-temperature selective catalytic reduction of NO with NH_3 over Mn_2O_3-doped Fe_2O_3 hexagonal micro sheets. ACS Appl Mater Interfaces 8:5224–5233

Lin S, Lu D, Liu Z (2012) Removal of arsenic contaminants with magnetic γ-Fe_2O_3 nanoparticles. Chem Eng J 211–212:46–52

Liu Z, Woo Ihl S (2006) Recent advances in catalytic $DeNO_x$ science and technology. Catal Rev 48:43–89

Liu F, Asakura K, He H, Shan W, Shi X, Zhang C (2011) Influence of sulfation on iron titanate catalyst for the selective catalytic reduction of NO_x with NH_3. Appl Catal B Environ 103:369–377

Liu C, Yang S, Ma L, Peng Y, Hamidreza A, Chang H, Li J (2013) Comparison on the performance of α-Fe_2O_3 and γ-Fe_2O_3 for selective catalytic reduction of nitrogen oxides with ammonia. Catal Lett 143:697–704

Liu Z, Su H, Chen B, Li J, Woo IS (2016) Activity enhancement of WO_3 modified Fe_2O_3 catalyst for the selective catalytic reduction of NO_x by NH_3. Chem Eng J 299:255–262

Pietrogiacomi D, Magliano A, Ciambelli P, Sannino D, Campa CM, Indovina V (2009) The effect of sulphation on the catalytic activity of CoO_x/ZrO_2 for NO reduction with NH_3 in the presence of O_2. Appl Catal B Environ 89:33–40

Roy S, Marimuthu A, Deshpande PA, Hegde MS, Madras G (2008) Selective catalytic reduction of NO_x: mechanistic perspectives on the role of base metal and noble metal ion substitution. Ind Eng Chem Res 47:9240–9247

Sing WSK, Everett HD, Haul WAR, Moscou L, Pierotti AR, Rouquerol J, Siemieniewska T (1985) Reporting physisorption data for gas/solid systems with special reference to the determination of surface area and porosity. Pure Appl Chem 57:603–619

Singh K, Ohlan A, Kotnala KR, Bakhshi KA, Dhawan KS (2008) Dielectric and magnetic properties of conducting ferromagnetic composite of polyaniline with γ-Fe_2O_3 nanoparticles. Mater Chem Phys 112:651–658

Wang H, Qu Z, Xie H, Maeda N, Miao L, Wang Z (2016) Insight into the mesoporous $Fe_xCe_{1-x}O_{2-\delta}$ catalysts for selective catalytic reduction of NO with NH_3: regulable structure and activity. J Catal 338:56–67

Wang T, Zhu C, Liu H, Xu Y, Zou X, Xu B, Chen T (2018) Performance of selective catalytic reduction of NO with NH_3 over natural manganese ore catalysts at low temperature. Environ Technol 39:317–326

Wu GZ, Gao FJ (2012) Synthesis of γ-Fe_2O_3 nanoparticles by homogeneous co-precipitation method. Micro Nano Lett 7:533

Xie G, Liu Z, Zhu Z, Liu Q, Ge J, Huang Z (2004) Simultaneous removal of SO_2 and NO_x from flue gas using a CuO/Al_2O_3 catalyst sorbent: II. Promotion of SCR activity by SO_2 at high temperatures. J Catal 224:42–49

Xiong BZ, Hu Q, Liu YD, Wu C, Zhou F, Wang ZY, Jin J, Lu MC (2016) Influence of partial substitution of iron oxide by titanium oxide on the structure and activity of iron-cerium mixed oxide catalyst for selective catalytic reduction of NO_x with NH_3. Fuel 165:432–439

Yadav D, Prasad R (2016) Low-temperature de-NO_x technology-a challenge for vehicular exhaust and its remediation: an overview. Procedia Technol 24:639–644

Yang YX, Li B, Sun L, Huang WZ, Cheng MX, Zhang WT, Xing F (2012a) Effect of the surface structure of α-Fe_2O_3 on the selective catalytic reduction of NO by NH_3. Acta Phys-Chim Sin 28(1):184–188

Yang S, Li J, Wang C, Chen J, Ma L, Chang H, Chen L, Peng Y, Yan N (2012b) Fe-Ti spinel for the selective catalytic reduction of NO with NH_3: mechanism and structure-activity relationship. Appl Catal B Environ 117–118:73–80

Yao G, Wang F, Wang X, Gui K (2010) Magnetic field effects on selective catalytic reduction of NO by NH_3 over Fe_2O_3 catalyst in a magnetically fluidized bed. Energy 35:2295–2300

Chapter 65
In Situ Visual Monitoring of Rotary Air Preheater Blockage: Setup and Image Analysis

Cheng Li, Qian Huang, Guanqing Liu, Xiao Sha, and Shuiqing Li

Abstract Blockage of air preheaters has become a common problem in coal-fired power plants due to the formation of sticky ammonium bisulfate. The lack of efficient monitoring technique hinders the efforts to prevent the problem. In this paper, we propose a novel in-situ visual monitoring system on the cold end of the air preheater. The system is inexpensive but powerful enough to reveal the temporal evolution and spatial distribution of blockage fractions. By virtual of 3D printing, we built a lab-scale test structure of the contaminated corrugate plate of the air preheater, and took images under simulating dark environments. We revealed that integrating Gaussian filtration for noise elimination and then K-means clustering for image segmentation exhibits the best performance for images from nearly true circumstances. Moreover, the convolutional neural network manifests its ability to learn the blockage fraction and thus its future applicability for image processing with a well-labeled image dataset.

Keywords Rotary air preheater · Blockage · Visualized monitoring · Image processing

65.1 Introduction

Blockage of the air preheater has become a common problem in China's coal-fired power plant (Xiao 2019). The adverse effect is clear: the efficiency of the exchanger, and thus the boiler, are reduced; the flow resistance grows remarkably and seriously threatens the safety of the unit. The problem arises with the wide application of selective catalytic reduction (SCR) systems for NO_x removal owing to the stricter demand for pollutant controls, as well as the tendency to spray ammonia excessively

C. Li · Q. Huang (✉) · X. Sha · S. Li
Key Laboratory for Thermal Science and Power Engineering of Ministry of Education, Department of Energy and Power Engineering, Tsinghua University, Beijing 100084, China
e-mail: huangqian@tsinghua.edu.cn

G. Liu
Huadian Electric Power Research Institute Co, Beijing 100084, China

J. Lyu and S. Li (eds.), *Clean Coal and Sustainable Energy*, Environmental Science and Engineering,
https://doi.org/10.1007/978-981-16-1657-0_65

for a sufficient denitrification efficiency (Yang and Ming 2007). This leads to the formation of ammonium bisulfate (abbreviated as ABS) in the flue gas, which can further condense onto the heat exchange element of the air preheater (Bruke and Johnson 1982). In a temperature range of 550–590 K, a viscous liquid form of ABS greatly promotes the capture of incoming fly ash and subsequently causes blockage of air preheaters (Menasha et al. 2011).

Several techniques have been proposed to prevent blockage. A strategy is to elevate the operating temperature of the air preheater. This can be achieved by either preheating the air using external heat sources or recirculating the flue gas (Shuangchen et al. 2010). Another one, which is more widely applied, is to introduce sootblowers for the air preheater (Shi et al. 2019). However, it remains a challenge to optimize the practical operation of the abovementioned devices; Even worse, sootblowing onto the clean surface can even damage the heat exchange element. This difficulty is mainly attributed to the lack of reliable indicators manifesting the real-time blockage degree of the air preheater.

A common and useful index is the pressure drop for the flue gas flowing through the air preheater (Teruela et al. 2005). Obtained conveniently at a relatively low cost, the pressure drop correlates positively with the degree of blockage at a constant flow rate. Since pressure is heavily fluctuated, the exponential moving average algorithm was used to reduce measurement errors, and efforts were paid to compare the pressure difference of varied loads of the unit by converting to a standard state (Zhang and Li 2012). It is also possible to derive the real-time heat transfer coefficient of the air preheater based on the measured inlet and outlet temperatures of the air preheater (Wang et al. 2000). Clearly, a reduced heat transfer coefficient implies the existence of internal blockage.

Although these methods are simple and inexpensive, they are severely limited by the number of measuring points, measurement inaccuracy (especially of pressure and flowrate), and the inability to reveal the spatial information of blockage sites. Therefore, we propose in this article the direct visual monitoring of the cold end of air preheaters. It is a feasible approach because blockage affects the shape of heat exchange elements and can be manifested from the image. Moreover, the cold end of the air preheater is the inlet of ash-free air at room temperature, making it possible for the monitoring setup. The images can be further processed by image segmentation algorithms to quantify the extent of blockage (Chen 2018). This promising method has not been reported before.

In this paper, we demonstrate the in situ visual monitoring approach and analyze the images obtaining from a 3D-printed air preheater blockage model. We determine a denoise procedure and a segmentation algorithm mostly suitable for this application by comparing the performance of several image analysis algorithms. The degree of blockage, and the spatial distribution of blocking sites of the cold end of the air preheater can thus be derived.

65.2 Methodology

65.2.1 Setup of the Visual Monitoring System

Figure 65.1 shows schematic of the visual monitoring system. The key devices are several dust-proof industrial cameras radially aligned in the secondary air compartment of the cold end of the air preheater. These cameras are vertically placed to take photographs of the cold end of the air preheater. The distance between the cameras and the preheater cold end is less than 1 m when considering both the image quality and the actual deployment environment. And this further decides the number of cameras needed by incorporating the field of view of each camera and the ideal requirement to cover the whole radial region. One would thus expect at least 7–8 cameras in this setup. With the rotation of the air preheater, the cold end can be fully covered by the cameras. The camera exposure time should be less than 150 ms to avoid significant smears in the photos. Hence, DC fill-in lights are needed to improve the internal lighting conditions, and the brightness should be greater than 100 lx.

The cameras are connected to the industrial control computer, which can remotely control the camera shooting frequency, exposure time, etc. Besides, the captured images are transmitted to the industrial computer on which the blockage fractions are real-time solved by the image segmentation algorithm. The result, together with the temporal-spatial information of clogs, can be further transmitted to the control center for operational feedbacks.

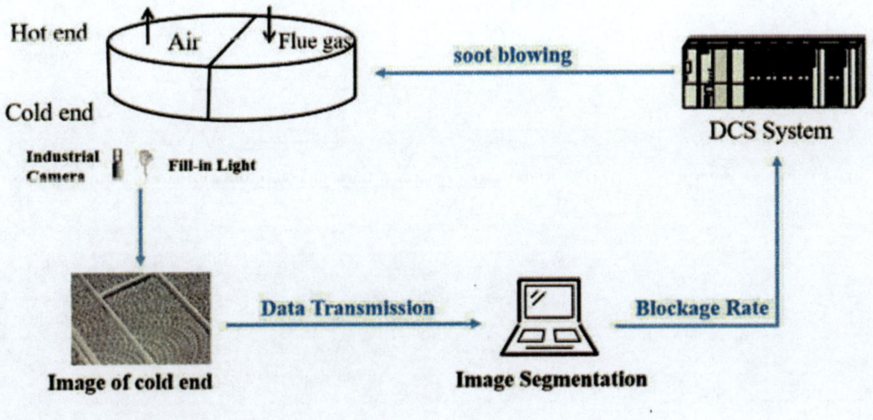

Fig. 65.1 Setup of the visual monitoring system

65.2.2 3D-Printed Model Simulating the Blockage Morphology

To demonstrate the visual monitoring approach, and to facilitate exploring image processing algorithms, we used a 3D-printed model to simulate the blockage morphology. In the air preheater, the corrugated plates are arranged in specific types to form heat exchange elements (Wang et al. 2018). In this work we chose a NF-type board (see details in Fig. 65.2b) as the template to design a 3D model (see Fig. 65.2c). A gray filament (type CR-PLA) was used for printing the 'deposit-free' structure, as shown in Fig. 65.2a. To further mimic the ash blockage, we adhered clay (type XCX-36) to the corrugated board to simulate the cold end morphology with different blockage fractions. An industrial camera (type MV-CA013-20GM/GC) was applied to take photos of the homemade model in the dark scene with the aid of a fill-in light. Again, this photography procedure simulates the real situation as much as possible. The resulting image, as shown in Fig. 65.3a, is believed to reproduce the industrial spot to the best extent.

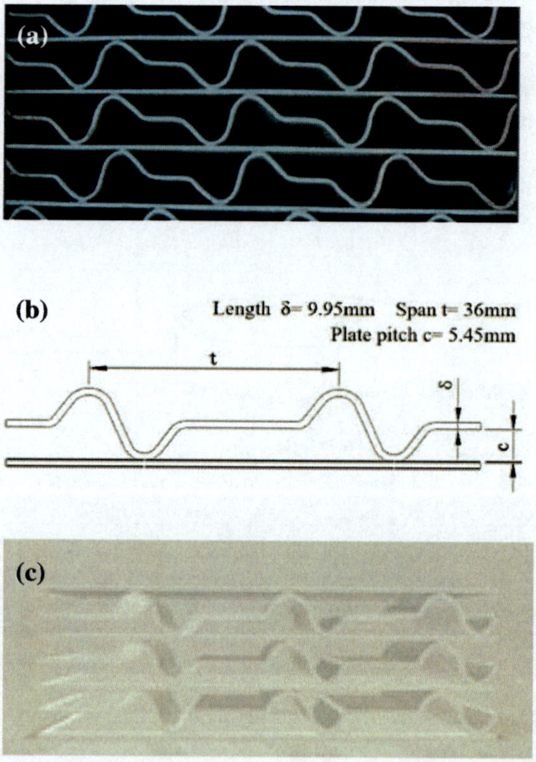

Fig. 65.2 3D-printed models of the corrugated plate

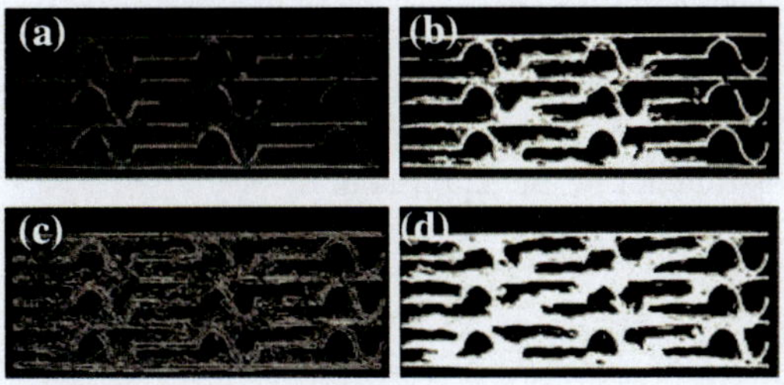

Fig. 65.3 Image segmentation results: **a** The original grayscale image; **b** OTSU result; **c** Adaptive threshold result; **d** K-means clustering result

65.2.3 Image Segmentation Algorithms

We investigated three image segmentation algorithms for quantification of the blockage fraction out of the images. All the algorithms divide the grayscale picture into a 'blocked' area (generally higher gray values) and an 'unblocked' area (with smaller gray values) by different criteria (Mobahi et al. 2011).

The *maximum between-class variance algorithm* (OTSU) finds an optimal gray threshold T for segmenting all pixels into two classes (gray levels greater or less than T) while maximizing a variance g between classes (Otsu 1979). For an image and any arbitrary threshold T, the variance g is defined as:

$$g = \omega_0(\mu_0 - \mu)^2 + \omega_1(\mu_1 - \mu)^2.$$

Here ω_i is the proportion of pixels belonging to class i, and μ_i is the average gray level of class i ($i = 1, 2$). Besides, $\mu = \omega_0\mu_0 + \omega_1\mu_1$ is the average gray level of the image.

Different from OTSU which uses a single threshold to classify every pixel of the image, the *adaptive threshold algorithm* determines a local threshold for each pixel as the weighted average of gray values of its neighboring points (Ma et al. 2009). This is expected to achieve better performance in real images with uneven illumination and large background gray variation. In this work, we used the Gaussian method, namely, a Gaussian-type convolution kernel is applied to weight the gray value in the selected area (specifically, 11 pixels $\times$ 11 pixels) so that the influence of the surrounding points on the center point is gradually weakened.

The third one, *K-means clustering algorithm*, is an unsupervised learning approach for classification (Zalik 2008). The procedure for our purpose is detailed as below. Each pixel is granted with a three-dimensional feature x_i ($I = 1, 2, 3$), where x_1 is its gray value, and x_2, x_3 are the mean gray value and the variance of the eight neighborhoods, respectively. To begin with, we randomly choose K points as the cluster centers (namely K classes). For each pixel, it belongs to class j if its distance to the jth center is the closest among the K distances. Here the distance is defined by the L^2-norm in the 3D feature space x_i. By this way we now have K clusters and therefore K renewed cluster centers, which in turn allows a second round of classification of all pixels. Repeat this procedure iteratively until the number of updated classification points in one iteration is less than 0.2%. With K clusters (classes) in hand, we assign the class whose center has the smallest gray value (x_1) as the 'unblocked' area. This is because the unblocked area usually features low gray values in the image due to the short exposure time. Clearly, K is an adjustable parameter in this approach.

65.2.4 Influence of Noise

The practical environment and data transmission can bring noises into the images, and the impact on image analysis needs further investigation. In this work we introduced two types of noises, Gaussian and salt-and-pepper, to the original image (Sun 2012). The mean value of Gaussian noise is 0, and the variance is 0.01; for salt-and-pepper noise, the ratio of salt and pepper is set to 1, and the total number of salt and pepper is 10 000. We further explored the effect of Gaussian filtering for noise reduction (Moncef et al. 1992). A two-dimensional Gaussian kernel is used as a convolution function to weight and average the gray values of pixels in the selected area (specifically, 9 pixels $\times$ 9 pixels) to eliminate noise.

65.2.5 Convolutional Neural Network

Convolutional neural network (CNN) is a popular and powerful deep learning framework for image processing, and it is desirable to test the performance of CNN in our applications. Briefly, CNN involves multiple steps of convolution and pooling to reduce the dimension of data features and hence the number of training parameters. The gradient descent algorithm is used to train the model parameters (Alex et al. 2017). As a supervised learning approach, it requires labeled samples for training. In this work, we used blockage fractions derived from the image segmentation algorithm as the label of images. This could help demonstrate the applicability of CNN in our case.

65.3 Results and Discussion

65.3.1 Image Segmentation Results

Different segmentation algorithms are used to process the images taken in the dark scene with a fill-in light, as shown in Fig. 65.3. It is seen from Fig. 65.3a that the left and right sides of the original image are darker (with smaller gray values) due to the uneven lighting condition. The corrugated plate is more reflective and thus higher in the gray value than the simulated ash and unblocked area. As for the segmentation results, OTSU (see Fig. 65.3b) apparently misidentifies the simulated ash on the left and right sides as 'unblocked' area. This underestimation of the blockage fraction results from the use of a single threshold for all pixels, which is easily biased by the uneven illumination. The adaptive threshold algorithm in Fig. 65.3c seems to overcome such problem by using localized threshold for each pixel. However, this approach incorrectly classifies many points in the blocked area with relatively low grey values into the 'unblocked' regime, resulting in a non-fully-connected blocked area and an inaccurate estimation of blockage fraction. In contrast, the K clustering method in Fig. 65.3d generates the most satisfactory segmentation result with interconnected blocked areas while properly eliminate the uneven lighting effects.

65.3.2 Noise Influence

Figure 4 shows the image segmentation results after the addition of Gaussian noise, which is specified in Sect. 65.2.4. The noise addition has little effect on the original greyscale image because the disturbed grey value is generally small; see Fig. 65.4a. Similarly, the influence on the OTSU result in Fig. 65.4b is negligible. As for the adaptive threshold algorithm, it is able to eliminate Gaussian noise after weighted average of gray values in the neighborhood, and the blocked area is still not fully connected; see Fig. 65.4c. The K-means clustering algorithm in Fig. 65.4d, however, is greatly affected by the noise because it classifies most of the positive Gaussian error points as blocked point due to the low gray value of the image background.

Figure 65.5 shows the image segmentation results after the addition of salt-and-

Fig. 65.4 Image segmentation after Gaussian noise addition: **a** original greyscale image with Gaussian noise; **b** OTSU result; **c** Adaptive threshold result; **d** K-means clustering result

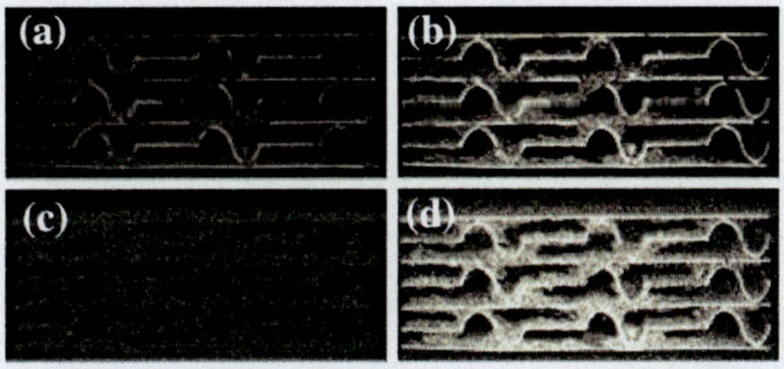

Fig. 65.5 Image segmentation after salt-and-pepper noise addition: **a** original greyscale image with salt-and-pepper noise; **b** OTSU result; **c** Adaptive threshold result; **d** K-means clustering result

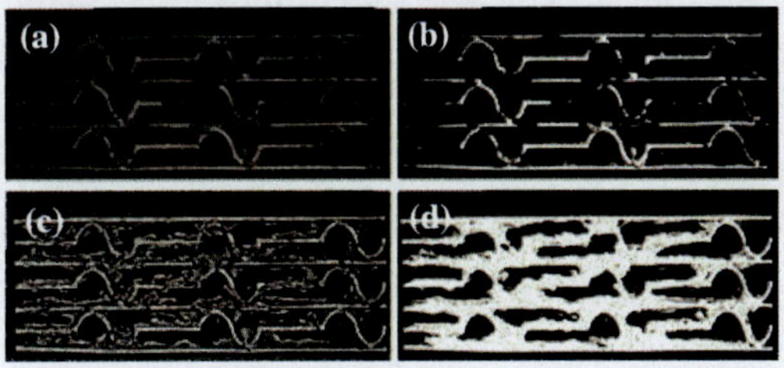

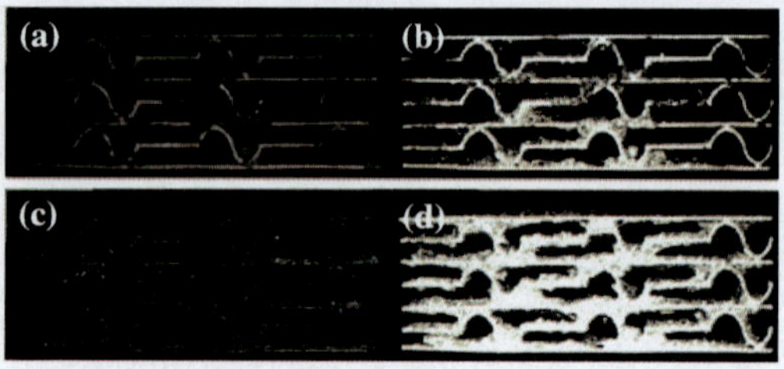

Fig. 65.6 Image segmentation after Gaussian noise addition and then Gaussian filtering: **a** original greyscale image with Gaussian noise; **b** OTSU result; **c** Adaptive threshold result; **d** K-means clustering result

pepper noise, which is specified in Sect. 65.2.4. As compared to Gaussian noise in which most perturbations are small in grey value, the salt-and-pepper noise introduces 0 (pepper) and 255 (salt) into the gray value distribution, and results in a higher OTSU threshold and much poorer segmentation performance; see Fig. 65.5b. For the adaptive threshold algorithm, it always classifies the salt noise points into the blocked area and causes large inaccuracy in Fig. 65.5c. The K-means clustering method, however, provides the chance to adjust the number of class K. By setting $K = 3$, the resulting class whose center has the smallest gray value is classified as the unblocked area, while the class with brightest center is the salt noise. In this way the majority of pixels can be properly segmented, as shown in Fig. 65.5d, and an accurate blockage fraction can be derived.

We further apply Gaussian filtering method to the perturbed image with Gaussian noise in Fig. 65.4a. Figure 65.6a shows the original greyscale image after filtering. The performance of different segmentation algorithm is varied. Gaussian filtering effectively improves the segmentation results of OTSU (see Figs. 65.6b and 4b) and K-means clustering (see Fig. 65.6d and 65.4d). However, the problem caused by uneven light in the OTSU method cannot be completely resolved. In addition, after filtering, the difference between pixels is reduced, and the segmentation effect of the adaptive threshold algorithm becomes worse; see Fig. 65.6c.

65.3.3 CNN Performance

In this section we establish a CNN model to learn the blockage fraction based on the image segmentation method. We feed the images of 'deposited' 3D-printed structures to the CNN model, and the label (namely, the blockage fraction) is derived by the K-means clustering algorithm which exhibits the best performance. In order to expand the size of the picture set, we randomly cut the original image (1024×1024) into multiple 64×64 sub-images for CNN model training. This procedure helps reduce the complexity of the CNN model. The structure of the CNN used is plotted in Fig. 65.7, which includes 2 pooling layers, 1 convolutional layer (10 channels, convolution kernel 5×5), and 1 fully- connected layer.

For each input image, the accuracy of the CNN result can be characterized by:

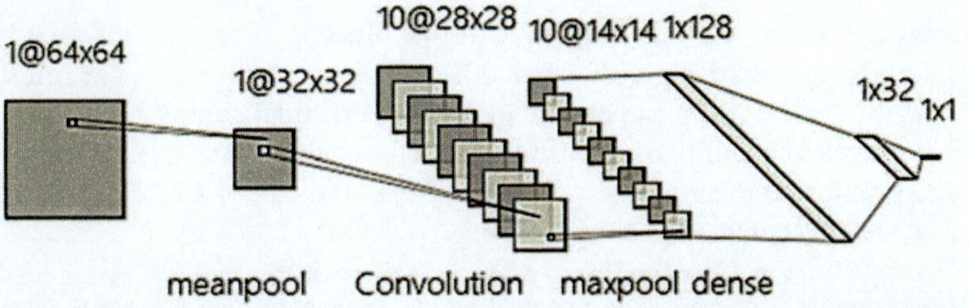

Fig. 65.7 CNN model structure

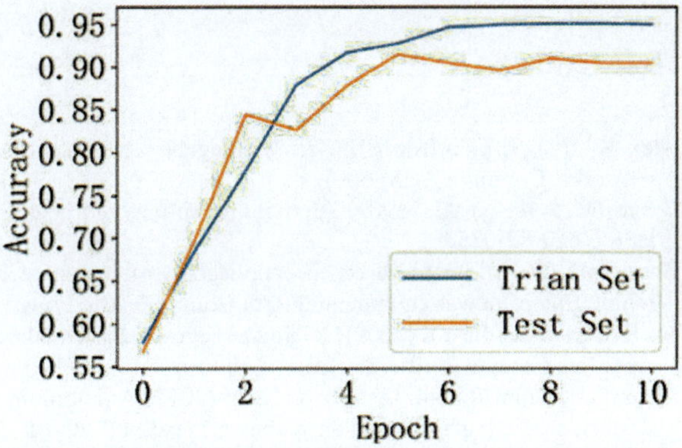

Fig. 65.8 CNN performance in learning and predicting the blockage fraction

$$\text{Accuracy} - 1 - \frac{|\text{CNN\_ Result} - \text{K\_ means\_ Result}|}{\text{K\_ means\_ Result}}.$$

The overall accuracies of CNN results on the training and test sets are derived by averaging the accuracy of each image. Figure 65.8 shows the evolution of prediction accuracy with the training epoch on both the training and test sets. The accuracy reaches ~90% on the test set, indicating the ability of CNN model to learn the blockage fraction from the K-means clustering algorithm.

65.4 Conclusions

This paper proposes a novel, in-situ visual monitoring system for the blockage of air preheaters in coal-fired power plants. The system on the cold end of the air preheater is practically feasible, low-cost, easy to operation, and can reveal the temporal-spatial information of clogs which is beyond the reach of traditional monitoring techniques. Based on the proposal, we built a 3D-printed model to mimic the clean and/or heavily-fouled corrugated plate of the air preheater, and took photos under a simulating dark environment. We delved into the performance of several image processing algorithms for deriving the real-time blockage fraction. It is found that a process involving Gaussian filtration (for noise removal) and then K-means clustering segmentation exhibits the best performance under practical complex conditions. We also demonstrate the ability of convolutional neural network (CNN) for learning the blockage fraction, indicating the potential of massive use of CNN once a labeled image dataset is available for our purpose.

Acknowledgements This work was supported by the National Natural Science Foundation of China (Grants No. 51906122 and 51725601).

References

Alex K, Ilya S, Geoffrey EH (2017) ImageNet classification with deep convolutional neural networks. Commun ACM 60(6)

Bruke JM, Johnson KL (1982) Ammonium sulfate and bisulfate formation in air preheaters. US EPA 600/7-82-025a

Chen WH (2018) Research on the application of image recognition technology in the online monitoring of power equipment. North China Electric Power University (Chinese)

Ma BG, Qiao LL, Jia YB (2009) Cell image segmentation method based on local adaptive threshold. Appl Res Comput 26(2):755–756

Menasha J, Dunn-Rankin D, Muzio L et al (2011) Ammonium bisulfate formation temperature in a bench-scale single-channel air preheater. Fuel 90(7):2445–2453

Moncef G, Edward J, Coyle C et al (1992) An overview of median and stack filtering. Circ Syst Signal Process 11(1):8–45

Mobahi H, Rao SR, Yang AY et al (2011) Segmentation of natural images by texture and boundary compression. Int J Comput Vis 95(1):86–98

Otsu N (1979) A threshold selection method from gray-level histograms. IEEE 1:62–66

Shuangchen M, Xin J, Sun YX et al (2010) The formation mechanism of ammonium bisulfate in SCR flue gas denitrification process and control thereof. Thermal Power Gener 39(08):12–17 (Chinese)

Shi Y, Wen J, Cui F et al (2019) An optimization study on soot-blowing of air preheaters in coal-fired power plant boilers. Energies 12:958

Sun HY (2012) Study on image Gaussian noise and Pepper and salt noise denoising algorithm. Fudan University (Chinese)

Teruela E, Cortés C, Ignacio Díez L et al (2005) Monitoring and prediction of fouling in coal-fired utility boilers using neural networks. Eng Sci 60:5035–5048

Wang E, Li K, Mao J et al (2018) Experimental study of flow and heat transfer in rotary air preheaters with honeycomb ceramics and metal corrugated plates. Appl Therm Eng 130(5):1549–1557

Wang JG, Xu ZM, Yang SR (2000) On-line monitoring model of ash deposits on air preheater. Proc CSEE 20(7):37–39 (Chinese)

Xiao C (2019) Analysis of air preheater blockage in 1000 MW coal—fired boiler. Sci Technol Vis 12:179–180 (Chinese)

Yang D, Ming Xu (2007) Discussion on the application of the SCR technology in coal-fired power plants. ELectr Power Environ Protect 23(1):48–53 (Chinese)

Zhang XA, Li YH (2012) Calculation of cleanness factor for sootblowing optimization of air preheaters. Power Equip 26(5):320–322 (Chinese)

Zalik KR (2008) An efficient K-means clustering algorithm. Pattern Recogn Lett 29(9):1385–1391

Chapter 66
Effect of Chemical Composition on Adsorption and Agglomeration Characteristics of Ash Particles After Sulfuric Acid Adsorption

Jiahao Jiang, Yu Yan, Jin Guo, Yuesheng Li, Lei Deng, and Defu Che

Abstract The low-low temperature electrostatic precipitator has been applied to enhance the efficiency of power plants. In this paper, the influences of chemical constitutions on the adsorption properties of particulate matters with vitriolic acid and the agglomeration after adsorption are researched. Experiments are conducted on a one-dimensional adsorption system, which simulates the working environment of low-low-temperature electrostatic precipitator. The consequences display that the agglomeration between particulate matters in scanning electron microscopy are loose pellets agglomeration together, adhesion of loose pellets to massive particles, and massive particles agglomeration together. For Na and K, the amounts of elements added have little effect on the adsorption of the sulfuric acid by ash particles. For Mg, Fe, and Ca, when the oxide is used as an additive, the adsorption capacity of ash samples to the vitriolic acid was enhanced and the agglomeration phenomenon of the ash granules was more remarkable with increasing elements amounts. In addition, the change of Mg has great impact on vitriolic acid adsorption and granules agglomeration.

Keywords Chemical constitution · Vitriolic acid adsorption · Particles agglomeration · Low-low-temperature electrostatic precipitator

---

J. Jiang · Y. Yan · L. Deng (✉) · D. Che
State Key Laboratory of Multiphase Flow in Power Engineering, School of Energy and Power Engineering, Xi'an Jiaotong University, Xi'an 710049, China
e-mail: leideng@mail.xjtu.edu.cn

J. Guo
Guangdong Institue of Special Equipment Inspection and Research, Foshan 528251, China

Y. Li
Guangdong Institue of Special Equipment Inspection and Research, Shunde Institue of Inspection, Foshan 528300, China

J. Lyu and S. Li (eds.), *Clean Coal and Sustainable Energy*, Environmental Science and Engineering,
https://doi.org/10.1007/978-981-16-1657-0_66

66.1 Introduction

China's economy has developed rapidly and energy consumption has increased in recent years. Coal-based energy consumption has brought about serious environmental pollution problems (Yuan et al. 2018). Air pollution caused by particulate matter, especially PM10 and PM2.5, has caused widespread concern in society (Rohde 2015). The Chinese government issued the "Action Plan for Coal Energy Saving and Emission Reduction and Reconstruction (2014–2020)" to protect the atmosphere (Lei 2015). The plan put forward new requirements for the emission of atmospheric pollutants from coal-fired power plants: NO_x, SO_2, dust, Hg emission concentration is lower than 50, 35, 10, 0.03 mg m^{-3}(reference oxygen content is 6%). In this case, many ultra-clean emission technologies were developed to decrease pollutants emission in coal-fired power plants. Among them, low-low-temperature electrostatic precipitator (LLT ESP) has the advantages of high dust removal efficiency and synergistic removal of SO_3, which has received extensive attention (Jianguo et al. 2014; Haibao et al. 2014; Jun 2013; Zhifu et al. 2017).

Low-low-temperature electric dust removal technology is a new type of efficient pollutants control technology, derived from wet electrostatic precipitators technology (Jianguo et al. 2014; Haibao et al. 2014). The difference between LLT ESP and traditional ESP is the temperature of flue gas at ESP inlet. It reduces the temperature to 90 °C below the acid dew point. As the temperature of flue gas decreasing, the SO_3 condenses into vitriolic acid and is adsorbed on the surface of granules, which reduces the specific resistance of the dust (Back 2009). The reduction of specific resistance effectively prevents the anti-corona phenomenon of dust, thereby improving the efficiency of electric dust removal. The SO_3 adsorbed on the dust is removed together with the dust (Jianguo et al. 2014). The decrease of flue gas temperature will increase the electric field breakdown voltage in ESP and improve the efficiency of dust removal. In addition, the decrease of exhaust gas temperature will result the amount of flue gas decreasing, which will increase the dust collecting area. The more dust collecting area is more conducive to the dust trapping by the electrode plate. In a typical 600 MW unit which was equipped LLT ESP in Ningde, the dust emission value of LLT ESP outlet can reach 20 mg m^{-3} while the operation temperature of ESP was about 95 °C. In addition, the SO_3 removal efficiency was 74%. Moreover, the dust concentration at the outlet of desulfurization tower can generally be reduced to about 5 mg m^{-3} when the unit was equipped with wet desulfurization technology (Qingliang 2013).

However, there also exists a series of problems in operation such as reentrainment of dust, deposition, blockage and corrosion (Jun 2013; Pingyang, et al. 2013; Huifeng 2013). In view of the low temperature corrosion and the deposition and blockage in ash bucket that may exist during the operation of LLT ESP. Many studies investigated the characteristics of granules agglomeration and vitriolic acid adsorption. Yu et al. (2017), Yubo et al. (2019) studied the adsorption of particulate matters to vitriolic acid mist and the agglomeration of granules during the raction. They were more concerned about the influence of operation temperature, reaction time and particles diameter. In addition, Guojie et al. (2009), Yong et al. (2009), Hanxiao et al. (2015) pointed out that liquid bridge force played a significance role in the agglomeration in the research of ash particles humidification. Liqiang (2011) believed that the presence of SO_3 promoted the formation of liquid film on the surface of ash particles. There are few studies on the effects of chemical constitution of particulate matters on the operation of LLT ESP.

In this study, the ash granules collected from the economizer of a typical 660 MW coal-fired boiler were prepared as the research object. The content of related elements in the ash samples were changed by adding corresponding compounds. The experiments of sulfuric acid adsorption were carried out in a fixed bed furnace at 90 °C to study the influence of chemical constitutions on the properties of adsorption and agglomeration of ash granules.

66.2 Experimental Section

The ash samples used in the experiments were collected from the economizer of a typical 660 MW coal-fired boiler. The adsorption experiments were carried out on a fixed bed system, as shown in Fig. 66.1. The temperature control of the experiments was carried out by using three resistance furnaces, which corresponded to three parts of gas preheating, sulfuric acid decomposition and adsorption reaction (200, 500, 90 °C). Nitrogen was used as the carrier gas, and the pre-heated nitrogen gas carried the decomposed sulfuric acid into the adsorption reaction zone to react with ash samples. The exhaust gas was treated by the NaOH solution.

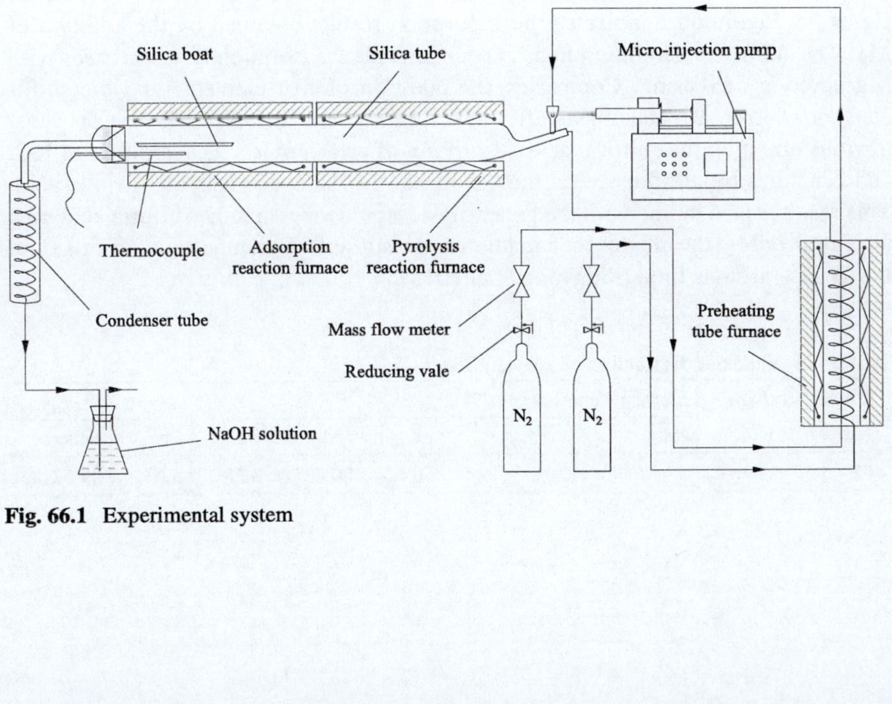

Fig. 66.1 Experimental system

The flow rate of vitriolic acid was 150 μL min^{-1}, and the flow rate of nitrogen was 2500 mL min^{-1}. The adsorption reaction temperature was controlled at 90 °C, which was nearby the operation condition of the LLT ESP. The raction time was 20 min. In the experiments, compounds of K, Na, Ca, Mg and Fe were used as additives, and the chemical components in ash samples were changed by direct addition. The addition amount of each compound was determined according to the content of each chemical component in the ash. The test samples before and after adsorption were analyzed by sulfur element analysis and scanning electron microscope (SEM).

66.3 Results and Discussion

66.3.1 Effect of Chemical Constitution on Adsorption

In order to determine the amount of various chemical components addition, the constitutions of ash samples were analyzed. Table 66.1 shows the content of elements in ash samples testing by XRF. Among these, the amount of Na, Mg, and K in ash particles is less than 1% (wt), and the fraction of Fe and Ca is nearby 3–5% (wt). Thereby, the elements addition amounts of Ca and Fe are 4–16% (wt), and the element addition amounts of Na, K and Mg are 2–8% (wt). The selected element additions are oxides of each element, wherein the oxides of Na and K are too active to be preserved and added, so NaCl and KCl are selected as additions.

The results of adsorption are shown in Figs. 66.2 and 66.3. Figure 66.2 shows the S adsorptive amounts of ash samples with Ca or Fe addition. It can be seen that the S concentration in specimens increased after reaction with the addition of Ca or Fe. Figure 66.3 indicates the adsorption results obtained by the addition of Mg. The sulfur concentration in ash specimens after adsorption also increases with Mg addition increasing. Comparing the addition of five elements and the sulfur concentration in ash specimens after adsorption, the addition of Mg has the most obvious effect on the sulfuric acid adsorption of ash samples. The addition of KCl and NaCl has little influence on the adsorption of ash specimens on vitriolic acid. This result is also in line with the generally accepted view that the sulfuric acid mist is neutralized by the alkalecence matter on the surface of granules and adsorbed on the surface at same time (Pingyang et al 2013).

Table 66.1 Elements distribution of ash samples

Particles size/μm	Elements content (wt)/%					
	Na	K	Mg	Ca	Fe	Others
<68	0.41	0.63	0.43	3.28	5.63	89.62

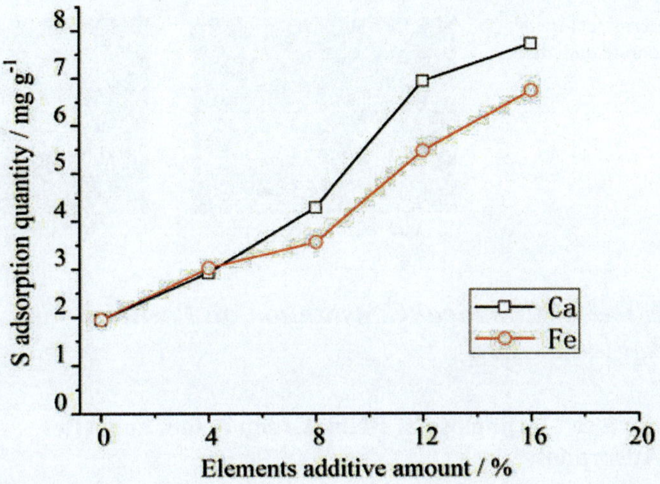

Fig. 66.2 Effect of Ca, Fe addition

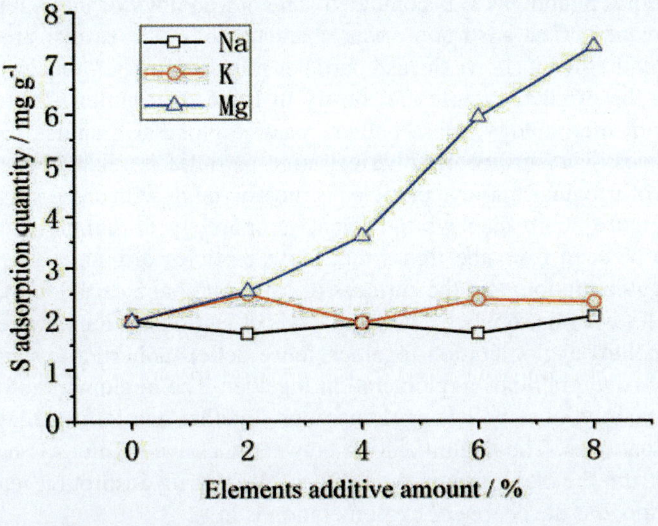

Fig. 66.3 Effect of Mg, Na, K addition

Fig. 66.4 Microstructure of blank group before and after adsorption

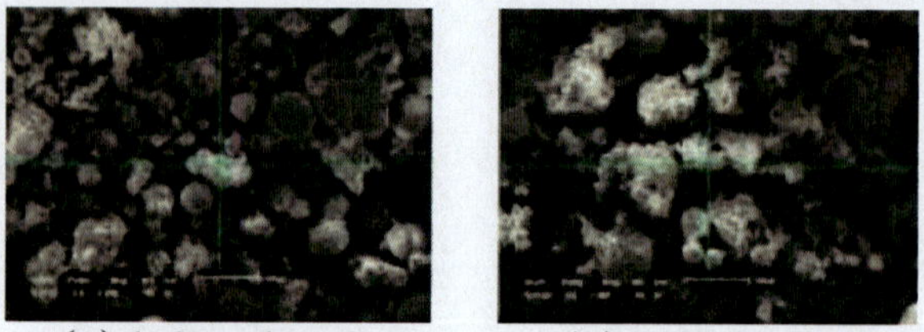

(a) before adsorption (b) after adsorption

66.3.2 *Effect of Chemical Constitution on Particles Agglomeration*

66.3.2.1 Surface Morphology of Blank Group Before and After Adsorption

Scanning electron microscope was used to characterize the particles before and after raction, and the agglomeration phenomenon of ash samples with different elements addition amounts was compared. The morphology of the blank ash specimen before and after adsorption was characterized. The results are shown in Fig. 66.4. Figure 66.4a shows the ash particles morphology before adsorption. Ash particles in the original sample are mostly in loose distribution. There are three main kinds of morphology: loose pellets, massive globular granules and irregular shape particles. The surface of spherical large particles is relatively smooth, and the surface of irregular massive particles is mostly rough with some small particles adhering. Figure 66.4b displays the micro morphology of samples after adsorption. It can be seen from the figure that loose particles originally scattered have occurred agglomeration, and the surfaces of irregular shape particles and spherical large particles become rougher. The main kinds of agglomeration between granules are: loose pellets agglomeration together, loose pellets adhering to massive granules and massive granlules agglomeration together. The agglomeration exits after reaction is mainly loose pellets agglomeration together and loose pellets adhering to massive particles. The agglomeration between massive granules is not obvious. Therefore, from the blank group, the ash particles during adsorption reation exists agglomeration, but the degree of agglomeration is low.

66.3.2.2 Surface Morphology of Ash Particles After Adsorption with Mg Addition

Figure 66.5 exhibits the SEM results of ash specimens after adsorbing vitriolic acid with disparate addition amounts of Mg. In Fig. 66.5a which the added amount of Mg is 2%, the agglomeration of adsorbed ash granules still exits as loose pellets agglomeration together and loose pellets adhering to massive particles. It is similar to the condition of blank group. However, the adsorbed specimens which the Mg addition

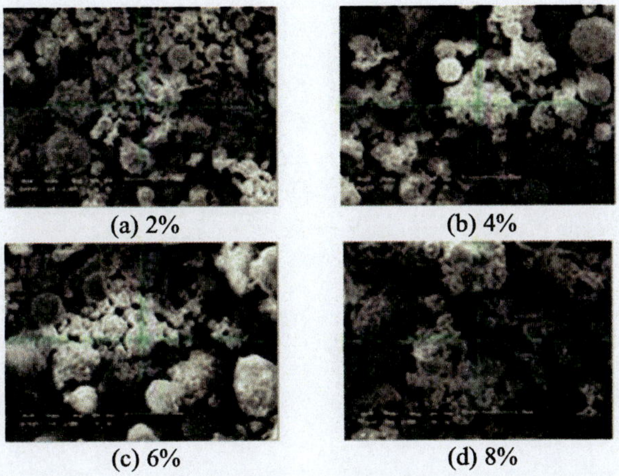

Fig. 66.5 Microstructure of adsorbed granules with Mg addition

amounts are 4–8% show different agglomeration phenomena from the specimens that without addition. The agglomeration between ash granules is more obvious. In addition, the agglomeration of pellets still exists, and the loose small pellets are reduced. The agglomeration phenomenon of loose pellets adhering to the surface of massive granules increases, and agglomeration between massive granules begin to appear with the increase of the addition amount. Especially, when the addition amount is 8%, the agglomeration phenomenon between massive granules is most prominent. MgO is an alkaline substance that can react with sulfuric acid and SO_3 in the flue gas. The content of magnesium oxide in the samples increases, which enhances the ability to adsorb sulfuric acid mist of granules, so that the surface of ash particles becomes wet, and the viscosity of granules are increased which makes the agglomeration of ash granules more obvious.

66.3.2.3 Surface Morphology of Ash Particles After Adsorption with Ca Addition

Figure 66.6 exhibits the SEM results of ash specimens after adsorbing vitriolic acid with addition of Ca. In Fig. 66.6a which the added amount of Ca is 4%, the agglomeration of adsorbed ash granules is analogical to the blank group. As the Ca additive increasing, the agglomeration phenomenon appearing between granules becomes more distinct. The main kind of agglomeration changed from between loose pellets agglomeration together to massive granules agglomeration together. The influence of CaO and MgO addition is analogical. The addition of calcium oxide increases the content of basic compounds in samples and enhances the ability of particles to absorb vitriolic acid mist and SO_3. Subsequently, the increased viscosity of ash granules facilitates the agglomeration between granules.

EDS scanning was applied to detect the elements content of adsorbed ash specimens which the addition amount of Ca is 8%. The location of dot selection is in the

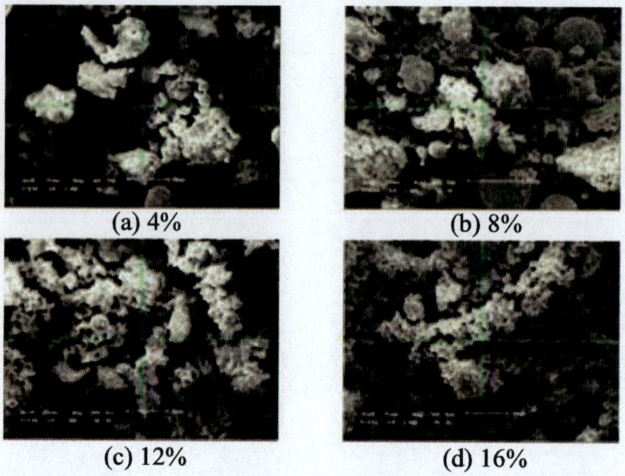

Fig. 66.6 Microstructure of adsorbed granules with Ca addition

region where the agglomeration phenomenon of granules is concentrated, as shown in Fig. 66.7. The distribution of elements detected by scanning are shown in Table 66.2. The mass fraction of S is about 5% and the mass fraction of Ca is about 18% at the position that the agglomeration is prominent. Both Ca and S in this area exceed the average content of ash particles. It can be inferred that the added calcium oxide promotes samples to adsorb vitriolic acid and SO3 in the flue gas (Table 66.3).

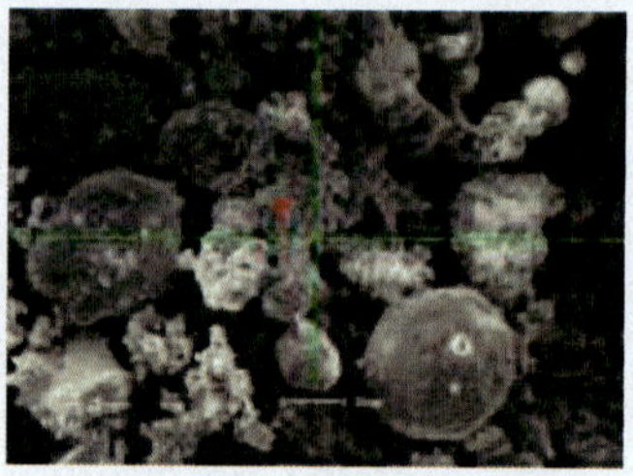

Fig. 66.7 Scanning dot position

Table 66.2 Adsorption sample with 8% Ca addition EDS results

Elements content (wt)/%						
O	Na	Si	S	Al	Ca	Other
61.65	0.47	6.96	5.31	1.18	17.85	6.58

Table 66.3 Adsorption sample with 16% Fe addition EDS results

Elements content (wt)/%					
O	Si	S	Al	Fe	Other
57.52	15.01	1.07	6.89	17.19	2.32

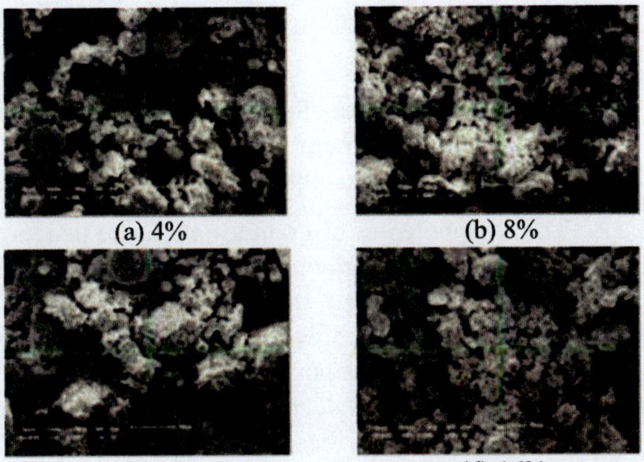

Fig. 66.8 Microstructure of adsorbed granules with Fe addition

66.3.2.4 Surface Morphology of Ash Particles After Adsorption with Fe Addition

The SEM results of the morphology of ash granules after adsorbing vitriolic acid with disparate addition amounts of Fe adsorption is shown in Fig. 66.8. The morphology of the granules of the sample changes little after adsorbing acid with 4% Fe addition comparing with blank group. The kinds of agglomeration are shown in loose pellets agglomeration together and loose pellets adhering to massive granules. When the added mass fraction of Fe is 8%, the microstructure of the specimens after adsorbing vitriolic acid changes more distinctly, and the agglomeration between massive granules is more significant. When the addition amount is further increased to 12 and 16%, the main kind of agglomeration changed from between loose pellets agglomeration together to massive granules agglomeration together.

The adsorbed ash specimens that the added mass faction of Fe additive was 16% was detected by EDS. Figure 66.9 shows the scanned region which is selected according to the agglomeration degree. The mass fration of various elements of granules is shown in Table 66.2. The mass fraction of S is 1.07%, and the mass fraction of Fe is 17.19%. Both Fe and S in this area exceed the average content of ash particles. The added Fe_2O_3 promotes samples to adsorb vitriolic acid and SO_3 in the flue gas.

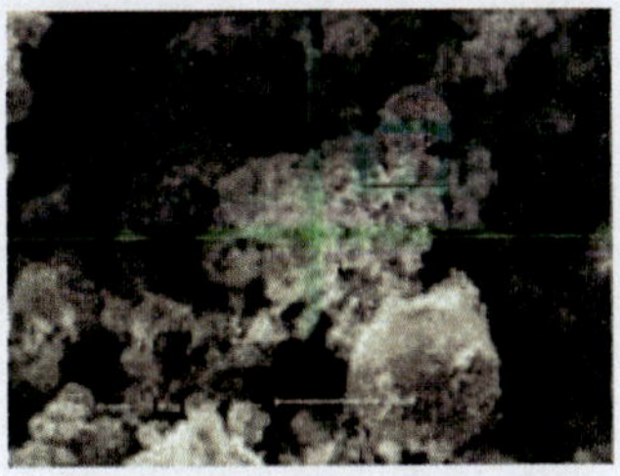

Fig. 66.9 Scanning position

66.4 Conclusions

In this study, the influence of chemical constitution of ash granules on the properties of adsorption and agglomeration after vitriolic acid adsorption was carried out. The experimental research has obtained the following conclusions:

(1) The morphologies of granules are loose pellets, irregular shape particles, and massive globular granules. The phenomenon of agglomeration after adsorption reaction mainly manifests in three kinds: loose pellets agglomeration together, adhesion of loose pellets to massive particles, and maasive particles agglomeration together.
(2) The influence of NaCl and KCl additives on the ash granules adsorbing vitriolic acid is not obvious. The addition of MgO, Fe_2O_3 and CaO facilitates the granules adsorbing vitriolic acid to different extents, and the influence of Mg addition is greater.
(3) The addition of MgO, Fe_2O_3 and CaO promotes the agglomeration of the particles to varying degrees and promotes the agglomeration between loose pellets and loose pellets transforms into the agglomeration between massive granules.

Acknowledgements This work was supported by the National Key R&D Program of China (2017YFB0602102).

References

Back A (2009) Enhancing ESP efficiency for high resistivity fly ash by reducing the flue gas temperature. In: 11th international conference on electrostatic precipitation
Guojie Q, Yong D, Lin C et al (2009) Review of submicron particles humidifying agglomeration. Chem Ind Eng Peogress 28(5):745–749 (in Chinese)
Huifeng W (2013) Application of the lower temperature electrostatic precipitator. In: Proceedings of 15th conference of ESP (in Chinese)
Haibao Z, Jianguo L, Yuzhong H et al (2014) Research and application on low-low temperature electrostatic precipitator technology. China Electr Power 47(10):117–121 (in Chinese)
Hanxiao L, Jianguo Y, Zhuhai L et al (2015) Mechanism analysis and research methods of particle coagulation. Electr Energy 36(1):107–111 (in Chinese)
Jun J (2013) Application of low-low temperature electrostatic precipitator and its problems. In: Proceedings of 15th conference of ESP (in Chinese)
Jianguo L, Zhuhai L, Yuzhong H et al (2014) Research and application on electric. Precipitation technology with low-low temperature. China Environ Prot Ind 03:28–34 (in Chinese)
Lei M (2015) Study on the technical route for ultra-low emission transformation of coal-fired power plants. Sci Technol Commun 7(24):98–100 (in Chinese)
Liqiang Q, Yongtao Y, Yawei S (2011) Influence of SO_3 on electrostatic precipitation of fine particles in flue gas. J Chin Soc Power Eng 31(7):539–543 (in Chinese)
Pingyang M, Zhu M, Jinwu D et al (2013) Application of low-low temperature and moving plate ESP technology in improving efficiency of ESP. In: Proceedings of 15th conference of ESP (in Chinese)

Qingliang X (2013) Application of low-low temperature ESP technology in 600 MW unit of ningde power plant. In: Proceedings of 15th conference of ESP
Rohde RA, Muller RA (2015) Air pollution in China: mapping of concentrations and sources. Plos One 10(8):e0135749
Yuan F, Wanyi Z, Jiawen C et al (2018) Analysis on the current situation and development trend of energy resources in China. Conserv Util Miner Resour 04:34–42 (in Chinese)
Yu Y, Ke S, Yun S et al (2017) Adsorption and agglomeration characteristics of ash particles after reducing flue gas temperature below the acid dew point. Energy Procedia 142(33):01–06
Yubo Z, Yu Y, Fangfang H et al (2019) Experimental study on agglomeration characteristics of ash particles in low-low temperature flue gas system. Thermal Power Gener 48(1):35–42 (in Chinese)
Yong D, Guojie Q, Lin C et al (2009) Analysis on agglomeration of humidified particles in CFD-FGD process. J Power Eng 29(7):671–675 (in Chinese)
Zhifu Q, Jingwei L, Dongli W et al (2017) Particulate matter emission characteristics and removal efficiencies of a low-low temperature electrostatic precipitator. Energy Fuels 31(2):1741–1746

Chapter 67
Characteristics of Hg^0 Re-emission Caused by Sulfite in a Wet Flue Gas Desulfurization System

Jialing Xu, Jingjing Bao, Jiguo Tang, Min Du, Zhengyu Mo, and Licheng Sun

Abstract Re-emission of Hg^0 caused by sulfite and metal ion in a wet flue gas desulfurization (WFGD) system was investigated experimentally in this study. The effects of $SO_3{}^{2-}$ concentration, slurry temperature, pH, O_2 concentration, and metal ions on Hg^{2+} reduction and Hg^0 re-emission were studied. The mechanisms of Hg^0 re-emission were explored based on the kinetics of mercury reduction and the equilibrium of mercuric-sulfite complexes. The results indicated that Hg^{2+} was reduced by $SO_3{}^{2-}$, which caused a re-emission of Hg^0. It is worth mentioning that increasing $SO_3{}^{2-}$ concentration showed an inhibitory effect on Hg^0 re-emission because of the formation of $Hg(SO_3)_2{}^{2-}$. The reduction of Hg^{2+} has a positive correlation with slurry temperature. The rate constant of Hg^{2+} reduction increased by approximately 11 times with temperature increasing from 40 to 60 °C. A decrease in pH can affect the equilibrium of mercuric-sulfite complexes, thus intensifying Hg^0 re-emission. The re-emission amount of Hg^0 increased from 0.4 to 23.8 μg with pH decreasing from 6.0 to 4.0. The calculation reveals that the stabilities of $HgSO_3$ and $Hg(SO_3)_2{}^{2-}$ are weakened at lower pH values, which promotes their decomposition to Hg^0. Although a secondary release of Hg^0 was observed in the presence of O_2, Hg^0 re-emission was inhibited with increasing O_2 concentration. In addition, adding Cu^{2+} or Fe^{2+} ions is found to enhance Hg^0 re-emission because of the reducing ability of metal ions.

Keywords Hg^{2+} reduction · Hg^0 re-emission · Mechanisms · Wet flue gas desulfurization system

67.1 Introduction

It is well known that mercury is harmful to the environment and human health due to its high toxicity, volatility, bioaccumulation, and long-distance transmission. A large amount of mercury is released to the atmosphere from anthropogenic sources, which

J. Xu · J. Bao (✉) · J. Tang · M. Du · Z. Mo · L. Sun
State Key Laboratory of Hydraulics and Mountain River Engineering, College of Water Resource & Hydropower, Sichuan University, Chengdu 610065, China
e-mail: baojj247@scu.edu.cn

J. Lyu and S. Li (eds.), *Clean Coal and Sustainable Energy*, Environmental Science and Engineering,
https://doi.org/10.1007/978-981-16-1657-0_67

mainly include coal fired power plants and ferrous metals manufacturing facilities (Pirrone and Cinnirella 2010; Guan et al. 2017). Although plenty of efforts have been made to reduce mercury emissions from anthropogenic activities, the accumulation of mercury in the environment has continued to increase recently (Jaishankar and Tseten 2014). The Minamata Convention on Mercury was signed in 2013 to protect human health and the environment from mercury pollution. Thereafter, the control of mercury emissions from coal-fired power plants has become more urgent.

There are three forms of mercury species in coal-fired flue gas: elemental mercury (Hg^0), oxidized mercury (Hg^{2+}), and particle-bound mercury (Hg_p) (Galbreath and Zygarlicke 2000). These mercury and mercury compounds can be removed by electrostatic precipitators (ESPs), selective catalytic reduction (SCR) systems, wet flue gas desulfurization (WFGD) systems, and other air pollution control devices (APCDs) in power plants. It should be noted that, as the last unit of APCD, the WFGD system has been used to efficiently remove SO_2 and Hg^{2+}, but can not remove insoluble Hg^0. The reductive ions in the desulfurization slurry (such as $SO_3{}^{2-}$ and $HSO_3{}^{-}$) can reduce Hg^{2+} to Hg^0, resulting in poor mercury removal performance. Since the WFGD system is considered an effective way to remove Hg^{2+}, Hg^{2+} reduction and Hg^0 re-emission become a hot topic in mercury removal by wet scrubbing (Liu and Wang 2015). It is determined that the sulfite concentration, temperature, and the pH of slurry are the main factors affecting Hg^0 re-emission (Omine and Romero 2012; Van Loon and Mader 2000; Chang and Zhao 2017). Among them, $SO_3{}^{2-}$ were confirmed to play an important role in Hg^0 re-emission (Omine and Romero 2012). The intramolecular redox reaction of $HgSO_3$ was proved to be the reason for Hg^{2+} reduction (Loon and Mader 2000). However, owing to the formation of a stable complex $Hg(SO_3)_2{}^{2-}$, excess $SO_3{}^{2-}$ inhibits the re-emission of Hg^0 (Chang and Zhao 2017). In addition, higher temperatures were reported to accelerate Hg^{2+} reduction and lead to more re-emission of Hg^0 (Ochoa-González and Díaz-Somoano 2013; Wang and Liu 2010; Xu and Bao 2018). Wo and Zhang (2009) and Díaz-Somoano and Unterberger (2007) performed a series of experiments, showing that the increase in pH results in an inhibition of Hg^0 re-emission, while Omine and Romero (2012) proposed the opposite conclusion. Obviously, the experimental results obtained in different studies are still controversial. And the mechanisim of pH affecting Hg^0 re-emission has been rarely mentioned.

The effects of other coexisting ions on Hg^0 re-emission have also been extensively studied to explore the mechanisms of Hg^{2+} reduction and Hg^0 re-emission in WFGD systems. Halide ions, as common ions coexisted in desulfurization liquid, have been thoroughly studied in previous studies (Ochoa-González and Díaz-Somoano 2013b; Liu and Wang 2011). Stable complexes such as $HgCl_n{}^{2-n}$, $HgSO_3Cl^-$, and $HgSO_3Cl_2{}^{2-}$ are probably formed in the presence of halide ions, which can promote Hg^{2+} retention in the liquid phase and suppress Hg^0 re-emission. The metal ions in the desulfurization liquid have also attracted some attention recently (Tang and Xu 2010; Hou and Lu 2012; Peng and Liu 2017). Tang et al. (2010) claimed that Pb^{2+} and Cu^{2+} can compete with Hg^{2+} to react with additives, thus the mercury removal by additives was weakened after adding them. A consistent phenomenon was also observed that the metal ions could inhibit Hg^{2+} precipitation by dithiocarbamate type chelating resin (DTCR), following the order $Cu^{2+} > Ni^{2+} > Pb^{2+} > Zn^{2+}$ (Hou and Lu 2012). Peng and Liu (2017) specially studied the effect of copper ions on Hg^0 re-emission. The results showed that Cu^{2+} first transferred to Cu^+, which then reduced Hg^{2+} to Hg^0. Cu^+ was also considered a catalyst for the decomposition of $Hg(SO_3)_2{}^{2-}$. However, $CuCl_2$ was used as the copper source in this study. Since Cl^- may influence the process of Hg^{2+} reduction and Hg^0 re-emission, the investigation is supposed to be further discussed in halogen-free conditions. Moreover, previous work mainly focused on the effect of metal ions on mercury removal by additives. Very few studies have evaluated the re-emission of Hg^0 in the presence of metal ions. Therefore, it is important to assess the effects of metal ions for a deep understanding of the mechanism of Hg^0 re-emission in WFGD systems.

In this study, the influences of sulfite concentration, slurry temperature, pH, O_2 concentration, and metal ions on Hg^{2+} reduction and Hg^0 re-emission were investigated in a simulated WFGD system. A set of experiments was carried out in a bubble reactor, and the reduction ratio of Hg^{2+} during the desulfurization process was also studied. In addition, the mechanisms of Hg^{2+} reduction and Hg^0 re-emission were discussed according to the kinetics of Hg^{2+} reduction and chemical equilibrium.

67.2 Experimental

67.2.1 Experimental Apparatus and Method

To investigate Hg^{2+} reduction and Hg^0 re-emission in a wet flue gas desulfurization system, a lab-scale device was set up as shown in Fig. 67.1. It comprises a gas generation unit, a bubble reactor, a testing part, and an exhaust gas treatment section. The simulated flue gas, including N_2 and O_2, was supplied by compressed steel

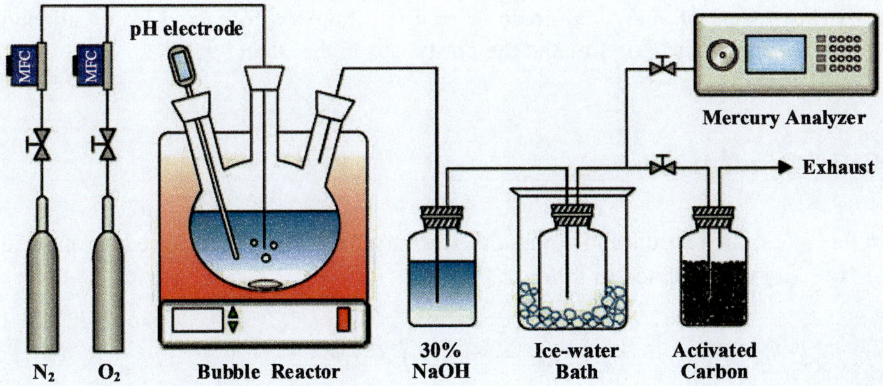

Fig. 67.1 Schematic of experimental apparatus

cylinders and controlled by mass flow controllers. The total flow rate of the simulated was 2 L/min. The bubble reactor mainly concludes a water bath and a three-necked glass flask reactor with a capacity of 1L. The water bath was used to maintain the temperature of the desulfurization liquid at the desired level. A pH electrode was inserted into the liquid to obtain the pH of the slurry. In the experiments, Na_2SO_3 was first added to the reactor as the source of sulfite. NaOH and HNO_3 were then used to adjust the pH of the slurry. Thereafter, 60 μg/L Hg^{2+} was quickly added into the liquid. A magnetic rotor was used to make a full mix of the gas and liquid phases during the desulfurization process. It should be noted that the O_2 remained off before the experiments began because sulfite is prone to oxidization and protonation. In addition, to prevent the influence of the condensation of water vapor and the adsorption of mercury on the internal surface of Teflon pipes on the reliability of results, the temperature of flue gas was kept at approximately 120 °C using auxiliary heat tapes. In the testing part, 30% NaOH and ice water bath were used to remove acid gas and retain water vapor, respectively. After that, the Hg^0 concentration was monitored using a VM3000 mercury analyzer (Mercury Instruments GmbH, Germany). The concentration of residual mercury in the liquid phase was detected using an atomic fluorescence spectrometer (RGF-6800, Beijing Ruiguang, China). In the exhaust gas treatment section, the activated carbon was employed to adsorb mercury from the exhaust gas to ensure environmental safety.

67.2.2 Experimental Materials

The nitrogen and oxygen gases were supplied by Chengdu Dongfeng Gas Co. Ltd. Anhydrous sodium sulfite, anhydrous calcium sulfate, nitric acid, sodium hydroxide and potassium dihydrogen phosphate were obtained from Chengdu Kelong Chemical Co., Ltd. Disodium hydrogen phonphare dodecahydrate and actived carbon were purchased from Chengdu Jinshan Chemical Reagent Co., Ltd. All the chemicals above were of analytical grade. Mercuric nitrate was obtained from Guizhou Tongrenhu Chemical Co., Ltd and the purity was higher than 99.9%.

67.2.3 Calculation

In this work, the re-emission amount of mercury (m_{Hg^0}, μg) and the reduction ratio of Hg^{2+} (α) were defined as follows:

$$m_{Hg^0} = 10^{-3} L \int C_{Hg^0} dt \tag{67.1}$$

$$\alpha = \frac{10^{-3} L \int C_{Hg^0} dt}{10^{-3} L \int C_{Hg^0} dt + C_{Hg^{2+}} V} \times 100\% \quad (67.2)$$

where C_{Hg^0} is the concentration of Hg^0 in the flue gas (μg/m^3), $C_{Hg^{2+}}$ is the concentration of Hg^{2+} in the desulfurization liquid (μg/L), L is the flow rate of the flue gas (L/min), V is the volume of the desulfurization liquid (L) and t is the desulfurization time (min).

67.3 Results and Discussion

67.3.1 Effect of Sulfite

In order to investigate the effect of sulfite on Hg^{2+} reduction and Hg^0 re-emission, a series of experiments were carried out with SO_3^{2-} concentrations ranging from 1.0 to 7.5 mmol/L. The pH and temperature of the desulfurization liquid were set to 5.0 and 50 °C, respectively. The results are illustrated in Fig. 67.2. Hg^0 re-emission occurred in the presence of SO_3^{2-} but the re-emission amount of Hg^0 decreased with increasing SO_3^{2-} concentration. As shown in Reactions (67.3) and (67.4), the reduction of Hg^{2+} by SO_3^{2-} may be the reason for Hg^0 re-emission. The standard electrode potentials of Hg^{2+}/Hg^0 and SO_3^{2-}/SO_4^{2-} are +0.85 V and −0.93 V, respectively (Surhone and Tennoe 2013). Thus the redox reactions between Hg^{2+} and SO_3^{2-} easily occurred in the desulfurization liquid, resulting in the re-emission of Hg^0. However, the results of Fig. 67.2b show that the reduction ratio of Hg^{2+} is less than 5% in all tests. This means that only a small amount of Hg^{2+} is reduced to Hg^0 in the experiments. Actually, the concentration of Hg^{2+} added to the liquid was much lower than that of SO_3^{2-}. In this case, Reaction (67.5) may play a dominan role in the desulfurization process and most of the mercury was stabilized as the form of $Hg(SO_3)_2^{2-}$. As a result, Hg^{2+} reduction was suppressed evidently and the re-emission amount of Hg^0 was quite low.

$$Hg^{2+} + SO_3^{2-} \longleftrightarrow HgSO_3 \quad (67.3)$$

$$HgSO_3 + H_2O \longrightarrow Hg^0 + SO_4^{2-} + 2H^+ \quad (67.4)$$

$$Hg^{2+} + 2SO_3^{2-} \longleftrightarrow Hg(SO_3)_2^{2-} \quad (67.5)$$

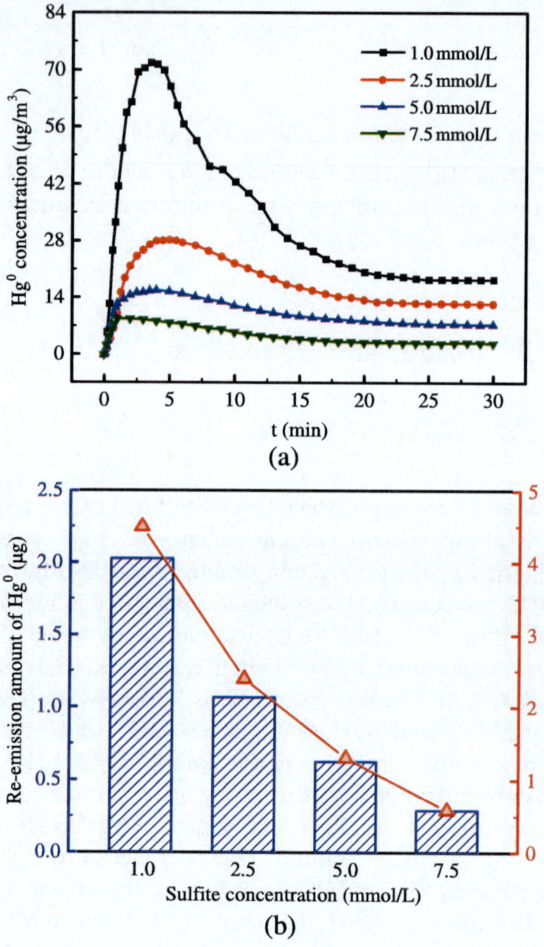

Fig. 67.2 Effect of sulfite concentration on Hg^0 re-emission. **a** Concentrations of Hg^0 re-emission at different sulfite concentrations **b** re-emission amounts of Hg^0 and reduction ratios of Hg^{2+} at different sulfite concentrations

67.3.2 Effect of Temperature

The slurry temperature in a typical WFGD system generally ranges from 40 to 60 °C, thus the experiments were conducted under the same temperature conditions. The pH of the liquid was maintained at 5.0 and sulfite concentration was 1 mmol/L. It can be seen from Fig. 67.3 that higher temperature intensified the reduction of Hg^{2+}. When the temperature increased from 40 to 60 °C, the re-emission amount and the reduction ratio of Hg^{2+} increased by approximately 10.8 and 12.2 times, respectively. This indicates that the rate constant of Hg^{2+} reduction is highly dependent on temperature. Based on the rate constants measured by Feinberg and Kurien (2015), the rate constants under certain conditions can be obtained using the Eyring-Polanyi equation (Loon and Mader 2001). The Eyring-Polanyi equation is defined as:

Fig. 67.3 Effect of temperature on Hg^0 re-emission **a** Concentrations of Hg^0 re-emission at different temperatures **b** re-emission amounts of Hg^0 and reduction ratios of Hg^{2+} at different temperatures

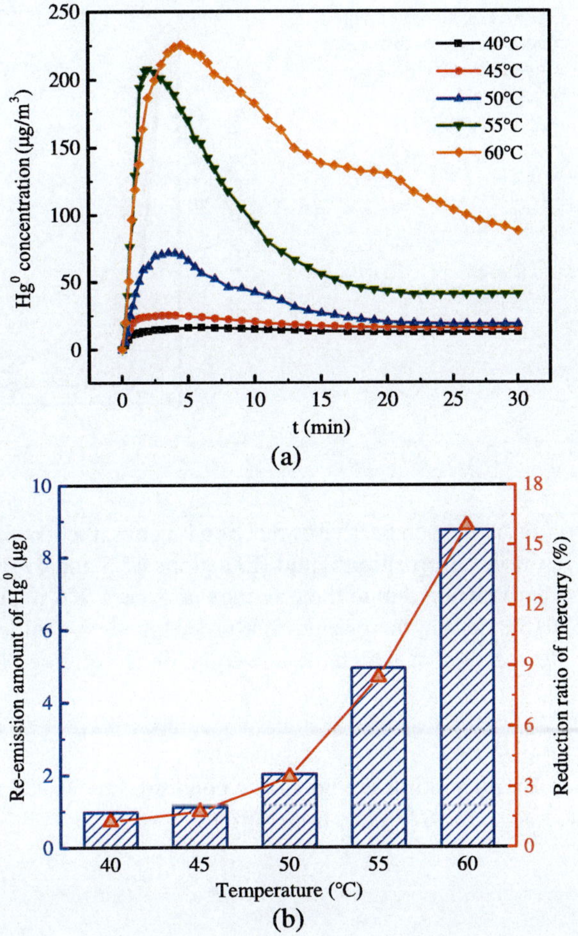

$$\ln \frac{k}{T} = -\frac{\Delta H}{RT} + \ln \frac{k_B}{h} + \frac{\Delta S}{R} \quad (67.6)$$

where k is the reaction rate constant, k_B is the Boltzmann constant (1.3806×10^{-23} JK^{-1}), R is the universal gas constant (8.314 $Jmol^{-1}$ K^{-1}), h is the Planck constant (6.626×10^{-34} Js), T is the absolute temperature (K), ΔH is the enthalpy of activation and ΔS is the entropy of activation.

The calculation shows that the rate constants for 40 °C, 45 °C, 50 °C, 55 °C, 60 °C are 7.5×10^{-2}, 13.98×10^{-2}, 25.56×10^{-2}, 45.91×10^{-2} and 81.05×10^{-2} s^{-1}, respectively. It suggests that the rate constant has an obviously positive correlation with temperature, which verifies the experimental results that Hg^0 re-emission was intensified with increasing temperature.

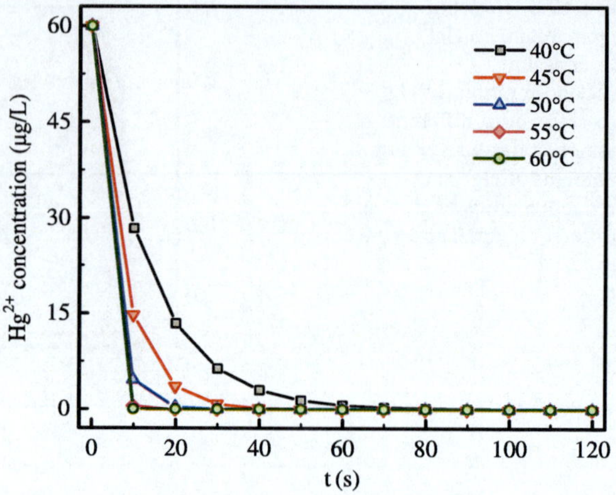

Fig. 67.4 Calculated Hg^{2+} concentrations at different temperatures ($[Hg^{2+}] =$ 60 μg/L, pH = 5)

From the perspective of reaction kinetics, the reaction of Hg^{2+} reduction occurred in the desulfurization liquid (Reactions 67.3 and 67.4) is considered a pseudo-first-order reaction due to the presence of excess $SO_3{}^{2-}$ and H_2O (Feinberg and Kurien 2015). Herein, the pseudo-first-order kinetic model is used to calculate the rate for this reaction, of which the rate equation is expressed as:

$$r = k \times C_{Hg^{2+}} \tag{67.7}$$

By integrating the boundary condition t = 0, $C_{Hg^{2+}} = C_{Hg^{2+},0}$; t = t, $C_{Hg^{2+}} = C_{Hg^{2+},t}$, Eq. (67.6) can transform to

$$C_{Hg^{2+},t} = C_{Hg^{2+},0} \times e^{-kt} \tag{67.8}$$

where r is the rate for Hg^{2+} reduction, k is the rate constant for Hg^{2+} reduction (s^{-1}), $C_{Hg^{2+}}$ is the concentration of Hg^{2+} in the liquid phase (mol/L) and t is the reaction time (s).

Assuming that Hg^{2+} is completely reduced to Hg^0 by sulfite, the calculated Hg^{2+} concentrations in the liquid at the temperature of 40–60 °C are illustrated in Fig. 67.4. The results show that increasing temperature can accelerate Hg^{2+} reduction, which is consistent with the experimental results shown in Fig. 67.3.

67.3.3 Effect of pH

The effect of pH on Hg^{2+} reduction and Hg^0 re-emission is shown in Fig. 67.5. The pH ranged from 4.0 to 6.0. The slurry temperature was set to 50 °C and sulfite concentration was 1 mmol/L. It can be seen that Hg^0 re-emission was suppressed

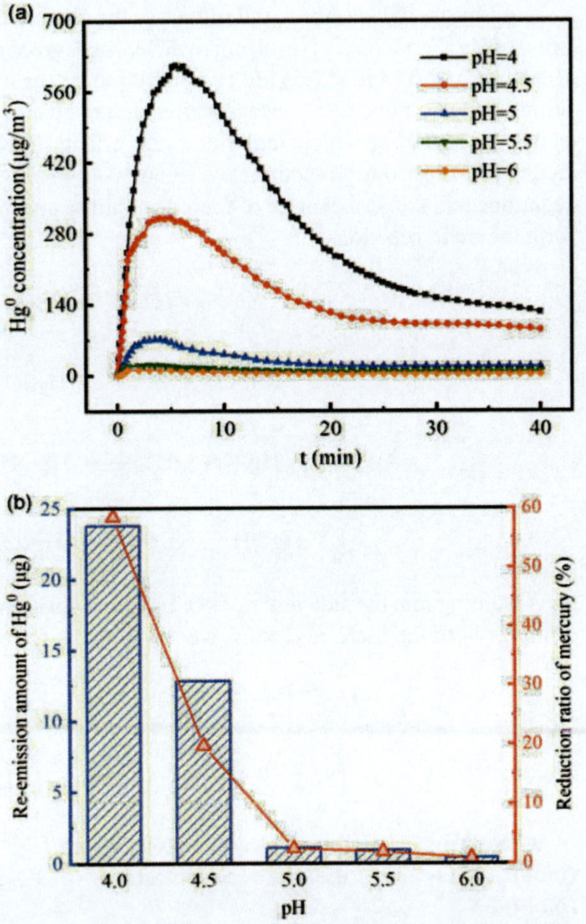

Fig. 67.5 Effect of pH on Hg^0 re-emission **a** Concentrations of Hg^0 re-emission at different pH values **b** re-emission amounts of Hg0 and reduction ratios of Hg^{2+} at different pH values

markedly with an increase in pH. The reduction ratio of Hg^{2+} also significantly decreased from 58.5 to 0.8% as pH increased from 4.0 to 6.0. The probable reason was reported that the pH variation had an obvious impact on the equilibrium of sulfite species (Omine and Romero 2012). ${SO_3}^{2-}$ would transform to ${HSO_3}^-$ when pH is less than 5.0, where $HgSO_3H^+$ complex is prone to form (shown in Reactions 67.9 and 67.10). Since $HgSO_3H^+$ is not as stable as $HgSO_3$, the enhancement of Hg^0 re-emission was observed.

$$Hg^{2+} + HSO_3^- \longleftrightarrow HgSO_3H^+ \quad (67.9)$$

$$HgSO_3H^+ + H_2O \longrightarrow Hg^0 \uparrow + SO_4^{2-} + 3H^+ \quad (67.10)$$

In addition, pH variation will influence the chemical equilibriums of the complexation of Hg^{2+} and $SO_3{}^{2-}$, resulting in different degrees of Hg^0 re-emission. Generally, Reactions (67.3) and (67.5) are considered to be the main reactions of the complexation of Hg^{2+} and $SO_3{}^{2-}$. Meanwhile, there are side reactions during the process of Hg^{2+} reduction, which include acidic effects (Reactions 67.11 and 67.12) and hydrolysis reactions (Reactions 67.13 and 67.14) (Loon and Mader 2001). The side reactions will cause a change of mercuric-sulfite complex equilibrium by competing with the main reactions.

$$SO_3^{2-} + H^+ \longleftrightarrow HSO_3^- \tag{67.11}$$

$$HSO_3^- + H^+ \longleftrightarrow H_2SO_3 \tag{67.12}$$

$$Hg^{2+} + OH^- \longleftrightarrow HgOH^- \tag{67.13}$$

$$HgOH^- + OH^- \longleftrightarrow Hg(OH)_2 \tag{67.14}$$

Assuming that the side reactions of mercuric-sulfite complexation are permitted to compete with the main reactions, we arrive at the following mass balance reactions:

$$\left[Hg^{2+\prime}\right] = \left[Hg^{2+}\right] + \left[HgOH^-\right] + \left[Hg(OH)_2\right] \tag{67.15}$$

$$\left[SO_3^{2-\prime}\right] = \left[SO_3^{2-}\right] + \left[HSO_3^-\right] + [H_2SO_3] \tag{67.16}$$

where $\left[Hg^{2+\prime}\right]$ is the total concentrations of Hg^{2+} not complexed with $SO_3{}^{2-}$ (mol/L), $[SO_3{}^{2-\prime}]$ is the total concentrations of $SO_3{}^{2-}$ not complexed with Hg^{2+} (mol/L).

Thus, the side reaction coefficients of hydrolysis reaction and acidic effect can be respectively expressed as follows:

$$\alpha_{Hg^{2+}(OH)} = \frac{[Hg^{2+\prime}]}{[Hg^{2+\prime}]} = \frac{[Hg^{2+}] + [HgOH^-] + [Hg(OH)_2]}{[Hg^{2+\prime}]} \tag{67.17}$$

$$\alpha_{SO_3^{2-}(OH)} = \frac{[SO_3^{2-\prime}]}{[SO_3^{2-}]} = \frac{[SO_3^{2-}] + [HSO_3^-] + [H_2SO_3]}{[SO_3^{2-}]} \tag{67.18}$$

Thereby, the absolute stability canstants of $HgSO_3$ (K_1) and $Hg(SO_3)_2{}^{2-}$ (K_2) are corrected to conditional stability constants $K_1{}'$ and $K_2{}'$. The calculations are listed below:

$$K\prime_1 = \frac{[HgSO_3]}{[Hg^{2+\prime}][SO_3^{2-\prime}]} = \frac{K_1}{\alpha_{Hg^{2+}(OH)}\alpha_{SO_3^{2-}(H)}} \tag{67.19}$$

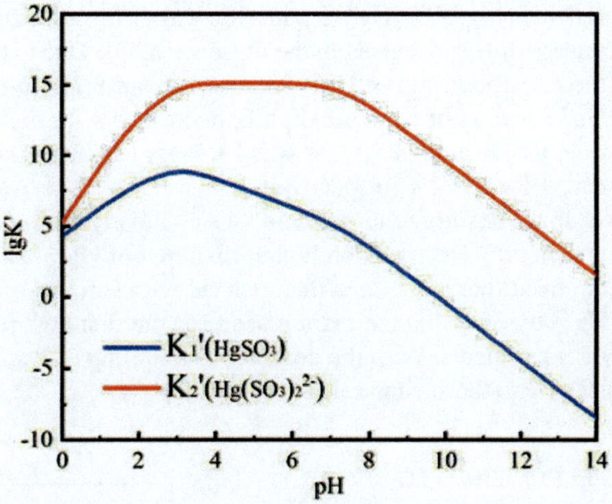

Fig. 67.6 Effect of pH on the conditional stability constants of $HgSO_3$ and $Hg(SO_3)_2^{2-}$ (T = 25°C)

$$K'_2 = \frac{[Hg(SO_3)_2^{2-}]}{[Hg^{2+'}][SO_3^{2-'}]^2} = \frac{K_2}{\alpha_{Hg^{2+}(OH)}\alpha^2_{SO_3^{2-}(H)}} \quad (67.20)$$

The conditional stability constants of $HgSO_3$ and $Hg(SO_3)_2^{2-}$ with pH are illustrated in Fig. 67.6. When pH is less than 4.0, both K_1' and K_2' increase with increasing pH. While at pH of 4.0–7.0, K_1' decreases gradually but K_2' maintains at a fixed value. It is speculated that lower pH can promote the decomposition of mercuric-sulfite complexes and causes an enhancement of Hg^0 re-emission. In the pH range of 4.0–7.0, SO_3^{2-} is the major sulfite species in the liquid and its concentration increases evidently as pH increases (Omine and Romero 2012). In this case, mercury mainly exists in the desulfurization liquid as the form of $Hg(SO_3)_2^{2-}$, and the stability of $Hg(SO_3)_2^{2-}$ affected by pH is very important. As shown in Fig. 67.6, the conditional stability constant of $Hg(SO_3)_2^{2-}$ keeps constant in the pH range of 4.0–6.0. This means the side reactions have no impact on Hg^0 re-emission. However, sulfite concentration increases with increasing pH value (Omine and Romero 2012), where stable $Hg(SO_3)_2^{2-}$ forms easily and Hg^0 re-emission is further inhibited.

67.3.4 Effect of O_2 Concentration

Sulfite ions are easily oxidized by the oxygen introduced into the simulated WFGD system, thus it is necessary to evaluate the effect of O_2 concentration on Hg^0 re-emission. Various oxygen concentrations (6–12%) were added to the reactor and N_2 is continuously suppiled as a balanced gas. The slurry temperature was 50°C and the sulfite concentration was 1 mmol/L. A buffer solution was used to maintain the pH at 5.0 and avoid experimental bias (Stergaršek and Horvat 2008). The results are shown in Fig. 67.7. Hg^0 re-emission was observed with or without O_2 but a secondary release of Hg^0 occurred in the presence of O_2. Due to the second re-emission of Hg^0, the reduction ratio of Hg^{2+} increased from 8.7% to more than 15%. However, the Hg^0 re-emission amount slightly decreased with increasing O_2 concentration. The re-emission amount of Hg^0 was 13.4 μg, 12.6 μg, 10.4 μg with O_2 concentration of 6%, 9%, and 12%, respectively.

In the desulfurization liquid, $SO_3{}^{2-}$ mostly exists in a free state since the concentration of $SO_3{}^{2-}$ is much higher than that of Hg^{2+}. During the experiments, O_2 first oxidized free $SO_3{}^{2-}$ and then reacted with $Hg(SO_3)_2{}^{2-}$. Therefore, the reduction of Hg^{2+} by $SO_3{}^{2-}$ is the main reason for the first Hg^0 re-emission within ten minutes of experiments. With the continuous bubbling of O_2, $Hg(SO_3)_2{}^{2-}$ will break down, leading to the second release of Hg^0.

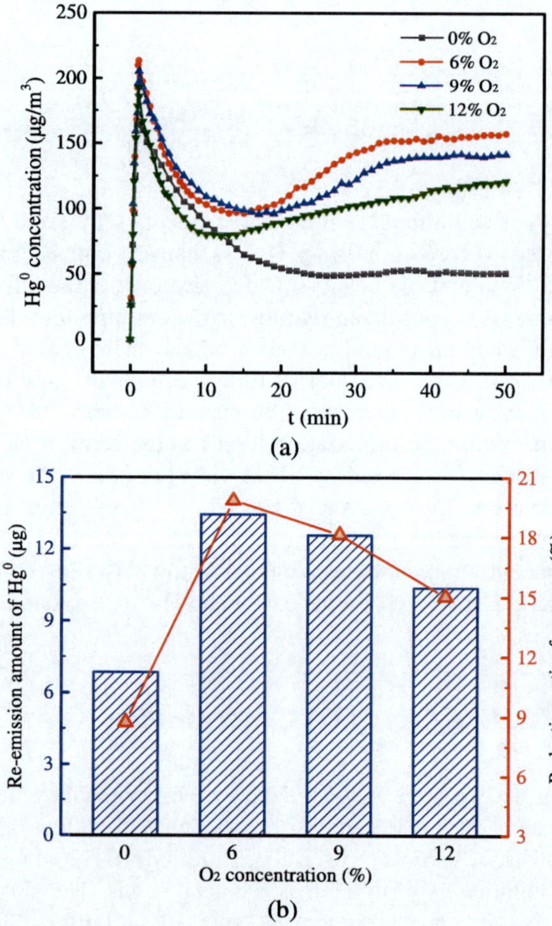

Fig. 67.7 Effect of O_2 concentration on Hg^0 re-emission **a** Concentrations of Hg^0 re-emission at different O_2 concentrations **b** re-emission amounts of Hg0 and reduction ratios of Hg^{2+} at different O_2 concentrations

Furthermore, the increase in O_2 concentration promotes the conversion from SO_3^{2-} to SO_4^{2-} and the formation of $HgSO_4$ and $HgSO_3SO_4^{2-}$ (Loon and Mader 2000). The reactions are shown in Reactions (67.21)–(67.23). $HgSO_3SO_4^{2-}$ can be stabilized in the desulfurization liquid, but $HgSO_4$ will be further extensively hydrolyzed to $Hg_3O_2(SO_4)$, which is readily dissolved in dilute mineral acids and may mainly exists in these solutions as undissociated molecules (Acuña-Caro 2009). Therefore, at higher content of O_2, a larger amount of mercury was retained in the liquid phase.

$$2SO_3^{2-} + O_2 \longleftrightarrow 2SO_4^{2-} \quad (67.21)$$

$$Hg^{2+} + SO_4^{2-} \longleftrightarrow HgSO_4 \quad (67.22)$$

$$2Hg(SO_3)_2^{2-} + O_2 \longleftrightarrow 2HgSO_3SO_4^{2-} \quad (67.23)$$

67.3.5 *Effect of Metal Ions*

Copper and Ferrous ions, as two typical heavy metal ions, exist in various flue gas scrubbing solutions (Guan and Ni 2009). The presence of metal ions may have a great impact on Hg^0 re-emission. The experiments were conducted in a simulated WFGD system. 0.5% $CaSO_4$ was added as the component of desulfurization slurry. The sulfite concentration was 1 mmol/L. The slurry temperature and pH were maintained at 50°C and 5.0. Cu^{2+} and Fe^{2+} ions were respectively added to investigate the influence of metal ions on Hg^0 re-emission. The results are shown in Figs. 67.8 and

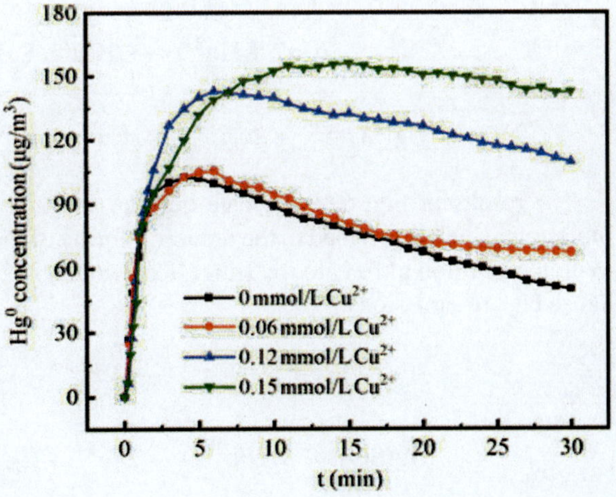

Fig. 67.8 Effect of copper ions on Hg^0 re-emission

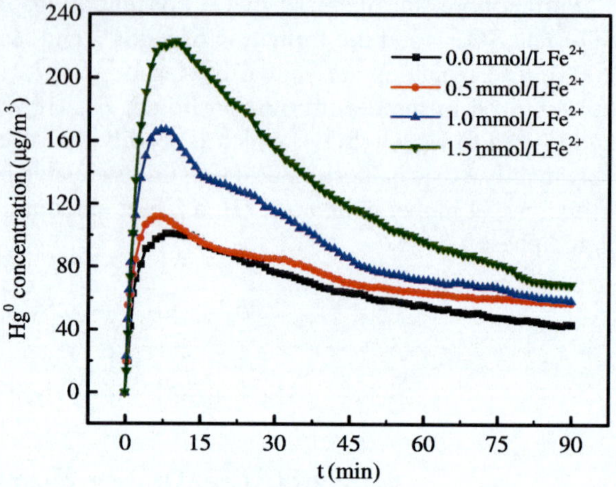

Fig. 67.9 Effect of ferrous ions on Hg^0 re-emission

67.9. The enhancement of Hg^0 re-emission was observed with the addition of Cu^{2+} or Fe^{2+}. Meanwhile, increasing the concentration of metal ions strengthened Hg^{2+} reduction and Hg^0 re-emission.

It has been reported that Cu^{2+} ions can be reduced to Cu^+ in the presence of $SO_3{}^{2-}$ (Khmeleva and Beattie 2003; Shen and Fornasiero 2001). And the electron exchanges between Cu^+ and Cu^{2+} ions in an acidic solution (McConnell and Weaver 1956). Due to the electron transfer, the reduction of Hg^{2+} and the decomposition of $Hg(SO_3)_2{}^{2-}$ can easily occur. In addition, Peng and Liu (2017) found that Cu^+ is the key to catalyzing the decomposition of $Hg(SO_3)_2{}^{2-}$. The more Cu^{2+} exists in the desulfurization liquid, the more Cu^+ is formed. As a result, the reduction of Hg^{2+} by Cu^+ is accelerated and Hg^0 re-emission is intensified. The reactions are listed below:

$$SO_3^{2-} + 2Cu^{2+} + H_2O \longleftrightarrow 2Cu^+ + SO_4^{2-} + 2H^+ \tag{67.24}$$

$$2Cu^+ + Hg^{2+} \longrightarrow 2Cu^{2+} + Hg^0 \uparrow \tag{67.25}$$

$$Hg(SO_3)_2^{2-} + 2Cu^+ \longrightarrow 2Cu^{2+} + Hg^0 \uparrow + 2SO_3^{2-} \tag{67.26}$$

The results of Fig. 67.9 indicate that the addition of Fe^{2+} can enhance Hg^0 re-emission, which is ascribed to the reducing ability of Fe^{2+} (Senior and Sarofim 2000). With the addition of Fe^{2+} to the slurry, Reactions (67.27) and (67.28) may occur and cause Hg^0 re-emission.

$$2Fe^{2+} + Hg^{2+} \longrightarrow 2Fe^{3+} + Hg^0 \uparrow \tag{67.27}$$

$$Hg(SO_3)_2^{2-} + 2Fe^{2+} \longrightarrow 2Fe^{3+} + Hg^0 \uparrow + 2SO_3^{2-} \tag{67.28}$$

67.4 Conclusion

The effects of sulfite concentration, slurry temperature, pH, O_2 concentration, and metal ions on Hg^{2+} reduction and Hg^0 re-emission were investigated in a simulated WFGD system. The mechanisms of temperature and pH affecting Hg^0 re-emission were also proposed based on the experimental and calculated results. The experimental results show that Hg^{2+} can be reduced to Hg^0 in the presence of $SO_3{}^{2-}$, whereas excess $SO_3{}^{2-}$ can stabilize mercury in the desulfurization liquid as the form of $Hg(SO_3)_2{}^{2-}$. Hg^0 re-emission was enhanced with increasing temperature and decreasing pH, which is mainly attributed to the acceleration of Hg^{2+} reduction. The kinetic results indicate that the rate constant of Hg^{2+} reduction increases with increasing temperature, leading to an increase in Hg^0 re-emission. As the pH decreases, the unstable $HgHSO_3{}^+$ is easily formed and then decomposed to Hg^0. Due to the side reactions, a decrease in pH also can influence the equilibrium of mercuric-sulfite complexes, and thus weakening the stabilities of these complexes. Compared with an N_2 atmosphere, a secondary release of Hg^0 was observed with the addition of O_2. Increasing O_2 concentration inhibited Hg^0 re-emission to some extent. In addition, both Cu^{2+} and Fe^{2+} show the ability to reduce Hg^{2+}. Therefore, Hg^0 re-emission was intensified with increasing concentration.

Acknowledgements This work was supported by the National Natural Science Foundation of China (51506099, 51676101 and 51606130) and the Natural Science Foundation of Jiangsu Province (BK20161558).

References

Acuña-Caro C (2009) Brechtel K: The effect of chlorine and oxygen concentrations on the removal of mercury at an FGD-batch reactor. Fuel 88(12):2489–2494

Chang L, Zhao Y (2017) Effect of sulfite on divalent mercury reduction and re-emission in a simulated desulfurization aqueous solution. Fuel Process Technol 165:138–144

Díaz-Somoano M, Unterberger S (2007) Mercury emission control in coal-fired plants: the role of wet scrubbers. Fuel Process Technol 88(3):259–263

Feinberg AI, Kurien U (2015) The kinetics of aqueous mercury (II) reduction by sulfite over an array of environmental conditions. Water Air Soil Pollut 226(4):119

Galbreath KC, Zygarlicke CJ (2000) Mercury transformations in coal combustion flue gas. Fuel Process Technol 65:289–310

Guan B, Ni W (2009) Removal of Mn (II) and Zn (II) ions from flue gas desulfurization wastewater with water-soluble chitosan. Sep Purif Technol 65(3):269–274

Guan Q, Liu Z, Fang X et al (2017) Experimental study on dense-phase pneumatic conveying of coal powder at high pressures. Clean Energy 1(1):50–67

Hou J, Lu R (2012) Effect of heavy metals on the stabilization of mercury (II) by DTCR in desulfurization solutions. J Hazard Mater 217:224–230

Jaishankar M, Tseten T (2014) Toxicity, mechanism and health effects of some heavy metals. Interdisc Toxicol 7(2):60–72

Khmeleva TN, Beattie DA (2003) Surface study of the effect of sulphite ions on copper-activated pyrite pre-treated with xanthate. Miner Eng 16(7):601–608

Van Loon L, Mader E (2000) Reduction of the aqueous mercuric ion by sulfite: UV spectrum of $HgSO_3$ and its intramolecular redox reaction. J Phys Chem A 104(8):1621–1626
Van Loon LL, Mader EA (2001) Sulfite stabilization and reduction of the aqueous mercuric ion: kinetic determination of sequential formation constants. J Phys Chem A 105(13):3190–3195
Liu Y, Wang Y (2011) A mechanism study of chloride and sulfate effects on Hg^{2+} reduction in sulfite solution. Fuel 90(7):2501–2507
Liu Y, Wang Y (2015) A study on removal of elemental mercury in flue gas using fenton solution. J Hazard Mater 292:164–172
McConnell HM, Weaver HE Jr (1956) Rate of electron exchange between cuprous and cupric ions in hydrochloric acid solutions by nuclear magnetic resonance. J Chem Phys 25(2):307–311
Omine N, Romero CE (2012) Study of elemental mercury re-emission in a simulated wet scrubber. Fuel 91(1):93–101
Ochoa-González R, Díaz-Somoano M (2013a) The capture of oxidized mercury from simulated desulphurization aqueous solutions. J Environ Manage 120:55–60
Ochoa-González R, Díaz-Somoano M (2013b) Effect of anion concentrations on Hg^{2+} reduction from simulated desulphurization aqueous solutions. Chem Eng J 214:165–171
Pirrone N, Cinnirella S (2010) Global mercury emissions to the atmosphere from anthropogenic and natural sources. Atmos Chem Phys 10(13):5951–5964
Peng B, Liu Z (2017) Effect of copper ions on the mercury re-emission in a simulated wet scrubber. Fuel 190:379–385
Senior CL, Sarofim AF (2000) Gas-phase transformations of mercury in coal-fired power plants. Fuel Process Technol 63(2–3):197–213
Shen WZ, Fornasiero D (2001) Flotation of sphalerite and pyrite in the presence of sodium sulfite. Int J Miner Process 63(1):17–28
Stergaršek A, Horvat M (2008) The role of flue gas desulphurisation in mercury speciation and distribution in a lignite burning power plant. Fuel 87(17–18):3504–3512
Surhone LM, Tennoe MT (2013) Standard electrode potential
Tang T, Xu J (2010) Enhanced Hg^{2+} removal and Hg^0 re-emission control from wet fuel gas desulfurization liquors with additives. Fuel 89(12):3613–3617
Wang Y, Liu Y (2010) Experimental study on the absorption behaviors of gas phase bivalent mercury in Ca-based wet flue gas desulfurization slurry system. J Hazard Mater 183(1–3):902–907
Wo J, Zhang M (2009) Hg^{2+} reduction and re-emission from simulated wet flue gas desulfurization liquors. J Hazard Mater 172(2–3):1106–1110
Xu J, Bao J (2018) Characteristics and inhibition of Hg^0 re-emission in a wet flue gas desulfurization system. Energy Fuels 32(5):6111–6118

Chapter 68
Study of the Concentration of Mercury in Coal Used in Combustion, in an Area of Boyacá, Colombia, South America

Sonia Guerra L, Manuel Romero, and Daniel Ballen

Abstract According to research of the Servicio Geológico Colombiano (SGC by Spanish initials), the Only Plan of Mercury and the Strategic Sector Plan for the elimination the mercury use, a project that studies the occurrence of mercury in the coal in an area of Boyacá is developed as an integral part in the generation of geoscientist knowledge of the subsoil of the national territory. In this context, and in order to establish a baseline of the mercury content in the coals of the area Sogamoso—Tunja and its effect on combustion processes, Characterization and Processing of Minerals and Coals Group carried out 84 samples in front of mines of 16 municipalities of the area, characterizing samples in 24 parameters, physic-chemical and petrographic, obtaining an important input in aspects of: classification, research in combustion, mercury in the combustion cycle, evaluation environmental, social and economic. It includes an analysis of mercury and its relations with the other measured parameters, given its importance in the production of coal energy and the associated environmental impact. The average content of mercury in dried base in analyzed samples is 127 μg/kg. 89% of the coals have values less than 300 μg/kg. According to the results, an acceptable significant direct correlation was identified between the mercury content with Ash, sulfur, pyritic sulfur, and no correlation with chlorine content, which reacts in the combustion.

Keywords Coal · Characterization · Mercury · Combustion

S. Guerra L (✉) · M. Romero · D. Ballen
Dirección de Laboratorios, Servicio Geológico Colombiano, Bogotá, D.C. 111321, Colombia
e-mail: sguerra@sgc.gov.co

M. Romero
e-mail: mhromero@sgc.gov.co

D. Ballen
e-mail: vballen@sgc.gov.co

J. Lyu and S. Li (eds.), *Clean Coal and Sustainable Energy*,
Environmental Science and Engineering,
https://doi.org/10.1007/978-981-16-1657-0_68

68.1 Methodology

- To carry out the complete characterization of the coals sampled in coal mines used in combustion, in an area of Boyacá (Fig. 68.1) with the energy company that uses them.
- Mapping the quality of the coal basins of the department of Boyacá with the mercury content of geological origin based on the coal quality database.
- To analyze mercury in the complete carbon combustion cycle, from the mine to the emissions of mercury on the direct influence of the coal-fired power plant.
- Fully evaluate the results obtained to potentiate the use of coal and give added value.
- Conducting samplings to elaborate stratigraphic columns
- To relate mercury with other elements that are present in coal to contribute to the development of a sustainable model that mitigates environmental impacts.

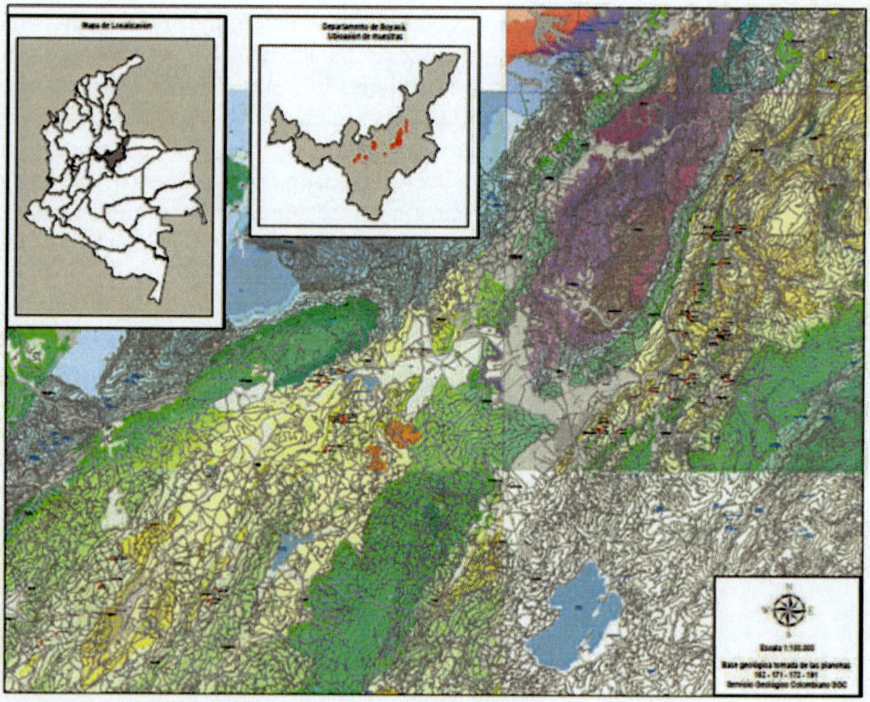

Fig. 68.1 Map of sampling zones in the Deparment of Boyaca. The dots are showing the areas that were sampled for this study

68.2 Results

Coal as a heterogeneous mineral contains mercury in different concentrations and ways of combination. The concentrations of mercury in the coals of the Boyaca's area vary between 8 and 486 μg/kg. In this coal zone, the average content of mercury is lower than the earth's crust content (Fig. 68.2).

Sulphides are generally form by Mercury, so its content is directly related to the content of sulphides. According to the results, an acceptably significant direct correlation was identified between the mercury content with: ash, sulfur, pyritic sulfur and mineral material and not linear correlation with the chlorine content with which it reacts in combustion.

It is important to analyze the concentration of chlorine since it is the halogen that helps to increase the retention of mercury in precipitators and desulfurizers by reacting with it in the combustion gases of coal (Kolker 2011).

A database generated with the characteristics of the coals used in combustion to contribute to the quality map of Colombian coals.

To Contribute to the proposal of a sustainable model to mitigate the impacts generated in the combustion process.

It was sampling and identifying nineteen (19) samples of coal, and rock from the top and bootom of the coal bed to make six (6) stratigraphic columns of the areas where the sampled mines are located. According to the stratigraphic information, the geology of the region is of great complexity and is characterized by a dominantly compressive environment related to the uplift of the northern Andes. The area has a set of faults and folds that frame the arrangement of carbon mantles for the study (Porras 2019). In this regard, the major structures highlighted are:

- Sinclinorio de Paz de Río (Ulloa et al. 2003)
- Soapaga failure (Ulloa et al. 1981, 2003)
- Boyacá failure (Ulloa et al. 1981, 2003).

Based on the above, the laboratory group considers carrying out other studies of the complete cycle of the mercury present in coal, from its exploitation to its use in other processes such as the production of bricks, steel and coking.

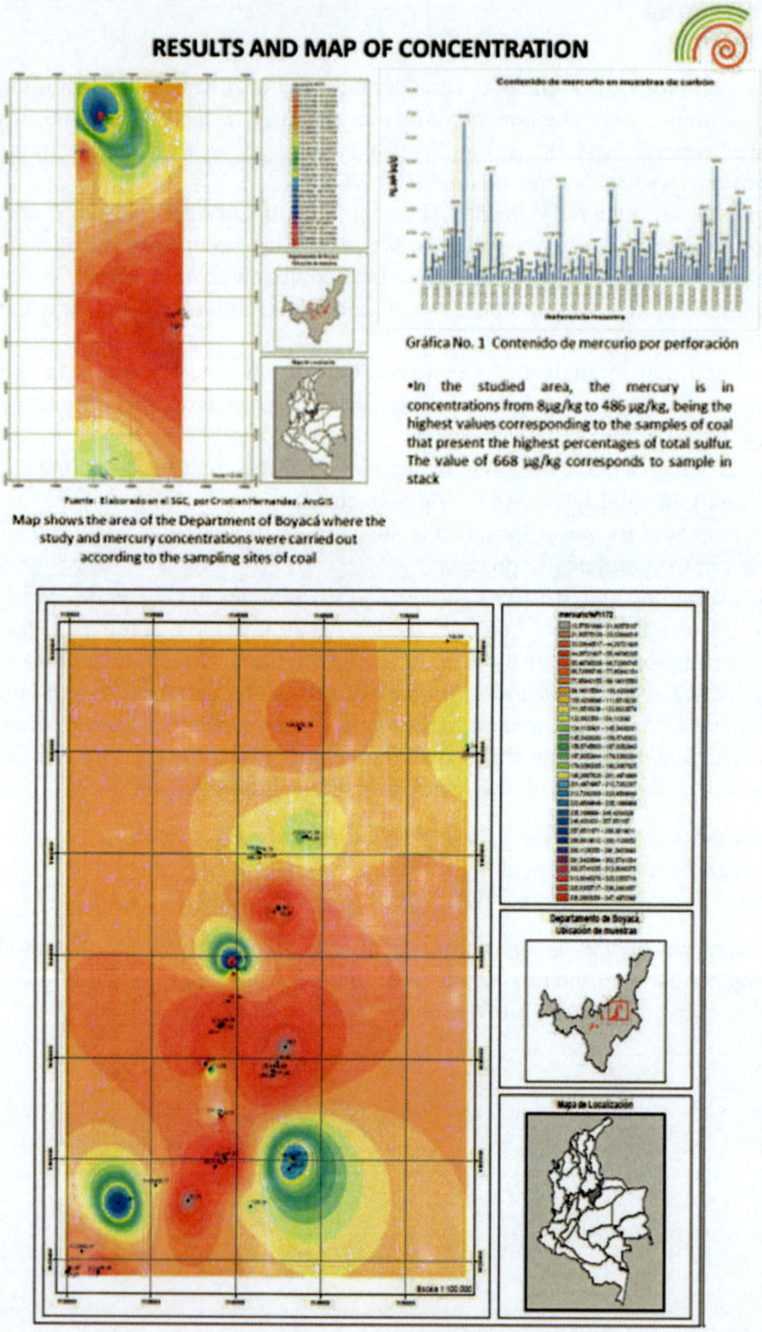

Fig. 68.2 Concetrations of Mercury. Maps show the area of the Department of Boyacá where the study and mercury and sulfur concentrations were carried out according to the sampling sites of coal. *Source* Prepared in the SGC, by Cristian Hernandez. ArcGIS

Acknowledgements This work was supported by the Characterization and Processing of Minerals and Coal Group, Laboratory Direction, SGC and the Special with the collaboration of Power Plant Fire coal Paipa I, II y III of the group GENSA.

References

Kolker AJC (2011) Mercury and Halogens in coal—their role in determining mercury emissions from coal combustion. USGS United Stated

Porras P (2019) Análisis de carbones térmicos de Boyacá para evaluar su potencial metalífero (Pregrado). Universidad Nacional, Bogotá

Ulloa C, Rodríguez E, Renzi G (1981) Geología de la Plancha 171 Duitama, 191 Tunja, 211 Tauramena y 230 Monterrey Cuadrángulos J – 12 Y K – 13. Memoria Explicativa. Boletin Geológico 24(1976):55. Retrieved from http://recordcenter.sgc.gov.co/B4/13010010024364/documento/pdf/0101243641101000.pdf

Ulloa C, Rodríguez E, Rodríguez G (2003) Geología de la Plancha 172, Paz de Río, Memoria explicativa. Geología De La Plancha 172 Paz De Río, 111

Chapter 69
An Experimental Study on Ash Deposition Problem of Low-Low Temperature Flue Gas System

Yu Yan, Jiahao Jiang, Jin Guo, Yuesheng Li, Lei Deng, and Defu Che

Abstract In order to achieve the ultra-clean emission standards, many power plants in China have completed the installation of low-low temperature flue gas system. However, during the operation of this system, many problems have emerged, such as the ash deposition problem in the heat exchanger and dry ESP (electrostatic precipitator). In this paper, the fixed adsorption system and entrained flow experimental system were built to study the ash deposition characteristics in low-low temperature flue gas system under different temperatures and particle sizes. The repose angle was measured to evaluate the flowability and adhesion capacity of ash particles. The results show that as the increase of adsorption time, the particle size increased gradually, and the agglomeration effect is attenuated after the particles have reached adsorption saturation. The time at which the ash particles began to agglomerate with each other was defined as the agglomeration threshold. As the adsorption temperature decreased, the particle size growth rate increased. The smaller the size of ash particles, the more significant the agglomeration effect. Large particles were easy to adsorb small particles, but the agglomeration effect between large particles was weak. In addition, in the initial stage of particle agglomeration, it was mainly included the adhesion between smaller particles and the adhesion of smaller particles on the surface of large particles. The followability of the ash particles was greatly reduced after the adsorption and agglomeration process. The repose angle of the ash particles increased as the temperature of the flue gas decreased. The smaller the particle size, the larger the repose angle changes, and the more pronounced the repose angle is influenced by the adsorption temperature. The flue gas temperature cannot be

---

Y. Yan · J. Jiang · L. Deng · D. Che (✉)
State Key Laboratory of Multiphase Flow in Power Engineering, School of Energy and Power Engineering, Xi'an Jiaotong University, Xi'an 710049, China
e-mail: dfche@mail.xjtu.edu.cn

J. Guo
Guangdong Institue of Special Equipment Inspection and Research, Foshan 528251, China

Y. Li
Guangdong Institue of Special Equipment Inspection and Research, Shunde Institue of Inspection, Foshan 528300, China

J. Lyu and S. Li (eds.), *Clean Coal and Sustainable Energy*, Environmental Science and Engineering,
https://doi.org/10.1007/978-981-16-1657-0_69

blindly dropped in pursuit of high sulfur oxide removal efficiency and dust removal efficiency.

Keywords Ash deposition · Low-low temperature flue gas system · Agglomeration · Repose angle

69.1 Introduction

With the support of China's sustainable energy policy, new energy sources are developing rapidly at present. However, due to the feature of the resource endowment, coal is still the main energy resource of China (Xunzhang et al. 2017; Zhang et al. 2017; Ren and Xia 2017). In the year 2017, China's total coal consumption reached 2.7 billion tons of standard coal equivalent (tce), of which coal for power generation accounted for 47.5% (National Bureau of Statistics of China 2018 2018; Xuehui et al. 2018). The large-scale use of coal in power plant has caused many serious environment problems, such as the air pollution and the acid rain (Quan et al. 2014; Hu et al. 2016; Chen et al. 2017). Thus it urges coal-fired power plants to carry out ultra-low emission retrofits to achieve the deep removal of fine particles and sulfur oxides (Chen et al. 2018).

Low-low temperature flue gas system is developed by Mitsubishi Corporation. It can achieve multi pollutant deep removal such as SO_2, SO_3, PM even $PM_{2.5}$ with high efficiency and low cost (Jianguo et al. 2014). In this system, a tubular heat exchanger installed upstream of electrostatic precipitator (abbreviated as ESP) reduces flue gas temperature below the acid dew point. This system is suitable for low sulfur coal and has remarkable application effects in Japan. According to Nakayama et al. (2011), the SO_3 concentration of FGD outlet is reduced to lower than 0.1 ppm, and the PM collecting efficiency of ESP is almost over 99% in Japanese power plant. However, the sulfur content of coal in China is generally higher and the types of coal used in China's power plant are changeable (Yunfei et al. 2005; Yuegang et al. 2015). After adsorbing H_2SO_4 mist, the serious agglomeration of ash particles happens, which finally leads to the ash deposition problem of the heat exchanger in the applications of low-low temperature flue gas system in China. Therefore, to solve the ash deposition problem, it is necessary to study the agglomeration characteristics of fly ash during the flue gas cooling process in the exchanger.

In this paper, the fixed bed adsorption experimental system was used to studied the agglomeration effect of ash particles. In this part, laser diffraction particle size analyzer (Mastersizer 3000) was used to analyze the size distribution of ash particles. The influences of adsorption temperature and particle size were studied. The entrained flow adsorption experimental system was built to studied the flowability of ash particles after adsorption reaction. The repose angle was measured to evaluate the flowability and adhesion capacity of ash particles. The results of this paper can help alleviate the ash deposition problem of the heat exchanger in low-low temperature flue gas system.

69.2 Experimental Section

69.2.1 Ash Samples and Experimental Condition

The experimental coal ash samples were collected from the inlet and outlet of the low-temperature economizer of a 660 MW heat power plant, respectively. At the inlet of the economizer, the temperature of flue gas was higher than the acid dew point. Hence, the ash collected form the inlet was considered as original sample without adsorption reaction with H_2SO_4 mist. Correspondingly, the ash samples from the outlet were the ash after the adsorption reaction. In order to guarantee the samples were uncontaminated, all samples were stored in the Ziploc bags for transportation.

The adsorption temperatures were set to 80, 90, 100, 130 °C to simulate the actual operating temperature of low-low temperature flue gas system. The ash samples were dried and sieved into four size (≤68, >68~75, >75~91, and >91~125 μm) before the experiment. Because the ash particles were not a perfect spherical shape, especially the ash samples after adsorption and agglomeration, a characterization method was need to represent the size of the ash particles. In this experiment, D (National Bureau of Statistics of China 2018; Ren and Xia 2017) was chosen for it was more sensitive to large particles and irregular particle than D (Ren and Xia 2017; Zhang et al. 2017) and $D_y(x)$.

69.2.2 Fixed Bed Adsorption Experimental System

The fixed bed adsorption experimental system consisted of four parts: gas preheating part, SO_3 generation part, adsorption reaction part and sulfuric acid collection part. The system is shown in Fig. 69.1. The 20% H_2SO_4 solution was filled in the glass syringe and injected into the silica tube by the micro-injection pump at a speed of 0.15 mL min^{-1}. A stream of N_2 heated by a spiral tube carried the H_2SO_4 droplets into the pyrolysis reaction furnace, in which the H_2SO_4 would decompose into gaseous SO_3 and H_2O. The temperature of the mixed gas dropped in the furnace and the adsorption reaction occurred in the adjacent furnace. A silica boat was placed on the tail section of the silica tube, where a thermocouple was fixed to control the adsorption temperature. The ash sample was evenly sprinkled on the silica boat by a sieve. After the experiment, the size distribution of ash particles was tested by laser diffraction particle size analyzer.

69.2.3 The Entrained Flow Adsorption Experimental System

The experiment was conducted on the entrained flow adsorption system, which was designed to simulate the cooling process of flue gas in heat exchanger. As shown in

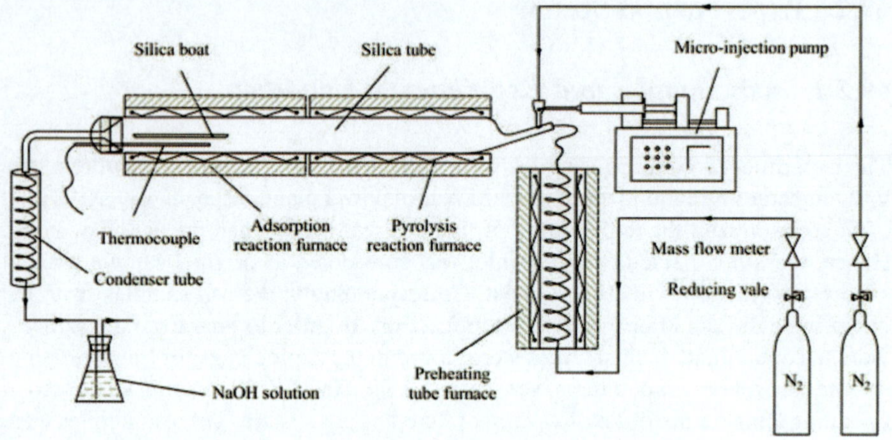

Fig. 69.1 Schematic diagram of the fixed bed adsorption experimental system

Fig. 69.2, the system consisted of four parts: SO_3 generation part, coal ash feeding part, adsorption reaction part and ash collection part. There were a micro-injection pump, a tube furnace and a silica tube in the SO_3 generation part. 20% H_2SO_4 solution was filled in a glass syringe on a micro-pump. The tube furnace was set to 550 °C before the experiment. When the experiment began, the H_2SO_4 solution was injected into the silica tube by the pump at a speed of 0.15 mL min^{-1}. At the same time, coal ash was blew into the adsorption reaction tube by a micro-ash feeder at a speed of 4 L min^{-1}. With the drop of the mixed gas temperature, the adsorption between ash particles and H_2SO_4 would occurred. After the adsorption, the ash was collected by

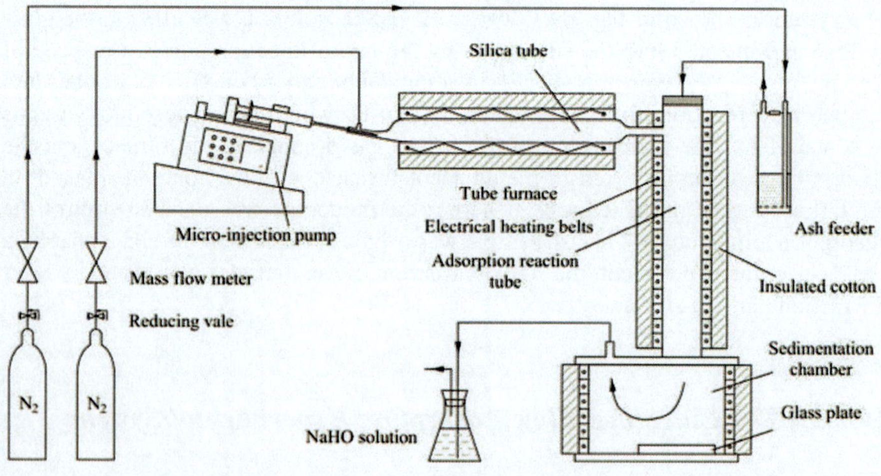

Fig. 69.2 Schematic diagram of the entrained flow adsorption experimental system

Table 69.1 The calibration of SO_3 content of the flue gas

Injection speed (mL min^{-1})	SO_3 content (μL L^{-1})			
	Group 1	Group 2	Group 3	Mean value
0.100	379.41	389.16	385.15	384.57
0.125	579.73	576.33	585.08	580.38
0.150	692.51	709.46	686.93	696.30

a glass plate placed on the bottom of a sedimentation chamber. The repose angle of the ash sample was obtained by a funnel and an ash stacking table. After stacking, the ash sample was photographed and the angle was measured by CAD on a computer.

69.2.4 Calibration Tests of the Entrained Flow Adsorption Experimental System

Calibration tests were taken before the experiment on the entrained flow adsorption system. SO_3 content of the flue gas under different injection speed and the ash feeding speed were calibrated. Manual controlled condensation method (CCM) was used to collect SO_3 and the content was analyzed by the ultraviolet spectrophotometer. When flue gas flow rate was 6 L min^{-1} the tests results are shown in Table 69.1. Three groups of repeated experiments were carried out and the difference between these results was within the error range indicating that the experimental system is practicable and the results are credible.

The results of calibration experiments of ash feeder system are shown in Fig. 69.3, which demonstrate the ash feeding speed of different ash particle sizes. As shown in the figure, the ash feed speed is linear and accurate under different conditions, which indicates that the ash feed system is effective to guarantee appropriate feed speed of different ashes.

69.3 Result and Discussion

69.3.1 The Agglomeration Effect of Ash Particles

The agglomeration effect of ash particles was represented by particle size growth rate. The particle size growth rate under different adsorption time is shown in Fig. 69.4. The two ordinates in the figure represent the absolute and relative increase in particle size of the ash sample after the adsorption reaction, and the selected particle size is 68–75 μm.

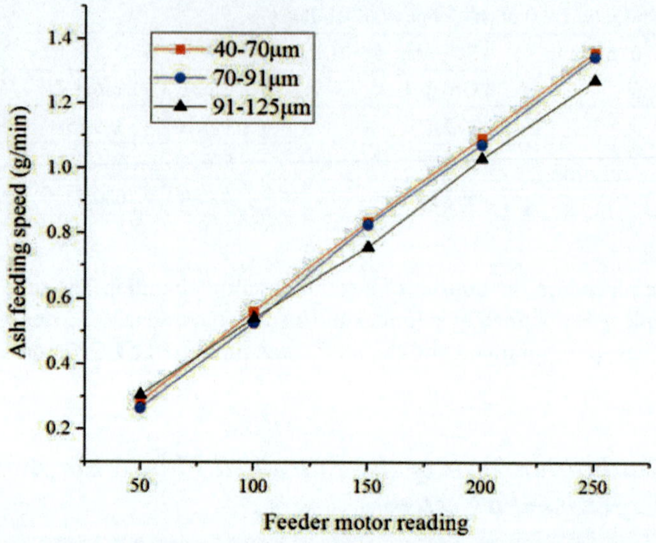

Fig. 69.3 Calibration of ash feeding speed of different particle sizes

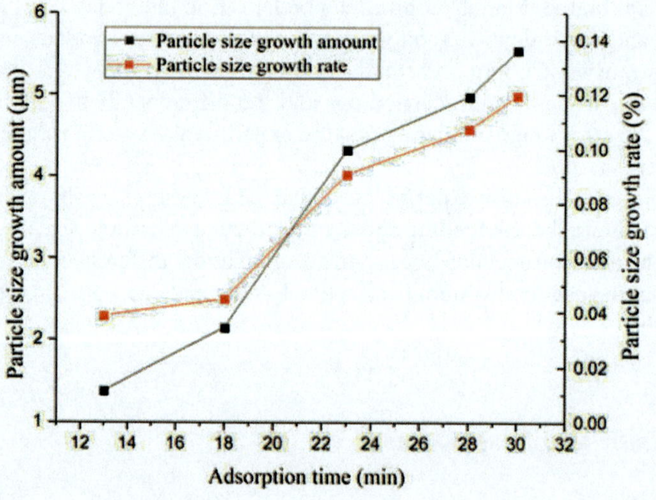

Fig. 69.4 The particle size growth rate of 68–75 μm ash

It can be seen from Fig. 69.4 that as the increase of adsorption time, the particle size of the ash particles increased gradually, and the agglomeration became remarkable. Before 23 min, the particle size increase rate was higher, and then significantly slowed down. The reason may be that the ash particles was in the rapid adsorption phase before 23 min. The amount of adsorbed sulfuric acid mist was large at this phase, so agglomeration effect was better. After 23 min, the adsorption reaction of the ash

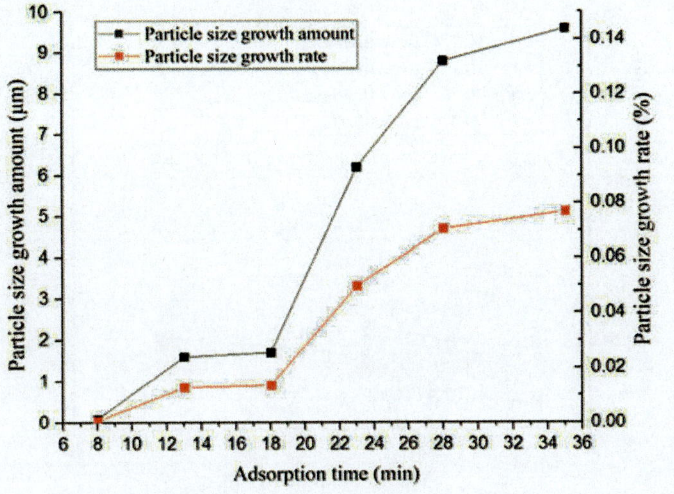

Fig. 69.5 The particle size growth rate of 91–125 μm ash

particles gradually approaches saturation, and the amount of H_2SO_4 mist adsorbed at this stage is less. Hence, the increase rate of ash particle size was dropped. In other words, the effect of agglomeration was weakened. The relative increase of particle size was less than 14%. That is because the increase rate was obtained by the average particle size, and a part of ash particles didn't agglomeration with each other.

The size increase rate of 91–125 μm ash particles is shown in Fig. 69.5. As the adsorption time increased, the particle size of the ash particles increased. And before 28 min, the particle size increase rate was higher, and then significantly slowed down. There was almost no change in particle size before 8 min. From this phenomenon, it can be inferred that during the adsorption process, the agglomeration of ash particles didn't occur at the beginning. The time at which the ash particles began to agglomerate with each other was defined as the agglomeration threshold. Before the agglomeration threshold, although the ash particles had been adsorbed with the H_2SO_4 mist, no agglomeration occurs between them. After the adsorption time over the agglomeration threshold, agglomeration begins to occur between the particles.

69.3.2 The Influence of Adsorption Temperature on Agglomeration Effect

In this study, the influence of adsorption temperature on agglomeration effect was investigated at 90 and 100 °C. The selected particle size rang was 68–75 μm. The experimental results are shown in Fig. 69.7. It can be inferred from Fig. 69.6 that as the adsorption temperature decreased, the particle size growth rate increased. That is, the lower the adsorption temperature, the more obvious the agglomeration

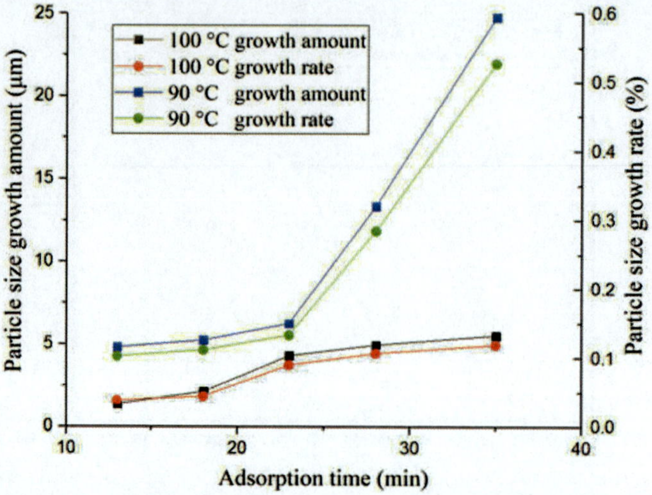

Fig. 69.6 The influence of adsorption temperature on agglomeration effect

effect. When the temperature was lowered by 10 °C, the agglomeration effect of the particles was greatly improved. The reason for this phenomenon was that as the adsorption temperature dropped, the humidity of the H_2SO_4 mist increases. After the adsorption reaction, the liquid film on the surface of the particulate matter became thick. The liquid film on the surface of the particles could promote the adhesion of the particles after collision. So the particles after collision with each other were not easily separated, and the reaction between the particles promoted the agglomeration between ash particles. Finally, the fine particle aggregated into larger particles.

69.3.3 The Influence of Particle Size on Agglomeration Effect

Based on the study of the influence of adsorption temperature on the agglomeration effect, the influence of particle size was investigated. The experimental temperature was chosen to be 100 °C, and the particle size of the ash sample was 68–75, 75–91, and 91–125 μm. The experimental results are shown in Fig. 69.7. It can be seen from the figure, that the smaller the size of ash particles, the more significant the agglomeration effect. Large particles were easy to adsorb small particles, and the agglomeration effect between large particles was weak. The agglomeration phenomenon of particulate matter after adsorption of sulfuric acid mist is beneficial to the removal of fine particles. However, the agglomeration of ash particles also may cause ash deposition problem in heat exchangers. In addition, in the initial stage of agglomeration of ash particles, the agglomeration phenomenon was mainly included the adhesion between smaller particles and the adhesion of smaller particles on the

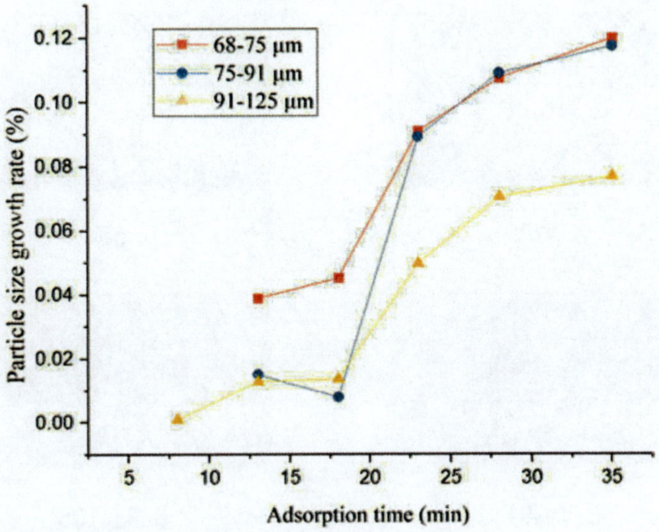

Fig. 69.7 Effect of particle size on agglomeration effect

surface of large particles. As the adsorption time prolongs, agglomeration began to occur between large particles.

69.3.4 The Flowability of Ash Particles

The ash samples taken from the inlet of economizer were under the condition that flue gas temperature was higher than acid dew point and the adsorption reaction didn't occur. After testing, the repose angle of the ash samples at each particle size is shown in Fig. 69.8 which is about the particle size of ≤68, >68~75, >75~91, and >91~125 μm.

It can be seen from Fig. 69.8 that the economizer coal ash was dry and loose. There was no obvious agglomeration. The repose angle of the four particle sizes of the dry coal ash particles was in the range of 50–57°, and when the particle size was less than 68 μm, the deposition angle was 56.32°. As the particle size increased, the coal ash particle accumulation angle gradually decreased. When the particle size was 91–125 μm, the deposition angle decreased to 51.65°.

The repose angle of the dry ash particles decreased with the increase of the particle size, indicating that the smaller the particle size, the lower the flowability. The reason for this phenomenon may be that as the particle size of the coal ash decreases, the molecular gravitational force and electrostatic attraction between the particles gradually increase, reducing the flowability of the ash particles. Secondly, the smaller the particle size, the more likely the adsorption and agglomeration occur between the particles, resulting in an increase in the angle of repose and a decrease in flowability. In

Fig. 69.8 The repose angle of ash sample of the economizer inlet

addition, as the particle size of the coal ash decreases, the particles tend to form close packing, resulting in a decrease in gas permeability and an increase in compression ratio.

The ash samples taken from the outlet of economizer was under the condition that after the adsorption and agglomeration process. After testing, the repose angle of the ash samples at each particle size is shown in Fig. 69.9 which is about the particle size of ≤68, >68~75, >75~91, and >91~125 μm.

After the flue gas temperature is dropped, the ash particles after absorption obviously agglomerated, and the adhesion of the ash particles was enhanced. Compared with the inlet samples, the upper portion of the ash particle pile was sharper and the repose angle was significantly increased. It can be seen that the ash particle repose angle varied greatly with the increase of the particle size. When the particle size is less than 68 μm, the deposition angle was 72.10°, which was much larger than the angle of the inlet ash particles under the same particle size. As the particle size increases, the coal ash particle repose angle gradually decreased. When the particle size was 91–125 μm, the repose angle decreased to 57.37°. However, it still lager than the repose angle of inlet ash samples, indicating that the followability of the ash particles was greatly reduced after the adsorption and agglomeration process. Therefore, the ash deposition problem occurred in the economizer. In addition, when the particle size is small, the repose angle was very large, much larger than the 91–125 μm particles. It indicates that the agglomeration is more likely to occur between small particles.

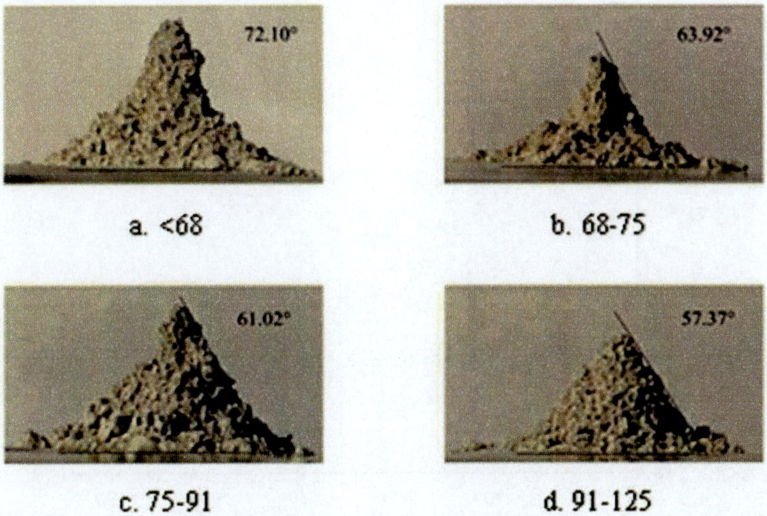

Fig. 69.9 The repose angle of ash sample of the economizer outlet

69.3.5 The Influence of Flue Gas Temperature on Ash Particle Flowability

To study the influence of flue gas temperature on the flowability of ash particles, adsorption experiments of the inlet ash samples were carried out in the entrained flow adsorption experimental system. The experimental temperatures selected to be 80, 90, 100 and 130 °C, covering the actual selected temperature parameters in the power plants. Since when the temperature is higher than 100 °C, it has little influence on the flowability of the particles, the experiments under 110 and 120 °C were skipped. The experiment results are shown in the following figure.

As can be seen from Fig. 69.10, the repose angle of the ash particles increased as the temperature of the flue gas decreased. Secondly, the smaller the particle size, the larger the repose angle changes, and the more pronounced the repose angle is influenced by the adsorption temperature. In addition, when the temperature was dropped below 100 °C, the repose angle increased sharply and the flowability of ash particles reduced greatly, especially for the smaller particles. Therefore, in the application of the low-low temperature flue gas system, the flue gas temperature cannot be blindly dropped in pursuit of high sulfur oxide removal efficiency and dust removal efficiency. Or it may cause serious ash deposition problem of the heat exchanger and the dust remover. Further, when the flue gas temperature was dropped, the agglomeration effect between large particles was low, and the decrease in the temperature did not significantly affect the agglomeration and the flowability of the large particles.

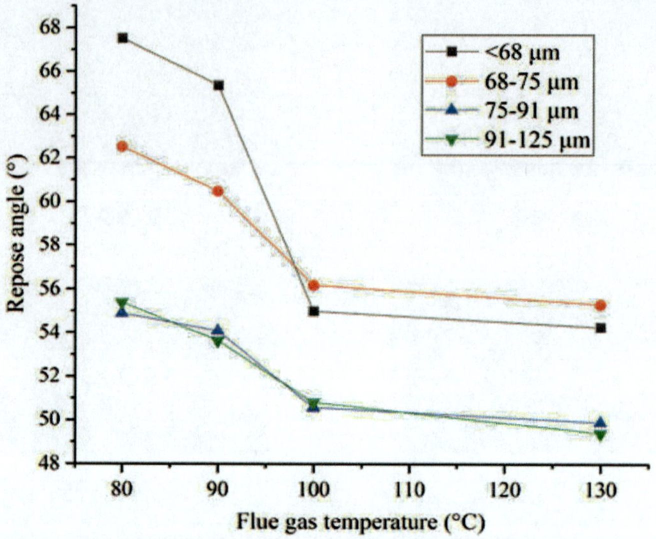

Fig. 69.10 The repose angle of ash samples after agglomeration

69.4 Conclusions

In this paper, the agglomeration effect of ash particles and the flowability of ash particles after adsorption reaction were studied on the fixed bed adsorption experimental system and the entrained flow adsorption experimental system, respectively. The laser diffraction particle size analyzer was used, and the repose angle of different ash samples were measured. The influences of adsorption temperature and particle size were studied.

The results show that as the increase of adsorption time, the particle size increased gradually, and the agglomeration effect is attenuated after the particles have reached adsorption saturation. The time at which the ash particles began to agglomerate with each other was defined as the agglomeration threshold. As the adsorption temperature decreased, the particle size growth rate increased. The smaller the size of ash particles, the more significant the agglomeration effect. Large particles were easy to adsorb small particles, but the agglomeration effect between large particles was weak. In addition, in the initial stage of particle agglomeration, it was mainly included the adhesion between smaller particles and the adhesion of smaller particles on the surface of large particles. The followability of the ash particles was greatly reduced after the adsorption and agglomeration process. The repose angle of the ash particles increased as the temperature of the flue gas decreased. The smaller the particle size, the larger the repose angle changes, and the more pronounced the repose angle is influenced by the adsorption temperature. The flue gas temperature cannot be blindly dropped in pursuit of high sulfur oxide removal efficiency and dust removal efficiency.

Acknowledgements This work was supported by the National Key R&D Program of China (2017YFB0602102).

References

Chen X, Shao S, Tian Z (2017) Impacts of air pollution and its spatial spillover effect on public health based on China's big data sample. J Clean Prod 142:915–925

Chen J, Chen C, Cao Y (2018) Review on the latest developments in modified vanadium-titanium-based SCR catalysts. Chin J Catal 39(8):1347–1365

Hu J, Duan F, He K (2016) Characteristics and mixing state of S-rich particles in haze episodes in Beijing. Front Env Sci Eng **10**(5)

Jianguo Z, Zhuhai Z, Yuzhong H (2014) Research and application of low-temperature and low-voltage electric dust removal technology. China's Environ Protect Indus 3:28–34 (in Chinese)

National Bureau of Statistics of China 2018 (2018) China Statistics Press, Beijing (in Chinese)

Quan J, Tie X, Zhang Q (2014) Characteristics of heavy aerosol pollution during the 2012–2013 winter in Beijing. China Atmos Environ 88:83–89

Ren F, Xia L (2017) Analysis of China's primary energy structure and emissions reduction targets by 2030 based on multiobjective programming. Math Probl Eng 2017:1–8

Xuehui L, Dan S, Xinwei N (2018) China's energy supply situation and potential risk analysis. China Energy 07:25–30 (in Chinese)

Xunzhang P, Wenying C, Clarke LE (2017) China's energy system transformation towards the 2 °C goal: implications of different effort-sharing principles. Energ Policy 103:116–126

Nakayama Y, Nakamura S, Takeuchi Y, Itoh M, Okino S (2011) MHI high efficiency system—proven technology for multi pollutant removal. Research & Development Center, Nishi-ku

Yuegang T, Xin H, Aiguo C (2015) Distribution characteristics and sedimentation control of sulfur content in coals in China. J Coal 40(09):1977–1988 (in Chinese)

Yunfei L, Wenhua L, Ying J (2005) Study on distribution characteristics of sulfur in coal in China. Coal Conversion 03:14–18 (in Chinese)

Zhang D, Wang J, Lin Y (2017) Present situation and future prospect of renewable energy in China. Renew Sustain Energy Rev 76:865–871

Chapter 70
Condensation Characteristic of Sulfuric Acid Vapor on Low-Temperature Surface of Tube Heat Exchanger

Ke Sun, Yu Yan, Fangfang Hu, Lei Deng, and Defu Che

Abstract In the low-low temperature flue gas system, a part of the H_2SO_4 vapor is adsorbed by the ash particles, and the other part will condense on the surface of the heat exchanger, resulting in low-temperature corrosion. In this paper, a numerical model for the condensation of H_2SO_4 vapor on low-temperature surface is constructed. This model combines the thermodynamic phase equilibrium theory, the multicomponent transport theory and the heat transfer theory. The effects of flue gas properties, the size and arrangement of heat exchanger are studied. The results show that, the condensation rate is mainly related to the concentration gradient of H_2SO_4 vapor near the surface. As the flue gas or surface temperature increases, condensation rate decreases. When the flue gas velocity increases, the mass transfer resistance is reduced, thereby the condensation rate is increased. As the content of H_2SO_4 in the flue gas increases, the concentration gradient of H_2SO_4 vapor near the surface increases, and the condensation rate of increases linearly. But the content of water vapor has little effect on the H_2SO_4 condensation rate. For the tube heat exchanger, when the working conditions keep the same, the smaller the outer diameter of the tube, the greater the condensation rate of the H_2SO_4 vapor. Compared with the inline arrangement, when the tubes are staggered, the disturbance of flue gas is stronger, and the condensation rate is larger.

Keywords Condensation characteristic · Low-low temperature flue gas system · Corrosion · Tube heat exchanger

---

K. Sun · Y. Yan · F. Hu · L. Deng · D. Che (✉)
State Key Laboratory of Multiphase Flow in Power Engineering, School of Energy and Power Engineering, Xi'an Jiaotong University, Xi'an 710049, China
e-mail: dfche@mail.xjtu.edu.cn

K. Sun
Huadian Electric Power Research Institute, Hangzhou 310030, China

F. Hu
China Energy Engineering Group Shaanxi Electric Power Design Institute Co., Ltd, Xi'an 710054, China

J. Lyu and S. Li (eds.), *Clean Coal and Sustainable Energy*,
Environmental Science and Engineering,
https://doi.org/10.1007/978-981-16-1657-0_70

Abbreviations

Nomenclature

$\bar{a}_i\bar{a}$	Activity coefficients in liquid phases
c_p	Specific heat of the fluid, J kg^{-1} K^{-1}
C_{μ}, $C_{2\varepsilon}$	Constant in the k-ε turbulent current model
$D_{i,m}$	Mass diffusivity coefficient, m^2 s^{-1}
f_i	Fugacity
f_i^0	Fugacity under 1 atmosphere, atm^{-1}
G_i	Gibbs free energy
$\Delta\Delta \underline{G_i}^v$	The change in Gibbs free energy during the vaporization of water or sulfuric acid, kJ mol^{-1}
k	Thermal conductivity of the fluid, W m^{-2} K^{-1}
K_0, K_1	Modified equilibrium constant
$\bar{L}_i^1$	Partial molar enthalpy, J mol^{-1}
p	Pressure, Pa
p_i	Partial pressure of sulfuric acid molecule, Pa
$p_{i,0}$	Apparent partial pressure of sulfuric acid, Pa
Sc_t	Turbulent Schmidt number
T	Temperature of the fluid, K
U	Fluid velocity, m s^{-1}
y_i	Mass fraction

Greek symbols

μ	Dynamic viscosity, N s m^{-2}
μ_t	Turbulent eddy viscosity
μ_i	Chemical potential
ρ	Density, kg m^{-3}
σ_k, σ_ε	Constant in the k-ε turbulent current model
$\phi_{i,0}$	Apparent fugacity coefficient

Subscripts

i	Species i
g	Gas
l	Liquid

70.1 Introduction

Although alternative energy developed rapidly these years, coal is still the main energy resource of China (Ren and Xia 2017; Mi et al. 2018; Zhang et al. 2017). Among various uses of coal in China, the use for power generation accounts for the main part (National Bureau of Statistics of China 2018). As a consequence, a large amount of pollutant has been emitted into the atmosphere and has caused serious environment problems, such as the smog and hazy weather appeared in northern China in recent winters. To control the pollutant emissions, many ultra-low emission reforms have been carried out in coal-fired power plants (Liu et al. 2018), including low-low temperature flue gas system.

Low-low temperature flue gas system is first proposed by Mitsubishi corporation. Due to its high pollutant removal efficiency and low operating costs (Li et al. 2014; Hanxiao et al. 2018; Qing 2018), it has been widely applied in China's power plants (Sihua et al. 2018; Yanwu et al. 2018). In this system, a tubular heat exchanger is installed before the electrostatic precipitator (ESP) and the flue gas temperature is reduced below the acid dew point. Then the condensed sulfuric acid mist will be adsorbed by ash particles and be removed in the dry ESP (Nakayama et al. 2011). However, the application effect of this system in China is not as good as it in Japan (Hanxiao et al. 2018; Ketao et al. 2019; Guangcai et al. 2018). The sulfur content of coal used in China is generally higher and the coal types are changeable (Jun 2005; Yunfei et al. 2005). A portion of the sulfuric acid mist that is not adsorbed by ash particles will condense and deposit on the surface of the heat exchanger, causing corrosion problem. Hence, it is important to study the condensation characteristic of sulfuric acid vapor on low-temperature surface of heat exchanger.

Currently, most studies focus on the condensation of water vapor on the low-temperature surface. For the investigations of sulfuric acid condensation, the most of the present works use empirical estimation formula methods (Cen 1994), without considering the geometric characteristics and the influence of operating conditions. Dietz and Marchal (1977) and Wilson (1989) modeled the deposition of sulfuric acid by using mass transfer theory or fugacity equations, but the calculations of these methods were particularly complex and were not suitable for the calculation of different objects in the engineering project. Baosheng (2006), Zhiyong et al. (2005) and Hui et al. (2013) studied the deposition rate of sulfuric acid vapor under different working conditions by Fluent. However, in these studies, the sulfuric acid vapor concentration at the wall surface was assumed constant and the effects of the size and arrangement of heat exchanger were neglected. Therefore, to better predict the condensation and deposition of sulfuric acid vapor on the surface of the heat exchanger, the condensation model of sulfuric acid vapor needs to be improved, and the effects of heat exchanger size and arrangement need to be studied.

In this paper, numerical simulation of the condensing process of sulfuric acid vapor was performed by applying the vapor–liquid equilibrium of sulfuric acid solution and modified fugacity theory. The main effects of flue gas properties, size and arrangement form of smooth tube exchanger on the deposition rate were analyzed.

70.2 Physical and Numerical Models

70.2.1 Physical Models

In this study, the bare tube heat exchanger was taken as the study object. The effects of flue gas properties, heat exchanger size and arrangement on the deposition rate of sulfuric acid vapor were investigated. The schematic grid diagram of the inline bare tube heat exchanger is shown in Fig. 70.1. Two rows of tubes were taken along the flow direction. The inlet and outlet sections were each extended by twice the length of the tube pitch, respectively. Local grid refinement near the tube wall surface was carried out. Furthermore, the verify of grid independence was performed. The specific geometry and calculation conditions of the heat exchanger are shown in Table 70.1.

In the simulation process, the deposition model is established on the basis of the following assumptions: (1) The flue gas is considered to be a ternary mixture of air, sulfuric acid vapor and water vapor. (2) Only the condensation of sulfuric acid vapor on the tube wall surface is considered. The condensation in the mainstream of flue gas was neglected. (3) The heat transfer resistance and mass transfer resistance of the condensate on the surface of the heat exchanger are neglected. (4) The wall temperature of the heat exchanger is assumed constant, and the latent heat released during the condensation of sulfuric acid is neglected. (5) The temperature of the acid condensate is constant equal to the wall temperature of the heat exchanger. (6) The secondary evaporation effect of acid condensate in the flue gas is neglected.

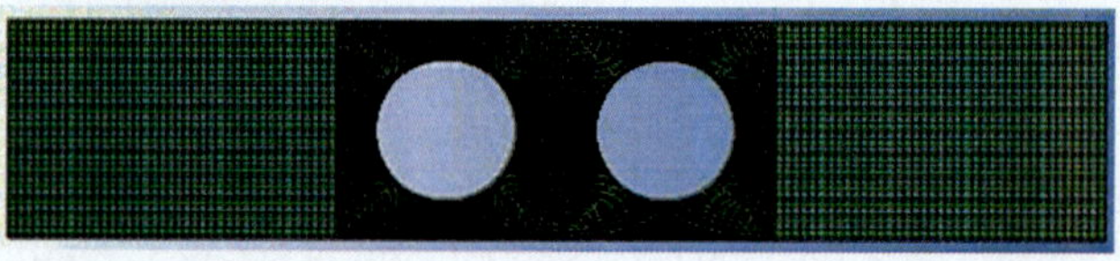

Fig. 70.1 The schematic grid diagram of the inline bare tube

Table 70.1 The specific geometry and calculation conditions of the heat exchanger

Item	Parameter
Outer diameter of the tube (mm)	35~55
The horizontal pitch between adjacent tube rows (mm)	70
The longitudinal pitch between adjacent tube rows (mm)	70
Flue gas temperature (K)	380~430
Flue gas flow rate (m s^{-1})	6~10
Sulfuric acid vapor volume fraction (10^{-6})	5~30
Water vapor volume fraction (%)	3~18
Wall temperature of the tube (K)	340~365

Different calculation methods of acid dew point were compared and analyzed in this study. The empirical formula proposed by A. G. Okkes (Gençay 1987) is adopted. This formula is obtained is obtained by Muller data regression, taking into account the influence of SO_3 and water vapor contents in the flue gas. Verified by many researchers, the acid dew point temperature estimated by this formula matches well with the experimental data. Therefore, this formula is used to estimate the flue gas acid dew point. It is considered that when the heat exchanger wall temperature is lower than the estimated value of acid dew point, the sulfuric acid vapor will condense, otherwise condensation will not occur.

70.2.2 Governing Equation

The governing equations for solving the problem of sulfuric acid vapor condensation in flue gas mainly include continuity equation, momentum conservation equation, energy conservation equation and component transport equation. The steady-state forms of each control equation are as follows:

$$\nabla \cdot U = 0 \tag{70.1}$$

$$\rho U \cdot \nabla U = -\nabla p + \mu \nabla^2 U \tag{70.2}$$

$$\rho U \cdot \nabla T = \frac{k}{c_{\mathrm{p}}} \nabla^2 T \tag{70.3}$$

$$\rho U \cdot \nabla y_i = \nabla \cdot \left[\left(\rho D_{i,\mathrm{m}} + \frac{\mu_{\mathrm{t}}}{Sc_{\mathrm{t}}} \right) \nabla y_i \right] \tag{70.4}$$

The Reynolds number of this study ranged from 12,000 to 32,000. Therefore, considering the influence of turbulent pulsation, the Standard k-ε two equation model was used. This model includes the kinetic energy equation and the turbulent energy dissipation rate equation. The specific forms are as follows:

$$\frac{\partial}{\partial x_i}(\rho k u_i) = \frac{\partial}{\partial x_j}\left[\left(\mu + \frac{\mu_{\mathrm{t}}}{\sigma_{\mathrm{k}}} \right) \frac{\partial k}{\partial x_j} \right] - \rho \varepsilon \tag{70.5}$$

$$\frac{\partial}{\partial x_i}(\rho \varepsilon u_i) = \frac{\partial}{\partial x_j}\left[\left(\mu + \frac{\mu_{\mathrm{t}}}{\sigma_{\varepsilon}} \right) \frac{\partial \varepsilon}{\partial x_j} \right] - C_{2\varepsilon} \rho \frac{\varepsilon^2}{k} \tag{70.6}$$

In the calculation process, the inlet area adopted the speed boundary condition, and the velocity of the inlet fluid was uniform. The inlet temperature and the volume fraction of each component were given. In the calculation region, the periodic boundary was used in the direction parallel to the fluid flow, and the symmetrical boundary was adopted perpendicular to the fluid flow direction. The outlet was free to flow. The wall temperature of the bare tube was set to be constant. The surface of the tube wall was in the no-slip boundary condition.

The calculation formula of sulfuric acid vapor and water vapor deposition rate is:

$$m_i = -\left(\rho D_{i,\mathrm{m}} + \frac{\mu_\mathrm{t}}{Sc_\mathrm{t}}\right)\nabla y_i \tag{70.7}$$

70.2.3 Vapor–Liquid Equilibrium and Interface Boundary Conditions

As the boundary condition for the simulation calculation, the concentration of sulfuric acid vapor around the wall surface of the heat exchanger tube is the key factor affecting the mass transfer process. It can be obtained from the vapor–liquid phase equilibrium data of the sulfuric acid solution. Abel (1948,2002) proposed an fugacity equation of sulfuric acid solution for estimating the vapor composition at the vapor–liquid interface from the solution concentration. Gmitro and Vermeulen (1964) made some modifications to the fugacity equation using the updated thermodynamic parameters. In this study, this equation was used to calculate the concentration of sulfuric acid vapor and water vapor at the vapor–liquid interface. The derivation process of the fugacity equation is as follows.

When the sulfuric acid vapor and the condensed acid solution reached a two-phase equilibrium, the chemical potential μ of the water or sulfuric acid component in both vapor and liquid phases should be equal. That is:

$$\underline{G}_i^{\mathrm{g0}} + RT\ln\left(\frac{\overline{f}_i^{\mathrm{g}}}{f_i^{\mathrm{g0}}}\right) = \underline{G}_i^{\mathrm{l0}} + RT\ln\left(\frac{\overline{f}_i^{\mathrm{l}}}{f_i^{\mathrm{l0}}}\right) \tag{70.8}$$

Then the fomula $\Delta\underline{G}_i^{\mathrm{v}}$ were substituted into the above Eq. (70.8) to get:

$$\ln f_i^{\mathrm{g}} = \frac{-\Delta\underline{G}_i^{\mathrm{v}}}{RT} + \ln\overline{a}_i \tag{70.9}$$

Bring the liquid phase activity definition $\overline{a}_i = \overline{f}_i^{\mathrm{l}}/f_i^{\mathrm{l0}}$ into Eq. (70.8):

$$\underline{G}_i^{\mathrm{g0}} + RT\ln\left(\frac{\overline{f}_i^{\mathrm{g}}}{f_i^{\mathrm{g0}}}\right) = \underline{G}_i^{\mathrm{l0}} + RT\ln\overline{a} \tag{70.10}$$

According to the phase equilibrium theory, the fugacity of the sulfuric acid or water component in the solution is equal to that at the vapor–liquid interface. That is $\overline{f}_i^{\mathrm{g}} = \overline{f}_i^{\mathrm{l}}$. Bring it into the Eq. (70.10) and simplify as:

$$\ln \overline{f}_i^{\mathrm{l}} = \frac{-\Delta G_i^{\mathrm{v}}}{RT} + \ln \overline{a} \tag{70.11}$$

Bring the Gibbs–Helmholtz equation in to (11) and integrate over the entire temperature range to get:

$$\ln \overline{a}_i = \ln \overline{a}_i(298) + \frac{\overline{L}_i^{\mathrm{l}}(298)}{R}\left[\frac{1}{T} - \frac{1}{298}\right] + \int_{298}^{T} \frac{1}{RT^2}\left[\int_{298}^{T} \left(\underline{C}_{\mathrm{p}i}^{\mathrm{l0}} - \overline{C}_{\mathrm{p}i}^{\mathrm{l}}\right) dT\right] dT \tag{70.12}$$

Using Kirchhoff's law and the definition of free energy, transform the Gibbs–Helmholtz formula and integrated from 298 K to the system temperature. The final fugacity equation can be obtained:

$$\frac{\Delta \underline{G}_i^{\mathrm{v}}}{RT} = \frac{-\Delta \underline{S}_i^{\mathrm{v}}(298)}{R} + \frac{\Delta \underline{H}_i^{\mathrm{v}}(298)}{RT} - \int_{298}^{T} \frac{1}{RT^2}\left[\int_{298}^{T} \left(\underline{C}_{\mathrm{p}i}^{\mathrm{g}} - \underline{C}_{\mathrm{p}i}^{\mathrm{l}}\right) dT\right] dT \tag{70.13}$$

Bring the Eq. (70.12) and (70.13) into (11):

$$\ln \overline{f}_i^{\mathrm{l}} = \frac{\Delta \underline{S}_i^{\mathrm{v}}(298)}{R} - \frac{\Delta \underline{H}_i^{\mathrm{v}}(298)}{RT} + \ln \overline{a}_i(298) + \frac{\overline{L}_i^{\mathrm{l}}(298)}{R}\left[\frac{1}{T} - \frac{1}{298}\right] + \int_{298}^{T} \frac{1}{RT^2}\left[\int_{298}^{T} \left(\underline{C}_{\mathrm{p}}^{\mathrm{g}} - \overline{C}_{\mathrm{p}}^{\mathrm{l}}\right) \mathrm{d}T\right] dT \tag{70.14}$$

In the above equation, C_p can be expanded into:

$$\overline{C}_{\mathrm{p}}^{\mathrm{g}} = a + bT + cT^2 \tag{70.15}$$

$$\overline{C}_{\mathrm{p}}^{\mathrm{l}} = \overline{C}_{\mathrm{p}(298)}^{\mathrm{l}} + \alpha(T - 298) \tag{70.16}$$

Bring Eq. (70.15) and (70.16) into (14) and then integrate the obtained equation to get:

$$\ln \overline{f}_i^{\mathrm{l}} = aA + bB + cC + \Delta \underline{H}_i^{\mathrm{v}} D + \Delta \underline{S}_i^{\mathrm{v}} E$$

$$+\overline{C}_{pi}^{l} F + \overline{L}_i^{l} G + \alpha_i H + \ln \overline{a_i}(298) \tag{70.17}$$

$$A = \frac{1}{R}\left[\ln\left(\frac{T}{298}\right) + \frac{298}{T} - 1\right]$$

$$B = \frac{1}{R}\left[\frac{298^2}{2T} + \frac{T}{2} - 298\right]$$

$$C = \frac{1}{R}\left[\frac{298^3}{3T} - \frac{298^2}{2} + \frac{T^2}{6}\right]$$

$$D = -\frac{1}{RT}$$

$$E = \frac{1}{R}$$

$$F = \frac{1}{R}\left[\ln\left(\frac{298}{T}\right) - \frac{298}{T} + 1\right]$$

$$G = \frac{1}{R}\left[\frac{1}{T} - \frac{1}{298}\right]$$

$$H = \frac{1}{R}\left[298 ln(\frac{T}{298}) + \frac{298^2}{2T} - \frac{T}{2}\right]$$

In the simulation process, it was assumed that the gas was a rational gas. Then, the Eq. (70.17) can be converted into the following partial pressure forms:

$$\begin{aligned}\ln p_i =& aA + bB + cC + \Delta \underline{H}_i^{v} D + \Delta \underline{S}_i^{v} E \\ &+ \overline{C}_{pi}^{l} F + \overline{L}_i^{l} G + \alpha_i H + \ln \overline{a_i}(298)\end{aligned} \tag{70.18}$$

The left side of the above equation is the vapor partial pressure of the species i at the vapor–liquid interface. The first five terms on the right side are the relevant thermodynamic parameters of species i, and the other terms are some parameters controlled by the concentration. The thermodynamic parameters of pure water proposed by Kelley (1960) are used in this study, as well as the parameters of the fugacity equations of different concentrations of sulfuric acid solution proposed by Giaugue et al. (1961).

Equation (70.18) is an equation for the relationship among the partial pressure of sulfuric acid or water vapor at the vapor–liquid interface, the molar concentration of the condensed acid solution, and the temperature of the condensed acid solution. As long as two of them are given, the other amount can be obtained. In order to make the calculation accuracy of the sulfuric acid concentration higher, it is necessary to introduce a fugacity correction coefficient to correct the value obtained in (18):

$$p_{i,0} = \frac{p_i}{\phi_{i,0}} \tag{70.19}$$

$$\phi_{i,0} = \frac{\overline{f}^{1}_{w}}{\overline{f}^{1}_{w} + K_0 + K_1(\overline{f}^{1}_{w})^2} \tag{70.20}$$

When looking for the concentration of sulfuric acid vapor at the wall, the following iterative method was uesd:

(I) In order to ensure that the calculation can converge, it is assumed that the water vapor concentration near the wall is equal to the water vapor concentration in the mainstream. The concentration of the condensed acid near the wall is obtained from the Eq. (70.18) according to the wall temperature.

(II) It is assumed that the initial value of the molar concentration of the condensed acid at the wall is obtained in step (I). The concentration of sulfuric acid vapor and water vapor at the wall are calculated from the Eq. (70.18) and the fugacity equations of water vapor and sulfuric acid vapor, respectively.

(III) The mass deposition rate of sulfuric acid vapor and water vapor calculated from Eq. (70.7).

(IV) The concentration of the condensed sulfuric acid is calculated from the obtained mass deposition rate of sulfuric acid vapor and water vapor. And then correct the initial concentration of the condensed acid in step (II).

(V) Repeat steps (II)~(V) until the calculation results converge.

70.3 Physical and Numerical Models

70.3.1 Vapor–Liquid Equilibrium and Interface Boundary Conditions

In order to verify the reliability of the model constructed in this study, the calculation results of this model were verified using the experimental results reported in previous study (Zhang 1991). This study focused on the flow characteristics of flue gas in a bare tube. The length and the inner diameter of the tube were 0.315 m and 0.01 m, respectively. The sulfuric acid vapor content in the flue gas was 3.3 μL L^{-1}. The water vapor content was 10%. The flue gas inlet flow rate was 4.5 m s^{-1}. The flue temperature and the tube wall temperature were 414 K and 350 K, respectively. The comparison between the experimental results and the simulated data is shown in Fig. 70.2.

As can be seen from the figure, the deposition rate of sulfuric acid vapor at the inlet of the tube is largest. The condensate deposition rate drops rapidly along the direction of mainstream. This is because the concentration of sulfuric acid vapor at the inlet is the largest. Hence, the concentration gradient of sulfuric acid vapor near the wall is the largest. As the condensation process occurred, the concentration of sulfuric

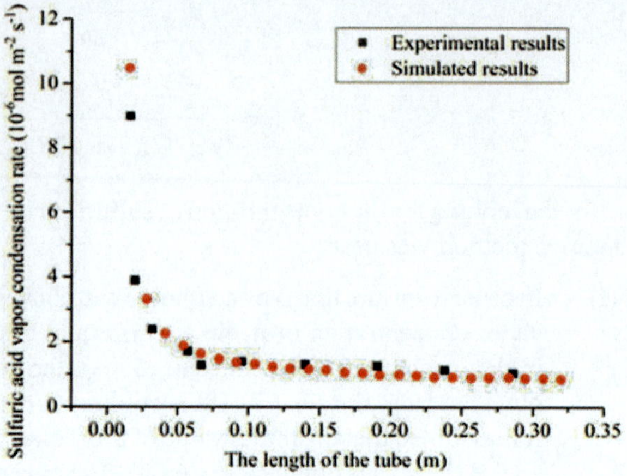

Fig. 70.2 The comparison between the experimental results and the simulated data

acid vapor in the flue gas decreases in the downstream area of the tube, reducing the concentration gradient of the acid vapor at the wall surface. So the condensation deposition rate drops. The simulation results are very close to the experimental data, indicating that the numerical calculation model constructed in this study can be used for the prediction of sulfuric acid condensation deposition on the surface of heat exchangers.

70.3.2 *Effect of Flue Gas Temperature and Velocity*

Figure 70.3 illustrates the simulation result of sulfuric acid deposition rate under different flue gas temperatures and velocities, where the concentration of sulfuric acid

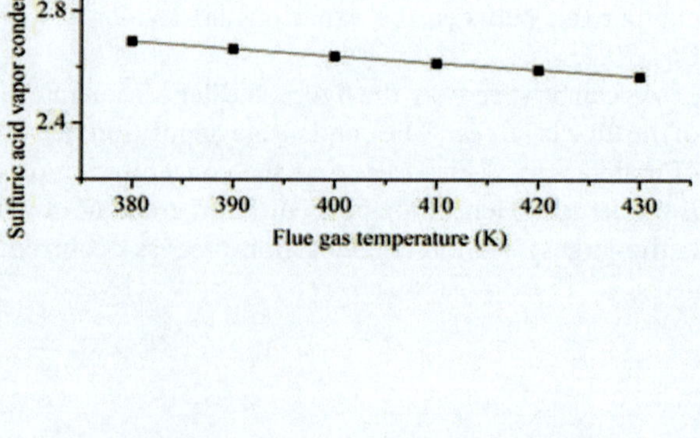

Fig. 70.3 Effect of flue gas temperature and velocity on sulfuric acid deposition rate

in flue gas is 20 μL L^{-1}, the water vapor volume fraction is 9%. As shown in the figure, when the flue gas temperature changes from 380 to 430 K, the deposition rates of sulfuric acid are reduced by 4.53%, 4.41% and 4.34% for the conditions with a flue gas velocity of 6 m s^{-1}, 8 m s^{-1} and 10 m s^{-1}, respectively. The result suggests that with the increase of the flue gas temperature, sulfuric acid deposition rate decreases. It can be explained as follow. When temperature increasing, the condensation rates of water vapor and acid vapor decline, while the decline degree of the former is higher than that of the latter, since water vapor has a lower dew point temperature. Consequently, the concentration of condensation acid on the wall surface increases, the saturated partial pressure of sulfuric acid vapor near the wall increases correspondingly, and the driving force of mass transfer decreases, resulting the condensation deposition rate of sulfuric acid decreases.

The figure also shows that the higher flow velocity of flue gas result in the higher condensation deposition rate of acid under consistent condensation condition. For example, when the flue gas temperature is 380 K and the flue gas velocity changes from 6 to 10 m s^{-1}, sulfuric acid deposition rate changes from 2.69×10^{-6} to 3.37×10^{-6} kg m^{-2} s^{-1}, within a 25.46% difference. This is because the raising of flue gas' flow velocity does not only increase the turbulence intensity of flue gas, but also thin the boundary layer of the fluid, which reduces the mass transfer resistance of flue gas and lead to the increase of acid deposition rate. The result suggests that compared with smoke temperature, flue gas velocity has a greater influence on sulfuric acid deposition rate.

70.3.3 *Effect of Sulfuric Vapor Content in Flue Gas*

Figure 70.4 illustrates the simulation results of sulfuric acid deposition rate changing

Fig. 70.4 Effect of sulfuric acid content in flue gas on sulfuric acid deposition rate

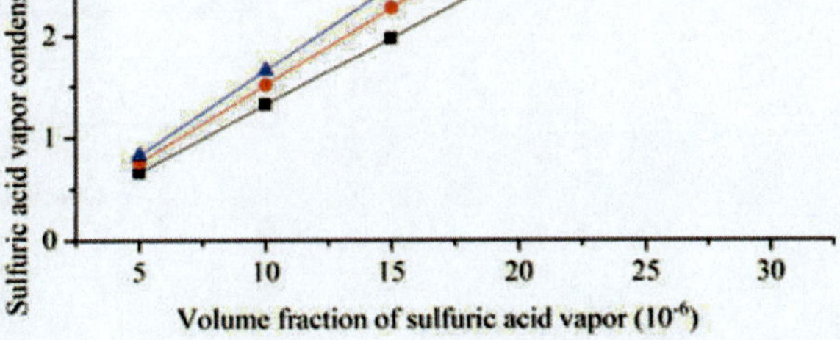

with the sulfuric vapor content of flue gas. As shown in the figure, with the increase of the volume fraction of acid vapor in flue gas, sulfuric acid deposition rate increases nearly linearly, and the higher the flue gas velocity, the faster the condensation rate increases. This is because when the sulfuric vapor content in the flue gas increases, the concentration gradient between the sulfuric acid vapor in the flue gas and that on the wall increase, which is conducive to the mass transfer process and leads to the increase of sulfuric acid's deposition rate. When the volume fraction of acid vapor increases from 5×10^{-6} to 3×10^{-5}, the acid deposition rate increases by about 5 times under all the three flue gas velocities, indicating that the content of sulfuric vapor in the flue gas has a significant influence on sulfuric acid deposition rate.

70.3.4 *Effect of Water Vapor Content in Flue Gas*

Figure 70.5 illustrates the simulation results of sulfuric acid deposition rate under different water vapor contents in flue gas. The figure shows that when water vapor content changes, sulfuric acid deposition rate is almost consistent. Theoretically, the increase of water vapor content in the flue gas enhances the water vapor mass transfer process, and decreases the concentration of the condensate acid on the wall accordingly. The saturated acid vapor partial pressure decreases correspondingly, resulting sulfuric acid deposition rate increases. However, since the partial pressure of sulfuric acid vapor near the surface of the heat exchanger is much lower than that of acid vapor in the mainstream flue gas, when the partial pressure of sulfuric acid vapor near the wall is reduced, sulfuric acid deposition rate will not change significantly.

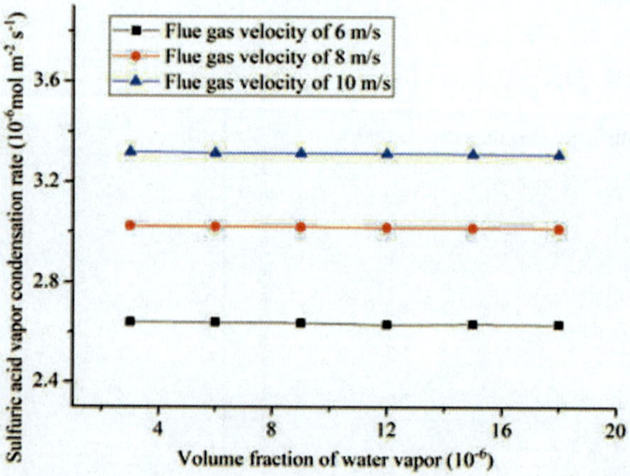

Fig. 70.5 Effect of water vapor content in flue gas on sulfuric acid deposition rate

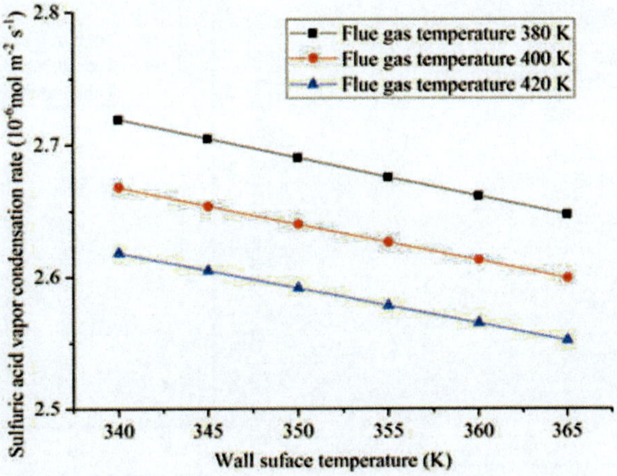

Fig. 70.6 Effect of wall temperature on sulfuric acid deposition rate

70.3.5 Effect of Wall Temperature

The influence of heat exchanger's wall temperature on sulfuric acid deposition rate is shown in Fig. 70.6. With the increase of wall temperature, sulfuric acid deposition rate decreases. The effcet machinism is similar with that between flue gas velocity and sulfuric acid deposition rate. The condensation rates of water vapor and acid vapor decline with the increase of wall temperature, and the decline degree of the former is higher than that of the latter. Consequently, the concentration of condensation acid on the wall surface increases, the saturated partial pressure of sulfuric acid vapor near the wall increases, and the driving force of mass transfer decreases, resulting the condensation deposition rate of sulfuric acid decreases. When the wall temperature rises from 340 to 365 K, the deposition rates of sulfuric acid all decrease by about 2.5% under the three different flue gas temperatures. When the flue gas temperature rises from 380 to 400 K, the deposition rates of sulfuric acid all decrease by about 2% under different wall temperatures. Both the wall temperature and flue gas temperature influence sulfuric acid's deposition rate by changing the concentration of condensation acid on the wall.

70.3.6 Effect of Heat Exchanger Size

Figure 70.7 illustrates the simulation results of sulfuric acid deposition rate with different heat exchanger sizes, reflecting that the smaller the outer diameter of the naked tube, the higher deposition rate of sulfuric acid. To explore the underlying causes, distributions of sulfuric acid vapor concentration on the surface of naked tube heat exchangers with different outer diameters are drawn in Fig. 4.8. The condensation conditions stay consistent.

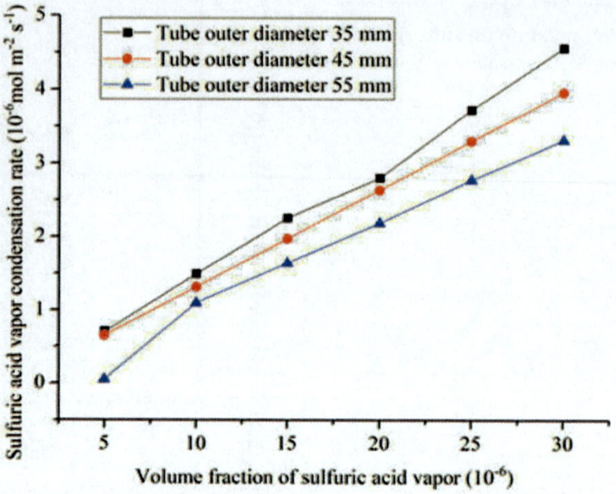

Fig. 70.7 Effect of heat exchanger size on sulfuric acid deposition rate

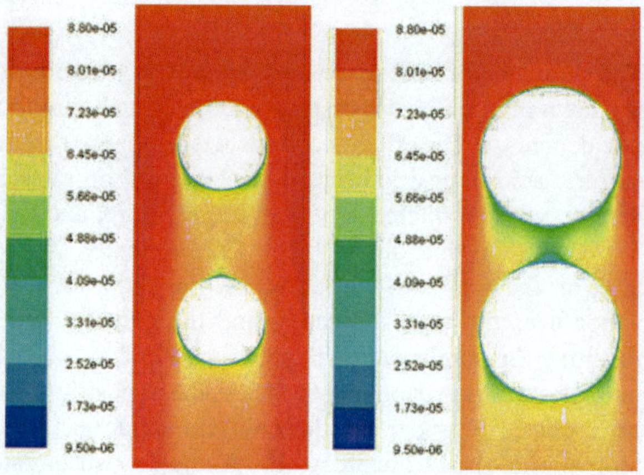

Fig. 70.8 Effect of heat exchanger size on sulfuric acid deposition rate

As Fig. 70.8 shown, with the same longitudinal pitch, the smaller the outer diameter of the tubes is, the larger the relative pitch between the tubes is, the more fully the mainstream smoke develops, and the higher concentration of sulfuric acid vapor between the tubes is. Compared with the tube with an external diameter of 55 mm, the sulfuric acid vapor concentration between the tubes with an external diameter of 35 mm is significantly greater, so the mass transfer driving force is stronger, and sulfuric acid deposition rate is higher.

70.3.7 *Effect of Heat Exchanger Arrangement*

Figure 70.9 shows the change of sulfuric acid deposition rate with acid vapor content

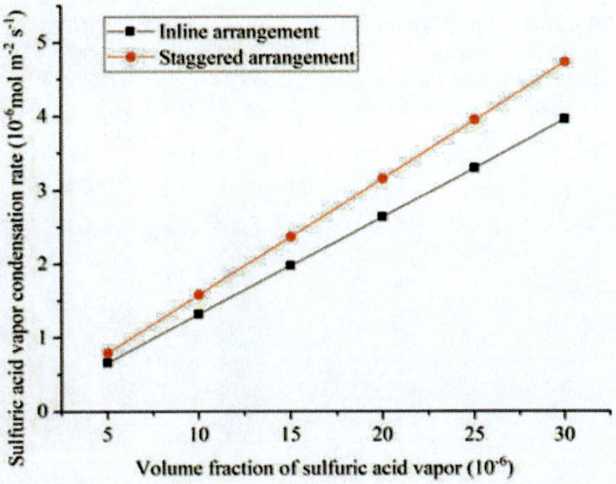

Fig. 70.9 Effect of heat exchanger arrangement on sulfuric acid deposition rate

under different heat exhchanger arrangements. The flue gas temperature is 400 K, wall temperature is 350 K, water vapor volume fraction is 9%, and flue gas velocity is 6 m s^{-1}.

As shown in the figure, sulfuric acid deposition rate in tubes with staggered arrangement is obviously higher than that in tubes with in-line arrangement. When the volume fraction of sulfuric acid vapor is 2×10^{-5}, sulfuric acid deposition rate in the in-line arrangement is 2.64 kg m^{-2} s^{-1}, while in the staggered arrangement, sulfuric acid deposition rate increases by 19.5% to 3.15 kg m^{-2} s^{-1}. When the naked tube is staggered, the disturbance of flue gas and the mass transfer process are enhanced, resulting a higher deposition rate of sulfuric acid.

Figure 70.10 and 70.11 show the distribution of gas flue velcity and sulfuric acid vapor concentration on the surface of naked tubes with different arrangements,

Fig. 70.10 Distribution of gas flue velcity on the surface of naked tubes with different arrangements

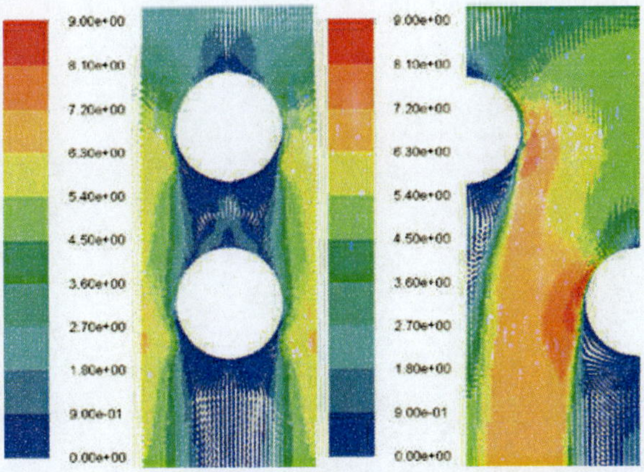

(a) In-line arrangement; (b) Staggered arrangement

Fig. 70.11 Distribution of sulfuric acid vapor concentration on the surface of naked tubes with different arrangements

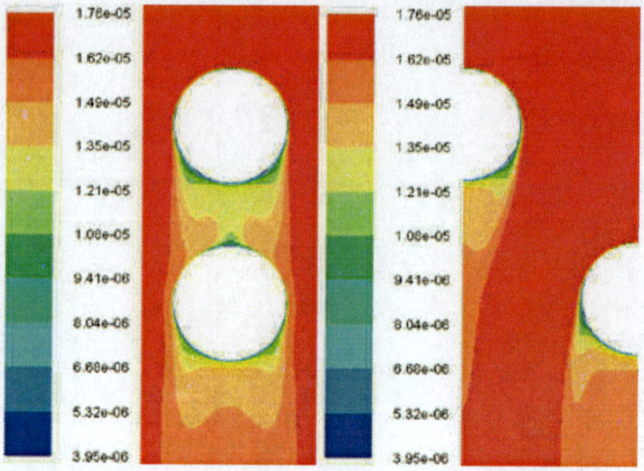

(a) In-line arrangement; (b) Staggered arrangement

respectively. The outer diameter of the bared-tube heat exchanger is 45 mm. The flue gas velocity and concentration of sulfuric acid vapor between the tubes with in-line arrangement are relatively small. When the arrangement of tubes changes from in-line to staggered, the leeward side of the upper naked tubes is scoured by flue gas as well, the velocity of flue gas and concentration of sulfuric acid vapor increase, resulting the mass transfer driving force and sulfuric acid deposition rate increase.

70.4 Conclusions

In this study, numerical simulation of the condensing process of sulfuric acid vapor was performed by applying the vapor–liquid equilibrium of sulfuric acid solution and modified fugacity theory. The main effects of flue gas properties, size and arrangement form of smooth tube exchanger on the deposition rate were analyzed.

The results show that, the condensation rate is mainly related to the concentration gradient of H_2SO_4 vapor near the surface. As the flue gas or surface temperature increases, condensation rate decreases. When the flue gas velocity increases, the mass transfer resistance is reduced, thereby the condensation rate is increased. As the content of H_2SO_4 in the flue gas increases, the concentration gradient of H_2SO_4 vapor near the surface increases, and the condensation rate of increases linearly. But the content of water vapor has little effect on the H_2SO_4 condensation rate. For the tube heat exchanger, when the working conditions keep the same, the smaller the outer diameter of the tube, the greater the condensation rate of the H_2SO_4 vapor. Compared with the inline arrangement, when the tubes are staggered, the disturbance of flue gas is stronger, and the condensation rate is larger.

Acknowledgements This work was supported by the National Key R&D Program of China (2017YFB0602102).

References

Abel E (1948) On the experimental bases for the calculation of the sulfuric acid vapor pressure above the sulfuric acid-water, system. J Phys Colloid Chem 52(5):908–914

Abel E (2002) The vapor phase above the system sulfuric acid–water. J Phys Chem 50(3):260–283

Baosheng J, Zhiyong T, Keqin S, Zhaoping Z (2006) Numerical prediction of condensate deposition rate of sulfuric acid in chimney of coal-fired power station. J China Electr Eng 09:40–44 (in Chinese)

Cen K (1994) Principles and calculations for prevention of ash, slagging, wear and corrosion in boilers and heat exchangers. Science Press (in Chinese)

Dietz R, Marchal D (1977) Attempt to predict quantitatively the acid deposits formed downstream of furnaces operating on heavy fuel oil. Int Chem Eng (United States) 17:4(1–2): 80–86

Gençay A (1987) Get Acid Dew Point of Flue Gas 66:7

Giauque W, Hornung E, Kunzler J, Rubin T (1961) Additions and corrections-the thermodynamic properties of aqueous sulfuric acid solutions and hydrates from 15 to 300 degrees K. J Am Chem Soc 83(24):5047

Gmitro JI, Vermeulen T (1964) Vapor-liquid equilibria for aqueous sulfuric acid. AICHE J 10(5):740–746

Guangcai Z, Zhixiang Z, Hongwei C, Nan L, Mingxi X (2018) Analysis of corrosion causes of flue gas reheater and countermeasures. Thermal Power Gener 47(04):110–115 (in Chinese)

Hanxiao L, Yuping Y, Jianguo L, Yuzhong H, Dongxu X, Lin Z, Shengqing Z (2018) Application of electrostatic precipitator technology integrated in high-ash coal ultra-low emission engineering. Environ Sci Technol 41(04):151–159 (in Chinese)

Hanxiao L, Yuping Y, Jianguo P, Yuzhong H, Lin Z, Xu D, Luo J, Zhao S, Wu J (2018) Analysis of the challenges facing the development of electrostatic precipitator technology. China's Environ Protect Indus (02):18–22. (in Chinese)

Hui H, Yaling H, Guihua T (2013) Numerical study on the deposition characteristics of sulfuric acid vapor on the surface of heat exchanger. J xi'an Jiaotong Univ 47(11):7–13 (in Chinese)

Jun H (2005) Distribution characteristics and causes of sulfur in coal in China. Coal Conversion 04:5–10 (in Chinese)

Kelley KK (1960) Contributions to the data on theoretical metallurgy: XIII. High temperature heat content, heat capacity and entropy data for the elements and inorganic compounds. Procedia Soc Behav Sci 110(5):259–266

Ketao X, Junfei M, Zhang Y (2019) Analysis of operation status and optimization measures of water-based flue gas heat exchangers. China Power 52(01):118–123

Li J, Li Z, He Y, Zhao H, Shunli Yu (2014) Research and application of low-low temperature and low temperature electric dust removal technology. China's Environ Protect Indus 3:28–34 (in Chinese)

Liu X, Zhao Y, Sun Y, Xiaofu Hu, Chen F (2018) Research progress on SO_3 control and removal technology in coal-fired power plants. Power Sci Eng 34(2):56–62 (in Chinese)

Mi Z, Zheng J, Meng J, Shan Y, Zheng H, Ou J, Guan D, Wei Y (2018) China's energy consumption in the new normal. Earth's Future 6(7):1007–1016

Nakayama Y, Nakamura S, Takeuchi Y, Itoh M, Okino S (2011) MHI high efficiency system—proven technology for multi pollutant removal. Research & Development Center, Nishi-ku

National Bureau of Statistics of China 2018 (2018) China Statistics Press, Beijing. (in Chinese)

Qing H (2018) Application research of double-effect low-temperature economizer technology for coal-fired boilers. Thermal Power Eng 33(11):75–80

Ren F, Xia L (2017) Analysis of China's primary energy structure and emissions reduction targets by 2030 based on multiobjective programming. Math Probl Eng 2017:1–8

Sihua Z, Su W, Zhensong T, Cunyi S, Junjie L (2018) Research status and engineering application of ultra-low emission control technology for coal-fired power plants. Modern Chem Indus 38(02):162–165 (in Chinese)

Wilson RW (1989) Condensation of sulfuric acid and water mixtures in laminar flow. D, Lehigh University, Bethlehem

Yanwu G, Peng L, Hongbin C (2018) Application of low temperature and low temperature electric dust removal system in ultra-supercritical unit. Electr Energy 39(01):98–100

Yunfei L, Wenhua L, Ying J, Xiangfei B (2005) Study on distribution characteristics of sulfur in coal in China. Coal Convers 03:14–18 (in Chinese)

Zhang Z (1991) Mass transfer processes in flow passages of rotating regenerative air preheaters. Lehigh University

Zhang D, Wang J, Lin Y, Si Y, Huang C, Yang J, Huang B, Li W (2017) Present situation and future prospect of renewable energy in China. Renew Sustain Energy Rev 76:865–871

Zhiyong T, Baosheng J, Keqin S, Zhaoping Z (2005) Numerical simulation of sulfuric acid corrosion in power plant chimney. J Southeast Univ (Natural Science Edition) 05:103–106. (in Chinese)

Chapter 71
Partitioning Behavior of Arsenic in an Ultra-Supercritical Coal-Fired Power Plant Equipped with APCDs for Ultra-Low Emission

Zhipeng Shi, Zhijun Huang, Wei Hua, and Lunbo Duan

Abstract To study the partitioning behavior of arsenic (As) in an ultra-supercritical coal-fired power plant equipped with air pollution control devices (APCDs) for ultra-low emission, the US EPA Method 29 was used to simultaneously sample flue gas before and after selective catalytic reduction (SCR), electrostatic precipitator (ESP), and wet flue gas desulfurization (WFGD). The solid and liquid samples including feed coal, bottom ash, ESP ash, limestone slurry, flush water, WFGD gypsum and wastewater were simultaneously collected in step with flue gas sampling. The mass distribution of As was obtained based on mass balance calculation. Results indicate that the mass balance rate of As is in the range of 78.40–117.68%, which is acceptable. At the outlet of air-preheater, particulate-bound As is the dominant species in the flue gas, which accounts for over 98%. Most of As were migrated to ESP ash and bottom ash, accounting for 95.19% and 4.28%, respectively. Both As^{5+} and As^{3+} existed in the feed coal and bottom ash, while As^{5+} was found to be the major form in the ESP ash, WFGD gypsum and wastewater. The removal efficiency of As by ESP and WFGD is 99.3–99.4%, while that by ESP is 99.87–99.90% and that by WFGD is 26.32–46.77%. The new APCD (mercury absorption device) can oxidize As^{3+} to As^{5+} in the flue gas, but somewhat increase the emission concentration of As. As tends to enrich in ESP ash (relative enrichment index = 1.28) but dilute in bottom ash (relative enrichment index = 0.23).

Keywords Coal-fired power plant · Ultra-low emission · Arsenic · Emission characteristics · Speciation

Z. Shi · L. Duan (✉)
Key Laboratory of Energy Thermal Conversion and Control, Ministry of Education, Southeast University, Nanjing 210096, China
e-mail: duanlunbo@seu.edu.cn

Z. Huang · W. Hua
Environmental Technology Center, Jiangsu Fangtian Power Technology Company, Nanjing 211102, China

J. Lyu and S. Li (eds.), *Clean Coal and Sustainable Energy*,
Environmental Science and Engineering,
https://doi.org/10.1007/978-981-16-1657-0_71

71.1 Introduction

Arsenic (As) is expected to be one of the most hazardous trace elements due to its high toxicity and carcinogenicity (Xu and Zheng 2013; Zhao and Zhang 2008; Kang and Liu 2011). In recent years, increasing emission of As has caused serious damages to the environment and human health (Duan and Tan 2013; Zhou and Liu 2017). With the huge consumption of coal annually, coal-fired power plant is recognized as the major anthropogenic emission source of As (Tian and Lu 2013; Zhao and Wang 2008; Tian and Wang 2011). As bound in coal can be released and redistributed into combustion by-products such as bottom ash, fly ash, and the flue gas during combustion processes (Tang and Liu 2012).

Actually, the toxicity of As is highly related to its speciation, and As^{3+} is more toxic than As^{5+} (Hughes 2002; Duker and Carranza 2005) Therefore, analysis of the speciation of As in combustion by-products of coal is beneficial for their subsequent treatment. Numerous studies have revealed that the As in these by-products is mainly in the form of arsenite (As^{3+}) and arsenate (As^{5+}) (Huggins and Senior 2007; Goodarzi and Huggins 2001; Shah and Strezov 2008; Zhao and Duan 2018). Shah and Strezov (2007) used X-ray absorption fine structure spectrometry (XAFS) to determine the speciation of As in bottom ash and fly ash from different power plants, which found that majority As in bottom ash and fly ash was in the less toxic As^{5+} form. Zhao and Duan (2018) also found $As^{5+}$was the main As form in fly ash from a coal-fired power plant by high-performance liquid chromatography coupled with inductively coupled plasma-mass spectrometry (HPLC-ICP-MS). However, previous studies mainly focused on the bottom ash and fly ash, while gypsum and wastewater from desulfurization systems have rarely been investigated.

To reduce the emission of the conventional pollutants, many air pollution control devices (APCDs) suchasselectivecatalyticreduction(SCR), electrostatic precipitator (ESP) or fabric fiber (FF), and wet flue gas desulfurization (WFGD) have been used increasingly in coal-fired power plants. The application of APCDs can also affect the partitioning and emission characteristics of As (Martinello and Oliveira 2014; Yue and Chen 2005; Zhao and Duan 2017a). Duan and Cui (2017) investigated the partitioning behavior of As in several circulating fluidized bed boilers, which found the removal efficiencies of As by APCDs (SNCR + ESP/FF + WFGD) are in the range of 99.2–99.6%. Zhao and Duan (2017b) also carried out a similar test in a pulverized boiler equipped with APCDs (SCR + ESP + WFGD), results showed that 99.96% of As can be removed by APCDs. Their findings all show that current APCDs in power plants are beneficial for the removal of As. However, research on large-scale units such as 1000 MW ultra- supercritical units is still insufficient. Moreover, to reduce Hg emissions, the new APCD mercury absorption device (MAD) has been developed and piloted, and the power plant in this study was first equipped with a MAD in China. Due to the difference in chemical properties of Hg and As, the influence of the new APCD on As emission and migration is uncertain.

Therefore, the main objectives of this study are to determine: (1) the mass balance and distribution of As in the entire system; (2) concentration and speciation of As in combustion by-products; (3) removal efficiency of As by APCDs; and (4) emission characteristics of As.

71.2 Materials and methods

71.2.1 Boiler Description

The study was conducted at a 1000 MW pulverized coal-fired power plant in Jiangsu province. To achieve ultra-low emission, SCR, ESP and WFGD were employed to reduce the emission of NOx, particle matters (PM) and SOx. Besides, this power plant was first equipped with a MAD in China. The MAD sprays HBr-modified fly ash into the flue gas to oxidize and absorb the Hg^0, which can remove more than 90% of the total Hg in the flue gas. Under the normal operation of the boiler unit, the emission concentration of NOx, SOx and PM are 44.4, 29.84 and 1.61 mg/ Nm^3, which meets the ultra-low emission standards (Shi and Yang 2016).

71.2.2 Sampling Procedure

To study the partitioning behaviors of As in the entire power plant and the influence of the new APCD (MAD) on the As in the flue gas, the experiments were carried out under the conditions of the MAD opening and the MAD closing, respectively. All sampling tests were carried out twice and the load of boiler was 80 ± 5% during the sampling procedure.

For the flue gas, the simultaneous sampling locations are provided in the Fig. 71.1. According to US EPA method 29, the sampling procedure was carried out by US Apex instrument, as shown in Fig. 71.2, which is consisted of a heated probe, a heated filter, glass impingers, ice bath box, vacuum pump and flow meter. To reduce the condensation loss of As, the probe and filter were heated to above 120 °C during the sampling procedure. The particulate As was collected by the filter while gaseous As was absorbed by the glass impingers with 5% (v./v.) nitric acid (HNO3) and 10% (v./v.) hydrogen peroxide (H2O2). For each test, the sampling process lasted for one hour. The solid and liquid samples including feed coal, bottom ash, ESP ash, limestone slurry, flush water, WFGD gypsum and wastewater were simultaneously collected in step with flue gas sampling. All samples collected were stored in specific containers and transported to the laboratory immediately for As analysis.

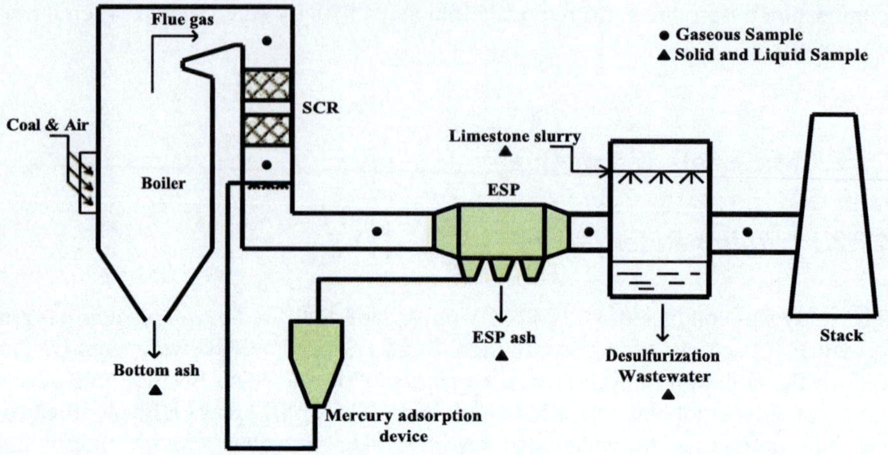

Fig. 71.1 Sampling sites in the coal-fired power plant

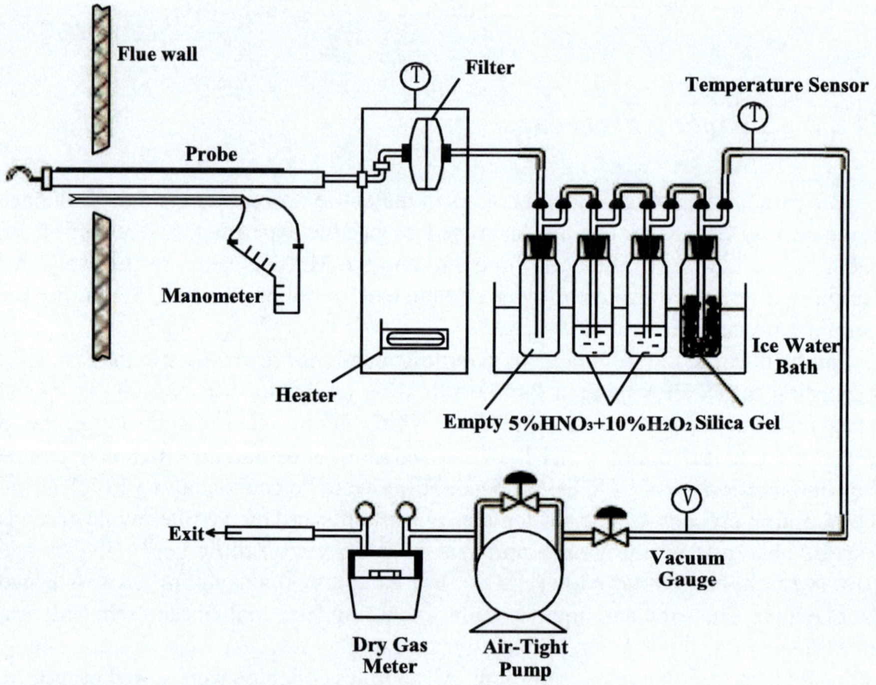

Fig. 71.2 Schematic configuration of sampling device of EPA Method 29

71.2.3 Analysis Methods

71.2.3.1 Element Analysis

The concentration of As in the liquid samples including absorption solution and flush water was directly measured by ICP-MS after filtration. The solid samples such as feed coal, bottom ash, ESP ash and WFGD gypsum were first air-dried, separated by quartering split, and then milled and sieved through a 200-mesh sieve (75 μm) for chemical pretreatment. Then the dry samples were digested by a mixture of acid (HNO_3: HCL: HF = 3:1:1) in a microwave oven.

Finally, the content of As in the digestion solution can be directly determined using ICP-MS. For the limestone slurry and WFGD wastewater, which are turbid liquids, the solid and liquid portions were separated through filtration and drying, and then the total concentration of As was obtained by calculating the As content in solid and liquid portions, respectively. All tests were carried out for three times and the relative standard deviation of parallel results is less than 5%. The showing results in this paper are average values.

71.2.3.2 Speciation Analysis

The speciation of As in the solid and liquid samples were determined by high-performance liquid chromatography coupled with ICP-MS (HPLC-ICP- MS). The solid samples such as feed coal, bottom ash and ESP ash were digested using the microwave acid extraction procedure. About 10 mg dry samples were weighed into HDPE bottles and 10 ml solution of 0.1% (v./v.) HNO_3 was added. Then the digest solution was centrifuged at 7000 rpm for 10 min and filtered through 0.25 μm PVDF filter. The content of As^{3+} and As^{5+} in the clear solution or liquid samples can be directly determined by HPLC-ICP-MS. The accuracy of As^{3+} and As^{5+} content in samples was determined through a blending solution of GBW08666 and GBW08667 as a standard reference. The chromatogram of As^{3+} and As^{5+} obtained from the GBW08666 and GBW08667 is shown in Fig. 71.3. All tests were carried out for three times and the relative standard deviation of parallel results is less than 5%.

71.3 Results and Discussion

71.3.1 Samples Analysis Results

71.3.1.1 Coal Analysis

The proximate and element analysis of the feed coal are shown in Table 71.1. The

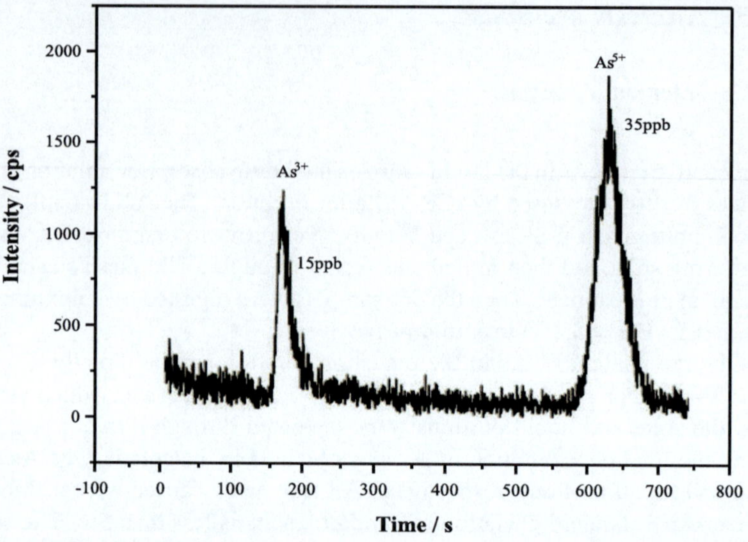

Fig. 71.3 Chromatogram of As^{3+} and As^{5+} based on mixed solution of GBW08666 and GBW08667

Table 71.1 Proximate and element analysis of the coal[a]

Proximate analysis				Qnet, ar MJ/kg	Element analysis					
M	A	V	FC		C	H	O	N	S	As
%	%	%	%		%	%	%	%	%	mg/kg
1.32	25.96	26.77	46.87	22.89	57.56	3.53	10.45	0.84	0.63	1.82

[a]All the value is based on as received basis

coal belongs to lean coal according to the National Coal Classification Standard of China (GB/T5751-2009). Some researchers found that flue gas components such as SO2 can affect the speciation of trace elements (Gao and Liu 2005). It was reported that the content of S in Chinese coal is in the range of 0.2–8% (Ketris and Yudovich 2009). This coal belongs to low sulfur coal, the content of S in which is 0.63%. Furthermore, the concentration of As in the feed coal is 1.82 mg/kg, which is lower than the average value of both Chinese coal (3.79 mg/kg) (Dai 2012) and world's coal (8.3 mg/kg) (Lin and Huang 2012). The difference may due to their different surrounding condition and coal formation process.

71.3.1.2 Concentration and Speciation of As in Samples

To study the content of As in the by-products of coal combustion, the bottom ash, the ESP ash, the WFGD gypsum and wastewater were simultaneously sampled and analyzed. The results are presented in Table 71.2. As shown, the concentration of As in the bottom ash (1.62 mg/kg) was lower than that in the feed coal, whereas it in the ESP ash (9.01 mg/kg) was comparatively higher, which indicates that As is readily volatilized during the combustion process and is inclined to condense on the fly ash particles. Moreover, As is also found in the WFGD gypsum and wastewater, which proves that the WFGD system can partly remove the As in the flue gas.

Also, speciation analysis was conducted because the toxicity of As is highly related to its speciation. The results are presented in Table 71.2. Both As^{3+} and As^{5+} exist in the bottom ash, accounting for 26.54 and 61.11%, respectively. (Duker and Carranza 2005) have found that As^{5+} compounds are more stable under oxidizing conditions while As^{3+} compounds are more stable under reducing conditions. Therefore, the reducing atmosphere of the combustion zone in furnace may be the reason for the high proportion of As^{3+} in bottom ash than that in ESP ash. Compared to the bottom ash, the proportion of As^{5+} in the ESP ash is higher, accounting for 97.23% while no As^{3+} was detected. Zhao and Duan (2018) carried out some tests to determine As speciation in the ESP ash from a coal-fired power plant by HPLC-ICP- MS, which found the ratio of As^{5+} to total As was 98.63%. Huggins and Senior (2007) confirmed that As was found predominantly to be in the form of As^{5+} based on the speciation analysis of As in 10 fly ash from full-scale utility plants by XAFS. Their findings all show a good agreement with the results in this study. Actually, under the action of high voltage electric field plasma discharge in ESP, H2O, O_2, and HCl in the flue gas will ionize and form active oxidants such as OH, O and O_3, which have strong oxidizing ability (Zhang and Wang 2015). These active components may oxidize As^{3+} to As^{5+}. In addition, the HBr-modified fly ash sprayed by MAD may also oxidize As^{3+}. For WFGD gypsum and wastewater, As^{5+} is the dominant speciation while As^{3+} was not detected, which may due to the oxidation air sent in the oxidation zone of the WFGD system.

Table 71.2 Concentration and speciation of As in coal and combustion by-products

Samples	$As3+$	$As5+$	Total As	Proportion of As^{5+} (%)
	mg/kg (mg/L)	mg/kg (mg/L)	mg/kg (mg/L)	
Coal	0.27	1.48	1.82	81.32
Bottom ash	0.43	0.99	1.62	61.11
ESP ash	n.d.[b]	8.76	9.01	97.23
Gypsum	n.d.[b]	0.12	0.13	92.31
Wastewater	n.d.[b]	0.03	0.04	75.00

[b]Not detected

71.3.2 *Mass Balance Rate and Distribution of As Across the Power Plant*

71.3.2.1 Mass Balance Rate

The mass balance rate is often used to verify data reliability and credibility for trace element test in power plant, which is defined as the following Eq. (71.1):

$$R = \frac{M_{out}}{M_{in}} \tag{71.1}$$

where *R* represents the mass balance rate of As. *Min* is the total amount of As entering the system during sample process while *Mout* is the total amount of As leaving the system during sample process. For the whole system, The input sources of As include feed coal, limestone slurry and flush water. The output sources of As include bottom ash, ESP ash, WFGD gypsum, WFGD wastewater and flue gas emitted to the atmosphere. Similarly, the mass balance rate of each APCD can be calculated through this formula.

Take all parameters above into Eq. (71.1), detailed Eq. (71.2) is as following:

$$\begin{aligned} R_{system} = C_{ba} * m_{ba} + C_{fa} * m_{fa} + C_g * m_g + C_{ww} * m_{ww} \\ + C_{eg} * m_{eg} C_f * m_f + C_{ls} * m_{ls} + C_{fw} * m_{fw} \end{aligned} \tag{71.2}$$

where *Rsystem* represents the mass balance rate of As across the whole system *Cf, Cls, Cfw, Cba, Cfa, Cg, Cww, Ceg* represent the concentration of As in feed coal, limestone slurry flush water bottom ash, ESP ash, WFGD gypsum, WFGD wastewater and flue gas emitted to the atmosphere, μg/kg or μg/m^3. The parameters *mf, mls, mfw, mba, mfa, mg, mww, meg* represents the flow rate of feed coal, limestone slurry flush water bottom ash, ESP ash, WFGD gypsum, WFGD wastewater and flue gas emitted to the atmosphere, kg/h or m^3/h. As shown in Fig. 71.4, the mass balance rate of As for the entire system is 107.41% while this for each APCDs are in the range of 78.40–117.68%. Actually, some factors like operating parameter fluctuation, error in analysis may affect the mass balance rate. Thus, the values in range of 70–130% are acceptable, as reported by other researchers previously (Reddy and Basha 2005; Quick and Irons 2002; Zhang and Zhang 2016). The values in this study are in this range, which suggests that the experimental results are reliable.

71.3.2.2 Mass Distribution of As

The As in the feed coal will migrate to the bottom ash, the ESP ash, the WFGD gypsum, the WFGD wastewater and the flue gas emitted to the atmosphere. Analysis of the mass distribution of As in combustion by-products is beneficial for their

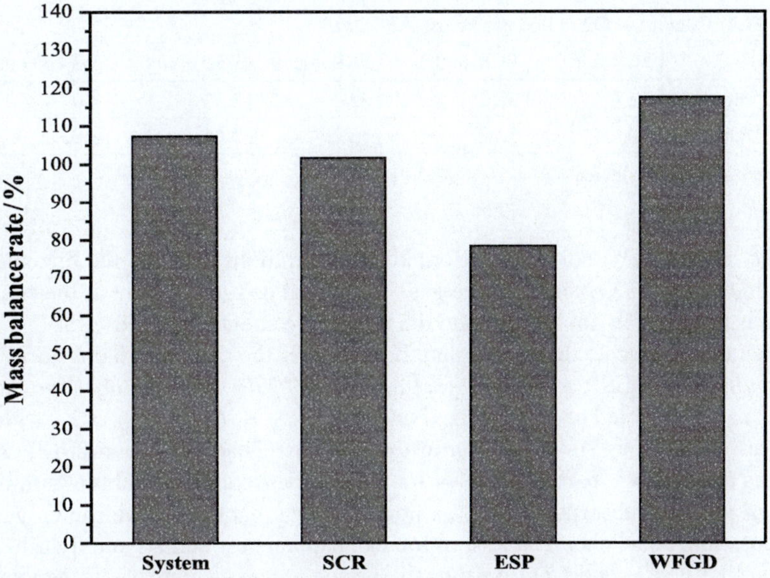

Fig. 71.4 Mass balance rates of As of the entire system and each APCD

Table 71.3 Mass distribution proportions of As (%)

	Bottom ash	ESP ash	Gypsum	Wastewater	Stack
Proportion	4.28	95.19	0.33	0.09	0.11

following treatment. Here, the results are shown in Table 71.3. The majority of As is distributed in the bottom ash and the ESP ash, which accounts for 95.19% and 4.28%, respectively.

The proportions of As in the WFGD gypsum and wastewater are comparatively lower, only accounting for 0.33% and 0.09%, respectively. The mass of As in the flue gas emitted to the atmosphere is little, accounting for 0.11%. The results agree with that of Zhao and Duan (2017a), which indicate that more attention should be paid to the treatment of the bottom ash and the ESP ash.

71.3.3 *Concentration, Forms, and Removal Efficiency of As in Flue Gas Across APCDs*

71.3.3.1 Concentration and Forms of As in Flue Gas

To study the influence of the new APCD (MAD) on the As in the flue gas, this work was carried out separately under the conditions of the MAD opening and the MAD

Table 71.4 Content of O2 in flue gas across APCDs (%)

	SCR inlet	SCR outlet	ESP inlet	ESP outlet	WFGD outlet
MAD [c] open	2.37	3.15	2.94	3.53	4.07
MAD [c] close	2.41	3.02	3.22	3.66	3.83

[c]Mercury absorption device

closing, respectively. For comparison, all concentrations of As in the flue gas have been unified to 6% O_2 and dry flue gas. The actual oxygen content at the sampling points is provided in Table 71.4, and the results are shown in Table 71.5.

The results indicate that particulate-bound As is the mainly existing form in the flue gas before the ESP, accounting for 98.45% ~ 99.37%. In the combustion process, As in coal is expected to be vaporized and form As_2O_3, As_2O and AsO_2, which are the possible gaseous As speciation in the flue gas (Zhao and Duan 2018). As the flue gas cools down, part of gaseous As will condense and recombine with fly ash through physical absorption such as homogeneous nucleation and heterogeneous condensation. Also, chemical absorption can happen between As and fly ash. Some oxides, like Fe_2O_3 and CaO on fly ash surface can react with As to generate As compounds (Huang and Jin 2003). Thus, particulate-bound As is the predominant form in the flue gas, which is consistent with the results reported by Zhao and Duan (2017a).

When the MAD opened, the concentration of As increased. This phenomenon can be explained as follows: on the one hand, the MAD is used to remove gaseous Hg, which is the predominant form of Hg in the flue gas. The HBr-modified ESP ash sprayed into the flue gas by the MAD can oxidize gaseous Hg to make it more readily be absorbed, while it has almost none influences on the particulate-bound Hg or As. Therefore, the MAD will not remove As due to the extremely high proportion of particulate-bound As in the flue gas. On the other hand, Álvarez-Ayuso and Querol (2006) studied the relative enrichment degree of different trace elements in the fly ash, which found As is prone to enrich in the ESP ash while Hg tends to dilute. Thus, the ESP ash sprayed into the flue gas will make the As enriched in the ash back to the flue gas, resulting in an increase in the concentration of As.

71.3.3.2 Removal Efficiency of As Across APCDs

SCR and MAD have little effect on the removal of As. To better describe the removal effect of ESP and WFGD, the removal efficiency of As is defined as follows Eq. (71.3):

$$\eta = \frac{C_{in} - C_{out}}{C_{in}} \times 100\% \qquad (71.3)$$

Table 71.5 Concentration and forms of As in flue gas across APCDs [d]

Condition		SCR inlet		SCR outlet		ESP inlet		ESP outlet		WFGD outlet	
		μg/Nm^3	%	μg/Nm^3	%	μg/Nm^3	%	μg/Nm^3	%	μg/Nm^3	%
MAD open	Asg	3.13	0.95	5.21	1.55	3.12	0.68	0.08	12.90	0.09	27.27
	Asp	324.97	99.05	331.19	98.45	457.04	99.32	0.54	87.10	0.24	72.73
	AsT	328.10	100.00	336.40	100.00	460.16	100.00	0.62	100.00	0.33	100.00
MAD close	Asg	3.82	1.18	4.18	1.27	2.39	0.63	0.03	7.89	0.13	46.43
	Asp	319.65	98.82	324.86	98.73	379.48	99.37	0.35	92.11	0.15	53.57
	AsT	323.47	100.00	329.04	100.00	381.87	100.00	0.38	100.00	0.28	100.00

[d] Asg: gas phase As; Asp: particulate-bound As; AsT: total As

where, η represents the removal efficiency. *Cin* and *Cout.* represent the total concentration of As in flue gas at the inlet and outlet of the ESP or WFGD, $\mu g/m^3$.

The removal efficiency of As across APCDs are shown in Fig. 71.5. ESP has the greatest removal efficiency (99.87–99.90%) because particulate-bound As is the predominant existing form. Particulate-bound As can be removed simultaneously with the deposition of particulate matter (PM) by the ESP. After the WFGD, both particulate-bound As and gaseous As reduced, which may due to the flushing process of the limestone slurry. Studies have revealed that the limestone slurry can react with As and form $Ca_3(AsO_4)_2$, which can remove some As in the flue gas (Córdoba and Ochoa-Gonzalez 2012; Zhu and Zhuo 2013; Shuang and Shi 2014). In this study, the removal efficiency of WFGD is 26.32–46.77%. Furthermore, the total removal efficiency of ESP + WFGD is 99.93–99.94%, which agrees with the results of Zhao and Duan (2017b). The results indicate that current APCDs are beneficial for the removal of As besides the capture of conventional air pollutants.

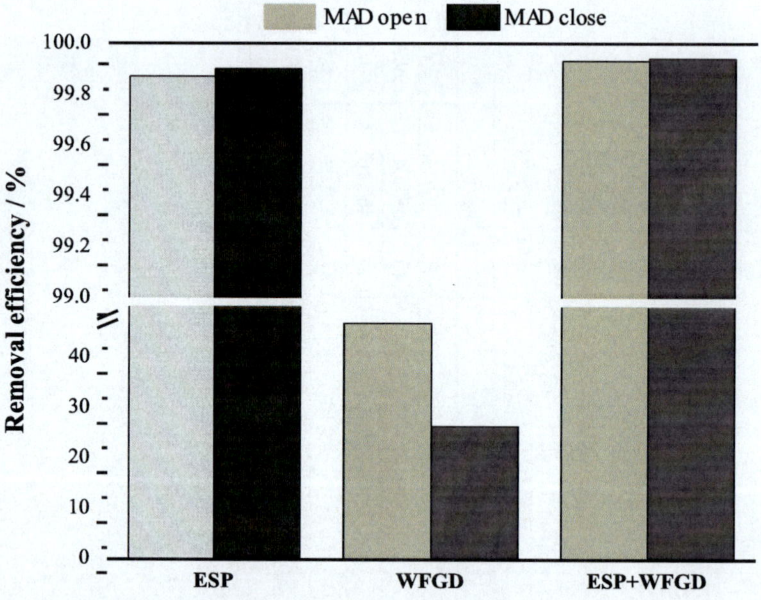

Fig. 71.5 Removal efficiency of As by APCDs

71.3.4 Relative Enrichment Index and Emission Factor of As

71.3.4.1 Relative Enrichment Index of As in Bottom Ash and ESP Ash

To study the enrichment characteristics of As in bottom ash and ESP ash. The relative enrichment index is defined as follows:

$$REI = \frac{C_{i,ba/fa} * A_{ba/fa}}{C_{i,coal}} \quad (71.4)$$

where *REI* represents the relative enrichment index of As in bottom ash or ESP ash. $Ci,ba/fa$ and $Ci,coal$ represent the concentration of As in bottom ash or ESP ash and feed coal, μg/kg. *Aba/fa* represents the bottom ash or ESP ash content in coal, %.

Actually, the larger the *REI* is, the more likely As is enriched in ashes, and the smaller the *REI* is, the more possible As is diluted or depleted in ashes (Tang and Liu 2012; Bhangare and Ajmal 2011).

In this study, the *REI* of As in bottom ash and ESP ash are 0.05 and 1.02, respectively, which indicates that As was inclined to enrich in the ESP ash but dilute in the bottom ash. This is mainly because most of the As in coal vaporizes during combustion. As the flue gas cool down, the As will condense and recombine with fly ash through physical and chemical absorption, which can finally be removed by ESP. Thus, the As will accumulate in the ESP ash.

71.3.4.2 Emission Factor of As

The concentrations of As emitted to the atmosphere from stack under two conditions (MAD open/close) are 0.33 μg/Nm^3 and 0.28 μg/Nm^3, respectively. The concentrations are truly low, which indicates that current APCDs can effectively remove the As in the flue gas. To compare emission characteristics of As between different conditions or different power plants, the emission factor is defined as follows:

$$EF = \frac{M_s}{M_f * LHV} \quad (71.5)$$

where *EF* represents the emission factor of As, g/TJ. *Ms* represents the mass amount of As emitted to the atmosphere per hour, mg/h. *Mf* represents the mass amount of feeding coal per hour, t/h. *LHV* represents the low calorific value of feed coal, MJ/kg.

In this study, the emission factor of As is 0.09 g/TJ when the MAD is open, while this value is 0.08 g/TJ when the MAD is open, while this value is 0.08 g/TJ when the MAD is closed. Thus, the use of the MAD will somewhat increase the emission amount of As.

71.4 Conclusions

Partitioning behavior and emission characteristic of As at different operating conditions was investigated on a 1000 MW pulverized coal-fired power plant equipped with SCR, MAD, ESP and WFGD. The conclusions were drawn as follows.

1. The mass balance rates of As for the entire system and each APCD are in the acceptable range of 70–130%;
2. The majority of As is distributed in bottom ash and ESP ash, accounting for 95.19% and 4.28%, respectively;
3. Both As^{5+} and As^{3+} existed in the feed coal and bottom ash, while As^{5+} was found to be the major species in the ESP ash, WFGD gypsum and wastewater;
4. Particulate-bound As is the main form in the flue gas. The removal efficiency of As by ESP + WFGD is 99.3–99.4%, while that by ESP is 99.87–99.90% and that by WFGD is 26.32–46.77%;
5. As tends to enrich in ESP ash ($REI = 1.28$) but dilute in bottom ash ($REI = 0.23$);
6. The new APCD (MAD) can help oxidize As^{3+} to As^{5+} in the flue gas, but somewhat increase the emission concentration of As.

Acknowledgements This project was funded by the National Natural Science Foundation of China through Grant No. 51661125011.

References

Álvarez-Ayuso E, Querol X (2006) Environmental impact of a coal combustion-desulphurisation plant: Abatement capacity of desulphurisation process and environmental characterisation of combustion by- products. Chemosphere 65(11):2009–2017

Bhangare RC, Ajmal PY (2011) Distribution of trace elements in coal and combustion residues from five thermal power plants in India. Int J Coal Geol 86(4):349–356

Córdoba P, Ochoa-Gonzalez R (2012) Partitioning of trace inorganic elements in a coal-fired power plant equipped with a wet flue gas desulphurisation system. Fuel 92:145–157

Dai SF (2012) Geochemistry of trace elements in Chinese coals: A review of abundances, genetic types, impacts on human health, and industrial utilization. Int J Coal Geol 94(3):3–21

Duan LB, Cui J (2017) Partitioning behavior of Arsenic in circulating fluidized bed boilers co-firing petroleum coke and coal. Fuel Process Technol 166:107–144

Duan J, Tan J (2013) Atmospheric heavy metals and Arsenic in China: Situation, sources and control policies. Atmos Environ 74(2):93–101

Duker AA, Carranza EJM (2005) Arsenic geochemistry and health. Environ Int 31(5):0–641

Gao LF, Liu GJ (2005) The study of sulfur geochemical in Chinese coal. Bulletin of Mineralogy Petrology & Geochemistry 24(1):79–87 (in Chinese)

Goodarzi F, Huggins FE (2001) Monitoring the species of arsenic, chromium and nickel in milled coal, bottom ash and fly ash from a pulverized coal-fired power plant in western Canada. J Environ Monit 3(1):1–6

Huang YJ, Jin BS (2003) Study on groups of trace elements during coal combustion. J Southeast Univ 33(2):148–152 (in Chinese)

Huggins FE, Senior CL (2007) Selenium and Arsenic Speciation in Fly Ash from Full-Scale Coal-Burning Utility Plants. Environ Sci Technol 41(9):3284–3289
Hughes MF (2002) Arsenic toxicity and potential mechanisms of action. Toxicol Lett 133(1):1–16
Kang Y, Liu GJ (2011) Arsenic in Chinese coals: distribution, modes of occurrence, and environmental effects. Sci Total Environ 412–413:1–13
Ketris MP, Yudovich YE (2009) Estimations of clarkes for carbonaceous biolithes: world averages for trace element contents in black shales and coals. Int J Coal Geol 78(2):135–148
Lin CW, Huang SH (2012) From electrostatic precipitation to nanoparticle generation. J Aerosol Sci 51:57–65
Martinello K, Oliveira ML (2014) Direct identification of hazardous elements in ultra-fine and nanominerals from coal fly ash produced during diesel co-firing. Sci Total Environ 470–471(2):444–452
Quick WJ, Irons RMA (2002) Trace element partitioning during the firing of washed and untreated power station coals. Fuel 81(5):665–672
Reddy MS, Basha S (2005) Evaluation of the emission characteristics of trace metals from coal and fuel oil fired power plants and their fate during combustion. J Hazard Mater 123(1–3):242–249
Shah P, Strezov V (2007) Speciation of arsenic and selenium in coal combustion products. Energy Fuels 21(2):506–512
Shah P, Strezov V (2008) Speciation of As, Cr, Se and Hg under coal fired power station conditions. Fuel 87(10–11):1859–1869
Shi WZ, Yang MM (2016) Ultra-low emission technical route of coal-fired power plants and the cooperative removal. Proceeding of the CSEE 36(16):4308–4318
Shuang D, Shi Y (2014) Emission characteristics of Cd, Pb and Mn from coal combustion: Field study at coal- fired power plants in China. Fuel Process Technol 126(10m):469–475
Tang Q Liu GJ (2012) Distribution and fate of environmentally sensitive elements (arsenic, mercury, stibium and selenium) in coal-fired power plants at Huainan, Anhui, China. Fuel 95:334–339
Tian HZ, Lu L (2013) A review of key hazardous trace elements in chinese coals: abundance, occurrence, behavior during coal combustion and their environmental impacts. Energy Fuels 27(2):601–614
Tian HZ, Wang Y (2011) Atmospheric emissions estimation of Hg, As, and Se from coal-fired power plants in China, 2007. Sci Total Environ 409(16):3078–3081
Xu SJ, Zheng N (2013) Geochemistry and health risk assessment of arsenic exposure to street dust in the zinc smelting district. Northeast China. Environmental Geochemistry & Health 35(1):89–99
Yue Y, Chen L (2005) Experimental study on characteristics of particulate matter size distribution and trace elements enrichment in emissions from a pulverized coal-fired boiler. Proceeding of the CSEE 25(18):74–79 (in Chinese)
Zhang L, Wang SX (2015) Mercury transformation and speciation in flue gases from anthropogenic emission sources: A critical review. Atmos Chem Phys 16(4):32889–32929
Zhang KH, Zhang DX (2016) Capture of gas-phase arsenic by ferrospheres separated from fly ashes. Energy Fuels 30:8746–8752
Zhao SL Duan YF (2017a) Study on emission of hazardous trace elements in a 350MW coal-fired power plant. Part 2. Arsenic, chromium, barium, manganese, lead. Environmental Pollution. 226:404–411
Zhao SL, Duan YF (2017b) Partitioning and emission of hazardous trace elements in a 100MW coal-fired power plant equipped with SCR, ESP, wet FGD. Energy Fuels 31(11):12383–12389
Zhao SL, Duan YF (2018) Distribution and speciation transformation of hazardous trace element arsenic in particulate matter of a coal-fired power plant. Energy Fuels 32:56049–56055
Zhao Y, Wang SX (2008) Primary air pollutant emissions of coal-fired power plants in China: Current status and future prediction. Atmos Environ 42(36):8442–8452
Zhao YC, Zhang JY (2008) Arsenic emission during combustion of high arsenic coals from Southwestern Guizhou. China. Energy Convers Manage 49(4):615–624
Zhou CC, Liu GJ (2017) Effect of ash composition on the partitioning of arsenic during fluidized bed combustion. Fuel 204:91–97

Zhu ZW, Zhuo YQ (2013) Trace element distribution during wet flue gas desulphurization system. Tsinghua Univ Tech 53:330–335 (in Chinese)

Chapter 72
A New Scheme for Synergetic Removal of NH_3 and SO_3 and Particulate Matter in the Flue Gas of Coal-Fired Boiler

Limin Wang, Dechao Li, Yan Yu, Chunli Tang, Lei Deng, and Defu Che

Abstract In China, coal-fired boilers generally utilize the selective catalytic reduction (SCR) denitration technology to achieve low NO_x emission. The SCR catalyst will inevitably convert some of the SO_2 in the flue gas into SO_3. NH_3 escaping from the SCR denitration system can react with SO_3 generated in the flue gas to form $(NH_4)_2SO_4$ and NH_4HSO_4. NH_4HSO_4 is highly hygroscopic, viscous and corrosive and easy to induce ash deposition on the heat transfer surfaces at the temperature below its dew point. Since the condensation temperature of NH_4HSO_4 is just in the operating temperature range of the rotary air preheater (RAPH), the RAPH suffered more serious sticky ash deposition problems seriously impacting the economical and safe operation of the boiler. In this study, a new scheme was proposed for synergetic removal of NH_3 slipped from SCR, SO_3 and particulate matters in flue gas. The flue gas outlet temperature of the RAPH is raised to the dew point temperature of NH_4HSO_4 by reducing the heat transfer area, and then NH_3 and SO_3 in the flue gas can be used as the flue gas conditioning agent to modify the adhesion property and specific resistance of the particles. Therefore, by using this new scheme, not only the particulate matters but SO_3 and NH_3 escaped from SCR in the flue gas can be removed in the electrostatic precipitator, and the ash deposition corrosion problem on the low temperature heating surface can be greatly alleviated. The key to this new system is the proper parameters design to achieve sufficient adsorption for NH_3 and SO_3 and the appropriate agglomeration of particulate matters. Therefore, the adsorption and agglomeration characteristics of ash particles were experimentally studied. The results showed that the agglomeration of ash particles can be significantly enhanced by the adsorption reaction, which can increase the mean size of particle matters. The molar ratio between the absorbed N and S contents always maintained in the range of 1.2–1.8. The adsorption of the SO_3 and NH_3 would interact with the ash particles leading to the morphology changes. Besides, the adsorption of NH_3 and SO_3 could

L. Wang · D. Li · Y. Yu · L. Deng · D. Che (✉)
School of Energy and Power Engineering, Xi'an Jiaotong University, Xi'an 710049, China
e-mail: dfche@mail.xjtu.edu.cn

C. Tang
School of Human Settlements and Civil Engineering, Xi'an Jiaotong University, Xi'an 710049, China

J. Lyu and S. Li (eds.), *Clean Coal and Sustainable Energy*,
Environmental Science and Engineering,
https://doi.org/10.1007/978-981-16-1657-0_72

strengthen the agglomeration of fine ash particles, and produces highly cohesive and relatively large particles, resulting in high collection efficiency due to the reduced rapping losses and re-entrainment. The experimental results initially proved that the proposed system would be a promising approach for the alleviation of the sticky ash deposition and for the synergetic removal of NH_3 and SO_3 and particulate matters. Further more detailed study would be made to examine the effects of the ratio of NH_3 and SO_3 in the gas, the adsorption temperature and the particle size on the adsorption rates and adsorption ratio of NH_3 and SO_3, and the specific resistance of the particles, and the particle agglomeration, in order to obtain proper operating parameters for the synergetic removal system.

Keywords Particulate matter · Adsorption · Agglomeration · Synergetic removal · Parameter design

72.1 Introduction

The world today is facing increasingly serious environmental pollution problem, and China is facing more severe emission reduction situation due to the coal-dominated energy structure (International Energy Agency 2018). In recent years, a wide range of continuous smog has hit many cities in China especially in winter, which adversely impacted the health of people. China has implemented more strict contaminant emission standard, NO_x < 50 mg/Nm^3, SO_2 < 35 mg/Nm^3 and particulate matter (PM) < 5 mg/Nm^3 (Ministry of Environmental Protection et al. 2016).

In order to achieve low NO_x emission, coal-fired boilers in China generally utilize the selective catalytic reduction (SCR) denitration technology (Fang and Lu 2018). The SCR catalyst inevitably converts some of the SO_2 in the flue gas into SO_3 while promoting the NO_x reduction reaction. NH_3 escaping from the SCR denitration system can react with SO_3 generated in the flue gas to form $(NH_4)_2SO_4$ and NH_4HSO_4. NH_4HSO_4 is highly hygroscopic, viscous and corrosive and easy to induce ash deposition on the heat transfer surfaces at the temperature below its dew point (Wang 2017; Navarrete 2015). Since the condensation temperature of NH_4HSO_4 is just in the operating temperature range of the rotary air preheater (RAPH), the RAPH suffered more serious sticky ash deposition problems (Jones et al. 1982), which not only increases the cost of soot blowing but leads to unplanned outage seriously impacting the economical and safe operation of the boiler. Although many researchers have conducted related investigations to optimize the structure of SCR reactor to reduce ammonia slip and to optimize the structure of rotary air preheater, the problem can still not be well solved (Chen 2017; Muzio 2017; Menasha 2010; Chothani and Morey 2008).

In addition, in order to further reduce PM emissions, the flue gas conditioning technology is often used on the electrostatic precipitators of coal-fired boilers because flue gas conditioning technology is more efficient and more flexible than other technologies (Li 2018). Flue gas conditioning technology is to spray a certain amount of conditioning agent (usually including SO_3, NH_3, H_2SO_4, NH_4HSO_4 or special adhesive agent and resistance-reducing agent) into the flue gas to change the adhesion and the specific resistance of ash particles at the inlet of the electrostatic precipitator to improve PM removal efficiency (Li 2018; Yu 2015; Baldrey 2001). SO_3 can react with water vapor in the flue gas to form H_2SO_4 which can adhere to the surface of ash particles (Shou et al. 2016). NH_3 can react with SO_3 in flue gas to form NH_4HSO_4 and $(NH_4)_2SO_4$. H_2SO_4 and NH_4HSO_4 with electrolyte properties adhere to the surface of ash particles can reduce their specific resistance (Navarrete 2015; Bian 2015). Besides, NH_4HSO_4 attached to the surface of ash particles has strong adhesion, which can improve PM removal efficiency and reduce $PM_{2.5}$ emissions by promoting the agglomeration of the fine particles and integrating the fine particles into large particles (Yu 2015; Bian 2015).

In this study, a new scheme was proposed for synergetic removal of NH_3 slipped from SCR, SO_3 and particulate matter in flue gas. The flue gas outlet temperature of the RAPH is raised to the dew point temperature of NH_4HSO_4 by reducing the heat transfer area, and then NH_3 and SO_3 in the flue gas can be used as the flue gas conditioning agent to modify the adhesion property and specific resistance of the particles. Therefore, by using this new scheme, not only the particulate matters but SO_3 and NH_3 escaped from SCR in the flue gas can be removed in the electrostatic precipitator, and the ash deposition corrosion problem on the low temperature heating surfaces can be greatly alleviated. In addition, experiments are carried out on a laboratory-scale fixed bed adsorption reaction system to study the effects of SO_3 and NH_3 on the adsorption and agglomeration characteristics of ash particles.

72.2 The New System

As shown in Fig. 72.1, a new system was proposed in our study for synergetic removal of NH_3 and SO_3 and particulate matter in the flue gas of the coal-fired power plant. After passing through the SCR denitrification system, the flue gas entered the rotary air preheater and was cooled down from about 380 to about 200 °C close to the dew point of the NH_4HSO_4. Thus, there would not be plugging in the RAPH caused by the condensation of NH_4HSO_4. Besides, the NH_3 and SO_3 in the flue gas could be used to conduct the flue gas conditioning, which would not only increase the dust removal efficiency of the ESP but also remove the NH_3 and SO_3 from the flue gas in the ESP. A continuous emission monitoring system (CEMS) was installed in the downstream of the RAPH to monitor the flowrate, the temperature and the species concentrations of the flue gas from the RAPH in real time. And the air conditioning auxiliary system could help to adjust the concentrations of conditioning agent (i.e., NH_3 and SO_3) according to the data of CEMS in order to obtain an optimum efficiency

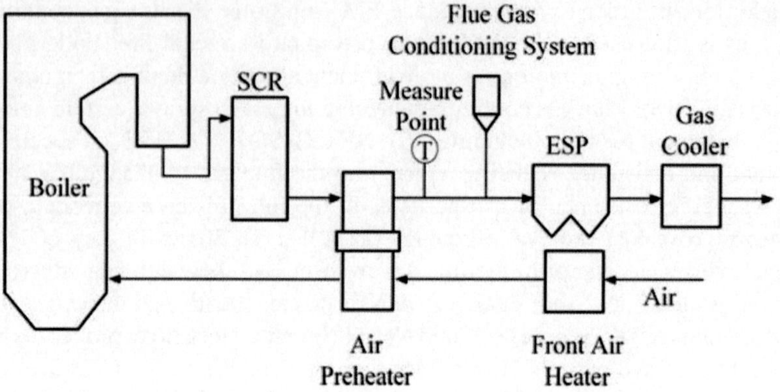

Fig. 72.1 A scheme for synergetic removal of NH_3 and SO_3 and particulate matter

of ESP. After the flue gas passed the ESP, the NH_3, SO_3 and particulate matter could be greatly reduced. As the result, the ash deposition and corrosion problems caused by H_2SO_4 and NH_4HSO_4 would rarely happen in the gas deep cooler, and thus the gas deep cooler could utilize more compact heat exchange areas, and flue gas temperature can be further reduced to achieve deep recovery of flue gas waste heat. The gas deep cooler can be used to heat feed water or to heat the combustion air combining with the front air heater. The front air heater can also heat the combustion air by using the extraction steam. Compared with the conventional system, the flue gas heat exchangers in this system would not only improve the removal rate of fine particles avoiding serious ash deposition and corrosion, but also recover the waste heat of flue gas efficiently.

72.3 Experimental System and Procedure

As shown in Fig. 72.2, the adsorption experiments were conducted on a fixed bed system which consisted of flue gas part, SO_3 generation part, gas preheating part and adsorption reaction part. The NH_3, SO_2, N_2, CO_2 and O_2 are mixed in the mixing chamber and then blew into the tube furnace. The SO_3 generation part consisted of a micro-injection pump, a porous core and a tube furnace. The glass syringe on the pump was filled with 5% sulfuric acid solution and the temperature of tube furnace was set to 530° before the experiment starting. Then the sulfuric acid solution injected into the silica tube by the micro-injection pump would decompose into gaseous SO_3 and H_2O at the high temperature. The gas from the tube furnace can simulate the flue gas at the outlet of the SCR system in the application.

The coal ash was flatly lain on the porcelain boat which was placed in the adsorption reaction tube. The temperature in the adsorption reaction tube was set to 180 °C which was lower than the dew point of the NH_4HSO_4, which covered the operating

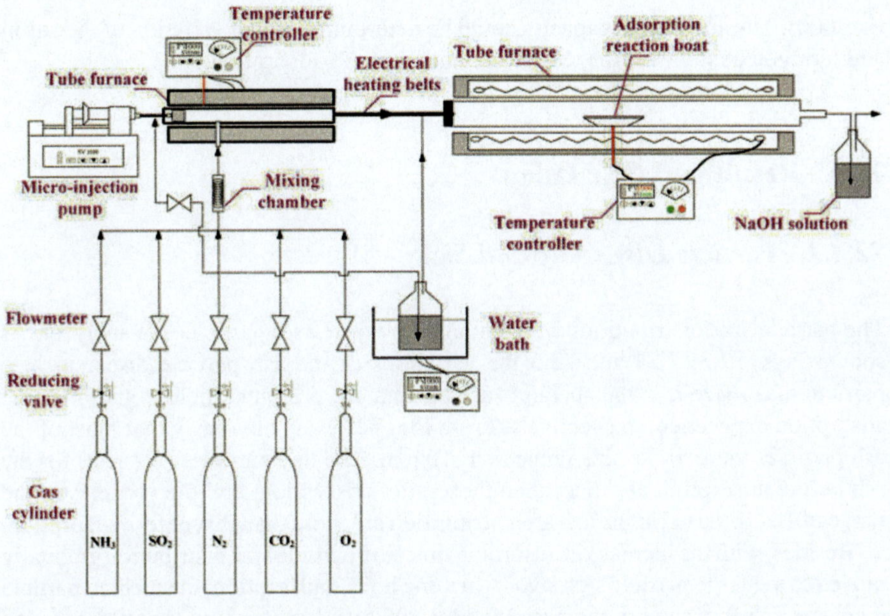

Fig. 72.2 Schematic diagram of the experimental system

temperature range of the new synergetic removal system. The calibration tests were conducted to ensure the stability of the experimental system and the accuracy of the results. The diameter of the test section is 33 mm, and the gas velocity passing the ash is 0.05 m/s. Table 72.1 presents the flue gas composition in the experiment.

The particle size distributions were tested by the Laser Particle Size Analyzer (BT-9300H), and the particle dispersity, mean diameter and the specific surface area were analyzed. The ash morphologies were obtained by the Scanning Electron Microscope (SEM-SU3500) magnifying two thousand times, and the variation of the agglomeration characteristics were obtained. The sulfur and nitrogen contents of the ashes were analyzed by the Automatic Elemental Analyzer (EUROVECTOR

Table 72.1 The volume fraction of each composition

Composition	Volume fraction
CO_2	11.4%
H_2O	11.7%
O_2	3.3%
N_2	73.5%
SO_2	0.1%
SO_3	80 ppm
NH_3	80 ppm

EA3000). The adsorption capacity could be determined by the variation of the sulfur and nitrogen contents of the ash before and after the adsorption.

72.4 Result and Discussion

72.4.1 Particle Dispersity and Size

The particle size distributions were obtained by the Laser particle size analyzer. As shown Figs. 72.3, 72.4 and 72.5 the variations of the ash particle dispersity, the particle mean size and the specific surface area are presented before and after the adsorption experiment, respectively. From Fig. 72.3, it is obviously that most of the ash particles were in the size range of 1–10 μm. The fine particles (<1 μm) for the ash before adsorption are more than those after adsorption, and the particles in the range of 10–100 μm for the ash after adsorption are more than that before adsorption.

Besides, with the increase in adsorbing time, the particle size distribution gradually moved to the large particle size area. After one hour's adsorption, the median particle diameter (D50) of the ash increased from 3.895 to 5.348 μm, and the volume mean diameter increased from 7.743 to 9.556 μm, and the surface mean diameter increased from 1.543 to 2.089 μm. In addition, the specific surface area gradually decreased from 1.254 to 0.926 m^2/g with the adsorbing time. Due to the adsorption, the ash particles aggregated into larger particles which were easier to be removed in the ESP.

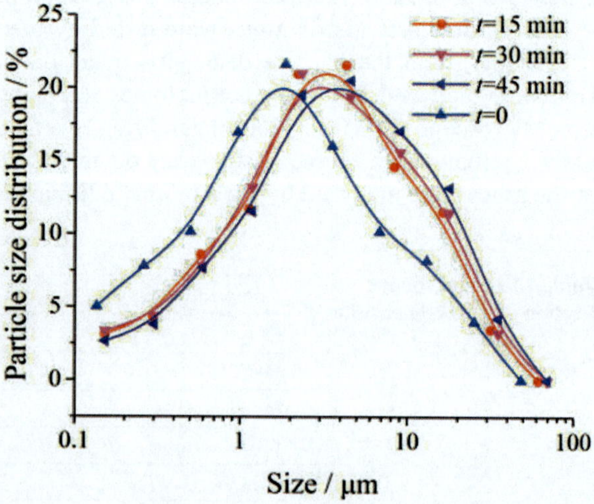

Fig. 72.3 The ash particle dispersity with absorbing time

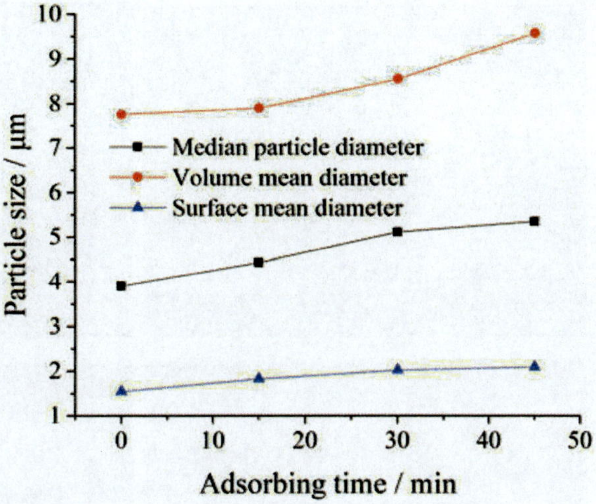

Fig. 72.4 The average diameter of the ash particles

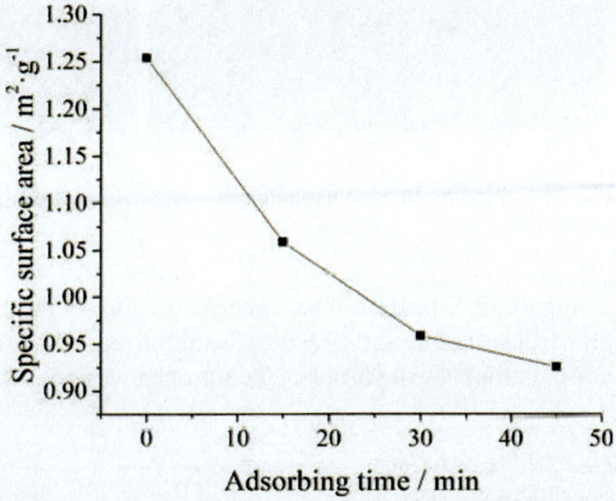

Fig. 72.5 The variation of the specific surface area

72.4.2 Agglomeration Characteristics

The ash particle surface morphologies were obtained by SEM magnifying two thousand times in order to analyze the agglomeration characteristics. Figure 72.6 shows the changes of particles before and after absorption process. As shown in this figure, the original ash particles were finer and more crushed and had irregular shapes even under 2,000× focal length. By contrast, the ash after adsorption were characterized by much relatively larger irregular shaped particles which were formed by the adhesion or agglomeration between fine particles. As the adsorption time increases, the agglomeration of particles gradually become more intense.

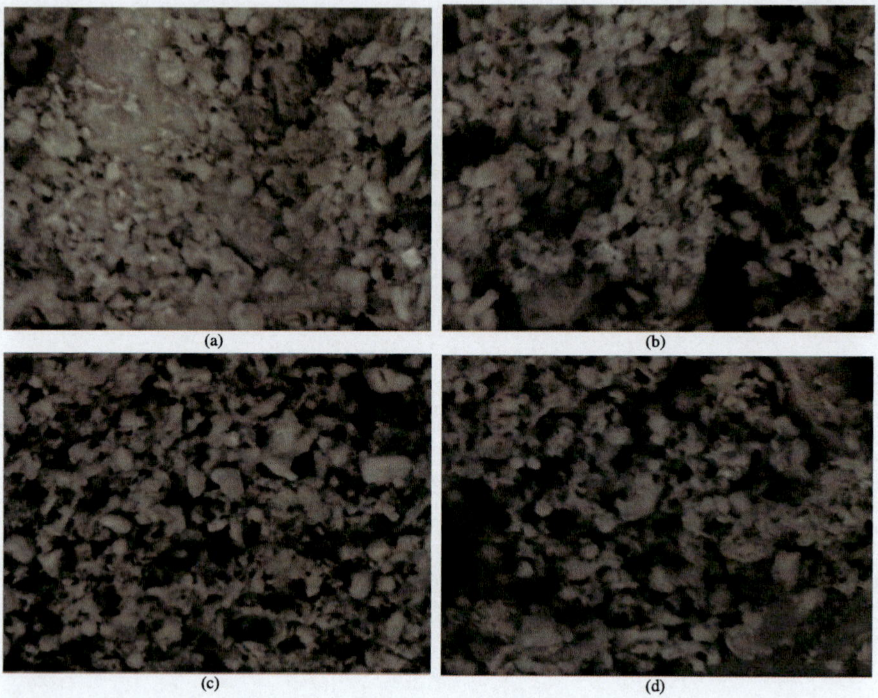

Fig. 72.6 Particle surface morphologies **a** original ash and **b–d** ashes after absorbing for 15, 30 and 45 min

Figure 72.7 illustrates the content variations of the nitrogen and sulfur elements in the particles before and after the adsorption reaction by automatic elemental analyzer. For the original ash particles, the nitrogen was not detected, and the sulfur element

Fig. 72.7 The elemental composition of ash particles

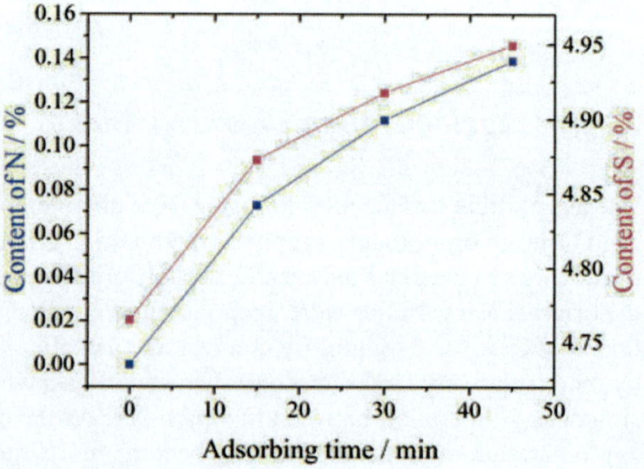

accounted 4.765% of the total mass. As absorbing time increased to 45 min, both of the contents of nitrogen and sulfur elements gradually increases and finally reached 0.143 and 4.949%. The molar ratio between the absorbed N and S contents maintained in the range of 1.2–1.8. It indicates that the adsorption of the SO_3 and NH_3 would interact with the ash particles leading to the morphology changes. The ammonia reacts with the sulfur trioxide and produces ammonium bisulfate (Trivedi and Phadke 2008; Brendel et al. 2000).

$$NH_3 + SO_3 + H_2O = NH_4HSO_4 \tag{72.1}$$

Ammonium bisulfate exists as a corrosive melt between the formation temperature and the melting point. It becomes a semi-liquid, acting like a glue when adsorbed by the fly ash particles, when the flue gas temperature is above 150 °C. And it exists as a very hygroscopic solid and can also be very corrosive below the melting point. This could strengthen the agglomeration of fine ash particles, and produces highly cohesive and relatively large particles, resulting in high collection efficiency due to the reduced rapping losses and re-entrainment (Baldrey 2001; Trivedi and Phadke 2008; Shanthakumar et al. 2008).

72.5 Conclusion

(1) A new scheme was proposed for synergetic removal of NH_3 slipped from SCR, SO_3 and particulate matters in the flue gas, and the NH_3 and SO_3 adsorption characteristics and the agglomeration characteristics of ash particles were investigated.

(2) The agglomeration of ash particles can be significantly enhanced by the adsorption reaction, which can increase the mean size of particle matters.

(3) The molar ratio between the absorbed N and S contents always maintained in the range of 1.2–1.8. It indicates that the adsorption of the SO_3 and NH_3 would interact with the ash particles leading to the morphology changes.

(4) The adsorption of NH_3 and SO_3 could strengthen the agglomeration of fine ash particles, and produces highly cohesive and relatively large particles, resulting in high collection efficiency due to the reduced rapping losses and re-entrainment.

(5) The results initially proved the proposed system would be a promising approach for the alleviation of the sticky ash deposition and for the synergetic removal of NH_3 and SO_3 and particulate matters. The detailed research will be carried out in our further study to examine the effects of the ratio of NH_3 and SO_3 in the gas, the adsorption temperature and the particle size on the adsorption rates and adsorption ratio of NH_3 and SO_3, and the specific resistance of the particles, and the particle agglomeration, in order to obtain proper operating parameters for the synergetic removal system.

Acknowledgements This work was supported by the China Postdoctoral Science Foundation Grant (No. 2019M653625) and by the Fundamental Research Funds for the Central University, China (Grant No. XJJ2018071).

References

Baldrey KE (2001) Advanced flue gas conditioning as a retrofit upgrade to enhance pm collection from coal-fired electric utility boilers. Office of scientific & technical information technical reports

Bian D (2015) Application of double dust (NH3+SO3) flue gas quenching and tempering technology for 600 MW unit. Industry and technology forum, pp 58–60

Brendel G et al (2000) Investigation of ammonia adsorption on fly ash due to installation of selective catalytic reduction systems. West Virginia University Research Corporation (US)

Chen H et al (2017) Corrosion and viscous ash deposition of a rotary air preheater in a coal-fired power plant. Appl Therm Eng 113:373–385

Chothani C, Morey R (2008) Ammonium bisulfate (ABS) measurement for SCR NOx control and air heater protection. Carnegie: Breen energy solution, pp 1–13

Fang C, Lu C (2018) Research on ultra-low emission operation optimization technology of SCR denitration system. China Electric Power 51:143–148

International Energy Agency (2018) Energy efficiency 2018. https://www.iea.org/topics/energy-efficiency

Jones GD et al (1982) Impact of NOx selective catalytic reduction processes on flue gas cleaning systems. Final report Oct 80–Oct 81

Li X (2018) Analysis of the effect of flue gas quenching and tempering on the efficiency of electric precipitator in coal-fired units. Coal Technol 37

Menasha J (2010) Characterization of ammonium bisulfate formation in a bench-scale single-channel air preheater. Dissertations & theses-gradworks

Ministry of Environmental Protection et al (2016) Full implementation of ultra-low emission and energy-saving renovation work plan for coal-fired power plants. Energy Conserv Environ Protect 32–32

Muzio L et al (2017) Ammonium bisulfate formation and reduced load SCR operation. Fuel 206:180–189

Navarrete B et al (2015) Effect of flue gas conditioning on the cohesive forces in fly ash layers in electrostatic precipitation. Environ Prog Sustain Energy 34(5):1379–1383

Shanthakumar S, Singh DN, Phadke RC (2008) Flue gas conditioning for reducing suspended particulate matter from thermal power stations. Prog Energy Combust Sci 34(6):685–695

Shou C, Zhai Z, Xie Y, Zou Z, Chunhong L, Min L, Jingwei L, Xiaodong L, Wenhua L (2016) Engineering application test study on particle removal characteristics of low-temperature electrostatic precipitator. China Electr Eng J 36:4326–4332

Trivedi SN, Phadke RC (2008) Flue gas conditioning. In: International conference on electrostatic precipitation

Wang Y et al (2017) Study of ash fouling on the blade of induced fan in a 330MW coal-fired power plant with ultra-low pollutant emission. Appl Therm Eng 118:283–291

Yu H (2015) Research on compound flue gas quenching and tempering technology, North China Electric Power University

Chapter 73
A Population Balance Model for Fine Particle Removal Inside the Electrostatic Precipitator

Lu Duan, Qian Huang, and Shuiqing Li

Abstract The electrostatic precipitators (ESP), as the major dust removal devices, requires more advanced designing tools to effectively reduce the fine particulate matter (PM). In this paper a multidimensional population balance model (PBM) is proposed to describe the mechanisms including particle mitigation, coagulation/breakage, nucleation/condensation and deposition/re-suspension inside ESPs. The governing equation is solved after discretizations of space, particle size and velocity. The numerical investigation of a mechanistic lab-scale ESP at Tsinghua shows reasonably good agreements with experimental measured particle size distribution (PSD) and collection efficiencies. Therefore, the model is able to benefit the state-of-the-art design and retrofit of ESPs.

Keywords Electrostatic precipitator (ESP) · Population balance model (PBM) · Particle charging · Particle size distribution (PSD) · Collection efficiency

73.1 Introduction

The electrostatic precipitator (ESP) has become a major dust removal device in industry, especially in coal-fired power plants, for over a hundred years. Although ESP is capable of removing most of the total PM mass, it may be less efficient in collecting micro-sized PM, a major source of air pollutant in many countries (Park and Chun 2002). Hence the design and retrofit of ESPs require quantitative models to evaluate the performance of collecting fine PMs.

However, at the present stage, industrial ESPs are still designed, to a large extent, empirically. Using the well-known Deutch model (Tianchi and Hongxing 1999;

L. Duan · Q. Huang · S. Li (✉)
Department of Energy and Power Engineering, Tsinghua University, Beijing 100084, China
e-mail: lishuiqing@tsinghua.edu.cn

L. Duan
China Coal Research Institute Company of Energy Conservation, Beijing 100013, China

J. Lyu and S. Li (eds.), *Clean Coal and Sustainable Energy*,
Environmental Science and Engineering,
https://doi.org/10.1007/978-981-16-1657-0_73

Liyan 2008; Batmunkh and Battogtokh 2007; Xinlin et al. 2009), the 'lumped' mitigation velocity is determined from design parameters of the unit and is then used for the model selection of ESPs. But this single-valued mitigation velocity can be seriously biased for fine particles. In recent years, the CFD-assisted ESP design (Long and Yao 2010; Guo et al. 2013) plays an increasingly important role by revealing the flow field and particle trajectory inside the device. Nevertheless, it is still a major challenge for ESP designers to accurately evaluate the removal efficiency of polydispersed fly ash particles through CFD. This is mainly attributed to the complex particle transport process inside the ESP, during which charged particles are driven by the electric field and fluid flow, while experiencing a dynamic coagulation-breakage process. If the inlet temperature of flue gas is sufficiently low, nucleation and condensation of SO_3/H_2SO_4 could make the process even more complicated (Xiao et al. 2015).

In order to describe the complex process mentioned above, the framework of population balance equation (PBE) stands out as it models the evolution of particle size distribution (PSD) and can naturally incorporate the various mechanisms. PBM has been widely used in atmospheric science (Marshall and Li 2014; Friedlander 2000), industrial synthesis of nano-particles and combustion of solid fuels (Qian 2017). Recent applications of PBM to ESP are mostly built on the spatial homogeneous assumption, deriving size-dependent deposition kernel functions onto the plate (Xuhui 2015; Shuhao et al. 2017; Haibo 2007; Renjie 2007) and further studying the dynamic evolution of SO_3/H_2SO_4 (Xuhui 2015). A detailed investigation of particle mitigation and removal, with the spatial distribution of particle concentration resolved, is still rarely reported in the literature.

To address this issue, a multidimensional PBE-based model incorporating particle charging, mitigation, deposition and re-suspension was developed. The spatial distribution of PSD, removal rate and collection efficiency of fine particles were solved out of the model. A comparison with the mechanistic experiments validated the model accuracy and demonstrated its potential for a powerful assistant tool to ESP designs.

73.2 Theoretical

In this work, the typical wire-plate ESP is simplified to a 2D structure, as depicted in Fig. 73.1. Fly ash particles evolve under the mechanisms of charging, migration, deposition and re-suspension. Gravity is not included here. The particle concentration $n = n(t, x, y, d_p, \mathbf{u}_p)$ is related to the residence time, spatial position, particle size and velocity. A first discretization of the particle size yields N_1 variables $n_k = n_k(t, x, \mathbf{u}_p)$ (k = 1, ..., N_1) which are governed by:

$$\frac{\partial n_k}{\partial t} + \nabla \cdot (n_k \mathbf{u}_p) = \left[\frac{dn_k}{dt}\right]_{dep} + \left[\frac{dn_k}{dt}\right]_{re} \tag{73.1}$$

Fig. 73.1 Grid mesh

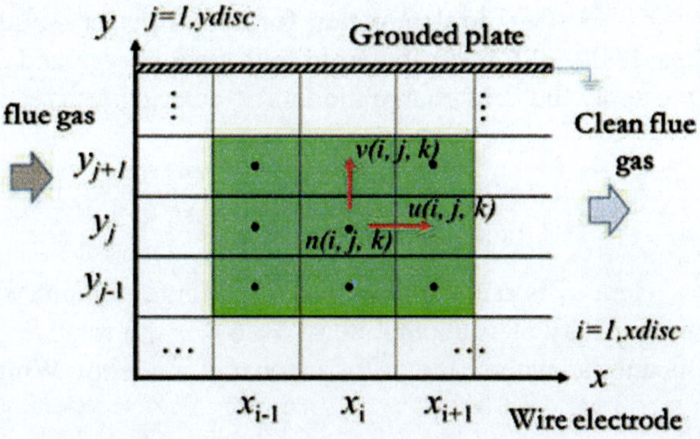

Here the second term on the LHS models the particle migration. The first term on the RHS represents particle deposition onto the plate, having the form

$$\left[\frac{dn_k}{dt}\right]_{dep} = -(1-\alpha)\frac{\partial n_k}{\partial n}dA \tag{73.2}$$

where $\frac{\partial n_k}{\partial n}$ is the particle concentration flux, dA is the cell area of the collected plate, and α is the coefficient of resuspension. The second term on the RHS represents particle resuspension, which can be written as

$$\left[\frac{dn_k}{dt}\right]_{dep} = \alpha\frac{\partial n_k}{\partial n}dA \tag{73.3}$$

In the migration term, the velocity of charged particle is affected by the electrostatic and fluid drag forces, namely.

$$m_p\frac{d\mathbf{u}_p}{dt} = \mathbf{F}_D + \mathbf{F}_E \tag{73.4}$$

Here $\mathbf{F}_D$ is the fluid drag force having the form.

$$\mathbf{F}_D = \frac{1}{8}\pi d_p^2\rho_f C_D(\mathbf{u}_f - \mathbf{u}_p)|\mathbf{u}_f - \mathbf{u}_p| \tag{73.5}$$

in which d_p is the particle diameter, ρ_f is the fluid density, $\mathbf{u}_f$ is the fluid velocity, and C_D is the drag coefficient. The C_D is a function of the particle Reynold number $\mathrm{Re}_p = \frac{\rho_p d_p^2 U_f}{18\mu_f L}$

$$C_D = \begin{cases} \frac{24}{\mathrm{Re}_p}(1 + 0.15\mathrm{Re}_p^{0.687}), & Re_p < 800 \\ 0.44, & Re_p \geq 800 \end{cases} \tag{73.6}$$

$\mathbf{F}_E = q\mathbf{E}$ is the electrostatic force, and q is the amount of charge. Particle charging inside the ESP is attributed to both field charge and diffusion charge. In this work we apply the field charge model for spheric particles in White (1951):

$$q_s = \frac{3\pi \epsilon_0 \epsilon_p}{\epsilon_p + 2} d_p^2 E \tag{73.7}$$

Here q_s is the saturated particle charge amount, $\epsilon_0 = 8.85 \times 10^{-12}$F/m is the permittivity of vacuum, and $\epsilon_p = 6.0$ is the relative permittivity of particle. In the meantime, we use the diffusion charge model by White (Pauthenier et al. 1932):

$$q_D = \frac{2\pi \epsilon_0 d_p k T}{e} \ln\left(1 + t/\tau^*\right) \tag{73.8}$$

in which q_D is the diffusion charge, $e = 1.6 \times 10^{-19}$C is the electron charge, $k = 1.38 \times 10^{-23}$J/K is the Boltzmann constant, T is the flue gas temperature, and τ^* is the time constant of diffusion charge.

It is believed that the field charge becomes saturated in $\sim 10^{-2}$s (Pauthenier et al. 1932), far less than the typical residence time of particles inside the ESP. Hence in this work we assume that particles are fully charged according to Eq. (73.6), whereas the diffusion charge saturates at $t/\tau^* = 500$. The overall particle charge is assumed to be additive, namely $q_p = q_s + q_D$.

73.3 Experimental

Experiments in a lab-scale ESP operated at room temperature (Jin 2013), sketched in Fig. 73.2, were used for the model validation. The inlet velocity of the particle-laden flow was 0.3 m/s. A direct voltage of −20 kV was exerted on the wire electrode, and the current was 2 mA. In this mechanistic rig, fly ash particles fed by the scraper feeder had a concentration similar to the inlet of typical power plant ESPs. Particle concentrations were measured by Dekati Electric Low Pressure Impactor (ELPI).

The fluid flow field $\mathbf{u}_f$ and the electric field $\mathbf{E}$ were simulated using COMSOL 5.3, so that the population balance model Eq. (73.1) can be solved under the assumption of one-way coupling of particles. To do that, the variable n_k was further discretized in each spatial cell, and a list of discrete velocities was used in the numerical solution.

73.4 Results and Discussion

Figure 73.3a shows the simulated 2D fluid field. The downward regions of nodes are featured with small velocity magnitudes. In Fig. 73.3b, c, elliptical electric potential

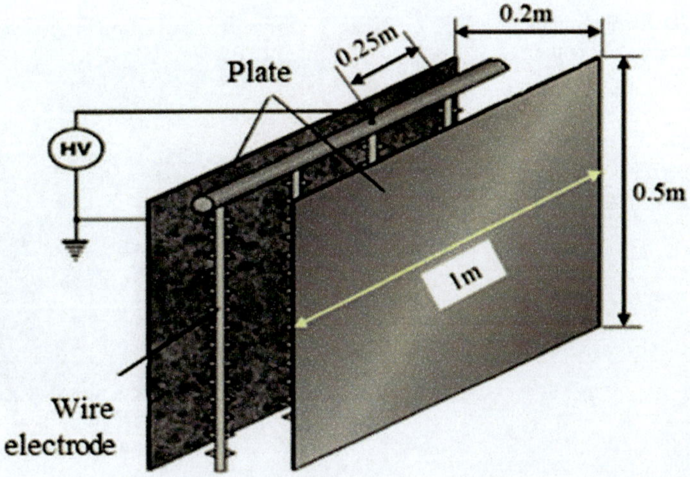

Fig. 73.2 Schematic of the electrostatic precipitator

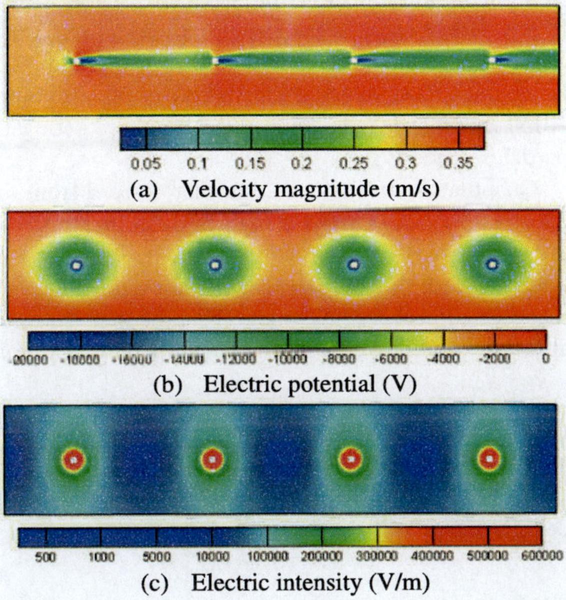

Fig. 73.3 Distribution of the flow field and electric field. **a** Velocity magnitude (m/s). **b** Electric potential (V). **c** Electric intensity (V/m)

contours are formed around each node, and the electric intensity decreases rapidly away from the node.

Using the field simulation results, the size-discretized PBE Eq. (73.1) was solved for the spatial distribution of particles inside the ESP after a large duration time. Figure 73.4 shows the dimensionless particle concentration normalized by the inlet

Fig. 73.4 2D distribution of particle concentration in the ESP

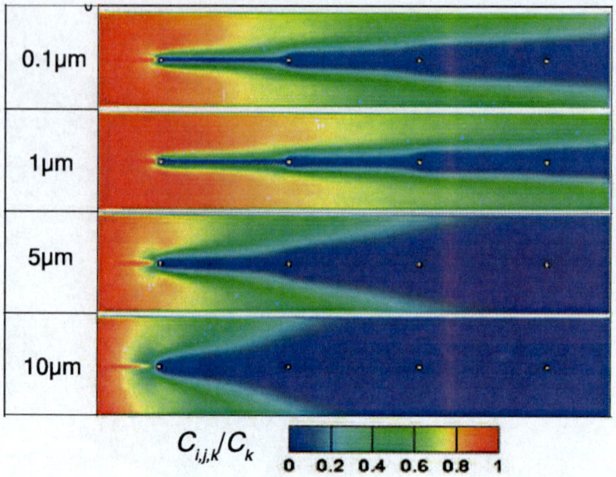

concentration for a variety of particle sizes inside the ESP. It is shown that particle concentrations decrease along the flow, and the removal rates of particles gradually decrease in the flow direction. Generally, larger particles are more easily collected by the plate because of the greater mitigation velocities. However, 1-μm particle has a larger outlet concentration than that of 0.1-μm due to a less charge capacity and hence mitigation velocity. This is consistent with other experimental results (Xuhui 2015). Hence, this model is capable of predicting the dynamic evolution of fine particles inside ESPs.

Quantitative results can be further derived from the simulation. The cumulative particle size distributions at the ESP inlet and outlet are presented in Fig. 73.5, and the overall mass concentrations of PM_1, $PM_{2.5}$ and PM_{10} are plotted in Fig. 73.6. Firstly, we see that the size discretization procedure ensures an accurate initial PSD reconstruction at the ESP inlet. Then, the numerical results of PSD at the ESP outlet agree reasonably well with the experiment, expect for that of particles smaller than

Fig. 73.5 Numerical and experimental results of particle mass concentration. The inset shows the grade efficiency of the ESP

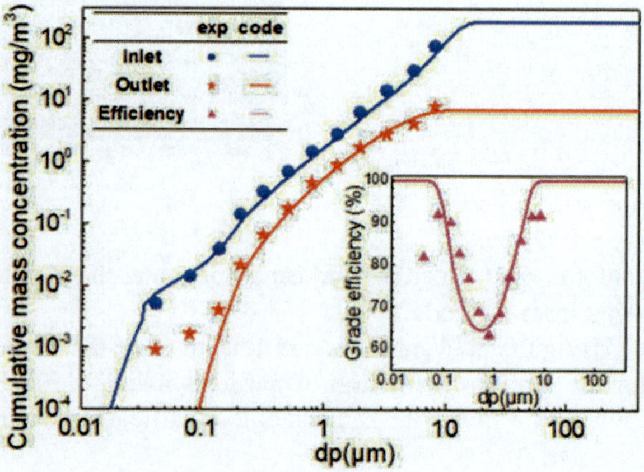

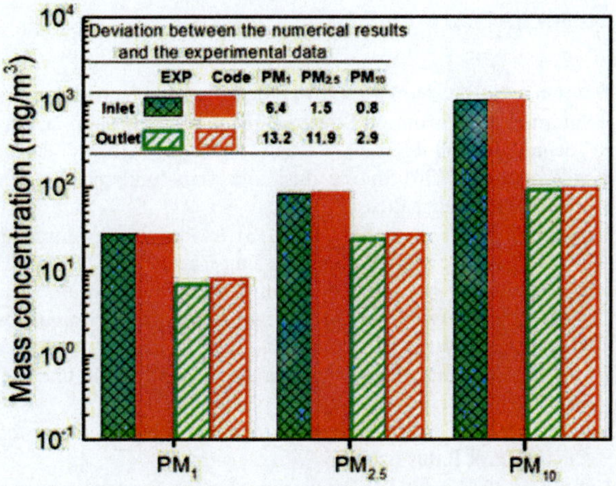

Fig. 73.6 Numerical and experimental results of particle mass concentration of PM_1, $PM_{2.5}$ and PM_{10}

0.1 μm ($PM_{0.1}$). This may be attributed to the inaccuracy of particle charging model in the ultrafine size regime. In the inset of Fig. 73.5, the total collection efficiency for each ELPI stage is also well predicted, except that the model makes a lower prediction in the size range of 1–10 μm. This may be caused by a lack of diffusion mechanism in the model and thus an underestimation of particle transport. In the micro-size range where diffusion is not significant, the modeling results match well with the experiments. Finally, the predicted emitted mass concentrations of PM_1, $PM_{2.5}$ and PM_{10} differ from the experimental results by 13.2%, 11.9% annd 2.6% respectively.

73.5 Conclusion

A 2D population balance model was established to investigate the dynamic evolution of fine particles inside the electrostatic precipitator (ESP). The model was validated with experimental results from a mechanistic lab-scale ESP. Spatial distributions of differently-sized particles were solved from the model and clearly reveal the particle removal characteristics inside the ESP. Furthermore, the model predictions on the collection efficiency and emission concentration both match quantitatively well with experiments. Future work on exploring a wider range of operation parameters would further benefit the model to become a powerful assistant tool for the state-of-the-art ESP design.

Acknowledgements This work was supported by the National Key Research and Development Prigram of China (2017YFB0603203).

References

Batmunkh S, Battogtokh Z (2017) Modeling an electrostatic precipitator (ESP) with ESPVI 4.0W. International Forum on Strategic Technology, Mongolia, Mongolian University of Science and Technology 650–653

Friedlander SK (200) Smoke, dust, and haze–fundamentals of aerosol dynamics, 2nd. New York, Oxford University Press

Guo BY, Yang SY, Xing M et al (2013) Toward the development of an integrated multiscale model for electrostatic precipitation. Ind Eng Chem Res 52(33):11129–11282

Haibo Z (2007) Stochastic solution of population balance modeling and the research on high-efficiency removal of particulate matter from coal combustion. Huazhong University of Science and Technology

Jin X (2013) Research on the capture technology of fine particles in electrostatic precipitator. Tsinghua University

Liyan C (2008) Design & analysis on electrical precipitator for million kW generating set. China Environ Prot Indus 6:50–55

Long ZW, Yao Q (2010) Evaluation of various particle charging models for simulating particle dynamics in electrostatic precipitators. J Aerosol Sci 41(7):702–718

Marshall JS, Li SQ (2014) Adhesive particle flow—a discrete element approach. Cambridge University Press, New York

Park JH, Chun CH (2002) An improved modelling for prediction of grade efficiency of electrostatic precipitators with negative corona. J Aerosol Sci 33:673–694

Pauthenier M, Moreau-Hanot M, (1932) La charge des particles spheriques dans un champ ionize. J Phys Radium 3:590–615

Qian H (2017) Effect of mineral on the particulate formation and ash deposition during pulverized coal combustion. Tsinghua University

Renjie L (2007) Numerical simulation of bipolarly charged electrostatic precipitator for collecting inhalable particles. Huazhong University of Science and Technology

Shuhao W, Shuiqing L, Duan L (2017) Study on the coagulation of vapor condensates and fine particulates in a phase-change agglomerator. Proceedings of the CSEE 37(24):7230–7236

Tianchi M, Hongxing Z (1999) Discuss on the electrostatic precipitation design and modify. Huazhong Electric Power 5:27–30

White HJ (1951) Particle charging in electrostatic precipitation. Trans Am Inst Electric 70(2):1186–1191

Xiao G, Wang XH, Yang G et al (2015) An experimental investigation of electrostatic precipitation in a wire–cylinder configuration at high temperatures. Powder Technol 269:166–177

Xinlin S, Jin G, Wanmu Z (2009) Development professional software of electrostatic precipitators for HFCDS system. Cement Technol 6:88–92

Xuhui Z (2015) Studies on synergetic removal of fine particulates and SO_3 by an extra cold-side electrostatic precipitator. Tsinghua University

Chapter 74
Effect of Calcination Atmosphere on High Temperature H_2S Removal of Mn_xO_y/Al_2O_3 Sorbent in Synthesis Gas

Li Haifeng, Su Sheng, Liu Lijun, Xu Kai, Hu Song, Wang Yi, and Xiang Jun

Abstract The Mn_xO_y/Al_2O_3 for high temperature H_2S removal were prepared by wet impregnation method with various calcination atmosphere, including N_2, H_2 and air. The desulfurization and regeneration process were conducted at 850 °C. The result showed that the sorbent calcined in N_2 (MN) and sorbent calcined in H_2 (MH) had higher sulfur capacity than that of the sorbent calcined in air (MA) during successive desulfurization-regeneration cycles. The characterization result showed that part of Mn_3O_4 dissolving into the inner of Al_2O_3 cannot involve the reaction with H_2S, which maybe the reason of durability decreasing of desulfurization performance during successive desulfurization-regeneration cycles. In successive desulfurization-regeneration cycles, the tissue structure stability of MN and MH were better than that of the MA, and more active component MN-based oxides were reacted with Al_2O_3 in MN and MH, the product was beneficial for the H_2S removal. IIowever, more Mn_3O_4 dissolved into the inner of Al_2O_3 in MA, which cannot react with H_2S. These phenomena may be the reason that the higher sulfur capacity of MN and MH than that of MA during successive desulfurization-regeneration cycles.

Keywords H_2S removal · Mn_xO_y/Al_2O_3 · Calcination atmosphere

74.1 Introduction

Manganese oxides as the sorbent are used by researcher for H_2S removal in gasification gas under high temperature, because that Mn-based oxides have high desulfurization ability, fast initial reaction rate with H_2S, and high mechanical stability

L. Haifeng · S. Sheng (✉) · L. Lijun · X. Kai (✉) · H. Song · W. Yi · X. Jun
State Key Laboratory of Coal Combustion, School of Energy and Power Engineering, Huazhong University of Science and Technology, Wuhan 430074, China
e-mail: susheng@mail.hust.edu.cn

X. Kai
e-mail: shajia0711@126.com

Jianghan University, Wuhan 430056, China

J. Lyu and S. Li (eds.), *Clean Coal and Sustainable Energy*, Environmental Science and Engineering,
https://doi.org/10.1007/978-981-16-1657-0_74

in the temperature range of 400–1000 °C (Chytil et al. 2017, 2015; Liu et al. 2016; Bakker et al. 2003). Meanwhile, when the temperature is lower than 1200 °C, the Mn-based oxides are reduced to MnO rather than metal Mn though gasification gas containing high concentration reduction atmosphere (mainly CO and H_2). Al_2O_3 as the supporter in sorbent can improve the distribution of active components with high manganese content to improve the sulfur capacity, and Al_2O_3 is stable and has a good catalytic effect for H_2S decomposition under high-temperature (Xia et al. 2017). So Al_2O_3-supported manganese sorbent attracted more attention of researchers (Wang et al. 2013a, b).

Many researchers (Bai et al. 2009; Abdel-Mageed et al. 2013; Chand et al. 2013) reported that the particle sizes, defect concentrations, valence states and phase structures of samples are different with various calcination atmospheres, which may give rise to variations in sample properties. So, the calcination atmosphere in preparation process of sorbent may have effect on the desulfurization performance of sorbent, and it not clear that the effect of different physical and chemical properties of sorbent with various calcination atmospheres on the durability of desulfurization performance of sorbent during successive desulfurization-regeneration cycles.

This work synthesized a series of Mn_xO_y/Al_2O_3 sorbents using incipient wet impregnation method with various calcination atmospheres. The chemical and tissue structures of the sorbent samples were characterized with scanning electron microscopy (SEM), Nitrogen Isothermal Adsorption, and X-ray powder diffraction (XRD), and the desulfurization performances of the sorbents at high temperature (850 °C) were investigated. The desulfurization performance durability of sorbents during the successive desulfurization–regeneration cycles were discussed by analyzing the relationship between characterization results and desulfurization performance.

74.2 Experiment

74.2.1 Sorbents

2 mol L^{-1} $Mn(NO_3)_2$ was as the metal precursors to be used to prepare the Mn_xO_y/Al_2O_3 sorbents with repeated wet impregnation method. The wet impregnation process was conducted in the oscillator for 24 h at 40 °C. and then the sample was dried under 80 °C for 24 h. The dried sample was conducted with further wet impregnation process without intermediate calcination. Samples were obtained after five repeated impregnations. Finally, the samples were calcined at 900 °C in the air, pure N_2 and pure H_2, respectively. The sample were denoted as the MN, MH, MA according to the calcination atmosphere with N_2, H_2 and Air, respectively.

74.2.2 Characterization of Sorbents

The ASAP 2020 analyzer (Micromeritics Instrument Corporation, America) was used to characterize the tissue structure at low temperature (77 K). The sorbents were degassed in vacuum under 350 °C for 4 h before analysis with the N_2 adsorption. The surface area, pore volume and average pore diameter were calculated using adsorption isotherm while pore size distribution was estimated by Barrett, Joyner and Halena (BJH) method, and the surface area was measured by the Brunauer–Emmett–Teller (BET) equation (Ren et al. 2016).

The XRD characterization were performed with the Bruker AXS D8 Advanced X-ray diffractometer (Cu Kα, k = 0.15406 nm, 40 kV, 40 mA). The step scanning mode with 2θ between 20° and 80°, and a step size is 0.02° (0.24 s per step). The crystalline phases were identified by reference to powder diffraction data (PDF).

The morphologies characteristics of these Mn-based sorbents were characterized using the SU8010 filed-emission scanning electron microscopy (FE-SEM), accelerating voltage at 15 kV, the resolution is 1 nm (Li et al. 2018).

74.2.3 Desulfurization and Regeneration Tests for the Sorbents

The test included the desulfurization process and regeneration process in every cycle, and the whole test repeats five cycles. A vertically oriented quartz tube was used as the reactor of desulfurization and regeneration process, and the inner diameter was 6 mm. The desulfurization and regeneration reactor were heated with an electrical furnace, and 0.5 g of the sorbents were packed in the reactor and placed with in the constant temperature zone.

The test temperature increased to 850 °C with the rate of 5 °C under the N_2 atmosphere. The desulfurization process and regeneration process were conducted under atmospheric pressure. The gas mixture included 2000 mg m^{-3} H_2S, 10% H_2 and N_2 balance (More gas composition would lead to complex problems for the research and analysis, which may make the research unclearly, so the research just selective H_2 to make the reduction condition for H_2S removal, which is the principal property of synthesis gas). The weight hour space velocity (WHSV) of desulfurization process is 12 L h^{-1} g^{-1}. The concentration of H_2S in inlet gas and tail gas was analyzed by Clarus 500 gas chromatograph (GC). The breakthrough time in desulfurization process was defined as the time when the H_2S concentration in tail gas reached 50 mg m^{-3}. The sulfur capacity of sorbent at breakthrough time point was denoted as the breakthrough sulfur capacity (BSC) of sorbent, which can be calculated by the following formula:

$$BSC = WHSV \times \frac{M_S}{Vm} \times \frac{22.4}{M_{H_2S}} \times \left[\int_0^t (C_{\text{in}} - C_{\text{out}})dt \right] \times 10^{-3}$$

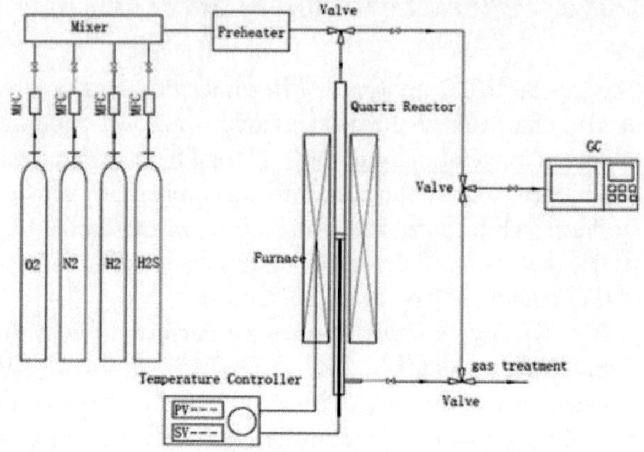

Fig. 74.1 Schematic diagram of desulfurization-regeneration process over Mn_xO_y/Al_2O_3

where BSC is the effective sulfur capacity of sorbent (mg $S \cdot g^{-1}$ sorbent); WHSV is weight hour space velocity ($L \cdot h^{-1}\ g^{-1}$); Ms is the molar weight of S (32.06 g mol^{-1}), and M_{H2S} and V_m is the molar weight of H_2S (34.06 g mol^{-1}) and the molar volume of H_2S at 1 atm and 25 °C (24.5 L mol^{-1}), respectively; t is the breakthrough time for desulfurization (h) when $C_{out} = 50$ mg m^{-3}; C_{in} and C_{out} are the inlet and tai gas concentration (mg m^{-3}) of H_2S, respectively (Xia et al. 2017).

When the desulfurization process finished, pure N_2 flushed the system for 30 min, then the sulfide sorbent was regenerated with a 5 vol% O_2/95 vol% N_2 mixture under 850 °C, and the WHSV of regenerated gas was 24 L h^{-1} g^{-1} the regeneration process finished when there was no detection of SO_2 in the tail gas with the $KMnO_4$ solution (Fig. 74.1).

At the end of regeneration process, pure N_2 was used to flush the whole system for 30 min, then the next desulfurization-regeneration cycle continued to conduct, or pure N_2 was used to purge the whole system until the temperature of bed was cooled down to the room temperature.

74.3 Result and Discussion

74.3.1 *H_2S Absorption Capacity Over Mn_xO_y/Al_2O_3 Sorbent*

Effect of calcination atmospheres on desulfurization performance and durability of the Mn_xO_y/Al_2O_3 samples during several successive desulfurization-regeneration cycles are shown in Fig. 74.2. It is observed that the BSC of the fresh sorbents obtained with various calcination atmospheres are obviously different. The BSC of MH is highest in all sorbents, the BSC of MN is higher than BSC of MA. It indicates that the calcination atmosphere has effect on the desulfurization performance of fresh sorbents.

Fig. 74.2 Breakthrough curves and BSC of the samples with various calcination atmosphere at 900 °C in five successive desulfurization–regeneration cycles (**a** MH; **b** MN; **c** MA)

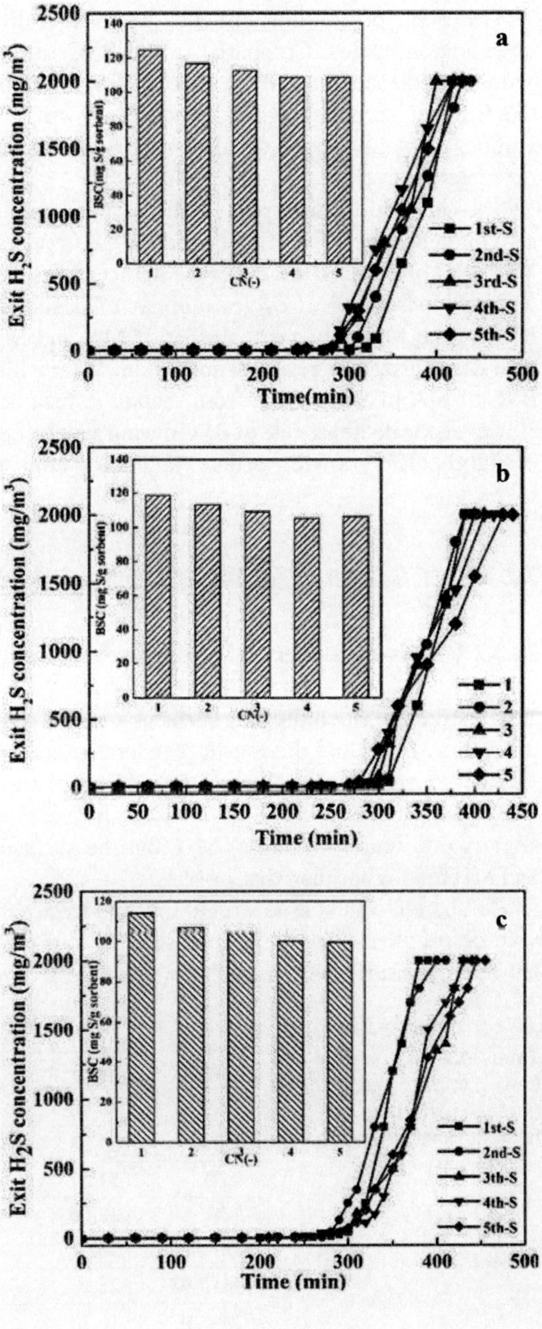

The BSC of all the sorbents decrease following with the desulfurization-regeneration cycles. Compared to the BSC of fresh sorbents, the BSC of all the samples decrease seriously in second desulfurization-regeneration cycle. However, the BSC of sorbent slightly change following with the increasing of the cycle number from the second desulfurization-regeneration cycle to fifth desulfurization-regeneration cycle.

The calcination atmosphere not only has effect on the BSC of fresh sorbents, but also influence the BSC of sorbent after the first desulfurization-regeneration cycles. The BSC of sorbent are different with various calcination atmospheres following with the increasing of cycles number of desulfurization-regeneration cycles. The BSC of the MH is better than that of MN and MA in every cycle from second to fifth desulfurization-regeneration cycle. Meanwhile, the BSC of MH is better than BSC of MA in every cycle from second to fifth desulfurization-regeneration cycle. However, the decrease rate of BSC during several desulfurization-regeneration cycles is slightly different with various calcination atmosphere.

74.3.2 Characterization of the Sorbent Samples

74.3.2.1 Tissue Structural of the Sorbent Samples

Tissue structures including the surfaces area, pore volumes and average pore diameter of fresh sorbents and the fourth regenerated sorbents are shown in Table 74.1. The surfaces area and pore volumes of the sorbents are different with various calcination atmospheres. The result shows that the surfaces area and pore volumes of MN are slightly different with that of MH, but the surfaces area and pore volumes of MN and MH are higher than that of MA.

Compared to the tissue structure of the fresh samples, the surfaces area and pore volumes of all the samples decrease after four desulfurization-regeneration cycles. However, the surfaces area and pore volumes of MN-4C and MH-4C are still similar,

Table 74.1 BET surface area (S_{BET}), average pore diameter (D_e) and pore volume (V_t) of fresh sorbents and fourth regenerated sorbents

Sample	S_{BET} ($m^2\ g^{-1}$)	V_t($cm^3\ g^{-1}$)	D_e (nm)
PA900	80.00	0.46	23.72
MN	56.25	0.26	18.23
MH	51.57	0.26	20.14
MA	30.26	0.18	23.99
MN-4C	26.51	0.21	10.24
MH-4C	25.13	0.19	12.27
MA-4C	16.78	0.15	34.61

−4C: the sorbent after four desulfurization-regeneration cycles; **PA900**: the pure Al_2O_3 calcined at 900 °C and that are higher than the surfaces area and pore volumes of MA-4C. It indicates that the tissue structures of MN and MH are more stable than the structure of MA during successive desulfurization-regeneration cycles.

74.3.2.2 SEM Analysis of the Sorbent Samples

The SEM of the fresh sorbent and fourth regenerated sorbent are shown in Fig. 74.3. it can be seen that the surface of MN and MH sorbent exist more pore structure than that of MA. It is agreement with the BET result of fresh sorbent shown in Table 74.1. The pore structure of MN and MH are beneficial for the gas diffusion in desulfurization process. After the four desulfurization-regeneration cycles, the surface become rough and less pore exist in the surface of MN-4C and MH-4C compared to the fresh sorbent. Though the surface of MA-4C is rough after four desulfurization-regeneration cycles, which can improve the H_2S adsorption in the surface, the total surface area and pore volume of MA-4C are lower than that of the fresh sorbent, and slightly lower than that of MN-4C and MH-4C, which is shown in Table 74.1.

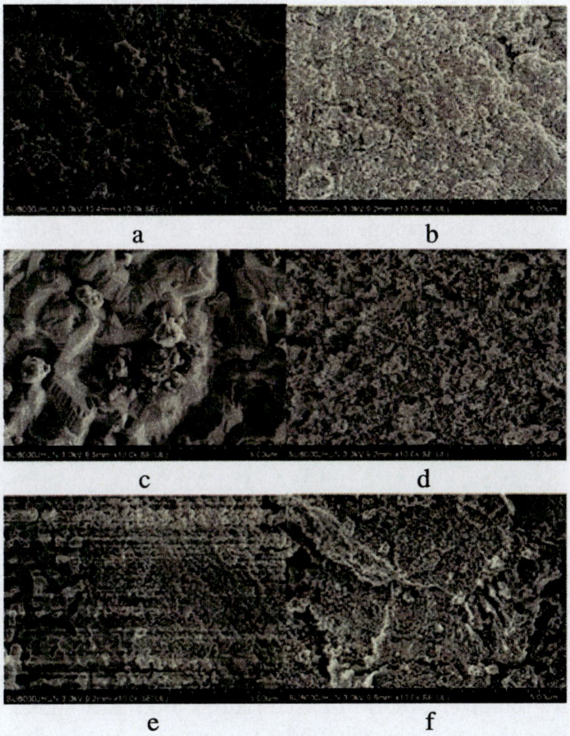

Fig. 74.3 The SEM of fresh sorbent and fourth regenerated sorbent (**a** MN; **b** MH; **c** MA; **d** MN-4C; **e** MH-4C; **f** MA-4C)

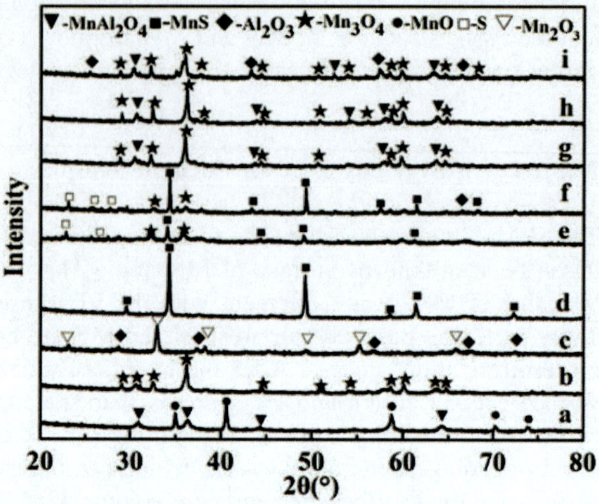

Fig. 74.4 XRD patterns of sorbents (**Fresh sorbent**: **a** MH; **b** MN; **c** MA. **Sulfide product**: **d** MH-S; **e** MN-S; **f** MA-S. **Fourth regenerated sorbent**: **g** MH-4C; **h** MN-4C; **i** MA-4C)

74.3.2.3 XRD of the Sorbent Samples

The XRD result of the fresh sorbents, sulfide sorbent and fourth regenerated sorbents are shown in Fig. 74.4. It can be observed in Fig. 74.4 that the phase of Mn-based oxides are different with various calcination atmospheres. As shown in Fig. 74.4, the phase of Mn_2O_3 can be seen at the diffraction peak of $2\theta = 23.13°$, 32.95°, 38.24°, 45.18°, 49.35°, 55.19°, 64.08°, 65.81°, and the phase of α-Al_2O_3 appears at $2\theta =$ 25.51°, 37.72°, 52.47°, 57.41°, 68.10° when the calcination atmosphere of sorbents is air.

The dominant peaks at $2\theta = 28.93°$, 32.40°, 60.77°, 64.65° over the MN sorbent are described to the phase Mn_3O_4. According to the report of Xia et al. (2015), the formation of Mn_3O_4 may because the low oxygen atmosphere during the regeneration process. This XRD result indicates that the phase of Mn_3O_4 also can be obtained from the precursor without oxygen. The XRD pattern of the sorbent calcined in the H_2 atmosphere appears the phase of MnO, which is observed at 34.96°, 40.59°, 58.75°, 70.22°, 73.84°. the MnO phase is the reduction product of Mn-oxides by H_2.

The fresh sorbent calcined in H_2 atmospheres appears the phase of $MnAl_2O_4$ while the absence in MN and MA. The $MnAl_2O_4$ is formed in the H_2 atmosphere during the prepare process, the existence of MnO in MH sorbent may indicate the MnO is necessary to obtain $MnAl_2O_4$, the conclusion is agreement with the conclusion in the report (Chytil et al. 2015).

Figure 74.4d–f show the XRD pattern of sulfide product in desulfurization process of fresh sorbent. It can be seen that MnS exists in the sorbent, which the reaction product between Mn-based oxides and H_2S. Because the gas includes higher concentration of H_2, the Mn-based oxides are firstly reduced into the MnO at short time, then the reduce product-MnO reacts with H_2S to form the MnS (Chytil et al. 2015).

The element sulfur in Fig. 74.4e, f appears in the MN and MA sorbents, it may be formed with the decomposition of H_2S under high temperature (Xia et al. 2015).

However, the part of Mn_3O_4 appears in the sulfide products of MN and MA, indicating that not all Mn-based oxides reacts with H_2S in gas. However, the sulfide product of MH cannot find Mn-based oxides, indicating that all the Mn-based oxide involve into the reaction with H_2S. It may be the reason that the MH sorbent has higher BSC than other sorbents.

Figure 74.4g–i shows that the XRD results of the sorbents calcined for different atmosphere at 900 °C after four successive desulfurization-regeneration cycles. The phase Mn_3O_4 and $MnAl_2O_4$ appear on all the sorbents, whose diffraction peak at $2\theta = 44.44°$, 50.71°, 56.01°, 58.51°, 59.84°, 64.65°, 74.15° and 30.58°, 36.01°, 43.81°, 57.97°, 63.69° respectively. It is observed that the phase of $MnAl_2O_4$ also appears in MA sorbent, indicating that the interaction of Mn-based oxides and Al_2O_3 substrate generated Mn-Al alloy during the desulfurization-regeneration cycle.

The sulfide product of sorbent could be regenerated completely during regeneration process because $MnSO_4$ and MnS peaks are not detected in regenerated sorbent. $MnSO_4$ is less likely formed with lower oxygen concentrations under the higher regeneration temperatures. The result is agreement with the former report, which shows that MnS could be oxidized completely at 700 °C with lower O_2 concentration.

74.3.3 Mechanism Analysis

The BSC of MH is highest in all the sorbents, and the BSC of MN is higher than BSC of MA in every desulfurization-regeneration cycle. The XRD result shows that the part of Mn_3O_4 is shown in sulfide product of MN and MA, which cannot react with H_2S during desulfurization process. It may because that the part of Mn-based oxides dissolve into the inner of Al_2O_3, and then they are covered by Al_2O_3 in preparation process and desulfurization process, which cannot involve the reaction with H_2S. The phenomenon may be caused by the phase transformation of Al_2O_3 under high temperature and the transformation between Al_2O_3 and $MnAl_2O_4$ (Li et al. 2018). Though the $MnAl_2O_4$ phase are detected in MH sorbent, which form with the reaction between Al_2O_3 with the part of Mn-based oxides dissolving into the inner of Al_2O_3 in preparation process, the $MnAl_2O_4$ is the active component for H_2S removal. So, the initial BSC of MH is higher than other sorbents.

During the desulfurization-regeneration cycles, the mutual conversion between Mn-based oxides and MnS, phase transformation of Al_2O_3, the transformation of Al_2O_3 and $MnAl_2O_4$ and shrinkage and disappearance of pores would cause more Mn_3O_4 to be covered by Al_2O_3. It would be the reason that the BSC of sorbents decrease during successive desulfurization-regeneration cycles. As shown in Table 74.1, the BET surface and pore volume of MN-4C and MH-4C is higher than MA-4C, which may cause less Mn-based oxides to be covered by Al_2O_3. So, the MA-4C has lower BSC than other sorbents in fifth desulfurization process.

74.3.4 Conclusions

1. The Mn_xO_y/Al_2O_3 is prepared with various calcination atmosphere (H_2, N_2 and air). MH sorbent has the highest initial BSC in all sorbent, and the BSC of MN is higher than that of MA. The same result can also find in every subsequent desulfurization-regeneration cycle.
2. The calcination atmosphere has effect on the phase of Mn-based oxides in sorbents. MnO and $MnAl_2O_4$ exists in MH, Mn_2O_3 exists in MA and Mn_3O_4 exists in MN. The BET surface and pore volume of MN and MH are higher than that of MA, the same result is found in the regenerated sorbents after four desulfurization-regeneration cycles.
3. The part of Mn-based oxides are covered by Al_2O_3 in MA and MH, which lead less initial BSC of MN and MA than that of MH. The more Mn_3O_4 are covered by Al_2O_3 during successive desulfurization-regeneration cycles, which lead the less BSC of MA than that of MH and MN.

Acknowledgements This work was supported by the Strategic International Scientific and Technological Innovation Cooperation Funds of National Key R&D Program of China (2016YFE0204000) and the National Natural Science Foundation of China (NO. 51576086, 51576081). Technical support from the Analytical and Testing Center of Huazhong University of Science and Technology are highly appreciated.

References

Abdel-Mageed AM, Eckle S, Anfang HG, Behm RJ (2013) Selective CO methanation in CO_2-rich H_2 atmospheres over a Ru/zeolite catalyst: the influence of catalyst calcination. J Catal 298:148–160

Bai L, Zhou Y, Zhang Y, Liu H, Sheng X, Duan Y (2009) Effect of calcination atmosphere on the catalytic properties of PtSnNaMg/ZSM-5 for propane dehydrogenation. Catal Commun 10(15):2013–2017

Bakker WJW, Kapteijn F, Moulijn JA (2003) A high capacity manganese-based sorbent for regenerative high temperature desulfurization with direct sulfur production. Chem Eng J 96(1–3):223–235

Chand R, Obuchi E, Katoh K, Luitel HN, Nakano K (2013) Effect of transition metal doping under reducing calcination atmosphere on photocatalytic property of TiO_2 immobilized on SiO_2 beads. J Environ Sci 25(7):1419–1423

Chytil S, Kure M, Lødeng R, Blekkan E (2017) A: Performance of Mn-based H_2S sorbents in dry, reducing atmosphere-Manganese oxide support effects. Fuel 196:124–133

Chytil S Kure M., Lødeng R, Blekkan E. A: On the initial deactivation of Mn_xOy-Al_2O_3 sorbents for high temperature removal of H_2S from producer gas. Fuel Processing Technology, 2015, 133, 183–194.

Li H, Su S, Hu S, Xu K, Jiang L, Wang Y, Xu J, Xiang J (2018) Effect of preparation conditions on Mn_xO_y/Al_2O_3 sorbent for H_2S removal from high-temperature synthesis gas. Fuel 223:115–124

Liu D, Zhou W, Wu J (2016) CeO_2-MnOx/ZSM-5 sorbents for H_2S removal at high temperature. Chem Eng J 284:862–871

Ren X, He Q, Yang Z, Li T, Chang L, Bao W (2016) Effect of temperature-programmed pyrolysis pre-treatment on desulfurization performance of Zn-based sorbent prepared by lignite as support. Process Saf Environ Prot 100:142–149

Wang J, Liang B, Parnas R (2013a) Manganese-based regenerable sorbents for high temperature H_2S removal. Fuel 107:539–546

Wang J, Liu L, Han L, Hu Y, Chang L, Bao W (2013b) Alumina-supported manganese oxide sorbent prepared by sub-critical water impregnation for hot coal gas desulfurization. Fuel Process Technol 110:235–241

Xia H, Zhang F, Zhang Z, Liu B (2015) Synthesis of functional xLayMn/KIT-6 and features in hot coal gas desulphurization. Phys Chem Chem Phys 17(32):20667–20676

Xia H, Liu B, Li Q, Huang Z, Cheung ASC (2017) High capacity Mn-Fe-Mo/FSM-16 sorbents in hot coal gas desulfurization and mechanism of elemental sulfur formation. Appl Catal B 200:552–565

Part VII
Design and Operating Experiences

Chapter 75
A Dataset Analysis of Particular Matter Removal Techniques in China's Coal-Fired Power Plant

Xiaoyu Li, Yu Ni, and Hui Long

Abstract We perform a dataset analysis of particulate matter (PM) removal techniques in China's coal-fired power plant. The database we built contains 164 units in all regions of China. It covers many typical coal samples and lasts from 2005 to 2019 when the PM emission limit largely reduces from 200 to 10 (or even 5) mg/m^3 in China. The most popular primary PM removal technique is the low–low temperature ESP. It is revealed that the specific collection area and inlet flue gas temperature keep adjusting to satisfy the stricter regulation within this period. Usually the use of secondary PM removal technique is necessary for ultra-low emission, with WFGD and demister dominants. The WESP increases the PM removal efficiency, but is always companied with huge electrical costs in operation. We then propose a technology roadmap for the purpose of ultra-low PM emission of coal-fired power plants under different conditions.

Keywords Particulate matter removal · Coal-fired power plant · Electrostatic precipitator · Database · Techno-economic analysis

75.1 Introduction

Coal severs as the largest primary energy source in China mainly due to the lack of other fossil fuels like oil and natural gas. Coal combustion raises many serious environmental concerns including the SO_2 and NO_x emissions. Besides, the particulate matter (PM) stands out in recent years as a human health hazard because it easily penetrates into the lung (Siegmann 2000). Faced with these adverse effects, Chinese government has been increasingly strengthening its regulation on emission control for coal-fired units. Table 75.1 lists the emission standards for PM, SO_2

X. Li (✉) · Y. Ni · H. Long (✉)
China Power Engineering Consulting Group Co., Ltd., Xicheng District, Beijing 100120, China
e-mail: xyli@cpecc.net

H. Long
e-mail: hlong@cpecc.net

J. Lyu and S. Li (eds.), *Clean Coal and Sustainable Energy*, Environmental Science and Engineering,
https://doi.org/10.1007/978-981-16-1657-0_75

Table 75.1 Emission standard for coal-fired unit in China, units: mg/m^3 when vol% of O_2 is 6%

Emission standards	PM	SO_2	NO_x
GB13223-1996 (Ministry of Environmental Protection of the People's Republic of China 1996)	200	1200	650
GB13223-2003 (Ministry of Environmental Protection of the People's Republic of China 2003)	100 (50)	400	450
GB13223-2011 (Ministry of Environmental Protection of the People's Republic of China 2011)	30 (20)	100 (50)	100
Ultra-Low Emission 2015 (Ministry of Environmental Protection, National Development and Reform Commission, National Energy Administration 2015)	10 (5)	35	50

Number in parentheses is the standard for key areas of China

and NO_x in China from 1996 to 2015. Notably, from 2015, the PM emission limit reduces to 10 mg/m^3 and even to 5 mg/m^3 in some urban regions. Together with the ultra-low emission limits for SO_2 and NO_x, the 2015 rule brought about a new wave of improving the emission control equipments in coal-fired power plants, a lasting endeavor in the past two decades.

After 4–5 years, most coal-fired power plants were retrofitted to meet the PM standards. And we are now in a position to overview the situation and evaluate different technical approaches. In general, the primary PM removal technique includes the electrostatic precipitator (ESP) and bag filter (Yao et al. 2009). The further improvement of ESP involves rotating electrode, high-frequency power, pulsed power and simply extending electrical fields. Other types of primary techniques, including the low–low temperature ESP (Liu et al. 2019) and electrostatic fabric precipitator (Yao et al. 2009), have been employed to further enhance the efficiency of fine PM removal. It is worth mentioning that the low–low temperature ESP has a synergetic effect on removing SO_3 in the flue gas (Liu et al. 2019; Huang 2019; Chen et al. 2015). By the year of 2011, the PM standard (30 or 20 mg/m^3) can be satisfied by most power plants equipped with only primary PM removal devices. However, to meet the 2015 ultra-low emission line, coal-fired power plants must resort to secondary PM removal techniques including wet flue gas desulfuration (WFGD) (Zhu and Yang 2020), high-efficiency demister (Shi et al. 2016), and wet-ESP (WESP) (Shi et al. 2016; Yang et al. 2015). It is the synergetic effect in these equipments that helps further ramp up the PM removal efficiency.

Owing to the vast diversity in China's power plants, many different technical routes have been used for the purpose of ultra-low PM emission while aiming at optimal techno-economical feasibility. The optimal technical route relies on the unit parameter, the coal property, and even the emission standard it intends to meet. It is believed that, after so many years' trial-and-error, insights from China's experience and lessons can be beneficial to the application of PM removal techniques in other regions of the world. However, to our best knowledge, there is currently no public overview on the PM control technology equipped in China's power plants.

The objective of this paper is to build an ESP database for coal-fired units in all regions of China. To do this, we collected design and operating data from leading ESP manufactures, environmental research institutes, and coal-fired power plants. Putting together coal properties, ESP structures, flue gas parameters and PM concentrations, we are able to perform a thorough techno-economic analysis of the prevalent techniques for ultra-low PM emission in the coal-fired system.

75.2 ESP Database

In this work, we base on an ESP database that we build for coal-fired power plants in all regions of China, containing both design/check and actual operating data. The database was detailed in Li and Long (2019). Briefly, the data were collected from leading ESP manufacturers, environmental research institutes, and coal-fired plants. The database is composed of datasets of coal property, flue gas, ESP structure and PM concentrations. At present, 164 kinds of coal samples and operating data from 164 units are included, and the database continues to grow and renew. In particular, 62 out of 164 units are located in East China, the more developed regions; only 6 units are located in South China, while 15–25 units are included for other regions of China. In terms of the capacity, 24 units are greater han 1000 MW, 52 units in the range 500–700 MW, 72 units in the range 300–400 MW, and 16 units less than 300 MW. Notice that the data are collected for different years, and only a part of the units are collected after 2015 when the ultra-low emission is achieved.

A major advantage of the database is the ability to reveal the PM removal capacity under real operating conditions. Figure 75.1 shows the PM penetration rate of the ESPs in 70 coal-fired units. Here we define the penetration rate ε as below:

$$\varepsilon = 1 - \eta = \frac{\text{outlet PM concentration}}{\text{inlet PM concentration}} \tag{75.1}$$

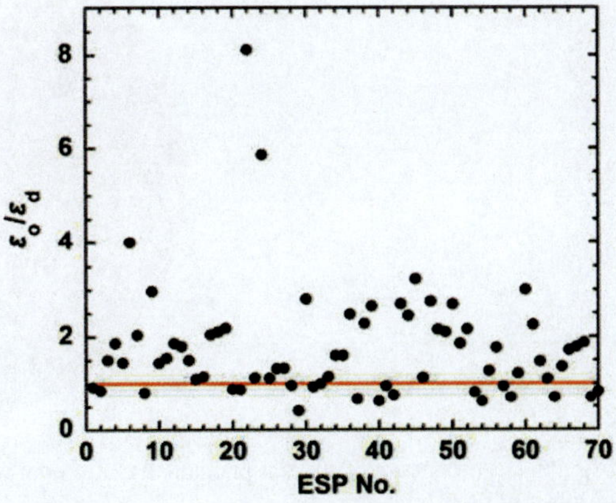

Fig. 75.1 The ratio of PM penetration rate in actual operation over the desined value of ESPs in 70 coal-fired units

Here η is the PM removal efficiency of the ESP. The y-axis of Fig. 75.1 is the ratio of ε under operating conditions (denoted ε_o) over designed values (denoted ε_d). It is seen that for most ESPs, the operational penetration rate of PM is 1–3 times of the desined value, and some unit the ratio even tops 6–8. This clearly reflects the need to improve the design and operation of PM removal devices in coal-fired units.

75.3 Results and Discussion

Based on the ESP database, the coal properties were detailed in Li and Long (2019). Briefly, the utility power coal has a large diversity in ash content. For the 164 coal samples, 23% are low ash ($A_{ar} \leq 18$ wt%), 35% are medium ash ($18 < A_{ar} \leq 25$ wt%), 30% are medium-to-high ash ($25 < A_{ar} \leq 35$ wt%), and 12% are high ash ($A_{ar} >$ 35 wt%). The high-ash coals are mainly burnt in Southwest China, and the low-ash coals are mainly used in the coastal region. Furthermore, the utility coals in China are generally low or medium in sulfur content, with 90% of the samples in the dataset are low-sulfur (S < 1 wt%) and medium-sulfur ($1 < S \leq 2$ wt%). The high-sulfur coals are mainly burnt in Southwest China, the medium-sulfur coals mainly burnt in North China, and most other regions use low-sulfur coals.

Figure 75.2 then shows the prevalence of primary PM removal technology equipped in the 164 units that have all been retrofitted to achieve ultra-low emission. We see that 31% of the units apply the high-frequency power for the ESP. The

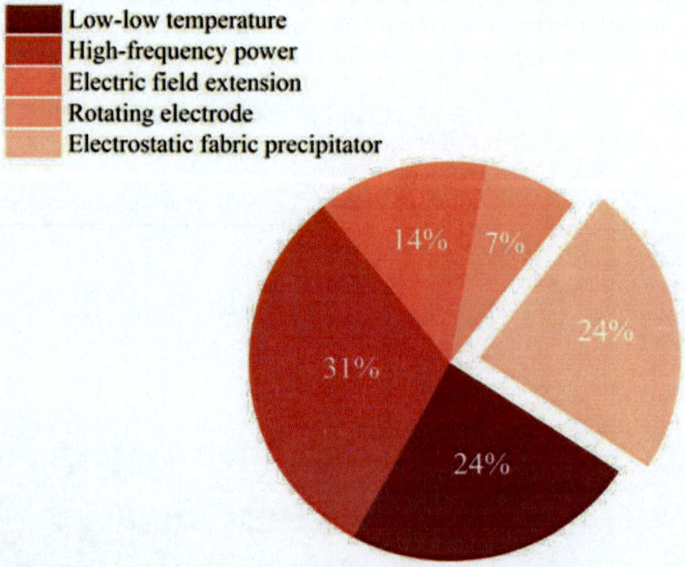

Fig. 75.2 Percentage of several primary PM removal techniques used by the power plants for the purpose of ultra low emission

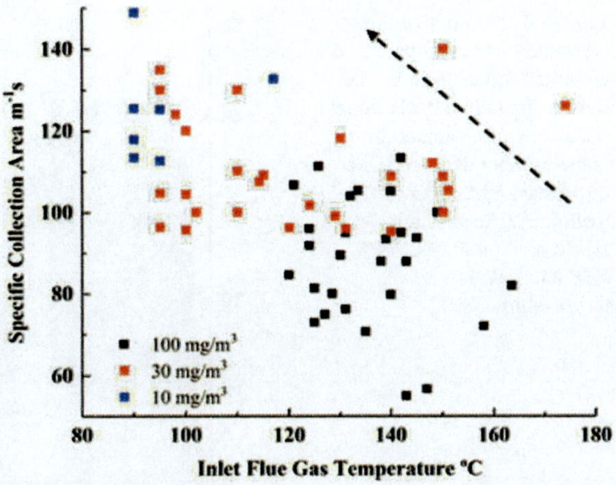

Fig. 75.3 Specific collection area versus Inlet flue gas temperature for 80 units using ESP in China. Different colors represent the parameters for varied emission standards from 100 (black), 30 (red) to 10 (blue) mg/m^3

low–low temperature ESP and electrostatic fabric precipitator both account for 24%. During the retrofit, 14% of the units chose to extend the electric field of the ESP. Only 7% of the units resort to the rotating electrode in the ESP.

With the lower and lower emission standard over the past two decades, the evolution and/or development of ESPs in China are revealed in Fig. 75.3, which scatter plots the specific collection area and the inlet flue gas temperature of the ESP. Different colors represent the goals of progressive emission limits. When the limit is as high as 100 mg/m^3 (from 2003 to 2011), the specific collection area usually falls within 60–100 m^{-1} s, and the inlet flue gas is at 120–150 °C. Then from 2011 to 2015, the limit drops to 30 mg/m^3. During this period, the specific collection area increases to 90–130 m^{-1} s, while the inlet flue gas widely ranges from 90–150 °C. Indeed, the low inlet temperature (<100 °C) is mainly attributed to the use of the low–low temperature ESP. When the ultra-low limit takes effect from 2015, we see from the relatively few cases that the application of low–low temperature ESP is dominant, and the specific collection area further increases to 110–150 m^{-1} s. Therefore, the trend of increasing specific collection area and decreasing flue gas temperature is clearly revealed from the data of different historical periods, and it again highlights the importance to match the parameters with emission goals.

The prevalence of secondary PM removal techniques is shown in Fig. 75.4. Note that only 72 units are equipped with secondary devices. Also shown is the number of primary techniques for each combination. We see that 42 units (58%) have installed WFGD and demister, 20 (28%) use WFGD and WESP, and only 10 (14%) use all the three techniques. For all three cases, the most popular primary technique is the low–low temperature ESP, followed by the conventional ESP and ESFP.

It is doubtless that the incorporation of PM removal equipments induces higher cost so that a careful economical analysis is critical. Here we perform a case study of a power plant retrofit. The original ESP has 2 chambers 4 electric fields, and its specific collection area is 96 m^{-1} s. The goal of retrofit is to decrease the PM emission from

Fig. 75.4 Number of three secondary PM removal technical routes used by the power plants in our database. Also shown in each column is the number of three kinds of primary PM removal techniqes (Electrostatic fabric precipitator (ESFP), ESP and low–low temperature ESP)

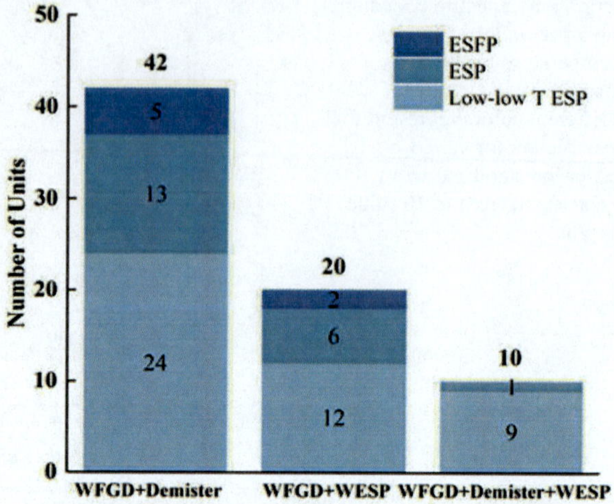

30 mg/m^3 to 5 mg/m^3. To do this, there are 2 routes. Route I combination of low–low temperature ESP, WFGD and demister. It works because the PM concentration at the outlet of low–low temperature ESP can be 20 mg/m^3, and the efficiency of combining WFGD and demister is higher than 75%. Route II uses ESP, WFGD and WESP. The PM concentration at the outlet of ESP is 35 mg/m^3, and the efficiency of combining WFGD and WESP can be higher than 90%. Thus, both routes can meet the standard, but the costs are different. Figure 75.5 shows the comparison of equipment cost, dedusting electric cost and total annual cost of the two routes. The cost of water, maintenance and financial are neglected. It is seen that although Route I is more costly in the equipments, its electric cost is much lower than the other route,

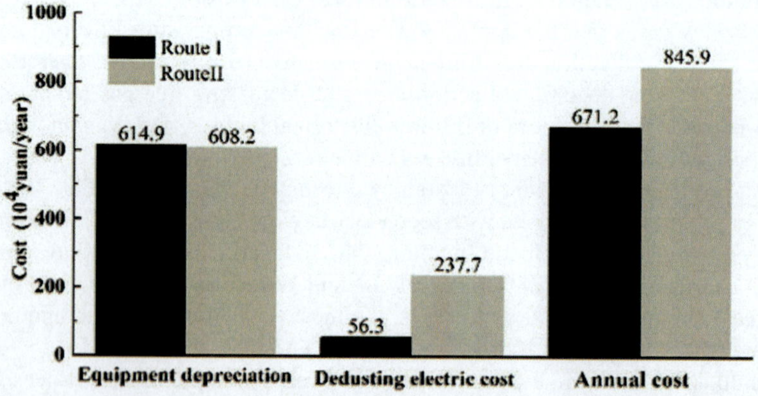

Fig. 75.5 Cost evaluation of the two retrofit routes in CNY (Unit: 10^4 Yuan per year). Route I is a combination of low–low temperature ESP, WFGD and demister, while Route II uses ESP, WFGD and WESP. The total annual cost is made up of the equipment depreciation and the dedusting electric cost

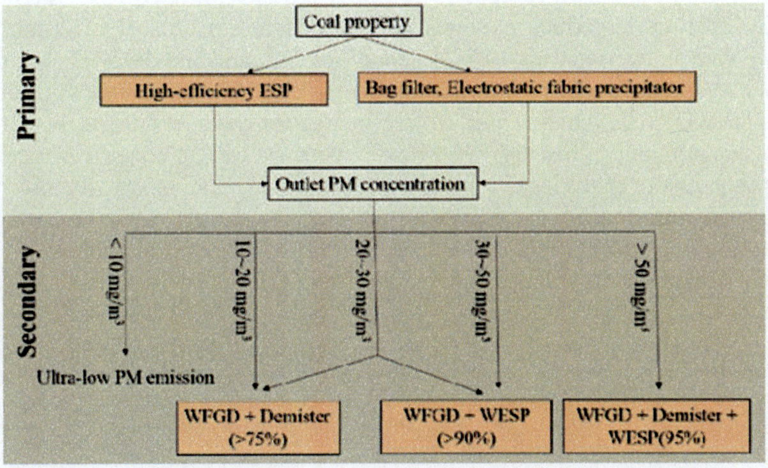

Fig. 75.6 Technical routes for primary and secondary PM equipments under different conditions

resulting in a total annual cost 1.747 million CNY less than that of route II. We may thus conclude that route1 is better than route II.

Based on the practices recorded in our dataset, we can propose a technical route for ultra-low PM emission in the coal-fired power unit, as shown in Fig. 75.6. The main focus is an optimal combination for different senarios. The first key factor is the coal property, and the two most popular primary PM control techniques can be considered, namely the high-efficiency ESP and the bag filter (or the electrostatic fabric precipitator). Then the outlet PM concentration acts as another major factor. If the value drops below 10 mg/m^3, then the secondary PM control device is unnecessary; if the value lies between 10 and 20 mg/m^3, a combinaton of WFGD and demister (with the total efficiency >75%) is recommended; if the value is 30–50 mg/m^3, a combination of WFGD and WESP (with the total efficiency >90%) is recommended; if the value is 20–30 mg/m^3, 'WFGD + demister' can be used in case the variation of coal property is small, and 'WFGD + WESP' should be used otherwise; if the value is greater than 50 mg/m^3, the full combination of WFGD, demister and WESP (with the total efficiency >95%) needs to be employed. This route can serve as a basic 'roadmap' for power-plant decision makers to balance costs and technical capacity for ultra-low emission control.

75.4 Conclusions

In this work we perform a thorough analysis on the existing status of PM removal technologies deployed in China's coal-fired power plants. We build a database of 164 units including both design and operating conditions. The data we collected covers all regions of China, includes many representative coal properties, and lasts from

2005 to 2019 during which the emission limit decreases gradually. Based on the data we give an overview on the prevalence, combination and development of both primary and secondary PM control technologies. We elucidate how technical indicators of the ESP, mainly the specific collection area and the inlet flue gas temperature, improve continuously to satisfy the growing demand on PM control. A case study of the economical analysis shows the total cost per year can be heavily affected by the electric cost, and the use of WESP is especially costly in this sense. Finally, we propose a technical route for evaluating several PM removal approaches under different conditions in order to achieve the best performance both technically and economically.

Acknowledgements This work is supported by National Key Research and Development Program of China (No. 2017YFB0603202).

References

Chen M, Hu YQ, Gui B (2015) Synergistic control technology for ultra-low PM emission from coal-fired power plants and its application. Electr Power 48(9):146–151 (in Chinese)

Huang YM (2019) Removal performance of PM2.5 and SO3 by low-low temperature electric dust removal technology. Chin J Environ Eng 13(12):2924–2933 (in Chinese)

Li XY, Long H (2019) Characteristics of power coal and recommendations on dust removal techniques for power plants based on the database of electrostatic precipitator and coal property. Electr Power 52(10):171–176 (in Chinese)

Liu HX, Li JG, Yao YP et al (2019) Study on SO3 removal performance of low-low temperature electrostatic precipitator system. Power Gener Technol 40(4):2096–4528 (in Chinese)

Ministry of Environmental Protection of the People's Republic of China (1996) Emission standard of air pollutants for thermal power plants: GB13223-1996. China Environmental Science Press, Beijing (in Chinese)

Ministry of Environmental Protection of the People's Republic of China (2003) Emission standard of air pollutants for thermal power plants: GB13223-2003. China Environmental Science Press, Beijing (in Chinese)

Ministry of Environmental Protection of the People's Republic of China (2011) Emission standard of air pollutants for thermal power plants: GB13223-2011. China Environmental Science Press, Beijing (in Chinese)

Ministry of Environmental Protection, National Development and Reform Commission, National Energy Administration (2015) Notice on the issuance of the work plan for the comprehensive implementation of ultra-low emissions and energy-saving upgrading of coal-fired power plants. http://www.mee.gov.cn/gkml/hbb/bwj/201512/t20151215_319170.htm

Shi WZ, Yang MM, Zhang XH et al (2016) Ultra-low emission technical route of coal-fired power plants and the cooperative removal. 36(16):4308–4318 (in Chinese)

Siegmann K (2000) Soot formation in flames. J Aerosol Sci 31(S1):S217–S218

Yang ZD, Chang QY, Yue T et al (2015) Experimental study on simultaneous control of SO_2 and PM by wet electrostatic precipitato. J Eng Thermophys 6:1365–1370 (in Chinese)

Yao Q, Li SQ, Xu HW et al (2009) Studies of formation and control of combustion particulate matter in China: a review. Energy 34(9):1296–1309

Zhu Y, Yang YL (2020) Influence analysis on haze control of ultra-low emission wet flue gas desulfurization of coal-fired power plants. Power Gener Technol 41(3):295–300 (in Chinese)

Chapter 76
Dynamic Characteristics of a Boiler with Low-NO_X Combustion

Jinjing Li, Lin Yang, Shi Yang, Zhenning Zhao, Yuanyuan Li, and Qingfeng Zhang

Abstract Due to the strictest nitric oxide (NO_x) emission limit in the world, a large number of thermal power plants in China have reformed their pulverized coal (PC) boilers with the 3rd generation low-NO_x combustion technology which integrates bias combustion burners with separated over fire air (SOFA) nozzles. The 3rd generation low-NO_x combustion technology effectively reduces the NO_x concentration in flue gas at furnace outlet, from more than 350 mg/m^3 to 100–150 mg/m^3 in steady states when burning bituminous coal. However, as the utilization of the 3rd generation low-NO_x combustion technology, the dynamic process of PC boiler becomes extremely slower than it used to be, on the countrary to the demands of power grid, who expects that the coal fired power plant changes its power load as rapidly as possible. In order to accelerate load changing rate in PC boilers with the 3rd generation low-NO_x combustion technology, operating optimization tests, including combustion optimizations in steady states and study on dynamic characteristics during load changing processes, were conducted on a 600 MW commercial boiler. The former works reduce NO_x concentration, from 130 to 100 mg/m^3, in flue gas at furnace outlet. In addition, the latter works accelerate boiler load changing rate from 1% ECR to 2% ECR, narrowing down the range of steam temperature fluctuation. All these works not only enhance the security of boiler and turbine, but also speed up power load changing process, especially benifit to the flexibility of coal-fired thermal power plant.

Keywords Low-NO_x combustion · Deeply air staging combustion · Boiler load changing characteristic · Steam temperature control · Commercial boiler operating

---

J. Li (✉) · Z. Zhao · Y. Li · Q. Zhang
North China Electric Power Research Institute Co., Ltd., Beijing 100045, China
e-mail: Liau@139.com

L. Yang
Inner Mongolia Shangdu Power Generation Co., Ltd., Xilinguole 027200, China

S. Yang
China Coal Research Institute Company of Energy Conservation CCRI, Beijing 100013, China

J. Lyu and S. Li (eds.), *Clean Coal and Sustainable Energy*,
Environmental Science and Engineering,
https://doi.org/10.1007/978-981-16-1657-0_76

76.1 Introduction

Since the year 2012, the strictest NO_x emission limit in the world has been implemented in China. According to this emission standard (Ministry of Environmental Protection of the People's Republic of China 2011), nitric oxide emission of thermal power plants must be limited within 100 mg/m^3 (dry basis, 6% O_2 in flue gases). Assuming the removal efficiency of selective catalytic reduction (SCR) system is 70%, the NO_x concentration in flue gas at SCR inlet/furnace outlet, should be less than 330 mg/m^3. This is far below the NO_x emission level of conventional pulverized coal combustion technologies. Researchers reported that the conventional tangentially fired furnace discharged about 500–700 mg/m^3 (Dıez et al. 2008) NO_x in its flue gas, that the conventional wall fired furnace discharged about 600–800 mg/m^3 (Ying et al. 2011) NO_x in its flue gas, and that the conventional down fired furnace discharged about 1500 mg/m^3 (Kuang et al. 2014) or even more NO_x in its flue gas. In other words, The conventional pulverized coal combustion technologies did not feed the NO_x emission reduction demands of coal fired power plants.

A large number of thermal power plants in China have reformed their coal fired boilers with the 3rd generation low-NO_x combustion technology. It integrates bias combustion (Cao et al. 2004; Jing et al. 2011) burners with separated over fire air (SOFA) nozzles (Zhou et al. 2011; Li et al. 2015). The bias combustion burner prevents nitrogen in fuel from contacting with oxygen in air too earlier. The SOFA nozzle constructs a NO_x reduction zone in the space between the burners and the SOFA nozzles. The NO_x is reduced by carbon monoxide or other amino agents produced in the burner zone, because of few air supply in reduction zone. In a 600 MW tangentially fired boiler, the height of reduction zone is usually greater than 7 m and NO_x concentration at tangentially fired furnace outlet canbe reduced to 100–200 mg/m^3 in steady states when burning bituminous coal. However, the reduction zone divides the furnace into two separated combustion zone along height, the burner zone and the SOFA zone. The the reduction zone seperates the traditional flame core in furnace, by lowering the maximum temperature and rebuilding two smaller high temperature zones. Consequently, it is difficult to maintain the steam temperature stable when the boiler load changes drastically. So the dynamic process of PC boiler with low-NO_x combustion technology reform becomes extremely slower than that without reformation.

In power grid, thermal power plants are usually regarded as the major power suppliers and load regulator, whose load rate is adjusted following grid load in real time dynamically. In China, the power plant load changing rate should be no less than 2% of rated capacity, otherwise they would have to pay an economical penalty. Moreover, power plant managers expect to change their load as quickly as possible, pursuing a considerable bonus from the power grid. Consequently, an operation technology is urgently needed to match the low-NO_x combustion technology, enhancing boiler load changing rate.

76.2 The Tangentially Fired Boiler

A tangentially fired boiler of thermal power plant with a unit capacity of 600 MW was built in 2006, and reformed with the 3rd generation low-NO_x combustion technology in 2012. It is a drum boiler with forced water circulation. The boiler performance and fuel analysis are shown in Table 76.1.

As shown in Fig. 76.1, all the burners and SOFA nozzles are installed on four corners of the furnace. The burners are designed based on vertical bias combustion technology (Fig. 76.2), where dense phase of coal-air mixture from every two adjacent burners are arranged together to build a fuel rich zone. The burners are swayed vertically to adjust the reheated steam temperature. The secondary air rotates clockwise in furnace, while the primary air rotates anticlockwise. The included angle between the directions of the primary and the secondary air is 7°, forming the air staging in cross section. Oxygen is insufficient in the burner zone, and the total excess air coefficient is in a range of 0.75–0.80. Above the reduction zone, there are seven layers of SOFA nozzles, providing the residual air for fuel combustion. The SOFA

Table 76.1 Boiler performances and fuel analysis

Content	Unit	Value
Boiler performance in boiler maximum continuous rating		
Superheat steam flow	t/h	2028
Superheat steam exit pressure	MPa	17.48
Superheat steam exit temperature	°C	541
Reheat steam flow	t/h	1677.7
Reheat steam entrance pressure	MPa	3.84
Reheat steam exit pressure	MPa	3.66
Reheat steam entrance temperature	°C	323
Reheat steam exit temperature	°C	541
Economizer water entrance temperature	°C	278.3
Proximate analysis of coal		
Moisture (as received)	%	14.0
Volatile matter (dry and ash free basis)	%	33.19
Ash (as received)	%	7.04
Lower heating value (as received)	kJ/kg	23.390
Ultimate analysis of coal		
Carbon (as received)	%	63.25
Hydrogen (as received)	%	3.40
Oxygen (as received)	%	11.18
Nitrogen (as received)	%	0.64
Sulphur (as received)	%	0.50

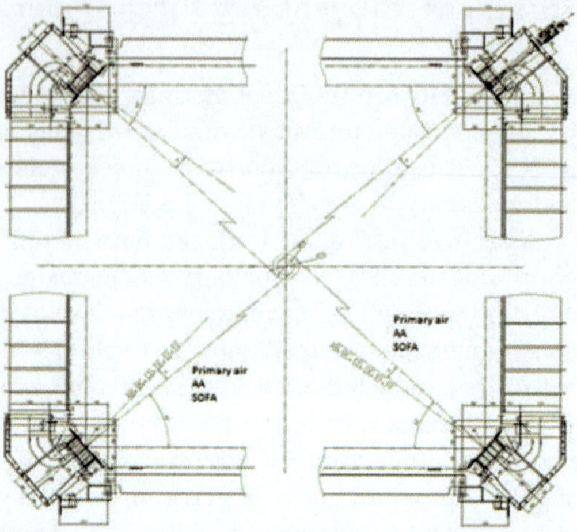

Fig. 76.1 Configuration of low-NO_x coal combustion system

nozzles can be swayed vertically to control reheated steam temperature as well, and also can be swayed horizontally to eliminate the temperature deviation of lateral flue gas at furnace outlet.

Three fuel rich zones are respectively located between the layers of burner B and C, the layers of burner D and E, and the layers of burner F and SOFA1. Wall air (WA) nozzles are installed in every fuel rich zone, where air injects into the furnace in a direction parallel to the water wall, to prevent it from high temperature corrosion.

As shown in Fig. 76.3, the steam reheating system is consisting of wall reheater, platen reheater and final reheater in steam pass. The water spray attemperator works only in accident. In the furnace, the flame is tightly surrounded by the wall reheater and platen reheater, where gas temperature is extremely high and sensitive to the location of flame core. When the burners/SOFA nozzles are nose up, the flame core rises up, transferring more heat to platen reheater and final reheater and increasing reheated steam temperature. As two reheated steam temperature adjusting methods, it is more economical to sway burners/SOFA nozzles vertically than to spray water. Therefore, the operators usually adjust reheated steam temperature by swaying burners and SOFA nozzles in steady state.

76.3 Method

In general, dynamic characteristics are researched basing on the step response curves and approximate transfer functions (Alzaharnah et al. 2013; Bhambare et al. 2007). These two methods are usually very effective on transient analysis. In this article, the boiler load changing process, which continues a quite long time, was studied. Dozens of operation items are carried out during the dynamic process, effecting many changes in different degrees on the steam pressure, steam temperature and

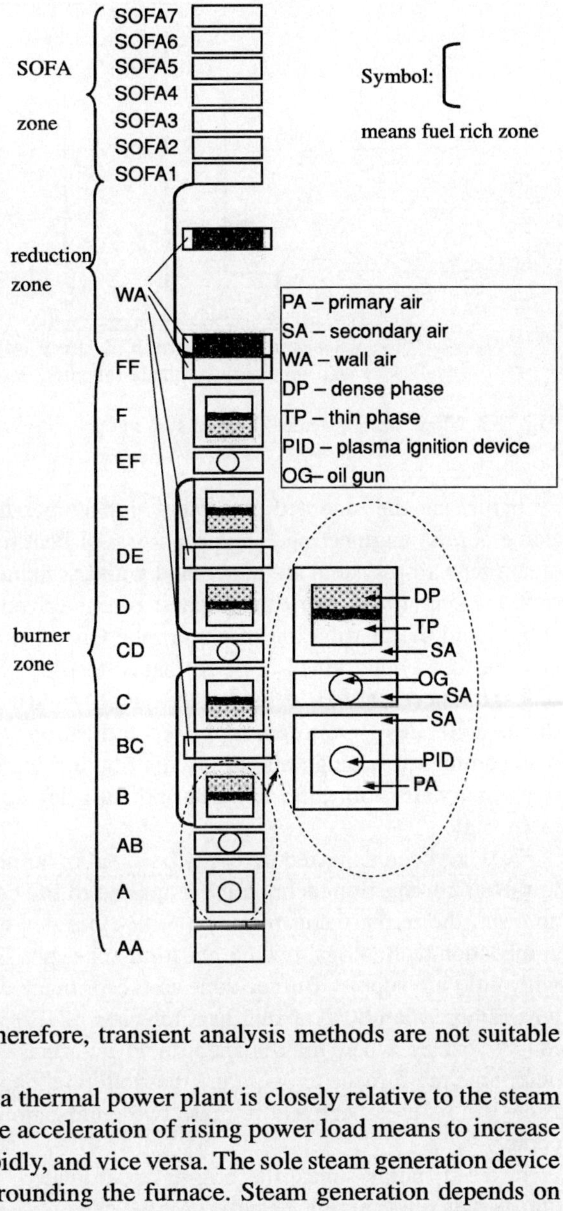

Fig. 76.2 Configuration of low-NO_x coal combustion system

NO_x emission for example. Therefore, transient analysis methods are not suitable for this work.

The electical power load of a thermal power plant is closely relative to the steam generating rate of its boiler. The acceleration of rising power load means to increase boiler steam generation rate rapidly, and vice versa. The sole steam generation device in boiler is the water wall surrounding the furnace. Steam generation depends on the heat transfer of water wall, mainly influenced by the temperature distribution in furnace and the ash fouling of water pipe. Operators adjust the former factor by operations optimizing combustion, and the later factor by ash blowing. As ash blowing is a periodical operation item independent on load changing, combustion optimization is the only controllable way to accelerate boiler steam generating rate.

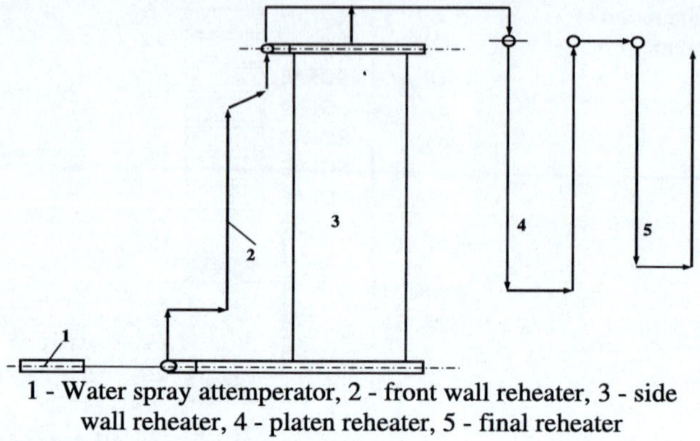

Fig. 76.3 Flow chart of steam reheating system

In furnace, the zone with extremely high temperature (1500–1600 K) is called the flame core in engineering. The proportion of heat transfer between water wall and steam reheating system are concerned with the flame core height, which is effected by three operation factors: the fineness of pulverized coal, the centralization of coal supply and the distribution of air supply. Once the optimal fineness of pulverized coal has been determined in performance test, operators scarcely change its status unless it is necessary to do that. As the quantity and capacity of pulverizer are definite due to boiler design, the coal supply centralization is barely changeable. In practice, it is common to change air supply distribution via adjusting the gross air supply, the primary air supply, the air baffles opening degree and the burners/SOFA nozzles pitch angle.

Fuel has been expected to totally burn out in burner zone until the 3rd generation low-NO_x combustion technology is applied to PC boiler. As the dividual supply of air from the burners and from the SOFA nozzles in the 3rd generation low-NO_x combustion technology, two combustion zones are separated by the reduction zone with little air supply. Burner zone is the primary combustion zone near burners, where more than 80% of fuel heat releases. SOFA zone is the burn out zone near SOFA nozzles, where the remanent fuel heat releases. These two combustion zones determine the furnace temperature distribution along height. Two high temperature zones are built respectively in these two combustion zones, regarded as two flame cores.

In the dynamic states, the original coordinated control system (CCS) changes unit power load relying heavily on coal feeding control. For example, the gross coal feeding rate (GCFR) increases excessively in load up process, and decreases excessively in load down process. Excess/insufficiency of fuel heat results a drastic change of steam temperature in load changing process, shown in Fig. 76.4. The faster unit power load changes, the faster steam temperature rises or falls. Especially when load changing speed is greater than 1.5% of rated capacity, the steam temperature probably exceeds its safety limits. For example, the steam temperature at rear platen

Fig. 76.4 Boiler dynamic performances under traditional control

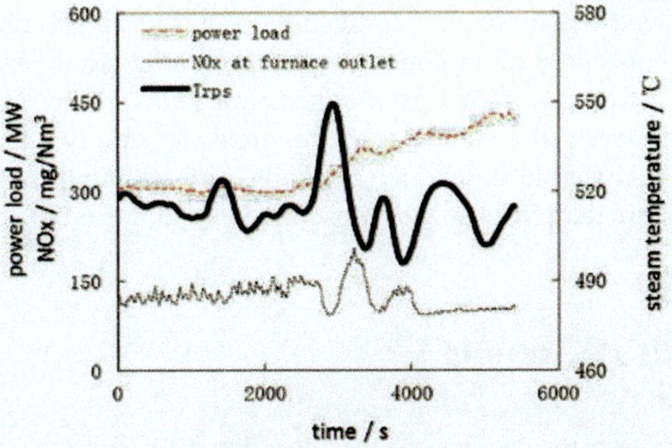

Load up process

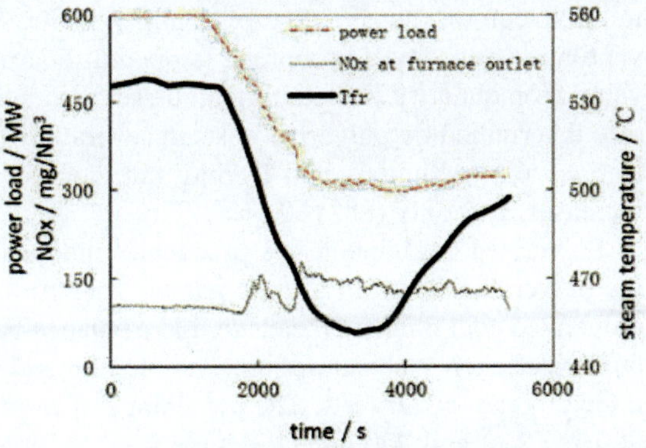

Load down process

superheater outlet (Trps) rises to 550 °C, over its high alarm value (540 °C) in load up process, causing risk of the rear platen superheater overheating. Inversely, the steam temperature at final reheater outlet (Tfr) drops to 450 °C or even lower in load down process, causing water hammer risk of the turbine blades.

Combustion optimization consists of two works: (1) matching steam generating rate with power load in static states, (2) balancing steam temperature with NO_x emission in dynamic states. Numerous key parameters are selected to be concerned, according to different combustion optimization states. Three typical power loads of the unit, 600, 450 and 300 MW, are selected as the static states for combustion optimization. In these static states, the key parameters are boiler efficiency, water spray flowrate and NO_x concentration in flue gas at furnace outlet. Two dynamic combustion optimization states are those the unit power load rises from 300 to 600 MW (load up process), or declines from 600 to 300 MW (load down process). On the one hand, the steam temperature at rear platen superheater outlet zooms in load up process. On the other hand, the steam temperature at final reheater outlet drops steeply in load down process. In dynamic states, corresponding steam temperature and NO_x concentration in flue gas at furnace outlet are the key parameters instead.

In static states, four adjustment items are executed, the the primary air supply, pulverized coal fineness, the gross air supply and the air baffles opening degree. In dynamic states, two methods, redistributing air supply along height and putting burners/SOFA nozzles nose up/down, are tried to stabilize steam temperature.

76.4 Results

The primary air supply is correlated to coal feeding rate. The optimal ventilation quantity in pulverizer satisfies not only pneumatic transmission and desiccation of pulverized coal but also NO_x reduction in furnace. Pulverized coal pneumatic transmission requires more ventilation quantity, while NO_x reduction in furnace needs less ventilation quantity. Appropriate temperature at pulverizer outlet prefers the fitting ventilation quantity. Accounting all these factors, the optimal ventilation quantity is also determined by pulverizer tests in several different coal feeding rates, as shown in Fig. 76.5. Although coal feeding rate doubles (30–60 t/h), ventilation quantity increases 25% only (88–110 t/h).

Pulverized coal fineness is practically independent of unit power load, because the pulverizers work in a quite narrow region of ventilation quantity. Optimizing pulverized coal fineness means to find a balance between combustion efficiency and pulverizer energy consumption. The recommended pulverized coal fineness determined by pulverizer tests. The percentage of residue on sieve with the mesh size of 90 μm (R_{90}) and 200 μm (R_{200}) are used to indicate the pulverized coal fineness. As every four burners correspond to one pulverizer, the R_{90} and R_{200} in each burner are measured. The average value of R_{90} is recommended no more than 20%. The maximum value of R_{200} is recommended no more than 5% in addition.

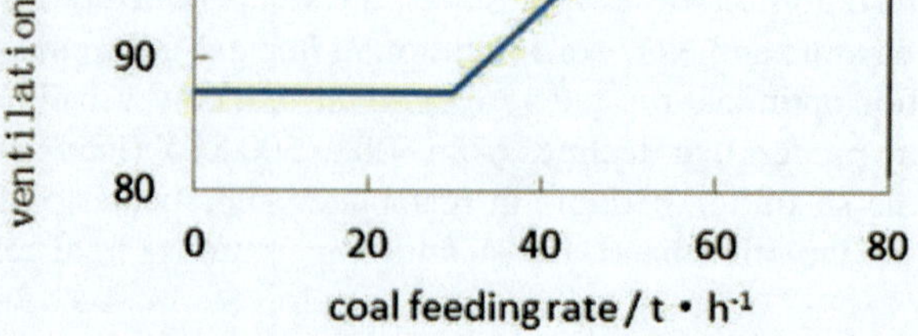

Fig. 76.5 Recommended ventilation quantity in pulverizer

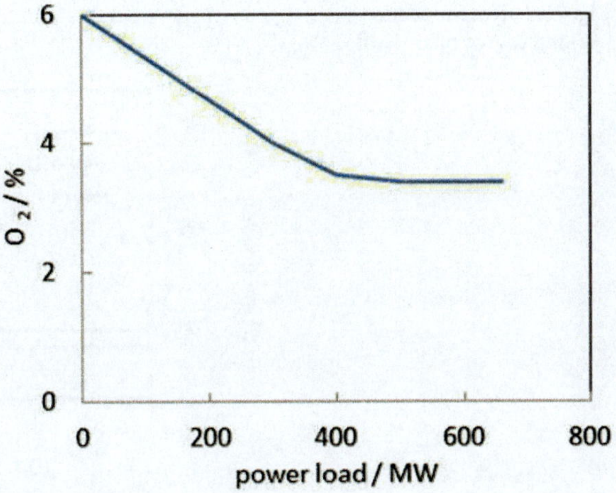

Fig. 76.6 Recommended oxygen content in flue gas

In the work of optimizing gross air supply and air baffle opening degree, the CO and NO_x content are measured at furnace outlet synchronously, representing the combustion efficiency and pollutant production respectively. Boiler operators monitor flue gas oxygen content (O_2) at furnace outlet to determine an appropriate gross air supply. On one hand, higher O_2 increases NO_x concentration in flue gas, contributing to the air pollution. On the other hand, lower O_2 increases CO concentration in flue gas, reducing combustion efficiency. The researchers change O_2 constantly to obtain an optimal value of O_2, while the unit power load is kept stable. The optimal values of O_2 contributes both high combustion efficiency and relatively low NO_x emission. Plotting optimal values of O_2 with unit power load, a recommended O_2 curve is graphed out, shown as Fig. 76.6. The greater unit power load is, the closer gross air supply approaches to the theoretical air requirement of fuel.

Uniform opening degrees of air baffles are not suitable for this low-NO_x combustion technology, because of its design of fuel rich zones. Air baffle opening degrees of every layer should be adjusted accurately and precisely to obtain a right value in each case of stable unit power load, balancing the CO and NO_x content in flue gas. The recommended air baffle opening degree curves are shown in Fig. 76.7. The locations of corresponding burners/SOFA nozzles are shown in Fig. 76.2. The opening degrees of secondary air baffles barely change except the layers AB, CD and FF. The baffles of separated over fire air open layer by layer, while the unit power load increases. An exception is SOFA4, opening finally and incompletely in order to construct the further air staging in SOFA zone.

All these recommended curves are implanted in the automatic control system of the power plant, resulting boiler a better static performance shown in Fig. 76.8. The NO_x at furnace outlet decreases 20.7 mg/m^3 without lowering boiler efficiency. The water spray flowrates decrease as well, which means that the water wall gains more heat in furnace. Consequently, the power plant becomes able to change its load in

Fig. 76.7 Recommended opening degree of air baffle

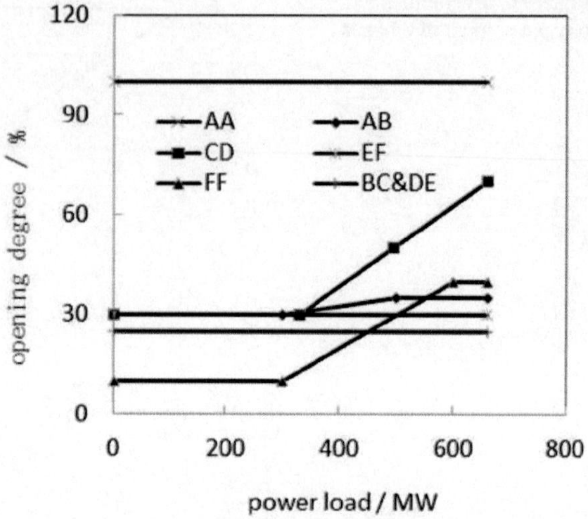

Secondary air baffle

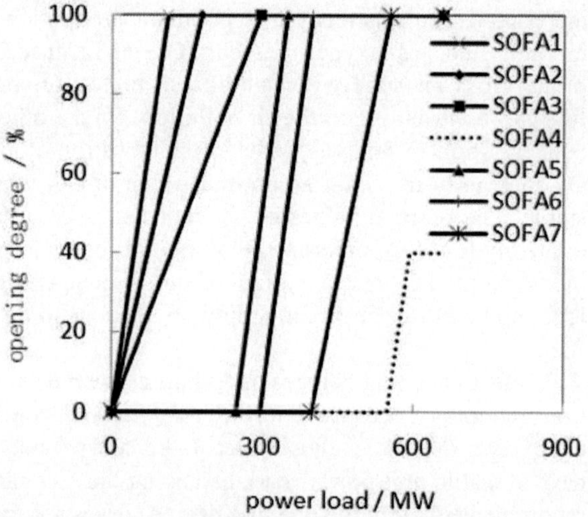

Separated over fire air baffle

a speed no less than 2% rated capacity, because combustion optimization matches steam generating rate with power load better in static states. This is quite important for the dynamic combustion optimization.

The dynamic combustion optimization, named air redistribution, was conducted in load changing process with a speed of 2% of rated unit capacity. In load up process, operators add air supply in burner zone, as well as minus that in SOFA zone in load up process, based on air baffle opening degree shown in Fig. 76.7. It results in a greater combustion proportion in burner zone, strengthening the lower flame

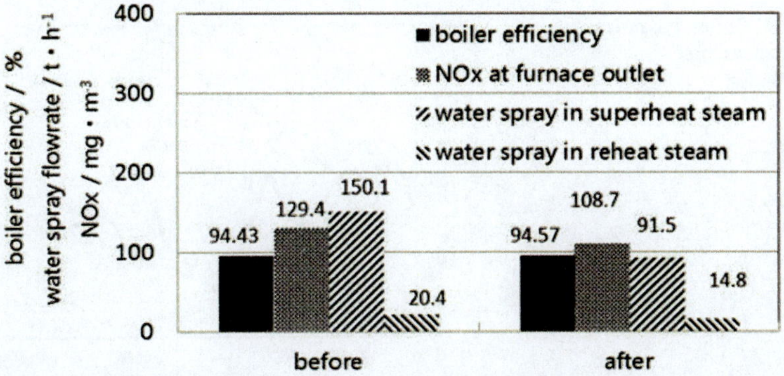

Fig. 76.8 Optimized static performances of boiler combustion and heat transfer

core and wakening the upper flame core. Meanwhile, they put the burners/SOFA nozzles nose down to move flame cores downward. As a result of these operation, heat transfer in water wall increases faster, and steam generation grows faster as well. In load down process, the air redistribution operations are supposed converse, including to add air supply in SOFA zone, to minus air supply in burners zone and to put the burners/SOFA nozzles nose up. These works make the upper flame core growing, the lower flame core shrinking and both of them moving upward, then the steam generation decreases more rapidly.

The results of air redistribution in dynamic states are shown in Fig. 76.9. As these works effectively reduces GCFR change, the Trps rises up to no more than 530 °C in load up process, and the Tfr declines down to no less than 480 °C in load down process. These works not only improve the security of boiler and turbine, but also speed up the power load changing process of coal fired thermal power unit. Though the NO_x at furnace outlet increases over 200 mg/m^3 for about 3 min, its maximum value is less than 300 mg/m^3, in these load changing process. It means that the NO_x emission can still feed the standard limit (100 mg/m^3) with the assistance of selective catalytic reduction (SCR) system.

76.5 Conclusions

To satisfy the demand of power grid, thermal power unit should change its power load as rapid as possible in dynamic process. To accelerate the unit power load change means to increase/decrease boiler steam flow rapidly. Excessive GCFR changing causes unsafety steam temperature fluctuation, in traditional control strategy. Combustion optimization is the other controllable way to accelerate changing of boiler steam flow, avoiding unsafety steam temperature. It consists of two works, matching boiler steam generating rate with unit power load in static states, and balancing steam temperature fluctuations with NO_x emissions in dynamic states.

Fig. 76.9 Boiler dynamic performances after air redistribution

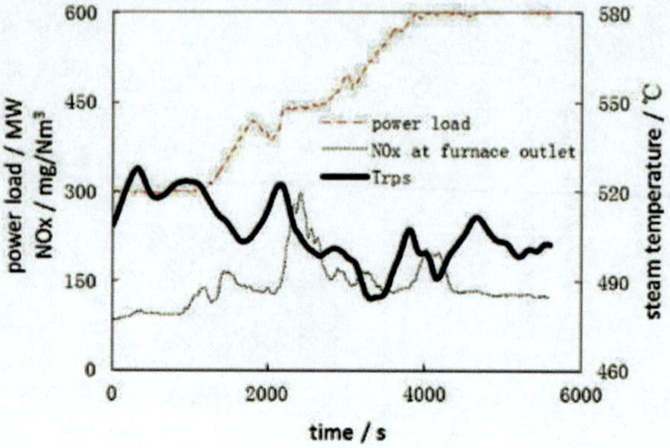

Load up process

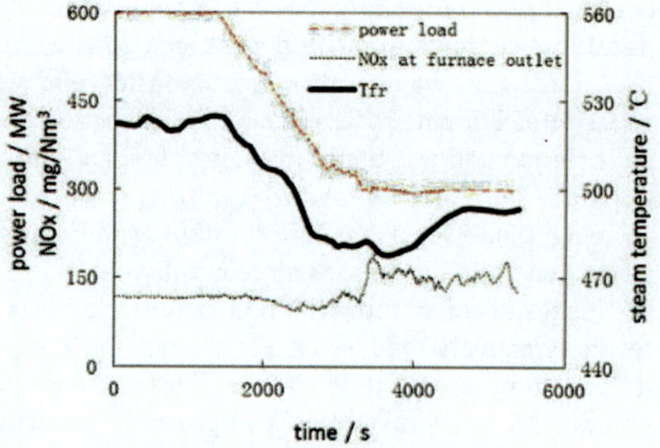

Load down process

Works in static states improve boiler static performance. In addition, the thermal unit becomes able to change its power load in a speed of no less than 2% rated capacity. Works of air redistribution are quite effective for steam temperature control especially in dynamic process with great load changing speed, as they effectively reduces GCFR change with an acceptable NO_x increasement at furnace outlet. These works are benefit to the flexibility of coal-fired thermal power plant.

Acknowledgements This work was supported by the Key Project of Natural Science Foundation of China (No. 29936090) and the Special Funds for Major State Basic Research projects (G1999022102).

References

Alzaharnah I, Habib MA, Elshafei M et al (2013) Control of the boiler swing rate for NO emission minimization. Energy Fuels 27(10):6079–6086

Bhambare KS, Mitra SK, Gaitonde UN (2007) Modeling of a coal-fired natural circulation boiler. J Energy Resour Technol Trans ASME 129(2):159–167

Cao H-J, Tang B-G, Xu H-B (2004) Influence of coal bias combustion mode on NOx emission from power station boiler. Proc Chin Soc Electr Eng 24(8):229–232

Dıez LI, Cortes C, Pallares J (2008) Numerical investigation of NO_x emissions from a tangentially-fired utility boiler under conventional and overfire air operation. Fuel 87:1259–1269

Jing J, Li Z, Zhu Q et al (2011) Influence of primary air ratio on flow and combustion characteristics and NOx emissions of a new swirl coal burner. Energy 36:1206–1213

Kuang M, Li Z, Ling Z et al (2014) Characterization of coal combustion and steam temperature with respect to staged-sir angle in a 600 MWe down-fired boiler. Energy Fuels 28:4199–4205

Li M, Wang X, Sun S et al (2015) Influence of overfire air jet form on low NO_x retrofit effect of an opposed firing boiler. J Chin Soc Power Eng 35(4):263–369

Ministry of Environmental Protection of the People's Republic of China & General Administration of Quality Supervision, Inspection and Quarantine of the People's Republic of China (2011) Emission standard of air pollutants for thermal power plants

Ying M, Dai C, Hu W et al (2011) Low NO_x burner retrofit and operation adjustment of 600 MW opposed wall firing boiler. Electr Power 44(4):55–58

Zhou H, Mo G, Si D et al (2011) Numerical simulation of the NO_x emissions in a 1000 MW tangentially fired pulverized coal boiler: influence of the multi-group arrangement of the separated over fire air. Energy Fuels 25(5):2004–2012

Chapter 77
Study on Modeling and Control Strategy for Combustion Optimization of Pulverized Coal Boiler

Miao Liu, Gengda Li, Xin Wang, and Baowei Chen

Abstract Pulverized coal boilers are often in a state of frequent changes in load, coal type, combustion and other parameters. The generation and removal of nitrogen oxides are closely related to the operating conditions of coal-fired units. Therefore, it is of great economic and environmental significance to study the generation and removal characteristics of nitrogen oxides under full operating conditions of coal-fired units. In this paper, combined with the formation mechanism of nitrogen oxides in coal-fired boilers, the long-term short-term memory neural network model is used to predict the formation of nitrogen oxides in 660 MW front-wall and back-wall pulverized coal fired boilers. Then the optimization of burner switching in variable load process is studied.

Keywords NO_X · Mechanism analysis · Long short term memory · Boiler combustion optimization · Burner switching optimal

77.1 Introduction

In recent years, with the increasingly serious environmental problems, a large number of researchers have carried out relevant research on the generation characteristics of NO_X in coal-fired units. The study of NO_X generation characteristics of coal-fired units can be carried out through boiler test and numerical simulation respectively. In the initial stage of the study, the influence of combustion parameters on the characteristics of NO_X formation was mainly studied through experimental study. The experimental study can directly and effectively obtain reliable experimental data, such as the combustion adjustment test of coal-fired units. However, due to the characteristics of high ash content and high combustion temperature in the flue gas of coal-fired units, it is difficult for the test and measurement equipment to run for a long time because of the harsh measuring environment. At the same time, the general

M. Liu · G. Li (✉) · X. Wang · B. Chen
Guodian New Energy Technology Research Institute Co., Ltd, 2 Street, Technology Park North, Changping District, Beijing 102209, China
e-mail: gengda.li@chnenergy.com.cn

J. Lyu and S. Li (eds.), *Clean Coal and Sustainable Energy*, Environmental Science and Engineering,
https://doi.org/10.1007/978-981-16-1657-0_77

test and measurement results are also local point measurements, so it is impossible to measure the parameters of the furnace as a whole, and it is difficult to test and analyze directly in some extreme conditions to ensure the safety of the equipment. The numerical simulation research can obtain detailed data of combustion related physical quantities, and gradually become an effective means to study the generation characteristics of NO_X in power plant boilers. At the same time, it can simulate some extreme conditions and analyze combustion conditions in detail. Previous studies on NO_X generation in furnaces mainly focus on two aspects: the NO_X generation model and the influence of boiler combustion operating conditions on NO_X generation.

Choi and Chang (2009) established a full-scale boiler combustion, heat transfer, flow and NO_X generation model considering superheater, reheater and economizer for 500 MW tangentially fired boilers, and studied the effects of air staged combustion on the generation characteristics of thermal NO_X and fuel-type NO_X. Dez et al. (2008) By comparing the temperature field in the furnace with the NO concentration field and the experimental values, the effects of exhaust air on combustion and NO_X formation in the furnace were studied. Yaming et al. (2015) simulated the NO_X generation characteristics and fly ash carbon content of 660 MW impact coal-fired boiler, and studied the effects of burnout rate and burner mode on NO_X formation and carbon content in fly ash. Zhengyang et al. (2009) studied the effects of excess air ratio, burnout air speed, combustor combination and boiler load on the generation characteristics of NO_X. Jun et al. (2009) studied the influence of coal quality and pulverized coal particle size on the formation characteristics of NO_X.

On the other hand, coal-fired units are often in the state of frequent changes in load and combustion parameters, so it becomes a new research topic to study the influence of load fluctuation or real-time changes of combustion parameters on the generation characteristics of NO_X. Because of the large size of the boiler structure, the complex combustion process and the long duration of the transient process (minute level), it takes too long to simulate the generation of NO_X in CFD transient state, which makes it difficult to simulate the generation characteristics of NO_X in real time. In terms of time scale, the process of NO_X generation in combustion is much smaller than that in transient state. Therefore, real-time NO_X generation can be studied by the existing methods of data collation and analysis, and the prediction of NO_X emission in the whole process can be obtained NO_X generation data analysis method in furnace is mainly carried out by two methods at present, one is artificial neural network and modified network method, the other is support vector machine method.

Tronci et al. (2002) analyzed the operation data of a 4.8 MW circulating fluidized bed boiler using an artificial neural network model, and studied the effects of boiler load, fuel, air volume and exhaust air rate on pollutant emissions. Zhou et al. (2004) predicted and studied the emission characteristics of NO_X from tangentially fired pulverized coal boilers using artificial neural network method, and optimized the combustion control parameters. Tan et al. (2016) used artificial neural network method and genetic algorithm to predict NO_X emission of 210 MW pulverized coal combustion boiler, and optimized combustion control parameters. Si et al. (2009) analyzed the relationship between operation parameters and NO_X emission of 700 MW pulverized coal fired boiler by using extreme learning machine method. The results of this method are more accurate than those of boiler operation data.

Zhou et al. (2010) used support vector model to predict NO_X emission from boiler combustion and genetic algorithm to optimize boiler combustion operation parameters. By improving the support vector model, Zhou et al. (2012) predicted and studied the relationship between boiler NO_X emission and combustion conditions. Yang et al. (2016; Liukkonen et al. 2011) used support vector model to predict NO_X emission of 300 MW boiler, and used particle swarm optimization to optimize combustion operation parameters. Smrekar et al. (2013) monitored the NO_X emission in the transient process of CFB boilers by self-organizing graph method, and predicted the NO_X emission in the transient process of CFB boilers by linear model and non-linear model. The predicted results were in good agreement with the measured results.

In addition to the above two main analysis methods, some scholars also use other methods to predict the generation of NO_X. Ilamathi et al. (2013) proposed a generalized clustering neural network method to predict and study the influence of combustion parameters on NO_X emission in 300 MW pulverized coal fired boiler. Zheng et al. (2009) used autoregressive model, neural network model and vector regression model to analyze the prediction performance of NO_X emission characteristics of coal-fired boilers.

In summary, the predecessors have done a lot of research on the generation of NO_X in boilers, but it can be found that the research mainly focuses on the generation characteristics of NO_X in single furnace. Coal-fired unit is a huge and complex system. The change of NO_X generation in unit will lead to the change of NO_X content at the entrance of SCR denitrification, which will affect the denitrification characteristics. However, previous studies on the generation and removal of NO_X from coal-fired units are relatively scarce. For a 660 MW front and rear wall hedge boiler, this paper studies the generation characteristics of NO_X from two aspects: steady state and transient process.

77.2 Formation Mechanism of NO_X in Pulverized Coal Combustion Process

The pulverized coal combustion process of coal-fired units includes a series of complex physical and chemical processes. The nitrogen oxides produced by combustion mainly refer to NO and NO_2, in addition to a small amount of NO (Yaoji 2011). Among them, NO accounted for more than 95%, NO 25%,N_2O only accounted for about 1% (Shien and Zhengning 2008; Weiping 2008). According to the source and formation of nitrogen, NO_X produced by pulverized coal combustion can be divided into three types: thermal type, fuel type and fast type.

(1) Thermal type NO_X

Thermal type NO_X is mainly caused by the destruction of N≡N bond and oxidation of N_2 in air at high temperature. Its formation process can be described by the Zeldovich mechanism (Zeldovitch 1964) proposed by Zeldovich, a former Soviet scientist.

$$O_2 + M \Leftrightarrow 2O + M \tag{77.1}$$

$$N_2 + O \Leftrightarrow N + NO \tag{77.2}$$

$$N + O_2 \Leftrightarrow NO + O \tag{77.3}$$

When the temperature is higher than 1800 K, thermal NO_X is produced in large quantities, and the relationship between the formation rate of thermal NO_X and temperature can be described by Arrhenius law (Kefa 2002):

$$\frac{d[NO]}{dT} = 3 \times 10^{14} e^{-542000/(RT)} [O_2]^{1/2} [N_2] \tag{77.4}$$

where: $[NO]$ is the concentration of NO (mol · m^{-3}); R is the gas molar constant (J · mol^{-1} · K^{-1}); T is the reaction thermodynamic temperature (K); $[O_2]$ is the concentration of O_2 (mol · m^{-3}); $[N_2]$ is the concentration of N_2 (mol · m^{-3}).

In the fuel-rich state, the following reactions occur during the formation of thermodynamic NO_X:

$$N + OH \Leftrightarrow NO + H \tag{77.5}$$

At present, the combustion of coal-fired units in the main combustion zone of boilers adopts air staged combustion to ensure low NO_X generation. Near the main combustion zone, the fuel-rich state is considered. The NO generation path from reaction (77.1), (77.2) to (77.5) is considered. At this time, the NO generation value is:

$$\frac{d[NO]}{dT} = 3 \times 10^{16} T^{1/2} e^{-69090/T} [O_2][N_2] \tag{77.6}$$

From the above formula, it can be seen that the formation of thermodynamic NO_X has nothing to do with the N in fuel. The combustion temperature and the concentration of N_2 and O_2 in the furnace are the main influencing factors. Under different working conditions of coal-fired units, the N_2 concentration in the furnace is basically unchanged, while the O_2 concentration and furnace temperature change significantly. Under variable operating conditions, the thermal NO_X generation of coal-fired units is mainly affected by the concentration of oxygen in the main combustion zone and combustion temperature. From formula (77.4) and formula (77.6), we can see that temperature has a more significant effect on it, which shows an exponential relationship. Therefore, the formation of thermal NO_X in coal-fired units can be controlled by reducing the concentration or temperature of O_2 in the main combustion zone of the furnace, such as staged combustion and flue gas recirculation.

(2) Fast type NO_X

Fast type NO_X mainly occurs under fuel-rich conditions. The hydrocarbon ion group CHi produced by coal combustion impacts N_2 molecules to form HCN and CN, then reacts to form NHi, and then oxidizes to form fast NO_X (Baitao 2014). For coal-fired units, the proportion of fast-type NO_X is very small compared with thermal and fuel-type NO_X, which only accounts for less than 5% of the total generated NO_X.

(3) Fuel type NO_X

Fuel type NO_X is mainly formed by oxidizing nitrogen compounds in coal. The formation of fuel type NO_X occurs mainly in the two stages of combustion of pulverized coal and coke. Most of the N-containing organic compounds in the volatile components precipitate with the volatile components, such as HCN and NH_3. These N-containing organic compounds are subsequently oxidized to form NO_X, with these two substances as intermediate products.

Nitrogen in fuel mainly exists in the form of $C{\equiv}N$ (bond energy 791 kJ $\cdot$ mol^{-1}) and N–C (bond energy 450 kJ $\cdot$ mol^{-1}). It is easier to decompose and oxidize than N–N (bond energy 945 kJ $\cdot$ mol^{-1}) in air. This is also the reason why the formation rate of fuel-type NO_X is much higher than that of thermodynamic type.

It is found that among the three nitrogen oxides produced during pulverized coal combustion, thermal type NO_X accounts for 15–20% of total nitrogen oxides, fast type NO_X accounts for 0–5% and fuel type NO_X accounts for 75–80% (Kefa 2004). Therefore, the research on the generation characteristics of NO_X in coal-fired units can mainly focus on the thermal type NO_X and fuel type NO_X.

77.3 NO_X Generation in Transient Process of Coal-fired Units

In order to eliminate load fluctuation in power grid, coal-fired units frequently fluctuate over a long period of time, which has a significant impact on the generation of NO_X. In this section, aiming at the study of the characteristics of NO_X generation in the transient process of coal-fired units, the prediction model of NO_X generation in the transient process is established, and the prediction of NO_X generation in the variable load process is carried out.

77.3.1 Long Short-Term Memory

The mechanism analysis and Long Short-Term Memory (LSTM) are combined to analyze the dynamic characteristics of boiler emission and boiler efficiency, and a non-linear dynamic model of thermal power unit including dynamic description of boiler combustion adjustment is established. Based on this model, a coordinated multi-objective predictive control strategy for boiler combustion optimization with low NO_X emission and high efficiency is proposed. The simulation results illustrate that the control strategy can effectively improve the operational flexibility of thermal power units and achieve the goal of energy saving and emission reduction for coal-fired boilers.

Figure 77.1 shows the operation process of LSTM structure. In the forward propagation process of LSTM, as an intermediate state LSTM unit, the input data includes not only new input, but also the output of the previous moment. Firstly, the input data is pretreated as input layer $x(t)$, and the hidden state $h(t-1)$ and input sequence data $x(t)$ of the input hidden layer of the forgetting gate are updated. The output $f(t)$ of the forgetting gate is obtained through an activation function, commonly sigmoid. Since the output $f(t)$ of sigmoid is between [0, 1], the output $f(t)$ here represents the probability of forgetting the hidden cell state of the upper layer. Mathematical expressions are as follows:

$$f(t) = \sigma(W_f[h(t-1), x(t)] + b_f) \tag{77.7}$$

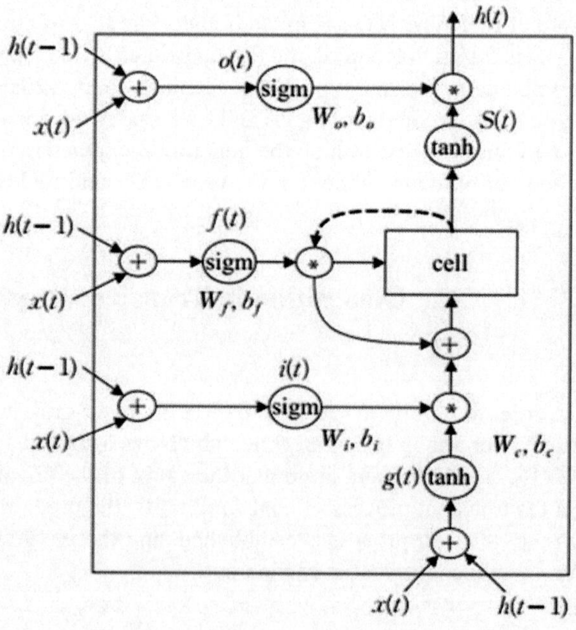

Fig. 77.1 LSTM algorithm flow

where, W_f, b_f are coefficients and biases of linear relationship, σ is sigmoid activation function.

77.3.2 NO_X Production Prediction Model for Unit Transient Process

In theory, the transient simulation of NO_X generation in the transient process of coal-fired units can be carried out by means of CFD numerical simulation. However, due to the complex physical and chemical reactions involved in the pulverized coal combustion problem, the time-consuming numerical simulation calculation is enormous. At the same time, the transient process of coal-fired units is generally minute in time scale, and the time step of transient process numerical simulation method is usually milliseconds. It is difficult to simulate the characteristics of NO_X generation in the whole transient process by CFD method.

The input parameters of the network are input to the various parameters needed for combustion, including the coal feed rate of each burner, the primary air flow rate, the secondary air flow rate and the separated overfire air flow rate. The functional relationship between combustion input parameters and output parameters is established by the neural network model. The value of NO_X generation in the transient process is predicted by the variation of combustion input parameters.

The time scale of NO_X generation process is much smaller than that of transient process. Transient process can be simulated by global quasi-steady state process. This paper collects and analyses the relevant operation data of the coal-fired boiler NO_X generation for 50 h (once per minute). The input coal quantity, primary air quantity, secondary air quantity and exhausted air quantity of each burner are arranged as the input parameters of the model. At the same time, the corresponding measured NO_X concentration at each time point is trained as the output parameters to ensure the training model. In this paper, the first two-thirds of the collected data are used for model training, and the last third of the data are used for model validation. The calculation results are shown in Fig. 77.2. It can be found that the training network is reliable as a whole, and the training part and the calibration part have high follow-up and accuracy. At the same time, the trend of NO_X generation and prediction in the transient process is consistent with the trend of acquisition value. The specific error analysis of the collected discrete values shows that the error of training data is less than 10%, and the error of calibration part is between −18%~+16%, which has high accuracy.

Fig. 77.2 LSTM model verification

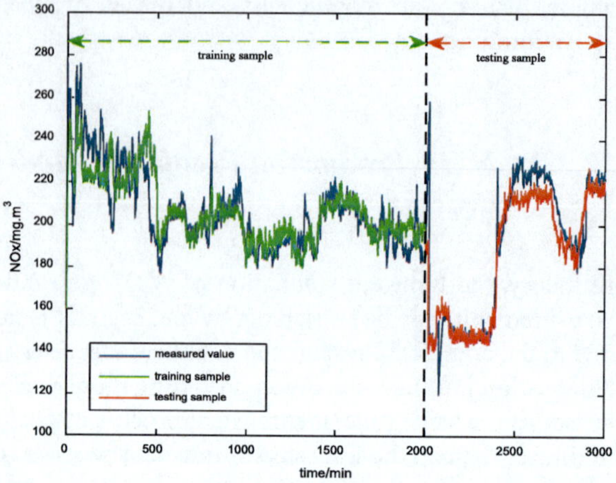

77.4 Optimal Control of Burner

77.4.1 Load Function Fitting

In order to ensure stable combustion and low nitrogen combustion of coal-fired units, the number of burner layers should be adjusted according to load fluctuation. In this paper, the boiler is equipped with six layers of burner. From bottom to top, which are named A layer, F layer, C layer, E layer, D layer and B layer respectively, as shown in Fig. 77.3 Fire detection and surveillance. They are equipped with two layers and

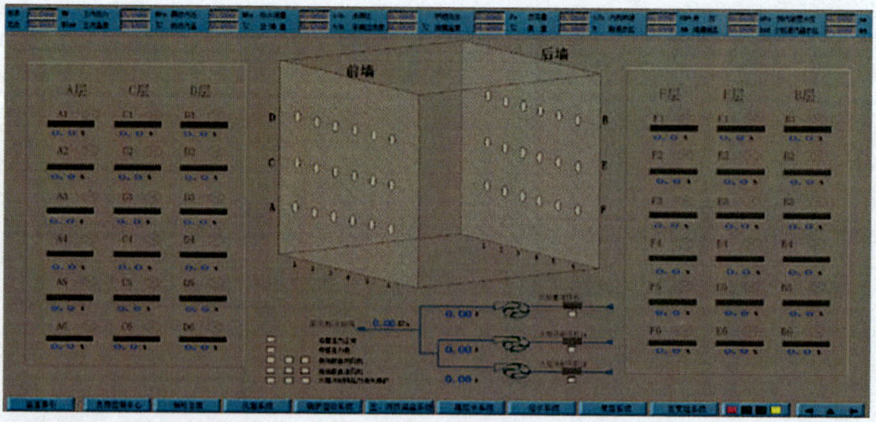

Fig. 77.3 Fire detection and surveillance

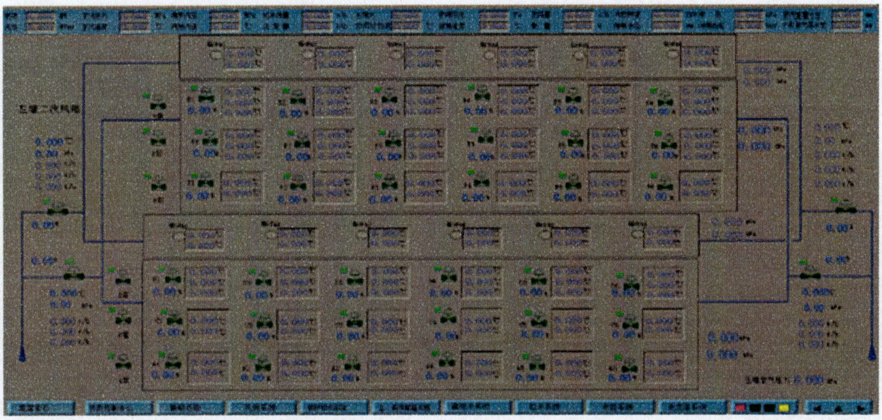

Fig. 77.4 Burner auxiliary throttle control

six SOFA air vents on the top of burner. They are named Left-SOFA-I, Left-SOFA-II, Left-SOFA-III, Right-SOFA-I, Right-SOFA-II, Right-SOFA-III respectively, as shown in Fig. 77.4 Burner auxiliary throttle control.

According to the data collected in the distributed control system (DCS), the quadratic function fitting is carried out for the relationship between coal feed and load, primary air and load, secondary air and load, and separated overfire air (SOFA) and load respectively.

(1) Coal feeder:

As shown in Fig. 77.5, the scatter points represent the distribution of coal quantity and load of A-layer burner, and the blue solid line is the quadratic function curve of coal quantity and load of A-layer burner.

The coal quantity and load function of each layer burner are as follows:

$$f_A(x) = 0.0001735x^2 - 0.1105x + 52.45$$

$$f_B(x) = 0.0002643x^2 - 0.1992x + 67.6$$

$$f_C(x) = 0.0001489x^2 - 0.08741x + 46.2$$

Fig. 77.5 Coal load distribution curve

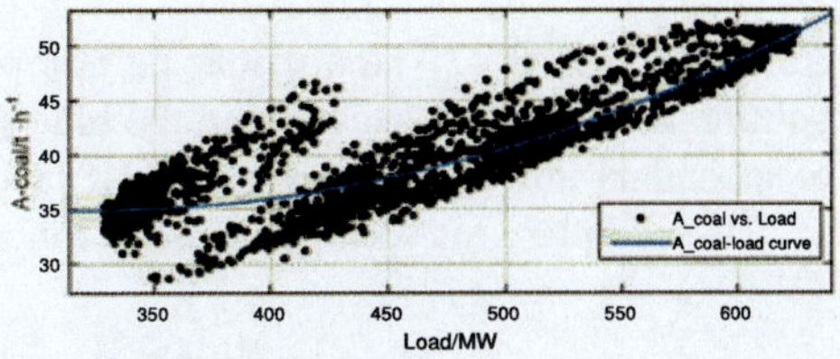

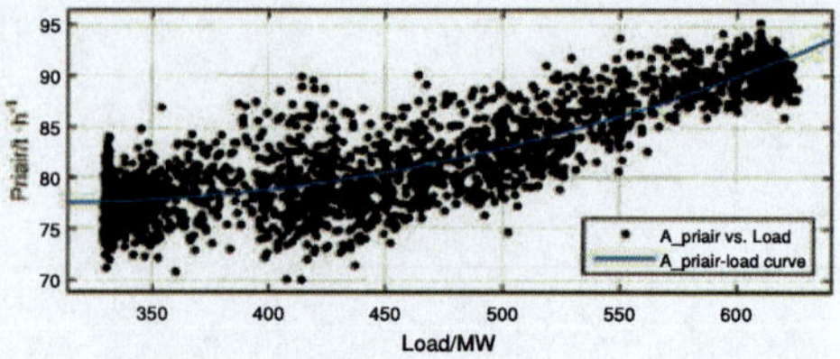

Fig. 77.6 Primary air load distribution curve

$$f_D(x) = 0.0000538x^2 + 0.04842x - 5.041$$

$$f_E(x) = 0.0001568x^2 - 0.09257x + 46.95$$

$$f_F(x) = 0.0001862x^2 - 0.1224x + 54.31$$

(2) Primary air:

As shown in Fig. 77.6, the scatter points represent the distribution of primary air volume and load in Layer A. The blue solid line is a quadratic function curve of primary air volume and load in Layer A.

The primary air volume and load function of each layer are as follows:

$$f_A(x) = 0.0001423x^2 - 0.08677x + 90.75$$

$$f_B(x) = 0.0001804x^2 - 0.1464x + 108.9$$

$$f_C(x) = 0.0001366x^2 - 0.08424x + 91.32$$

$$f_D(x) = 0.0001423x^2 - 0.1237x + 112.2$$

$$f_E(x) = 0.00004857x^2 - 0.0009976x + 72.18$$

$$f_F(x) = 0.00002381x^2 - 0.01143x + 75.53$$

(3) Secondary air:

As shown in Fig. 77.7, the scatter point represents the distribution of secondary air volume and load in L1 layer, and the blue solid line is the quadratic function curve of secondary air volume and load in L1 layer.

The secondary air volume and load function of each layer are as follows:

$$f_{L1}(x) = 0.000543x^2 + 0.0239x + 292.4$$

Fig. 77.7 Secondary air load distribution curve

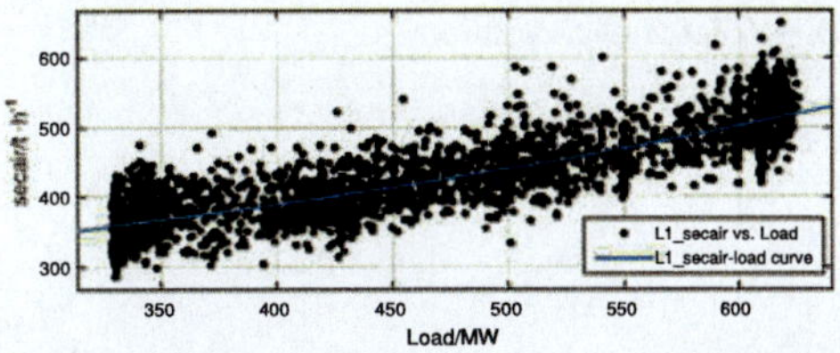

$$f_{L2}(x) = 0.0003956x^2 + 0.1147x + 331.1$$

$$f_{L3}(x) = 0.0004751x^2 + 0.08047x + 282$$

$$f_{R1}(x) = 0.00022x^2 + 0.7628x + 161.9$$

$$f_{R2}(x) = 0.0001223x^2 + 0.923x + 90.18$$

$$f_{R3}(x) = -0.00002701x^2 + 1.181x - 63.89$$

(4) SOFA air:

As shown in Fig. 77.8, the scatter point represents the distribution of SOFA air volume and load in L1 layer, and the blue solid line is the quadratic function curve of SOFA air volume and load in L1 layer.

$$f_{L1}(x) = 0.0005328x^2 + 0.1116x + 9.038$$

$$f_{L2}(x) = 0.0004745x^2 + 0.1918x - 22.6$$

$$f_{L3}(x) = 0.0004691x^2 + 0.201x - 25.19$$

$$f_{R1}(x) = -0.0008335x^2 + 1.471x - 309.4$$

Fig. 77.8 SOFA air load distribution curve

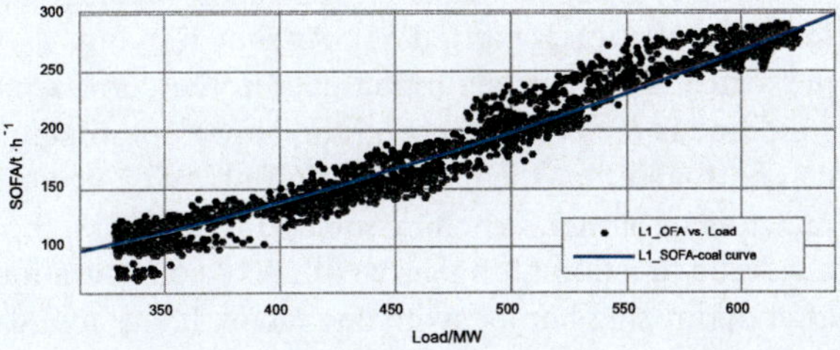

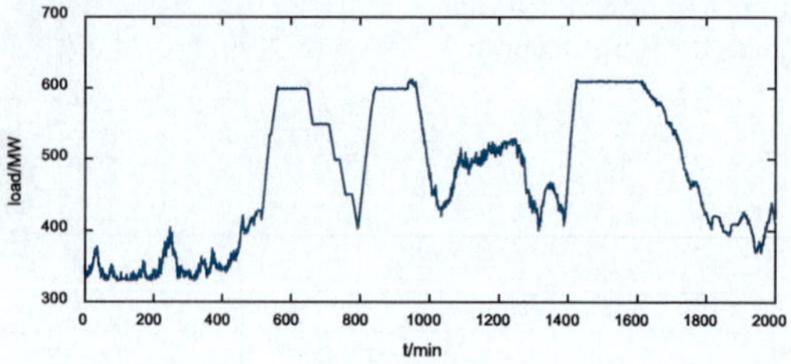

Fig. 77.9 Load variation

$$f_{R1}(x) = -0.0008015x^2 + 1.413x - 291.6$$

$$f_{R3}(x) = -0.0007691x^2 + 1.392x - 285.3$$

77.4.2 Optimal Scheme of Burner Switching in Variable Load Process

Load variation is shown in the Fig. 77.9. In order to ensure stable combustion and low nitrogen combustion, primary and secondary air of coal-fired units change according to fitting function, and the number of burner layers needs to be adjusted accordingly with load fluctuation.

Six-layer burner and two-layer SOFA air are put into operation at 90% BMCR.

Under 75% BMCR condition, the upper five-layer burner is put into operation. The number of operation layers is F layer, C layer, E layer, D layer and B layer burners, and the number of SOFA air operation vents are Left-SOFA-II, Left-SOFA-III, Right-SOFA-I, Right-SOFA-II, Right-SOFA-III.

Under 60% BMCR condition, the upper four layers of burner are put into operation. The number of layers of burner is C layer, E layer, D layer and B layer. The number of vents of SOFA is Left-SOFA-II, Left-SOFA-III, Right-SOFA-II, Right-SOFA-III.

Under 50% BMCR condition, the upper three layers of burners are put into operation. The number of layers of burners is E, D and B. The number of vents of SOFA is Left-SOFA-III, Right-SOFA-II, Right-SOFA-III.

Through the established long and short memory neural network model, the simulation calculation is carried out. As shown in Fig. 77.10, the blue dot is divided into the actual NO_X concentration data in the corresponding time period, and the red solid line is the NO_X concentration after optimizing the burner switching mode. It can be found that the concentration of NO_X decreases as a whole after the optimization of burner switching mode. Figure 77.11 shows the scatter distribution and quadratic function fitting curve of NO_X concentration varying with load before and after optimizing burner switching mode. It was found that the average concentration

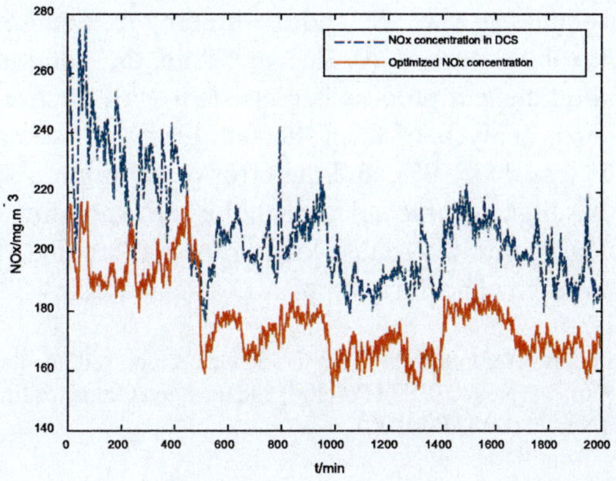

Fig. 77.10 Comparison of NO_X concentration before and after optimization

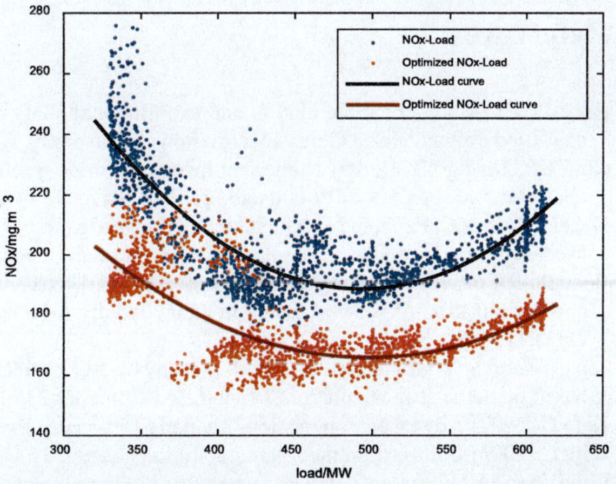

Fig. 77.11 Comparison of NO_X concentration under different loads

of NO_X decreased by 13.87% after optimization.

$$f_L(x) = 0.0004745x^2 + 0.1918x - 22.6$$

$$f_R(x) = -0.0008015x^2 + 1.413x - 291.6$$

77.5 Summary

In this study, we take a 660 MW front and rear wall pulverized coal boiler as the research object, the Long Short Term Memory neural network model is used to predict the generation of nitrogen oxides, it can be found that the training network is reliable as a whole, and the training part and the calibration part have high follow-up and accuracy. At the same time, the trend of NO_X generation and prediction in the transient process is consistent with the trend of acquisition value. The specific error analysis of the collected discrete values shows that the error of training data is less than 10%, and the error of calibration part is between −18%~+16%, which has high accuracy. And then the corresponding optimization of the burner switching load point in variable load process is studied. The simulation results show that the average concentration of NO_X decreased by 13.87% after optimization.

Acknowledgements This work was supported by the China Energy Science and Technology Project (No. 2018E2TP00200) and the Seed Funds for Jinqiao Project of Beijing Science Association (No. 2018B3TP00100).

References

Baitao (2014) Numerical simulation and experimental study of low NOx combustion system for coal-fired boilers. North China Electric Power University, Beijing

Choi CR, Chang NK (2009) Numerical investigation on the flow, combustion and NOx emission characteristics in a 500 MW tangentially fired pulverized-coal boiler. Fuel 88(9):1720–1731

Díez LI, Cortés C, Pallarés J (2008) Numerical investigation of NOx emissions from a tangentially-fired utility boiler under conventional and overfire air operation. Fuel 87(7):1259–1269

Ilamathi P, Selladurai V, Balamurugan K et al (2013) ANN-GA approach for predictive modeling and optimization of NOx emission in a tangentially fired boiler. Clean Technol Environ Policy 15(1):125–131

Jun L, Weiping Y, Chunyan L et al (2009) Study on NO x release from pulverized coal fired boilers based on numerical calculation. J Electr Eng China 29(23):13–19

Kefa C (2002) Advanced combustion. Zhejiang University Press, Hangzhou

Kefa C (2004) Combustion theory and pollution control. Machinery Industry Press, Beijing

Liukkonen M, Hiltunen T, Hälikkä E et al (2011) Modeling of the fluidized bed combustion process and NOx emissions using self-organizing maps: an application to the diagnosis of process states. Environ Modelling Softw 26(5):605–614

Shien H, Zhengning Z (2008) Clean utilization of coal and pollution control. China Electric Power Press, Beijing

Si F, Romero CE, Yao Z et al (2009) A new approach for function approximation in boiler combustionoptimization based on modified structural AOSVR. Expert Syst Appl 36(4):8691–8704

Smrekar J, Potočnik P, Senegačnik A (2013) Multi-step-ahead prediction of NOx emissions for a coal-based boiler. Appl Energy 106(11):89–99

Tan P, Xia J, Zhang C et al (2016) Modeling and reduction of NOx emissions for a 700 MW coal-fired boiler with the advanced machine learning method. Energy 94:672–679

Tronci S, Baratti R, Servida A (2002) Monitoring pollutant emissions in a 4.8MW power plant through neural network. Neurocomputing 43(1):3–15

Weiping Y (2008) Clean coal power generation technology. China Electric Power Press, Beijing

Yaming L, Fangyong L, Qisheng X et al (2015) Numerical simulation of NOx emission characteristics of 600 MW counter-fired boilers. J Power Eng 35(05):341–347

Yang T, Cui C, Shen Y et al (2016) A novel denitration cost optimization system for power unit boilers. Appl Therm Eng 96:400–410

Yaoji C (2011) W-type flame boiler burning anthracite with low NOx combustion mechanism and modular experimental study. Zhejiang University, Hangzhou

Zeldovitch J (1964) The oxidation of nitrogen in combustion and explosions. Acta Physicochimi 4:21

Zheng LG, Zhou H, Cen KF et al (2009) A comparative study of optimization algorithms for low NOx combustion modification at a coal-fired utility boiler. Expert Syst Appl 36(2):2780–2793

Zhengyang G, Wei S, Lijun F et al (2009) Numerical simulation of NO emission characteristics of double tangential circle boilers for 1000MW ultra supercritical units. J Electr Eng China 29(32):12–18

Zhou H, Cen K, Fan J (2004) Modeling and optimization of the NOx emission characteristics of a tangentially fired boiler with artificial neural networks. Energy 29(1):167–183

Zhou H, Zheng L, Cen K (2010) Computational intelligence approach for NOx emissions minimization in a coal-fired utility boiler. Energy Convers Manage 51(3):580–586

Zhou H, Pei Zhao J, Gang Zheng L et al (2012) Modeling NOx emissions from coal-fired utility boilers using support vector regression with ant colony optimization. Eng Appl Artif Intell 25(1):147–158

Chapter 78
Characteristics of RB Control Loop of Large Lignite Power Plant Boiler and Analysis and Comparison

Y. Zhang, L. Cheng, Q. Zhang, Z. Zhao, and M. Gao

Abstract In recent years, lignite has been widely used as an important reserve energy in China. In view of the characteristics of lignite, such as high moisture content, high volatile content and low calorific value, in order to achieve the best expected effect, a plant has carried out many RB experiments. Reasonable use of the boiler itself heat storage, through constant adjustment of water, coal and other inertial links, so that the unit safety and stability to low load. According to the characteristics of lignite boiler, a series of RB logic rationalization control suggestions are put forward.

Keywords Lignite boiler · RB · Thermal automatic control

The purpose of RB test is to test the adaptability of unit and control system under fault condition, and to test the operation ability of unit under fault condition and the performance and function of control system. The realization of RB function provides safety guarantee for unit under highly automated operation mode, makes all parts of unit coordinate quickly and steadily, ensures the safe, stable and economic operation of unit, and avoids it. The impact of load on power grid (Yan 2003). Reasonable design and normal operation of unit RB function are of great significance to the safe and stable operation of power plant and even the whole power grid (Wang et al. 2011).

In our country, RB tests of different units are analyzed and studied for fuel, feed water and steam pressure regulation methods (Liu 2010; Cai et al. 2003; Zhang et al. 2011; Hao 2012). The research done by foreign advanced scholars on the methods of improving RB reliability of auxiliary equipment of once-through circulation boiler has provided us with a lot of valuable experience. The RB test of 660 MW lignite boiler was carried out three times. The first condition was single lignite condition, the second and third condition was lignite blending combustion mode. The parameters were compared and the optimization space was found, which provided the parameter optimization space for better accident treatment of the unit.

Y. Zhang · L. Cheng · Q. Zhang · Z. Zhao (✉) · M. Gao
North China Electric Power Research Institute Co., Ltd, Beijing 100045, China
e-mail: zhaozn@sina.com

J. Lyu and S. Li (eds.), *Clean Coal and Sustainable Energy*,
Environmental Science and Engineering,
https://doi.org/10.1007/978-981-16-1657-0_78

Table 78.1 Main technical parameters of boiler

Item	Unit	BMCR
Superheated steam flow	t/h	2117
Superheated steam pressure	MPa, g	29.40
Superheated steam temperature	°C	605
Feed water temperature	°C	309
Reheated steam flow rate	t/h	1748
Reheated steam inlet pressure	MPa, g	5.789
Reheated steam outlet pressure	MPa, g	5.589
Reheat steam inlet temperature	°C	359
Reheated steam outlet temperature	°C	623

78.1 Systematic Introduction

In a coal-electricity integration project of a plant, 2 × 660 MW ultra-supercritical air-cooled unit, the boiler type is B&WB-2117/29.4-M. Boiler design of ultra-supercritical parameters, variable pressure once-through furnace, single furnace, front and rear wall hedging combustion. There are three layers of front wall (A/B/C grinding respectively from bottom to top), four layers of rear wall (D/E/F/G grinding respectively from bottom to top), and five burners in each layer. There are 7 OFA nozzles in front and rear walls (14 in total). The main design parameters of the boiler are shown in Table 78.1 below.

Each boiler is equipped with two ANT-2884/1400B secondary movable blade adjustable axial flow induced draft fans produced by Horton Luxury Engineering Co., Ltd. Two Hawton Luxury Engineering Co., Ltd. are manufactured with ANT-2240/1400F secondary movable blade adjustable axial flow primary fan. Two ANN-2450/1400N movable blade adjustable axial-flow blowers manufactured by Luxury Engineering Limited.

The designed and checked coal types of this project are lignite. The specific coal quality is shown in the Table 78.2.

78.2 RB Main Circuit

After the RB trigger, the main loop of related parameters control is:

(1) CCS controlled excision, cut to TF mode, TF forbidden increase 10S.
(2) Sliding pressure mode of unit operation to constant pressure.
(3) The target pressure decreases at a certain rate according to the sliding pressure curve.
(4) RB control 40S, give target coal quantity according to target load, reduce coal quantity according to load reduction rate.

Table 78.2 Analysis of designed and checked coal types for boilers

Item	Symbols	Units	Coal Types
Receipt of base carbon	Car	%	40.89
Receipt of base hydrogen	Har	%	2.77
Receipt of base oxygen	Oar	%	9.87
Received base nitrogen	Nar	%	0.52
Received base total sulfur	St, ar	%	0.21
Receiving base ash	Aar	%	10.24
Receiving base moisture	Mt	%	35.5
Air drying base moisture	Mad	%	9.32
Dry ashless base volatiles	Vdaf	%	48.26
Receiving base low calorific value	Qnet, ar	MJ/kg	14.69

(5) RB retains four coal mills (G-C-F Mill Tripping in turn), with an interval of 7S and a coal reduction rate of 10 t/s.
(6) Temperature at the middle point should be prohibited from increasing or decreasing by 30S, and then adjusted.
(7) Inertia time of setting value for main control of water supply.
(8) Close the overheated, reheated and desuperheated water regulating door (pulse 30 s) and cut it manually. Close 20S.
(9) Setting the upper limit of the speed of the air supply fan, induced draft fan, primary fan regulating door and feed pump.
(10) When MFT or primary fan RB, according to the main steam flow rate and the number of operation units of the induced draft fan, the induced draft fan's movable blade overshoots and closes one opening.
(11) When RB occurs, water supply control, air supply control, air induction control, fuel control and primary air control are all excised for large adjustment deviation protection.

78.3 RB Test Process and Problems Encountered

The plant has carried out three times of Rb test of draft fan and induced draft fan. During the first test, the same kind of coal was burned in the plant (condition 1). The second (condition 2) and third (condition 3) boilers were burned with lignite blending. The bottom grinding (A.B.D.E) used one kind of coal and the upper grinding (C.F.G) used one kind of coal. The coal quality during the test period is shown in the table below.

The test process parameters are as Table 78.3

Table 78.3 Condition 1/2/3 coal quality

Item	Symbols	Condition1	Condition 2		Condition 3	
			C.F.G	A.B.D.E	C.F.G	A.B.D.E
Total moisture content	Mt (%)	32.75	32.41	32	41.19	35.14
Air drying base moisture	Mad (%)	17.06	16.3	15.72	16.88	19.56
Aad	Aad (%)	18.2	19.43	24.12	8.8	17.74
Receiving base ash	Aar (%)	14.75	15.69	19.46	6.23	14.3
Vad	Vad (%)	27.44	28.71	27.32	31.14	27.72
Dry ashless base volatiles	Vdaf (%)	42.38	44.67	45.41	41.9	44.21
Air-drying based total sulfur	St, ad (%)	1.38	1.1	1.2	1.43	0.93
Dry-based total sulfur	St, d (%)	1.69	1.31	1.42	1.72	1.16
Received base total sulfur	St, ar (%)	1.03	0.89	0.97	1.01	0.75
Air Drying Base Hydrogen Value	Had (%)	3.08	3.17	3	3.5	3.07
Calorific value of air-dried base cartridge	Qb, ad (J/g)	18.69	18.99	17.47	22.04	18.38
High calorific value of drying base	Qgr, d (MJ/kg)	18.54	22.54	20.57	26.32	22.72
Receiving base low calorific value	Qnet, ar (MJ/kg)	22.35	13.96	12.76	14.02	13.42
Receiving base low calorific value	Qnet, ar (cal/g)	3291	3339	3050	3352	3208

78.3.1 Condition 1

The set value of unit feed water control is two first-order inertia. The water supply instruction begins to decrease obviously in 7 s. The fastest rate of water supply instruction decline is 1250 t/h. The actual water supply flow begins to decrease after 10 s, and the fastest rate is 1000 t/h. 1 min 32 s feed water flow control is stable, flow rate is 840 t/h (Fig. 78.1).

The outlet temperature of separator began to decrease after 43 s, reached the lowest value 413 C after 1 min 47 s, began to rise after 2 min, and reached the highest value 502 C (8 min 38 s) in the later period.

Table 78.4 The test process parameters are as follows

Item	Units	Condition1		Condition 2		Condition 3	
		Initial value	Process value	Initial value	Process value	Initial value	Process value
Coal quantity	t/h	396	197.8, 27 s	439	240, 23 s	408	229, 21 s
Feed water flow rate	t/h	1874	839, 1 min 32 s	2060	991, 2 min 26 s	1932	1106, 2 min 22 s
Outlet temperature of separator	°C	420	413 (min), 1 min 47 s; 502 (max), 8 min 38 s	421	402 (min) 2 min 26 s; 425 (before MFT) 5 min 5 s	420	407 (min), 2 min 22 s
Superheat	°C	46	40 (min), 1 min 47 s; 127.9 (max), 8 min 38 s	48	29 (min) 2 min 26 s; 50 (before MFT), 5 min 5 s	–	–
Superheated steam temperature	°C	592	574 (min), 3 min; 593 (max), 7 min	552	503 (min), 4 min; 508 (before MFT), 5 min 5 s	586	551 (min), 3 min 22 s; 571, 8 min 22 s
Reheated steam temperature	°C	598	557 (min), 13 min	587	578 (min), 5 min 5 s	593	558 (min), 36 min 52 s
Load	MW	660	288.48 (min), 3 min 13 s; 313, 4 min 16 s	656	331, 3 min 36 s; Min 176, 5 min 8 s	660	632, 26 s, 532; 427, 3 min 15 s
Integrated valve position	%	72.4	38.2 (min), 3 min 13 s; 40, 4 min 16 s	78.8	41, 3 min 36 s; 29, 5 min 8 s	72.1	69.4, 26 s; 58.88, 1 min 57 s; 47.5, 3 min 15 s; 47.7, 4 min 10 s
Main steam pressure setting/actual value	Mpa	28.1	27.133/26.24, 3 min 13 s; 26.29/26.31, 4 min 16 s	27.86	26.01/25.095, 3 min 36 s; 25.668/24.927, 5 min 8 s	27.71	27.71/27.47 (26 s); 26.347/25.812, 1 min 57 s; 25.95/25.29, 3 min 15 s; 25.21/25.21, 4 min 10 s

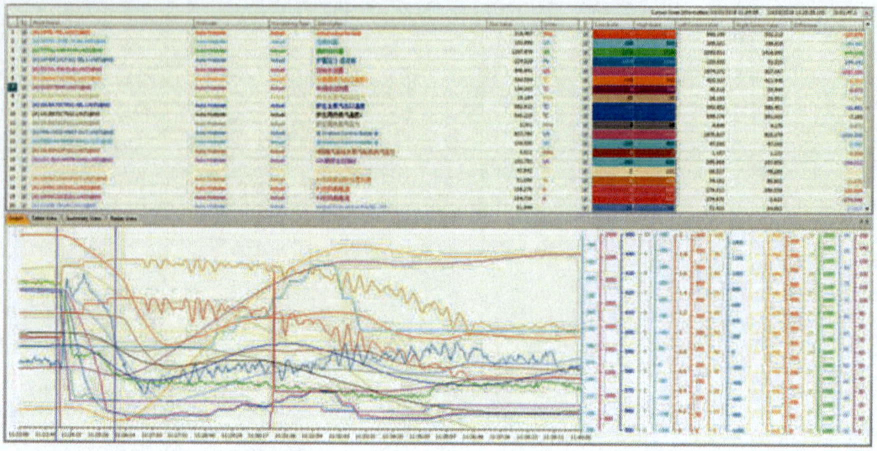

Fig. 78.1 Test curve of condition 1

Load of unit is lowest to 288 MW and comprehensive valve position is lowest to 38%.

RB action, first order inertia (10 s) of set value of main steam pressure. After RB action 2:35, the actual value of main steam pressure of the unit is stable 26.18 MPa. After 4 min of action, the set value of pressure coincides with the actual value.

After 40 s, the main steam temperature of the unit began to drop, reaching the lowest value of 574/560 C in 2 min and 50 s, and rising to 593/577 C in the later period. The lowest reheated steam temperature is 557.

78.3.2 Condition 2

On the basis of condition 1, condition 2 reduces the water reduction rate. The outlet temperature of the separator drops slightly and then turns back. The outlet temperature of the separator does not exceed the temperature (up to 425 C). After RB triggering, the superheated steam temperature drops rapidly and drops 49 C for the maximum 4 min (Fig. 78.2).

After RB triggering, the main steam pressure decreases rapidly, the pressure instruction is always higher than the actual value of the main steam pressure, the comprehensive valve position of the turbine is always closed down, the load of the unit decreases rapidly, and the load is lowest to 176 mw. Because the small turbine steam source is only supplied by four-stage extraction, the steam source pressure at the inlet of the steam pump decreases, the steam pump can not maintain the feed water flow. Boiler tripping, the first feed water flow is low.

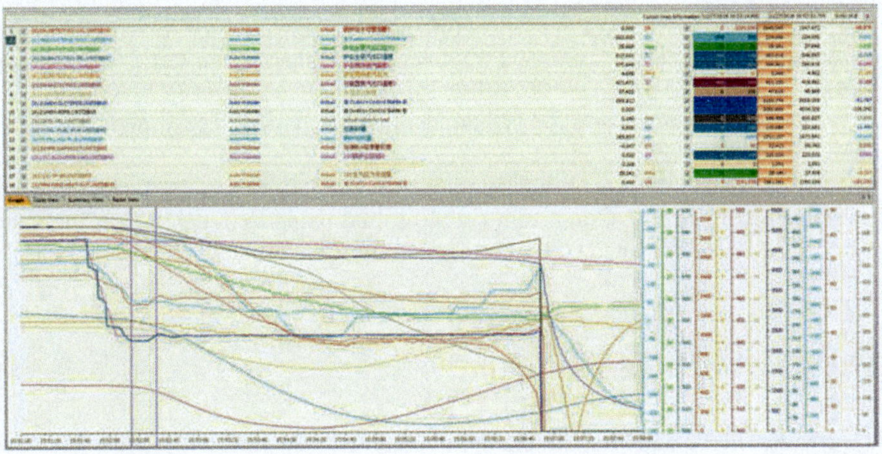

Fig. 78.2 Test curve of condition 2

78.3.3 Condition 3

The set value of unit feed water control is two first-order inertia. The water supply instruction begins to decrease obviously in 7 s. The fastest rate of decline of water supply instruction is 600 t/h. The actual water supply flow begins to decrease after 15 s, and the fastest rate is 600 t/h. When RB is reset, the feed water flow is stable at 1100 t/h (Fig. 78.3).

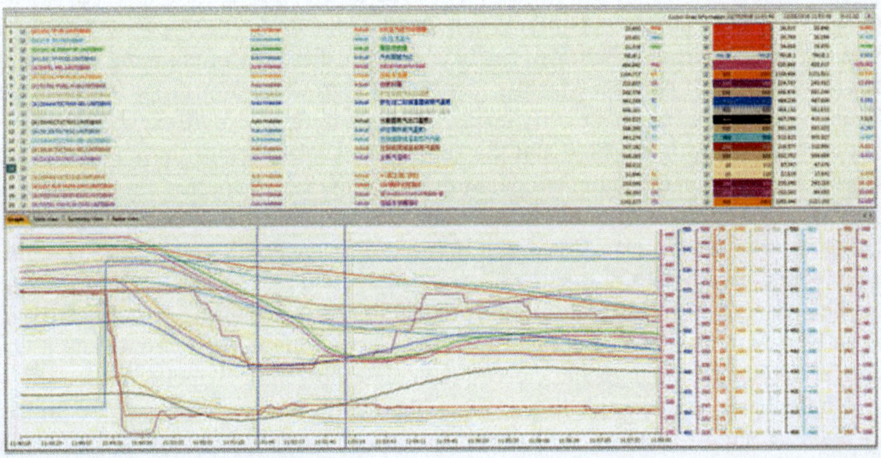

Fig. 78.3 Test curve of condition 3

The outlet temperature of the separator began to decrease after 46 s, reaching the lowest value 407 C after 2 min and 22 s, and the highest value 431 C in the later period.

Load of unit is lowest to 411 MW and comprehensive valve position is lowest to 47%.

After 33 s, the main steam temperature of the unit began to drop, reaching the lowest value of 551/542 C in 3 min and 40 s, and rising to 570/555 C in the later period. The lowest reheated steam temperature is 563.

78.3.4 Result Analysis

(1) Water Inertia and Fuel Matching Regulation

Boiler combustion process belongs to the balance of input and output energy, and the mass balance between feed water and evaporation. The main input parameters are feed water flow, coal flow and total air flow, and the main output energy is steam pressure, temperature, unit load and boiler heat storage. When RB triggers under high load, one-sided auxiliary machine trips and the unit needs emergency load reduction. The unit is reduced from a high load balance point to a low load balance point. Due to the different heat storage of units under different loads, it is necessary to fully release the heat storage from the furnace during the process of unit decreasing from high load to low load. How to safely and steadily drop to another equilibrium point needs to fully adjust the rate of decline between feed water and fuel. The rapid decrease of fuel volume directly leads to the lower load of the unit and the faster decrease of superheated steam temperature. When the unit load drops, the super-steam pump without reliable steam source will easily lead to the difficulty of pumping water and the low tripping of boiler feed water flow. If the feed water flow drops too fast, the boiler heat storage can not be fully released in a short time, resulting in the separator outlet temperature is too high and the superheated steam temperature is too high. The unit trips when the outlet temperature of separator is too high.

The main problems of working condition one are: after RB operation, the water reduction rate of RB test of feed and induced draft fans is relatively too fast, the coal reduction rate does not match the water reduction rate seriously, coupled with the heat storage of the boiler itself, the outlet temperature of the separator decreases slightly at first, and then rises rapidly to 502 C, resulting in the intermediate point temperature being too high and close to the unit temperature. Tripping value.

On the basis of working condition one, the inertia time of feed water control is increased from 15 to 33 s, and the temperature of the middle point of the unit is increased to 425 °C (Table 78.4).

(2) The influence of coal quality on RB test process

Coal quality has great influence on boiler combustion during RB test. Different RB process parameters should be adopted for different coal types. Because the blending burning mode is adopted in the second working condition, because the calorific value of the bottom grinding coal is lower than that of the upper grinding during the blending burning process, the first working condition after RB triggering and the second working condition target coal quantity correspond to 330 MW. Because of the high calorific value of the grinding tripping after RB triggering, the actual bottom grinding coal calorific value can not maintain 330 MW operation, the unit load and excess. Hot steam pressure and superheated steam temperature decrease seriously. The target coal quantity is defined as the coal quantity corresponding to 350 MW load in the third working condition.

(3) Effect of Pressure Curve and Pressure Reduction Rate on Unit Load

After RB triggering, the operation mode of the unit is cut to follow the turbine, and the turbine valve is adjusted to track the target pressure at a certain pressure-reducing rate. The actual pressure of the unit has a great influence on the fuel drop rate of the unit, the underwater drop rate of the unit and the regulation of the turbine regulating valve. The pressure target value can be determined by the corresponding value of sliding pressure curve and load or RB sliding pressure value. After triggered by RB, the actual pressure decreases rapidly, the actual pressure decreases rapidly corresponding to the target load, the sliding pressure curve pressure of the target load decreases gradually, and the steam turbine regulating valve closes rapidly to reduce the pressure drop rate. When the actual pressure of the unit decreases slowly, the actual pressure is higher than the target pressure, and the turbine regulating door will open to maintain the stable operation of the unit. The latter sets the sliding pressure rate of RB and the target value of sliding pressure, which is generally 17–19 MPa. After the trigger of RB, the sliding pressure can be reduced to the target pressure at a certain rate to maintain the stable operation of the current pressure. This unit adopts the first method mentioned above, and adopts the step-by-step pressure regulation method, that is, the pressure drop rate is regulated in two stages: the pressure drop rate is 0.7 MPa/min in the first 100 s and 0.5 MPa/min in the second 100 s after RB, which fully alleviates the influence of excessive pressure drop before RB triggering and the actual pressure drop slowing down in the later period. Because of the great change of blended coal quality and the rapid drop of actual pressure, the third condition is adjusted on the basis of the second condition. Finally, the pressure reduction rate is changed to RB trigger, which is 1.1 MPa/min in the first three minutes and 0.9 MPa/min after three minutes. After 4 min and 10 s, the target pressure is close to the actual pressure.

Acknowledgements The purpose of RB test process of unit is to ensure long-term safe and stable operation of unit. Therefore, the relevant parameters of RB test should consider the condition of long-term operation of unit. Therefore, the parameter adjustment and optimization before RB test and the more stable coal type should be adopted in RB test process. Different RB parameter loops are adjusted according to different coal types and different combustion modes.

References

Cai Y, Xu Z, Zhao W, Qu Y (2003) Runback performance test and study of 350 MW direct-fired boiler. Electr Power 10:80–85 (in Chinese)

Hao Y (2012) Research on RB control strategy of 600MW power unit. Master thesis, North China Electric Power University (in Chinese)

Liu M (2010) Influence of units' characters upon the run back test. Therm Power Gener 39(06):44–47+50 (in Chinese)

Wang D, Xia Y, Qi Z (2011) Design and implementation of runback control strategy in 600MW supercritical once-through unit. Instrumentation. 18(03):81–83 (in Chinese)

Yan G (2003) Research on the coordinated control system of the 300MW unit. Master thesis, Taiyuan University of Technology (in Chinese)

Zhang X, Wu S, Chen W (2010) RB test and common issues of 670 MW supercritical units. Shandong Electr Power 04:74–77 (in Chinese)

Chapter 79
Power Plant Boiler Operation Optimization System Based on CO Control

Y. P. Sun, H. J. Cao, Q. F. Zhang, and C. Y. Liu

Abstract Based on the characteristics of CO emission from power plant boilers, a CO online monitoring system suitable for power plant boilers was determined through discussion of four aspects: technical principle, flue gas treatment mode, system stability and economy, system installation location. Based on the CO emission value and the relevant operating parameters of the boiler, the boiler economic evaluation indicators were established, including boiler efficiency, superheating and reheat steam temperature, desuperheating water and auxiliary mechanical electrical consumption. According to the evaluation result, the CO emission at the air preheater inlet is controlled within a reasonable range by adjusting the secondary air volume. The application results at a 600 MW power plant show that compared with the original oxygen control method, the power plant boiler operation optimization system based on CO control can reduce the coal consumption of power supply by 2 g/(kW h).

Keywords Power plant boiler · CO control · Economic evaluation · Air volume

Foreword

At present, the operation of power station boilers faces two major pressures: under the premise of changing coal quality, the pressure of energy saving and consumption reduction is increasing; with the implementation of new environmental protection standards, the technical transformation of power station boilers is inevitable, and the pressure of emission reduction during boilers operation is getting bigger and bigger. Due to the change of coal quality, the amount of oxygen controlled in boiler operation tends to deviate from the optimal oxygen amount, which tends to cause low thermal efficiency and high NOx concentration. Moreover, in order to reduce NOx emission concentration, low-oxygen combustion is adopted by many power plants, which is easy to cause the increase of CO emission concentration and the reduce of the boiler thermal efficiency.

Y. P. Sun (✉) · H. J. Cao · Q. F. Zhang · C. Y. Liu
North China Electric Power Research Institute CO Ltd., Beijing 100045, China

J. Lyu and S. Li (eds.), *Clean Coal and Sustainable Energy*, Environmental Science and Engineering,
https://doi.org/10.1007/978-981-16-1657-0_79

Compared to combustion operation optimization based on oxygen control, combustion optimization based on CO control in flue gas has proven to be a more efficient optimization method (Xu et al. 1991; Deguchi et al. 2002; Cheng 2003; Zhao et al. 2004; Qiu and Zhang 2004). The CO concentration has a certain relationship with the fly ash combustibles, the heat loss of the flue gas, etc. The monitoring of the CO concentration can keep the boiler running under a low excess air while avoiding slagging and high temperature corrosion caused by the lack of oxygen in the furnace. Even the unevenness of combustion in the furnace was found in time. Therefore, real-time monitoring of CO concentration in boiler flue gas is an effective method to monitor boiler operating status in real time and improve boiler efficiency, which is of great significance for improving the safety and economy of boiler operation.

The application of online CO monitoring technology in power plant is analyzed from three aspects: technical principles, flue gas treatment mode, system stability and economy. Then an online CO monitoring system suitable for power station boilers is established; According to the CO emission concentration of the power station boiler and the boiler related operating parameters, a set of operation economic analysis and air volume adjustment method for the power station boiler is established. Finally, the application of this method in boiler operation optimization of a 660 MW power station unit is introduced.

79.1 Power Plant Boiler CO Monitoring Technology

On-line CO monitoring technology for power plant boilers includes non-dispersive infrared absorption method and semiconductor laser absorption spectroscopy. The two techniques are compared and analyzed from three aspects: technical principle, flue gas treatment mode, system stability and economy.

79.1.1 Technical Principle

The non-dispersive infrared absorption method is based on the principle that CO selectively absorbs radiation at 4.67 and 4.72 wavelengths, and this method is also used by the standard HJ/T44-1999 issued by the State Environmental Protection Administration of China in the CO determination of the exhaust of solid pollution sources (HJ/T 44-1999). This method is one of the most widely used methods because it is stable and reliable, has strong anti-interference ability and can be used for a long time. Semiconductor laser absorption spectroscopy (DLAS) is actually derived from non-dispersive infrared absorption method, which replaces infrared light with a laser beam of a specific wavelength (Innami et al. 2011). It is not affected by interference of background gas, ash and window pollution due to the very small spectral width of the laser beam. It can automatically correct the effects of temperature and pressure on the measurement, which enable more accurate online gas analysis.

79.1.2 Flue Gas Treatment Mode

The flue gas detection technology is divided into two types: the extraction type and the in situ type according to the different treatment methods of the flue gas. The extraction method is to continuously extract the measured flue gas, and the flue gas needs to be filtered, heated, condensed, dehydrated or diluted before entering the gas analyzer. The system is complicated, and the detection result may be affected and lagging behind by the treatment of the flue gas. The CO on-line monitoring system based on non-dispersive infrared absorption method generally adopts the extraction type flue gas treatment method; while the in-situ type does not need to extract the flue gas from the flue, no sample transmission is needed, and no pretreatment of flue gas is needed. So the system is relatively simple, there is no time lag problem of component detection. The CO on-line monitoring system based on semiconductor laser absorption spectroscopy adopts on-beam in-situ technology, that is, the transmitter and receiver instruments are installed on both sides of the flue. The component concentration is detected based on the weakening of a light beam in specific wavelength by the component concentration of the flue gas. Xi and Liu (2007) shows that the in-situ technique based on semiconductor laser absorption spectroscopy is 44 s faster than the extraction system based on non-dispersive infrared absorption method.

79.1.3 System Stability and Economy

From the aspect of stability, the laser-on-beam detection technology has the problem of small laser alignment angle between the transmitting end and the receiving end, and the large flue width will also affect the laser detection, but the technology fundamentally solves the problems of traditional extraction gas analysis technology based on non-spectral infrared absorption method such as response lag, frequent maintenance, easy to block and leak, more wearing parts and high operating cost.

From the aspect of economy, the on-beam in-situ gas analysis system based on laser absorption spectrum is more expensive. The laser detection system can generally only detect one gas concentration, while the extraction gas analysis system based on non-spectroscopic infrared absorption method can generally detect up to five gas concentrations simultaneously. The development of this technology is more mature, and more manufacturers can be selected.

In summary, with the continuous development and application of laser technology, the advantages of on-beam in-situ technology based on semiconductor laser absorption spectroscopy will be more obvious. It is a worthwhile on-line CO monitoring technology for power plant boilers.

79.1.4 System Installation Location

It is also important to determine a reasonable system installation location after selecting the appropriate on-line CO detection technology for the power plant boiler. There are several main installation locations for power plant boiler systems:

(1) Upper part of the furnace burning zone (Deguchi et al. 2002) and furnace exit (Innami et al. 2011). The two locations are close to the combustion area, and the detected CO concentration can timely and accurately reflect the combustion situation in the furnace. However, due to the large volume of the power station boiler and the complicated flow field in the furnace, a large number of monitoring instruments need to be arranged. The system is in a high ambient temperature and requires frequent maintenance;
(2) Precipitator outlet (Xi and Liu 2007). In this flue the requirements on the equipment are low and the maintenance is convenient due to the low flue gas temperature, uniform flow field and the low soot concentration, but this place is too far away from the CO production area, and the detected CO concentration cannot accurately reflect the combustion in the furnace in real time;
(3) Air preheater inlet. According to the parameters of the flue gas operating conditions (temperature, concentration, flow field), the real-time monitoring, the number of system installations and the operating conditions of the equipment, the air preheater inlet is the ideal location for installing an online CO monitoring system.

79.2 Power Plant Boiler Operation Optimization Method and System Based on CO Control

The power plant boiler operation optimization system based on CO control includes CO online monitoring system, boiler economic evaluation system and air volume optimization system. The system diagram is shown in Fig. 79.1. The CO online monitoring system is used to monitor the CO concentration in the flue gas in real time, and the boiler system economy is evaluated online according to the monitoring value of the CO and the relevant operating parameters of the boiler, such as the exhaust gas temperature and the oxygen content of the flue gas. The adjustment and optimization of air volume are carried out based on the boiler economy result, which can achieve precise control of boiler air volume and reduction of coal consumption of unit to improve the safety and economy of boiler operation.

The power plant boiler flue gas CO on-line monitoring system monitors the CO concentration in the flue gas by means of a in-situ gas analysis device based on semiconductor laser absorption spectroscopy installed at the inlet of the air preheater.

The economic evaluation system of the power plant boiler uses the boiler operating parameters in the DCS system of the power plant, such as the flue gas temperature,

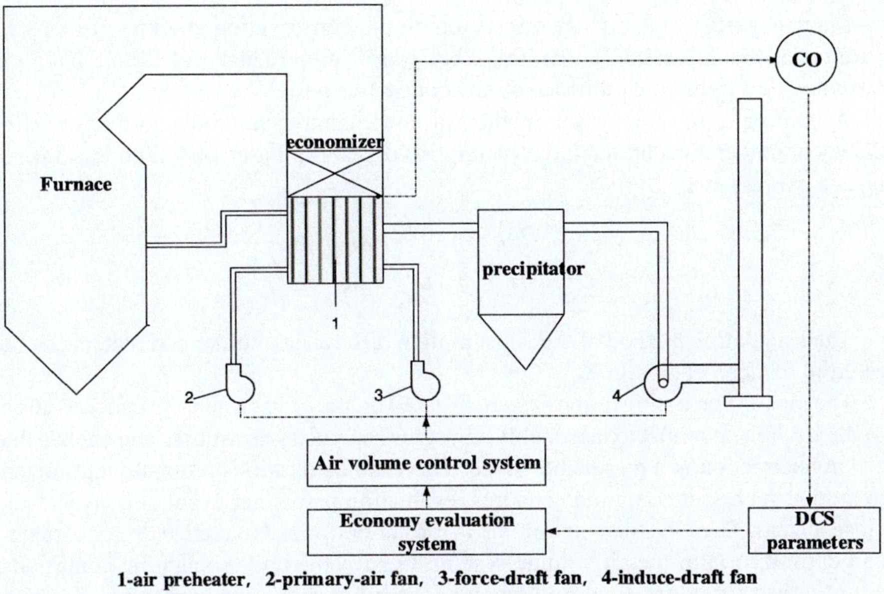

Fig. 79.1 The boiler operation optimization system based on CO control

the oxygen content of the flue gas, the carbon content of the fly ash, the CO concentration, etc., to calculate the boiler thermal efficiency and the coal consumption of the unit in real time (GB, T10184-2015). The consumption difference analysis of the parameters that affect the boiler operation economics is established to determine the key parameters. Many experts have conducted research on the real-time evaluation of boiler operation economics (Yan et al. 2001), but generally do not consider the impact of chemical incomplete combustion loss on boiler thermal efficiency. The main reason is that the power plant lacks real-time monitoring of chemical incomplete combustion products such as CO. This paper will use the monitored CO concentration to calculate the chemical incomplete combustion loss.

A boiler operation economic evaluation parameter ω is proposed, taking into account the effect on the coal consumption of the unit of factors that affect the economics of boiler operation, such as the exhaust gas temperature, the oxygen content of the exhaust gas, the carbon content of the fly ash, the CO concentration and the power consumption of the fan.

$$\omega = \frac{\partial b_g}{\partial \theta_{py}} + \frac{\partial b_g}{\partial CO} + \frac{\partial b_g}{\partial O_2} + \frac{\partial b_g}{\partial C_{fh}} + \frac{\partial b_g}{\partial W_{fj}} \tag{79.1}$$

In the formula, $\frac{\partial b_g}{\partial \theta_{py}}$, $\frac{\partial b_g}{\partial CO}$, $\frac{\partial b_g}{\partial O_2}$, $\frac{\partial b_g}{\partial C_{fh}}$, $\frac{\partial b_g}{\partial W_{fj}}$ are the change rate of coal consumption caused by exhaust gas temperature, CO concentration, oxygen content of exhaust gas, carbon content of fly ash and power consumption of fan.

Many experts have conducted research on the consumption analysis of relevant parameters that affect the economics of boiler operation (Chen et al. 2002), but they have not considered the influence of CO concentration.

According to the calculation method of unit coal consumption and thermal efficiency of power plant boiler, the consumption difference based on CO concentration can be expressed as:

$$\frac{\partial b_g}{\partial CO} = \frac{\partial b_g}{\partial \eta_{gl}}\left(\frac{\partial \eta_{gl}}{\partial CO} + \frac{\partial \eta_{gl}}{\partial \alpha}\frac{\partial \alpha}{\partial CO}\right) \quad (79.2)$$

The calculation method of the consumption difference of other parameters can be referred to Chen et al. (2002).

The air volume optimization system is based on the oxygen and CO concentration of the air preheater inlet to reasonably adjust the secondary air volume and control the CO concentration at a reasonable value. The wind coal ratio is rationally optimized to reduce the boiler operation economic evaluation parameter ω continuously, even close to zero. Then the total air volume is further optimized to realize the adjustment and optimization of the air volume system. Finally, the boiler operating conditions are observed for a long time. In particular, attention should be paid to the slagging phenomenon in the boiler furnace. If there is obvious slagging phenomenon, the air volume entering the furnace can be appropriately increased under the premise that the evaluation parameter ω is small.

79.3 Optimized Operation of Power Plant Boiler Based on CO Control

A power plant boiler is a sub-critical parameter, natural circulation, front and rear wall hedging combustion mode, an intermediate reheat, single furnace, balanced ventilation, solid-state slagging steam drum boiler. The boiler has a long-term problem of improper operation air volume setting, resulting in high CO concentration in the flue gas, large deviation of oxygen and CO concentration on the left and right sides of the inlet flue of the air preheater. Table 79.1 shows the comparison of boiler thermal efficiency test data of the boiler before and after overhaul.

It can be seen from the table that the flue gas CO concentration is very high before and after overhaul. The incomplete combustion heat loss of the boiler gas calculated based on the CO concentration is 0.528% and 1.062% respectively, which has strong impact on the thermal efficiency of the boiler. So it is very necessary to consider the influence of CO emission concentration in boiler thermal efficiency calculation. The oxygen content of flue gas only reduces from 3.91 to 3.84% before and after overhaul, while the CO emission concentration after the overhaul is one times higher than before the overhaul, resulting in an incomplete combustion loss of boiler gas increased by 0.534%. Therefore, the change of oxygen content in the flue gas can not

Table 79.1 The comparison of boiler thermal efficiency test data of the boiler before and after overhaul

Items	Unit	Before overhaul	After overhaul
Unburned carbon of fly ash	%	2.35	4.20
Unburned carbon of slag	%	4.72	2.30
O_2 in flue gas	%	3.91	3.84
CO in flue gas	ppm	1322	2717
Flue gas temperature	°C	131.1	131.30
Excess air coefficient	–	1.224	1.2142
Flue gas heat loss	%	5.559	5.818
Unburned combustible loss of gas	%	0.528	1.062
Unburned combustible loss of solid	%	1.256	1.903
Heat dissipation loss	%	0.331	0.348
Physical heat loss of ash slag	%	0.233	0.234
Thermal efficiency	%	92.09	90.64

fully reflect the change of boiler thermal efficiency, and the traditional optimization of the operating air volume based on the oxygen is one-sided. In response to this problem, a laser spectroscopic gas on-line analyzer was installed on both sides of the inlet flue of the air preheater to monitor the O_2 and CO concentrations, whose model is NEO LaserGas II, and to optimize the boiler operating air volume.

By monitoring the CO concentration, the boiler thermal efficiency and coal consumption can be more accurately reflected. Therefore, the boiler efficiency and coal consumption can be ensured at a reasonable level by controlling the CO emission concentration of the boiler. The selection of the CO emission concentration control range is mainly based on the calculation of the consumption difference in the boiler economic evaluation method, and the control range does not take into account the change of the boiler load. The reason is, on the one hand, according to the conclusion in the literature 1, the reasonable control range of CO emission concentration is basically independent of the boiler load, on the other hand, if the load range is added to the feedback of CO emission, it will not only increase the time of feedback control response, but also increase the complexity and instability of thermal logic. On the basis of considering the factors such as boiler load, boiler efficiency loss and boiler auxiliary electromechanical consumption, the upper limit of CO emission concentration control is selected as 600 mg/Nm^3, and the lower limit control of the control is 100 mg/Nm^3 (Fig. 79.2).

The calculation results of thermal efficiency of power plant boilers before and after input of power plant boiler operation optimization method based on CO control are given in Table 79.2.

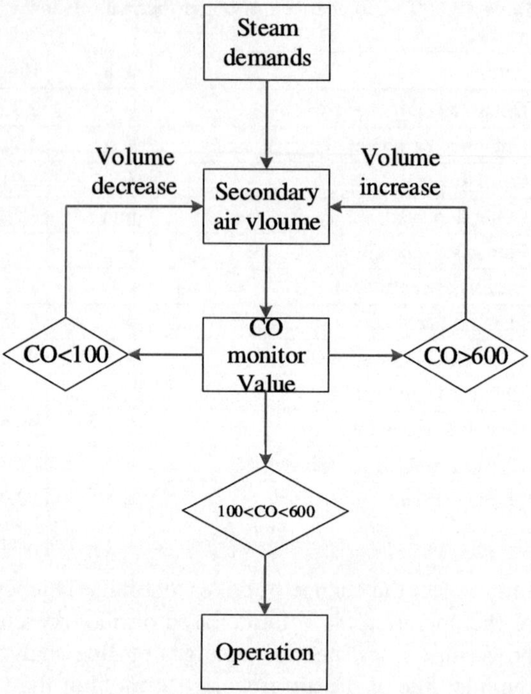

Fig. 79.2 Schematic diagram of boiler air volume optimization control system based on CO

Table 79.2 Thermal efficiency results of power plant boilers before and after Operation optimization

Items	Unit	Before optimization	After optimization
Unburned carbon of fly ash	%	2.36	2.02
Unburned carbon of slag	%	2.08	2.24
O_2 in flue gas	%	3.61	4.42
CO in flue gas	ppm	3697	520
Flue gas temperature	°C	143.0	142.8
Excess air coefficient	–	1.195	1.262
Flue gas heat loss	%	5.848	6.170
Unburned combustible loss of gas	%	1.444	0.204
Unburned combustible loss of solid	%	1.149	1.003
Heat dissipation loss	%	0.353	0.358
Physical heat loss of ash slag	%	0.244	0.244
Thermal efficiency	%	90.96	92.02

Table 79.3 Changes in power supply coal consumption with different parameters after operation optimization

Items	Boiler thermal efficiency	Desuperheating water of superheater	Desuperheating water of reheater	Auxiliary electromechanical consumption	Total
Changes in power supply coal consumption	−2.94	−0.29	−0.03	0.34	−2.93

From the table, it can be seen that through the operation optimization of power plant boiler based on CO control, the CO concentration in flue gas decreases obviously, from 3697 to 520 ppm, and the boiler thermal efficiency is increased by 1.06%.

In addition to the change of boiler thermal efficiency, the operation optimization will also change the desuperheating water amount of superheater and reheater, and auxiliary electromechanical consumption, which will cause the change of power supply coal consumption. The change of power supply coal consumption of the unit after optimization is given in Table 79.3.

It can be seen from the table that when the consumption differences of boiler efficiency, desuperheating water and auxiliary electromechanical consumption are considered comprehensively, the coal consumption of the boiler can be reduced by at least 2 g/(kW h) through the optimal adjustment of the boiler air volume, which is a great improvement for the economy of the boiler.

79.4 Conclusion

At present, the operation of power station boilers faces the dual pressures of energy saving and emission reduction and coal quality change. The operation optimization of power plant boilers based on CO control has higher sensitivity, less impact from air leakage, and appropriate coal quality than that based on O_2 control.

The application of CO detection technology based on non-dispersive infrared absorption method and laser absorption spectroscopy were compared and analyzed from three aspects: technical principle, flue gas treatment mode and system stability and economy. The installation location of the system for large power plant boilers was also discussed, and then the online CO monitoring system for the power plant boilers was finally determined: the in-situ CO detection technology based on laser absorption spectroscopy was adopted at the inlet of the air preheater.

On the basis of monitoring the CO concentration in the flue gas of power plant boilers, a method for economic evaluation of power plant boilers was proposed, which considers CO concentration and combines with the consumption difference analysis of multiple parameters. The reasonable control of the air volume in the furnace and flue gas CO concentration achieve the purpose of accurately controlling the air volume of the power plant boiler and reducing the coal consumption of the unit. It is applied in a 660 MW unit power plant boiler. The results show that the method can reduce the coal consumption of the power supply by 2 g/(kW h) at the full load condition.

References

Chen HW, Li Y, Zhen ZH et al (2002) Relationship between controllable parameter changes and coal consumption of thermal power units. Thermal Power Gen 2002(4):29–33, 46

Cheng B (2003) Combustion control system based on O2 and CO monitor. J ANHUI Electric Power Univ Staff 01:10–13

Deguchi Y, Noda M, Abe M et al (2002) Improvement of combustion control through real-time measurement of O2 and CO concentrations in incinerators using diode laser absorption spectroscopy. Proc Combust Inst 29(1):147–153

GB/T10184-2015. Performance test code for utility boiler. 2016-07-01

HJ/T 44-1999. Stationary source emission−determination of carbon monoxide−non-dispersive infrared absorption method. 1999-08-18

Innami Y, Murata A, Yuki Y et al (2011) Real-time CO measurement in a coal fired boiler with a TDLS analyzer. In: 2011 Proceedings of SICE annual conference, pp 92–96

Qiu LCH, Zhang JH (2004) Coal fired boilers combustion optimization technology based on CO. Zhejiang Electric Power 2004(03):23–25

Xi W, Liu X (2007) LDS6 laser gas analyzer's application for real time monitoring of converter gas. In: Automation expert meeting, Kunming

Xu CHK, Jia CHQ, Yang X et al (1991) Experimental research on O_2/CO coordinated control combustion optimization technology. Thermal Power Gen 01:7–19

Yan WP, Liang XJ, Chen BK (2001) The study of a method for the real-time calculation of utility boiler efficiency based on on-line detected data. J Eng Thermal Energy Power 16(2):189–190

Zhao N, Cao HT, Li XJ et al (2004) Optimizing combustion control system based on O2 and CO signals. Electric Power Sci Eng 03:23–25

Chapter 80
Technical Measures in Design and Operation of the 1000 MW Supercritical Boiler Burning High-Slagging-Propensity Coal

Zhou Lyu, Qian Huang, Yuanping Yang, and Shuiqing Li

Abstract Centralized coal utilization in large-scale units is still an indispensable part of the energy structure of many developing countries to maintain grid stability. This work is focused on the technical considerations on 1000 MW boiler design and operation for burning high-slagging propensity coals. The design and check coal samples are relatively low in heating values, and exhibit high to severe slagging propensity based on several existing slagging indices. Then the key issues for boiler design are discussed. The use of opposite firing has the advantage of slagging mitigation. Details are provided for the furnace size, burners layout, furnace exit flue gas temperature, furnace arch structure, tube array pitches, and the sootblower deployment. Besides the design considerations, operational optimization is required to reduce the risk of slagging, in which combustion control, slag monitor and effective sootblowing stand out.

Keywords 1000 MW coal-fired boiler · Slagging propensity · Boiler design · Furnace volumetric heat release rate · Furnace exit flue gas temperature

80.1 Introduction

While low carbon energy becomes more prevalent worldwide, many developing countries like China still have to rely heavily on coal for a long period of time; for instance, in 2019 coal accounts for about 60% of the primary energy in China (BP Statistical Review of World Energy 2019). The role that coal plays in the energy structure is not only a cost effective resource, but also the most important approach to maintain grid stability with increasing capacities of renewable energy. The last few

Z. Lyu · Q. Huang (✉) · Y. Yang · S. Li
Key Laboratory for Thermal Science and Power Engineering of Ministry of Education, Department of Energy and Power Engineering, Tsinghua University, Beijing 100084, China
e-mail: huangqian@tsinghua.edu.cn

Z. Lyu
China Energy Longyuan Power Technology & Engineering Co., Ltd., Beijing 100039, China

J. Lyu and S. Li (eds.), *Clean Coal and Sustainable Energy*, Environmental Science and Engineering,
https://doi.org/10.1007/978-981-16-1657-0_80

decades have seen the vast development of coal fired power plants, as the dominant concentrated energy supply, towards greater capacities and higher steam parameters for the higher cycle efficiency and the ability to purify the flue gas through centralized processes (Grace 2010). At the present stage, 600–1000 MW units with ultralow pollutant emission are becoming more and more common in China, especially for newly built ones.

For power plant operators, it is the eternal theme to pursue safe, efficient, clean and economic running of the unit, even when flexibility is increasingly focused on in recent years. Ash-related slagging and fouling problems are among the most serious concerns, especially when considering the diversified coal property the boiler could face for the reason of cost (Bryers 1996). Some well-known examples are the alkali-rich coals exploited in Australia and Northwest China which cause severe problems in both countries (Li 2004; Li et al. 2015). There are numerous works to explore ash deposition mechanism, prediction and countermeasures (Bryers 1996; Li et al. 2015; Baxter 1993; Huang et al. 2018). One practical strategy in operation is to manipulate the coal and/or ash property through additives or co-combustion techniques (Huang et al. 2018). In the meantime, the industry community is always interested in the improvement of boiler design for burning coals more prone to slag and/or foul. Progress in this direction is able to help achieve the techno-economic feasibility of coal-fired power plant with more options for coals.

This paper provides discussions on the technical measures of large-scale boiler design for burring high-slagging propensity coals through a case study. The 1000 MW unit is currently under construction in Northern China. The coal properties and the boiler design details are thoroughly discussed, providing insights to future centralized utilization of coals prone to slagging problems.

80.2 Coal Properties

Table 80.1 lists the properties of coal burnt in the 1000 MW unit, including a design coal and two kinds of check coal. Generally, the coal samples contain 10–16 wt% of moisture and 9–18 wt% of ash. The gross heat values range from 17–19 MJ/kg. Regarding the elemental compositions, Fig. 80.1 presents a coalification diagram of the coal samples (Baxter 1993). Also included in some other coal samples in the literature (Huang et al. 2017). It is demonstrated in this plot that lower values of atomic O/C and higher values of atomic H/C are featured with higher heating value (Baxter 1993). This is consistent with our case, in which the coal samples all lie in the lower right region with lower heating values as compared with other coals.

Then, the ash fusion temperature and ash compositions give clues to the slagging propensity of the coal samples. There are some existing slagging indices in the literature based on the coal property data, as listed in Table 80.2 (Degereji et al. 2012). Some indices imply a more severe slagging propensity when taking larger values, while others exhibit opposite features. The quantitative indication of slagging potential is referred to (Degereji et al. 2012). Figure 80.2 shows the slagging (and

Table 80.1 Coal properties

	Design coal	Check coal #1	Check coal #2
Proximate analysis (wt%)			
Moisture (air dried)	10.77	10.58	16.78
Volatile (dried ash free)	38.39	40.39	39.52
Ash (as received)	15.35	18.24	9.72
HHV (MJ/kg, as received)	19.08	17.56	19.13
Ultimate analysis (wt%, as received)			
C	49.73	46.38	48.22
H	2.81	2.62	3.19
N	0.64	0.61	0.60
S_{total}	1.27	0.97	1.94
Cl	0.007	0.005	0.012
O	9.70	10.38	10.33
Ash fusion temperatures (°C)			
DT	1270	1250	1110
ST	1280	1260	1120
HT	1300	1280	1130
FT	1330	1300	1140
Ash composition (wt% in ash)			
SiO_2	37.88	41.66	23.01
Al_2O_3	21.85	22.32	5.91
Fe_2O_3	11.76	7.60	23.03
CaO	10.68	11.49	18.84
MgO	2.30	2.37	5.07
SO_3	9.75	8.05	19.13
K_2O	1.40	2.01	0.56
Na_2O	2.96	2.74	3.59
P_2O_5	0.020	0.020	0.002
TiO_2	0.91	1.23	0.09

fouling) propensity of the coal samples predicted by the indices. According to the mostly-used B/A ratio, all the coal samples are high in the slagging potential, with the design coal #2 being the most remarkable one. This is consistent with most other indices' predictions. Both silica/alumina ratio and silica ratio manifest the slagging tendency of SiO_2 is greater than alumina, but lower than the basic components. And both identify the check coal #2 as the most problematic. Notably, if alkali contents are highlighted, the equivalent of Na in Fig. 80.2d indicates more severe slagging propensity of the design coal and the check coal #1 than the check coal #2. These

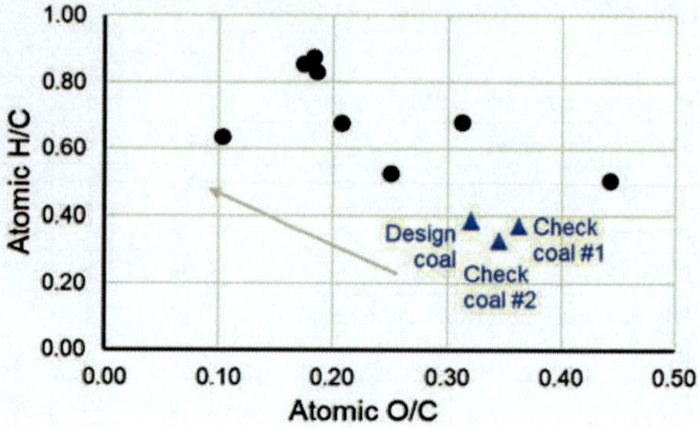

Fig. 80.1 Coalification diagram of the design and check coal samples. Also shown are some other coal samples reported in the literature (Huang et al. 2017)

Table 80.2 Coal slagging indices (Degereji et al. 2012)

Basis	Slagging index
B/A ratio[a]	$B/A = \frac{Fe_2O_3+Cao+Mgo+K_2O+Na_2O}{SiO_2+Al_2O_3+TiO_2}$
Silica/alumina ratio[a]	$R_{S/A} = SiO_2/Al_2O_3$
Silica ratio[b]	$SR = 100 \times SiO_2/(SiO_2 + Fe_2O_3 + CaO + MgO)$
Equivalent of Na[a]	$E_{Na} = (Na_2O + 0.659K_2O) \times A_d/100$
Fusibility[b]	$AF = 0.8\ DT + 0.2\ HT$
Sulphur content index[a]	$Index = (B/A) \times S$

[a]Higher value indicates higher slagging propensity
[b]Higher value indicates lower slagging propensity

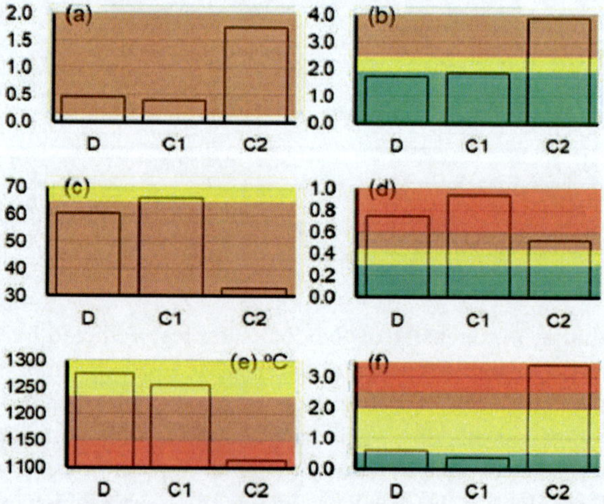

Fig. 80.2 Slagging and fouling propensity of the coal samples (D: design coal; C1: check coal #1; C2: check coal #2) based on various criteria (Degereji et al. 2012). **a** B/A ratio, **b** silica/alumina ratio, **c** silica ratio, **d** equivalent of Na, **e** fusibility (Unit: °C), **f** sulphur content index. In all the panels, green corresponds to low (ash deposition propensity), yellow to medium, orange to high, and red to severe

indices thus clearly reveal medium or high slagging propensities of the coal samples, which calls for careful considerations in the design and operation of the boiler.

80.3 Technical Measures

Figure 80.3 presents the general view of the boiler design, and the unit is under construction in Northern China. Some key parameters are given in Table 80.3. The 1000 MW ultra-supercritical unit has a Π-type once-through single-furnace opposite-firing boiler and a single-reheat cycle. The steam parameters at the boiler exit are 29.4 MPa/605 °C/623 °C. In what follows we shall discuss the major technical measures in designing the boiler to reduce the risk of severe slagging.

First of all, we note the opposite firing of the boiler with multilayer positioned burners ensures a more uniform distribution of heat flux than the tangential fired boiler. The heat flux difference, defined as the ratio of maximal heat flux over the minimum on the walls, is less than 1.15, while for tangentially-fired boilers this value is usually 1.3–1.5. As a result, we expect smaller temperature differences for both the steam and flue gas at the furnace exit, a benefit to slagging mitigation and the heat transfer tube safety. The swirl bunrer has also more freedom to adjust the flame length and stiffness so that no overheating or reducing atmosphere is present near the waterwalls. It helps prevent tube overtemperature and corrosion. Moreover, by adjusting the number of operating burners, the boiler can be more flexible to meet the underload condition down to 25%.

It is then critical to determine the furnace size and other related parameters. The key is the volumetric heat release rate, which is designed as 72.8 kW/m^3 for this

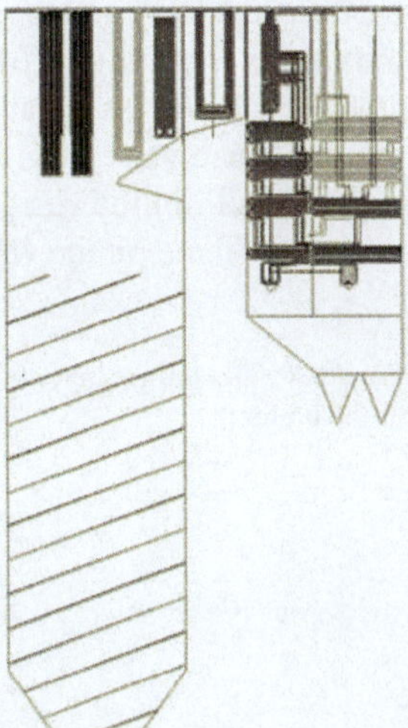

Fig. 80.3 General view of the boiler design

Table 80.3 Key boiler parameters[a]

Furnace cross section and height	mm	33,128.7 × 15,728.7 × 68,500
Furnace cavity	m^3	30,718
Effective radiant heating area of the furnace	m^2	6713
Heating area of the burner section	m^2	1490
Distance between the upmost burner and the platen bottom	m	28.4
Furnace volumetric heat release rate	kW/m^3	72.8
Furnace sectional thermal load	MW/m^2	4.436
Wall heat flux of the burner section	MW/m^2	1.556
Furnace exit flue gas temperature	°C	967
Flue gas temperature at platen superheater bottom	°C	1339
Steam flowrate out of the superheater	t/h	2980
Excess air coefficient at the furnace exit	–	1.14
Boiler efficiency (based on LHV)	%	94.45

[a]Based on the BMCR (boiler maximum continuous rating) condition and the design coal

boiler. The oppositely-fired boiler can use the heat transfer area more effectively without the need to tilt the burner, as compared with the tangentially-fired boiler. Therefore, it is possible and indeed practical to choose a slightly higher value of the volumetric heat release rate and hence a smaller furnace size, while preventing the formation of massive slags.

Based on the whole furnace structure, it comes to the optimization of burner layout. As shown in Fig. 80.4, the boiler is equipped with three layers of burners in either the front or rear wall, and each layer includes eight swirl burners. The distance between the OFA and the top layer is 7000 mm, and the distance between each layer of burners is chosen as 6100 mm, resulting in a wall heat flux of the burner section as 1.556 MW/m^2; see Table 80.3. It is regarded safe to prevent slagging in the burner region. Meanwhile, it is beneficial to increase the distance between the burner to the side wall. In the current design, this distance is set as 3964 mm to eliminate the impact of flame on the wall.

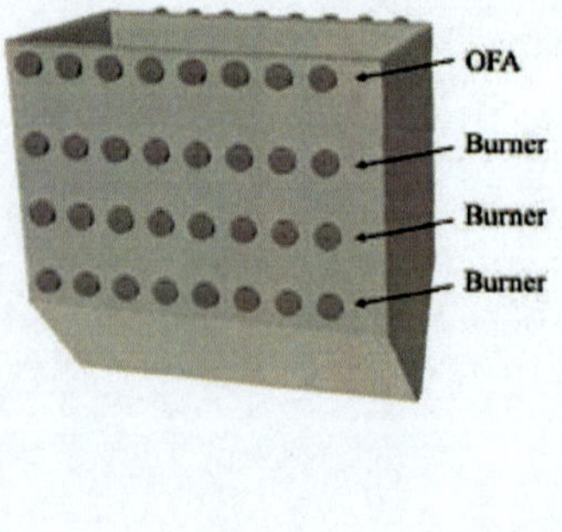

Fig. 80.4 The burner layout of the boiler

Because the slagging propensity, especially the sticking probability, is closely related to the temperature, the design of furnace exit flue gas temperature is essential. Figure 80.5 shows the designed BMCR flue gas temperature at different boiler locations for the design and check coal samples. It is seen that the furnace exit flue gas temperature (G) is substantially lower than the deformation and soft temperatures of the coal ashes. However, the flue gas temperature at the platen superheater inlet is higher than the ash fusion temperature, indicating a full melting of ash particles and greater sticking probabilities. Therefore, it provides important implications for the boiler operation to avoid platen superheater slagging; that is, to ensure a sufficient

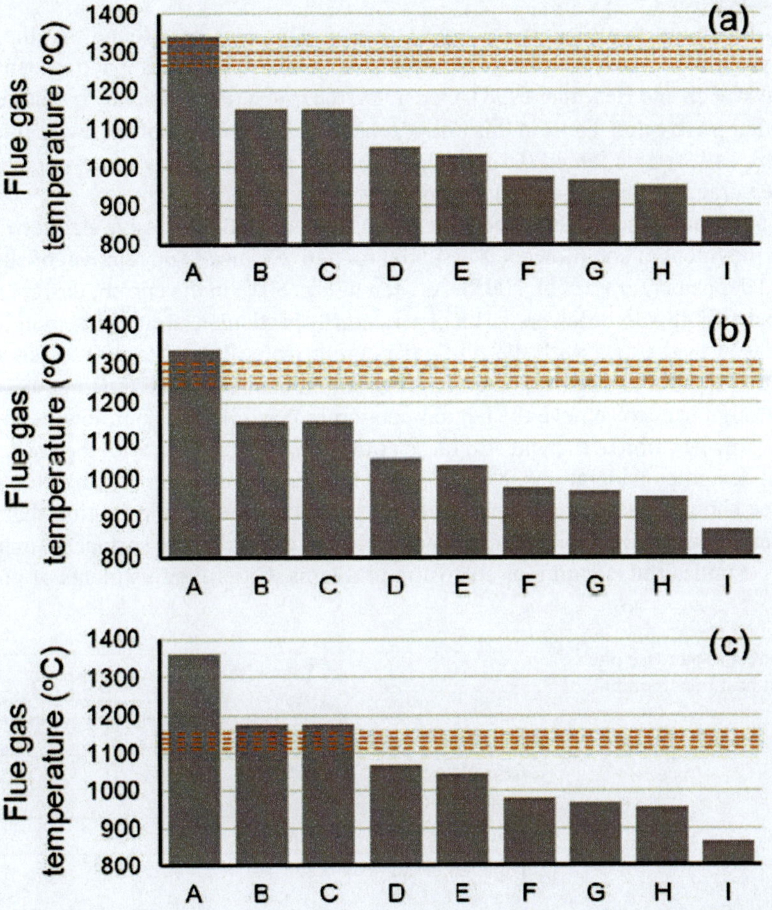

Fig. 80.5 Designed BMCR flue gas temperatures at different boiler locations for **a** the design coal **b** check coal #1 and **c** check coal #2. Dashed lines in each panel are the ash fusion tempertures of the coal samples. Location A: panel superheat inlet; B: platen superheater outlet; C: rear platen superheater inlet; D: rear platen superheater outlet; E: final superheater inlet; F: final superheater outlet; G: furnace exit; H: high temperature reheater inlet; I: high temperature reheater outlet

burnout at this site, and to closely monitor and effectively sootblowing the high-risk region.

Another strategy of the boiler design is to use a greater ratio of the furnace arch depth over the furnace depth. The idea manifests itself as an efficient approach to increase the local flue gas velocity, enhance mixing and spatial uniformity, and to reduce the thermal shock imposed on the rear platen and final superheaters. On the other hand, it is beneficial to slightly enlarge the distance between the rear platen superheater and the upper side of the furnace arch when burning high-slagging-propensity coals. With adequate sootblowers placed in the rear platen superheater region, the current design is helpful to prevent slagging, fouling or even blockage of the furnace arch.

As for fouling problem in the convective tube bundles, it is found that the spanwise pitch ratio has a prominent blockage effect on the fly ash impacting behaviors (Zdravkovich and Bearman 1998). Generally, the impaction efficiency increases with a smaller pitch ratio. Thus, in this boiler relatively large values of spanwise pitch are chosen, as listed in Table 80.4, which eliminates the possibility of severe slagging or even clogging in the horizontal and vertical flue ducts.

In addition to the structural considerations for preventing massive slag formation, the deployment of sootblowers is essential for active control and removal of slag and fouled deposits (Teruel et al. 2005). As seen in Fig. 80.6, in the current design, a total number of 148 sootbolowers is incorporated. In particular, around 68 sootblowers are placed in the area with the flue gas ranging from 400 °C to 900 °C, in which a serious fouling is often expected for high-alkali coal combustion (Li et al. 2015). This design tries to achieve the techno-economic feasibility of sootblowing.

Future operation of the unit still faces critical challenges even with the abovementioned design considerations. Most importantly, the combustion organization in the furnace should maintain a proper high temperature zone. Any deviation that leads to overheating or reducing atmosphere near the heat transfer surface is expected to cause undesired slagging or corrosion problems. Careful adjustments of coal-air

Table 80.4 Spanwise pitch ratio of heat transfer tube arrays

	Spanwise pitch (mm)	Tube outer diameter (mm)
Platen superheater	1275	51
Rear platen superheater	600	51
Final superheater	600	51
High temperature reheater	225	51
Low temperature reheater	112.5	60
Low temperature superheater	112.5	51
Economizer	112.5	60

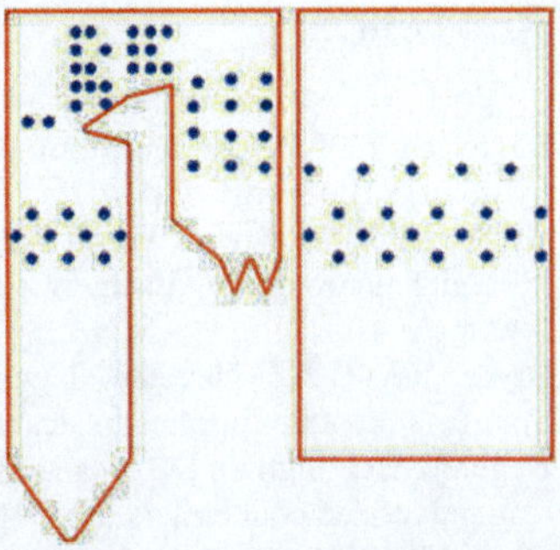

Fig. 80.6 Positions of sootblowers in the boiler

ratios are much required for that purpose. Moreover, intensive monitoring and effective sootblowing are always among the necessary measures during boiler operations. Both can benefit from the development of intelligent power plant equipped with more advanced monitoring and analysis techniques.

80.4 Conclusions

In this paper, we discuss the technical measures in the design of 1000 MW boiler for burning high-slagging-propensity coal. The design and check coal samples are relatively low in heating values, and exhibit high to severe slagging propensity based on several existing slagging indices. Then the key issues for boiler design are discussed. The use of opposite firing has the advantage of slagging mitigation. Details are provided for the furnace size, burners layout, furnace exit flue gas temperature, furnace arch structure, tube array pitches, and the sootblower deployment. Besides, operational optimization is required to reduce the risk of slagging, in which combustion control, slag monitor and effective sootblowing stand out. With the unit currently under construction, future work can be directed to test the slagging propensity of the coal samples under real combustion conditions, and to develop a practical and efficient procedure to monitor slagging behaviors in the boiler during operation.

Acknowledgements This work was supported by the National Natural Science Foundation of China (Grant No. 51906122 and 51725601).

References

Baxter LL (1993) Ash deposition during biomass and coal combustion: A mechanistic approach. Biomass Bioenergy 4:85–102

BP Statistical Review of World Energy 2019. https://www.bp.com/content/dam/bp/business-sites/en/global/corporate/pdfs/energy-economics/statistical-review/bp-stats-review-2020-full-report.pdf.

Bryers RW (1996) Fireside slagging, fouling, and high-temperature corrosion of heat-transfer surface due to impurities in steam-raising fuels. Prog Energy Combust Sci 22:29–120

Degereji MU, Ingham DB, Ma L (2012) Numerical assessment of coals/blends slagging potential in pulverized coal boilers. Fuel 102:345–353

Grace CT (2010) Coal combustion research. Nova Science Publishers, New York

Huang Q, Li S, Li G, Yao Q (2017) Mechanisms on the size partitioning of sodium in particulate matter from pulverized coal combustion. Combust Flame 182:313–323

Huang Q, Zhang Y, Yao Q, Li S (2018) Mineral manipulation of Zhundong lignite towards fouling mitigation in a down-fired combustor. Fuel 232:519–529

Li CZ (2004) Advances in the science of Victorian brown coal. Elsevier. Boston, Amsterdam

Li G, Li S, Huang Q, Yao Q (2015) Fine particulate formation and ash deposition during pulverized coal combustion of high-sodium lignite in a down-fired furnace. Fuel 143:430–437

Teruel E, Cortes C, Diez LI, Arauzo I (2005) Monitoring and prediction of fouling in coal-fired utility boilers using neural networks. Chem Eng Sci 60:5035–5048

Zdravkovich MM, Bearman PW (1998) Flow around circular cylinders – Volume 1: Fundamentals. Oxford Science Publications

Chapter 81
The Synergistic Performance of Heat and Electricity Studied from the Heating Demand Side and the Supply Side

Pan Zhang, Weiliang Wang, and Junfu Lyu

Abstract As the increasingly fast development of cogeneration of heat and power and renewable energy, it is necessary to review the traditional technical route, and develop a more flexible and efficient cogeneration system. The thermal power units on the heating side and the heating users on the thermal demand side, including the whole heat network transmission system, have a great impact on the power grid dispatching. Exploring the regulating potential of the whole heating system can effectively improve the peak regulation of the power grid and the absorption capacity of new energy. Based on the minimum stable combustion load of the boiler, the peak shaving capacity of the cogeneration unit can be increased by 36 million kW compared with the pure condensation operation. In the social energy consumption system, the cogeneration unit can deeply peak-shaving to promote new energy consumption, and save 2.58 million tons of standard coal per year. In addition, different response and delay characteristics of electricity and heat can be utilized to effectively cooperate with frequency modulation operation of power grid and improve power supply quality and operation stability of power grid. At the same time, on the demand side, the adoption of heat storage method can well adjust the peak and valley difference of the power grid.

Keywords Peak shaving · Thermoelectric synergy · Thermoelectric decoupling · Heat storage

---

P. Zhang · W. Wang (✉) · J. Lyu
Key Laboratory for Thermal Science and Power Engineering of Ministry of Education, Department of Energy and Power Engineering, Tsinghua University, Beijing 100084, China
e-mail: 13810659647@163.com

P. Zhang
Shengyuan Hi-Tech (Beijing) Technology Co. Ltd, Beijing 100081, China

J. Lyu and S. Li (eds.), *Clean Coal and Sustainable Energy*, Environmental Science and Engineering,
https://doi.org/10.1007/978-981-16-1657-0_81

81.1 Background

By the end of 2018, China's installed capacity of power generation was 1.9 billion kW, thermal power capacity was 1.14 billion kW, and renewable energy capacity was 710 million kW, accounting for 37.4% of the total installed capacity (National Bureau of Statistics 2018). During the period of energy transformation, the basic position of coal in primary energy structure remains unchanged, the dominant function of coal-fired power generation in energy structure remains unchanged, and the key role of coal in energy structure adjustment remains unchanged. Under the new energy situation, coal-fired power generation is developing towards more efficient, more flexible, cleaner, lower carbon, more circular economy and more wisdom.

At the same time, with the slowdown of the growth of electricity demand and the large-scale development of renewable energy, China's electricity market has shown new and obvious characteristics, that is, the peak-valley difference of power load in the power system is getting bigger and bigger, and the duration is also increasing. Only through the peak-shaving of power grid can meet the current situation Therefore, new peak-shaving power sources was needed urgently. In order to make the grid more receptive to new energy sources such as wind power, thermal power units should actively participate in peak shaving. Flexibility modification of thermal power units will make a great contribution to new energy consumption. The 13th Five-Year Plan of Electric Power Development defines the goal of flexible transformation of coal-fired power in China as 220 million kW of installed capacity of peak shaving units by 2020. About 154 billion kWh (Long et al. 2018) of new energy generation is expected to be absorbed (Fig. 81.1).

In recent years, the installed capacity of cogeneration units and the scale of centralized heating have increased rapidly. In 2017, cogeneration units were installed at 550 million kilowatts, accounting for 50% of the total thermal plants (Huang 2019). 1.4 billion GJ/year heat was used for heating, and 700 million GJ/year heat was used for industrial production. The inherent concept of "fixing electricity by heat" limits the peak load regulation capability of cogeneration units to a certain extent (Fig. 81.2).

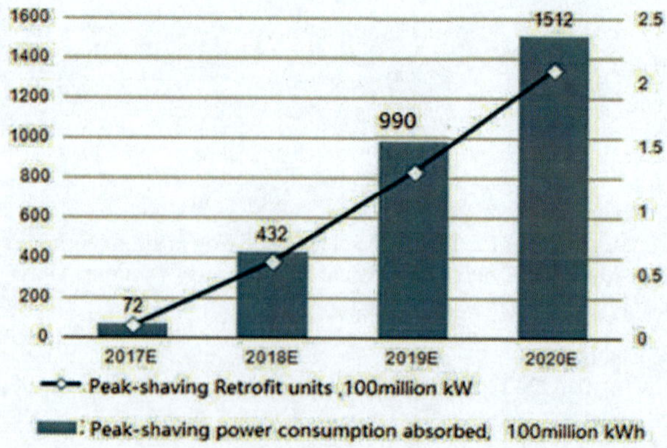

Fig. 81.1 Relationship between peak shaving capacity of thermal power units and new energy consumption

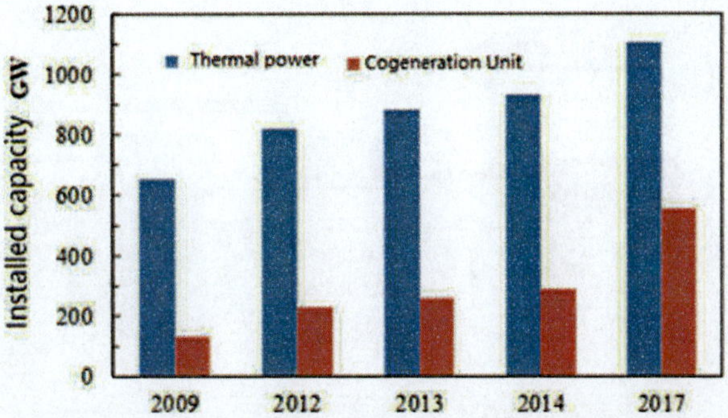

Fig. 81.2 Development and capacity of Cogeneration Units in China

At the same time, the demand for heating is growing rapidly. By 2018, the national heating area will be about 23.6 billion square meters, of which cogeneration will bear 3.5 billion square meters. Hot water pipe network is about 200,000 km and steam pipe network is about 12,000 km. The demand for clean heating is increasing. The standard coal consumption for central heating has reached nearly 200 million tons. On the one hand, it needs to meet the rapid growth of heating demand, on the other hand, it needs to replace the traditional way of heating (International Energy Agency 2018).

It is generally believed that the peak shaving capacity of cogeneration units will be limited by heat supply, especially for units with large heat supply. In order to find some ways of "thermoelectric decoupling" to release the peak-shaving capacity of thermal power units, such as low-pressure cylinder cutting technology, electric boiler, various heat storage technology have been applied in heating units in recent years, and also played a positive role. In the whole energy structure, we should discover and excavate the role of thermal power units in peak shaving and new energy consumption. At the same time, the analysis of thermal power units should be more comprehensive and systematic. The source, the pipeline network and the user side should be regarded as an organic whole system, and the potential of peak regulation and frequency regulation of heat supply units should be brought into full play.

81.2 Peak Shaving Potential Analysis of Cogeneration Units

For general thermal power units, peak shaving capacity is evaluated on the basis of the minimum non-oil stable combustion load of the boiler. Although the minimum steady burning load of the unit can be reduced by oxygen-enriched combustion technology, it is not easy to popularize this technology on a large scale because of the high cost of oxygen production, and the operation of denitrification system at low load will be limited by too low temperature. Based on the minimum evaporation of the boiler, the influence of heating unit on peak shaving is analyzed. Figure 81.3

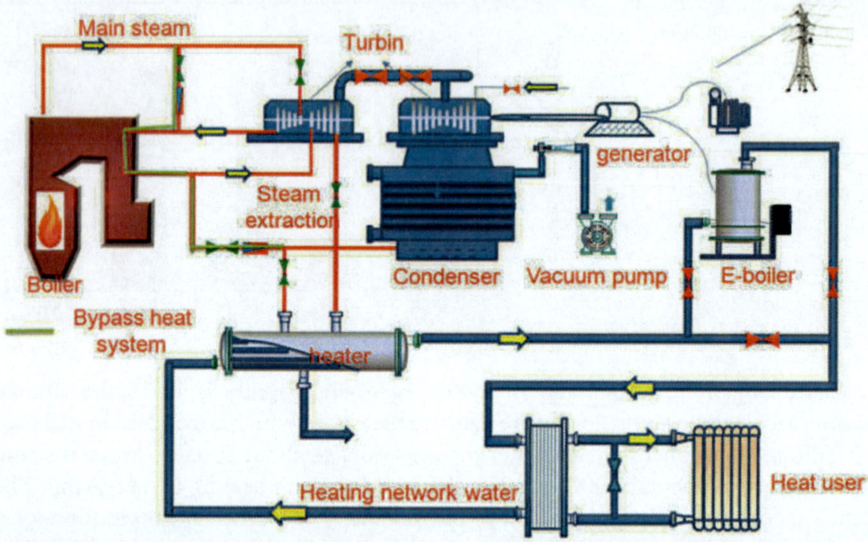

Fig. 81.3 A schematic diagram of a thermoelectric decoupling system

shows a thermoelectric decoupling system diagram. This thermoelectric decoupling system mainly includes the following regulation systems: A. Steam turbine bypass heating system, through high pressure and low bypass of steam turbine to realize the supplementary heating of the boiler to the heating system at low load, using the existing bypass design capacity, can increase the heating capacity at low load. If the bypass heating system is expanded and reformed, the heating capacity can be improved (Xue et al. 2018). B. There are many cases of steam turbine LP cylinder cutting retrofit in heat supply units in the past two years, which effectively improves the heat supply capacity of units under low load. The increment of heating capacity basically utilizes the minimum flow rate of low-pressure cylinder, which is generally 100–150 t/h. After cylinder cutting operation, the extraction heat supply capacity of the unit can be increased by 70–100 MW (Chen et al. 2019; Liu et al. 2017). C. In order to achieve deeper peak shaving, a certain capacity of electric boiler and regenerative system can be equipped to assist peak shaving. However, the capacity of electric boilers should not be too large, and cannot be used as a regular regulation mode. Electricity is a kind of high-grade energy, and the conversion of electricity into heat is an inefficient way of utilization. D. Peak shaving capacity of generating units can be further increased through energy storage regulation of heating network. For example, the water storage capacity of 10 km heat supply pipeline reaches 30 000 t. If the temperature of supply and return water is increased by 10 C at the same time during the period of high electric load, it is equivalent to storing 10 C heat in 30 000 t water, which is equivalent to a large-scale regenerative tank of 8 000 m^3 in the power plant.

For heating units, the ability of thermoelectric decoupling of units should be brought into full play by using cylinder cutting and bypass heating. Then deep peak shaving is carried out through various heat storage methods and electric boiler regulation. From another point of view, in the case of meeting the minimum evaporation flow rate of the boiler, the heating unit can bring lower load because of heating. If part of the steam is supplied by bypass instead of working in the steam turbine, the power generated by the heating steam in the steam turbine is the increased peak shaving capacity. In the next analysis, based on the decoupling mode of bypass heating, the peak shaving potential of heating units is quantitatively analyzed when the minimum stable combustion of boilers is satisfied.

When participating in the consultation and research project of China Academy of Engineering, the strategic research project of energy saving and emission reduction of coal-fired generating units in China, the total heat supply of existing cogeneration units in China was counted. By 2018, the heat supply scale of China's cogeneration will be about 2.1 billion GJ (including steam supply), and the incremental part that can be developed in recent years will be 600 million GJ. The total annual heat supply will reach about 2.7 billion GJ. The heat that can be supplied by 1t main steam (including heat absorber in reheater) is calculated as 3.5GJ. 2.7 billion GJ can be converted to 700 million tons of main steam; the average hours of annual heating are calculated in 4000 h, equivalent to 192,000 tons of main steam consumption per hour. The power generation capacity of 10 MW corresponding to the average 32 t/h main steam is calculated. Considering the minimum steady combustion load, the peak load regulation capacity of 60 million kW can be increased compared with the pure condensation unit because of the bypass heating of the heating unit. Considering that the actual heating load of some units exceeds the main steam heating load under the minimum steady combustion load of the units. Multiplied by 0.6 coefficient, the peak shaving capacity of heating unit is 36 million kW higher than that of pure condensation unit, which is absolutely helpful to the consumption of new energy.

81.3 Analysis of Social Energy Conservation Caused by Deep Peak-Shaving of Heating Units

Based on the above analysis, the heating unit is beneficial to peak shaving, especially by-pass heating can give full play to the deep peak shaving ability of the heating unit. However, this decoupling method is equivalent to the boiler heating directly, steam is not used to generate electricity, there is no "cogeneration" effect, so intuitively compared with the original cogeneration method, it is sure that energy consumption has increased. However, using this mode of operation, itself is for peak shaving as a "way out" for new energy. So the analysis of energy consumption should be placed in the energy supply system of the whole society for analysis. For heating units, because of the different heating parameters and modes, it is more complicated to analyze the energy consumption of heating in detail. A simplified model is presented

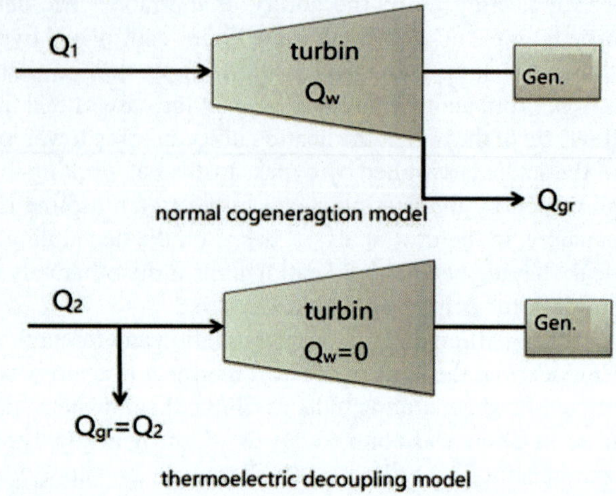

Fig. 81.4 Thermoelectric decoupling energy consumption analysis model

for analysis and comparison. As shown in the following figure: under the general cogeneration mode, the heat supply is supplied by steam extraction mode, the heat output of the boiler is Q_1, the power generation is Q_w and the heat supply is Q_{gr}; after adopting the thermoelectric decoupling heating mode, the heat supply Q_{gr} remains unchanged, and the heat output of the boiler is used for heating. Power generation load Q_w of steam turbine is absorbed by new energy generation. In the analysis of the whole energy supply system, under the condition that both the power generation Q_w and the heat supply Q_{gr} remain unchanged, the reduction of the heat output of the boiler under the thermoelectric decoupling mode is energy saving. According to the simple heat balance, it can be concluded that the output heat reduction of the boiler is the power generation Q_w of the steam turbine. The statistics of heat supply Q_{gr} are known. The total social energy saving can be obtained by deducing the relationship between Q_w and Q_{gr} (Fig. 81.4).

The following analysis deduces the relationship between electricity and heat: the total heat entering the steam turbine is Q_1, in which the power generation in the steam turbine is Q_w and the external heat supply is Q_{gr}; when not heating, the efficiency of steam in the steam turbine is n, the average heat consumption rate of the steam turbine is 8200 kJ/kWh, then n is 43.9%, the heat input into the steam turbine is Q_1, and the power generation is Q_{w1}; heating affects the work done by the steam in the steam turbine. It is Q_{w2}; The standard coal consumption for heating is b_{gr}, and the boiler efficiency is 91%, and the calculation results show that b_{gr} is 37.5 kg/GJ; the average standard coal consumption for heating is $\overline{b}_{gr}$. Through the research of the subject, the average coal consumption of heating and heating can reach 15 kg/GJ and that of industrial steam can reach 28 kg/GJ. Then there are the following equations among the variables:

$$\begin{cases} Q_1 = Q_w + Q_{gr} \\ Q_{w1} = Q_1\eta \\ Q_{w1} = Q_w + Q_{w2} \\ \frac{Q_{w2}}{Q_{w1}} = \frac{\bar{b}_{gr}}{b_{gr}} \end{cases}$$

By substituting the relevant values, it can be concluded that for heating, $Q_w/Q_{gr} = 0.356$; for industrial steam supply, $Q_w/Q_{gr} = 0.125$. The annual heat supply of cogeneration is 2.7 billion GJ, including 1.8 billion GJ for heating and 900 million GJ for industrial heating. The annual effective decoupling peak-shaving time of heating units is calculated by 500 h, which is equivalent to 75.62 million GJ of less heat input of steam turbines, 91% of boiler efficiency, and 2.58 million tons of standard coal can be saved annually.

81.4 Improve the Peak-Shaving and Frequency-Regulating Ability of Cogeneration Units by Cooperative Operation

Above is a combined thermoelectric system. The power supply system consists of conventional thermal power units, gas combined cycle units, hydropower and nuclear power units, wind and photovoltaic new energy generating units, and cogeneration units. The connection between power and thermal system is that on the one hand, the power grid supplies power to the terminal heating equipment, and on the other hand, the cogeneration unit supplies heat to users through the central heating network of the city. The supply side of the power system is equipped with some storage facilities, such as hydropower stations, lithium batteries and so on. The supply side of cogeneration is also equipped with some storage equipment, such as thermal storage tanks. The thermoelectric combined system is an inseparable organic whole. The general analysis is mainly based on the "static" model, without considering the heat storage capacity and thermal inertia characteristics of the whole heating system. If the combined thermoelectric system is considered as an organic whole system, the peak shaving potential of thermoelectric units can be further explored. Electric energy has the characteristics of easy transmission and difficult storage, while thermal energy has the characteristics of easy storage and difficult transmission. Composition of thermoelectric combined system makes the two advantages complement each other. Using the characteristics of large inertia of thermal system, thermal load can be shifted or peak-shaving and valley-filling (Fig. 81.5).

The basic principle is that the heating temperature of the end user fluctuates within a reasonable range above the red line temperature. According to the demand of peak-shaving capacity in full-time period of power grid, under the premise of meeting the heating requirements, the thermal storage capacity and delay characteristics of

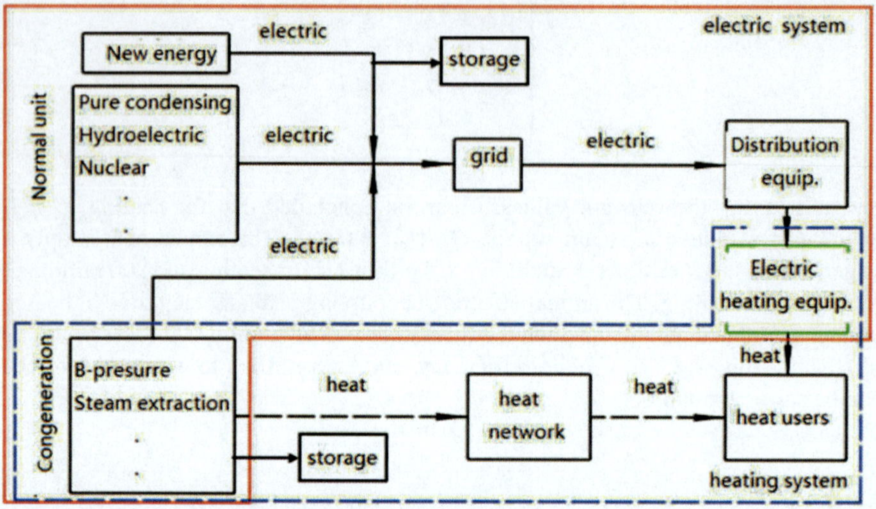

Fig. 81.5 Composition of thermoelectric combined system

thermal system are fully utilized to make cogeneration units respond to the peak-shaving requirements of power grid to the greatest extent. Figure 81.6 is the result of (Li et al. 2017) optimal model for combined dispatching of 200 MW heat-supply units, which utilizes the thermal dynamic characteristics of buildings and heating networks to improve the peak shaving capacity of units. It can be seen from the blue curve in the figure that the power output of the cogeneration unit is strictly restricted by the heat load in the operation mode of "fixing electricity by heat" before considering the thermal dynamic characteristics of the building and the heating network, which is consistent with it all the time. Therefore, the unit's electric output is affected by the "thermoelectric coupling" characteristics, which can only change

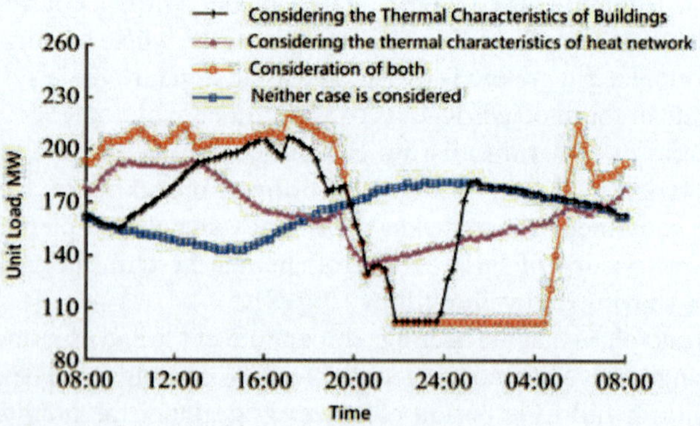

Fig. 81.6 Power output of cogeneration units considering dynamic characteristics of buildings and heating networks

in a very small range, and the peak shaving ability is low. Considering the thermal dynamic characteristics of building and heating network, the thermal output of the unit is no longer strictly restricted by the thermal load. The thermal output cannot keep synchronization with the actual demand of heating load, but has a strong correlation with the dispatching curve of the power grid. Thus, the range of the unit's output is widened up and down, and the peak shaving ability of the unit is improved.

Because the heat network has the ability of heat storage, at the same time, the heat inertia of the heat network is very strong. Changing the operating state of the heat source end in a short time, such as reducing or increasing the amount of steam extracted from heating in a short time, will not affect the quality of heating. At the same time, the frequency modulation rate of the unit can be increased by utilizing the energy storage characteristics of the heat network system. Although the power output of low pressure cylinder is insufficient during heating, the load response speed is reduced. Especially, the primary frequency regulation ability of the unit decreases with the increase of the extraction capacity of the heating steam turbine (Qi 2018). If the heat network is used to store energy, a coordinated control scheme of heat supply units with feedforward heating is adopted, that is, the heat load of working steam is stored in the heat network or released from the heat network in a short time, thus increasing the variable load rate of the unit. Field experiments in LPS thermal power plant show that the actual maximum variable load rate of the unit can reach 4% Pc/min (Deng 2016). This rate is much higher than the 1.5% Pc/min stipulated by the power grid, which can effectively relieve the pressure of peak load regulation and frequency regulation brought by large-scale grid-connected new energy power. Under the coordinated regulation mode of thermal system, it is necessary to break the current contract paradigm and assessment mode between heating power plants and heating companies, and participate in peak-load and frequency-regulation of power grid, so as to achieve common benefits. If the heating unit can bring its own FM potential into full play, it will play an important role in regulating the quality of power supply, increasing the proportion of new energy connected to the network, and trans regional power transmission.

At the same time, in addition to centralized heating, quite a number of heating buildings in China adopt distributed heating, such as single or centralized air conditioning, gas wall-mounted furnace, electric heating and other ways. Especially in the past two years, the policy of coal-to-electricity conversion has been encouraged and promoted, and the electric energy heating has developed rapidly. In order to encourage the promotion of coal-to-electricity conversion, the peak-valley tariff policy has been basically adopted in all regions of northern China, which effectively promotes the development of the heating mode of electric energy storage. Electric heat storage is to convert electricity into heat and store it in the heat storage medium by using lower electricity price during peak-shaving Valley period, and release heat from the heat storage device for heating in other periods. If this method of electric heat storage also participates in peak shaving operation of power grid, it will be more conducive to peak shaving of power grid. Taking a 1 million square meters electric heat storage as an example, the required heat storage power during peak-shaving valley period is 100 MW. If the heating area of electric heat storage reaches 100 million square meters, the heat storage power can reach 10 million kW. If the users of electric heat storage can effectively participate in peak shaving and electricity trading, it is undoubtedly very good for peak shaving of power grid. Therefore, the power grid should encourage end-users to participate in electricity market transactions in a "packaged" way, and enjoy peak-shaving benefits equally with the supply side.

81.5 Conclusion

4.1 Based on the minimum stable combustion load of the boiler, the peak shaving capacity of the cogeneration unit can be increased by 36 million kW compared with the pure condensation operation. In the social energy consumption system, the cogeneration unit can deeply peak-shaving to promote new energy consumption, and save 2.58 million tons of standard coal per year. Whether from the increase of peak shaving capacity or from the energy saving, it is advantageous for cogeneration units to participate in peak shaving, so the peak shaving capacity of cogeneration units should be fully released. It is necessary to strengthen the guidance of cogeneration units to participate in deep peak shaving from policies and markets.

4.2 Utilizing the heat storage and thermal inertia characteristics of the thermal pipe network system and buildings, the peak shaving characteristics of the heating unit can be further released under the condition of meeting the heating demand of the end users. The variable load rate of cogeneration units can be effectively increased by using thermal pipe network system to store and release heat in a short time. Through the coordinated operation of the thermal system, the advantages of peak load regulation and frequency regulation of cogeneration units can be further brought into play, which can effectively improve the quality of power regulation, increase the proportion of new energy grid-connected and promote power transmission across regions.

4.3 Distributed electric heating users, especially the terminal heating users using electric heat storage, will help the grid peak-shaving if they can effectively participate in the electricity peak-shaving market transactions. The peak-shaving potential released by 100 million square meters of thermal storage users is 10 million kWh, and the response rate of electric load increase and decrease is fast.

Acknowledgements This work was supported by the National Key Research and Development Program of China (2016YFB0600205), and Science & Technology Research Project of the State Grid Co., Ltd. (52020119000Q)

References

Chen YH, Bao WW, Zhang M, Li JH, Zhang XH, Yuan JL, Zhou Y (2019) Thermodynamic analysis of cooling steam for low pressure cylinder cutting technology of supercritical 350 mw unit. Therm Power Gener 05:133–138

Deng TY (2016) Energy storage characteristic analysis and fast variable load control of heating units. North China Electric Power University (Beijing)

Huang QL (2019) The Way of Green Energy Development. Distrib Source 4(02):1–7

International Energy Agency (IEA). (2018) Research report on the development of clean heating in China

Li P, Wang HX, Wang Y, Han Y, Li WD (2017) Improving peak shaving capacity of cogeneration units by utilizing thermal dynamic characteristics of buildings and heat networks. Power Syst Autom 41(15):26–33

Liu L, Yang JG, Wang D, Liu C (2017) Research on thermoelectric decoupling technology for thermal turbine units. Ningxia Electric Power 06:62–66

Long H, Huang JJ (2018) Analysis of the development direction of coal-fired power generation design technology in the 13th five-year plan. Power Gener Technol 39(01):13–17

National Bureau of Statistics (2018) Statistical bulletin on national economic and social development 2018

Qi W (2018) Analysis of the influence of different extraction conditions on frequency modulation capability of heating units. North China Electric Power University

Xue CY, Yang RZ, Wang T, Gu WW, Gao Q, Zhang YH (2018) Application of combined high-low bypass heating of steam turbine on supercritical 350 MW units. Therm Power Gener 47(05):101–105

Chapter 82
Effect of Recirculated Flue Gas on 660 MW Double Reheated Boiler

H. Xiao, Y. X. Wu, L. L. Feng, C. W. Meng, H. Zhang, M. Zhang, and Z. Chai

Abstract To investigate effect of recirculated flue gas (RFG) on combustion, heat transfer and pollutant emission of double reheated boiler, a three dimension CFD model was established and a scalar equation of mass fraction of RFG was involved to analyze the dispersion of RFG in the furnace. Realizable k-ε model, Euler-Lagrange method, DO model and finite-rate/eddy-dissipation were applied to solve the turbulent flow, dispersion of pulverized coal, radiation and gas-phase turbulent combustion in the furnace. The results show that when the RFG ratio is low, recirculated flue gas was mainly distributed in the central zone of the furnace and spiraled upward in the combustion zone. When the amount of RFG increases, it was diffused around the furnace from the central furnace to the vicinity of the wall, which effectively reduced the temperature. Moreover, NOx emission decreases with the increase of RFG ratio but the carbon burnout decreases apparently when the RFG ratio is larger than 10%. The results show that a RFG ratio of 6–10% was the best suitable rate for 50% THA condition.

Keywords Double reheated boiler · Flue gas recirculation · Combustion · Heat distribution · NOx

H. Xiao · Y. X. Wu (✉) · L. L. Feng · H. Zhang · M. Zhang
Key Laboratory for Thermal Science and Power Engineering of the Ministry of Education, Department of Energy and Power Engineering, Tsinghua University, Beijing 100084, China
e-mail: wuyx09@tsinghua.edu.cn

C. W. Meng
Sibley School of Mechanical and Aerospace Engineering, Cornell University, Ithaca, NY 14853, USA

Z. Chai
Shanghai Boiler Works Co., Ltd., Shanghai 200240, China

J. Lyu and S. Li (eds.), *Clean Coal and Sustainable Energy*, Environmental Science and Engineering,
https://doi.org/10.1007/978-981-16-1657-0_82

82.1 Introduction

At present, thermal power generation is still the main method of electric power production in China (McHugh 2018). Improving steam parameters is the most direct means to improve the efficiency of the unit (Bugge et al. 2014). However, due to the limitation of high temperature resistant materials, it is difficult to achieve a great increase in steam temperature to 700 centigrade (Zhu and Zhou 2009; Abe 2015). Therefore, double reheated boiler has become a good choice to improve the efficiency of the unit (Li et al. 2014). For example, the double reheated boiler in Taizhou has become one of the most efficient boilers in the world (Yong 2015). However, the difficulty of energy distribution and temperature regulation of boiler heating surface is increased due to more heating surfaces (Cai et al. 2016), especially under wide load. And the uncontrollability of temperature regulation of double reheated boiler is greatly improved (Zhang et al. 2013). Therefore, the key problem in the application of double reheated boiler is to find flexible means of regulating the steam temperature of the secondary reheating boiler. And flue gas damp, flue gas recirculation, and tilting burner are the common methods to control the temperature. Previous studies have shown that flue gas damp and tilting burner have some limitations (Chuansheng 2016; He 2015; Jinying et al. 2016) except their advantages so flue gas recirculation is an important supplement for double reheated boiler (Zhang et al. 2015). And it's of great significance to comprehend the effect of RFG on combustion, heat transfer and pollutant emission of double reheated boiler because RFG may be unfavorable to economy and reliability of the unit, while little such research about the RFG's effect of on combustion, heat transfer and pollutant emission of double reheated boiler was reported (Zhi-Yun and Bao-Jun 2018; Ming et al. 2018).

82.2 Geometry Model and Meshing

The boiler is tangentially fired and RFG is injected into the furnace on both sides of the ash hopper, there are two groups of flue gas recirculation nozzles, six nozzles in each group. There are six groups of burner nozzles in the boiler, which are recorded as A, B, C, D, E, F from lowest to highest and two groups of burnout air nozzles. They are all installed around the furnace. The burner nozzle includes primary air nozzle and secondary air nozzle.

Because the structure of burner nozzle is very complex, it is simplified to some extent. In order to reduce the pseudo-diffusion error caused by the angle between the flow direction and the grid line, the grid line is rotated by 45° and the mesh near the burner nozzle is radiant (Mallinson and Davis 1973), similar to the actual flow direction of the fluid. Three sets of grids with similar structure and different densities are divided, and the calculation is carried out under cold conditions. The velocity field in the furnace area is observed. It is found that in more than 1.8 million grids,

Fig. 82.1 Schematic diagram of mesh generation

the grid density no longer has a significant impact on the calculation results. Finally, the number of grids is determined (Fig. 82.1).

82.3 Numerical Model

The turbulence information was calculated through Realizable k-ε model (Manić et al. 2017). Discrete-Ordinates model (Wacławiak and Kalisz 2012) was adopted to calculate the radiation heat transfer, diffusion dynamically coupled model was used to calculate coke reaction (Chenwei et al. 2015) and finite-rate/eddy-dissipation model (Achim et al. 2009; Khaldi et al. 2016) was used to solve the coupling of turbulence and chemical reactions.

In order to quantitatively analyze the distribution characteristics of RFG in the furnace, the mass fraction of recirculating flue gas is defined as follows:

$$\frac{\partial \rho \phi_z}{\partial t} + \frac{\partial}{\partial x_i}\left(\rho u_i \phi_z - \left(\rho D_z + \frac{\mu_t}{Sc_t}\right)\frac{\partial \phi_z}{\partial x_i}\right) = S_{\phi_z}$$

Dz is the mass transfer coefficient of recycling flue gas, which is approximately equal to the mass transfer coefficient of oxygen in nitrogen; Sc_t is the turbulent Schmidt number, which is 0.75; μ_t is the turbulent viscosity coefficient, which is calculated according to the corresponding turbulence model; $S_{\Phi z}$ is the source term of RFG, because the RFG belongs to the external input scalar, its source term $S_{\Phi k}$ = 0. Based on the above definition of recirculating flue gas, the entrance conditions of each boundary can be determined as follows: the scalar Φ_z at the entrance of recirculating flue gas is defined as 1, and the other entrance scalar Φ_z is defined as 0. By solving the transfer equation in the computational domain, the ratio of RFG in the boiler can be obtained, and the diffusion process of RFG can be mastered.

Table 82.1 Industrial and elemental analysis of coal

Industrial analysis	M_t/%	A_{ar}/%	V_{ar}/%	FC_{ar}/%	LHV/(MJ/kg)
	11.9	28.25	22.86	36.98	18.4
Elemental analysis	C_{ar}/%	H_{ar}/%	N_{ar}/%	O_{ar}/%	$S_{t,ar}$/%
	47.4	3.24	0.8	7.21	1.2

82.4 Boundary Conditions

In order to facilitate the analysis of particles, according to the requirement that coal particles meet R90 = 18–20%, the particles are divided into five grades and the industrial and elemental analysis of pulverized coal is shown in Table 82.1.

When the volatilization of pulverized coal is removed, the decomposition reaction takes place. According to the data of elemental analysis of coal, based on the assumption of element conservation and proportion between components, the calculation can be determined. The release ratio of each component in the process of volatilization can be determined as shown in the following formula.

$$m_{Vol} \rightarrow 0.032N_2 + 0.043H_2S + 1.315H_2 + 0.256CO + 0.128CO_2$$
$$+0.241C_2H_2 + 0.12CH_4$$

Inlets of flue gas, primary and secondary air were set to be mass flow inlet and the outlet was pressure outlet. The temperature of different heating surfaces is set as a constant according to the first boundary condition according to the thermodynamic calculation table.

82.5 Effect of Load Reduction on Double Reheated Boiler

Under BMCR condition, primary air was injected through nozzle A, B, C, D and E, and the rate of RFG was 4%. When the load was reduced to 75% THA, nozzle A was closed and nozzle B was closed under 50% THA, while the rate of RFG was not changed.

Under different loads, the temperature field tends to be high around and low at the center. The highest temperature in the furnace area is located near the burner nozzle. As for the overall temperature, with the decrease of load, the average temperature in the furnace decreases considerably. In the burner area in operation, there is still a distinct high temperature area. As the load decreases, the low temperature zone in the center of the furnace begins to expand outward as Fig. 82.2.

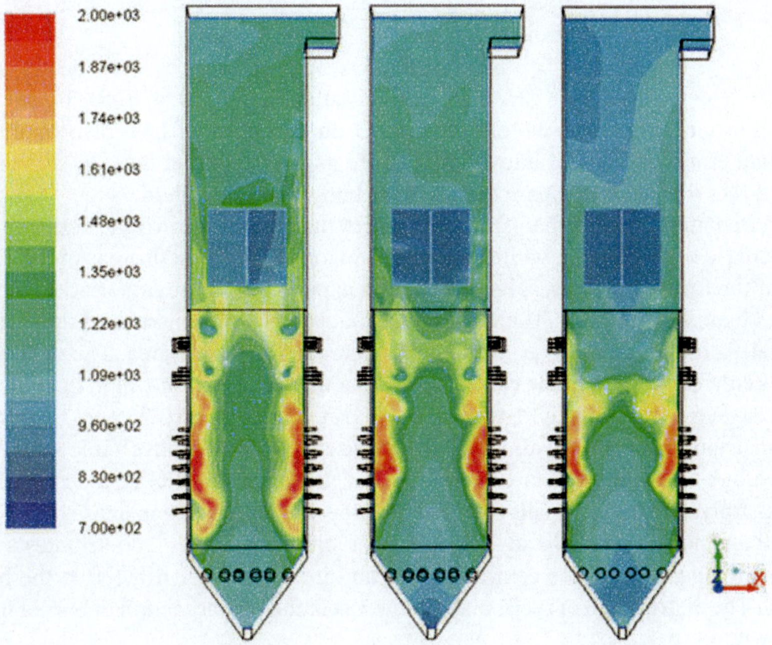

Fig. 82.2 Temperature of central section of furnace under three different loads (K)

Besides, the burnout rate of pulverized coal under three different loads was all equal to 100%, which shows that the a flue gas recirculation rate of 4% has no effect on the normal burnout of pulverized coal.

However, while the load decreases, the absolute heat absorption of each heating surface decreases significantly, but at the same time, the heating surface of furnace water wall, which is mainly radiation heating surface, its heat absorption decreases slightly, and the heat absorption of the rear heating surface decreases greatly. Therefore, it is necessary to adjust the temperature in order to maintain the stability of the endothermic ratio.

The absorptive heat of furnace water wall and tail heating surface (primary reheater high temperature section, secondary reheater high temperature section and high temperature superheater) under three kinds of loads and their decreasing range are shown in the Table 82.2.

Table 82.2. Three kinds of loads and their decreasing range

	Water wall (10^8 W)	Drop rate, %	Tail heating surface (10^8 W)	Drop rate, %
BMCR	5.7	0	0.92	0
75%THA	4.9	14.0	0.62	32.6
50%THA	4.1	28.1	0.38	58.7

82.6 Effect of RFG

Under the low load of 50% THA, six gas recirculation rates (0, 4, 6, 10, 14 and 20%) were selected to analyze the effect of RFG on combustion, heat distribution and pollutant emission. RFG is a low temperature gas in the burner area. Its distribution has a direct impact on the temperature distribution and flow field.

Recirculated flue gas mainly distributes in the central area of the furnace. With the increase of its content, while the proportion of recirculated flue gas in the central area of the furnace increases, its distribution expands from the center of the furnace to the vicinity of the wall. The largest distribution area of recirculated flue gas also gradually diffuses from the cold ash hopper area to the burner area. The distribution of recirculated flue gas in the central section of the furnace is shown in the Fig. 82.3.

At the burner nozzle height, when the RFG rate is less than 10%, there is almost no RFG distribution near the wall. The RFG in the central region plays a role in lowering the average temperature. Starting from 10%, the share of RFG in the central area rises sharply, and the expansion of RFG near the wall causes the increase of the share of wall and nozzle. For 14 and 20% flue gas recirculation rate, it contributes a lot to the low temperature in the central region, and it is widely distributed in the hearth region. The distribution of recirculating flue gas in the height section of burner nozzle is shown in Fig. 82.4.

And the temperature field is significantly affected by the RFG. On the central section of the furnace, with the increase of flue gas recirculation rate, the low temperature zone of the cold ash hopper area gradually expands to the main combustion zone

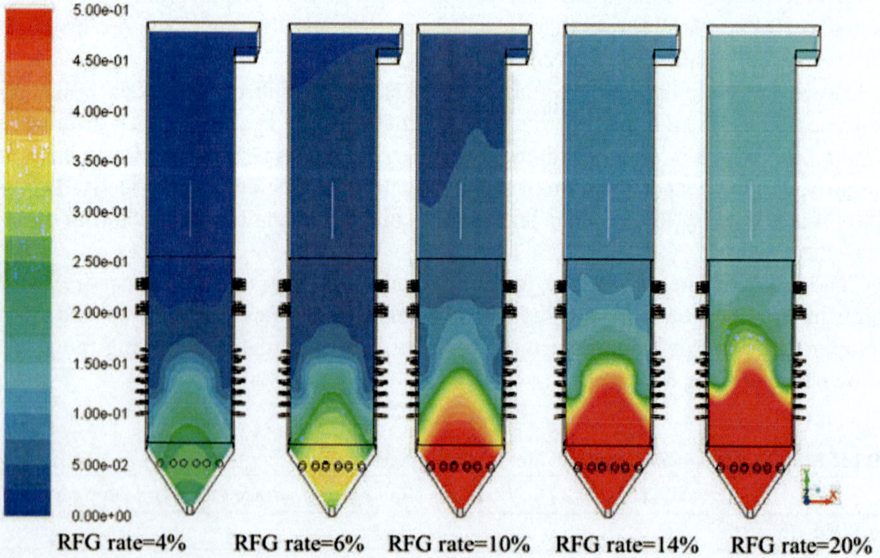

Fig. 82.3 Distribution of RFG in central section of furnace

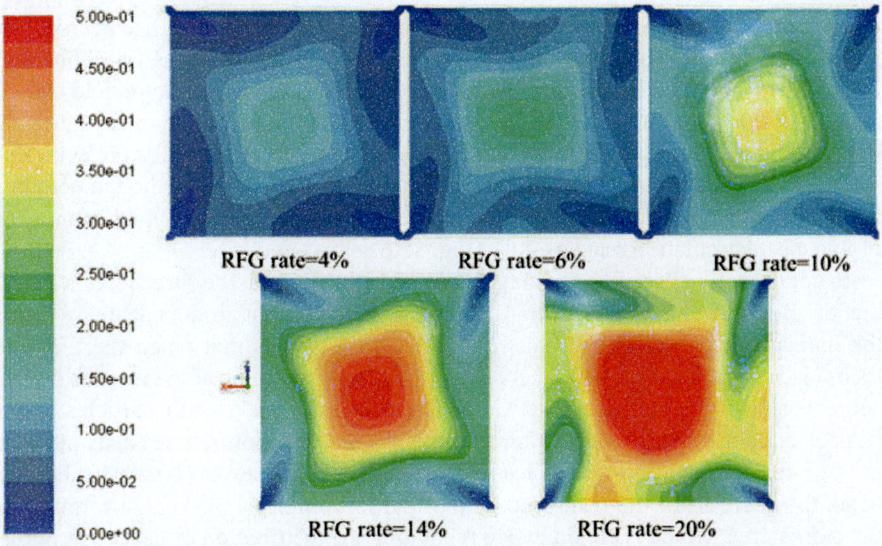

Fig. 82.4 Distribution of recirculating flue gas in the first layer burner nozzle height section

of the furnace, which reduces the temperature level of the main combustion zone and maintains a higher combustion temperature near the furnace wall. The temperature field of the central section of the furnace at various flue gas recirculation rates is shown in Fig. 82.5.

Under 50% THA load, it can be found that the temperature distribution still shows the trend of low center and high around. With the increase of flue gas recirculation rate, the temperature near the wall began to maintain a higher temperature, but when

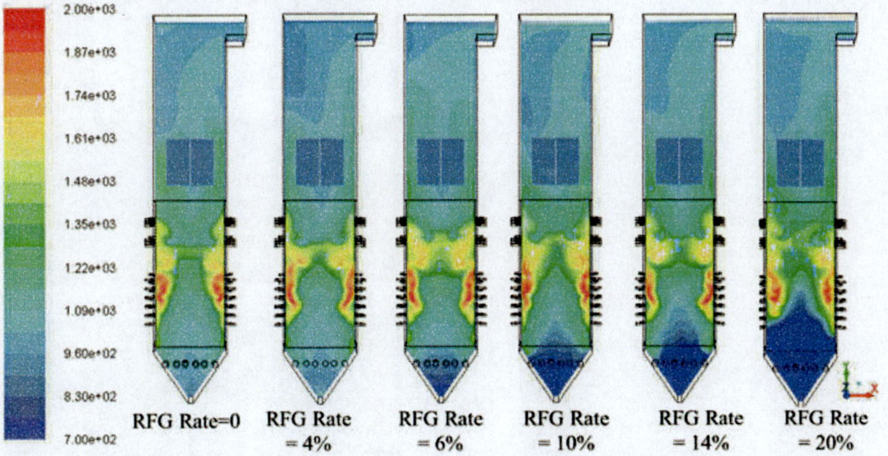

Fig. 82.5 Central section temperature of furnace under different flue gas recirculation rates (k)

its content increased to 20%, on the one hand, the central temperature had been below 1000 K, on the other hand, the temperature near the wall also decreased significantly. As for the symmetry of temperature field, the distribution of temperature field is very symmetrical without the addition of flue gas recirculation. With the introduction of RFG, the temperature field of front and rear walls, left and right walls are centrally symmetrical. When the rate of flue gas recirculation rises to 10%, the temperature field is irregular, and the shape of tangential circle can hardly be seen when the rate of flue gas recirculation reaches 20% (Fig. 82.6).

In addition, by observing the change of residual carbon in the furnace, it is found that the Burnout Situation in the burner area and the burnout air area decreases with the increase of recycling flue gas. The calculation shows that when the flue gas recirculation rate reaches 10%, there is still 1% particle residue at the furnace outlet. Moreover, with the increase of RFG, the average residence time of particles in the furnace continues decreasing. After the flue gas recirculation rate exceeds 6%, the average residence time of particles in the furnace decreases by 10%. On the one hand, the decrease of the residence time of particles inhibits the heat absorption in the burner area, which is conducive to regulating the distribution of heat absorption, on the other hand, it has an impact on the burnout of particles. The residence time and burnout rate of particles at different flue gas recirculation rates is shown in Fig. 82.7.

With the increase of flue gas recirculation rate, the heat absorption share of water wall in furnace decreases gradually, and the heat absorption share of tail heating surface increases gradually, which has a more obvious effect when the flue gas recirculation rate reaches 4%. When the flue gas recirculation rate reaches 20%, the distribution of endothermic ratio is basically consistent with the rated load. The heat

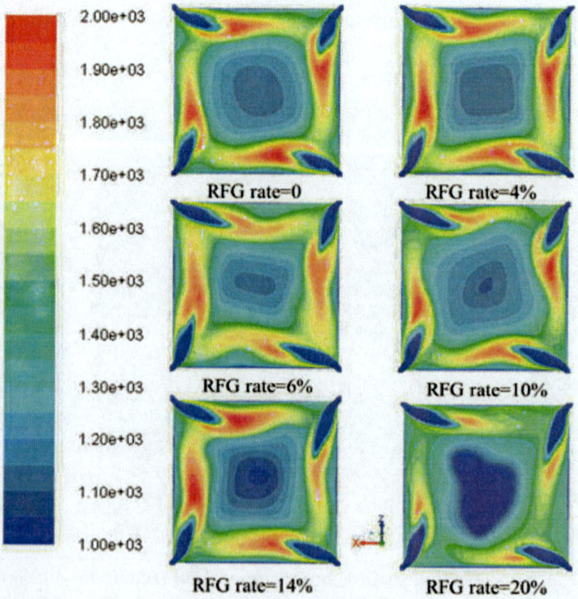

Fig. 82.6 Temperature distribution at nozzle height section of first layer burner (K)

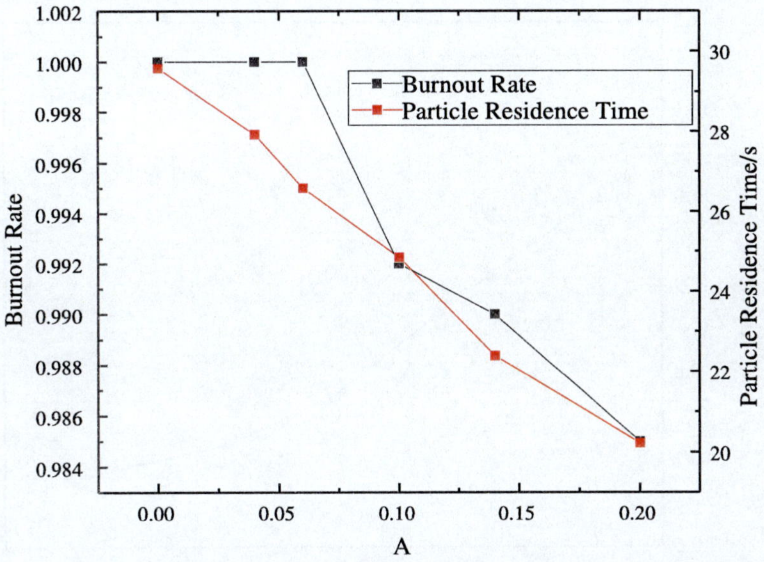

Fig. 82.7 Residence time and burnout rate of particles at different flue gas recirculation rates

absorption and decrease range under different flue gas recirculation rates, of furnace water wall are shown in Table 82.3.

With the increase of RFG content, the concentration of NOx decreased gradually, but the effect of RFG on the change of NOx was weakened in the range above 10%. The principle of reducing NOx by flue gas recirculation is mainly to reduce the temperature of the furnace area. Without the RFG, the furnace area maintains a very high temperature level. When the recirculation flue gas is added, the temperature level in the furnace area is lowered, resulting in a significant decrease in the content of NOx. On the other hand, as a reducing gas, with the increase of the recycling gas, the reducing atmosphere in the furnace gradually strengthens, which can also play a role in weakening the formation of NOx (Fig. 82.8).

Table 82.3 Heat absorption and descender under different flue recirculation rate

RFG rate (%)	Heat absorption of water wall ($\times 10^8$ W)	Descender (%)
0	4.23	0
4	4.14	2.12
6	4.00	5.43
10	3.97	6.14
14	3.91	6.85
20	3.40	19.6

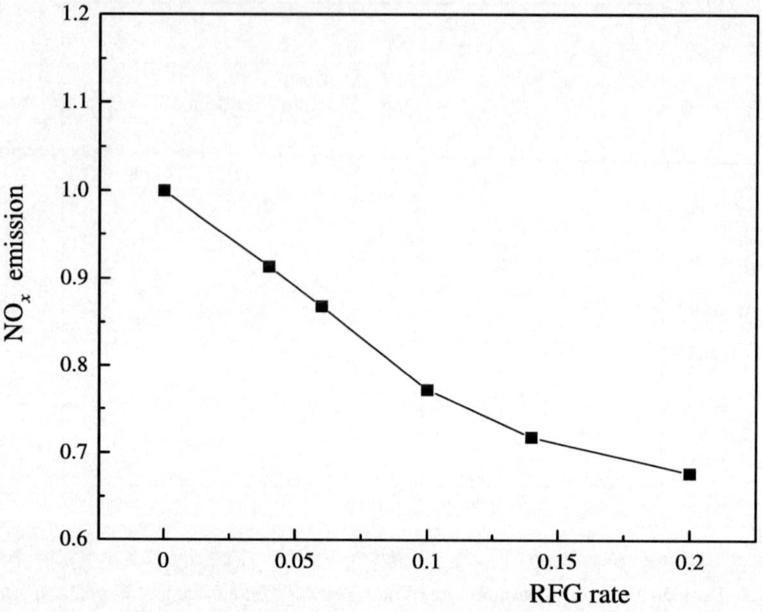

Fig. 82.8 NOx emission under different RFG rate (Normalization)

82.7 Conclusion

With the decrease of boiler load, when the amount of flue gas recirculation remains unchanged, the proportion of heat absorbed by furnace water wall increases, and the share of heat absorbed by tail convection heating surface decreases, which is the main reason for the decrease of steam temperature. At 50% THA load, when the RFG rate is low, it mainly distributes in the central area of the furnace and spirals upward in the combustion area. With the increase of the RFG rate, the RFG gradually diffuses into wall area of the furnace and affects the combustion and flame characteristics of the fuel. When the proportion of RFG exceeds 10%, the maximum flame temperature decreases significantly, and the average cross-section temperature decreases to 80 °C. At the same time, the decrease of flame temperature leads to a significant decrease in radiation heat absorption of the water wall of the furnace and increases the heat absorption share of the convective heating surface. Regarding the generation of NOx, RFG can reduce the emission of NOx. When the proportion of NOx exceeds 10%, the influence on NOx is not significant, and the burnout rate of coke decreases significantly. Comprehensive analysis shows that the optimum RFG rate for double reheated boiler under low load ranges from 6 to 10% of the total flue gas volume.

Acknowledgements This work was supported by the National Key R&D Program of China (2017YFB0602102).

References

Abe F (2015) Research and development of heat-resistant materials for advanced USC Power plants with steam temperatures of 700 °C and above. Engineering 1(2):211–224

Achim D, Naser J, Morsi YS et al (2009) Numerical investigation of full scale coal combustion model of tangentially fired boiler with the effect of mill ducting. Heat Mass Transf 46(1):1–13

Bugge J, Kjær S, Blum R (2014) High-efficiency coal-fired power plants development and perspectives. Energy 31(10):1437–1445

Cai B, Gao H, Wang J et al (2016) Dynamic process characteristic analysis and control strategy verification on double-reheat ultra-supercritical coal-fired power units. In: Proceedings of the CSEE, 19

Chenwei M, Yuxin W, Hai Z et al (2015) Investigation of coal reactivity effects on a 1000 MW tangential coal-firing boiler. In: International symposium on coal combustion

Chuansheng LI (2016) Engineering application of reheat steam temperature control mode for double-reheat tower boilers. Therm Power Gener

Yong DJ (2015) 1000 MW Supercritical double reheat boiler technology. Value Eng

He W (2015) Match characteristics and influence of Dongfang double reheat heating surface. Chongqing University

Ming JG, Xin G, Hui-Yi LI (2018) Numerical simulation on NOx distribution of ultra Supercritical double reheating pulverized coal boiler. Power Syst Eng

Jinying Z, Qiusheng Z, Jianhong LV et al (2016) Analysis on the application of reheat steam temperature control strategy in ultra supercritical units with secondary reheat cycle. Electr Power

Khaldi N, Chouari Y, Mhiri H et al (2016) CFD investigation on the flow and combustion in a 300 MWe tangentially fired pulverized-coal furnace. Heat Mass Transf 52(9):1881–1890

Li Y, Zhou L, Gang X et al (2014) Thermodynamic analysis and optimization of a double reheat system in an ultra-supercritical power plant. Energy 74(2):202–214

Mallinson GD, Davis GDV (1973) The method of the false transient for the solution of coupled elliptic equations. J Comput Phys 12(4):435–461

Manić NG, Jovanović VV, Stojiljković DD et al (2017) Application of different turbulence models for improving construction of small-scale boiler fired by solid fuel. Therm Sci 809–823

McHugh L (2018) Reducing emissions from coal: a role for the World Bank. World Coal Magaz

Wacławiak K, Kalisz S (2012) A practical numerical approach for prediction of particulate fouling in PC boilers. Fuel 97:38–48

Zhang G, Xu W, Wang X et al (2015) Analysis and optimization of a coal-fired power plant under a proposed flue gas recirculation mode. Energy Convers Manage 102:161–168

Zhang FW, Liu YY, Tan HZ et al (2013) Study of secondly reheat technique of supercritical fire power generators. Electric Power Surv Des

Zhi-Yun ZL, Bao-Jun BS (2018) Influence of flue gas recirculation rate on temperature of double-reheater unit. Power Syst Eng

Zhu BT, Zhou RC (2009) Problems paid attention to raise the ultra supercritical units steam parameters. Proceedings of the CSEE 29:95–100

Chapter 83
Furnace Outlet Temperature Prediction Model of a 350 MW Ultra-Supercritical Boiler

Tianyu Zhang and Zhenning Zhao

Abstract Ultra-supercritical boiler has been widely applied due to its advantages of high combustion efficiency and low pollutant emission. Furnace outlet temperature can indicate the boiler operating status, because of which, related investigations draw various researchers' attention. Based on a 350 MW design data, this paper established a model for predicting furnace outlet temperature. Comparison of design data and predicted result validates the accuracy of present model. Furnace outlet temperature of different operating parameter (e.g. excess air coefficient) is predicted, which can provide reference for actual operation

Keywords Ultra-supercritical boiler · Temperature prediction

83.1 Introduction

Ultra-supercritical boiler has been widely applied due to its advantages of high combustion efficiency and low pollutant emission. Furnace outlet temperature can indicate the boiler operating status, because of which, related investigations draw various researchers' attention. Based on a 350 MW design data, this paper established a model for predicting furnace outlet temperature. Comparison of design data and predicted result validates the accuracy of present model. Furnace outlet temperature of different operating parameter (e.g. excess air coefficient) is predicted, which can provide reference for actual operation.

The requirements of high combustion efficiency, low pollutant emission and operating flexibility drive the investigations on fossil-fire plant with high parameter. In recent decades, an increasing number of once-through boilers with ultra-supercritical parameter are designed and put into services, and many researchers focus on the heat transfer, fluid flow and construction of boiler. However, due to the complexity of combustion process and heat transfer in furnace, investigation may not be possible to involve all the whole process of actual operating condition.

T. Zhang (✉) · Z. Zhao
North China Electric Power Research Institute, Beijing 100032, China

J. Lyu and S. Li (eds.), *Clean Coal and Sustainable Energy*,
Environmental Science and Engineering,
https://doi.org/10.1007/978-981-16-1657-0_83

In present paper, a model of classical zero-dimensional model is established to cast light on the heat transfer capacity, heat flux distribution (radiative and convective), temperature of theoretic combustion and furnace outlet at different operation condition with varies in unit load and excess air coefficient. Relative result may provide reference and perspective for graduates, operator and researchers.

83.2 Model and Governing Equations

See Table 83.1.

In order to utilize the zero-dimension model, several assumptions should be made:

1. Fossil combustion and heat transfer in furnace proceeds separately.
2. Combustion process finishes immediately and combustion products diffuse into whole furnace with uniform physical character.
3. The influence of platen superheater on heat transfer within furnace is neglected.

Based on the assumptions above, the flue temperature at furnace outlet Tout (K) can be calculated as:

$$T_{out} = \frac{T_a}{M\left(\frac{\sigma_0 \Psi F_1 a_1 T_a^3}{\varphi B_j V_{cpj}}\right)^{0.6} + 1} \quad (83.1)$$

where

M Parameter related with height on which the highest flame temperature lies;
σ_0 Absolute blackbody radiation coefficient, 5.67×10^{-11} kJ/(m^2 s K^4);
Ψ Thermal Effectiveness Coefficient;
F_1 Heating surfaces area in furnace;
a_1 Furnace blackness;
T_a theoretical combustion temperature;
φ Heat preservation coefficient;

Table 83.1 Boiler main parameter

Item	Unit	BMCR	75% THA	50% THA	30% THA
Main steam flow	t/h	1172	748	486	352
Main steam pressure	MPa g	25.4	23.5	15.7	11.5
Main steam temperature	°C	571	571	571	571
Reheat steam flow	t/h	962	631	420	307
Reheat steam pressure	MPa g	5.22	4.84	2.25	1.60
Reheat steam temperature	°C	569	569	569	544
Feedwater temperature	°C	295	267	243	226

B_j Calculated fuel consumption;
V_{cpj} Mean specific heat.

The calculation of parameter involved in Eq. 2-1 can be referred in literature (Thermodynamic calculation of boiler unit 1973). After getting the theoretical combustion temperature and furnace outlet temperature, the radiative and convective heat transfer flux (Q_f and Q_d) can be calculated as:

$$Q_f = a_1 \Psi \sigma_0 T_x^4 F_1$$

$$Q_d = \alpha_d (T_x - T_b) F_1$$

$$\frac{T_x}{T_a} = \left(\frac{T_{out}}{T_a}\right)^n$$

$$T_b = T_s + \frac{s}{\lambda} \Psi a_1 \sigma_0 T_x^4$$

where

T_x Mean temperature of flame;
T_b Fouling surface temperature;
α_d Heat transfer coefficient, calculation method varies, present study take advantage in Miu (1996).

Based on the model above, the theoretical combustion temperature Ta, furnace outlet temperature Tout, radiative heat transfer capacity, convective heat transfer capacity can be obtained. In present study, a validation is made by comparing the calculated furnace outlet temperature with design data, then working condition of different excess air coefficients is investigated.

83.3 Calculation Condition, Validation and Result Analysis

Four typical working conditions of BMCR, 75% THA, 50% THA and 30% BMCR are in scope of present investigation (Table 83.2).

Table 83.2 Main calculating parameter setup

	BMCR	75%	50%	30%
Coal feed (t/h)	187.0	129.5	89.2	67.5
Excess air coefficient	1.15	1.20	1.32	1.45
Hot wind temperature (°C)	330	310	293	282

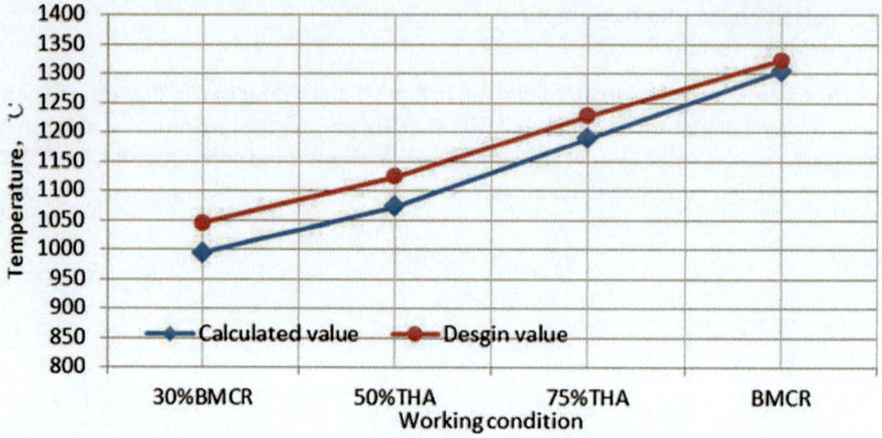

Fig. 83.1 Validation: comparison of calculated and design temperature

Hence, the calculated furnace outlet temperature and design data can be compared as shown Fig. 83.1.

As can be seen from the figure above, the deviations between the calculated and design temperature at four typical working conditions are limited within 50 °C, and this value is generally accepted in furnace thermal calculation, which validates the accuracy of present model, including relative parameter setup.

The theoretical combustion temperature can indicate the stability of combustion to a certain extent, and is determined by coal and excessive air coefficient. Figure 83.2 shows the influence of boiler load and operating excess air coefficient on theoretical combustion temperature. It can be seen obviously that as the load increases and excess air coefficient decreases, the theoretical combustion temperature rises accordingly. One reason low load requires high air coefficient is that though the combustion

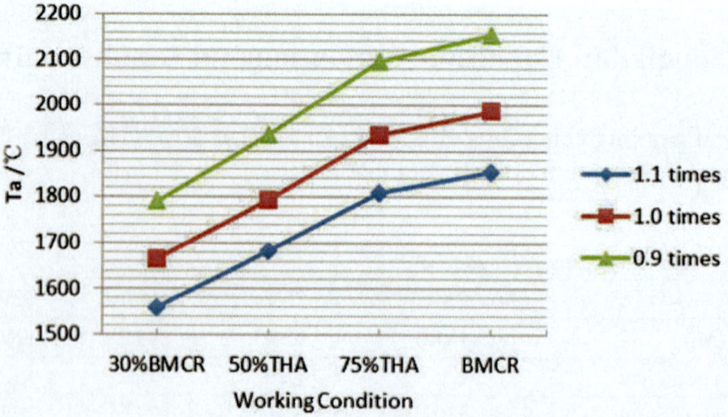

Fig. 83.2 Theoretical combustion temperature under different load and excess air coefficient

period lies in dynamic period, excess air is needed to promote the burn out of the coal particle.

In terms of radiative and convective heat transfer capacity, the four typical working conditions exhibit an inclination: with boiler load increases, the amount of heat transfer in radiative form increases while the form in conduction decreases, as shown in Fig. 83.3. And calculated result also accords with the assumption of neglect of convection in furnace in some furnace thermal calculation model.

In actual operating status, change in air coefficient occurs frequently, resulting in variation of combustion. Hence, it is important to get the changing rules of critical parameter in heat transfer—furnace outlet temperature. Figure 83.4 shows the deviation from design condition. When the excess air coefficient in furnace decreases, the outlet temperature increases. However, by comparing the theoretical combustion temperature and furnace outlet temperature, it can be noticed that changing amplitude in theoretical combustion temperature is much more than furnace outlet temperature. It is mainly because the rise in combustion temperature leads to more furnace heat absorption, especially in form of radiation. According result can be seen

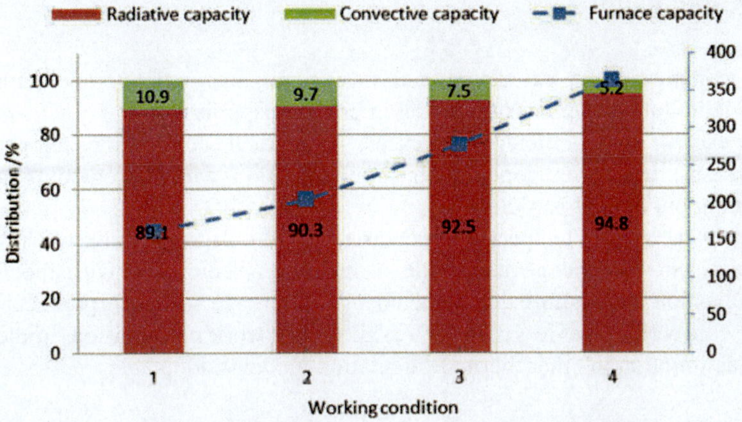

Fig. 83.3 Furnace overall capacity and distribution of radiation and convection

Fig. 83.4 Excess air coefficient's influence on furnace outlet temperature

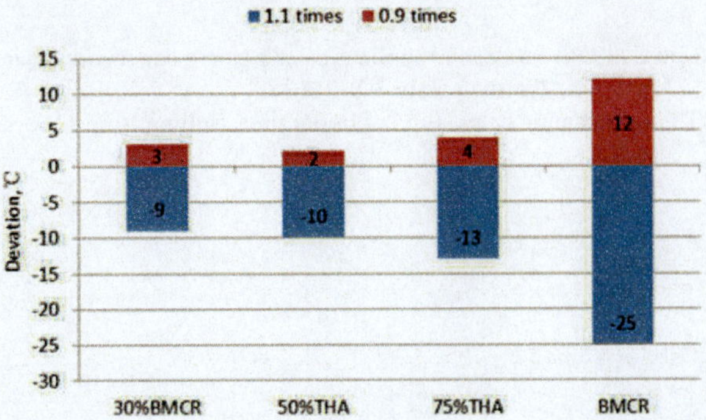

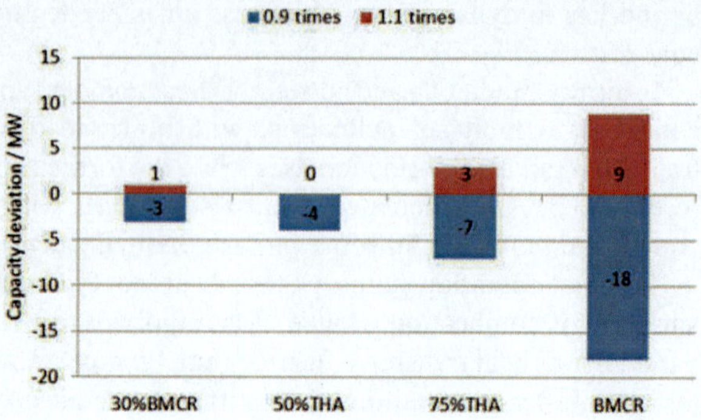

Fig. 83.5 Excess air coefficient's influence on furnace overall capacity

in Fig. 83.5, and the change in BCMR condition is the most in the four compared working condition.

83.4 Conclusion

1. By comparing the calculated and design furnace outlet temperature, the validation of zero-dimension model in present paper is made.
2. As excess air coefficient decreases, theoretical combustion temperature increases, indicating good combustion condition without regarding to the burning out of coal particle.
3. The furnace outlet temperature increases with the decrease in excess air coefficient, but the change amplitude is much less compared with theoretical combustion temperature due to the change in furnace heat absorption capacity.
4. The convective takes less than 10% in all studied working conditions, indicating the assumption in other thermal calculation model is adoptable.

References

Miu Z (1996) Model and application of particle convection heat transfer in pulverized coal furnace. J Shanghai Jiaotong Univ 7:100–105

Thermodynamic calculation of boiler unit. Beijing Boiler Works (1973)

Chapter 84
Study on the Stability of High Pressure Pneumatic Transport of the Mixture of Pulverized Coal and Extraction Residue of Direct Coal Liquefaction Residue

B. Z. Peng, X. H. Fang, H. Q. An, Z. Liu, Y. Li, and Z. Y. Feng

Abstract The use of direct liquefaction residue of coal has been the focus of attention. Due to the liquefaction residue contains more organic compounds, it is necessary to recover these organic compounds by extraction, and then the extraction residue is considered as a feedstock, which mixed with coal into a powder and enter a dry-fed entrained flow coal gasifier for gasification. However, the stability of high pressure pneumatic transport of the mixture of pulverized coal and extraction residue is crucial to the safe and stable operation of dry-fed entrained flow coal gasifiers. In this paper, A set of up to 6.0 MPag high pressure dense phase transport experiments was built, and high pressure pneumatic conveying experiments were carried out using a high-volatile bituminous coal and its mixture with 20 wt% extraction residue of direct coal liquefaction residue (DCLR) in pipes with diameters of 25 mm and 15 mm, respectively, and at back pressures of 1.0–4.0 MPag. By comparing and calculating the relative standard deviations (RSDs) of different signals on the transport stability, the RSD of solid flow rate was selected as an indicator for conveying stability. The result show that the RSDs of both transport stability signals of coal and its mixture with 20 wt% extraction residue of DCLR is less than 4%, which indicated that good stability was found in both the coal and its mixture with 20 wt% extraction residue of DCLR.

Keywords Pneumatic transport · Pulverized coal · Coal direct liquefaction residue · Extraction residue · Stability

84.1 Introduction

Coal is still the main energy in china so far and coal gasification is a key technology for coal chemical field. Coal direct liquefaction (CDL) is a process for convering coal

B. Z. Peng (✉) · X. H. Fang · H. Q. An · Z. Liu · Y. Li · Z. Y. Feng
National Institute of Clean and Low Carbon Energy, Beijing 102211, China
e-mail: baozi.peng@chnenergy.com.cn

J. Lyu and S. Li (eds.), *Clean Coal and Sustainable Energy*, Environmental Science and Engineering,
https://doi.org/10.1007/978-981-16-1657-0_84

to the liquid fuels and the remain residue is up to about 30% of the total raw coal (Shu et al. 2003). Therefore, the utilization of direct coal liquefaction residue (DCLR) has become a focus topic. Many methods for dealing with DCLR were presented, such as combustion, gasification (Lv et al. 2015), co-pyrolysis with low-rank coals (Li et al. 2015), directly use as asphalt or asphalt modifiers (Ji et al. 2018) and so on. Among all these methods, gasification is considered to be the best method for dealing with large scale DCLR. However, due to the DCLR contains more organic compounds, it is necessary to recover these organic compounds by extraction (Li et al. 2012), and then the remaining extraction residue that mixed with coal is used as a gasification feedstock. However, there is still no any relevant report on high pressure pneumatic transport of the mixture of pulverized coal and extraction residue of DCLR so far.

In the gas flow dry bed feed gasification process, the stable transport of pulverized coal is the key to ensure the safe and stable operation of the gasifier. In recent years, many researchers have carried out some research on the stable transport of pulverized coal by using signal statistical analysis methods. Such as Lin et al. (2014) studied the steady flow of pulverized coal flow in the Shell gasifier, which the result showed the relative standard deviation of pulverized coal flow fluctuations in the stable operation of the Shell gasifier was 2–3%, and the maximum fluctuation range was 14–17%. Cong et al. (2013) used the relative standard deviation of the pressure signal to characterize the transport stability and presented a coal flow stability criterion. It was indicated that the flow was stable during the relative standard deviation of the pressure signal is less than 0.1, which the Fr number of the pipeline is greater than 10. Based on this, the minimum gas velocity stably delivered on pilot plant and Shell gasification industrial plant was also predicted. Liang et al. (2012) investigated the influences of operating parameters and material properties on conveying characteristics, and used Wavelet transform and Shannon entropy analysis to obtain stability criterion. In addition, Liang et al. (2015) also used the relative standard deviation of the pulverized coal flow signal to characterize the transport stability and analyzed the variation of the transport stability with the total pressure difference of the transport. Xie et al. (2013) studied the transport stability by using the maximum peak power spectral density and relative standard deviation of the pressure signal, which defined the minimum pressure drop speed at the local minimum value of the above left parameters as the boundary of the stable transport, and showed that the gas velocity was below the minimum pressure drop speed of about 1–1.5 m/s operating conditions. Mallick and Wypych (2009) used the fluctuation in the transport experiment and the "non-linearity" of the receiving tank feed as the criteria for conveying instability, who proposed that the boundary of the pulverized coal conveying stability was $Fr = 5.7$. He et al. (2014) used the Hurst exponent analysis of the differential pressure signal of the tube length of the horizontal tube to characterize the transport stability. When the superficial gas velocity of the pipe was less than 1.5 times the deposition rate, the pulverized coal transport stability became worse. Ma et al. (2015) believe that the fluctuation of coal powder mass flow in a GSP gasifier is less than 5% (maximum amplitude of 10%) is stable.

It can be seen that the pulverized coal transport stability in the industry is mainly concerned with the fluctuation of the pulverized coal flow rate, while the experimental research on the pulverized coal transport stability mainly uses the signal statistical analysis method. Due to the high measurement accuracy and low cost of the pipeline pressure signal or differential pressure signal, most of the studies are based on statistical analysis and qualitative or quantitative stability criteria. The conditions for stable transportation are mostly related to the pipeline gas velocity or the number of pipeline Fr. However, the stability index of pulverized coal transportation is not uniform, and there are no comparative analyses between different researchers suggesting the differences between stability indicators and conditions. In this paper, high pressure pneumatic conveying experiments were carried out using a high-volatile bituminous coal and its mixture with 20 wt% extraction residue of coal direct liquefaction residue (DCLR) in pipes with diameters of 25 mm and 15 mm, respectively. The relative standard deviation of solid flow rate was used as an indicator for investigate the transport stability by using a dense-phase pneumatic conveying experiment.

84.2 Experimental Section

84.2.1 Experimental Facility and Procedure

The flow diagram of the high pressure dense phase pneumatic conveying system is shown in Fig. 84.1. High pressure nitrogen flows from a nitrogen tank installed in a truck to a storage vessel through pressure reducing valves, and is then fed into a delivery vessel (receiving vessel). Pressurizing gas is used to pressurize the vessels, fluidizing gas is used to fluidize coal powder for easier transportation, and supplementary gas is used to suppress solid deposition and to adjust gas velocity and particle concentration in the pipe. The pressure of vessels is controlled by electric control

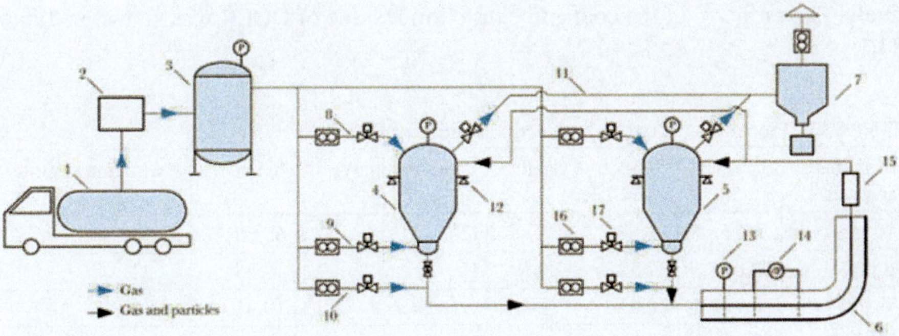

Fig. 84.1 Flow diagram of the dense-phase conveying system

valves installed at pressurizing gas pipe and relieving gas pipe. The volume flow rate of fluidizing gas is adjusted as the pressure of the vessel varies until coal powder inside the bottom of the vessel reaches optimal fluidization status. The fluidized coal powder is discharged from the bottom of the delivery vessel, mixing with the supplementary gas, and flows into the conveying line through a coal distributor. The conveying line consists of three parallel branches of different pipe diameters. Several valves are installed at the entrance and exit of the branches for switching between these branches. The three parallel branches join together through another coal distributor, following which is a solid mass flowmeter installed in a vertical downward pipe. The gas solid mixture eventually flows into the receiving vessel, where gas and solid particles are separated, and carrier gas is vented after a bag-filter. The delivery vessel and receiving vessel are exchangeable, and coal powder is circulated inside the conveying system. New coal material is fed into the delivery vessel by the screw feeder system, and used coal material can be discharged from the bottom of the bag-filter.

The flow rate of supplementary gas is measured by vortex flowmeters, and the flow rates of pressurizing gas and fluidizing gas are measured by mass flowmeters. Seven pressure transducers and eight differential pressure transducers are installed along the conveying line. The weights of vessels are measured by load cells and are used to calculate the average solid flow rate in a period of time. The instantaneous solid flow rate is measured by the solid mass flowmeter of Thermo Ramsey, which consists of a velocity sensor, a concentration sensor, and an integrator. The output signals of measuring instruments are converted by A/D converters, and then recorded by a PLC control system at a frequency of 2 Hz. The detailed experimental facility and procedure can be refered in Guan et al. (2017).

84.2.2 Material

SH coal and its mixture with 20 wt% extraction residue of DCLR as the conveying material are carried out the high pressure pneumatic transport experiment, respectively. The property of the coal and extraction residue of DCLR was shown in Table 84.1.

Table 84.1 Properties of coal and extraction residue of DCLR

Proximate analysis	SH coal	Extraction residue of DCLR	Ultimate analysis	SH coal	Extraction residue of DCLR
M_{ad}/%	6.31	1.64	C	67.75	53.75
A_{ad}/%	10.02	37.52	H	3.85	2.22
V_{ad}/%	30.75	21.01	O	10.81	0.46
FC_{ad}/%	52.92	39.83	N	0.85	0.41
			S	0.41	4.0

84.3 Results and Discussion

84.3.1 Pressure Fluctuation

In the process of the high pressure pneumatic transport of the pulverized coal, the pressure stability is a key parameter for conveying stability. According to the method mentioned above, the pneumatic transport experiments of two kind of conveying material were measured at back pressure 4.0 MPa and in pipes with diameters of 25 mm (solid/gas mass ratio 14) and 15 mm (solid/gas mass ratio 10), respectively. The results that pressure fluctuation in horizontal flow, vertical up flow and vertical down flow were shown in Figs. 84.2 and 84.3. Due to the limited weight of pulverized coal, the conveying time in pipe is short and only the stability time about 300 s was shown in Figs. 84.2 and 84.3.

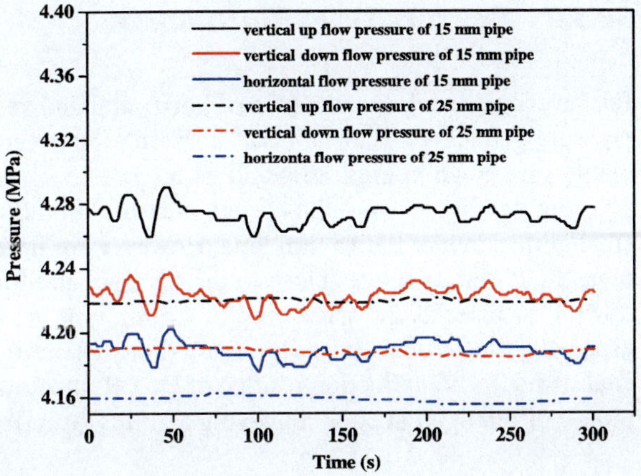

Fig. 84.2 Variation of pipe pressure with time for SH coal in pipes with diameters of 25 and 15 mm at back pressure 4.0 MPa and differential pressure $\Delta P = 0.50$ MPa

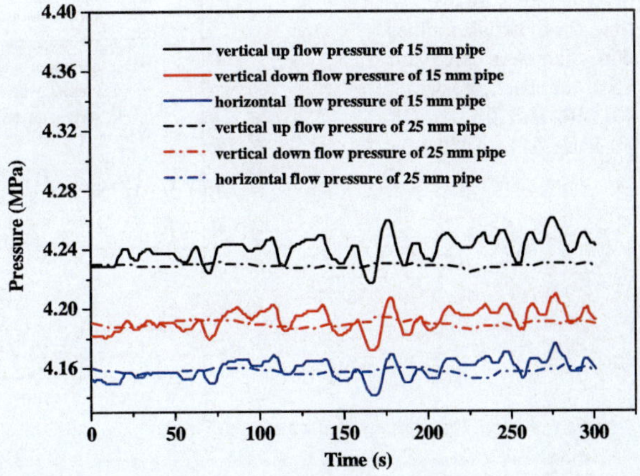

Fig. 84.3 Variation of pipeline pressure with time for SH coal contain 20 wt% extraction residue of DCLR in pipes with diameters of 25 and 15 mm at back pressure 4.0 MPa and differential pressure $\Delta P = 0.50$ MPa

As shown in Figs. 84.2 and 84.3, it can be seen that the variation of pipe pressure with time show a good agreement for the solid horizontal flow, vertical up flow and vertical down flow of whether the diameter 25 mm or 15 mm pipe. For 15 mm pipe, the pressure fluctuation is larger because of its smaller convey space, which a small amount of gas filling the pipe can cause large pressure fluctuation. In addition, it should be seen that there is a large deviation of convey pressure in 15 mm pipe compared with 25 mm pipes for SH coal. The reason may be caused by the different bulk density of SH coal and extraction residue of DCLR, which the bulk density of SH coal is 0.86 g/m^3 and the bulk density of extraction residue of DCLR is 0.97 g/m^3. As the bulk density of solid is larger, the corresponding carrier gas volume in pipe is smaller, and so the effect of carrier gas on conveying pressure would be reduced. Overall, both of them have less pressure fluctuation in 15 and 25 mm pipe for SH coal and its mixture with 20 wt% extraction residue of DCLR.

84.3.2 Pressure Drop Fluctuation

For investigate the pressure drop of horizontal and vertical pipe, several differential pressure gauges were installed at horizontal and vertical section of the pipe. The result was shown in Figs. 84.4 and 84.5.

From the Figs. 84.4 and 84.5, the pressure drop fluctuation in the 25 mm pipe is smaller than that of the 15 mm pipe. However, for horizontal pipe, the pressure drop mean in 25 mm pipe was close to the 15 pipe, and for vertical up flow and vertical down flow pipe, the pressure drop in 25 mm pipe is larger than that of 15 mm pipe, especially the vertical up flow pressure drop. Therefore, regardless of SH coal or SH coal with 20 wt% extraction residue of DCLR, the flow pattern in 25 mm and 15 mm pipe is different and better conveying stability occurred in the 25 mm pipe.

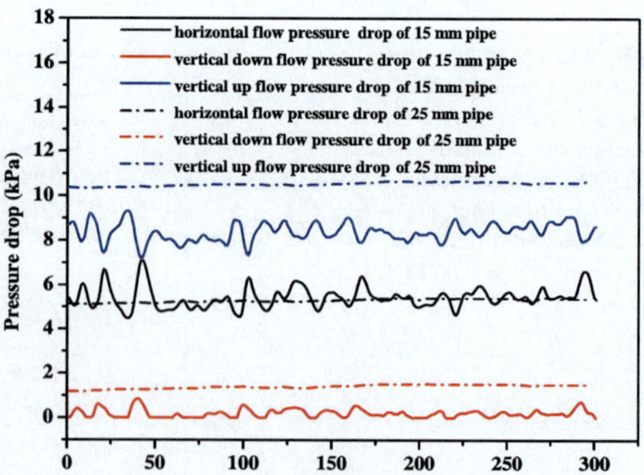

Fig. 84.4 Variation of pipeline pressure drop with time for SH coal in pipes with diameters of 25 and 15 mm at back pressure 4.0 MPa and differential pressure $\Delta P = 0.50$ MPa

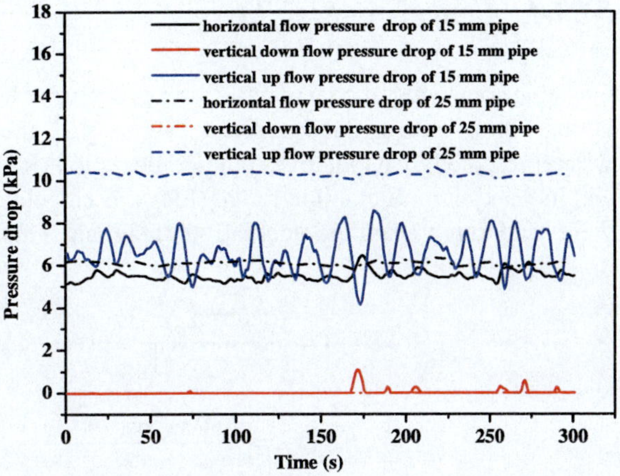

Fig. 84.5 Variation of pipeline pressure drop with time for SH coal contain 20 wt% extraction residue of DCLR in pipes with diameters of 25 and 15 mm at back pressure 4.0 MPa and differential pressure $\Delta P =$ 0.50 MPa

84.3.3 Pulverized Coal Flow Rate, Velocity and Concentration Fluctuation

The pulverized coal flow rate is an important data in industrial gasification process. Because of the high cost of pulverized coal flow meter, there is only one that was installed in 25 mm pipe. The results were shown in Fig. 84.6.

As shown in Fig. 84.6, it is very clear that the concentration of SH coal in pipe is larger than that SH coal with 20 wt% extraction residue of DCLR. At the same convey drive force, it is certain that larger bulk density solid corresponding to lower convey concentration. At the same time, the flow rate and velocity will also become lower. Additionally, it can also be seen that there is a sudden change for SH coal with 20 wt% extraction residue of DCLR in the position of about 40 s in Fig. 84.6. That is because the amount of carrier gas sudden reduced due to a mistake operation. After the mistake operation was corrected, the flow rate, velocity and concentration become more stable.

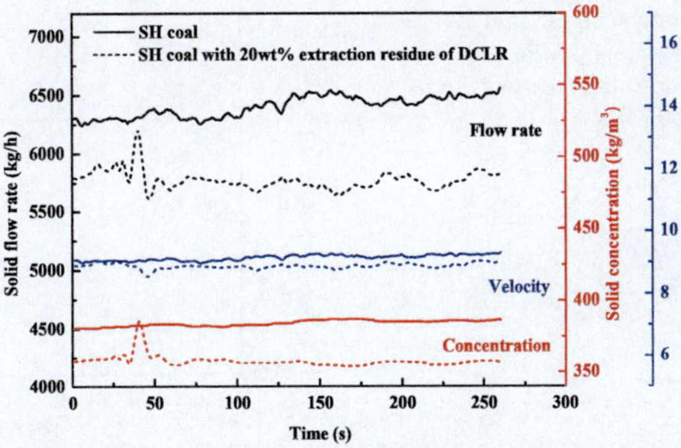

Fig. 84.6 Variation of flow rate, velocity and concentration of pulverized coal with time for SH coal and its mixture contain 20 wt% extraction residue of DCLR in pipes with diameters of 25 mm at back pressure 4.0 MPa and differential pressure $\Delta P =$ 0.50 MPa

84.3.4 Conveying Stability

Unstable conveying of coal powder always occurred with simultaneous signal fluctuations. For discrete signal data series $\{y_{1,} y_{2,\dots} y_N\}$, the average fluctuation amplitude was represented by its relative standard deviation (RSD) as calculated by Eq. (84.1), and its maximum fluctuation range (FR) was calculated by Eq. (84.2). RSD and FR were used to represent the fluctuation of signals (Guan et al. 2017).

$$RSD(y) = \frac{\sqrt{\frac{1}{N-1}\sum_{i=1}^{N}(y_i - \overline{y})^2}}{\overline{y}}, \overline{y} = \frac{1}{N}\sum_{i=1}^{N} y_i \tag{84.1}$$

$$FR(y) = \frac{\max(y_i) - \min(y_i)}{\overline{y}} \times 100\% \tag{84.2}$$

For investigating the conveying stability, three representative signals including the pressure of a horizontal pipe P_{h}, the pressure drop per unit length of a horizontal pipe $\Delta P_{\mathrm{h}}/\Delta L$, and solid flow rate M_s were used to analyze the fluctuation. The result was shown in Figs. 84.7 and 84.8.

As we can see from Figs. 84.7 and 84.8 that the value of RSD(M_s) and FR(M_s) have less fluctuation with increasing apparent gas velocity. For SH coal, the RSD(M_s) value is less than 1.5% and FR(M_s) value is less than 5.3%. For SH coal with 20 wt% extraction residue of DCLR, the RSD(M_s) value is less than 1.2% and FR(M_s) value is less than 4.8%. This result indicated that both of two conveying material show good conveying stability. At the extent, the conveying stability of SH coal with 20 wt% extraction residue of DCLR was better than that of SH coal because of the lower RSD(M_s) and FR(M_s) value. In addition, the relative standard deviation of flow rate, velocity and concentration for SH coal and its mixture with 20 wt% extraction residue of DCLR were also calculated. The result was shown in Table 84.2. From

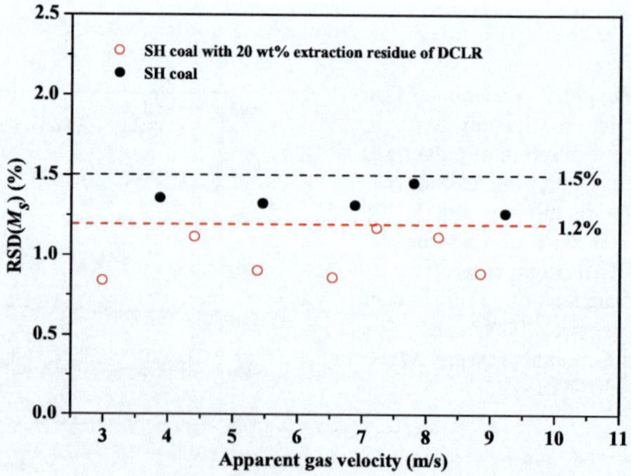

Fig. 84.7 Variation of RSD(M_s) with apparent gas velocity in pipes with diameters of 25 mm at back pressure 4.0 MPa and differential pressure $\Delta P =$ 0.50 MPa

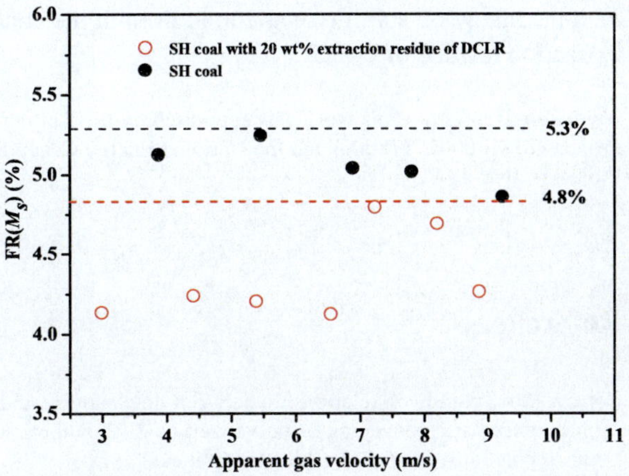

Fig. 84.8 Variation of FR(M_s) with apparent gas velocity in pipes with diameters of 25 mm at back pressure 4.0 MPa and differential pressure $\Delta P =$ 0.50 MPa

Table 84.2 Signal relative standard deviation for two conveying material

RSD and FR	SH coal	SH coal with 20 wt% extraction residue of DCLR
RSD(Ms) × 10^2	1.3158	1.3562
RSD(Vs) × 10^2	1.7536	1.1429
RSD(Cs) × 10^2	1.6248	0.9053
FR(Ms) × 10^2	0.093	0.0918
FR(Vs) × 10^2	3.6195	2.6595
FR(Cs) × 10^2	0.0566	0.0865

Table 84.2, it can be seen that the maximum relative standard deviation is less than 4%. It means that the conclusion which both of two conveying material have good conveying stability was further confirmed.

84.4 Conclusion

High pressure pneumatic conveying experiments for SH coal and its mixture with 20 wt% extraction residue of DCLR was carried out in pipes with diameters of 25 mm and 15 mm, respectively. The variation of pressure, pressure drop, solid flow rate, velocity and concentration with time was analyzed. The result indicate that although both of them have good conveying stability, but the conveying stability of conveying material with larger bulk density is relatively better. By comparing and calculating the relative standard deviations (RSDs) of different signals on the conveying stability, the RSD of solid flow rate was selected as an indicator for conveying stability. The result show that the RSDs of both conveying stability signals of SH coal and its mixture with 20 wt% extraction residue of DCLR is less than 4%, which further demonstrate good stability was found in both the coal and its mixture with 20 wt% extraction residue of DCLR.

Acknowledgements This work was supported by the Beijing Science and Technology Planning Project (Z181100005118006) and the National Energy Group Science and Technology Innovation Project (GJNY-18-72).

References

Cong X (2013) Study on relationship between flow patterns and pipeline pressure signals in dense-phase pneumatic conveying of pulverized coal (PhD thesis). East China University of Science and Technology, Shanghai, China (in Chinese)

Guan QL, Liu Z, Fang XH et al (2017) Experimental study on dense-phase pneumatic conveying of coal powder at high pressures. Clean Energy 1(1):50–67

He C, Shen X, Zhou H (2014) Stability analysis of dense phase pneumatic conveying of pulverized coal at high pressure. J Chem Ind Eng (china) 65(03):820–828 (in Chinese)

Ji J, Wang D, Suo Z, Xu Y, Xu SF (2018) Study on direct coal liquefaction residue influence on mechanical properties of flexible pavement. Int J Pave Res Tech 11:355–362

Li Y, Zhang XP, Lai SY, Dong HF, Chen XL, Wang XL, Nie Y, Sheng Y, Zhang SJ (2012) Ionic liquids to extract valuable components from direct coal liquefaction residues. Fuel 94:617–619

Li XH, Xue YL, Feng J, Yi Q, Li WY, Guo XF, Liu K (2015) Co-pyrolysis of lignite and Shendong coal direct liquefaction residue. Fuel 144:342–348

Liang C, Xu P, Chen XP, Zhao CS (2012) Flow characteristics and stability of dense-phase pneumatic conveying of pulverized coal under high pressure. Exp Therm Fluid Sci 41:149–157

Liang C, Shen L, Chen X et al (2015) Conveying characteristics and resistance properties in high-pressure dense-phase pneumatic conveying of anthracite and petroleum coke. J Chem Eng Jpn 48(3):163–174

Lin W, Wang Y, Gong X et al (2014) Stability analysis of pneumatic conveying of pulverized coal in Shell coal gasification plant. Chem Eng (china) 42(01):50–54 (in Chinese)

Lv DM, Yuchi W, Bai ZQ, Bai J, Kong LX, Guo ZX, Yan JC, Li W (2015) An approach for utilization of direct coal liquefaction residue: blending with low-rank coal to prepare slurries for gasification. Fuel 145:143–150

Ma Y, Luo C, Jing Y et al (2015) Conveying problem of GSP gasification pulverized coal and optimizing measures. China Powder Sci Technol 21(03):96–99 (in Chinese)

Mallick SS, Wypych PW (2009) Minimum transport boundaries for pneumatic conveying of powders. Powder Tech 194(3):181–186

Shu GP, Shi SD, Li KJ (2003) Coal liquefaction technology. China Coal Industry Publishing House, Beijing, p 179

Xie K, Guo X, Cong X et al (2013) Minimum pressure drop velocity and stability analysis in pneumatic conveying of pulverized coal in commercial-scale horizontal pipe. J Chem Ind Eng (china) 64(06):1969–1975 (in Chinese)

Chapter 85
Experimental Study of the Flat-Flame Pulverized Coal Gasification Technology

H. Q. An, Z. Liu, X. H. Fang, Z. Y. Feng, B. Z. Peng, Y. Li, and W. H. Li

Abstract A flat-flame pulverized coal gasification technology is presented and its test results on a 3 t/d gasification test facility are discussed. The test results show that the performance of flat-flame gasification technology is better than the traditional top-placed single-burner technologies, and the test facility is running stably during the test periods. It is revealed that, the flat-flame burner enables higher turbulent intensity and enhances the turbulent mixing between coal particles and oxidant. The overall gasification reaction rate is increased by abovementioned effects. When the mass ratio of O_2/coal is 0.8, the carbon conversion, volume fraction of efficient syngas and cold gas efficiency of flat-flame burner were respectively 96.7%, 88.6% and 82.1%, which are completely better than single-burner gasification. Preliminary test results and analysis validates the technical feasibility of flat-flame gasification technology and it is ready for scale-up for industry-scale demonstrations.

Keywords Flat-flame pulverized coal gasification · Burner · Test facility · Flow field

85.1 Introduction

Coal gasification is a determinately efficient way to converts coal into gaseous products, which are widely used for power generation, chemical products and H_2 production (Liu et al. 2019). Entrained flow gasification technology is generally known as large-scale, clean and high efficiency (Brown et al. 1986; Kong et al. 2014),which is an important part of the clean coal conversion technology in China. Comparing with entrained flow gasification with coal-water-slurry, entrained flow gasification with pulverized coal has advantages of low water consumption, high carbon conversion and heat efficiency, better adaptability (Higman et al.).

H. Q. An (✉) · Z. Liu · X. H. Fang · Z. Y. Feng · B. Z. Peng · Y. Li · W. H. Li
National Institute of Clean and Low Carbon Energy, Beijing 102211, China
e-mail: anhaiquan@nicenergy.com

J. Lyu and S. Li (eds.), *Clean Coal and Sustainable Energy*, Environmental Science and Engineering,
https://doi.org/10.1007/978-981-16-1657-0_85

In recent years, many entrained flow gasification technologies have been developed in China (Wu et al. 2008; An et al. 2017a; Chao et al. 2012), all these technologies attempt to manufacture larger gasifier for more coal capacity and higher efficiency. However, lager gasifier means larger investment and running cost. Therefore, the flat-flame gasification technology is proposed as high efficiency with smaller gasifier.

When a coal particle was introduced into the gasifier, it would be pyrolysed in 50 ms first, then the volatiles would combust extremely quickly, and the fixed-carbon would gasify for 0.5–1 s. However, the residence time of coal particles in the gasifier was about 5–10 s with conventional gasification technologies. Possible reasons are the engineering design margin and diffusion and reaction between coal particle and gasifying agent could not reach ideal condition. As a matter of fact, during the gasification temperature of 1300–1500 °C, the gasification overall reaction rate was mainly controlled by the mass diffusion, instead of chemical reaction temperature (An et al. 2017b; Dai et al. 2017). Thus, the non-premixed impinging-type burner (Abraham et al. 2007) and the MILD combustion concept (Cavaliere and Joannon 2004) were employed, and a novel flat flame gasification technology was introduced in this paper. A flat flame burner includes several coal nozzles, and three streams of gasifying agent were impinged with each coal nozzle. It helps increasing the mass diffusion between coal particles and gasifying agent, leading to a uniform distribution in the high temperature region of gasifier. Consequently, high carbon conversion would be approached in a smaller reaction zone.

It is expected the flat flame gasification technology would achieve comparable performance as conventional gasification technologies within a much smaller reaction chamber and reduce the cost. Extensive research had been conducted to develop this technology in previous studies, including the flow field characterization of a cold gasifier module under atmospheric pressure (Fang et al. 2016), dense phase pneumatic conveying of pulverized coal at high pressure (Fang et al. 2017) and pilot scale tests. In this paper, a 3 t/d-scale gasification test facility was introduced and the test results are presented and discussed.

85.2 Pilot Scale Test Facilities

The block flow diagram and a photograph of the pilot scale of flat-flame pressurized entrained flow gasification system can be seen in Fig. 85.1. The flat-flame gasifier is a pulverized coal-fed, pilot scale (3 ton per day) entrained flow reactor operated at 0.5–1.0 MPa. It is a cylindrical and refractory lined reactor with an inside diameter of 300 mm and a height of 1500 mm. The vertically oriented gasifier has a top-down flow design with an integrated quencher (another 2350 mm of height) at the bottom.

The test procedure is as follows: at the start phase the pulverized coal was loaded into the atmospheric storage vessel, then dropped into the high pressure feed hopper. The gasifier was initially heated up to around 1000 °C with a natural gas burner at atmospheric pressure for several hours, so that the refractory material could hold

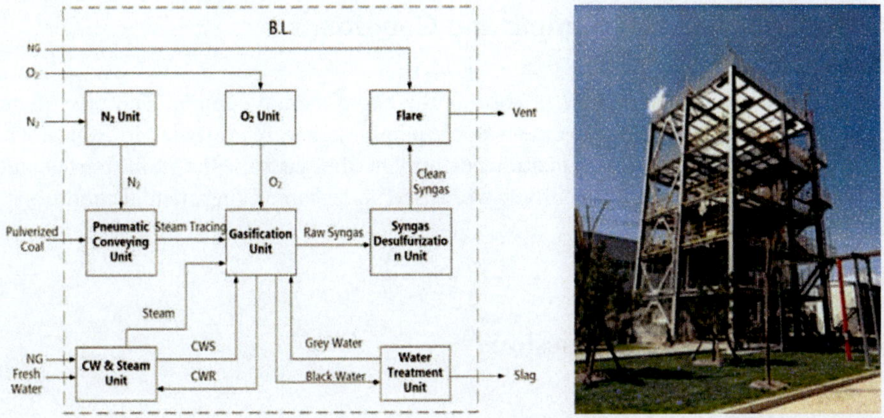

Fig. 85.1 The block flow diagram and picture of gasification test facility

enough quantity of heat. After that, the natural gas burner was replaced with a process burner in a short time period. The pulverized coal was conveyed into the gasifier by high pressure nitrogen, and the appropriate proportion of oxygen was fed into the gasifier at the same time. Finally, the coal to oxygen ratio can be adjusted gradually to the desired value.

The operation pressure was measured by a pressure gauge and controlled by the syngas outlet. The temperature in the reactor was measured by 16 thermoelectric couples, which were evenly distributed at the reactor wall. Besides, the syngas components were detected by a gas analyzer and an infrared spectrometer.

The flat flame burner was shown as Fig. 85.2, it consisted of three vertical coal unit nozzles, which distributed evenly around a circle centered on main coal tube. Each coal unit nozzle was matched up with three same streams of gasifying agent. These three streams were evenly distributed centered on coal unit nozzle and impinged with coal stream at a same angle with vertical. The flow of coal is equally divided three streams by the coal distributor.

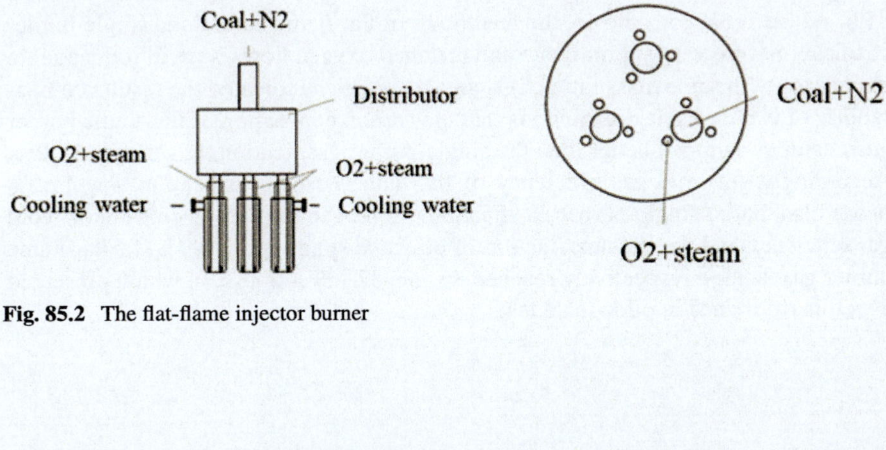

Fig. 85.2 The flat-flame injector burner

85.3 Experimental Sample and Conditions

A kind of conventional coal was used as the experimental sample. The proximate analysis, ultimate analysis and other sample parameters were presented in Table 85.1.

In order to compare the gasification performance between the flat-flame burner and traditional single burner, both burners were used under same experimental conditions, as the Table 85.2 shown.

85.4 Results and Discussion

85.4.1 Running Conditions

The stable running conditions of gasification system were recorded and shown as Fig. 85.3. The representative parameters for running stability were shown as Fig. 85.3a, which were temperature, pressure, flow of coal and oxygen, respectively. It was obvious that the temperature and pressure in the gasifier kept stable during the gasification. The flow of coal, which was controlled by the differential pressure between coal high pressure delivery tank and gasifier, kept basically stable and had no major impact on gasification system.

The volume fractions of CO and H_2 in the dry syngas were shown in Fig. 85.3b. At the beginning, the volume fractions of CO and H_2 were fluctuated slightly, which was probably caused by the fluctuation of oxygen flow. Then, they gradually stabilized when the gasification was stable. The volume fraction of ($CO + H_2$) was over 88%, which indicated the flat-flame gasification with well results.

85.4.2 Gasification Results

The gasification results of single-burner and flat-flame burner were shown as Table 85.3. As the Table 85.3 shown, the feed flows of flat-flame burner and single-burner gasification were very approximate and matched oxygen flows were introduced into the reactor with same mass ratio of O_2 and coal, which indicated the results comparability of two cases. It was obvious that the carbon conversion of flat-flame burner gasification was much better than the single-burner gasification, larger with 14.2%. Accordingly, the cold gas efficiency of flat-flame burner gasification was 11.6% larger than that of single-burner gasification. Moreover, the carbon conversion, cold gas efficiency and the volume fraction of effective syngas ($CO + H_2$) of flat-flame burner gasification respectively reached 96.7%, 82.1% and 88.6%, which presented a well performance in pilot-scale test.

Table 85.1 Coal analysis

Proximate analysis /wt%				LHV	Ash fusion temperature/°C
M_{ar}	V_{daf}	FC_{ar}	Ash_{ar}	MJ/kg	Flow temperature
1.42	36.43	55.96	10.55	27.93	1210

Ultimate analysis /wt%(daf)					Particle size distribution /wt%		
C	H	O	N	S	<50 μm	<75 μm	<90 μm
84.54	5.06	8.91	1.08	0.42	75.76	91.32	97.99

Table 85.2 Main operation parameters of the gasification facility

Operation parameter	Units	Value
Coal feeding	kg/h	~200
Pressure	MPaG	0.5
Temperature	°C	1300 –1400
Oxygen/Coal ratio	kg/kg	0.8

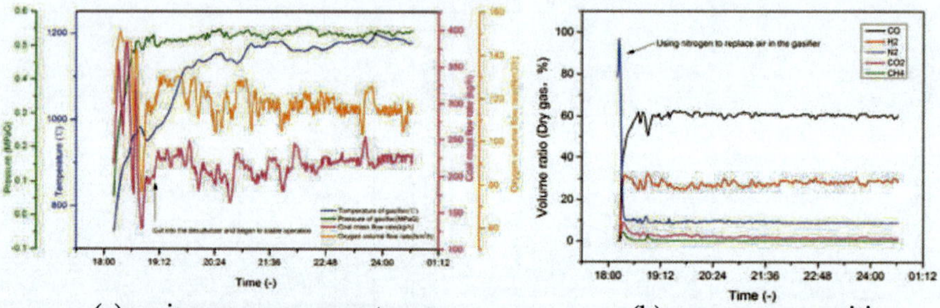

(a) main process parameters (b) syngas composition

Fig. 85.3 The flat-flame pulverized coal gasifier performance

Table 85.3 Gasification results of single and flat-flame burner

	Units	Single burner	Flat-flame burner
Feed flow	kg/h	221	212
Ratio of oxygen and coal	kg/kg	0.8	0.8
Gasification pressure	MPa	0.5	0.5
Carbon conversion	%	82.5	96.7
Cold gas efficiency	%	70.5	82.1
Syngas composition, dry basis			
CO	vol.%	56.98	60.47
H_2		29.78	28.09
CO_2		2.11	2.00
N_2		10.53	9.16
CH_4		0.58	0.28

In order to compare with other gasification technologies, primary gasification results of flat-flame burner and other gasification technologies from pilot-scale experiments can be seen in Table 85.4.

Table 85.4 Different gasification technology results comparison

Gasification technology	O_2/coal (kg/kg)	Gasification pressure (MPa)	Carbon conversion (%)	$CO + H_2$ (dry basis, vol.%)	Cold gas efficiency (%)
Dry feed + refractory brick (Gong et al. 2014)	0.84	2.0	98.8	89.1	\
2-stage entrained flow dry powder gasification (Xu et al. 2014)	0.63	3.0	\	93.3	79.9
Dry feed + refractory brick (Ünlü et al. 2017)	0.67	0.1	\	82.8	72
Coal water slurry + refractory brick (An et al. 2017a)	0.65	0.1	93.2	80.4	\
Single-burner gasification	0.8	0.5	82.5		70.5
Flat-flame gasification	0.8	0.5	96.7	88.6	82.1

Comparing with other gasification technologies, the results of flat-flame burner gasification also can obtain a well performance. The different coal type decided different mass ratio of O_2/coal, it directly influenced the gasification temperature. Under stable running conditions, the primary gasification results were related to the coal particle gasification rate and residence time. When the gasification pressure was increasing, the gas flow rate was decreasing by the gas volume was compressed. The coal particle residence time was longer because of slower carrier gas rate. Thus, the cases under high gasification pressure can obtain higher carbon conversion and volume fraction of efficient syngas, as the references shown. In relatively short reaction time, the flat-flame gasification results presented little difference with case results under high pressure, which indicated the gasification rate of flat-flame was much higher. In addition, it was also proved by the single-burner gasification results, which obtained dissatisfactory results with similar coal particle residence time.

85.4.3 *Gasification Temperature*

The temperature field in the flat-flame burner gasifier was improved by MILD combustion technology. Backflow of vast high temperature syngas could heat the reactants and dilute the oxidant concentration. Thus, the gasification was enhanced and the combustion was weakened. Wünning and Wünning (1997) defined the entrained rate as K_v to express the quantity of backflow, the equation was presented as:

$$K_v = \frac{M_E}{M_F + M_A} \tag{85.1}$$

M_E, M_F and M_A are the mass flow of backflow, fuel and reacting gas, respectively. It indicated a stronger refluxing with a larger K_V. The temperature decline was smaller and peak temperature was lower at the same time. Particularly, in the flat-flame burner gasifier, a distinctly higher K_V was shown adjacent the nozzle by CFD simulation as Fig. 85.4. Correspondingly, in the flat-flame burner gasifier the temperature distributed evenly and the highest temperature is lower than that in the single burner gasifier (Tu et al. 2015).

The experimental results of temperature also proved the simulation. The temperature field measured by 12 high temperature thermocouples was presented in Fig. 85.5.

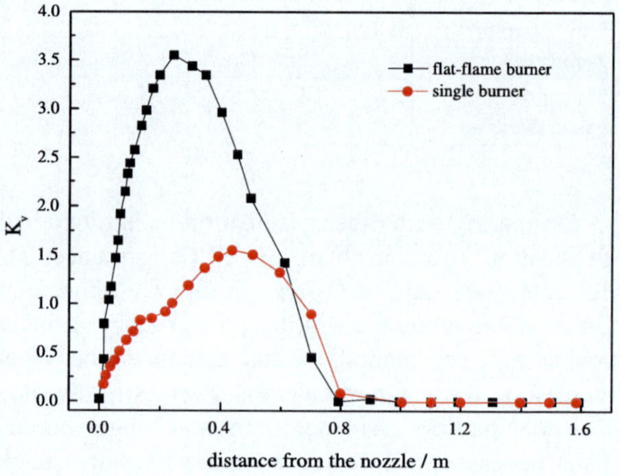

Fig. 85.4 K_V distribution with flat-flame burner and single burner

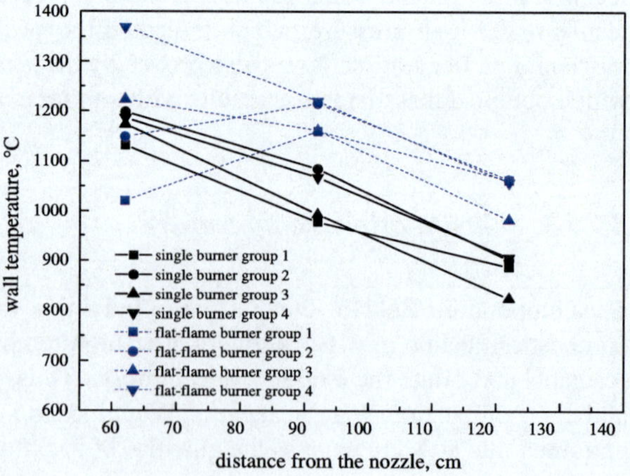

Fig. 85.5 temperature distribution with flat-flame burner and single burner (Experimental results)

Comparing with the temperature distribution in single-burner gasifier, it distributed evenly in the flat-flame burner gasifier, which became convictive evidence for stronger backflow by the flat-flame burner.

85.4.4 Flow Field Analysis

As many investigations (Dai et al. 2017; Keller et al. 2018; Irfan et al. 2011) suggested, the overall gasification rate of coal char with CO_2 was controlled by both chemical reaction and pore diffusion when the temperature was higher than 1200 °C. Thus, the mass diffusion rate became a key impact on influencing overall gasification rate. The flat-flame burner could improve the mass transfer between coal surface and gasifying agent by impinging streams from multiple nozzles. Moreover, turbulence in the gasifier was also enhanced and reactants were heated by strong entrainment of syngas with high temperature.

Impinging stream is proposed and experimentally verified by Elperin, which indicates a good promotion on heat and mass transfer (Abraham et al. 2007). In detail, two equivalent streams of gas and solid flow along the same axis in opposite direction, collides on the impinging point. The impinged stream would create a narrow high turbulent zone, and provide a well condition for enhancing heat and mass transfer. The impinged stream by the flat-flame burner brought coal particle back to the gas stream from the burner, which enhanced the turbulence intensity. The multiple back mixing helped generating an obviously more sever turbulent zone than single burner. In addition, the coal particle oscillation caused by impinging stream could help extending the residence time of coal in the oxygen-rich zone, which was helpful to achieve higher carbon conversion. For flat-flame burner, the high concentration of coal and high concentration of oxygen had a well overlapped zone, which was conductive to increase the gas–solid contact probability and improve the oxygen diffusion to the coal particle surface. Thus, the overall gasification reaction rate in flat-flame burner gasifier was faster.

85.5 Conclusion

Flat-flame pulverized coal gasification technology is preliminarily validated on a 3 t/d gasification test facility. The gasification results of flat-flame burner and conventional single burner were compared and discussed. As the experimental results shown, the carbon conversion, volume fraction of efficient syngas and cold gas efficiency of flat-flame burner were respectively 96.7, 88.6 and 82.1%, which were obviously better than the single-burner. Comparing with other gasification technology, the flat-flame burner gasification can achieve well result with shorter particle residence time.

Based on the MILD combustion and impinging stream theory, the flow field adjacent the nozzle was studied. In detail, the turbulence intensity adjacent the flat-flame burner was stronger than that of the single burner, which caused by the stronger impinging stream with flat-flame. The stronger turbulence intensity could promote the backflow of high temperature syngas and heated the injected reactant, which helped creating a more uniform temperature distribution and a favorable flow field. Moreover, it also helped the high coal concentration zone overlapped with high oxygen concentration zone on flat-flame burner. It was proven that the overall gasification reaction rate in flat-flame burner gasifier was faster because of enhanced mass diffusion effect.

In conclusion, the flat-flame burner gasification technology has been proved as a promising gasification technology with better performance. It could be predicted that the gasification performance of flat-flame burner would achieve better performance at the industrial scale with reduced cost of equipment investment.

Acknowledgements This work was supported by the National Energy Group Science and Technology Innovation Project (GJNY-18-72), National Key Research and Development Program (2017YFB0602603) and Beijing Science and Technology Planning Project (Z181100005118006).

References

Abraham Z, Luzzatio K, Tamir A (2007) Application of free impinging streams to the combustion of gas and pulverized coal. Combust Sci Technol 60(1):31–44

An H, Yu J, Fan J et al (2017a) Experiment study on entrained flow gasification technology with dry slag by second-stage water supply. Powder Technol 306:10–16

An H, Yu J, Jiang Y et al (2017b) Kinetics of steam and CO_2 gasification with high ash fusion temperature coal char under elevated pressure. Energy Sour Part A: Recov Util Environ Effects 39(24):2188–2194

Brown BW, Smoot LD, Hedman PO (1986) Effect of coal type on entrained gasification. Fuel 65(5):673–678

Chao L, Dai Z, Li W et al (2012) 3d numerical study of particle flow behavior in the impinging zone of an opposed multi-burner gasifier. Powder Technol 225

Cavaliere A, Joannon MD (2004) Mild combustion. Prog Energy Combust Sci 30(4):329–366

Dai B, Hoadley A, Zhang L (2017) Characteristics of high temperature C-CO_2 gasification reactivity of Victorian brown coal char and its blends with high ash fusion temperature bituminous coal. Fuel 202:352–356

Fang X, Gong Z, Feng Z et al (2016) Research on flow field of gasifier with seven nozzles based on PIV technology. Clean Coal Technol 22(6):34–39

Fang X, Guan Q, Gong Z et al (2017) Construction and commissioning of high pressure dense phase conveying device for pulverized coal. Coal Process Compr Util 2:68–72

Gong X, Lu W, Guo X et al (2014) Pilot-scale comparison investigation of different entrained-flow gasification technologies and prediction on industrial-scale gasification performance. Fuel 129:37–44

Higman C, Burgt VD et al (2007) Gasification, 2nd ed

Irfan MF, Usman MR, Kusakabe K (2011) Coal gasification in Co atmosphere and its kinetics since 1948: a brief review. Energy 36(1):12–40

Keller F, Küster F, Meyer B (2018) Determination of coal gasification kinetics from integral drop tube furnace experiments with steam and CO_2. Fuel

Kong L, Bai J, Bai Z et al (2014) Improvement of ash flow properties of low-rank coal for entrained flow gasifier. Fuel 120:122–129
Liu M, Shen Z, Liang Q et al (2019) Experimental studies on two dimensional particle swarm gasification of different coal chars and petroleum coke at high temperature. Fuel 241:973–984
Tu Y, Liu H, Chen S et al (2015) Numerical study of combustion characteristics for pulverized coal under oxy-mild operation. Fuel Process Technol 135:80–90
Ünlü A, Kayahan U, Argönül A et al (2017) Pilot scale entrained flow gasification of Turkish lignites. J Energy Inst 90(1):159–165
Wu Y, Zhang JS, Yue GX et al (2008) Analysis of the gasification performance of a staged entrained flow gasifier by presumed Pdf model. Proc CSEE 28(26):29–34
Wünning J, Wünning J (1997) Flameless oxidation to reduce thermal no-formation. Prog Energy Combust Sci 23(1):81–94
Xu S, Ren Y, Wang B et al (2014) Development of a novel 2-stage entrained flow coal dry powder gasifier. Appl Energy 113:318–323

Chapter 86
Research on Evaluation of Pulverized Coal Flow Stability in Dilute Phase Pneumatic Conveying Based on Pressure Fluctuation of Resistance Components

Yang Shi

Abstract By several empirical methods, including pressure drop measurement and electrical capacitance tomography (ECT), flow pattern, pressure drop and pressure fluctuation characteristics of the horizontal bend and venture pipes were investigated. The results showed that, pressure drop signals of venturi is appropriate to evaluate the flow stability. Statistical evaluation indexes, based on effective pressure drop signal of venturi pipe, was constructed to evaluate the powder flow stability. The results showed that, pressure drop signals of venturi is appropriate to evaluate the powder flow stability in different operation conditions.

Keywords Pulverized coal · Flow stability · Pressure fluctuation · Resistance components · Evaluation

86.1 Introduction

Pneumatic conveying has been widely utilized in numerous production fields. In some particular production process, powder mass flow must be maintained stable. For improving the flow stability, besides technology upgrade of pneumatic transport equipment, an effective and easy operation method for evaluating the flow stability must be established. By this method, the technical reform effect can be quantitative assessmed and the feeding stability of different techniques can be compared.

The evaluation method of flow stability can be divided into two species: (1) direct weighing, by this method, the flow rate can be obtained by subtracting the real-time value of banker weight. (2) indirect measurement, the flow rate is measured by

Y. Shi (✉)
China Coal Research Institute Company of Energy Conservation, Beijing 100013, China
e-mail: 1773005065@qq.com

State Key Laboratory of High Efficient Mining and Clean Utilization of Coal Resources, Beijing 100013, China

National Energy Technology & Equipment Laboratory of Coal Utilization and Emission Control, Beijing 100013, China

J. Lyu and S. Li (eds.), *Clean Coal and Sustainable Energy*, Environmental Science and Engineering,
https://doi.org/10.1007/978-981-16-1657-0_86

tomography, ultrasonic or other technologies. To apply the former method, weighing equipment with high range and high accuracy must be used, which is not unfit for industrial field. For the second method, expensive and sophisticated technology measuring instruments will be utilized. The reliability of these instruments is poor in the harsh environment.

The pressure drop signals of transportation pipe, containing abundant gas–solid flow information, is the comprehensive reflection of flow behavior, pipe structure. A large number of researchers have established powder flow measurement method by analysing pressure drop signals of resistance components (Jin et al. 1999; Zhiyao et al. 1995; Zongmin et al. 2001; Wu and Xie 2007, Shouguang 2015, Wang et al. 2016). Although the accuracy of these methods is deficiency, the principle can be used for evaluating powder flow stability.

To evaluate the powder flow stability by pressure drop of pipe, proper type of resistance components must be selected. In this article, the pressure characteristics, flow pattern and fluctuation characteristics of the pressure signals of the horizontal elbow and venture pipes are studied on the pneumatic conveying test bench. According to the gas–solid flow characteristics of these two resistance components, the proper type has been selected. The atmospheric pressure dilute phase pneumatic conveying test bench was set up as the real pulverized coal storage system on industrial boiler, shown as Fig. 86.1.

Pulverized coal in the bin is transported by rotary valve into the mixer, where mixed with primary air. The air-coal mixture flow through the pipe with diameter of 1 mm, finally regurgitate into the bin with weighing device.

The length horizontal pipe is 20 m, includes straight pipe, venturi pipe and 90° bend pipe. The pressure-measuring-points and ECT poins are arranged along the horizontal pipeline, shown as Fig. 86.2.

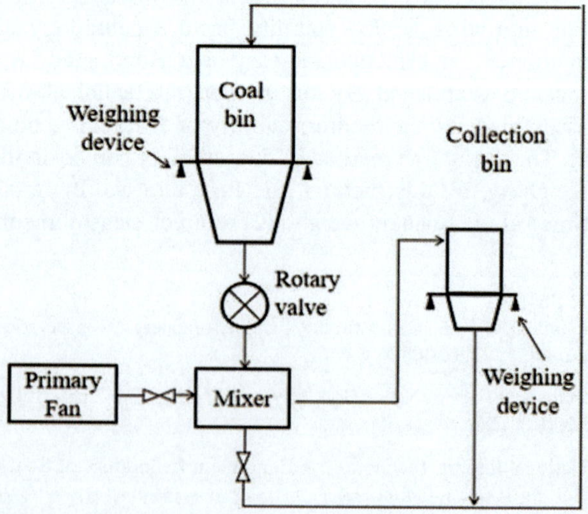

Fig. 86.1 Schematic diagram of test bench

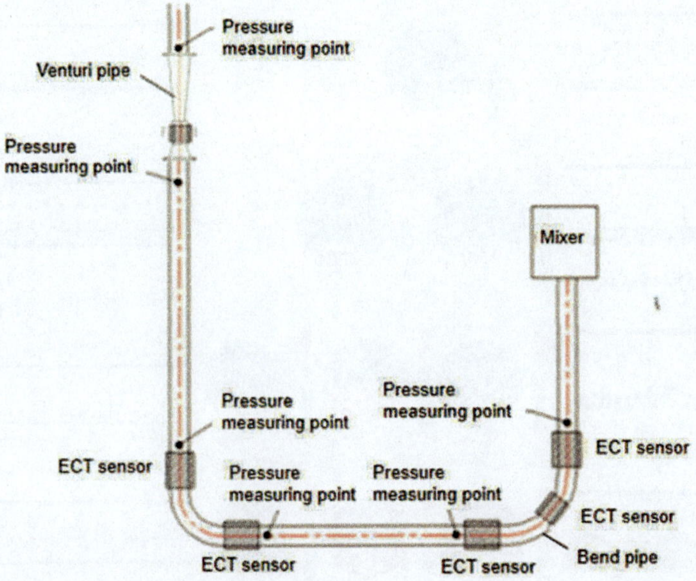

Fig. 86.2 Schematic diagram of measuring point

The primary air flow rate and pulverized coal feed rate are controlled by inverters. The primary air flow rate is measured by V-cone flow meter with the accuracy of 0.5%. The pressure difference of venturi and bend pipes are measured by Keller PD-23 pressure sensors, with the accuracy of 0.3%.

To observe the flow behavior of pulverized coal inside the pipeline, ECT measuring poins are arranged on the straight pipe, venturi pipe and 90° bend pipe.

Shenfu bituminous coal, used in actual production, is selected as experimental powder in this research. The average particle size and repose angle of pulverized coal are 76 μm and 48°,the bulk density and tap density are 459 and 766 kgm^{-3} respectively. The pulverized coal moisture content is about 5%.

86.2 Results and Discussion

86.2.1 Pressure Drop Characteristic

The pressure drop of bend and venturi pipes in different operation conditions are shown in Figs. 86.3 and 86.4. As illustrated in these two figures, in constant flow conditions, the pressure drop of bend pipe significantly increases with increasing of primary air flow. When the air flow rate reaches a certain value, the pressure drop rise rate decreases. This trend between bend pipe and primary air flow is consistent with

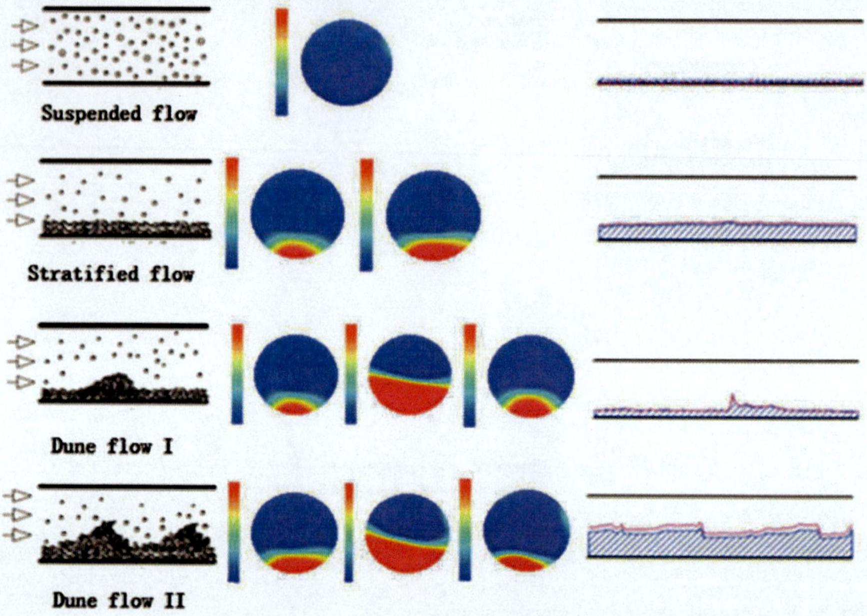

Fig. 86.3 ECT images corresponding to different flow patterns

Literature (Liu et al. 2012). While the relationship between venturi pipe pressure drop and primary air flow is approximate linear increasing.

The classic formula for calculatinghorizontal bend pressure drop (Chi et al. 2000) is illustrated as formula (86.1):

$$P = (1 + \mathrm{K}m)\xi U_{\mathrm{g}}^2 = \xi U_{\mathrm{g}}(U_{\mathrm{g}} + \mathrm{K}\frac{\mathrm{G_s}}{\rho_{\mathrm{g}}\mathrm{A}}) \tag{86.1}$$

ξ is drag coefficient when gas flow through the bend, which is related to the bend size and structure. m is solid–gas mass ratio.U_{g} is superficial gas velocity, m/s. K is empirical coefficient.G_{s} is pulverized coal mass flow rate, kg s^{-1}. ρ_{g} is gas density, kg m^{-3}. A is cross-sectional area of the pipe, m^2.

The pressure drop of venturi pipe can be expressed as formula (86.2):

$$\begin{aligned} P &= a\beta^b\theta_1^c\theta_2^d L^e(G_{\mathrm{p}}U_g - f) \\ &= a\beta^b\theta_1^c\theta_2^d L^e(\frac{G_{\mathrm{s}}}{\mathrm{A}}U_g - f) \end{aligned} \tag{86.2}$$

β is throttle ratio. θ_1 is contraction angle, °. θ_2 is expansion angle, °. L is the length of throat, mm. G_{p} is the mass flux of pulverized coal, kg m^{-2} s^{-1}. G_{s} is mass flow rate of pulverized coal, kgs^{-1}. A is flow area, m^2. U_{g} is superficial gas velocity, m/s. a, b, c, d, e, f are empirical coefficients.

Fig. 86.4 Flow patterns characteristics under different operating conditions

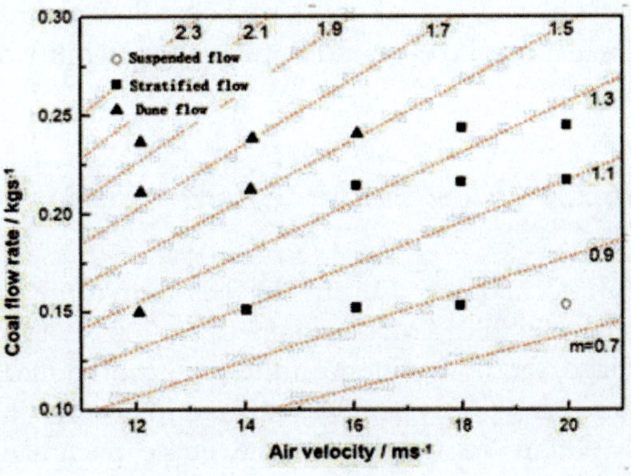

(A) Bend pipe

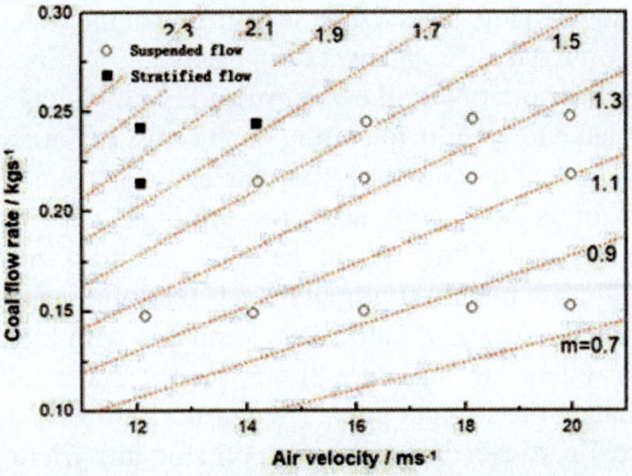

(B) Venturi pipe

From formula (86.1) and (86.2), the pressure drop of horizontal bend is the quadratic function of superficial gas velocity, and is the linear function of mass flow rate of pulverized coal. While the pressure drop of venturi is the linear function of superficial gas velocity and mass flow rate of pulverized coal. Therefore, the press drop of horizontal bend increases with the gas velocity nonlinearly, and the press drop of venturi increases with the gas velosity linearly.

As pressure drop of bend is the quadratic function of gas velocity, minor variation in air velocity will result in acute change in pressure drop which covering the pressure variation caused by the instantaneous change of the pulverized coal flow. The pressure drop of venturi is linear function of gas velocity and pulverized coal mass flow as the stability of gas velocity is largely superior to that of pulverized coal mass flow, It can be approximated that the pressure variation is caused by the instantaneous change of the pulverized coal flow. So the pressure variation of venturi pipe is linear with the variation of pulverized coal flow and it is more suitable to use the pressure fluctuation of the venturi to reflect the stability of pulverized coal flow.

86.2.2 Flow Pattern Characteristic

ECT technology can effectively observe the gas–solid flow state in the pipeline (Jaworski 2001; Madhaudana et al. 2001). According to the ECT frame-by-frame image, the flow pattern in the pipe can be clearly distinguished. In this article, 8 electrode ECT system was utilized to observe the gas–solid flow state.

Within the range of dilute phase pneumatic conveying operating conditions included in this article, suspended flow, stratified flow, dune flow I, dune flow II (fast move) will appear, shown as Fig. 86.3.

The flow pattern in the horizontal bend and the venturi under different operations is shown in Fig. 86.4. With the decrease of the superficial gas velocity or the increase of the pulverized coal flow rate, suspended, stratified, and dune flows occur successively in the horizontal elbow. Besides the solid–gas mass ratio, the flow pattern is also related to specific operating conditions. In the condition of same solid–gas ratio, but different other operating parameters, gas–solid flow pattern may be different as well.

In the bend pipe, when the solid–gas mass ratio is less than 0.8,the flow pattern is suspended flow. When the solid–gas mass ratio is in the range of 0.8–1.3,the flow pattern is mainly stratified flow. If the solid–gas mass ratio is larger than 1.3, the flow pattern is mainly dune flow. Compared with literature (Zhuxin 2007; Fangbo 2007), stratified flow and dune flow occur at lower solid–gas mass ratio in the bend pipe than in the straight pipe. The reason for that is the kinetic energy loss of pulverized coal particles caused by the collision and friction between particles and bend pipe wall (Wang et al. 2013; Pan et al. 1999). As a result, pulverized coal deposition is more inclined to happen in the bend pipe than in the straight pipe.

In the venturi pipe, when the solid–gas mass ratio is less than 1.7, the flow pattern is suspended flow. When it is larger than 1.7, the flow pattern is stratified flow. Dune flow dosen't occur in the venturi pipe within the operating conditions accomplished in this article. The reason for that is the "suction effect" to gas and solid of the venturi pipe (Kai 2015). At the same, the increase of gas velocity also improves the powder carrying ability of the gas flow. As a result, pulverized coal deposition is more inclined to happen in the bend pipe than in the straight pipe. So near the venturi pipe zone, pulverized coal does not tend to deposit.

If the gas–solid flow pattern is dune flow, the actual flow area of gas is effect by the fluctuation of pulverized coal. So the fluctuation of pressure drop signal is much more vioment in dune flow condition than that in suspended and stratified flow condition. Due to the flow pattern in the venture pipe maintains suspended and stratified flow even in the condition of high gas–solid mass ratio, the pressure drop of venture pipe is more stable and suitable for reflecting the stability of pulverized coal.

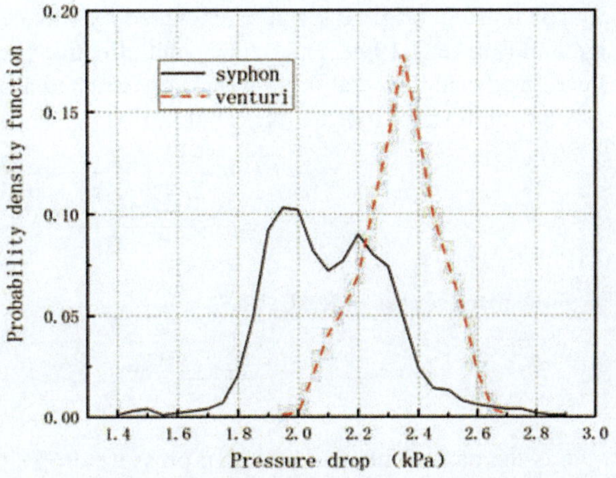

Fig. 86.5 Probability density distribution of pressure drop signals in bend and venturi

86.2.3 Statistical Analysis of Pressure Drop Signal

For comparing the pressure fluctuation of the bend and the venturi pipe, the pressure drop signal of these two pipes in the same operating condition is analyzed statistically. The probability density function (PDF) distribution feature of the bend and the venturi pipe pressure signal are illustrated in Fig. 86.5.

From Fig. 86.5, the range of PDF of the bend pipe pressure signal, with high dispersion and bimodality, is 1–3 kPa. While the range of PDF of the venturi pipe pressure signal, with low dispersion and unimodality, is 1.8–2.7 kPa.

When the primary air is 16 ms^{-1}, the coal flow rate is 0.24 kgs^{-1}, the gas–solid flow pattern in the bend pine is dune flow, while that in the venturi pipe is suspension flow. According to literature (Chen et al. 2004), the dispersion of PDF increases with the complexity and disturbing degree of gas–solid flow. At the same operating condition, the complexity and disturbing degree of gas–solid flow in the venturi pipe are lower than those in the bend pipe. And the flow state in the venturi pipe is more stable.

As a result, the utilization of pressure drop signal in the venturi pipe to evaluate the flow stability of pulverized coal, which can avoid the disturbance of dune flow, is feasible theoretically.

86.3 Application of the Utilization of Pressure Drop Signal to Evaluate the Flow Stability

According to the investigation of the pressure drop and flow pattern characteristics of the bend and the venturi pipe, the venturi pipe pressure signal is selected in this article.

The noise in original signal is excluded by wavelet spatial correlation denoising method (Fusheng 1999; Fang 2009),and effective pressure drop signal is obtained. Pulverized coal flow stability index is constructed as follows:

Flow uniformity index J (Xie et al. 2013):

$$\mathrm{J} = \frac{\sqrt{\frac{1}{n}\sum_{i=1}^{n}(\mathrm{P}_i - \mathrm{P_a})^2}}{\mathrm{P_a}} \tag{86.3}$$

Flow discreteness index L:

$$\mathrm{L} = \frac{\mathrm{P_{max}} - \mathrm{P_{min}}}{\mathrm{P_a}} \tag{86.4}$$

n is the data number of effective pressure drop signal. Pi is the value of the ith date. Pa is the mathematical expectation of effective signal. Pmax is the maximum value in effective signal. Pmin is the minimum value in effective signal.

The smaller uniformity index J and discreteness index L, the more uniform and the narrower the variation range of the instantaneous coal flow during testing time are.

In this article, change rate of coal bin weight is also used to reflect the mass flow of pulverized coal, and the stability of coal flow can be evaluated by analyzing the weight change signal as well.

Comparison between uniformity index J, discreteness index L obtained by analyzing venturi pressure signal and standard deviation coefficient, range coefficient of weight change signal of coal bin are illustrated in Fig. 86.6.

The smaller standard deviation coefficient, range coefficient of weight change signal of coal bin, the higher the stability of coal flow is. From Fig. 86.6, all the indexes, whether obtained from the pressure signal or the weight change signal,

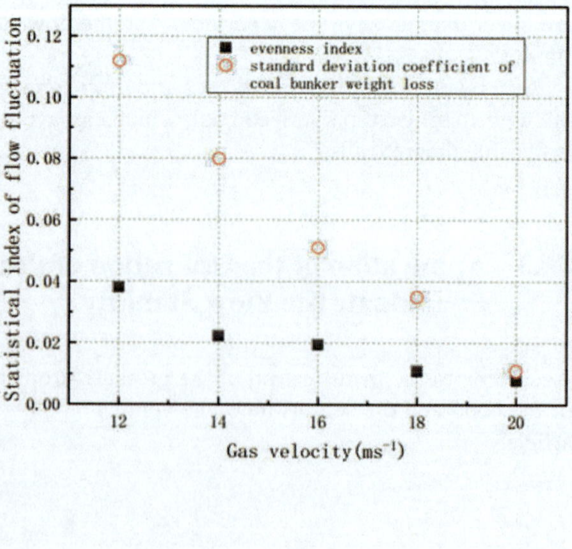

Fig. 86.6 Statistical index with gas velocity

decrease with increasing of gas velosity. That means, at a certain coal feed rate, the higher primary air velocity, the more stable the coal flow is. This is consistent with the previous analysis. The consistency between indexes obtained from the pressure signal and the weight change signal demonstrates that, the pulverized coal flow stability can be effectively evaluated by pressure drop of venturi pipes.

86.4 Conclusion

(1) The pressure drop of both bend and venturi increase with the increase of primary air velocity and pulverized coal feed rate. In the condition of constant pulverized coal feed rate, as primary air velocity increases, the pressure drop of bend increases significantly first and than decreases gradually. While the pressure drop of ventur increases linearly with the increase of primary air velocity.
(2) As the collision and friction between particles and bend pipe wall, the kinetic energy of pulverized coal particles is attenuated significantly and pulverized coal deposition is more inclined to happen in the bend pipe than in the straight pipe. As the "suction effect" to gas and solid in the venturi pip and the powder carrying ability of gasflow is improved by the increase of gas velocity, coal particles do not tend to deposit near the venturi pipe zone.
(3) PDF distribution of the bend pipe pressure signal is dispersed and bimodal. While PDF distribution of the venturi pipe pressure signal is concentrated and unimodal. At the same operating condition, the complexity and disturbing degree of gas–solid flow in the venturi pipe are lower than those in the bend pipe. And the flow state in the venturi pipe is more stable.
(4) According to the investigation of the pressure drop and flow pattern characteristics of the bend and the venturi pipe, the venturi pipe pressure signal is suitable to evaluate the coal flow stability in dilute phase pneumatic conveying. Statistical indexes are constructed and they can be used to reflect the coal flow stability effectively.

Acknowledgements This work was supported by China Coal Research Institute Innovation Fund project (No. 2018JC01, No. 2017CX02), and China Coal Science and Technology Group Innovation Fund Project (No.2018MS003)

References

Chi Z, Zhang L, Pan W et al (2000) Experiment and calculation of pressure drop of pulverized-coal in vertical pipes and circular bends. J Col 6(4):315–319
Chen Z, Zhao Q, He L (2004) Statistical investigation on transient characteristics of gas flow rate increasing under slug flow in horizontal pipeline. CIESC J 55(8):1274–1279

Fang W (2009) Research and application of signal de-noising based on wavelet analysis. Xihua University, Chengdu
Fusheng Y (1999) Engineering analysis and application of wavelet analysis. Beijing Science Press, Beijing, pp 1–145
Jaworski A and Dyakowski T (2001) application of electrical capacitance tomography for measurement of gas–solids flow characteristics in a pneumatic conveying system. Meas Sci Technol 12:1109–1119
Jin F, Liu R, Xia Ji et al (1999) Gas/solid two-phase mass flow rate measurement by using differential pressure-concentration method. J Northeast Univ (Natural Science) 20(5),461-463
Kai L (2015) Study on the characteristics and laws of gas-coal mixture through venturi under high concentration. East China University of Science and Technology, Shanghai
Liu Z, Yue Y, Zhang X et al (2001) Gas–solid measurement and numerical study using an extended-throat venturimeter. J Jinan Univ 15(4):296–298
Liu X, Duan G, Liu Z (2012) Experiment study and numerical simulation on gas-solid two-phase flow in horizontal elbow pipe. J Jinan Univ (Sci and Tech) 26(3):230–235
Madhaudana S, Zhu K, Wang C et al (2001) Electrical capacitance tomography measurements on the pneumatic conveying of solids. Ind Eng Chem Res 2001(40):4216–4226
Omine and Romero (2012) Study of elemental mercury re-emission in a simulated wet scrubber. Fuel 91(1):93–101
Pan W, Chi Z, Cen K (1999) Study on the numerical simulation to primary air flow in bend of the boiler. Power Eng 19(6):438–444
Shouguang C (2015) The soft sensor method for double-elbow solid mass flow measurement. Hebei United University, Tangshan
Wu Z, Xie F (2007) Optimization of Venturi tube design for pipeline pulverized coal flow measurements. J-Tsinghua Univ (Sci & Tech) 47(5):670–673
Wang C, Sun X, Zhang W (2013) Investigation on characteristics of dilute gas-solid two phase flow within 90° circular-cross bend. Chin Quart Mech 34(3):487–493
Xie K, Guo X, Cong X et al (2013) Minimum pressure drop velocity and stability analysis in pneumatic conveying of pulverized coal in commercial-scale horizontal pipe. CIESC J 64(6):1969–1975
Zhu F (2007) Investigation of the mass fraction of pulverized coal for the power plant by Micro-wave technique. University of Science and Technology, Nanjing
Zhuxin L (2007) Design and applied research of the real-time monitoring system for coal concentration of boilers of power plant. North China Electric Power University, Berjing
Zhiyao H, Zekui Z, Haiqing Li (1995) The method based on the differential pressure and the powder concentration for on-line measurement of powder flow rate in pneumatic conveying. J Chem Eng Chin Univ. 9(3):239–243

Chapter 87
Research on Dense Phase Pneumatic Conveying of the Mixture of Pulverized Coal and Extract Residue of Coal Liquefaction Residue at High Pressure

X. H. Fang, Z. Liu, H. Q. An, Z. Y. Feng, B. Z. Peng, Y. Li, and Y. G. Wang

Abstract In order to study on the high pressure dense phase conveying characteristics of extract residue of coal liquefaction residue (hereinafter referred to ER) blending to pulverized coal, the pulverized coal (M1) and the pulverized coal blending 20wt.% extraction residue (M2) were prepared as experimental samples. The conveying experiments were conduct at back pressures (receiver pressure) of 2 and 4Mpag in a pipeline with 25 mm inner diameter. The results showed that at the same superficial gas velocity, there is no significant difference in particles velocity when conveying M1 and M2 samples. The increasing of back pressure reduced the velocity difference between local superficial gas velocity and particles velocity. The conveying phenomena showed that both the mass flow rate and the particle concentration of these two samples increased with the increasing of flow rate of the fluidizing gas. When the fluidization number increased to 2.5, the fluidization effect were good enough to keep the mass flow rate of M1 and M2 stable; when back pressure decreased from 4 to 2Mpag, blending ER to pulverized coal will lead to a decrease in mass flow rate of the mixture powder. The effect of the blending ER to pulverized coal on particle concentration was mainly appeared in the low superficial gas velocity region, in the high gas velocity region, the particle concentration of M1 and M2 was similar during the conveying processes.

Keywords Dense-phase pneumatic conveying · Extract residue of direct coal liquefaction residue · Fluidizing gas · Supplementary gas · Superficial gas velocity

X. H. Fang (✉) · Y. G. Wang
China University of Mining & Technology, Beijing 100083, China
e-mail: fangxinhui@nicenergy.com

X. H. Fang · Z. Liu · H. Q. An · Z. Y. Feng · B. Z. Peng · Y. Li
National Institute of Clean and Low Carbon Energy, Beijing 102211, China

J. Lyu and S. Li (eds.), *Clean Coal and Sustainable Energy*,
Environmental Science and Engineering,
https://doi.org/10.1007/978-981-16-1657-0_87

87.1 Introduction

Direct coal liquefaction (Zhenzhen et al. 2017; Weike et al. 2017) can produce 30 wt.% liquefaction residue. Extracting heavy oil and asphalt from coal liquefaction residue is a way of deep processing for the coal liquefaction residue. The extracting process produce about 50 wt.% ER. ER contains the unreacted coal, minerals and external catalyst (Yan et al. 2015). The suitable treatment of ER has not yet been found. On one hand, ER has the characteristics of high percentage of carbon, heat value and low percentage of moisture, theoretically it can be considered as a raw material for coal gasification; on the other hand, ER contains a high percentage of ash, sulfur, and iron-based catalyst. Putting ER in gasifiers directly will result in a very high sulfur content in the syngas and also increasing the grey water treatment capacity, reducing the yield of effective gas, which may have devastating effects on gasifiers and subsequent devices, therefore, blending ER into the pulverized coal for co-gasification is a way to deal with large-scale resource utilization.

Dense phase conveying of pulverized coal is the upstream system of coal gasification (Xiangyu and Yao 2015), the stability and reliability of the conveying process is an important premise for the operation of coal gasification unit (Wei et al. 2011; Huijun and Jie 2015; Akanksha et al. 2018). The properties of ER, such as density and surface structure are significantly different from those of pulverized coal, and since the extracting process uses coal liquefied oil or coal tar as the extracting agent, a small amount of solvent oil remains in ER (Hongyan et al. 2017), resulting in a certain cohesiveness of the mixture powder, which may affect the flow characteristics of the mixture powder. Since the large-scale industrial extracting devices have not been constructed, there is no research reported on the pneumatic conveying characteristics of pure ER powder or pulverized coal blending ER.

In this paper, an experimental study was carried out on the pulverized coal blending ER at the pressures of 2Mpag and 4Mpag. The high-pressure dense phase conveying characteristics of the mixture powder were systematically investigated. This is of great significance for designing a more advanced gasification process.

87.2 Experimental Part

The experimental facility of high pressure dense-phase pneumatic conveying of coal powder is consists of a nitrogen supply system, screw feeder system, coal delivery system, conveying and measurement system, coal receiving system and data acquisition and control system. The flow diagram of the dense phase conveying system is shown in Fig. 87.1. High pressure nitrogen flows from a nitrogen tank installed in a truck to a storage vessel through pressure-reducing valves and are then fed into a delivery vessel (receiving vessel). Pressurizing gas is used to pressurize the vessels, fluidizing gas is used to fluidize coal powder for easier transportation and supplementary gas is used to suppress solid deposition and to adjust gas velocity and particle

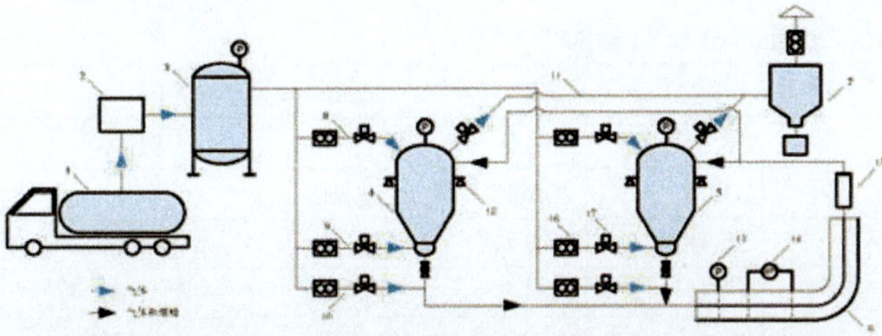

1-nitrogen tanker, 2-pressure reducing valve group, 3-gas storage tank, 4-transmission tank, 5-receiver tank, 6-measurement pipeline, 7-bag dust collector, 8-charged gas, 9-fluidization Gas, 10-Supply Gas, 11-Distributed Gas, 12-Weighing Sensor, 13-Pressure Sensor, 14-Differential Pressure Sensor, 15-Pulverized Coal Flow Meter, 16-Gas Flow Meter, 17-Electric Regulating Valve

Fig. 87.1 Brief diagram of pulverized coal high pressure dense phase transport experimental system

concentration in the pipe. The pressure of vessels is controlled by electric control valves installed at the pressurizing gas pipe and relieving gas pipe. The volume flow rate of fluidizing gas is adjusted as the pressure of the vessel varies until coal powder inside the bottom of the vessel reaches the optimal fluidization status. The fluidized coal powder is discharged from the bottom of the delivery vessel, mixing with the supplementary gas, and flows into the conveying line through a coal distributor. The conveying line consists of three parallel branches of different pipe diameters. Several valves are installed at the entrance and exit of the branches for switching between these branches. Conveying distance can be changed by altering the length of the horizontal conveying line using a set of valves. The three parallel branches join together through another coal distributor, following which is a solid mass flowmeter installed in a vertical downward pipe. The gas–solid mixture eventually flows into the receiving vessel, where gas and solid particles are separated, and carrier gas is vented after passing through a bag filter. The delivery vessel and receiving vessel are interchangeable, and coal powder is circulated inside the conveying system. New coal material is fed into the delivery vessel by the screw feeder system, and used coal material can be discharged from the bottom of the bag filter.

The flow rate of supplementary gas is measured by vortex flowmeters, and the flow rates of pressurizing gas and fluidizing gas are measured by mass flowmeters. Seven pressure transducers and eight differential pressure transducers are installed along the conveying line. The weights of vessels are measured by load cells and are used to calculate the average solid flow rate during a period of time. The instantaneous solid flow rate is measured by the solid mass flowmeter of Thermo Ramsey, which consists of a velocity sensor, a concentration sensor and an integrator. The output signals of measuring instruments are converted by A/D converters and then recorded by a PLC control system at a frequency of 2 Hz.

There are two kinds of raw materials for the high pressure dense phase pneumatic conveying experiments. One is the low-grade bituminous pulverized coal, which is defined as M1, and the other is a mixture powder consists of 20 wt.% ER and 80 wt.%

Table 87.1 Property data of M1 and M2

Parameter	M1	M2	Parameter	M1	M2
M_{ar}/%	3.35	3.25	C_{daf}/%	81.05	82.51
V_d/%	31.71	29.64	H_{daf}/%	4.65	4.45
FC_d/%	55.18	52.24	O_{daf}/%	12.70	10.31
A_d/%	13.11	18.12	N_{daf}/%	1.03	0.96
V_{daf}/%	36.50	36.20	S_{daf}/%	0.57	1.77
$Q_{net, ar}$/(MJ/kg)	26.44	25.83			

Table 87.2 Physical property data of M1 and M2

	Average particle size μm	True density kg/m^3	Tap density kg/m^2	Bulk density kg/m^3	Compression %
M1	33	1449	725	535	26
M2	35	1649	790	496	37

pulverized coal, which is defined as M2. The property data of these two conveying samples are as shown as Tables 87.1 and 87.2. It can be seen that the moisture and the average particle size are similar between M1 and M2. The main differences in properties of these two samples are the true density, compressibility.

87.3 Results and Discussion

87.3.1 *Effect of Superficial Gas Velocity on Particle Velocity*

Figure 87.2 show that the effect of operating parameters on the particle velocity in the conveying pipeline. As shown as Fig. 87.2, it can be seen that the particle velocity is mainly affected by the local superficial gas velocity. When M1 and M2 samples were conveying, at the same local superficial gas velocity, there was no significant difference for the particle velocity of the two samples. The particle velocity increased with the superficial gas velocity increasing, but the particle velocity was lower than the local superficial gas velocity. When the superficial gas velocity of the pipeline was high, the flow pattern in the pipeline was a kind of dilute phase suspension flow. At higher gas velocities, the particle–wall friction force was larger because of larger particle velocities, and the gas–solid drag force was also larger according to the force balance of particles in the dilute phase, therefore, the velocity difference between gas and solid should be larger to drive the particles. That trend was similar to the results predicted by the correlation developed by Arastoopour and Gidaspow (Hamid and Dimitri 1979).

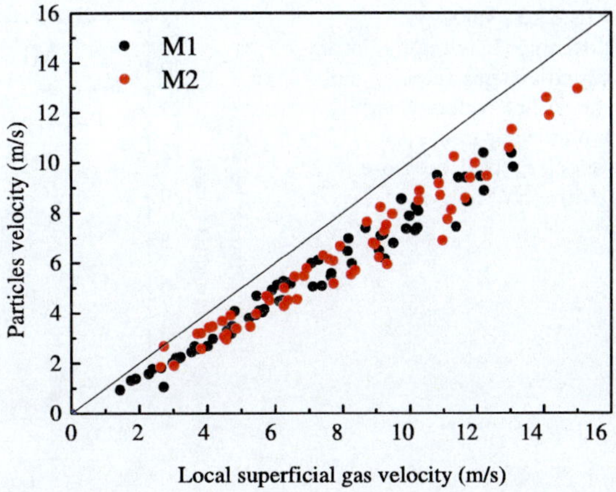

Fig. 87.2 Particle velocity varies with the superficial gas velocity at the position of mass flow rate flowmeter

Further analysis of the velocity difference between the local superficial gas velocity and particle velocity at different back pressures of 2Mpag and 4Mpag was conducted, as Fig. 87.3 shown, when the back pressure increased from 2Mpag to 4Pag, the velocity difference reduced for the reason of pressure increasing, from Fig. 87.3a, it can be seen that the velocity difference increased with the increasing of superficial gas velocity, and the maximum velocity difference reached 4.5 m/s, while under high back pressure 4Mpag, as shown as Fig. 87.3b, the velocity difference does not show an obviously increasing with the superficial gas velocity increasing, and the maximum velocity difference is not more than 3.5 m/s. These were mainly because the viscosity of the gas phase increased as the back pressure increasing. The improvement of gas viscosity leads to enhancing in the ability of carrying particles, which caused the velocity between the apparent gas velocity and the particle velocity decreasing.

87.3.2 Effect of Fluidizing Gas on Conveying Characteristics

In order to maintain the stability and controllability of the dense phase conveying process, it is necessary to determine the flow rate of the fluidizing gas flowing into fluidization chamber of the delivery vessel. For the purpose of maintaining the same fluidization effect under different operation conditions, the following formula is used to calculate the fluidized gas flow rate:

$$Q_f = U_f A_f = \frac{U_f}{U_{mf}} U_{mf} A_f$$

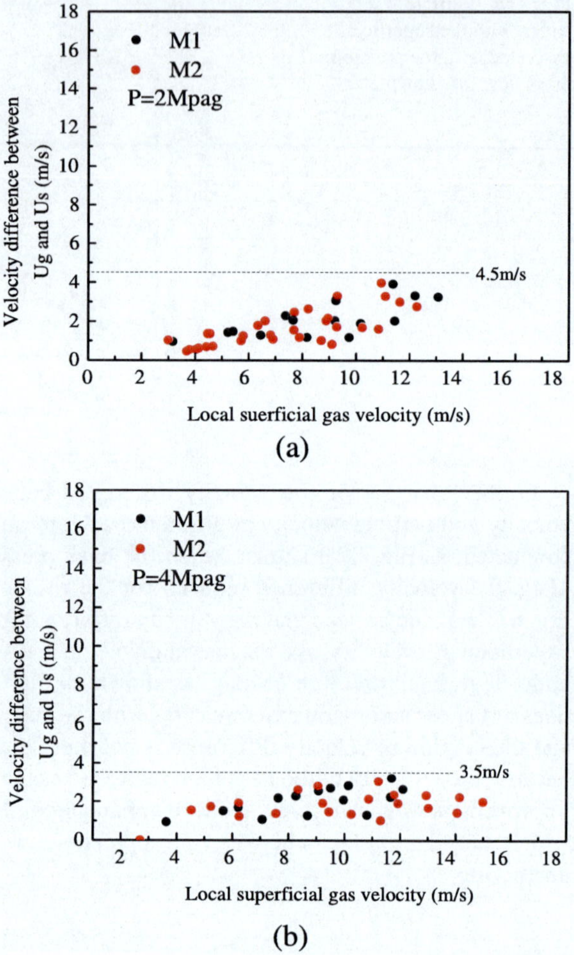

Fig. 87.3 Velocity difference between the local superficial gas velocity and the particle velocity varies with the apparent gas velocity **a** back pressure 2Mpag **b** 4Mpag

where U_f is the average gas velocity in the fluidized chamber, U_{mf} is the minimum fluidizing gas which is calculated by Wen and Yu (1966), $\frac{U_f}{U_{mf}}$ is the fluidization number,A_f is the area of fluidization plate.

In this paper, U_{mf} was calculated firstly according to the structure of the fluidization chamber, and the flow rate of fluidization gas adjusted by changing fluidization number. As Fig. 87.4 shown, it can be seen that the mass flow rate of M1 and M2 increased significantly with fluidization number increasing at the low fluidization number. When the fluidization number of M1 was more than 1.5, the mass flow rate of M1 is basically unchanged. When the fluidization number of the M2 was more than 2.5, the mass flow rate of M2 was trend to be stable. As Fig. 87.5 shown, with the fluidization number increasing, the mass ratio of solid–gas was gradually decreasing when conveying M1 and M2. The decreasing of mass ratio of solid–gas indicated

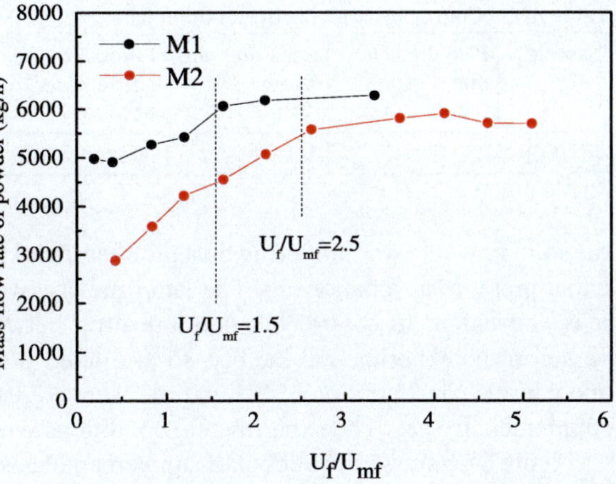

Fig. 87.4 Mass flow rate of powder varies with the fluidization number

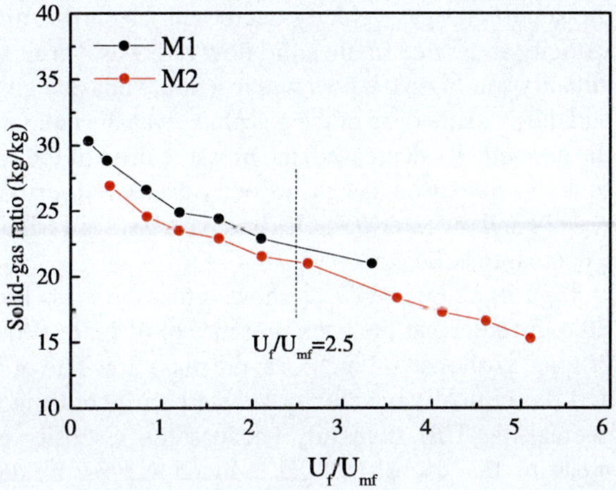

Fig. 87.5 Mass ratio of solid–gas varies with the fluidization number

that the particle concentration in pipeline is decreasing, which gone against the dense phase conveying process. Therefore, the fluidization number should be greater than 2.5 which ensure that the fluidization effect is good and the mass ratio of solid–gas was above 20 when conveying M1 and M2.

87.3.3 Effect of Supplemental Gas on Conveying Characteristics

Diagram of conveying was widely used to investigate the characteristics of high pressure dense-phase conveying. In the diagram, the pressure drop per unit length was plotted against the superficial gas velocity with the solid flow rate as a parameter, or

Table 87.3 Conveying experiment condition table

Samples	Pipe diameter, mm	Back pressure, MPag	Pressure difference, Mpa	Mass flow rate, kg/h	Concentration, kg/m^3
M1/M2	25	4/2	0.75/0.5/0.25	2000~9000	120~520

the solid flow rate was plotted against the superficial gas velocity with the total differential pressure as a parameter. The latter method was chosen in this paper because it is convenient to control the pressure drop between the delivery and receiving vessel in the experimental facility, so at a fixed pressure drop, the mass flow rate and particle concentration of M1 and M2 were adjusted by changing the flow rate of supplementary gas. The experimental conditions were listed as Table 87.3 shown.

Figure 87.6 showed the effect of supplemental gas on the characteristics of dense-phase conveying for M1 and M2. At the same back pressure and total differential pressure, the mass flow rate of powder increased to a maximum and decreased as the superficial gas velocity decreased, as shown in Fig. 87.6a. The superficial gas velocity at the maximum solid flow rate was a critical velocity. When superficial gas velocity was high, the flow was in a dilute phase region with low solid concentrations, and the pressure drop of the gas phase was the main part of the total pressure drop. As the gas velocity decreased, the pressure drop of the gas phase decreased and the solid flow rate increased. As gas velocity decreased further, solid concentration increased rapidly and the inter-particles and particle–wall friction resistance, which resulted in a decreased solid flow rate.

Figures 87.6a and 87.7a showed that the mass flow rate of M1 and M2 showed a little difference at the back pressure of 4Mpag. When the back pressure reduced to 2Mpag, as shown as Fig. 87.7a, the mass flow rate of M2 showed obvious decreasing, and the critical gas velocity showed an increasing trend due to the back pressure decreasing. This is mainly because the viscosity of the gas was affected by the pressure, the viscosity of gas reduced as pressure decreasing from 4 to 2Mpag. For the weak carrying ability of conveying gas at low back pressure, the higher superficial gas velocity was needed to make sure that the powder mass flow rate achieved the maximum value.

Figure 87.6b with Fig. 87.7b showed that in the low superficial gas velocity region, which the superficial gas was lower than the critical velocity, the particle concentration of M2 was lower than M1. When the superficial gas velocity increasing more than critical gas velocity, the particle concentration of M1 and M2 were almost similar, this indicated that the ER blending mainly affected particle concentration in the low superficial gas velocity region. For the low superficial gas velocity, the carrying ability of the conveying gas is not enough to fully suspend the pulverized particles in pipeline. Since the true density of ER is greater than the pulverized coal, some of ER particles will depose at the bottom of the pipeline. The deposition of ER particles caused reducing of particles number of the mixture powder when passing through the section of density measurement, which resulted in the lower particle concentration when conveying M1 and M2.

Fig. 87.6 Mass flow rate, particle concentration varies with local superficial gas velocity at back pressure of 4Mpag **a** mass flow rate **b** particle concentration

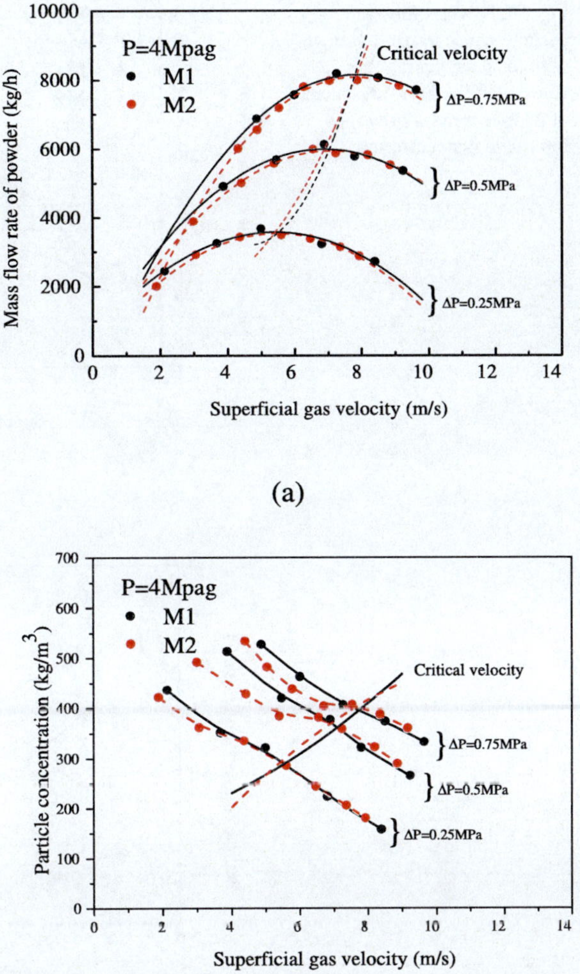

(a)

(b)

87.3.4 *Effect of Pressure Drop on Conveying Characteristics*

For the purpose of reducing the influence of supplemental gas on the mass flow rate, in this paper, changed the mass flow rate of powder by adjusting the pressure drop between delivering and receiving vessel at the condition of the flow rate of supplementary gas was zero. The pressure drop was the main power source for conveying process. As Fig. 87.8 shown, the mass flow rate of M1 and M2 increased significantly with the increasing of pressure drop. When pressure drop increased

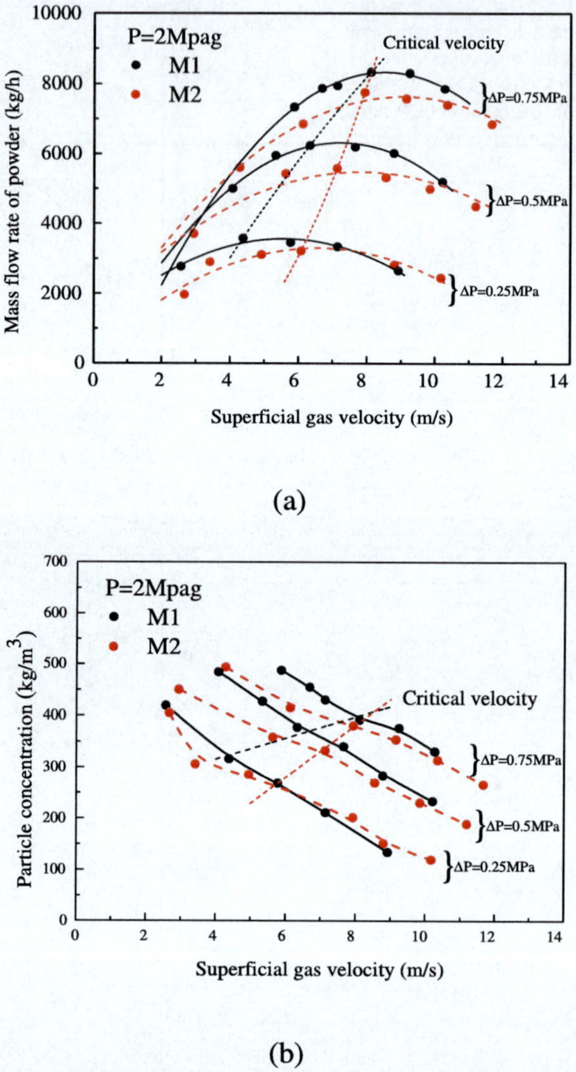

Fig. 87.7 Mass flow rate, particle concentration varies with local superficial gas velocity at the back pressure of 2Mpag **a** mass flow rate **b** particle concentration

from 0.25 to 1.1 MPa, the mass flow rate of M1 and M2 increased from 2000 to over 8000.

It can be seen from Fig. 87.8a, b that at the same pressure drop, the mass flow rate of M2 was smaller than those of M1, mainly because the pure density of ER contained in M2 was much more than pulverized coal, a part of ER particles depositing at the bottom of pipeline in conveying process resulted in the decreasing of mass flow rate of M2.

Fig. 87.8 Mass flow rate varies with pressure drop **a** P = 4Mpag, **b** P = 2Mpag

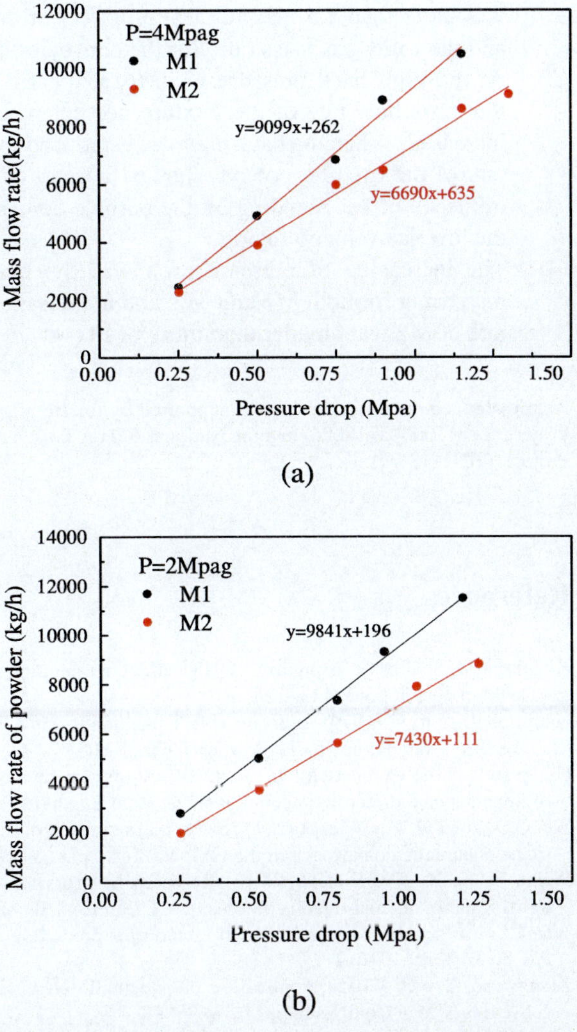

87.4 Summary

Through the experimental study on the dense phase conveying of two kinds of samples M1 and M2, the following conclusions are obtained:

(1) At the same superficial gas velocity, there is no significant difference between the particle velocity and local superficial velocity when conveying M1 and M2, and the increasing of back pressure is the main reason of reducing the velocity difference.

(2) With the increasing of fluidizing gas, the mass flow rate and solid–gas mass ratio decreased when conveying M1 and M2. When the fluidization number

was more than 2.5, the fluidization effect in fluidization chamber was good, and the solid–gas mass ratio for the conveying process was above 20.

(3) At the high back pressure of 4Mpag, the ER blending had little effect on the mass flow rate of the mixture powder when the superficial gas velocity increased. When the back pressure is reduced from 4 to 2Mpag, the mass flow rate of the mixture powder showed a trend of significantly decreasing. The influence of ER blending on the particle concentration is mainly reflected in the low gas velocity region.

(4) The increasing of pressure drop was the main reason of mass flow rate increasing for both M1 and M2, and at the same pressure drop, the mass flow rate of M2 was smaller than those of M1

Acknowledgements This work was supported by the Beijing Science and Technology Planning Project (Z181100005118006) and the National Energy Group Science and Technology Innovation Project (GJNY-18-72).

References

Akanksha M, Shalini G, Tripurari S (2018) Effect of operating parameters on coal gasification. Int J Coal Sci Technol 5(2):113–125

Hamid A, Dimitri G (1979) Analysis of IGT pneumatic conveying data and fast fluidization using a thermohydrodynamic model. Powder Technol 22(1):77–87

Hongyan C, Shifu C, Guodong W et al (2017) Study and development on extraction and utilization of Shenhua coal direct liquefaction residue. Coal Eng 49(S1):61–66

Wei G, Xiaolei G, Yue W et al (2011) Study on the impact of the flowability of the pulverized coal to the pneumatic conveyance of the GSP gasifier. Coal Chem Indus 39(05):9–11

Weike C, Jie X, Wei S et al. (2017) Research progress and prospect of sulfur migration and transformation in coal liquefaction residue. Clean Coal Technol 23(03):1–6

Wen CY, Yu YH (1966) A generalized method for predicting the minimum fluidization velocity. AIChE J 12(3):610–612

Xiangyu Z, Yao X (2015) Application of pneumatic conveying technology in pulverized coal conveying. Clean Coal Technol 21(02):85–88

Yan G, Tianbao W, Haolei D et al (2015) Extractant recycle from raffinate of direct coal liquefaction residue. Clean Coal Technol 21(06):67–71

Huijun Y, Jie C (2015) Brief analysis of stable operation of pulverized coal conveying system for GSP gasification process and its influence on gasification process. Chem Fertilizer Indus 42(06):51–54

Zhenzhen S, Ming S, Ye H et al. (2017) Modified asphalt with the extract fractions of Shenhua direct coal liquefaction residue. Chem Indus Eng Progress 36(09):3273–3279

Chapter 88
The Process of Self-heating Sludge Incineration Based on Granular Heat Carrier

Kong Hao, Miao Miao, Yang Hairui, Zhang Xuyi, and Zhang Man

Abstract Among the current sludge treatment methods, drying and incineration is one of the most frequently adopted for sludge processing since it works efficiently and can reduce pollution caused by sludge. However, sludge with high moisture content needs a lot of heat for drying before stable burning in the furnace, which increases its cost and causes the stink. This paper introduces a novel process of self-heating sludge incineration based on granular heat carrier as shown in Fig. 88.1. Self-heat incineration means that the process uses self-generating heat for drying and incineration to reduce heat consumption. Free falling particles curtain is used to quickly cool down the flue gas temperature to below 300 °C, to avoid the temperature 300–800 °C that may cause regeneration of dioxin. The particles heated works as heat carrier to dry the wet sludge with water content over 70% in the fluidized bed dryer. The dried sludge is incinerated in the insulated fluidized bed furnace to provide heat to burn other wet sludge directly injected into the furnace. A 240 t/d unit with such process was manufactured and operated in 2017, and the performance is satisfied. SO_2 emission is less than 30 mg/Nm^3, NO_x emission is less than 100 mg/Nm^3, PCDD/Fs emission is less than 0.034 ng-TEQ/Nm^3 and the investment of this process is less than 250,000 yuan/t/d.

Keywords Granular heat carrier · Self heating · Sludge incineration

88.1 Introduction

With the rapid development of industry and the increasing population of the city, the discharge of urban sewage has been increasing unprecedently. Sludge, as one of the products in the sewage treatment process, has also been gradually increasing. It is predicted that by 2020, the output of urban sludge and industrial sludge in China will exceed 80 million tons/year (Xue et al. 2018). Sludge is a complex heterogeneous

K. Hao · M. Miao · Y. Hairui · Z. Xuyi · Z. Man (✉)
Key Laboratory for Thermal Science and Power Engineering of Ministry of Education, State Key Laboratory of Power Systems, Department of Energy and Power Engineering, Tsinghua University, Beijing 100084, China

J. Lyu and S. Li (eds.), *Clean Coal and Sustainable Energy*,
Environmental Science and Engineering,
https://doi.org/10.1007/978-981-16-1657-0_88

body composed of bacterial micelles and persistent organic pollutants (Hudson and Lowe 1996; Davis 1996). So, if sludge is not treated effectively, it will cause serious damage to the environment and can also cause infectious diseases. The treatment of sludge tends to be a tough problem of the society (Praise 2009; Chi et al. 2011; Chen et al. 2014). In this paper, a novel treatment process of sludge named as The Process of Self-heating Sludge Incineration based on Granular Heat Carrier is introduced, and the advantages of this process compared to the traditional process and the current shortcomings is clarified. Finally, the application demonstration of this process is introduced.

88.2 The Process of Sludge Incineration

At present, the treatment of sludge includes: landfill, recycling/composing and incineration. Among these methods, incineration can treat sludge effectively and harmlessly, and meets the basic objectives of sludge treatment and disposal, so it has been widely used (Wolski et al. 2003; Deng et al. 2008). There are two main methods for sludge incineration: mixed burning with other fuel and incineration after drying (Gao 2014).

88.2.1 Mixed Burning with Other Fuel

The main methods of mixed burning with other fuel is mixing a small amount of sludge in waste incineration and coal-fired power plants. This method is simple and the amount of mixed sludge is small, so it does not affect the original combustion process. Generally, there are no new technical difficulties (Ninomiya et al. 2003). In Europe, there are more than 100 power plants using this method to treat biosolid waste including sludge. However, the distribution of general sewage treatment plants is relatively dispersed, and the daily sludge treatment is relatively small, and the power plants are often far away from the sludge source, resulting in too long a sludge transportation distance and an increase in cost.

88.2.2 Incineration After Drying

Another method of the treatment of sludge is incineration after drying. It is because if the sludge is directly incinerated, because of the high moisture in the sludge, not only the sludge of heat can't be obtained, but also a large amount of auxiliary fuel is consumed (Vaxelaire et al. 2000; Youssef and Kahil 2016; Bein 1994). It is necessary to dry sludge to reduce moisture to a certain extent before incineration so that the sludge can keep burning without adding auxiliary fuel (Stasta et al. 2005; Vaxelaire et al. 1999). Direct thermal drying, indirect thermal drying and solar drying are the most widely used three kinds of methods for industrial sludge treatment. Although new drying technologies such as microwave drying and frying drying have appeared, they only stay in the experimental stage, are temporarily not applied to the actual sludge treatment (Zhou et al. 2012). Direct thermal drying means that heat transfer is directly contacted by hot sludge and hot gas, and the heat of the intake air provides the latent heat required for liquid evaporation in the sludge. This part of heat is always provided by fuel. Indirect heat drying means that the heat transfer is completed by the contact of the wet sludge with the hot surface. The heating medium is separated from the metal wall surface, and the evaporated liquid is removed from the heating medium. This part of heat is always provided by waste heat. Solar drying which is using inexhaustible clean solar, is currently the lowest energy consumption in all drying processes, and has the advantages of simple structure and low technical requirements, which is conducive to large-scale promotion.

However, the core technology of sludge drying currently used is still the traditional process decades ago. Direct thermal drying can reduce the sludge volume at one time, eliminating the trouble of sludge subsequent treatment, but it also requires a large amount of heat. The high-temperature drying is also not conducive to large-scale promotion. Indirect thermal drying can solve energy consumption problems by utilizing various types of waste heat, but the drying efficiency is low. It is necessary to build a large number of sludge drying plants to meet the growing demand for sludge drying. And solar drying also has obvious shortcomings, due to the dispersion of solar energy, the drying temperature of the sludge is low, and the drying efficiency is difficult to compare with direct thermal drying and indirect thermal drying.

88.2.3 Sludge Drying and Incineration Integration

If these traditional dry incineration methods are used, it will usually cause bad odor and the drying energy consumption is high, resulting in low energy efficiency. Therefore, it is invented a new process named "Sludge drying and incineration integration".

This process combines the heat drying process with the utilization of waste heat which is generated by the sludge combustion, it is not necessary to set the heat drying process separately, and the waste heat generated during sludge incineration can also be utilized, thereby improving energy efficiency. But there are also two main issues with this process. The first issue is how to achieve sludge self-sustained burning (Li and Qi 2002; Li et al. 2001; Wang et al. 2013; Hai and Ma 2012; Liao et al. 2016), the other issue is how to reduce dioxin emissions.

88.3 The Process of Self-heating Sludge Incineration Based on Granular Heat Carrier

88.3.1 Process Introduction

In response to the two key issues mentioned above: how to achieve self-sustaining combustion of sludge with high moisture and how to achieve low/ultra-low emissions of pollutants (especially dioxin) during combustion. A novel process named "the process of self-heating sludge incineration based on granular heat carrier" was developed and using this novel process an industrial application of 240 t/d was built, which process flow is shown in Fig. 88.1.

This process adopts an approximately adiabatic fluidized combustion mode. The fuel is proportionally mixed with wet sludge and dry sludge, and wet sludge is dried by multi-stage drying of the granular heat carrier in the down-flow bed and the fluidized bed. The ratio of the dry sludge and wet sludge is adjusted to achieve self-sustained combustion. The heat of the solid particulate heat carrier used to dry the sludge is derived from the falling cloud bed for counter current flue gas heating. In addition, heat exchange using granular heat carrier realizes gas–solid direct contact heat transfer and rapid cooling of flue gas. The rapid cooling of flue gas can fast pass through the range of 300–800 °C, in such temperature the metal oxide in the hot ash acts as a catalyst, and the carbon in the hot ash, free chlorine in the flue gas, and hydrogen chloride as reactants can cause dioxin synthetize, which may account for 70–90% in total emissions. Besides, the rapid cooling of flue gas can also achieve a function of slowing metal corrosion. At the same time, the heated surface is arranged

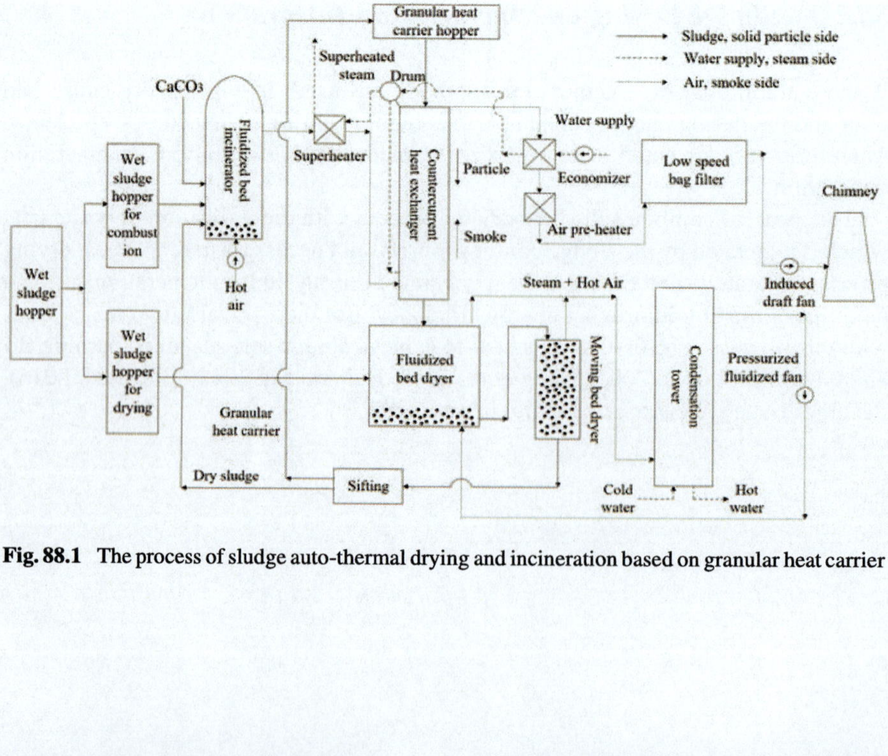

Fig. 88.1 The process of sludge auto-thermal drying and incineration based on granular heat carrier

at an appropriate position, and the residual heat of the flue gas can be continuously utilized to generate superheated steam at a specific pressure temperature.

88.3.2 Power Plant Based on the Process and Measurement Data

Two years ago, Shaanxi Xuanyi Environmental Protection Technology Liability Limited, in Xi'an Second Wastewater Treatment Plant, built a project based on the process of self-heating sludge incineration based on granular heat carrier, and completed all commissioning. And at the end of 2017, a series of tests on this project was carried out, the results proved that the project has achieved good performance. Firstly from the perspective of project cost, the investment of the project is less than 250,000 yuan/ton/day. The investment costs of other power plants are as follows: The investment costs of Shanghai Shidongkou, which is using German equipment, is about 550,000 yuan/ton/day. The investment costs of Suzhou, which is using French technology, is about 400,000 yuan/ton/day. And the investment costs of Wenzhou, which is using Japanese equipment, is about 350,000 yuan/ton/day. It is obvious that the investment cost of the project is much lower than other project costs.

Secondly from the perspective of pollutant emissions, this project activates desulfurization by adding limestone desulfurization and tail section moisture addition induced activation to make SO_2 emission is less than 30 mg/Nm^3. For reducing nitrogen oxide emission, it is achieved by keeping the furnace temperature around 900 °C and the furnace outlet O_2 concentration less than 6%. It is measured that NO_x emission is less than 100 mg/Nm^3. And for emission of dioxin, which has attracted more and more attention from the public. The detailed measurements on the emissions of dioxins are listed in Table 88.1.

It is also measured that CO concentration is in the range of 300–1000 mg/Nm^3, it can be concluded that there is no direct correlation between dioxin emissions and CO emissions. And for condition 3, some PVC was added into the furnace, and there is no significant change in the emission of dioxins. Therefore, we believe that taking appropriate measures, chlorine-containing components will not increase dioxin emissions.

Table 88.1 Measurement of dioxin emission data under three conditions

Date	Measured component	Condition 1 (ng/Nm^3)	Condition 2 (ng/Nm^3)	Condition 3 (ng/Nm^3)
2017.5	Smoke	0.075	0.035	0.032
2017.12	Smoke	0.034	0.034	0.0059

88.3.3 Advantages and Disadvantages

In addition to the advantages of low investment costs and low pollutant emissions, this project also has other advantages. For example, this project uses a fluidized bed as a sludge incineration device, which has high efficiency and good fuel suitability. Using the particle heat carrier technology and counter current heat exchanger can improve the utilization efficiency of flue gas waste heat. And the project used a two-stage dryer for drying sludge, which can improve the utilization of heat during sludge drying and make the sludge drying better.

The process also has some disadvantages that it needs to add a small part of the sludge filter press equipment for ignition start. In the next work, it will continue to improve the disadvantages to achieve better results.

88.4 Conclusion

The process of sludge auto-thermal drying and incineration based on granular heat carrier is introduced. It is a low-cost process that enables low emissions of pollutants, recycling of ash and efficient use of flue gas heat. Though it needs to add a small part of the sludge filter press equipment for ignition start, it is a process with market potential.

Acknowledgements Financial support of this work by the National Natural Science Foundation of China (U1810126) are gratefully acknowledged.

References

Bein DJ (1994) Two-stage sludge drying apparatus and method. Patent: 5365676

Chen W, Deng M, Luo H et al (2014) Study on the characteristics of malodor and volatile organic compounds produced by direct drying of sludge. Environ Sci 35(8):2897–2902

Chi Y, Xue C, Xi L et al (2011) The nature and inactivation method of pathogenic microorganisms in sewage sludge. J Tianjin Urban Construct Univ 17(1):48–54

Davis RD (1996) The impact of EU and UK environmental pressures on the future of sludge treatment and disposal. Water Environ J 10(1):65–69

Deng W, Yan J, Li X et al (2008) Drying of paper sludge and pollutant discharge characteristics of incineration system. Combust Sci Technol 14(6):545–550

Gao Y (2014) Analysis of the ratio of blending and burning of municipal modified sludge incineration in waste-assisted semi-drying. Tsinghua University

Hai L, Ma X (2012) Simulation of co-incineration of sewage sludge with municipal solid waste in a grate furnace incinerator. Waste Manage 32(3):561–567

Hudson JA, Lowe P (1996) Current technologies for sludge treatment and disposal. Water and Environ J 10(6):436–441

Li X, Qi Y (2002) Research and application of sludge fluidized bed incineration technology. J Combust Sci Technol 02:159–162

Li A, Wang Z, Wei L, Cai J (2001) Experimental study on combustion characteristics of sewage sludge in fluidized bed. J Shenyang Inst Aeronaut Eng 04:43–45
Liao J, Li W, Shan Y, Li Z (2016) Engineering design and commissioning operation of Wenzhou sludge concentrated drying incineration plant. China Water Wastewater 32(22):73–77
Ninomiya Y, Zhang L, Sakano T, Kanaoka C, Masui M (2003) Transformation of mineral and emission of particulate matters during co-combustion of coal with sewage sludge. Fuel 83(6):751–764
Praise (2009) Study on benzene series and phenol in sludge and its release characteristics. Zhejiang University, Hangzhou
Stasta P, Boran J, Bebar L, Stehlik P, Oral J (2005) Thermal processing of sewage sludge. Appl Therm Eng 26(13):1420–1426
Vaxelaire J, Bongiovanni JM, Puiggali JR (1999) Mechanical dewatering and thermal drying of residual sludge. Environ Technol 20(1):29–36
Vaxelaire J, Bongiovanni JM, Mousques P, Puiggali JR (2000) Thermal drying of residual sludge. Water Res 34(17):4318–4323
Wang C, Yu Q, Gao Z, Sheng S, Wu C, Wang K (2013) Removal of pollutants from 600t/d new sludge spray drying-incineration demonstration project. J Environ Eng 7(08):3196–3202
Wolski N, Maier J, Hein KRG (KRG) Fine particle formation from co-combustion of sewage sludge and bituminous coal. Fuel Process Technol 85(6):673–686 (2003)
Xue C, Kong X, Wang S et al (2018) The status quo, development and incentive policy requirements of urban sludge treatment and disposal in China. Water Purificat Technol 37(12):33–39
Youssef AS, Kahil MA (2016) Solar sludge drying for Medina Al-Munawarah sewage treatment plant in the Kingdom of Saudi Arabia. J Environ Eng 142(12):0501–6006
Zhou D, Wei W, Zhou X, Chao Z, Liying Z, Rui Qi (2012) Development and engineering application of municipal sludge hydrothermal drying system. China Water Wastewater 28(19):4–7